JACARANDA MATHS QUEST
SPECIALIST MATHEMATICS 11

VCE UNITS 1 AND 2 | SECOND EDITION

JACARANDA MATHS QUEST SPECIALIST MATHEMATICS 11

VCE UNITS 1 AND 2 | SECOND EDITION

RAYMOND ROZEN

JENNIFER NOLAN

GEOFF PHILLIPS

CONTRIBUTING AUTHORS

Romina Norello

Douglas Scott

Beverly Langsford Willing

Second edition published 2023 by
John Wiley & Sons Australia, Ltd
155 Cremorne Street, Cremorne, Vic 3121

First edition published 2016

Typeset in 10.5/13 pt TimesLTStd

ISBN: 978-1-119-87665-6

The covers of the *Jacaranda Maths Quest VCE Mathematics* series are the work of Victorian artist Lydia Bachimova.

Lydia is an experienced, innovative and creative artist with over 10 years of professional experience, including five years of animation work with Walt Disney Studio in Sydney. She has a passion for hand drawing, painting and graphic design.

Illustrated by diacriTech and Wiley Composition Services

Typeset in India by diacriTech

A catalogue record for this book is available from the National Library of Australia

Printed in Singapore
M121040_230822

Contents

About this resource

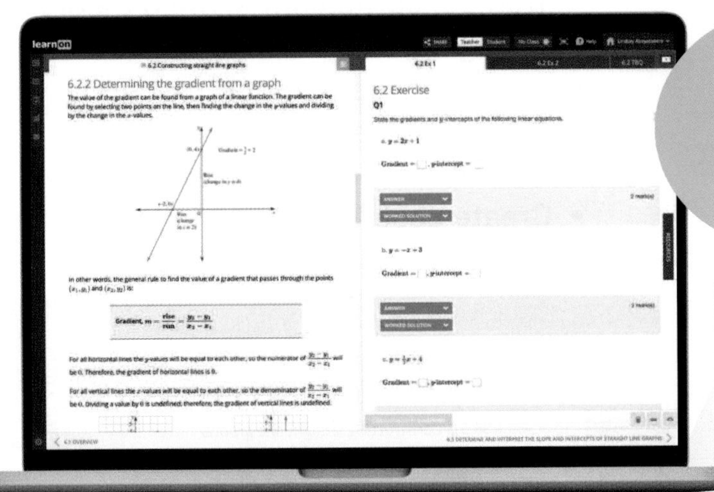

Everything you need for your students to succeed

JACARANDA MATHS QUEST
SPECIALIST MATHEMATICS 11 VCE UNITS 1 AND 2 | SECOND EDITION

Developed by expert Victorian teachers for VCE students

Tried, tested and trusted. The NEW Jacaranda VCE Mathematics series continues to deliver curriculum-aligned material that caters to students of all abilities.

Completely aligned to the VCE Mathematics Study Design

Our expert author team of practising teachers ensures 100 per cent coverage of the new VCE Mathematics Study Design (2023–2027).

Everything you need for your students to succeed, including:

- NEW! Access targeted question sets including exam-style questions and all relevant past VCAA exam questions since 2013. Ensure assessment preparedness with practice Mathematical investigations.
- NEW! Be confident your students can get unstuck and progress, in class or at home. For every question online they receive immediate feedback and fully worked solutions.
- NEW! Teacher-led videos to unpack concepts, plus VCAA exam questions and exam-style questions to fill learning gaps after COVID-19 disruptions.

Learn online with Australia's mos

Everything you need for each of your lessons in one simple view

- **Trusted, curriculum-aligned theory**
- **Engaging, rich multimedia**
- **All the teacher support resources you need**
- **Deep insights into progress**
- **Immediate feedback for students**
- **Create custom assignments in just a few clicks**

Practical teaching advice and ideas for each lesson provided in teachON.

Each lesson linked to the Key Knowledge (and Key Skills) from the VCE Mathematics Study Design.

Reading content and rich media including embedded videos and interactivities.

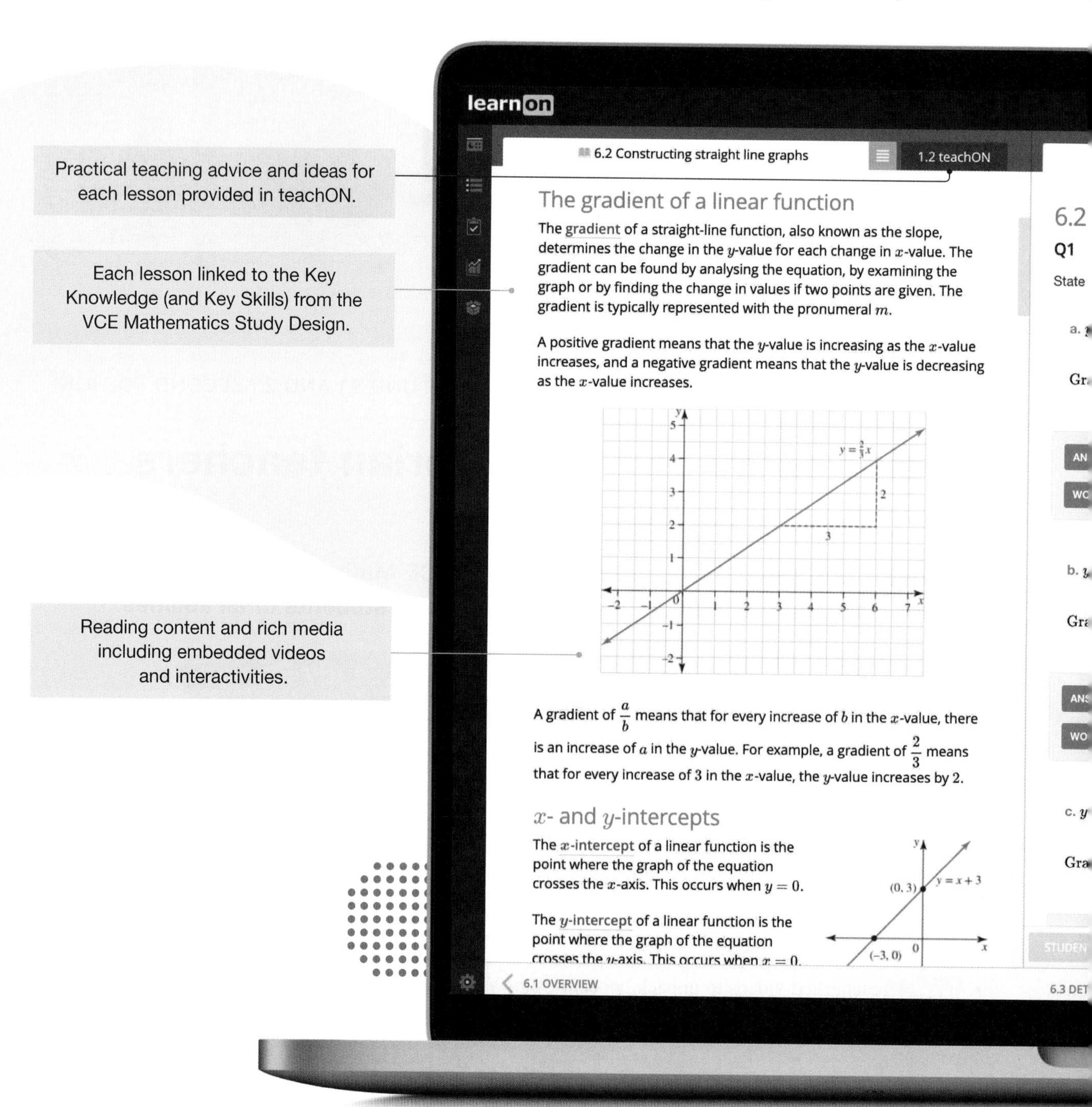

powerful learning tool, learnON

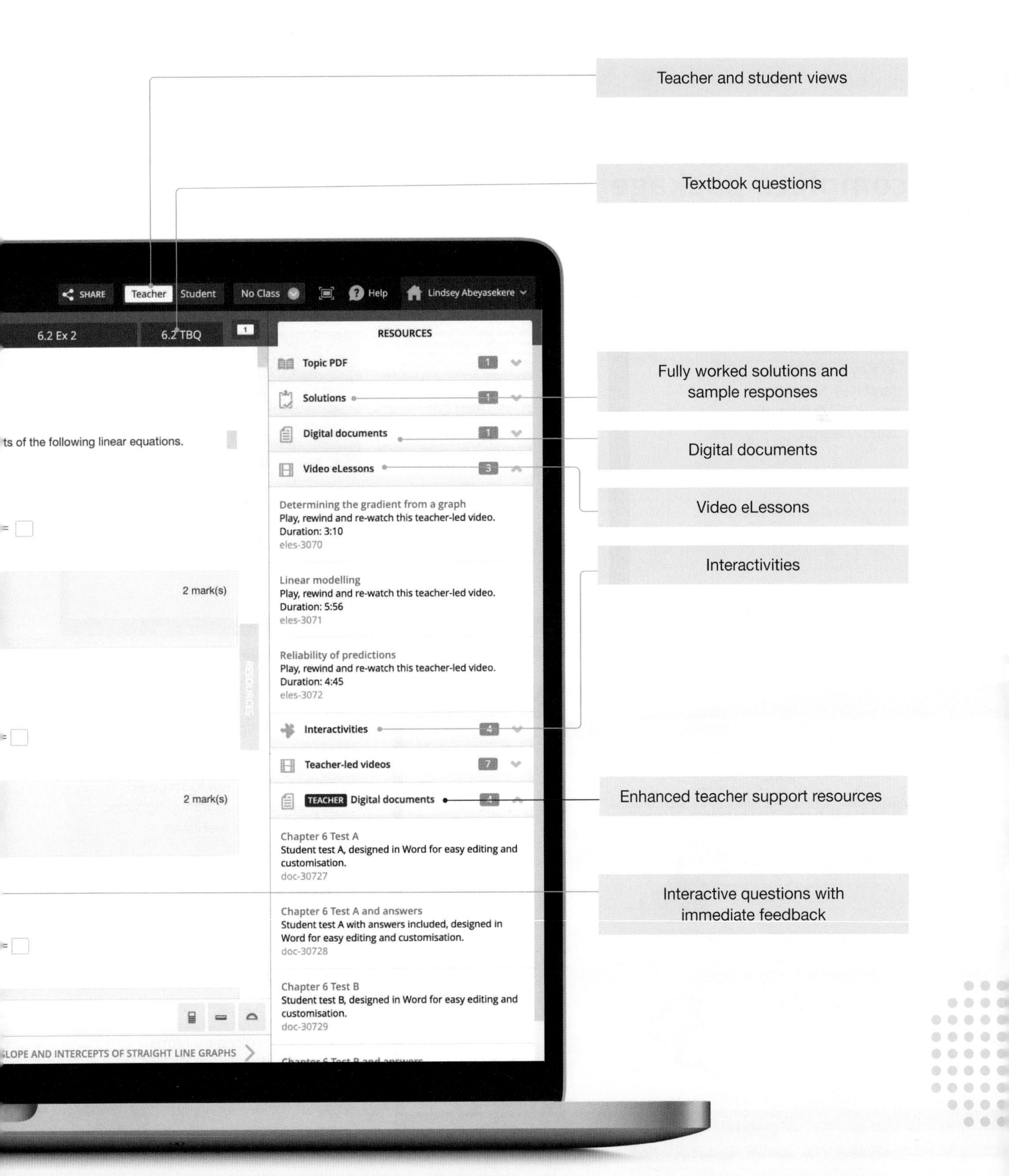

Get the most from your online resources

Online, these new editions are the complete package

Trusted Jacaranda theory, plus tools to support teaching and make learning more engaging, personalised and visible.

Each topic is linked to Key Knowledge (and Key Skills) from the VCE Mathematics Study Design.

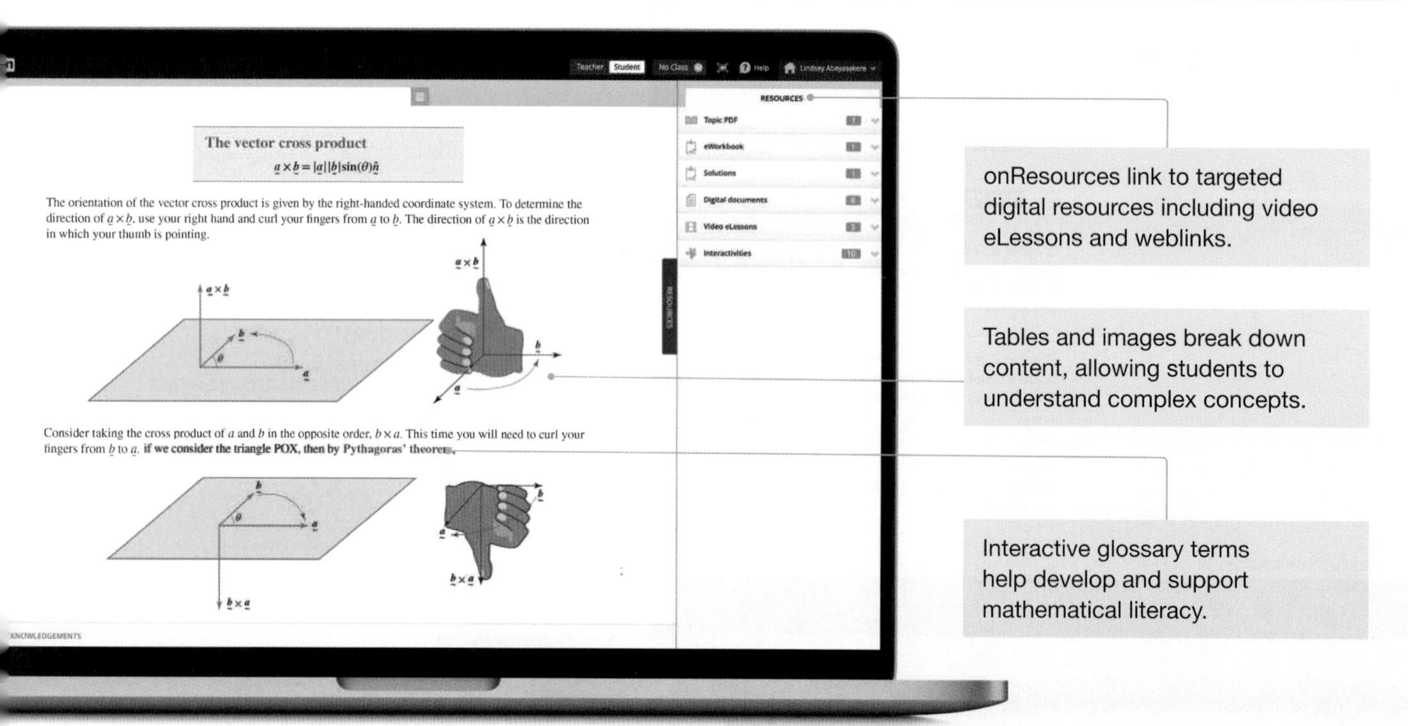

onResources link to targeted digital resources including video eLessons and weblinks.

Tables and images break down content, allowing students to understand complex concepts.

Interactive glossary terms help develop and support mathematical literacy.

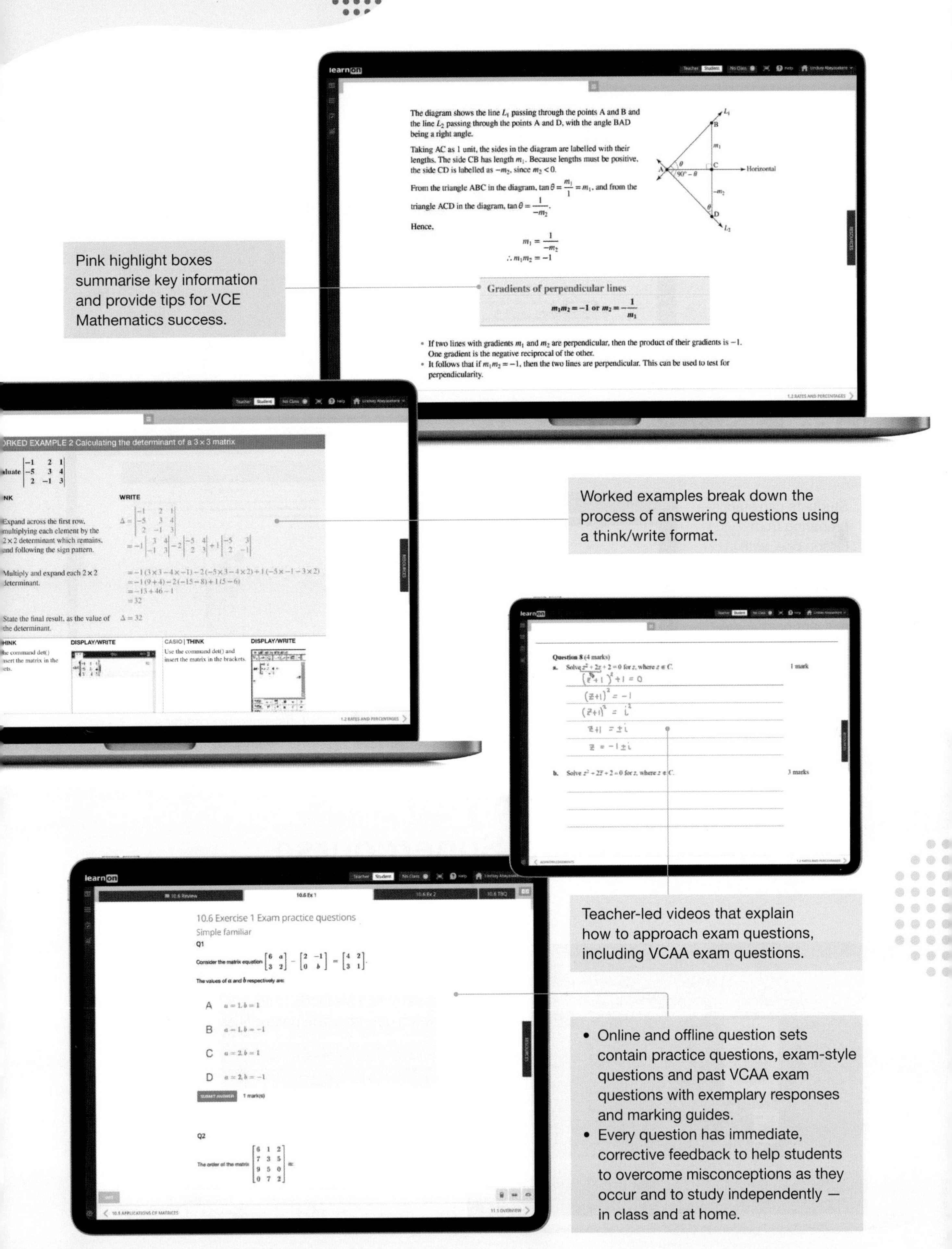

Pink highlight boxes summarise key information and provide tips for VCE Mathematics success.
Gradients of perpendicular lines
$m_1 m_2 = -1$ or $m_2 = -\frac{1}{m_1}$
Worked examples break down the process of answering questions using a think/write format.
Teacher-led videos that explain how to approach exam questions, including VCAA exam questions.
10.6 Exercise 1 Exam practice questions
Simple familiar
• Online and offline question sets contain practice questions, exam-style questions and past VCAA exam questions with exemplary responses and marking guides.
• Every question has immediate, corrective feedback to help students to overcome misconceptions as they occur and to study independently — in class and at home.

Topic reviews

Topic reviews include online summaries and topic-level review exercises that cover multiple concepts. Topic-level exam questions are structured just like the exams.

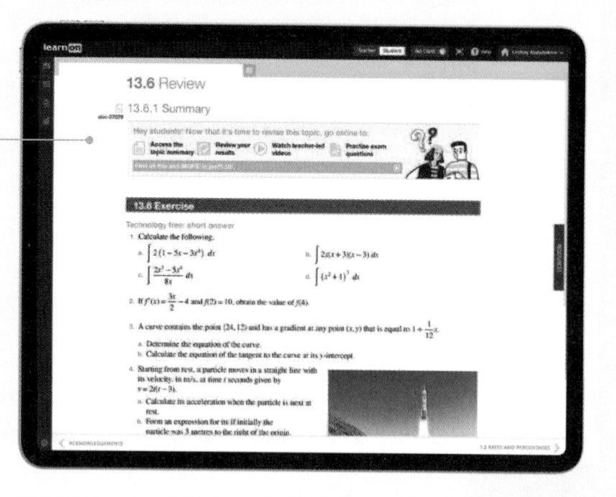

End-of-topic exam questions include relevant past VCE exam questions and are supported by teacher-led videos.

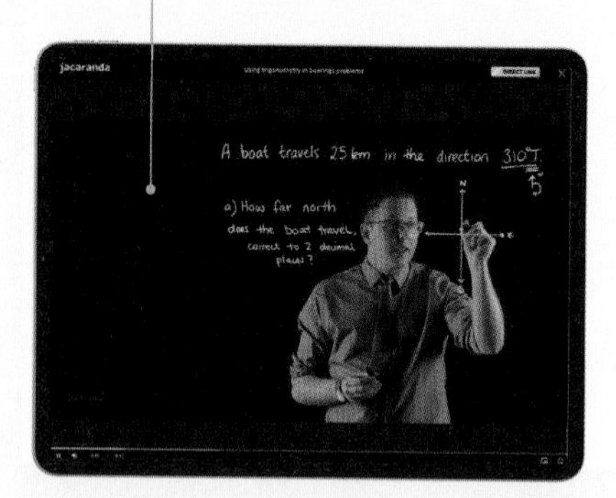

Get exam-ready!

Students can start preparing from lesson one, with exam questions embedded in every lesson — with relevant past VCAA exam questions since 2013.

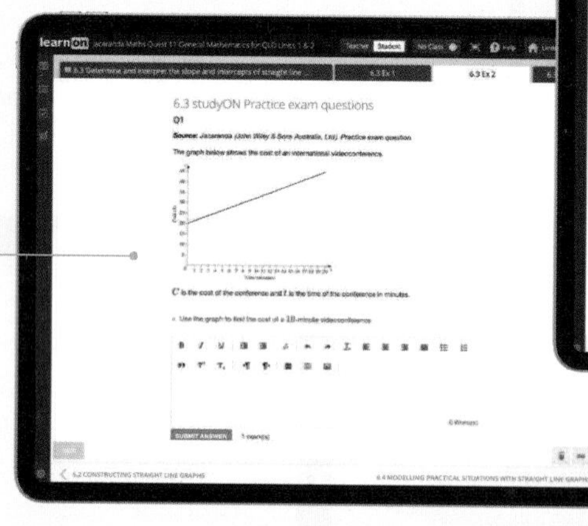

Customisable practice Mathematical investigations available to build student competence and confidence.

Combine units flexibly with the Jacaranda Supercourse

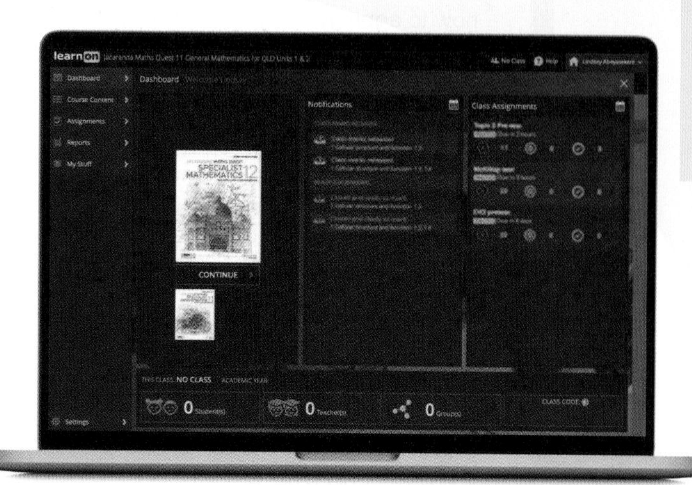

Build the course you've always wanted with the Jacaranda Supercourse. You can combine all Specialist Mathematics Units 1 to 4, so students can move backwards and forwards freely. Or combine General and Methods Units 1 & 2 for when students switch courses. The possibilities are endless!

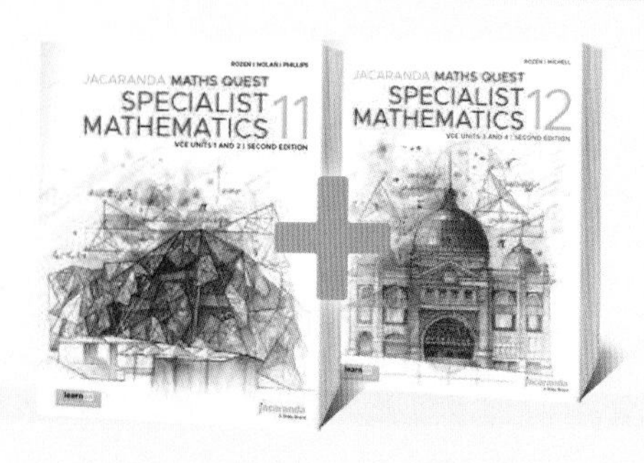

A wealth of teacher resources

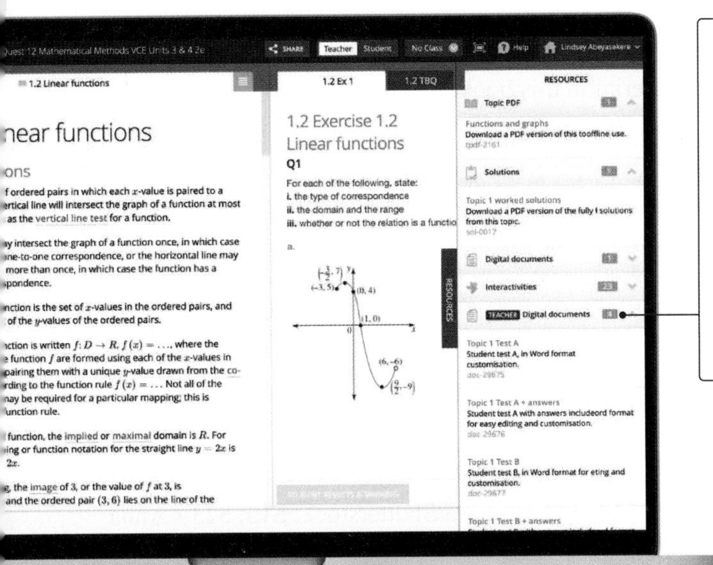

Enhanced teacher support resources, including:

- work programs and curriculum grids
- teaching advice and additional activities
- quarantined topic tests (with solutions)
- quarantined Mathematical investigations (with worked solutions and marking rubrics)

Customise and assign

A testmaker enables you to create custom tests from the complete bank of thousands of questions (including past VCAA exam questions).

Reports and results

Data analytics and instant reports provide data-driven insights into performance across the entire course.

Show students (and their parents or carers) their own assessment data in fine detail. You can filter their results to identify areas of strength and weakness.

Acknowledgements

The authors and publisher would like to thank the following copyright holders, organisations and individuals for their assistance and for permission to reproduce copyright material in this book.

Selected extracts from the VCE Mathematics Study Design (2023–2027) are copyright Victorian Curriculum and Assessment Authority (VCAA), reproduced by permission. VCE® is a registered trademark of the VCAA. The VCAA does not endorse this product and makes no warranties regarding the correctness and accuracy of its content. To the extent permitted by law, the VCAA excludes all liability for any loss or damage suffered or incurred as a result of accessing, using or relying on the content. Current VCE Study Designs and related content can be accessed directly at www.vcaa.vic.edu.au. Teachers are advised to check the VCAA Bulletin for updates.

Images

• © Andrew Rich/E+/Getty Images: **499** • © Science History Images/Alamy Stock Photo: **537** • robert mobley/Shutterstock: **588** • © frankies/Shutterstock: **503, 535, 544, 549** • © Orla/Shutterstock: **515, 523** • Anna Kireieva/Shutterstock: **165** • ArCaLu/Shutterstock: **368** • Brilliantist Studio/Shutterstock: **164** • Elena Veselova/Shutterstock: **746** • enciktepstudio/Shuterstock: **159** • GreenArt/Shutterstock: **149** • Kaewta Mungkung/Shutterstock: **154** • © 3d_illustrator/Shutterstock: **280** • © A and N photography/Shutterstock: **97** • © A.RICARDO/Shutterstock: **235** • © acid2728k/Shutterstock: **216** • © Action Sports Photography/Shutterstock: **69** • © Africa Studio/Shutterstock: **586** • © agsandrew/Shutterstock: **667** • © Aleksandar Todorovic/Shutterstock: **384** • © Alhovik/Shutterstock: **15** • © Alita Bobrov/Shutterstock: **606** • © Anatoliy Babiy/iStockphoto/Getty Images: **14** • © Andreas Nilsson/Shutterstock: **64** • © angelo gilardelli/Shutterstock: **113** • © Ann Kosolapova/Shutterstock: **451, 452, 466** • © Anton Balazh/Shutterstock.com: **449** • © artefacti/Shutterstock: **609** • © Asisyaj/Shutterstock: **2** • © BCFC/Shutterstock: **424** • © Berezovska Anastasia/Shutterstock: **405** • © blackboard1965/Shutterstock: **755** • © bonchan/Shutterstock: **773** • © brize99/Shutterstock: **425** • © Bruce Rolff/Shutterstock: **509** • © Byron W.Moore/Shutterstock: **118** • © Cbenjasuwan/Shutterstock: **467** • © cigdem/Shutterstock: **218** • © Corepics VOF/Shutterstock: **30** • © David Marchal/iStockphoto/Getty Images: **250** • © De Visu/Shutterstock: **450** • © Diana Taliun/Shutterstock: **621** • © Dima Sidelnikov/Shutterstock: **21** • © djumandji/Shutterstock: **210** • © donatas1205/Shutterstock: **29** • © DR Travel Photo and Video/Shutterstock: **469** • © Dudarev Mikhail/Shutterstock: **586** • © dvoevnore/Shutterstock: **627** • © Elegor/Shutterstock: **120** • © Elena Pominova/Shutterstock: **433** • © encikAn/Shutterstock: **103** • © Ewa Studio/Shutterstock: **439** • © Fasttailwind/Shutterstock: **458** • © file404/Shutterstock: • © FiledIMAGE**99**/Shutterstock: **283** • © filippo giuliani/Shutterstock: **25** • © Find the beauty/Shutterstock: **279** • © Florin Stana/Shutterstock: **384** • © Francesco Bonino/Shutterstock: • © g-stockstudio**63**/Shutterstock: **52** • © Goodluz/Shutterstock: **376** • © GrAl/Shutterstock: **612** • © graphit/Shutterstock: **770118** • © ImageFlow • © Image 100: /Shutterstock: **259** • © In-Finity/Shutterstock: **194** • © INTERFOTO/Personalities/Alamy Stock Photo: **190** • © Jessica L Archibald/Shutterstock: **581** • © Joe Belanger/Shutterstock: • © Joel Shawn**91**/Shutterstock: **580** • © Jon Tyson/Unsplash.com: • © Joseph M. Arseneau**111**/Shutterstock: **491** • © KalpanaBS/Shutterstock: **440** • © kentoh/Shutterstock: **613** • © Keo/Shutterstock: **600** • © Kkulikov/Shutterstock: **216** • © kovalto1/Shutterstock: **287, 300** • © Laralova/Shutterstock: • © Lenscap Photography/Shutterstock: **234** • © Lightspring/Shutterstock: **621** • © Liu zishan/Shutterstock: **578, 619** • © Lorna Roberts/Shutterstock: **1** • © MAGNIFIER/Shutterstock: **745** • © Maridav/Shutterstock: **46** • © Mariusz Niedzwiedzki/Shutterstock: **207** • © Maryna Kulchytska/Shutterstock: **420** • © Melanie Lemahieu/Shutterstock: **46** • © Michelangelus/Shutterstock: **269** • © Monkey Business Images/Shutterstock: **33** • © monkeybusinessimages/Getty Images: **11** • © mstanley/Shutterstock: • © muratart**112**/Shutterstock: **189** • © Neale Cousland/Shutterstock: **241** • © Nelson Charette Photo/Shutterstock: **497, 243** • © No formal credit line required: • © Olga Popova/Shutterstock: **406** • © PA Images/Alamy Stock Photo: **53** • © Palau/Shutterstock: **367, 378** • © Papik/Shutterstock: • © Paul

Bradbury/Getty Images: **14** • © Paulista/Adobe Stock: **628** • © Paulo Vilela/Shutterstock: **469** • © Pinky Rabbit/Shutterstock: **577** • © Pressmaster/Shutterstock: **31** • © pzUH/Shutterstock: • © raigvi/Shutterstock: **233** • © Raimundas/Shutterstock: **498** • © Rawpixel/Shutterstock: **768** • © Ruth Black/Shutterstock: **772** • © Sami Sarkis/PhotoDisc: Inc/Getty Images, **168** • © Santanor/Shutterstock: **400** • © schankz/Shutterstock: **78** • © serg_dibrova/Shutterstock: **13** • © Sergey Nivens/Shutterstock: **115** • © Sharlotta/Shutterstock: **421** • © silvae/Shutterstock: **489** • © simez78/Shutterstock: **54** • © Stacey Newman/Shutterstock: **397** • © svariophoto/Shutterstock: **238** • © Sveta/E+/Getty Images: **608** • © Syda Productions/Shutterstock: **123** • © teekid/iStock.com/Getty Images: **618** • © Thomas La Mela/Shutterstock: **96** • © tonymax/Getty Images: **584** • © urfin/Shutterstock: **747** • © Valerii Honcharuk/Adobe Stock: **389** • © Vasyl Shulga/Shutterstock: **168** • © Viewfinder Australia Photo Library: **166** • © Vorobyeva/Shutterstock: **773** • © worldswildlifewonders/Shutterstock: **100** • © Yuriy Belmesov/Shutterstock: **415, 434, 441** • © Zenphotography/Shutterstock: **433**

Every effort has been made to trace the ownership of copyright material. Information that will enable the publisher to rectify any error or omission in subsequent reprints will be welcome. In such cases, please contact the Permissions Section of John Wiley & Sons Australia, Ltd.

1 Combinatorics

LEARNING SEQUENCE

Fully worked solutions for this topic are available online.

1.1 Overview

Hey students! Bring these pages to life online

 Watch videos  Engage with interactivities  Answer questions and check results

Find all this and MORE in jacPLUS

1.1.1 Introduction

Ever since you were a small child, even before you started school, you have had the experience of counting. These simple counting techniques are built upon to develop very sophisticated ways of counting, in a field of mathematics known as combinatorics. Permutations and combinations allow us to calculate the number of ways objects belonging to a finite set can be arranged.

Have you encountered the pigeon-hole principle before? It is a deceptively simple statement that can be used to identify patterns in huge amounts of data, such as in DNA analysis.

Have you encountered Pascal's triangle before? It spans the mathematical fields of combinations, probability, the binomial theorem, Fibonacci numbers and the bell-shaped normal distribution. In developing his triangle, Blaise Pascal (1623–1662) made a fundamental contribution to the field of combinatorics.

In this topic you will apply combinatorics to determine, for example, the number of ways a team of 5 players can be chosen from a group of 10. Consider its usefulness in developing rosters for staff or flow charts for projects. Just as Pascal's triangle spans mathematical fields, combinatorics spans industries as varied as gambling, internet information transfer and security, communication networks, computer chip architecture, logistics and DNA modelling. It has applications in any field where different choices mean different efficiencies.

KEY CONCEPTS

This topic covers the following key concepts from the VCE Mathematics Study Design:

- the pigeon-hole principle and its use in solving problems and proving results
- the inclusion-exclusion principle for the union of two sets and the union of three sets
- permutations and combinations and their use in solving problems involving arrangements and selections with or without repeated elements
- derivation and application of simple combinatorial identities.

1.2 Counting techniques

LEARNING INTENTION

At the end of this subtopic you should be able to:
- apply the inclusion–exclusion, multiplication and addition principles to determine combinations and probabilities.

1.2.1 Review of set notation

A **set**, S, is a collection of objects. The objects in a set are referred to as the **elements** of the set.

A set can be written in a variety of ways. Consider the following. Let the sample space be the set of numbers between 1 and 20, that is $\{1, 2, 3, 4, 5, 6, 7, 8, 9, 10, 11, 12, 13, 14, 15, 16, 17, 18, 19, 20\}$.

Let the set S be the set of even numbers between 1 and 20 inclusive.

S can be:
- written as a list: $S = \{2, 4, 6, 8, 10, 12, 14, 16, 18, 20\}$ (in any order)
- written as a rule: $S = \{n : n = 2r \text{ for } 1 \leq r \leq 10\}$
- shown in a Venn diagram.

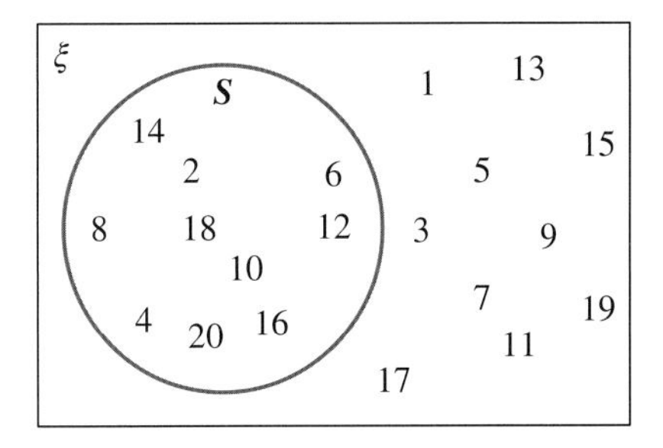

The complement of S, written as S', is the set of all things **not** in S. In this example, $S' = \{1, 3, 5, 7, 9, 11, 13, 15, 17, 19\}$.

The complete set of objects being considered is called the **universal set**, ξ. It is represented by the rectangle in the Venn diagram, and is abbreviated with the Greek letter ξ (pronounced 'ksi'). In this example, $\xi = \{1, 2, 3, 4, 5, 6, 7, 8, 9, 10, 11, 12, 13, 14, 15, 16, 17, 18, 19, 20\}$.

Now consider a second set, T, which is the set of numbers that are multiples of 3 between 1 and 20; that is, $T = \{3, 6, 9, 12, 15, 18\}$. The sets S and T can be combined in various ways.

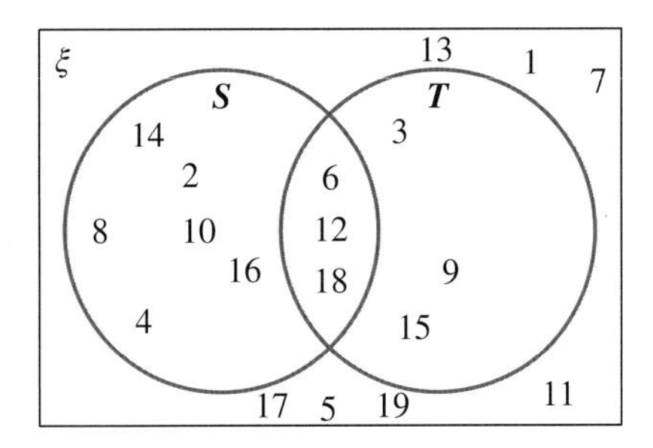

The **union** of S and T is all the elements in either S or T or both. It is shown as follows.

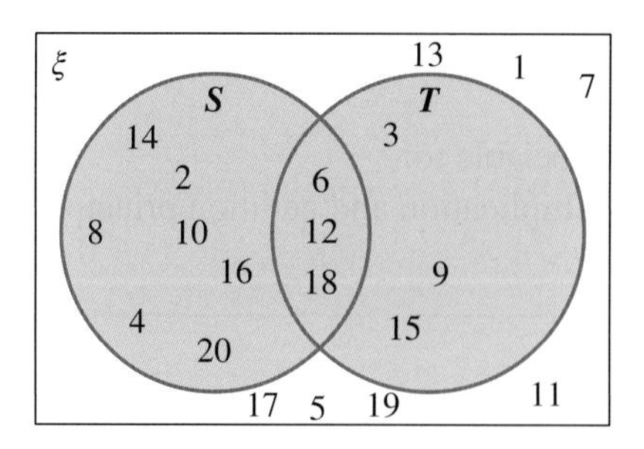

$$S \cup T = \{2, 3, 4, 6, 8, 9, 10, 12, 14, 15, 16, 18, 20\}$$

The **intersection** of S and T is all the elements that are in both S and T.

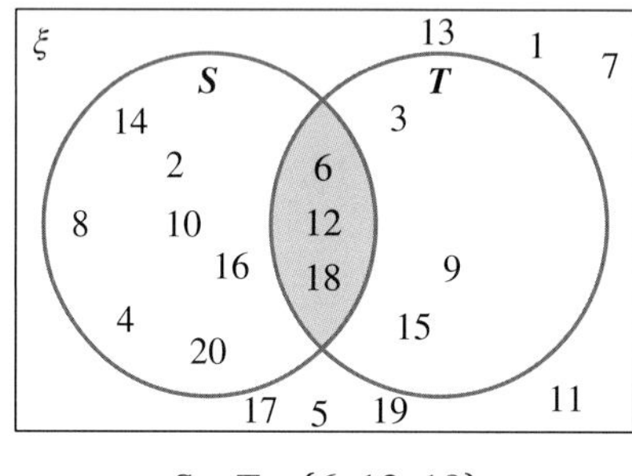

$$S \cap T = \{6, 12, 18\}$$

1.2.2 The inclusion–exclusion principle

Continuing the example from subsection 1.2.1, we have the two sets $S = \{2, 4, 6, 8, 10, 12, 14, 16, 18, 20\}$ and $T = \{3, 6, 9, 12, 15, 18\}$.

If we calculate the number of elements in the two sets, we obtain $n(S) = 10$ and $n(T) = 6$.

In general, for two sets S and T, $n(S \cup T) = n(S) + n(T) - n(S \cap T)$.

This is known as the **inclusion–exclusion principle**.

$n(S \cup T)$ is the number of elements in the union of S and T (i.e. the number of elements in S or T or both). This is equal to $n(S)$ (the number of elements in S) plus $n(T)$ (the number of elements in T) minus $n(S \cap T)$ (the number of elements in both S and T, as these have already been counted in sets S and T).

For our example above:

$$n(S \cup T) = n(S) + n(T) - n(S \cap T)$$

$$n(S) = 10,\ n(T) = 6,\ n(S \cap T) = 3$$

So, $n(S \cup T) = 13$, which can be confirmed by counting all the elements that occur in S, T or both.

If three sets are involved, the inclusion–exclusion principle becomes:

$$n(S \cup T \cup R) = n(S) + n(T) + n(R) - n(S \cap T) - n(T \cap R) - n(S \cap R) + n(S \cap T \cap R)$$

The inclusion–exclusion principle

For 2 sets, S and T:

$$n(S \cup T) = n(S) + n(T) - n(S \cap T)$$

For 3 sets, S, T and R:

$$n(S \cup T \cup R) = n(S) + n(T) + n(R) - n(S \cap T) - n(T \cap R) + n(S \cap T \cap R)$$

WORKED EXAMPLE 1 Applying the inclusion–exclusion principle

Let R be the set of natural numbers between 15 and 30 inclusive that are divisible by 2.
Let S be the set of natural numbers between 15 and 30 inclusive that are divisible by 3.
Let T be the set of natural numbers between 15 and 30 inclusive that are divisible by 5.

a. **Construct a Venn diagram to represent R, S and T.**
b. **Use this diagram to evaluate $n(R \cup S \cup T)$.**
c. **Recall the inclusion–exclusion principle to compute $n(R \cup S \cup T)$.**

THINK

a. 1. R is the set of natural numbers between 15 and 30 divisible by 2, so:
$R = \{16, 18, 20, \ldots 30\}$

WRITE

a.

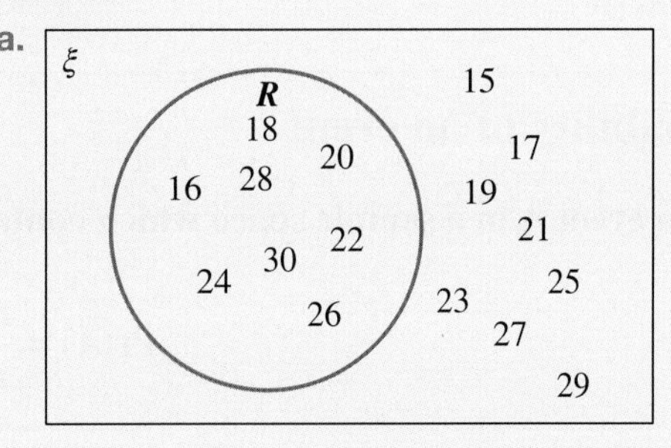

2. S is the set of natural numbers between 15 and 30 divisible by 3, so:
$S = \{15, 18, 21, 24, 27, 30\}$
$R \cap S$ are the numbers that occur in both S and R, so:
$R \cap S = \{18, 24, 30\}$

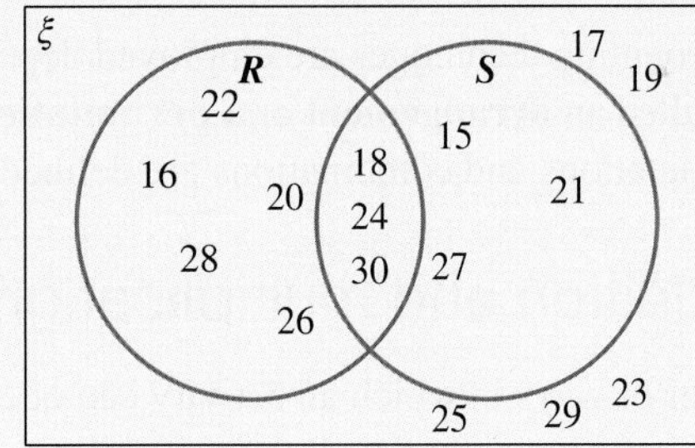

3. T is the set of natural numbers between 15 and 30 divisible by 5, so:
$T = \{15, 20, 25, 30\}$
$R \cap T = \{20, 30\}$
$S \cap T = \{15, 30\}$
$R \cap S \cap T = \{30\}$

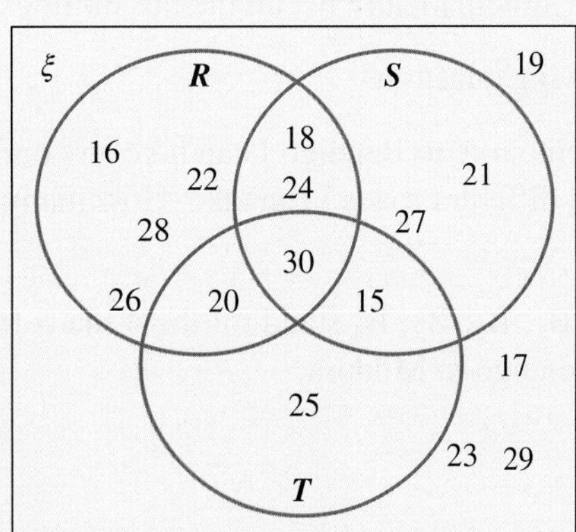

b. $n(R \cup S \cup T)$ is the number of elements in sets R, S and T. Count the number of elements within each portion of each circle.	**b.** $n(R \cup S \cup T) = 4+2+1+2+1+1+1$ $= 12$
c. **1.** Recall the inclusion–exclusion principle formula.	**c.** $n(S \cup T \cup R) = n(S) + n(T) + n(R) - n(S \cap T) - n(T \cap R) - n(S \cap R) + n(S \cap T \cap R)$
2. The number of elements in each set can be substituted into the formula.	$n(S \cup T \cup R) = 6+4+8-2-2-3+1$ $= 12$ This agrees with the calculation from the Venn diagram in part **b**.

1.2.3 Types of counting techniques

Counting techniques allow us to determine the number of ways an activity can occur. This in turn allows us to calculate the **probability** of an event. Recall from your earlier probability studies that the theoretical probability of event A, $\Pr(A)$, can be determined by counting the number of elements in A and dividing by the total number in the sample space, ξ. Note that this only applies in cases where each outcome in the sample space is equally likely.

Probability of an event

For an event A in a sample space which contains outcomes which are equally likely:

$$\Pr(A) = \frac{n(A)}{n(\xi)}$$

Different types of counting techniques are employed depending on whether order is important. When order is important, this is called an **arrangement** or a **permutation**; when it is not important, it is called a **selection** or a **combination**. Permutations and combinations are defined more formally in sections 1.3 and 1.5.

1.2.4 The addition and multiplication principles

To count the number of ways in which an activity can occur, first make a list. Let each outcome be represented by a letter and then systematically list all the possibilities.

Consider the following question:

In driving from Melbourne to Bendigo I can take any one of 4 different roads *and* in driving from Bendigo to Mildura there are 3 different roads I can take. How many different routes can I take in driving from Melbourne to Mildura?

To answer this, let B_1, B_2, B_3, B_4 stand for the 4 roads from Melbourne to Bendigo and M_1, M_2, M_3 stand for the 3 roads from Bendigo to Mildura.

Use the figure to systematically list the roads:

B_1M_1, B_1M_2, B_1M_3

B_2M_1, B_2M_2, B_2M_3

B_3M_1, B_3M_2, B_3M_3

B_4M_1, B_4M_2, B_4M_3

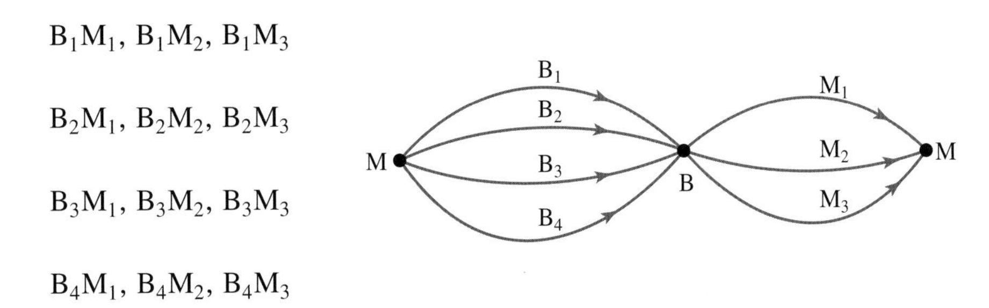

Hence, there are 12 different ways I can drive from Melbourne to Mildura.

In the above example it can be argued logically that if there are 4 ways of getting from Melbourne to Bendigo and 3 ways of getting from Bendigo to Mildura, then there are 4×3 ways of getting from Melbourne to Mildura.

This idea is formalised in the **multiplication principle**.

The multiplication principle should be used when there are operations or events (say, A and B), where one event is followed by the other – that is, when order is important.

The multiplication principle

If there are *n* ways of performing operation A and *m* ways of performing operation B, then there are $n \times m$ ways of performing A *and* B in the order AB.

A useful technique for solving problems based on the multiplication principle is to use boxes. In the example above we would write the following.

1st	2nd
4	3

The value in the '1st' column represents the number of ways the first operation — the trip from Melbourne to Bendigo — can be performed.

The value in the '2nd' column stands for the number of ways the second operation — the trip from Bendigo to Mildura — can be performed.

To apply the multiplication principle you multiply the numbers in the lower row of boxes.

WORKED EXAMPLE 2 Applying the multiplication principle (1)

Two letters are to be chosen from the set of 5 letters. A, B, C, D and E, where order is important.
a. Recall how to list all the different ways that this may be done.
b. Use the multiplication principle to calculate the number of ways that this may be done.
c. Determine the probability the first letter will be a C.

THINK	WRITE
a. 1. Begin with A in first place and make a list of each of the possible pairs.	a. AB AC AD AE

2. Make a list of each of the possible pairs with B in the first position.

BA BC BD BE

3. Make a list of each of the possible pairs with C in the first position.

CA CB CD CE

4. Make a list of each of the possible pairs with D in the first position.

DA DB DC DE

5. Make a list of each of the possible pairs with E in the first position.
Note: AB and BA need to be listed separately as order is important.

EA EB EC ED

b. The multiplication principle could have been used to determine the number of ordered pairs.

1. Rule up two boxes that represent the pair.

b.

5	4

2. Write down the number of letters that may be selected for the first box. That is, in first place any of the 5 letters may be used.
3. Write down the number of letters that may be selected for the second box. That is, in second place, any of the 4 letters may be used.
Note: One less letter is used to avoid repetition.
4. Evaluate.

$5 \times 4 = 20$ ways

5. Answer the question.

There are 20 ways in which 2 letters may be selected from a group of 5 where order is important.

c. 1. Recall the probability formula. The total number in the set $n(\xi)$ was determined in part **b**.

c. $\Pr(A) = \dfrac{n(A)}{n(\xi)}$

$n(\xi) = 20$

2. Let A be the event that the pair starts with a C. Draw a table showing the requirement imposed by the first letter to be C.

1	

3. Complete the table. Once the first letter has been completed, there are 4 choices for the second letter. Use the multiplication principle to determine the number of combinations starting with C.

1	4

There are $1 \times 4 = 4$ possible combinations beginning with C.
So, $n(A) = 4$.

4. Use the probability formula to answer the question.

$$\Pr(A) = \frac{n(A)}{n(\xi)} = \frac{4}{20} = \frac{1}{5}$$

This is confirmed by examining the answer to part **a**.

WORKED EXAMPLE 3 Applying the multiplication principle (2)

a. Use the multiplication principle to calculate how many ways an arrangement of 5 numbers can be chosen from {1, 2, 3, 4, 5, 6}.

b. Determine the probability of the number ending with 4.

THINK

WRITE

a. 1. Instead of listing all possibilities, draw 5 boxes to represent the 5 numbers chosen.
Label each box on the top row as 1st, 2nd, 3rd, 4th and 5th.
Note: The word 'arrangement' implies order is important.

a.

2. Fill in each of the boxes showing the number of ways a number may be chosen.
In the 1st box there are 6 choices for the first number. In the 2nd box there are 5 choices for the second number as 1 number has already been used. In the 3rd box there are 4 choices for the third number as 2 numbers have already been used. Continue this process until each of the 5 boxes is filled.

1st	2nd	3rd	4th	5th
6	5	4	3	2

3. Use the multiplication principle as this is an *'and'* situation.

$$\begin{aligned}\text{No. of ways} &= 6 \times 5 \times 4 \times 3 \times 2 \\ &= 720\end{aligned}$$

4. Answer the question.

An arrangement of 5 numbers may be chosen 720 ways.

b. 1. Recall the probability formula.
The total number of arrangements, $n(\xi)$, was determined in part a.

b. $\Pr(A) = \dfrac{n(A)}{n(\xi)}$

$n(\xi) = 720$

2. Let A be the event that the number ends with 4.
Draw a table showing the requirement imposed by the last letter to be 4.

1st	2nd	3rd	4th	5th
				1

3. Complete the table. Once the last number has been completed, there are 5 choices for the number in the first position and 4 choices for the next number. Continue this process until each of the 5 columns has been filled. Use the multiplication principle to determine the number of combinations ending with 4.

1st	2nd	3rd	4th	5th
5	4	3	2	1

There are $5 \times 4 \times 3 \times 2 \times 1 = 120$ possible combinations ending with 4.
So, $n(A) = 120$.

4. Use the probability formula to answer the question.

$$\begin{aligned}\Pr(A) &= \frac{n(A)}{n(\xi)} \\ &= \frac{120}{720} \\ &= \frac{1}{6}\end{aligned}$$

This is confirmed by examining the answer to part a.

Now consider a different situation, one in which the two operations do not occur one after the other.

I am going to travel from Melbourne to either Sydney *or* Adelaide. There are 4 ways of travelling from Melbourne to Sydney and 3 ways of travelling from Melbourne to Adelaide.

How many different ways can I travel to either Sydney *or* Adelaide?

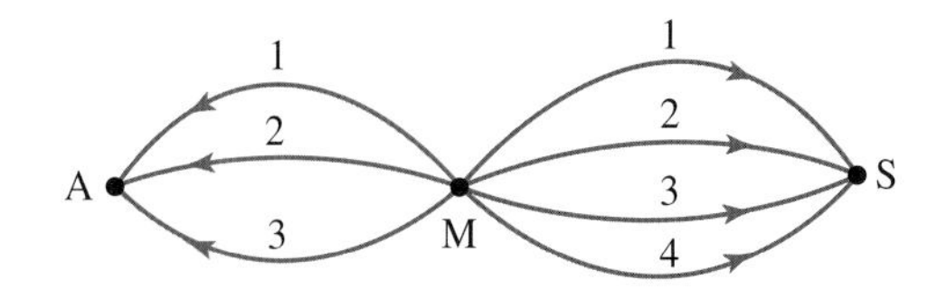

It can be seen from the figure that there are $4 + 3 = 7$ ways of completing the journey. This idea is summarised in the **addition principle**.

The addition principle should be used when two distinct operations or events occur in which one event is not followed by another — that is, when the events are mutually exclusive.

The addition principle

If there are n ways of performing operation A and m ways of performing operation B, then there are $n + m$ ways of performing A *or* B.

WORKED EXAMPLE 4 Applying the addition principle

One or two letters are to be chosen from the set of 6 letters A, B, C, D, E, F. Assuming order is important, use the multiplication principle and the addition principle to calculate:

a. **the number of ways to choose 2 letters**

b. **the number of ways to choose 1 or 2 letters.**

THINK	WRITE
a. 1. Determine the number of ways of choosing 1 letter.	a. Number of ways of choosing 1 letter $= 6$
2. Rule up two boxes for the first and second letters.	1st: 6, 2nd: 5
3. Determine the number of ways of choosing 2 letters from 6. In the 1st box there are 6 choices for the first letter. In the 2nd box there are 5 choices for the second letter as 1 letter has already been used.	
4. Use the multiplication principle (as this is an '*and*' situation) to evaluate the number of ways of choosing 2 letters from 6.	Number of ways of choosing 2 letters $= 6 \times 5$ $= 30$
5. Answer the questions.	There are 30 ways of choosing 2 letters.
b. 1. Determine the number of ways of choosing 1 or 2 letters from 6 letters. Use the addition principle as this is an '*or*' situation.	b. The number of ways of choosing 1 or 2 letters is $6 + 30 = 36$.
2. Answer the question.	There are 36 ways of choosing 1 or 2 letters from 6.

The multiplication and addition principles can be used to count the number of elements in the union of two or three sets in the same way as for one set. Remember, the multiplication principle is used with '*and*' situations, and the addition principle is used with '*or*' situations.

WORKED EXAMPLE 5 Application of the addition and multiplication principles

Oscar's cafe offers a choice of 3 starters, 9 main courses and 4 desserts.

a. **Determine how many choices of 3-course meals (starter, main, dessert) are available.**

b. **Determine how many choices of starter and main course meals are offered.**

c. **Determine how many choices of meals comprising a main course and dessert are offered.**

d. **Determine how many choices of 2- or 3-course meals are available (assuming that a main course is always ordered).**

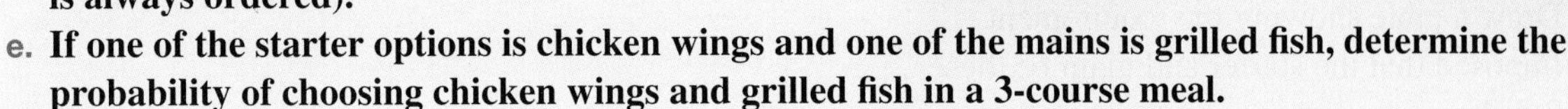

e. **If one of the starter options is chicken wings and one of the mains is grilled fish, determine the probability of choosing chicken wings and grilled fish in a 3-course meal.**

THINK / **WRITE**

a. 1. Consider each course as separate sets and rule up 3 boxes to represent each course — starter, main, dessert. Label each box on the top row as S, M and D.

2. Determine the number of ways of choosing each meal: starter = 3, main = 9, dessert = 4.

a.

S	M	D
3	9	4

3. Use the multiplication principle (as this is an '*and*' situation) to evaluate the number of choices of 3-course meals.

$$\text{Number of choices} = 3 \times 9 \times 4$$
$$= 108$$

4. Answer the question.

There are 108 choices of 3-course meals.

b. 1. Rule up 2 boxes to represent each course — starter, main. Label each box on the top row as S and M.

2. Determine the number of ways of choosing each meal: starter = 3, main = 9.

b.

S	M
3	9

3. Use the multiplication principle (as this is an '*and*' situation) to evaluate the number of choices of starter and main courses.

$$\text{Number of choices} = 3 \times 9$$
$$= 27$$

4. Answer the question.

There are 27 choices of starter and main course.

c. 1. Rule up 2 boxes to represent each course — main and dessert. Label each box on the top row as M and D.

2. Determine the number of ways of choosing each meal: main = 9, dessert = 4.

c.

M	D
9	4

3. Use the multiplication principle (as this is an '*and*' situation) to evaluate the number of choices of main course and dessert.

$$\text{Number of choices} = 9 \times 4$$
$$= 36$$

4. Answer the question.	There are 36 choices of main course and dessert.
d. 1. Determine the number of ways of choosing 2- or 3-course meals, assuming that a main course is always ordered. Use the addition principle as this is an '*or*' situation.	**d.** The number of ways of choosing 2- or 3-course meals, assuming that a main course is always ordered, is: $108 + 27 + 36 = 171$
2. Answer the question.	There are 171 ways of choosing 2- or 3-course meals, assuming that a main course is always ordered.
e. 1. Recall the probability formula. The total number in the set $n(\xi)$ was determined in part **a**.	**e.** $\Pr(A) = \dfrac{n(A)}{n(\xi)}$ $n(\xi) = 108$
2. Let A be the event that the meal contains chicken wings and grilled fish. Draw a table showing the requirement imposed that the starter and main be these 2 dishes.	S | M | D 1 | 1 |
3. Complete the table — there are 4 dessert options. Use the multiplication principle to determine the number of combinations of the meal.	S | M | D 1 | 1 | 4 There are $1 \times 1 \times 4 = 4$ possible combinations for the meal. So, $n(A) = 4$.
4. Use the probability formula to answer the question.	$\Pr(A) = \dfrac{n(A)}{n(\xi)} = \dfrac{4}{108} = \dfrac{1}{27}$ The probability of choosing chicken wings and grilled fish as part of the 3-course meal is $\dfrac{1}{27}$.

1.2 Exercise

Students, these questions are even better in jacPLUS

Receive immediate feedback and access sample responses

Access additional questions

Track your results and progress

Find all this and MORE in jacPLUS

Technology free

1. WE1 Let R be the set of natural numbers between 30 and 45 inclusive that are divisible by 2.
   Let S be the set of natural numbers between 30 and 45 inclusive that are divisible by 3.
   Let T be the set of natural numbers between 30 and 45 inclusive that are divisible by 5.
   a. Construct a Venn diagram to represent R, S and T.
   b. Use this diagram to evaluate $n(R \cup S \cup T)$.
   c. Recall the inclusion–exclusion principle to compute $n(R \cup S \cup T)$.

2. Recall the inclusion–exclusion principle to calculate the number of cards in a deck of 52 that are either red or even or a 4.
3. Student Services has the following data on Year 11 students and their sport commitments:
   - 18 play no sport.
   - 16 play netball (and possibly other sports).
   - 24 play football.
   - 20 are involved in a gym program.
   - 7 play netball and football.
   - 6 play netball and are in the gym program.
   - 15 play football and are involved in the gym program.
   - 5 students do all three activities.

   Determine how many students there are in Year 11.
4. WE2 Two letters are to be chosen from A, B and C, where order is important.
   a. Recall how to list all the different ways that this may be done.
   b. Use the multiplication principle to calculate the number of ways that this may be done.
   c. Determine the probability the last letter will be a B.
5. List all the different arrangements possible for a group of 2 colours to be chosen from B (blue), G (green), Y (yellow) and R (red).
6. List all the different arrangements possible for a group of 3 letters to be chosen from A, B and C.
7. a. WE3 Use the multiplication principle to calculate how many ways an arrangement of 2 letters can be chosen from A, B, C, D, E, F and G.
   b. Calculate how many ways an arrangement of 3 letters can be chosen from 7 different letters.
   c. Calclate how many ways an arrangement of 4 letters can be chosen from 7 different letters.
   d. Calculate how many different arrangements of 5 letters can be made from 7 letters.
   e. Determine the probability of the letters starting with an E.

8. a. A teddy bear's wardrobe consists of 3 different hats, 4 different shirts and 2 different pairs of pants. Determine how many different outfits the teddy bear can wear.
   b. A surfboard is to have 1 colour on its top and a different colour on its bottom. The 3 possible colours are red, blue and green. Caclulate how many different ways the surfboard can be coloured.
   c. A new phone comes with a choice of 3 cases, 2 different-sized screens and 2 different storage capacities. With these choices, determine how many different arrangements are possible.
   d. Messages can be sent by placing 3 different coloured flags in order on a pole. If the flags come in 4 colours, determine how many different messages can be sent.

9. a. **WE4** One or two letters are to be chosen in order from the letters A, B, C, D, E, F and G. Use the multiplication principle and the addition principle to calculate the number of ways can this be done.
   b. Two or three letters are to be chosen in order from the letters A, B, C, D, E, F and G. Determine how many ways this can be done.

10. Manish is in a race with 7 other runners. If we are concerned only with the first, second and third placings, determine how many ways Manish can finish first or second or third.

11. **WE5** Hani and Mary's restaurant offers its patrons a choice of 4 entrees, 10 main courses and 5 desserts.
   a. Determine how many choices of 3-course meals (entree, main, dessert) are available.
   b. Determine how many choices of entree and main course are offered.
   c. Determine how many choices of meals comprising a main course and dessert are offered.
   d. Determine how many choices of 2- or 3-course meals are available (assuming that a main course is always ordered).
   e. Determine the probability of choosing vegetable soup for entree and roast for main in a 3-course meals.

12. Jake is able to choose his work outfits from the following items of clothing: 3 jackets, 7 shirts, 6 ties, 5 pairs of trousers, 7 pairs of socks and 3 pairs of shoes.
   a. Calculate how many different outfits are possible if he wears one of each of the above items. (He wears matching socks and matching shoes.)
   b. If Jake has the option of wearing a jacket or not, but he must wear one of each of the above items, determine how many different outfits are possible. Justify your answer.

13. The local soccer team sells 'doubles' at each of their games to raise money. A 'double' is a card with 2 digits on it representing the score at full time. The card with the actual full time score on it wins a prize. If the digits on the cards run from 00 to 99, determine how many different tickets there are.

14. Jasmin has a phone that has a 4-digit security code. She remembers that the first number in the code was 9 and that the others were 3, 4 and 7 but forgets the order of the last 3 digits. Determine how many different trials she must make to be sure of unlocking the phone.

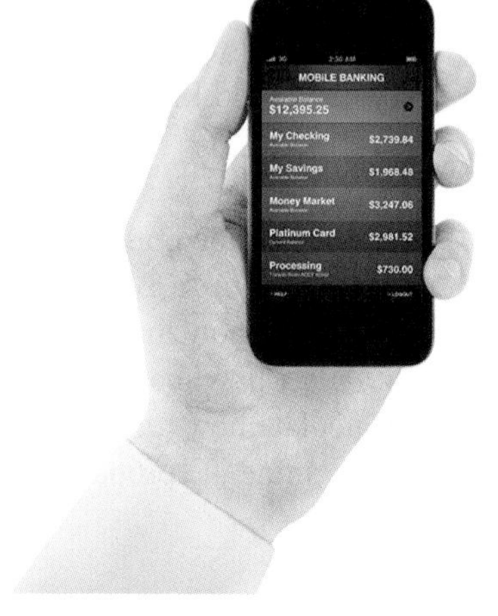

15. Julia has a banking app that has two 4-digit codes. She remembers that she used the digits 1, 3, 5 and 7 on the first code and 2, 4, 6 and 8 on the second code, but she cannot remember the order. Determine the maximum number of trials she would need to make before she has opened both codes. (Assume that she can try an unlimited number of times and once the first code is correct, she can try the second code.)

16. Determine how many different 4-digit numbers can be made from the numbers 1, 3, 5 and 7 if the numbers can be repeated (that is 3355 and 7777 are valid).

17. Determine how many 4-digit numbers can be made from the numbers 1, 3, 5, 7, 9 and 0 if the numbers can be repeated. (Remember — a number cannot start with 0.)

18. A combination lock has 3 digits each from 0 to 9.

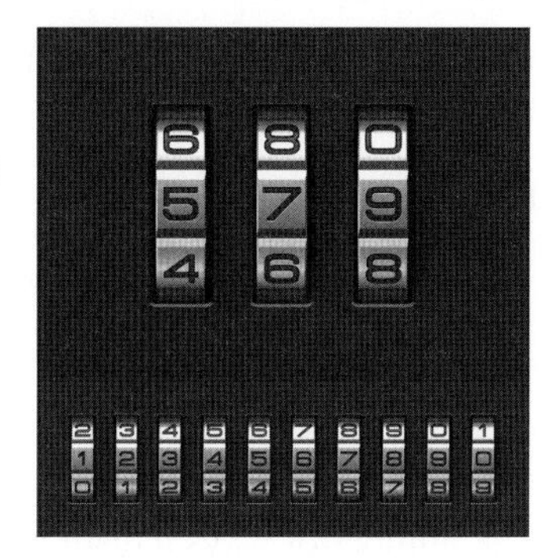

a. Determine how many combinations are possible.

The lock mechanism becomes loose and will open if the digits are within one either side of the correct digit. For example, if the true combination is 382, then the lock will open on 271, 272, 371, 493 and so on.

b. Determine how many combinations would unlock the safe.

c. List the possible combinations that would open the lock if the true combination is 382.

1.2 Exam questions

Question 1 (1 mark) TECH-ACTIVE

MC $A=\{3, 4, \pi, 7\}, B=\{1, 4, 5, 9\}, C=\{n: n=3r \text{ for } 1 \leq r \leq 4, r \in Z\}$. $(A \cap B) \cup C$ equals

A. ϕ **B.** $\{4\}$ **C.** $\{3,6,9\}$ **D.** $\{3,4,6,9\}$ **E.** $\{3,4,6,9,12\}$

Question 2 (1 mark) TECH-FREE

Michael is buying a sound system for his music studio. He has a choice of 6 pairs of speakers, 4 amplifiers, 3 DJ controllers and 2 turntables. Calculate the number of different systems possible if he chooses one of each type of component.

Question 3 (1 mark) TECH-FREE

Janelle must choose 1 first semester unit from 3 units of geography and 2 units of mathematics and 1 second semester unit from 2 units of English, 3 units of science and 3 units of IT. Determine how many different 2-units courses are possible.

More exam questions are available online.

1.3 Factorials and permutations

LEARNING INTENTION

At the end of this subtopic you should be able to:
- calculate permutations using factorials
- calculate the number of arrangements of a group of objects arranged in a circle.

1.3.1 Factorials

The Physical Education department is to display 5 new trophies along a shelf in the school foyer and wishes to know in how many ways this can be done.

Using the multiplication principle from the previous section, the display may be done in the following way:

Position 1	Position 2	Position 3	Position 4	Position 5
5	4	3	2	1

That is, there are $5 \times 4 \times 3 \times 2 \times 1 = 120$ ways.

Depending on the number of items we have, this method could become quite time consuming.

In general when we need to multiply each of the integers from a particular number, n, down to 1, we write $n!$, which is read as n **factorial**.

Hence:

$$\begin{aligned} 6! &= 6 \times 5 \times 4 \times 3 \times 2 \times 1 \\ &= 720 \\ 8! &= 8 \times 7 \times 6 \times 5 \times 4 \times 3 \times 2 \times 1 \\ &= 40\,320 \\ n! &= n \times (n-1) \times (n-2) \times (n-3) \times \ldots \times 3 \times 2 \times 1 \end{aligned}$$

The number of ways n distinct objects may be arranged is $n!$ (n factorial).

Factorials

$$n! = n \times (n-1) \times (n-2) \times (n-3) \times \ldots \times 3 \times 2 \times 1$$

That is, $n!$ is the product of each of the integers from n down to 1.

A special case of the factorial function is $0! = 1$.

WORKED EXAMPLE 6 Evaluating factorials

Evaluate the following factorials.

a. $4!$ b. $7!$ c. $\frac{8!}{5!}$ d. $\frac{(n-1)!}{(n-3)!}$

THINK	WRITE
a. 1. Write 4! in its expanded form and evaluate.	a. $4! = 4 \times 3 \times 2 \times 1$ $= 24$
2. Verify the answer obtained using the factorial function on a calculator.	
b. 1. Write 7! in its expanded form and evaluate.	b. $7! = 7 \times 6 \times 5 \times 4 \times 3 \times 2 \times 1$ $= 5040$
2. Verify the answer obtained using the factorial function on a calculator.	
c. 1. Write each factorial term in its expanded form.	c. $\frac{8!}{5!} = \frac{8 \times 7 \times 6 \times 5 \times 4 \times 3 \times 2 \times 1}{5 \times 4 \times 3 \times 2 \times 1}$
2. Cancel down like terms.	$= 8 \times 7 \times 6$
3. Evaluate.	$= 336$
4. Verify the answer obtained using the factorial function on a calculator.	
d. 1. Write each factorial term in its expanded form.	d. $\frac{(n-1)!}{(n-3)!} = \frac{(n-1)(n-2)(n-3)(n-4) \times \ldots \times 3 \times 2 \times 1}{(n-3)(n-4) \times \ldots \times 3 \times 2 \times 1}$
2. Cancel like terms.	$= (n-1)(n-2)$

TI | THINK

c. On a Calculator page complete the entry line as shown.

DISPLAY/WRITE

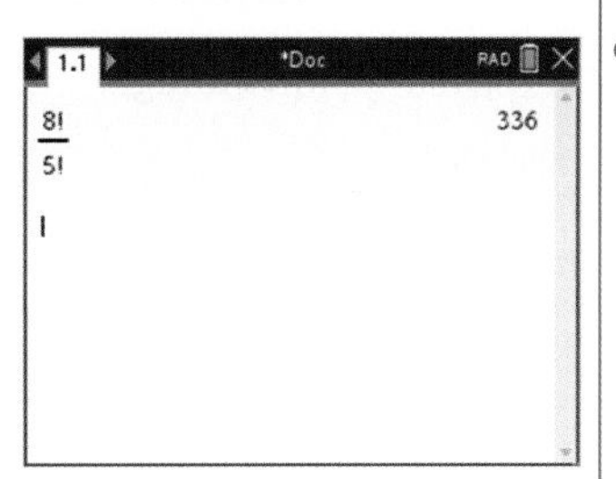

CASIO | THINK

c. From the Advance palette on the Keyboard, complete the entry line as shown by selecting !

DISPLAY/WRITE

$\frac{8!}{5!}$ 336

In parts **c** and **d** of Worked example 6, there was no need to fully expand each factorial term.

The factorial $\frac{8!}{5!}$ could have first been simplified to $\frac{8 \times 7 \times 6 \times 5!}{5!}$ and then the 5! terms cancelled.

The factorial $\frac{(n-1)!}{(n-3)!}$ could have first been simplified to $\frac{(n-1)(n-2)(n-3)!}{(n-3)!}$ and then the $(n-3)!$ terms cancelled.

1.3.2 Permutations

The term permutation is often used instead of the term arrangement, and in this section we begin by giving a formal definition of permutation.

Previously, we learned that if you select 3 letters from 7 where *order is important*, the number of possible arrangement is:

1st	2nd	3rd
7	6	5

$$\begin{aligned}\text{The number of arrangements} &= 7 \times 6 \times 5 \\ &= 210\end{aligned}$$

This value may also be expressed in factorial form: $7 \times 6 \times 5 = \frac{7 \times 6 \times 5 \times 4!}{4!} = \frac{7!}{4!}$

Using more formal terminology we say that in choosing 3 things from 7 things where order is important, the number of permutations is ${}^7P_3 = 7 \times 6 \times 5$. The letter P is used to remind us that we are finding permutations.

The number of ways of choosing ***r*** things from ***n*** distinct things is given by the following rule.

Permutations

$$\begin{aligned}{}^n\mathbf{P}_r &= n \times (n-1) \times \ldots \times (n-r+1) \\ &= \frac{n \times (n-1) \times \ldots \times (n-r+1)(n-r)!}{(n-r)!} \\ &= \frac{n!}{(n-r)!}\end{aligned}$$

The definition of ${}^n\mathrm{P}_r$ may be extended to the cases of ${}^n\mathrm{P}_n$ and ${}^n\mathrm{P}_0$.

${}^n\mathrm{P}_n$ represents the number of ways of choosing n objects from n distinct things.

$$\begin{aligned}{}^n\mathrm{P}_n &= n \times (n-1) \times (n-2) \times \ldots \times (n-n+1)\\ &= n \times (n-1) \times (n-2) \times \ldots \times 1\\ &= n!\end{aligned}$$

From the definition:

$$\begin{aligned}{}^n\mathrm{P}_n &= \frac{n!}{(n-n)!}\\ &= \frac{n!}{0!}\end{aligned}$$

Therefore, equating both sides, we obtain: $n! = \frac{n!}{0!}$.

This can occur only if $0! = 1$.

$$\begin{aligned}{}^n\mathrm{P}_0 &= \frac{n!}{(n-0)!}\\ &= \frac{n!}{n!}\\ &= 1\end{aligned}$$

Special cases of permutations

$$\mathbf{{}^n P_n = n!}$$

$$\mathbf{{}^n P_0 = 1}$$

WORKED EXAMPLE 7 Calculating permutations

a. Calculate the number of permutations for ${}^6\mathrm{P}_4$ by expressing it in expanded form.
b. Write ${}^8\mathrm{P}_3$ as a quotient of factorials and hence evaluate.

THINK	WRITE
a. 1. Write down the first 4 terms beginning with 6.	a. ${}^6\mathrm{P}_4 = 6 \times 5 \times 4 \times 3$
2. Evaluate.	$= 360$
b. 1. Recall the rule for permutations.	b. ${}^n\mathrm{P}_r = \frac{n!}{(n-r)!}$
2. Substitute the given values of n and r into the permutation formula.	${}^8\mathrm{P}_3 = \frac{8!}{(8-3)!}$ $= \frac{8!}{5!}$
3. Use a calculator to evaluate 8! and 5!	$= \frac{40\,320}{120}$
4. Evaluate.	$= 336$

TI | THINK

a. On a Calculator page, press MENU, then select:
5: Probability
2: Permutations.
Complete the entry line as nPr(6,4).

DISPLAY/WRITE

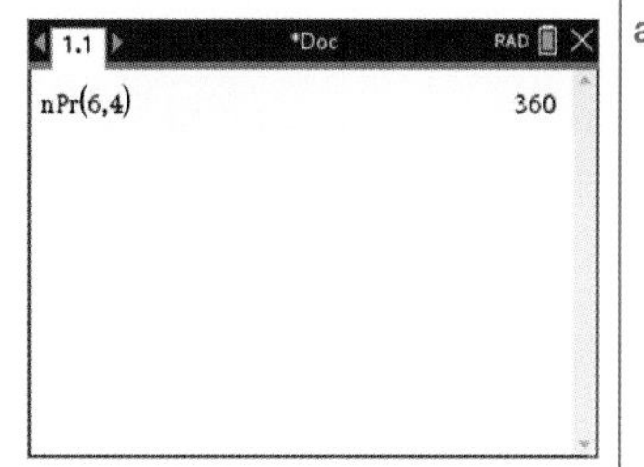

CASIO | THINK

a. From the Advance palette on the Keyboard, select nPr. Complete the entry line as shown.

DISPLAY/WRITE

WORKED EXAMPLE 8 Using permutations to calculate probabilities

The netball club needs to appoint a president, secretary and treasurer. From the committee 7 people have volunteered for these positions. Each of the 7 nominees is happy to fill any one of the 3 positions.

a. Determine how many different ways these positions can be filled.

b. For three years, the same 7 people volunteer for these positions. Determine the probability one of them is president 3 years in a row.

THINK

a. 1. Recall the rule for permutations.
Note: Order is important, so use permutations.

2. Substitute the given values of n and r into the permutation formula.

3. Use a calculator to evaluate 7! and 4!

4. Evaluate.

5. Answer the question.

b. 1. In three years of the president's position, each year this could be awarded to one of the 7 people.

2. Let A be the event that the same person fills the position three years in a row.

WRITE

a. $^{n}\text{P}_{r} = \dfrac{n!}{(n-r)!}$

$^{7}\text{P}_{3} = \dfrac{7!}{(7-3)!}$

$= \dfrac{7!}{4!}$

$= \dfrac{5040}{24}$

$= 210$

There are 210 different ways of filling the positions of president, secretary and treasurer.

b.

7	7	7

The total number of ways the president's position could be filled is $7 \times 7 \times 7$.

$n(\xi) = 7 \times 7 \times 7$
$= 343$

There are 7 choices of the same person to fill the positions three years in a row.

$n(A) = 7$

3. Calculate the probability that the same person fills the president's position three years in a row.

$$P(A) = \frac{n(A)}{n(\xi)} = \frac{7}{7 \times 7 \times 7} = \frac{1}{49}$$

The probability that one of the 7 volunteers, fills the president's position three years in a row is $\frac{1}{49}$.

1.3.3 Arrangements in a circle

Consider this problem: In how many different ways can 7 people be seated, 4 at a time, on a bench?

By now you should quickly see the answer: ${}^{7}P_{4} = 840$.

Let us change the problem slightly: in how many different ways can 7 people be seated, 4 at a time, at a circular table?

The solution must recognise that when people are seated on a bench, each of the following represents a different arrangement:

ABCD BCDA CDAB DABC

However, when the people are sitting in a circle, each of these represents the *same* arrangement. It is important to note that in a circle arrangement we do not consider positions on the circle as different — it does not matter where the circle starts.

In each case B has A on the left and C on the right.

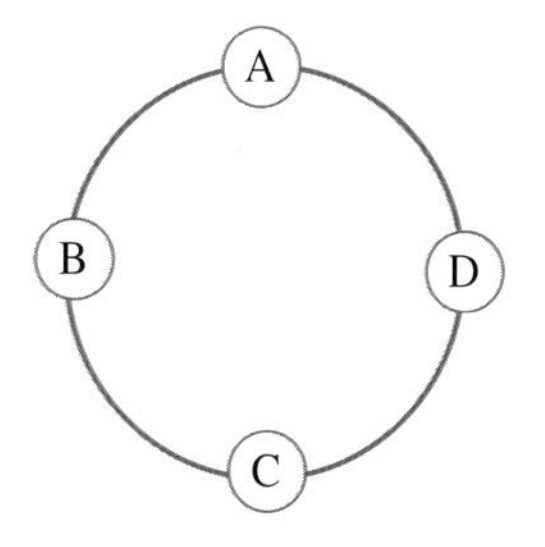

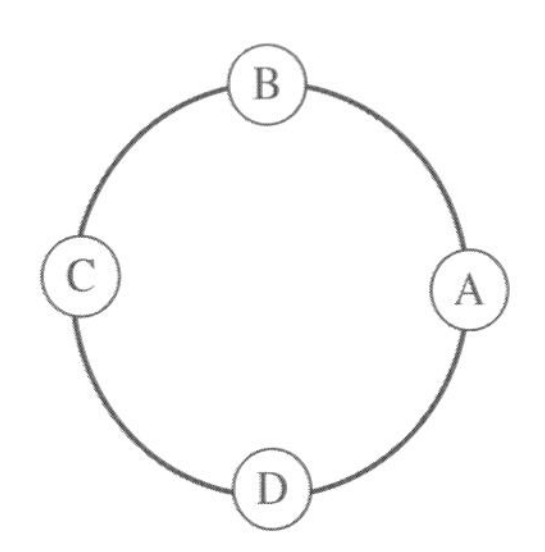

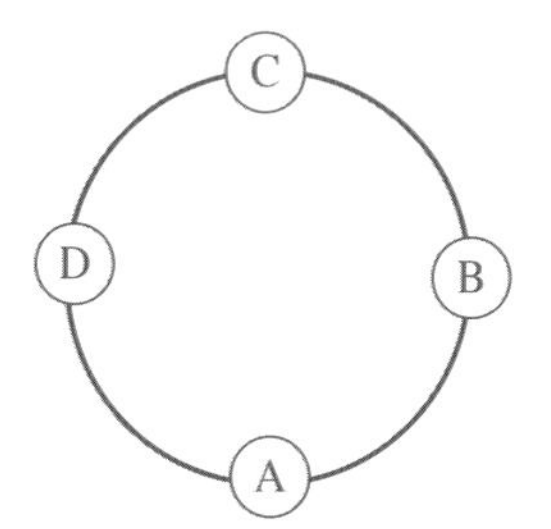

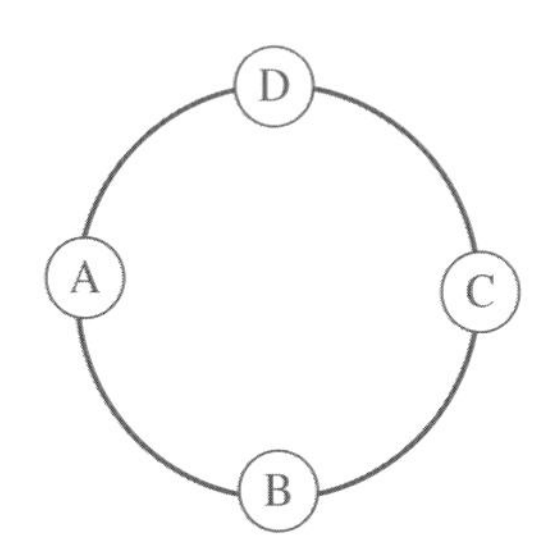

We conclude that the number ${}^{7}P_{4}$ gives 4 times the number of arrangements of 7 people in a circle 4 at a time.

Therefore, the number of arrangements is $\frac{{}^{7}P_{4}}{4} = 210$.

Arrangements in a circle

In general, the number of different ways n objects can be arranged, r at a time, in a circle is:

$$\frac{{}^{n}P_{r}}{r}$$

WORKED EXAMPLE 9 Calculating arrangements in a circle

a. By recalling the appropriate formula, give an expression for the number of different arrangements if, from a group of 8 people, 5 are to be seated at a round table.
b. Evaluate this expression.
c. Each table receives one lucky door prize and one lucky seat prize. Determine the probability of the same person at one table winning both.

THINK	WRITE
a. 1. Write down the rule for the number of arrangements in a circle.	a. $\frac{{}^{n}\text{P}_r}{r}$
2. Substitute the given values of n and r into the formula.	$= \frac{{}^{8}\text{P}_5}{5}$
3. Answer the question.	The number of ways of seating 5 people from a group of 8 people at a round table is given by the expression $\frac{{}^{8}\text{P}_5}{5}$.
b. 1. Use a calculator to evaluate ${}^{8}\text{P}_5$.	b. ${}^{8}\text{P}_5 = \frac{6720}{5}$
2. Evaluate.	$= 1344$
3. Answer the question.	The number of ways of seating 5 from a group of 8 people at a round table is 1344.
c. 1. There are 2 prizes, and each prize can be won by any one of the 5 people.	c. | 5 | 5 | The total number of ways the prizes could be won is 5×5. $n(\xi) = 5 \times 5$ $= 25$
2. Let A be the event that the same person wins both prizes.	There are 5 choices of the same person to win both prizes. $n(A) = 5$
3. Calculate the probability that the same person wins both.	$\text{P}(A) = \frac{n(A)}{n(\xi)}$ $= \frac{5}{5 \times 5}$ $= \frac{1}{5}$ The probability of the same person winning both prizes is $\frac{1}{5}$.

1.3 Exercise

Students, these questions are even better in jacPLUS

Receive immediate feedback and access sample responses

Access additional questions

Track your results and progress

Find all this and MORE in jacPLUS

Technology free

1. Recall the definition of $n!$ and write each of the following in expanded form.

   a. 4! b. 5! c. 6! d. 7!

2. WE6c Evaluate the following factorials.

   a. $\frac{9!}{5!}$ b. $\frac{10!}{4!}$ c. $\frac{7!}{3!}$ d. $\frac{6!}{0!}$

3. WE6d Simplify the following factorials, in terms of n.

   a. $\frac{n!}{(n-5!)}$ b. $\frac{(n+3)!}{(n+1)!}$ c. $\frac{(n-3)!}{n!}$ d. $\frac{(n-2)!}{(n+2)!}$

Technology active

4. WE6a, b Evaluate the following factorials.

   a. 10! b. 14! c. 9! d. 16!

5. WE7a Calculate each of the following by expressing it in expanded form.

   a. ${}^{8}P_{2}$ b. ${}^{7}P_{5}$ c. ${}^{8}P_{7}$

6. WE7b Write each of the following as a quotient of factorials and hence evaluate.

   a. ${}^{9}P_{6}$ b. ${}^{5}P_{2}$ c. ${}^{18}P_{5}$

7. Use your calculator to determine the value of each of the following.

   a. ${}^{20}P_{6}$ b. ${}^{800}P_{2}$ c. ${}^{18}P_{5}$

8. WE8 A soccer club will appoint a president and a vice-president. Eight people have volunteered for either of the two positions.
   a. Determine how many different ways these positions can be filled.
   b. For three years the same 8 people volunteered for these positions. Determine the probability that one of them is president for both years.

9. There are 26 players in an online game. Determine how many different results for 1st, 2nd, 3rd and 4th can occur.

10. A rowing crew consists of 4 rowers who sit in a definite order. Determine how many different crews are possible if 5 people try out for selection.

11. The school musical needs a producer, director, musical director and script coach. Nine people have volunteered for any of these positions. Determine how many different ways the positions can be filled. (*Note*: One person cannot take on more than 1 position.)

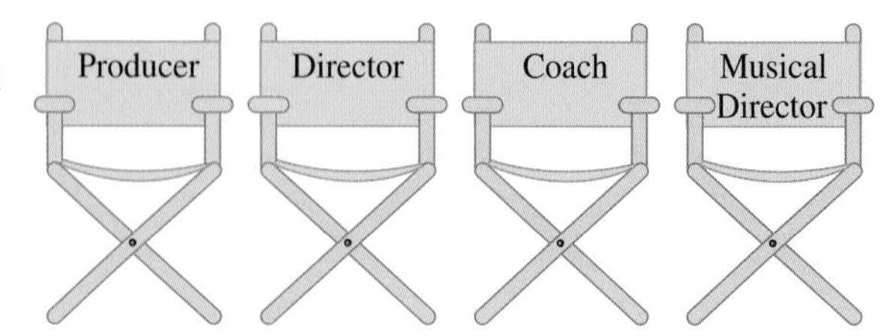

12. There are 14 swimmers in a race. Determine how many different ways the 1st, 2nd and 3rd positions can be filled.

13. WE9 a. By recalling the appropriate formula, give an expression for the number of different arrangements if, from a group of 15 people, 4 are to be seated at a round table.
    b. Evaluate this expression.
    c. Each table receives a lucky door prize, a lucky seat prize and a best-dressed prize. Determine the probability of the same person at one table winning all 3 prizes.

14. A round table seats 6 people. From a group of 8 people, give an expression for, and hence calculate, the number of ways 6 people can be seated at the table.

15. At a dinner party for 10 people all the guests were seated at a circular table. Determine how many different arrangements were possible.

16. At one stage in the court of Camelot, King Arthur and 12 knights would sit at the round table. If each person could sit anywhere, determine how many different arrangements were possible.

1.3 Exam questions

Question 1 (1 mark) TECH-ACTIVE

MC If ${}^{2n}P_n = 1680$ then n is equal to

A. 3 **B.** 4 **C.** 5 **D.** 6 **E.** 7

Question 2 (2 marks) TECH-FREE

A child has wooden letters spelling out their name GRACE on their wardrobe. Determine how many names can be created by rearranging these wooden letters in any combination that is more than 2 letters long.

Question 3 (1 mark) TECH-FREE

Determine how many ways the Mathematics, Science, Physics and Chemistry prizes can be awarded from 25 students if no student can win more than one prize.

More exam questions are available online.

1.4 Permutations with restrictions

LEARNING INTENTION

At the end of this subtopic you should be able to:

- calculate permutations for situations involving like objects and restrictions.

1.4.1 Like objects

A 5-letter word is to be made from 3 As and 2 Bs. How many different permutations or arrangements can be made?

If the 5 letters were all different, it would be easy to calculate the number of arrangements. It would be $5! = 120$. Perhaps you can see that when letters are repeated, the number of different arrangements will be less than 120. To analyse the situation let us imagine that we can distinguish one A from another. We will write A_1, A_2, A_3, B_1 and B_2 to represent the 5 letters.

As we list some of the possible arrangements we notice that some are actually the same, as shown in the table.

$A_1A_2B_1A_3B_2$ $A_1A_3B_1A_2B_2$ $A_2A_1B_1A_3B_2$ $A_2A_3B_1A_1B_2$ $A_3A_1B_1A_2B_2$ $A_3A_2B_1A_1B_2$	$A_1A_2B_2A_3B_1$ $A_1A_3B_2A_2B_1$ $A_2A_1B_2A_3B_1$ $A_2A_3B_2A_1B_1$ $A_3A_1B_2A_2B_1$ $A_3A_2B_2A_1B_1$	Each of these 12 arrangements is the same — AABAB — if $A_1 = A_2 = A_3$ and $B_1 = B_2$.
$B_2A_1A_2B_1A_3$ $B_2A_1A_3B_1A_2$ $B_2A_2A_1B_1A_3$ $B_2A_2A_3B_1A_1$ $B_2A_3A_1B_1A_2$ $B_2A_3A_2B_1A_1$	$B_1A_1A_2B_2A_3$ $B_1A_1A_3B_2A_2$ $B_1A_2A_1B_2A_3$ $B_1A_2A_3B_2A_1$ $B_1A_3A_1B_2A_2$ $B_1A_3A_2B_2A_1$	Each of these 12 arrangements is the same — BAABA — if $A_1 = A_2 = A_3$ and $B_1 = B_2$.

The number of repetitions is 3! for the A s and 2! for the Bs. Thus, the number of different arrangements in choosing 5 letters from 3 As and 2 Bs is $\frac{5!}{3! \times 2!}$.

Permutations with like objects

The number of different ways of arranging n objects made up of groups of repeated (identical) objects, n_1 in the first group, n_2 in the second group and so on, is:

$$\frac{n!}{n_1!\,n_2!\,n_3!\ldots n_r!}.$$

***Note*:** If there are elements of the group that are not duplicated, then they can be considered as a group of 1. It is not usual to divide by 1!; it is more common to show only those groups that have duplications.

WORKED EXAMPLE 10 Calculating permutations with like objects

Determine how many different permutations of 7 counters can be made from 4 black and 3 white counters.

THINK	WRITE
1. Write down the total number of counters.	There are 7 counters in all; therefore, $n = 7$.
2. Write down the number of times any of the coloured counters are repeated.	There are 3 white counters; therefore, $n_1 = 3$. There are 4 black counters; therefore, $n_2 = 4$.
3. Write down the rule for arranging groups of like things.	$\frac{n!}{n_1!\,n_2!\,n_3!\ldots n_r!}$
4. Substitute the values of n, n_1 and n_2 into the rule.	$= \frac{7!}{3! \times 4!}$
5. Expand each of the factorials.	$= \frac{7 \times 6 \times 5 \times 4 \times 3 \times 2 \times 1}{3 \times 2 \times 1 \times 4 \times 3 \times 2 \times 1}$
6. Simplify the fraction.	$= \frac{7 \times 6 \times 5}{6}$
7. Evaluate.	$= 35$
8. Answer the question.	Thirty-five different arrangements can be made from 7 counters, of which 3 are white and 4 are black.

1.4.2 Restrictions

Sometimes restrictions are introduced so that a smaller number of objects from the original group need to be considered. This results in limiting the numbers of possible permutations.

WORKED EXAMPLE 11 Calculating permutations with restrictions (1)

A rowing crew of 4 rowers is to be selected, in order from the first seat to the fourth seat, from 8 candidates. Determine how many different arrangements are possible if:

a. **there are no restrictions**
b. **Jason or Kris must row in the first seat**
c. **Jason must be in the crew but he can row anywhere in the boat**
d. **Jason is not in the crew.**

THINK	WRITE
a. 1. Write down the permutation formula. ***Note*:** 4 rowers are to be selected from 8 and the order is important.	a. ${}^{n}P_{r} = \frac{n!}{(n-r)!}$
2. Substitute the given values of n and r into the permutation formula.	${}^{8}P_{4} = \frac{8!}{(8-4)!}$ $= \frac{8!}{4!}$
3. Expand the factorials or use a calculator to evaluate 8! and 4!.	$= \frac{8 \times 7 \times 6 \times 5 \times \cancel{4} \times \cancel{3} \times \cancel{2} \times \cancel{1}}{\cancel{4} \times \cancel{3} \times \cancel{2} \times \cancel{1}}$ $= 8 \times 7 \times 6 \times 5$
4. Evaluate.	$= 1680$
5. Answer the question.	There are 1680 ways of arranging 4 rowers from a group of 8.
b. 1. Apply the multiplication principle since two events will follow each other; that is, Jason will fill the first seat and the remaining 3 seats will be filled in $7 \times 6 \times 5$ ways or Kris will fill the first seat and the remaining 3 seats will be filled in $7 \times 6 \times 5$ ways. | J | 7 | 6 | 5 | or | K | 7 | 6 | 5 |	b. No. of arrangements = no. of ways of filling the first seat × no. of ways of filling the remaining 3 seats. $= 2 \times {}^{n}P_{r}$
2. Substitute the values of n and r into the formula and evaluate.	$= 2 \times {}^{7}P_{3}$ $= 2 \times 210$ $= 420$

3. Answer the question.

There are 420 ways of arranging the 4 rowers if Jason or Kris must row in the first seat.

c. 1. Apply the addition principle, since Jason must be in either the first, second, third or fourth seat. The remaining 3 seats will be filled in $7 \times 6 \times 5$ ways each time.

J	7	6	5	+	7	J	6	5
7	6	J	5	+	7	6	5	J

c. No. of arrangements
= no. of arrangements with Jason in seat 1
+ no. of arrangements with Jason in seat 2
+ no. of arrangements with Jason in seat 3
+ no. of arrangements with Jason in seat 4.

2. Substitute the values of n and r into the formula.

No. of arrangements

$$= 1 \times {}^7P_3 + 1 \times {}^7P_3 + 1 \times {}^7P_3 + 1 \times {}^7P_3$$
$$= 4 \times {}^7P_3$$
$$= 4 \times 210$$

3. Evaluate.

$$= 840$$

4. Answer the question.

There are 840 ways of arranging the 4 rowers if Jason must be in the crew of 4.

d. As Jason is not in the crew, there are only 7 candidates. Four rowers are to be chosen from 7 and order is important.

d. $${}^7P_4 = \frac{7!}{(7-4)!}$$
$$= \frac{7!}{3!}$$
$$= \frac{7 \times 6 \times 5 \times 4 \times 3 \times 2 \times 1}{3 \times 2 \times 1}$$
$$= 7 \times 6 \times 5 \times 4$$
$$= 840$$

There are 840 ways of arranging the crew when Jason is not included.

WORKED EXAMPLE 12 Calculating permutations with restrictions (2)

a. Calculate the number of permutations of the letters in the word COUNTER.
b. Calculate the number of permutations in which the letters C and N appear side by side.
c. Calculate the number of permutations in which the letters C and N appear apart.
d. Determine the probability of the letters C and N appearing side by side.

THINK

a. 1. Count the number of letters in the given word.

WRITE

a. There are 7 letters in the word COUNTER.

2. Determine the number of ways the 7 letters may be arranged.

The 7 letters may be arranged $7! = 5040$ ways.

3. Answer the question.

There are 5040 permutations of letters in the word COUNTER.

b. 1. Imagine the C and N are 'tied' together and are therefore considered as 1 unit. Determine the number of ways C and N may be arranged: CN and NC.

b. Let C and N represent 1 unit. They may be arranged $2! = 2$ ways.

2. Determine the number of ways 6 things can be arranged. ***Note*:** There are now 6 letters: the 'CN' unit along with O, U, T, E and R.	Six things may be arranged $6! = 720$ ways.
3. Determine the number of permutations in which the letters C and N appear together.	The number of permutations $= 2 \times 6!$ $= 2 \times 720$ $= 1440$
4. Answer the question.	There are 1440 permutations in which the letters C and N appear together.
c. 1. Determine the total number of arrangements of the 7 letters.	**c.** Total number of arrangements $= 7!$ $= 5040$
2. Write down the number of arrangements in which the letters C and N appear together, as obtained in **a**.	Arrangements with C and N together $= 1440$
3. Determine the difference between the values obtained in steps 1 and 2. ***Note*:** The number of arrangements in which C and N are apart is the total number of arrangements less the number of times they are together.	The number of arrangements $= 5040 - 1440$ $= 3600$
4. Answer the question.	The letters C and N appear apart 3600 times.
d. 1. Recall the probability formula.	**d.** $\Pr(A) = \dfrac{n(\xi)}{n(A)}$
2. State the number of elements in the set ξ, that is $n(\xi)$.	From part **a**, there are 5040 permutations of letters in the word COUNTER. $n(\xi) = 5040$
3. Determine the number of elements in the set A, that is the number of arrangements in which the letters C and N appear side by side.	From part **b**, there are 1440 permutations in which the letters C and N appear side by side. $n(A) = 1440$
4. Calculate the answer.	$\Pr(A) = \dfrac{n(A)}{n(\xi)}$ $= \dfrac{1440}{5040}$ $= \dfrac{2}{7}$ The probability of the letters C and N appearing side by side is $\dfrac{2}{7}$.

WORKED EXAMPLE 13 Calculating permutations with restrictions (3)

Consider the two words 'PARALLEL' and 'LINES'.

a. Calculate how many arrangements of the letters of the word LINES have the vowels grouped together.

b. Calculate how many arrangements of the letters of the word LINES have the vowels separated.

c. Calculate how many arrangements of the letters of the word PARALLEL are possible.

d. Determine the probability that in a randomly chosen arrangement of the word PARALLEL the letters A are together.

THINK	WRITE
a. 1. Group the required letters together.	a. There are two vowels in the word LINES. Treat these letters, I and E, as one unit.
2. Arrange the unit of letters together with the remaining letters.	Now there are four groups to arrange: (IE), L, N, S. These arrange in 4! ways.
3. Use the multiplication principle to allow for any internal rearrangements.	The unit (IE) can internally rearrange in 2! ways. Hence, the total number of arrangements is: $4! \times 2! = 24 \times 2$ $= 48$
b. 1. State the method of approach to the problem.	b. The number of arrangements with the vowels separated is equal to the total number of arrangements minus the number of arrangements with the vowels together.
2. State the total number of arrangements.	The five letters of the word LINES can be arranged in $5! = 120$ ways.
3. Calculate the answer.	From part **a**, there are 48 arrangements with the two vowels together. Therefore, there are $120 - 48 = 72$ arrangements in which the two vowels are separated.
c. 1. Count the letters, stating any identical letters.	c. The word PARALLEL contains 8 letters of which there are 2 As and 3 Ls.
2. Recall the rule $\frac{n!}{n_1!\, n_2! \ldots}$ and state the number of distinct arrangements.	There are $\frac{8!}{2! \times 3!}$ arrangements of the word PARALLEL.
3. Calculate the answer.	$\frac{8!}{2! \times 3!} = \frac{8 \times 7 \times 6 \times 5 \times \cancel{4}^{2} \times \cancel{3!}}{\cancel{2} \times \cancel{3!}}$ $= 3360$ There are 3360 arrangements.
d. 1. State the number of elements in the sample space.	d. There are 3360 total arrangements of the word PARALLEL, so $n(\xi) = 3360$ or $\frac{8!}{2! \times 3!}$.
2. Group the required letters together.	For the letters A to be together, treat these two letters as one unit. This creates seven groups: (AA), P, R, L, L, E, L, of which three are identical Ls.

3. Calculate the number of elements in the event.

The seven groups arrange in $\frac{7!}{3!}$ ways. As the unit (AA) contains two identical letters, there are no distinct internal rearrangements of this unit that need to be taken into account. Hence, $\frac{7!}{3!}$ is the number of elements in the event.

4. Calculate the required probability.
Note: It helps to use factorial notation in the calculations.

The probability that the As are together

$$= \frac{\text{number of arrangements with the As together}}{\text{total number of arrangements}}$$
$$= \frac{7!}{3!} \div \frac{8!}{2! \times 3!}$$
$$= \frac{7!}{3!} \times \frac{2! \times 3!}{8 \times 7!}$$
$$= \frac{2}{8}$$
$$= \frac{1}{4}$$

1.4 Exercise

Students, these questions are even better in jacPLUS

Receive immediate feedback and access sample responses

Access additional questions

Track your results and progress

Find all this and MORE in jacPLUS

Technology active

1. Recall the appropriate formula and calculate the number of different arrangements can be made using the 6 letters of the word NEWTON, assuming:
   a. the first N is distinct from the second N
   b. there is no distinction between the 2 Ns.
2. Determine how many different permutations can be made using the 11 letters of the word ABRACADABRA.
3. **WE10** Determine how many different arrangements of 5 counters can be made using 3 red and 2 blue counters.
4. Determine how many different arrangements of 9 counters can be made using 4 black, 3 red and 2 blue counters.
5. A collection of 12 books is to be arranged on a shelf. The books consist of 3 copies of *Great Expectations*, 5 copies of *Catcher in the Rye* and 4 copies of *Huntin', Fishin' and Shootin'*. Determine how many different arrangements of these books are possible.

6. A shelf holding 24 cans of dog food is to be stacked using 9 cans of Yummy and 15 cans of Ruff for Dogs. Determine how many different ways the shelf can be stocked.

7. **WE11** A cricket team of 11 players is to be selected, in batting order, from 15. Writing your answers in standard form correct to 3 significant figures, determine how many different arrangements are possible if:
   a. there are no restrictions
   b. Arjun must be in the team at number 1
   c. Arjun must be in the team but he can be anywhere from 1 to 11
   d. Arjun is not in the team.

8. The Student Council needs to fill the positions of president, secretary and treasurer from 6 candidates. Each candidate can fill only one of the positions. Determine how many ways can this be done if:
   a. there are no restrictions
   b. Tan must be secretary
   c. Tan must have one of the 3 positions
   d. Tan is not in any of the positions.

9. The starting 5 in a basketball team is to be picked, in order, from the 10 players in the squad. Determine how many ways can this be done if:
   a. there are no restrictions
   b. Jamahl needs to be player number 5
   c. Jamahl and Anfernee must be in the first 5 players (starting 5)
   d. Jamahl is not in the team.

10. **WE12** a. Calculate the number of permutations of the letters in the word MATHS.
   b. Calculate the number of permutations in which the letters M and A appear together.
   c. Calculate the number of permutations in which the letters M and A appear apart.
   d. Determine the probability of the letters M and A appearing apart.

11. A rowing team of 4 rowers is to be selected in order from 8 rowers.
   a. Calculate how many different ways this can be done.
   b. Calculate the number of these ways in which 2 rowers, Jane and Lee, can sit together in the boat.
   c. Calculate the number of ways in which the crew can be formed without using Jane or Lee.
   d. Determine how many ways the crew can be formed if it does not contain Jane.

12. A decathlon has 12 runners.
   a. Calculate how many ways 1st, 2nd and 3rd can be filled.
   b. Calculate how many ways 1st, 2nd and 3rd can be filled if Najim finishes first.

13. **WE13** Consider the words SIMULTANEOUS and EQUATIONS.
   a. Calculate how many arrangements of the letters of the word EQUATIONS have the letters Q and U grouped together.
   b. Calculate how many arrangements of the letters of the word EQUATIONS have the letters Q and U separated.
   c. Calculate the number of possible arrangements of the letters of the word SIMULTANEOUS.
   d. Determine the probability that in a randomly chosen arrangement of the word SIMULTANEOUS both the letters U are together.

14. The clue in a crossword puzzle says that a particular answer is an anagram of STOREY. An anagram is another word that can be obtained by rearranging the letters of the given word.
   a. Determine the number of possible arrangements of the letters of STOREY.
   b. The other words in the crossword puzzle indicate that the correct answer is O__T__. Calculate how many arrangements are now possible and identify the word.

1.4 Exam questions

Question 1 (1 mark) TECH-FREE

Determine the total number of arrangements of the letters of the word FACTORIAL, which starts and ends with A.

Question 2 (1 mark) TECH-ACTIVE

MC The number of ten–digit mobile phone numbers having at least one of their digits repeated is:

A. 36 288 **B.** 3 628 800 **C.** 99 963 712 **D.** 36 288 000 **E.** 9 996 371 200

Question 3 (2 marks) TECH-ACTIVE

Determine how many ways the letters of the word CALCULUS can be arranged in a row

a. If there are no restrictions. **(1 mark)**

b. If the two C's are separated. **(1 mark)**

More exam questions are available online.

1.5 Combinations

LEARNING INTENTION

At the end of this subtopic you should be able to:
- calculate the number of possible combinations in situations where order is not important
- calculate probabilities in situations involving combinations.

1.5.1 When order does not matter

A group of things chosen from a larger group where order is not important is called a combination. In previous sections we performed calculations of the number of ways a task could be done where order is important — permutations or arrangements. We now examine situations where *order does not matter*.

Suppose 5 people have nominated for a committee consisting of 3 members. It does not matter in what order the candidates are placed on the committee, it matters only whether they are there or not. If order was important we know there would be ${}^5P_3$, or 60, ways in which this could be done. Here are the possibilities:

ABC ACB BAC BCA CAB CBA
ABD ADB BAD BDA DAB DBA
ABE AEB BAE BEA EAB ABA
ACE AEC CAE CEA EAC ECA
ACD ADC CAD CDA DAC DCA
ADE AED DAE DEA EAD EDA
BCD BDC CBD CDB DBC DCB
BCE BEC CBE CEB EBC ECB
BDE BED DBE DEB EBD EDB
CDE CED DCE DEC ECD EDC

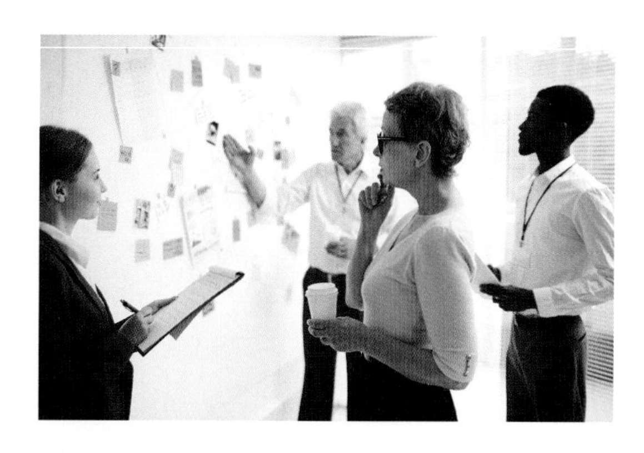

The 60 arrangements are different only if we take order into account; that is, ABC is different from CAB and so on. You will notice in this table that there are 10 distinct committees corresponding to the 10 distinct rows. Each row merely repeats, in a different order, the committee in the first column. This result (10 distinct committees) can be arrived at logically:

1. There are ${}^{5}P_{3}$ ways of choosing or selecting 3 from 5 in order.
2. Each choice of 3 is repeated 3! times.
3. The number of distinct selections or combinations is ${}^{5}P_{3} \div 3! = 10$.

This leads to the general rule of selecting r objects from n objects.

Combinations

The number of ways of choosing or selecting r objects from n distinct objects, where order is not important, is given by ${}^{n}C_{r}$:

$$ {}^{n}C_{r} = \frac{{}^{n}P_{r}}{r!} $$

C is used to represent combinations.

WORKED EXAMPLE 14 Evaluating combinations

Write these combinations as statements involving permutations, then calculate them.

a. ${}^{7}C_{2}$

b. ${}^{20}C_{3}$

THINK	WRITE
a. 1. Recall the rule for ${}^{n}C_{r}$.	a. ${}^{n}C_{r} = \frac{{}^{n}P_{r}}{r!}$
2. Substitute the given values of n and r into the combination formula.	${}^{7}C_{2} = \frac{{}^{7}P_{2}}{2!}$
3. Simplify the fraction.	$= \frac{\left(\frac{7!}{5!}\right)}{2!}$ $= \frac{7!}{5!} \div 2!$ $= \frac{7!}{5!} \times \frac{1}{2!}$ $= \frac{7 \times 6 \times 5!}{5! \times 2 \times 1}$
4. Evaluate.	$= \frac{7 \times 6}{2 \times 1}$ $= \frac{42}{2}$ $= 21$
b. 1. Write down the rule for ${}^{n}C_{r}$.	b. ${}^{n}C_{r} = \frac{{}^{n}P_{r}}{r!}$

2. Substitute the values of n and r into the formula.

$$^{20}C_3 = \frac{^{20}P_3}{3!}$$

3. Simplify the fraction.

$$= \frac{\left(\frac{20!}{17!}\right)}{3!}$$
$$= \frac{20!}{17!} \div 3!$$
$$= \frac{20!}{17!} \times \frac{1}{3!}$$
$$= \frac{20 \times 19 \times 18 \times 17!}{17! \times 3 \times 2 \times 1}$$

4. Evaluate.

$$= \frac{20 \times 19 \times 18}{3 \times 2 \times 1}$$
$$= \frac{6840}{6}$$
$$= 1140$$

WORKED EXAMPLE 15 Applying combinations

Apply the concept of ^nC_r to calculate the number ways a basketball team of 5 players be selected from a squad of 9 if the order in which they are selected does not matter.

THINK

1. Recall the rule for ^nC_r.
Note: Since order does not matter, use the ^nC_r rule.

WRITE

$$^nC_r = \frac{^nP_r}{r!}$$

2. Substitute the values of n and r into the formula.

$$^9C_5 = \frac{^9P_5}{5!}$$

3. Simplify the fraction.

$$= \frac{\left(\frac{9!}{4!}\right)}{5!}$$
$$= \frac{9!}{4!} \div 5!$$
$$= \frac{9!}{4!} \times \frac{1}{5!}$$
$$= \frac{9!}{4!\ 5!}$$
$$= \frac{9 \times 8 \times 7 \times 6 \times 5!}{4 \times 3 \times 2 \times 1 \times 5!}$$

4. Evaluate.

$$= \frac{9 \times 8 \times 7 \times 6}{4 \times 3 \times 2 \times 1}$$
$$= \frac{3024}{24}$$
$$= 126$$

The formula we use to determine the number of ways of selecting r objects from n distinct objects, where order is not important, is useful but needs to be simplified.

$$^{n}C_{r} = \frac{^{n}P_{r}}{r!}$$
$$= \frac{\frac{n!}{(n-r)!}}{r!}$$
$$= \frac{n!}{r!\,(n-r)!}$$

Alternative form of the combinations formula

$$^{n}C_{r} = \frac{n!}{r!\,(n-r)!} = \binom{n}{r}$$

$0 \leq r \leq n$ where r and n are non-negative integers.

The formula for $^{n}C_{r}$ is exactly that for the binomial coefficients used in the binomial theorem, which is explored with reference to Pascal's triangle in section 1.7.

WORKED EXAMPLE 16 Evaluating combinations (2)

Determine the value of the following.

a. $^{12}C_{5}$

b. $\binom{10}{2}$

THINK

a. 1. Recall the formula for $^{n}C_{r}$.

2. Substitute the given values of n and r into the combination formula.

3. Simplify the fraction.

4. Evaluate.

WRITE

a. $^{n}C_{r} = \frac{n!}{(n-r)!\,r!}$

$$^{12}C_{5} = \frac{12!}{(12-5)!\,5!}$$
$$= \frac{12!}{7!\,5!}$$
$$= \frac{12 \times 11 \times 10 \times 9 \times 8 \times 7!}{7! \times 5 \times 4 \times 3 \times 2 \times 1}$$
$$= \frac{12 \times 11 \times \cancel{10} \times \cancel{9}^{3} \times \cancel{8}^{2}}{\cancel{5} \times \cancel{4} \times \cancel{3} \times \cancel{2} \times 1}$$
$$= 12 \times 11 \times 3 \times 2$$
$$= 792$$

b. 1. Recall the rule for $\binom{n}{r}$.

b. $\binom{n}{r} = {}^nC_r$
$= \frac{n!}{(n-r)!\,r!}$

2. Substitute the given values of n and r into the combination formula.

$\binom{10}{2} = \frac{10!}{(10-2)!\,2!}$
$= \frac{10!}{8!\,2!}$

3. Simplify the fraction.

$= \frac{10 \times 9 \times 8!}{8! \times 2 \times 1}$
$= \frac{10 \times 9}{2 \times 1}$

4. Evaluate.

$= \frac{90}{2}$
$= 45$

1.5.2 Probability calculations

The combination formula is always used in selection problems. Most calculators have a ${}^nC_r$ key to assist with the evaluation when the figures become large.

Both the multiplication and addition principles apply and are used in the same way as for permutations.

The calculation of probabilities from the rule $\Pr(A) = \frac{n(A)}{n(\xi)}$ requires that the same counting technique is used for the numerator and denominator. We have seen for permutations that it can assist calculation to express numerator and denominator in terms of factorials and then simplify. Similarly for combinations, express the numerator and denominator in terms of the appropriate combinatoric coefficients and then carry out the calculations.

WORKED EXAMPLE 17 Calculating probabilities using combinations

A committee of 5 students is to be chosen from 7 boys and 4 girls. Use ${}^nC_r$ and the multiplication and addition principles to answer the following.

a. **Calculate how many committees can be formed.**
b. **Calculate how many of the committees contain exactly 2 boys and 3 girls.**
c. **Calculate how many committees have at least 3 girls.**
d. **Determine the probability of the oldest and youngest students both being on the committee in part a.**

THINK

a. 1. As there is no restriction, choose the committee from the total number of students.

WRITE

a. There are 11 students in total from whom 5 students are to be chosen. This can be done in ${}^{11}C_5$ ways.

2. Use the formula ${}^{n}C_r = \dfrac{n!}{r! \times (n-r)!}$ to calculate the answer.

$$\begin{aligned}{}^{11}C_5 &= \frac{11!}{5! \times (11-5)!} \\ &= \frac{11!}{5! \times 6!} \\ &= \frac{11 \times 10 \times 9 \times 8 \times 7 \times 6!}{5! \times 6!} \\ &= \frac{11 \times 10 \times 9 \times 8 \times 7}{5 \times 4 \times 3 \times 2 \times 1} \\ &= 462\end{aligned}$$

There are 462 possible committees.

b. 1. Select the committee to satisfy the given restriction.

b. The 2 boys can be chosen from the 7 boys available in ${}^{7}C_2$ ways. The 3 girls can be chosen from the 4 girls available in ${}^{4}C_3$ ways.

2. Use the multiplication principle to form the total number of committees.
Note: The upper numbers on the combinatoric coefficients sum to the total available, $7+4=11$, while the lower numbers sum to the number that must be on the committee, $2+3=5$.

The total number of committees that contain two boys and three girls is ${}^{7}C_2 \times {}^{4}C_3$.

3. Calculate the answer.

$$\begin{aligned}{}^{7}C_2 \times {}^{4}C_3 &= \frac{7!}{2! \times 5!} \times 4 \\ &= \frac{7 \times 6}{2!} \times 4 \\ &= 21 \times 4 \\ &= 84\end{aligned}$$

There are 84 committees possible with the given restriction.

c. 1. List the possible committees that satisfy the given restriction.

c. As there are 4 girls available, at least 3 girls means either 3 or 4 girls. The committees of 5 students that satisfy this restriction have either 3 girls and 2 boys, or they have 4 girls and 1 boy.

2. Write the number of committees in terms of combinatoric coefficients.

3 girls and 2 boys are chosen in ${}^{4}C_3 \times {}^{7}C_2$ ways.
4 girls and 1 boy are chosen in ${}^{4}C_4 \times {}^{7}C_1$ ways.

3. Use the addition principle to state the total number of committees.

The number of committees with at least three girls is ${}^{4}C_3 \times {}^{7}C_2 + {}^{4}C_4 \times {}^{7}C_1$.

4. Calculate the answer.

$$\begin{aligned}{}^{4}C_3 \times {}^{7}C_2 + {}^{4}C_4 \times {}^{7}C_1 &= 84 + 1 \times 7 \\ &= 91\end{aligned}$$

There are 91 committees with at least 3 girls.

d. 1. State the number in the sample space.

d. The total number of committees of 5 students is ${}^{11}C_5 = 462$ from part **a**.

2. Form the number of ways the given event can occur.

Each committee must have 5 students. If the oldest and youngest students are placed on the committee, then 3 more students need to be selected from the remaining 9 students to form the committee of 5. This can be done in ${}^{9}C_3$ ways.

3. State the probability in terms of combinatoric coefficients.

Let A be the event the oldest and the youngest students are on the committee.

$$\Pr(A) = \frac{n(A)}{n(\xi)} = \frac{^{9}C_{3}}{^{11}C_{5}}$$

4. Calculate the answer.

$$\begin{aligned}\Pr(A) &= \frac{9!}{3! \times 6!} \div \frac{11!}{5! \times 6!} \\ &= \frac{9!}{3! \times 6!} \times \frac{5! \times 6!}{11!} \\ &= \frac{1}{3!} \times \frac{5!}{11 \times 10} \\ &= \frac{5 \times 4}{110} \\ &= \frac{2}{11}\end{aligned}$$

The probability of the committee containing the youngest and the oldest students is $\frac{2}{11}$.

WORKED EXAMPLE 18 Identifying patterns in combination calculations

Evaluate the following using your calculator and comment on your results.

a. $^{9}C_{3}$ **b.** $^{9}C_{6}$ **c.** $^{15}C_{5}$ **d.** $^{15}C_{10}$ **e.** $^{12}C_{7}$ **f.** $^{12}C_{5}$

THINK

a–f. Use your calculator to evaluate the listed combinations.

WRITE

a. $^{9}C_{3} = 84$
b. $^{9}C_{6} = 84$
c. $^{15}C_{5} = 3003$
d. $^{15}C_{10} = 3003$
e. $^{12}C_{7} = 792$
f. $^{12}C_{5} = 792$

Comment on your results.

So $^{9}C_{3} = {^{9}C_{6}}$, $^{15}C_{5} = {^{15}C_{10}}$ and $^{12}C_{7} = {^{12}C_{5}}$.

It appears that when, for example, $^{12}C_{p} = {^{12}C_{q}}$ $p + q = 12$.

TI | THINK

a. On a Calculator page, press MENU, then select:
5: Probability
3: Combinations.
Complete the entry line as nCr(9,3).

DISPLAY/WRITE

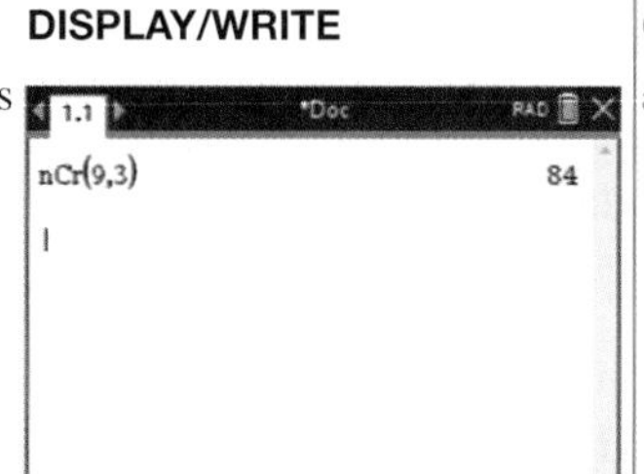

CASIO | THINK

a. From the Advance palette on the Keyboard, select nCr. Complete the entry line as shown.

DISPLAY/WRITE

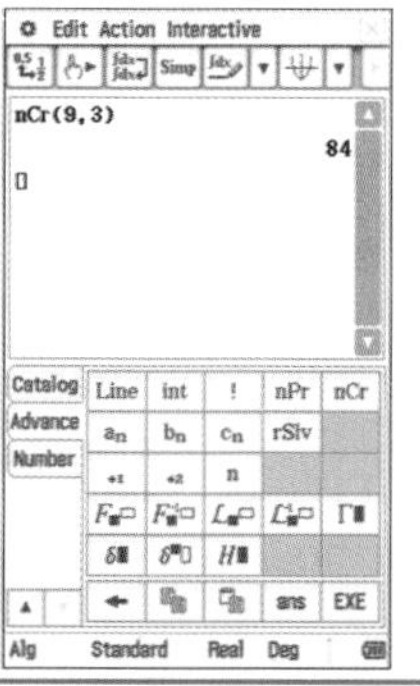

For each of the preceding examples, it can be seen that ${}^{n}\mathrm{C}_{r} = {}^{n}\mathrm{C}_{n-r}$. This may be derived algebraically:

$$\begin{aligned}
{}^{n}\mathrm{C}_{n-r} &= \frac{{}^{n}\mathrm{P}_{n-r}}{(n-r)!} \\
&= \frac{\left(\frac{n!}{[n-(n-r)]!}\right)}{(n-r)} \\
&= \frac{\left(\frac{n!}{r!}\right)}{(n-r)!} \\
&= \frac{n!}{r!} \times \frac{1}{(n-r)!} \\
&= \frac{n!}{r!\,(n-r)!} \\
&= \frac{n!}{(n-r)!\,r!} \\
&= \frac{{}^{n}\mathrm{P}_{r}}{r!} \\
&= {}^{n}\mathrm{C}_{r}
\end{aligned}$$

Drawing on our understanding of combinations, we have:

- ${}^{n}\mathrm{C}_{r} = {}^{n}\mathrm{C}_{r-1}$, as choosing r objects must leave behind $(n-r)$ objects and vice versa
- ${}^{n}\mathrm{C}_{0} = 1 = {}^{n}\mathrm{C}_{n}$, as there is only one way to choose none or all of the n objects
- ${}^{n}\mathrm{C}_{1} = n$, as there are n ways of choosing 1 object from a group of n objects.

1.5 Exercise

Technology free

1. **WE14** Write each of the following as statements in terms of permutations.

   a. ${}^{8}\mathrm{C}_{3}$ b. ${}^{19}\mathrm{C}_{2}$ c. ${}^{1}\mathrm{C}_{1}$ d. ${}^{5}\mathrm{C}_{0}$

2. Write each of the following using the notation ${}^{n}\mathrm{C}_{r}$.

   a. $\frac{{}^{8}\mathrm{P}_{2}}{2!}$ b. $\frac{{}^{9}\mathrm{P}_{3}}{3!}$ c. $\frac{{}^{8}\mathrm{P}_{0}}{0!}$ d. $\frac{{}^{10}\mathrm{P}_{4}}{4!}$

3. **WE15** Apply the concept of ${}^{n}\mathrm{C}_{r}$ to calculate the number of ways three types of ice-cream can be chosen in any order from a supermarket freezer if the freezer contains the following number of types of ice-cream.

   a. 3 types b. 6 types c. 10 types d. 12 types

4. A mixed netball team must have 3 women and 4 men in the side. If the squad has 6 women and 5 men wanting to play, determine how many different teams are possible.

5. A *quinella* is a bet made on a horse race that pays a win if the punter selects the first 2 horses in any order. Determine how many different quinellas are possible in a race that has the following number of horses in it.

   a. 8 horses b. 16 horses

Technology active

6. A cricket team of 11 players is to be chosen from a squad of 15 players. Determine how many ways can this be done.

7. A basketball team of 5 players is to be chosen from a squad of 10 players. Determine how many ways can this be done.

8. WE16 Determine the value of each of the following.

 a. ${}^{12}C_4$ b. ${}^{11}C_1$ c. ${}^{12}C_{12}$

 d. $\binom{21}{15}$ e. $\binom{100}{1}$ f. $\binom{17}{14}$

9. From a pack of 52 cards, a hand of 5 cards is dealt.
 a. Calculate how many different hands there are.
 b. Determine how many of these hands contain only red cards.
 c. Determine how many of these hands contain only black cards.
 d. Determine how many of these hands contain at least one red and at least one black card.

10. WE17 A committee of 5 students is to be chosen from 6 boys and 8 girls. Use ${}^nC_r$ and the multiplication principle to answer the following.
 a. Calculate how many committees can be formed.
 b. Calculate how many of the committees contain exactly 2 boys and 3 girls.
 c. Calculate how many committees have at least 4 boys.
 d. Determine the probability of neither the oldest nor the youngest student being on the committee.

11. A music collection contains 32 albums. Determine how many ways 5 albums can be chosen from the collection.

Questions 12–14 refer to the following information.

The Maryborough Tennis Championships involve 16 players. The organisers plan to use 3 courts and assume that each match will last on average 2 hours and that no more than 4 matches will be played on any court per day.

12. In a 'round robin' each player plays every other player once.
 a. If the organisers use a round robin format, determine how many games will be played in all.
 b. Determine how long the tournament lasts.

13. The organisers split the 16 players into two pools of 8 players each. After a 'round robin' within each pool, a final is played between the winners of each pool.
 a. Determine how many matches are played in the tournament.
 b. Determine how long the tournament lasts.

14. A 'knock out' format is one in which the loser of every match drops out and the winners proceed to the next round until there is only one winner left.
 a. If the game starts with 16 players, determine how many matches are needed before a winner is obtained.
 b. Determine how long the tournament lasts.

15. Lotto is a gambling game played by choosing 6 numbers from 45. Gamblers try to match their choice with those numbers chosen at the official draw. No number can be drawn more than once and the order in which the numbers are selected does not matter.
 a. Calculate how many different selections of 6 numbers can be made from 45.
 b. Suppose the first numbers drawn at the official draw are 42, 3 and 18. Determine how many selections of 6 numbers will contain these 3 numbers.

 Note: This question ignores supplementary numbers. Lotto is discussed further in the next section.

16. a. WE18 Calculate the value of:

i. ${}^{12}C_3$ and ${}^{12}C_9$ ii. ${}^{15}C_8$ and ${}^{15}C_7$ iii. ${}^{10}C_1$ and ${}^{10}C_9$ iv. ${}^{8}C_3$ and ${}^{8}C_5$

b. Describe what you notice in your results for part a. Give your answer as a general statement such as 'The value of ${}^nC_r$ is...'.

1.5 Exam questions

Question 1 (1 mark) TECH-ACTIVE

MC Four marbles are to be selected from a bag containing six green and three purple marbles. Determine how many different ways this can be done if there are to be two purple marbles in the selection.

A. ${}^6C_2 \times {}^3C_2$ B. ${}^9C_2 \times {}^7C_2$ C. ${}^6C_2 + {}^3C_2$ D. ${}^9C_2 + {}^7C_2$ E. $\dfrac{{}^9C_2 \times {}^7C_2}{21}$

Question 2 (1 mark) TECH-FREE

At Marina's café, sandwiches can be made with cos lettuce, carrot, avocado, capsicum, tomato and red onion. Determine how many different sandwiches are possible.

Question 3 (2 marks) TECH-FREE

A rugby union squad has 12 forwards and 10 backs in training. A team consists of 8 forwards and 7 backs. Determine how many different teams can be chosen from the squad.

More exam questions are available online.

1.6 Applications of permutations and combinations

LEARNING INTENTION

At the end of this subtopic you should be able to:

- apply your knowledge of permutations and combinations to solve problems in real world scenarios.

1.6.1 Permutations and combinations in the real world

Counting techniques, particularly those involving permutations and combinations, can be applied in gambling, logistics and various forms of market research. In this section we investigate when to use permutations and when to use combinations as well as examining problems associated with these techniques.

Permutations are used to count when order is important. Some examples are:

- the number of ways the positions of president, secretary and treasurer can be filled
- the number of ways a team can be chosen from a squad *in distinctly different positions*
- the number of ways the first three positions of a race can be filled.

Combinations are used to count when order is not important. Some examples are:

- the number of ways a committee can be chosen
- the number of ways a team can be chosen from a squad
- the number of ways a hand of 5 cards can be dealt from a deck.

These relatively simple applications of permutations and combinations are explored in the worked examples that follow. However, it is important to be mindful that the modern world relies on combinatorial algorithms. These algorithms are important for any system that benefits from finding the fastest ways to operate. Examples include communication networks, molecular biology, enhancing security and protecting privacy in internet information transfer, data base queries and data mining, computer chip design, simulations and scheduling.

WORKED EXAMPLE 19 Applications of permutations and combinations (1)

a. Ten points are marked on a page and no three of these points are in a straight line. Determine how many triangles can be drawn joining these points.

b. Determine how many different 3-digit numbers can be made using the digits 1, 3, 5, 7 and 9 without repetition.

THINK	WRITE
a. 1. ***Note*****:** A triangle is made by choosing 3 points. It does not matter in what order the points are chosen, so ${}^{n}C_{r}$ is used. Recall the rule for ${}^{n}C_{r}$.	**a.** ${}^{n}C_{r} = \frac{n!}{(n-r)!\,r!}$
2. Substitute the given values of n and r into the combination formula.	${}^{10}C_{3} = \frac{10!}{(10-3)!\,3!}$ $= \frac{10!}{7!\,3!}$
3. Simplify the fraction.	$= \frac{10 \times 9 \times 8 \times 7!}{7! \times 3 \times 2 \times 1}$ $= \frac{10 \times \cancel{9}^{3} \times \cancel{8}^{4}}{\cancel{3} \times \cancel{2} \times 1}$
4. Evaluate.	$= 10 \times 3 \times 4$ $= 120$
5. Answer the question.	120 triangles may be drawn by joining 3 points.
6. Verify the answer obtained by using the combination function on a calculator.	
b. 1. ***Note*****:** Order is important here. Recall the rule for ${}^{n}P_{r}$.	**b.** ${}^{n}P_{r} = \frac{n}{(n-r)!}$
2. Substitute the given values of n and r into the permutation formula.	${}^{5}P_{3} = \frac{5!}{(5-3)}$ $= \frac{5!}{2!}$
3. Evaluate.	$= \frac{5 \times 4 \times 3 \times 2}{2!}$ $= 5 \times 4 \times 3$ $= 60$
4. Answer the question.	Sixty 3-digit numbers can be made without repetition from a group of 5 numbers.
5. Verify the answer obtained by using the permutation function on a calculator.	

WORKED EXAMPLE 20 Applications of permutations and combinations (2)

Jade and Kelly are 2 of the 10 members of a basketball squad. Calculate how many ways can a team of 5 be chosen if:

a. **both Jade and Kelly are in the 5**

b. **neither Jade nor Kelly is in the 5**

c. **Jade is in the 5 but Kelly is not.**

THINK	WRITE
a. 1. ***Note***: Order is not important, so ${}^{n}C_r$ is used. Recall the rule for ${}^{n}C_r$.	a. ${}^{n}C_r = \frac{n!}{(n-r)!\ r!}$
2. ***Note***: If Jade and Kelly are included, then there are 3 positions to be filled from the remaining 8 players. Substitute the given values of n and r into the combination formula.	${}^{8}C_3 = \frac{8!}{(8-3)!\ 3!}$ $= \frac{8!}{5!\ 3!}$
3. Simplify the fraction.	$= \frac{8\times7\times6\times5}{5!\times3\times2\times1}$ $= \frac{8\times7\times\cancel{6}}{\cancel{3}\times\cancel{2}\times1}$
4. Evaluate.	$= 8\times7$ $= 56$
5. Answer the question.	If Jade and Kelly are included, then there are 56 ways to fill the remaining 3 positions.
b. 1. ***Note***: Order is not important, so ${}^{n}C_r$ is used. Recall the rule for ${}^{n}C_r$.	b. ${}^{n}C_r = \frac{n!}{(n-r)!\ r!}$
2. ***Note***: If Jade and Kelly are not included, then there are 5 positions to be filled from 8 players. Substitute the given values of n and r into the combination formula.	${}^{8}C_5 = \frac{8!}{(8-5)!\ 5!}$ $= \frac{8!}{3!\ 5!}$
3. Simplify the fraction.	$= \frac{8\times7\times6\times5!}{3\times2\times1\times5!}$ $= \frac{8\times7\times\cancel{6}}{\cancel{3}\times\cancel{2}\times1}$
4. Evaluate.	$= 8\times7$ $= 56$
5. Answer the question.	If Jade and Kelly are not included, then there are 56 ways to fill the 5 positions.

c. 1. ***Note*:** Order is not important, so ^nC_r is used. Recall the rule for ^nC_r.

c. $^nC_r = \frac{n!}{(n-r)!\,r!}$

2. ***Note*:** If Jade is included and Kelly is not, then there are 4 positions to be filled from 8 players.
Substitute the given values of n and r into the combination formula.

$^8C_4 = \frac{8!}{(8-4)!\,4!}$
$= \frac{8!}{4!\,4!}$

3. Simplify the fraction.

$= \frac{8 \times 7 \times 6 \times 5 \times 4!}{4 \times 3 \times 2 \times 1 \times 4!}$
$= \frac{\cancel{8} \times 7 \times \cancel{6}^2 \times 5}{\cancel{4} \times \cancel{3} \times \cancel{2} \times 1}$

4. Evaluate.

$= 7 \times 2 \times 5$
$= 70$

5. Answer the question.

If Jade is included and Kelly is not, then there are 70 ways to fill the 4 positions.

6. Verify each of the answers obtained by using the combination function on a calculator.

1.6.2 Lotto systems

An interesting application of combinations as a technique of counting is a game that Australians spend many millions of dollars on each week — lotteries. There are many varieties of lottery games in Australia. To play Saturday Gold Lotto in Queensland, a player selects 6 numbers from 45 numbers. The official draw chooses 6 numbers and 2 supplementary numbers. Depending on how the player's choice of 6 numbers matches the official draw, prizes are awarded in different divisions.

Division 1: 6 winning numbers

Division 2: 5 winning numbers and one of the supplementary numbers

Division 3: 5 winning numbers

Division 4: 4 winning numbers

Division 5: 3 winning numbers and one of the supplementary numbers

If the official draw was:

Winning numbers						Supplementaries	
13	42	6	8	20	12	2	34

A player who chose:

8 34 13 12 20 45

would win a Division 4 prize and a player who chose:

8 34 13 12 22 45

would win a Division 5 prize.

A player may have 7 lucky numbers — 4, 7, 12, 21, 30, 38 and 45 — and may wish to include all possible combinations of these 7 numbers in a 6 numbers lotto entry.

This can be done as follows:

4	7	12	21	30	38
4	7	12	21	30	45
4	7	12	21	38	45
4	7	12	30	38	45
4	7	21	30	38	45
4	12	21	30	38	45
7	12	21	30	38	45

The player does not have to fill out 7 separate entries to enter all combinations of these 7 numbers 6 at a time. Instead, they can complete a 'System 7' entry by marking 7 numbers on the entry form.

A System 9 consists of all entries of 6 numbers from the chosen 9 numbers.

WORKED EXAMPLE 21 Calculating winning combinations in a lottery

Use the information on lottery systems given above.
A player uses a System 8 entry with the numbers 4, 7, 9, 12, 22, 29, 32 and 36.
The official draw for this game was 4, 8, 12, 15, 22, 36 with supplementaries 20 and 29.

a. **Determine how many single entries are equivalent to a System 8.**
b. **List 3 of the player's entries that would have won Division 4.**
c. **Determine how many of the player's entries would have won Division 4.**

THINK	WRITE
a. 1. ***Note*:** Order is not important, so ${}^nC_r$ is used. Recall the rule for ${}^nC_r$.	a. ${}^nC_r = \dfrac{n!}{(n-r)!\ r!}$
2. ***Note*:** A System 8 consists of all entries consisting of 6 numbers chosen from 8. Substitute the given values of n and r into the combination formula.	${}^8C_6 = \dfrac{8!}{(8-6)!\ 6!}$ $= \dfrac{8!}{2!\ 6!}$
3. Simplify the fraction.	$= \dfrac{8 \times 7 \times 6!}{2 \times 7 \times 6!}$ $= \dfrac{\cancel{8}^4 \times 7}{\cancel{2} \times 1}$
4. Evaluate.	$= 4 \times 7$ $= 28$
5. Answer the question.	A System 8 is equivalent to 28 single entries.
6. Verify each of the answers obtained by using the combination function on a calculator.	
b. ***Note*:** Division 4 requires 4 winning numbers. The player's winning numbers are 4, 12, 22 and 36. Any of the other 4 numbers can fill the remaining 2 places. List 3 of the player's entries that would have won Division 4.	b. Some of the possibilities are: 4 12 22 36 7 9 4 12 22 36 7 29 4 12 22 36 7 32

c. 1. ***Note*: Order is not important, so ^nC_r is used. Recall the rule for ^nC_r.	**c.** $^nC_r = \frac{n!}{(n-r)!\,r!}$
2. ***Note*: To win Division 4 the numbers 4, 12, 22 and 36 must be included in the entry. The other 2 spaces can be filled with any of the other 4 numbers in any order. Substitute the given values of n and r into the combination formula.	$^4C_2 = \frac{4!}{(4-2)!\,2!}$ $= \frac{4!}{2!\,2!}$
3. Simplify the fraction.	$= \frac{4 \times 3 \times 1!}{\not{2} \times 1 \times 2!}$ $= \frac{\not{4}^2 \times 3}{\not{2} \times 1}$
4. Evaluate.	$= 2 \times 3$ $= 6$
5. Answer the question.	Six of the player's entries would have won Division 4.
6. Verify each of the answers obtained by using the combination function on a calculator.	

1.6 Exercise

Students, these questions are even better in jacPLUS

Receive immediate feedback and access sample responses

Access additional questions

Track your results and progress

Find all this and MORE in jacPLUS

Technology active

1. WE19 Determine how many ways there are:
   a. to draw a line segment between 2 points on a page with 10 points on it
   b. to make a 4-digit number using the digits 2, 4, 6, 8 and 1 without repetition
   c. to choose a committee of 4 people from 10 people
   d. for a party of 15 people to shake hands with one another.

2. Determine how many ways there are:
   a. for 10 horses to fill 1st, 2nd and 3rd positions
   b. to choose a team of 3 cyclists from a squad of 5
   c. to choose 1st, 2nd and 3rd speakers for a debating team from 6 candidates
   d. for 20 students to seat themselves in a row of 20 desks, write your answer in standard form correct to 2 significant figures.

3. The French flag is known as a tricolour flag because it is composed of 3 bands of colour. Determine how many different tricolour flags can be made from the colours red, white, blue and green if each colour can be used only once in one of the 3 bands and order is important.
4. In a taste test a market research company has asked people to taste 4 samples of coffee and try to identify each as one of four brands. Subjects are told that no 2 samples are the same brand. Determine how many different ways the samples can be matched to the brands.
5. **WE20** A volleyball team of 6 players is to be chosen from a squad of 10 players. Calculate how many ways can this be done if:
   a. there are no restrictions
   b. Stephanie is to be in the team
   c. Stephanie is not in the team
   d. two players, Stephanie and Alison, are not both in the team together.

6. A cross-country team of 4 runners is to be chosen from a squad of 9 runners. Determine how many ways this can be done if:
   a. there are no restrictions
   b. Cecily is to be one of the 4
   c. Cecily and Michael are in the team
   d. either Cecily or Michael but not both are in the team.

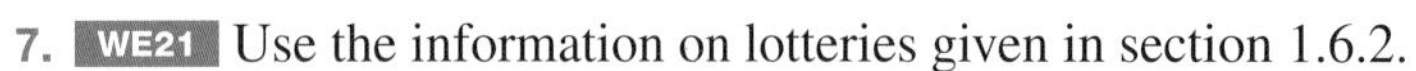

7. **WE21** Use the information on lotteries given in section 1.6.2.

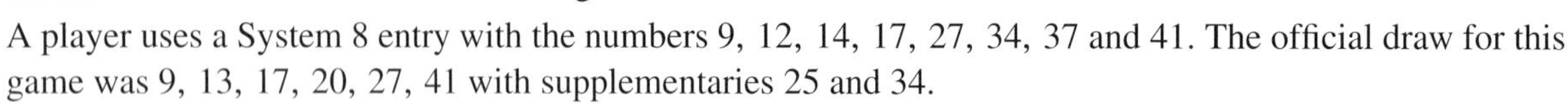

   A player uses a System 8 entry with the numbers 9, 12, 14, 17, 27, 34, 37 and 41. The official draw for this game was 9, 13, 17, 20, 27, 41 with supplementaries 25 and 34.
   a. Determine how many single entries are equivalent to a System 8.
   b. List 3 of the player's entries that would have won Division 4.
   c. Determine how many of the player's entries would have won Division 4.
8. Use the information on lotteries given in section 1.6.2.
   A player uses a System 9 entry with the numbers 7, 10, 12, 15, 25, 32, 35, 37 and 41. The official draw for this game was 7, 11, 15, 18, 25, 39 with supplementaries 23 and 32.
   a. Determine how many single entries are equivalent to a System 9.
   b. List 3 of the player's entries that would have won Division 5.
   c. Determine how many of the player's entries would have won Division 5.
9. In the gambling game roulette, if a gambler puts \$1 on the winning number he will win \$35. Suppose a gambler wishes to place five \$1 bets on 5 different numbers in one spin of the roulette wheel. If there are 36 numbers in all, determine how many ways the five bets can be placed.
10. A soccer team of 11 players is to be chosen from a squad of 17. If one of the squad is selected as goalkeeper and any of the remaining players can be selected in any of the positions, determine how many ways can this be done if:
    a. there are no other restrictions
    b. Karl is to be chosen
    c. Karl and Andrew refuse to play in the same team
    d. Karl and Andrew are either both in or both out.

Questions* 11 *and* 12 *refer to the following information.

Keno is a popular game in clubs and pubs around Australia. In each round a machine randomly generates 20 winning numbers from 1 to 80. In one entry a player can select up to 15 numbers.

11. Suppose a player selects an entry of 6 numbers.
    a. Determine how many ways an entry of 6 numbers can contain 6 winning numbers.
    Suppose an entry of 6 numbers has exactly 3 winning numbers in it.
    b. Determine how many ways the 3 winning numbers can be chosen.
    c. Determine how many ways the 3 losing numbers can be chosen.
    d. Determine how many entries of 6 numbers contain 3 winning numbers and 3 losing numbers.

12. Suppose a player selects an entry of 20 numbers.
    a. Determine how many ways an entry of 20 numbers can contain 20 winning numbers.
    b. Suppose an entry of 20 numbers has exactly 14 winning numbers in it.
        i. Determine how many ways the 14 winning numbers can be chosen.
        ii. Determine how many ways the 6 losing numbers can be chosen.
        iii. Determine how many entries of 20 numbers contain 14 winning numbers and 6 losing numbers.
        iv. Determine how many entries of 20 numbers contain no winning numbers.

1.6 Exam questions

Question 1 (1 mark) TECH-FREE

To win a racing quinella you have to select the first two runners across the finishing line in any order. At $0.50 a bet, determine how much it would cost to cover all quinella possibilities in a 24-hours race.

Question 2 (1 mark) TECH-FREE

Determine how many 4-digit numbers that are divisible by 10 can be formed from the numbers 3, 5, 7, 8, 9, 0 such that no number repeats.

Question 3 (1 mark) TECH-FREE

The official version of Oz Lotto requires you to select seven numbers from a total of 45 numbers. A standard game consists of 7 randomly chosen numbers and costs $1.45. Determine how much it would cost you to ensure that you won first prize.

More exam questions are available online.

1.7 Pascal's triangle and the pigeon-hole principle

LEARNING INTENTION

At the end of this subtopic you should be able to:
- evaluate combinations by using Pascal's triangle
- calculate numbers in Pascal's triangle using combinations
- use the pigeon-hole principle to evaluate statements about shared properties.

1.7.1 Pascal's triangle

Combinations are useful in other areas of mathematics, such as probability and binomial expansions. If we analyse the ${}^nC_r$ values closely, we notice that they produce the elements of any row in **Pascal's triangle** or each of the coefficients of a particular binomial expansion.

The triangle shown was named after the French mathematician Blaise Pascal. He was honoured for his application of the triangle to his studies in the area of probability.

Each new row in Pascal's triangle is obtained by first placing a 1 at the beginning and end of the row and then adding adjacent entries from the previous row.

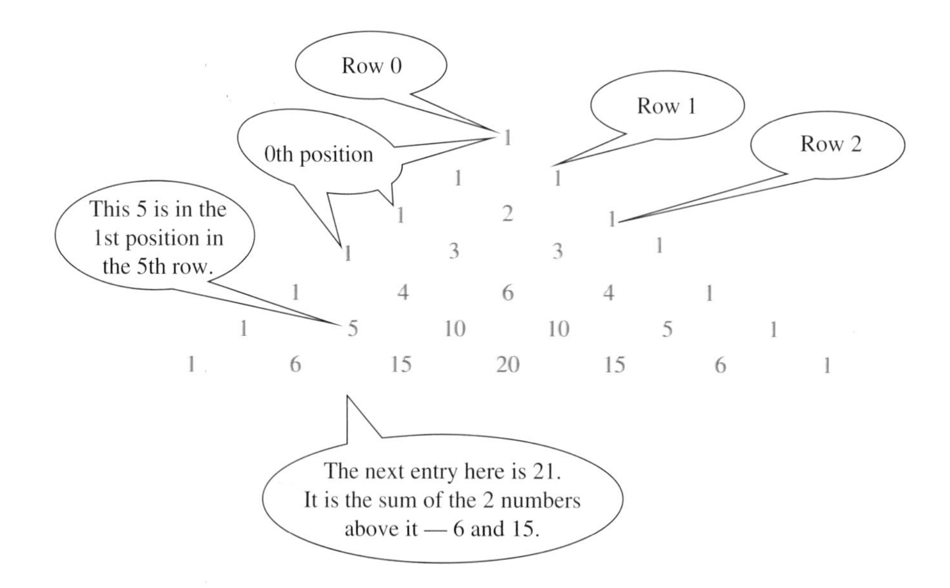

Each element in Pascal's triangle can be calculated using combinations. For example, 10 is the 2nd element in the 5th row of Pascal's triangle; that is, ${}^5C_2 = 10$ (assuming 1 is the zeroth (0th) element). Hence, the triangle can be written using $\binom{n}{r}$ or ${}^nC_r$ notation.

$n = 0$:	1	${}^0C_0$
$n = 1$:	1 1	${}^1C_0$ ${}^1C_1$
$n = 2$:	1 2 1	${}^2C_0$ ${}^2C_1$ ${}^2C_2$
$n = 3$:	1 3 3 1	${}^3C_0$ ${}^3C_1$ ${}^3C_2$ ${}^3C_3$
$n = 4$:	1 4 6 4 1	${}^4C_0$ ${}^4C_1$ ${}^4C_2$ ${}^4C_3$ ${}^4C_4$
$n = 5$:	1 5 10 10 5 1	${}^5C_0$ ${}^5C_1$ ${}^5C_2$ ${}^5C_3$ ${}^5C_4$ ${}^5C_5$

Note that the first and last number in each row is always 1.

Each coefficient is obtained by adding the two coefficients immediately above it.

The binomial expansion can therefore be generalised using combinations.

Pascal's triangle shows that the rth element of the nth row of Pascal's triangle is given by ${}^nC_r$. It is assumed that the 1 at the beginning of each row is the 0th element.

This gives **Pascal's identity** as follows.

Pascal's identity

$$\mathbf{{}^nC_r = {}^{n-1}C_{r-1} + {}^{n-1}C_r \text{ for } 0 < r < n}$$

The relationship between Pascal's triangle and combinations can be extended to the **binomial theorem**. This theorem gives a rule for expanding an expression such as $(a+b)^n$. Expanding expressions such as this may become quite difficult and time consuming using the usual methods of algebra. Consider the coefficients of the binomial expansion.

$$(x+y)^0 = 1$$
$$(x+y)^1 = x+y$$
$$(x+y)^2 = x^2+2xy+y^2$$
$$(x+y)^3 = x^3+3x^2y+3xy^2+y^3$$
$$(x+y)^4 = x^4+4x^3y+6x^2y^2+4xy^3+y^4$$
$$(x+y)^5 = x^5+5x^4y+10x^3y^2+10x^2y^3+5xy^4+y^5$$

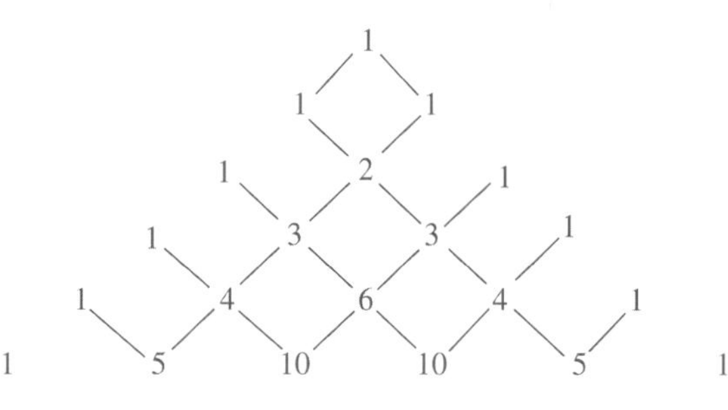

The coefficients in the binomial expansion are equal to the numbers in Pascal's triangle.

These relationships are summarised in the following table.

$(x+y)^0 = 1$	Coefficient: 1	${}^0C_0$
$(x+y)^1 = x+y$	Coefficients: 1 1	${}^1C_0, {}^1C_1$,
$(x+y)^2 = x^2+2xy+y^2$	Coefficients: 1 2 1	${}^2C_0, {}^2C_1, {}^2C_2$
$(y+x)^3 = y^3+3x^2y+3xy^2+x^3$	Coefficients: 1 3 3 1	${}^3C_0, {}^3C_1, {}^3C_2, {}^3C_3$
$(x+y)^4 = x^4+4x^3y+6x^2y^2+4xy^3+y^4$	Coefficients: 1 4 6 4 1	${}^4C_0, {}^4C_1, {}^4C_2, {}^4C_3, {}^4C_4$
$(x+y)^5 = x^5+5x^4y+10x^3y^2+10x^2y^3+5xy^4+y^5$	Coefficients: 1 5 10 10 5 1	${}^5C_0, {}^5C_1, {}^5C_2, {}^5C_3, {}^5C_4, {}^5C_5$

The binomial expansion can therefore be generated using combinations.

Generating the binomial expansion using combinations

$$\begin{aligned}(x+y)^n &= {}^nC_0x_n + {}^nC_1x^{n-1}y + {}^nC_2x^{n-2}y^2 + \ldots + {}^nC_rx^{n-r}y^r + \ldots + {}^nC_ny^n \\ &= x^n + {}^nC_1x^{n-1}y + {}^nC_2x^{n-2}y^2 + \ldots + {}^nC_rx^{n-r}y^r + \ldots + y^n\end{aligned}$$

WORKED EXAMPLE 22 Identifying numbers in Pascal's triangle

Refer to Pascal's triangle above and answer the following questions.

a. Determine the number in the 4th position in the 6th row.

b. Complete the 7th row in Pascal's triangle.

c. The numbers 7 and 21 occur side by side in the 7th row. Determine what element in the 8th row occurs below and in between these numbers.

THINK	WRITE
a. 1. Locate the 6th row and the 4th position. ***Note*:** Remember the 0th row is 1 and the first row is 1 1. In the 6th row the 1 on the left is in the 0th position.	a. 6th row ⇒ 1 6 15 20 **15** 6 1
2. Answer the question.	The number in the 4th position in the 6th row is 15.
b. 1. Write down the elements of the 6th row.	b. 6th row ⇒ 1 6 15 20 **15** 6 1

2. Obtain the 7th row.
   a. Place the number 1 at the beginning of the row.
   b. Add the first 2 adjacent numbers from the 6th row (1 and 6).
   c. Place this value next to the 1 on the new row and align the value so that it is in the middle of the 2 numbers (directly above) that created it.
   d. Repeat this process with the next 2 adjacent numbers from the 6th row (6 and 15).
   e. Once the sums of all adjacent pairs from the sixth row have been added, place a 1 at the end of the row.

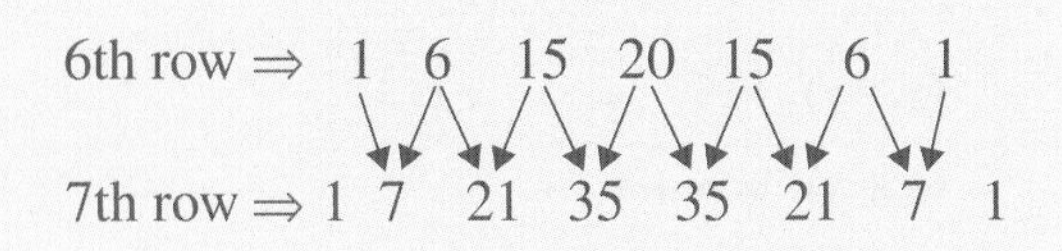

3. Answer the question.

The 7th row is
1 7 21 35 35 21 7 1.

c. 1. Add the numbers 7 and 21 in order to obtain the element in the 8th row that occurs below and in between these numbers.

c. 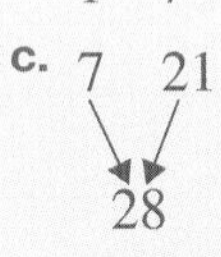

2. Answer the question.

The element in the 8th row that occurs below and in between 7 and 21 is 28.

WORKED EXAMPLE 23 Identifying numbers in Pascal's triangle using combinations

Use combinations to calculate the number in the 5th position in the 9th row of Pascal's triangle.

THINK	WRITE
1. Write down the combination rule.	${}^{n}C_{r}$
2. Substitute the values for n and r into the rule. ***Note***: The row is represented by $n = 9$. The position is represented by $r = 5$.	${}^{9}C_{5} = 126$
3. Evaluate using a calculator.	
4. Answer the question.	The value of the number in the 5th position in the 9th row is 126.

WORKED EXAMPLE 24 Expanding a binomial using combinations (1)

Use the binomial theorem to expand $(a + 2)^4$.

THINK	WRITE
1. Recall the rule for the binomial theorem.	$(x+y)^n = x^n + {}^{n}C_{1}x^{n-1}y^1 + \ldots + {}^{n}C_{r}x^{n-r}y^r + \ldots y^n$
2. Substitute the values for a, b and n into the rule: $x = a, y = 2$ and $n = 4$.	$(a+2)^4 = a^4 + {}^{4}C_{1}a^3 2^1 + {}^{4}C_{2}a^2 2^2 + {}^{4}C_{3}a^1 2^3 + 2^4$
3. Simplify.	$= a^4 + 4 \times a^3 \times 2 + 6 \times a^2 \times 4 + 4 \times a \times 8 + 16$ $= a^4 + 8a^3 + 24a^2 + 32a + 16$

WORKED EXAMPLE 25 Expanding a binomial using combinations (2)

Determine the 4th term in the expansion of $(x+y)^7$.

THINK	WRITE
1. Recall that the rule for the 4th term can be obtained from the binomial theorem: $(x+y)^n = x^n + {}^nC_1 x^{n-1}y^1 + \dots {}^nC_r x^{n-r}y^r + \dots y^n$ Write down the rule for the rth term.	rth term $= {}^nC_r x^{n-r}y^r$
2. Substitute the values for n and r into the rule: $n=7$ and $r=4$.	$= {}^7C_4 x^{7-4}y^4$
3. Simplify. *Note:* The 0th term corresponds to the first element of the expansion.	$= 35x^3y^4$
4. Answer the question.	The 4th term is equal to $35x^3y^4$.

1.7.2 Pigeon-hole principle

The **pigeon-hole principle** is a useful principle which can be used to make statements about the number of things within a set which share a particular property.

The pigeon-hole principle

If there are $(n+1)$ pigeons to be placed in n pigeon-holes, then there is at least one pigeon-hole with at least two pigeons in it.

Notes:
- Note the precise use of language in this statement, in particular the importance of the phrase 'at least'.
- Some may view the pigeon-hole principle as an obvious statement, but used cleverly it is a powerful problem-solving tool.

WORKED EXAMPLE 26 Applying the pigeon-hole principle (1)

Show that in a group of 13 people there are at least 2 whose birthday falls in the same month.

THINK	WRITE
1. Think of each person as a pigeon and each month as a pigeon-hole.	There are 12 months and 13 people.
2. If there are 13 pigeons to be placed in 12 holes, at least one hole must contain at least two pigeons.	Using the pigeon-hole principle: 13 people to be assigned to 12 months. At least one month must contain at least two people. That is, at least two people have birthdays falling in the same month.

Generalised pigeon-hole principle

If there are $(nk+1)$ pigeons to be placed in n pigeon-holes, then there is at least one pigeon-hole with at least $(k+1)$ pigeons in it.

WORKED EXAMPLE 27 Applying the pigeon-hole principle (2)

Show that in a group of 37 people there are at least 4 whose birthdays lie in the same month.

THINK	WRITE
1. Think of each person as a pigeon and each month as a pigeon-hole.	There are 12 months and 37 people.
2. Recall the generalised pigeon-hole principle.	Using the generalised pigeon-hole principle, 37 people are to be assigned to 12 months.
3. $(nk + 1)$ pigeons to be allocated to n holes; $n = 12 \rightarrow k = 3$	The value of n is 12 and k is 3. So at least one month has at least $(k + 1)$ or 4 people in it. That is, at least 4 people have birthdays falling in the same month.

WORKED EXAMPLE 28 Applying the pigeon-hole principle (3)

On resuming school after the Christmas vacation, many of the 22 teachers of Eastern High School exchanged handshakes. Mr Yisit, the Social Science teacher, said, 'Isn't that unusual — with all the handshaking, no two people shook hands the same number of times.'

Not wanting to spoil the fun, the Mathematics teacher, Mrs Pigeon, said respectfully, 'I am afraid you must have counted incorrectly. What you say is not possible.'

Explain how Mrs Pigeon can make this statement.

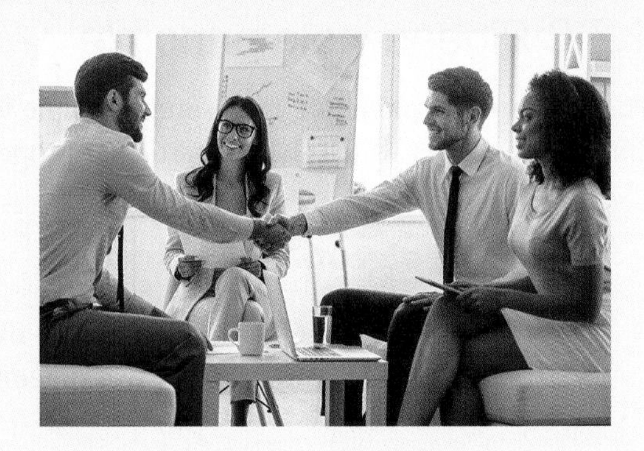

THINK	WRITE
1. Think of the possible number of handshakes by a person as a pigeon-hole.	For each person there are 22 possible numbers of handshakes; that is, 0 to 21.
2. If two or more people have 0 handshakes, the problem is solved. Consider the cases where there is 1 person with 0 handshakes or 0 persons with 0 handshakes.	1 person with 0 handshakes: If there is 1 person with 0 handshakes, there can be no person with 21 handshakes. Thus, there are 21 people to be assigned to 20 pigeon-holes. Therefore, there must be at least one pigeon-hole with at least two people in it. 0 people with 0 handshakes: If there is no person with 0 handshakes, there are 22 people to be assigned to 21 pigeon-holes. Therefore, there must be at least one pigeon-hole with at least two people in it (at least two people have made the same number of handshakes).
3. Conclude using a sentence.	Thus, there are at least two people who have made the same number of handshakes.

1.7 Exercise

Students, these questions are even better in jacPLUS

Receive immediate feedback and access sample responses

Access additional questions

Track your results and progress

Find all this and MORE in jacPLUS

Technology free

1. Write the first 8 rows in Pascal's triangle.

2. WE22 Refer to Pascal's triangle in section 1.7.1 and answer the following questions.
   a. Determine the number in the 4th position in the 8th row.
   b. Complete the 9th row in Pascal's triangle.
   c. If 9 and 36 occur side by side in the 9th row, determine what element in the 10th row occurs below and in between these numbers.

Technology active

3. WE23 Use combinations to:
   a. calculate the number in the 7th position of the 8th row of Pascal's triangle
   b. calculate the number in the 9th position of the 12th row of Pascal's triangle
   c. generate the 10th row of Pascal's triangle.

4. WE24 Use the binomial theorem to expand:
   a. $(x+y)^2$
   b. $(n+m)^3$
   c. $(a+3)^4$

5. WE25 a. Determine the 4th term in the expansion of $(x+2)^5$.
   b. Determine the 3rd term in the expansion of $(p+q)^8$.
   c. Determine the 7th term in the expansion of $(x+2)^9$.

6. a. In Pascal's triangle, calculate the sum of all elements in the:
      i. 0th row
      ii. 1st row
      iii. 2nd row
      iv. 3rd row
      v. 4th row
      vi. 5th row.
   b. i. Describe what you notice in your results from part a.
      ii. Complete the statement: 'The sum of the elements in the nth row of Pascal's triangle is...'

7. Use your statement result from question 6 to deduce a simple way of calculating:

$$^6C_0 + {^6C_1} + + {^6C_2} + {^6C_3} + {^6C_4} + {^6C_5} + {^6C_6}$$

8. WE26 In a cricket team consisting of 11 players, show that there are at least 2 whose phone numbers have the same last digit.

9. WE27 A squad of 10 netballers is asked to nominate when they can attend training. They can choose Tuesday only, Thursday only, or Tuesday and Thursday. Show that there is at least one group of at least 3 players who agree with one of these options.

10. J&L lollies come in five great colours — green, red, brown, yellow and blue. Determine how many J&Ls I need to select to be sure I have 6 of the same colour.

11. The new model WBM roadster comes in burgundy, blue or yellow with white or black trim. That is, the vehicle can be burgundy with white or burgundy with black and so on. Determine how many vehicles need to be chosen to ensure at least 3 have the same colour combination.

12. Explain whether it is possible to show that in a group of 13 people, there are at least 2 whose birthdays fall in February.

13. WE28 Nineteen netball teams entered the annual state championships. However, it rained frequently and not all games were completed. No team played the same team more than once. Mrs Organisit complained that the carnival was ruined and that no two teams had played the same number of games. Show that she is incorrect in at least part of her statement and that at least two teams played the same number of games.

14. Prove that in any group of 6 people either at least 3 are mutual friends or at least 3 are strangers.

1.7 Exam questions

Question 1 (1 mark) TECH-ACTIVE

MC On one evening in Australia, 1.4 million people watched the national news. We can be certain that at least x people from the same state/territory watched the news. Determine the maximum value of x.

- **A.** 139 999
- **B.** 174 999
- **C.** 175 000
- **D.** 200 000
- **E.** 233 333

Question 2 (1 mark) TECH-FREE

There are 21–30 people swimming at the local pool. The pool has 10 lanes. Show that there is at least 1 lane with 3 or more people in it.

Question 3 (2 marks) TECH-FREE

In an all-boys school of 393 students, every students must wear the school uniform comprising of the following options: short-sleeve shirt, long-sleeve shirt, jumper, shorts and pants. A students must wear 1 type of shirt, shorts or pants and has the option of wearing the jumper.

Show that there are at least 50 students wearing the exact same uniform.

More exam questions are available online.

1.8 Review

1.8.1 Summary

doc-37044

Hey students! Now that it's time to revise this topic, go online to:

Access the topic summary | Review your results | Watch teacher-led videos | Practise exam questions

Find all this and MORE in jacPLUS

1.8 Exercise

Technology free: short answer

1. a. State the inclusion–exclusion principle for 3 sets, R, S and T.
   b. Use the inclusion–exclusion principle to calculate the number of cards in a deck of 52 cards that are either red, a jack or a court card (king, queen or jack).
2. a. State the multiplication principle.
   b. One or two letters are to be chosen from the letters A, B, C, D, E and F. Determine how many different ways can this be done without replacement, if order is important.
3. a. State the definition of ${}^n\text{P}_r$.
   b. Without a calculator, compute the value of ${}^{10}\text{P}_3$.
   c. Prove that ${}^n\text{P}_r = \dfrac{n!}{(n-r)!}$.
4. A free-style snowboard competition has 15 entrants. Determine how many ways the first, second and third places can be filled. You may wish to use technology to answer this question.
5. a. Suppose 5 people are to be seated. Explain why there are fewer ways of seating 5 people at a circular table compared with seating the group on a straight bench.
   b. Determine how many ways can 5 people be seated at a round table.
6. The main cricket ground in Brisbane is called the Gabba. It is short for Woolloongabba. Determine how many different arrangements of letters can be made from the word WOOLLOONGABBA. You may wish to use technology to answer the question.
7. Apply the concept of ${}^n\text{C}_r$ to calculate the number of ways 12 different ingredients can be chosen from a box of 30 different ingredients. Describe what you can conclude about the ingredients left behind. Do not use algebra to explain this.
8. A committee of 5 men and 5 women is to be chosen from 8 men and 9 women. Determine how many ways this can be done.

Technology active: multiple choice

9. **MC** There are 12 people on the committee at the local football club. Determine how many ways can a president and a secretary be chosen from this committee.

   A. 2 B. 23 C. 132 D. 144 E. 12!

10. **MC** A TV station runs a cricket competition called *Classic Catches*. Six catches, A to F, are chosen and viewers are asked to rank them in the same order as the judges. The number of ways in which the six catches can be ranked is

    A. 1 B. 6 C. 30 D. 120 E. 720

11. **MC** Identify which one of the following permutations cannot be calculated.

A. ${}^{1000}P_{100}$ B. ${}^{1}P_{0}$ C. ${}^{8}P_{8}$ D. ${}^{11}P_{10}$ E. ${}^{4}P_{8}$

12. **MC** The result of 100! is greater than 94!.
Identify which of the following gives the best comparison between these two numbers.

A. 100! is 6 more than 94!
B. 100! is 6 times bigger than 94!
C. 100! is about 10 000 more than 94!
D. 100! is ${}^{100}P_6$ times bigger than 94!
E. 100! is about 1 000 000 more than 94!

13. **MC** If the answer is 10, identify which of the following options best matches this answer.

A. The number of ways 1st and 2nd can occur in a race with 5 entrants
B. The number of distinct arrangements of the letters in NANNA
C. The number of permutations of the letters in POCKET where P and O are together
D. ${}^{10}P_2 \div {}^{4}P_2$
E. The number of ways to select the top 3 horses in a 6 horse race

14. **MC** If the answer is 240, identify which of the following options best matches this answer.

A. The number of ways 1st and 2nd can occur in a race with 5 entrants
B. The number of distinct arrangements of the letters in NANNA
C. The number of permutations of the letters in POCKET where P and O are apart
D. ${}^{10}P_2 \div {}^{4}P_2$
E. The number of ways to select the top 3 horses in a 6 horse race

15. **MC** At a party there are 40 guests and they decide to have a toast. Each guest 'clinks' glasses with every other guest. Determine how many clinks there are in all.

A. 39
B. 40
C. 40!
D. 780
E. 2048

16. **MC** On a bookshelf there are 15 books — 7 geography books and 8 law books. Abena selects 5 books from the shelf — 2 geography books and 3 law books. Determine how many different ways she can make this selection.

A. ${}^{15}C_2 \times {}^{15}C_3$ B. ${}^{15}C_7 \times {}^{15}C_8$ C. ${}^{7}C_2 \times {}^{8}C_3$ D. ${}^{7}C_2 + {}^{8}C_3$ E. ${}^{5}C_2 \times {}^{5}C_3$

17. **MC** A netball team consists of 7 different positions: goal defence, goal keeper, wing defence, centre, wing attack, goal attack and goal shooter. The number of ways a squad of 10 players can be allocated to these positions is:

A. 10!
B. 7!
C. $\dfrac{10!}{7!}$
D. ${}^{10}P_7$
E. ${}^{10}C_7$

18. **MC** $16x^3$ is a term in the binomial expansion of:

A. $(x+2)^3$ B. $(x+4)^3$ C. $(x+2)^4$ D. $(x+4)^4$ E. $(x+16)^3$

Technology active: extended response

19. a. A school uses identification (ID) cards that consist of two letters from A to D followed by 3 digits chosen from 0 to 9. Each digit may be repeated but letters cannot be repeated. If the school receives about 800 new students each year, determine how many years it will take for the school to run out of unique ID numbers.

    b. For a scene in a movie, five boy–girl couples are needed. If they are to be selected from 10 boys and 12 girls, determine how many ways can this be done. (Assume the order of the couples does not matter.)

20. A plane is covered in points at 1-unit spacing. Each point on the plane is coloured red, blue or white. Show there are three points of the same colour at a maximum distance $2\sqrt{2}$ from each other.

1.8 Exam questions

Question 1 (2 marks) TECH-FREE

Determine how many ways the letters of the word DIFFERENTIAL can be arranged in a row

a. if there are no restrictions. **(1 mark)**

b. if the F's are separated. **(1 mark)**

Question 2 (1 mark) TECH-ACTIVE

MC Old MacDonald had a farm E-I-E-I-O.

The number of different arrangements of the letters E-I-E-I-O in a straight line is

A. $5!$

B. 5

C. $\frac{5!}{2!}$

D. $\frac{5!}{2!\,2!}$

E. $5!\ 2!\ 2!$

Question 3 (2 marks) TECH-ACTIVE

A netball team of 7 players is to be chosen from a squad of 11 players. Suppose and member can play any position. Determine how many ways this can be done

a. if each player is chosen to play a particular position. **(1 mark)**

b. If players have no particular position. **(1 mark)**

Question 4 (3 marks) TECH-ACTIVE

Determine how many ways five men and five women can be arranged in a row if

a. there are no restrictions. **(1 mark)**

b. the men and women occupy alternate positions. **(1 mark)**

c. all the men are next to each other. **(1 mark)**

Question 5 (3 marks) TECH-ACTIVE

In the expansion of $(ax+b)^4$ two of the terms are $-540x^3$ and $-1500x$. Determine the possible values of a and b.

More exam questions are available online.

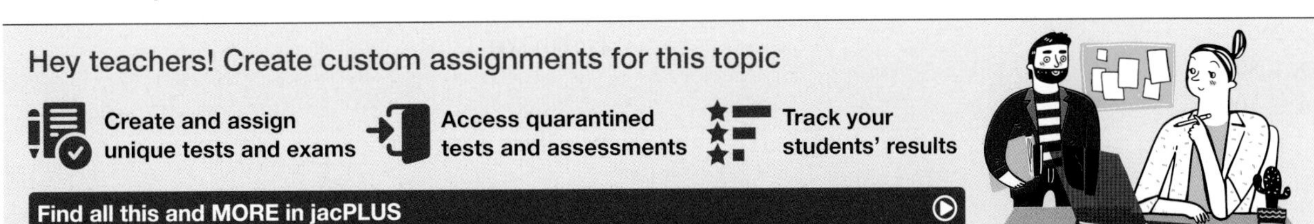

Answers

Topic 1 Combinatorics

1.2 Counting techniques

1.2 Exercise

1. a.

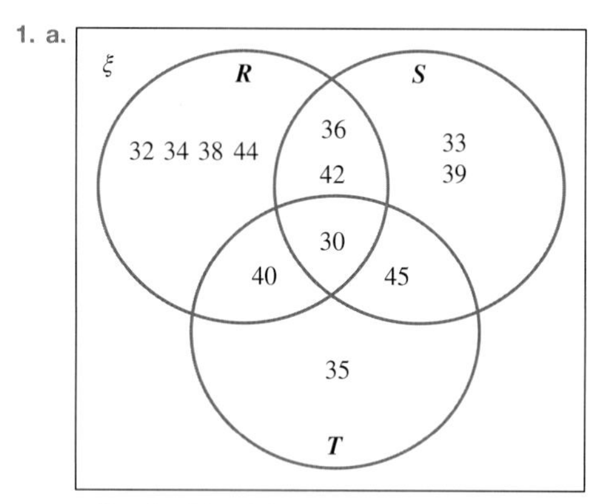

   b. $n(R \cup S \cup T) = 12$
   c. $n(R \cup S \cup T) = 12$
2. n(red, even or 4) = 36
3. 55 students
4. a. AB BA CA
   AC BC CB
   b. 6
   c. $\frac{1}{3}$
5. BG GB YB RB
   BY GY YG RG
   BR GR YR RY
6. ACB BAC CAB
   ABC BCA CBA
7. a. 42 b. 210 c. 840 d. 2520 e. $\frac{1}{7}$
8. a. 24 b. 6 c. 12 d. 24
9. a. 49 b. 252
10. 126
11. a. 200 b. 40 c. 50 d. 290 e. $\frac{1}{40}$
12. a. 13 230
    b. 17 640
    Jake may wear 13 230 outfits with a jacket or 4410 outfits without a jacket. Therefore, he has a total of 17 640 outfits to choose from. The assumption made with this problem is that no item of clothing is exactly the same; that is, none of the 7 shirts are exactly the same.
13. 100
14. 6
15. 48
16. 256
17. 1080
18. a. 1000
    b. 27
    c. 271 371 471
    272 372 472
    273 373 473
    281 381 481
    282 382 482
    283 383 483
    291 391 491
    292 392 492
    293 393 493

1.2 Exam questions

Note: Mark allocations are available with fully worked solutions online.

1. E
2. 144
3. 40

1.3 Factorials and permutations

1.3 Exercise

1. a. $4 \times 3 \times 2 \times 1$
   b. $5 \times 4 \times 3 \times 2 \times 1$
   c. $6 \times 5 \times 4 \times 3 \times 2 \times 1$
   d. $7 \times 6 \times 5 \times 4 \times 3 \times 2 \times 1$
2. a. 3024 b. 151 200 c. 840 d. 720
3. a. $n(n-1)(n-2)(n-3)(n-4)$
   b. $(n+3)(n+2)$
   c. $\dfrac{1}{n(n-1)(n-2)}$
   d. $\dfrac{1}{(n+2)(n+1)n(n-1)}$
4. a. 3 628 800 b. 87 178 291 200
   c. 362 880 d. 20 922 789 888 000
5. a. $8 \times 7 = 56$
   b. $7 \times 6 \times 5 \times 4 \times 3 = 2520$
   c. $8 \times 7 \times 6 \times 5 \times 4 \times 3 \times 2 = 40\,320$
6. a. $\dfrac{9!}{3!} = 60\,480$ b. $\dfrac{5!}{3!} = 20$ c. $\dfrac{18!}{13!} = 1\,028\,160$
7. a. 27 907 200 b. 639 200 c. 1 028 160
8. a. 56 b. $\frac{1}{8}$
9. 358 800
10. 120
11. 3024
12. 2184
13. a. $\dfrac{{}^{15}P_4}{4}$ b. 8190 c. $\dfrac{1}{16}$
14. 3360
15. 362 880
16. 479 001 600

1.3 Exam questions

Note: Mark allocations are available with fully worked solutions online.

1. B
2. 300
3. 13 800

1.4 Permutations with restrictions

1.4 Exercise

1. a. ${}^6P_6 = 720$ b. $\frac{{}^6P_6}{2} = 360$
2. 83 160
3. 10
4. 1260
5. 27 720
6. 1 307 504
7. a. 5.45×10^{10} b. 3.63×10^9 c. 4.00×10^{10} d. 1.45×10^{10}
8. a. 120 b. 20 c. 60 d. 60
9. a. 30 240 b. 3024 c. 6720 d. 15 120
10. a. 120 b. 48 c. 72 d. $\frac{3}{5}$
11. a. 1680 b. 180 c. 360 d. 840
12. a. 1320 b. 110
13. a. 80 640 b. 282 240 c. 119 750 440 d. $\frac{1}{6}$
14. a. 720 b. 24, OYSTER

1.4 Exam questions

Note: Mark allocations are available with fully worked solutions online.

1. 5040
2. E
3. a. 5040 b. 3780

1.5 Combinations

1.5 Exercise

1. a. $\frac{{}^8P_3}{3!}$ b. $\frac{{}^{19}P_2}{2!}$ c. $\frac{{}^1P_1}{1!}$ d. $\frac{{}^5P_0}{0!}$
2. a. ${}^8C_2$ b. ${}^9C_3$ c. ${}^8C_0$ d. ${}^{10}C_4$
3. a. 1 b. 20 c. 120 d. 220
4. 100
5. a. 28 b. 120
6. 1365
7. 252
8. a. 495 b. 11 c. 1 d. 54 264 e. 100 f. 680
9. a. 2 598 960 b. 65 780 c. 65 780 d. 2 467 400
10. a. 2002 b. 840 c. 126 d. $\frac{36}{91}$
11. 201 376
12. a. 120 b. 10 days
13. a. 57 b. 4 days 6 hours
14. a. 15 b. 1 day 4 hours
15. a. 8 145 060 b. 11 480
16. a. i. 220, 220 ii. 6435, 6435 iii. 10, 10 iv. 56, 56
    b. The value of ${}^nC_r$ is the same as ${}^nC_{n-r}$.

1.5 Exam questions

Note: Mark allocations are available with fully worked solutions online.

1. A
2. 63
3. 59400

1.6 Applications of permutations and combinations

1.6 Exercise

1. a. 45 b. 120 c. 210 d. 105
2. a. 720 b. 10 c. 120 d. 2.4×10^{18}
3. 24
4. 24
5. a. 210 b. 126 c. 84 d. 140
6. a. 126 b. 56 c. 21 d. 70
7. a. 28
   b. Any combination with 9, 17, 27 and 41.
   Sample responses include: 9 17 27 41 12 14
   9 17 27 41 12 37 9 17 27 41 12 34
   c. 6
8. a. 84
   b. Any combination with 7, 15, 25 and 32.
   Sample responses include: 7 15 25 32 10 12
   7 15 25 32 10 35 7 15 25 32 10 37
   c. 10
9. 376 992
10. a. 8008 b. 5005 c. 5005 d. 4004
11. a. 38 760 b. 1140 c. 34 220 d. 39 010 800
12. a. 1
    b. i. 38 760 ii. 50 063 860 iii. 1 940 475 213 600 iv. 4 191 844 505 805 495

1.6 Exam questions

Note: Mark allocations are available with fully worked solutions online.

1. $138
2. 60
3. $65 800 449

1.7 Pascal's triangles and the pigeon-hole principle

1.7 Exercise

1. See table at the bottom of the page*
2. a. 70
   b. 1 9 36 84 126 126 84 36 9 1
   c. 45
3. a. 8
   b. 220
   c. ${}^{10}C_0\ {}^{10}C_1\ {}^{10}C_2\ {}^{10}C_3\ {}^{10}C_4\ {}^{10}C_5\ {}^{10}C_6\ {}^{10}C_7\ {}^{10}C_8\ {}^{10}C_9\ {}^{10}C_{10}$
   1 10 45 120 210 252 210 120 45 10 1
4. a. $x^2 + 2xy + y^2$
   b. $n^3 + 3n^2m + 3nm^2 + m^3$
   c. $a^4 + 12a^3 + 54a^2 + 108a + 81$
5. a. $80x$ b. $56p^5q^3$ c. $4608x^2$
6. a. i. 1 ii. 2 iii. 4
   iv. 8 v. 16 vi. 32
   b. i. The sum of the elements in each row of Pascal's triangle is a power of 2:

Row	Sum
0	$2^0 = 1$
1	$2^1 = 2$
2	$2^2 = 4$
3	$2^3 = 8$
4	$2^4 = 16$
5	$2^5 = 32$

   ii. The sum of the elements in the nth row of Pascal's triangle is 2^n.
7. $2^6 = 64$

8, 9. Sample responses can be found in the worked solutions in the online resources.

10. 26
11. 13

12–14. Sample responses can be found in the worked solutions in the online resources.

1.7 Exam questions

Note: Mark allocations are available with fully worked solutions online.

1. C

2, 3. Sample responses can be found in the worked solutions in the online resources

1.8 Review

1.8 Exercise

Technology free: short answer

1. a. $n(R \cup S \cup T) = n(R) + n(S) + n(T) - n(R \cap S) - n(R \cap T) - n(S \cap T) + n(R \cap S \cap T)$
   b. n (red, court card or jack) = 32
2. a. If there are n ways of performing operation A and m ways of performing operation B, then there are $n \times m$ ways of performing A *and* B in the order AB.
   b. 64 ways
3. a. ${}^nP_r$ is the number of ways of choosing r objects from n distinct things, when order is important. ${}^nP_r = \dfrac{n!}{(n-r)!}$
   b. ${}^{10}P_3 = 720$
   c. Sample responses can be found in the worked solutions in the online resources.
4. ${}^{15}P_3 = 2730$
5. a. When people are sitting in a circle, we cannot tell the difference between arrangements such as between $\{A, B, C, D, E\}$ and $\{C, D, E, A, B\}$. In a circle, these represent the same arrangement. Therefore, there are fewer ways to arrange people in a circle than in a straight line.
   b. 24
6. 32 432 400
7. ${}^{30}C_{12} = {}^{30}C_{30-12} = {}^{30}C_{18} = 86\,493\,225$.
   Each separate time we choose r things from n distinct things, we also leave $n - r$ objects. (We can interchange taking and leaving.) Hence, ${}^nC_r = {}^nC_{n-r}$.
8. 7056

Technology active: multiple choice

9. C
10. D
11. E

*1.

Row																	
0									1								
1								1		1							
2							1		2		1						
3						1		3		3		1					
4					1		4		6		4		1				
5				1		5		10		10		5		1			
6			1		6		15		20		15		6		1		
7		1		7		21		35		35		21		7		1	
8	1		8		28		56		70		56		28		8		1

12. D

13. B

14. C

15. B

16. D

17. D

18. D

Technology active: extended response

19. a. 15 years b. 199 584

20. Sample responses can be found in the worked solutions in the online resources.

1.8 Exam questions

Note: Mark allocations are available with fully worked solutions online.

1. a. 59 875 200 b. 49 896 000

2. D

3. a. 1 663 200 b. 330

4. a. 3 628 800 b. 28 800 c. 86 400

5. $a = -3,\ b = 5$ or $a = 3,\ b = -5$

2 Sequences and series

LEARNING SEQUENCE

Fully worked solutions for this topic are available online.

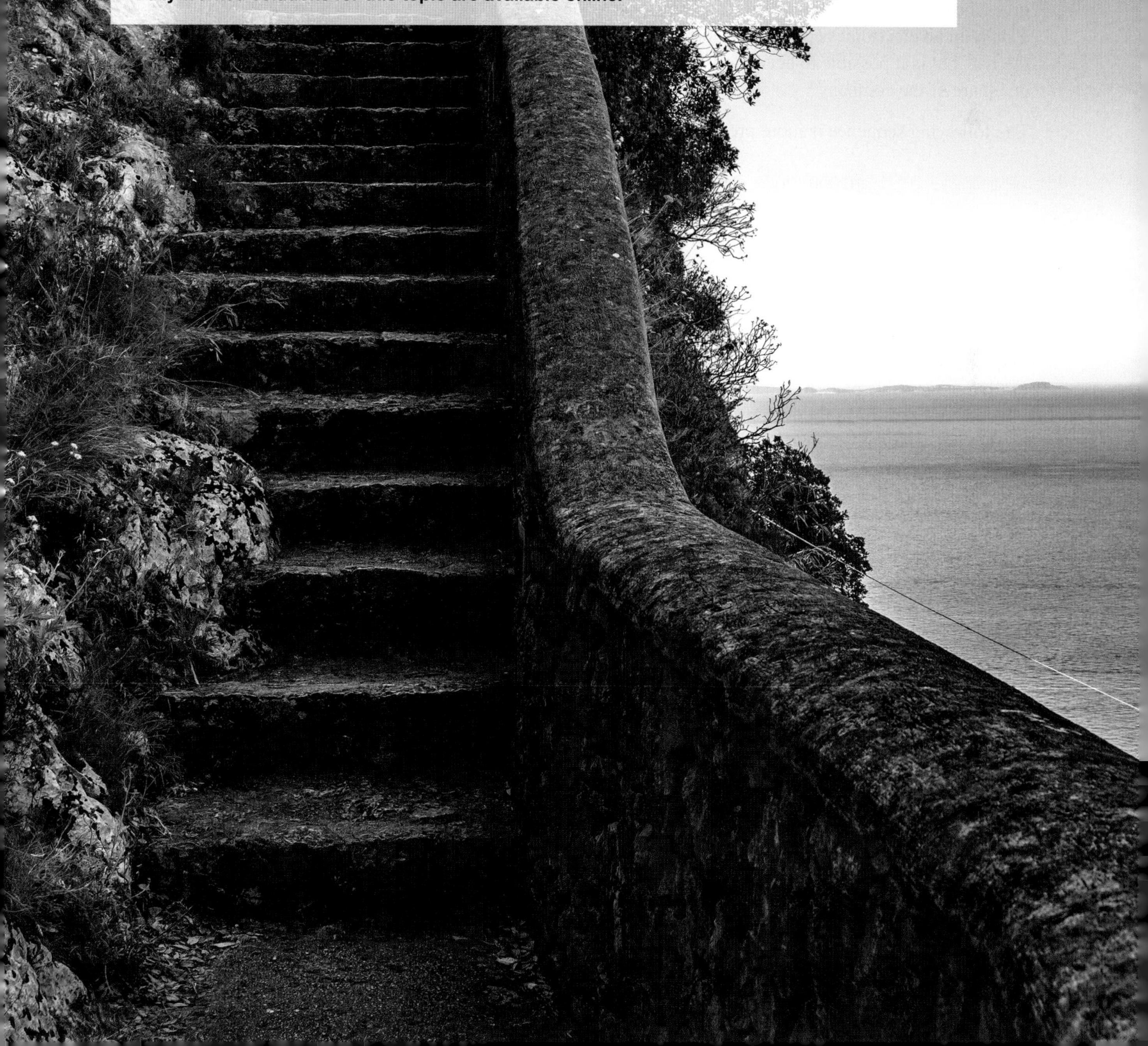

2.1 Overview

Hey students! Bring these pages to life online

Watch videos

Engage with interactivities

Answer questions and check results

Find all this and MORE in jacPLUS

2.1.1 Introduction

Sequences of numbers play an important part in our everyday life. For example, the following sequence:

2.25, 2.37, 2.58, 2.57, 2.63, ...

gives the end-of-day trading price (for 5 consecutive days) of a share in an electronics company. It looks like the price is on the rise, but is it possible to accurately predict the future price per share of the company?

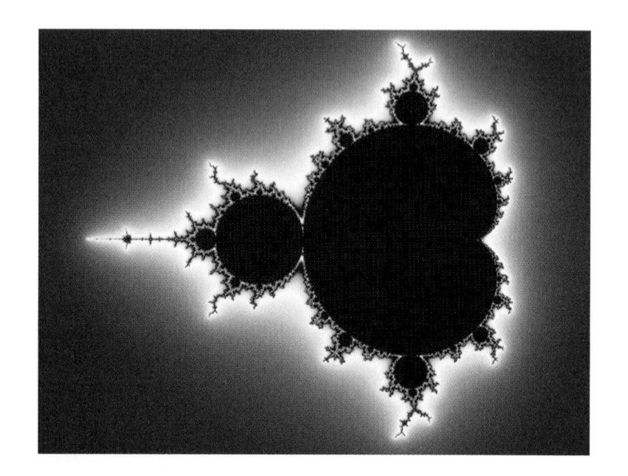

The following sequence is more predictable:

10 000, 9000, 8100, ...

This is the estimated number of radioactive decays of a medical compound each minute after administration to a patient. The compound is used to diagnose tumours. In the first minute, 10 000 radioactive decays are predicted; during the second minute, 9000, and so on. Can you predict the next number in the sequence? You're correct if you said 7290. Each successive term here is 90% of, or 0.90 times, the previous term.

Sequences are strings of numbers. They can be finite in number or infinite. Number sequences may follow an easily recognisable pattern or they may not. A great deal of recent mathematical work has gone into deciding whether certain strings follow a pattern (in which case subsequent terms could be predicted) or whether they are random (in which case subsequent terms cannot be predicted). This work forms the basis of chaos theory, speech recognition software for computers, weather prediction and stock market forecasting, to name but a few uses. The list is almost endless. The image above is a visual representation of a sequence of numbers called a Mandelbrot set.

KEY CONCEPTS

This topic covers the following key concepts from the VCE Mathematics Study Design:

- definitions of sequences and series, arithmetic and geometric sequences and their partial sums
- the limiting behaviour as $n \to \infty$ of the terms t_n in a geometric sequence and dependence on the value of the common ratio
- sequences generated by recursion
- solution of first order linear recurrence relations of the form $t_{n+1} = at_n + b,\ a \neq 0,$ with constant coefficients and their application to financial problems and population modelling.

2.2 Describing sequences

LEARNING INTENTION

At the end of this subtopic you should be able to:

- define sequences by listing them sequentially, using a functional definition or using a recursive definition
- examine the behaviour of logistic sequences as n increases.

2.2.1 Describing sequences

Sequences that follow a pattern can be described in a number of different ways. They may be listed in sequential order; they may be described as a functional definition; or they may be described in an iterative definition.

Listing in sequential order

Consider the sequence of numbers t: $\{5, 7, 9, \ldots\}$. The numbers in sequential order are firstly 5 then 7 and 9, with the indication that there are more numbers to follow. The symbol t is the name of the sequence, and the first three terms in the sequence shown are $t_1 = 5, t_2 = 7$ and $t_3 = 9$. The fourth term, t_4, if the pattern were to continue, would be the number 11. In general, t_n is the nth term in the sequence. In this example, the next term is simply the previous term with 2 added to it, with the first term being the number 5.

Another possible sequence is t: $\{5, 10, 20, 40, \ldots\}$. In this case it appears that the next term is twice the previous term. The fifth term here, if the pattern continued, would be $t_5 = 80$. It can be difficult to determine whether or not a pattern exists in some sequences. Can you find the next term in the following sequence?

$$t: \{1, 1, 2, 3, 5, 8, \ldots\}$$

Here the next term is the sum of the previous *two terms*, hence the next term would be $5 + 8 = 13$, and so on. This sequence is called the Fibonacci sequence and is named after its discoverer Leonardo Fibonacci, a thirteenth century mathematician.

Here is another sequence; can you find the next term here?

$$t: \{7, 11, 16, 22, 29, \ldots\}$$

In this sequence the difference between successive terms increases by 1 for each pair. The first difference is 4, the next difference is 5 and so on. The sixth term is thus 37, which is 8 more than 29.

Functional definition

A **functional definition** of a sequence of numbers is expressed in the form: $t_n = f(n)$.

An example could be: $t_n = 2n - 7,\ n \in \{1, 2, 3, 4, \ldots\}$

Using this definition the nth term can be readily calculated. For this example $t_1 = 2 \times 1 - 7 = -5$, $t_2 = 2 \times 2 - 7 = -3$, $t_3 = 2 \times 3 - 7 = -1$ and so on. We can readily calculate the 100th term, $t_{100} = 2 \times 100 - 7 = 193$, simply by substituting the value $n = 100$ into the expression for t_n.

Look at the following example:

$$d_n = 4.9n^2,\ n \in \{1, 2, 3, \ldots\}$$

For this example, in which the sequence is given the name d, $d_1 = 4.9 \times 1^2 = 4.9$, and $d_2 = 4.9 \times 2^2 = 19.6$. Listing the sequence would yield d: $\{4.9, 19.6, 44.1, 78.4, \ldots\}$. The 10th term would be $4.9 \times 10^2 = 490$.

Recursive definition

A sequence can be generated by the repeated use of an instruction. This is known as **recursion**. Term n is represented by t_n; the next term after this one is represented by t_{n+1}, while the term before t_n is t_{n-1}. For these sequences, the first term must be stated.

Look at the following example:

$$t_{n+1} = 3t_n - 2; t_1 = 6.$$

The first term, t_1, is 6 (this is given in the definition), so the next term, t_2, is $3 \times 6 - 2 = 16$, and the following term is $3 \times 16 - 2 = 46$. In each and all cases, the next term is found by multiplying the previous term by 3 and then subtracting 2. We could write the sequence out as a table:

n	t_n	Comment
1	$t_1 = 6$	Given in the definition
2	$t_2 = 3t_1 - 2$ $= 3 \times 6 - 2$ $= 16$	Using t_1 to find the next term, t_2
3	$t_3 = 3t_2 - 2$ $= 3 \times 16 - 2$ $= 46$	Using t_2 to find the next term, t_3
4	$t_4 = 3t_3 - 2$ $= 3 \times 46 - 2$ $= 136$	Using t_3 to find the next term, t_4

An example of this sequence using notation found in a spreadsheet would be:

A1 = 6 (the first term is equal to 6)

A2 = 3 × A1 − 2 (the next term is 3 times the previous term minus 2).

You could then apply the **Fill Down** option in the **Edit** menu of the spreadsheet from cell A2 downwards to generate as many terms in the sequence as required. This would result in the next cell down being three times the previous cell, less 2. The recursive definition finds a natural use in a spreadsheet environment and consequently is used often. A drawback is that you cannot find the nth term directly as in the functional definition, but an advantage is that more complicated systems can be successfully modelled using recursive descriptions.

WORKED EXAMPLE 1 Determining terms in sequences

a. **Determine the next three terms in the sequence:** $\left\{14, 7, \frac{7}{2}, \ldots\right\}$.

b. **Determine the 4th, 8th and 12th terms in the following sequence:** $e_n = n^2 - 3n, n \in \{1, 2, 3, \ldots\}$.

c. **Determine the second, third and fifth terms for the following sequence:** $k_{n+1} = 2k_n + 1, k_1 = -0.50$.

THINK	WRITE
a. In this example the sequence is listed and a simple pattern is evident. From inspection, the next term is half the previous term and so the sequence would be $14, 7, \frac{7}{2}, \frac{7}{4}, \frac{7}{8}, \frac{7}{16}$.	a. The next three terms are $\frac{7}{4}, \frac{7}{8}, \frac{7}{16}$.

b. 1. This is an example of a functional definition. The nth term of the sequence is found simply by substitution into the expression $e_n = n^2 - 3n$.

b. $e_n = n^2 - 3n$

2. Determine the fourth term by substituting $n = 4$.

$$e_4 = 4^2 - 3 \times 4 \\ = 4$$

3. Determine the eighth term by substituting $n = 8$.

$$e_8 = 8^2 - 3 \times 8 \\ = 40$$

4. Determine the 12th term by substituting $n = 12$.

$$e_{12} = 12^2 - 3 \times 12 \\ = 108$$

c. 1. This is an example of a recursive definition. We can find the second, third and fifth terms for the sequence $k_{n+1} = 2k_n + 1, k_1 = -0.50$ by recursion.

c. $k_{n+1} = 2k_n + 1,$
$k_1 = -0.50$

2. Substitute $k_1 = -0.50$ into the formula to determine k_2.

$$k_2 = 2 \times -0.50 + 1 \\ = 0$$

3. Continue the process until the value of k_5 is found.

$$k_3 = 2 \times 0 + 1 \\ = 1$$

$$k_4 = 2 \times 1 + 1 \\ = 3$$

$$k_5 = 2 \times 3 + 1 \\ = 7$$

4. Write the answer.

Thus $k_2 = 0, k_3 = 1$ and $k_5 = 7$.

2.2 Exercise

Students, these questions are even better in jacPLUS

Receive immediate feedback and access sample responses

Access additional questions

Track your results and progress

Find all this and MORE in jacPLUS

Technology free

1. **WE1** a. Determine the next three terms in the sequence: $\left\{3, \frac{3}{2}, \frac{3}{4}, \ldots\right\}$.
   b. Determine the second, fourth and sixth terms in the following sequence: $t_n = 4 \times 3^{n-2}, n \in \{1, 2, 3, \ldots\}$.
   c. Determine the second, third and fifth terms for the following sequence: $k_{n+1} = k_n + 2, k_1 = -5$.
2. a. Determine the next three terms in the sequence: $\{2, -5, 8, -11, 14, \ldots\}$.
   b. Determine the 4th, 8th and 12th terms in the following sequence: $t_n = n^2 - n + 41, n \in \{1, 2, 3, \ldots\}$.
   c. Determine the second, third and fourth terms for the following sequence: $k_{n+1} = -\left(k_n^{\ 2}\right) - 2, k_1 = 3$.

3. For each of the following sequences, write a rule for obtaining the next term in the sequence and hence evaluate the next three terms.

a. $\{1, 4, 7, \ldots\}$
b. $\{1, 0, -1, -2, \ldots\}$
c. $\{1, 4, 16, 64, \ldots\}$
d. $\{2, 5, 9, 14, 20, \ldots\}$

4. For each of the following sequences, write a rule for obtaining the next term in the sequence and hence evaluate the next three terms.

a. $\{3, 4, 7, 11, 18, \ldots\}$
b. $\{2a - 5b, a - 2b, b, -a + 4b, \ldots\}$
c. $\{1, 0, -1, 0, 1, \ldots\}$
d. $\{1.0, 1.1, 1.11, \ldots\}$

5. Determine the first, fifth and tenth terms in the following sequences.

a. $t_n = 2n - 5, n \in \{1, 2, 3, \ldots\}$
b. $t_n = \dfrac{n}{n+1}, \in \{1, 2, 3, \ldots\}$
c. $t_n = 17 - 3.7n, n \in \{1, 2, 3, \ldots\}$
d. $t_n = 5 \times \left(\dfrac{1}{2}\right)^n, n \in \{1, 2, 3, \ldots\}$

6. Determine the first, fifth and tenth terms in the following sequences.

a. $t_n = 5 \times \left(\dfrac{1}{2}\right)^{(3-n)}, n \in \{1, 2, 3, \ldots\}$
b. $t_n = (-1)^n + n, n \in \{1, 2, 3, \ldots\}$
c. $t_n = 3^n 2^{-n}, n \in \{1, 2, 3, \ldots\}$
d. $t_n = a + (n - 1)d, n \in \{1, 2, 3, \ldots\}$

Technology active

7. Using technology, determine the third, eighth and tenth terms in the following sequences.

a. $t_{n+1} = -2t_n, t_1 = -3$
b. $t_{n+1} = t_n - 7, t_1 = 14$
c. $t_{n+1} = -t_n + 2, t_1 = 3$
d. $t_{n+1} = t_n + (-1)^n t_n, t_1 = 3$

8. For the sequences in question 7, use technology to generate their graphs. Place the term number on the horizontal axis and the value of the term on the vertical axis.

9. Study the pattern in each of the following sequences and where possible write the next two terms in the sequence, describing the pattern that you use.

a. $5, 6, 8, 11, \ldots$
b. $4, 9, 12, 13, 12, 9, \ldots$
c. $9, 8, 9, 0, \ldots$

10. Study the pattern in each of the following sequences and where possible write the next two terms in the sequence, describing the pattern that you use.

a. $6, 12, 12, 6, 1\frac{1}{2}, \ldots$
b. $5, 8, 13, 21, \ldots$
c. $1, 3, 7, 15, \ldots$

11. Write the iterative definition for each of the following sequences.

a. $\{7, 5, 3, 1, -1, \ldots\}$
b. $\{12, 6, 3, 1.5, \ldots\}$
c. $\{12, 12.6, 13.2, \ldots\}$

12. Write the iterative definition for each of the following sequences.

a. $\{2, 11, 56, 281, \ldots\}$
b. $\{4, -12, 36, \ldots\}$
c. $\{2, 4, 16, 256, \ldots\}$

2.2 Exam questions

Question 1 (1 mark) TECH-FREE

Specify a rule for the sequence 2, 5, 10, 17, 26...

Question 2 (1 mark) TECH-FREE

Determine the next three terms in the sequence $a^2, -a, 1, \ldots$

Question 3 (1 mark) TECH-FREE

Write down the first five terms in the sequence $t_n = \dfrac{(n+1)^3}{n}$.

More exam questions are available online.

2.3 Arithmetic sequences

LEARNING INTENTION

At the end of this subtopic you should be able to:
- identify and define arithmetic sequences
- plot graphs of arithmetic sequences.

2.3.1 Identifying and defining arithmetic sequences

At a racetrack a new prototype racing car unfortunately develops an oil leak. Each second, a drop of oil hits the road. The driver of the car puts her foot on the accelerator and the car increases speed at a steady rate as it hurtles down the straight. The diagram below shows the pattern of oil drops on the road with the distances between the drops labelled.

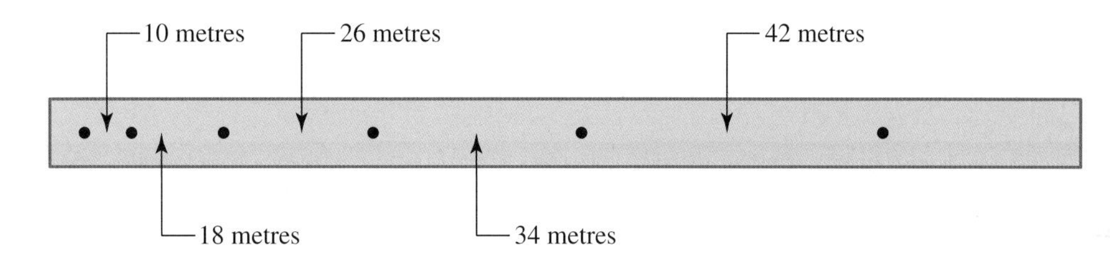

The sequence of distances travelled in metres each second is $\{10, 18, 26, 34, 42, \ldots\}$. The first term in the sequence, t_1, is 10, and as you can see, each subsequent term is 8 more than the previous term. This type of sequence is given a special name — an **arithmetic sequence**.

Definition of an arithmetic sequence

An arithmetic sequence is a sequence where there is a common difference between any two successive terms.

The difference between successive terms in an arithmetic sequence is called the common difference, d. They can be increasing (if $d > 0$), decreasing (if $d < 0$) or constant (if $d = 0$). The first term in an arithmetic is denoted by a.

Some examples of arithmetic sequences are given below.

Arithmetic sequence	First term	Common difference
$\{5, 7, 9, 11, 13, \ldots\}$	$a = 5$	$d = 2$
$\{30, 20, 10, 0, -10, \ldots\}$	$a = 30$	$d = -10$
$\{4, 4, 4, 4, 4, \ldots\}$	$a = 4$	$d = 0$
$\{-77, -66, -55, -44, -33, \ldots\}$	$a = -77$	$d = 11$
$\{3.1, 5.4, 7.7, 10, 12.3, \ldots\}$	$a = 3.1$	$d = 2.3$

The common difference and the first term provide enough information to define any particular arithmetic sequence. We can therefore describe any arithmetic sequence using a functional definition.

As the term number, n, increases by 1, the value of the term increases by the common difference, d, and when $n = 1$, the value of the term is a. The functional definition for an arithmetic sequence is therefore: $t_n = a + (n - 1)d$, $n \in N$. This definition can also be written as $t_n = (a - d) + nd$, $n \in N$. This functional definition is sometimes called the rule for the nth term as the value of the nth term can be easily determined by substituting the value of n into the formula.

Looking at the sequence $\{5, 7, 9, 11, 13, ...\}$ we have $a = 5$ and $d = 2$. Substituting these values into the functional definition gives a rule of $t_n = (5 - 2) + n \times 2$, $n \in N$ which simplifies to $t_n = 2n + 3$, $n \in N$.

Arithmetic sequences

Arithmetic sequences are defined by the formulas:

$$t_n = a + (n - 1)d,\ n \in N$$
$$t_n = (a - d) + nd,\ n \in N$$

Where a is the value of the first term and d is the common difference.

WORKED EXAMPLE 2 Identifying and defining arithmetic sequences

State which of the following are arithmetic sequences by calculating the difference between successive terms. For those that are arithmetic, determine the next term in the sequence, t_4, and consequently determine the functional definition for the nth term for the sequence, t_n.

a. $t: \{4, 9, 15, \ldots\}$
b. $t: \{-2, 1, 4, \ldots\}$

THINK	WRITE
a. 1. To check that a sequence is arithmetic, see if a common difference exists.	a. $9 - 4 = 5$ $15 - 9 = 6$
2. There is no common difference as $5 \neq 6$.	Since there is no common difference the sequence is not arithmetic.
b. 1. To check that a sequence is arithmetic, see if a common difference exists.	b. $1 - -2 = 3$ $4 - 1 = 3$
2. The common difference is 3.	The sequence is arithmetic with the common difference $d = 3$.
3. The next term in the sequence, t_4, can be calculated by adding 3 to the previous term, t_3.	$t_4 = t_3 + 3$ $= 4 + 3$ $= 7$
4. To determine the functional definition, write the formula for the nth term of the arithmetic sequence.	$t_n = a + (n - 1) \times d$ $= (a - d) + nd$
5. Identify the values of a and d.	$a = -2$ and $d = 3$
6. Substitute $a = -2$ and $d = 3$ into the formula and simplify.	$t_n = (-2 - 3) + n \times 3$ $t_n = 3n - 5$

The common difference and first term of an arithmetic sequence can be determined from any two terms, t_a and t_b. First, determine the common difference by calculating the difference between the term values and then dividing by the difference in the term numbers. The formula $d = \frac{t_a - t_b}{a - b}$ describes this process. Once the common difference is known, the first term can be determined by solving the rule for the nth term for a.

WORKED EXAMPLE 3 Determining missing terms in an arithmetic sequence

Determine the missing terms in the arithmetic sequence {41, *e*, 55, *f*, ...}.

THINK

1. The terms $t_1 = 41$ and $t_3 = 55$ are known. Use the formula to determine the common difference.

WRITE

$$\begin{aligned} d &= \frac{t_3 - t_1}{3 - 1} \\ &= \frac{55 - 41}{2} \\ &= \frac{14}{2} \\ &= 7 \end{aligned}$$

2. Use the common difference to determine the missing terms.

$$\begin{aligned} e &= t_2 \\ &= t_1 + d \\ &= 41 + 7 \\ &= 48 \end{aligned} \qquad \begin{aligned} f &= t_4 \\ &= t_3 + d \\ &= 55 + 7 \\ &= 62 \end{aligned}$$

WORKED EXAMPLE 4 Determining the rule for an arithmetic sequence from two terms

Determine 16th and *n*th terms in the arithmetic sequence with 4th term 15 and 8th term 37.

THINK

1. The terms $t_4 = 15$ and $t_8 = 37$ are known. Use the formula to determine the common difference.

WRITE

$$\begin{aligned} d &= \frac{t_8 - t_4}{8 - 4} \\ &= \frac{37 - 15}{4} \\ &= \frac{22}{4} \\ &= 5.5 \end{aligned}$$

2. Solve the formula $t_n = a + (n - 1)\,d$ for a using the known value of t_4.

$$\begin{aligned} t_n &= a + (n - 1) \times d \\ t_4 &= a + (4 - 1) \times d \\ 15 &= a + 3 \times 5.5 \\ 15 &= a + 16.5 \\ a &= -1.5 \end{aligned}$$

3. State the rule for the nth term.

$$\begin{aligned} t_n &= (a - d) + nd \\ &= (-1.5 - 5.5) + n \times 5.5 \\ &= 5.5n - 7 \end{aligned}$$

4. To calculate the 16th term, substitute $n = 16$ into the formula established in the previous step.

$$\begin{aligned} t_{16} &= 5.5 \times 16 - 7 \\ &= 88 - 7 \\ &= 81 \end{aligned}$$

2.3.2 Graphing an arithmetic sequence

WORKED EXAMPLE 5 Graphing arithmetic sequences

Consider the arithmetic sequence 2, 4, 6, 8, 10, ...

a. **Draw up a table showing the term number with its value.**
b. **Graph the values in the table.**
c. **From your graph, determine the value of the tenth term in the sequence.**

THINK

a. Draw up a table to show the term number and value.

WRITE/DRAW

a.

Term number	1	2	3	4	5
Term value	2	4	6	8	10

b. The value of the term depends on the term number, so 'value' is graphed on the y-axis. Draw up a suitable scale on both axes, plot the points.

b.

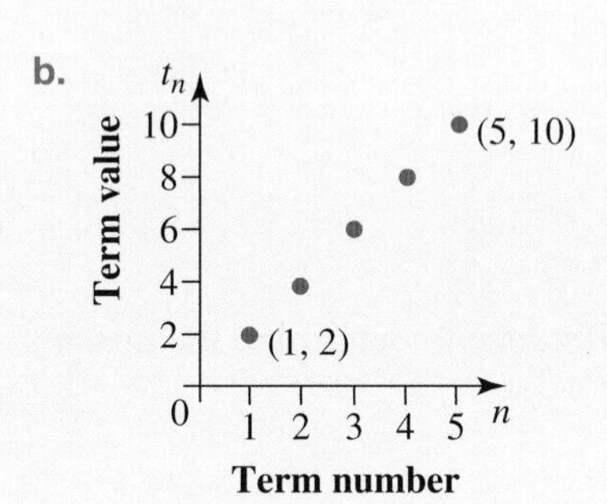

c. 1. To determine a later value in the sequence draw a straight line through the points that were plotted in part **b.** and see where this line intersects the vertical line representing term number 10. Identify the term value of this point.

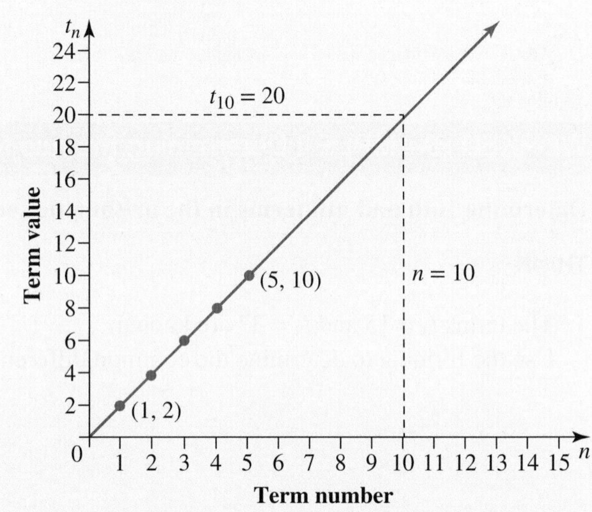

2. Write the answer.

c. The tenth number in this sequence is 20

2.3 Exercise

Students, these questions are even better in jacPLUS

Receive immediate feedback and access sample responses

Access additional questions

Track your results and progress

Find all this and MORE in jacPLUS

Technology free

For this exercise you may use a scientific calculator to aid in calculations.

1. WE2 State which of the following are arithmetic sequences by calculating the difference between successive terms. For those that are arithmetic, determine the next term in the sequence, t_4, and consequently determine the functional definition for the nth term for the sequence, t_n.
   a. t: $\{3, 6, 12, \ldots\}$
   b. t: $\{-3, 0, 3, \ldots\}$
2. State which of the following are arithmetic sequences by calculating the difference between successive terms. For those that are arithmetic, determine the next term in the sequence, t_4, and consequently determine the functional definition for the nth term for the sequence, t_n.
   a. t: $\{4, 7, 11, \ldots\}$
   b. t: $\{-2, -6, -10, \ldots\}$
3. Determine the term given in brackets for each of the following arithmetic sequences.
   a. $\{4, 9, 14, \ldots\}$, (t_{21})
   b. $\{-2, 10, 22, \ldots\}$, (t_{58})
   c. $\{-27, -12, 3, \ldots\}$, (t_{100})
   d. $\{2, -11, -24, \ldots\}$, (t_{2025})
4. Determine the functional definition for the nth term of the arithmetic sequence:
   a. where the first term is 5 and the common difference is -3
   b. where the first term is 2.5 and the common difference is $\frac{1}{2}$
   c. where the first term is -3 and the common difference is 3
   d. where the first term is $2x$ and the common difference is $5x$.
5. WE3 Determine the missing terms in this arithmetic sequence: $\{16, m, 27, n\}$.
6. Determine missing terms in this arithmetic sequence: $\{33, x, 61, y\}$.
7. WE4 Calculate the fourth term and nth term in the arithmetic sequence whose first term is 6 and whose seventh term is -10.
8. Calculate the eighth and nth terms in an arithmetic sequence with the third term 3.44 and the 12th term 5.42.
9. Determine the nth term in the arithmetic sequence where the first term is 6 and the third term is 10.
10. Determine the nth term in the arithmetic sequence where the first term is 3 and the third term is 13.
11. If $t_{10} = 100$ and $t_{15} = 175$, determine the first term, the common difference and hence the nth term for the arithmetic sequence.
12. If $t_{10} = \frac{-1}{2}$ and $t_{13} = \frac{3}{4}$, determine the first term, the common difference and hence the nth term for the arithmetic sequence.

13. For the arithmetic sequence $\{22, m, n, 37, \ldots\}$, determine the values for m and n.

14. The first three terms in an arithmetic sequence are 37, 32 and 27, and the kth term is -3. Determine the value for k.

15. Determine the value of x such that the following forms an arithmetic progression: $\ldots, x, 3x+4, 10x-7, \ldots$

16. **WE5** Consider the arithmetic sequence $4, 8, 12, 16, 20, \ldots$

a. Draw up a table showing the term number with its value.
b. Graph the values in the table.
c. From your graph, determine the value of the tenth term in the sequence.

17. Consider the arithmetic sequence $10, 13, 16, 19, 22, \ldots$

a. Draw up a table showing the term number with its value.
b. Graph the values in the table.
c. From your graph, determine the value of the tenth term in the sequence.

18. For the following arithmetic sequences, determine the recursive definition and use technology to generate the first 50 numbers in the sequence.

a. t_n: $\{3, 7, 11, \ldots\}$
b. t_n: $\{-3, 0, 3, \ldots\}$
c. t_n: $\left\{\frac{2}{7}, \frac{11}{14}, \frac{9}{7}, \ldots\right\}$

19. For the following arithmetic sequences, determine the recursive definition and use technology to generate the first 50 numbers in the sequence.

a. t_n: $\left\{\frac{3}{4}, \frac{3}{2}, \frac{9}{4}, \ldots\right\}$
b. t_n: $\left\{\frac{1}{4}, \frac{-3}{2}, \frac{-13}{4}, \ldots\right\}$
c. t_n: $\{2\pi+3, 4\pi+1, 6\pi-1, \ldots\}$

20. The ratio between the first term and the second term in an arithmetic sequence is $\frac{3}{4}$. The ratio between the second term and the third term is $\frac{4}{5}$.

a. Calculate the ratio of the third term to the fourth term.
b. Determine the ratio of the nth and the $(n+1)$th term in the sequence.

2.3 Exam questions

Question 1 (2 marks) TECH-FREE

Determine the common difference of the arithmetic sequence given by $t_n = (n+1)^2 - n(n+3) + 4,\ n \in N$.

Question 2 (1 mark) TECH-FREE

In an arithmetic sequence, $t_{16} = 5t_3$ and the common difference is 4. Determine the value of the first term.

Question 3 (2 marks) TECH-FREE

p, q, r, s are four consecutive terms in an arithmetic sequence. Show that $qr - ps = 2(r-s)^2$.

More exam questions are available online.

2.4 Arithmetic series

LEARNING INTENTION

At the end of this subtopic you should be able to:
- calculate the sum of terms in arithmetic sequences.

2.4.1 Definition of a series

In many cases we are interested, not only in the values of terms in a sequence, and the sequence's long term behaviour, but also in the sum of the terms in a sequence. For an example, we return to the oil drops on the racetrack from the start of the previous section on arithmetic sequences. The distance covered by the car each second illustrated the concept of an arithmetic sequence.

The total distance covered by the car is the sum of the individual distances covered each second. So after one second the car has travelled 10 m, after two seconds the car has travelled $10 + 18\text{ m} = 28\text{ m}$, after three seconds the car has travelled a total distance of $10 + 18 + 26\text{ m} = 54\text{ m}$, and so on.

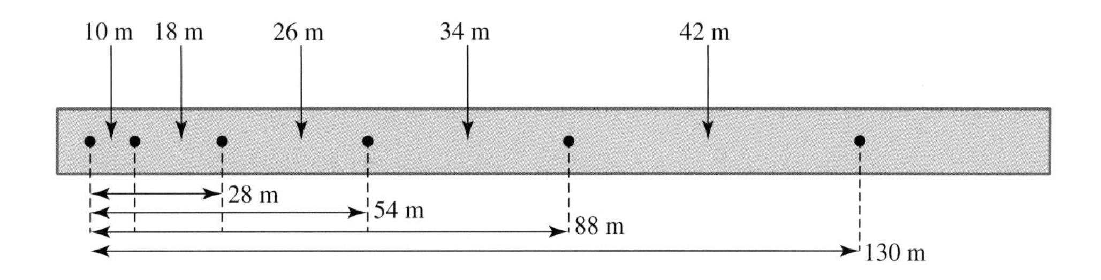

Definition of a series

A series, S_n, is the sum of a sequence of n terms $t_1 + t_2 + t_3 + \ldots + t_n$.

Thus:

$$\begin{aligned}S_1 &= t_1\\ S_2 &= t_1 + t_2\\ S_3 &= t_1 + t_2 + t_3\\ S_n &= t_1 + t_2 + t_3 + \ldots + t_{n-2} + t_{n-1} + t_n\end{aligned}$$

2.4.2 Arithmetic series

Since arithmetic sequences have a first term of $t_1 = a$ and a common difference of d, we can express the sum of their terms neatly. An arithmetic series with n terms S_n is defined as follows.

$$\begin{aligned}S_n &= t_1 + t_2 + t_3 + \ldots + t_{n-2} + t_{n-1} + t_n\\ &= a + (a + d) + (a + 2d) + \ldots + \big(a + (n-3)d\big) + \big(a + (n-2)d\big) + \big(a + (n-1)d\big)\end{aligned}$$

If we denote the last term (t_n) in the arithmetic sequence by l, we can express the series as follows:

$$S_n = a + (a + d) + (a + 2d) + \ldots + (l - 2d) + (l - d) + l$$

Writing the series in reverse order:

$$S_n = l + (l - d) + (l - 2d) + \ldots + (a + 2d) + (a + d) + a$$

Notice that if we add these two expressions for S_n the result is the sum of n groups of $(a + l)$.

$$\begin{aligned} 2S_n &= (a + l) + (a + l) + (a + l) + \ldots \quad (n \text{ times}) \\ &= n(a + l) \end{aligned}$$

Dividing both sides of this equation by 2 gives an expression for the arithmetic series S_n.

$$S_n = \frac{n}{2}(a + l)$$

Recall that l is the nth term of the arithmetic sequence, $l = a + (n - 1)d$, so this expression for the series S_n can also be expressed as $S_n = \frac{n}{2}[2a + (n - 1)d]$.

These expressions allow us to calculate the sum of an arithmetic sequence without having to add up all of the individual terms.

Arithmetic series

The sum of the first n terms in an arithmetic series is given by:

$$S_n = \frac{n}{2}(a + l) \text{ or } S_n = \frac{n}{2}[2a + (n - 1)d]$$

Where a is the first term of the arithmetic sequence, l is the last term of the arithmetic sequence and d is the common difference.

It is worthwhile also to note that $S_{n+1} = S_n + t_{n+1}$. This tells us that the next term in the series S_{n+1} is the present sum, S_n, plus the next term in the sequence, t_{n+1}. This result is useful in spreadsheets where one column gives the sequence and an adjacent column is used to give the series.

WORKED EXAMPLE 6 Calculating the sum of an arithmetic sequence

Calculate the sum of the first 20 terms in the sequence t_n: {12, 25, 38, ...}.

THINK	WRITE
1. Write the formula for the sum of the first n terms in the arithmetic sequence.	$S_n = \frac{n}{2}(2a + (n - 1)d)$
2. Identify the variables.	$a = 12, d = 25 - 12, n = 20$ $= 13$
3. Substitute values of a, d and n into the formula and evaluate.	$S_{20} = \frac{20}{2}(2 \times 12 + 19 \times 13)$ $S_{20} = 2710$

2.4 Exercise

Students, these questions are even better in jacPLUS

Receive immediate feedback and access sample responses

Access additional questions

Track your results and progress

Find all this and MORE in jacPLUS

Technology free

1. WE6 Calculate the sum of the first 20 terms in the sequence $t_n : \{1, 3, 5, \ldots\}$.
2. Calculate the sum of the first 50 terms in the sequence $t_n = 3n + 7, n \in \{1, 2, 3, \ldots\}$
3. a. Calculate the sum of the first 50 positive integers.
   b. Calculate the sum of the first 100 positive integers.
4. a. Calculate the sum of all the half-integers between 0 and 100.
   Note: The sequence of half-integers is $\left\{\frac{1}{2}, 1\frac{1}{2}, 2\frac{1}{2}, 3\frac{1}{2}, \ldots\right\}$.
   b. Compare your answer with that for question 3b.
5. Calculate the sum of the first 12 terms of an arithmetic sequence in which the second term is 8 and 13th term is 41.
6. A sequence of numbers is defined by t_n: $\{15, 9, 3, -3, \ldots\}$.
   a. Calculate the sum of the first 13, 16 and 19 terms in the sequence.
   b. Calculate the sum of all the terms between and including t_{10} and t_{15}.
7. A sequence of numbers is defined by $t_n = 2n - 7$, $n \in \{1, 2, 3, \ldots\}$. Calculate:
   a. the sum of the first 20 terms
   b. the sum of all the terms between and including t_{21} and t_{40}
   c. the average of the first 40 terms. *Hint:* You need to calculate the sum first.
8. Determine the equation that gives the sum of the first n positive integers.
9. a. Show that the sum of the first n odd positive integers is equal to the perfect square n^2.
   b. Show that the sum of the first n even positive integers is equal to $n^2 + n$.

Technology active

10. A sequence is $5, 7, 9, 11, \ldots$ Determine how many consecutive terms need to be added to obtain a sum of 357.
11. Consider the sum of the first n positive integers. Determine the value of n when the sum will first exceed 1000.
12. The first term in an arithmetic sequence is 5, and the sum of the first 20 terms is 1240. Determine the common difference, d.
13. The sum of the first four terms of an arithmetic sequence is 58, and the sum of the next four terms is twice that number. Determine the sum of the following four terms.
14. The sum of a series is given by $S_n = 4n^2 + 3n$. Use the result that $t_{n+1} = S_{n+1} - S_n$ to prove that the sequence of numbers, t_n, whose series is $S_n = 4n^2 + 3n$ is arithmetic. Determine both the functional and recursive equations for the sequence, t_n.

2.4 Exam questions

Question 1 (2 marks) TECH-FREE

Calculate the sum of all natural numbers between 1 and 100 that are divisible by three.

Question 2 (2 marks) TECH-FREE

Calculate the sum of the first 15 terms of the sequence 5, 9, 13, 17, ...

Question 3 (2 marks) TECH-FREE

Consider the sequence $t_n = pn + q$. Calculate the sum of the first 24 terms in the sequence by first showing that the sequence is arithmetic.

More exam questions are available online.

2.5 Geometric sequences

LEARNING INTENTION

At the end of this subtopic you should be able to:

- identify and define geometric sequences
- plot graphs of geometric sequences.

2.5.1 Geometric sequences

A farmer is breeding worms that he hopes to sell to local shire councils to decompose waste at rubbish dumps. Worms reproduce readily and the farmer expects a 10% increase per week in the mass of worms that he is farming. A 10% increase per week would mean that the mass of worms would increase by a constant factor of $\left(1 + \frac{10}{100}\right) = 1.1$.

He starts off with 10 kg of worms. By the beginning of the second week he will expect $10 \times 1.1 = 11$ kg of worms, by the start of the third week he would expect $11 \times 1.1 = 10 \times (1.1)^2 = 12.1$ kg of worms, and so on. This is an example of a **geometric sequence**.

Definition of a geometric sequence

A geometric sequence is a sequence where each term is obtained by multiplying the preceding term by a certain constant factor.

The multiplicative factor between successive terms in a geometric sequence is called the common ratio, r, and it can be determined by dividing any two successive terms: $r = \frac{t_{n+1}}{t_n}$. The first term of a geometric sequence is denoted by a.

All geometric sequences can therefore be written in terms of a and r as follows: $\left\{a, ar, ar^2, ar^3, \ldots, ar^{(n-1)}\right\}$.

Some examples of geometric sequences are given below.

Geometric sequence	First term	Common ratio
{10, 11, 12.1, 13.31, 14.641, ...}	$a = 10$	$r = 1.1$
{1, 2, 4, 8, 16, ...}	$a = 1$	$r = 2$
{15, 15, 15, 15, 15, ...}	$a = 15$	$r = 1$
{128, 64, 32, 16, 8, ...}	$a = 128$	$r = \frac{1}{2}$
{3, −6, 12, −24, 48, ...}	$a = 3$	$d = -2$

Geometric sequences

The nth term of a geometric sequence is given by:

$$t_n = ar^{(n-1)}$$

where a is the first term of the sequence, and r is the common ratio, given by:

$$r = \frac{t_{n+1}}{t_n}.$$

WORKED EXAMPLE 7 Determining whether a sequence is geometric

State whether the sequence t_n: {2, 6, 18, ...} is geometric by calculating the ratio of successive terms. If it is geometric, determine the next term in the sequence, t_4, and the nth term for the sequence, t_n.

THINK	WRITE
1. Calculate the ratio $\frac{t_2}{t_1}$.	$\frac{t_2}{t_1} = \frac{6}{2}$ $= 3$
2. Calculate the ratio $\frac{t_3}{t_2}$.	$t_3 = \frac{18}{6}$ $= 3$
3. Compare the ratios and make your conclusion.	Since $\frac{t_2}{t_1} = \frac{t_3}{t_2} = 3$, the sequence is geometric with the common ratio $r = 3$.
4. Because the sequence is geometric, determine the fourth term by multiplying the preceding (third) term by the common ratio.	$t_4 = t_3 \times r$ $= 18 \times 3$ $= 54$
5. Write the general formula for the nth term.	$t_n = ar^{n-1}$
6. Identify the values of a and r.	$a = 2; r = 3$
7. Substitute the values of a and r into the general formula.	$t_n = 2 \times 3^{n-1}$
8. Check the value for t_4.	$t_4 = 2 \times 3^{4-1} = 2 \times 27 = 54$

The expression for the nth term of a geometric sequence can be used to determine the values of a and r for geometric sequences when we are given two terms in the sequence.

To determine these values, substitute the known values from the given terms into the expression for the nth term to form equations which can be solved simultaneously.

WORKED EXAMPLE 8 Determining the rule for a geometric sequence

Determine the nth term and the tenth term in the geometric sequence where the first term is 3 and the third term is 12.

THINK	WRITE
1. Write the general formula for the nth term in the geometric sequence.	$t_n = ar^{n-1}$
2. State the value of a (the first term in the sequence) and the value of the third term.	$a = 3;\ t_3 = 12$
3. Substitute all known values into the general formula.	$12 = 3 \times r^{3-1}$ $= 3 \times r^2$
4. Solve for r (note that there are two possible solutions).	$r^2 = \frac{12}{3}$ $= 4$ $r = \pm\sqrt{4}$ $= \pm 2$
5. Substitute the values of a and r into the general equation. Because there are two possible values for r, you must show both expressions for the nth term of the sequence.	So $t_n = 3 \times 2^{n-1}$, or $t_n = 3 \times (-2)^{n-1}$
6. Determine the tenth term by substituting $n = 10$ into each of the two expressions for the nth term.	When, $n = 10$, $t_{10} = 3 \times 2^{10-1}$ (using $r = 2$) $= 3 \times 2^9$ $= 1536$ or $t_{10} = 3 \times (-2)^{10-1}$ (using $r = -2$) $= 3 \times (-2)^9 = -1536$

WORKED EXAMPLE 9 Determining the rule for a geometric sequence given two terms

The fifth term in a geometric sequence is 14 and the seventh term is 0.56. Determine the common ratio, r, the first term, a, and the nth term for the sequence.

THINK	WRITE
1. Write the general rule for the nth term of a geometric sequence.	$t_n = ar^{n-1}$
2. Use the information about the fifth term to form an equation. Label it [1].	When $n = 5$, $t_n = 14$ $14 = a \times r^{5-1}$ $14 = a \times r^4$ [1]

3. Similarly, use information about the seventh term to form an equation. Label it [2].	When $n = 7, t_n = 0.56$ $0.56 = a \times r^{7-1}$ $0.56 = a \times r^6$ [2]
4. Solve the equations simultaneously: divide equation [2] by equation [1] to eliminate a.	$\frac{[2]}{[1]}$ gives $\frac{ar^6}{ar^4} = \frac{0.56}{14}$
5. Solve for r.	$r^2 = 0.04$ $r = \pm\sqrt{0.04}$ $= \pm 0.2$
6. Because there are two solutions, we have to perform two sets of computations. Consider the positive value of r first. Substitute the value of r into either of the two equations, say equation [1], and solve for a.	If $r = 0.2$ Substitute r into [1]: $a \times (0.2)^4 = 14$ $0.0016a = 14$ $a = 14 \div 0.0016$ $= 8750$
7. Substitute the values of r and a into the general equation to determine the expression for the nth term.	The nth term is: $t_n = 8750 \times (0.2)^{n-1}$
8. Now consider the negative value of r. Substitute the value of r into either of the two equations, say equation [1], and solve for a. (Note that the value of a is the same for both values of r.)	If $r = -0.2$ Substitute r into [1] $a = (-0.2)^4 = 14$ $0.0016a = 14$ $a = 14 \div 0.0016$ $= 8750$
9. Substitute the values of r and a into the general formula to determine the second expression for the nth term of the sequence.	The nth term is: $t_n = 8750 \times (-0.2)^{n-1}$
10. State the possible solutions.	The first term is $a = 8750$. The common ratio is $r = \pm 0.2$. The two possible rules for the nth term are: $t_n = 8750 \times (0.2)^{n-1}$ or $t_n = 8750 \times (-0.2)^{n-1}$

2.5.2 Graphs of geometric sequences

Unlike graphs of arithmetic sequences, which are always linear, the shape of the graph of a geometric sequence depends on the value of the common ratio, r.

Graphs of geometric sequences with $0 < r < 1$

When $0 < r < 1$, the points in the sequence get closer to the origin as the term number increases. Sequences that get closer and closer to a particular value as the term number increases are said to **converge** to that particular value. These graphs are curved and get flatter as the term number increases.

The graph of the geometric sequence with $a = 128$ and $r = \frac{1}{2}$ is shown below.

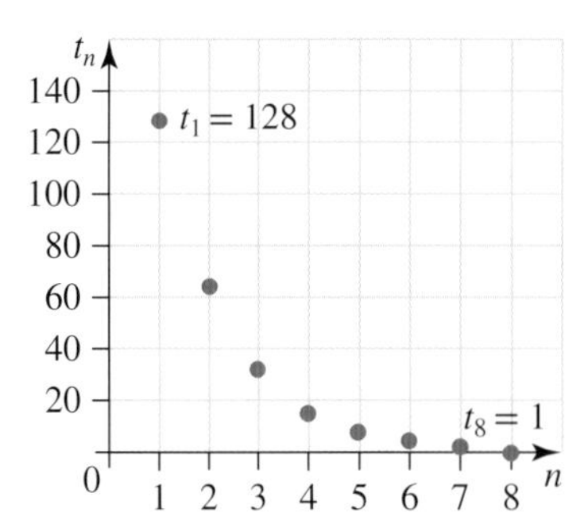

Graphs of geometric sequences with $r > 1$

When $r > 1$, the points lie on a curve that increases in magnitude as the term number increases. Sequences like this, which do not converge to any particular value are said to **diverge**.

The graph of the geometric sequence with $a = -1$ and $r = 2$ is shown below.

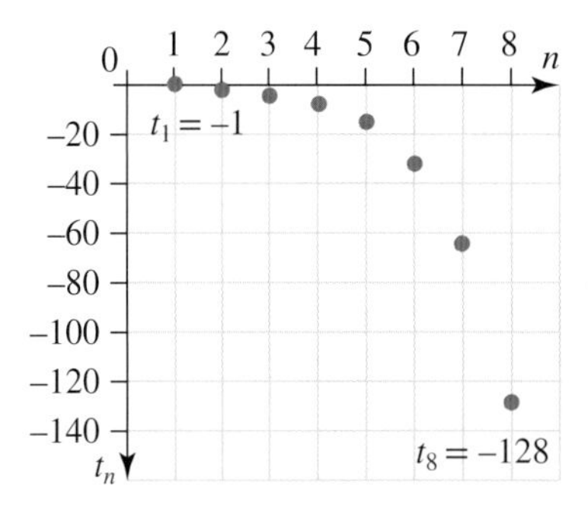

Graphs of geometric sequences with $r < 0$

When $r < 0$ the values of the terms continually flip from positive to negative as the term number increases. The graphs of these sequences therefore oscillate on either side of the horizontal axis.

When $-1 < r < 0$, the sequence converges to 0 as the term number increases.

For example, the graph of the geometric sequence with $a = 128$ and $r = -\frac{1}{2}$ is shown below.

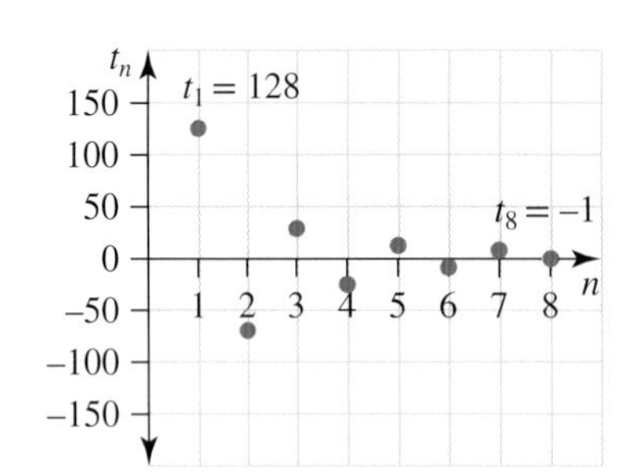

When $r < -1$ the sequence gets further away from the horizontal axis as the term number increases (diverges).

For example, the graph of the geometric sequence with $a = 1$ and $r = -2$ is shown below.

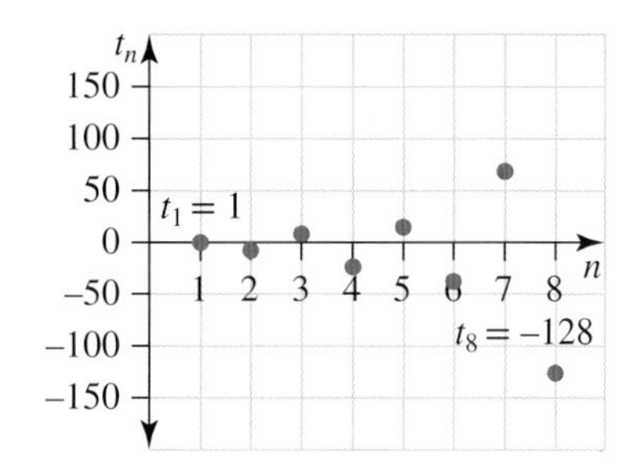

Graphs of geometric sequences with $r = 1$ and $r = -1$

When $r = 1$, the sequence is constant and so the graph of the sequence horizontal.

For example, the graph of the geometric sequence with $a = 5$ and $r = 1$ is shown below.

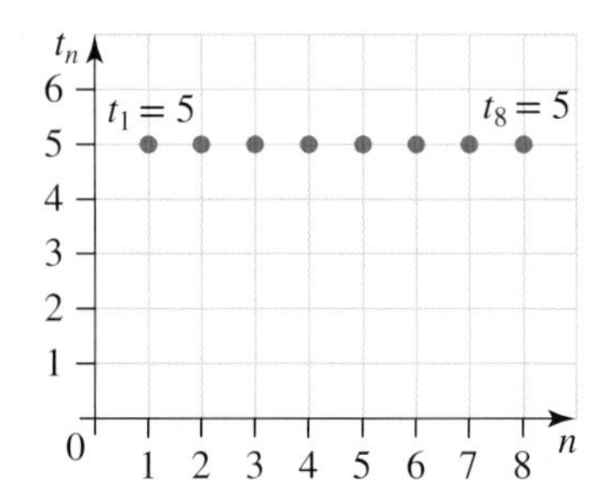

When $r = -1$, the sequence oscillates between the negative and positive of the first term, a.

For example, the graph of the geometric sequence with $a = 3$ and $r = -1$ is shown below.

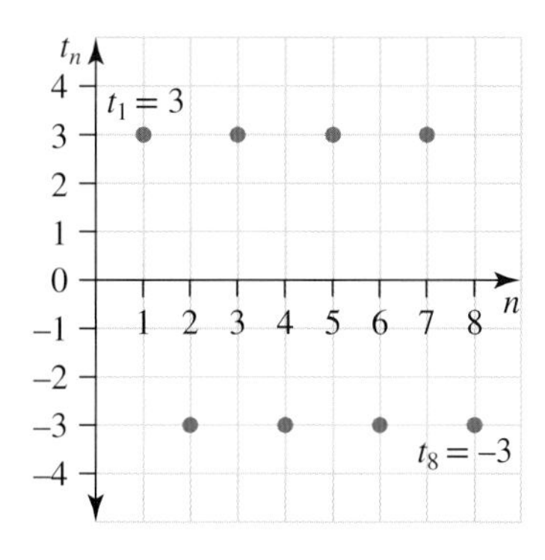

WORKED EXAMPLE 10 Graphing a geometric sequence

Consider the geometric sequence 2, 4, 8, 16, 32, ...

a. Draw up a table showing the term number and its value.

b. Graph the entries in the table.

c. Comment on the shape of the graph.

THINK	WRITE/DRAW
a. Draw up a table showing the term number and its corresponding value.	a. (see table below)

Term number	1	2	3	4	5
Term value	2	4	8	16	32

b. The value of the term depends on the term number, so 'Term value' is graphed on the y-axis. Draw a set of axes with suitable scales and plot the points.

b.

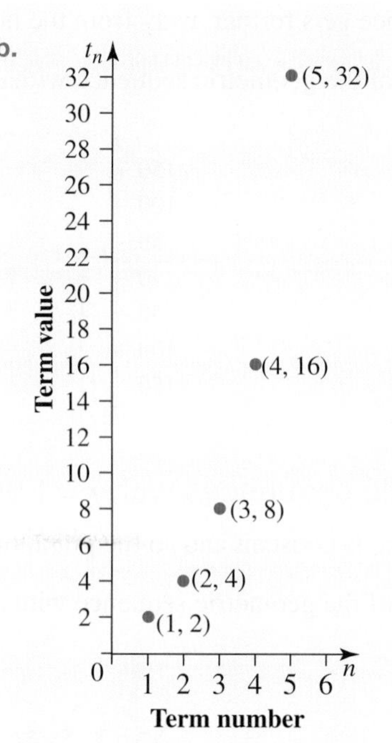

c. Comment on the shape of the curve.

c. The points lie on a smooth curve which increases rapidly. Because of this rapid increase in value, it would be difficult to use the graph to predict future values in the sequence with any accuracy.

2.5 Exercise

Technology free

1. **WE7** State whether the sequence is geometric by calculating the ratio of successive terms for t_n: $\{3, 6, 12, \ldots\}$. If the sequence is geometric, determine the next term in the sequence, t_4, and the nth term for the sequence, t_n.

2. State whether the sequence is geometric by calculating the ratio of successive terms for t_n: $\left\{-3, 1, \frac{-1}{3}, \ldots\right\}$.
If the sequence is geometric, determine next term in the sequence, t_4, and the nth term for the sequence, t_n.

3. For each of the following:
   i. show that the sequence is geometric
   ii. determine the nth term and consequently the sixth and tenth terms.

   a. t: $\{5, 10, 20, \ldots\}$
   b. t: $\{2, 5, 12.5, \ldots\}$
   c. t: $\{1, -3, 9, \ldots\}$
   d. t: $\{2, -4, 8, \ldots\}$

4. For each of the following:
   i. show that the sequence is geometric
   ii. determine the nth term and consequently the sixth and tenth terms.

   a. $t: \left\{\frac{1}{2}, 1, 2, \ldots\right\}$
   b. $t: \left\{\frac{1}{3}, \frac{1}{12}, \frac{1}{48}, \ldots\right\}$
   c. $t: \{x, 3x^4, 9x^7, \ldots\}$
   d. $t: \left\{\frac{1}{x}, \frac{2}{x^2}, \frac{4}{x^3}, \ldots\right\}$

5. **WE8** Determine the nth term and the tenth term in the geometric sequence where the first term is 2 and the third term is 18.

6. Determine the nth term and the tenth term in the geometric sequence where the first term is 1 and the third term is 4.

7. Determine the nth term and the tenth term in the geometric sequence where:
   a. the first term is 5 and the fourth term is 40
   b. the first term is -1 and the second term is 2
   c. the first term is 9 and the third term is $\frac{1}{81}$. (State why there are two possible answers.)

8. Calculate the fourth term in the geometric sequence where the first term is 6 and the seventh term is $\frac{3}{32}$.

Technology active

9. Determine the nth term in the geometric sequence where the first term is 3 and the fourth term is $6\sqrt{2}$.

10. **WE9** The third term in a geometric sequence is 100 and the fifth term is 400. Determine the common ratio, r, the first term, a, and the nth term for the sequence.

11. The fourth term in a geometric sequence is 48 and the eighth term is 768. Determine the common ratio, r, the first term, a, and the nth term for the sequence.

12. For the geometric sequence t: $\{3, m, n, 192, \ldots\}$, determine the values for m and n.

13. Consider the geometric sequence t: $\{16, m, 81, n, \ldots\}$. Determine the values of m and n, if it is known that both are positive numbers.

14. For the geometric sequence t: $\{a, 15, b, 0.0375, \ldots\}$, determine the values of a and b, given that they are positive numbers.

15. If $t_2 = \frac{1}{2}$ and $t_5 = \frac{27}{16}$, determine the first term, a, the common factor, r, and hence the nth term for the geometric sequence.

16. Determine the value of x such that the following sequence forms a geometric progression: $x - 1, 3x + 4, 6x + 8$.

17. Determine the three missing terms in the sequence: 8, ____, ____, ____, $\frac{1}{32}$ such that the sequence of numbers is geometric.

18. **WE10** Consider the geometric sequence $1, 3, 9, 27, 81, \ldots$
   a. Draw up a table showing the term number and its value.
   b. Graph the entries in the table.
   c. Comment on the shape of the graph.

19. Consider the geometric sequence $-20, 10, -5, 2.5, -1.25, \ldots$
   a. Draw up a table showing the term number and its value.
   b. Graph the entries in the table.
   c. Comment on the shape of the graph.

20. The difference between the first term and the second term in a geometric sequence is 6. The difference between the second term and the third term is 3.

 a. Calculate the difference between the third term and the fourth term.

 b. Determine the nth term in the sequence.

21. The first two terms in a geometric sequence are 120 and 24, and the kth term is 0.0384. Determine the value for k.

2.5 Exam questions

Question 1 (1 mark) TECH-ACTIVE

MC For the geometric sequence $m, -\frac{n}{m}, \frac{n^2}{m^3}, ...$, the common ration and the tenth term are

A. $-\frac{n}{m}, -\frac{n^9}{m^8}$ B. $-\frac{n}{m}, \frac{n^9}{m^8}$ C. $-\frac{n}{m^2}, -\frac{n^9}{m^{17}}$ D. $-\frac{n}{m^2}, \frac{n^9}{m^{17}}$ E. $-\frac{n^2}{m^2}, -\frac{n^{18}}{m^{17}}$

Question 2 (1 mark) TECH-FREE

Insert one term between 2 and 8 to form a geometric sequence.

Question 3 (2 marks) TECH-FREE

Determine the number which forms a geometric sequence when added to each of the numbers 11, 17, 25.

More exam questions are available online.

2.6 Geometric series

LEARNING INTENTION

At the end of this subtopic you should be able to:
- calculate the sum of terms in geometric sequences
- calculate the sum to infinity of geometric sequences which have $|r| < 1$.

2.6.1 Geometric series

When we add up or sum the terms in a sequence we get the series for that sequence. If we look at the geometric sequence {2, 6, 18, 54, ...}, where the first term $t_1 = a = 2$ and the common ratio is 3, we can quickly calculate the first few terms in the series of this sequence.

$$S_1 = t_1 = 2$$
$$S_2 = t_1 + t_2 = 2 + 6 = 8$$
$$S_3 = t_1 + t_2 + t_3 = 2 + 6 + 18 = 26$$
$$S_4 = t_1 + t_2 + t_3 + t_4 = 2 + 6 + 18 + 54 = 80$$

In general the sum of the first n terms is:

$$S_n = t_1 + t_2 + t_3 + ... + t_{n-2} + t_{n-1} + t_n$$

For a geometric sequence the first term is a, the second term is ar, the third term is ar^2 and so on up to the nth term, which is ar^{n-1}. Thus:

$$S_n = a + ar + ar^2 + \ldots + ar^{n-3} + ar^{n-2} + ar^{n-1} \qquad [1]$$

If we multiply equation [1] by r we get:

$$rS_n = ar + ar^2 + ar^3 + \ldots + ar^{n-2} + ar^{n-1} + ar^n \qquad [2]$$

Note that on the right-hand side of equations [1] and [2] all but two terms are common, namely the first term in equation [1], a, and the last term in equation [2], ar^n. If we take the difference between equation [2] and equation [1] we get:

$$rS_n - S_n = ar^n - a \qquad [2]-[1]$$

$$\therefore \quad (r-1)S_n = a(r^n - 1)$$

$$\therefore \quad S_n = \frac{a(r^n - 1)}{r-1}; r \neq 1 \qquad (r \text{ cannot equal } 1)$$

We now have an equation that allows us to calculate the sum of the first n terms of a geometric sequence.

Geometric series

The sum of the first n terms of a geometric sequence is given by:

$$S_n = \frac{a(r^n - 1)}{r-1}; r \neq 1$$

where a is the first term of the sequence and r is the common ratio.

WORKED EXAMPLE 11 Calculating the sum of terms in geometric sequences

Calculate the sum of the first five terms (S_5) of these geometric sequences.

a. t_n: $\{1, 4, 16, \ldots\}$

b. $t_n = 2(2)^{n-1}, n \in \{1, 2, 3, \ldots\}$

c. $t_{n+1} = \frac{1}{4}t_n, t_1 = \frac{-1}{2}$

THINK	WRITE
a. 1. Write the general formula for the sum of the first n terms of the geometric sequence.	a. $S_n = \frac{a(r^n - 1)}{r-1}$
2. Write the sequence.	t_n: $\{1, 4, 16, \ldots\}$
3. Identify the variables: a is the first term; r can be established by calculating the ratio; n is known from the question.	$a = 1$; $r = \frac{4}{1} = 4$; $n = 5$

4. Substitute the values of a, r and n into the formula and evaluate.

$$S_5 = \frac{1(4^5 - 1)}{4 - 1} = \frac{1024 - 1}{3} = 341$$

b. 1. Write the sequence.

2. Compare the given rule with the general formula for the nth term of the geometric sequence $t_n = ar^{n-1}$ and identify values of a and r; the value of n is known from the question.

b. $t_n = 2(2)^{n-1},\ n \in \{1, 2, 3, \ldots\}$

$a = 2; r = 2; n = 5$

3. Substitute values of a, r and n into the general formula for the sum and evaluate.

$$S_5 = \frac{2(2^5 - 1)}{2 - 1} = 62$$

c. 1. Write the sequence.

c. $t_{n+1} = \frac{1}{4}t_n, t_1 = \frac{-1}{2}$

2. This is an iterative formula, so the coefficient of t_n is r; $a = t_1$; n is known from the question.

$r = \frac{1}{4}$; $a = \frac{-1}{2}$; $n = 5$

3. Substitute values of a, r and n into the general formula for the sum and evaluate.

$$S_5 = \frac{\frac{-1}{2}\left[\left(\frac{1}{4}\right)^5 - 1\right]}{\frac{1}{4} - 1} = \frac{\frac{-1}{2} \times \left(\frac{1}{1024} - 1\right)}{\frac{-3}{4}} = \frac{-341}{512}$$

2.6.2 The infinite sum of a geometric sequence where $r < 1$

When the constant ratio, r, is less than 1 and greater than -1, that is, $\{r: -1 < r < 1\}$, each successive term in the sequence gets closer to zero. This can readily be shown with the following two examples.

$$g: \left\{2, -1, \frac{1}{2}, \frac{-1}{4}, \ldots\right\} \text{ where } a = 2 \text{ and } r = \frac{-1}{2}$$

$$h: \left\{40, \frac{1}{2}, \frac{1}{160}, \ldots\right\} \text{ where } a = 40 \text{ and } r = \frac{1}{80}$$

In both the examples, successive terms approach zero as n increases. In the second case the approach is more rapid than in the first, and the first sequence alternates positive and negative. A simple investigation whereby the first few terms are plotted will quickly reveal that for geometric sequences with the size or magnitude of $r < 1$, the series eventually settles down to a near constant value. We say that the series converges to a value S_∞, which is the sum to infinity of all terms in the geometric sequence. We can find the value S_∞ by recognising that as $n \to \infty$ the term $r^n \to 0$, provided r is between -1 and 1. We write this technically as $-1 < r < 1$ or $|r| < 1$. The symbol $|r|$ means the magnitude or size of r. Using our equation for the sum of the first n terms:

$$S_n = \frac{a(r^n - 1)}{r - 1};\ r \neq 1$$

Taking -1 as a common factor from the numerator and denominator:

$$S_n = \frac{a(1-r^n)}{1-r}$$

As $n \to \infty$, $r^n \to 0$ and hence $1 - r^n \to 1$. Thus the top line or numerator will equal a when $n \to \infty$:

$$S_\infty = \frac{a}{1-r}; \quad |r| < 1$$

We now have an equation that allows us to calculate the sum to infinity, S_∞, of a geometric sequence.

The sum to infinity of a geometric sequence

The sum to infinity, S_∞, of a geometric sequence is given by:

$$S_\infty = \frac{a}{1-r}; |r| < 1$$

where a is the first term of the sequence and r is the common ratio whose magnitude is less than one.

WORKED EXAMPLE 12 Calculating the sum to infinity of geometric sequences

a. Calculate the sum to infinity for the sequence t_n: $\{10, 1, 0.1, \ldots\}$.

b. Determine the fourth term in the geometric sequence whose first term is 6 and whose sum to infinity is 10.

THINK	WRITE
a. 1. Write the formula for the nth term of the geometric sequence.	a. $t_n = ar^{n-1}$
2. From the question we know that the first term, a, is 10 and $r = 0.1$.	$a = 10$, $r = 0.1$
3. Write the formula for the sum to infinity.	$S_\infty = \frac{a}{1-r}; \ |r| < 1$
4. Substitute $a = 10$ and $r = 0.1$ into the formula and evaluate.	$S_\infty = \frac{10}{1-0.1}$ $S_\infty = \frac{10}{0.9} = \frac{100}{9}$ $= 11\frac{1}{9}$
b. 1. Write the formula for the sum to infinity.	b. $S_\infty = \frac{a}{1-r}; \ |r| < 1$
2. From the question we know that the infinite sum is equal to 10 and that the first term, a, is 6.	$a = 6$; $S_\infty = 10$
3. Substitute known values into the formula.	$10 = \frac{6}{1-r}$
4. Solve for r.	$10(1-r) = 6$ $10 - 10r = 6$ $r = 0.4$

5. Write the general formula for the nth term. $t_n = ar^{n-1}$

6. To determine the fourth term substitute $a = 6, n = 4$ and $r = 0.4$ into the formula and evaluate.

$$t_4 = 6 \times (0.4)^3 = 0.384$$

2.6 Exercise

Students, these questions are even better in jacPLUS

Receive immediate feedback and access sample responses

Access additional questions

Track your results and progress

Find all this and MORE in jacPLUS

Technology active

1. WE12 Calculate the sum of the first five terms (S_5) of these geometric sequences.
   a. t_n: $\{1, 2, 4, \ldots\}$
   b. $t_n = 3(-2)^{n-1}, n \in \{1, 2, 3, \ldots\}$
   c. $t_{n+1} = 2t_n, t_1 = \dfrac{3}{2}$
2. Calculate the sum of the first five terms (S_5) of these geometric sequences.
   a. t_n: $\{1, 3, 9, \ldots\}$
   b. $t_n = -4(1.2)^{n-1}, n \in \{1, 2, 3, \ldots\}$
   c. $t_{n+1} = \dfrac{1}{2}t_n, t_1 = \dfrac{-2}{3}$
3. Consider the following geometric sequences and determine the terms indicated.
   a. The first term is 440 and the 12th term is 880. Determine S_6.
   b. The fifth term is 1 and the eighth term is 8. Determine S_1, S_{10}, S_{20}.
4. Determine what minimum number of terms of the series $2 + 3 + 4\frac{1}{2} + \ldots$ must be taken to give a sum in excess of 100.
5. WE12 a. Calculate the sum to infinity for the sequence $t_n : \left\{1, \dfrac{1}{2}, \dfrac{1}{4}, \ldots\right\}$.
   b. Determine the fourth term in the geometric sequence whose first term is 4 and whose sum to infinity is 6.
6. a. Calculate the sum to infinity for the sequence $t_n : \left\{1, \dfrac{2}{3}, \dfrac{4}{9}, \ldots\right\}$.
   b. Determine the fourth term in the geometric sequence whose first term is 1 and whose sum to infinity is $\dfrac{3}{5}$.
7. For the infinite geometric sequence $\left\{1, \dfrac{1}{4}, \dfrac{1}{16}, \ldots\right\}$, Calculate the sum to infinity. Consequently, determine what proportion each of the first three terms contributes to this sum as a percentage.

8. A sequence of numbers is defined by $t_n = 3\left(\frac{1}{2}\right)^{n-1}, n \in \{1, 2, 3, \ldots\}$.

a. Calculate the sum of the first 20 terms.
b. Calculate the sum of all the terms between and including t_{21} and t_{40}.
c. Calculate the sum to infinity, S_∞.

9. A sequence of numbers is defined by t_n: $\{9, -3, 1, \ldots\}$.

a. Calculate the sum of the first nine terms.
b. Calculate the sum of all the terms between and including t_{10} and t_{15}.
c. Calculate the sum to infinity, S_∞.

10. The first term of a geometric sequence is 5 and the fourth term is 0.078 125. Calculate the sum to infinity.

11. The sum of the first four terms of a geometric sequence is 30 and the sum to infinity is 32. Determine the first three terms of the sequence if the common ratio is positive.

12. For the geometric sequence $\sqrt{5}+\sqrt{3}, \sqrt{5}-\sqrt{3}, \ldots$, determine the common ratio, r, and the sum of the infinite series, S_∞.

13. If $1 + 3x + 9x^2 + \ldots = \frac{2}{3}$, determine the value of x.

14. If the common ratio for a geometric sequence is 0.99 and the sum to infinity is 100, determine the value of the first and second terms in the sequence.

15. a. Show that $x^n - 1$ has a factor $(x-1)$ for $n \in \{1, 2, 3, 4, 5\}$.
b. Is there a pattern in the results? Make an educated guess about the way that you could factorise $x^n - 1$.

16. A student stands at one side of a road 10 metres wide, and walks halfway across. The student then walks half of the remaining distance across the road, then half the remaining distance again and so on.

a. Comment on whether the student will ever make it *past* the other side of the road.
b. Comment on if the width of the road affects your answer.

2.6 Exam questions

Question 1 (1 mark) TECH-FREE

Calculate the sum of the first 10 terms of the geometric series −2, +4, −8, +15, −32, ...

Question 2 (1 mark) TECH-FREE

The sum of the first 12 terms of a geometric sequence is 12 285. If the first term is 3, determine the common ratio.

Question 3 (3 marks) TECH-FREE

For a particular geometric sequence $S_{2n} = 5S_n$. Show that $r = \sqrt[n]{4}$.

More exam questions are available online.

2.7 Applications of sequences and series

LEARNING INTENTION

At the end of this subtopic you should be able to:
- apply your knowledge of sequences and series to real-life problems.

2.7.1 Solving real-life problems using sequences and series

This section consists of a mixture of problems where the work covered in the first five exercises is applied to a variety of situations.

The following general guidelines can assist you in solving the problems.
1. Read the question carefully.
2. Decide whether the information suggests an arithmetic or geometric sequence. Check to see if there is a constant difference between successive terms or a constant ratio. If there is neither, look for a simple number pattern such as the difference between successive terms changing in a regular way.
3. Write the information from the problem using appropriate notation. For example, if you are told that the fifth term is 12, write $t_5 = 12$. If the sequence is arithmetic, you then have an equation to work with, namely: $a + 4d = 12$. If you know the sequence is geometric, then $ar^4 = 12$.
4. Define what you have to calculate and write an appropriate formula or formulas. For example, if you have to find the tenth number in a sequence that you know is geometric, you have an equation: $t_{10} = ar^9$. This can be calculated if a and r are known or can be established.
5. Use algebra to determine what is required in the problem.

WORKED EXAMPLE 13 Applications of geometric sequences

In 1970 the cost of 1 megabyte of computer memory was $2025. In 1980 the cost for the same amount of memory had reduced to $45, and by 1990 the cost had dropped to $1.

a. **Assuming the pattern continues through the years, determine the cost of 1 megabyte of memory in the year 2000.**

b. **Calculate how much memory, in megabytes, you could buy for $10 in the year 2010 based on the trend.**

THINK

a. 1. Present the given information in a table.

2. Study the table. The information suggests a geometric sequence for the cost at each ten-year interval. Verify this by checking for a constant ratio between successive terms.

3. To calculate the cost in the year 2000, determine the fourth term in the sequence by multiplying the preceding (third) term by the common ratio.

4. Interpret the result and clearly answer the question.

WRITE

a.

Year	1970	1980	1990	2000	2010
Cost ($)	2025	45	1	?	?

$45 \div 2025 = \frac{1}{45}$ and $1 \div 45 = \frac{1}{45}$, so the three terms form a geometric sequence with common ratio $r = \frac{1}{45}$.

$t_4 = t_3 \times r$

$t_4 = 1 \times \frac{1}{45}$

$= \frac{1}{45}$

$= 0.022\ldots$

In the year 2000 you would have paid about 2 cents for a megabyte of memory.

b. 1. If the cost of 1 megabyte can be found in the year 2010, then the amount of memory purchased for $10 can be determined. To calculate the predicted cost in the year 2010, the fifth term in the sequence needs to be determined.

b. $t_5 = t_4 \times r$

$= \frac{1}{45} \times \frac{1}{45}$

$= \frac{1}{2025}$ of a dollar per megabyte

2. Take the reciprocal of t_5 to get the amount of memory per dollar.

The amount of memory per dollar is 2025 megabytes.

3. Calculate the amount of memory that can be purchased for $10.

So $10 would buy $10 \times 2025 = 20250$ megabytes.

Expressing recurring decimals as fractions using sequences

It is possible to express any recurring decimal as a fraction by breaking it up into a sequence of repeated sections added together. The following worked example demonstrates this process.

WORKED EXAMPLE 14 Expressing a recurring decimal as a fraction

Express the recurring decimal 0.131 313 13 ... as a proper fraction.

THINK	WRITE
1. Express the given number as a geometric series.	$0.131\,313\ldots = 0.13 + 0.0013 + 0.000\,013\ldots$
2. State the values of a and r.	$a = 0.13$ and $r = \frac{0.0013}{0.13} = 0.01$
3. Calculate the sum to infinity, S_∞. Write the formula for the sum to infinity.	$S_\infty = \frac{a}{1-r}$
4. Substitute values of a and r into the formula and simplify.	$S_\infty = \frac{0.13}{1-0.01}$ $S_\infty = \frac{0.13}{0.99}$
5. Multiply both numerator and denominator by 100 to get rid of the decimal point.	$S_\infty = \frac{13}{99}$

2.7.2 Linear recurrence relations

Linear recurrence relations are sequences of the form $t_{n+1} = at_n + b, \quad a \neq 0$, these are also known as difference equations and two special cases of these have already been studied.

Case (1): When $a = 1$ the recurrence relation becomes $t_{n+1} = t_n + b$ or, equivalently, $t_{n+1} - t_n = d$, which is an arithmetic progression with common difference d.

Case (2): when $b = 0$ it becomes $t_{n+1} = at_n$ or, equivalently, $\frac{t_{n+1}}{t_n} = r$, which is a geometric progression with common ratio r.

Applications of linear recurrence relations to financial problems

These two special types linear recurrence relations have applications to financial problems, being straight line flat depreciation and compound interest.

WORKED EXAMPLE 15 Linear recurrence relations – special cases

a. A car has an initial value of \$20 000 and depreciates each year by \$2000. Write a recurrence relation for the value of the car, C_n, at the end of the nth year and determine the value of the car after 9 years.

b. Sam invests \$2000 into an account that pays compound interest at 5% per annum, interest paid annually. Let V_n be the value of the investment at the end of the nth year. Determine the value of the investment after 4 years.

THINK	WRITE
a. 1. Write down the first term.	a. The value of the car, at the end of the first year is $C_1 = 20\,000 - 2000 = \$18\,000$
2. Each year the car depreciates by \$2000.	$C_{n+1} = C_n - 2000$
3. This is an arithmetic sequence or arithmetic progression.	$C_n = a + (n-1)d, \quad a = 18\,000, \quad d = -2000$ $C_n = 18\,000 - 2000\,(n-1), \quad C_1 = 18\,000$
4. Determine the value of the car after 9 years.	$C_9 = 18\,000 - 2000\,(9-1)$ $C_9 = 18\,000 - 8 \times 2000$ $C_9 = \$2000$
***Note*:** After the 10th year the value of the car is zero, as we can't have negative values.	The sequence $C_{n+1} = C_n - 2000$ is only valid for $n = \{1, 2, 3, 4, 5, 6, 7, 8, 9, 10\}$
b. 1. Determine the value at the end of the first year	$V_1 = 2000 \times \left(1 + \frac{5}{100}\right) = \$2100$
2. At the end of each year, the value increases by 5%, of the previous year.	$V_{n+1} = \left(1 + \frac{5}{100}\right) V_n$ $V_{n+1} = 1.05V_n$
3. This is a geometric sequence or progression.	$V_n = ar^{n-1}, \quad a = 2100, \quad r = 1.05$ $V_n = 2100 \times 1.05^{n-1}, \quad V_1 = 2100$
4. Calculate the value after 4 years.	$V_4 = 2100 \times 1.05^3 = \$2431.01$

General linear recurrence relations

The general linear recurrence relation is of the form $t_{n+1} = at_n + b, \;\; t_1 = u, \;\; a \neq \{0, 1\}, \;\; b \neq 0$. It is not an arithmetic sequence or a geometric sequence.

Since the recurrence relation states a rule for how to move from term to term, we can determine the values of the terms of the sequence systematically by computing each term in order.

Financial applications of linear recurrence relations

One application of linear recurrence relations is in financial situations such as investments and loan repayments. In both of these cases, a is dependent on the interest rate, $r\%$, per time period and is given by $a = 1 + \frac{r}{100}$. The value of b relates to the amount that is being added to the investment or paid off the loan at each time period. In the case of an investment, money is usually added to the balance each month or year, so b is positive, and in the case of a loan repayment, each payment reduces the balance of the loan, so b is negative.

WORKED EXAMPLE 16 Modelling financial situations using linear recurrence relations

Nigel is saving money for a car and opens a savings account which pays 4% interest per month. He opens the account with a balance of $5000 and plans to deposit $200 into the account at the start of each month.

a. **Write a recurrence relation to model the balance of the account at the start of each month.**

b. **The car that Nigel is planning on purchasing costs $8000. Calculate how long it will take for Nigel to be able to purchase the car.**

THINK	WRITE
a. 1. Determine the values of a and b. Note that since Nigel is depositing money into the account each month, b will be positive.	a. $a = 1 + \frac{4}{100} = 1.04$ $b = 200$
2. State the value of the account at the start of the first month.	$t_1 = 5000$
3. Using these values, write the recurrence relation that models the balance of the account at the start of each month.	$t_{n+1} = 1.04t_n + 200,\ t_1 = 5000$
b. 1. Use the recurrence relation to calculate the balance at the start of the second month.	b. $t_2 = 1.04t_1 + 200$ $= 1.04 \times 5000 + 200$ $= \$5400$
2. Use the recurrence relation to calculate the balance at the start of the third month.	$t_3 = 1.04t_2 + 200$ $= 1.04 \times 5400 + 200$ $= \$5816$
3. Continue using the recurrence relation to determine the balance at the end of the next months, stopping once the balance reaches $8000.	$t_4 = 1.04t_3 + 200$ $= 1.04 \times 5816 + 200$ $= \$6248.64$ $t_5 = 1.04t_4 + 200$ $= 1.04 \times 6248.64 + 200$ $= \$6698.59$ $t_6 = 1.04t_5 + 200$ $= 1.04 \times 6698.59 + 200$ $= \$7166.53$ $t_7 = 1.04t_6 + 200$ $= 1.04 \times 7166.53 + 200$ $= \$7653.19$ $t_8 = 1.04t_7 + 200$ $= 1.04 \times 7653.19 + 200$ $= \$8159.32$
4. State the answer.	Nigel will be able to purchase the car at the start of the 8th month.

Modelling population growth or decline using linear recurrence relations

Another application of linear recurrence relations is population modelling based on a constant growth rate, $k\%$, (proportional to the current population) and a constant number joining or leaving the population at regular time intervals.

In this case, a relates to the growth rate and is given by $a = 1 + \frac{k}{100}$, and b corresponds to the number joining or leaving the population at each time period.

WORKED EXAMPLE 17 Modelling population growth using linear recurrence relations

The number of people P_n in millions in a city at the start of the nth year is given by $P_{n+1} = 1.2P_n - l$. At the start of the first year there were one million people.

a. If the population of the city is maintained at one million people, determine the value of l.

b. If $l = 0.02$, determine the number of people in the city at the start of the fourth year.

c. When $l = 0.02$, the population at the start of the nth year is given by: $P_n = 0.9 \times (1.2)^{n-1} + 0.1$. Verify that this formula works by calculating the population of the city at the start of the fourth year, and comparing it to your answer from part b.

THINK	WRITE
a. To maintain a constant population of $P_n = 1$, the number of people leaving each year, l, must equal the increase in population due to the 20% growth rate.	a. $P_1 = 1$ $P_2 = 1.2P_1 - l$ $= 1.2 - l$ For $P_2 = P_1$: $1 = 1.2 - l$ $l = 0.2$
b. 1. Use the recurrence relation to calculate the population at the start of the second year.	b. $P_{n+1} = 1.2P_n - 0.02, \quad P_1 = 1$ $P_2 = 1.2P_1 - 0.02$ $P_2 = 1.2 \times 1 - 0.02$ $P_2 = 1.18$ There are 1.18 million people at the start of the second year.
2. Use the recurrence relation to calculate the population at the start of the third year.	$P_3 = 1.2P_2 - 0.02$ $P_3 = 1.2 \times 1.180 - 0.02$ $P_3 = 1.396$ There are 1.396 million people at the start of the third year.
3. Use the recurrence relation to calculate the population at the start of the fourth year.	$P_4 = 1.2P_3 - 0.02$ $P_4 = 1.2 \times 1.396 - 0.02$ $P_4 = 1.6552$ There are 1.6552 million people at the start of the fourth year.

c. 1. Use the formula $P_n = 0.9 \times (1.2)^{n-1} + 0.1$ to calculate the population at the start of the fourth year.	c. $P_n = 0.9 \times (1.2)^{n-1} + 0.1$ $P_4 = 0.9 \times (1.2)^3 + 0.1$ $= 1.6552$
2. Compare this to the answer calculated in part **b**.	The results are exactly the same. The population of the city at the start of the fourth year is 1.6552 million.

2.7 Exercise

Students, these questions are even better in jacPLUS

Receive immediate feedback and access sample responses

Access additional questions

Track your results and progress

Find all this and MORE in jacPLUS

Technology active

1. **WE13** In 1970 the Smith family purchased a small house for $60 000. Over the following years, the value of their property rose steadily. In 1975 the value of the house was $69 000 and in 1980 it reached $79350.
   a. Assuming that the pattern continues through the years, calculate (to the nearest dollar) the value of the Smiths' house in i 1985 and ii 1995.
   b. Determine by what factor the value of the house will have increased by the year 2015, compared to the original value.
2. An accountant working with a company commenced on a salary of $58 000 and has received a $4200 increase each year.
   a. Determine how much she earned in her 15th year of employment.
   b. Determine how much she has earned from the company altogether in those 15 years.
3. A chemist has been working with the same company for 15 years. She commenced on a salary of $28 000 and has received a 4% increase each year.
   a. State what type of sequence of numbers her annual income follows.
   b. Determine how much she earned in his 15th year of employment.
   c. Calculate how much she has earned from the company altogether.
   d. Determine his increase in salary at the end of i his 1st and ii his 14th year of employment.
4. A biologist is growing a tissue culture in a Petri dish. The initial mass of the culture was 20 milligrams. By the end of the first day the culture had a mass of 28 milligrams.
   a. Assuming that the daily growth is *arithmetic*, determine the mass of the culture after the second, third, tenth and *n*th day.
   b. Determine on what day the mass of the culture will first exceed 200 milligrams if its growth is arithmetic.
   c. Assuming that the daily growth is *geometric*, determine the mass of the culture after the second, third, tenth and *n*th day.
   d. Determine on what day the mass of the culture will first exceed 200 milligrams if its growth is geometric.

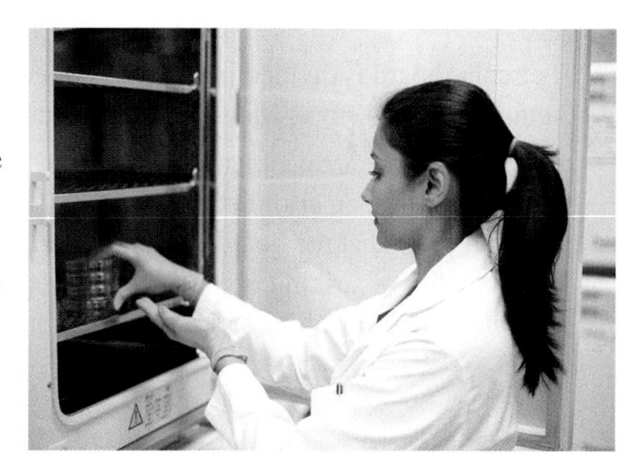

5. Logs of wood can be stacked so that there is one more log on each descending layer than on the previous layer. The top row has 6 logs and there are 20 rows.
   a. Calculate how many logs are in the stack altogether.
   b. The logs are to be separated into two equal piles. They are separated by removing logs from the top of the pile. Determine how many rows down the workers will take away before they remove half the stack.
6. Kind-hearted Kate has 200 movie tickets to give away to people at the shopping centre. She gives the first person one ticket, the next person two tickets, the third person four tickets and so on following a geometric progression until she can no longer give the nth person 2^{n-1} tickets. Determine how many tickets the last lucky person received and calculate how many tickets Kate had left.
7. The King of Persia, so the story goes, offered Xanadu any reward to secure the safety of his kingdom. As his reward, Xanadu requested a chessboard with one grain of rice on the first square, two grains on the second, four on the third and so on until the 64th square had its share of rice deposited.
   a. Determine the total number of grains of rice that the king needed to supply.
   b. If each grain of rice weighs 0.10 grams, calculate how many kilograms of rice this represents. (*Note*: There are 1000 grams in 1 kilogram.)
8. As legend has it, the King of Constantinople offered Xanadu's cousin Yittrius any reward to secure the safety of his city. This Yittrius accepted: she requested a chessboard with one grain of rice on the first square, three grains of rice on the second square, five grains of rice on the third square and so on until the 64th square had its share of rice deposited.
   a. Determine the total number of grains of rice that the king needed to supply.
   b. If each grain of rice weighs 0.10 grams, calculate how many kilograms of rice this represents. Write your answer correct to 2 decimal places. (*Note*: There are 1000 grams in 1 kilogram.)
9. A hiker walks 36 km on the first day and $\frac{2}{3}$ that distance on the second. Every day thereafter she walks $\frac{2}{3}$ of the distance she walked on the day before. Determine if the will hiker cover the distance of 100 km to complete the walk. If so, calculate which day she will complete the walk.
10. **WE14** Express the recurring decimal 0.1111 . . . as a proper fraction.
11. Express the recurring decimal 0.575 757 . . . as a proper fraction.
12. Determine the fraction equivalent of the following recurring decimal numbers by writing the decimal number as a sum of infinite terms.
   a. 0.333 333 333 . . .
   b. 2.343 434 . . .
   c. 3.142 142 142 . . .
   d. 21.2121 . . .
   e. 16.666 . . .
13. A circular board is divided into a series of concentric circles of radius 1 cm, 2 cm, 3 cm and 4 cm as shown in the diagram.
   a. Determine the areas of each of the successive shaded regions and show that they form an arithmetic progression.
   b. A dart is fired at the board at random and hits the board. Determine the probability of striking each of the four regions of the board.
   (*Note*: The probability of striking a region = area of region ÷ total area.)

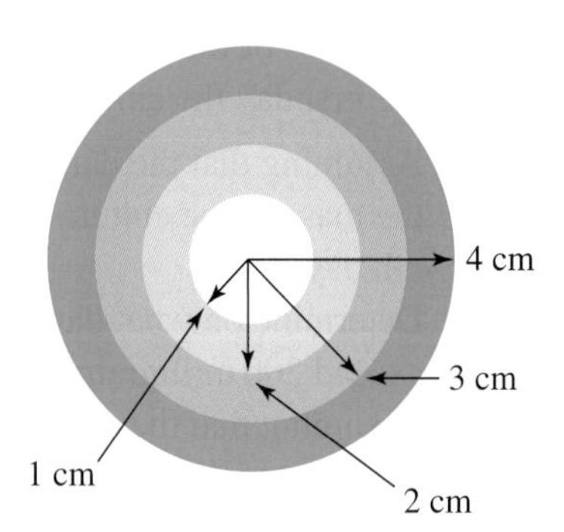

14. A bullet is fired vertically up into the air. In the first second it has an average speed of 180 m/s; that is, it travels 180 m up into the air during the first second. Each second its average speed diminishes by 12 m/s. Thus during the 2nd second the bullet has an average speed only 168 m/s and accordingly travels 168 m further up into the air.

a. Determine an equation for the average speed of the bullet for the nth second that it is in the air.
b. Determine the time when the average speed of the bullet is equal to zero.
c. Determine the maximum height of the bullet above where it was fired.

15. Coffee cools according to Newton's Law of Cooling, in which the temperature of coffee *above* room temperature drops by a constant fraction each unit of time. The table below shows the temperature of a cup of coffee in a room at 20°C each minute after it was made.

Time (min)	Temp. (°C)
1	80.0
2	74.0
3	68.6

Remember to subtract the room temperature from the temperature of the coffee before you do your calculations.
The person who made the coffee will drink it only if it has a temperature in excess of 50°C. Calculate the minimum time after the cup of coffee has been made before it becomes undrinkable.

16. Two arithmetic sequences, t_n and u_n, are multiplied together. That is, each term is multiplied by the other to form a new term.

$$t_n = 2n - 3, n \in \{1, 2, 3, \ldots\} \text{ and}$$
$$u_n = 3n, n \in \{1, 2, 3, \ldots\}$$

Show that the new sequence of numbers $t_1 \times u_1, t_2 \times u_2, t_3 \times u_3, \ldots$ is an arithmetic series and hence determine the arithmetic sequence for that new series.
(*Hint:* For a sequence a_n with a series $A_n, a_n = A_n - A_{n-1}$.)

17. **WE15** a. A machine has an initial value of $30 000 and depreciates each year by $3000. Write a recurrence relation for the value of the machine, M_n, at the end of the nth year and determine the value of the machine after 6 years.

b. Peter invests $3000 into an account that pays compound interest at 3% per annum, interest paid annually. Let V_n be the value of the investment at the end of the nth year. Determine the value of the investment after 8 years.

18. **WE16** Indah is saving money for her retirement and opens a savings account which pays 6% interest per year. She opens the account in 2023 with a balance of $60 000 and plans to deposit $3500 into the account at the start of each year (starting in 2024).

a. Write a recurrence relation to model the balance of the account at the start of each month.
b. Indah is planning on retiring once she has saved $100 000. Determine the year in which Indah will retire.

19. Arturo is taking out a loan of $1200 to buy a new laptop. The bank is charging 3% interest per month and Arturo is paying the loan off with monthly payments of $200, which are taken off the balance after interest is applied. Determine how many months it will take Arturo to pay off the loan.

20. WE17 The number of fish F_n in a trout farm at the start of the nth month is given by $F_{n+1} = 1.1F_n - c$. At the start of the first month there were 1000 fish.

a. If the number of fish is to be maintained at 1000, determine the value of c.

b. If $c = 90$, determine the number of fish in the trout farm at the start of the fourth month.

c. When $c = 90$, the number of fish in the trout farm at the start of the nth month is given by $F_n = 100 \times (1.1)^{n-1} + 900$. Verify that this formula works by calculating the number of fish in the trout farm at the start of the fourth month, and comparing it to your answer from part b.

2.7 Exam questions

Question 1 (2 marks) TECH-ACTIVE

Using compound interest, calculate the future value of \$1000 invested for 10 years 8% per annum.

Question 2 (2 marks) TECH-FREE

The planet Angloopa has a population of 6.6 billion nargs in the year 2094 AP. If the population of nargs increases at a rate of 10% per Angloopa year (approx. 1.75 Earth years), determine during which year, AP, the population will have tripled.

Question 3 (3 marks) TECH-FREE

The number of koalas K_n on a plantation at the start of the nth year is given by $K_{n+1} = 0.98K_n - 5$ and at the start of the first year there were 200 koalas. (*Note:* round all decimal answers down to the nearest integer when stating answers.)

a. Determine the number of koalas on the plantation at the start of the fourth year. **(1 mark)**

b. The number of koalas at the start of the nth year is given by the formula $K_n = 450 \times 0.98^{n-1} - 250$. Use this formula to determine the year in which the number of koalas falls below 100. **(2 marks)**

More exam questions are available online.

2.8 Review

2.8.1 Summary

doc-37045

Hey students! Now that it's time to revise this topic, go online to:

Access the topic summary

Review your results

Watch teacher-led videos

Practise exam questions

Find all this and MORE in jacPLUS

2.8 Exercise

Technology free: short answer

1. Write the recursive definition for each of the following sequences.
   a. $\{7, 11, 19, 35, 67, \ldots\}$
   b. $\{-2, 5, 26, 677, \ldots\}$

2. For the arithmetic sequence where $t_3 = 10$ and $t_6 = 478$, determine:
   a. the functional rule for the nth term in the sequence
   b. the recursive rule for the sequence.

3. A car at a racetrack starts from rest and travels 0.5 m in the 1st second and 1.0 m in the 2nd second, following an arithmetic progression in the distances covered each subsequent second.
   a. Determine how far it will travel during the 10th second.
   b. After 10 seconds of motion, calculate how far it will have travelled in total.
   c. To the nearest whole second, determine how long will it take to travel 1000 m (1 km). *Note:* you may use a scientific calculator in this question.

4. At Bugas Heights a radiation leak in a waste disposal tank potentially exposes staff to a 1000 milli-rem dose on the first day of the accident, an 800 milli-rem dose on the second day after the accident and a 640 milli-rem dose on the third day after the accident. *Note:* you may use a scientific calculator in this question.
   a. Assuming a geometric sequence, determine the amount of potential exposure dose on the 10th day.
   b. Calculate the total potential exposure dose in the first 5 days.

5. Calculate the sum of each of the following expressions.
   a. $1 + \frac{1}{4} + \frac{1}{16} + \frac{1}{64} \ldots$
   b. $1 - \frac{2}{3} + \frac{4}{9} - \frac{8}{27} \ldots$

6. Determine the fraction equivalent to each of the following recurring decimals.
   a. 0.222 222
   b. 2.454 545 454

Technology active: multiple choice

7. **MC** Consider the sequence $t_{n+1} = 2t_n + 4; t_3 = 12$. The second term in the sequence is
   A. 10 B. 6 C. 28 D. 4 E. 8

8. **MC** A series is listed as 3, 10, 21, 36, … The next term in the series is
   A. 51 B. 52 C. 53 D. 54 E. 55

9. MC The 23rd term in the sequence of numbers $\{7, 3, -1, \ldots\}$ is

A. -88 B. -81 C. -74 D. -83 E. 90

10. MC Consider the arithmetic sequence $\{52, a, 41, b\}$. The numerical value of the expression $a - 3b$ is

A. -60 B. $-64\frac{1}{2}$ C. $-67\frac{1}{2}$ D. -71 E. $72\frac{1}{2}$

11. MC Consider the arithmetic sequence $\{x-2y, 3x-4y, 4x-7y, \ldots\}$. An expression for y in terms of x is

A. $y=x$ B. $y=-x$ C. $y=-2x$ D. $y=2x$ E. $y=-3x$

12. MC A car is accelerating such that in the 1st second it travels 2.0 metres, in the 2nd second it travels 3.5 metres, in the 3rd second it travels 5.0 metres, and so on for a total of 15 seconds. The total distance travelled by the car is

A. 630 m B. 93.75 m C. 187.5 m D. 375 m E. 315 m

13. MC The sum of the first four terms in an arithmetic sequence is 70. The sum of the first six terms is 63. The sixth term of the sequence is equal to

A. -14 B. -7 C. 0 D. 7 E. 14

14. MC For a geometric sequence, the fourth term is 5 and the seventh term is -625. The second term in the sequence is

A. -2.5 B. -1.25 C. 0.25 D. -0.25 E. 0.20

15. MC The sum of an infinite geometric sequence is 5.6 with the common ratio equal to 0.20. The sum of the first four terms of the geometric sequence is closest to

A. 5.0 B. 5.2 C. 5.4 D. 5.6 E. 5.8

16. MC The sum of the first 10 terms of a geometric sequence is 400. The next term in the sequence is 3 times the previous term. The first term in the sequence is

A. $\frac{17}{731}$ B. $\frac{400}{1473}$ C. $\frac{100}{7381}$ D. $\frac{200}{781}$ E. $\frac{10}{387}$

Technology active: extended response

17. Consider a square of side length 2 units.
   a. Determine the perimeter of the square.
   b. Each of the four midpoints form the vertices of a new square inscribed within the original square. Determine the perimeter of this new square.
   c. Repeat the process to determine the perimeter of a third square inscribed within the second.
   d. Give an expression for the perimeter of the nth square.

18. Consider the following iterative definitions.
   a. $t_{n+1} = t_n - \frac{3}{4},\ t_1 = \frac{1}{8}$
   b. $t_{n+1} = at_n,\ t_1 = b^2$
   c. $t_{n+1} = 3t_n^2 - 1.5,\ t_1 = 0.5$

   If each of these definitions is used to generate a sequence of numbers:
   i. decide whether the sequence is arithmetic, geometric or neither
   ii. determine its fourth term.

19. Two gymnasiums in the same area of country Victoria offer yearly memberships, with an increase each year due to the increasing cost of staff.

Year	Cost of membership each year ($)	
	Looking Good Gym	Feel Fit Gym
2019	550	550
2020	575	566.50
2021	600	583.50
2022	625	601
2023	650	619.03

a. Assume that 2019 is year 1.
   i. Write down the cost of a membership at Looking Good Gym in year n.
   ii. Assuming that Feel Fit Gym follows a geometric progression, write down the cost of a membership in year n.
   iii Write down the amount of money, correct to 2 decimal places, that each gym would charge in the 15th year.
   iv. Calculate which year the cost of Feel Fit Gym will first be greater than Looking Good Gym.
   v. Given that you pay membership for 25 years, calculate which gymnasium would be cheaper and by how much.

b. A third gymnasium, Bodywork, is also in the area and its membership costs follow the iterative formula $C_{n+1} = 1.04C_n$, where $C_1 = c$.
   i. If the initial membership in 2019 is the same as for the other two clubs, write down the value of c.
   ii. Write down the costs of membership at Bodywork for the first four years, rounding answers to 2 decimal places.

20. On an island in the Pacific Ocean the population of a species of insect (species A) is increasing geometrically with a population of 10 000 in 1990 and an annual growth rate of 12.0%. Another species of insect (species B) is also increasing its population, but it is increasing arithmetically with numbers 15 000 in 1990 and an annual increment of 1000 per annum.

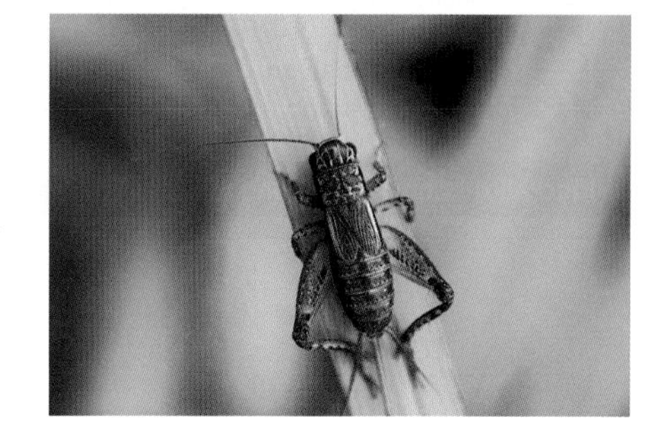

a. Using technology, determine the difference in the numbers of the two species during the last decade of the twentieth century (that is, up to 1999).
b. Determine in what year the first species will be greater in number than the second species, assuming that growth rates remain fixed.

A scientist has a mathematical model that indicates the species can cohabit provided that they have equal numbers in the year 2000.

c. If the growth rate in species A is to remain unchanged, calculate what the annual increment in species B would need to be to achieve this.
d. If the annual increment in species B is to remain unchanged, calculate what the growth rate in species A would need to be to achieve this.

2.8 Exam questions

Question 1 (1 mark) TECH-FREE

A pendulum starts its swing through an angle of 60° from one extreme to the other.

Each oscillation of the pendulum swings out an angle which is 60% of the preceding oscillation. Determine the angle that the pendulum swings through in its 10th oscillation.

Question 2 (2 marks) TECH-FREE

Two friends, Thanh and Maria, start work at Blogg Brothers Emporium. Initially they are paid the same annual salary of $25 000.

Thanh's salary increases at the end of each year at the flat rate of $1250 per annum while Maria's salary increases at the end of each year at 4% per annum.

Determine who earns the greater salary at the end of the sixth year.

Question 3 (2 marks) TECH-FREE

The nth term of a geometric series is $6(0.25)^n$. Calculate the sum to infinity.

Question 4 (2 marks) TECH-FREE

A ball drops from a height of 10 metres. After bouncing, it reaches 70% of its original height. Assuming a geometric sequence, determine on which bounce the ball passes a rebound height of 1 metre for the last time.

Question 5 (3 marks) TECH-FREE

The sum of three consecutive terms of an arithmetic series is 36 and the sum of the next three terms is 45. If the first term is a, calculate the sum of the first 12 terms in terms of a.

More exam questions are available online.

Answers

Topic 2 Sequences and series

2.2 Describing sequences

2.2 Exercise

1. a. $\frac{3}{8}, \frac{3}{16}, \frac{3}{32}$ b. 4, 36, 324 c. $-3, -1, 3$

2. a. $-17, 20, -23$
   b. 53, 97, 173
   c. $-11, -123, -15\,131$

3. a. Add 3 (to the previous term); 10, 13, 16.
   b. Subtract 1 (from the previous term); $-3, -4, -5$.
   c. Multiply by 4; 256, 1024, 4096.
   d. The difference between the terms increases by 1 for each pair; 27, 35, 44.

4. a. Add the preceding two terms; 29, 47, 76.
   b. Add $3b - a$; $-2a + 7b, -3a + 10b, -4a + 13b$.
   c. Many possible answers — assume the sequence repeats; $0, -1, 0$.
   d. Append 1 to the decimal expansion of the preceding term; 1.111, 1.1111, 1.111 11.

5. a. $-3, 5, 15$ b. $\frac{1}{2}, \frac{5}{6}, \frac{10}{11}$
   c. $13.3, -1.5, -20$ d. $\frac{5}{2}, \frac{5}{32}, \frac{5}{1024}$

6. a. $\frac{5}{4}, 20, 640$ b. 0, 4, 11
   c. $\frac{3}{2}, \frac{243}{32}, \frac{59\,049}{1024}$ d. $a,\ a + 4d,\ a + 9d$

7. a. -12, 384, 1536
   b. $0, -35, -49$
   c. $3, -1, -1$
   d. 0, 0, 0

8. a.

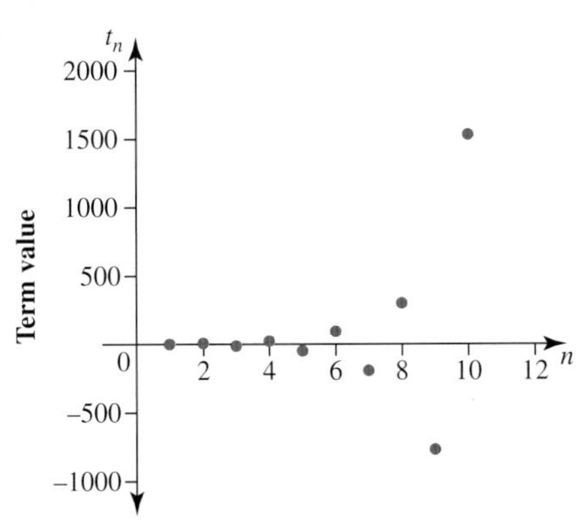

b.

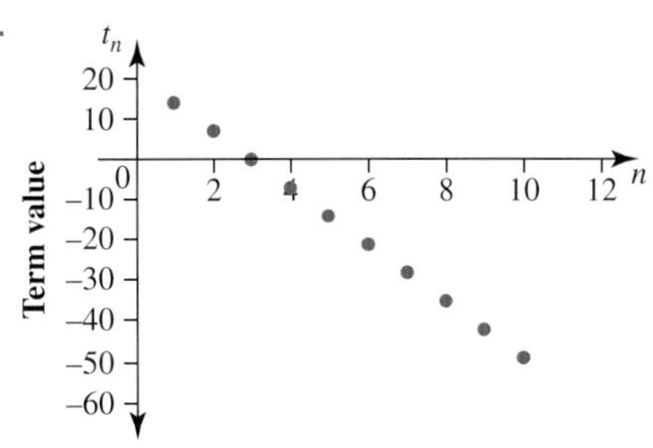

c.

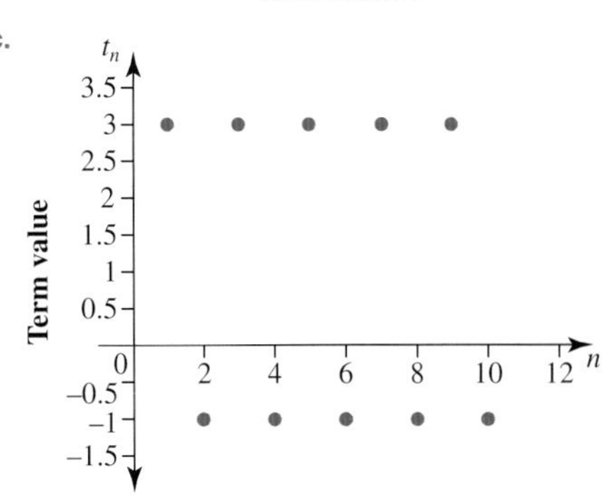

d.

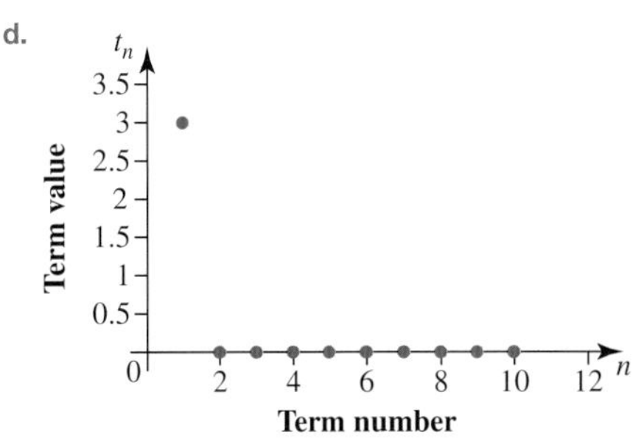

9. a. 15, 20; the difference between subsequent terms increases by 1.
   b. There are many possible answers. A possible pattern is the addition of 5, then 3, then 1, then -1. The next two terms are $4, -3$. Here the difference between successive terms follows an arithmetic sequence.
   c. Many possible answers as there is no obvious pattern. It could be the start of a telephone number.

10. a. Each successive term is multiplied by an increasing factor of $\frac{1}{2}$, starting with $\left(\frac{1}{2}\right)^{-1} = 2$, then $\left(\frac{1}{2}\right)^{0} = 1$, and then $\left(\frac{1}{2}\right)^{1}$ followed by $\frac{1}{4}; \frac{3}{16}, \frac{3}{256}$.
    b. 34, 55; each subsequent term is the sum of the preceding two terms.
    c. 31, 63; terms are 1 less than powers of 2.

11. a. $t_{n+1} = t_n - 2, t_1 = 7$
    b. $t_{n+1} = t_n \div 2, t_1 = 12$
    c. $t_{n+1} = t_n + 0.6, t_1 = 12$

12. a. $t_{n+1} = t_n \times 5 + 1, t_1 = 2$
    b. $t_{n+1} = -3t_n, t_1 = 4$
    c. $t_{n+1} = (t_n)^2, t_1 = 2$

2.2 Exam questions

Note: Mark allocations are available with the fully worked solutions online.

1. $t_n = n^2 + 1,\ n \in N$

2. $-\frac{1}{a}, \frac{1}{a^2}, -\frac{1}{a^3}$

3. $8, \frac{27}{2}, \frac{64}{3}, \frac{125}{4}, \frac{216}{5}$

2.3 Arithmetic sequences

2.3 Exercise

1. a. Not arithmetic
   b. Arithmetic, difference $= 3; t_4 = 6, t_n = -6 + 3n$
2. a. Not arithmetic
   b. Arithmetic, difference $= -4; t_4 = -14, t_n = 2 - 4n$
3. a. 104 b. 682
   c. 1458 d. $-26\,310$
4. a. $t_n = 8 - 3n, n = 1, 2, 3, \ldots$
   b. $t_n = 2 + \frac{n}{2}, n = 1, 2, 3, \ldots$
   c. $t_n = -6 + 3n,\ n = 1, 2, 3, \ldots$
   d. $t_n = -3x + 5nx, n = 1, 2, 3, \ldots$
5. $m = 21.5,\ n = 32.5$
6. $x = 47,\ y = 75$
7. $-2, \frac{26}{3} - \frac{8}{3}n$
8. $4.54, t_n = 2.78 + 0.22n$
9. $t_n = 4 + 2n, n = 1,\ 2,\ 3, \ldots$
10. $t_n = 5n - 2, n = 1,\ 2,\ 3, \ldots$
11. $-35;\ 15;\ t_n = 15n - 50,\ n = 1,\ 2,\ 3\ldots$
12. $-4\frac{1}{4};\ \frac{5}{12};\ t_n = -4\frac{2}{3} + \frac{5}{12}n,\ n = 1,\ 2,\ 3\ldots$
13. $m = 27, n = 32$
14. 9
15. 3
16. a.

Term number	1	2	3	4	5
Term value	4	8	12	16	20

b.

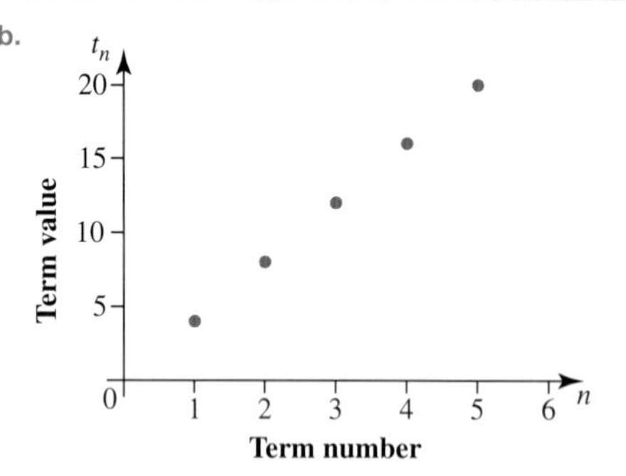

c. $t_{10} = 40$

17. a.

Term number	1	2	3	4	5
Term value	10	13	16	19	22

b.

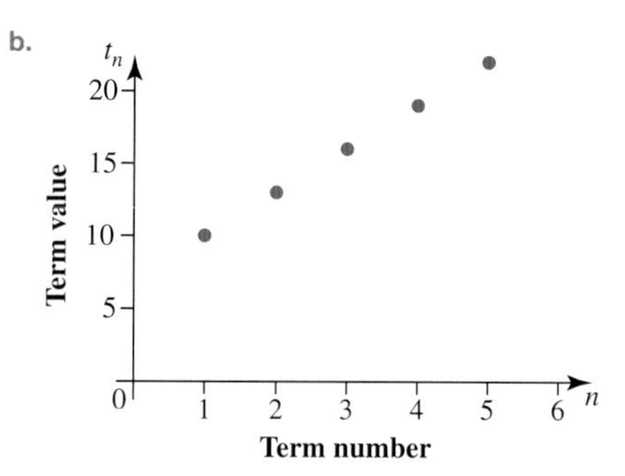

c. $t_{10} = 37$

18. a. $t_{n+1} = t_n + 4; t_1 = 3$
    b. $t_{n+1} = t_n + 3; t_1 = -3$
    c. $t_{n+1} = t_n + \frac{1}{2};\ t_1 = \frac{2}{7}$
19. a. $t_{n+1} = t_n + \frac{3}{4};\ t_1 = \frac{3}{4}$
    b. $t_{n+1} = t_n - \frac{7}{4};\ t_1 = \frac{1}{4}$
    c. $t_{n+1} = t_n + 2\pi - 2;\ t_1 = 2\pi + 3$
20. a. $\frac{5}{6}$ b. $\frac{n+2}{n+3}$

2.3 Exam questions

Note: Mark allocations are available with the fully worked solutions online.

1. $d = -1$
2. $a = 5$
3. Sample responses can be found in the worked solutions in the online resources.

2.4 Arithmetic series

2.4 Exercise

1. 400
2. 4175
3. a. 1275 b. 5050
4. a. 5000
   b. Each of the 100 terms is $\frac{1}{2}$ less than its corresponding term in question 3b. There are 100 terms, so the answer to this question is 50 less than in question 3b.
5. 258
6. a. $-273, -480, -741$
   b. -324
7. a. 280 b. 1080 c. 34
8. $\frac{n(n+1)}{2}$
9. Sample responses can be found in the worked solutions in the online resources.
10. 17
11. 45
12. 6

13. 174

14. The iterative equation is $t_{n+1} = t_n + 8, t_1 = 7$. The functional equation is $t_n = 8n - 1, n = 1, 2, 3, \ldots$

2.4 Exam questions

Note: Mark allocations are available with the fully worked solutions online.

1. 1683
2. 495
3. $12(25p + 2q)$

2.5 Geometric sequences

2.5 Exercise

1. Geometric, ratio $= 2; t_4 = 24; t_n = 3 \times 2^{n-1}$
2. Geometric, ratio $= -\frac{1}{3}; t_4 = \frac{1}{9}; t_n = (-3)^{2-n}$
3. a. $t_n = 5 \times 2^{n-1}, t_6 = 160, t_{10} = 2560$
   b. $t_n = 2 \times 2.5^{n-1}, t_6 = 195.31, t_{10} = 7629.39$
   c. $t_n = 1 \times (-3)^{n-1}, t_6 = -243, t_{10} = -19\,683$
   d. $t_n = 2 \times (-2)^{n-1}, t_6 = -64, t_{10} = -1024$
4. a. $t_n = \frac{1}{2} \times 2^{n-1}, t_6 = 16, t_{10} = 256$
   b. $t_n = \frac{1}{3} \times \left(\frac{1}{4}\right)^{n-1}, t_6 = \frac{1}{3072}, t_{10} = \frac{1}{786\,432}$
   c. $t_n = x \times (3x^3)^{n-1}, \; t_6 = 243x^{16}, \; t_{10} = 19\,683x^{28}$
   d. $t_n = \frac{1}{x} \times \left(\frac{2}{x}\right)^{n-1}, t_6 = \frac{32}{x^6}, t_{10} = \frac{512}{x^{10}}$
5. There are two possible answers because the ratio could be -3 or 3. The nth term is $t_n = 2 \times 3^{n-1}$ or $t_n = 2 \times (-3)^{n-1}$, $t_{10} = \pm 39\,366$.
6. There are two possible answers because the ratio could be -2 or 2. The nth term is $t_n = 2^{n-1}$ or $t_n = (-2)^{n-1}$, $t_{10} = \pm 512$.
7. a. The nth term is $t_n = 5 \times 2^{n-1}, t_{10} = 2560$.
   b. The nth term is $t_n = -1 \times (-2)^{n-1}, t_{10} = 512$.
   c. There are two possible answers because the ratio could be $-\frac{1}{27}$ or $\frac{1}{27}$. The nth term is $t_n = 3^{5-3n}$ or $t_n = (-3)^{5-3n}, \; t_{10} = \pm 3^{-25}$.
8. $\pm \frac{3}{4}$
9. $3 \times 2^{\left(\frac{n-1}{2}\right)}$
10. $t_1 = 25, \; r = \pm 2, \; t_n = 25 \times 2^{n-1}$ or $t_n = 25 \times (-2)^{n-1}$
11. $a = \pm 6, r = \pm 2, t_n = 6 \times 2^{n-1}$ or $t_n = -6 \times (-2)^{n-1}$
12. $m = 12, n = 48$
13. $m = 36, \; n = \frac{729}{4}$
14. $a = 300, b = 0.75$
15. $t_1 = \frac{1}{3}, \; r = \frac{3}{2}, t_n = 3^{n-2}2^{1-n}$
16. -6
17. $2, \frac{1}{2}, \frac{1}{8}$, or $-2, \frac{1}{2}, -\frac{1}{8}$
18. a.

Term number	1	2	3	4	5
Term value	1	3	9	27	81

b.

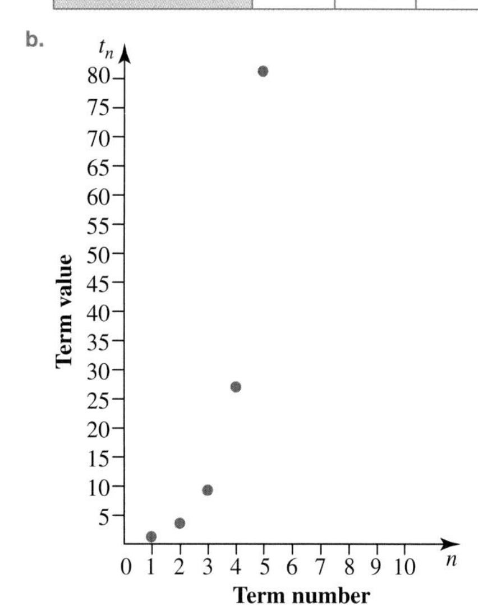

c. The points lie on a smooth curve which increases rapidly. Because of this rapid increase in value, it would be difficult to use the graph to predict future values in the sequence with any accuracy.

19. a.

Term number	1	2	3	4	5
Term value	−20	10	−5	2.5	−1.25

b.

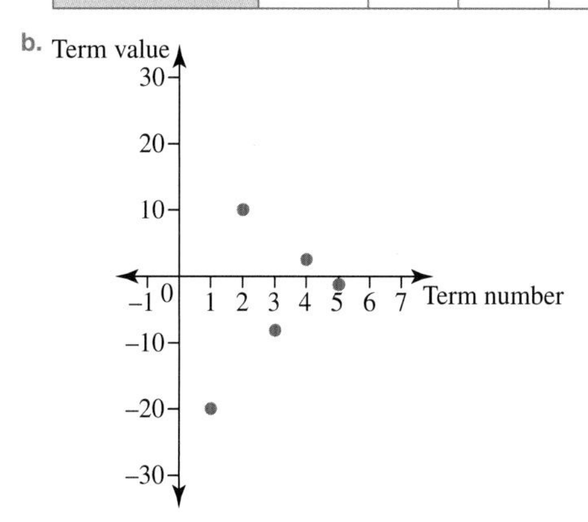

c. The points are oscillating and converge on a point.

20. a. $\frac{3}{2}$ b. $\frac{24}{2^n}$

21. $k = 6$

2.5 Exam questions

Note: Mark allocations are available with the fully worked solutions online.

1. C
2. The missing term is 4 or -4
3. 7

2.6 Geometric series

2.6 Exercise

1. a. 31 b. 33 c. 46.5
2. a. 121 b. -29.8 c. $-\frac{31}{24}$
3. a. 3108 b. $\frac{1}{16}, 63\frac{15}{16}, 66\ 535\frac{15}{16}$
4. 9
5. a. 2 b. $\frac{4}{27}$
6. a. 3 b. $-\frac{8}{27}$
7. $\frac{4}{3}$; 75%, 18.75%, 4.6875%
8. a. $6\left[1-\left(\frac{1}{2}\right)^{20}\right] = 5.999\ 994\ 278$
   b. 5.722×10^{-6}
   c. 6
9. a. $\frac{27}{4}\left[1-\left(\frac{-1}{3}\right)^{9}\right] = 6.750\ 343$
   b. -3.425×10^{-4}
   c. $6\frac{3}{4}$
10. $6\frac{2}{3}$
11. 16, 8, 4
12. $4-\sqrt{15}, \frac{\left(\sqrt{3}+\sqrt{5}\right)}{\left(\sqrt{15}-3\right)} = \frac{\left(4\sqrt{3}+3\sqrt{5}\right)}{3}$
13. $-\frac{1}{6}$
14. 1, 0.99
15. Sample responses can be found in the worked solutions in the online resources.
16. a. Mathematically, the student will never make it past the other side of the road. After each attempt, the distance remaining is halved, and this result is the extra distance walked at the next attempt. Thus the distance travelled across the road approaches but never reaches 10 metres.
    b. As shown in part a, the extra distance travelled at each attempt is equal to half the remaining distance from the previous attempt. Given that there will always be an amount remaining to travel, only half this amount can be achieved on the next attempt, regardless of the width of the road.

2.6 Exam questions

Note: Mark allocations are available with the fully worked solutions online.

1. 682
2. $r = 2$
3. Sample responses can be found in the worked solutions in the online resources.

2.7 Applications of sequences and series

2.7 Exercise

1. a. i. $91 253 ii. $120 681
   b. 3.518 times
2. a. $116 800 b. $1 311 000
3. a. Geometric
   b. $48 487
   c. $560 660
   d. i. $1120
      ii. $1865
4. a. 36 mg, 44 mg, 100 mg, $20 + 8n$ mg
   b. 23rd day
   c. 39 mg, 55 mg, 579 mg, $28 \times (1.4)^{n-1}$ or 20×1.4^n
   d. Seventh day
5. a. 310
   b. The workers must remove 12 full rows and 17 logs from the 13th row.
6. The last person received 64 tickets and Kate had 73 left.
7. a. 1.8×10^{19} grains of rice
   b. 1.8×10^{15} kg
8. a. 4096 grains of rice
   b. 0.41 kg
9. Yes, seventh day
10. $\frac{1}{9}$
11. $\frac{57}{99}$
12. a. $\frac{1}{3}$ b. $2\frac{34}{99} = \frac{232}{99}$ c. $3\frac{142}{999} = \frac{3139}{999}$
    d. $21\frac{7}{33} = \frac{700}{33}$ e. $16\frac{2}{3} = \frac{50}{3}$
13. a. $\pi, 3\pi, 5\pi, 7\pi$ — arithmetic progression with $a = \pi$ and $d = 2\pi$
    b. $\frac{1}{16}, \frac{3}{16}, \frac{5}{16}, \frac{7}{16}$
14. a. $192 - 12n$ m/s
    b. During the 16th second
    c. 1440 m
15. After 7 minutes the coffee has cooled to below 50°C.
16. The sequence for the arithmetic series $t_n u_n$ is $12n - 15$, $n \in \{1, 2, 3, \ldots\}$
17. a. $12 000 b. $3800.31
18. a. $V_{n+1} = 1.06V_n + 3500,\ V_1 = \$60\,000$
    b. 2028
19. 6 months
20. a. $c = 100$ b. 1033 c. 1033

2.7 Exam questions

Note: Mark allocations are available with the fully worked solutions online.

1. $2158.92
2. 2105
3. a. 173 b. 13th year

2.8 Review

2.8 Exercise

Technology free: short answer

1. a. $t_{n+1} = 2t_n - 3; t_1 = 7$
   b. $t_{n+1} = (t_n)^2 + 1; t_1 = -2$
2. a. Functional rule: $t_n = 156n - 458$
   b. Recursive: $t_{n+1} = t_n + 156; t_1 = -302$
3. a. 5 m b. 27.5 m c. 63 s
4. a. 134.2 milli-rem
   b. 3361.6 milli-rem
5. a. $\frac{4}{3}$ b. $\frac{3}{5}$
6. a. $\frac{2}{9}$ b. $2\frac{5}{11}$

Technology active: multiple choice

7. D
8. E
9. B
10. A
11. B
12. C
13. B
14. E
15. D
16. C

Technology active: extended response

17. a. 8
   b. $4\sqrt{2}$
   c. 4
   d. $P_n = 8\left(\frac{\sqrt{2}}{2}\right)^{n-1}$
18. a. i. Arithmetic
   ii. $t_4 = -2\frac{1}{8}$
   b. i. Geometric
   ii. $t_4 = a^3b^2$
   c. i. Neither
   ii. $t_4 = -1.394\,531\,25$
19. a. i. $L_n = 525 + 25n$
   ii. $F_n = 550 \times 1.03^{n-1}$
   iii. Looking Good Gym: \$900.00
   Feel Fit Gym: \$831.92
   iv. 2047
   v. Feel Fit would be cheaper by \$1197.40
   b. i. $c = 550$
   ii. 2019: \$550
   2020: \$572
   2021: \$594.88
   2022: \$618.68
20. a.

Year	n	Pop A growth rate = 1.12	Pop B annual increment = 1000	Difference
1990	1	10 000	15 000	5000
1991	2	11 200	16 000	4800
1992	3	12 544	17 000	4456
1993	4	14 049	18 000	3951
1994	5	15 735	19 000	3265
1995	6	17 623	20 000	2377
1996	7	19 738	21 000	1262
1997	8	22 107	22 000	−107
1998	9	24 760	23 000	−1760
1999	10	27 731	24 000	−3731
2000	11	31 058	25 000	−6058
2001	12	34 785	26 000	−8785

   b. During 1997
   c. Annual increment of 1606 insects.
   d. Annual growth rate of 9.596%

2.8 Exam questions

Note: Mark allocations are available with the fully worked solutions online.

1. 0.60°
2. Thanh
3. 2
4. On the 6th bounce.
5. $12a + 66$

3 Logic and algorithms

LEARNING SEQUENCE

Fully worked solutions for this topic are available online.

3.1 Overview

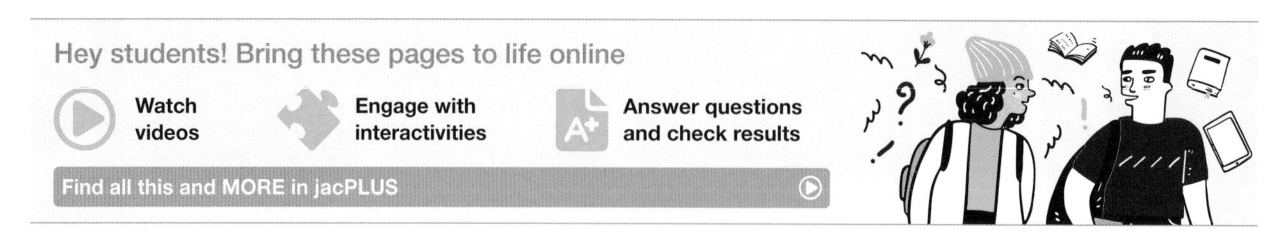

3.1.1 Introduction

Older than calculus (17th century CE), algebra (9th century CE) and even geometry (300 years BCE) is the study of logic. Some of the material described in this topic was developed by Aristotle, one of the most famous of the ancient Greek philosophers, yet it is still used today by people as diverse as mathematicians, lawyers, engineers and computer scientists. All of our modern digital technology owes its birth to the application of the principles of logic; every meaningful computer program ever written has relied on the principles you will learn in this topic.

Furthermore, logic can be seen as the study of **argument**. You will be able to analyse logically the arguments of teachers, politicians and advertisers to determine if they should convince you of their ideas, programs and products.

KEY CONCEPTS

This topic covers the following key concepts from the VCE Mathematics Study Design:

- propositions, connectives, truth values, truth tables and Karnaugh maps
- Boolean algebra
- tautologies, validity and proof patterns and the application of these to proofs in natural language and laws and properties of Boolean algebra, the algebra of sets and propositions
- logic gates and circuits, and simplification of circuits
- definition of an algorithm and the fundamental constructs needed to describe algorithms: sequence, decision (selection, choice, if … then … blocks) and repetition (iteration and loops)
- construction and implementation of basic algorithms incorporating the fundamental constructs using pseudocode.

3.2 Statements (propositions), connectives and truth tables

LEARNING INTENTION

At the end of this subtopic you should be able to:
- create truth tables to determine the truth of compound statements.

3.2.1 Statements

A **statement** is a sentence that is either true or false. For example, 'This book is about mathematics' is a true (T) statement, whereas 'The capital of Australia is Perth' is a false (F) statement.

Some sentences are not statements at all. 'Go to the store' is an instruction; 'How old are you?' is a question; 'See you later!' is an exclamation, and 'You should see the latest Spielberg movie' is a suggestion. 'The bus is the best way to get to work' is an opinion.

To determine whether a sentence is a statement, put the expression 'It is true that ...' (or 'It is false that ...') at the front of the sentence. If it still makes sense, then it is a statement.

Beware of some 'near-statements' such as 'I am tall' or 'She is rich' because these are relative sentences; they require more information to be complete. They can be turned into statements by saying 'I am tall compared to Ismaya' or 'She is rich compared to Cooper'.

In some textbooks, statements are called **propositions**.

WORKED EXAMPLE 1 Classifying statements

Classify the following sentences as either statements, instructions, suggestions, questions, opinions, exclamations or 'near-statements'. If they are statements, indicate whether they are true, false or indeterminate without further information.

a. **Germany won World War II.**
b. **Would you like to read my new book?**
c. **The most money that Mary can earn in one day is $400.**
d. **When it rains, I wear rubber boots.**
e. **Hello!**
f. **You will need to purchase a calculator in order to survive Year 11 Mathematics.**
g. **Do not run in the hallways.**
h. **You should read this book.**
i. **I am short.**

THINK	WRITE
a. Put the phrase 'It is true that ...' in front of the sentence. If the new sentence makes sense, it is classed as a statement.	a. This is a (false) statement.

b. Put the phrase 'It is true that ...' in front of the sentence. If the new sentence makes sense, it is classed as a statement.

b. This is a question.

c. Put the phrase 'It is true that ...' in front of the sentence. If the new sentence makes sense, it is classed as a statement.

c. This is a statement. We cannot at this time determine if it is true or false without further information.

d. Put the phrase 'It is true that ...' in front of the sentence. If the new sentence makes sense, it is classed as a statement.

d. This is a (presumably true) statement.

e. Put the phrase 'It is true that ...' in front of the sentence. If the new sentence makes sense, it is classed as a statement.

e. This is an exclamation.

f. Put the phrase 'It is true that ...' in front of the sentence. If the new sentence makes sense, it is classed as a statement.

f. This is a (true) statement.

g. Put the phrase 'It is true that ...' in front of the sentence. If the new sentence makes sense, it is classed as a statement.

g. This is an instruction.

h. Put the phrase 'It is true that ...' in front of the sentence. If the new sentence makes sense, it is classed as a statement.

h. This is a suggestion.

i. Put the phrase 'It is true that ...' in front of the sentence. If the new sentence makes sense, it is classed as a statement.

i. This is a near-statement because it requires additional information to be complete. It can be turned into a statement by saying 'I am shorter than Karen'.

3.2.2 Connectives and truth tables

Two (or more) statements can be combined into compound statements using a connective. For example, the statement 'The book is new and about mathematics' is a compound of the single statements 'The book is new' and 'The book is about mathematics'.

Notice the **connective**, 'and', that is used to join the two statements. Two main connectives — 'and' and 'or' — are used in compound sentences. Other connectives are 'not', 'if ... then ...' and 'if and only if ...'.

The truth of a compound statement is determined by the truth of the separate single statements. These separate single statements are sometimes called atomic sentences.

Considering this example, there are four cases:

Case 1: 'The book is new' is true. 'The book is about mathematics' is true.

Case 2: 'The book is new' is true. 'The book is about mathematics' is false.

Case 3: 'The book is new' is false. 'The book is about mathematics' is true.

Case 4: 'The book is new' is false. 'The book is about mathematics' is false.

This list can be summarised using a truth table.

Let $p =$ 'The book is new' and

$q =$ 'The book is about mathematics'.

What about the third column? This represents the **truth value** of the compound statement 'p and q'. To determine this truth value we need to examine the logical definition of the connective 'and'. For the compound statement to be true, both single statements must be true. If either is false, then the whole statement is false. Therefore, we can complete the truth table for 'and' (using the common symbol ∧ to represent 'and').

p	q	$p \wedge q$
T	T	T
T	F	F
F	T	F
F	F	F

Similarly, the truth table for 'or', using the symbol ∨, is shown. The implication here is that it takes only one (or both) of the statements to be true for a statement such as 'Mary went to the store or the library' to be true. If she went to the store, then certainly she went to the store or the library. Similarly, if she went to the library in this example, the statement would be true.

p	q	$p \vee q$
T	T	T
T	F	T
F	T	T
F	F	F

There are some compound statements where it is not possible for both statements to be true at the same time. For example: 'John is 15 or 16 years old'. Clearly, in this case John cannot be both 15 and 16. This is an example of 'exclusive or'.

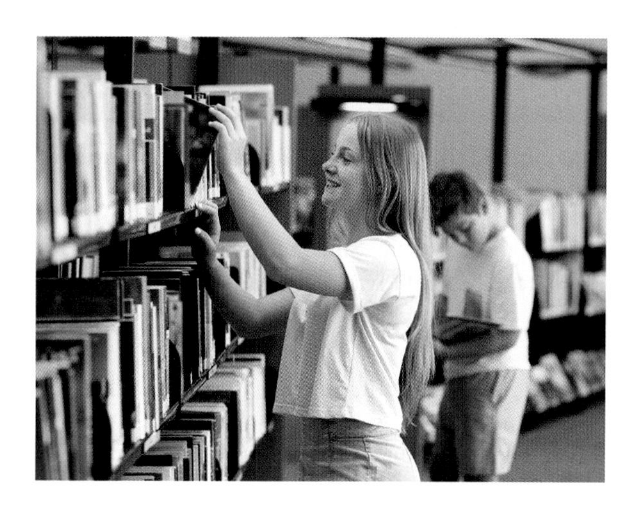

Also be careful not to confuse the logical use of 'and' with the common English usage. For example, the sentence 'Boys and girls are allowed in the swimming pool after 6 pm' is made up of the compound sentences 'Boys are allowed …' and 'Girls are allowed …'. In reality, what is being said is that either boys *or* girls or both are allowed, so logically the sentence should be 'Both boys and girls are allowed in the swimming pool after 6 pm'.

In some textbooks 'and' is called the **conjunction** and 'or' is called the **disjunction**.

WORKED EXAMPLE 2 Creating a truth table for a statement

Determine the truth table for the compound statement:
'The suspect wore black shoes or was a female wearing a skirt.'

THINK	WRITE
1. Identify and label the individual statements.	p = 'The suspect wore black shoes.' q = 'The suspect was female.' r = 'The suspect wore a skirt.'

2. Form a compound statement. Clearly p is separate from q and r.

p or (q and r)

$p \vee (q \wedge r)$

Note: Use brackets to indicate the separation.

3. Create a truth table. Because there are three statements and each can have two values (T or F), there are $2 \times 2 \times 2 = 8$ rows in the table. The $(q \wedge r)$ column is completed by looking at the q and r columns.
4. The last column is completed by looking at just the p column and at the $(q \wedge r)$ column.

p	q	r	$(q \wedge r)$	$p \vee (q \wedge r)$
T	T	T	T	T
T	T	F	F	T
T	F	T	F	T
T	F	F	F	T
F	T	T	T	T
F	T	F	F	F
F	F	T	F	F
F	F	F	F	F

As can be observed in the last column above, p 'dominates' the table. Regardless of the truth of q and r, the entire statement is true if p is true (rows 1–4). Otherwise, if p is false then both q and r must be true (row 5).

Negation

Another connective is the negation, or 'not', which is denoted by the symbol $\neg$. This is merely the opposite of the original statement. For example, if $p =$ 'It is raining', then $\neg p =$ 'It is not raining'.

Be careful when negating English sentences. For example, the negation of 'I am over 21' isn't 'I am under 21', but 'I am not over 21'. Can you see the difference?

WORKED EXAMPLE 3 Completing a truth table

Complete the truth table for the compound statement $p \vee \neg p$.

THINK

Set up a truth table. Since there is only one statement here (p), we need only two rows, either p is true or p is false.

WRITE

p	$\neg p$	$p \vee \neg p$
T	F	T
F	T	T

Note: The compound statement in Worked example 3 is always true! An English sentence equivalent to this statement could be 'I will be there on Monday or I will not be there on Monday'.

Karnaugh maps

Karnaugh maps are an alternate way of displaying the information in a truth table.

Instead of using columns to represent each proposition, Karnaugh maps display the propositions in a table, with one proposition along the top row and the other along the left column.

For example, consider an empty truth table for two propositions:

p	q	
T	T	
T	F	
F	T	
F	F	

The empty Karnaugh map for two propositions looks like:

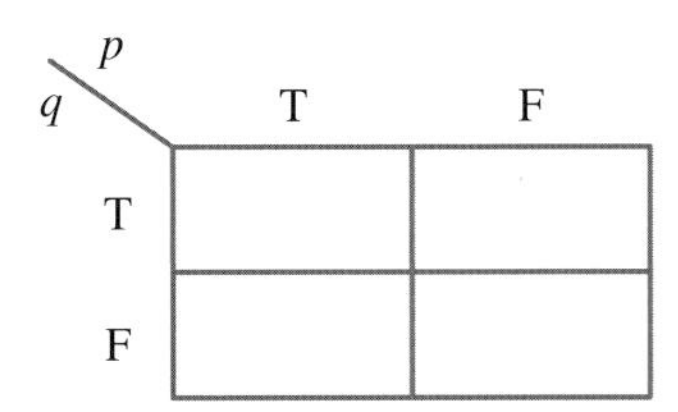

The left column corresponds to p being true, and the right column corresponds to p being false. Similarly, the top row corresponds to q being true, and the bottom row corresponds to q being false.

The true/false values that fill the cells in the Karnaugh map correspond to the values in the third column of a truth table.

For example, consider the propositions p, q and $p \vee \neg q$.

The truth table and Karnaugh map for these propositions are shown below.

p	q	$p \vee \neg q$
T	T	T
T	F	T
F	T	F
F	F	T

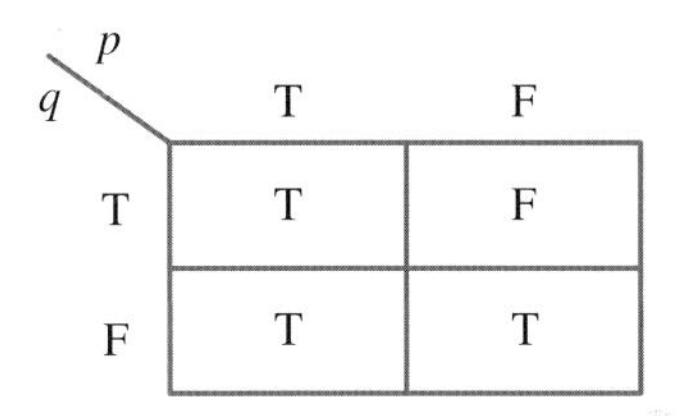

Note that these are equivalent; the third column of the truth table corresponds to the values in the cells of the Karnaugh map.

Equivalent statements

Two statements are equivalent if their truth tables are identical. Each row of the truth tables must match. If there is even one difference, then the statements are not equivalent. The symbol $\leftrightarrow$ is used to indicate equivalence, as in $p \leftrightarrow q$. This is read as 'p is true if and only if q is true'.

Note that the equivalence operators, $p \leftrightarrow q$, have a truth table of their own, as shown.

This clearly demonstrates that $p \leftrightarrow q$ is true when the truth value of p equals the truth value of q; that is, if both p and q are true or if both p and q are false.

p	q	$p \leftrightarrow q$
T	T	T
T	F	F
F	T	F
F	F	T

WORKED EXAMPLE 4 Completing a truth table

By completing truth tables, show that $\neg(p \wedge q) \leftrightarrow (\neg p \vee \neg q)$.

THINK

1. Set up a truth table. Because there are two statements, we need $2 \times 2 = 4$ rows.

WRITE

p	q	$\neg p$	$\neg q$	$(p \wedge q)$	$\neg(p \wedge q)$	$(\neg p \vee \neg q)$
T	T	F	F	T	F	F
T	F	F	T	F	T	T
F	T	T	F	F	T	T
F	F	T	T	F	T	T

2. Complete the $\neg p$ and $\neg q$ columns by negating p and q separately.

3. Complete the $(p \wedge q)$ column.
4. Negate the $(p \wedge q)$ column.
5. Write the column for $(\neg p \vee \neg q)$ using columns 3 and 4.
6. Observe that the final two columns are equal in every row. Since their truth tables are identical, the two statements are equivalent.

3.2 Exercise

Students, these questions are even better in jacPLUS

Receive immediate feedback and access sample responses

Access additional questions

Track your results and progress

Find all this and MORE in jacPLUS

Technology free

1. WE1 Classify the following sentences as statements (propositions), instructions, suggestions, exclamations, opinions or 'near-statements'. If they are statements, then indicate whether they are true, false or indeterminate without further information.
   a. That was the best Hollywood movie this year.
   b. That movie had the most Oscar nominations this year.
   c. When the power fails, candles are a good source of light and heat.
   d. Why did you use that candle?
   e. Mary is tall for her age.

2. Break up the following compound statements into individual single statements.
   a. The car has 4 seats and air conditioning.
   b. The Departments of Finance and Defence were both over budget last year.
   c. Bob, Carol, Ted and Alice went to the hotel.
   d. To be a best-seller a novel must be interesting and relevant to the reader.
   e. Either Sam or Nancy will win the trophy.
   f. You can choose from ice-cream or fruit for dessert. We have vanilla or strawberry ice-cream.
   g. There are some statements which cannot be proved to be true or false.
   h. Most of my friends studied Mathematics, Physics, Engineering or Law and Arts.
3. Convert the following pairs of simple sentences into a compound sentence. Be sure to use 'and' and 'or' carefully.
   a. John rode his bicycle to school. Mary rode her bicycle to school.
   b. The book you want is in row 3. The book you want is in row 4.
   c. The weather is cold. The weather is cloudy.
   d. Many people read novels. Many people read history.
   e. In a recent poll, 45% preferred jazz. In a recent poll, 35% preferred classical music.
   f. Two is an even number. Two is a prime number.

4. As you saw in Worked example 4, if there is a compound statement with two single statements, p and q, then there are $2 \times 2 = 4$ rows in the truth table. List all the different rows for compound statements made up of:

   a. 3 single statements
   b. 4 single statements
   c. 5 single statements.

   You should be able to develop a pattern of completing the Ts and Fs in a logical sequence.

5. **WE2** Write the following compound sentences in symbolic form (p, q, r), and determine the truth table.
   a. John, Zia and David passed General Mathematics.
   b. Either Alice and Renzo, or Carla, will have to do the dishes. (Note the use of commas.)
   c. The committee requires two new members. One must be a female, the other must be either a student or a professor.

6. **WE3** Complete the truth tables for the following compound statements.
   a. $p \wedge \neg q$
   b. $\neg p \wedge \neg q$
   c. $(p \wedge q) \wedge r$

7. Complete the truth tables for the following compound statements.
   a. $p \vee \neg q$
   b. $\neg p \vee \neg q$
   c. $(p \vee q) \vee r$

8. Let $p =$ 'It is raining', $q =$ 'I bring my umbrella'. Write a sentence for the following compound statements.
   a. $p \wedge q$
   b. $p \vee q$
   c. $\neg p \wedge q$

9. Let $p =$ 'Peter likes football', and $q =$ 'Quentin likes football'. Write a sentence for the following compound statements.
   a. $p \wedge q$
   b. $p \vee q$
   c. $p \wedge \neg q$

10. **WE4** By completing truth tables, show that $\neg(p \vee q) \leftrightarrow (\neg p \wedge \neg q)$.

11. Determine whether the compound statement $\neg(p \vee q)$ is equivalent to $\neg p \vee \neg q$.

12. Determine whether the following compound statement pairs are equivalent.
    a. $(p \wedge q) \vee \neg p$
       $(p \vee q) \wedge \neg p$
    b. $(p \vee q) \vee \neg p$
       $p \vee \neg p$

13. Determine if the brackets in an expression alter the truth table by comparing $(p \wedge q) \vee r$ with $p \wedge (q \vee r)$.

14. Repeat question 13 with the following statement pairs.
    a. i. Compare $(p \wedge q) \wedge r$ with $p \wedge (q \wedge r)$.
       ii. Compare $(p \vee q) \vee r$ with $p \vee (q \vee r)$.
    b. Based on the results of questions 13 and 14, state what you might conclude about the effect of brackets on a compound expression.

3.2 Exam questions

Question 1 (1 mark) TECH-FREE

Complete the truth table for the compound statement $p \wedge \neg p$.

Question 2 (1 mark) TECH-ACTIVE

MC p and q, or p and r can be written as

A. $(p \wedge q) \wedge (p \wedge r)$
B. $(p \wedge q) \vee (p \wedge r)$
C. $(p \vee q) \wedge (p \vee r)$
D. $(p \vee q) \vee (p \vee r)$
E. $(p \wedge q) \neg (p \wedge r)$

Question 3 (2 marks) TECH-FREE

Determine the missing connectives between p, q and r in the last column of the following truth table.

p	q	r	$p\,?\,q\,?\,r$
T	T	T	T
T	T	F	T
T	F	T	T
F	T	T	T
T	F	F	F
F	T	F	F
F	F	T	T
F	F	F	F

More exam questions are available online.

3.3 Valid and invalid arguments

LEARNING INTENTION

At the end of this subtopic you should be able to:

- determine the validity of arguments by looking at the premises and seeing whether the conclusion follows logically from them.

The purpose of the logical connectives 'and', 'or' and 'not' is to form statements, true or false, in order to evaluate the truth, or otherwise, of something called an argument. An argument is a set of one or more propositions (statements). Before we can evaluate arguments, we need one more connective: the implication (or conditional) statement.

3.3.1 Implication

Consider the following 'classical' statement: 'If it is raining, then I bring my umbrella.' This is the combination of the two statements 'It is raining' and 'I bring my umbrella', connected by two words: 'if' and 'then'. Each of the two statements has individual truth values; either could be true or false. The first statement is called the **antecedent**, the second is called the **consequent**, and in symbolic form this is written as $p \to q$.

This is called **implication** because the first statement implies the second; it is also called **conditional**, because the outcome of the second statement is conditional on the first.

How can we determine the truth table of $p \to q$? This is not as simple as employing a mere definition.

Referring to our example, consider the question 'Under what conditions would $p \to q$ be a lie?'

1. If it *is* indeed raining and I bring my umbrella then, clearly $p \to q$ is true.
2. If it is raining and I *don't* bring my umbrella, then I lied to you! Thus, $p \to q$ is false.
3. What if it is *not* raining? I have told you *nothing* about what I would do in that case. I might either bring my umbrella, or I might not. In either case you *cannot* say I lied to you, so $p \to q$ is true.

p	q	$p \to q$
T	T	T
T	F	F
F	T	T
F	F	T

The truth table can be constructed to summarise this situation.

This leads us immediately to ask the question: Is $p \to q$ the same as $q \to p$?

WORKED EXAMPLE 5 Using a truth table to determine equivalence

Using a truth table, determine whether $(p \to q) \leftrightarrow (q \to p)$.

THINK

1. Set up a truth table for p, q and $p \to q$. $p \to q$ is shown in the 3rd column.
2. Exchange the roles of p and q to determine the truth table for $q \to p$. This is shown in the last column.
3. Clearly, they are not equivalent.

WRITE

p	q	$p \to q$	$q \to p$
T	T	T	T
T	F	F	T
F	T	T	F
F	F	T	T

This is a most important result; it is a result that people who think they are arguing logically often mistake for a valid statement. In this example, 'If I bring my umbrella, then it is raining' says a very different thing from the original statement and is called its **converse**. It seems to be making the argument that my bringing the umbrella can control the weather!

3.3.2 Converse, contrapositive and inverse

As we have just seen, there are alternative forms of $p \to q$, such as the converse. These and their relationships with $p \to q$ are shown in the table.

Relationships to $p \to q$

Name	Symbol	Relationship to $p \to q$
Implication	$p \to q$	**(assumed) True**
Converse	$q \to p$	**False**
Contrapositive	$\neg q \to \neg p$	**True**
Inverse	$\neg p \to \neg q$	**False**

Often the contrapositive is a more realistic way of stating an implication than the original statement is. Be careful, however, not to use the converse or inverse as they are (generally) false when $p \to q$ is true.

3.3.3 Arguments

An **argument** is a series of statements divided into two parts — the premises and the conclusion. The **premises** are a series of statements intended to justify the **conclusion**. For example, consider the following argument:

A terrier is a breed of dog.	Premise
Rover is a terrier.	Premise
Therefore, Rover is a dog.	Conclusion

Generally, an argument will have only one conclusion and (usually) two premises.

Conclusion and premise indicators

To help identify the conclusion, look for words or phrases such as: therefore, accordingly, hence, thus, consequently, it must be so, so, it follows that, implies that.

What follows one of these conclusion indicators is the conclusion; by default everything else is a premise. There are also premise indicators: because, given that, since, seeing that, may be inferred from, owing to, for, in that.

In a formal argument, the conclusion comes after the premises.

WORKED EXAMPLE 6 Identifying premises and conclusions

Identify the premises and conclusions for each of the following arguments.

a. **A Commodore is a model of a Holden car.**
   My car is a white Commodore.
   Therefore, my car is a Holden.

b. **Military defence depends upon adequate government funding.**
   Adequate government funding depends on a healthy economy.
   A healthy economy depends upon an intelligent fiscal policy.
   Military defence depends upon an intelligent fiscal policy.

c. **Pregnant mothers should not smoke.**
   Cigarettes can harm the foetus.

THINK	WRITE	
a. Examine each sentence looking for the conclusion indicators, or examine the sequence of the sentences.	a. A Commodore is a model of a Holden car.	Premise
	My car is a white Commodore.	Premise
	Therefore, my car is a Holden.	Conclusion
b. Note how the sequence of statements connects one with the next. The last is therefore the conclusion.	b. Military defence depends upon adequate government funding.	Premise
	Adequate government funding depends on a healthy economy.	Premise
	A healthy economy depends upon an intelligent fiscal policy.	Premise
	Military defence depends upon an intelligent fiscal policy.	Conclusion
c. In this case the sentences have been reversed. This is a common mistake.	c. Pregnant mothers should not smoke.	Conclusion
	Cigarettes can harm the foetus.	Premise

In some textbooks, statements are called **propositions** and arguments are called **inferences**.

3.3.4 Categorical propositions and the deductive argument

The standard argument consists of two premises and a conclusion:

All dogs are mammals.	Premise
Rover is a dog.	Premise
Therefore, Rover is a mammal.	Conclusion

Note: Observe the use of the key word 'All'. Beware of arguments that use the key word 'some', as in 'Some journalists are hard-working'. This is a weaker form of argument, the study of which is beyond the scope of this course.

The first premise is called a **categorical statement** or **proposition**, and this form of argument can be called the classical **deductive argument**. However, as we shall see, there are many cases where we will not have a **valid** deductive argument, even if everything looks correct; these situations are called **fallacies**. As an example, consider the following argument:

All dogs are mammals.	Premise
Rover is a mammal.	Premise
Therefore, Rover is a dog.	Conclusion

Clearly, no one should be convinced by this argument. Both premises might be true, but the conclusion does not follow logically from them, and we would say that this is an **invalid** argument. This is an example of a formal, or structural, fallacy.

Some categorical propositions can be turned into implications. For instance, the statement 'All dogs are mammals' can be written as 'If it is a dog, then it is a mammal'. This says exactly the same thing.

Beware of statements such as 'If it is sunny tomorrow, I will go to the beach'. This is not the same as saying 'On all sunny days I will go to the beach'. The key word here is 'tomorrow' — this restricts the statement so that the key word 'all' cannot be used. However, the implication can still be used in a valid argument:

If it is sunny tomorrow, I will go to the beach.

After checking the weather tomorrow:

It is sunny.

I will go to the beach.

This is certainly a valid argument. At this point, we can define a symbolic form for this kind of deductive argument.

Symbolic form of an argument

The symbolic form of an argument is:

$$\frac{\textbf{Premise 1} \\ \textbf{Premise 1}}{\textbf{Conclusion}}$$

***Note:* There may be more than 2 premises; in general, all premises are above the horizontal line and the conclusion is below it.**

Putting the argument from above into this form we have:

$$\begin{array}{l} p \to q \\ \underline{p \quad} \\ q \end{array}$$

In this case, p is 'It is sunny tomorrow' and q is 'I go to the beach tomorrow'. Given the premises $p \to q$ and p are true, it follows that q is true. This argument is therefore valid. Note that this is only one form of (potentially) valid argument.

WORKED EXAMPLE 7 Determining whether arguments are valid

Determine whether the following arguments are valid.

a. **All mathematics books are interesting.**
This is a book about mathematics.
Therefore, this book is interesting.

b. **If I study hard, I will pass Physics.**
I passed Physics.
I must have studied hard.

c. **Some history books are boring.**
This book is about history.
Therefore, this book is boring.

d. **If I don't study, I will fail Physics.**
I didn't study.
I will fail Physics.

THINK	WRITE
a. 1. Change the first statement to: 'If … then …'	a. If it is a mathematics book, then it is interesting.
2. (a) Assign each statement a symbol.	p = It is a mathematics book. q = It is interesting.
(b) Put the argument into symbolic form.	$p \to q$ $\underline{p \quad}$ q
(c) Determine whether the conclusion follows logically from the premises.	If p is true and $p \to q$ is true, then q must also be true.
3. Determine whether it is a valid form.	Yes, this is a valid form for an argument.
b. 1. (a) Assign each statement a symbol.	b. p = I study hard. q = I will pass Physics.
(b) Put the argument into symbolic form.	$p \to q$ $\underline{q \quad}$ p
(c) Determine whether the conclusion follows logically from the premises.	If q is true and $p \to q$ is true, p is not necessarily true.
2. Determine whether it is a valid form.	No, this is not a valid form for an argument.
c. Consider the first statement. Note the use of the word 'some'.	c. The use of the word 'some' means that the statement cannot be put into this form. Thus, the entire argument is not valid.
d. 1. (a) Assign each statement a symbol.	d. p = I don't study. q = I will fail Physics.
(b) Put the argument into symbolic form.	$p \to q$ $\underline{p \quad}$ q
(c) Determine whether the conclusion follows logically from the premises.	If p is true and $p \to q$ is true, then q must also be true.
2. Determine whether it is a valid form.	Yes, this is a valid form for an argument.

Note: Even if the statements are expressed in negative form: 'I don't study' … 'I will fail Physics', it is still possible to have a valid argument. Can you devise a 'positive' argument that is the equivalent to the one in part **d** of Worked example 7?

Valid and sound arguments

It is important to note that an argument may be valid even if the truth of the component statements cannot be established. Consider the following (nonsense) argument:

All fribbles are granches.

A hommie is a fribble.

Therefore, a hommie is a granch.

We certainly cannot establish the truth of the two premises (let alone know what fribbles, granches or hommies are), but presuming they are true, the argument is valid. Furthermore, consider the argument:

If it is a dog, then it can do algebra.

Rover is a dog.

Therefore, Rover can do algebra.

This is a valid form of argument, but one (or more) of the premises is (are) false. Despite it's validity, this argument is unlikely to convince anyone of the mathematical ability of dogs.

In cases such as this one, where the logic of the argument is valid but one or more of the premises are false we say that the argument is valid, but not **sound**. A sound argument is a valid argument which has premises that are true.

3.3.5 Valid forms of argument

There are many valid forms of argument. We shall limit our discussion to the most important ones, five of which are tabulated below.

Argument form and name	Example
$p \rightarrow q$ p ——— q *Modus ponens*	If Mary is elected, then she must be honest. Mary was elected. ——— Mary must be honest. This is our standard form.
$p \vee q$ $\neg p$ ——— q Disjunctive syllogism	Either John or Jemma was born in Canada. John was not born in Canada. ——— Jemma was born in Canada. Note that the roles of p and q can be interchanged here.
$p \rightarrow q$ $q \rightarrow r$ ——— $p \rightarrow r$ Hypothetical syllogism	If it is raining, I will bring my umbrella. If I bring my umbrella,then I will not get wet. ——— If it is raining I will not get wet. Many statements (p, q, r, …) can be linked together this way to form a valid argument.
$p \rightarrow q$ $\neg q$ ——— $\neg p$ *Modus tollens*	If I study hard, I will pass Physics. I did not pass Physics. ——— I did not study hard. This is a valid form of a negative argument.

$p \to q \wedge r \to s$ $p \wedge r$ $q \wedge s$ Constructive dilemma	If we holiday in France, we will have to practise speaking French, and if we holiday in Germany, we will have to practise German. We will holiday in France and Germany. We will have to practise speaking French and German.

There are several other forms more complex than these that are beyond the scope of this course.

3.3.6 Proving the validity of an argument form

It may not be satisfactory to merely declare that the five arguments in the previous table are automatically valid. There is a way to mathematically establish their validity using a truth table. The procedure is as follows.

Proving the validity of an argument

Step 1. Set up a single truth table for all the premises and for the conclusion.

Step 2. Examine the row (or rows) in the table where *all* the premises are true.

Step 3. If the conclusion is true in each of the cases in step 2, then the argument is valid. Otherwise it is invalid.

WORKED EXAMPLE 8 Establish the validity of arguments (1)

Establish the validity of the *modus ponens* argument, namely:

$$\begin{array}{l} p \to q \\ p \\ \hline q \end{array}$$

THINK

1. Set up a truth table for each of the premises, namely p and $p \to q$, and the conclusion q. Note that p and q are set up first in the usual way, and $p \to q$ is completed from them.

WRITE

p	q	$p \to q$
T	T	T
T	F	F
F	T	T
F	F	T

2. Determine the rows where all the premises are true.

The premises are all true in the 1st row only.

3. Compare with the conclusion column (q).

The conclusion is also true, so the argument is valid.

WORKED EXAMPLE 9 Establish the validity of arguments (2)

Show that the following argument is invalid.

$$\begin{array}{l} p \rightarrow q \\ \underline{q} \\ p \end{array}$$

THINK

1. Set up a truth table for each of the premises, namely q and $p \rightarrow q$, and the conclusion p. Note that p and q are set up first in the usual way, and $p \rightarrow q$ is completed from them.

WRITE

p	q	$p \rightarrow q$
T	T	T
T	F	F
F	T	T
F	F	T

2. Determine the rows where all the premises are true.

The premises are all true in the 1st row and 3rd row.

3. Compare with the conclusion column (p).

The conclusion is true in the 1st row but false in the 3rd, so the argument is invalid.

The argument shown in Worked example 9 is a common error in logical argument and is called **affirming the consequent**.

In conclusion, if an argument fits exactly one of the five given forms, then it is immediately assumed to be valid; otherwise it must be established to be valid using truth tables.

3.3.7 Tautologies

In logic, a tautology is a formula or assertion involving propositions that is true in every possible interpretation.

Tautologies involving 2 propositions

WORKED EXAMPLE 10 Showing a tautology involving 2 propositions

Use a truth table to show that $\neg(p \wedge q) \leftrightarrow (\neg p \vee \neg q)$ is a tautology.
This is one of De Morgan's laws.

THINK

1. Set up the truth table. Since there are two propositions, we need four rows.
2. In the third column complete true and false for $p \wedge q$.
3. In the fourth, fifth and sixth columns negate $p \wedge q$, p and q respectively.
4. In the seventh column complete true and false for $\neg p \vee \neg q$.
5. In the last (eighth) column form $\neg(p \wedge q) \leftrightarrow (\neg p \vee \neg q)$.
Since the fourth and seventh columns are the same, the last column contains only true values T; it is proved.

WRITE

p	q	$p \wedge q$	$\neg(p \wedge q)$	$\neg p$	$\neg q$	$\neg p \vee \neg q$	$(p \wedge q) \leftrightarrow (\neg p \vee \neg q)$
T	T	T	F	F	F	F	T
T	F	F	T	F	T	T	T
F	T	F	T	T	F	T	T
F	F	F	T	T	T	T	T

Tautologies involving 3 propositions

WORKED EXAMPLE 11 Showing a tautology involving 3 propositions

Use a truth table to show that $(p \to q) \wedge (q \to r) \to (p \to r)$ is a tautology.
Showing this is equivalent to establishing the law of hypothetical syllogism.

THINK

1. Set up the truth table. Since there are three propositions, we need $2^3 = 8$ rows.
2. In the fourth column form $p \to q$.
3. In the fifth column form $q \to r$.
4. In the sixth column form $(p \to q) \wedge (q \to r)$.
5. In the seventh column form $p \to r$.
6. In the last column form $(p \to q) \wedge (q \to r) \to (p \to r)$. Since it contains only true values it is proved.

WRITE

p	q	r	$p \to q$	$q \to r$	$(p \to q) \wedge (q \to r)$	$p \to r$	$(p \to q) \wedge (q \to r) \to (p \to r)$
T	T	T	T	T	T	T	T
T	T	F	T	F	F	F	T
T	F	T	F	T	F	T	T
T	F	F	F	T	F	F	T
F	T	T	T	T	T	T	T
F	T	F	T	F	F	T	T
F	F	T	T	T	T	T	T
F	F	F	T	T	T	T	T

3.3 Exercise

Students, these questions are even better in jacPLUS

Receive immediate feedback and access sample responses

Access additional questions

Track your results and progress

Find all this and MORE in jacPLUS

Technology free

1. WE5 Using truth tables, determine whether $(p \to q) \leftrightarrow (\neg q \to \neg p)$.
2. Establish the truth table for the inverse; namely, show that $(p \to q)$ is not equivalent to $(\neg p \to \neg q)$.
3. Let $p =$ 'It is bread' and $q =$ 'It is made with flour'. Write out the implication, converse, contrapositive and inverse in sentences.
4. WE6 Identify the premises and conclusion in the following arguments.
   a. All cats are fluffy.
      My pet is a cat.
      My pet is fluffy.
   b. Two is the only even prime number.
      Prime numbers are divisible by themselves and 1.
      All even numbers are divisible by themselves and by 2.
   c. Growing apples depends on good water.
      Growing apples depends on good irrigation.
      Good water depends on good irrigation.

5. **WE7** Determine which of the following are valid arguments.

 a. If you are a mathematician, you can do algebra.
 You are a mathematician.
 You can do algebra.

 b. All footballers are fit.
 David is not a footballer.
 David is not fit.

 c. If it is a native Australian mammal, then it is a marsupial.
 A wombat is a native Australian mammal.
 A wombat is a marsupial.

 d. Some TV shows are boring.
 Neighbours is a TV show.
 Neighbours is boring.

 e. All musicians can read music.
 Louise can read music.
 Louise is a musician.

6. Look again at the arguments in question 5 which were not valid. If possible, turn them into valid arguments. Assume that the first statement in each argument is always correct.

7. **WE8** Establish the validity of the disjunctive syllogism argument, namely:

$$\begin{array}{l} p \vee q \\ \underline{\neg q} \\ q \end{array}$$

8. Establish the validity of the three remaining valid forms of argument, namely:

 a. hypothetical syllogism:
 $$\begin{array}{l} p \rightarrow q \\ \underline{q \rightarrow r} \\ p \rightarrow r \end{array}$$

 b. *modus tollens*:
 $$\begin{array}{l} p \rightarrow q \\ \underline{\neg q} \\ \neg p \end{array}$$

 c. constructive dilemma:
 $$\begin{array}{l} p \rightarrow q \wedge r \rightarrow s \\ \underline{p \wedge r \qquad\quad} \\ q \wedge s \end{array}$$

9. The following are valid arguments. Determine which of the five forms of argument were used.

 a. Either you clean up your room or you will not watch any television tonight. You did not clean up your room.
 Therefore you will not watch any television tonight.

 b. If you help your mother with the dishes, I will take you to the football game tomorrow. I didn't take you to the football game.
 Therefore you didn't help your mother with the dishes.

 c. If you study statistics, then you will understand what standard deviation means. You studied statistics.
 Therefore you will understand what standard deviation means.

10. **WE9** Determine the validity of the following argument:

$$\begin{array}{l} p \rightarrow q \\ \underline{\neg p} \\ \neg q \end{array}$$

11. Show that the following is an example of the argument in question 10 above.
 If elected with a majority, my government will introduce new tax laws.
 My government was not elected with a majority.
 Therefore, my government will not introduce new tax laws.

12. A common argument is of the form:
 If you work hard, then you will become rich.
 You don't work hard.
 Therefore, you will not become rich.

 a. Put this argument in symbolic form.

 b. Show that it is an invalid form of argument. (This is called 'denying the antecedent'.)

13. Determine the validity of the following arguments.

a. $p \rightarrow q$
$r \rightarrow \neg q$
———
$p \rightarrow \neg r$

b. $\neg p \wedge \neg q$
$r \rightarrow q$
———
r

c. $\neg p \rightarrow \neg q$
p
———
q

14. Determine the validity of the following arguments.

a. All dogs have five legs.
All five-legged creatures are called chickens.
Therefore, all dogs are chickens.

b. All dogs have five legs.
All chickens have five legs.
Therefore, all dogs are chickens.

15. Determine the validity of the following arguments.

a. If you deposit money in the bank, then you will earn interest.
You didn't earn any interest.
Therefore, you didn't deposit any money in the bank.

b. If I wanted an easy course to study, I would choose Human Development and if I wanted an interesting course to study, I would choose General Mathematics.
I can choose an easy course, and an interesting one.
Therefore, I will study Human Development and General Mathematics.

16. **WE10** Use a truth table to show that $\neg (p \vee q) \leftrightarrow (\neg p \wedge \neg q)$ is a tautology.

17. Use a truth table to show that $(p \wedge q) \rightarrow p$ is a tautology.

18. Use a truth table to show that $\left[(p \rightarrow q) \wedge \neg q\right] \rightarrow \neg p$ is a tautology; that is, establish the law of *modus tollens*.

19. **WE11** Use a truth table to show that $\left[(p \vee q) \vee r\right] \leftrightarrow \left[p \vee (q \vee r)\right]$ is a tautology. This is showing the associative law for $\vee$.

20. Use a truth table to show that $\left[p \vee (q \wedge r)\right] \leftrightarrow \left[(p \vee q) \wedge (p \vee r)\right]$ is a tautology. This is showing the distributive law.

3.3 Exam questions

Question 1 (2 marks) TECH-FREE

Examine the validity of the following argument.

The supplement of an obtuse angle is an acute angle.

Angle A is not obtuse.

Hence, the supplement of angle A is not acute.

Question 2 (2 marks) TECH-FREE

Use a truth table to determine whether $(p \rightarrow q) \leftrightarrow (\neg q \rightarrow \neg p)$.

Question 3 (2 marks) TECH-FREE

Use tautology to verify the validity or otherwise of the following *modus tollens* argument.

$\neg p \rightarrow \neg q$
q
———
p

More exam questions are available online.

3.4 Boolean algebra and digital logic

LEARNING INTENTION

At the end of this subtopic you should be able to:
- convert between Boolean expressions, truth tables and logic circuits
- determine a truth table for a given logic circuit
- draw a logic circuit for a given Boolean expression.

3.4.1 Boolean algebra

In ordinary algebra we use the letters x, y to denote any real numbers, $x, y \in R$.

In Boolean algebra, we use the letters A, B to denote the values of 0 or 1 only, $A, B \in \{0, 1\}$. Note that a Boolean variable must be either 0 or 1, it cannot be both at the same time. Since this is a new type of algebra, we need to define new operations for Boolean variables. There are only three Boolean operations: NOT, AND and OR. These operations are denoted as NOT $A = A'$, A AND $B = A \bullet B$, and A OR $B = A + B$. They are defined as shown by the truth tables below.

NOT	
A	A'
0	1
1	0

AND		
A	B	$A \cdot B$
0	0	0
0	1	0
1	0	0
1	1	1

OR		
A	B	$A + B$
0	0	0
0	1	1
1	0	1
1	1	1

Note that the plus sign works very differently in Boolean algebra than it does for addition in regular algebra. For example, in Boolean algebra $1 + 1 = 1$. Be careful not to confuse the Boolean + with the + used for addition.

Laws of Boolean algebra

Idempotent laws	$A \cdot A = A$ $A + A = A$
Complement laws	$A \cdot A' = 0$ $A + A' = 1$
Identity laws	$A \cdot 1 = 1$ $A \cdot 0 = 0$ $A + 0 = A$ $A + 1 = 1$
Commutative laws	$A \cdot B = B \cdot A$ $A + B = B + A$
Associative laws	$(A + B) + C = A + (B + C)$ $(A \cdot B) \cdot C = A \cdot (B \cdot C)$
Distributive laws	$A \cdot (B + C) = A \cdot B + A \cdot C$ $A + (B \cdot C) = (A + B) \cdot (A + C)$

WORKED EXAMPLE 12 Using truth tables to verify Boolean expressions

Use a truth table to show that $A + A \cdot B = A$.

THINK

1. Draw up the truth table. We need four rows: in the first two columns the values of A and B; in column three, A AND B.

WRITE

A	B	$A \cdot B$	$A + A \cdot B$
0	0	0	0
0	1	0	0
1	0	0	1
1	1	1	1

2. Since the first and last columns are identical, the result follows. This result is known as the first absorption law.

$A + A \cdot B = A$

WORKED EXAMPLE 13 Using the laws of Boolean algebra to simplify Boolean expressions

Using the laws of Boolean algebra, simplify each of the following Boolean expressions.

a. $A \cdot B + A \cdot B'$

b. $A \cdot B \cdot C \cdot \left(A + B' + C\right)$

THINK

a. 1. Use the first distributive law.

2. Use the second complement and identity laws.

b. 1. Use the distributive laws.

2. Use the idempotent, complement and commutative laws.

WRITE

a. $A \cdot B + A \cdot B' = A \cdot (B + B')$

$= A \cdot 1$

$= A$

b. $A \cdot B \cdot C \cdot (A + B' + C)$

$= A \cdot B \cdot C \cdot A + A \cdot B \cdot C \cdot B' + A \cdot B \cdot C \cdot C$

$= A \cdot B \cdot C + 0 + A \cdot B \cdot C$

$= A \cdot B \cdot C$

3.4.2 Digital truth values

The contribution of logic and Boolean algebra to the design of digital computers is immense. All digital circuits rely on the application of the basic principles we have learned in this topic. Computer software is constructed using **logic gates** based on some of the rules of logic laid down by Aristotle.

Digital circuits consist of electrical current flowing through wires that connect the various components. The computer recognises the presence of current as 'True' and the absence of current as 'False'.

Furthermore, it is the accepted convention that we denote the presence of current by 1 and the absence by 0. (In some systems the value of 1 is given to positive current and 0 to negative current.) Thus, we have the basic conversion rule that we will apply here as shown in the table below.

Logical value	Digital value	Spoken value
False	0	Off
True	1	On

The so-called 'on–off' values come from the notion of a switch: if the current flows through, the switch is on; otherwise it is off, just like a light switch.

3.4.3 Gates

A gate is an electrical component that controls the flow of current in some way. It is similar to a gate on a farm, which sometimes lets the sheep through and sometimes doesn't. The simplest possible gate is the switch itself. It has two states, on and off, as shown in the figures. When drawing a switch on a diagram it is conventional to show the 'off' position. By combining switches in certain configurations, we can create simple logic circuits.

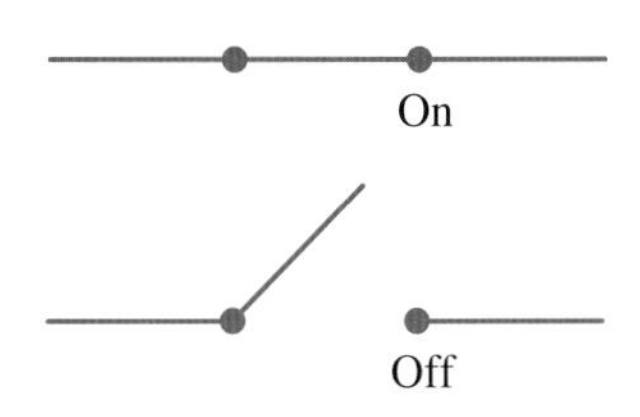

WORKED EXAMPLE 14 Truth tables for switches

Consider the pair of switches arranged (in parallel) as shown in the figure. Assume there is electricity at *P*. Determine which positions of the two switches, *x* and *y*, will allow a current to flow through.

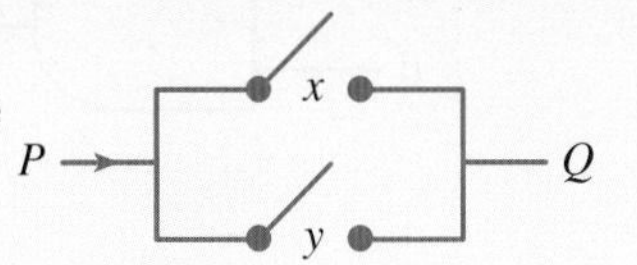

THINK

1. List the possible positions for each switch. Switch x can be either off or on (0 or 1) independently of y, so there are $2 \times 2 = 4$ possible positions.
2. Consider $x = 0$, $y = 0$. There will be no current at Q. Otherwise, if $x = 1$ there will be a current at Q. Similarly, if $y = 1$ there will be a current at Q. If both $x = 1$ and $y = 1$ there will be a current at Q.
   Note: We can consider this as the 'truth table' for this circuit. Because of the similarity of this truth table to the Boolean operator '+' ('or'), we can symbolise this circuit as $Q = x + y$.

WRITE

x	y	Q
0	0	0
0	1	1
1	0	1
1	1	1

In theory, a computer could be constructed from nothing more than thousands (millions, billions …) of switches. However, the design of a logic circuit would be a long, time-consuming process. Furthermore, it is not clear 'who' turns the switches on or off. Hence, more complex logic gates were constructed as 'black box' components that could be combined quickly to perform relatively complex operations.

A logic gate consists of one or two inputs and one output. The inputs are 'wires' that are either off (0) or on (1). Similarly, the output is either 0 or 1. Inputs require a continuous source of electric current in order to remain at either 0 or 1.

The following table shows the gates we will use. Note that inputs are always on the left; output is always on the right.

<table>
<tr><th>Name</th><th>Symbol</th><th>Truth table</th><th>Comments</th></tr>
<tr><td>NOT</td><td></td><td><table><tr><th>Input</th><th>Output</th></tr><tr><td>0</td><td>1</td></tr><tr><td>1</td><td>0</td></tr></table></td><td>Equivalent to Boolean 'not'</td></tr>
<tr><td>OR</td><td>A
B</td><td><table><tr><th>Input A</th><th>Input B</th><th>Output</th></tr><tr><td>0</td><td>0</td><td>0</td></tr><tr><td>0</td><td>1</td><td>1</td></tr><tr><td>1</td><td>0</td><td>1</td></tr><tr><td>1</td><td>1</td><td>1</td></tr></table></td><td>Equivalent to Boolean 'or'</td></tr>
</table>

<table>
<tr><td>NOR</td><td></td><td><table><tr><th>Input A</th><th>Input B</th><th>Output</th></tr><tr><td>0</td><td>0</td><td>1</td></tr><tr><td>0</td><td>1</td><td>0</td></tr><tr><td>1</td><td>0</td><td>0</td></tr><tr><td>1</td><td>1</td><td>0</td></tr></table></td><td>Equivalent to Boolean ‘or’ followed by ‘not’
Note that the shape is the same as the OR gate, with the open pink circle representing NOT in this symbol.</td></tr>
<tr><td>AND</td><td></td><td><table><tr><th>Input A</th><th>Input B</th><th>Output</th></tr><tr><td>0</td><td>0</td><td>0</td></tr><tr><td>0</td><td>1</td><td>0</td></tr><tr><td>1</td><td>0</td><td>0</td></tr><tr><td>1</td><td>1</td><td>1</td></tr></table></td><td>Equivalent to Boolean ‘and’</td></tr>
<tr><td>NAND</td><td></td><td><table><tr><th>Input A</th><th>Input B</th><th>Output</th></tr><tr><td>0</td><td>0</td><td>1</td></tr><tr><td>0</td><td>1</td><td>1</td></tr><tr><td>1</td><td>0</td><td>1</td></tr><tr><td>1</td><td>1</td><td>0</td></tr></table></td><td>Equivalent to Boolean ‘and’ followed by ‘not’
Note that the shape is the same as the AND gate, with the open pink circle representing NOT in this symbol.</td></tr>
</table>

NAND and NOR gates, although they lack equivalent Boolean expressions, are convenient ways of combining AND or OR with NOT. For example, a NAND gate is equivalent to the combination shown.

Very sophisticated circuits can be constructed from combinations of these five gates, and the truth table of the output for all possible inputs can be determined.

WORKED EXAMPLE 15 Truth tables for gates

Determine the truth table for the output Q in terms of the inputs a, b and c.

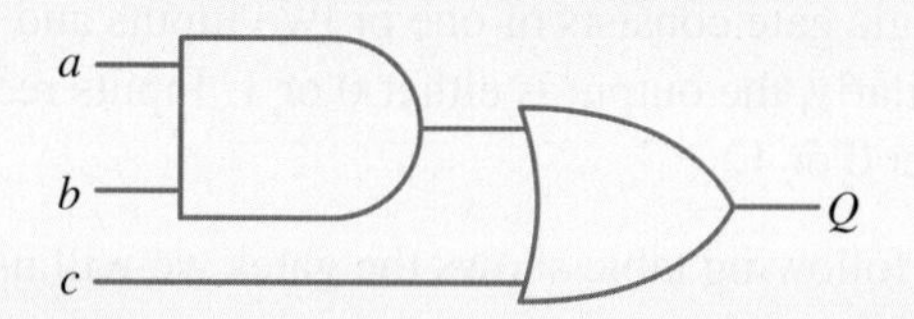

THINK

1. Working from left to right, determine the truth table for the output d in terms of inputs a and b.

WRITE

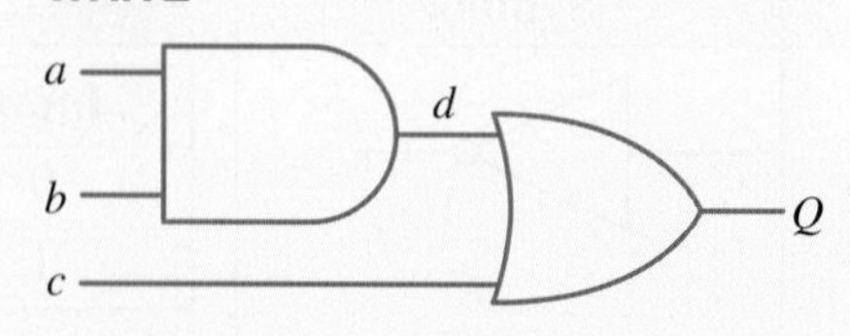

2. Use the truth table for an AND gate.

a	b	d
0	0	0
0	1	0
1	0	0
1	1	1

3. Now consider the output d to be the input to the OR gate, combined with c to determine the truth table at Q.
Note that the first four rows correspond to step 2 for the case of $c = 0$, and the second four rows correspond to step 2 for the case of $c = 1$.

a	*b*	*d*	*c*	*Q*
0	0	0	0	0
0	1	0	0	0
1	0	0	0	0
1	1	1	0	1
0	0	0	1	1
0	1	0	1	1
1	0	0	1	1
1	1	1	1	1

An alternative approach is to start with all inputs (a, b and c) and lay out a 'blank' truth table for these three inputs. Add columns for each gate as required.

A 'blank' truth table for three inputs.

a	*b*	*c*
0	0	0
0	0	1
0	1	0
0	1	1
1	0	0
1	0	1
1	1	0
1	1	1

The completed truth table.

a	*b*	*c*	*d*	*Q*
0	0	0	0	0
0	0	1	0	1
0	1	0	0	0
0	1	1	0	1
1	0	0	0	0
1	0	1	0	1
1	1	0	1	1
1	1	1	1	1

It should be clear that this truth table is equivalent to the one in step 3 of Worked example 15, with the rows in different order. Furthermore, this circuit of an AND and an OR gate is logically equivalent to the statement $(a \wedge b) \vee c$, or in Boolean algebra terms $(a \cdot b) + c$.

3.4.4 Simplifying logic circuits

In some cases an apparently complex circuit can be reduced to a simpler one.

WORKED EXAMPLE 16 Simplifying logic circuits

Determine a circuit equivalent to the one shown.

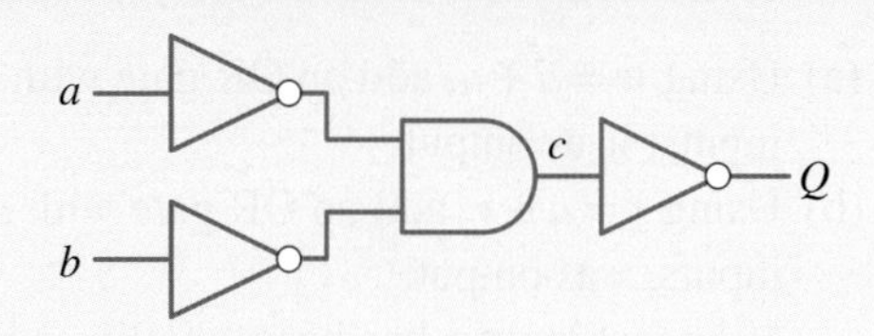

THINK

1. Determine the truth table of the circuit. Start by determining the output at c. Note that the inputs to the AND gate are 'inverted' by the two NOT gates.

WRITE

a	*b*	*c*
0	0	1
0	1	0
1	0	0
1	1	0

Boolean expression = $(a' \cdot b')$

2. Complete the truth table by determining the output at Q. This is just the negation of c.

a	b	c	Q
0	0	1	0
0	1	0	1
1	0	0	1
1	1	0	1

3. Write out the Boolean expression for Q by working backwards from Q.

$Q = c'$ (but $c = a' \cdot b'$)
$= (a' \cdot b')'$

4. Simplify, using the rules for Boolean algebra.

$Q = a'' + b''$ — 2nd De Morgan's law
$= a + b$ — 'double' negative

5. Create the equivalent circuit. In this case it is a single OR gate.

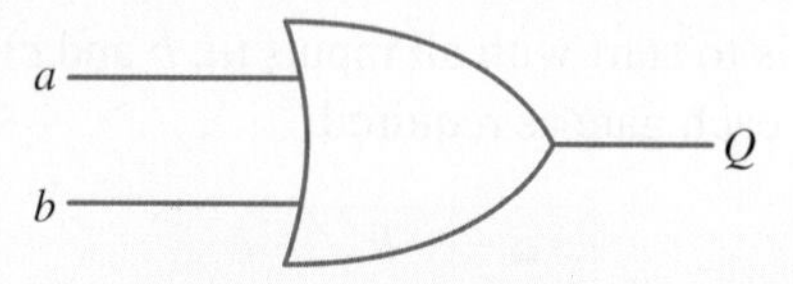

The original, more complicated circuit might have been used because of availability or cost of components. Otherwise, it would be advantageous to use the circuit in step 5.

Often, one has to design or draw a logic circuit given a Boolean expression.

WORKED EXAMPLE 17 Drawing a logic circuit for a Boolean expression

Draw the logic circuit for the Boolean expression $Q = (a + b') \cdot (a + c')$.

THINK

1. Determine the number of 'independent' inputs.
2. Reduce the original Boolean expression to simpler component parts. This last expression is as simple as possible.
3. Begin with the last, 'simplest' expression. This is an AND gate with w and x as inputs, Q as output.
4. (a) Using $w = a + u$, add an OR gate with a and u as inputs, w as output.
   (b) Using $x = a + v$, add an OR gate with a and v as inputs, x as output.
   Note that input a has been 'duplicated' for each OR gate.

WRITE

1. There are three inputs: a, b and c.
2. Let $u = b'$ and $v = c'$
$Q = (a + u) \cdot (a + v)$
Let $w = a + u$
Let $x = a + v$
$Q = w \cdot x$
3. 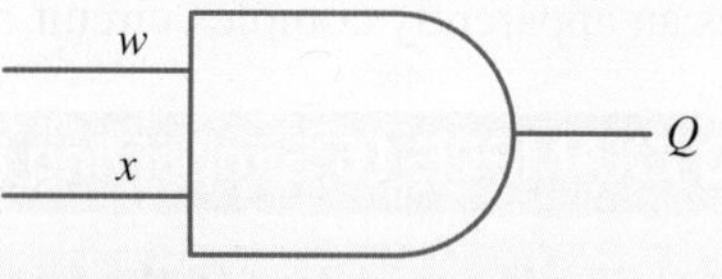

4. 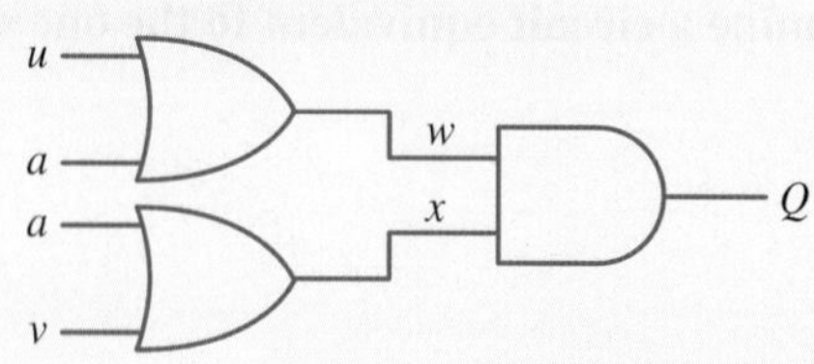

5. (a) Using $u = b'$ and $v = c'$, add two NOT gates to complete the circuit.
(b) The two a inputs must be connected.

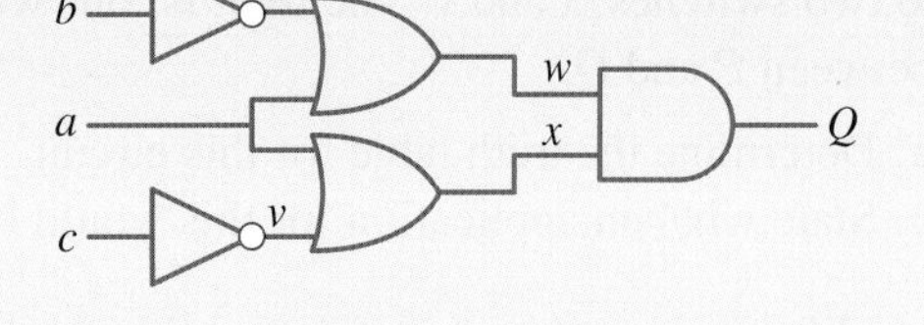

6. Draw the completed circuit with inputs as specified in the Boolean expression.

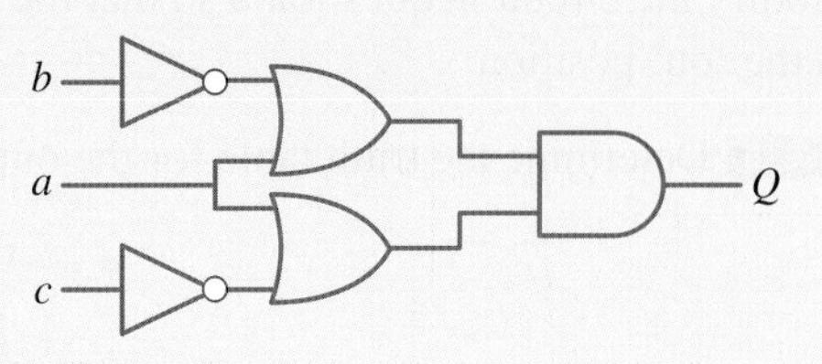

3.4 Exercise

Technology free

1. WE12 Use a truth table to show that $A + A' \cdot B = A + B$.

2. a. Use a truth table to show that $(A \cdot B)' = A' + B'$.
b. Use a truth table to show that $(A + B)' = A' \cdot B'$.

3. WE13 Using the laws of Boolean algebra, simplify each of the following Boolean expressions.
a. $A' \cdot B' + A' \cdot B$
b. $A \cdot B \cdot C' \cdot (A' + B + C)$

4. Using the laws of Boolean algebra, simplify each of the following Boolean expressions.
a. $A \cdot B + A \cdot B' + A' \cdot B$
b. $A' \cdot B \cdot C \cdot (A + B' + C)$

5. Using the laws of Boolean algebra, simplify each of the following Boolean expressions.
a. $(A + B) \cdot (A + B')$
b. $(A' + B') \cdot (A + B')$

6. Using the laws of Boolean algebra, simplify each of the following Boolean expressions.
a. $A \cdot B + A' \cdot B + A \cdot B' + A' \cdot B'$
b. $(A + B) \cdot (A' + B) \cdot (A + B') \cdot (A' + B')$

7. WE14 Consider the pair of switches arranged (in series) as shown. Assuming that there is electricity at P, determine when there is current at Q for various positions of the switches x and y.

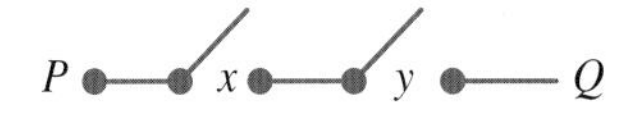

8. Consider the three switches arranged as shown.
a. Assuming that there is electricity at P, determine when there is current at Q, for various positions of the switches x, y and z.
b. Write a Boolean expression equivalent to this circuit.

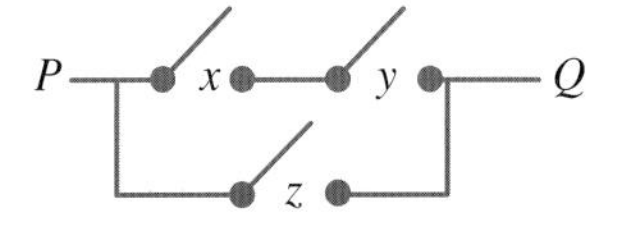

9. Consider the circuit shown which represents a light fixture in a hallway connected to two switches, x and y. The light is 'on' whenever there is a direct connection between P and Q.

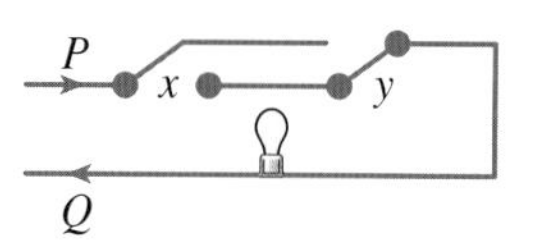

a. Determine the truth table for this circuit.
b. State what an application for this would be.

10. Modify the circuit in question 9 so that the light comes on only when either (or both) of the two switches is in the 'on' position.

11. **WE15** Determine the truth table for the output Q, in terms of the inputs a, b and c for the circuit shown.

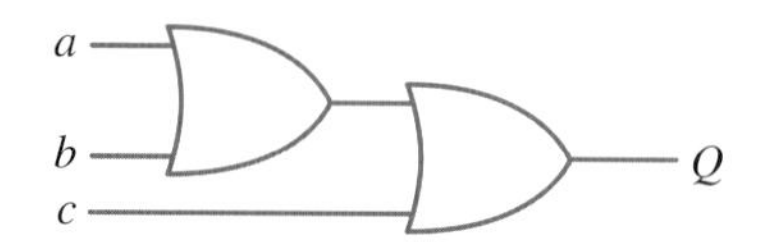

12. Determine the truth table for the output Q, in terms of the inputs a, b and c for the circuit shown.

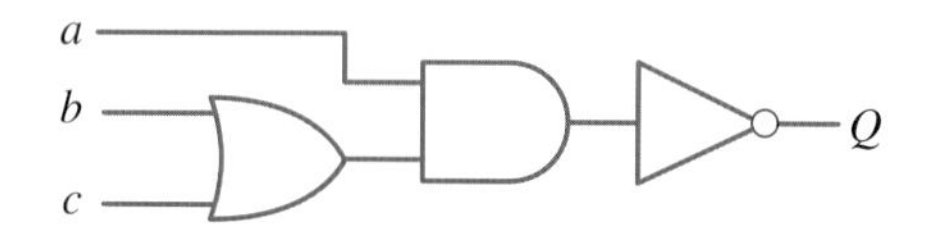

13. **WE16** Consider the circuit shown.

a. Determine the truth table for the output Q, in terms of the inputs a, b and c.
b. Hence, show that this circuit is equivalent to the one in question 12.

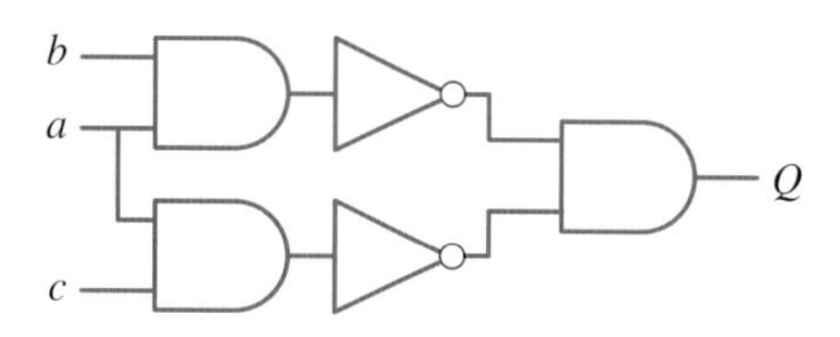

14. For the circuit shown, determine the truth table for the output Q, in terms of the inputs a, b, c and d.

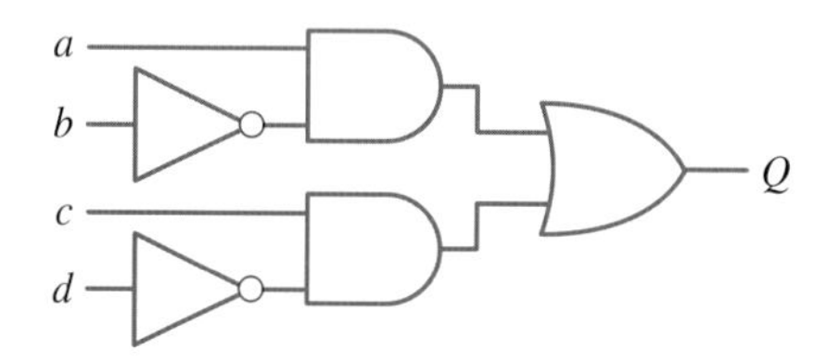

15. Consider the following circuit which shows an alarm system used to protect a safe.
If there is a '1' at R, the alarm rings.
If there is a '1' at Q, the safe can be unlocked.

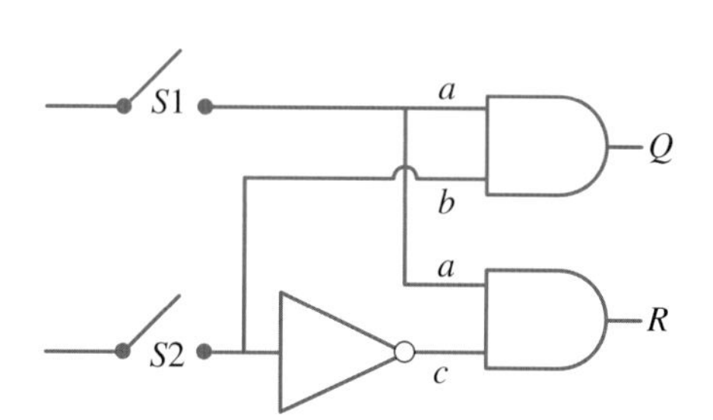

a. Determine the truth table for this circuit.
b. Hence, describe the operation of this circuit.

16. a. Use De Morgan's laws to show that $a \cdot b = (a' + b')'$.
b. Hence, construct a logic circuit equivalent to an AND gate.

17. Show how a single NAND gate can be the equivalent of a NOT gate.

18. **WE17** Determine the logic circuit for the Boolean expression $Q = a \cdot (b + c)'$.

19. The designer of the circuit in question 18 does not have any NOT gates available.
Re-design the circuit using NOR and/or NAND gates to replace any NOT gates.

20. Design a logic circuit for the Boolean expression $Q = a \cdot b + a' \cdot c$, without using any NOT gates.

3.4 Exam questions

Question 1 (2 marks) TECH-FREE

Using the laws of Boolean algebra, show that $A + A \cdot B = A$.

Question 2 (2 marks) TECH-FREE

Determine the Boolean equations for this electrical circuit.

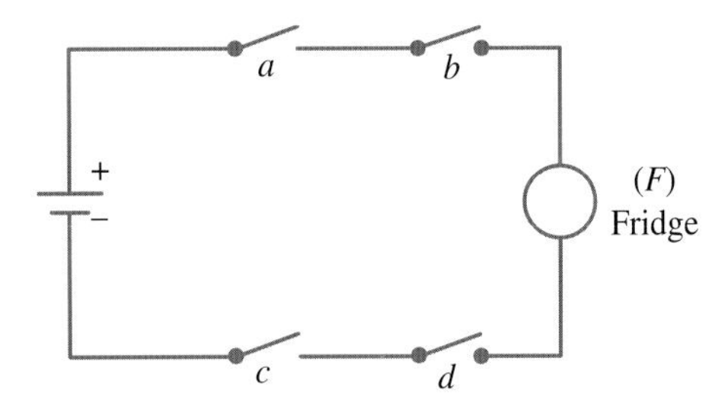

Question 3 (3 marks) TECH-FREE

Use truth tables to show that the second logic circuit is a simplified version of the first logic circuit.

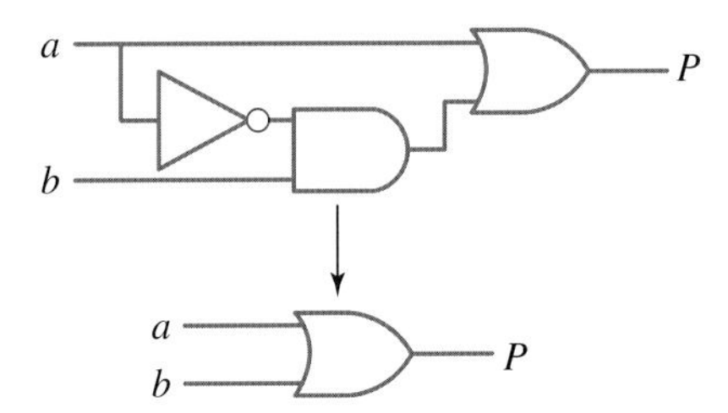

More exam questions are available online.

3.5 Sets and Boolean algebra

LEARNING INTENTION

At the end of this subtopic you should be able to:

- use Boolean symbols, set symbols and logic symbols to prove that logical statements are equivalent.

3.5.1 Sets and their properties

As covered in Topic 1, a set is a collection of objects (or members) that have something in common.

Sets can be **finite**, containing a fixed number of members, such as the set $A = \{1, 2, 3, 4, 5, 6, 7, 8, 9, 10\}$ with 10 members, or **infinite**, such as the set of positive integers, $N = \{1, 2, 3, ...\}$.

Implicit in sets is the concept that there are objects in the set and objects not in the set. If an object x is in set A, we write $x \in A$, and if object y is not in set A, we write $y \notin A$.

Remember that sets can be displayed visually using a Venn diagram, as shown. The area inside the circle represents the set with its members $A = \{2, 4, 6, 8\}$. The white area outside the circle represents all objects *not* in the set. In future we will not generally show the members in the set, but state its 'rule'. What could be the rule for the set in this figure?

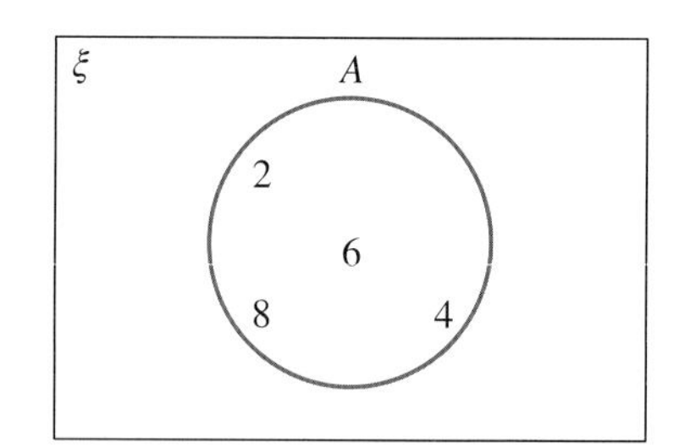

The rectangle itself represents the **universal** set, the set of all possible members (some are in A, some are not), and is denoted by the symbol ξ. In this example the universal set could be *all* the integers.

As in arithmetic, there are a series of operations and properties that enable us to manipulate sets. Consider two sets, A and B, and the possible operations on them.

Intersection: Symbol:	The area in common between two sets is known as the intersection and is shown here in blue. $A \cap B$ 'A intersection B' or 'in both A and B'	
Union: Symbol:	The area in either A or B is the union and is shown here in blue. $A \cup B$ 'A union B' or 'in either A or B or both'	
Negation: Symbol:	The area not in A is the negation or complement and is shown here in blue. A' 'Complement of A' or 'A-prime' or 'not in A'	

Given these operations, we can now look at the rules of sets, comparing them to the rules of arithmetic. For some laws we will need three sets.

Name	Symbolic form	Description	Corresponding arithmetic
Commutative Law	1. $A \cup B = B \cup A$ 2. $A \cap B = B \cap A$	Order of a single operation is not important.	$a + b = b + a$ $a \times b = b \times a$
Identity sets	1. $A \cup \varnothing = A$ 2. $A \cap \xi = A$	The null set ($\varnothing$) has no effect on 'union'; the universal set has no effect on 'intersection'.	$a + 0 = a$ $a \times 1 = a$

Complements	1. $A \cup A' = \xi$ 2. $A \cap A' = \varnothing$	Inverse	$a + (-a) = 0$ $a \times \frac{1}{a} = 1,\ a \neq 0$
Associative Law	1. $A \cup (B \cup C) = (A \cup B) \cup C$ 2. $A \cap (B \cap C) = (A \cap B) \cap C$	The placement of brackets has no effect on the final result when the operations are the same.	$a + (b + c) = (a + b) + c$ $a \times (b \times c) = (a \times b) \times c$
Distributive Law	1. $A \cap (B \cup C) = (A \cap B) \cup (A \cap C)$ 2. $A \cup (B \cap C) = (A \cup B) \cap (A \cup C)$	Bracketed expressions can be expanded when different operations are involved.	$a \times (b + c) = a \times b + a \times c$ (Note that only the first of these laws applies to arithmetic.)
Closure	Consider sets A, B and S. If $A, B \subset S$, then $A \cup B \subset S$ $A \cap B \subset S$	Performing operations on a set will create a result that still belongs to the same class of sets (S).	If a and b are real numbers, then: $a + b$ is a real number $a \times b$ is a real number. *Note:* Closure also applies to addition and multiplication of integers, rational numbers and complex numbers.

It is important to note that union (∪) acts similarly to addition, and intersection (∩) is similar to multiplication, except in the complements, where their roles are reversed.

Although the commutative laws are self-evident, the remaining laws can be demonstrated using Venn diagrams. Closure is a concept that, for now, will have to be taken for granted. For example, closure applies for integers with the operations of addition and multiplication. It does not apply for division, for example $\frac{1}{2}$, as the result (0.5) is not an integer, even though 1 and 2 are.

3.5.2 Boolean algebra

By replacing the set symbols with Boolean ones, we get the laws of Boolean algebra, which are exactly the same as those for sets.

Set name	Set symbol	Boolean name	Boolean symbol
Intersection	$\cap$	and	$\cdot$
Union	$\cup$	or	$+$
Complement	$'$	not	$'$
Universal set	ξ	'everything'	I
Null set	$\varnothing$	'nothing'	O

Thus the set laws can be restated as Boolean laws:

Name	Set law	Boolean law
Commutative Law	1. $A \cup B = B \cup A$ 2. $A \cap B = B \cap A$	$A + B = B + A$ $A \cdot B = B \cdot A$
Identity	1. $A \cup \varnothing = A$ 2. $A \cap \xi = A$	$A + O = A$ $A \cdot I = A$

Complements	1. $A \cup A' = \xi$ 2. $A \cap A' = \varnothing$	$A + A' = I$ $A \cdot A' = O$
Associative Law	1. $A \cup (B \cup C) = (A \cup B) \cup C$ 2. $A \cap (B \cap C) = (A \cap B) \cap C$	$A + (B + C) = (A + B) + C$ $A \cdot (B \cdot C) = (A \cdot B) \cdot C$
Distributive Law	1. $A \cup (B \cap C) = (A \cup B) \cap (A \cup C)$ 2. $A \cap (B \cup C) = (A \cap B) \cup (A \cap C)$	$A + (B \cdot C) = (A + B) \cdot (A + C)$ $A \cdot (B + C) = (A \cdot B) + (A \cdot C)$
Closure	Whatever applies to sets also applies to Boolean algebra.	

Only the first distributive law may require some explanation. Do not confuse the Boolean '+' sign with addition!

WORKED EXAMPLE 18 Using Venn diagrams to establish the distributive law

Establish the distributive law, namely $A \cup (B \cap C) = (A \cup B) \cap (A \cup C)$, using Venn diagrams.

THINK

WRITE

1. Consider the left-hand side term $(B \cap C)$, which is the *intersection* of B and C.

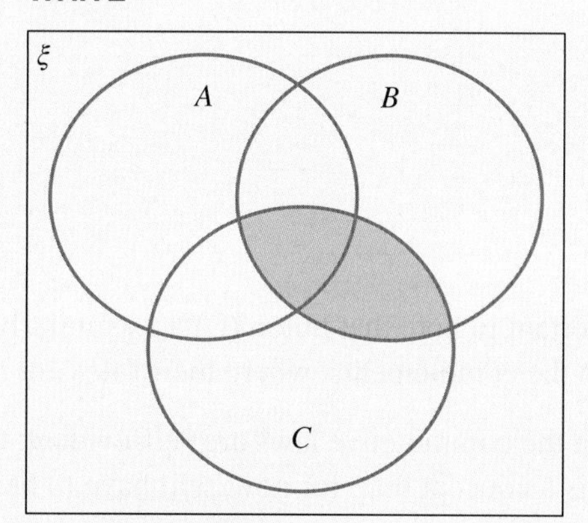

2. Now, create the *union* with A, namely $A \cup (B \cap C)$. In this figure, the red shading shows the 'new' area added. The final result is the region that has either colour.

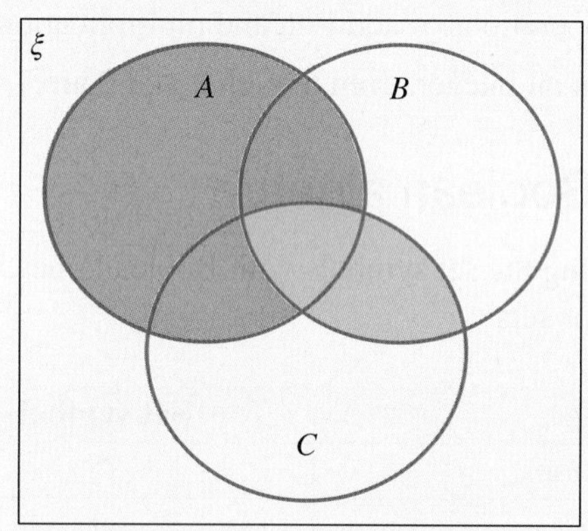

3. Now, consider the first 'term' of the right-hand side, namely $(A \cup B)$.

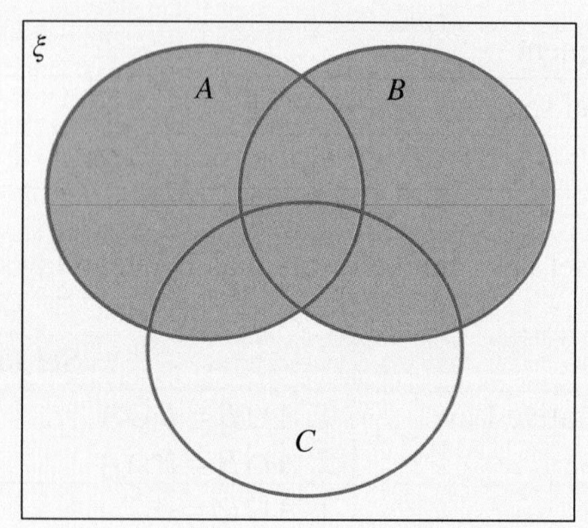

4. Now, consider the second term of the right-hand side, namely $(A \cup C)$.

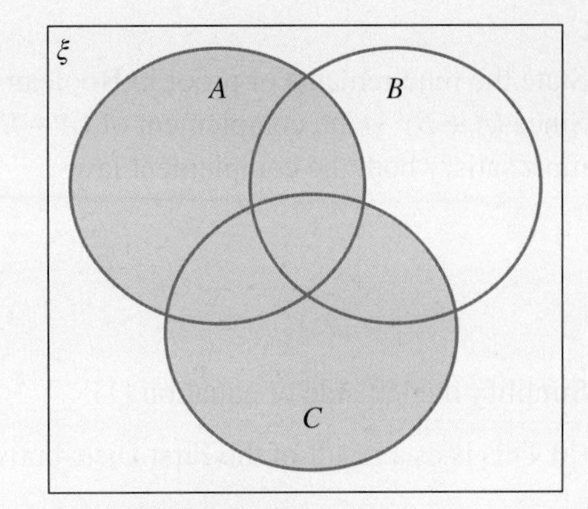

5. Now, consider the intersection of the two regions in steps 3 and 4, which produces the region $(A \cup B) \cap (A \cup C)$. The purple area is the resultant region.

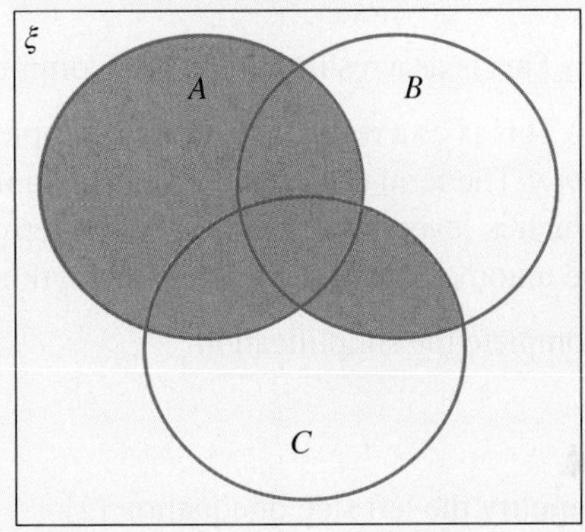

6. Compare the two results.

Clearly the area in step 2 equals the area in step 5; thus, $A \cup (B \cap C) = (A \cup B) \cap (A \cup C)$.

3.5.3 De Morgan's laws and additional results

There are two further important results in Boolean algebra involving the negation of the union and intersection operations. These rules, called De Morgan's laws, can be proved using the results from Boolean algebra, or can be demonstrated using Venn diagrams.

De Morgan's First Law states:

$$(A + B)' = A' \cdot B'$$

De Morgan's Second Law states:

$$(A \cdot B)' = A' + B'$$

These laws can be interpreted as saying that 'the complement of union is intersection' and 'the complement of intersection is union'.

WORKED EXAMPLE 19 Proving the first of De Morgan's laws

Prove the first of De Morgan's laws, namely that the complement of the union of two sets is the intersection of their complements using:

a. **the rules of Boolean algebra**

b. **Venn diagrams.**

THINK | **WRITE**

a. 1. State the requirements of proof in Boolean algebra terms. Since $(A+B)'$ is the complement of $(A+B)$, then $A'\cdot B'$ must satisfy both the complement laws.

a. If $(A+B)' = A'\cdot B'$, then the two complement laws must be satisfied.
Therefore, we must show that:
$(A+B)+(A'\cdot B') = I$ — 1st Complement Law [1]
$(A+B)\cdot(A'\cdot B') = O$ — 2nd Complement Law [2]

2. Simplify the left side of equation [1].

1st Complement Law

(a) This is as a result of the First Distributive Law.

$\text{LHS} = (A+B)+(A'\cdot B')$
$= (A+B+A')\cdot(A+B+B')$

(b) This is as a result of the First Commutative Law.

$= (A+A'+B)\cdot(A+B+B')$

(c) This is as a result of the First Complement Law.
Note: The term $(I+B)$ represents the union of B with I, which is 'everything'. Similarly, the term $(A+I)$ represents the union of A with I, which is 'everything'.

$= (I+B)\cdot(A+I)$

3. Complete the simplification.

$= I\cdot I$
$= I$
$= \text{RHS}$ QED

4. Simplify the left side of equation [2].

2nd Complement Law

(a) This is as a result of the Second Commutative Law.

$\text{LHS} = (A+B)\cdot(A'\cdot B')$
$= (A'\cdot B')\cdot(A+B)$

(b) This is as a result of the Second Distributive Law.

$= A'\cdot B'\cdot A + A'\cdot B'\cdot B$

(c) This is as a result of the Second Commutative Law.
Note: The term $A'\cdot A$ is the intersection of A and its complement, which is 'nothing' or O. Similarly, $B'\cdot B = O$.

$= A'\cdot A\cdot B' + A'\cdot B'\cdot B$

5. Complete the simplification. Note that the intersection and the union of O with any set must be O, since there is nothing in O.

$= O\cdot B' + A'\cdot O$
$= O + O$
$= O$
$= \text{RHS}$ QED

b. 1. Draw a Venn diagram representing the left-hand side of the equation, that is $(A\cup B)'$.
(a) Draw a rectangle with two large, partly intersecting circles. Label one of the circles as A and the other as B.
(b) Identify the portion required.
Note: $A\cup B$ represents the portion inside the two circles. Therefore, its complement $(A\cup B)'$ is represented by the portion outside the two circles.
(c) Shade the required portion.

b. $(A\cup B)'$

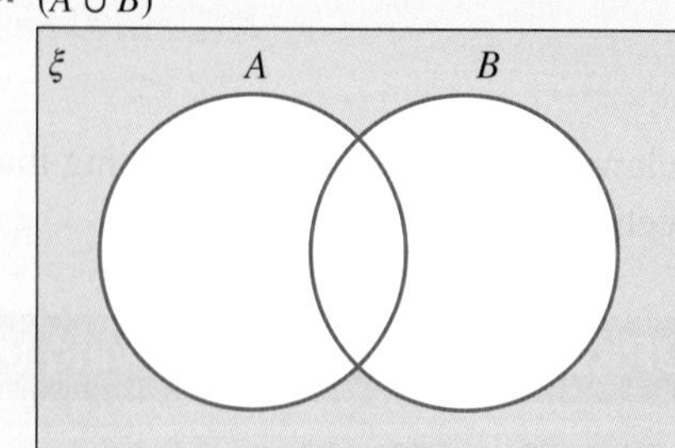

2. Draw a Venn diagram representing the right-hand side of the equation, that is $A' \cap B'$.

(a) Draw a rectangle with two large, partly intersecting circles. Label one of the circles as A and the other as B.

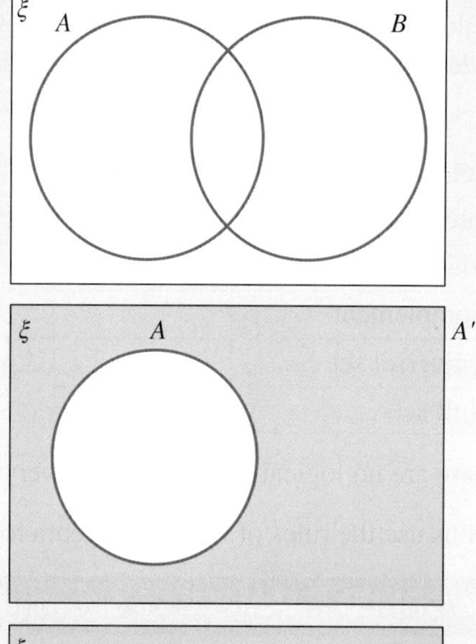

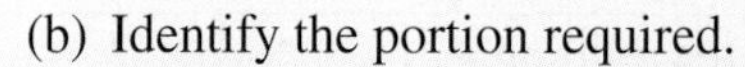

(b) Identify the portion required.
Note: A', the complement of A, represents the portion outside the two circles and the non-intersecting part of circle B, B', the complement of B, represents the portion outside the two circles and the non-intersecting part of circle A. $A' \cap B'$ is represented by the common shaded portion, that is the portion outside the two circles.

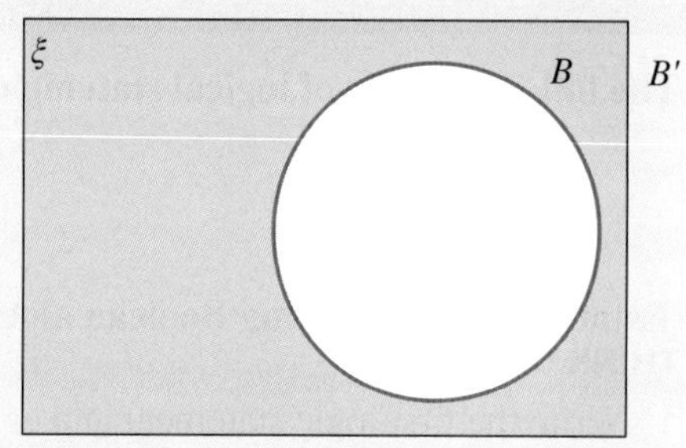

(c) Shade the required portion.

$A' \cap B'$

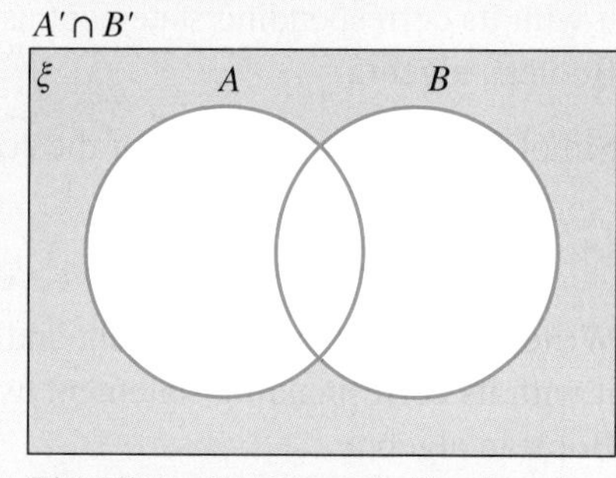

3. Comment on the Venn diagrams obtained.

The Venn diagrams obtained are identical; therefore, De Morgan's First Law, $(A \cup B)' = A' \cap B'$, holds true.

The results in Worked example 19 establish the first of De Morgan's laws. The second law can be proved in a similar fashion.

Based on the rules for Boolean algebra, some important additional results can be tabulated.

Set rule	Boolean rule	Explanation
$A \cup A = A$	$A + A = A$	The union of any set with itself must still be itself.
$A \cap A = A$	$A \cdot A = A$	The intersection of any set with itself must still be itself.
$A \cup \xi = \xi$	$A + I = I$	The union of any set with 'everything' must be 'everything', I.
$A \cap \phi = \phi$	$A \cdot O = O$	The intersection of any set with 'nothing' must be 'nothing', O.
$A \cap (A \cup B) = A$	$A \cdot (A + B) = A$	Consider that the only part of $(A + B)$ that intersects with A must be just A itself.
$A \cup (A \cap B) = A$	$A + (A \cdot B) = A$	Consider the fact that $A \cdot B$ is within A if $B \subset A$, or is A if $A' \subset B$, so that its union with A must be just A itself.

These results are easily established with Venn diagrams and are left as an exercise.

At this point it is worth noting that the key operations of sets and Boolean algebra are intimately related to those of deductive logic. These can be summarised by adding columns to an earlier table.

Set name	Set symbol	Logic name	Logic symbol	Boolean name	Boolean symbol
Intersection	$\cap$	and	$\wedge$	and	$\cdot$
Union	$\cup$	or	$\vee$	or	$+$
Complement	$'$	not	$\neg$	not	$'$
Universal set	ξ			'everything'	I
Null set	$\varnothing$			'nothing'	O

There are no logical equivalents to 'everything' or 'nothing'.

Let us use the rules of Boolean algebra to prove an earlier result.

WORKED EXAMPLE 20 Showing that a pair of logical statements are equivalent

The following pair of logical statements are equivalent:

$$(p \vee q) \vee \neg p$$
$$p \vee \neg p$$

Establish this fact using Boolean algebra.

THINK	WRITE	
1. Write the first logic statement and equate it with its corresponding statement using Boolean algebra.	$(p \vee q) \vee \neg p = (P + Q) + P'$	
2. Simplify the right-hand side of the equation.	$= (P + P') + Q$	1st Commutative Law
	$= I + Q$	Identity Law
	$= I$	
3. Write the second logic statement and equate it with its corresponding statement using Boolean algebra.	$p \vee \neg p = P + P'$	Complements
4. Simplify the right-hand side of the equation.	$= I$	
5. Comment on the results obtained.	The two statements are both equal to I and therefore equivalent to each other. QED	

3.5 Exercise

Students, these questions are even better in jacPLUS

Receive immediate feedback and access sample responses

Access additional questions

Track your results and progress

Find all this and MORE in jacPLUS

Technology free

1. WE17 Demonstrate the 2nd Associative Law, namely:
$A \cap (B \cap C) = (A \cap B) \cap C$ using Venn diagrams.
2. Demonstrate the 2nd Distributive Law, namely:
$A \cap (B \cup C) = (A \cap B) \cup (A \cap C)$ or $A \cdot (B + C) = A \cdot B + A \cdot C$ using Venn diagrams.
3. Write the following sets using the notation $A = \{...\}$.
   a. $A =$ the set of all even positive integers less than 20
   b. $B =$ the set of all positive integers divisible by 4
   c. $C =$ the set of all even prime numbers
   d. $D =$ the set of court cards in a deck of playing cards
   e. $E =$ the set of integers, less than 0, which are square numbers
   f. $F =$ the set of integers less than 10
4. State which of the sets in question 3 are finite.
5. Demonstrate, using a Venn diagram, the intersection of the following two sets:
$A =$ the set of two-digit positive odd numbers
$B =$ the set of two-digit square numbers.
List the members of the intersection on the diagram.
6. Demonstrate, using a Venn diagram, the intersection of the following two sets:
$A =$ the set of two-digit positive even numbers
$B =$ the set of two-digit palindromes (numbers which are the same backwards and forwards).
List the members of the intersection on the diagram.
7. Demonstrate on a Venn diagram the regions defined by:
   a. $A \cap B'$
   b. $A' \cap B'$
   c. $A' \cap (B \cap C)$.
8. The laws of sets can be demonstrated with specific sets.
Let $A = \{1, 2, 3, 4, 5, 6, 7, 8, 9, 10\}$, $B = \{2, 4, 6, 8, 10\}$, $C = \{1, 4, 9\}$.
Consider the 1st Distributive Law: $A \cup (B \cap C) = (A \cup B) \cap (A \cup C)$.
   a. Determine the set represented by the expression $(B \cap C)$.
   b. Determine the set represented by $A \cup (B \cap C)$.
   c. Determine the set represented by $(A \cup B)$.
   d. Determine the set represented by$(A \cup C)$.
   e. Determine the set represented by $(A \cup B) \cap (A \cup C)$ and show that this is the same set as that in the answer to part b.
9. Let $A = \{1, 2, 3, 4, 5, 6, 7, 8, 9, 10\}$, $B = \{2, 4, 6, 8, 10\}$, $C = \{1, 4, 9\}$.
   a. Determine the set represented by the expression $(B \cup C)$.
   b. Determine the set represented by $A \cap (B \cup C)$.
   c. Determine the set represented by $(A \cap C)$.
   d. Determine the set represented by $(A \cap B)$.
   e. Determine the set represented by $(A \cap B) \cup (A \cap C)$ and show that this is the same set as that in the answer to part b.

10. WE19a Using the rules for Boolean algebra, prove the 2nd of De Morgan's Laws: $(A \cdot B)' = A' + B'$

11. Simplify the following logical expressions, using the rules of Boolean algebra.
 a. $A + A' \cdot B + A \cdot B$ b. $(A + B + A') + B'$ c. $A + A' \cdot B$ d. $A \cdot B \cdot (A + C)$

12. WE19b Using Venn diagrams, show that the following statements are true.
 a. $(A \cup B) \cap A = A$ b. $(A \cup B) \cap B' = A$ c. $A \cup B \cap A' = A \cup B$

13. WE20 Determine, using Boolean algebra, if the following two statements are equivalent.

$$(p \wedge q) \wedge \neg p$$
$$p \wedge \neg p$$

14. Prove the following using Boolean algebra.
 a. $A + B + A' + B' = I$ b. $(A + B) \cdot A' \cdot B' = O$
 c. $(A + B) \cdot (A + B') = A$ d. $A \cdot B + C \cdot (A' + B') = A \cdot B + C$

15. A conditional circuit: Up until now, we have not seen a digital equivalent, or even a Boolean equivalent, of the important logical expression $a \rightarrow b$. The truth table for the conditional statement is shown.

a	b	$a \rightarrow b$
0	0	1
0	1	1
1	0	0
1	1	1

 a. From the following list of statements, determine which one has the same truth table as $a \rightarrow b$.
 i. $a \cdot b'$ ii. $a' \cdot b$ iii. $a' + b$ iv. $(a + b') \cdot a \cdot b' \cdot b$
 b. Design a logic circuit equivalent to $a \rightarrow b$.
 c. Design a logic circuit equivalent to $b \rightarrow a$.
 d. Determine a Boolean statement equivalent to $(a \rightarrow b) \cdot a$.
 e. Determine a Boolean statement equivalent to the modus ponens argument, namely:

$$\begin{array}{c} a \rightarrow b \\ \underline{a \quad\;\;} \\ b \end{array}$$

 and simplify, as much as possible, using the result from part d.
 f. Design a circuit equivalent to the Boolean statement from part e, and show that the output is always 1. Thus, you have established the validity of the modus ponens argument.

3.5 Exam questions

Question 1 (3 marks) TECH-FREE

Determine using Boolean algebra if $(p \wedge q) \vee \neg q$ and $p \vee \neg q$ are equivalent.

Question 2 (1 mark) TECH-ACTIVE

MC If $\xi = \{1, 2, 3, 4, 5, 6, 7, 8, 9\}$, $B = \{1, 3, 5, 7, 9\}$, $C = \{3, 5, 7, 8\}$, $(B \cap C) \cup B'$ equals
 A. $\{2, 4, 6, 8\}$ B. $\varnothing$ C. $\{2, 3, 4, 5, 6, 7, 8\}$
 D. $\{6\}$ E. $\{2, 3, 4, 5, 6, 7, 8, 8\}$

Question 3 (4 marks) TECH-FREE

If $A = \{p, q, r\}$, $B = \{p, q, r, s\}$, $I = \{p, q, r, s, t, u, v\}$, verify De Morgan's Law $(A + B)' = A' \cdot B'$.

More exam questions are available online.

3.6 Algorithms and pseudocode

LEARNING INTENTION

At the end of this subtopic you should be able to:
- understand how algorithms can be used to solve problems
- write pseudocode to perform simple calculations and tasks.

3.6.1 What is an algorithm?

An algorithm is a set of rules or instructions used to solve a particular problem. In mathematics this may be a set of processes or calculations to solve a particular problem such as the method we use to solve a long division problem, but algorithms occur everywhere in day to day life. For example, the recipe for baking a cake can be thought of as an algorithm.

Complex problems are better solved using computers, but to we need to instruct the computer on what to do by writing computer programs. To make sure the computer can understand the program it must be written in a specific computer language. There are many computer languages around, with names such as Python, Java, C++, Lua and some older languages such as Pascal, Fortran, Cobol and Basic. In this topic we will not delve into any particular language. Instead, we will use pseudocode which avoids the specific rules and notations used in each of these computer languages.

3.6.2 What is a pseudocode?

Pseudocode is a plain English language description of the steps in solving an algorithm. Pseudocode often uses some of the conventions of programming language, but is intended for human reading rather than machine reading, and is independent of the particular syntax, that is the rules and requirements of a specific programming language.

3.6.3 Writing pseudocode

While there is no agreed conventions or standards for writing pseudocode, throughout this topic we will follow the following conventions:
- Pseudocodes will start with the word 'begin' and finish with the word 'end'.
- All statements will be written on a single line and will be carried out in sequential order.
- We will not be concerned with punctuation or whether the words are upper or lowercase. Punctuation will be used solely for the purpose of making the pseudocode readable.
- Variable names cannot contain spaces, or any of the key words used in the pseudocode language (such as *numeric* or *display*).
- We will choose names that are appropriate (for example, we can use the variable 'length' to represent the length).
- We will need to define our variables used in the code as follows: the word 'numeric' indicates that a variable will be used as a real number, the word 'char' indicates that a variable will be used to represent characters such as letters of the alphabet or names (in programming languages these are called strings). For example, the following line shows how to define three numeric variables in a pseudocode:
  numeric length, width, height.

- Once a variable is defined we can assign a value to it. To prompt the user to assign a value to a variable we will use the word 'input' followed by the variable name.
- When we store a value to a variable, we will use the notation $x = 3$. What we mean here is that the variable named x now has the value of 3. Note that since there is no set language for pseudocode, there are alternative notations for variable storage. Examples include, $x \leftarrow 3$ and $x := 3$.
- The left-hand side is the variable name and the right-hand side is the new value of the variable. The right-hand side can be an expression involving variables that have had values assigned to them, for example via input statements or previous assignments.
- The word 'display' will be used to display an output to the user. Outputs will usually contain some text (written in quotation marks in the pseudocode) and some output variables (usually these are defined in terms of variables that have been assigned values via the 'input').

Structure of pseudocode

begin

statement 1

statement 2

.....

end

WORKED EXAMPLE 21 Calculating the perimeter and area of a rectangle

Write a pseudocode to calculate the perimeter and area of a rectangle.

THINK	WRITE
1. Start the code with begin. Determine which input variables and which output variables are needed. In this situation we need input variables *length* and *width* and output variables *perimeter* and *area*. These are all numeric variables, so write them out on the second line, separated by commas as shown.	begin numeric length, width, perimeter, area
2. We need the user to input the values of the input variables. Use the 'display' to give some context to the user about which variable they are inputting.	display 'Enter the length' input length display 'Enter the width' input width
3. Perform the required calculations. Calculate the perimeter and area of the rectangle from the inputs. Note we use * to indicate multiplication.	perimeter = 2 * length + 2 * width area = length * width
4. We need to output the results of our calculated output values.	display 'The perimeter of the rectangle is', perimeter display 'The area of the rectangle is', area
5. End the code with end.	end

The complete pseudocode is below, note the indentation to make the blocks of code easier to read.

```
begin
    numeric length, width, perimeter, area
    display 'Enter the length'
    input length
    display 'Enter the width'
    input width
    perimeter = 2 * length + 2 * width
    area = length * width
    display 'The perimeter of the rectangle is', perimeter
    display 'The area of the rectangle is', area
end
```

3.6.4 Selection or conditional constructs

For Worked example 21, the statements are executed line by line in order. Often in programming or pseudocode we need to decide which set of statements or blocks of statements are selected or repeated. One way of achieving this will depend on whether a statement which is a proposition which evaluates to a Boolean expression of 'true' or 'false'. This is done using the if.. (Boolean expression).. then.. endif construct

Selection constructs

if (Boolean expression) then

{some lines of code which will be executed if the Boolean expression in the above line is true}

endif

Note: If the Boolean expression is false, the lines of code are not executed and the pseudocode moves onto the following lines.

WORKED EXAMPLE 22 Determining the greatest of two numbers

Write a pseudocode to determine which of two numbers is greater.

THINK	WRITE
1. Start the code with begin. Only two input variables are need in this problem, no other output variables. We will call them num1 and num2.	begin numeric num1, num2
2. Display and prompt for the two input variables.	display 'Enter the first number' input num1 display 'Enter the second number' input num2

3. Use the if statement to decide if the first number is the largest of the two numbers.	if (num1 > num2) then
	display 'The first number', num1, 'is the largest'
	endif

4. Use the if statement to decide if the second number is the largest of the two numbers.	if (num2 > num1) then
	    display 'The second number', num2, 'is the largest'
	endif

5. If neither of the above two conditions are met, test to check if the two numbers are equal. Note the two equal signs which are used to test if the values are equal, not to assign the value of num1 to the value of num2.	if (num1 == num2) then
	    display 'The two numbers are equal'
	endif

6. End the code with end.	end

The complete pseudocode is below.

```
begin
    numeric num1, num2
    display 'Enter the first number'
    input num1
    display 'Enter the second number'
    input num2
    if (num1 > num2) then
        display 'The first number', num1, 'is the largest'
    endif
    if (num2 > num2) then
        display 'The second number', num2, 'is the largest'
    endif
    if (num1 == num2) then
        display 'The two numbers are equal'
    endif
end
```

3.6.5 Extended if then elseif

Sometimes there will be more than one case to consider, or the cases are mutually exclusive, that is if one occurs then the other can't, in these situations we use the if.. then.. elseif.. else.. endif constructs. Also, we can combine Boolean expressions using 'and', 'or' to get true or false values.

if then elseif

if (Boolean expression) then

    {execute these lines of pseudocode if the above Boolean expression is true}

elseif (Boolean expression)

    {execute these lines of pseudocode if the above Boolean expression is true}

elseif (Boolean expression)

{execute these lines of pseudocode if the above Boolean expression is true}

else

{execute these lines of pseudocode if none of the above Boolean expressions were true}

endif

Note that pseudocode is not necessarily unique; there are many ways to write code and to achieve the same outputs. For example, the pseudocode for Worked example 22 could also be have written as:

```
begin
    numeric num1, num2
    display 'Enter the first number'
    input num1
    display 'Enter the second number'
    input num2
    if (num1 == num2) then
        display 'The numbers are equal'
    elseif (num1 > num2) then
        display 'The first number', num1, 'is the largest'
    else
        display 'The second number', num2, 'is the largest'
    endif
end
```

WORKED EXAMPLE 23 Determining the maximum of three numbers

Write a pseudocode to determine the maximum of three numbers.

THINK	WRITE
1. Start the code with begin. Three input variables are needed in this problem, and one output variable to store the information about which variable is the maximum. We will call the input variables num1, num2 and num3 and the output variable maxx. (Sometimes in programming languages the word max is a reserved word and cannot be used as a variable name.)	begin numeric num1, num2, num3, maxx
2. Display and prompt for the three input variables.	display 'Enter the first number' input num1 display 'Enter the second number' input num2 display 'Enter the third number' input num3

3. Use the if statement to decide the maximum of the first two numbers and assign the variable maxx to one of these.

```
if (num1 < num2) then
        maxx = num2
else
        maxx = num1
endif
```

4. Use the if statement to decide if the third number is greater than the maximum of the other two numbers.

```
if (num3 > maxx) then
        maxx = num3
endif
```

5. Output the largest of the three numbers.

```
display 'The maximum of the three numbers is', maxx
```

6. End the code with end.

```
end
```

The complete pseudocode is below.

```
begin
        numeric num1, num2, num3, maxx
        display 'Enter the first number'
        input num1
        display 'Enter the second number'
        input num2
        display 'Enter the third number'
        input num3
        if (num1 < num2) then
                maxx = num2
        else
                maxx = num1
        endif
        if (num3 > maxx) then
                maxx = num3
        endif
        display 'The maximum of the three numbers is', maxx
end
```

WORKED EXAMPLE 24 Displaying different messages based on the time

Write a pseudocode to take the time from a 24 clock and output 'Good morning' between 12 am and 12 pm, 'Good afternoon' between 12 pm and 6 pm, 'Have a good night' between 6 pm and 10 pm, or 'Time to go to bed' between 10 pm and 12 am.

THINK	WRITE
1. Start the code with begin. Only one input is needed in this problem, the current time.	begin numeric time
2. Display and prompt for the time.	display 'Using a 24-hour clock, enter the time as a decimal. For example, 13:27 = 13.27' input time
3. Use the if statement to decide whether the time is between 12 am and 12 pm.	if ((time ≥ 0) and (time < 12)) then display 'Good morning'
4. Use the elseif statement to decide whether the time is between 12 pm and 6 pm.	elseif ((time ≥ 12) and (time < 18)) then display 'Good afternoon'
5. Use the elseif statement to decide whether the time is between 6 pm and 10 pm.	elseif ((time ≥ 18) and (time < 22)) then display 'Have a good night'
6. Use the elseif statement to decide whether the time is between 10 pm and 12 am.	elseif ((time ≥ 22) and (time < 24)) then display 'Time to go to bed'

7. If none of the above conditions are met, the time must be invalid. End the if..then..elseif block with endif.	else
	display 'Invalid time'
	endif

8. End the code with end.	end

The complete pseudocode is below.

```
begin
    numeric time
    display 'Using a 24-hour clock, enter the time as a decimal. For example, 13:27 = 13.27'
    input time
if ((time ≥ 0) and (time < 12)) then
    display 'Good morning'
elseif ((time ≥ 12) and (time < 18)) then
    display 'Good afternoon'
elseif ((time ≥ 18) and (time < 22)) then
elseif ((time ≥ 22) and (time < 24)) then
    display 'Have a good night'
    display 'Time to go to bed'
else
    display 'Invalid time'
endif
end
```

3.6.6 Programming using CAS calculators

CAS calculators are capable of creating and running programs such as the ones we have been modelling throughout this topic. The language and methodology differs depending on the CAS calculator used; however, they all use similar steps to these we have been using in the pseudocodes throughout this topic. The following Worked example is a repeat of the previous one, using the TI-Nspire and CASIO ClassPad CAS calculators.

WORKED EXAMPLE 25 Determine the maximum of three numbers using CAS

Write a program on your CAS calculator to determine the maximum of three numbers.

TI | THINK — **DISPLAY/WRITE**

1. Open a new TI-Nspire document and choose 9: Add Program Editor and 1: New... to open a new program.

2. Type in a name for the program, call it 'maxof3' and tab and press OK.

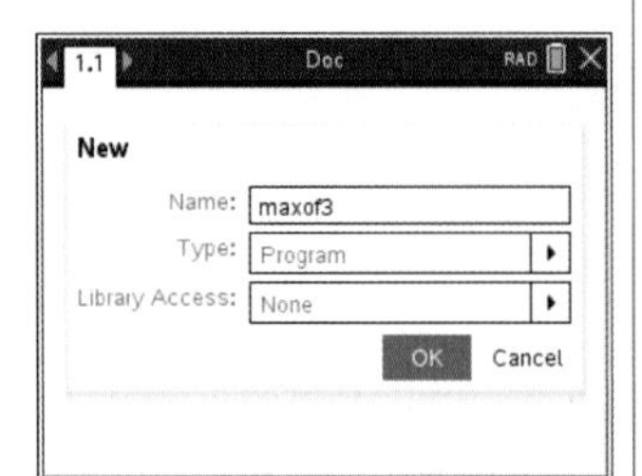

3. Some of the code is already present. Press MENU option 3: Define Variables and 1: Local.

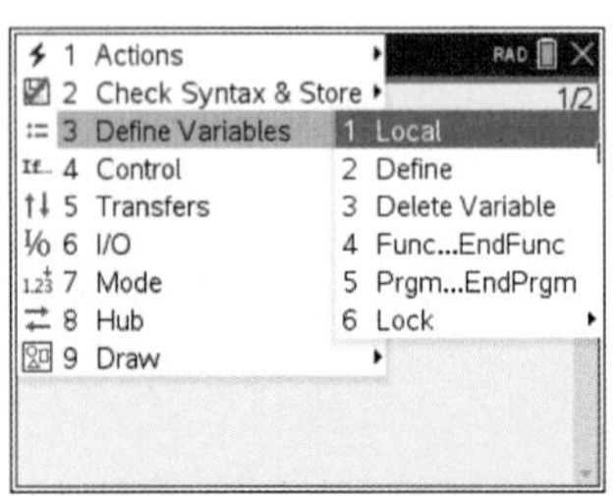

4. Type in the variable names we will use. Notice the keywords are coloured.

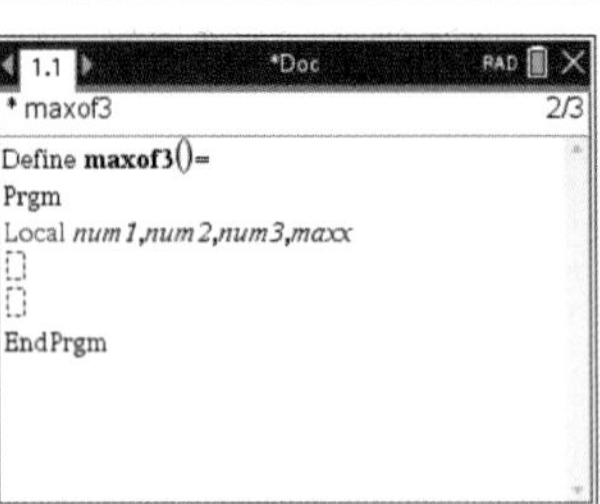

CASIO | THINK — **DISPLAY/WRITE**

1. Open the Program application found on the second page of the Menu.

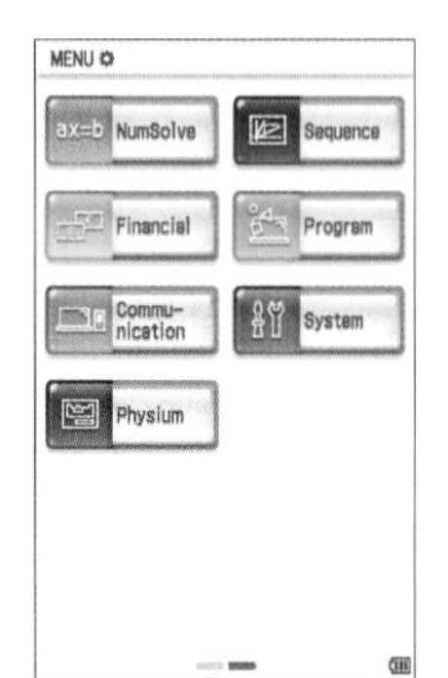

2. Select Edit in the toolbar and then choose New File.

3. Type in a name for the program. This example will be called 'maxof3'. Tap OK.

4. Define the variables to use by first selecting Misc > Variable > Local and then typing in the variable names.

5. To obtain the input press MENU option 6: I/O and 3: Request.

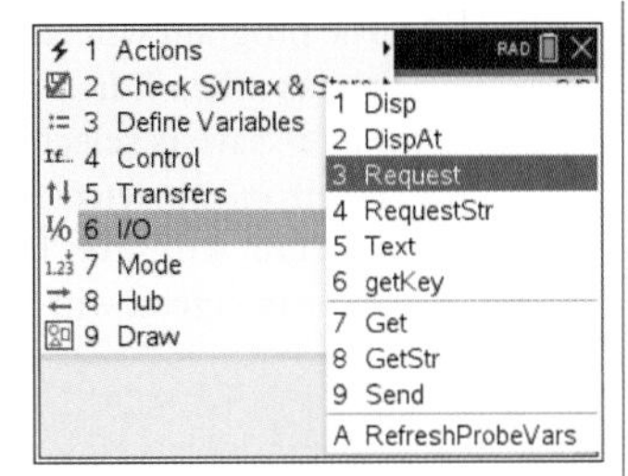

6. Complete and repeat for the I/O (input output) for the other variables as shown.

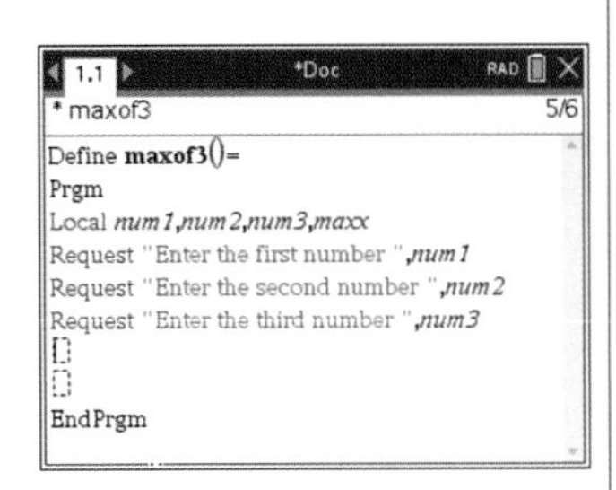

7. To obtain other constructs to change the flow of the program, press MENU option 4: Control and 3: If…Then…Else…Endif.

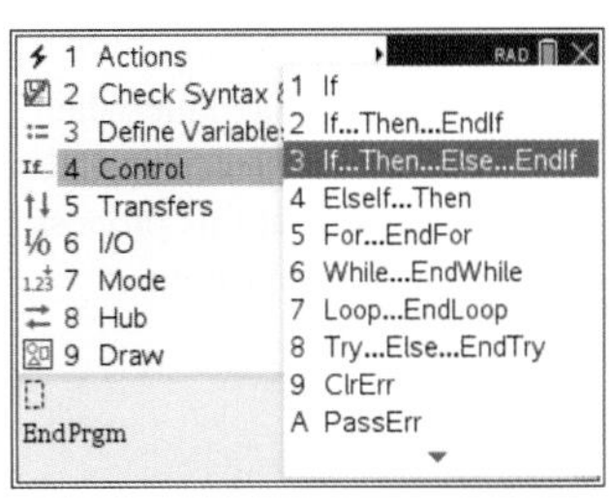

8. Repeat for If…Then…EndIf and type in the rest of the code as shown. Notice again the colour coding and the automatic indentation.

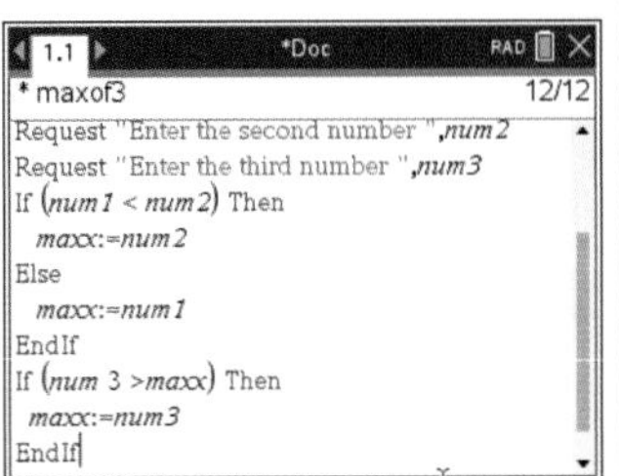

9. For the output press MENU option 6: I/O and 1: Disp.

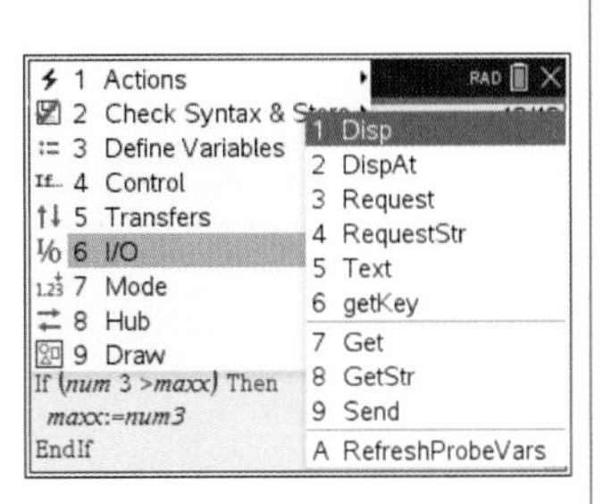

5. To enable the numbers to be entered by the user, select the Input command in the I/O sub-menu and complete the coding as shown.

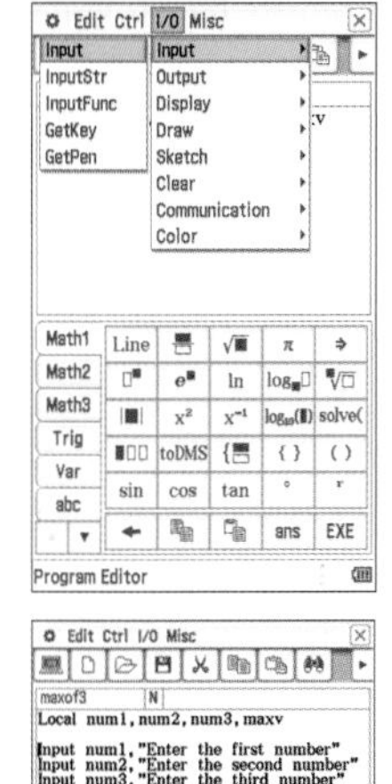

6. To obtain other command constructs to change the flow of the program, use the options available in the Ctrl menu. In this example the items of If…Then…IfEnd will be used.

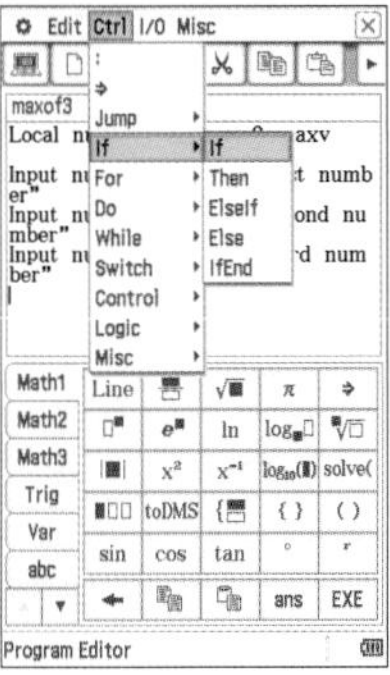

7. Type in the code as shown. Note the use of the colon to enter multiple commands on the same line.

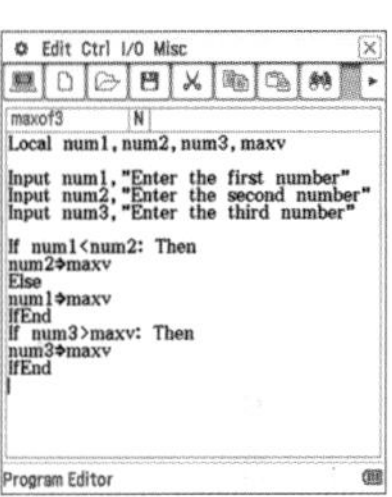

8. For the output display select the menu I/O. Commonly used options include ClrText (clears any previous output from the screen) and Print (prints text and results of the program).

9. Complete the program as shown. This example will print a statement, the three numbers that were used and then the maximum of these numbers.

10. Complete the program as shown.

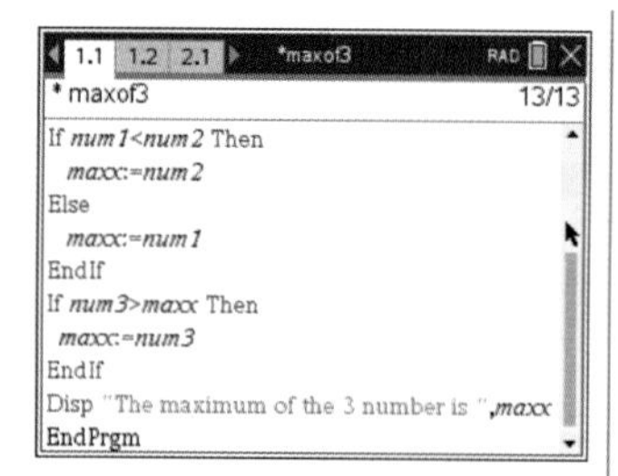

11. To check the syntax press MENU option 2: Check Syntax & Store and 1: Check Syntax & Store (or the shortcut key Ctrl + B).
Provided there are no errors, that is, the code is syntax free, you will get a message stating 'maxof3' stored successfully.

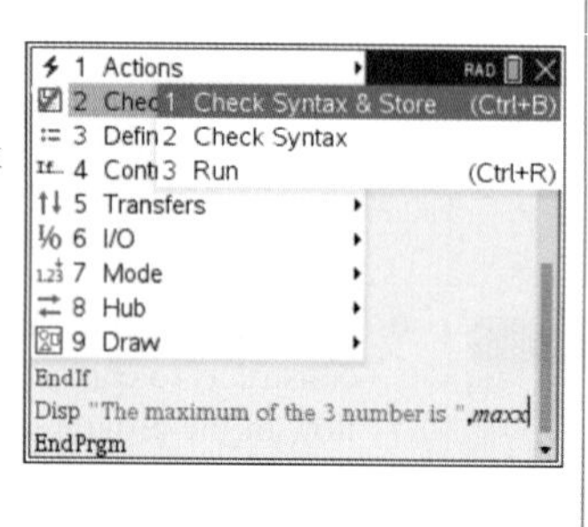

12. To run the program press MENU option 2: Check Check Syntax & Store and 3: Run (or the shortcut key Ctrl + R).
Press enter, then enter in some values, tab to OK, press enter and repeat for the other variables.

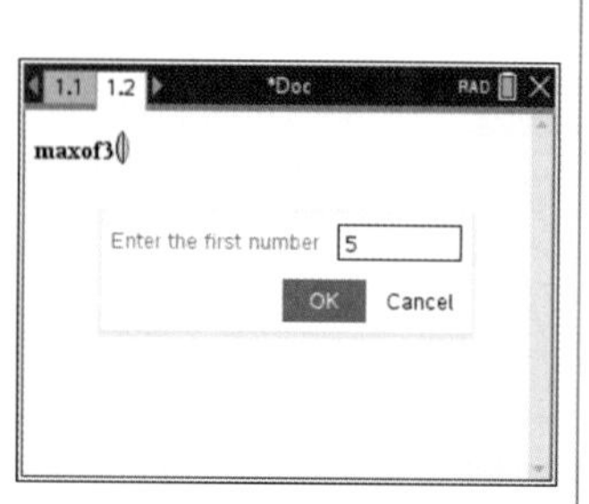

13. The outputs are shown on a Calculator page.
You can also save the document and use the TI-Nspire calculator to check your code for many of the other worked examples and pseudocode that you can write, remembering there will be subtle changes.

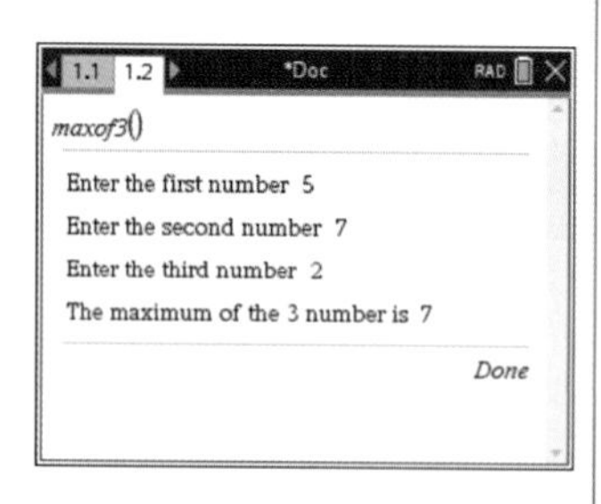

10. The program will test the syntax when Edit > Save File > Save is selected. If no errors are detected, the program will save. If not, an error screen will appear.

11. To run the program, select the icon to open the Program Loader screen.
Select the program to load by tapping the Name: down arrow button and selecting 'maxof3'.
Tap or Run > Run Program.

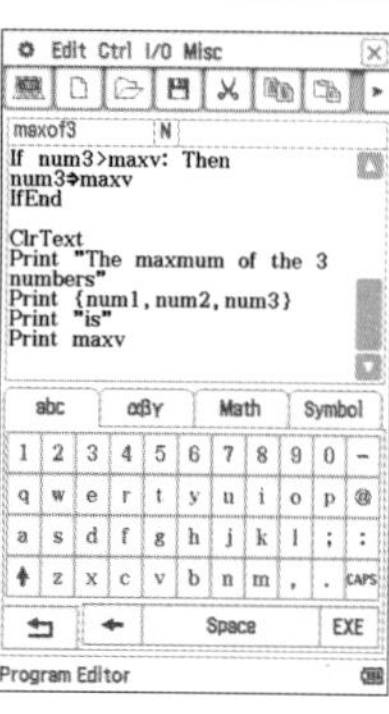

12. Enter a value when prompted. Tap OK.
Repeat this process for the other prompts.

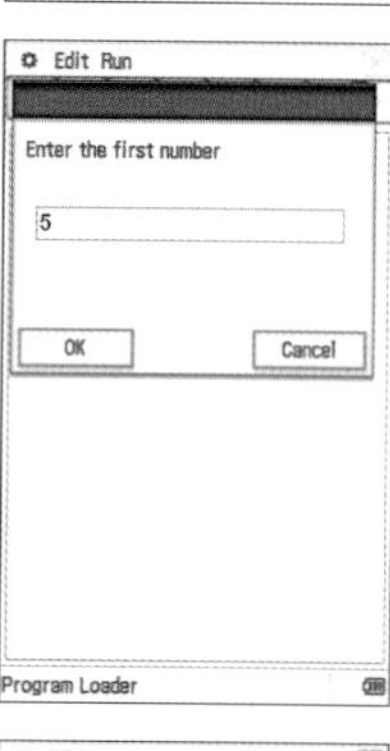

13. The outputs are shown on the screen.
You can use the ClassPad calculator to check your code for many of the other worked examples and pseudocode that you can write, remembering there will be subtle changes.

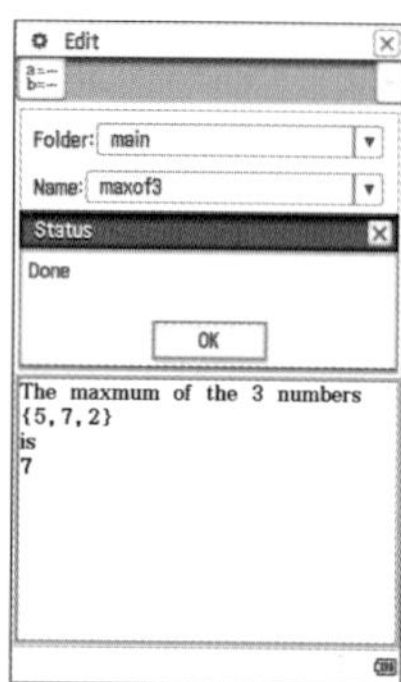

3.6.7 Case constructs

Often the 'if then else' constructs can become quite complicated. An alternative approach is case constructs. A case construct tests specific cases for a variable and once it comes across one case that is true it executes the following lines of code and then breaks out of the case construct without testing the remaining cases.

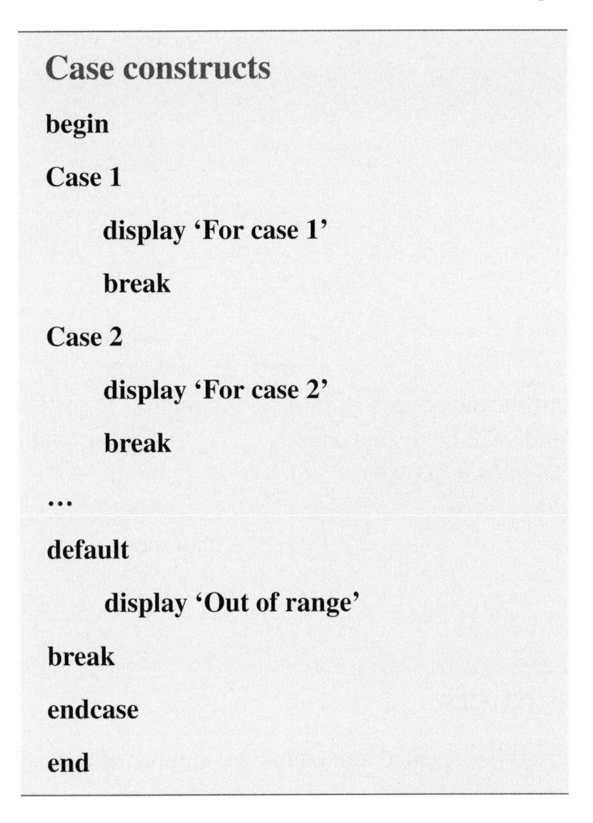

Case constructs

```
begin
Case 1
      display 'For case 1'
      break
Case 2
      display 'For case 2'
      break
...
default
      display 'Out of range'
break
endcase
end
```

The case command is not available on the TI-Nspire CAS programming.

WORKED EXAMPLE 26 Writing a pseudocode using a case command

Write pseudocode to list the day of the week, with Monday being 1 and Sunday being 7, and the day number entered as input.

THINK	WRITE
1. Start the code with begin. Only one input variable is needed in the problem, no output variables. We will call it daycode.	begin numeric daycode
2. Display and prompt for the daycode.	display 'Enter the day code' input daycode

3. As 1 represents Monday, use the case statement to display the corresponding week day name and break out of the case.	begin
	case 1
	    display 'Monday'
	    break

4. As 2 represents Tuesday, use the case statement to display the corresponding week day name.	case 2
	    display 'Tuesday'
	    break

5. Repeat for the other days of the week.

```
case 3
    display 'Wednesday'
    break
case 4
    display 'Thursday'
    break
case 5
    display 'Friday'
    break
case 6
    display 'Saturday'
    break
case 7
    display 'Sunday'
    break
```

6. Generate an error message if the daycode is invalid. The message 'Value not valid' will be displayed if the number entered is not an integer between 1 and 7 inclusive.

```
default
    display 'Value not valid'
    break
```

7. End the cases.

```
endcase
```

8. End the code with end.

```
end
```

3.6.8 Repetition constructs

When a block of statements needs to be repeated a fixed known number of times we use the for loop. These repetition constructs are also known as iterations. A for loop has the form:

for (index variable name = initial value, final value, increment value)

The index variable name that is commonly used for this variable is i, j or counter, the initial value is usually 1, and the incremental value again if usually 1, and if omitted will be assumed to be 1. Note that again we will use indentation and finish the block with endfor to indicate all of these statements within this loop will executed and repeated.

Repetition

for (index variable name = initial value, final value, increment value)

    statement 1

    statement 2

    …

endfor

WORKED EXAMPLE 27 Creating a list of numbers and their squares (1)

Write a pseudocode using a for loop to print out the first 10 natural numbers and their squares.

THINK	WRITE
1. Start the code with begin. No input variables are needed in the problem, only output variables which are the counter, and the value squared.	begin numeric counter, squaredvalue
2. We need a heading indicating what will be displayed.	display 'Number Square'
3. Write the for loop.	for (counter = 1, 10)
4. We need to calculate and display the output, and end the for loop.	squaredvalue = counter * counter display counter, squaredvalue endfor
5. End the code with end.	end

The complete pseudocode is below

```
begin
    display 'Number Square'
    numeric counter, valuesquared
    for (counter = 1, 10)
        square = counter * counter
        display counter, square
    endfor
end
```

Note that the output from this code will be:

```
Number, Square
1     1
2     4
3     9
4     16
5     25
6     36
7     49
8     64
9     81
10    100
```

In the Worked example above the statements are executed a fixed number of times that is known in advance. We could have an input to determine the number of times the loop would be executed. Sometimes we may need to make decisions made when a variable value has changed.

Now in mathematics a statement such as $x = x + 1$ is meaningless and has no solution, but in coding it is often used. What it is really saying, is increment or increase the value of x by 1. Using this we can use the while (Boolean expression is true).. endwhile construct as a loop which is pre-tested at the beginning of the block. When the conditional Boolean expression is true, repeat the block of statements, up to the end… while, but when the condition is false, the block of statements are not executed. Within the block of code we need to change a value so that the Boolean expression which is tested every time must eventually become false, otherwise the loop and the code will never terminate. Note that if the condition of the while loop is initially false, then the while loop will not be executed at all.

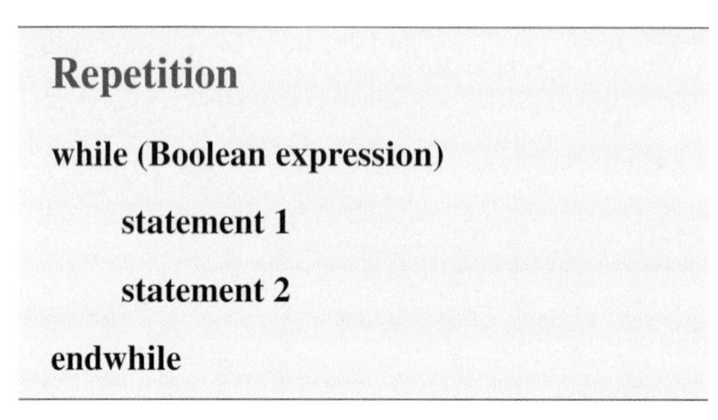

Often code can be written in many ways, in the next example we repeat the last example using the while.. endwhile block.

WORKED EXAMPLE 28 Creating a list of numbers and their squares (2)

Write a pseudocode using a while loop to print out the first 10 natural numbers and their squares.

THINK	WRITE
1. Start the code with begin. No input variables are needed in the problem, only output variables which are the counter, and the value squared.	begin numeric counter, squaredvalue
2. We need to initialise the value of counter, before the while loop starts.	counter = 1
3. We need a heading indicating what will be displayed.	display 'Number Square'
4. Write the while loop.	while (counter <=10)
5. We need to calculate and display the output, and increment the value of counter.	squaredvalue = counter * counter display counter, squarevalue counter = counter + 1
6. End the while loop.	endwhile
7. End the code with end.	end

The complete pseudocode is below and the output is the same as from Worked example 27.

The statements inside the while loop are executed and tested, as soon as the variable counter becomes greater than 10, the condition is false, and the loop is terminated.

```
begin
    numeric counter, squaredvalue
    counter = 1
    display 'Number Square'
    while (counter <=10)
        squaredvalue = counter * counter
        display counter, squarevalue
        counter = counter + 1
    endwhile
end
```

3.6.9 Using functions

Often when writing pseudocode we may need to use other built in mathematical functions and mathematical notations, some of these are described in this section.

For example, we will assume the value of pi (π) is built in and has the value 3.14159.

When raising number to powers with indices, we write for example x^n. In pseudocode we can write this as x**n, and for square roots $\sqrt{x}$ we can write sqrt(x). Often we may need the Boolean expression 'not equal to', for which we will use !=.

Operations on integers are useful as well, for example $\text{mod}(14, 3) = 2$, so that mod gives the remainder of the division. Other functions to obtain random numbers include randInt(1, n), which gives a random integer between 1 and n inclusive. We will also assume the values sin(x) and cos(x) etc are available and can be used when required.

While the above notes are only an introduction to writing pseudocode, the core of writing pseudocode (and programming in general) is the ability to represent the algorithms using constructs such as 'sequence', 'if … then … else', 'case', 'for loop' and 'while'. These constructs are also called keywords and are used to describe the control flow of the algorithm. Note that variable names cannot contain any of these special words.

3.6 Exercise

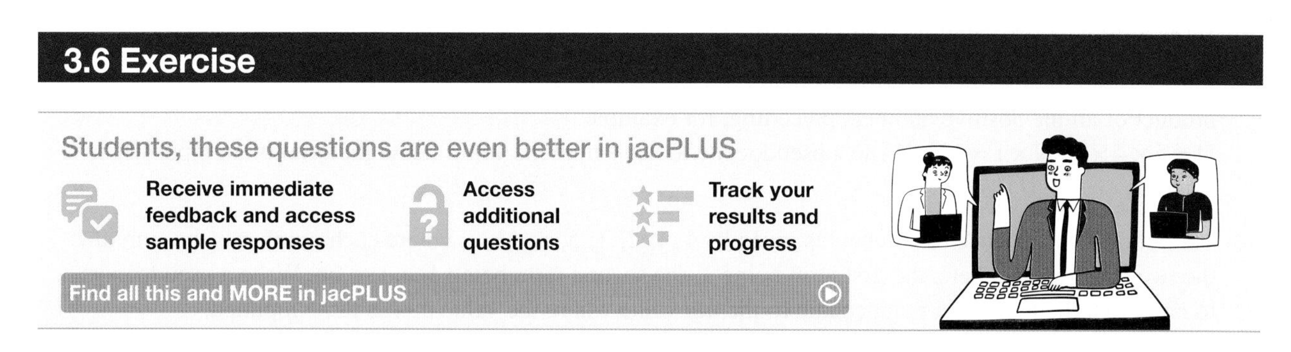

Technology free

1. WE21 Write a pseudocode to calculate the area and circumference of a circle.
2. Write a pseudocode to calculate the area and perimeter of a square.
3. Write a pseudocode to calculate the average of two numbers.

4. Write a pseudocode to calculate the volume of a cone.
5. Write a pseudocode to calculate the area of a parallelogram.
6. **WE22** Write a pseudocode to output a 'pass' if a student scores 50% or more on a test, or a 'fail' if a student scores less than 50% on the test, both the students mark and total marks for the test will need to be given as inputs.
7. **WE23** Write a pseudocode to determine the minimum of three numbers.
8. Write a pseudocode to determine whether a number is even or odd.
9. **WE25** Write a pseudocode to give the grades of students results, that is output 'A' for marks 80% or greater, 'B' for marks 70–79, 'C' for marks 60–69, 'D for marks 50–59 and 'E' for marks less than 50. Assume the value entered is a percentage rounded to the nearest whole number.
10. Write a pseudocode to input your age in years and if you are less than 12 years old, output 'Only a child', if your age is between 13 and 19, output 'Just a teenager', if your age is 20 to 29, output 'In your twenties', if your age is 30 to 39, output 'In your thirties', if your age is 40 to 49, output 'In your forties', if your age is 50 to 59, output 'In your fifties', if your age is 60 to 99, output 'You can retire', otherwise output 'Wow over 100'.
11. **WE26** Write a pseudocode to list the months of the year according to the number value (assume that January corresponds to 1 and December corresponds to 12).
12. Write a pseudocode to decide if a letter of the alphabet is a vowel.
13. Write a pseudocode to add, subtract, multiply or divide two numbers entered, when prompted for the operation, if 1 is addition, 2 is subtraction, 3 is multiplication and 4 is division.
14. **WE27** Write a pseudocode using for loops to print out the first 10 natural numbers and their square roots.

15. Write a pseudocode using for loops to sum all the numbers from 1 up to 100 inclusive.
16. Write a pseudocode using for loops to sum all the odd numbers from 1 up to n, when n is entered as input.
17. The factorial for a positive number or integer, which is denoted by n, denoted or represented as $n!$, is the product of all the positive numbers preceding, for example $5! = 5 \times 4 \times 3 \times 2 \times 1 = 120$. Write a pseudocode to determine the value of the factorial of an input when entered.
18. The Fibonacci sequence of numbers is given by 1, 1, 2, 3, 5, 8, 13, ... where each number is the sum of the two previous numbers, the first term being 1, the second term being 1 and so on. Write a pseudocode to output all the Fibonacci numbers, up to the nth value where the value of n is to be inputted.
19. **WE28** Write a pseudocode using while loops to print out the first 10 natural numbers and their square roots.
20. Write a pseudocode using while loops to sum all the numbers from 1 up to 100 inclusive.
21. Write a pseudocode using while loops to sum all the odd numbers from 1 up to n, when n is entered as input.
22. Write a pseudocode using while loops to determine the value of a factorial.

23. Write a pseudocode using while loops to determine the mean of a list of numbers which are inputted, the inputs continue until the number -1 is entered, this is a trigger to stop inputting numbers.

24. 'Guess the number game'. Write a pseudocode to generate a random number between 1 and 100 inclusive, and let a user try to guess the number, outputs are 'Too high', or 'Too low' or 'Correct number', depending on the guess. Guesses continue until the correct number is guessed, the code then states the total number of guesses needed to guess the number.

25. Write a pseudocode to sort three numbers in ascending order.

3.6 Exam questions

Question 1 (2 marks) TECH-FREE

Write a pseudocode to determine the volume of a cylinder.

Question 2 (5 marks) TECH-FREE

The following equation shows how to convert from degrees Fahrenheit to degrees Celsius:
$C = 5/9\ (F - 32)$.

a. Write pseudocode to convert a temperature entered as Fahrenheit to its Celsius equivalent. **(2 marks)**

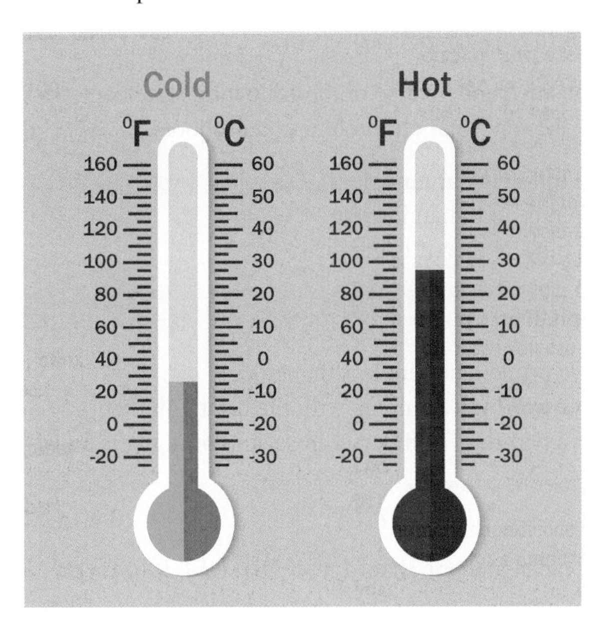

b. Write pseudocode to tabulate temperatures in Celsius to their Fahrenheit equivalence for values of Celsius from 0 to 40 in steps of 2 degrees. **(3 marks)**

Question 3 (4 marks) TECH-FREE

Write a pseudocode to solve a quadratic equation, where the coefficient of the quadratic term is non-zero.

More exam questions are available online.

3.7 Review

3.7.1 Summary

doc-37046

Hey students! Now that it's time to revise this topic, go online to:

Access the topic summary

Review your results

Watch teacher-led videos

Practise exam questions

Find all this and MORE in jacPLUS

3.7 Exercise

Technology free: short answer

1. Write the following compound statements in symbolic form.
   a. Melpomeni and Jacques purchased new bicycles.
   b. Either it is cold or it is warm and sunny.
   c. The dinner was late, expensive and poorly cooked.

2. Determine the truth table for $(p \wedge q) \vee (\neg p \wedge \neg q)$.

3. Write the converse, contrapositive and inverse of the following statement:
   If a politician is intelligent, she sends her children to good schools.

4. Establish the validity of the following argument:
   If the bicycle is not red then it is an Italian bicycle.
   If a bicycle is not an Italian bicycle then it is green.
   My bicycle is red.
   Therefore the bicycle is not Italian.

5. Let A = the set of all positive prime numbers less than 100.
   Let B = the set of all positive two-digit numbers with the digit 1 in them.
   Let C = the set of all positive two-digit numbers whose sum of digits is equal to 7.
   List the following sets:
   a. $A \cap B$
   b. $A \cup (B \cap C)$
   c. $A \cap B \cap C$

6. Prove, using the rules of Boolean algebra, that $(A + A' \cdot B) \cdot (B + B \cdot C) = B$.

7. Design a logic circuit equivalent to the Boolean expression $Q = [A \cdot (B' \cdot C')] + [A \cdot (B \cdot C)]$.

Technology active: multiple choice

8. **MC** The sentences 'The capital of Australia is Canberra', 'Australia is part of the Southern Hemisphere' and 'Australia's population is over 20 million' are examples of
   A. Statements
   B. Instructions
   C. Suggestions
   D. Exclamations
   E. Near-statements

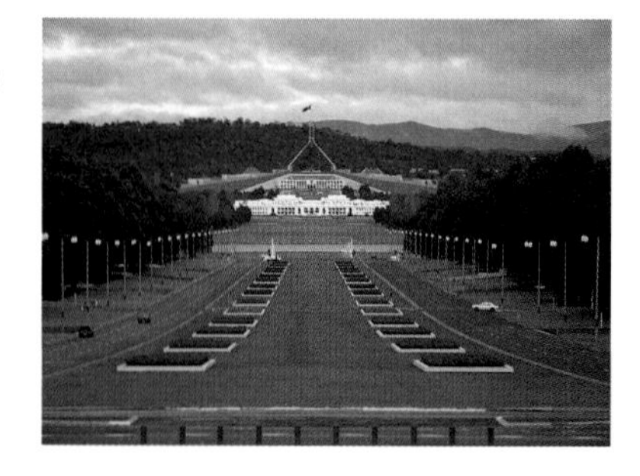

9. **MC** If there is a compound statement with 6 single statements; p, q, r, s, t and v, then the number of rows there will be in the truth table is

A. 6 **B.** 8 **C.** 12 **D.** 36 **E.** 64

10. **MC** The truth table shown represents

p	q	?
T	T	T
T	F	T
F	T	F
F	F	T

A. $p \wedge q$ **B.** $p \vee q$ **C.** $p \vee \neg q$ **D.** $\neg p \vee q$ **E.** $p \wedge \neg q$

11. **MC** The sentence 'I like either ham or steak with eggs for breakfast' can be symbolised as

A. $h \vee (s \vee e)$ **B.** $h \wedge (s \wedge e)$ **C.** $h \wedge (s \vee e)$ **D.** $h \vee (s \wedge e)$ **E.** $h \rightarrow (s \wedge e)$

12. **MC** The inverse statement to: 'If I buy a new coat then I am happy' is

A. If I don't buy a new coat then I am not happy.
B. If I am not happy then I won't buy a new coat.
C. If I am happy then I will buy a new coat.
D. If I don't buy a new coat then I am happy.
E. If I am happy then I won't buy a new coat.

13. **MC** The following argument is an example of which valid form?
If I study hard I will pass my exams.
I did not pass my exams.
I did not study hard.

A. *Modus ponens* **B.** Disjunctive syllogism **C.** Hypothetical syllogism
D. *Modus tollens* **E.** Constructive dilemma.

14. **MC** The shaded area in the figure shown represents

A. $A \cup B$
B. $A' \cup B$
C. $A \cup B'$
D. $A \cap B'$
E. $(A \cup B)'$

15. **MC** If $I =$ the 'universal set', $O =$ the 'empty set' and $A =$ any set, then $A \cdot I$ is

A. A **B.** I **C.** O
D. A' **E.** none of these

16. **MC** The Boolean expression equivalent to the circuit shown is

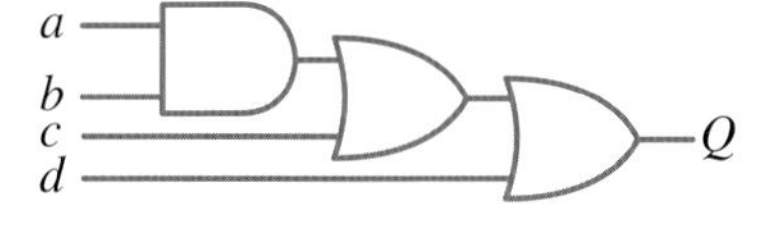

A. $Q = (a \cdot b) + (c \cdot d)$ **B.** $Q = (a \cdot b) \cdot (c + d)$
C. $Q = [(a + b) \cdot c + d]$ **D.** $Q = [(a \cdot b) + c] + d$
E. $Q = a \cdot [b + (c + d)]$

17. **MC** The Boolean equivalent to the circuit shown is

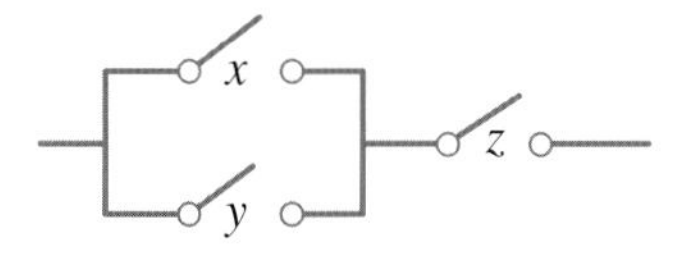

A. $(x + y) + z$ **B.** $(x + y) \cdot z$ **C.** $x \cdot (y + z)$
D. $x \cdot (y \cdot z)$ **E.** $x \cdot (y + z)$

Technology active: extended response

18. In addition to the valid forms of argument introduced in this topic, there are several others, including the *destructive dilemma*. Consider the following argument.

If we want to reduce greenhouse gases, we should use more nuclear power, and if we wish to reduce nuclear accidents we should use conventional power.

We will either not use nuclear power or not use conventional power.

Therefore, we will either not reduce greenhouse gases or we will not reduce the risk of nuclear accidents.

a. Put each statement into symbolic form.
b. Set up the truth table for the three statements.
c. Determine if the argument is valid by determining the rows in the truth table where all premises are true and comparing them with the conclusion.
d. Use this technique to determine the validity of the following argument.

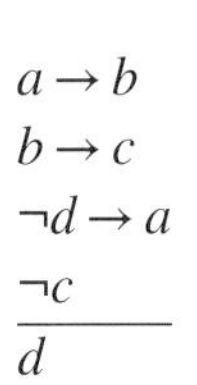

$$\begin{array}{l} a \to b \\ b \to c \\ \neg d \to a \\ \underline{\neg c \quad\quad} \\ d \end{array}$$

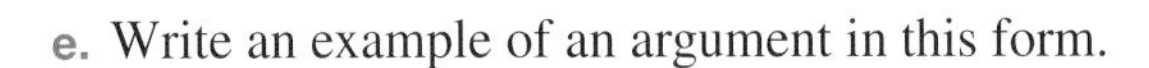

e. Write an example of an argument in this form.

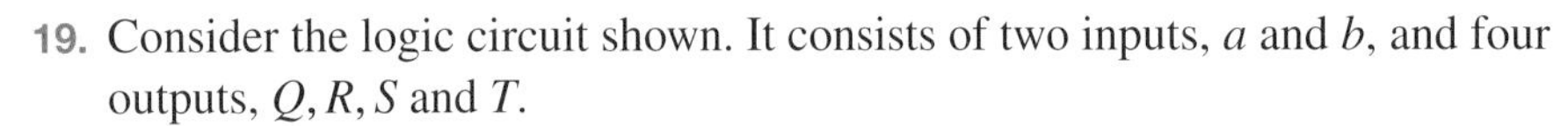

19. Consider the logic circuit shown. It consists of two inputs, a and b, and four outputs, Q, R, S and T.

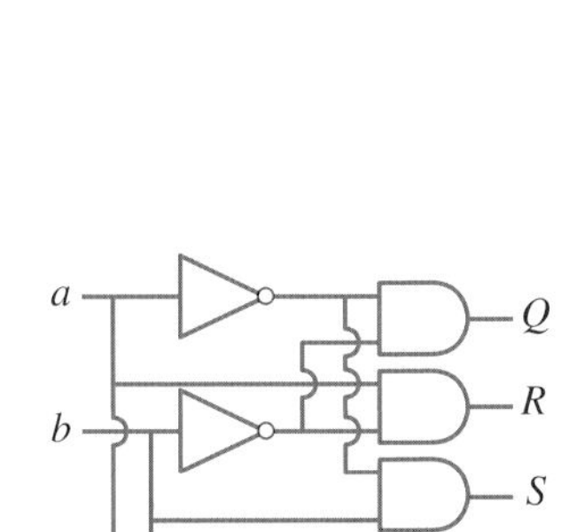

a. Determine the outputs when $a = b = 0$.
b. Determine the outputs when $a = 0$ and $b = 1$.
c. Repeat for the remaining possible values of a and b.
d. Show that $Q = 1$ only when $a = b = 0$.
e. Describe the pattern for this circuit, which is called a 2-bit decoder.
f. The circuit has two NOT gates and four AND gates for the 2-bit decoding of a and b. Determine how many gates would be required for a 3-bit, 4-bit and n-bit decoder.

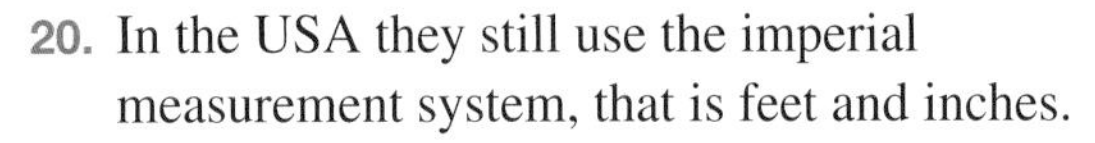

20. In the USA they still use the imperial measurement system, that is feet and inches.

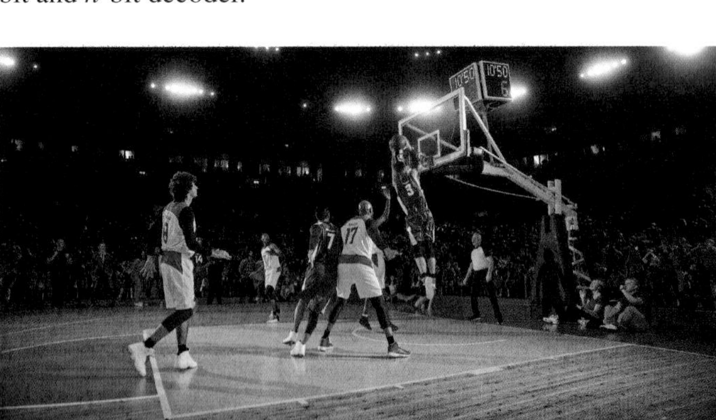

If we hear a basketball player is 6 feet 4 inches, we may wish to know what this is in centimetres.

Given the conversions 1 inch = 2.54 cm and 12 inches = 1 foot. Write pseudocode for each of the following.

a. To convert a height given in feet and inches into centimetres.
b. To convert a height given in centimetres to feet and inches.
c. To tabulate heights from 4 feet 6 inches to 6 feet 8 inches into centimetres using step sizes of two inches.

3.7 Exam questions

Question 1 (1 mark) TECH-ACTIVE

MC Consider the following valid argument.

If John plays for us on Saturday, then we will win.

If we win on Saturday, then we will come in first place on the ladder.

If we come in first place on the ladder, then we play our first final at home.

Therefore, if John plays for us on Saturday, then we play our first final at home.

This is an example of

A. *modus ponens*
B. disjunctive syllogism
C. hypothetical syllogism
D. *modus tollens*
E. constructive dilemma

Question 2 (1 mark) TECH-ACTIVE

MC Premise: All argans are flims. Premise: All flimps are doinks.

The conclusion is

A. all doinks are flimps.
B. all doinks are argans.
C. all argans are doinks.
D. all flimps are argans.
E. doinks are a subset of argans.

Question 3 (1 mark) TECH-ACTIVE

MC Consider the following logic circuit

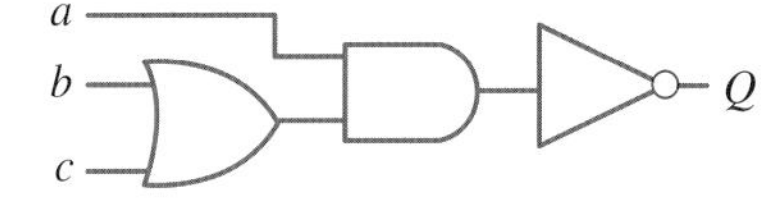

The Boolean equivalent to this circuit is:

A. $[a \cdot (b+c)]'$
B. $a' + (b' \cdot c')$
C. $(a' + b') \cdot (a' + c')$
D. Both **A** and **B** are equivalent to the circuit
E. **A**, **B** and **C** are all equivalent to the circuit

Question 4 (2 marks) TECH-FREE

Use truth tables to determine whether $(p \to q) \leftrightarrow (q \to \neg p)$.

Question 5 (8 marks) TECH-FREE

By considering the elements in the various sets below, verify each of the following statements.

$$\xi = \{1, 2, 3, 4, 5, 6, 7, 8, 9, 10, 11, 12, 13, 14, 15\}$$
$$A = \{1, 2, 3, 4, 5\}$$
$$B = \{4, 5, 6, 7, 8, 9\}$$
$$C = \{5, 8, 9, 10, 11\}$$

a. $(A \cup B)' = A' \cap B'$ **(2 marks)**
b. $(A \cap B)' = A' \cup B'$ **(2 marks)**
c. $(A \cup B) \cap C = (A \cap C) \cup (B \cap C)$ **(2 marks)**
d. $(A \cap B \cap C)' = A' \cup B' \cup C'$ **(2 marks)**

More exam questions are available online.

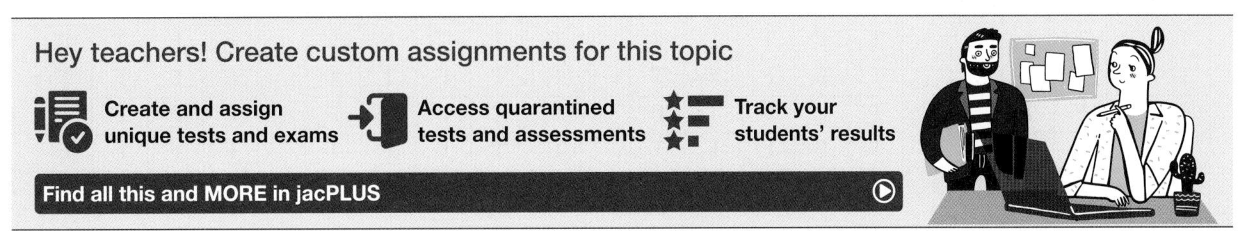

Answers

Topic 3 Logic and algorithms

3.2 Statements (propositions), connectives and truth tables

3.2 Exercise

1. a. Opinion
   b. Indeterminate without further information
   c. True statement
   d. Question
   e. Near-statement
2. a. The car has 4 seats.
   The car has air conditioning.
   b. The Department of Finance was over budget last year.
   The Department of Defence was over budget last year.
   c. Bob went to the hotel.
   Carol went to the hotel.
   Ted went to the hotel.
   Alice went to the hotel.
   d. To be a best-seller a novel must be interesting to the reader.
   To be a best-seller a novel must be relevant to the reader.
   e. Sam will win the trophy.
   Nancy will win the trophy.
   f. You can choose vanilla ice-cream for dessert.
   You can choose strawberry ice-cream for dessert.
   You can choose fruit for dessert.
   g. There are some statements that cannot be proved to be true.
   There are some statements that cannot be proved to be false.
   h. Most of my friends studied Mathematics.
   Most of my friends studied Physics.
   Most of my friends studied Engineering.
   Most of my friends studied Law.
   Most of my friends studied Arts.
3. a. John and Mary rode their bicycles to school.
   b. The book you want is in row 3 or 4.
   c. The weather is cold and cloudy.
   d. Many people read novels or history.
   e. In a recent poll 80% preferred jazz or classical music.
   f. Two is an even prime number.
   Two is the only even prime number (alternative answer).
4. a.

p	q	r
T	T	T
T	T	F
T	F	T
T	F	F
F	T	T
F	T	F
F	F	T
F	F	F

8 ways

b.

p	q	r	s
T	T	T	T
T	T	T	F
T	T	F	T
T	T	F	F
T	F	T	T
T	F	T	F
T	F	F	T
T	F	F	F

p	q	r	s
F	T	T	T
F	T	T	F
F	T	F	T
F	T	F	F
F	F	T	T
F	F	T	F
F	F	F	T
F	F	F	F

16 ways

c. See the table at the bottom of the page.*

5. a. $p =$ John passed.
$q =$ Zia passed.
$r =$ David passed.
$p \wedge q \wedge r$
$2^3 = 8$ rows

p	q	r	$p \wedge q \wedge r$
T	T	T	T
T	T	F	F
T	F	T	F
T	F	F	F
F	T	T	F
F	T	F	F
F	F	T	F
F	F	F	F

*4. c.

p	q	r	s	t
T	T	T	T	T
T	T	T	T	F
T	T	T	F	T
T	T	T	F	F
T	T	F	T	T
T	T	F	T	F
T	T	F	F	T
T	T	F	F	F

p	q	r	s	t
T	F	T	T	T
T	F	T	T	F
T	F	T	F	T
T	F	T	F	F
T	F	F	T	T
T	F	F	T	F
T	F	F	F	T
T	F	F	F	F

p	q	r	s	t
F	T	T	T	T
F	T	T	T	F
F	T	T	F	T
F	T	T	F	F
F	T	F	T	T
F	T	F	T	F
F	T	F	F	T
F	T	F	F	F

p	q	r	s	t
F	F	T	T	T
F	F	T	T	F
F	F	T	F	T
F	F	T	F	F
F	F	F	T	T
F	F	F	T	F
F	F	F	F	T
F	F	F	F	F

32 ways

b. p = Alice does the dishes.
q = Renzo does the dishes.
r = Carla does the dishes.
$(p \wedge q) \vee r$
$2^3 = 8$ rows

p	q	r	$(p \wedge q) \vee r$
T	T	T	T
T	T	F	T
T	F	T	T
T	F	F	F
F	T	T	T
F	T	F	F
F	F	T	T
F	F	F	F

c. p = female member
q = student
r = professor
$2^3 = 8$ rows
$p \wedge (q \vee r)$

p	q	r	$p \wedge (q \vee r)$
T	T	T	T
T	T	F	T
T	F	T	T
T	F	F	F
F	T	T	F
F	T	F	F
F	F	T	F
F	F	F	F

6. a.

p	q	$p \wedge \neg q$
T	T	F
T	F	T
F	T	F
F	F	F

b.

p	q	$\neg p \wedge \neg q$
T	T	F
T	F	F
F	T	F
F	F	T

c.

p	q	r	$(p \wedge q) \wedge r$
T	T	T	T
T	T	F	F
T	F	T	F
T	F	F	F
F	T	T	F
F	T	F	F
F	F	T	F
F	F	F	F

7. a.

p	q	$p \vee \neg q$
T	T	T
T	F	T
F	T	F
F	F	T

b.

p	q	$\neg p \vee \neg q$
T	T	F
T	F	T
F	T	T
F	F	T

c.

p	q	r	$(p \vee q) \vee r$
T	T	T	T
T	T	F	T
T	F	T	T
T	F	F	T
F	T	T	T
F	T	F	T
F	F	T	T
F	F	F	F

8. a. It is raining and I bring my umbrella.
   b. It is raining or I bring my umbrella.
   c. It is not raining and I bring my umbrella.
9. a. Peter and Quentin like football.
   b. Peter or Quentin like football.
   c. Peter likes football and Quentin does not like football.

10.

p	q	$\neg(p \vee q)$	$(\neg p \wedge \neg q)$
T	T	F	F
T	F	F	F
F	T	F	F
F	F	T	T

11.

p	q	$\neg(p \vee q)$	$\neg p \wedge \neg q$
T	T	F	F
T	F	F	T
F	T	F	T
F	F	T	T

Not equivalent

12. a.

p	q	$(p \wedge q)$	$\neg p$	$(p \wedge q) \vee \neg p$	$(p \vee q)$	$(p \vee q) \wedge \neg p$
T	T	T	F	T	T	F
T	F	F	F	F	T	F
F	T	F	T	T	T	T
F	F	F	T	T	F	F

Not equivalent

b.

p	q	$(p \vee q)$	$\neg p$	$(p \vee q) \vee \neg p$	$(p \vee \neg q)$
T	T	T	F	T	T
T	F	T	F	T	T
F	T	T	T	T	T
F	F	F	T	T	T

Equivalent

13.

p	q	r	$(p \wedge q) \vee r$	$p \wedge (q \vee r)$
T	T	T	T	T
T	T	F	T	T
T	F	T	T	T
T	F	F	F	F
F	T	T	T	F
F	T	F	F	F
F	F	T	T	F
F	F	F	F	F

Not equivalent

14. a. i.

p	q	r	$(p \wedge q) \wedge r$	$p \wedge (q \wedge r)$
T	T	T	T	T
T	T	F	F	F
T	F	T	F	F
T	F	F	F	F
F	T	T	F	F
F	T	F	F	F
F	F	T	F	F
F	F	F	F	F

Equivalent

ii.

p	q	r	$(p \vee q) \vee r$	$p \vee (q \vee r)$
T	T	T	T	T
T	T	F	T	T
T	F	T	T	T
T	F	F	T	T
F	T	T	T	T
F	T	F	T	T
F	F	T	T	T
F	F	F	F	F

Equivalent

b. Brackets have no effect on expressions with a single $\vee$ or $\wedge$ operator, but they do have an effect if they are mixed up together.

3.2 Exam questions

Note: Mark allocations are available with the fully worked solutions online.

1.

p	$\neg p$	$p \wedge \neg p$
T	F	F
F	T	F

2. B

3. $p \wedge q \vee r$

3.3 Exercise

1.

p	q	$\neg p$	$\neg q$	$(p \to q)$	$(\neg q \to \neg p)$
T	T	F	F	T	T
T	F	F	T	F	F
F	T	T	F	T	T
F	F	T	T	T	T

Equivalent.

2.

p	q	$\neg p$	$\neg q$	$(p \to q)$	$(\neg p \to \neg q)$
T	T	F	F	T	T
T	F	F	T	F	T
F	T	T	F	T	F
F	F	T	T	T	T

Not equivalent.

3. Implication: If it is bread then it is made with flour.
Converse: If it is made with flour then it is bread.
Contrapositive: If it is not made with flour then it is not bread.
Inverse: If it is not bread then it is not made with flour.

4. a. Premise: All cats are fluffy.
Premise: My pet is a cat.
Conclusion: My pet is fluffy.

b. Premise: All even numbers are divisible by themselves and 2.
Premise: Prime numbers are divisible by themselves and 1.
Conclusion: Two is the only even prime number.

c. Premise: Growing apples depends on good water.
Premise: Good water depends on good irrigation.
Conclusion: Growing apples depends on good irrigation.

5. **a** and **c** are valid.

6. b. All footballers are fit.
David is not fit.
David is not a footballer.

d. Cannot be made into a valid argument.

e. All musicians can read music.
Louise is a musician.
Louise can read music.

7.

p	q	$p \vee q$	$\neg p$
T	T	T	F
T	F	T	F
F	T	T	T
F	F	F	T

Conclusion is true whenever all premises are true (3rd row), thus a valid argument.

8. a.

p	q	r	$p \to q$	$q \to r$	$p \to r$
T	T	T	T	T	T
T	T	F	T	F	F
T	F	T	F	T	T
T	F	F	F	T	F
F	T	T	T	T	T
F	T	F	T	F	T
F	F	T	T	T	T
F	F	F	T	T	T

Conclusion is true whenever all premises are true (1st, 5th, 7th and 8th rows), thus a valid argument.

b.

p	q	$p \rightarrow q$	$\neg q$	$\neg p$
T	T	T	F	F
T	F	F	T	F
F	T	T	F	T
F	F	T	T	T

Conclusion is true whenever all premises are true (4th row), thus a valid argument.

c.

p	q	r	s	$p \rightarrow q$	$r \rightarrow s$	$p \rightarrow q \wedge r \rightarrow s$	$p \wedge r$	$q \wedge s$
T	T	T	T	T	T	T	T	T
T	T	T	F	T	F	F	T	F
T	T	F	T	T	T	T	F	T
T	T	F	F	T	T	T	F	F
T	F	T	T	F	T	F	T	F
T	F	T	F	F	F	F	T	F
T	F	F	T	F	T	F	F	F
T	F	F	F	F	T	F	F	F
F	T	T	T	T	T	T	F	T
F	T	T	F	T	F	F	F	F
F	T	F	T	T	T	T	F	T
F	T	F	F	T	T	T	F	F
F	F	T	T	T	T	T	F	F
F	F	T	F	T	F	F	F	F
F	F	F	T	T	T	T	F	F
F	F	F	F	T	T	T	F	F

Conclusion is true whenever all premises are true (1st row), thus a valid argument.

9. a. Disjunctive syllogism

b. *Modus tollens*

c. *Modus ponens*

10.

p	q	$p \rightarrow q$	$\neg p$	$\neg q$
T	T	T	F	F
T	F	F	F	T
F	T	T	T	F
F	F	T	T	T

Since premises are both true in row 3 and 4, but conclusion is false in row 3 then this is not a valid argument.

11. $p \rightarrow q$

$\underline{\neg p}$

$\neg q$

$p =$ If elected with a majority

$q =$ My government will introduce new tax laws

12. a. $p \rightarrow q$
$\underline{\neg p}$
$\neg q$

b.

p	q	$\neg p$	$\neg q$	$p \rightarrow q$
T	T	F	F	T
T	F	F	T	F
F	T	T	F	T
F	F	T	T	T

Premises are both true in rows 3 and 4, but conclusion is false in row 3 therefore this is not a valid argument.

13. a.

p	q	r	$p \rightarrow q$	$r \rightarrow \neg q$	$p \rightarrow \neg r$
T	T	T	T	F	F
T	T	F	T	T	T
T	F	T	F	T	F
T	F	F	F	T	T
F	T	T	T	F	T
F	T	F	T	T	T
F	F	T	T	T	T
F	F	F	T	T	T

Valid argument.

b.

p	q	r	$\neg p \wedge \neg q$	$r \rightarrow p$
T	T	T	F	T
T	T	F	F	T
T	F	T	F	T
T	F	F	F	T
F	T	T	F	F
F	T	F	F	T
F	F	T	T	F
F	F	F	T	T

Invalid argument.

c.

p	q	$\neg p \rightarrow \neg q$
T	T	T
T	F	T
F	T	F
F	F	T

Invalid argument.

14. a. Valid - hypothetical syllogism
b. Invalid

15. a. Valid – *modus tollens*
b. Valid – constructive dilemma

16. $\neg(p \vee q) \leftrightarrow (\neg p \wedge \neg q)$

p	q	$p \vee q$	$\neg(p \vee q)$	$\neg p$	$\neg q$	$\neg p \wedge \neg q$	$\neg(p \vee q) \leftrightarrow (\neg p \wedge \neg q)$
T	T	T	F	F	F	F	T
T	F	T	F	F	T	F	T
F	T	T	F	T	F	F	T
F	F	F	T	T	T	T	T

17. $(p \wedge q) \rightarrow p$

p	q	$p \wedge q$	$(p \wedge q) \rightarrow p$
T	T	T	T
T	F	F	T
F	T	F	T
F	F	F	T

18. $[(p \rightarrow q) \wedge \neg q] \rightarrow \neg p$

p	q	$p \rightarrow q$	$\neg q$	$(p \rightarrow q) \wedge \neg q$	$\neg p$	$[(p \rightarrow q) \wedge \neg q] \rightarrow \neg p$
T	T	T	F	F	F	T
T	F	F	T	F	F	T
F	T	T	F	F	T	T
F	F	T	T	T	T	T

19. $[(p \vee q) \vee r] \leftrightarrow [p \vee (q \vee r)]$

p	q	r	$p \vee q$	$(p \vee q) \vee r$	$(q \vee r)$	$p \vee (q \vee r)$	$[(p \vee q) \vee r] \leftrightarrow [p \vee (q \wedge r)]$
T	T	T	T	T	T	T	T
T	T	F	T	T	T	T	T
T	F	T	T	T	T	T	T
T	F	F	T	T	F	T	T
F	T	T	T	T	T	T	T
F	T	F	T	T	T	T	T
F	F	T	F	T	T	T	T
F	F	F	F	F	F	F	T

20. $[p \vee (q \wedge r)] \leftrightarrow [(p \vee q) \wedge (p \vee r)]$

p	q	r	$q \wedge r$	$p \vee (q \wedge r)$	$p \vee q$	$p \vee r$	$(p \vee q) \wedge (p \vee r)$	$[p \vee (q \wedge r)] \leftrightarrow [(p \vee q) \wedge (p \vee r)]$
T	T	T	T	T	T	T	T	T
T	T	F	F	T	T	T	T	T
T	F	T	F	T	T	T	T	T
T	F	F	F	T	T	T	T	T
F	T	T	T	T	T	T	T	T
F	T	F	F	F	T	F	F	T
F	F	T	F	F	F	T	F	T
F	F	F	F	F	F	F	F	T

3.3 Exam questions

Note: Mark allocations are available with the fully worked solutions online.

1. The argument is valid.
2. The statements are equivalent.
3. Sample responses can be found in the worked solutions in the online resources.

3.4 Boolean algebra and digital logic

3.4 Exercise

1.

A	B	A'	$A' \cdot B$	$A + A' \cdot B$	$A + B$
0	0	1	0	0	0
0	1	1	1	1	1
1	0	0	0	1	1
1	1	0	0	1	1

2. a.

A	B	$A \cdot B$	$(A \cdot B)'$	A'	B'	$A' + B'$
0	0	0	1	1	1	1
0	1	0	1	1	0	1
1	0	0	1	0	1	1
1	1	1	0	0	0	0

b.

A	B	$A + B$	$(A + B)'$	A'	B'	$A' \cdot B'$
0	0	0	1	1	1	1
0	1	1	0	1	0	0
1	0	1	0	0	1	0
1	1	1	0	0	0	0

3. a. A' b. $A \cdot B \cdot C'$

4. a. $A + B$ b. $A' \cdot B \cdot C$

5. a. A b. B'

6. a. 1 b. 0

7.

x	y	Q
0	0	0
0	1	0
1	0	0
1	1	1

8. a.

x	y	z	Q
0	0	0	0
0	0	1	1
0	1	0	0
0	1	1	1
1	0	0	0
1	0	1	1
1	1	0	1
1	1	1	1

$Q = (x \cdot y) + z$

9. a.

x	y	Q
0	0	1
0	1	0
1	0	0
1	1	1

b. Used where there are 2 people who can activate the light separately.

10.

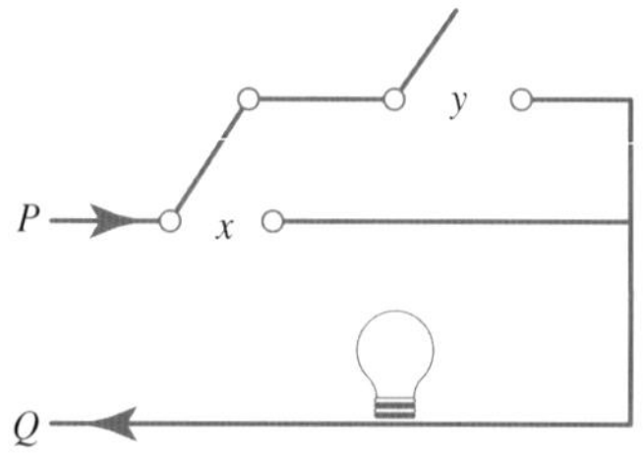

P
x
y
Q

11.

a	b	c	Output
0	0	0	0
0	0	1	1
0	1	0	1
0	1	1	1
1	0	0	1
1	0	1	1
1	1	0	1
1	1	1	1

12.

a	b	c	Output
0	0	0	1
0	0	1	1
0	1	0	1
0	1	1	1
1	0	0	1
1	0	1	0
1	1	0	0
1	1	1	0

13. a.

a	b	c	Output
0	0	0	1
0	0	1	1
0	1	0	1
0	1	1	1
1	0	0	1
1	0	1	0
1	1	0	0
1	1	1	0

b. Same truth table as question 12

14.

a	*b*	*c*	*d*	Output
0	0	0	0	0
0	0	0	1	0
0	0	1	0	1
0	0	1	1	0
0	1	0	0	0
0	1	0	1	0
0	1	1	0	1
0	1	1	1	0
1	0	0	0	1
1	0	0	1	1
1	0	1	0	1
1	0	1	1	1
1	1	0	0	0
1	1	0	1	0
1	1	1	0	1
1	1	1	1	0

15. a.

S1	*S2*	*a*	*b*	*c*	*Q*	*R*
0	0	0	0	1	0	0
0	1	0	1	0	0	0
1	0	1	0	1	0	1
1	1	1	1	0	1	0

b. When $S1 = 0$, the system is disabled; the safe can't be opened.
When $S1 = 1$ and $S2 = 0$, the alarm rings.
When $S1 = 1$ and $S2 = 1$, the safe can be opened.

16. a. Sample responses can be found in the worked solutions in the online resources.

b. 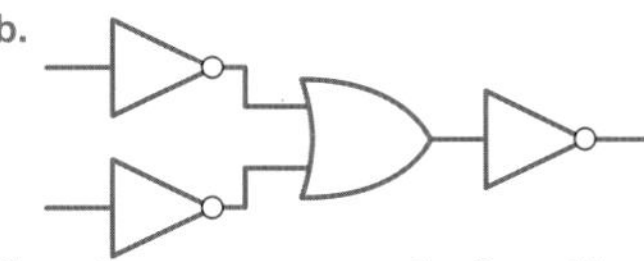

17. Sample responses can be found in the worked solutions in the online resources.

18. 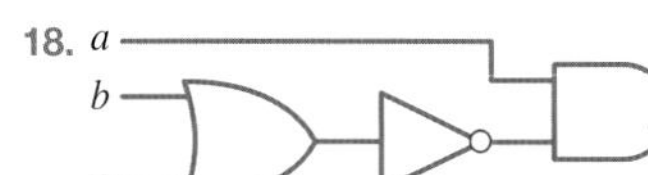

19. 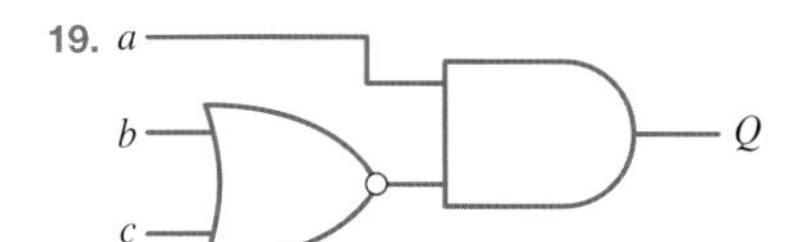

20. 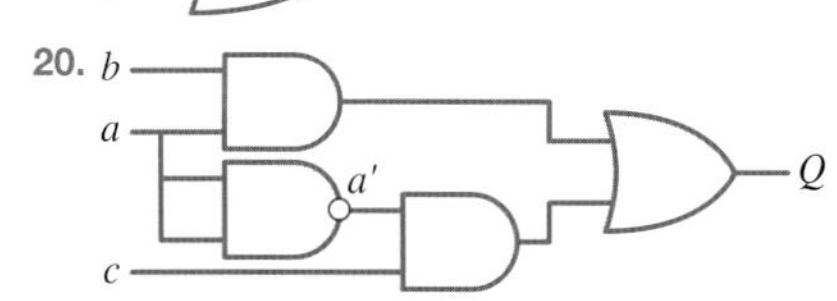

3.4 Exam questions

Note: Mark allocations are available with the fully worked solutions online.

1. Sample responses can be found in the worked solutions in the online resources.
2. $F = (a \cdot b) + (c \cdot d)$
3. Sample responses can be found in the worked solutions in the online resources.

3.5 Sets and Boolean algebra

3.5 Exercise

1. Sample responses can be found in the worked solutions in the online resources.
2. Sample responses can be found in the worked solutions in the online resources.
3. a. $A = \{2, 4, 6, 8, 10, 12, 14, 16, 18\}$
   b. $B = \{4, 8, 12, 16, \ldots\}$
   c. $C = \{2\}$
   d. $D = \{\text{Jack, Queen, King}\}$
   e. $\text{E} = \varnothing$
   f. $F = \{9, 8, 7, 6, \ldots\}$
4. A, C, D, E
5. 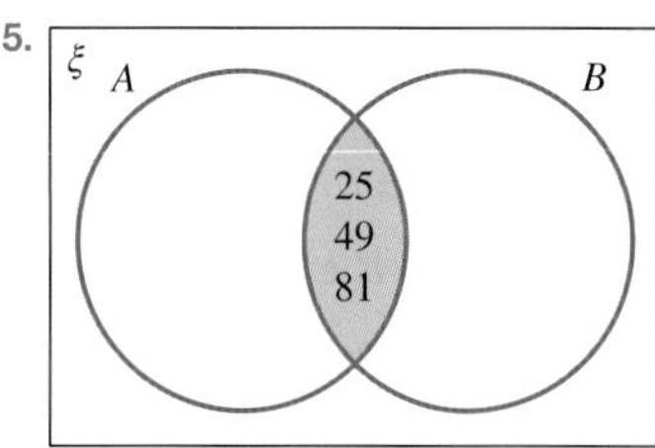

6. 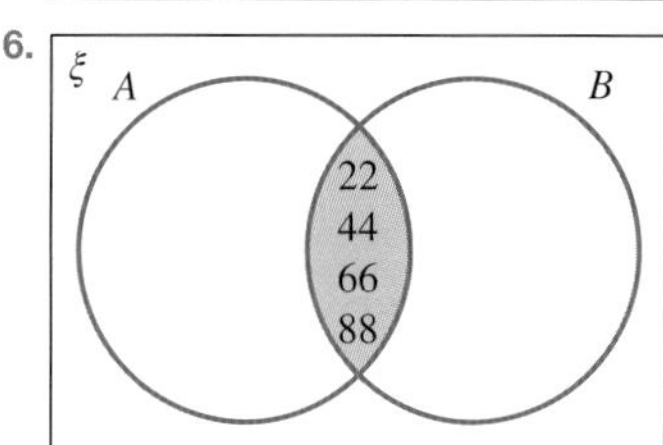

7. a. 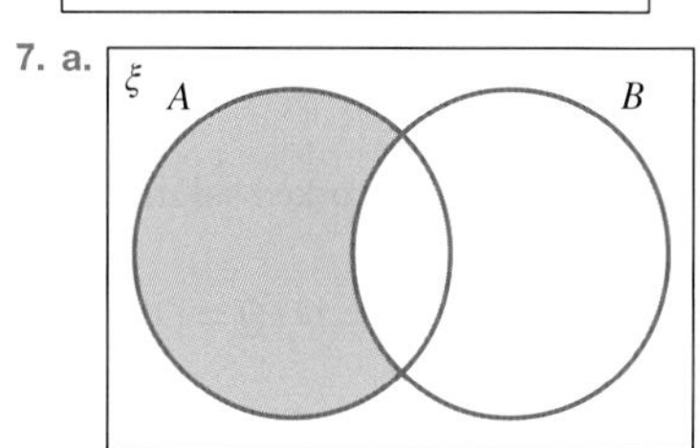

   b. 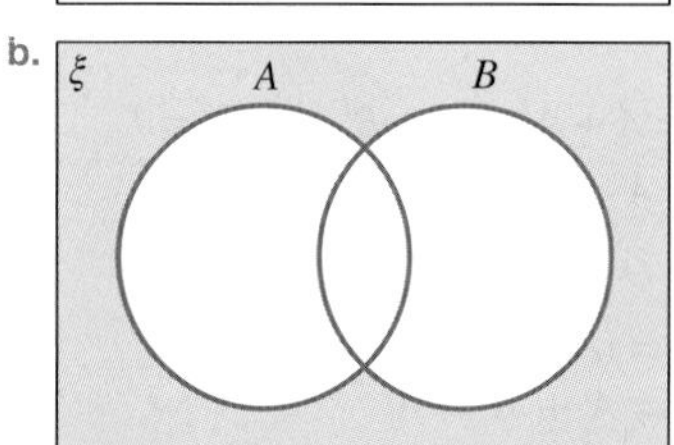

c. 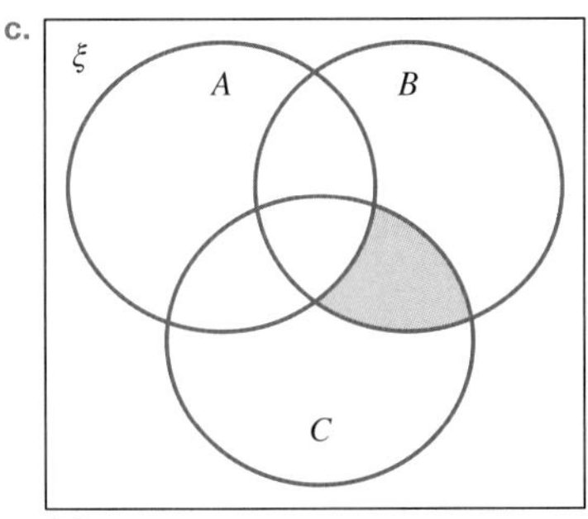

8. a. $\{4\}$
   b. $\{1, 2, 3, 4, 5, 6, 7, 8, 9, 10\}$
   c. $\{1, 2, 3, 4, 5, 6, 7, 8, 9, 10\}$
   d. $\{1, 2, 3, 4, 5, 6, 7, 8, 9, 10\}$
   e. $\{1, 2, 3, 4, 5, 6, 7, 8, 9, 10\}$
9. a. $(\text{B} \cup \text{C}) = \{1, 2, 4, 6, 8, 9, 10\}$
   b. $\text{A} \cap (\text{B} \cup \text{C}) = \{1, 2, 4, 6, 8, 9, 10\}$
   c. $(\text{A} \cap \text{C}) = \{1, 4, 9\}$
   d. $(\text{A} \cap \text{B}) = \{2, 4, 6, 8, 10\}$
   e. $(\text{A} \cap \text{B}) \cup (\text{A} \cap \text{C}) = \{1, 2, 4, 6, 8, 9, 10\}$
10. Part 1: Show that $(A \cdot B) + (A' + B') = I$.

$$\begin{aligned}(A \cdot B) + (A' + B') &= (A + A' + B') \cdot (B + A' + B') \\ &= (I + B') \cdot (A' + I) \\ &= (I) \cdot (I) \\ &= I \qquad \text{QED}\end{aligned}$$

Part 2: Show that $(A \cdot B) \cdot (A' + B') = O$.

$$\begin{aligned}(A \cdot B) \cdot (A' + B') &= A \cdot B \cdot A' + A \cdot B \cdot B' \\ &= O \cdot B + A \cdot O \\ &= O + O \\ &= O \qquad \text{QED}\end{aligned}$$

11. a. $A + B$
    b. I
    c. $A + B$
    d. $A \cdot B$
12. Sample responses can be found in the worked solutions in the online resources.
13. $(p \wedge q) \wedge \neg p = (P \cdot Q) \cdot P' = (P \cdot P') \cdot Q = O \cdot Q = O$

$$p \wedge \neg p = P \cdot P' = O \qquad \text{QED}$$

14. a. $A + B + A' + B' = A + A' + B + B' = I + I = I$
    b. $(A + B) \cdot A' \cdot B' = A \cdot A' \cdot B' + B \cdot A' \cdot B'$
    $= O \cdot B' + O \cdot A' = O + O = O$
    c. $(A + B) \cdot (A + B') = (A + B) \cdot A + (A + B) \cdot B'$
    $= A + A = A$
    d. $A \cdot B + C \cdot (A' + B') = A \cdot B + C \cdot (A \cdot B)'$
    $= A \cdot B + C$

15. a. iii. $a' + b$
    b. 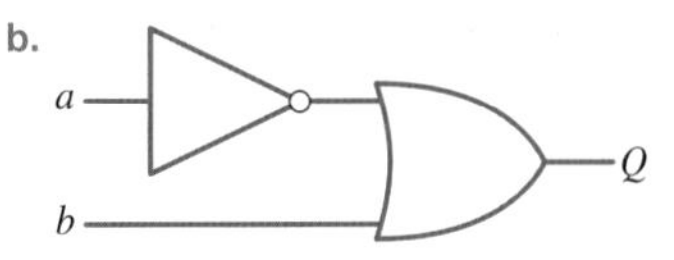

c. 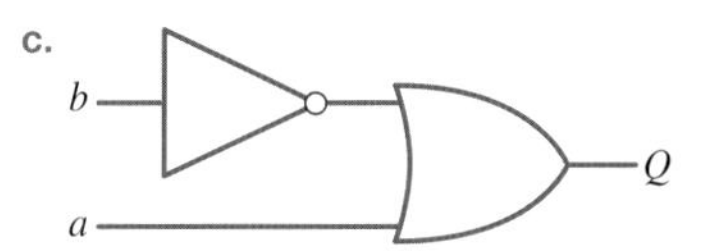

d. $a \cdot b$
e. $(a \cdot b)' + b$
f. 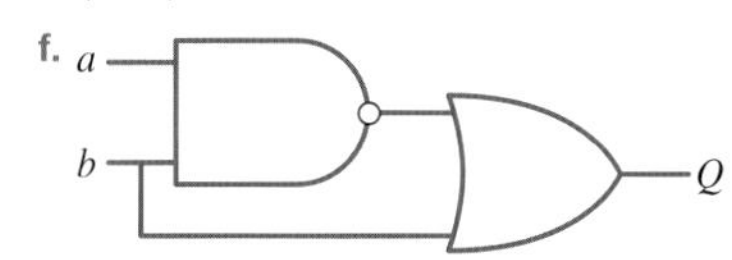

a	b	$(a \cdot b)'$	$(a \cdot b)' + b$
0	0	1	1
0	1	1	1
1	0	1	1
1	1	0	1

3.5 Exam questions

Note: Mark allocations are available with the fully worked solutions online.

1. They are equivalent.
2. C
3. Sample responses can be found in the worked solutions in the online resources.

3.6 Algorithms and pseudocode

3.6 Exercise

1. begin
    numeric radius, circumference, area
    display 'Enter the radius'
    input radius
    circumference = 2 * Pi * r
    area = Pi * r * r
    display 'The circumference of the circle is', circumference
    display 'The area of the circle is', area
   end
2. begin
    numeric length, area, perimeter
    display 'Enter the length'
    input length
    perimeter = 4 * length
    area = length * length
    display 'The perimeter of the square is', perimeter
    display 'The area of the square is', area
   end
3. begin
    numeric num1, num2, avg
    display 'Enter the first number'
    input num1
    display 'Enter the second number'
    input num2
    avg = (num1 + num2) / 2
    display 'The average of the two numbers is', avg
   end

4. begin
```
    numeric height, radius, volume
    display 'Enter the height of the cone'
    input height
    display 'Enter the radius of the cone'
    input radius
    volume = Pi * r * r * h / 3
    display 'The volume of the cone is', volume
end
```
5. begin
```
    numeric length, base, height, area
    display 'Enter the base length of the parallelogram'
    input base
    display 'Enter the height of the parallelogram'
    input height
    area = base * height
    display 'The area of the parallelogram is', area
end
```
6. begin
```
    numeric mark, totalmarks, grade
    display 'Enter the total marks of the test'
    input totalmarks
    display 'Enter the mark obtained on the test'
    input mark
    grade = mark / totalmarks * 100
    if (grade >= 50) then
        display 'Pass'
    endif
    if (grade < 50) then
        display 'Fail'
    endif
end
```
7. begin
```
    numeric num1, num2, num3, minx
    display 'Enter the first number'
    input num1
    display 'Enter the second number'
    input num2
    display 'Enter the third number'
    input num3
    if (num1 < num2) then
        minx = num1
    else
        minx = num2
    endif
    if (num3 < minx) then
        minx = num3
    endif
    display 'The minimum of the three numbers is', minx
end
```
8. begin
```
    numeric num
    display 'Enter an integer'
    input num
    if (mod(num2) == 0 ) then
        display 'The number is even'
    else
        display 'The number is odd'
    endif
end
```
9. begin
```
    numeric grade
    display 'Enter your grade'
    input grade
    if (grade >= 80) then
        display 'The result is A'
    elseif (grade >= 70) then
        display 'The result is B'
    elseif (grade >= 60) then
        display 'The result is C'
    elseif (grade >= 50) then
        display 'The result is D'
    else
        display 'The result is E'
    endif
end
```
10. begin
```
    numeric age
    display 'Enter your age in years'
    input age
    if (age <= 12) then
        display 'Only a child'
    elseif (age <= 19) then
        display 'Just a teenager'
    elseif (age <= 29) then
        display 'In your twenties'
    elseif (age <= 39) then
        display 'In your thirties'
    elseif (age <= 49) then
        display 'In your forties'
    elseif (age <= 59) then
        display 'In your fifties'
    elseif (age <= 99) then
        display 'You can retire'
    else
        display 'Wow over 100'
    endif
end
```
11. begin
```
    numeric monthcode
    disp 'Enter a number 1-12 to represent the month of the
year'
    input monthcode
    disp 'The month is'
    case 1
        display 'January'
        break
    case 2
        display 'February'
        break
    case 3
        display 'March'
        break
    case 4
        display 'April'
        break
    case 5
        display 'May'
        break
```

```
        case 6
            display 'June'
            break
        case 7
            display 'July'
            break
        case 8
            display 'August'
            break
        case 9
            display 'September'
            break
        case 10
            display 'October'
            break
        case 11
            display 'November'
            break
        case 12
            display 'December'
            break
        default
            display 'Not a valid month value'
            break
    end
```

12.
```
begin
        char letter
        disp 'Enter a letter'
        input letter
        case 'A' or 'a'
            display 'Yes it is a vowel'
            break
        case 'E' or 'e'
            display 'Yes it is a vowel'
            break
        case 'I' or 'i'
            display 'Yes it is a vowel'
            break
        case 'O' or 'o'
            display 'Yes it is a vowel'
            break
        case 'U' or 'u'
            display 'Yes it is a vowel'
            break
        default
            display 'Not a vowel'
            break
    end
```

13.
```
begin
        numeric num1, num2, sum, difference, product,
    quotient, operation
        disp 'Enter two numbers'
        input num1, num2
        disp 'Enter a number 1 for addition of the two numbers'
        disp 'Enter a number 2 for subtraction of the two
    numbers'
        disp 'Enter a number 3 for the product of the two
    numbers'
        disp 'Enter a number 4 for the quotient of the two
    numbers'
        input operation
        case 1
            sum = num1 + num2
            display 'The sum of the two numbers is', sum
            break
        case 2
            difference = num1 - num2
            display 'The difference of the two numbers is',
    difference
            break
        case 3
            product = num1 * num2
            display 'The product of the two numbers is', product
            break
        case 4
            quotient = num1 / num2
            display 'The quotient of the two numbers is',
    quotient
            break
        default
            display 'Not a valid operation code'
            break
    end
```

14.
```
begin
        numeric counter,
        disp 'Value Square Root'
        for (counter = 1, 10)
    begin
            display counter, sqrt(counter)
        endfor
    end
```

15.
```
begin
        numeric counter, number, sum
        sum = 0
        for (counter = 1, 100)
            sum = sum + counter
        endfor
        disp 'The sum of the first 100 numbers is', sum
    end
```

16.
```
begin
        numeric counter, number, finalvalue, sum
        sum = 0
        disp 'Enter a number'
        input finalvalue
        for (counter = 1, finalvalue, 2)
            sum = sum + counter
        endfor
        disp 'The sum of the odd numbers from 1 to',
    finalvalue, 'is', sum
    end
```

17.
```
begin
        numeric counter, number, fact
        fact = 1
        disp 'Enter a number'
        input number
        for (counter = 1, number)
            fact = fact * counter
        endfor
        disp 'The value of', number, 'factorial is', fact
    end
```

18. begin
```
    numeric nextfib, lastfib, beforefib
    lastfib = 1
    beforefib = 1
    disp lastfib, beforefib
    disp 'Enter a number'
    input number
    for (counter = 1, number)
        nextfib = lastfib + beforefib
        disp nextfib
        beforefib = lastfib
        lastfib = nextfib
    endfor
end
```
19. begin
```
        numeric counter
    counter = 1
    display 'Number Square Root'
    while (counter <= 10)
        display counter, sqrt(counter)
        counter = counter + 1
    endwhile
end
```
20. begin
```
    numeric counter, number, sum
    sum = 0
    counter = 0
    while (counter <= 100)
        sum = sum + counter
        counter = counter + 1
    endwhile
    disp 'The sum of the first 100 numbers is', sum
end
```
21. begin
```
    numeric counter, number, finalvalue, sum
    sum = 1
    disp 'Enter a number'
    input finalvalue
    counter = 1
    while (counter <= finalvalue)
        sum = sum + counter
        counter = counter + 2
    endwhile
    disp 'The sum of the odd numbers from 1 to',
finalvalue, 'is', sum
end
```
22. begin
```
    numeric counter, number, fact
    fact = 1
    counter = 1
    disp 'Enter a number'
    input number
    while (counter <= number)
        fact = fact * counter
        counter = counter + 1
    endwhile
    disp 'The value of', number, 'factorial is', fact
end
```
23. begin
```
    numeric counter, number, sum, average
    sum = 0
    number = 0
    counter = 0
    while (number != -1)
        disp 'Enter a number'
        input number
        sum = sum + number
        counter = counter + 1
    endwhile
    average = (sum + 1) / (counter - 1)
    disp 'The average of all the number is', average
end
```
24. begin
```
    numeric tries, guess, randnum
    tries = 0
    guess = 0
    randnum = randInt(1,100)
    while (guess != randnum)
        disp 'Enter a number'
        input guess
        tries = tries + 1
        if (guess > randnum) then
            disp 'Too high'
    endif
    if (guess < randnum) then
        disp 'Too low'
    endif
    endwhile
    disp 'Yes correct number you guessed it in', tries,
'guesses'
end
```
25. begin
```
    numeric x1, x2, x3
    disp 'Enter three numbers'
    input x1, x2, x3
    if (x1 < x2) and (x2 < x3) and (x1 < x3) then
        disp x1, x2, x3
    endif
    if (x1 < x2) and (x3 < x2) and (x1 < x3) then
        disp x1, x3, x2
    endif
    if (x2 < x1) and (x2 < x3) and (x1 < x3) then
        disp x2, x1, x3
    endif
    if (x2 < x1) and (x2 < x3) and (x3 < x1) then
        disp x2, x3, x1
    endif
    if (x1 < x2) and (x3 < x2) and (x3 < x1) then
        disp x3, x1, x2
    endif
    if (x2 < x1) and (x3 < x2) and (x3 < x1) then
        disp x3, x2, x1
    endif
end
```

3.6 Exam questions

Note: Mark allocations are available with the fully worked solutions online.

1.
```
begin
    numeric radius, height, volume
    display 'Enter the radius'
    input radius
    display 'Enter the height'
    input height
    volume = pi * r ** 2 * h
    display 'The volume of the cylinder is', volume
end
```

2. a.
```
begin
    numeric c, f
    display 'Enter the temperature in Fahrenheit'
    input f
    c = 5 / 9 * ( f – 32)
    display 'The equivalent temperature in Centigrade is', c
end
```

b.
```
begin
    numeric c, f
    display 'Centigrade Fahrenheit'
    for (c = 0, 40, 2)
    f = 32 + 9 * c / 5
    display c, f
    endfor
end
```

3.
```
begin
    numeric a, b, c, delta, x1, x2
    display 'Enter the values of a, b and c'
    input a, b, c
    delta = b * b – 4 * a * c
    if (delta > 0) then
        x1 = (-b + sqrt(delta)) / (2 * a)
        x2 = (-b - sqrt(delta)) / (2 * a)
        disp 'There are two solutions' x1, 'and', x2
    elseif (delta == 0 ) then
        x1 = -b / (2 * a)
        display 'There is one solution' x1
    else
        display 'There are no real solutions'
    endif
end
```

3.7 Review

3.7 Exercise

Technology free: short answer

1. a. $m \wedge j$

b. $c \vee (w \wedge s)$

c. $l \wedge e \wedge p$

2.

p	q	$(p \wedge q)$	$(\neg p \wedge \neg q)$	$(p \wedge q) \vee (\neg p \wedge \neg q)$
T	T	T	F	T
T	F	F	F	F
F	T	F	F	F
F	F	F	T	T

3. Converse: If she sends her children to good schools, the politician is intelligent.
Contrapositive: If she doesn't send her children to good schools, the politician is not intelligent.
Inverse: If a politician isn't intelligent she doesn't send her children to good schools.

4.

r	I	g	$\neg r \rightarrow I$	$\neg I \rightarrow g$	r	$\neg I$	
T	T	T	T	T	T	F	*
T	T	F	T	T	T	F	*
T	F	T	T	T	T	T	
T	F	F	T	F	T	T	
F	T	T	T	T	F	F	
F	T	F	T	T	F	F	
F	F	T	F	T	F	T	
F	F	F	F	F	F	T	

The conclusion (column 7) is false when the premises (columns 4, 5, 6) are all true (*). Invalid argument.

5. a. $\{11, 13, 17, 19, 31, 41, 61, 71, 91\}$
   b. $\{2, 3, 5, 7, 11, 13, 16, 17, 23, 29, 31, 37, 41, 43, 47, 53, 59, 61, 67, 71, 73, 79, 83, 89, 91, 97\}$
   c. $\{61\}$

6. $(A + A' \cdot B) \cdot (B + B \cdot C)$

$= (A + A' \cdot B) \cdot (B \cdot I + B \cdot C)$ $\{B \cdot I = B\}$

$= (A + A' \cdot B) \cdot B(I + C)$ {Distributive Law}

$= (A + A' \cdot B) \cdot B \cdot I$ $\{I + C = I\}$

$= A \cdot B \cdot I + A' \cdot B \cdot B \cdot I$ {Distributive Law}

$= A \cdot B + A' \cdot B$ $\{B \cdot I = B, B \cdot B = B\}$

$= (A + A') \cdot B$ {Distributive Law}

$= I \cdot B$ $\{A + A' = I\}$

$= B$ QED

7. 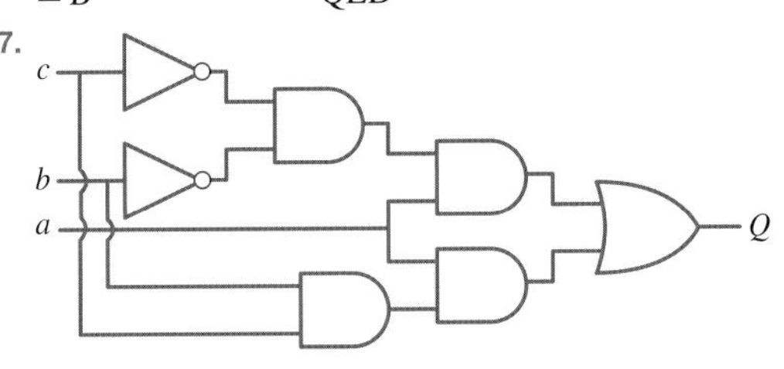

Technology active: multiple choice

8. A
9. E
10. C
11. D
12. A
13. D
14. C
15. A
16. D
17. B

Technology active: extended response

18. a. $S1 = (g \rightarrow n) \wedge (a \rightarrow c)$
   $S2 = \neg n \vee \neg c$
   $S3 = \neg g \vee \neg a$
   Premises (S1, S2) are all true in rows 4, 12, 13, 15, 16
   Conclusion (S3) is true in rows 3, 4, 7, 8, 9, 10, 11, 12, 13, 14, 15, 16
   Conclusion is true when premises are true, so this is a valid argument.

b.

g	n	a	c	$g \rightarrow n$	$a \rightarrow c$	$S1$	$S2 = \neg n \vee \neg c$	$S3 = \neg g \vee \neg a$
T	T	T	T	T	T	T	F	F
T	T	T	F	T	F	F	T	F
T	T	F	T	T	T	T	F	T
T	T	F	F	T	T	T	T	T
T	F	T	T	F	T	F	T	F
T	F	T	F	F	F	F	T	F
T	F	F	T	F	T	F	T	T
T	F	F	F	F	T	F	T	T
F	T	T	T	T	T	T	F	T
F	T	T	F	T	F	F	T	T
F	T	F	T	T	T	T	F	T
F	T	F	F	T	T	T	T	T
F	F	T	T	T	T	T	T	T
F	F	T	F	T	F	F	T	T
F	F	F	T	T	T	T	T	T
F	F	F	F	T	T	T	T	T

c. Premises $S1$ and $S2$ are all true in rows 4, 12, 13, 15 and 16. Conclusion is true when premises are true. Therefore a valid argument.

d.

a	b	c	d	$a \rightarrow b$	$b \rightarrow c$	$\neg d \rightarrow a$	$\neg c$
T	T	T	T	T	T	T	F
T	T	T	F	T	T	T	F
T	T	F	T	T	F	T	T
T	T	F	F	T	F	T	T
T	F	T	T	F	T	T	F
T	F	T	F	F	T	T	F
T	F	F	T	F	T	T	T
T	F	F	F	F	T	T	T
F	T	T	T	T	T	T	F
F	T	T	F	T	T	F	F
F	T	F	T	T	F	T	T
F	T	F	F	T	F	F	T
F	F	T	T	T	T	T	F
F	F	T	F	T	T	F	F
F	F	F	T	T	T	T	T
F	F	F	F	T	T	F	T

Argument is valid because conclusion is true when all premises are true (row 15).

e. Sample responses can be found in the worked solutions in the online resources.

19. a–d.

b	a	Q	R	S	T
0	0	1	0	0	0
0	1	0	1	0	0
1	0	0	0	1	0
1	1	0	0	0	1

e. This circuit 'decodes' or distinguishes the inputs, which are 00, 01, 10 and 11.

Only one of the outputs = 1 for each of the 4 possible inputs

f. 3-bit: 3 NOT, 8 AND
4-bit: 4 NOT and 16 AND
n-bit: n NOT and 2^n AND

20. a.
```
begin
    numeric feet, inches, cms
        disp 'Enter the height feet then inches'
        input feet, inches
        cms = (12 * feet + inches) * 2.54
    disp 'That height is', cms, 'cm'
end
```

b.
```
begin
    numeric feet, inches, cms
        disp 'Enter the height in cm'
        input cms
        inches = cms / 2.54
        feet = inches div 12
        inches = inches mod 12
    disp 'That height is', feet, 'feet', inches, 'inches'
end
```

c.
```
begin
    numeric counter, feet, inches, cms
    disp 'feet inches cms'
    inches = 4
    feet = 4
    counter = 0
    while (counter <= 12)
        counter = counter + 1
        inches = inches + 2
        if (inches == 12) then
    begin
        inches = 0
        feet = feet + 1
    endif
    cms = (12 * feet + inches) * 2.54
    disp feet, inches, cm
 endwhile
 end
```

3.7 Exam questions

Note: Mark allocations are available with the fully worked solutions online.

1. C
2. C
3. E
4. They are not equivalent.
5. Sample responses can be found in the worked solutions in the online resources.

4 Proof and number

LEARNING SEQUENCE

Fully worked solutions for this topic are available online.

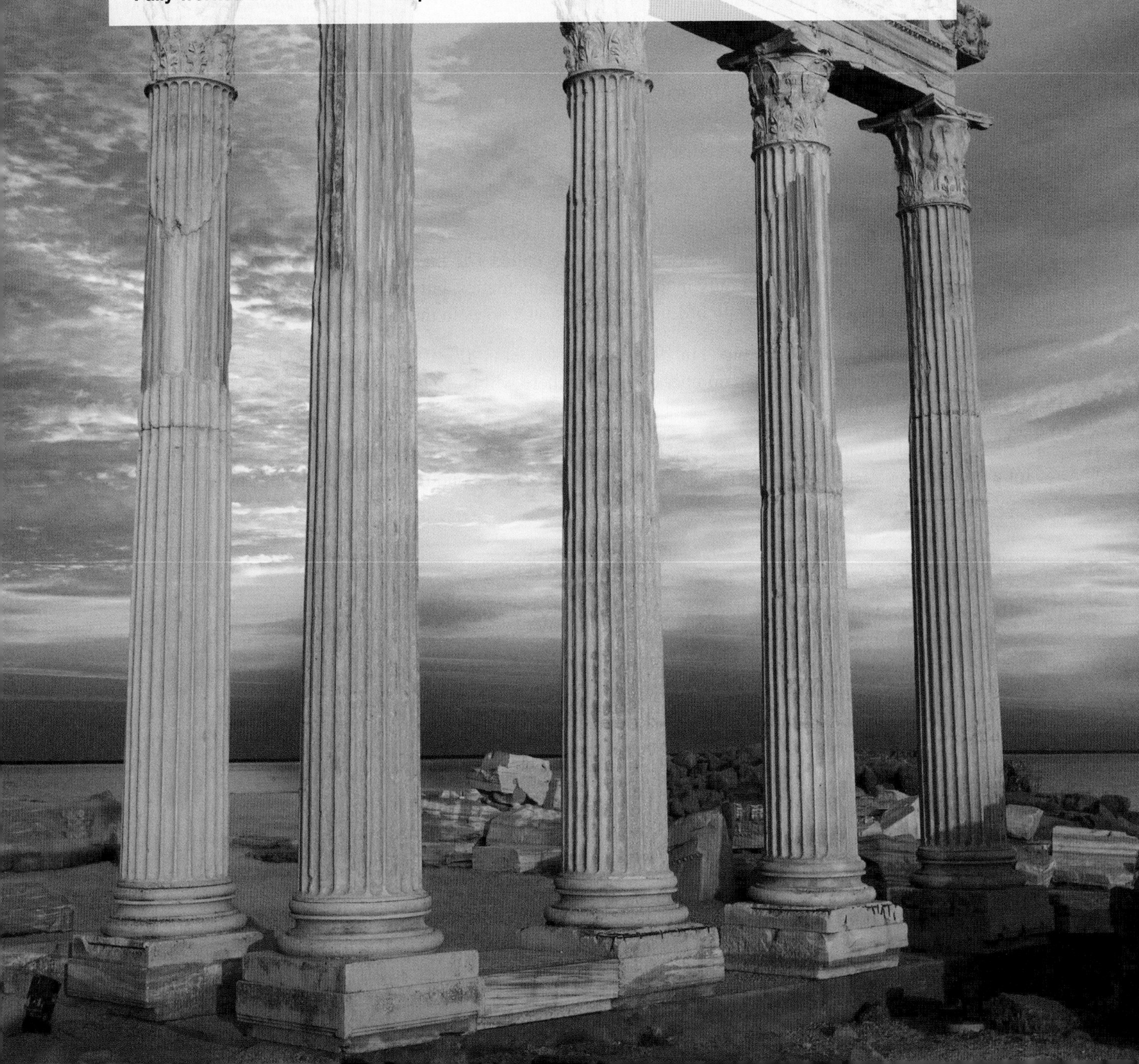

4.1 Overview

Hey students! Bring these pages to life online

Watch videos

Engage with interactivities

Answer questions and check results

Find all this and MORE in jacPLUS

In your study of mathematics, you have already used a number of rules, for example, Pythagoras's theorem. Mathematicians like to be certain that a rule is always true before they use it. They use proofs to demonstrate this. Once something is proven, it is certain to be true and it will always be true. Proofs make mathematics different to the sciences, because they are not like theories. (A theory is the best current explanation for our observations, and may be replaced by a better theory sometime in the future.)

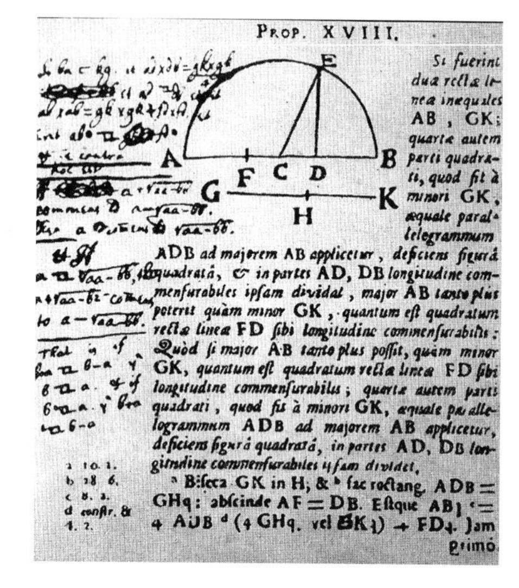

PROP. XVIII.

Si fuerint duæ rectæ lineæ inæquales AB, GK; quartæ autem parti quadrati, quod fit à minori GK, æquale parallelogrammum ADB ad majorem AB applicetur, deficiens figurâ quadratâ, & in partes AD, DB longitudine commensurabiles ipsam dividat, major AB tanto plus poterit quàm minor GK, quantum est quadratum rectæ lineæ FD sibi longitudine commensurabilis: Quòd si major AB tanto plus possit, quàm minor GK, quantum est quadratum rectæ lineæ FD sibi longitudine commensurabilis; quartæ autem parti quadrati, quod fit à minori GK, æquale parallelogrammum ADB ad majorem AB applicetur, deficiens figurâ quadratâ, in partes AD, DB longitudine commensurabiles ipsam dividet.

a Biseca GK in H; & b fac rectang. ADB = GHq: abscinde AF = DB. Estque ABq c = 4 ADB d (4 GHq. vel GKq) + FDq. Jam primò

a 10. 1.
b 28. 6.
c 8. 2.
d constr. & 4. 2.

In Greece during the fifth century BCE, philosophers developed the idea of proving that a mathematical statement or proposition was true. To do this they needed to agree on the definitions of some basic terms. Also, to have starting points for their arguments, they needed to agree that some basic statements, called axioms, were true. Once a statement was proven true using a rigorous logical argument based on the definitions and axioms, it was called a theorem. Once a theorem was proved, it could be used to prove other more complicated theorems. Euclid (325–263 BCE) collected these definitions, theorems and proofs into a series of 13 books called *The Elements* — a book so influential it has been used for over 2000 years, not only in the study of mathematics, but also the development of logic. The image shows Sir Isaac Newton's copy of Euclid with his own notes in the margin.

However, this certainty was challenged in 1931 by Kurt Gödel (1906–1978), who proved that in any complex mathematical system with a certain number of axioms, there will be some statements that can be neither proved nor disproved using the axioms.

If all theorems have already been proved, why is it important to learn how to construct a proof? The first reason for a proof is to convince someone that a statement is true. The second reason is that it helps us to understand why the statement is true.

KEY CONCEPTS

This topic covers the following key concepts from the VCE Mathematics Study Design:

- binary number systems
- number systems for the natural numbers, N, integers, Z, rational numbers, Q, real numbers, R, and complex numbers, C, and their fundamental properties and structure
- set notation and operations including element, intersection, union, complement sub-set and power set
- prime numbers, the fundamental theorem of arithmetic, and proof that there are infinitely many prime numbers
- conversion between fraction and decimal forms of rational numbers
- introduction to principles of proof including propositions and quantifiers, examples and counter-examples, direct proof, proof by contradiction, and proof using the contrapositive and mathematical induction
- simple proofs involving, for example, divisibility, sequences and series, inequalities and irrationality.

4.2 Number systems and mathematical statements

LEARNING INTENTION

At the end of this subtopic you should be able to:
- understand and use the language of mathematical logic to write and evaluate statements
- use quantifiers to write and evaluate the truth of mathematical statements
- convert numbers between decimal and binary forms.

4.2.1 The real number system

The set of **real numbers** consists of all the numbers that can be thought of as points on a number line. The set of real numbers is represented by the symbol R.

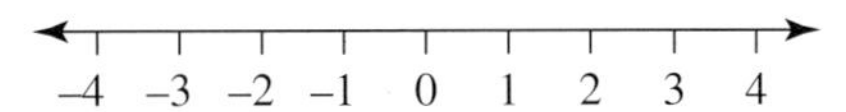

The set of real numbers has a number of subsets.

The set of **natural numbers** (or counting numbers) is represented by the symbol N, where $N = \{1, 2, 3, ...\}$.

The set of **integers** consists of all the positive and negative natural numbers and 0 (which is neither positive nor negative). Represented by the symbol Z, the set of integers can be divided into the subsets of positive integers (or natural numbers), represented by Z^+ (or N), and negative integers, represented by Z^-. That is:

$$Z = \{..., -3, -2, -1, 0, 1, 2, 3, ...\}$$
$$Z^+ = \{1, 2, 3, 4, 5, 6, ...\}$$
$$Z^- = \{-1, -2, -3, -4, -5, -6, ...\}$$

Using set notation, $Z = Z^+ \cup Z^- \cup \{0\}$.

Integers and natural numbers may be represented on the number line as illustrated below.

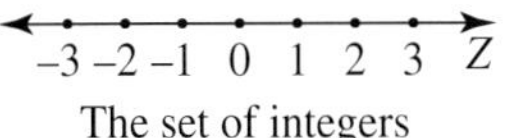

The set of integers

1 2 3 4 5 6 N

The set of positive integers or natural numbers

Z⁻ –6 –5 –4 –3 –2 –1

The set of negative integers

***Note*:** Integers on the number line are marked with a solid dot to indicate that they are the only points we are interested in.

When an integer is divided by another integer, the result is a **rational number**. More formally, a rational number can be written as the ratio of two integers, a and b, in the form $\frac{a}{b}$ where $b \neq 0$ and a and b do not have any common factors (except 1). The set of rational numbers is represented by the symbol Q.

In set notation,

$$Q = \left\{\frac{a}{b}, a, b \in Z, b \neq 0, \gcd(a, b) = 1\right\}$$

where $\gcd(a, b) = 1$ means that the greatest common divisor of a and b is 1.

If $b = 1$, then $\frac{a}{b} = a$. Therefore, integers are a subset of the set of rational numbers.

Some rational numbers may be expressed as terminating decimals (that is, they contain a specific number of digits). For example:

$$\frac{1}{2} = 0.5$$
$$\frac{1}{8} = 0.125$$
$$-\frac{9}{5} = -1.8$$

Other rational numbers may be expressed as recurring decimals (non-terminating or periodic decimals). For example:

$$\frac{1}{3} = 0.333\,333\,3\ \ldots = 0.\dot{3}$$
$$\frac{7}{11} = 0.636\,363\,63\ \ldots = 0.\dot{6}\dot{3} = 0.\overline{63}$$
$$\frac{53}{90} = 0.588\,888\,88\ \ldots = 0.5\dot{8}$$
$$\frac{15}{7} = 2.142\,857\,142\,857\ \ldots = 2.\dot{1}42\,85\dot{7} = 2.\overline{142\,857}$$

***Note*:** Recurring decimals with more than one digit can be shown by either a dot over the first digit and the last digit in the recurring sequence, or an overscore over the recurring sequence.

Rational numbers can be represented on a number line as illustrated below.

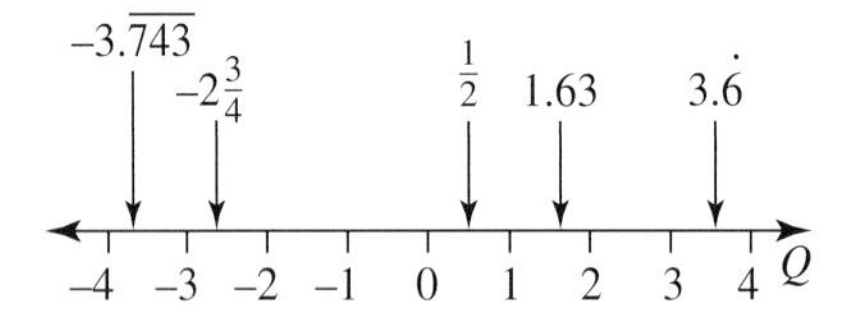

Any real numbers that are not rational numbers are called **irrational numbers**. These include **surds** (for example $\sqrt{2}$ and $\sqrt{10}$) and all other decimals that neither terminate nor recur, such as π and e. There is no common symbol for irrational numbers, but because they are the set of real numbers excluding the set of rational numbers, they can be represented by $R \setminus Q$.

Irrational numbers may be represented by decimals. For example,

$$\sqrt{0.03} = 0.173\,205\,080\ldots$$
$$\sqrt{18} = 4.242\,640\,68\ldots$$
$$-\sqrt{5} = -2.236.6797\ldots$$
$$-2\sqrt{7} = -5.291\,502\,62\ldots$$
$$\pi = 3.141\,592\,653\ldots$$
$$e = 2.718\,281\,828\ldots$$

These decimal approximations can be shown on a number line as illustrated below.

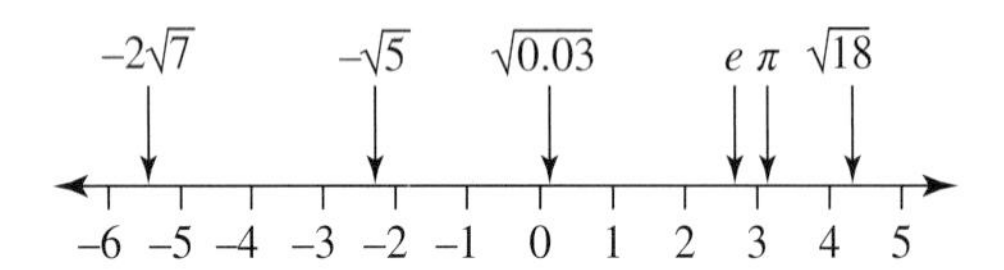

Using geometry, Pythagoras' **theorem** and a compass, irrational numbers in surd form can be represented on the number line exactly, as shown.

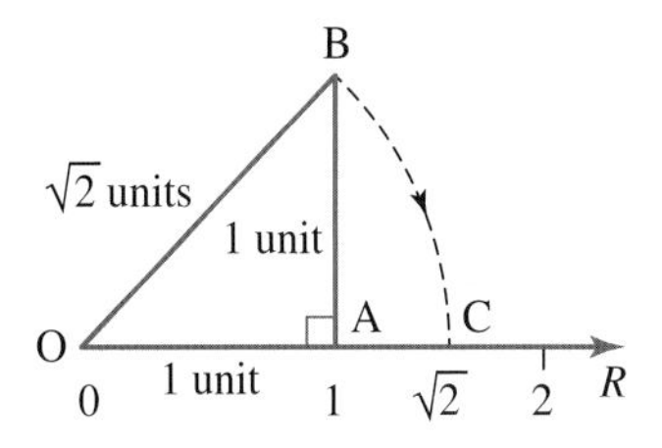

In summary, the set of real numbers can be divided into two main sets: rational and irrational numbers. These may be divided into further subsets as illustrated below.

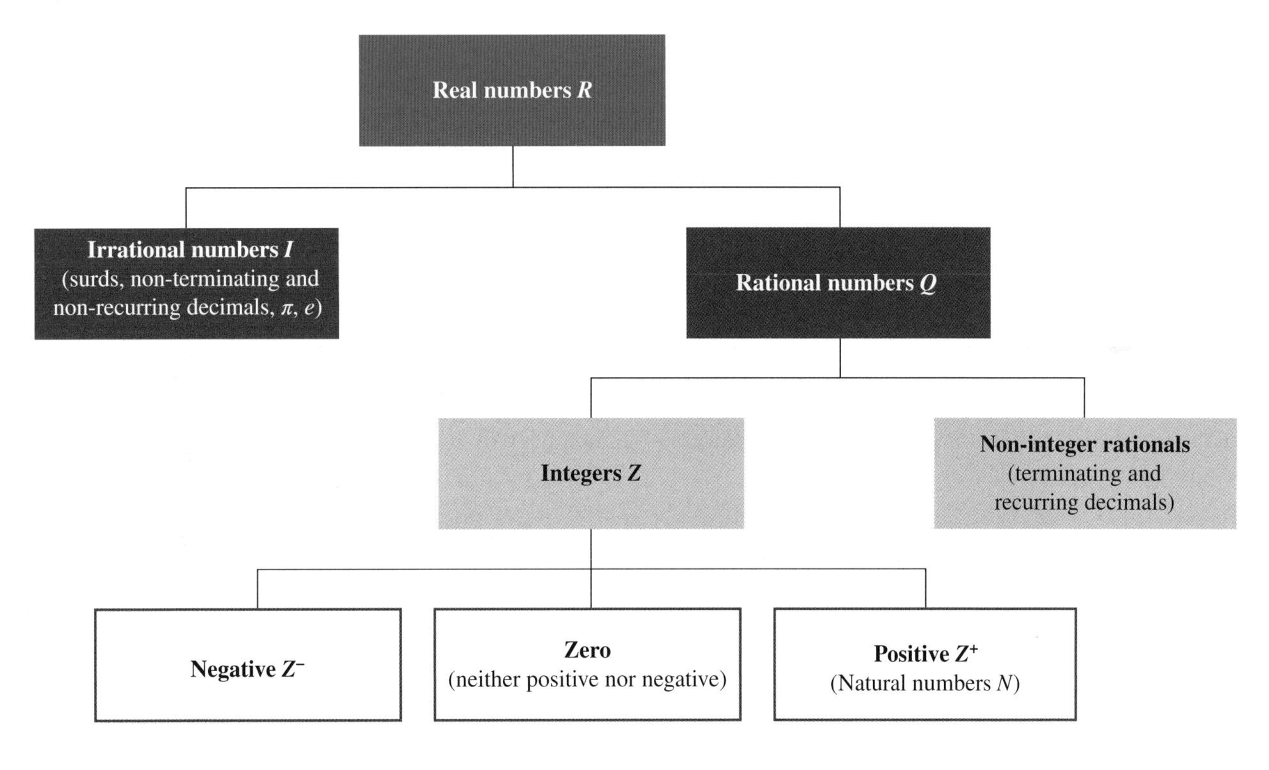

Definitions in mathematics are precise descriptions of mathematical terms, such as those given above for real numbers, integers, rational numbers and irrational numbers. Definitions make it easier for us to discuss and apply mathematical concepts.

WORKED EXAMPLE 1 Classifying numbers as rational or irrational

Identify the following numbers as either rational or irrational.

a. 7 b. $\frac{3}{11}$ c. 0.25

d. $0.010\,110\,111\ldots$ e. $3.\dot{7}$ f. $\sqrt{5}$

THINK	WRITE
a. 7 is an integer. All integers are rational.	a. Rational
b. $\frac{3}{11}$ is the quotient of two integers. It is rational.	b. Rational

c. 0.25 is a terminating decimal. It is rational. — c. Rational

d. 0.010 110 111 ... is neither terminating nor eventually repeating. It is irrational. — d. Irrational

e. $3.\dot{7}$ is a recurring decimal. It is rational. — e. Rational

f. $\sqrt{5}$ is a surd. It is irrational. — f. Irrational

4.2.2 Converting between fractions and decimals

Rational numbers can be expressed as the quotient of two integers or as numbers whose decimal expansions are either terminating or eventually recurring. It is necessary to be able to express each rational number in both formats.

Expressing common fractions as decimals

Consider the following decimal fractions and their common fraction equivalents.

$$0.1 = \frac{1}{10} \qquad 0.37 = \frac{37}{100} \qquad 0.163 = \frac{163}{100}$$

All terminating decimals can be expressed with a power of 10 as the denominator. As the only prime factors of 10 are 2 and 5, any common fraction whose denominator can be written as powers of 2 and/or 5 will be a terminating decimal.

Consider $\frac{1}{40}$. The denominator can be written as powers of 2 and 5: $40 = 2^3 \times 5$. Therefore, the number can be expressed as a terminating decimal. $\frac{1}{40} = \frac{1}{40} \times \frac{25}{25} = \frac{25}{1000} = 0.025$.

Consider $\frac{1}{3}$. As the denominator is not a multiple of either 2 or 5, the number will be a recurring decimal. This decimal can be calculated by division: $3\overline{)1.{}^{1}0{}^{1}0{}^{1}0}$ giving $0.333...$. Therefore, $\frac{1}{3} = 0.\dot{3}$.

Likewise, the rational number $\frac{5}{6}$ is a recurring decimal as $\frac{5}{6} = 0.8\dot{3}$, and can be calculated by division:

$6\overline{)5.{}^{5}0{}^{2}0{}^{2}0}$ giving $0.833...$. Therefore, $\frac{5}{6} = 0.8\dot{3}$.

WORKED EXAMPLE 2 Converting fractions to decimals

Express the following numbers as decimals.

a. $\frac{1}{50}$ b. $\frac{7}{9}$ c. $\frac{5}{11}$

THINK	WRITE
a. 1. The denominator can be written as powers of 2 and 5.	a. $50 = 5^2 \times 2$

2. Write the denominator as a power of 10. As $50 = 5^2 \times 2$, it needs to be multiplied by 2 in order to be written as a power of $10\left(10^2 = 100\right)$.

$$\frac{1}{50} = \frac{1}{5^2 \times 2} \times \frac{2}{2} = \frac{1 \times 2}{(5 \times 2)^2} = \frac{2}{10^2}$$

3. Write the result as the decimal fraction 2 hundredths.

$$\frac{1}{50} = \frac{2}{100} = 0.02$$

b. 1. The denominator cannot be written as powers of 2 and 5. The decimal will be recurring. Use division to determine the decimal equivalent.

b. $9\overline{)\,7.{}^{7}0{}^{7}0{}^{7}0}$ = 0. 7 7...

2. Write the result as a recurring decimal.

$$\frac{7}{9} = 0.\dot{7}$$

c. 1. The denominator cannot be written as powers of 2 and 5. The decimal will be recurring. Use division to determine the decimal equivalent.

c. $11\overline{)\,5.{}^{5}0{}^{6}0{}^{5}0{}^{6}0{}^{5}0}$ = 0. 4 5 4 5...

2. Write the result as a recurring decimal.

$$\frac{5}{11} = 0.\dot{4}\dot{5} \text{ (or } 0.\overline{45}\text{)}$$

TI | THINK	DISPLAY/WRITE	CASIO | THINK	DISPLAY/WRITE
c. On a Calculator page, complete the entry line as $\frac{5}{11}$. Press MENU then select: 2 Number 1 Convert to Decimal Then press ENTER. Alternatively, complete the entry line as $\frac{5}{11}$ and then press ctrl ENTER.	 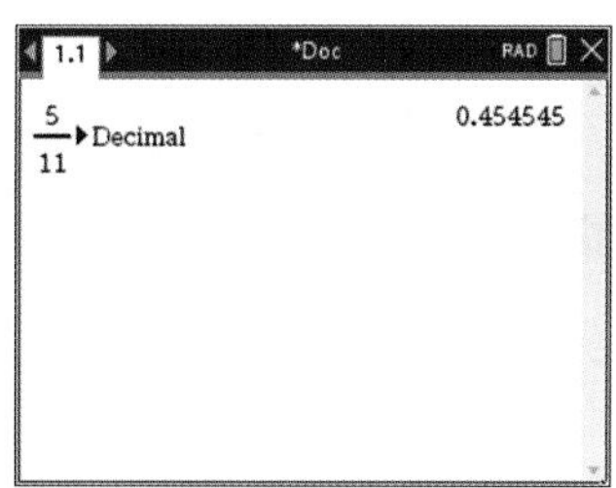	**c.** On a Main screen, complete the entry line as $\frac{5}{11}$ and click on the fraction/decimal conversion button in the top left corner.	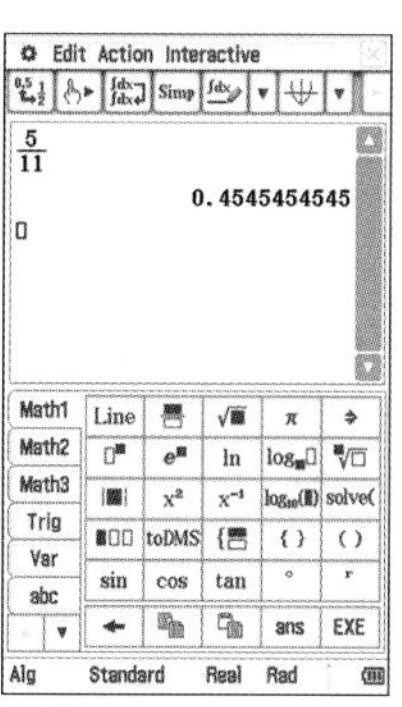

Expressing decimals as fractions

The fraction 0.04 can be expressed as $\frac{4}{100}$. This simplifies to $\frac{1}{25}$.

For recurring decimals, the process is demonstrated in the following example.

WORKED EXAMPLE 3 Expressing recurring decimals as fractions

Express the following recurring decimals as common fractions.

a. $\mathbf{0.\dot{2}}$

b. $\mathbf{3.2\dot{1}\dot{5}}$

THINK

a. 1. Let variable x equal the recurring decimal.

2. $10x$ and x will have identical recurring decimals.

WRITE

a. Let $x = 0.\dot{2}$
$= 0.2222\ldots$

$10x = 2.2222\ldots$

3. Subtracting x from $10x$ will result in a whole number.

$$10x - x = 2.\dot{2} - 0.\dot{2}$$
$$9x = 2$$

4. Solve for x.

$$x = \frac{2}{9}$$

5. Write your concluding statement.

$$0.\dot{2} = \frac{2}{9}$$

b. 1. Let variable x equal the recurring decimal.

b. Let $x = 3.2\dot{1}\dot{5}$.

2. Calculating $10x$ will result in a number with only the recurring part after the decimal.

$$10x = 32.\dot{1}\dot{5}$$

3. $10x$ and $1000x$ will have identical recurring decimals.

$$1000x = 3215.\dot{1}\dot{5}$$

4. Subtracting $10x$ from $1000x$ will result in a whole number.

$$1000x - 10x = 3215.\dot{1}\dot{5} - 32.\dot{1}\dot{5}$$
$$990x = 3183$$

5. Solve for x and simplify.

$$x = \frac{3183}{990}$$
$$= \frac{1061}{330}$$

6. Write your concluding statement.

$$3.2\dot{1}\dot{5} = \frac{1061}{330} \text{ or}$$
$$3.2\dot{1}\dot{5} = 3\frac{71}{330}$$

TI | THINK	DISPLAY/WRITE	CASIO | THINK	DISPLAY/WRITE
c. On a Calculator page, complete the entry line as 3.2151515151515. Press MENU then select: 2 Number 2 Approximate to Fraction Then press ENTER.	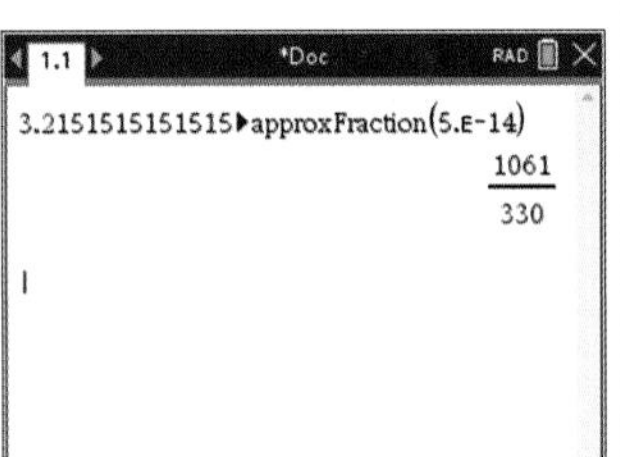	c. On a Main screen, complete the entry line as 3.2151515151515 and click on the fraction/decimal conversion button in the top left corner.	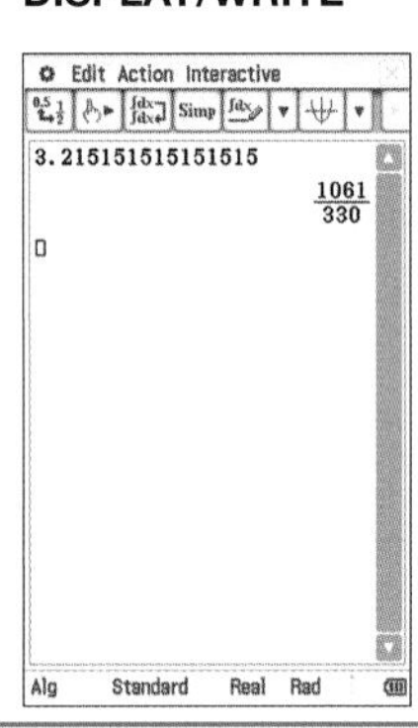

4.2.3 Number systems

The decimal number system

In the decimal number system, or base 10, the place value or positions of the digits is extremely important. Digits to the left of the decimal point are represented by ones, tens, hundreds, thousands, and so on, or expressed as powers of 10, 10^0, 10^1, 10^2, 10^3, and so on. For example

$$\begin{aligned}473 &= 400 + 70 + 3 \\ &= 4 \times 100 + 7 \times 10 + 3 \times 1 \\ &= 4 \times 10^2 + 7 \times 10^1 + 3 \times 10^0\end{aligned}$$

The binary number system

The binary number system is a base 2 system, which means it uses only 2 digits, 0 and 1. The place values in the binary number system correspond to powers of 2.

The first place value corresponds to $2^0 = 1$, the second corresponds to $2^1 = 2$, the third corresponds to $2^2 = 4$ and so on, with each place value corresponding to the next power of 2.

Powers of 2

2^{10}	2^9	2^8	2^7	2^6	2^5	2^4	2^3	2^2	2^1	2^0
1024	512	256	128	64	32	16	8	4	2	1

A binary number is equivalent to the sum of the powers of 2 that are represented by the digit 1. The digits 0 are equivalent to 0 times the power of 2 that the place value corresponds to.

For example, the binary number 10011 is equal to:

$$\begin{aligned} 1\times 2^4 + 0\times 2^3 + 0\times 2^2 + 1\times 2^1 + 1\times 2^0 &= 2^4 + 2^1 + 2^0 \\ &= 16 + 2 + 1 \\ &= 19 \end{aligned}$$

The table below shows the first 9 binary numbers, along with their decimal equivalent

Binary	Decimal
0	0
1	1
10	2
11	3
100	4
101	5
110	6
111	7
1000	8

Note that there are some situations where it may not be obvious whether a number is written in the binary or decimal number system. To counter this, we will use a subscript 10 or 2 at the end of each number to indicate which number system it corresponds to. For example, 100_{10} represents 100 in decimal and 100_2 represents 100 in binary (which is equivalent to 4 in the decimal system).

Converting decimal numbers to binary

To convert a number in decimal to binary we use successive division by 2. The remainder of each division is retained as a bit of the binary number with the first remainder represented as the least significant digit (rightmost), and the last remainder as the most significant digit (leftmost).

A nice way to visualise this is using a table. Consider the decimal number 57. Each row we divide the number by 2 and note the remainder. The process is repeated until the number reaches 0. The binary number is then the sequence of remainders, in reverse order ($57_{10} = 111001_2$).

Number	Remainder
57	
28	1
14	0
7	0
3	1
1	1
0	1

Binary digits from bottom to top

WORKED EXAMPLE 4 Converting between binary and decimal numbers

a. **Convert 11110011_2 to decimal.**
b. **Convert 178_{10} to binary.**

THINK	WRITE
a. 1. Write the binary number as powers of 2.	a. 11110011_2 $= 1\times2^7 + 1\times2^6 + 1\times2^5 + 1\times2^4 + 0\times2^3 + 0\times2^2 + 1\times2^1 + 1\times2^0$
2. Simplify the expression to remove the terms corresponding to 0's and then evaluate.	$= 2^7 + 2^6 + 2^5 + 2^4 + 2^1 + 2^0$ $= 128 + 64 + 32 + 16 + 2 + 1$ $= 243_{10}$
b. 1. Divide by 178 by 2	b. $\frac{178}{2} = 89$ and remainder 0
2. Divide by 89 by 2	$\frac{89}{2} = 44$ and remainder 1
3. Divide by 44 by 2	$\frac{44}{2} = 22$ and remainder 0
4. Divide by 22 by 2	$\frac{22}{2} = 11$ and remainder 0
5. Divide by 11 by 2	$\frac{11}{2} = 5$ and remainder 1
6. Divide by 5 by 2	$\frac{5}{2} = 2$ and remainder 1
7. Divide by 2 by 2	$\frac{2}{2} = 1$ and remainder 0
8. Divide by 1 by 2	$\frac{1}{2} = 0$ and remainder 1
9. Alternatively, continually dividing by 2 until zero is reached, read the remainders bottom up.	(see table below)

Number	Remainder
178	
89	0
44	1
22	0
11	0
5	1
2	1
1	0
0	1

The reverse sequence of remainders is 10110010.

10. State the result.	$178_{10} = 10110010_2$

4.2.4 Set notation and operations

In the previous topic, we learnt about set notation and operations. In this topic, we will apply these principles to make statements, deductions and proofs about sets of numbers.

The following table summarises the set notation and operations that were discussed in topic 3.

Or

A or B is true when A is true, or when B is true, or when both A and B are true. This is summarised by the truth table below. A or B can also be represented on a Venn diagram as the union, $A \cup B$.

A	B	A or B
T	T	T
T	F	T
F	T	T
F	F	F

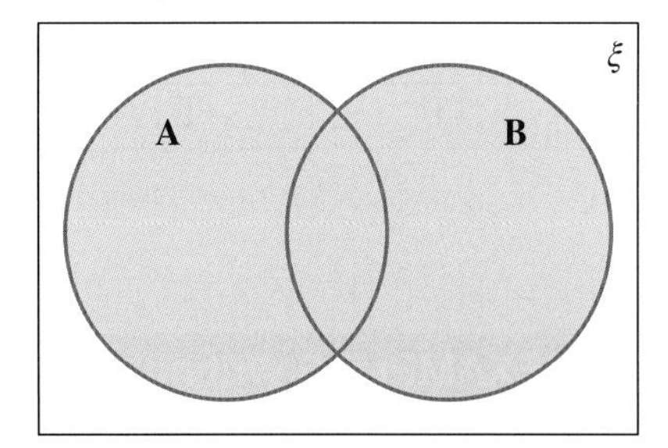

And

A and B is true only when both A and B are true. This is summarised by the truth table below. A and B can also be represented on a Venn diagram as the intersection, $A \cap B$.

A	B	A and B
T	T	T
T	F	F
F	T	F
F	F	F

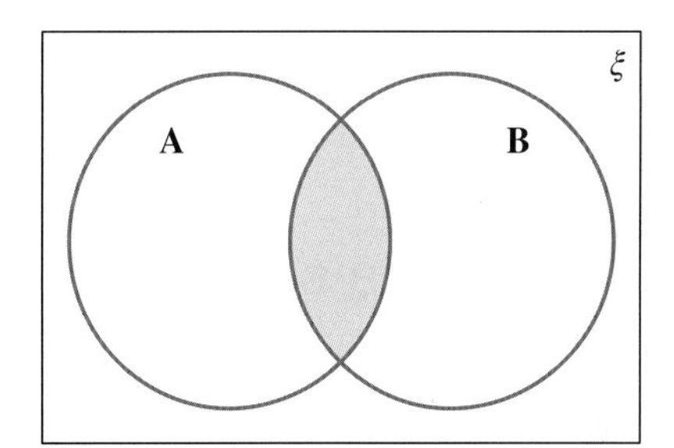

WORKED EXAMPLE 5 Classifying integers as even and/or prime

From the set of numbers {1, 2, 3, 4, 5, 6, 7, 8, 9}:
a. identify the numbers that are even or prime
b. identify the numbers that are even and prime.

THINK	WRITE
a. 1. Identify the even numbers.	a. Even numbers: {2, 4, 6, 8}
2. Identify the prime numbers.	Prime numbers: {2, 3, 5, 7}
3. Identify the numbers that are even or prime.	Even or prime numbers: {2, 3, 4, 5, 6, 7, 8}
b. Identify the numbers that are both even and prime.	b. Even and prime numbers: {2}

Negation

The negation of a statement is the opposite of the original statement. As a result, the negation of a true statement is false and the negation of a false statement is true. The negation of statement A is denoted ¬A. The truth table below summarises this. The negation can also be thought of as the complement in set theory, which is represented in a Venn diagram as A′.

A	¬A
T	F
F	T

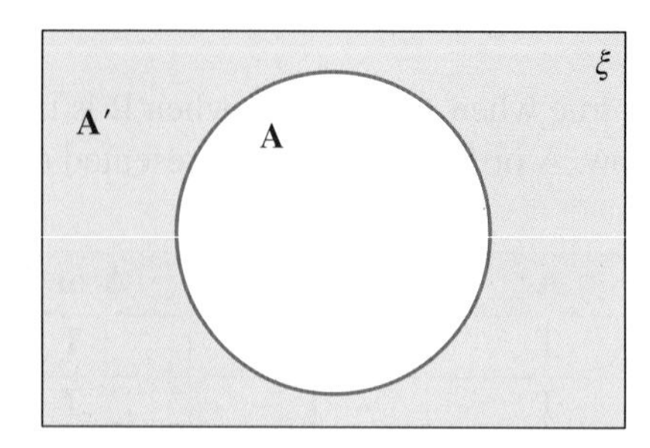

WORKED EXAMPLE 6 Writing negations of statements

Write the negation of the statement $5 > 7$.

THINK	WRITE
1. The negation of 'larger than' is 'not larger than'.	$5 \ngtr 7$
2. If a number is not larger than a number, it must be smaller than or equal to the number. This is a better way to write the statement.	$5 \leq 7$

4.2.5 Power of a set

The power of a set A is denoted by $P(\text{A})$ is defined as the set of all possible subsets including the null set and the set itself.

WORKED EXAMPLE 7 Listing the power of a set

Given the set $\text{A} = \{a, b, c\}$, list the power set.

THINK	WRITE
1. The set has 3 elements. List all sets with only one element from the set.	$n(\text{A}) = 3$ $\{a\}, \{b\}, \{c\}$
2. List all sets with two element from the set.	$\{a, b\}, \{b, c\}, \{a, c\}$
3. List the power set, including $\phi = \{\}$ the null set and the set itself.	$P(\text{A}) = \{\{a\}, \{b\}, \{c\}, \{a, b\}, \{b, c\}, \{a, c\}, \{a, b, c\}, \phi\}$

4.2.6 Set difference

The set difference of two sets A and B is denoted by A\B (the backslash) or in some books as A − B and is defined by $A \backslash B = \{x \in A, x \notin B\}$. This is shown visually for sets A and B below.

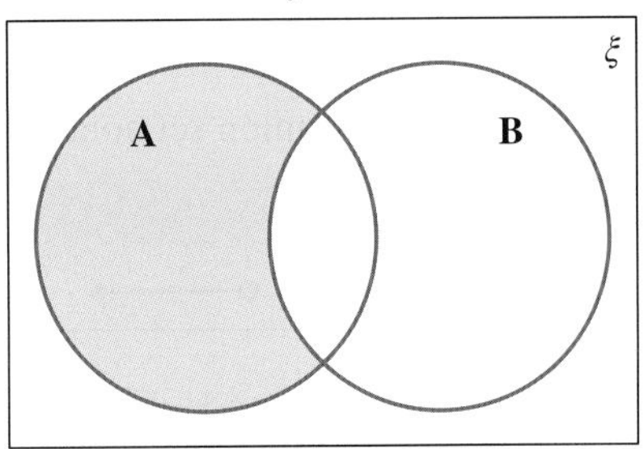

WORKED EXAMPLE 8 Listing the difference between two sets

Given the sets $A = \{1, 2, 3, 4, 5\}$ and $B = \{2, 4, 6, 8\}$, list the sets

a. **A\B**
b. **B\A**

THINK	WRITE
a. List the elements that in the set A which are not in the set B.	$A \backslash B = \{1, 3, 5\}$
b. List the elements that in the set B which are not in the set A.	$B \backslash A = \{6, 8\}$

4.2.7 Interval notation

Many sets will contain elements which are real numbers. In these situations, we can show the set on a number line using line segments to represent intervals. The endpoints of the line segments can either be included in the set or not included. To show that an endpoint is included in an interval we use a closed dot, whereas to show that an endpoint is not included in an interval, we use an open dot.

The following number lines show the four possible scenarios for intervals with open or closed endpoints.

The open interval: $\{x \in R : a < x < b\} = (a, b)$

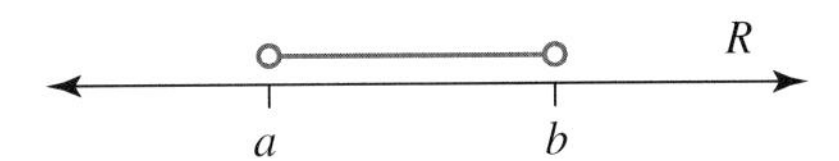

The closed interval $\{x \in R : a \le x \le b\} = [a, b]$

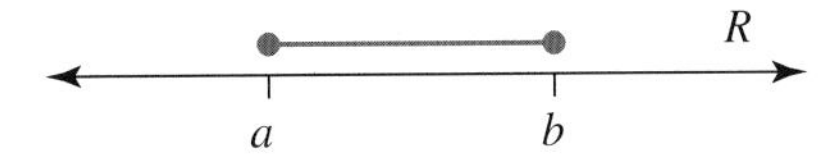

A half-open interval: $\{x \in R : a < x \le b\} = (a, b]$

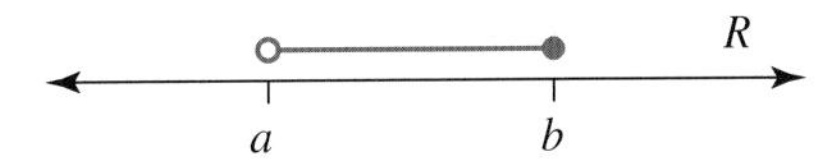

A half-open interval: $\{x \in R : a \leq x < b\} = [a, b)$

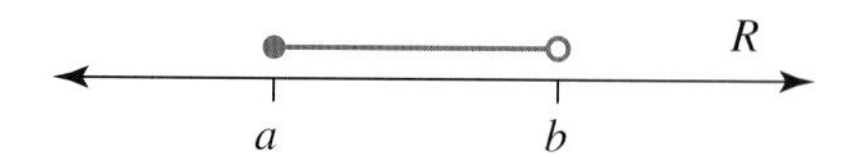

Sets can be made up of more than one interval using the union set notation. For example, the set shown on the number line below can be expressed as $(a, b) \cup (c, d]$.

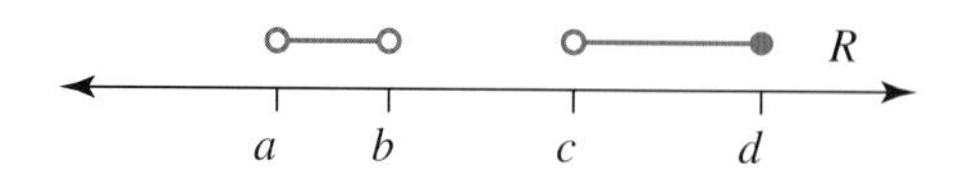

WORKED EXAMPLE 9 Expressing sets of real numbers in interval notation

Express the following sets using interval notation.

a. $R \backslash \{0, 1\}$

b. $R \backslash (0, 1)$

THINK	WRITE
a. The set of all real numbers except for the elements of the set, that is all real numbers except for 0 and 1.	$R \backslash \{0, 1\}$ $= (-\infty, 0) \cup (0, 1) \cup (1, \infty)$
b. The set of all real numbers, not including the real numbers in the interval between zero and one.	$R \backslash (0, 1)$ $= (-\infty, 0] \cup [1, \infty)$

4.2.8 Quantifiers

The proposition $9 > 7$ is a true statement, and the proposition $4 > 7$ is a false statement. However, $x > 7$ might be true or false depending on the value of x. A proposition that includes variables is known as a **propositional function**. Because it is a function, we can use function notation to name the function. In this case, we might call it P(x).

Quantifiers are used to give information about the values of the variables in propositional functions so that we can determine if the function is true or false.

The universal quantifier

The **universal quantifier,** *for all*, is written with the symbol $\forall$. This means that all possible values for the variable are considered.

Consider the propositional function $x > 7$. Using the universal quantifier, the statement becomes 'For all real numbers x, $x > 7$'. Written with symbols, the proposition is $\forall x \in R,\ x > 7$. As the proposition is not true for all possible values of x, it is a false statement.

WORKED EXAMPLE 10 Propositions using the universal quantifier

Consider the proposition $\forall x \in R,\ x^2 \geq 0$. Write the proposition in words and determine if it is true or false.

THINK	WRITE
1. The symbol $\forall$ means 'for all'. $x \in R$ means x is a member of the set of real numbers.	For all real numbers x, $x^2 \geq 0$.
2. If $x \geq 0,\ x^2 \geq 0$, and if $x < 0,\ x^2 > 0$. Therefore, the proposition is always true.	The proposition is true.

The existential quantifier

The **existential quantifier**, *there exists*, is written with the symbol $\exists$. This means that there is a value for the variable that would make the propositional function true.

Let us again consider the propositional function $x > 7$, but this time using the existential quantifier. The statement becomes 'There exists a real number, x, where $x > 7$'. Written in symbols, the proposition is $\exists x \in R,\ x > 7$. In this case, because there are values of x that make the proposition true, it is true.

WORKED EXAMPLE 11 Propositions using the existential quantifier

Consider the proposition $\exists x \in N,\ 2x + 1$ is a multiple of 3. Write the proposition in words and determine if it is true or false.

THINK	WRITE
The symbol $\exists$ means 'there exists'. $x \in N$ means x is a member of the set of natural numbers.	There exists a natural number x where $2x + 1$ is a multiple of 3. When $x = 1,\ 2x + 1 = 2 + 1 = 3$, which is a multiple of 3.
Try a simple case, such as $x = 1$.	The proposition is true for $x = 1$ and is therefore true.

Negating quantifiers

Consider the statement 'For all even numbers, x, x^2 is even'. This is a true statement. However, it would not be true if you could find one even number, x, where x^2 was not even.

Now consider the false statement 'All prime numbers are odd'. This can be rewritten as 'For all prime numbers, p, p is odd.' The negation of this false statement is the true statement 'There exists a prime number, p, where p is not odd'.

The negation of the statement $\forall x \in X,\ p(x)$ is the statement $\exists x \in X,\ \neg p(x)$

This also means that the negation of the statement $\exists x \in X,\ p(x)$ is the statement $\forall x \in X,\ \neg p(x)$.

Combining quantifiers

If a propositional function has more than one variable, then quantifiers are needed for all of the variables. If the quantifiers are the same type, the order that they are written in is immaterial. Also, if a quantifier is shown before a group of variables separated by commas, the quantifier applies to all of the variables in the group. For example, the following statements are equivalent.

$$\forall x \in R,\ \forall y \in R$$

$$\forall y \in R,\ \forall x \in R$$

$$\forall x,\ y \in R$$

If the quantifiers are different, then the order becomes important. The statement $\forall x,\ \exists y$ means that for each possible value of x, there exists a value for y that makes a proposition true. The statement $\exists x,\ \forall y$ means that there is a value for x that makes the statement true for any possible value of y.

WORKED EXAMPLE 12 Deciding whether a proposition involving quantifiers is true

For each of the propositions below, write the proposition in words and determine if it is true or false.

a. $\forall x \in R, \exists y \in R, y > x^2$

b. $\exists x \in R, \forall y \in R, y > x^2$

THINK	WRITE
a. 1. $\forall x$, $\exists y$ means that for each possible value of x, there exists a value for y that makes a proposition true.	a. For all real numbers x, there exists a real number y where $y > x^2$.
2. For every real number, there is another real number greater than the square of the original. Therefore, the proposition is true.	True
b. 1. $\exists x$, $\forall y$ means that there is a value for x that makes the statement true for any possible value of y.	b. There exists a real number x where for all real numbers y, $y > x^2$.
2. If y is a negative number, the inequality is never true. Therefore, the proposition is false.	False

4.2 Exercise

Students, these questions are even better in jacPLUS

Receive immediate feedback and access sample responses

Access additional questions

Track your results and progress

Find all this and MORE in jacPLUS

Technology free

1. WE1 Identify the following numbers as either rational or irrational.

   a. 6.2 b. 0.321 c. 0.25 d. 3.121 221 222 ...
   e. 156 f. $\sqrt{13}$

2. Identify the following numbers as either rational or irrational.

   a. $\sqrt{7}$ b. $\sqrt[3]{9}$ c. $\sqrt{81}$ d. π

3. WE2 Express the following numbers as decimals.

   a. $\frac{17}{20}$ b. $\frac{107}{125}$ c. $\frac{2}{3}$ d. $\frac{8}{15}$

4. Express the following numbers as decimals.

   a. $1\frac{3}{5}$ b. $\frac{9}{16}$ c. $\frac{22}{7}$ d. $\frac{13}{12}$

5. WE3 Express the following recurring decimals as common fractions.

   a. $0.\dot{5}$ b. $0.\dot{2}\dot{3}$ c. $3.0\dot{1}$ d. $6.\dot{1}\dot{6}$
   e. $7.1\dot{1}\dot{2}$ f. $0.\dot{7}0\dot{2}$

6. a. WE4 Convert 101011001_2 to decimal.
   b. Convert 183_{10} to binary.

7. a. Convert 111001000_2 to decimal.
   b. Convert 297_{10} to binary.

8. a. Convert 11000110_2 to decimal.
   b. Convert 154_{10} to binary.

9 **WE5** From the set of numbers {0, 3, 6, 9, 12, 15, 18, 21, 24}:
   a. identify the numbers that are odd or less than 12
   b. identify the numbers that are odd and less than12.

10 From the set of numbers {11, 22, 33, 44, 55, 66, 67, 68}:
   a. identify the numbers that are even or a multiple of 3
   b. identify the numbers that are even and a multiple of 3.

11. **WE6** Write the negation of the statement $11 < 20$.

12. Write the negation of the statement $40 \geq 39$.

13. **WE7** Given the set $A = \{2, 4, 6\}$ list the power set.

14. Given the set $A = \{a, b, c, d\}$ list the power set.

15. **WE8** Given the sets $A = \{1, 2, 3, 4, 5, 7, 8, 9, 10\}$ and $B = \{1, 3, 5, 7, 9, 11, 13, 15\}$, list the sets
   a. $A \backslash B$
   b. $B \backslash A$

16. Consider the everything set of natural numbers between 1 and 20 inclusive. Let A be set of numbers divisible by 3 and B be the set of numbers divisible by 4. List the sets:
   a. $A \backslash B$
   b. $B \backslash A$

17. **WE9** Express the following sets using interval notation.
   a. $R \backslash \{-1, 1\}$
   b. $R \backslash (-1, 1)$

18. Express the following sets using interval notation.
   a. $R \backslash \{-2, 3\}$
   b. $R \backslash [-2, 3]$

19. **WE10** For each of the propositions below, write the proposition in words and determine if it is true or false.
   a. $\forall x \in N,\ 2x$ is even.
   b. $\forall x \in N,\ 2x + 1$ is a multiple of 3.
   c. $\forall x \in N,\ x > 0$.

20. For each of the propositions below, write the proposition in words and determine if it is true or false.
   a. $\forall x \in R,\ x > 0$ or $x \leq 0$.
   b. $\forall x \in N,\ x$ is even.
   c. $\forall x \in R,\ x^2 + 1 \geq 0$.

21. **WE11** For each of the propositions below, write the proposition in words and determine if it is true or false.
   a. $\exists x \in Z,\ x + 5 = 7$.
   b. $\exists x \in R,\ x^2 < 0$.
   c. $\exists x \in N,\ x$ is even.

22. For each of the propositions below, write the proposition in words and determine if it is true or false.
   a. $\exists x \in R,\ x^2 + 1 = 0$.
   b. $\exists x \in R,\ x^2 - 1 = 0$.
   c. $\exists x \in Q,\ \sqrt{x} \in Q$.

23. **WE12** For each of the propositions below, write the proposition in words and determine if it is true or false.
   a. $\forall x \in R,\ \exists y \in R,\ y = x$.
   b. $\exists x \in R,\ \forall y \in R,\ y = x$.

24. For each of the propositions below, write the proposition in words and determine if it is true or false.
   a. $\forall x \in R,\ \exists y \in R,\ xy = 0$.
   b. $\exists x \in R,\ \forall y \in R,\ xy = 0$.

25. For each of the propositions below, determine if it is true or false.
   a. $\forall x \in R,\ \exists y \in R,\ x > y$.
   b. $\exists x \in R,\ \forall y \in R,\ x > y$.
   c. $\forall x \in R,\ \forall y \in R,\ x > y$.

4.2 Exam questions

Question 1 (1 mark) TECH-ACTIVE

MC The recurring decimal $0.\dot{2}0\dot{9}$ expressed as a fraction in the form $\frac{a}{b}$ is

A. $\frac{209}{1009}$
B. $\frac{209}{999}$
C. $\frac{209}{990}$
D. $\frac{209}{909}$
E. $\frac{29}{101}$

Question 2 (1 mark) TECH-FREE

Determine if the proposition $\forall x \in R, \forall y \in (x, \infty), x < y$ is true or false.

Question 3 (2 marks) TECH-FREE

Consider the statement '$\forall x \in N, x(x+1)$ is even'. Decide if this is a true statement and justify your decision.

More exam questions are available online.

4.3 Direct and indirect methods of proof

LEARNING INTENTION

At the end of this subtopic you should be able to:

- prove simple implications using a direct proof
- prove and disprove statements using indirect methods such as counter example, the contrapositive and contradiction.

4.3.1 Direct proofs

When trying to prove a statement directly you might think that we can simply show that the statement is true for a few randomly selected values, however regardless of how many cases a statement is shown to be true for, this does not prove that the statement is true in general.

A direct proof is a common method used to prove statements of the form 'if p then q' or 'p implies q', which is written mathematically as $p \rightarrow q$. The truth table below shows the possible truth values for the implication statement $p \rightarrow q$.

p	q	$p \rightarrow q$
T	T	T
T	F	F
F	T	T
F	F	T

The method of a direct proof is to take a statement p, which we assume to be true, and use it directly to show that q is true. Since we assume that p is true, only rows 1 and 2 of the table concern us, so for $p \rightarrow q$ to be true, all that is required is to show that q is true (as in row 1 of the above truth table).

Direct proof method

1. **Identify the statements p and q and assume that p is true.**
2. **Use the fact that p is true to directly show that q is true.**
3. **Therefore $p \rightarrow q$ is true. This completes the proof.**

Before we can start using direct proof to prove mathematical statements, we must know the exact definitions of the properties that we are wishing to prove.

Odd and even numbers

Definitions of odd and even numbers

An even number is an integer n such that $n = 2k$, where k is an integer.

An odd number is an integer n such that $n = 2k + 1$, where k is an integer.

For example, consider the number 14. Since $14 = 2 \times 7$ (and 7 is an integer), 14 is even. Alternatively, consider the number 15. Since $15 = 2 \times 7 + 1$ (and 7 is an integer), 15 is odd. Note that 15 is not even since there is no integer k such that $2k = 15$

All integers are either odd or even, so we could have defined odd numbers as integers which are not even.

WORKED EXAMPLE 13 Direct proof

Prove that the sum of two even integers is an even integer.

THINK	WRITE
1. Write an equivalent statement to be proved. p: a and b are even integers. q: their sum $a + b$ is even.	Proof: If a and b are even integers, then their sum $a + b$ is even.
2. Assume that p is true, that is a and b are two even integers. (Note that it would be incorrect to state $b = 2j$ as that would imply $a = b$.)	Let $a = 2j$ and $b = 2k$ where $j, k \in Z$.
3. Express the sum of a and b using the expressions formed in step 2. Recall that the sum of any two integers is an integer.	$a + b = 2j + 2k$ $= 2(j + k)$ $= 2i$ Where $i = j + k \in Z$
4. Use the definition of an even number to show that q is true.	$a + b = 2i,\ i \in Z$ Therefore $a + b$ is even.
5. We have shown that $p \rightarrow q$. State the conclusion.	The sum of two even integers is an even integer.

4.3.2 Indirect methods of proof

When attempting to construct a proof, sometimes the direct method is not the most efficient. Thankfully there are a variety of indirect methods of proof which can be applied in these situations.

Disproof by example/proof by counter example

Proofs in mathematics require more than just lots of examples to demonstrate that a conjecture is correct. Examples can help to convince us that a claim is likely to be true, but they cannot prove that a statement is true.

A single example is all that is needed to show that a conjecture is false. This is called a **counter example**.

To prove that $p \rightarrow q$ is false, it is necessary to find an example where p is true and q is not true.

WORKED EXAMPLE 14 Using a counter example to disprove a conjecture

Identify a counter example to disprove the conjecture 'All primes are odd numbers.'

THINK	WRITE
1. We need to think of an even number that is also prime.	The number 2 is both prime and even.
2. Write a concluding a statement.	Not all primes are odd numbers.

Contrapositive

The **contrapositive** of the conjecture 'If p, then q' is 'If not q, then not p.' Notice that both are p and q are negated and the order is reversed. The contrapositive of a true statement is also true.

Consider the true proposition 'If x is even, then $x + 1$ is odd.'

The contrapositive would be 'If $x + 1$ is not odd, then x is not even' or 'If $x + 1$ is even, then x is odd'. Notice that these propositions are also true.

Previously, we used a Venn diagram to demonstrate what $p \rightarrow q$ looks like when it is true. You can see on the diagram that if not q is true, then not p will also be true. Using symbols, if $p \rightarrow q$ is true, then $\neg q \rightarrow \neg p$ is also true.

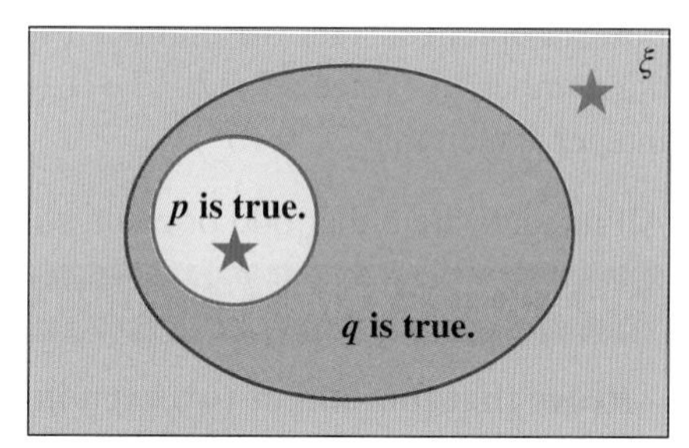

If $p \rightarrow q$ is true, then $\neg q \rightarrow \neg p$ is also true.
★ $p \rightarrow q$ is true.
★ q is not true $\rightarrow$ p is not true.

This can also be seen by comparing truth tables. $p \rightarrow q$ has the same truth table as $\neg q \rightarrow \neg p$.

We know that if a proposition is true, the contrapositive is also true. Therefore, if we prove that the contrapositive is true, the proposition is also true.

p	q	$p \rightarrow q$
T	T	T
T	F	F
F	T	T
F	F	T

p	q	$\neg q$	$\neg p$	$\neg q \rightarrow \neg p$
T	T	F	F	T
T	F	T	F	F
F	T	F	T	T
F	F	T	T	T

WORKED EXAMPLE 15 Proof using the contrapositive

Consider the statement 'If $n + m$ is odd, then $n \neq m$'.
a. Identify the propositions and write the contrapositive statement.
b. Prove the statement by proving the contrapositive.

THINK

a. Identify the propositions and their negatives. Write the contrapositive by negating both statements and reversing the order.

WRITE

a. p: $n + m$ is odd
$\neg p$: $n + m$ is even
q: $n \neq m$
$\neg q$: $n = m$
Contrapositive: If $n = m$, then $n + m$ is even.

THINK

b. 1. If $n = m$, then substitution can be used.

WRITE

b. If $n = m$, then
$$\begin{aligned} n + m &= n + n \\ &= 2n \end{aligned}$$

THINK

2. Write concluding statements.

WRITE

If $n = m$, then $n + m$ is even.
Therefore, if $n + m$ is odd, then $n \neq m$.

Proof by contradiction

To prove something by **contradiction**, we assume that what we want to prove is not true. The proof continues until the initial assumption is contradicted. As the initial assumption must be false, the alternative is true.

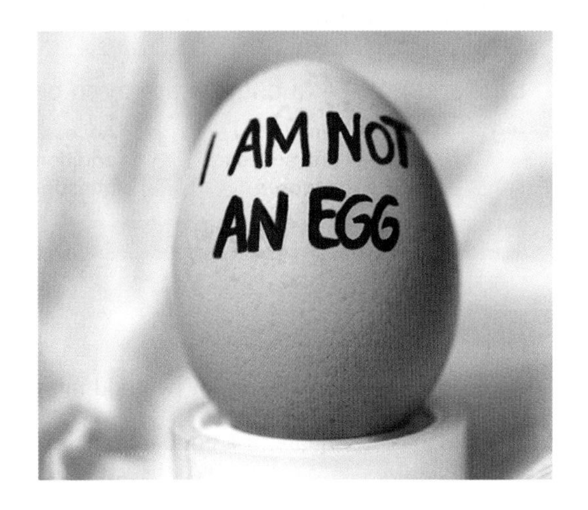

If you were trying to prove that $p \rightarrow q$ is true by contradiction, you would try to show the following:

1. Assume the proposition to be proved is false, that is, $p \rightarrow q$ is not true (if p is true and q is not true).
2. Show that the proposition contradicts the initial assumptions, that is, show that if p is true, q is true.
3. Conclude that as the solution does not meet the initial assumptions, the original assumption must be false, and hence the proposition is true: $p \rightarrow q$.

WORKED EXAMPLE 16 Proof by contradiction

Use proof by contradiction to prove the proposition 'There are no positive integer solutions to the Diophantine equation $x^2 - y^2 = 1$.'
***Note:* This could also have been written as 'If x and y are positive integers, then $x^2 - y^2 \neq 1$.'**

THINK	WRITE
1. Assume that the proposition is not true. This means that it is possible a solution exists where x and y are both positive integers.	Assume that $x^2 - y^2 = 1$, x, $y \in Z^+$.
2. Factorise the equation.	$(x+y)(x-y) = 1$
3. As x, $y \in Z^+$, $x+y$ is a positive integer and $x-y$ is an integer.	Two integers, $x+y$ and $x-y$, multiply to give 1. Both integers must equal 1 or -1. But, if x, $y \in Z^+$, then $x+y \in Z^+$. Therefore, $x+y \neq -1$. Thus, $x+y=1$ and $x-y=1$.
4. Use simultaneous equations solve the equations.	$x+y=1$ [1] $x-y=1$ [2] [1] + [2] : $2x = 2$ $x = 1$ Substituting $x=1$ into [1] results in $y=0$.
5. The initial assumption was that x, $y \in Z^+$; $y=0$ does not fit with this assumption.	$y=0$ contradicts the initial assumption that $y \in Z^+$.
6. As the only solution does not meet the original assumption, the original assumption must be false.	There are no positive integer solutions to the equation $x^2 - y^2 = 1$.

4.3 Exercise

Students, these questions are even better in jacPLUS

Receive immediate feedback and access sample responses

Access additional questions

Track your results and progress

Find all this and MORE in jacPLUS

Technology free

1. WE13 Prove that the sum of two odd integers is an even integer.
2. Prove that the square of an odd number is odd.
3. In the following list, prime numbers are marked in red.

1	2	3	4	5	6	7	8	9	10
11	12	13	14	15	16	17	18	19	20
21	22	23	24	25	26	27	28	29	30

   From the list, select counter examples to disprove each of the following statements.

   a. All numbers of the form $6n + 1$ are prime, $n \in N$.
   b. All numbers of the form $2^n + 1$ are prime, $n \in N$.

4. WE14 Identify counter examples to disprove each of the following conjectures.
   a. If $x^2 = 100$, then $x = 10$.
   b. If a quadrilateral has four congruent sides, then it is a square.
   c. If n is a multiple of 3, then it is odd.
   d. If n is a multiple of 5, then the last digit is 5.
5. Write the contrapositive statement to each of the following propositions.
   a. If today is Monday, then tomorrow is Tuesday.
   b. If A bisects a line segment, then A is the midpoint of the line segment.
6. If proof by contradiction was being used to justify that $\sqrt{ab} \leq \frac{a+b}{2}$, state what would the initial assumption be would be.
7. WE15 Consider the statement 'If $n > m$, then $n - m > 0$'.
   a. Write the contrapositive statement.
   b. Prove the statement by proving the contrapositive.
8. At the beginning of a proof by contradiction 'that x is a positive number', Ailsa wrote 'Assume x is a negative number.' State what Ailsa has overlooked.
9. WE16 Use proof by contradiction to prove the proposition 'There are no integer solutions to the Diophantine equation $4x^2 - y^2 = 1$'.
10. Use proof by contradiction to prove the proposition 'If $x, y \in Z$, then $2x + 4y \neq 7$'.

11. Analyse the following proof and identify the flaw in the logic.
Proposition: All integers are the same.
Assume that there are two integers a and b that are not the same.
Therefore, $\exists c$, $a = b + c$.
Multiplying both sides by $a - b$: $a(a-b) = (b+c)(a-b)$

Expanding the brackets: $a^2 - ab = ab - b^2 + ac - bc$

$$a^2 - ab - ac = ab - b^2 - bc$$
$$a(a - b - c) = b(a - b - c)$$

Therefore, $a = b$.

This contradicts the initial assumption.
Therefore, all integers are the same.

12. Determine if the proposition 'There are no integer solutions to the equation $25 - y^2 = x^2$' is true or false. Provide appropriate evidence to justify your claim.

13. Consider the statement 'If a and b are integers, there is no solution to the equation $a + b = \frac{1}{2}$'.
    a. Write the contrapositive statement.
    b. Prove the statement by proving the contrapositive.

14. Consider the statement 'If $x + y > 5$, then either $x > 2$ or $y > 3$, $x, y \in R$'. This statement can be justified by proving the contrapositive.
    a. Write the contrapositive statement.
    b. Justify the statement by proving the contrapositive.

15. Prove the statement 'There are no integer solutions to $x^2 - y^2 = 2$'.

16. Prove that statement 'For every $x \in Q^+$, there exists $y \in Q^+$ for which $y < x$'.

4.3 Exam questions

Question 1 (2 marks) TECH-FREE

Prove by counter example that the statement 'the cube root of x^3 is $\pm x$' is false.

Question 2 (1 mark) TECH-ACTIVE

MC The contrapositive of the conjecture 'If p, then q' is

A. 'If not p, then not q'.
B. 'If not q, then p'.
C. 'If not q, then not p'.
D. 'If q, then not p'.
E. 'If q, then p'.

Question 3 (3 marks) TECH-FREE

If you were trying to prove that $M \to N$ is true by contradiction, state what steps you would take. Explain the process.

More exam questions are available online.

4.4 Proofs with rational and irrational numbers

LEARNING INTENTION

At the end of this subtopic you should be able to:

- prove statements involving rational and irrational numbers using direct and indirect methods of proof.

Earlier in this chapter, you explored ways to construct proofs. In this section, you will use the skills that you developed to write proofs with rational and irrational numbers.

4.4.1 Proofs with consecutive numbers

If a series of numbers are consecutive, they can be written as n, $n+1$, $n+2$, ...

Sometimes, it is easier to call the middle number n. so the series would be ... $n-1$, n, $n+1$.

WORKED EXAMPLE 17 Proofs with consecutive numbers

Demonstrate that the sum of three consecutive natural numbers is equal to 3 times the middle number.

THINK	WRITE
1. As the example talks about the middle number, it might be easier to let the middle number be n, where n is a natural number.	Let the middle number be n, $n \in N$. The series is $n-1$, n, $n+1$.
2. Determine the sum of the three numbers.	$\text{Sum} = (n-1)+n+(n+1)$ $= 3n$
3. Write the concluding statement.	The sum of three consecutive natural numbers is equal to 3 times the middle number.

4.4.2 Prove that a number is irrational by contradiction

A proof by contradiction can also be used to prove that a number is irrational. Begin by assuming that the number is rational and work to prove that the assumption is false.

WORKED EXAMPLE 18 Proving that a number is irrational by contradiction

Prove that the following numbers are irrational.

a. $\sqrt{2}$

b. $\log_2(5)$

THINK	WRITE
a. 1. Use a proof by contradiction and assume that $\sqrt{2}$ is rational. This means that $\sqrt{2}$ can be written as the ratio of two integers a and b, where $b \neq 0$ and a and b have no common factors.	a. Assume $\sqrt{2}$ is rational. $\therefore \sqrt{2} = \dfrac{a}{b}$, $a, b \in Z$, $\gcd(a, b) = 1$, $b \neq 0$
2. Multiply by b and square both sides.	$\sqrt{2}b = a$ [1] $2b^2 = a^2$

3. This means that 2 divides a^2 evenly and therefore 2 divides a evenly.
Note: $a \mid b$ means that a divides into b evenly, or a is a factor of b.

$\therefore 2 \mid a^2$
$\therefore 2 \mid a$

4. If 2 divides a evenly, then a is a multiple of 2.

Let $a = 2m$, $m \in Z$. [2]

5. Substitute [2] in [1] and repeat the process to show that 2 will divide b evenly.

Substitute [2] in [1]:
$2b^2 = (2m)^2$
$2b^2 = 4m^2$
$b^2 = 2m^2$
$\therefore 2 \mid b^2$
$\therefore 2 \mid b$

6. 2 is a divisor of both a and b.

2 divides both a and b.
But $\gcd(a, b) = 1$.
The assumption is false.

7. Write your concluding statement.

$\sqrt{2}$ is irrational.

b. 1. Use proof by contradiction and assume that $\log_2(5)$ is rational.

b. Assume $\log_2(5) = \frac{a}{b}$,
$a, b \in Z$, $\gcd(a, b) = 1$, $b \neq 0$.

2. Rearrange the equation.

$2^{\frac{a}{b}} = 5$

$(2^a)^{\frac{1}{b}} = 5$

$2^a = 5^b$

3. The only case in which powers of 2 and 5 are equal is $2^0 = 5^0$.

As $\gcd(2, 5) = 1$, the only solution is $a = b = 0$.
But $b \neq 0$.
The assumption is false.

4. Write your concluding statement.

$\log_2(5)$ is irrational.

4.4.3 Proofs with odd and even numbers

If a proof involves odd or even numbers, you can use the fact that for any integer n, $2n$ will be even and $2n + 1$ will be odd.

WORKED EXAMPLE 19 Proofs with odd numbers

Prove that n^2 is odd if and only if n is odd.

THINK

WRITE

1. 'If and only if' means you need to prove that n^2 is odd if n is odd and also that n is odd if n^2 is odd. Begin by assuming that n is odd. This means that it is 1 more than an even number.

If n is odd, $n = 2a + 1$, $a \in Z$.

2. Determine an expression for n^2.

$n^2 = (2a + 1)^2$
$= 4a^2 + 4a + 1$

3. Odd numbers can be expressed as 1 more than an even number.	$= 2(2a^2 + 2a) + 1$ This is an odd number.
4. Write your concluding statement.	Therefore, if n is odd, then n^2 is odd.
5. Now assume that n^2 is odd, meaning that it can be written as 1 more than an even number.	If n^2 is odd, then $n^2 = 2b + 1, b \in Z$.
6. Consider $n^2 - 1$. This expression can be factorised as the difference of two squares.	$n^2 - 1 = (n + 1)(n - 1)$
7. Substitute $n^2 = 2b + 1$.	$(2b + 1) - 1 = (n + 1)(n - 1)$ $2b = (n + 1)(n - 1)$
8. If the product of two numbers is even, then one of the numbers must be even.	Either $n + 1$ is even, which means that n is odd, or $n - 1$ is even, which also means that n is odd.
9. Write your concluding statement.	Therefore, if n^2 is odd, then n is odd. Therefore, n^2 is odd if and only if n is odd.

4.4.4 Prove that a set of numbers is infinite

A proof by contradiction can be used to prove that a set of numbers is infinite (that is, it has an infinite number of members). Begin by assuming that the set is finite and work to prove that this assumption is false.

WORKED EXAMPLE 20 Proofs that a set of numbers is infinite

Prove that there are infinitely many prime numbers.

THINK	WRITE
1. Use a proof by contradiction. Begin by assuming that there are a finite number of primes.	Assume that $p_1 ... p_n$ are the only prime numbers.
2. There must be a number that is 1 more than the product of all of the primes. Call this number q. ***Note:*** The notation $\prod_{i=1}^{n} p_i$ indicates the product of all prime numbers between 1 and n.	Let $q = \prod_{i=1}^{n} p_i + 1$.
3. If we divide q by any of the primes, the remainder is 1. This means that q is not a product of any of the primes p_i.	If we divide q by any of the primes, the remainder is 1. Therefore, q is prime, or there is a prime number greater than p_n that is a factor of q.
4. There are more prime numbers than initially assumed.	This means that the assumption is false, as there must be a prime number greater than p_n.
5. Write your concluding statement.	There are infinitely many prime numbers.

4.4.5 Other proofs with real numbers

It can be possible to prove that an expression is larger or smaller than another expression. It can also be possible to prove that an expression is not equal to another expression.

WORKED EXAMPLE 21 Proofs involving equalities and inequalities

a. **Prove that for $a, b \in R, a^2 + b^2 \geq 2ab$.**

b. **Prove that $\frac{1}{x} + \frac{1}{y} \neq \frac{1}{x+y}$ for $x, y \neq 0$.**

THINK	WRITE
a. 1. The elements a^2, b^2 and $2ab$ are part of a perfect square expansion. A perfect square is always greater than or equal to 0.	**a.** $(a-b)^2 \geq 0$
2. Expand the brackets.	$a^2 - 2ab + b^2 \geq 0$
3. Rearrange the terms to form the proof.	$\therefore\ a^2 + b^2 \geq 2ab$
b. 1. Use a proof by contradiction and begin by assuming that the equation is true. Rewrite $\frac{1}{x} + \frac{1}{y}$ as a single fraction.	**b.** Assume that $\frac{1}{x} + \frac{1}{y} = \frac{1}{x+y}$ $\frac{y}{xy} + \frac{x}{xy} = \frac{1}{x+y}$ $\frac{y+x}{xy} = \frac{1}{x+y}$
2. Rearrange the equation.	$(x+y)^2 = xy$
3. $(x+y)^2$ will be positive. (It cannot equal zero as $x, y \neq 0$.)	$(x+y)^2 > 0$ $\therefore xy > 0$
4. Expanding $(x+y)^2$ to show an xy term.	$(x+y)^2 = xy$ $x^2 + 2xy + y^2 = xy$ $x^2 + y^2 = -xy$
5. $x^2 + y^2$ will be positive.	$x^2 + y^2 > 0$; therefore, $-xy > 0$ $xy < 0$
6. Identify the contradiction.	But earlier, we found that $xy > 0$.
7. State your conclusions.	Therefore, the assumption is false. $\frac{1}{x} + \frac{1}{y} \neq \frac{1}{x+y}$

4.4 Exercise

Technology free

1. WE17 Demonstrate that the sum of five consecutive natural numbers is equal to 5 times the middle number.

2. Demonstrate that if n and m are multiples of 3, then $n + m$ is a multiple of 3.

3. Demonstrate that if m, n and p are consecutive natural numbers, then $n^2 - mp = 1$.

4. An even number, n, can be written in the form $n = 2a,\ a \in Z$. Demonstrate that if m and n are even numbers, then $m^2 + n^2$ and $m^2 - n^2$ are both divisible by 4.

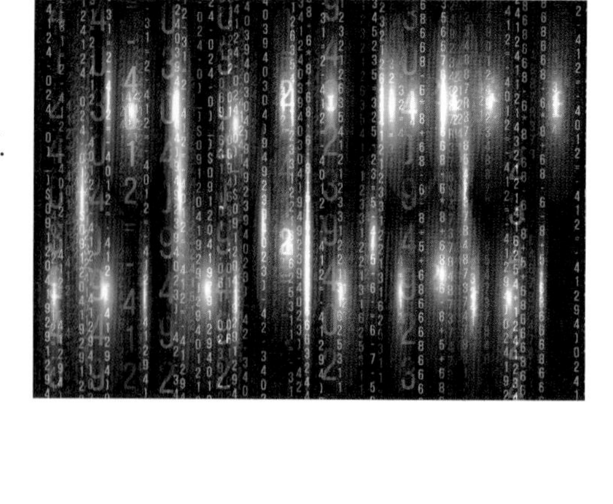

5. An odd number, n, can be written in the form $n = 2a + 1,\ a \in Z$. Demonstrate that the sum of two consecutive odd numbers is divisible by 4.

6. Consider four consecutive numbers, n_1, n_2, n_3, n_4. Demonstrate that $n_1 + n_2 + n_3 + n_4 = n_3 n_4 - n_1 n_2$.

7. WE18 Use a proof by contradiction to prove that $\sqrt{5}$ is irrational.

8. Use a proof by contradiction to prove that $\sqrt{3}$ is irrational.

9. Use a proof by contradiction for each of the following.
   a. Prove that $\log_3(7)$ is irrational.
   b. Prove that $\log_5(11)$ is irrational.

10. Prove that for any natural number n, $\dfrac{n(n+1)}{2}$ is a natural number.

11. Use your understanding of divisibility to show the following. *Note*: $a \mid b$ means that a divides into b evenly, or a is a factor of b.
    a. If $a \mid b$ and $b \mid c$, then $a \mid c$.
    b. If $a \mid b$ and $a \mid c$, then $a \mid (b + c)$.

12. WE19 Prove that n^2 is even if and only if n is even. *Hint:* When assuming that n^2 is even, consider factorising $n^2 - 1$.

13. Prove that all numbers of the form $n^3 - n,\ n \in Z$ are multiples of 6.

14. Prove that if x is rational and y is irrational, then $x + y$ is irrational.

15. WE20 Use a proof by contradiction for each of the following.
    a. Prove that there are infinitely many integers. (*Hint:* Assume that n is the largest integer.)
    b. Prove that there are infinitely many even numbers.

16. WE21 Prove that for $a > 0,\ a + \dfrac{1}{a} \geq 2$. *Hint:* Begin by noting that $(a - 1)^2 \geq 0$.

17. Prove that for $a, b \in R^+ \cup \{0\},\ \dfrac{1}{2}(a + b) \geq \sqrt{ab}$.

18. Two resistors have resistances of a ohms and b ohms. If the resistors are placed in series, the combined resistance is $R_S = a + b$. If the resistors are placed in parallel, the combined resistance is found using $\dfrac{1}{R_P} = \dfrac{1}{a} + \dfrac{1}{b}$. Justify that $R_S \geq R_P$.
    Hint: Use a proof by contradiction.

Technology active

19. Consider $\sqrt{n}$ where n is an integer. There are two irrational numbers, $\sqrt{2}$ and $\sqrt{3}$, between the rational numbers $\sqrt{1}=1$ and $\sqrt{4}=2$. There are four irrational numbers between the next rational pair, $\sqrt{4}=2$ and $\sqrt{9}=3$. Determine a rule for the number of irrational numbers of the form $\sqrt{n}$ between consecutive integers m and $m+1$. Prove that your rule works.

20. a. Arrange the integers 1 to 200 in a table similar to the one shown, marking all the primes and multiples of 6. You may like to use a spreadsheet or similar to assist with this.

1	2	3	4	5	6	7
	8	9	10	11	12	13
	14	15	16	17	18	19
	20	21	22	23	24	25
	26	27	28	29	30	31
	32	33	34	35	36	37
				Primes are in red.	Multiples of 6	Primes are in red.

Note that with the exception of the primes 2 and 3, the other primes are all 1 less than or 1 more than a multiple of 6. This means that primes can be expressed in the form $6n \pm 1$.

b. Sophie Germain primes, p, are prime numbers where $2p+1$ is also prime. This means that 2 is a Sophie Germain prime, because 5 is also prime. Determine the other Sophie Germain primes between 1 and 200. Note that, with the exception of 2 and 3, they are all 1 less than a multiple of 6. Justify why this is so.

4.4 Exam questions

Question 1 (1 mark) TECH-ACTIVE

MC Consider four consecutive numbers, a_1, a_2, a_3, a_4. Then $a_1+a_2+a_3+a_4$ equals

A. $a_3 \times a_4 - a_1 \times a_2$
B. $a_3 \times a_4 + a_1 \times a_2$
C. $a_1 \times a_2 - a_3 \times a_4$
D. $a_1 \times a_2 + a_3 \times a_4$
E. $a_2 \times a_4 - a_1 \times a_3$

Question 2 (1 mark) TECH-ACTIVE

MC If a is a factor of b and b is a factor of c, then

A. $\frac{a}{b}$ is a factor of $\frac{b}{c}$
B. ab is a factor of $\frac{c}{b}$
C. ab is a factor of c
D. a is a factor of c
E. a^3 is a factor of bc

Question 3 (3 marks) TECH-FREE

Prove that for all $x, y \in R$, $(x+1)^2+(y-1)^2 \geq 2x-2y-2xy+2$

More exam questions are available online.

4.5 Proof by mathematical induction

4.5.1 Proof by induction basics

LEARNING INTENTION

At the end of this subtopic you should be able to:
- prove mathematical statements using induction.

Imagine a series of dominoes, placed just close enough so that if the first one topples, then it will cause the next one to fall and so on. This process is how a **proof by mathematical induction** works.

Notice that for the dominoes to topple, they firstly need to be close enough together *and* the first domino needs to fall. Without both of these steps happening, some (or all) of the dominoes will remain standing.

The first step in a proof by induction is to prove that the formula works for $n = 1$. This is the equivalent of knocking over the first domino and is known as the **initial statement**. This step is important because it is necessary to prove that the formula works at least once.

The second step is to assume that since that formula works once, the formula will also be true when $n = k$. It is then necessary to prove that the formula is also true for $n = k + 1$. This is the equivalent of having the dominoes close enough together that they will topple and is known as the **inductive step**.

If the formula is true for $n = k$ also true for $n = 1$ and if it is also true for $n = k + 1$, this means it is also true for $n = 2$. And as it is true for $n = 2$, it is also true for $n = 3$ and so on. Therefore, the formula is true for all natural numbers.

WORKED EXAMPLE 22 Proof by induction (1)

Use mathematical induction to prove that $1 + 2 + 3 + \ldots + n = \dfrac{n(n+1)}{2}$ for $n \geq 1$.
Note that this was first proven by a form of mathematical induction by the Persian mathematician al-Karaji (953–1029) in the 10th century.

THINK	WRITE
1. Verify that the formula is true for $n = 1$.	If $n = 1$, LHS $= 1$. RHS $= \dfrac{1(1+1)}{2}$ $= \dfrac{1(2)}{2}$ $= 1$ True for $n = 1$.
2. There are some values of n that make the formula true. Write down the formula for $n = k$.	Assume that it is true for $n = k$. $1 + 2 + 3 + \ldots + k = \dfrac{k(k+1)}{2}$

3. Knowing that $1+2+3+...+k=\frac{k(k+1)}{2}$, add the next term $(k+1)$ to both sides.

When $n=k+1$, then next term in the series is $k+1$.

$$1+2+3+...+k+(k+1)=\frac{k(k+1)}{2}+(k+1)$$

4. Note that if $n=k+1$, then $\frac{n(n+1)}{2}=\frac{(k+1)[(k+1)+1]}{2}$ so we are trying to demonstrate that $\frac{k(k+1)}{2}+(k+1)=\frac{(k+1)[(k+1)+1]}{2}$.

Additionally, $k+1$ is a common factor and is part of the expression that we are working towards. Factorise and then simplify.

$$=(k+1)\left(\frac{k}{2}+1\right)$$

$$=(k+1)\left(\frac{k+2}{2}\right)$$

5. Write the expression so that it is clearly the same formula as when $n=k+1$.

$$=\frac{(k+1)[(k+1)+1]}{2}$$

6. Write a statement to explain what you have demonstrated and concluded.

If the statement is true for $n=k$, it is also true for $n=k+1$. The statement is true for $n=1$. Therefore, by mathematical induction, the statement is true for $n\geq 1$.

WORKED EXAMPLE 23 Proof by induction (2)

Use mathematical induction to prove that $2+4+8+...+2^n=2^{n+1}-2$, $n\geq 1$.

THINK / **WRITE**

1. Verify that the formula is true for $n=1$.

If $n=1$,
LHS $=2^1$ $=2$ RHS $=2^{1+1}-2$ $=2$
True for $n=1$.

2. There are some values of n that make the formula true. Write down the formula for $n=k$.

Assume that it is true for $n=k$.
$2+4+8+...+2^k=2^{k+1}-2$

3. Knowing that $2+4+8+...+2^k=2^{k+1}-2$, add the next term (2^{k+1}) to both sides.

When $n=k+1$, then next term in the series is 2^{k+1}.

$2+4+8+...+2^k+2^{k+1}$

4. Note that if $n=k+1$, then $2^{n+1}-2=2^{k+1+1}-2$ so we are trying to demonstrate that $2^{k+1}-2+2^{k+1}=2^{k+1+1}-2$.
$2^{k+1}+2^{k+1}$ can be rewritten as $2\times 2^{k+1}$.

$=2^{k+1}-2+2^{k+1}$
$=2^{k+1}+2^{k+1}-2$
$=2\times 2^{k+1}-2$

5. Use index laws to write the expression so that it is clearly the same as the formula when $n=k+1$.

$=2^{k+1+1}-2$

6. Write a statement to explain what you have demonstrated and concluded.

If the statement is true for $n = k$, it is also true for $n = k + 1$. The statement is true for $n = 1$. Therefore, by mathematical induction, the statement is true for $n \geq 1$.

4.5.2 Using $\sum$ notation

An alternative way to write the equation to be proved involves using the **summation sign**, $\sum$, the Greek letter $\sum$, (pronounced 'sigma'). In Worked example 24, the expression $1 + 2 + 3 + ... + n$ was used. An alternative way to write this is $\sum_{r=1}^{n} r$, which means to add up all the r values beginning with $r = 1$ and ending with $r = n$, that is $1 + 2 + 3 + ... n$.

In the next worked example, the expression $\sum_{r=1}^{n} 3r$ is used. When $r = 1$, $3r = 3$ and when $r = 2$, $3r = 6$. Therefore, $\sum_{r=1}^{n} 3r$ can be rewritten as $3 + 6 + 9 + ... + 3n$.

WORKED EXAMPLE 24 Proof by induction (3)

Use mathematical induction to prove that $\sum_{r=1}^{n} 3r = \frac{3n(1+n)}{2}$.

THINK

1. Verify that the formula is true for $n = 1$.

WRITE

If $n = 1$,

$$\begin{aligned}\text{LHS} &= 3 \times 1\\ &= 3\\ \text{RHS} &= \frac{3(1)(1+1)}{2}\\ &= \frac{3(2)}{2}\\ &= 3\end{aligned}$$

True for $n = 1$.

2. There are some values of n that make the formula true. Write down the formula for $n = k$.

Assume that it is true for $n = k$.

$$\sum_{r=1}^{k} 3r = \frac{3k(1+k)}{2}$$

3. Finding the sum of the first $k + 1$ terms is the same as finding the sum of the first k terms and then adding term $k + 1$.
In this instance, term $k + 1$ will be $3(k + 1)$ and we know that $\sum_{r=1}^{k} 3r = \frac{3k(1+k)}{2}$.

$$\begin{aligned}\sum_{r=1}^{k+1} 3r &= \sum_{r=1}^{k} 3r + 3(k+1)\\ &= \frac{3k(1+k)}{2} + 3(k+1)\end{aligned}$$

4. If the sum is true for the first $k+1$ terms, then the sum will equal $\frac{3(k+1)[1+(k+1)]}{2}$. Additionally, $k+1$ and 3 are common factors and are part of the expression that we are working towards. Factorise and then simplify.

$$= 3(k+1)\left(\frac{k}{2}+1\right)$$
$$= 3(k+1)\left(\frac{k+2}{2}\right)$$

5. Write the expression so that it is clearly the same as the formula when $n = k+1$.

$$= \frac{3(k+1)[1+(k+1)]}{2}$$

6. Write a statement to explain what you have demonstrated and concluded.

If the statement is true for $n = k$, it is also true for $n = k+1$. The statement is true for $n = 1$. Therefore, by mathematical induction, the statement is true for $n \geq 1$.

Proof by mathematical induction procedure

1. **Show the statement is true for $n = 1$**
2. **Assume the statement is true for $n = k$**
3. **Show the statement is true for $n = k+1$**
4. **State your conclusion.**

4.5 Exercise

Students, these questions are even better in jacPLUS

Receive immediate feedback and access sample responses

Access additional questions

Track your results and progress

Find all this and MORE in jacPLUS

Technology free

1. Consider the statement $2 + 4 + 6 + ... + 2n = n(n+1),\ n \geq 1$.
   a. Demonstrate that the statement is true for the first three possible values of n.
   b. If $2 + 4 + 6 + ... + 2k = k(k+1)$, demonstrate that $2 + 4 + 6 + ... + 2k + 2(k+1) = (k+1)(k+2)$.

2. Consider the statement $\sum_{r=1}^{n} 2^{r-1} = 2^n - 1$.
   a. Demonstrate that the statement is true for the first three possible values of n.
   b. If $\sum_{r=1}^{k} 2^{r-1} = 2^k - 1$, demonstrate that $\sum_{r=1}^{k+1} 2^{r-1} = 2^{k+1} - 1$.

3. Consider the statement $3 + 5 + 7 + ... (1 + 2n) = n(n+2),\ n \geq 1$.
   a. Demonstrate that the statement is true for the first possible value of n.
   b. Assume that the statement is true for $n = k$. Document this assumption by writing a suitable mathematical equation.

c. Use the assumption from part b to demonstrate that if the statement is true for $n = k$, it will also be true for $n = k + 1$.

d. Explain how you have demonstrated that $3 + 5 + 7 + ... (1 + 2n) = n(n + 2)$.

4. Consider the statement $\sum_{r=1}^{n} (2r - 1) = n^2$.

a. Demonstrate that the statement is true for the first possible value of n.

b. Assume that the statement is true for $n = k$. Document this assumption by writing a suitable mathematical equation.

c. Use the assumption from part b to demonstrate that if the statement is true for $n = k$, it will also be true for $n = k + 1$.

d. Explain how you have demonstrated that $\sum_{r=1}^{n} (2r - 1) = n^2$.

5. WE22 Use mathematical induction to prove that $5 + 6 + 7 + ... + (n + 4) = \frac{n(n + 9)}{2}, \quad n \geq 1$.

6. WE23 Use mathematical induction to prove that $3 + 6 + 12 + ... + 3 \times 2^{n-1} = 3(2^n - 1), \quad n \geq 1$.

7. WE24 Use mathematical induction to prove that $\sum_{r=1}^{n} 2^r = 2^{n+1} - 2$.

8. Use mathematical induction to prove that $\sum_{r=1}^{n} 2r = n(n + 1)$.

9. Use mathematical induction to prove that $\sum_{r=1}^{n} \frac{1}{r(r + 1)} = 1 - \frac{1}{n + 1}$.

10. Use mathematical induction to prove that $1^2 + 3^2 + 5^2 + ... + (2n - 1)^2 = \frac{1}{3}n(2n - 1)(2n + 1)$ for $n \geq 1$.

11. Use mathematical induction to prove that $a + ar + ar^2 + ... + ar^{n-1} = \frac{a(r^n - 1)}{r - 1}, \quad n \geq 1$.

12. Use mathematical induction to prove that $\sum_{r=1}^{n} r^2 = \frac{1}{6}n(n + 1)(2n + 1)$.

13. Use mathematical induction to prove that $\sum_{r=1}^{n} r^3 = \frac{1}{4}n^2(n + 1)^2$.

14. Use mathematical induction to prove that
$a + (a + d) + (a + 2d) + ... (a + [n - 1]d) = \frac{n}{2}[2a + (n - 1)d]$, for $n \geq 1$.

4.5 Exam questions

Question 1 (1 mark) TECH-ACTIVE

MC The first step in the proof by induction to prove $1 + 2 + 3 + 4 + ... + n = \frac{n(n + 1)}{2}$, is to prove the statement is true for

A. $n > 1$ **B.** $k > 1$ **C.** $n = 1$ **D.** $n = k$ **E.** $n = k + 1$

Question 2 (1 mark) TECH-ACTIVE

MC Consider the proposition $P(n)$ that

$$1^2 + 2^2 + 3^2 + \ldots\ldots + n^2 = \frac{1}{6}n(n+1)(2n+1)$$

If $n = k + 1$, the proposition would become

A. $1^2 + 2^2 + \ldots\ldots + (k+1)^2 = \frac{1}{6}(k+1)(k+2)(2k+1)$

B. $(k+1)^2 = \frac{1}{6}(k+1)(k+2)(2k+3)$

C. $1^2 + 2^2 + \ldots\ldots + \frac{1}{6}(k+1)^2 = k(k+1)(k+2)(2k+3)$

D. $1^2 + 2^2 + \ldots\ldots + (k+1)^2 = \frac{1}{6}(k+1)(k+2)(2k+3)$

E. $1^2 + 2^2 + \ldots\ldots + k^2 + (k+1)^2 = \frac{k}{6}(k+1)(2k+2)$

Question 3 (5 marks) TECH-FREE

Assuming the triangle inequality $|a+b| \leq |a| + |b|$ for $a, b \in R$.

Use mathematical induction to prove that $|x_1 + x_2 + \ldots + x_n| \leq |x_1| + |x_2| + \ldots + |x_n|, x_1, x_2, \ldots, x_n \in R$.

More exam questions are available online.

4.6 Proof of divisibility using induction

LEARNING INTENTION

At the end of this subtopic you should be able to:

- use induction to prove that an expression involving natural numbers is divisible by a particular integer.

4.6.1 Divisibility results

Proof by induction can also be used to prove that an expression is divisible by a certain number. An equation can be found by writing the expression as a multiple of that number. This means that if $n^3 - n$ is divisible by 3, then $n^3 - n = 3m$ (where m is an integer).

WORKED EXAMPLE 25 Divisibility proofs using induction (1)

Use a proof by induction to demonstrate that $n^3 - n$ is divisible by 3, $n > 1$.

THINK	WRITE
1. $n > 1$ means that the first value of n to check is $n = 2$. Verify that the formula is true for $n = 2$.	If $n = 2, n^3 - n = 8 - 2$ $= 6$ As 6 is divisible by 3, that claim is true for $n = 2$.
2. There are some values of n that make the claim true. If a number is divisible by 3, it can be written as $3m$, write this as a mathematical equation where $n = k$.	Assume that it is true for some $n = k$. $k^3 - k = 3m, \quad m \in Z$
3. Consider the statement when $n = k + 1$.	$(k+1)^3 - (k+1) = (k^3 + 3k^2 + 3k + 1) - (k+1)$

4. Substitute $k^3 - k = 3m$ to demonstrate that the statement can be written as a multiple of 3.

$$\begin{aligned} &= k^3 + 3k^2 + 3k + 1 - k - 1 \\ &= k^3 - k + 3k^2 + 3k \\ &= 3m + 3k^2 + 3k \\ &= 3(m + k^2 + k) \end{aligned}$$

5. Write a statement to explain that the statement is true for $n = k + 1$ if it is true for $n = k$.

The statement $k^3 - k = 3m,\ m \in Z$ is divisible by 3. Therefore if $k^3 - k$ is divisible by 3, then $(k+1)^3 - (k+1)$ is divisible by 3.

6. Write the concluding statement.

$n^3 - n$ is divisible by 3 if $n = 2$. It is true for $n = k + 1$ if it is true for $n = k$. Therefore, by mathematical induction, the statement is true $\forall\ n > 1$.

4.6.2 Using divisibility notation

When proving an expression is divisible by a certain number, an alternative way to write the equation to be proved involves using | **notation**. In Worked example 27, it was required to prove (R.T.P) that $n^3 - n$ is divisible by 3. An alternative way to write this is $3|(n^3 - n)$.

In the following worked example, the notation $5|(2^{4n+1} + 3)$ is used which is equivalent to $2^{4n+1} + 3$ is divisible by 5. In general terms, $a|b$ means b is divisible by a.

WORKED EXAMPLE 26 Divisibility proofs using induction (2)

Use a proof by induction to demonstrate that $5|\left(2^{4n+1} + 3\right)$, $n \geq 1$.

THINK / **WRITE**

1. Verify that the statement is true for $n = 1$.

If $n = 1$, $\quad 2^{4n+1} + 3 = 2^5 + 3$
$$\begin{aligned} &= 32 + 3 \\ &= 35 \end{aligned}$$

As 35 is divisible by 5, the statement is true for $n = 1$.

2. There are some values of n that make the claim true. If a number is divisible by 5, it can be written as $5m$, write this as a mathematical equation where $n = k$.

Assume that it is true for some $n = k$.
$2^{4k+1} + 3 = 5m$

3. Consider the statement when $n = k + 1$.

$2^{4(k+1)+1} + 3 = 2^{4k+4+1} + 3$

4. Rewrite using $2^{4k+4+1} = 2^{4k+1} \times 2^4$ and expand 2^4. As we want$16\left(2^{4k+1} + 3\right) = 16 \times 2^{4k+1} + 16 \times 3$ it is necessary also subtract 16×3 to maintain the equality. Substitute $2^{4k+1} + 3 = 5m$ to show that the claim is also true for $n = k + 1$.

$$\begin{aligned} 2^{4k+4+1} + 3 &= 2^{4k+1} \times 2^4 + 3 \\ &= 16 \times 2^{4k+1} + 3 \\ &= 16\left(2^{4k+1} + 3\right) - 3 \times 16 + 3 \\ &= 16\left(2^{4k+1} + 3\right) - 48 + 3 \\ &= 16 \times 5m - 45 \\ &= 5(16m - 9) \end{aligned}$$

Alternatively:

If $2^{4k+1}+3=5m$ then we can substitute $2^{4k+1}=5m-3$ to show that the claim is also true for $n=k+1$.

Alternatively:

$$\begin{aligned}2^{4(k+1)+1}+3 &= 2^{4k+4+1}+3\\ &= 2^{4k+1}\times 2^4+3\\ &= (5m-3)\times 16+3\\ &= 16\times 5m-3\times 16+3\\ &= 16\times 5m-3\times 15\\ &= 5(16m-9)\end{aligned}$$

5. Write a statement to explain that the statement is true for $n=k+1$ if it is true for $n=k$.

This is divisible by 5, therefore $2^{4(k+1)+1}+3$ is divisible by 5 if $2^{4k+1}+3$ is divisible by 5.

6. Write the concluding statement.

$2^{4n+1}+3$ is divisible by 5 if $n=1$. It is true for $n=k+1$ if it is true for $n=k$. Therefore, the statement is true $\forall\ n\geq 1$.

4.6 Exercise

Technology free

1. Consider the statement that 6^n+4 is divisible by 5, $n\in Z^+ \cup \{0\}$.
   a. Demonstrate that the statement is true for the first three possible values of n.
   b. If $6^k+4=5m,\ m\in Z$, demonstrate that $6^{k+1}+4$ is divisible by 5.
2. Consider the statement that 8^n-1 is divisible by 7, $n\in Z^+$.
   a. Demonstrate that the statement is true for the first three possible values of n.
   b. If $8^k-1=7m,\ m\in Z$, demonstrate that $8^{k+1}-1$ is divisible by 7.
3. Consider the statement: $3^{2n-1}+1$ is divisible by 4, $n\in Z^+$.
   a. Demonstrate that the statement is true for the first possible value of n.
   b. Assume that the statement is true for $n=k$. Document this assumption by writing a suitable mathematical equation.
   c. Use the assumption from part b to demonstrate that if the statement is true for $n=k$, it will also be true for $n=k+1$.
   d. Explain how you have demonstrated that $3^{2n-1}+1$ is divisible by 4, $n\in Z^+$.
4. Consider the statement: 10^n-4 is divisible by 12, $n\geq 2$.
   a. Demonstrate that the statement is true for the first possible value of n.
   b. Assume that the statement is true for $n=k$. Document this assumption by writing a suitable mathematical equation.
   c. Use the assumption from part b to demonstrate that if the statement is true for $n=k$, it will also be true for $n=k+1$.
   d. Explain how you have demonstrated that 10^n-4 is divisible by 12, $n\geq 2$.

5. **WE25** Use a proof by induction to demonstrate that $n^3 + 2n$ is divisible by 3, $n \geq 1$.

6. Use a proof by induction to demonstrate that $3^{4n} - 1$ is divisible by 80, $n \geq 1$.

7. Use a proof by induction to demonstrate that $5^n + 2 \times 11^n$ is divisible by 3, $n \geq 0$.

8. **WE26** Use a proof by induction to demonstrate that $9 | (4^n + 15n - 1)$, $n \geq 1$.

9. Prove, using a proof by induction, that $9 | (4^n + 5^n)$, n is odd.

10. Prove, using a proof by induction, that $5 | (2^{3n} - 3^n)$, $n > 0$.

11. Prove, using a proof by induction, that $21 | (4^{n+1} + 5^{2n-1})$, $n > 0$.

12. Prove, using a proof by induction, that $11 | (10^n - (-1)^n)$, $n \geq 1$.

13. Prove, using a proof by induction, that $19 \left| \left(\frac{5}{4} \times 8^n + 3^{3n-1} \right) \right.$, $n \in Z^+$.

14. Prove by induction that $9 | [(3n + 1)7^n - 1]$, $n \in Z^+$.

4.6 Exam questions

Question 1 (4 marks) TECH-FREE

Use proof by induction to prove that $12 \left| \left(3^{(2n+1)} - 3\right) \right.$, $n \in N$.

Question 2 (4 marks) TECH-FREE

Use proof by induction to prove that $n^4 - n^2$ is divisible by 2, $n \in N$

Question 3 (4 marks) TECH-FREE

Prove by induction that $2 \left| \left(5^{(n-3)} + 1\right) \right.$, where n is a natural number greater than 3.

More exam questions are available online.

4.7 Review

4.7.1 Summary

Hey students! Now that it's time to revise this topic, go online to:

Access the topic summary

Review your results

Watch teacher-led videos

Practise exam questions

Find all this and MORE in jacPLUS

4.7 Exercise

Technology free: short answer

1. Determine if the following are true statements.
   a. $\exists x \in N,\ x^2 < 1$
   b. $\forall x \in R,\ x > 3$
   c. $x > 4 \rightarrow x > 3$

2. For each of the following, write the converse and determine if implication ($\rightarrow$) or equivalence ($\leftrightarrow$) is the more appropriate symbol.
   a. If an animal is a monotreme, then it is a mammal.
   b. If a number is prime, then it has exactly two factors.
   c. If $x > 2$, then $x^3 > 8$.
   d. If $x = 9$, then $x^2 = 81$.

3. Write the negation of each of the following statements.
   a. $x < 4,\ x \in R$
   b. The number is odd or a multiple of 5.
   c. The number is an even perfect square.
   d. If x is an interesting number, then x has exactly 3 factors.

4. If Pravdeep was using proof by contradiction to prove that the diagonals of a trapezium do not bisect each other, state what her initial assumption would be.

5. Prove that the sum of four consecutive whole numbers is 2 less than a multiple of 4.

6. Prove by induction that $\sum_{r=1}^{n} (3r-2) = \frac{n(3n-1)}{2}$.

7. Prove by induction that $7 \mid \left(12^n + 2\left(5^{n-1}\right)\right),\ n \geq 1$.

Technology active: multiple choice

8. **MC** Expressed as a decimal, $\frac{2}{9}$ equals
   A. $0.\dot{2}$
   B. 0.2
   C. 0.22
   D. 0.222
   E. 0.222222

9. **MC** The negation of 'All plongs are umple' is
   A. No plongs are umple
   B. All plongs are not umple
   C. Some plongs are not umple
   D. Some plongs are umple
   E. No plongs are not umple

10. **MC** Let $p(x)$: $\forall x \in Z^{-}, x^2 > 0$ and $q(y)$: $\exists y \in Z, \dfrac{y}{2} < 0$ then

A. both p and q are true.
B. p is true and q is false.
C. p is false and q is true.
D. both p and q are false.
E. more information is required to determine if p and q are truth or false.

11. **MC** Three consecutive numbers are p, q and r but not necessarily in that order. If $r^2 = pq + 1$, the numbers in ascending order are

A. p, q, r B. q, p, r C. r, p, q D. r, q, p E. p, r, q

12. **MC** Expressed as a decimal rounded to 4 decimal places, $\dfrac{7}{15}$ equals

A. $0.466\dot{7}$ B. $0.466\dot{6}$ C. 0.4666 D. 0.4667 E. 0.4677

13. **MC** Within the set $\{1, 2, 3, 4, 5, 6, 7, 8, 9\}$, the set of numbers which are both odd and prime is

A. $\{1, 3, 5, 7, 9\}$ B. $\{3, 5, 7\}$ C. $\{3, 5, 7, 9\}$ D. $\{1, 3, 5, 7\}$ E. $\{1, 2, 3, 5, 7, 9\}$

14. **MC** The converse of the statement "If the sky is blue, then it is not raining" is

A. If it not raining then the sky is blue.
B. If it is not raining, then the sky is not blue.
C. If it is raining, then the sky is blue.
D. If it is raining, then the sky is not blue.
E. Either the sky is blue or it is not raining.

15. **MC** Consider the proposition $P(n)$ that 4 is a factor of $5n - 1$ for all natural numbers n. To prove this by mathematical induction, the first step would be to

A. assume $P(k)$ holds
B. consider $P(k+1)$
C. test if $P(1)$ holds
D. express $\left(5^{k+1} - 1\right)$ as $\left(5^{k+1} - 5 + 4\right)$
E. show that $P(k) \Rightarrow P(k+1)$

16. **MC** The quantity, $7^n - 1, n > 0$ is divisible by

A. 2 only B. 3 only C. 6 only D. 2 and 3 only E. 2, 3 and 6

17 **MC** If $P(n) = \sum_{i=1}^{n} 6i = 3n(n+1)$, then $P(k+1)$ is

A. $(3k+1)(k+2)$ B. $(3k+1)(2k+2)$ C. $(3k+3)(2k+2)$
D. $(3k+3)(k+2)$ E. $(3k+3)(2k+1)$

Technology active: extended response

18. If n_1 is a 2-digit number where the digit in the tens place is greater than the digit in the units place, the number n_2 is found by interchanging the two digits. Prove that $n_1 - n_2$ is a multiple of 9.

19. Consider the following conjecture: 'The sum of the squares of any pair of consecutive natural numbers is always 1 more than a multiple of 4.'

a. Use your calculator to explore examples and demonstrate that the conjecture is likely to be true.
b. Use algebra to prove the conjecture.

20. Use proof by induction to prove that $n^5 - n$ is divisible by 30 where $n \geq 2$.

4.7 Exam questions

Question 1 (1 mark) TECH ACTIVE

MC $\frac{22}{7}$, 3.14, 3.146, $\sqrt{2\frac{1}{4}}$, $\sqrt[3]{9}$, $\sqrt{(5^2+12^2)}$

Within this list of numbers, the irrational number is

A. $\frac{22}{7}$

B. 3.146

C. $\sqrt[3]{9}$

D. $\sqrt{(5^2+12^2)}$

E. $\sqrt{2\frac{1}{4}}$

Question 2 (1 mark) TECH ACTIVE

MC If $p=q, p, q \in N$, then $p+q$ is even.

The contrapositive is:

A. If $p \neq q, p, q \in N$, then $p+q$ is odd.

B. If $p=q, p, q \in N$, then $p+q$ is odd.

C. If $p+q$ is odd, $p, q \in N$, then $p \neq q$

D. If $p+q$ is odd, $p, q \in N$, then $p=q$

E. If $p=q,\ p, q \in N$, then $p-q$ is even.

Question 3 (5 marks) TECH-FREE

Prove by induction that $\sum_{i=1}^{n} 5^i = \frac{5}{4}(5^n - 1)$.

Question 4 (4 marks) TECH-FREE

Consider the following statement: $\exists x \in N,\ x(x-1)$ is odd. Is this a true statement? Show proof.

Question 5 (3 marks) TECH-FREE

Prove that for $a,\ b \in Z$, if $4 \mid (a^2+b^2)$, then a and b are not both odd.

More exam questions are available online.

Answers

Topic 4 Proof and number

4.2 Number systems and mathematical statements

4.2 Exercise

1. a. Rational b. Rational c. Rational
   d. Irrational e. Rational f. Irrational
2. a. Irrational b. Irrational
   c. Rational d. Irrational
3. a. 0.85 b. 0.856 c. $0.\dot{6}$ d. $0.5\dot{3}$
4. a. 1.6 b. 0.5625
   c. $3.\dot{1}42\,85\dot{7}$ or $3.\overline{142\,857}$ d. $1.08\dot{3}$
5. a. $\frac{5}{9}$ b. $\frac{23}{99}$
   c. $\frac{271}{90}$ or $3\frac{1}{90}$ d. $\frac{610}{99}$ or $6\frac{16}{99}$
   e. $\frac{2347}{330}$ or $7\frac{37}{330}$ f. $\frac{26}{37}$
6. a. 345 b. 10110111
7. a. 456 b. 100101001
8. a. 198 b. 10011010
9. a. $\{0, 3, 6, 9, 15, 21\}$ b. $\{3, 9\}$
10. a. $\{22, 33, 44, 66, 68\}$ b. $\{66\}$
11. $11 \geq 20$
12. $40 < 39$
13. $\{\{2\}, \{4\}, \{6\}, \{2, 4\}, \{4, 6\}, \{2, 6\}, \{2, 4, 6\}, \phi\}$
14. $\{\{a\}, \{b\}, \{c\}, \{d\}, \{a, b\}, \{a, c\}, \{a, d\}, \{b, c\}, \{b, d\}, \{c, d\}, \{a, b, c\}, \{a, b, d\}, \{a, c, d\}, \{b, c, d\}, \{a, b, c, d\}, \phi\}$
15. a. $A\backslash B = \{2, 4, 6, 8, 10\}$ b. $B\backslash A = \{11, 13, 15\}$
16. a. $A\backslash B = \{3, 6, 9, 15, 18\}$ b. $B\backslash A\, \{4, 8, 16, 20\}$
17. a. $(-\infty, -1) \cup (-1, 1) \cup (1, \infty)$
    b. $(-\infty, -1] \cup [1, \infty)$
18. a. $(-\infty, -2) \cup (-2, 3) \cup (3, \infty)$
    b. $(-\infty, -2] \cup [3, \infty)$
19. a. For all natural numbers, x, $2x$ is even. True
    b. For all natural numbers, x, $2x + 1$ is a multiple of 3. False
    c. For all natural numbers, x, $x > 0$. True
20. a. For all real numbers, x, either $x > 0$ or $x \leq 0$. True
    b. For all natural numbers, x, x is even. False
    c. For all real numbers, x, $x^2 + 1 \geq 0$. True
21. a. There is an integer, x, so that $x + 5 = 7$. True
    b. There is a real number, x, so that $x^2 < 0$. False
    c. There is a natural number, x, that is even. True
22. a. There is a real number, x, so that $x^2 + 1 = 0$. False
    b. There is a real number, x, so that $x^2 - 1 = 0$. True
    c. There is a rational number, x, so that $\sqrt{x}$ is also rational. True
23. a. For all real numbers, x, there exists a real number, y, so that $x = y$. True
    b. There exists a real number, x, so that for all real numbers, y, $x = y$. False
24. a. For all real numbers, x, there exists a real number, y, so that $xy = 0$. True
    b. There exists a real number, x, so that for all real numbers, y, $xy = 0$. True
25. a. True b. False c. False

4.2 Exam questions

Note: Mark allocations are available with the fully worked solutions online.

1. B
2. True
3. True

4.3 Direct and indirect methods of proof

4.3 Exercise

1, 2. Sample responses can be found in the worked solutions in the online resources.

3. a. 25 b. 9
4. a. $x = -10$
   b. Counter examples include: a rhombus
   c. Counter examples include: the numbers 6, 12, 18, ...
   d. Counter examples include: the numbers 10, 20, 30, ...
5. a. If tomorrow is not Tuesday, then today is not Monday.
   b. If A is not the midpoint of a line segment, then A does not bisect the line segment.
6. $\sqrt{ab} > \frac{a+b}{2}$
7. a. If $n - m \leq 0$, then $n \leq m$.
   b. Students will need to write their own proofs. Sample responses can be found in the worked solutions in the online resources.
8. Ailsa forgot that x could equal 0.

9, 10. Students will need to write their own proofs. Sample responses can be found in the worked solutions in the online resources

11. At the 'Expanding the brackets' step, $a(a - b - c) = b(a - b - c)$ only if $a - b - c = 0$. Therefore, the proof breaks down.
12. False. One solution is $x^2 = 9$ and $y^2 = 16$ (disproof by counter example).
13. a. There is a solution to $a + b = \frac{1}{2}$ if a or b is not an integer.
    b. Students will need to write their own proofs. A sample response can be found in the worked solutions in the online resources.
14. a. If $x \leq 2$ and $y \leq 3$, then $x + y \leq 5$.
    b. Students will need to write their own proofs. A sample response can be found in the worked solutions in the online resources.

15, 16. Students will need to write their own proofs. Sample responses can be found in the worked solutions in the online resources.

4.3 Exam questions

Note: Mark allocations are available with the fully worked solutions online.

1. Sample responses can be found in the worked solutions in the online resources.
2. C
3. Sample responses can be found in the worked solutions in the online resources.

4.4 Proofs with rational and irrational numbers

4.4 Exercise

1–19. Students will need to write their own proofs. Sample responses can be found in the worked solutions in the online resources.

20. a. See the table at the bottom of the page*

b. The Sophie Germain primes less than 200 are 2, 3, 5, 11, 23, 29, 41, 53, 83, 89, 113, 131, 173, 179 and 191. A sample response for the proof can be found in the worked solutions in the online resources.

4.4 Exam questions

Note: Mark allocations are available with the fully worked solutions online.

1. A
2. D
3. Sample responses can be found in the worked solutions in the online resources.

4.5 Proof by mathematical induction

4.5 Exercise

1. a. $n = 1$. LHS $= 2$. RHS $= 1(1 + 1) = 2$. LHS $=$ RHS
$n = 2$. LHS $= 2 + 4 = 6$. RHS $= 2(2 + 1) = 6$. LHS $=$ RHS
$n = 3$. LHS $= 2 + 4 + 6 = 12$. RHS $= 3(3 + 1) = 12$. LHS $=$ RHS

b. Sample responses can be found in the worked solutions in the online resources.

2. a. $n = 1$. $\sum_{r=1}^{n} 2^{r-1} = 2^0 = 1$, $2^n - 1 = 2^1 - 1 = 1$. True
$n = 2$. $\sum_{r=1}^{n} 2^{r-1} = 2^0 + 2^1 = 3$, $2^n - 1 = 2^2 - 1 = 3$. True
$n = 3$. $\sum_{r=1}^{n} 2^{r-1} = 2^0 + 2^1 + 2^2 = 7$, $2^n - 1 = 2^3 - 1 = 7$. True

b. Sample responses can be found in the worked solutions in the online resources.

3. a. $n = 1$. LHS $= 3$. RHS $= n(n + 2) = 1(3) = 3$. True

b. $3 + 5 + 7 + ... (1 + 2k) = k(k + 2)$

c. Sample responses can be found in the worked solutions in the online resources.

d. If the statement is true for $n = k$, it is also true for $n = k + 1$. The statement is true for $n = 1$. Therefore, the statement is true for $n \geq 1$.

4. a. $n = 1$. $\sum_{r=1}^{n} (2r - 1) = 2 - 1 = 1$. $n^2 = 1$. True

b. $\sum_{r=1}^{k} (2r - 1) = k^2$

c. Sample responses can be found in the worked solutions in the online resources.

d. If the statement is true for $n = k$, it is also true for $n = k + 1$. The statement is true for $n = 1$. Therefore, the statement is true for $n \geq 1$.

5–14. Sample responses can be found in the worked solutions in the online resources.

*20. a.

1	**2**	**3**	4	**5**	6	**7**
	8	9	10	**11**	12	**13**
	14	15	16	**17**	18	**19**
	20	21	22	**23**	24	25
	26	27	28	**29**	30	**31**
	32	33	34	35	36	**37**
	38	39	40	**41**	42	**43**
	44	45	46	**47**	48	49
	50	51	52	**53**	54	55
	56	57	58	**59**	60	**61**
	62	63	64	65	66	**67**
	68	69	70	**71**	72	**73**
	74	75	76	77	78	**79**
	80	81	82	**83**	84	85
	86	87	88	**89**	90	91
	92	93	94	95	96	**97**
	98	99	100			
				Primes are in **bold**.	Multiples of 6	Primes are in **bold**.

4.5 Exam questions

Note: Mark allocations are available with the fully worked solutions online.

1. C
2. D
3. Sample responses can be found in the worked solutions in the online resources.

4.6 Proof of divisibility using induction

4.6 Exercise

1. a. $n = 0$, $6^n + 4 = 1 + 4 = 5$. Divisible by 5.
   $n = 1$, $6^n + 4 = 6 + 4 = 10$. Divisible by 5.
   $n = 2$, $6^n + 4 = 36 + 4 = 40$. Divisible by 5.
   b. Sample responses can be found in the worked solutions in the online resources.
2. a. $n = 1$, $8^n - 1 = 8 - 1 = 7$. Divisible by 7.
   $n = 2$, $8^n - 1 = 64 - 1 = 63$. Divisible by 7.
   $n = 3$, $8^n - 1 = 512 - 1 = 511 = 73 \times 7$. Divisible by 7.
   b. Sample responses can be found in the worked solutions in the online resources.
3. a. $n = 1$. $3^{2n-1} + 1 = 3 + 1 = 4$. Divisible by 4.
   b. $3^{2k-1} + 1 = 4m,\ m \in Z$.
   c. Sample responses can be found in the worked solutions in the online resources.
   d. $3^{2n-1} + 1$ is divisible by 4 when $n = 1$. It is divisible by 4 for $n = k + 1$ if it is divisible by 4 for $n = k$. Therefore, it is divisible by 4 for $n \in Z^+$.
4. a. $n = 2$. $10^n - 4 = 100 - 4 = 96 = 12 \times 8$. Divisible by 12.
   b. $10^k - 4 = 12m,\ m \in Z$.
   c. Sample responses can be found in the worked solutions in the online resources.
   d. $10^n - 4$ is divisible by 12 when $n = 2$. It is divisible by 12 for $n = k + 1$ if it is divisible by 12 for $n = k$. Therefore it is divisible by 12 for $n \geq 2$.

5–14. Sample responses can be found in the worked solutions in the online resources.

4.6 Exam questions

Note: Mark allocations are available with the fully worked solutions online.

1. Sample responses can be found in the worked solutions in the online resources.
2. Sample responses can be found in the worked solutions in the online resources.
3. Sample responses can be found in the worked solutions in the online resources.

4.7 Review

4.7 Exercise

Technology free: short answer

1. a. False b. False c. True
2. a. If an animal is a mammal, then it is a monotreme. (→)
   b. If a number has exactly two factors, then it is prime. (↔)
   c. If $x^3 > 8$, then $x > 2$. (↔)
   d. If $x^2 = 81$, then $x = 9$. (→)
3. a. $x \geq 4, x \in R$
   b. The number is even and not a multiple of 5.
   c. The number is odd or not a perfect square.
   d. x is an interesting number and it does not have exactly 3 factors.
4. The diagonals of the trapezium bisect each other.

5–7. Students will need to write their own responses. Sample responses can be found in the worked solutions in the online resources.

Technology active: multiple choice

8. A
9. C
10. A
11. E
12. D
13. B
14. A
15. C
16. E
17. D

Technology active: extended response

18–20. Students will need to write their own responses. Sample responses can be found in the worked solutions in the online resources.

4.7 Exam questions

Note: Mark allocations are available with the fully worked solutions online.

1. C
2. C
3. Sample responses can be found in the worked solutions in the online resources.
4. The statement is not true. A sample proof can be found in the worked solutions in the online resources.
5. Sample responses can be found in the worked solutions in the online resources.

5 Matrices

LEARNING SEQUENCE

Fully worked solutions for this topic are available online.

5.1 Overview

Hey students! Bring these pages to life online

Watch videos

Engage with interactivities

Answer questions and check results

Find all this and MORE in jacPLUS

5.1.1 Introduction

Matrices were first used to solve systems of linear equations. Today they are used in many fields, including engineering, physics, economics and statistics. They are used primarily in the encoding and decoding of information. You have probably used matrix technology without realising it in encrypted messaging and internet banking.

The British mathematician Alan Turing led a team of mathematicians who were responsible for breaking the Germans' Enigma code during World War II. Their work helped the Allies to win the war in Europe and also led to the creation of the first computer. Turing built a machine to help with the decoding process. This was the start of the computer age, and now every computer-generated image is the result of matrix mathematics.

KEY CONCEPTS

This topic covers the following key concepts from the VCE Mathematics Study Design:

- matrix notation, dimension and the use of matrices to represent data
- matrix operations and algebra
- determinants and matrix equations, and simple applications.

5.2 Addition, subtraction and scalar multiplication of matrices

LEARNING INTENTION

At the end of this subtopic you should able to:
- add and subtract matrices
- multiply matrices by a scalar.

5.2.1 Introduction to matrices

The table shows the final medal tally for the top four countries at the 2020 Tokyo Olympic Games.

Country	Gold	Silver	Bronze
United States of America	39	41	33
People's Republic of China	38	32	18
Japan	27	14	17
Great Britain	22	21	22

This information can be presented in a matrix, without the country names, and without the headings for gold, silver and bronze:

$$\begin{bmatrix} 39 & 41 & 33 \\ 38 & 32 & 18 \\ 27 & 14 & 17 \\ 22 & 21 & 22 \end{bmatrix}$$

The data is presented in a rectangular array arranged in rows and columns, and is called a **matrix**. It conveys information such as that the second country won 38 gold, 32 silver and 18 bronze medals. This matrix has four rows and three columns. The numbers in the matrix, in this case representing the number of medals won, are called **elements of the matrix**.

Matrices

In general we enclose a matrix in square brackets and usually use capital letters to denote it. The size or **order of a matrix** is important, and is determined by the number of rows and the number of columns, strictly in that order.

Consider the following matrix A:

$$A = \begin{bmatrix} 3 & 5 \\ 4 & 7 \end{bmatrix}$$

A is a 2×2 matrix: it has two rows and two columns. When the number of rows and columns in a matrix are equal, it is called a **square matrix**.

$$B = \begin{bmatrix} 2 & -5 & -3 \\ -4 & 2 & -5 \\ 1 & 3 & 4 \end{bmatrix}$$

B is a 3×3 matrix: it has three rows and three columns.

$$C = \begin{bmatrix} 1 \\ -2 \\ 3 \end{bmatrix}$$

C is a 3×1 matrix as it has three rows and one column. If a matrix has only one column, it is also called a **column** or **vector matrix**.

$$D = \begin{bmatrix} 3 & -2 \end{bmatrix}$$

D is a 1×2 matrix as it has one row and two columns. If a matrix has only one row it is also called a **row matrix**.

$$E = \begin{bmatrix} 3 & 5 \\ -4 & 2 \\ -1 & 3 \end{bmatrix}$$

E is a 3×2 matrix as it has three rows and two columns.

$$F = \begin{bmatrix} 2 & 3 & 4 \\ 4 & -5 & -2 \end{bmatrix}$$

F is a 2×3 matrix as it has two rows and three columns.

Each element in a matrix can also be identified by its position in the matrix. For example, in matrix $A = \begin{bmatrix} 3 & 5 \\ 4 & 7 \end{bmatrix}$, the element 3 is in the first row and first column, so $a_{11} = 3$. The element 5 is in the first row and second column, so $a_{12} = 5$. The element 4 is in the second row and first column, so $a_{21} = 4$. Finally, the element 7 is in the second row and second column, so $a_{22} = 7$.

In general, we can write a 2×2 matrix as:

$$A = \begin{bmatrix} a_{11} & a_{12} \\ a_{21} & a_{22} \end{bmatrix}$$

A general matrix of order $m \times n$ can be written as:

$$B = \begin{bmatrix} b_{11} & b_{12} & b_{13} & \cdots & b_{1n} \\ b_{21} & b_{22} & b_{23} & \cdots & b_{2n} \\ b_{31} & b_{32} & b_{33} & \cdots & b_{3n} \\ . & . & . & \cdots & . \\ . & . & . & \cdots & . \\ . & . & . & \cdots & . \\ b_{m1} & b_{m2} & b_{m3} & \cdots & b_{mn} \end{bmatrix}$$

Here, b_{14} denotes the element in the first row and fourth column, b_{43} denotes the element in the fourth row and third column, and b_{ij} denotes the element in the ith row and jth column of the matrix B.

5.2.2 Operations on matrices

Operations include addition, subtraction and multiplication of two matrices. Division of matrices is not possible.

Equality of matrices

Two matrices are equal, if and only if, they have the same order and each of the corresponding elements are equal.

For example, if $\underset{3\times 1}{\begin{bmatrix} x \\ y \\ z \end{bmatrix}} = \underset{3\times 1}{\begin{bmatrix} 1 \\ -2 \\ 3 \end{bmatrix}}$, then $x=1$, $y=-2$ and $z=3$; if $\underset{2\times 2}{\begin{bmatrix} a & b \\ c & d \end{bmatrix}} = \underset{2\times 2}{\begin{bmatrix} 3 & 5 \\ 4 & 7 \end{bmatrix}}$, then $a=3$, $b=5$, $c=4$ and $d=7$.

Addition and subtraction of matrices

Only two matrices of the same order can be added or subtracted. To add or subtract two matrices, we add or subtract the elements in the corresponding positions. For example, if $P = \underset{2\times 2}{\begin{bmatrix} 2 & 3 \\ -1 & 5 \end{bmatrix}}$ and $Q = \underset{2\times 2}{\begin{bmatrix} 4 & -2 \\ 6 & 3 \end{bmatrix}}$, then:

$$P+Q=\begin{bmatrix} 2 & 3 \\ -1 & 5 \end{bmatrix}+\begin{bmatrix} 4 & -2 \\ 6 & 3 \end{bmatrix}=\begin{bmatrix} 2+4 & 3-2 \\ -1+6 & 5+3 \end{bmatrix}=\begin{bmatrix} 6 & 1 \\ 5 & 8 \end{bmatrix}$$

$$P-Q=P+(-Q)=\begin{bmatrix} 2 & 3 \\ -1 & 5 \end{bmatrix}-\begin{bmatrix} 4 & -2 \\ 6 & 3 \end{bmatrix}=\begin{bmatrix} 2-4 & 3+2 \\ -1-6 & 5-3 \end{bmatrix}=\begin{bmatrix} -2 & 5 \\ -7 & 2 \end{bmatrix}$$

Matrices can be added or subtracted only if they are of the same order.

To add matrices, simply add the elements in the corresponding positions. To subtract matrices, simply subtract the elements in the corresponding positions.

Addition and subtraction of matrices

For example, if $A = \begin{bmatrix} a_{11} & a_{12} & \cdots & a_{1n} \\ a_{21} & a_{22} & \cdots & a_{2n} \\ \vdots & \vdots & \ddots & \vdots \\ a_{m1} & a_{m2} & \cdots & a_{mn} \end{bmatrix}$ **and** $B = \begin{bmatrix} b_{11} & b_{12} & \cdots & b_{1n} \\ b_{21} & b_{22} & \cdots & b_{2n} \\ \vdots & \vdots & \ddots & \vdots \\ b_{m1} & b_{m2} & \cdots & b_{mn} \end{bmatrix}$**, then:**

$$A+B=\begin{bmatrix} a_{11}+b_{11} & a_{12}+b_{12} & \cdots & a_{1n}+b_{1n} \\ a_{21}+b_{21} & a_{22}+b_{22} & \cdots & a_{2n}+b_{2n} \\ \vdots & \vdots & \ddots & \vdots \\ a_{m1}+b_{m1} & a_{m2}+b_{m2} & \cdots & a_{mn}+b_{mn} \end{bmatrix}$$

and

$$A-B=\begin{bmatrix} a_{11}-b_{11} & a_{12}-b_{12} & \cdots & a_{1n}-b_{1n} \\ a_{21}-b_{21} & a_{22}-b_{22} & \cdots & a_{2n}-b_{2n} \\ \vdots & \vdots & \ddots & \vdots \\ a_{m1}-b_{m1} & a_{m2}-b_{m2} & \cdots & a_{mn}-b_{mn} \end{bmatrix}$$

Note that none of the matrices defined by

$$A=\underset{2\times 2}{\begin{bmatrix}3 & 5\\4 & 7\end{bmatrix}},\ B=\underset{3\times 3}{\begin{bmatrix}2 & -5 & -3\\-4 & 2 & -5\\1 & 3 & 4\end{bmatrix}},\ C=\underset{3\times 1}{\begin{bmatrix}1\\-2\\3\end{bmatrix}},\ D=\underset{1\times 2}{\begin{bmatrix}3 & -2\end{bmatrix}},\ E=\underset{3\times 2}{\begin{bmatrix}3 & 5\\-4 & 2\\-1 & 3\end{bmatrix}},\ F=\underset{2\times 3}{\begin{bmatrix}2 & 3 & 4\\4 & -5 & -2\end{bmatrix}},$$

can be added or subtracted from one another, as they are all of different orders.

WORKED EXAMPLE 1 Representing data in matrix form

At a football match one food outlet sold 280 pies, 210 hotdogs and 310 boxes of chips. Another food outlet sold 300 pies, 220 hotdogs and 290 boxes of chips. Represent each data set as a 1×3 matrix, and determine the total number of pies, hotdogs and chips sold by these two outlets.

THINK	WRITE
1. Use a 1×3 matrix to represent the number of pies, hotdogs and chips sold.	$\begin{bmatrix}\text{pies} & \text{hotdogs} & \text{chips}\end{bmatrix}$
2. Write the matrix for the sales from the first outlet.	$S_1=\begin{bmatrix}280 & 210 & 310\end{bmatrix}$
3. Write the matrix for the sales from the second outlet.	$S_2=\begin{bmatrix}300 & 220 & 290\end{bmatrix}$
4. Use the rules of addition of matrices to determine the sum of these two matrices.	$S_1+S_2=\begin{bmatrix}280+300 & 210+220 & 310+290\end{bmatrix}$ $=\begin{bmatrix}580 & 430 & 600\end{bmatrix}$

Scalar multiplication of matrices

A **scalar** is an entity with magnitude only; that is, it is a real number. To multiply any matrix by a scalar, we multiply every element in the matrix by the scalar.

$$\text{If } P=\begin{bmatrix}2 & 3\\-1 & 5\end{bmatrix}$$

then:

$$\begin{aligned}2P&=2\begin{bmatrix}2 & 3\\-1 & 5\end{bmatrix}\\&=\begin{bmatrix}2\times 2 & 2\times 3\\2\times(-1) & 2\times 5\end{bmatrix}\\&=\begin{bmatrix}4 & 6\\-2 & 10\end{bmatrix}\end{aligned}$$

WORKED EXAMPLE 2 Addition and scalar multiplication of matrices

Given the matrices $A = \begin{bmatrix} 3 \\ -5 \end{bmatrix}$, $B = \begin{bmatrix} -5 \\ 4 \end{bmatrix}$ **and** $C = \begin{bmatrix} 2 \\ y \end{bmatrix}$ **determine the values of** x **and** y **if** $xA + 2B = C$.

THINK	WRITE
1. Substitute for the given matrices.	$xA + 2B = C$ $x\begin{bmatrix} 3 \\ -5 \end{bmatrix} + 2\begin{bmatrix} -5 \\ 4 \end{bmatrix} = \begin{bmatrix} 2 \\ y \end{bmatrix}$
2. Apply the rules for scalar multiplication.	$\begin{bmatrix} 3x \\ -5x \end{bmatrix} + \begin{bmatrix} -10 \\ 8 \end{bmatrix} = \begin{bmatrix} 2 \\ y \end{bmatrix}$
3. Apply the rules for addition of matrices.	$\begin{bmatrix} 3x - 10 \\ -5x + 8 \end{bmatrix} = \begin{bmatrix} 2 \\ y \end{bmatrix}$
4. Apply the rules for equality of matrices.	$3x - 10 = 2$ $-5x + 8 = y$
5. Solve the first equation for x.	$3x = 12$ $x = 4$
6. Substitute for x into the second equation and solve this equation for y.	$-5x + 8 = y$ $y = 8 - 20$ $y = -12$
7. Answer the question.	$\therefore x = 4, \ y = -12$

5.2.3 Special matrices

The zero matrix

The 2×2 **null matrix** or **zero matrix** O, with all elements equal to zero, is given by $O = \begin{bmatrix} 0 & 0 \\ 0 & 0 \end{bmatrix}$.

When matrices $A + B = O$, matrix B is the **additive inverse** of A. Hence, $B = O - A$.

If $A = \begin{bmatrix} 2 & 4 \\ 5 & -3 \end{bmatrix}$ and $B = \begin{bmatrix} -2 & -4 \\ -5 & 3 \end{bmatrix}$,

$$A + B = \begin{bmatrix} 2 & 4 \\ 5 & -3 \end{bmatrix} + \begin{bmatrix} -2 & -4 \\ -5 & 3 \end{bmatrix} = \begin{bmatrix} 2 + (-2) & 4 + (-4) \\ 5 + (-5) & -3 + 3 \end{bmatrix} = \begin{bmatrix} 0 & 0 \\ 0 & 0 \end{bmatrix} = O.$$

Hence, B is the additive inverse of A.

The identity matrix

The 2×2 **identity matrix** I is defined by $I = \begin{bmatrix} 1 & 0 \\ 0 & 1 \end{bmatrix}$. This square matrix has ones on the leading diagonal and zeros on the other diagonal. An example of an identity matrix is shown in Worked example 3c.

WORKED EXAMPLE 3 Various operations on matrices

Given the matrices $A = \begin{bmatrix} 3 & 5 \\ 4 & 7 \end{bmatrix}$ **and** $B = \begin{bmatrix} -5 & -3 \\ 3 & 4 \end{bmatrix}$, **determine the matrix** X **if:**

a. $X = 2A - 3B$ b. $A + X = O$ c. $X = B + 2A - 3I$

THINK	WRITE
a. 1. Substitute for the given matrices.	**a.** $X = 2A - 3B$ $= 2\begin{bmatrix}3 & 5\\4 & 7\end{bmatrix} - 3\begin{bmatrix}-5 & -3\\3 & 4\end{bmatrix}$
2. Apply the rules for scalar multiplication.	$= \begin{bmatrix}6 & 10\\8 & 14\end{bmatrix} - \begin{bmatrix}-15 & -9\\9 & 12\end{bmatrix}$
3. Apply the rules for subtraction of matrices.	$= \begin{bmatrix}21 & 19\\-1 & 2\end{bmatrix}$
b. 1. Transpose the equation to make X the subject.	**b.** $A + X = O$ $X = O - A$
2. State the final answer.	$= \begin{bmatrix}0 & 0\\0 & 0\end{bmatrix} - \begin{bmatrix}3 & 5\\4 & 7\end{bmatrix}$ $= \begin{bmatrix}-3 & -5\\-4 & -7\end{bmatrix}$
c. 1. Substitute for the given matrices.	**c.** $X = B + 2A - 3I$ $= \begin{bmatrix}-5 & -3\\3 & 4\end{bmatrix} + 2\begin{bmatrix}3 & 5\\4 & 7\end{bmatrix} - 3\begin{bmatrix}1 & 0\\0 & 1\end{bmatrix}$
2. Apply the rules for scalar multiplication.	$= \begin{bmatrix}-5 & -3\\3 & 4\end{bmatrix} + \begin{bmatrix}6 & 10\\8 & 14\end{bmatrix} - \begin{bmatrix}3 & 0\\0 & 3\end{bmatrix}$
3. Apply the rules for addition and subtraction of matrices.	$= \begin{bmatrix}-5+6-3 & -3+10+0\\3+8-0 & 4+14-3\end{bmatrix}$
4. State the final answer.	$= \begin{bmatrix}-2 & 7\\11 & 15\end{bmatrix}$

TI | THINK — **DISPLAY/WRITE**

a. 1. On a Calculator page, press the template button and complete the entry line as:

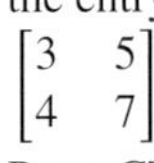

Press CTRL, then press VAR, then type a and press ENTER to store matrix A. Repeat this step to store matrix B.

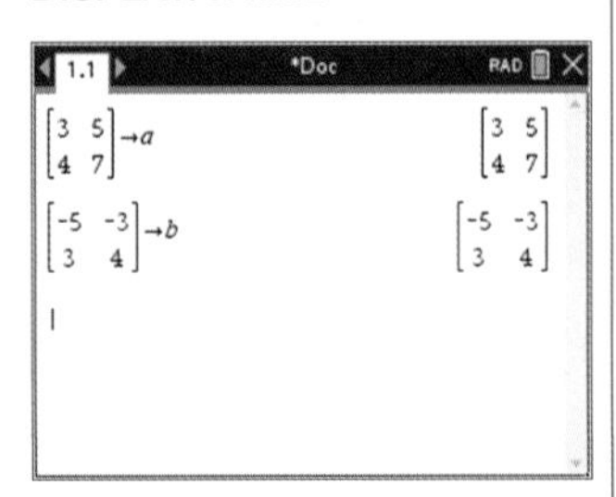

2. Complete the next entry line as $2a - 3b$ then press ENTER.

CASIO | THINK — **DISPLAY/WRITE**

a. Although it is possible to store matrices, the more efficient approach is to enter the expression to be evaluated as shown in the entry line.
The matrix templates are available from the Keyboard Math2 tab.
Select EXE to display the result.

5.2 Exercise

Students, these questions are even better in jacPLUS

Receive immediate feedback and access sample responses

Access additional questions

Track your results and progress

Find all this and MORE in jacPLUS

Technology free

1. **WE1** At football matches, commentators often quote player statistics. In one particular game, the top ranked player on the ground had 25 kicks, 8 marks and 10 handballs. The second ranked player on the same team on the ground had 20 kicks, 6 marks and 8 handballs, while the third ranked player on the same team on the ground had 18 kicks, 5 marks and 7 handballs. Represent each data set as a 1×3 matrix and determine the total number of kicks, marks and handballs by these three players.

2. At the end of a doubles tennis match, one player had 2 aces, 3 double faults, 25 forehand winners and 10 backhand winners, while his partner had 4 aces, 5 double faults, 28 forehand winners and 7 backhand winners. Represent this data as 2×2 matrices and determine the total number of aces, double faults, forehand and backhand winners for these players.

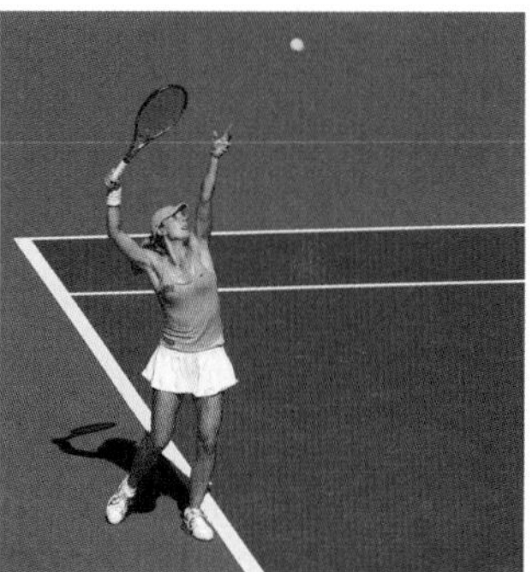

3. **WE2** Given the matrices $A = \begin{bmatrix} -3 \\ 4 \end{bmatrix}$, $B = \begin{bmatrix} 4 \\ 5 \end{bmatrix}$ and $C = \begin{bmatrix} 2 \\ y \end{bmatrix}$, determine the values of x and y if $xA + 2B = C$.

4. Given the matrices $A = \begin{bmatrix} 4 \\ -2 \\ 3 \end{bmatrix}$, $B = \begin{bmatrix} 3 \\ 5 \\ -1 \end{bmatrix}$ and $C = \begin{bmatrix} 6 \\ y \\ z \end{bmatrix}$, determine the values of x, y and z if $xA - 2B = C$.

5. **WE3** If $A = \begin{bmatrix} -2 & 4 \\ 3 & 5 \end{bmatrix}$ and $B = \begin{bmatrix} 2 & 4 \\ -1 & -3 \end{bmatrix}$, determine the matrix X given the following.

   a. $X = 3A - 2B$
   b. $2A + X = O$

6. If $A = \begin{bmatrix} a & b \\ c & d \end{bmatrix}$ and $B = \begin{bmatrix} 2 & 4 \\ -1 & -3 \end{bmatrix}$, determine the values of a, b, c and d given the following.

   a. $A + 2I - 2B = O$
   b. $3I + 4B - 2A = O$

7. If $A = \begin{bmatrix} -1 \\ 2 \end{bmatrix}$ and $B = \begin{bmatrix} 3 \\ 5 \end{bmatrix}$, determine the matrix C given the following.

   a. $C = A + B$
   b. $3A + 2C = 4B$

8. Consider these matrices: $A = \begin{bmatrix} 2 & 3 \\ -1 & 4 \end{bmatrix}$, $B = \begin{bmatrix} 4 & 5 \\ 2 & -3 \end{bmatrix}$ and $C = \begin{bmatrix} 1 & -2 \\ 5 & 4 \end{bmatrix}$.

   a. Determine the following matrices.
      i. $B + C$
      ii. $A + B$
   b. Verify the Associative Law for matrix addition: $A + (B + C) = (A + B) + C$.

9. If $A = \begin{bmatrix} 1 & 4 \\ -3 & 2 \end{bmatrix}$, $B = \begin{bmatrix} 4 & -2 \\ 3 & 5 \end{bmatrix}$ and $O = \begin{bmatrix} 0 & 0 \\ 0 & 0 \end{bmatrix}$, determine the matrix C given the following.

   a. $3A = C - 2B$
   b. $C + 3A - 2B = O$

10. Given the matrices $A=\begin{bmatrix}1 & 4\\-3 & 2\end{bmatrix}$, $B=\begin{bmatrix}4 & -2\\3 & 5\end{bmatrix}$, $O=\begin{bmatrix}0 & 0\\0 & 0\end{bmatrix}$ and $I=\begin{bmatrix}1 & 0\\0 & 1\end{bmatrix}$, determine the matrix C if the following apply.

a. $3A+C-2B+4I=O$

b. $4A-C+3B-2I=O$

11. If $A=\begin{bmatrix}x & -3\\2 & x\end{bmatrix}$ and $B=\begin{bmatrix}2 & y\\y & -3\end{bmatrix}$, determine the values of x and y given the following.

a. $A+B=\begin{bmatrix}7 & 4\\9 & 2\end{bmatrix}$

b. $B-A=\begin{bmatrix}-1 & 1\\-4 & -6\end{bmatrix}$

12. If $D=\begin{bmatrix}1 & 4 & 5\\-3 & 2 & -2\end{bmatrix}$ and $E=\begin{bmatrix}2 & -2 & 4\\1 & 4 & -3\end{bmatrix}$, determine the matrix C given the following.

a. $C=D+E$

b. $3D+2C=4E$

13. a. Given $A=\begin{bmatrix}2 & 3\\-1 & 4\end{bmatrix}$ write down the values of a_{11}, a_{12}, a_{21} and a_{22}.

b. State the 2×2 matrix B if $b_{11}=3$, $b_{12}=-2$, $b_{21}=-3$ and $b_{22}=5$.

14. a. Determine the 2×2 matrix A whose elements are $a_{ij}=2i-j$ for $j\neq i$ and $a_{ij}=ij$ for $j=i$.

b. Determine the 2×2 matrix A whose elements are $a_{ij}=i+j$ for $i<j$, $a_{ij}=i-j+1$ for $i>j$ and $a_{ij}=i+j+1$ for $i=j$.

15. The **trace of a matrix** A denoted by tr(A) is equal to the sum of leading diagonal elements. For 2×2 matrices, if $A=\begin{bmatrix}a_{11} & a_{12}\\a_{21} & a_{22}\end{bmatrix}$ then $\text{tr}(A)=a_{11}+a_{22}$. Consider the following matrices: $A=\begin{bmatrix}2 & 3\\-1 & 4\end{bmatrix}$, $B=\begin{bmatrix}4 & -2\\3 & 5\end{bmatrix}$ and $C=\begin{bmatrix}1 & -2\\5 & 4\end{bmatrix}$.

a. Calculate the following.

i. tr(A)

ii. tr(B)

iii. tr(C)

b. Determine if $\text{tr}(A+B+C)=\text{tr}(A)+\text{tr}(B)+\text{tr}(C)$.

c. Determine if $\text{tr}(2A+3B-4C)=2\text{tr}(A)+3\text{tr}(B)-4\text{tr}(C)$.

5.2 Exam questions

Question 1 (4 marks) TECH-FREE

Given the matrices $A=\begin{bmatrix}2 & 4\\5 & 8\end{bmatrix}$ and $B=\begin{bmatrix}-7 & -1\\2 & 3\end{bmatrix}$, determine the matrix X if $X=2B-3A+2I$.

Question 2 (1 mark) TECH-ACTIVE

MC If $\begin{bmatrix}7 & 5\\a & -2\end{bmatrix}-\begin{bmatrix}-2 & b\\3 & 7\end{bmatrix}=\begin{bmatrix}9 & 6\\-3 & -9\end{bmatrix}$, then the values of a and b are

A. $a=0$ and $b=1$.

B. $a=0$ and $b=-1$.

C. $a=-6$ and $b=-1$.

D. $a=-6$ and $b=1$.

E. $a=6$ and $b=1$

Question 3 (2 marks) TECH-FREE

Given the matrices $A=\begin{bmatrix}a & 3\\x & -6\end{bmatrix}$, $B=\begin{bmatrix}-2 & b\\4 & y\end{bmatrix}$ and $O=\begin{bmatrix}0 & 0\\0 & 0\end{bmatrix}$, determine the values of a, b, x and y given that $2A-3B=0$.

More exam questions are available online.

5.3 Matrix multiplication

LEARNING INTENTION

At the end of this subtopic you should be able to:
- identify whether the product of two matrices will be defined
- multiply matrices by one another, given that their matrix product is defined.

5.3.1 Multiplication of matrices

At the end of an AFL football match between Sydney and Melbourne the scores were as shown.

This information is represented in a matrix as:

$$\begin{array}{c} \\ \text{Sydney} \\ \text{Melbourne} \end{array} \begin{array}{c} \text{Goals Behinds} \\ \begin{bmatrix} 12 & 15 \\ 9 & 10 \end{bmatrix} \end{array}$$

One goal in AFL football is worth 6 points and one behind is worth 1 point.

This information is represented in a matrix as:

$$\begin{array}{c} \text{Goals} \\ \text{Behinds} \end{array} \begin{bmatrix} 6 \\ 1 \end{bmatrix}$$

To get the total points scored by both teams the matrices are multiplied.

$$\begin{bmatrix} 12 & 15 \\ 9 & 10 \end{bmatrix} \times \begin{bmatrix} 6 \\ 1 \end{bmatrix} = \begin{bmatrix} 12 \times 6 + 15 \times 1 \\ 9 \times 6 + 10 \times 1 \end{bmatrix} = \begin{bmatrix} 87 \\ 64 \end{bmatrix}$$

Multiplying matrices in general

Two matrices A and B may be multiplied together to form the product AB if the number of columns in the first matrix A, is equal to the number of rows in the second matrix B. If matrix A is of order $m \times n$ and matrix B is of order $n \times p$, then the product AB will be of order $m \times p$. The number of columns in the first matrix must be equal to the number of rows in the second matrix otherwise matrix multiplication is not defined.

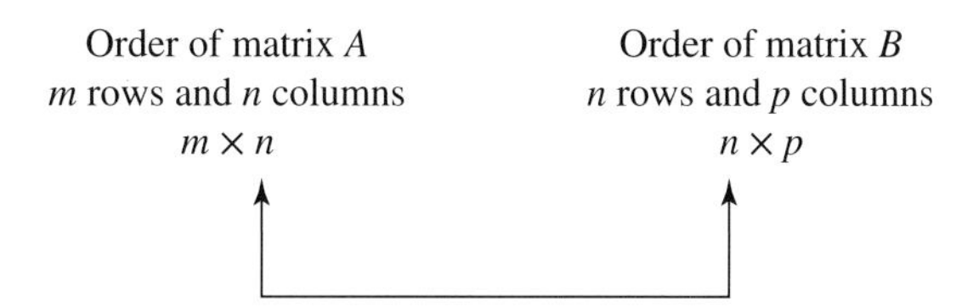

Therefore, matrix multiplication is defined and the order of the product AB will be $m \times p$ (the outside numbers).

The product is obtained by multiplying each element in each row of the first matrix by the corresponding elements of each column in the second matrix.

For an example of how this works in practise we will calculate the following matrix product:

$$\begin{bmatrix}1 & 2 & -1\\0 & 1 & 3\end{bmatrix}\times\begin{bmatrix}3 & 4\\2 & 0\\0 & -1\end{bmatrix}.$$

Multiply the first row by the first column. $(1\times3)+(2\times2)+(-1\times0)=7$ The value in row 1, column 1 is therefore 7.	$\begin{bmatrix}1 & 2 & -1\\0 & 1 & 3\end{bmatrix}\times\begin{bmatrix}3 & 4\\2 & 0\\0 & -1\end{bmatrix}=\begin{bmatrix}7 & \square\\\square & \square\end{bmatrix}$
Multiply the first row by the second column. $(1\times4)+(2\times0)+(-1\times-1)=5$ The value in row 1, column 2 is therefore 5.	$\begin{bmatrix}1 & 2 & -1\\0 & 1 & 3\end{bmatrix}\times\begin{bmatrix}3 & 4\\2 & 0\\0 & -1\end{bmatrix}=\begin{bmatrix}7 & 5\\\square & \square\end{bmatrix}$
Multiply the second row by the first column. $(0\times3)+(1\times2)+(3\times0)=2$ The value in row 2, column 1 is therefore 2.	$\begin{bmatrix}1 & 2 & -1\\0 & 1 & 3\end{bmatrix}\times\begin{bmatrix}3 & 4\\2 & 0\\0 & -1\end{bmatrix}=\begin{bmatrix}7 & 5\\2 & \square\end{bmatrix}$
Multiply the second row by the second column. $(0\times4)+(1\times0)+(3\times-1)=-3$ The value in row 2, column 2 is therefore −3.	$\begin{bmatrix}1 & 2 & -1\\0 & 1 & 3\end{bmatrix}\times\begin{bmatrix}3 & 4\\2 & 0\\0 & -1\end{bmatrix}=\begin{bmatrix}7 & 5\\2 & -3\end{bmatrix}$

Matrix multiplication

The product AB of two matrices A and B is defined if the number of columns in A is equal to the number of rows in B.

For example, if $A=\begin{bmatrix}a_{11} & a_{12} & \dots & a_{1n}\\a_{21} & a_{22} & \dots & a_{2n}\\\vdots & \vdots & \ddots & \vdots\\a_{m1} & a_{m2} & \dots & a_{mn}\end{bmatrix}$ **and** $B=\begin{bmatrix}b_{11} & b_{12} & \dots & b_{1p}\\b_{21} & b_{22} & \dots & b_{2p}\\\vdots & \vdots & \ddots & \vdots\\b_{n1} & b_{n2} & \dots & b_{np}\end{bmatrix}$**, then:**

$$AB=\begin{bmatrix}c_{11} & c_{12} & \dots & c_{1p}\\c_{21} & c_{22} & \dots & c_{2p}\\\vdots & \vdots & \ddots & \vdots\\c_{m1} & c_{m2} & \dots & c_{mp}\end{bmatrix}, \text{ where } c_{ij}=a_{i1}\times b_{1j}+a_{i2}\times b_{2j}+a_{i3}\times b_{3j}+\dots+a_{in}\times b_{nj}$$

WORKED EXAMPLE 4 Determining whether matrix multiplication is defined

Consider the following matrices:

$$A=\begin{bmatrix}1 & 3\\-1 & 5\\4 & 2\end{bmatrix}, B=\begin{bmatrix}2\\0\\-2\end{bmatrix}, C=\begin{bmatrix}3 & 5\\-1 & 6\end{bmatrix}, D=\begin{bmatrix}1 & 7 & 4\end{bmatrix}$$

a. **Determine if matrix multiplication is defined for each of the products given below.**
 i. AB ii. AC iii. DB iv. BD

b. **State the order of each product in part a if the matrix multiplication is defined.**

THINK	WRITE
a. 1. State the order of each of the matrices.	a. A: 3×2 matrix B: 3×1 matrix C: 2×2 matrix D: 1×3 matrix
2. Determine if the product is defined by checking that the number of columns of the first matrix is the same as the number of rows of the second matrix. ***Note:*** Check that the inside numbers are the same.	i. AB: A B 3×2 and 3×1 No, AB is not defined. ii. AC: A C 3×2 and 2×2 Yes, AC is defined. iii. DB: D B 1×3 and 3×1 Yes, DB is defined. iv. BD: B D 3×1 and 1×3 Yes, BD is defined.
b. If matrix multiplication is defined, the outside numbers give the order of the product.	b. i. Not defined ii. 3×2 iii. 1×1 iv. 3×3

WORKED EXAMPLE 5 Multiplication of 2×2 matrices

Given the matrices $A = \begin{bmatrix} 3 & 5 \\ 4 & 7 \end{bmatrix}$ and $B = \begin{bmatrix} -5 & -3 \\ 3 & 4 \end{bmatrix}$, determine the following matrices.

a. AB b. BA c. B^2

THINK	WRITE
a. 1. Substitute for the given matrices.	a. $AB = \begin{bmatrix} 3 & 5 \\ 4 & 7 \end{bmatrix} \begin{bmatrix} -5 & -3 \\ 3 & 4 \end{bmatrix}$
2. Check that matrix multiplication is defined. Since A and B are both 2×2 matrices, the product AB will also be a 2×2 matrix.	
3. Recall the rules for matrix multiplication and apply.	$= \begin{bmatrix} 3 \times (-5) + 5 \times 3 & 3 \times (-3) + 5 \times 4 \\ 4 \times (-5) + 7 \times 3 & 4 \times (-3) + 7 \times 4 \end{bmatrix}$
4. Simplify and give the final result.	$AB = \begin{bmatrix} 0 & 11 \\ 1 & 16 \end{bmatrix}$

b. 1. Substitute for the given matrices.

b. $BA = \begin{bmatrix} -5 & -3 \\ 3 & 4 \end{bmatrix} \begin{bmatrix} 3 & 5 \\ 4 & 7 \end{bmatrix}$

2. Check that matrix multiplication is defined. Since both A and B are 2×2 matrices, the product BA will also be a 2×2 matrix.

3. Recall the rules for matrix multiplication and apply.

$= \begin{bmatrix} (-5)\times 3 + (-3)\times 4 & (-5)\times 5 + (-3)\times 7 \\ 3\times 3 + 4\times 4 & 3\times 5 + 4\times 7 \end{bmatrix}$

4. Simplify and give the final result.

$BA = \begin{bmatrix} -27 & -46 \\ 25 & 43 \end{bmatrix}$

c. 1. $B^2 = B \times B$. Substitute for the given matrix.

c. $B^2 = \begin{bmatrix} -5 & -3 \\ 3 & 4 \end{bmatrix}^2 = \begin{bmatrix} -5 & -3 \\ 3 & 4 \end{bmatrix} \begin{bmatrix} -5 & -3 \\ 3 & 4 \end{bmatrix}$

2. Since B is a 2×2 matrix, B^2 will also be a 2×2 matrix. Apply the rules for matrix multiplication.

$= \begin{bmatrix} (-5)\times(-5) + (-3)\times 3 & (-5)\times(-3) + -3\times 4 \\ 3\times(-5) + 4\times 3 & 3\times(-3) + 4\times 4 \end{bmatrix}$

3. Simplify and give the final result.

$B^2 = \begin{bmatrix} 16 & 3 \\ -3 & 7 \end{bmatrix}$

TI | THINK — **DISPLAY/WRITE**

c. 1. On a Calculator page, press the template button and complete the entry line as:

$\begin{bmatrix} -5 & -3 \\ 3 & 4 \end{bmatrix}$

Press CTRL, then press VAR, then type b and press ENTER to store matrix B.

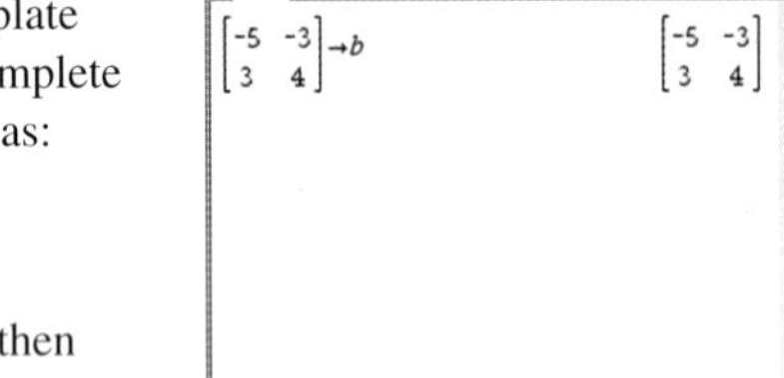

2. Complete the next entry line as b^2 then press ENTER.

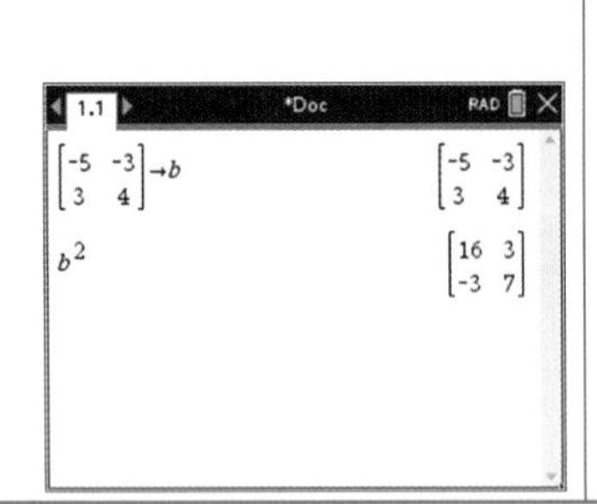

CASIO | THINK — **DISPLAY/WRITE**

c. From the Math2 keyboard template, select the 2×2 matrix and complete the entry line as shown. Select EXE to display the result.

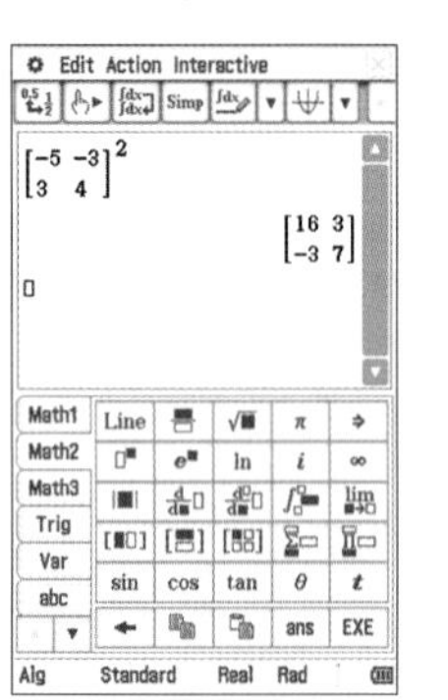

Worked example 6 shows that matrix multiplication in general is not commutative: $AB \neq BA$, although there are exceptions. It is also possible that one product is defined and the other is not defined, and that the products may have different orders. Note that squaring a matrix (when defined) is not the square of each individual element.

WORKED EXAMPLE 6 Multiplication of 3×2 and 2×3 matrices

Given the matrices $E = \begin{bmatrix} 3 & 5 \\ -4 & 2 \\ -1 & 3 \end{bmatrix}$ **and** $F = \begin{bmatrix} 2 & 3 & 4 \\ 4 & -5 & -2 \end{bmatrix}$, **determine the following matrices.**

a. *EF*
b. *FE*

THINK	WRITE
a. 1. Substitute for the given matrices.	a. $EF = \begin{bmatrix} 3 & 5 \\ -4 & 2 \\ -1 & 3 \end{bmatrix} \begin{bmatrix} 2 & 3 & 4 \\ 4 & -5 & -2 \end{bmatrix}$
2. Check that matrix multiplication is defined. Since *E* is a 3×2 matrix and *F* is a 2×3 matrix, the product *EF* will be a 3×3 matrix.	
3. Recall the rules for matrix multiplication and apply.	$= \begin{bmatrix} 3\times2+5\times4 & 3\times3+5\times(-5) & 3\times4+5\times(-2) \\ (-4)\times2+2\times4 & (-4)\times3+2\times(-5) & (-4)\times4+2\times(-2) \\ (-1)\times2+3\times4 & (-1)\times3+3\times(-5) & (-1)\times4+3\times(-2) \end{bmatrix}$
4. Simplify and give the final result.	$EF = \begin{bmatrix} 26 & -16 & 2 \\ 0 & -22 & -20 \\ 10 & -18 & -10 \end{bmatrix}$
b. 1. Substitute for the given matrices.	b. $FE = \begin{bmatrix} 2 & 3 & 4 \\ 4 & -5 & -2 \end{bmatrix} \begin{bmatrix} 3 & 5 \\ -4 & 2 \\ -1 & 3 \end{bmatrix}$
2. Check that matrix multiplication is defined. Since *F* is a 2×3 matrix and *E* is a 3×2 matrix, the product *EF* will be a 2×2 matrix.	
3. Recall the rules for matrix multiplication and apply.	$= \begin{bmatrix} 2\times3+3\times(-4)+4\times(-1) & 2\times5+3\times2+4\times3 \\ 4\times3+(-5)\times(-4)+2\times(-1) & 4\times5+(-5)\times2+(-2)\times3 \end{bmatrix}$
4. Simplify and give the final result.	$FE = \begin{bmatrix} -10 & 28 \\ 34 & 4 \end{bmatrix}$

5.3 Exercise

Technology free

1. State the order of each of the following matrices.

$$A=\begin{bmatrix}2 & 5\\ -1 & 3\\ 7 & -2\end{bmatrix},\ B=\begin{bmatrix}1\\ 5\\ 6\end{bmatrix},\ C=\begin{bmatrix}3 & 1\\ -1 & 6\end{bmatrix},\ D=\begin{bmatrix}8 & -1 & 1\end{bmatrix}$$

2. **WE4** Consider the following matrices:

$$A=\begin{bmatrix}2 & 5\\ -1 & 3\\ 7 & -2\end{bmatrix},\ B=\begin{bmatrix}1\\ 5\\ 6\end{bmatrix},\ C=\begin{bmatrix}3 & 1\\ -1 & 6\end{bmatrix},\ D=\begin{bmatrix}8 & -1 & 1\end{bmatrix}$$

a. Determine if matrix multiplication is defined for each of the products given below.

i. AB ii. BD iii. DA iv. CD v. DB vi. AC

b. State the order of each product in part a if the matrix multiplication is defined.

3. **WE5** Given the matrices $A=\begin{bmatrix}-2 & 4\\ 3 & 5\end{bmatrix}$ and $B=\begin{bmatrix}2 & 4\\ -1 & -3\end{bmatrix}$, determine the following matrices.

a. AB b. BA c. A^2 d. B^2

4. Given the matrices $A=\begin{bmatrix}-2 & 4\\ 3 & 5\end{bmatrix}$, $O=\begin{bmatrix}0 & 0\\ 0 & 0\end{bmatrix}$ and $I=\begin{bmatrix}1 & 0\\ 0 & 1\end{bmatrix}$, determine the following matrices.

a. AO b. OA c. AI d. IA

State what observations you can make from completing parts a to d.

5. a. Given the matrices $A=\begin{bmatrix}2 & 3\\ -1 & 4\end{bmatrix}$, $I=\begin{bmatrix}1 & 0\\ 0 & 1\end{bmatrix}$ and $O=\begin{bmatrix}0 & 0\\ 0 & 0\end{bmatrix}$, verify the following.

i. $AI=IA=A$ ii. $AO=OA=O$

b. Given the matrices $A=\begin{bmatrix}a & b\\ c & d\end{bmatrix}$ $I=\begin{bmatrix}1 & 0\\ 0 & 1\end{bmatrix}$ and $O=\begin{bmatrix}0 & 0\\ 0 & 0\end{bmatrix}$, verify the following.

i. $AI=IA=A$ ii. $AO=OA=O$

6. a. Given the matrices $A=\begin{bmatrix}2 & 3\\ -1 & 4\end{bmatrix}$ and $I=\begin{bmatrix}1 & 0\\ 0 & 1\end{bmatrix}$, verify that $(I-A)(I+A)=I-A^2$.

b. If $A=\begin{bmatrix}3 & 0\\ -4 & 0\end{bmatrix}$ and $B=\begin{bmatrix}0 & 0\\ 2 & -1\end{bmatrix}$, verify that $AB=O$ where $O=\begin{bmatrix}0 & 0\\ 0 & 0\end{bmatrix}$. Determine if $BA=O$.

c. If $A=\begin{bmatrix}a & 0\\ b & 0\end{bmatrix}$ and $B=\begin{bmatrix}0 & 0\\ x & y\end{bmatrix}$, verify that $AB=O$ where $O=\begin{bmatrix}0 & 0\\ 0 & 0\end{bmatrix}$. Determine if $BA=O$.

7. Given the matrices $A = \begin{bmatrix} a & b \\ c & d \end{bmatrix}$ and $X = \begin{bmatrix} x \\ y \end{bmatrix}$, determine the following matrices.

a. AX
b. XA

8. WE6 Given the matrices $D = \begin{bmatrix} 2 & -1 \\ -3 & 5 \\ -1 & -4 \end{bmatrix}$ and $E = \begin{bmatrix} 1 & 2 & -3 \\ 2 & -4 & 5 \end{bmatrix}$, determine the following matrices.

a. DE
b. ED

9. Given the matrices $C = \begin{bmatrix} 1 \\ -2 \end{bmatrix}$ and $D = \begin{bmatrix} 3 & -2 \end{bmatrix}$, determine the following matrices.

a. CD
b. DC

10. If $A = \begin{bmatrix} x & -3 \\ 2 & x \end{bmatrix}$ and $B = \begin{bmatrix} 2 & x \\ x & -3 \end{bmatrix}$, calculate the value of x given the following.

a. $AB = \begin{bmatrix} 3 & 18 \\ 13 & 3 \end{bmatrix}$
b. $BA = \begin{bmatrix} -16 & 10 \\ 10 & 24 \end{bmatrix}$
c. $A^2 = \begin{bmatrix} -2 & 12 \\ -8 & -2 \end{bmatrix}$
d. $B^2 = \begin{bmatrix} 8 & -2 \\ -2 & 13 \end{bmatrix}$

The following information relates to questions 11 and 12.

Consider the following matrices: $A = \begin{bmatrix} -1 \\ 2 \end{bmatrix}$, $B = \begin{bmatrix} 3 & -5 \end{bmatrix}$ and $C = \begin{bmatrix} 2 & 4 \\ -3 & 5 \end{bmatrix}$.

11. Calculate the following.

a. $A + B$
b. $A + C$
c. $B + C$
d. AB
e. BA
f. AC

12. Calculate the following.

a. CA
b. BC
c. CB
d. ABC
e. CBA
f. CAB

13. Given $D = \begin{bmatrix} 1 & -4 & 2 \\ -2 & 8 & -4 \end{bmatrix}$ and $E = \begin{bmatrix} 6 & 2 \\ 3 & -1 \\ 3 & -3 \end{bmatrix}$, determine the following matrices.

a. DE
b. ED
c. $E + D$
d. D^2

14. Consider the matrices $A = \begin{bmatrix} 2 & 3 \\ -1 & 4 \end{bmatrix}$, $B = \begin{bmatrix} 4 & 5 \\ 2 & -3 \end{bmatrix}$ and $C = \begin{bmatrix} 1 & -2 \\ 5 & 4 \end{bmatrix}$.

a. Verify the Distributive Law: $A(B + C) = AB + AC$.
b. Verify the Associative Law for Multiplication: $A(BC) = (AB)C$.
c. Determine if $(A + B)^2 = A^2 + 2AB + B^2$. Explain your answer.
d. Show that $(A + B)^2 = A^2 + AB + BA + B^2$.

15. a. If $P = \begin{bmatrix} -1 & 0 \\ 0 & 4 \end{bmatrix}$, determine the matrices P^2, P^3 and P^4, and deduce the matrix P^n.

b. If $Q = \begin{bmatrix} 2 & 0 \\ 0 & -3 \end{bmatrix}$, determine the matrices Q^2, Q^3 and Q^4, and deduce the matrix Q^n.

16. a. If $R=\begin{bmatrix}1 & 0\\3 & 1\end{bmatrix}$, determine the matrices R^2, R^3 and R^4, and deduce the matrix R^n.

b. If $S=\begin{bmatrix}0 & 3\\2 & 0\end{bmatrix}$, determine the matrices S^2, S^3 and S^4, and deduce the matrices S^8 and S^9.

17. If $A=\begin{bmatrix}2 & 3\\-1 & 4\end{bmatrix}$ and $I=\begin{bmatrix}1 & 0\\0 & 1\end{bmatrix}$, evaluate the matrix $A^2-6A+11I$.

18. a. If $B=\begin{bmatrix}4 & 5\\-2 & -3\end{bmatrix}$ and $I=\begin{bmatrix}1 & 0\\0 & 1\end{bmatrix}$, evaluate the matrix $B^2-B-22I$.

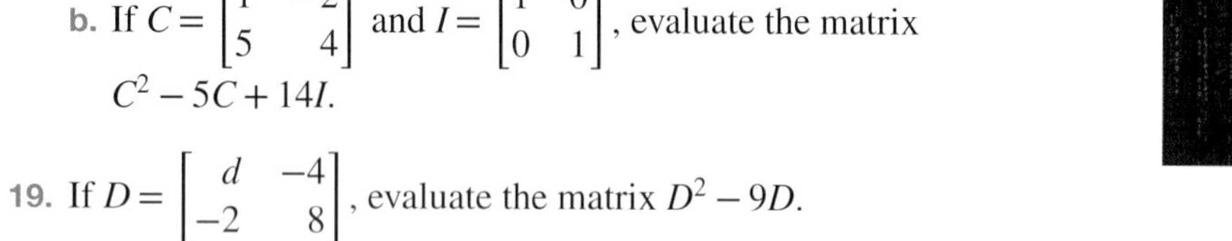

b. If $C=\begin{bmatrix}1 & -2\\5 & 4\end{bmatrix}$ and $I=\begin{bmatrix}1 & 0\\0 & 1\end{bmatrix}$, evaluate the matrix $C^2-5C+14I$.

19. If $D=\begin{bmatrix}d & -4\\-2 & 8\end{bmatrix}$, evaluate the matrix D^2-9D.

20. The trace of a matrix A, denoted by $\text{tr}(A)$, is equal to the sum of leading diagonal elements. For 2×2 matrices, if $A=\begin{bmatrix}a_{11} & a_{12}\\a_{21} & a_{22}\end{bmatrix}$, then $\text{tr}(A)=a_{11}+a_{22}$.

Consider the matrices $A=\begin{bmatrix}2 & 3\\-1 & 4\end{bmatrix}$ $B=\begin{bmatrix}4 & -2\\3 & 5\end{bmatrix}$ and $C=\begin{bmatrix}1 & -2\\5 & 4\end{bmatrix}$.

a. Calculate the following.

i. $\text{tr}(AB)$
ii. $\text{tr}(BA)$
iii. $\text{tr}(A)\text{tr}(B)$

b. Determine if $\text{tr}(ABC)$ is equal to $\text{tr}(A)\,\text{tr}(B)\,\text{tr}(C)$.

5.3 Exam questions

Question 1 (1 mark) TECH-ACTIVE

MC If A is a (2×3) matrix and B is a (3×1) matrix, then the product AB

A. is a (2×1) matrix.
B. is a (3×3) matrix.
C. is a (2×3) matrix.
D. does not exist.
E. is not defined.

Question 2 (1 mark) TECH-ACTIVE

MC The set of matrices that can be multiplied is

A. $\begin{bmatrix}4\\1\\4\end{bmatrix}$ and $\begin{bmatrix}5\\1\\3\end{bmatrix}$

B. $\begin{bmatrix}4 & 2\end{bmatrix}$ and $\begin{bmatrix}5 & 3\end{bmatrix}$

C. $\begin{bmatrix}3 & 1 & 2\end{bmatrix}$ and $\begin{bmatrix}5\\3\\1\end{bmatrix}$

D. $\begin{bmatrix}4 & 2\\7 & 3\end{bmatrix}$ and $\begin{bmatrix}9 & 1 & 5\\3 & 2 & 4\\8 & 1 & 3\end{bmatrix}$

E. $[7]$ and $\begin{bmatrix}3\\9\end{bmatrix}$

Question 3 (1 mark) TECH-ACTIVE

MC $\begin{bmatrix}1 & 4\\6 & 3\end{bmatrix}\times\begin{bmatrix}8 & 1 & 3\\0 & 9 & 4\end{bmatrix}$ is equal to

A. $\begin{bmatrix}8 & 19\\48 & 37\end{bmatrix}$

B. $\begin{bmatrix}8 & 37 & 19\\48 & 33 & 30\end{bmatrix}$

C. $\begin{bmatrix}48 & 37\\30 & 33\end{bmatrix}$

D. $\begin{bmatrix}8 & 48\\37 & 33\\19 & 30\end{bmatrix}$

E. $\begin{bmatrix}8 & 28 & 15\\48 & 42 & 34\end{bmatrix}$

More exam questions are available online.

5.4 Determinants and inverses

LEARNING INTENTION

At the end of this subtopic you should be able to:

- calculate the determinants of 2×2 and 3×3 matrices
- determine the inverses of 2×2 and 3×3 matrices.

5.4.1 Determinant of a 2×2 matrix

Associated with a square matrix is a single number called the **determinant of a matrix**. For a matrix A the determinant of the matrix A is denoted by $\det(A)$ or the symbol Δ. The determinant is represented not by the square brackets that we use for matrices, but by straight lines; that is,

if $A=\begin{bmatrix}a & b\\c & d\end{bmatrix}$, $\det(A)=\Delta=|A|=\begin{vmatrix}a & b\\c & d\end{vmatrix}$.

To evaluate the determinant, multiply the elements in the leading diagonal and subtract the product of the elements in the other diagonal.

Determinant of a 2×2 matrix

The determinant of a 2×2 matrix $A=\begin{bmatrix}a & b\\c & d\end{bmatrix}$ is defined as follows:

$$\det(A)=ad-bc$$

WORKED EXAMPLE 7 Calculating the determinant of a 2 × 2 matrix

Calculate the determinant of the matrix $F=\begin{bmatrix}3 & 5\\4 & 7\end{bmatrix}$.

THINK	WRITE
1. Apply the definition $\det(F)=ad-bc$ by multiplying the elements in the leading diagonal. Subtract the product of the elements in the other diagonal.	$F=\begin{bmatrix}3 & 5\\4 & 7\end{bmatrix}$ $\det(F)=3\times 7-5\times 4$ $=21-20$
2. State the value of the determinant.	$\det(F)=1$

5.4.2 Determinant of a 3×3 matrix

The determinant of a 3×3 matrix involves evaluating the determinants of three 2×2 matrices.

If $A = \begin{bmatrix} a & b & c \\ d & e & f \\ g & h & i \end{bmatrix}$, $\det(A) = a\begin{vmatrix} e & f \\ h & i \end{vmatrix} - b\begin{vmatrix} d & f \\ g & i \end{vmatrix} + c\begin{vmatrix} d & e \\ g & h \end{vmatrix}$.

The three sub-determinants are referred to as **minors**.

The coefficients of each sub-determinant are the elements of row 1 of the 3×3 matrix, that is, elements a, b and c.

The second coefficient, b, is given a negative sign.

Minors are formed by removing the row and column of each of these coefficients and using the remaining elements to give three 2×2 matrices.

For example,

$\begin{bmatrix} a & b & c \\ d & e & f \\ g & h & i \end{bmatrix}$ becomes $a\begin{bmatrix} e & f \\ h & i \end{bmatrix}$ $\quad$ $\begin{bmatrix} a & b & c \\ d & e & f \\ g & h & i \end{bmatrix}$ becomes $-b\begin{bmatrix} d & f \\ g & i \end{bmatrix}$ $\quad$ $\begin{bmatrix} a & b & c \\ d & e & f \\ g & h & i \end{bmatrix}$ becomes $c\begin{bmatrix} d & e \\ g & h \end{bmatrix}$

Determinant of a 3×3 matrix

If $A = \begin{bmatrix} a & b & c \\ d & e & f \\ g & h & i \end{bmatrix}$, $\det(A) = a\begin{vmatrix} e & f \\ h & i \end{vmatrix} - b\begin{vmatrix} d & f \\ g & i \end{vmatrix} + c\begin{vmatrix} d & e \\ g & h \end{vmatrix}$

Note: Although you may not be required to calculate the determinant of matrices of higher order than 2×2 without technology, the formula is included here to enhance your understanding. Worked example 9 demonstrates the calculation of higher-order determinants using technology.

WORKED EXAMPLE 8 Calculating the determinant of a 3 × 3 matrix

Evaluate $\begin{vmatrix} 2 & 1 & 3 \\ 1 & -1 & 2 \\ -1 & 2 & 0 \end{vmatrix}$.

THINK	WRITE
1. Use elements of row 1 as the coefficients of the minors.	$\begin{vmatrix} 2 & 1 & 3 \\ 1 & -1 & 2 \\ -1 & 2 & 0 \end{vmatrix}$ $= 2\begin{vmatrix} -1 & 2 \\ 2 & 0 \end{vmatrix} - 1\begin{vmatrix} 1 & 2 \\ -1 & 0 \end{vmatrix} + 3\begin{vmatrix} 1 & -1 \\ -1 & 2 \end{vmatrix}$
2. Evaluate the minors.	$= 2(-1 \times 0 - 2 \times 2) - 1(1 \times 0 - 2 \times -1) + 3(1 \times 2 - (-1) \times -1)$ $= 2(0 - 4) - 1(0 + 2) + 3(2 - 1)$ $= -8 - 2 + 3$ $= -7$

The determinant of a transpose matrix

A matrix, A, can be transposed to form a new matrix, A^T, by swapping the rows with the columns, as follows:

$$A = \begin{bmatrix} a & b & c \\ d & e & f \end{bmatrix}, \quad A^T = \begin{bmatrix} a & d \\ b & e \\ c & f \end{bmatrix}$$

For a square matrix, A, its transpose, A^T, is formed by flipping the other elements over the main diagonal. The determinants of these two matrices are equal.

Determinant of a transposed matrix

For square matrices:

$$A = \begin{bmatrix} a & b & c \\ d & e & f \\ g & h & i \end{bmatrix}, \quad A^T = \begin{bmatrix} a & d & g \\ b & e & h \\ c & f & i \end{bmatrix}$$

$$\det(A) = \det\left(A^T\right)$$

WORKED EXAMPLE 9 Calculating the determinant of a square matrix and its transpose

If $A = \begin{bmatrix} 1 & -3 & 2 \\ -1 & 4 & 3 \\ 0 & 2 & 5 \end{bmatrix}$, use technology to calculate the determinant of A and the determinant of the transpose of matrix A, A^T. State what conclusion you can make.

TI | THINK

1. On a Calculator page, press the template button and complete the entry line as:
$\begin{bmatrix} 1 & -3 & 2 \\ -1 & 4 & 3 \\ 0 & 2 & 5 \end{bmatrix}$
Press CTRL, then press VAR, then type a and press ENTER to store matrix A.

DISPLAY/WRITE

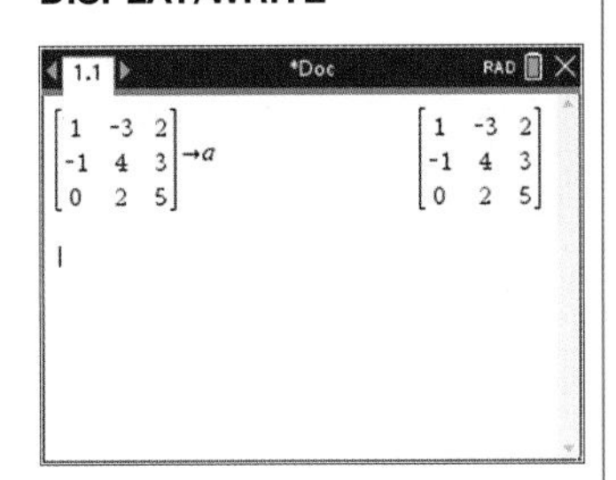

2. Press MENU then select:
7 Matrix & Vector
3 Determinant
Complete the entry line as det(a) then press ENTER.

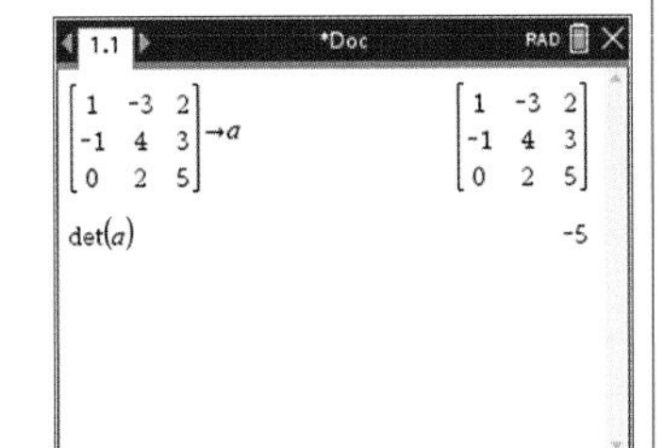

CASIO | THINK

1. From the Keyboard Math2 tab, select the 2×2 matrix format and select again to create the required 3×3 matrix.
Enter the element values as shown.

DISPLAY/WRITE

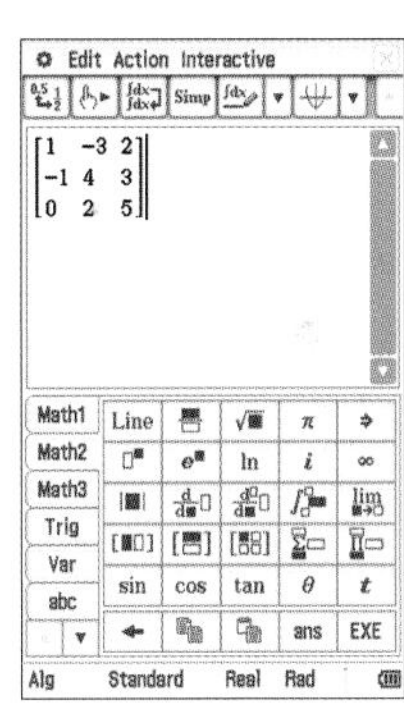

2. To highlight and select this completed 3×3 matrix, swipe across the matrix with a stylus.
3. Select Interactive > Matrix > Calculation > det.

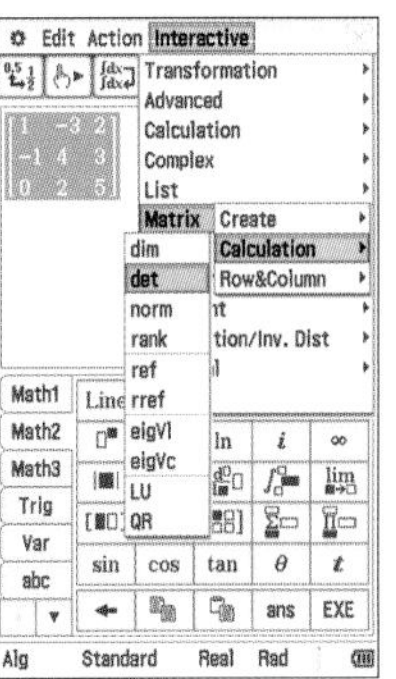

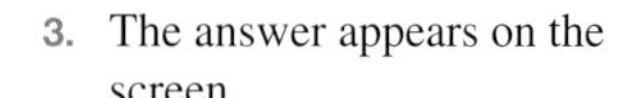

3. The answer appears on the screen.

$\det(A) = 5$

4. Press MENU then select:
7 Matrix & Vector
3 Determinant. Type a.
Press MENU then select:
7 Matrix & Vector
2 Transpose
and press ENTER.

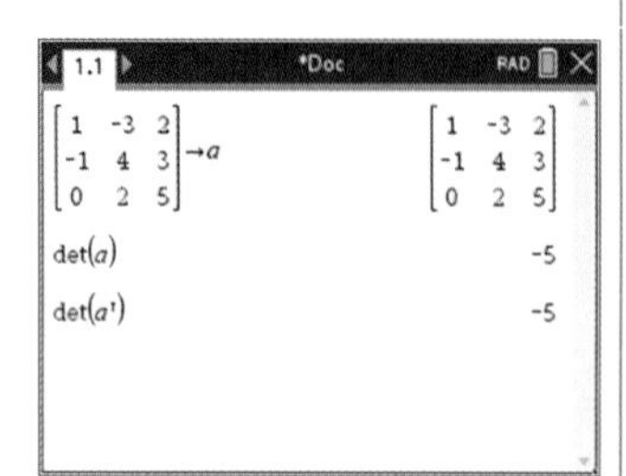

5. The answer appears on the screen.

$\det\left(A^{T}\right) = 5$

6. State your conclusion.

$\therefore \det(A) = \det\left(A^{T}\right)$

4. The answer appears on the screen.

5. Select the 3×3 matrix and drag it down to a new entry line.

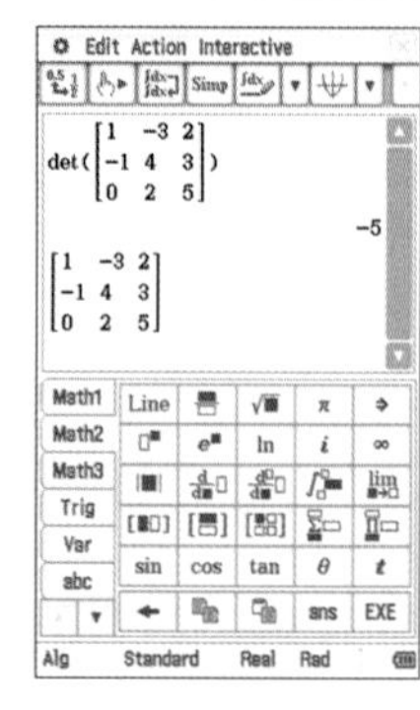

6. Highlight (select) the 3×3 matrix.
7. Select Interactive > Create > trn to create the transpose.

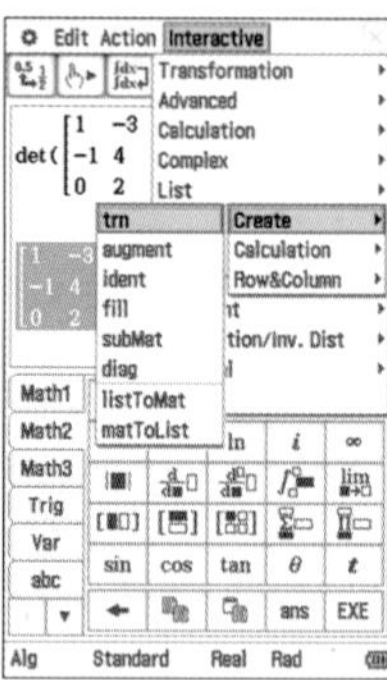

8. Highlight the already entered $\text{trn}\left(\begin{bmatrix}1 & -3 & 2\\-1 & 4 & 3\\0 & 2 & 5\end{bmatrix}\right)$.
9. Select Interactive > Matrix > Calculation > det.
10. The answer will appear on the screen.

11. State your conclusion

$\therefore \det(A) = \det\left(A^{T}\right)$

5.4.3 Inverse of a 2×2 matrix

The identity matrix I, defined by $I = \begin{bmatrix}1 & 0\\0 & 1\end{bmatrix}$, has the property that for a 2×2 non-zero matrix A, $AI = IA = A$.

When any square matrix is multiplied by its **multiplicative inverse**, the identity matrix I is obtained.

This is the same as multiplying 3 by its multiplicative inverse, $\frac{1}{3}$; the result is 1.

That is, $3 \times \frac{1}{3} = 1$ and $\frac{1}{3} \times 3 = 1$.

Thus, 3 is the multiplicative inverse of $\frac{1}{3}$ and $\frac{1}{3}$ is the multiplicative inverse of 3.

For a matrix, A, the multiplicative inverse is called the inverse matrix and is denoted by A^{-1}, and $AA^{-1} = A^{-1}A = I$. Note that $A^{-1} \neq \frac{1}{A}$ as division of matrices is not defined.

Consider the products of $A = \begin{bmatrix} 3 & 5 \\ 4 & 7 \end{bmatrix}$ and $A^{-1} = \begin{bmatrix} 7 & -5 \\ -4 & 3 \end{bmatrix}$

$$AA^{-1} = \begin{bmatrix} 3 & 5 \\ 4 & 7 \end{bmatrix}\begin{bmatrix} 7 & -5 \\ -4 & 3 \end{bmatrix} = \begin{bmatrix} 3\times 7 + 5\times(-4) & 3\times(-5) + 5\times 3 \\ 4\times 7 + 7\times(-4) & 4\times(-5) + 7\times 3 \end{bmatrix} = \begin{bmatrix} 1 & 0 \\ 0 & 1 \end{bmatrix}$$

$$A^{-1}A = \begin{bmatrix} 7 & -5 \\ -4 & 3 \end{bmatrix}\begin{bmatrix} 3 & 5 \\ 4 & 7 \end{bmatrix} = \begin{bmatrix} 7\times 3 + (-5)\times 4 & 7\times 5 + (-5)\times 7 \\ (-4)\times 3 + 3\times 4 & (-4)\times 5 + 3\times 7 \end{bmatrix} = \begin{bmatrix} 1 & 0 \\ 0 & 1 \end{bmatrix}$$

Now for the matrix $A = \begin{bmatrix} 3 & 5 \\ 4 & 7 \end{bmatrix}$, the determinant $\begin{vmatrix} 3 & 5 \\ 4 & 7 \end{vmatrix} = 3\times 7 - 5\times 4 = 1$.

$A^{-1} = \begin{bmatrix} 7 & -5 \\ -4 & 3 \end{bmatrix}$ is obtained from the matrix A by swapping the elements on the leading diagonal, and placing a negative sign on the other two elements.

In general, the inverse of a 2×2 matrix $A = \begin{bmatrix} a & b \\ c & d \end{bmatrix}$ can be determined in three simple steps.

Step 1: Evaluate $\frac{1}{\det(A)} = \frac{1}{ad - bc}$.

Step 2: Swap a with d and multiply b and c by -1 to form the matrix $\begin{bmatrix} d & -b \\ -c & a \end{bmatrix}$.

Step 3: Multiply the results of the previous steps together to form the inverse $A^{-1} = \frac{1}{ad - bc}\begin{bmatrix} d & -b \\ -c & a \end{bmatrix}$

Inverse of a 2×2 matrix

The inverse of a 2×2 matrix $A = \begin{bmatrix} a & b \\ c & d \end{bmatrix}$ is defined as:

$$A^{-1} = \frac{1}{ad - bc}\begin{bmatrix} d & -b \\ -c & a \end{bmatrix}$$

WORKED EXAMPLE 10 Determining the inverse of a 2×2 matrix

Determine the inverse of the matrix $P = \begin{bmatrix} 2 & 3 \\ -1 & 5 \end{bmatrix}$ and verify that $PP^{-1} = P^{-1}P = I$.

THINK	WRITE
1. Calculate the determinant. If $P = \begin{bmatrix} a & b \\ c & d \end{bmatrix}$, then $\|P\| = ad - bc$.	$\|P\| = \begin{vmatrix} 2 & 3 \\ -1 & 5 \end{vmatrix}$ $= 2\times 5 - 3\times(-1)$ $= 10 + 3$ $= 13$

2. To determine the inverse of matrix P, recall and apply the rule $P^{-1} = \frac{1}{ad-bc}\begin{bmatrix} d & -b \\ -c & a \end{bmatrix}$.

$$P^{-1} = \frac{1}{13}\begin{bmatrix} 5 & -3 \\ 1 & 2 \end{bmatrix}$$

3. Substitute and evaluate PP^{-1}.

$$PP^{-1} = \begin{bmatrix} 2 & 3 \\ -1 & 5 \end{bmatrix} \times \frac{1}{13}\begin{bmatrix} 5 & -3 \\ 1 & 2 \end{bmatrix}$$

4. Apply the rules for scalar multiplication and multiplication of matrices.

$$= \frac{1}{13}\begin{bmatrix} 2\times 5+3\times 1 & 2\times(-3)+3\times 2 \\ (-1)\times 5+5\times 1 & (-1)\times(-3)+5\times 2 \end{bmatrix}$$

5. Simplify the matrix product to show that $PP^{-1} = I$.

$$= \frac{1}{13}\begin{bmatrix} 13 & 0 \\ 0 & 13 \end{bmatrix} = \begin{bmatrix} 1 & 0 \\ 0 & 1 \end{bmatrix}$$

6. Substitute and evaluate $P^{-1}P$.

$$P^{-1}P = \frac{1}{13}\begin{bmatrix} 5 & -3 \\ 1 & 2 \end{bmatrix} \times \begin{bmatrix} 2 & 3 \\ -1 & 5 \end{bmatrix}$$

7. Use the rules for scalar multiplication and multiplication of matrices.

$$= \frac{1}{13}\begin{bmatrix} 5\times 2+(-3)\times(-1) & 5\times 3+(-3)\times 5 \\ 1\times 2+2\times(-1) & 1\times 3+2\times 5 \end{bmatrix}$$

8. Simplify the matrix product to show that $P^{-1}P = I$.

$$= \frac{1}{13}\begin{bmatrix} 13 & 0 \\ 0 & 13 \end{bmatrix} = \begin{bmatrix} 1 & 0 \\ 0 & 1 \end{bmatrix}$$

9. Answer the question.

$$PP^{-1} = P^{-1}P = \begin{bmatrix} 1 & 0 \\ 0 & 1 \end{bmatrix}$$

Singular matrices

If a matrix has a zero determinant then the inverse matrix does not exist, and the original matrix is termed a **singular matrix**.

$\frac{1}{\det(A)} = \frac{1}{0}$ is not defined as we cannot divide by zero.

Hence, if $\det(A) = 0$, then A^{-1} does not exist and matrix A is singular.

WORKED EXAMPLE 11 Verifying that a matrix is singular

Demonstrate that the matrix $\begin{bmatrix} -3 & 2 \\ 6 & -4 \end{bmatrix}$ is singular.

THINK

WRITE

1. Evaluate the determinant.

$$\begin{vmatrix} -3 & 2 \\ 6 & -4 \end{vmatrix} = (-3\times -4) - (2\times 6)$$
$$= 12 - 12$$
$$= 0$$

2. Since the determinant is zero, the matrix $\begin{bmatrix} -3 & 2 \\ 6 & -4 \end{bmatrix}$ is singular.

$$\begin{vmatrix} -3 & 2 \\ 6 & -4 \end{vmatrix} = 0$$

WORKED EXAMPLE 12 Applications of matrices

If $A = \begin{bmatrix} -2 & 4 \\ 3 & 5 \end{bmatrix}$ and $I = \begin{bmatrix} 1 & 0 \\ 0 & 1 \end{bmatrix}$, express the determinant of the matrix $A - kI$ in the form $pk^2 + qk + r$, stating the values of p, q and r. Hence evaluate the matrix $pA^2 + qA + rI$.

THINK	WRITE
1. Substitute to calculate the matrix $A - kI$. Recall and apply the rules for scalar multiplication and subtraction of matrices.	$A - kI = \begin{bmatrix} -2 & 4 \\ 3 & 5 \end{bmatrix} - k\begin{bmatrix} 1 & 0 \\ 0 & 1 \end{bmatrix} = \begin{bmatrix} -2-k & 4 \\ 3 & 5-k \end{bmatrix}$
2. Evaluate the determinant of the matrix $A - kI$.	$\det(A - kI) = \begin{vmatrix} -2-k & 4 \\ 3 & 5-k \end{vmatrix}$ $= (-2-k)(5-k) - 4 \times 3$
3. Simplify the determinant of the matrix $A - kI$.	$= -(2+k)(5-k) - 1$ $= -\left(10 + 3k - k^2\right) - 12$ $= k^2 - 3k - 22$
4. Equate the matrices and state the values of p, q and r.	$pk^2 + qk + r = k^2 - 3k - 22$ $\therefore p = 1; q = 3; r = -22$
5. Determine the matrix A^2.	$A^2 = \begin{bmatrix} -2 & 4 \\ 3 & 5 \end{bmatrix}\begin{bmatrix} -2 & 4 \\ 3 & 5 \end{bmatrix} = \begin{bmatrix} 16 & 12 \\ 9 & 37 \end{bmatrix}$
6. Substitute for p, q and r and evaluate the matrix $A^2 - 3A - 22I$.	$A^2 - 3A - 22I = \begin{bmatrix} 16 & 12 \\ 9 & 37 \end{bmatrix} - 3\begin{bmatrix} -2 & 4 \\ 3 & 5 \end{bmatrix} - 22\begin{bmatrix} 1 & 0 \\ 0 & 1 \end{bmatrix}$
7. Simplify by applying the rules for scalar multiplication of matrices.	$= \begin{bmatrix} 16 & 12 \\ 9 & 37 \end{bmatrix} - \begin{bmatrix} -6 & 12 \\ 9 & 15 \end{bmatrix} - \begin{bmatrix} 22 & 0 \\ 0 & 22 \end{bmatrix}$
8. Simplify and apply the rules for addition and subtraction of matrices.	$\therefore A^2 - 3A - 22I = \begin{bmatrix} 0 & 0 \\ 0 & 0 \end{bmatrix}$

5.4.4 Inverse of a 3 × 3 matrix

Calculators can be used to determine the inverse of a 3×3 matrix.

WORKED EXAMPLE 13 Determining the inverse of a 3 × 3 matrix using technology

Using technology, determine the inverse of the following matrix.

$$A = \begin{bmatrix} 3 & -1 & 3 \\ 1 & 2 & -1 \\ 5 & 3 & 4 \end{bmatrix}$$

TI | THINK	DISPLAY/WRITE
On a Calculator page, complete the entry line as shown.	$\begin{bmatrix}3 & -1 & 3\\1 & 2 & -1\\5 & 3 & 4\end{bmatrix}^{-1}$ $\begin{bmatrix}\frac{11}{21} & \frac{13}{21} & \frac{-5}{21}\\ \frac{-3}{7} & \frac{-1}{7} & \frac{2}{7}\\ \frac{-1}{3} & \frac{-2}{3} & \frac{1}{3}\end{bmatrix}$

CASIO | THINK	DISPLAY/WRITE
On a Main screen, complete the entry line as shown.	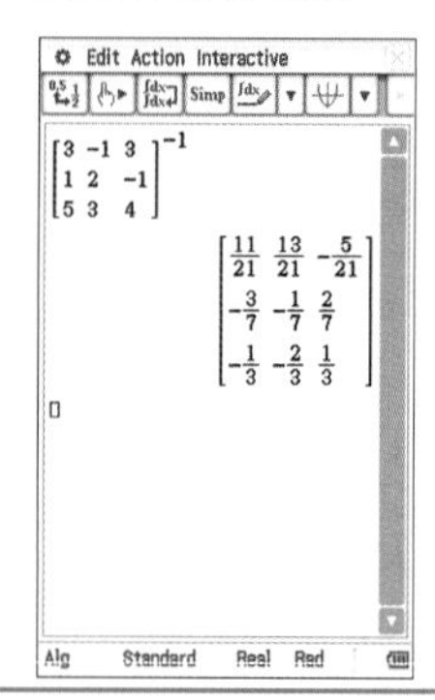

5.4 Exercise

Technology free

1. **WE7** Calculate the determinant of the matrix $G = \begin{bmatrix}-2 & 4\\3 & 5\end{bmatrix}$.

2. The matrix $\begin{bmatrix}x & 5\\3 & x+2\end{bmatrix}$ has a determinant equal to 9. Calculate the possible values of x.

3. **WE8** Evaluate the following.

   a. $\begin{vmatrix}2 & 1 & 3\\4 & -2 & 5\\-1 & 3 & 6\end{vmatrix}$

   b. $\begin{vmatrix}-1 & -1 & 0\\-3 & 4 & 2\\2 & 3 & 5\end{vmatrix}$

4. **WE10** Determine the inverse of the matrix $A = \begin{bmatrix}4 & -2\\5 & 6\end{bmatrix}$ and verify that $AA^{-1} = A^{-1}A = I$.

5. Determine the inverse matrix of each of the following matrices.

   a. $\begin{bmatrix}-1 & 0\\0 & 4\end{bmatrix}$

   b. $\begin{bmatrix}2 & 1\\0 & -3\end{bmatrix}$

   c. $\begin{bmatrix}2 & 0\\3 & 1\end{bmatrix}$

   d. $\begin{bmatrix}0 & -3\\2 & -1\end{bmatrix}$

6. The inverse of the matrix $\begin{bmatrix}2 & 3\\3 & 4\end{bmatrix}$ is $\begin{bmatrix}p & 3\\3 & q\end{bmatrix}$. Calculate the values of p and q.

7. **WE11** Demonstrate that the matrix $\begin{bmatrix}1 & -2\\-5 & 10\end{bmatrix}$ is singular.

8. Calculate the value of x if the matrix $\begin{bmatrix}x & 4\\3 & x+4\end{bmatrix}$ is singular.

9. Consider the matrix $P = \begin{bmatrix}6 & -2\\4 & 2\end{bmatrix}$.

   a. Calculate the following.

      i. $\det(P)$

      ii. P^{-1}

   b. Verify that $PP^{-1} = P^{-1}P = I$.

c. Calculate the following.

i. $\det\left(P^{-1}\right)$ ii. $\det(P)\det\left(P^{-1}\right)$

10. Calculate the value of x for each of the following.

a. $\begin{vmatrix} x & -3 \\ 4 & 2 \end{vmatrix} = 6$ b. $\begin{vmatrix} x & 3 \\ 4 & x \end{vmatrix} = 4$

11. Calculate the values of x if each of the following are singular matrices.

a. $\begin{bmatrix} x & -3 \\ 4 & 2 \end{bmatrix}$ b. $\begin{bmatrix} x & 3 \\ 4 & x \end{bmatrix}$

12. Given $A = \begin{bmatrix} -1 \\ 2 \end{bmatrix}$, $B = \begin{bmatrix} 3 & -5 \end{bmatrix}$ and $C = \begin{bmatrix} 2 & 4 \\ -3 & 5 \end{bmatrix}$, determine, if possible, the following matrices.

a. $(AB)^{-1}$ b. A^{-1} c. C^{-1} d. $(ABC)^{-1}$

13. **WE12** If $A = \begin{bmatrix} 2 & 3 \\ -1 & 5 \end{bmatrix}$ and $I = \begin{bmatrix} 1 & 0 \\ 0 & 1 \end{bmatrix}$, express the determinant of the matrix $A - kI$ in the form $pk^2 + qk + r$, stating the values of p, q and r. Hence, evaluate the matrix $pA^2 + qA + rI$.

14. If $A = \begin{bmatrix} 4 & -8 \\ -3 & 2 \end{bmatrix}$ and $I = \begin{bmatrix} 1 & 0 \\ 0 & 1 \end{bmatrix}$, determine the value of k for which the determinant of the matrix $A - kI$ is equal to zero.

15. If $A = \begin{bmatrix} 2 & -3 \\ -1 & -4 \end{bmatrix}$ and $I = \begin{bmatrix} 1 & 0 \\ 0 & 1 \end{bmatrix}$, express the determinant of the matrix $A - kI$, $k \in R$ in the form $pk^2 + qk + r$, stating the values of p, q and r. Hence, evaluate the matrix $pA^2 + qA + rI$.

Technology active

Questions 16 and 17 refer to the following matrices:

$$A = \begin{bmatrix} 2 & -3 \\ -1 & -4 \end{bmatrix}, B = \begin{bmatrix} 4 & 5 \\ 2 & 3 \end{bmatrix} \text{ and } C = \begin{bmatrix} 1 & -2 \\ 3 & 4 \end{bmatrix}$$

16. a. Calculate $\det(A)$, $\det(B)$ and $\det(C)$.
    b. Determine if $\det(AB) = \det(A)\det(B)$.
    c. Verify that $\det(ABC) = \det(A)\det(B)\det(C)$.

17. a. Calculate the matrices A^{-1}, B^{-1}, C^{-1}.
    b. Determine if $(AB)^{-1} = A^{-1}B^{-1}$.
    c. Determine if $(AB)^{-1} = B^{-1}A^{-1}$.
    d. Determine if $(ABC)^{-1} = C^{-1}B^{-1}A^{-1}$.

18. **WE9** If $A = \begin{bmatrix} 4 & 6 & 8 \\ -2 & 3 & 6 \\ -3 & 2 & -1 \end{bmatrix}$, use technology to calculate the determinants of A and the determinant of the transpose of matrix A, A^T. State what conclusion you can make.

19. **WE13** Use technology to determine the inverse of each of the following matrices.

a. $A = \begin{bmatrix} 5 & -1 & 2 \\ -2 & 3 & -1 \\ 6 & 4 & -4 \end{bmatrix}$ b. $B = \begin{bmatrix} 12 & 8 & 0 & 4 \\ 8 & 4 & 8 & 8 \\ 4 & 12 & 8 & 12 \\ 8 & 4 & 4 & 8 \end{bmatrix}$

20. Consider the matrices $B = \begin{bmatrix} -3 & 5 \\ -2 & 4 \end{bmatrix}$, $Q = \begin{bmatrix} 5 & 1 \\ 2 & 1 \end{bmatrix}$ and $I = \begin{bmatrix} 1 & 0 \\ 0 & 1 \end{bmatrix}$.
    a. Calculate the values of k for which the determinant of the matrix $B - kI = 0$.
    b. Determine the matrix $Q^{-1}BQ$.

5.4 Exam questions

Question 1 (1 mark) TECH-ACTIVE

MC Which of these matrices is singular?

A. $\begin{bmatrix} -1 & -2 \\ 2 & 1 \end{bmatrix}$ **B.** $\begin{bmatrix} 4 & -1 \\ 8 & 2 \end{bmatrix}$ **C.** $\begin{bmatrix} 5 & 1 \\ 1 & 5 \end{bmatrix}$ **D.** $\begin{bmatrix} 2 & 1 \\ 6 & 3 \end{bmatrix}$ **E.** $\begin{bmatrix} 5 & 3 \\ 4 & 2 \end{bmatrix}$

Question 2 (1 mark) TECH-ACTIVE

MC The inverse of the matrix $\begin{bmatrix} 4 & 7 \\ 2 & 9 \end{bmatrix}$ is:

A. $\begin{bmatrix} 9 & -7 \\ -1 & 2 \end{bmatrix}$ **B.** $\begin{bmatrix} \frac{9}{22} & \frac{-7}{22} \\ \frac{-1}{22} & \frac{2}{22} \end{bmatrix}$ **C.** 22 **D.** $\begin{bmatrix} \frac{4}{22} & \frac{7}{22} \\ \frac{2}{22} & \frac{9}{22} \end{bmatrix}$ **E.** $\begin{bmatrix} \frac{4}{22} & -\frac{7}{22} \\ -\frac{2}{22} & \frac{9}{22} \end{bmatrix}$

Question 3 (3 marks) TECH-FREE

If $A = \begin{bmatrix} 1 & 2 \\ 5 & 4 \end{bmatrix}$, determine the value(s) of k for which the determinant of the matrix $A - kI = 0$.

More exam questions are available online.

5.5 Matrix equations

LEARNING INTENTION

At the end of this subtopic you should be able to:

- represent simultaneous equations in matrix form
- use matrix algebra to solve matrix equations and systems of linear equations.

5.5.1 Introduction to matrix equations

Inverse matrices are used to solve matrix equations as division of matrices is not possible.

Consider the matrix equations $AX = B$ and $XA = B$, where A, B and X are matrices, and X needs to be found.

If $AX = B$, pre-multiply both sides by A^{-1}, the inverse of matrix A. Remember order of multiplication is important when multiplying matrices.

$$\Rightarrow A^{-1}AX = A^{-1}B$$

$$\text{Since } A^{-1}A = I,$$

$$IX = A^{-1}B \qquad \text{where } I = \begin{bmatrix} 1 & 0 \\ 0 & 1 \end{bmatrix}$$

$$\therefore X = A^{-1}B.$$

If $XA = B$, multiply both sides of the equation by A^{-1}, remembering order of multiplication is important.

$$\Rightarrow XAA^{-1} = BA^{-1}$$

$$\text{Since } AA^{-1} = I,$$

$$XI = BA^{-1} \qquad \text{where } I = \begin{bmatrix} 1 & 0 \\ 0 & 1 \end{bmatrix}$$

$$\therefore X = BA^{-1}.$$

Solving matrix equations

In general, if $AX = B$, then $X = A^{-1}B$, and if $XA = B$, then $X = BA^{-1}$.

WORKED EXAMPLE 14 Solving matrix equations

Given the matrices $A = \begin{bmatrix} 3 & -4 \\ 5 & -6 \end{bmatrix}$, $C = \begin{bmatrix} -1 \\ 2 \end{bmatrix}$ and $D = \begin{bmatrix} 3 & -2 \end{bmatrix}$, determine the matrix X if:

a. $AX = C$

b. $XA = D$.

THINK	WRITE
a. 1. If $AX = C$, pre-multiply both sides by the inverse matrix, A^{-1} and solve for X.	a. $AX = C$ $A^{-1}AX = A^{-1}C$ $IX = X = A^{-1}C$ $\therefore X = A^{-1}C$
2. Recall the determinant rule and apply to evaluate the determinant of the matrix A. $\det(A) = ad - bc$	$\det(A) = \begin{vmatrix} 3 & -4 \\ 5 & -6 \end{vmatrix} = 3 \times (-6) - (-4) \times 5 = 2$
3. Recall the inverse matrix formula to determine the inverse matrix A^{-1}. $A^{-1} = \frac{1}{ad - bc}\begin{bmatrix} d & -b \\ -c & a \end{bmatrix}$	$A^{-1} = \frac{1}{2}\begin{bmatrix} -6 & 4 \\ -5 & 3 \end{bmatrix}$
4. Substitute for the given matrices.	$X = \frac{1}{2}\begin{bmatrix} -6 & 4 \\ -5 & 3 \end{bmatrix} \times \begin{bmatrix} -1 \\ 2 \end{bmatrix}$
5. X is a 2×1 column matrix. Apply the rules to multiply the matrices.	$= \frac{1}{2}\begin{bmatrix} (-6) \times (-1) + 4 \times 2 \\ (-5) \times (-1) + 3 \times 2 \end{bmatrix}$
6. State the answer.	$= \frac{1}{2}\begin{bmatrix} 14 \\ 11 \end{bmatrix}$ $= \begin{bmatrix} 7 \\ \frac{11}{2} \end{bmatrix}$
b. 1. If $XA = D$, multiply both sides by the inverse matrix A^{-1}, and solve for X.	b. $XA = D$ $XAA^{-1} = DA^{-1}$ $XI = X = DA^{-1}$ $\therefore X = DA^{-1}$
2. Substitute for the given matrices.	$X = \begin{bmatrix} 3 & -2 \end{bmatrix} \times \frac{1}{2}\begin{bmatrix} -6 & 4 \\ -5 & 3 \end{bmatrix}$

3. X is a 1×2 matrix. Apply the rules to multiply the matrices.

$$X = \frac{1}{2}\begin{bmatrix} 3\times(-6)+(-2)\times(-5) & 3\times4+(-2)\times3 \end{bmatrix}$$

$$= \frac{1}{2}\begin{bmatrix} -8 & 6 \end{bmatrix}$$

4. State the answer.

$$= \begin{bmatrix} -4 & 3 \end{bmatrix}$$

TI | THINK

a. 1. On a Calculator page, press the template button and complete the entry line as:
$\begin{bmatrix} 3 & -4 \\ 5 & -6 \end{bmatrix}$
Press CTRL, then press VAR, then type a and press ENTER to store matrix A.
Repeat this step to store matrix C.

2. Complete the entry line as $a^{-1} \times c$ then press ENTER.

DISPLAY/WRITE

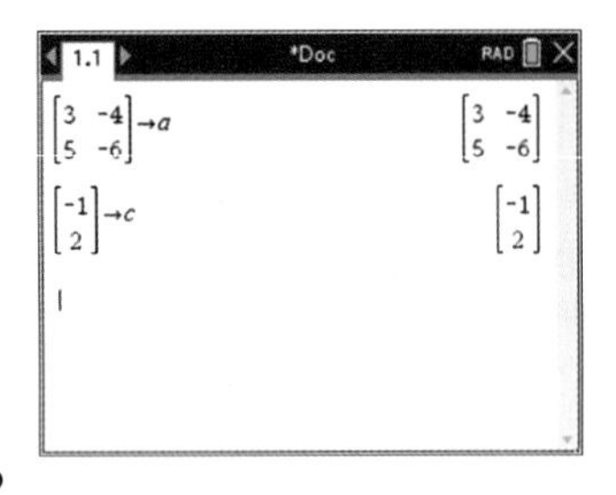

CASIO | THINK

a. From the Keyboard Math2 tab, select the 2×2 matrix format and enter the element values as shown. Select the index template to add the -1.
From the Keyboard Math2 tab, select the 2×1 matrix format and enter the elements as shown.
Select EXE to complete the calculation.

DISPLAY/WRITE

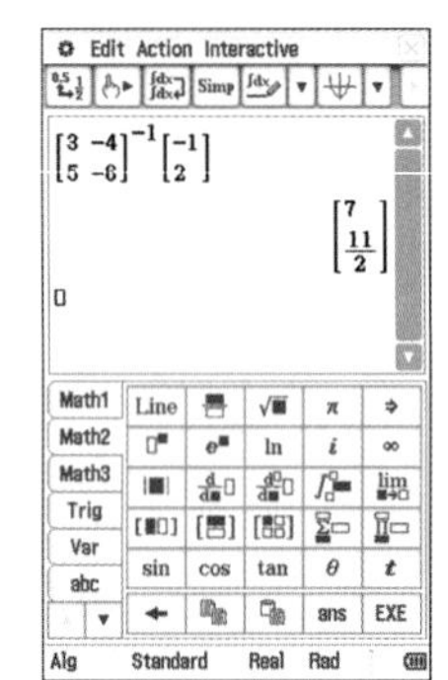

5.5.2 Solving 2 × 2 linear equations

Consider the two linear equations $ax + by = e$ and $cx + dy = f$.

These equations can be written in matrix form as follows:

$$\begin{bmatrix} a & b \\ c & d \end{bmatrix}\begin{bmatrix} x \\ y \end{bmatrix} = \begin{bmatrix} e \\ f \end{bmatrix}$$

If we let $A = \begin{bmatrix} a & b \\ c & d \end{bmatrix}$, $X = \begin{bmatrix} x \\ y \end{bmatrix}$ and $B = \begin{bmatrix} e \\ f \end{bmatrix}$, this equation is of the form $AX = B$, which can be solved for X, as $X = A^{-1}B$.

WORKED EXAMPLE 15 Solving simultaneous equations using inverse matrices

Solve for x and y using inverse matrices.

$$4x + 5y = 6$$
$$3x + 2y = 8$$

THINK

1. First rewrite the two equations as a matrix equation.

WRITE

$$\begin{bmatrix} 4 & 5 \\ 3 & 2 \end{bmatrix}\begin{bmatrix} x \\ y \end{bmatrix} = \begin{bmatrix} 6 \\ 8 \end{bmatrix}$$

2. Write down the matrices A, X and B.

$A = \begin{bmatrix} 4 & 5 \\ 3 & 2 \end{bmatrix}, X = \begin{bmatrix} x \\ y \end{bmatrix}$ and $B = \begin{bmatrix} 6 \\ 8 \end{bmatrix}$

3. Write as an equation and solve for X.

$AX = B$
$\therefore X = A^{-1}B$

4. Recall the determinant rule and calculate for matrix A.

$\Delta = \begin{vmatrix} 4 & 5 \\ 3 & 2 \end{vmatrix} = 4 \times 2 - 3 \times 5 = -7$

5. Recall the inverse matrix A^{-1} rule, and apply with rules for scalar multiplication to simplify this inverse.

$A^{-1} = \frac{1}{-7}\begin{bmatrix} 2 & -5 \\ -3 & 4 \end{bmatrix} = \frac{1}{7}\begin{bmatrix} -2 & 5 \\ 3 & -4 \end{bmatrix}$

6. The unknown matrix X satisfies the equation $X = A^{-1}B$. Write the equation in matrix form.

$X = \begin{bmatrix} x \\ y \end{bmatrix} = \frac{1}{7}\begin{bmatrix} -2 & 5 \\ 3 & -4 \end{bmatrix}\begin{bmatrix} 6 \\ 8 \end{bmatrix}$

7. Apply the rules for matrix multiplication. The product is a 2×1 matrix.

$= \begin{bmatrix} x \\ y \end{bmatrix} = \frac{1}{7}\begin{bmatrix} (-2) \times 6 + 5 \times 8 \\ 3 \times 6 + (-4) \times 8 \end{bmatrix}$

8. Apply the rules for scalar multiplication, and the rules for equality of matrices.

$X = \begin{bmatrix} x \\ y \end{bmatrix} = \frac{1}{7}\begin{bmatrix} 28 \\ -14 \end{bmatrix} = \begin{bmatrix} 4 \\ -2 \end{bmatrix}$

9. State the final answer.

$x = 4$ and $y = -2$

5.5.3 Geometrical interpretation of solutions

There are 3 possible cases for the solutions of systems of linear equations as follows:

- A unique solution
- No solution
- Infinitely many solutions

If the determinant is non-zero ($\Delta \neq 0$), then these two equations are consistent. Graphically, the two lines have different gradients and therefore they intersect at a unique point resulting in a *unique solution*.

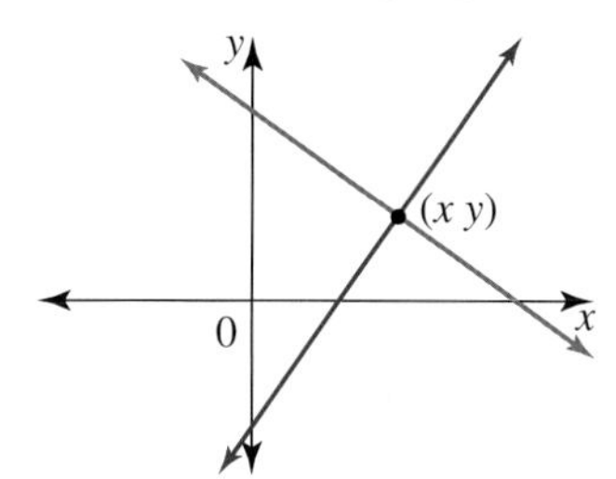

If the determinant is zero ($\Delta = 0$), then there are two possibilities.
The lines are parallel, which indicates that there is *no solution*. Graphically the two lines have the same gradient but different y-intercepts.

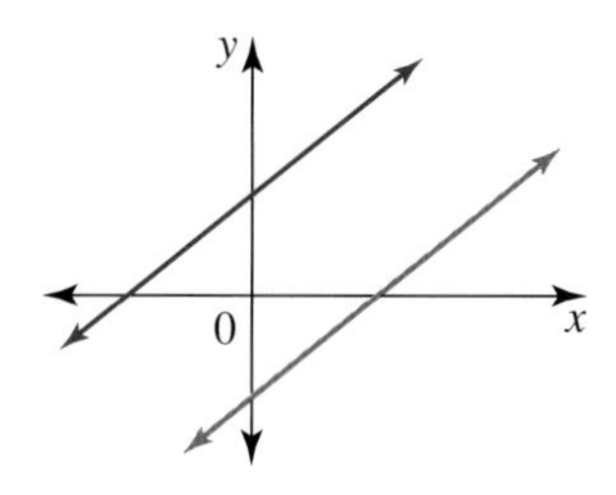

OR

The lines are multiples of one another, that is, they have the same gradient and the same y-intercept (they overlap). This indicates that there is an *infinite number of solutions*.

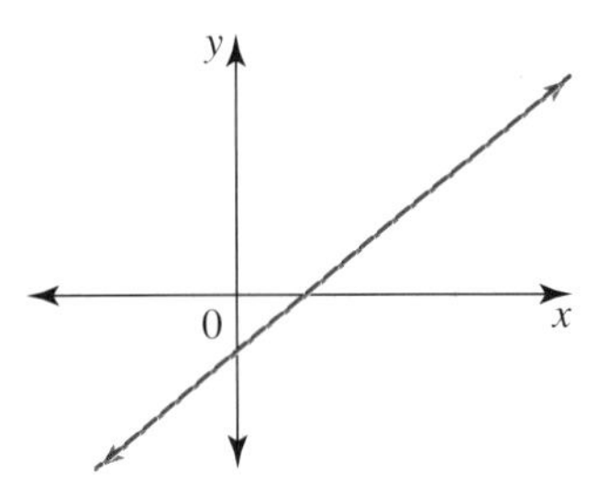

WORKED EXAMPLE 16 Interpreting the solutions of simultaneous equations (1)

Solve the following simultaneous linear equations using matrices and interpret the solution geometrically.

$$\begin{aligned} 3x - 2y &= 6 \\ -6x + 4y &= -10 \end{aligned}$$

THINK	WRITE
1. First write the two equations as a matrix equation.	$\begin{bmatrix} 3 & -2 \\ -6 & 4 \end{bmatrix}\begin{bmatrix} x \\ y \end{bmatrix} = \begin{bmatrix} 6 \\ -10 \end{bmatrix}$
2. Write down the matrices A, X and B.	$A = \begin{bmatrix} 3 & -2 \\ -6 & 4 \end{bmatrix}, X = \begin{bmatrix} x \\ y \end{bmatrix}$ and $B = \begin{bmatrix} 6 \\ -10 \end{bmatrix}$
3. Write as an equation and solve for X.	$AX = B$ $\therefore X = A^{-1}B$
4. Recall the determinant rule and calculate for matrix A.	$\Delta = \begin{vmatrix} 3 & -2 \\ -6 & 4 \end{vmatrix} = 3 \times 4 - (-2) \times (-6) = 0$
5. The inverse matrix A^{-1} does not exist. This method cannot be used to solve the simultaneous equations.	The matrix A is singular, as the determinant equals zero, which means the lines may be parallel or the exact same line.
6. Rearrange both equations into the form $y = mx + c$.	$3x - 2y = 6 \rightarrow y = \dfrac{3x - 6}{2} \rightarrow y = \dfrac{3}{2}x - 3$ $-6x + 4y = -10 \rightarrow y = \dfrac{6x - 10}{4} \rightarrow y = \dfrac{3}{2}x - \dfrac{5}{2}$

7. As the gradients are equal, the lines are parallel (y-intercepts are different). Alternatively:

The gradients of both lines are $\frac{3}{2}$, and the y-intercepts are -3 and $-\frac{5}{2}$. The lines are parallel.

8. Apply another method to solving simultaneous equations: the graphical method. Since both equations represent straight lines, determine the x- and y-intercepts.

Line [1]: $3x - 2y = 6$ crosses the x-axis at $(2, 0)$ and the y-axis at $(0, -3)$.
Line [2]: $-6x + 4y = -10$ crosses the x-axis at $\left(\frac{5}{3}, 0\right)$ and the y-axis at $\left(0, -\frac{5}{2}\right)$.

9. Sketch the graphs. Note that the two lines are parallel and therefore have no points of intersection.

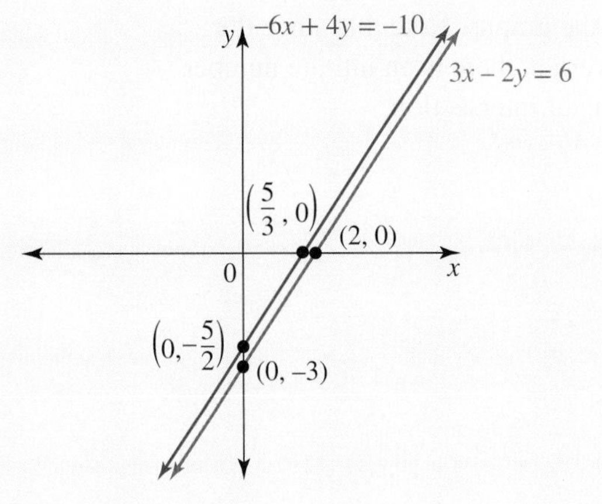

10. State the final answer.

There is no solution.

WORKED EXAMPLE 17 Interpreting the solutions of simultaneous equations (2)

Solve the following linear simultaneous equations for x and y, using matrices and interpret the solution geometrically.

$$\begin{aligned} 3x - 2y &= 6 \\ -6x + 4y &= -12 \end{aligned}$$

THINK

WRITE

1. First write the two equations as a matrix equation.

$$\begin{bmatrix} 3 & -2 \\ -6 & 4 \end{bmatrix} \begin{bmatrix} x \\ y \end{bmatrix} = \begin{bmatrix} 6 \\ -12 \end{bmatrix}$$

2. Write down the matrices A, X and B.

$A = \begin{bmatrix} 3 & -2 \\ -6 & 4 \end{bmatrix}$, $X = \begin{bmatrix} x \\ y \end{bmatrix}$ and $B = \begin{bmatrix} 6 \\ -12 \end{bmatrix}$

3. Write as an equation and solve for X.

$AX = B$
$\therefore X = A^{-1}B$

4. Recall the determinant rule and calculate for matrix A.

$$\Delta = \begin{vmatrix} 3 & -2 \\ -6 & 4 \end{vmatrix} = 3 \times 4 - (-2) \times (-6) = 0$$

5. The inverse matrix A^{-1} does not exist. This method cannot be used to solve the simultaneous equations.

The matrix A is singular, as the determinant equals zero, which means the lines may be parallel or the exact same line.

6. Rearrange both equations into the form $y = mx + c$.

$$3x - 2y = 6 \rightarrow y = \frac{3x - 6}{2} \rightarrow y = \frac{3}{2}x - 3$$

$$-6x + 4y = -12 \rightarrow y = \frac{6x - 12}{4} \rightarrow y = \frac{3}{2}x - 3$$

7. The lines have the same gradients and y-intercept. They are the same line. Alternatively:

Gradients $= \frac{3}{2}$
y-intercept $= -3$

8. Apply another method of solving simultaneous equations: the graphical method. Determine the x- and y-intercepts.

Line [1]: $3x - 2y = 6$ crosses the x-axis at $(2, 0)$ and the y-axis at $(0, -3)$.
Line [2]: $-6x + 4y = -12$ is actually the same line, since $[2] = -2 \times [1]$.

9. Sketch the graphs. Note that since the lines overlap, there is an infinite number of points of intersection.

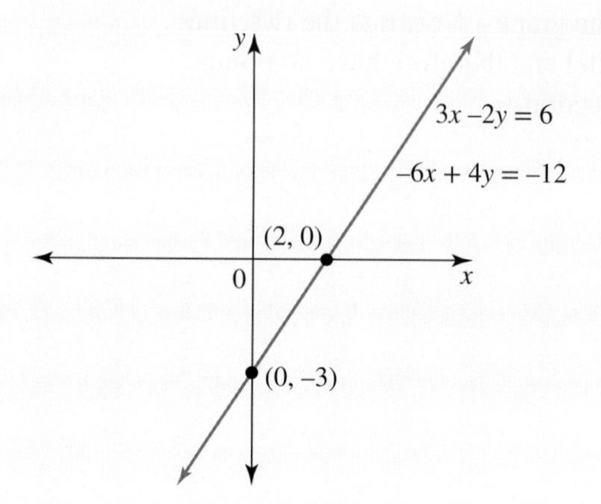

Since $3x - 2y = 6$, $\quad x = \frac{6 + 2y}{3}$.
If $y = 0$, $x = 2$: $\quad (2, 0)$
If $y = 1$, $x = \frac{8}{3}$: $\quad \left(\frac{8}{3}, 1\right)$
If $y = 2$, $x = \frac{10}{3}$: $\quad \left(\frac{10}{3}, 2\right)$
If $y = 3$, $x = 4$: $\quad (4, 3)$
In general, let $y = t$ so that $x = \frac{6 + 3t}{3}$.
As a coordinate: $\left(\frac{6 + 2t}{3}, \ t\right)$

10. State the final answer.

There is an infinite number of solutions of the form $\left(2 + \frac{2t}{3}, \ t\right)$ where $t \in R$.

WORKED EXAMPLE 18 Solving a parameter for the three possible cases

Determine the values of k for which the equations $kx - 3y = k - 1$ and $10x - (k + 1)y = 8$ have:

a. a unique solution

b. no solution

c. an infinite number of solutions.

(You are not required to determine the solution set.)

THINK

WRITE

1. First write the two equations as matrix equations.

$$\begin{bmatrix} k & -3 \\ 10 & -(k+1) \end{bmatrix} \begin{bmatrix} x \\ y \end{bmatrix} = \begin{bmatrix} k - 1 \\ 8 \end{bmatrix}$$

2. Write out the determinant, as it is the key to answering this question.

$$\Delta = \begin{vmatrix} k & -3 \\ 10 & -(k+1) \end{vmatrix}$$

3. Evaluate the determinant in terms of k.

$$\begin{aligned} \Delta &= -k(k+1)+30 \\ &= -k^2-k+30 \\ &= -\left(k^2+k-30\right) \\ \Delta &= -(k+6)(k-5) \end{aligned}$$

4. Let the determinant equal zero and solve for k.

$k=-6, k=5$

a. If $\Delta \neq 0$, the solution is unique; that is, there is a unique solution when $k \neq -6$ and $k \neq 5$.
There is either no solution or an infinite number of solutions when $\Delta = 0$.

a. There is a unique solution when $\Delta \neq 0$, that is, when $k \neq -6$ and $k \neq 5$, or $k \in R\backslash\{-6, 5\}$.

b. 1. Substitute $k=-6$ into the two equations.

b. $$\begin{aligned} -6x-3y=-7 &\Rightarrow 2x+y=\frac{7}{3} \\ &\Rightarrow y=-2x+\frac{7}{3} \\ 10x+5y=8 &\Rightarrow 2x+y=\frac{8}{5} \\ &\Rightarrow y=-2x+\frac{8}{5} \end{aligned}$$

2. The gradients are the same. The two equations represent parallel lines with different y-intercepts. Interpret the answer.

When $k=-6$ there is no solution, as the lines are parallel.

c. 1. Substitute $k=5$ into the two equations.

c. $$\begin{aligned} 5x-3y=4 &\Rightarrow y=\frac{5x-4}{3} \Rightarrow y=\frac{5}{3}x-\frac{4}{3} \\ 10x-6y=8 &\Rightarrow y=\frac{10x-8}{6} \Rightarrow y=\frac{5}{3}x-\frac{4}{3} \end{aligned}$$

2. The lines have the same gradient and y-intercept. They are the same line.

Gradient $=\frac{5}{3}$, y-intercept $=\frac{-4}{3}$

3. The two equations are multiples of one another. Interpret the answer.

When $k=5$ there are an infinite number of solutions.

5.5 Exercise

Students, these questions are even better in jacPLUS

Receive immediate feedback and access sample responses

Access additional questions

Track your results and progress

Find all this and MORE in jacPLUS

Technology free

1. **WE14** If $A = \begin{bmatrix} -2 & 4 \\ 3 & -5 \end{bmatrix}$, $C = \begin{bmatrix} -2 \\ 3 \end{bmatrix}$ and $D = \begin{bmatrix} 2 & -5 \end{bmatrix}$, determine matrix X given the following.

   a. $AX = C$
   b. $XA = D$

2. If $B = \begin{bmatrix} -5 & -3 \\ 3 & 4 \end{bmatrix}$, $C = \begin{bmatrix} -1 \\ 2 \end{bmatrix}$ and $D = \begin{bmatrix} 4 & 3 \end{bmatrix}$, determine matrix X given the following:

   a. $BX = C$
   b. $XB = D$

3. **WE15** Solve for x and y using inverse matrices.

$$\begin{aligned} 3x - 4y &= 23 \\ 5x + 2y &= 21 \end{aligned}$$

4. Solve for x and y using inverse matrices.

$$\begin{aligned} 2x + 5y &= -7 \\ 3x - 2y &= 18 \end{aligned}$$

5. Solve each of the following simultaneous linear equations using inverse matrices.

   a. $2x + 3y = 4$
      $-x + 4y = 9$
   b. $4x + 5y = -6$
      $2x - 3y = 8$
   c. $x - 2y = 8$
      $5x + 4y = -2$
   d. $-2x + 7y + 3 = 0$
      $3x + y + 7 = 0$

6. Consider the matrices $A = \begin{bmatrix} 1 & -2 \\ 5 & 4 \end{bmatrix}$, $B = \begin{bmatrix} 3 & 1 \\ -7 & 2 \end{bmatrix}$, $C = \begin{bmatrix} -5 \\ -19 \end{bmatrix}$ and $D = \begin{bmatrix} 7 & 14 \end{bmatrix}$.
   Calculate the matrix X in each of the following cases.

   a. $AX = C$
   b. $XA = B$
   c. $AX = B$
   d. $XA = D$

7. If $P = \begin{bmatrix} 1 & -2 \\ 3 & 4 \end{bmatrix}$, $Q = \begin{bmatrix} 2 & -1 \\ -3 & 6 \end{bmatrix}$ and $O = \begin{bmatrix} 0 & 0 \\ 0 & 0 \end{bmatrix}$, calculate the matrix X given the following.

   a. $XP - Q = O$
   b. $PX - Q = O$

8. **WE16** Solve the following simultaneous linear equations using matrices and interpret the solution geometrically.

$$\begin{aligned} 4x - 3y &= 12 \\ -8x + 6y &= -18 \end{aligned}$$

9. WE17 Solve the following simultaneous linear equations using matrices and interpret the solution geometrically.

$$4x - 3y = 12$$
$$-8x + 6y = -24$$

10. WE18 Determine the values of k for which the equations $(k+1)x - 2y = 2k$ and $-6x + 2ky = -8$ have:

a. a unique solution
b. no solution
c. an infinite number of solutions.

(You are not required to determine the solution set.)

11. Determine the value of k if the following simultaneous linear equations have no solution.

$$5x - 4y = 20$$
$$kx + 2y = -8$$

12. a. The line $\frac{x}{a} + \frac{y}{b} = 1$ passes through the points $(12, 6)$ and $(8, 3)$.

i. Write down two simultaneous equations that can be used to solve for a and b.
ii. Using inverse matrices, determine the values of a and b.

b. The line $\frac{x}{a} + \frac{y}{b} = 1$ passes through the points $(4, 5)$ and $(-4, -15)$.

i. Write down two simultaneous equations that can be used to solve for a and b.
ii. Using inverse matrices, determine the values of a and b.

13. Calculate the value of k if the following simultaneous equations for x and y have an infinite number of solutions.

$$5x - 4y = 20$$
$$kx + 2y = -10$$

14. Determine the values of k for which the following simultaneous linear equations have:

i. no solution
ii. an infinite number of solutions.

a. $x - 3y = k$
$-2x + 6y = 6$

b. $3x - 5y = k$
$-6x + 10y = 10$

15. Demonstrate that each of the following does not have a unique solution. Describe the solution set and solve if possible.

a. $x - 2y = 3$
$-2x + 4y = -6$

b. $2x - y = 4$
$-4x + 2y = -7$

16. Determine the values of k for which the following systems of equations have:

i. a unique solution
ii. no solution
iii. an infinite number of solutions
(You are not required to determine the solution set.)

a. $(k-2)x - 2y = k - 1$
$-4x + ky = -6$

b. $(k+1)x + 5y = 4$
$6x + 5ky = k + 6$

c. $(k-1)x - 3y = k + 2$
$-4x + 2ky = -10$

d. $2x - (k-2)y = 6$
$(k-5)x - 2y = k - 3$

17. Determine the values of p and q for which the following systems of equations has:
   i. a unique solution
   ii. no solution
   iii. an infinite number of solutions.
   (You are not required to determine the solution set.)

a. $\begin{aligned}-2x+3y&=p\\ qx-6y&=7\end{aligned}$ b. $\begin{aligned}4x-2y&=q\\ 3x+py&=10\end{aligned}$ c. $\begin{aligned}3x-py&=6\\ 7x-2y&=q\end{aligned}$ d. $\begin{aligned}px-y&=3\\ -3x+2y&=q\end{aligned}$

Technology active

18. Consider matrices $A=\begin{bmatrix}-2 & 3\\ 4 & 5\end{bmatrix}$, $B=\begin{bmatrix}2 & 19\\ 12 & -7\end{bmatrix}$, $C=\begin{bmatrix}3\\ 1\end{bmatrix}$ and $D=\begin{bmatrix}-1 & 3\end{bmatrix}$. Calculate the matrix X in each of the following cases.

a. $AX=C$ b. $XA=B$ c. $AX=B$ d. $XA=D$

19. a, b, c and d are all non-zero real numbers.
If $P=\begin{bmatrix}a & 0\\ 0 & d\end{bmatrix}$, determine P^{-1} and verify that $PP^{-1}=P^{-1}P=I$.

20. a, b, c and d are all non-zero real numbers.
   a. If $R=\begin{bmatrix}a & b\\ c & 0\end{bmatrix}$, determine R^{-1} and verify that $RR^{-1}=R^{-1}R=I$.
   b. If $S=\begin{bmatrix}0 & b\\ c & d\end{bmatrix}$, determine S^{-1} and verify that $SS^{-1}=S^{-1}S=I$.
   c. If $A=\begin{bmatrix}a & b\\ c & d\end{bmatrix}$, determine A^{-1} and verify that $AA^{-1}=A^{-1}A=I$.

5.5 Exam questions

Question 1 (4 marks) TECH-FREE

Express this system of simultaneous equations in matrix form and solve for x and y.

$$\begin{aligned}2x+5y&=-3\\ 7x-2y&=9\end{aligned}$$

Question 2 (1 mark) TECH-ACTIVE

MC Given matrices $A=\begin{bmatrix}3 & -1\\ 2 & -4\end{bmatrix}$ and $B=\begin{bmatrix}-7 & 7\\ -8 & -2\end{bmatrix}$, matrix X, such that $AX=B$, is

A. $\begin{bmatrix}-2 & 1\\ 3 & 2\end{bmatrix}$ **B.** $\begin{bmatrix}-2 & 3\\ 1 & 2\end{bmatrix}$ **C.** $\begin{bmatrix}20 & -30\\ 10 & -20\end{bmatrix}$ **D.** $\frac{1}{10}\begin{bmatrix}20 & -30\\ 10 & -20\end{bmatrix}$ **E.** $\frac{1}{5}\begin{bmatrix}-7 & -7\\ -18 & 7\end{bmatrix}$

Question 3 (1 mark) TECH-ACTIVE

MC Which of the following system of equations has a unique solution?

System I: $\begin{aligned}2x-y&=-3\\ 5x+y&=4\end{aligned}$ System II: $\begin{aligned}y&=5-2x\\ 4x+2y&=5\end{aligned}$ System III: $\begin{aligned}7x-5y&=1\\ 2x&=6+y\end{aligned}$

A. System I only
B. Systems I and II only
C. Systems I and III only
D. All systems
E. None of the systems

More exam questions are available online.

5.6 Review

5.6.1 Summary

c-37048

Hey students! Now that it's time to revise this topic, go online to:

Access the topic summary

Review your results

Watch teacher-led videos

Practise exam questions

Find all this and MORE in jacPLUS

5.6 Exercise

Technology free: short answer

1. Consider the following matrices.

$$A=\begin{bmatrix}5 & -2\\3 & 4\end{bmatrix},\ B=\begin{bmatrix}2 & -4 & 1\\3 & -5 & 2\\7 & -4 & 8\end{bmatrix},\ C=\begin{bmatrix}3 & -4\\1 & 5\\-7 & 2\end{bmatrix},\ D=\begin{bmatrix}2 & 1 & -5\\3 & -4 & 7\end{bmatrix},\ E=\begin{bmatrix}5\\-2\end{bmatrix}$$

a. Determine if any of the matrices be added to or subtracted from one another.
b. Determine all possible products.
c. Determine the order of the products which were possible in part b.

2. Calculate the values of x if each of the following is a singular matrix.

a. $\begin{bmatrix}x-2 & -2\\12 & 6\end{bmatrix}$
b. $\begin{bmatrix}x+1 & x-1\\4 & 3\end{bmatrix}$
c. $\begin{bmatrix}x+2 & 3\\5 & x\end{bmatrix}$
d. $\begin{bmatrix}x+3 & 5\\4 & x+2\end{bmatrix}$

3. Solve each of the following using inverse matrices.

a. $x+y=6a$
$4x-3y=3a+14b$

b. $3bx-2ay=0$
$bx+ay=5ab$

c. $x-y=6b$
$3x-4y=17b-a$

d. $\dfrac{x}{a}+\dfrac{y}{b}=2a+b$
$\dfrac{2x}{b}+\dfrac{3y}{a}=2a+6b$

4. Determine the values of p and q for which the following systems of equations have:

i. a unique solution
ii. no solution
iii. an infinite number of solutions.

a. $px+2y=5$
$-3x-5y=q$

b. $4x+3y=q$
$5x+py=7$

5. Let $A=\begin{bmatrix}a\\b\end{bmatrix}$ and $B=\begin{bmatrix}c & d\end{bmatrix}$.

a. Demonstrate that AB exists but $(AB)^{-1}$ does not.
b. Calculate BA and $(BA)^{-1}$.

6. A matrix A is called nilpotent if $A^n=O$, where $O=\begin{bmatrix}0 & 0\\0 & 0\end{bmatrix}$ for some positive integer n.

a. If $A=\begin{bmatrix}2 & -4\\1 & -2\end{bmatrix}$ show that $A^2=O$ and hence show that A is nilpotent.
b. Given the matrix $A=\begin{bmatrix}a & b\\c & d\end{bmatrix}$ where a, b, c and d are all non-zero real constants, determine how a, b, c and d must be related so that A is nilpotent.

7. a. If $A=\begin{bmatrix}2 & 1\\3 & -4\end{bmatrix}$, $I=\begin{bmatrix}1 & 0\\0 & 1\end{bmatrix}$ and $O=\begin{bmatrix}0 & 0\\0 & 0\end{bmatrix}$, express the determinant of the matrix $A-kI$ in the form pk^2+qk+r starting the values of p, q and r. Hence show that $pA^2+qA+rI=O$.

b. If $B=\begin{bmatrix}a & b\\c & d\end{bmatrix}$, $I=\begin{bmatrix}1 & 0\\0 & 1\end{bmatrix}$ and $O=\begin{bmatrix}0 & 0\\0 & 0\end{bmatrix}$, express the determinant of the matrix $B-kI$ in the form pk^2+qk+r stating the values of p, q and r. Hence, show that $pB^2+qB+rI=O$.

Technology active: multiple choice

8. **MC** If $A=\begin{bmatrix}1 & 2 & 3 & 4\end{bmatrix}$ and $B=\begin{bmatrix}3\\2\\1\\0\end{bmatrix}$ then the matrix product AB is equal to

A. $\begin{bmatrix}3 & 6 & 9 & 12\\2 & 4 & 6 & 8\\1 & 2 & 3 & 4\\0 & 0 & 0 & 0\end{bmatrix}$

B. $[36]$

C. $\begin{bmatrix}3\\4\\3\\0\end{bmatrix}$

D. $\begin{bmatrix}8\\12\\20\\24\end{bmatrix}$

E. $[10]$

9. **MC** If $P=\begin{bmatrix}1 & -1\\2 & 4\\3 & 5\end{bmatrix}$ and $Q=\begin{bmatrix}1 & 2\\1 & 2\end{bmatrix}$ then the matrix PQ is

A. $\begin{bmatrix}1 & -2\\2 & 8\\3 & 10\end{bmatrix}$ B. $\begin{bmatrix}1 & -2\\2 & 8\\3 & 6\end{bmatrix}$ C. $\begin{bmatrix}0 & 0\\6 & 12\\8 & 16\end{bmatrix}$ D. $\begin{bmatrix}1 & -2\\1 & -2\end{bmatrix}$ E. not defined

10. **MC** If $C=\begin{bmatrix}2 & 0\\-1 & -1\end{bmatrix}$ then the matrix C^3 is equal to

A. $\begin{bmatrix}8 & 0\\-3 & -1\end{bmatrix}$ B. $\begin{bmatrix}8 & 0\\-1 & -1\end{bmatrix}$ C. $\begin{bmatrix}6 & 0\\-3 & -3\end{bmatrix}$ D. $\begin{bmatrix}8 & 0\\-3 & -3\end{bmatrix}$ E. $\begin{bmatrix}-8 & 0\\1 & 1\end{bmatrix}$

11. **MC** Given the matrices $A=\begin{bmatrix}a_1 & a_2\\a_3 & a_4\\a_5 & a_6\end{bmatrix}$ and $B=\begin{bmatrix}b_1 & b_2 & b_3\end{bmatrix}$

The order of the matrix product BA is

A. 2×1 B. 3×1 C. 1×3 D. 1×2 E. not defined

12. **MC** Given the following matrices $A=\begin{bmatrix}a_1 & a_2\\a_3 & a_4\end{bmatrix}$ $B=\begin{bmatrix}b_1 & b_2\end{bmatrix}$ and $C=\begin{bmatrix}c_1\\c_2\end{bmatrix}$

The matrix product which does not exist is

A. AC B. BC C. CB D. CA E. BA

13. **MC** The simultaneous equations $x-3y=4$ and $2y-x=5$, can be represented in matrix form as

A. $\begin{bmatrix}1 & -3\\2 & -1\end{bmatrix}\begin{bmatrix}x\\y\end{bmatrix}=\begin{bmatrix}4\\5\end{bmatrix}$

B. $\begin{bmatrix}1 & -3\\2 & -1\end{bmatrix}\begin{bmatrix}x\\y\end{bmatrix}=\begin{bmatrix}5\\4\end{bmatrix}$

C. $\begin{bmatrix}-1 & 2\\1 & -3\end{bmatrix}\begin{bmatrix}x\\y\end{bmatrix}=\begin{bmatrix}5\\4\end{bmatrix}$

D. $\begin{bmatrix}1 & -3\\2 & -1\end{bmatrix}\begin{bmatrix}x & y\end{bmatrix}=\begin{bmatrix}4 & 5\end{bmatrix}$

E. $\begin{bmatrix}-1 & 2\\1 & -3\end{bmatrix}\begin{bmatrix}x & y\end{bmatrix}=\begin{bmatrix}5 & 4\end{bmatrix}$

14. **MC** If A is a 2×2 matrix and the determinant of the matrix A, $\det(A)=k$, then the determinant of the matrix A^2 is equal to

A. $2k$
B. k^2
C. $4k$
D. $4k^2$
E. Cannot be determined

15. **MC** Given the matrix $A=\begin{bmatrix}4 & b\\c & -1\end{bmatrix}$, then the statement which is true is

A. If $b=2$ and $c=-2$ the matrix is non-singular.
B. If $b=-2$ and $c=2$ the matrix is non-singular.
C. If $b=1$ and $c=-4$ the matrix is non-singular.
D. If $b=-3$ and $c=-1$ the matrix is singular.
E. If $b=\frac{1}{2}$ and $c=-8$ the matrix is singular.

16. **MC** If $B=\begin{bmatrix}2 & 1\\5 & 2\end{bmatrix}$ then B^{-1} is equal to

A. $\begin{bmatrix}-2 & 1\\5 & -2\end{bmatrix}$
B. $\begin{bmatrix}2 & -1\\-5 & 2\end{bmatrix}$
C. $\begin{bmatrix}\frac{1}{2} & -1\\\frac{1}{5} & \frac{1}{2}\end{bmatrix}$
D. $\frac{1}{9}\begin{bmatrix}2 & -1\\-5 & 2\end{bmatrix}$
E. $\begin{bmatrix}\frac{1}{2} & 1\\\frac{1}{5} & \frac{1}{2}\end{bmatrix}$

17. **MC** If $A^2+9I=O$ where O represents the null (zero) matrix and I represents the identity matrix, then the matrix A^{-1} is equal to

A. $\frac{1}{9}A$
B. $-\frac{1}{9}A$
C. $\frac{1}{3}I$
D. $-\frac{1}{3}I$
E. $-3I$

Technology active: extended response

18. a. Given the following matrices $A=\begin{bmatrix}2 & -1 & 3\\3 & 4 & 5\end{bmatrix}$ and $B=\begin{bmatrix}7 & 24 & 13\end{bmatrix}$, calculate the matrix X, if $XA=B$.

b. Given the following matrices $A=\begin{bmatrix}2 & -1\\-3 & 4\\5 & -2\end{bmatrix}$ $X=\begin{bmatrix}x & 4 & -2\\y & -2 & 3\end{bmatrix}$ and $B=\begin{bmatrix}-12 & 15\\13 & -10\end{bmatrix}$, determine the values of x and y if $XA=B$.

19. a. Given the following matrices $A=\begin{bmatrix}3 & -2\\-2 & 5\\4 & -1\end{bmatrix}$ $X=\begin{bmatrix}x & -2 & -1\\y & 3 & 5\end{bmatrix}$ and $B=\begin{bmatrix}12 & -17\\8 & 14\end{bmatrix}$, determine the values of x and y if $XA=B$.

b. Given the following matrices $A = \begin{bmatrix} 3 & -2 \\ 1 & 5 \end{bmatrix}$ and $B = \begin{bmatrix} 15 & -2 \\ -19 & -7 \end{bmatrix}$, calculate the matrix X if $A^4X + 80B = X$.

20. Given that b and c are non-zero real numbers and $Q = \begin{bmatrix} 0 & b \\ c & 0 \end{bmatrix}$:

a. calculate Q^{-1} and verify that $QQ^{-1} = Q^{-1}Q = I$.
b. if $n \in Z$, calculate Q^n.

5.6 Exam questions

Question 1 (4 marks) TECH-FREE

Express the following system of simultaneous equations in matrix form and solve for x and y.

$$3x - 2y = 7$$
$$5x + y = 3$$

Question 2 (1 mark) TECH-ACTIVE

MC The inverse of the matrix $\begin{bmatrix} 2 & 0 \\ 1 & 1 \end{bmatrix}$ is

A. $\begin{bmatrix} 1 & 0 \\ 0 & 1 \end{bmatrix}$
B. $\begin{bmatrix} 4 & -2 \\ -2 & 1 \end{bmatrix}$
C. $\frac{1}{2}\begin{bmatrix} 1 & 0 \\ -1 & 2 \end{bmatrix}$
D. not defined
E. none of these

Question 3 (6 marks) TECH-FREE

Using $A = \begin{bmatrix} 5 & 2 \\ 0 & 4 \end{bmatrix}$, $B = \begin{bmatrix} 2 & 4 \\ -1 & 3 \end{bmatrix}$ and $C = \begin{bmatrix} -1 & 0 \\ 0 & 4 \end{bmatrix}$, verify the Associative Law: $A(B + C) = AB + AC$.

Question 4 (4 marks) TECH-FREE

Given matrices $A = \begin{bmatrix} 1 & -2 \\ -3 & 4 \end{bmatrix}$ and $I = \begin{bmatrix} 1 & 0 \\ 0 & 1 \end{bmatrix}$, verify that $A^2I = IA^2 = A^2$.

Question 5 (4 marks) TECH-FREE

If $M^{-1} = \frac{1}{5}\begin{bmatrix} -3 & 2 \\ -4 & 1 \end{bmatrix}$, calculate M, and verify that $MM^{-1} = M^{-1}M = I$.

More exam questions are available online.

Answers

Topic 5 Matrices

5.2 Addition, subtraction and scalar multiplication of matrices

5.2 Exercise

1. $[63 \quad 19 \quad 25]$; 63 kicks, 19 marks, 25 handballs
2. $\begin{bmatrix} 6 & 8 \\ 53 & 17 \end{bmatrix}$; 6 aces, 8 double faults, 53 forehand winners, 17 backhand winners
3. $x = 2;\ y = 18$
4. $x = 3;\ y = -16;\ z = 11$
5. a. $\begin{bmatrix} -10 & 4 \\ 11 & 21 \end{bmatrix}$ b. $\begin{bmatrix} 4 & -8 \\ -6 & -10 \end{bmatrix}$
6. a. $a = 2;\ b = 8;\ c = -2;\ d = -8$
   b. $a = \frac{11}{2};\ b = 8;\ c = -2;\ d = -\frac{9}{2}$
7. a. $\begin{bmatrix} 2 \\ 7 \end{bmatrix}$ b. $\frac{1}{2}\begin{bmatrix} 15 \\ 14 \end{bmatrix}$
8. a. i. $\begin{bmatrix} 5 & 3 \\ 7 & 1 \end{bmatrix}$ ii. $\begin{bmatrix} 6 & 8 \\ 1 & 1 \end{bmatrix}$
   b. $\begin{bmatrix} 7 & 6 \\ 6 & 5 \end{bmatrix}$
   Sample response can be found in the worked solutions in the online resources.
9. a. $\begin{bmatrix} 11 & 8 \\ -3 & 16 \end{bmatrix}$ b. $\begin{bmatrix} 5 & -16 \\ 15 & 4 \end{bmatrix}$
10. a. $\begin{bmatrix} 1 & -16 \\ 15 & 0 \end{bmatrix}$ b. $\begin{bmatrix} 14 & 10 \\ -3 & 21 \end{bmatrix}$
11. a. $x = 5;\ y = 7$ b. $x = 3;\ y = -2$
12. a. $\begin{bmatrix} 3 & 2 & 9 \\ -2 & 6 & -5 \end{bmatrix}$ b. $\frac{1}{2}\begin{bmatrix} 5 & -20 & 1 \\ 13 & 10 & -6 \end{bmatrix}$
13. a. $a_{11} = 2;\ a_{12} = 3;\ a_{21} = -1;\ a_{22} = 4$
    b. $\begin{bmatrix} 3 & -2 \\ -3 & 5 \end{bmatrix}$
14. a. $\begin{bmatrix} 1 & 0 \\ 3 & 4 \end{bmatrix}$ b. $\begin{bmatrix} 3 & 3 \\ 2 & 5 \end{bmatrix}$
15. a. i. 6 ii. 9 iii. 5
    b. Yes
    c. Yes

5.2 Exam questions

Note: Mark allocations are available with the fully worked solutions online.

1. $X = \begin{bmatrix} -18 & -14 \\ -11 & -16 \end{bmatrix}$
2. B
3. $a = -3,\ b = 2,\ x = 6,\ y = -4$

5.3 Matrix multiplication

5.3 Exercise

1. $A = 3 \times 2$
   $B = 3 \times 1$
   $C = 2 \times 2$
   $D = 1 \times 3$
2. a. i. No ii. Yes
   iii. Yes iv. No
   v. Yes vi. Yes
   b. i. – ii. 3×3
   iii. 1×2 iv. –
   v. 1×1 vi. 3×2
3. a. $\begin{bmatrix} -8 & -20 \\ 1 & -3 \end{bmatrix}$ b. $\begin{bmatrix} 8 & 28 \\ -7 & -19 \end{bmatrix}$
   c. $\begin{bmatrix} 16 & 12 \\ 9 & 37 \end{bmatrix}$ d. $\begin{bmatrix} 0 & -4 \\ 1 & 5 \end{bmatrix}$
4. a. $\begin{bmatrix} 0 & 0 \\ 0 & 0 \end{bmatrix}$ b. $\begin{bmatrix} 0 & 0 \\ 0 & 0 \end{bmatrix}$
   c. $\begin{bmatrix} -2 & 4 \\ 3 & 5 \end{bmatrix}$ d. $\begin{bmatrix} -2 & 4 \\ 3 & 5 \end{bmatrix}$
   Observations:
   $AO = OA = O$
   $AI = IA = A$
5. a. i. $\begin{bmatrix} 2 & 3 \\ -1 & 4 \end{bmatrix}$ ii. $\begin{bmatrix} 0 & 0 \\ 0 & 0 \end{bmatrix}$
   b. i. $\begin{bmatrix} a & b \\ c & d \end{bmatrix}$ ii. $\begin{bmatrix} 0 & 0 \\ 0 & 0 \end{bmatrix}$
6. a. $\begin{bmatrix} 0 & -18 \\ 6 & -12 \end{bmatrix}$
   b. No, sample responses can be found in the worked solutions in the online resources.
   c. No, sample responses can be found in the worked solutions in the online resources.
7. a. $\begin{bmatrix} ax + by \\ cx + dy \end{bmatrix}$ b. Does not exist
8. a. $\begin{bmatrix} 0 & 8 & -11 \\ 7 & -26 & 34 \\ -9 & 14 & -17 \end{bmatrix}$ b. $\begin{bmatrix} -1 & 21 \\ 11 & -42 \end{bmatrix}$
9. a. $\begin{bmatrix} 3 & -2 \\ -6 & 4 \end{bmatrix}$ b. $[7]$
10. a. $x = -3$ b. $x = -4$
    c. $x = -2$ d. $x = 2$
11. a. Does not exist b. Does not exist
    c. Does not exist d. $\begin{bmatrix} -3 & 5 \\ 6 & -10 \end{bmatrix}$
    e. $[-13]$ f. Does not exist
12. a. $\begin{bmatrix} 6 \\ 13 \end{bmatrix}$ b. $[21 \quad -13]$
    c. Does not exist d. $\begin{bmatrix} -21 & 13 \\ 42 & -26 \end{bmatrix}$
    e. Does not exist f. $\begin{bmatrix} 18 & -30 \\ 39 & -65 \end{bmatrix}$

13. a. $\begin{bmatrix} 0 & 0 \\ 0 & 0 \end{bmatrix}$ b. $\begin{bmatrix} 2 & -8 & 4 \\ 5 & -20 & 10 \\ 9 & -36 & 18 \end{bmatrix}$

c. Does not exist d. Does not exist

14. a. $\begin{bmatrix} 31 & 9 \\ 23 & 1 \end{bmatrix}$

b. $\begin{bmatrix} 19 & -24 \\ -81 & -76 \end{bmatrix}$

c. $\text{LHS} = \begin{bmatrix} 44 & 56 \\ 7 & 9 \end{bmatrix}$

$\text{RHS} = \begin{bmatrix} 55 & 25 \\ 4 & -2 \end{bmatrix}$

No, since $BA \neq AB$.

d. $\begin{bmatrix} 44 & 56 \\ 7 & 9 \end{bmatrix}$

15. a. $P^2 = \begin{bmatrix} 1 & 0 \\ 0 & 16 \end{bmatrix}; P^3 = \begin{bmatrix} -1 & 0 \\ 0 & 64 \end{bmatrix}; P^4 = \begin{bmatrix} 1 & 0 \\ 0 & 256 \end{bmatrix};$

$P^n = \begin{bmatrix} (-1)^n & 0 \\ 0 & 4^n \end{bmatrix}$

b. $Q^2 = \begin{bmatrix} 4 & 0 \\ 0 & 9 \end{bmatrix}; Q^3 = \begin{bmatrix} 8 & 0 \\ 0 & -27 \end{bmatrix}; Q^4 = \begin{bmatrix} 16 & 0 \\ 0 & 81 \end{bmatrix};$

$Q^n = \begin{bmatrix} 2^n & 0 \\ 0 & (-3)^n \end{bmatrix}$

16. a. $R^2 = \begin{bmatrix} 1 & 0 \\ 6 & 1 \end{bmatrix}; R^3 = \begin{bmatrix} 1 & 0 \\ 9 & 1 \end{bmatrix}; R^4 = \begin{bmatrix} 1 & 0 \\ 12 & 1 \end{bmatrix};$

$R^n = \begin{bmatrix} 1 & 0 \\ 3n & 1 \end{bmatrix}$

b. $S^2 = \begin{bmatrix} 6 & 0 \\ 0 & 6 \end{bmatrix}; S^3 = \begin{bmatrix} 0 & 18 \\ 12 & 0 \end{bmatrix}; S^4 = \begin{bmatrix} 36 & 0 \\ 0 & 36 \end{bmatrix};$

$S^8 = \begin{bmatrix} 1296 & 0 \\ 0 & 1296 \end{bmatrix}; S^9 = \begin{bmatrix} 0 & 3888 \\ 2592 & 0 \end{bmatrix}$

17. $\begin{bmatrix} 0 & 0 \\ 0 & 0 \end{bmatrix}$

18. a. $\begin{bmatrix} -20 & 0 \\ 0 & -20 \end{bmatrix}$ b. $\begin{bmatrix} 0 & 0 \\ 0 & 0 \end{bmatrix}$

19. $\begin{bmatrix} d^2 - 9d + 8 & 4 - 4d \\ 2 - 2d & 0 \end{bmatrix}$

20. a. i. 39 ii. 39 iii. 54

b. No, sample responses can be found in the worked solutions in the online resources.

5.3 Exam questions

Note: Mark allocations are available with the fully worked solutions online.

1. A
2. C
3. B

5.4 Determinants and inverses

5.4 Exercise

1. -22

2. $x = 4, -6$

3. a. -53 b. -33

4. $\frac{1}{34}\begin{bmatrix} 6 & 2 \\ -5 & 4 \end{bmatrix}$

5. a. $\frac{1}{4}\begin{bmatrix} -4 & 0 \\ 0 & 1 \end{bmatrix}$ b. $\frac{1}{6}\begin{bmatrix} 3 & 1 \\ 0 & -2 \end{bmatrix}$

c. $\frac{1}{2}\begin{bmatrix} 1 & 0 \\ -3 & 2 \end{bmatrix}$ d. $\frac{1}{6}\begin{bmatrix} -1 & 3 \\ -2 & 0 \end{bmatrix}$

6. $p = -4,\ q = -2$

7. $\Delta = 0$

8. $x = -6, 2$

9. a. i. 20 ii. $\frac{1}{10}\begin{bmatrix} 1 & 1 \\ -2 & 3 \end{bmatrix}$

b. Sample responses can be found in the worked solutions in the online resources.

c. i. $\frac{1}{20}$ ii. 1

10. a. $x = -3$ b. $x = \pm 4$

11. a. $x = -6$ b. $x = \pm 2\sqrt{3}$

12. a. Does not exist

b. Does not exist

c. $\frac{1}{22}\begin{bmatrix} 5 & -4 \\ 3 & 2 \end{bmatrix}$

d. Does not exist

13. $p = 1,\ q = -7,\ r = 13;\ \begin{bmatrix} 0 & 0 \\ 0 & 0 \end{bmatrix}$

14. $k = -2,\ 8$

15. $p = 1,\ q = 2,\ r = -11;\ \begin{bmatrix} 0 & 0 \\ 0 & 0 \end{bmatrix}$

16. a. $-11;\ 2;\ 10$

b. Yes

c. Sample responses can be found in the worked solutions in the online resources.

17. a. $A^{-1} = -\frac{1}{11}\begin{bmatrix} -4 & 3 \\ 1 & 2 \end{bmatrix}$

$B^{-1} = \frac{1}{2}\begin{bmatrix} 3 & -5 \\ -2 & 4 \end{bmatrix}$

$C^{-1} = \frac{1}{10}\begin{bmatrix} 4 & 2 \\ -3 & 1 \end{bmatrix}$

b. No

c. Yes

d. Yes

18. $\det(A) = -140$

$\det\left(A^T\right) = -140$

$\det(A) = \det\left(A^T\right)$

19. a. $\begin{bmatrix} \frac{4}{39} & -\frac{2}{39} & \frac{5}{78} \\ \frac{7}{39} & \frac{16}{39} & -\frac{1}{78} \\ \frac{1}{3} & \frac{1}{3} & -\frac{1}{6} \end{bmatrix}$

b. $\begin{bmatrix} \frac{1}{16} & \frac{1}{16} & -\frac{1}{16} & 0 \\ \frac{1}{12} & 0 & \frac{1}{12} & -\frac{1}{6} \\ 0 & \frac{1}{4} & 0 & -\frac{1}{4} \\ -\frac{5}{48} & -\frac{3}{16} & \frac{1}{48} & \frac{1}{3} \end{bmatrix}$

20. a. $k = -1, 2$ b. $\begin{bmatrix} -1 & 0 \\ 0 & 2 \end{bmatrix}$

5.4 Exam questions

Note: Mark allocations are available with the fully worked solutions online.

1. D
2. B
3. $k = 6, -1$

5.5 Matrix equations

5.5 Exercise

1. a. $\begin{bmatrix} 1 \\ 0 \end{bmatrix}$ b. $\frac{1}{2}\begin{bmatrix} -5 & -2 \end{bmatrix}$

2. a. $\frac{1}{11}\begin{bmatrix} -2 \\ 7 \end{bmatrix}$ b. $\frac{1}{11}\begin{bmatrix} -7 & 3 \end{bmatrix}$

3. $x = 5,\ y = -2$

4. $x = 4,\ y = -3$

5. a. $x = -1,\ y = 2$ b. $x = 1,\ y = -2$
   c. $x = 2,\ y = -3$ d. $x = -2,\ y = -1$

6. a. $\frac{1}{7}\begin{bmatrix} -29 \\ 3 \end{bmatrix}$
   b. $\frac{1}{14}\begin{bmatrix} 7 & 7 \\ -38 & -12 \end{bmatrix}$
   c. $\frac{1}{14}\begin{bmatrix} -2 & 8 \\ -22 & -3 \end{bmatrix}$
   d. $\begin{bmatrix} -3 & 2 \end{bmatrix}$

7. a. $\frac{1}{10}\begin{bmatrix} 11 & 3 \\ -30 & 0 \end{bmatrix}$ b. $\frac{1}{10}\begin{bmatrix} 2 & 8 \\ -9 & 9 \end{bmatrix}$

8. No solution

9. $\left(3 + \frac{3t}{4}, t\right),\ t \in R$

10. a. $k \in R \backslash \{-3, 2\}$
    b. $k = -3$
    c. $k = 2$

11. $k = -\frac{5}{2}$

12. a. i. $\frac{12}{a} + \frac{6}{b} = 1,\ \frac{8}{a} + \frac{3}{b} = 1$
    ii. $a = 4,\ b = -3$
    b. i. $\frac{4}{a} + \frac{5}{b} = 1,\ -\frac{4}{a} - \frac{15}{b} = 1$
    ii. $a = 2,\ b = -5$

13. $k = -\frac{5}{2}$

14. a. i. $k \neq -3$ ii. $k = -3$
    b. i. $k \neq -5$ ii. $k = -5$

15. a. $(2t + 3, t),\ t \in R$ b. No solution

16. a. i. $k \in R \backslash \{-2, 4\}$ ii. $k = -2$
    iii. $k = 4$
    b. i. $k \in R \backslash \{-3, 2\}$ ii. $k = -3$
    iii. $k = 2$
    c. i. $k \in R \backslash \{-2, 3\}$ ii. $k = -2$
    iii. $k = 3$
    d. i. $k \in R \backslash \{1, 6\}$ ii. $k = 1$
    iii. $k = 6$

17. a. i. $q \neq 4,\ p \in R$ ii. $q = 4,\ p \neq -\frac{7}{2}$
    iii. $q = 4,\ p = -\frac{7}{2}$
    b. i. $p \neq -\frac{3}{2},\ q \in R$ ii. $p = -\frac{3}{2},\ q \neq \frac{40}{3}$
    iii. $p = -\frac{3}{2},\ q = \frac{40}{3}$
    c. i. $p \neq \frac{6}{7},\ q \in R$ ii. $p = \frac{6}{7},\ q \neq 14$
    iii. $p = \frac{6}{7},\ q = 14$
    d. i. $p \neq \frac{3}{2},\ q \in R$ ii. $p = \frac{3}{2},\ q \neq -6$
    iii. $p = \frac{3}{2},\ q = -6$

18. a. $\frac{1}{11}\begin{bmatrix} -6 \\ 7 \end{bmatrix}$ b. $\begin{bmatrix} 3 & 2 \\ -4 & 1 \end{bmatrix}$
    c. $\frac{1}{11}\begin{bmatrix} 13 & -58 \\ 16 & 31 \end{bmatrix}$ d. $\frac{1}{22}\begin{bmatrix} 17 & 3 \end{bmatrix}$

19. $P^{-1} = \begin{bmatrix} \frac{1}{a} & 0 \\ 0 & \frac{1}{d} \end{bmatrix}$

20. a. $R^{-1} = \begin{bmatrix} 0 & \frac{1}{c} \\ \frac{1}{b} & -\frac{a}{bc} \end{bmatrix}$ b. $S^{-1} = \begin{bmatrix} -\frac{d}{bc} & \frac{1}{c} \\ \frac{1}{b} & 0 \end{bmatrix}$
    c. $A^{-1} = \frac{1}{ad - bc}\begin{bmatrix} d & -b \\ -c & a \end{bmatrix}$

5.5 Exam questions

Note: Mark allocations are available with the fully worked solutions online.

1. $x = 1, y = -1$
2. B
3. C

5.6 Review

5.6 Exercise

Technology free: short answer

1. a. No, they all have different orders.

   b. AB does not exist.

   AC does not exist.

   $AD = \begin{bmatrix} 4 & 13 & -39 \\ 18 & -13 & 13 \end{bmatrix}$

   $AE = \begin{bmatrix} 29 \\ 7 \end{bmatrix}$

   BA does not exist.

   $BC = \begin{bmatrix} -5 & -26 \\ -10 & -33 \\ -39 & -32 \end{bmatrix}$

   BD does not exist.

   BE does not exist.

   $CA = \begin{bmatrix} 3 & -22 \\ 20 & 18 \\ -29 & 22 \end{bmatrix}$

   CB does not exist.

   $CD = \begin{bmatrix} -6 & 19 & -43 \\ 17 & -19 & 30 \\ -8 & -15 & 49 \end{bmatrix}$ $CE = \begin{bmatrix} 23 \\ -5 \\ -39 \end{bmatrix}$

   DA does not exist.

   $DB = \begin{bmatrix} -28 & 7 & -36 \\ 43 & -20 & 51 \end{bmatrix}$ $DC = \begin{bmatrix} 42 & -13 \\ -44 & -18 \end{bmatrix}$

   DE does not exist.

   EA does not exist.

   EB does not exist.

   EC does not exist.

   ED does not exist.

   c. D: 2×3, AE: 2×3, BC: 3×2, CA: 3×2, CD: 3×3, CE: 3×1, DB: 2×3, DC: 2×2

2. a. $x = -2$ b. $x = 7$

   c. $x = -5, 3$ d. $x = -7, 2$

3. a. $x = 3a + 2b, \ y = 3a - 2b$

   b. $x = 2a, \ y = 3b$

   c. $x = a + 7b, \ y = a + b$

   d. $x = ab, \ y = 2ab$

4. a. i. $p \neq \frac{6}{5}, \ q \in R$ ii. $p = \frac{6}{5}, \ q \neq -\frac{25}{2}$

   iii. $p = \frac{6}{5}, \ q = -\frac{25}{2}$

   b. i. $p \neq \frac{15}{4}, \ q \in R$ ii. $p = \frac{15}{4}, \ q \neq \frac{28}{5}$

   iii. $p = \frac{15}{4}, \ q = \frac{28}{5}$

5. a. $AB = \begin{bmatrix} ac & ad \\ bc & bd \end{bmatrix}$ b. $BA = [ac + bd]$

   $(BA)^{-1} = \left[\frac{1}{ac + bd}\right]$

6. a. Sample responses can be found in the worked solutions in the online resources.

   b. $a = -d, \ a^2 = d^2 = -bc$

7. a. $p = 1, q = 2, r = -11$. Sample responses can be found in the worked solutions in the online resources.

   b. $p = 1, q = -(a + d), r = ad - bc$. Sample responses can be found in the worked solutions in the online resources.

Technology active: multiple choice

8. E
9. C
10. A
11. D
12. D
13. C
14. B
15. E
16. A
17. B

Technology active: extended response

18. a. $\begin{bmatrix} -4 & 5 \end{bmatrix}$ b. $x = 5, \ y = -4$

19. a. $x = 4, \ y = -2$ b. $\begin{bmatrix} 3 & 4 \\ 2 & -1 \end{bmatrix}$

20. a. $Q^{-1} = \begin{bmatrix} 0 & \frac{1}{c} \\ \frac{1}{b} & 0 \end{bmatrix}$

    b. $Q^n = \begin{bmatrix} (bc)^{\frac{n}{2}} & 0 \\ 0 & (bc)^{\frac{n}{2}} \end{bmatrix}$ if n is even

    $Q^n = \begin{bmatrix} 0 & b^{\frac{n+1}{2}} c^{\frac{n-1}{2}} \\ b^{\frac{n-1}{2}} c^{\frac{n+1}{2}} & 0 \end{bmatrix}$ if n is odd

5.6 Exam questions

Note: Mark allocations are available with the fully worked solutions online.

1. $x = 1, \ y = -2$
2. C

3-5. Sample responses can be found in the worked solutions in the online resources.

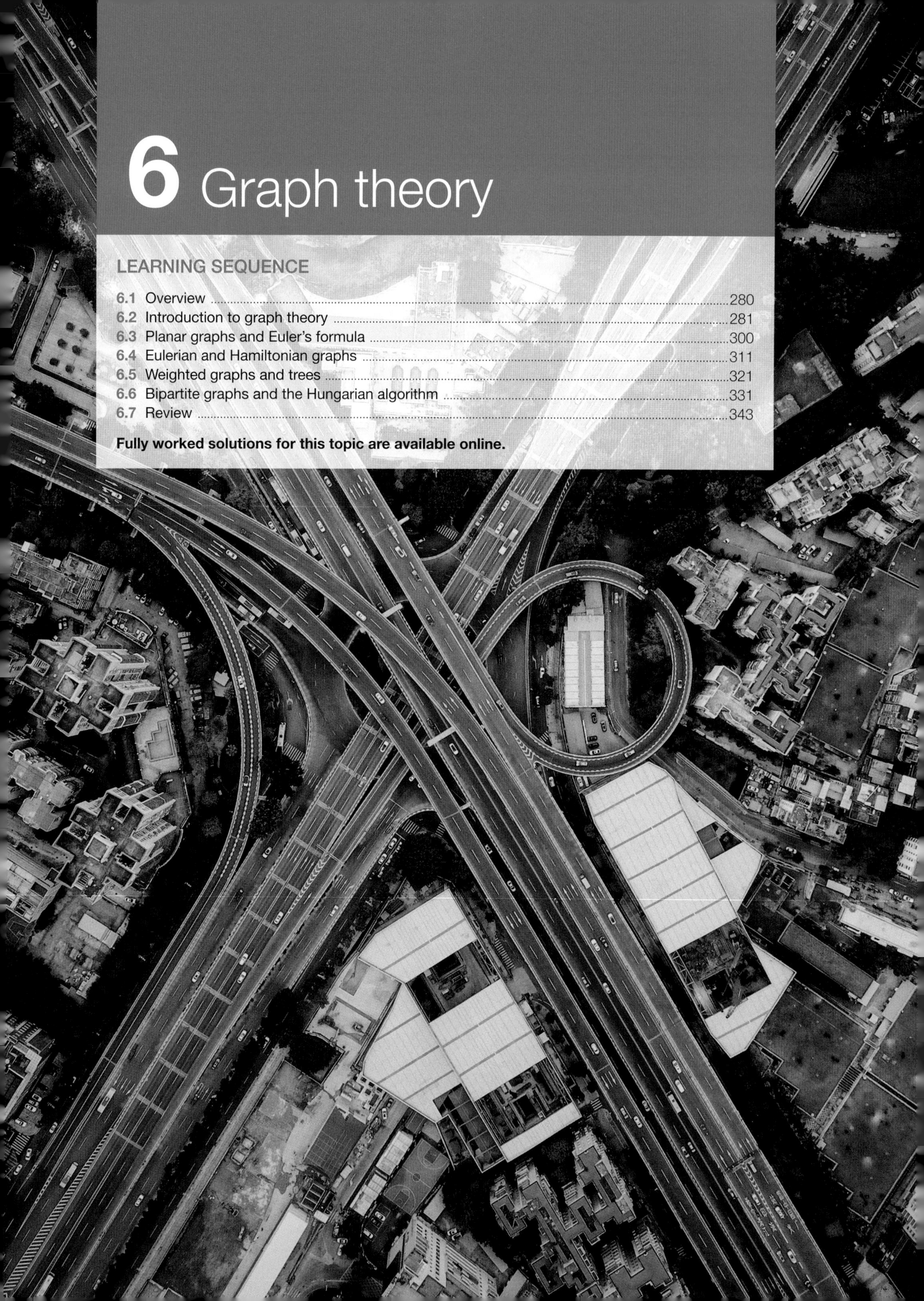

6 Graph theory

LEARNING SEQUENCE

Fully worked solutions for this topic are available online.

6.1 Overview

Hey students! Bring these pages to life online

 Watch videos

 Engage with interactivities

 Answer questions and check results

Find all this and MORE in jacPLUS

6.1.1 Introduction

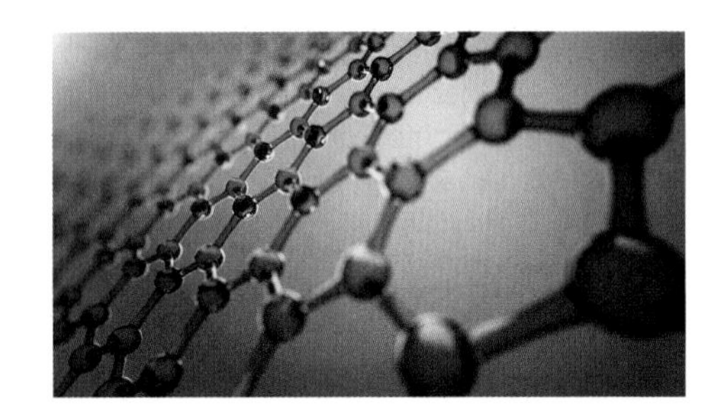

As you will have noticed in previous years, it is a common practice to draw diagrams and other visual and graphic representations when solving many mathematical problems. In the branch of mathematics known as graph theory, diagrams involving points and lines are used as a planning and analysis tool for systems and connections. Applications of graph theory include business efficiency, transportation systems, design projects, building and construction, food chains, communications networks, molecular structures, electrical circuits, and even social networks. The graphs referred to in graph theory are different from the graphs of a function and are called networks.

Mathematician Leonhard Euler (1707–83) is usually credited with being the founder of graph theory. He famously used it to solve a problem known as the 'Bridges of Königsberg'. For a long time it had been pondered whether it was possible to travel around the European city of Königsberg (now called Kaliningrad) in such a way that the seven bridges would only have to be crossed once each.

KEY CONCEPTS

This topic covers the following key concepts from the VCE Mathematics Study Design:

- vertices and edges for undirected graphs, including multiple edges and loops, and their representation using lists, diagrams and adjacency matrices
- examples of graphs from a range of contexts such as molecular structure, electrical circuits, social networks, utility connections and their use to discuss types of problems in graph theory including existence problems, construction problems, counting problems and optimisation problems
- the degree of a vertex and the result that the sum of all the vertex degrees is equal to twice the number of edges (the handshaking lemma)
- simple graphs, isomorphism, subgraphs, connectedness, complete graphs and the complement of a graph
- bipartite graphs, trees, and regular graphs (including the Platonic graphs)
- planar graphs and related proofs and applications such as:
  - Euler's formula $v+f-e=2$ for simple connected planar graphs
  - the complete graph K_n on n vertices has $\frac{n(n-1)}{2}$ edges
  - a regular graph with n vertices each of degree r has $\frac{nr}{2}$ edges
  - the planarity of various types of graphs, including all trees, complete graph on n vertices, if $n \leq 4$, and $K_{m,n}$, the complete bipartite graph with m, n vertices, if $m \leq 2$ or $n \leq 2$
- equivalent conditions for a simple graph with n vertices to be a tree
- trails and circuits, Euler circuits and Euler trails, Hamiltonian cycles and paths, and the Konigsberg bridge problem.

6.2 Introduction to graph theory

LEARNING INTENTION

At the end of this subtopic you should be able to:
- identify the vertices and edges of directed and undirected graphs, including multiple edges and loops
- determine the degree of a vertex
- identify simple, completed, connected and disconnected graphs, subgraphs, and isomorphism between two or more graphs
- be familiar with different representations of a graph including adjacency lists and matrices.

6.2.1 Graphs and networks

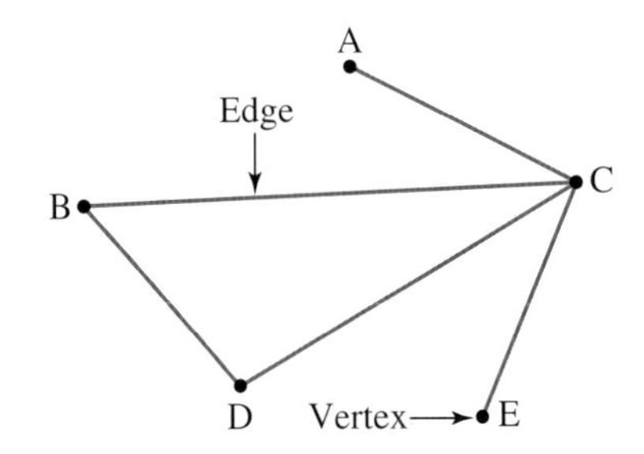

A **graph** or **network** is a series of points and lines that can be used to represent the connections that exist in various settings. The points that make up a graph are called vertices and the lines that join two vertices are called edges. Two vertices joined by an edge are said to be adjacent.

For example, the graph above has five vertices and five edges. Vertex A is adjacent to vertex C, vertex B is adjacent to vertices C and D, and so on.

Although edges are often drawn as straight lines, they don't have to be. As an example, a loop is an edge that connects a vertex to itself. It is also important to note that edges of a graph can intersect without there being a vertex.

An **isolated vertex** is one that is not connected by an edges to any other vertex. These may also be referred to as degenerate vertices.

6.2.2 Types of graphs

Simple graphs

A **simple graph** is one in which pairs of vertices are connected by one edge at most. The table below shows two simple graphs and one graph that is not simple. The third graph is not simple as the vertices A and D are connected by multiple edges.

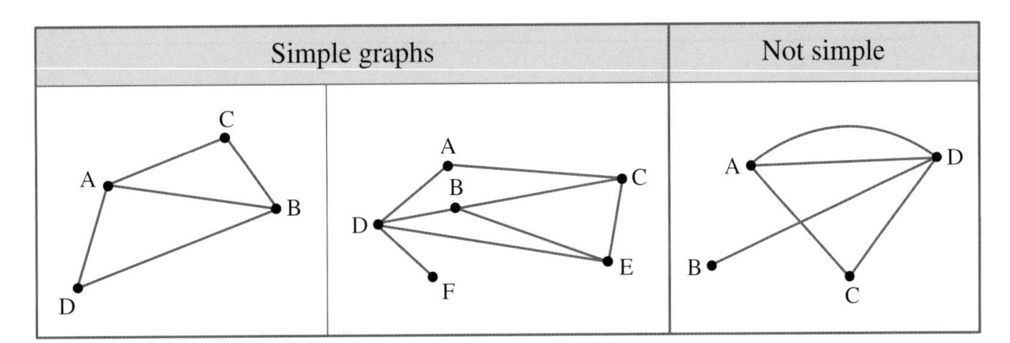

Connected graphs

A **connected graph** is a graph where it is possible to reach all vertices by moving along edges. A graph that is not connected is called a disconnected graph. A **bridge** is an edge in a connected graph that, if removed, would cause the graph to become disconnected. For example, the edge CE in the connected graph below is a bridge since the removal of that edge causes the graph to be disconnected.

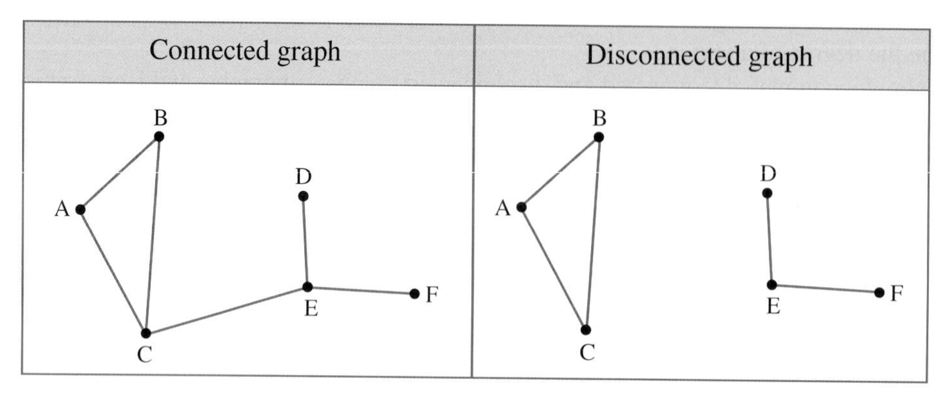

Complete graphs

A **complete graph** is a simple graph in which each vertex is connected to every other vertex. A complete graph is denoted by K_n where n is the number of vertices in the graph.

The number of edges in a complete graph is given by the formula:

$$\text{Number of edges of } K_n = \frac{n(n-1)}{2}$$

The graphs of K_4, K_5 and K_6 are shown below.

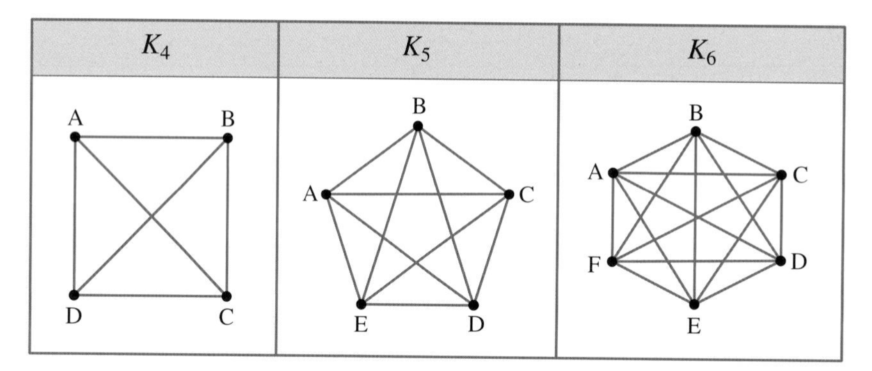

Directed graphs

All of the graphs described so far are called **undirected graphs** since it is possible to move along edges in both directions. **Directed graphs** are graphs in which you can only travel along each edge in one direction. The direction of edges is represented by an arrow on each edge pointing in the direction of allowed travel.

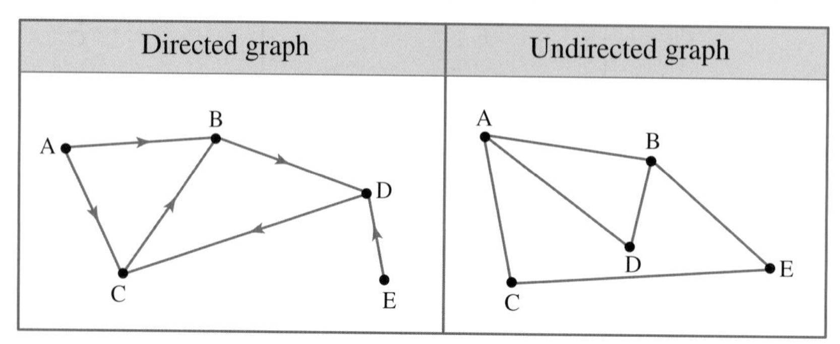

Null (degenerate) graphs

A graph with only isolated vertices (no edges) is called a **null graph** or a **degenerate graph**.

Null (degenerate) graphs	
A B C D E	A B C

6.2.3 Graphs representing paths or roads

Consider the road map shown.

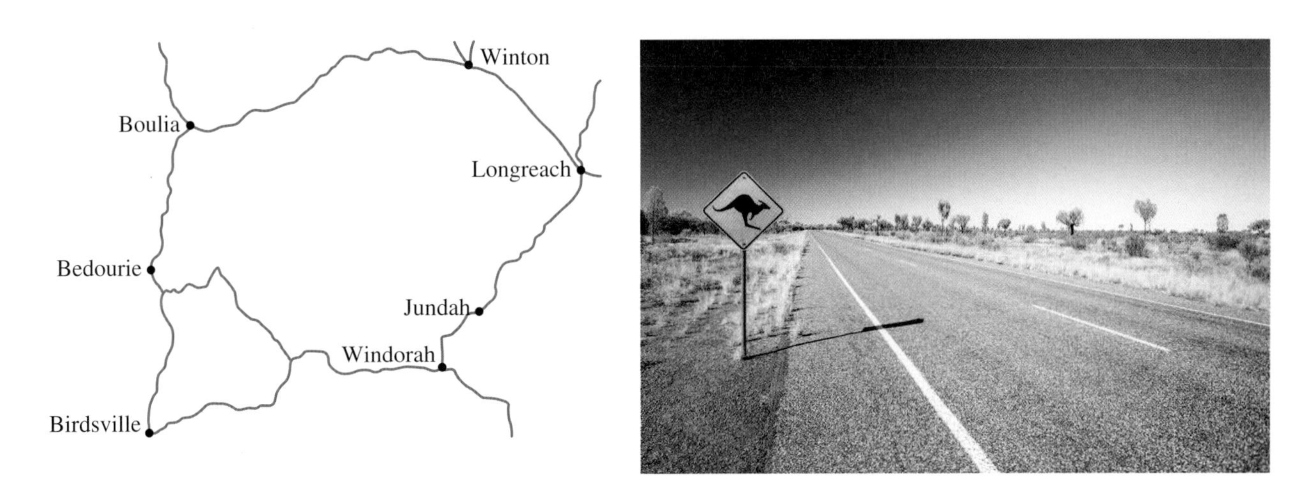

This map can be represented by the following graph or network.

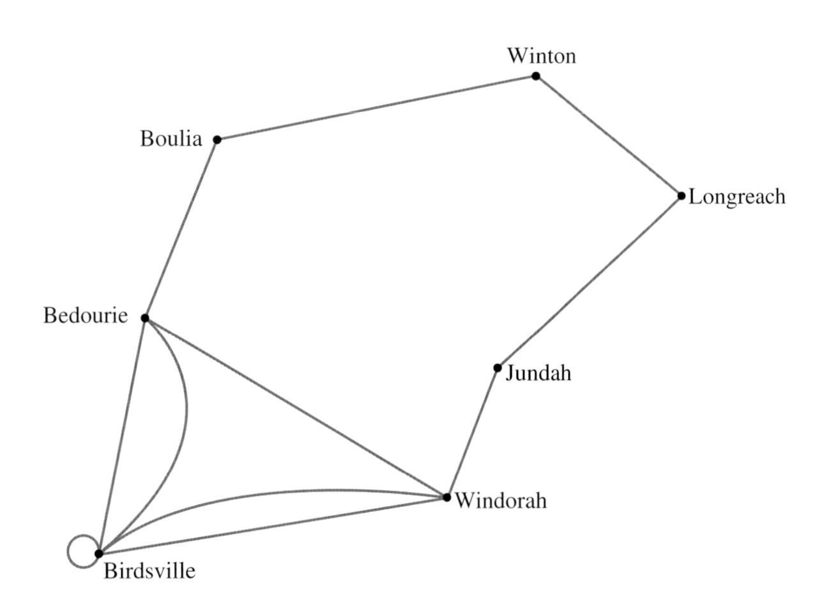

As there is more than one route connecting Birdsville to Windorah and Birdsville to Bedourie, they are each represented by an edge in the graph. The fact that some vertices are connected by multiple edges means that this is not a simple graph. Also, as it is possible to travel along a road from Birdsville that returns without passing through another town, this is represented by an edge. When this happens, the edge is called a **loop**.

WORKED EXAMPLE 1 Creating a graph to represent a system of paths

The diagram represents a system of paths and gates in a large park. Draw a graph to represent the possible ways of travelling to each gate in the park.

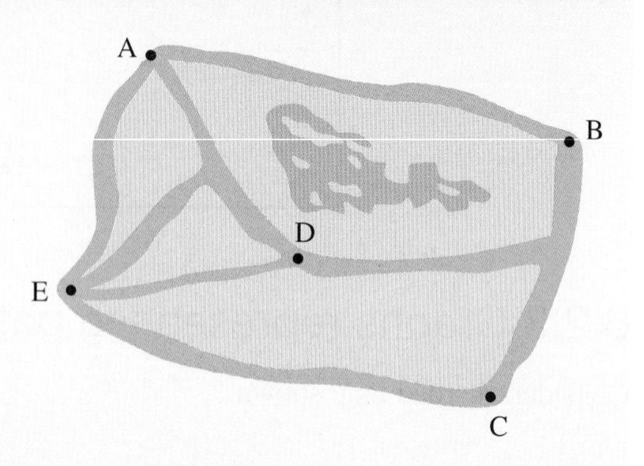

THINK	WRITE/DRAW
1. Identify, draw and label all possible vertices.	Represent each of the gates as vertices.
2. Draw edges to represent all the direct connections between the identified vertices.	Direct pathways exist for A–B, A–D, A–E, B–C, C–E and D–E.
3. Identify all the other unique ways of connecting vertices.	Other unique pathways exist for A–E, D–E, B–D and C–D.

A

B

D

E

C

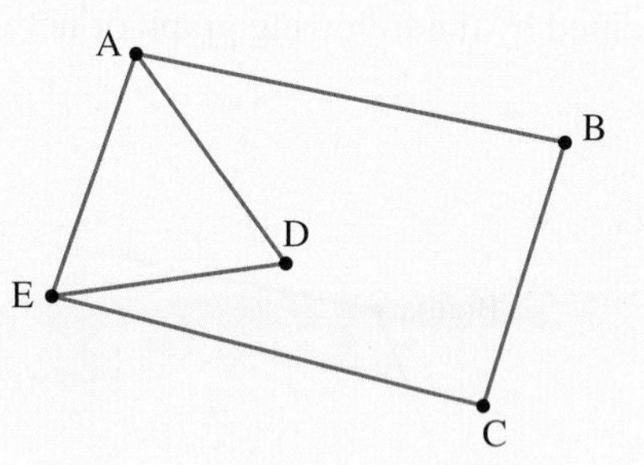

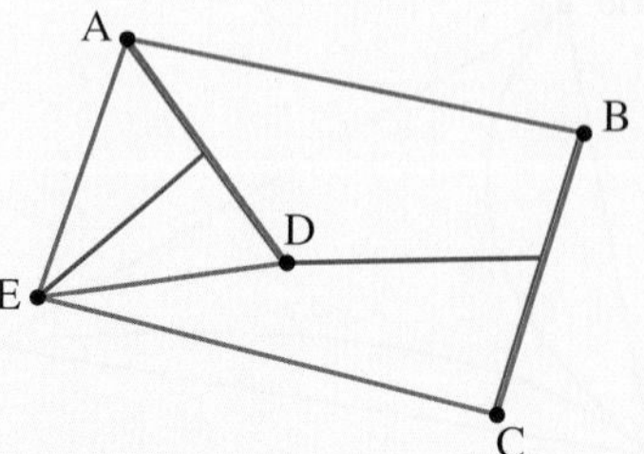

4. Draw the final graph.

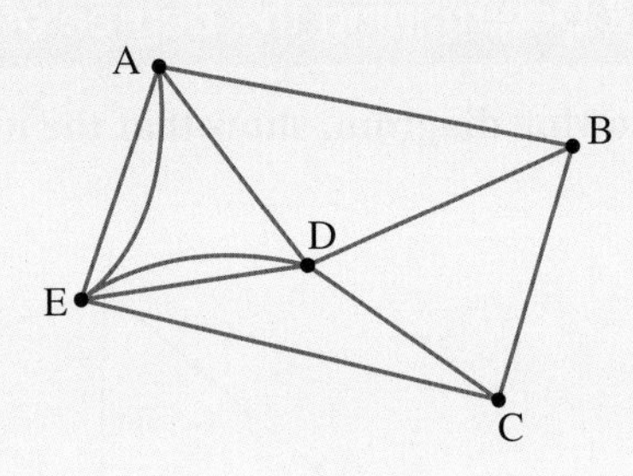

6.2.4 The degree of a vertex

When analysing the situation that a graph is representing, it can often be useful to consider the number of edges that are directly connected to a particular vertex. This is referred to as the **degree** of the vertex and is given the notation deg(V), where V represents the particular vertex. Note that a loop connected to a vertex adds 2 to the degree of that vertex. You can think of it as being connected to both ends of the edge that is the loop.

Since each edge has two ends, the sum of the degrees of each vertex in a graph will be double the number of edges.

The degree of a vertex

The degree of a vertex is equal to the number of edges that are directly connected to that vertex.

The sum of the degrees of the vertices in a graph is double the number of edges in the graph.

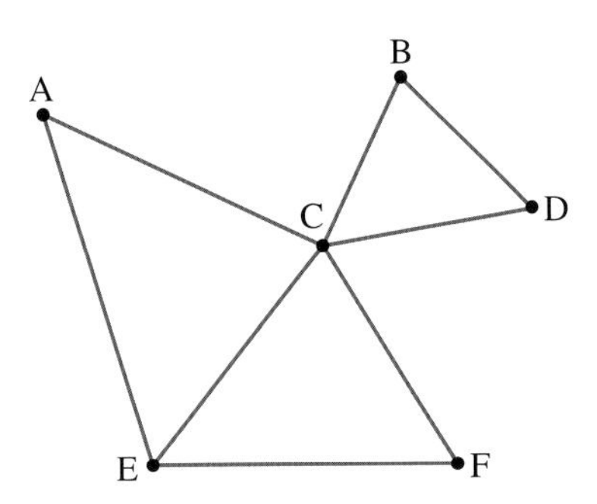

In the above diagram, $\deg(A) = 2$, $\deg(B) = 2$, $\deg(C) = 5$, $\deg(D) = 2$, $\deg(E) = 3$ and $\deg(F) = 2$. Note that the sum of the degrees in this graph is $2 + 2 + 5 + 2 + 3 + 2 = 16$, and there are 8 edges, which is half of the sum of the degrees of the vertices.

WORKED EXAMPLE 2 Determining the degree of each vertex in a graph

For the graph in the following diagram, show that the number of edges is equal to half the sum of the degree of the vertices.

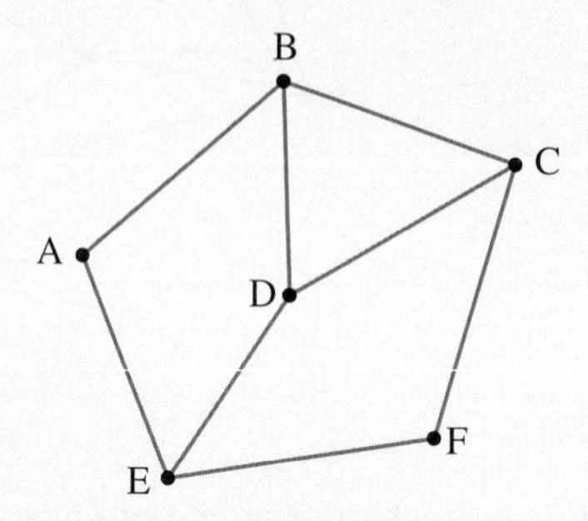

THINK	WRITE
1. Identify the degree of each vertex.	$\deg(A) = 2, \deg(B) = 3, \deg(C) = 3,$ $\deg(D) = 3, \deg(E) = 3,$ and $\deg(F) = 2$
2. Calculate the sum of the degrees for the graph.	The sum of the degrees for the graph $= 2 + 3 + 3 + 3 + 3 + 2$ $= 16$
3. Count the number of edges for the graph.	The graph has the following edges: A–B, A–E, B–C, B–D, C–D, C–F, D–E, E–F. The graph has 8 edges.
4. State the final answer.	The total number of edges in the graph is therefore half the sum of the degrees.

Regular graphs

If every vertex in a graph has the same degree, it is called a **regular graph**. If each vertex has degree r then the graph is regular of degree r or r-regular. Since the degree of every vertex in an r-regular graph is r, the sum of the degrees will be given by $n \times r$ where n is the number of vertices.

$$\begin{aligned}\sum \text{degrees of vertices} &= nr \\ &= 2 \times n\,(\text{edges})\end{aligned}$$

The number of edges in an r-regular graph is therefore equal to $\frac{nr}{2}$.

> **Regular graphs**
>
> **A regular graph is a graph where every vertex has the same degree.**
>
> **The number of edges in an r-regular graph is $\frac{nr}{2}$**
>
> **where n is the number of vertices and r is the degree of each vertex.**

It follows from this definition that every complete graph is also a regular graph, as the degree of every vertex in the complete graph K_n is $(n - 1)$. In other words, K_n is an $(n - 1)$-regular graph.

Another group of regular graphs that is of interest is the Platonic graphs, formed from the edges and vertices of the five platonic solids. The five platonic solids are five geometric solids whose faces are all identical, regular polygons that meet at the same angle.

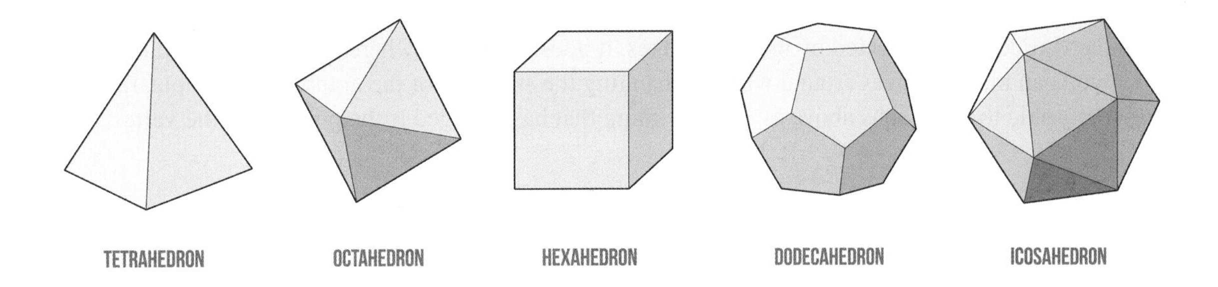

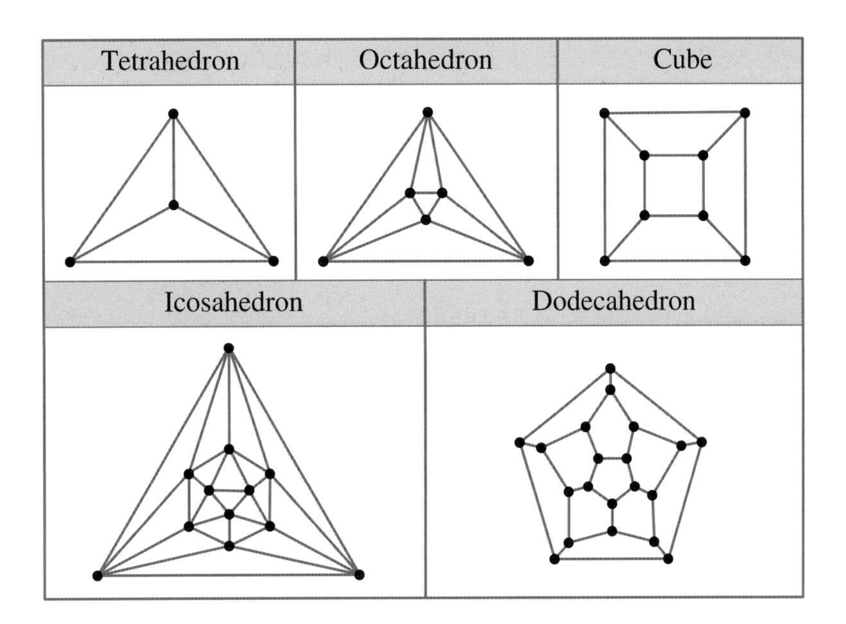

6.2.5 Isomorphism, subgraphs and the complement of a simple graph

Isomorphisms

In graph theory an isomorphism is a one-to-one correspondence between the vertices of two graphs, where the corresponding vertices in both graphs are connected by the same edges. Put simply, two graphs are **isomorphic** if their vertices and edges differ only by the way in which they are named. That is, the two graphs are structurally equivalent.

For example, consider the two graphs shown below. Both graphs have 5 vertices, and the same arrangement of edges between the vertices. You may notice that the vertices correspond to each other as A = I, B = J, C = H, D = F and E = G. A is connected to B and C just as I is connected to J and H, and so on for all the rest of the vertices.

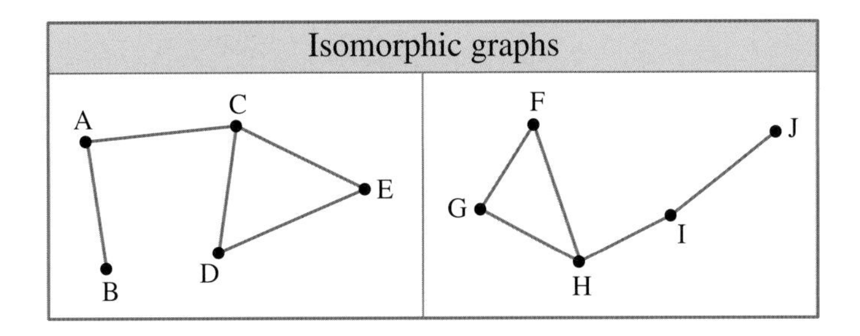

One helpful thing to note is that the way that a graph is drawn does not affect its inherent structure. That is, so long as the vertices are connected by the same edges, it does not matter where the vertices are placed. This means that you can move vertices around without changing the structure of the graph. For example, the graph below is isomorphic to the graphs above as the only thing that has changed is the position of the vertex B.

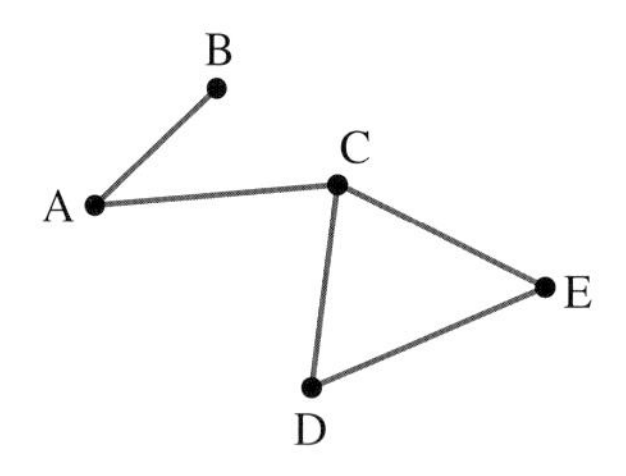

Isomorphic graphs

Two graphs are isomorphic if they have the same number of vertices, and corresponding vertices are connected by edges in the same way in both graphs. That is, they are structurally equivalent.

WORKED EXAMPLE 3 Determining isomorphism

Determine whether the two graphs below are isomorphic.

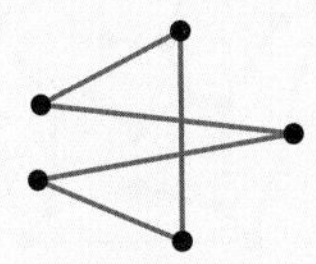

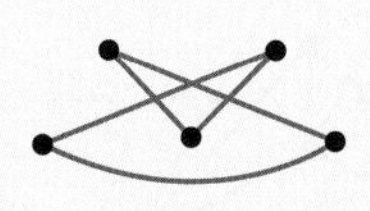

THINK	WRITE/DRAW
1. Are the number of vertices the same in each graph?	Both graphs have 5 vertices.
2. Are the number of edges the same in each graph?	Both graphs have 5 edges.
3. Check that the degrees of each vertex match for both graphs.	The degree of each vertex is 2 in the first graph. The degree of each vertex is 2 in the second graph.
4. Label each vertex on both graphs and check if there is a correspondence between the vertices. Starting with two vertices with the same number of edges create a correspondence between the two graphs. A → 1 By looking at the vertices adjacent to A we can see that: C → 4 D → 5 Since B is adjacent to D and E we get: B → 2 Since E is adjacent to B(2) and C (4) E → 3	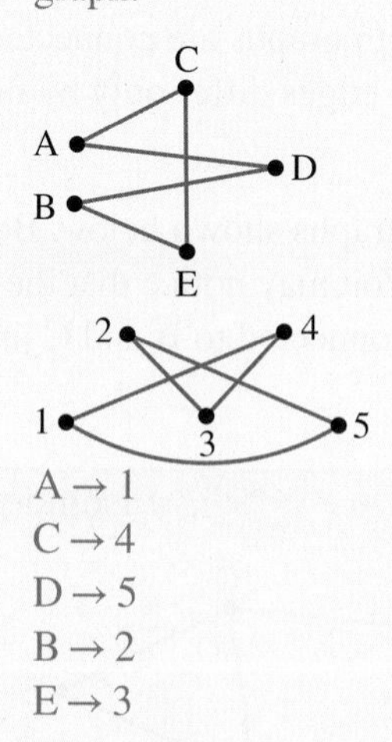 A → 1 C → 4 D → 5 B → 2 E → 3

5. Starting with two vertices with the same number of edges create a correspondence between the two graphs.
Look at the vertices adjacent to A.
B is adjacent to D and 2 is adjacent to 5.
E is adjacent to B and C and 3 is adjacent to 2 and 4.

A → 1
C → 4
D → 5
B → 2
E → 3

6. Since there is a one-to-one correspondence between the vertices of the two graphs they are isomorphic.

The two graphs are isomorphic since there is a one-to-one correspondence between the vertices.

Alternative method

1. Move vertex D to the left, vertex A up slightly and vertex B down slightly.

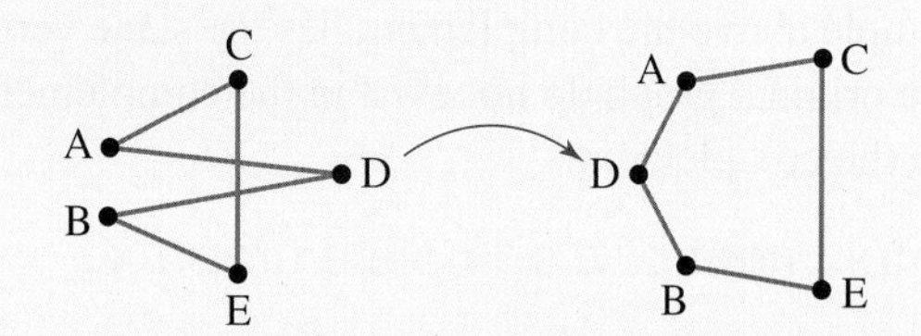

2. Move vertex 4 to the left, vertex 2 to the right and vertex 3 up.

3. It is now very clear that these graphs are isomorphic as they have the exact same structure. You can easily assign the vertex correspondence by looking at the graphs.

A → 1
C → 4
E → 3
B → 2
D → 5

Subgraphs

A **subgraph** is a graph whose vertices and edges are all contained within the original graph. A subgraph can be created by removing edges and vertices from the original graph.

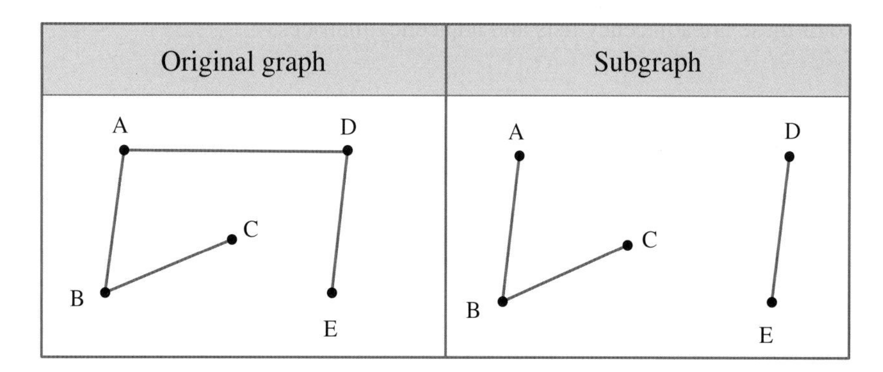

The complement of a graph

The **complement of a simple graph** is the graph that contains the same set of vertices and the complement set of edges. That is, any edges in the original simple graph are not edges in the graph of the complement, and any edges that do not exist in the original graph will exist in the complement.

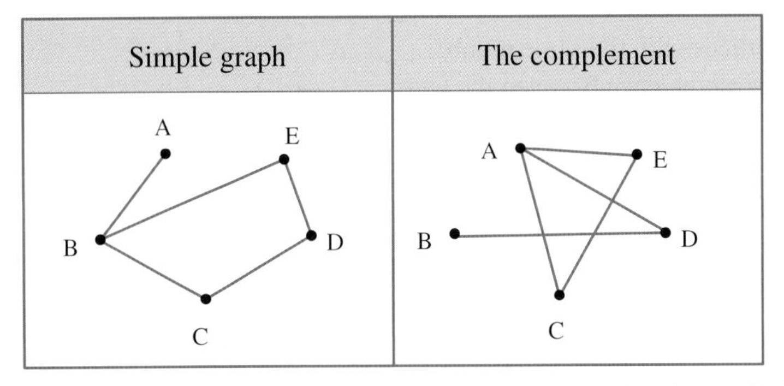

In the example above the complement has the same vertices as the original simple graph, but the edges that exist in the original graph do not exist in the complement and the edges that didn't exist in the original graph do exist in the complement.

The complement of a complete graph K_n

Since the complete graph on n vertices K_n contains every possible edge between vertices, the complement of the complete graph will have no edges. The complement of a complete graph is therefore a null graph.

K_5	Complement of K_5
B, A, C, E, D	B, A, C, E, D

6.2.6 Adjacency lists and matrices

While the diagram of graph with edges and vertices gives a clear visual representation of the structure of the system it is seeking to represent, this is not a particularly usable form. If we wanted to input a graph into a computer and run it through an algorithm to find the shortest distance between two points, we need other representations. Two of these are adjacency lists and adjacency matrices.

Adjacency lists

An **adjacency list** is a set of unordered lists, whereby each unordered list states the vertices that are adjacent to a particular vertex. Since the adjacency list contains the information regarding the number of vertices and how the vertices are connected by edges, it represents the overall graph.

Graph	Adjacency list
A, B, E, D, C	A ⟶ (B, D, D, E) B ⟶ (A, E) C ⟶ (C, D) D ⟶ (A, A, C) E ⟶ (A, B)

In each list the order of the successors in each list is irrelevant, all that matters is the number of occurrences of each successor. If a vertex appears multiple times in a specific list, that indicates multiple edges; if a vertex appears in its own list of successors then there is a loop.

Note: An empty list indicates a vertex with no edges connected to it.

Adjacency matrices

Matrices are often used when working with graphs. A matrix that represents the vertices and edges that connect the vertices of a graph is known as an **adjacency matrix**.

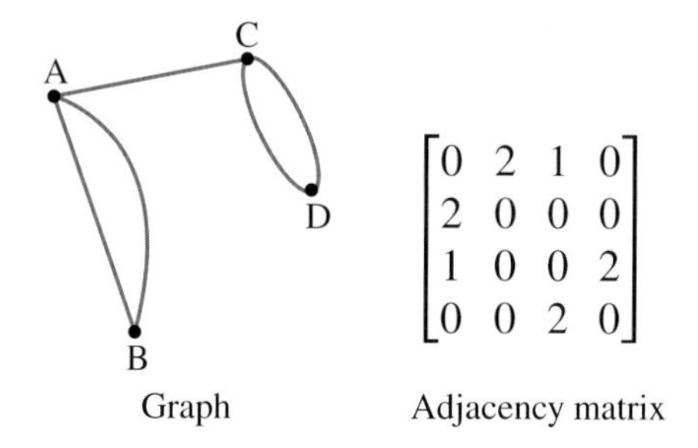

In the adjacency matrix, column 3 corresponds to vertex C and row 4 to vertex D. The '2' indicates the number of edges joining these two vertices.

$$\begin{array}{c} \\ A \\ B \\ C \\ \textcircled{D} \end{array} \begin{array}{c} \begin{array}{cccc} A & B & \textcircled{C} & D \end{array} \\ \begin{bmatrix} 0 & 2 & 1 & 0 \\ 2 & 0 & 0 & 0 \\ 1 & 0 & 0 & 2 \\ 0 & 0 & 2 & 0 \end{bmatrix} \end{array}$$

> **Adjacency matrices**
>
> **Each column and row of an adjacency matrix corresponds to a vertex of the graph. The numbers in the matrix indicate how many edges exist between these vertices.**

Characteristics of adjacency matrices

Adjacency matrices are square matrices with n rows and columns, where 'n' is equal to the number of vertices in the graph.

$$\begin{array}{l} \text{Column: } \; 1 \quad 2 \ldots n-1 \quad n \qquad \text{Row} \end{array}$$
$$\begin{bmatrix} 0 & 2 \ldots & 1 & 0 \\ 2 & 0 \ldots & 0 & 0 \\ \vdots & \vdots \ldots & \vdots & \vdots \\ 1 & 0 \ldots & 0 & 2 \\ 0 & 0 \ldots & 2 & 0 \end{bmatrix} \begin{array}{c} 1 \\ 2 \\ \vdots \\ n-1 \\ n \end{array}$$

Adjacency matrices are symmetrical around the leading diagonal.

$$\begin{bmatrix} 0 & 2 & 1 & 0 \\ 2 & 0 & 0 & 0 \\ 1 & 0 & 0 & 2 \\ 0 & 0 & 2 & 0 \end{bmatrix}$$

Any non-zero value in the leading diagonal will indicate the existence of a loop. For example, the 1 in the adjacency matrix below represents a loop at vertex B. ***Note:*** Although a loop contributes 2 to the degree of the vertex, it is only 1 edge, so is represented by a 1 in the adjacency matrix, not a 2.

$$\begin{bmatrix} 0 & 2 & 1 & 2 \\ 2 & \textcircled{1} & 0 & 0 \\ 1 & 0 & 0 & 2 \\ 2 & 0 & 2 & 0 \end{bmatrix}$$

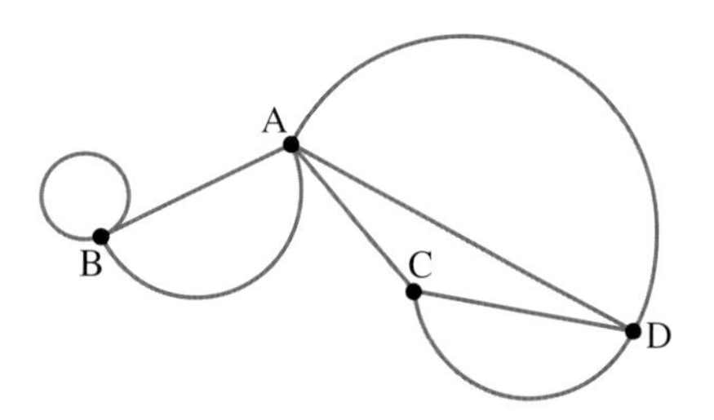

A row consisting of all zeros indicates an isolated vertex (a vertex that is not connected to any other vertex).

$$\begin{bmatrix} 0 & 0 & 0 & 1 \\ \boxed{0 \;\; 0 \;\; 0 \;\; 0} \\ 0 & 0 & 0 & 1 \\ 1 & 0 & 1 & 0 \end{bmatrix}$$

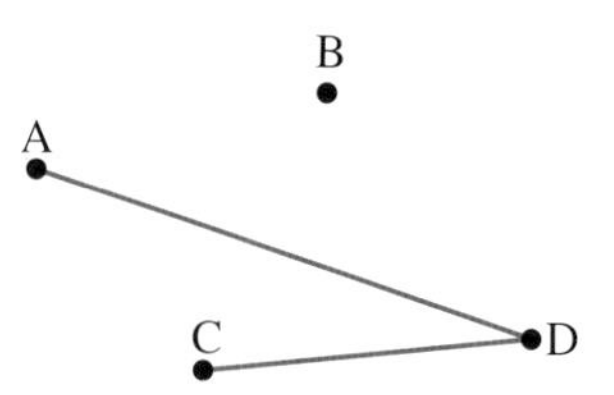

WORKED EXAMPLE 4 Creating an adjacency matrix from a graph

Construct the adjacency matrix for the given graph.

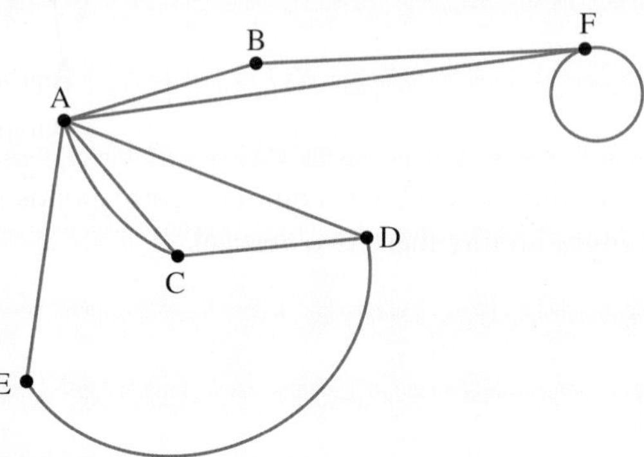

THINK

WRITE/DRAW

1. Draw up a table with rows and columns for each vertex of the graph.

	A	B	C	D	E	F
A						
B						
C						
D						
E						
F						

2. Count the number of edges that connect vertex A to the other vertices and record these values in the corresponding space for the first row of the table.

	A	B	C	D	E	F
A	0	1	2	1	1	1

3. Repeat step 2 for all the other vertices.

	A	B	C	D	E	F
A	0	1	2	1	1	1
B	1	0	0	0	0	1
C	2	0	0	1	0	0
D	1	0	1	0	1	0
E	1	0	0	1	0	0
F	1	1	0	0	0	1

4. Display the numbers as a matrix.

$$\begin{array}{c} \\ A \\ B \\ C \\ D \\ E \\ F \end{array} \begin{array}{c} \begin{array}{cccccc} A & B & C & D & E & F \end{array} \\ \begin{bmatrix} 0 & 1 & 2 & 1 & 1 & 1 \\ 1 & 0 & 0 & 0 & 0 & 1 \\ 2 & 0 & 0 & 1 & 0 & 0 \\ 1 & 0 & 1 & 0 & 1 & 0 \\ 1 & 0 & 0 & 1 & 0 & 0 \\ 1 & 1 & 0 & 0 & 0 & 1 \end{bmatrix} \end{array}$$

6.2 Exercise

Technology free

1. **WE1** The diagram shows the plan of a floor of a house. Draw a graph to represent the possible ways of travelling between each room of the floor.

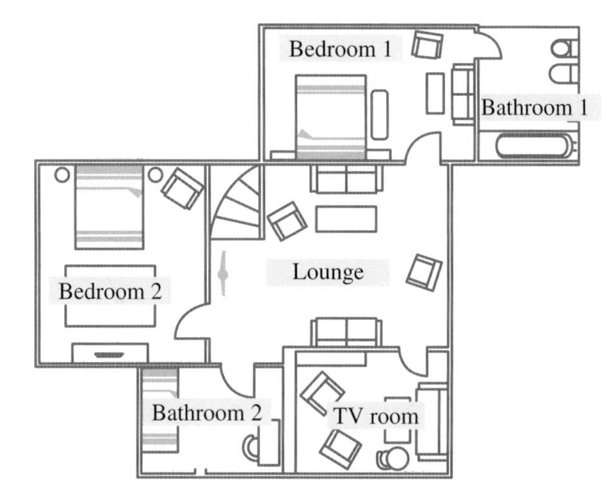

2. Draw a graph to represent the following tourist map.

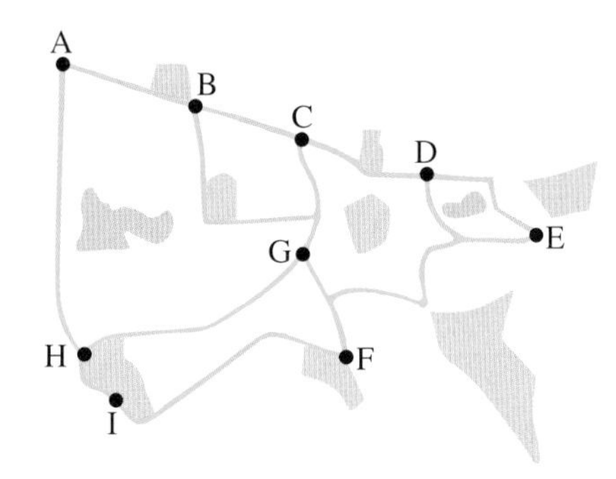

3. **WE2** For each of the following graphs, verify that the number of edges is equal to half the sum of the degree of the vertices.

a.

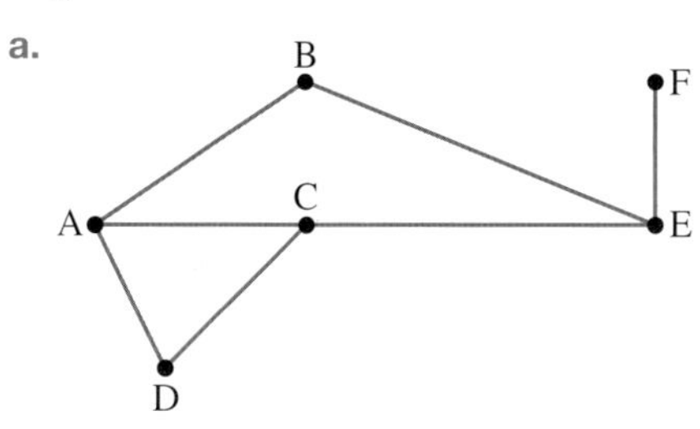

b.

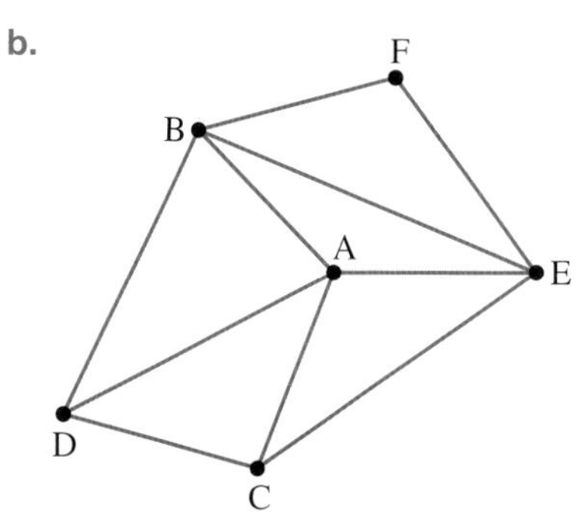

4. For each of the following graphs, verify that the number of edges is equal to half the sum of the degree of the vertices.

a. 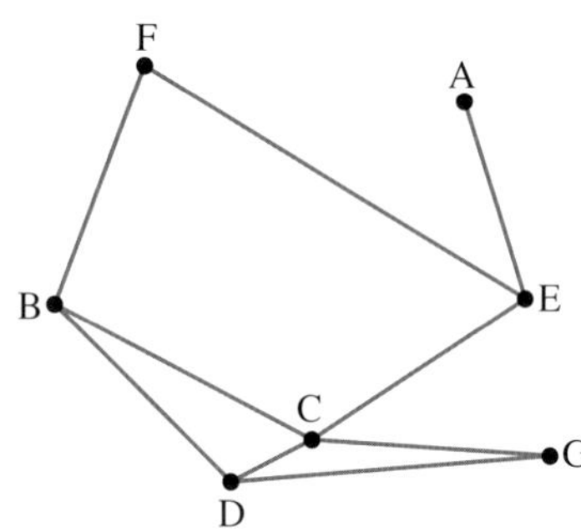

b. 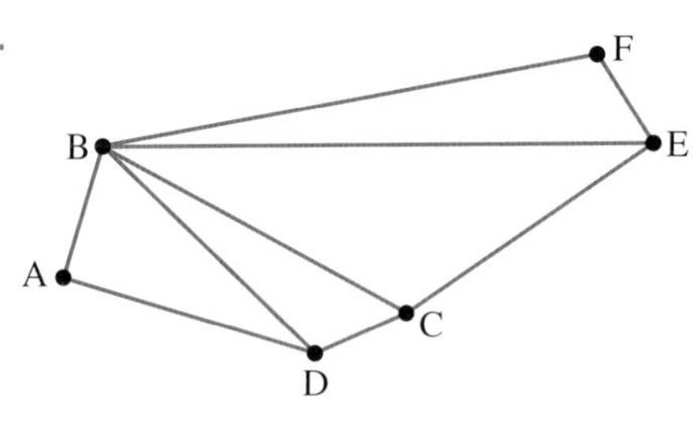

5. Identify the degree of each vertex in the following graphs.

a. 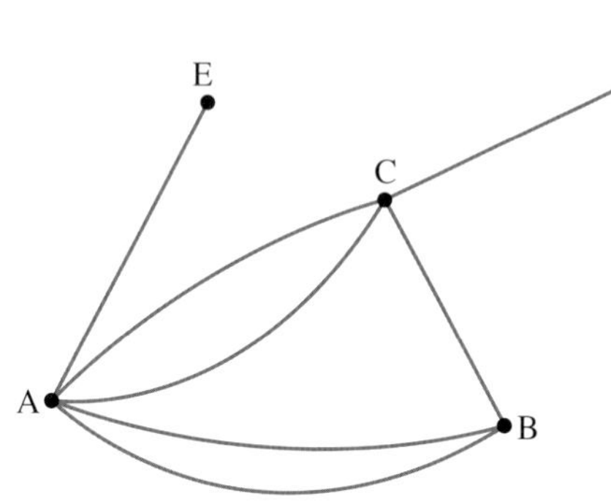

b. 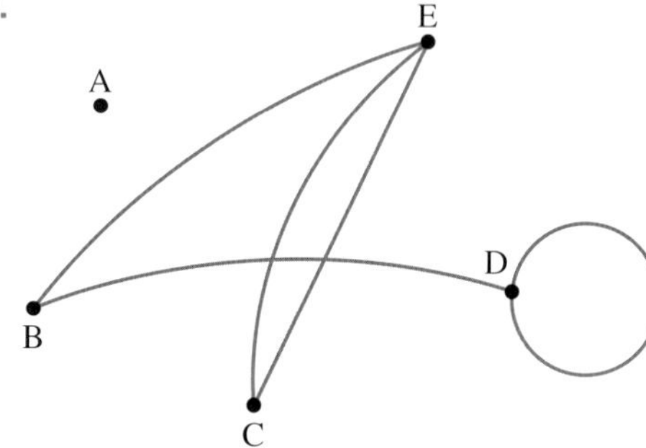

c. 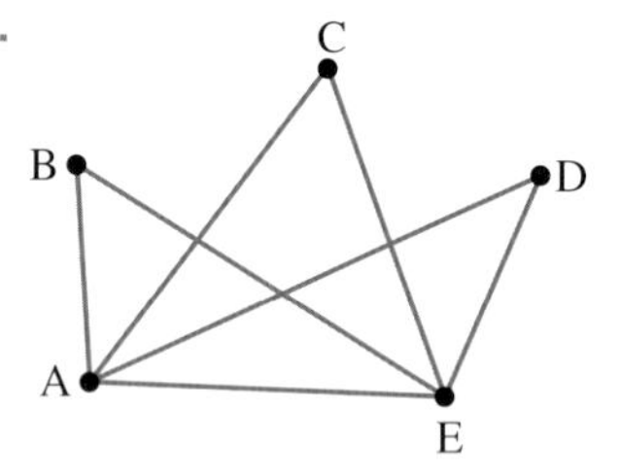

d. 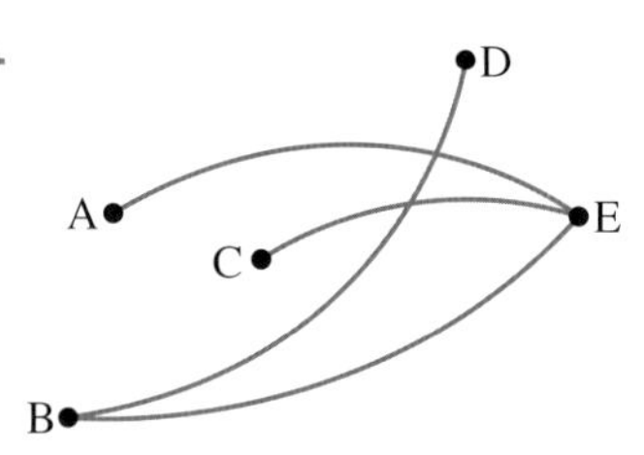

6. WE3 For each of the following pairs of graphs, determine whether the graphs are isomorphic.

a. 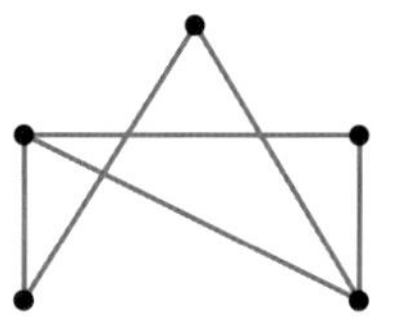

b. 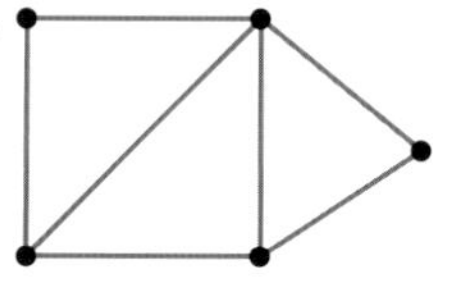

c. 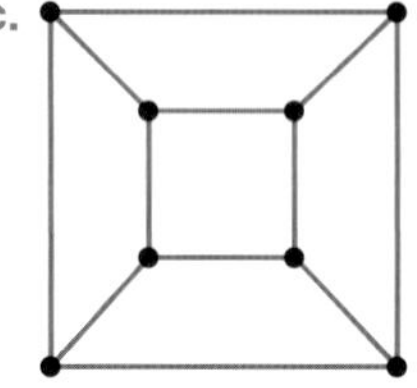

d. 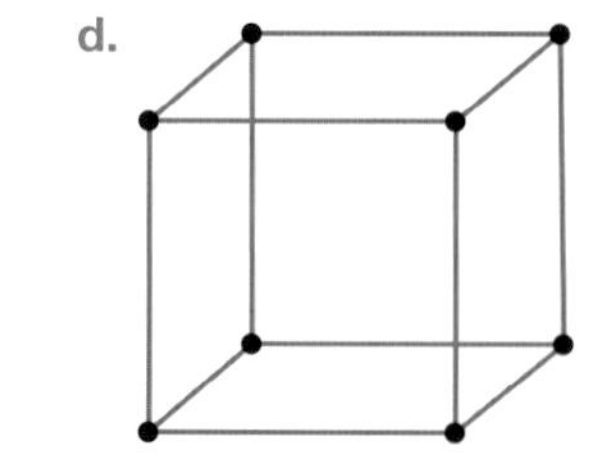

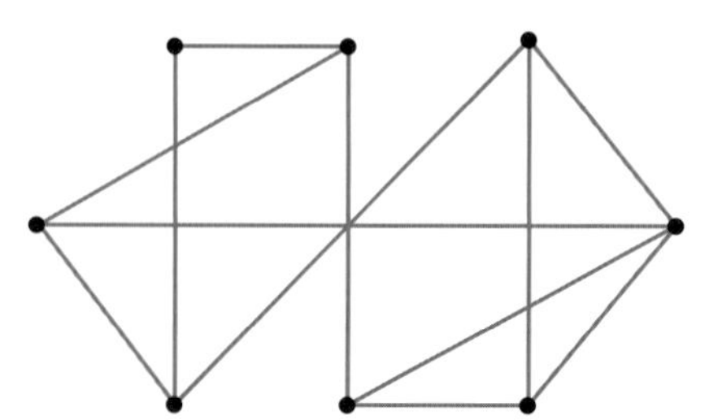

7. **WE4** Construct adjacency matrices for the following graphs.

a.

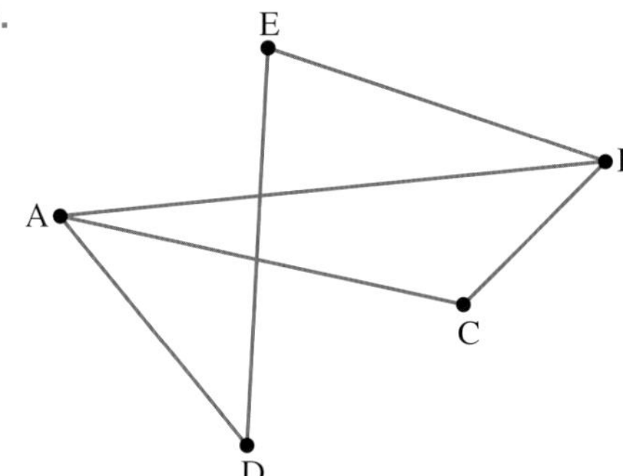

b.

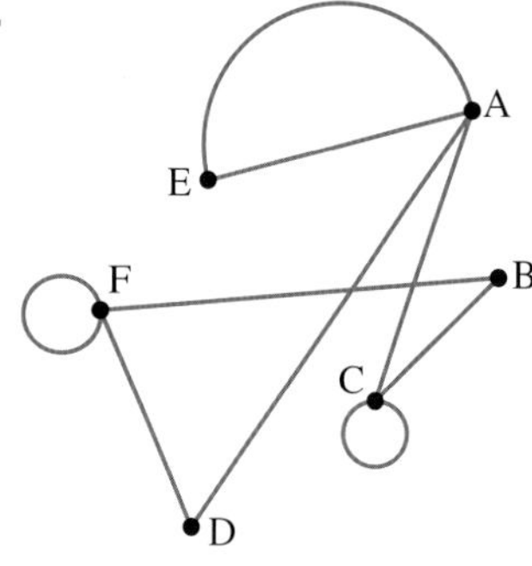

c.

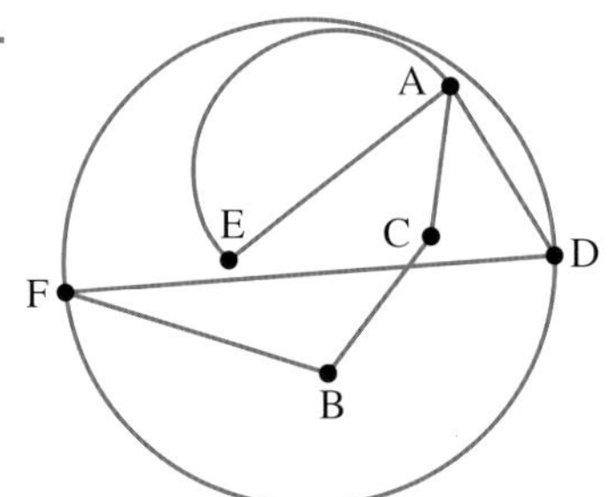

d.

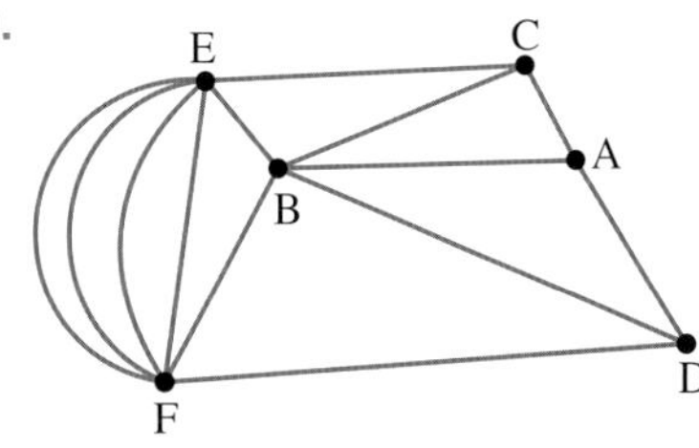

8. Construct adjacency lists for the graphs shown in question 7.

9. Draw graphs to represent the following adjacency matrices.

a. $\begin{bmatrix} 0 & 1 & 0 & 1 \\ 1 & 0 & 0 & 1 \\ 0 & 0 & 0 & 1 \\ 1 & 1 & 1 & 0 \end{bmatrix}$

b. $\begin{bmatrix} 1 & 0 & 2 & 0 \\ 0 & 0 & 0 & 1 \\ 2 & 0 & 0 & 2 \\ 0 & 1 & 2 & 0 \end{bmatrix}$

c. $\begin{bmatrix} 0 & 1 & 2 & 0 & 0 \\ 1 & 1 & 0 & 0 & 1 \\ 2 & 0 & 0 & 1 & 0 \\ 0 & 0 & 1 & 1 & 1 \\ 0 & 1 & 0 & 1 & 1 \end{bmatrix}$

d. $\begin{bmatrix} 2 & 0 & 1 & 1 & 0 \\ 0 & 0 & 0 & 0 & 0 \\ 1 & 0 & 1 & 0 & 2 \\ 1 & 0 & 0 & 0 & 1 \\ 0 & 0 & 2 & 1 & 0 \end{bmatrix}$

10. Complete the following adjacency matrices.

a. $\begin{bmatrix} 0 & 0 & \\ 0 & 2 & 2 \\ 1 & & 0 \end{bmatrix}$

b. $\begin{bmatrix} 2 & 1 & & 0 \\ & 0 & & \\ 0 & 1 & 0 & 1 \\ & 2 & & 0 \end{bmatrix}$

c. $\begin{bmatrix} 0 & & 1 & & 0 \\ 0 & 0 & & & 0 \\ & 0 & 0 & 0 & 2 \\ 1 & 0 & 0 & 0 & 1 \\ 0 & & & 1 & 0 \end{bmatrix}$

d. $\begin{bmatrix} 0 & 0 & 0 & 1 & 0 \\ 0 & 0 & 0 & 1 & 0 \\ & 0 & 0 & 0 & 1 \\ & & & 0 & 0 \\ 0 & & & 0 & 1 \end{bmatrix}$

11. Enter details for complete graphs in the following table.

Vertices	Edges
2	
3	
4	
5	
6	
n	

12. Copy and complete the following table for the graphs shown.

	Simple	Complete	Connected
Graph 1	Yes	No	Yes
Graph 2			
Graph 3			
Graph 4			
Graph 5			

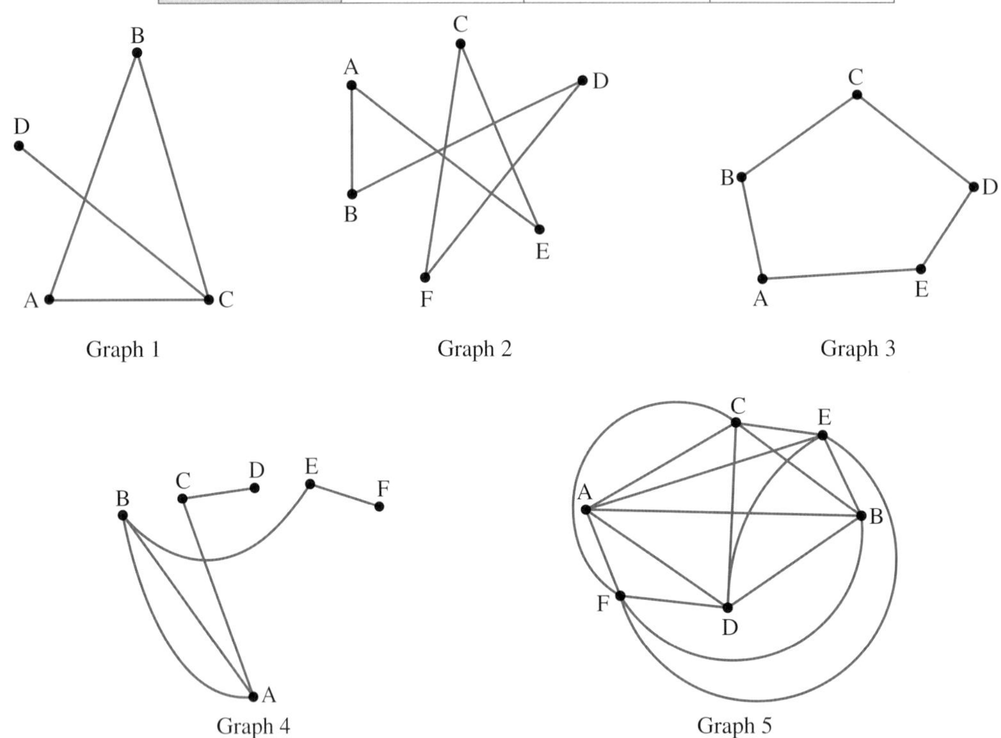

Graph 1 Graph 2 Graph 3 Graph 4 Graph 5

13. Construct the adjacency matrices for each of the graphs shown in question 12.

14. Draw a graph of:

a. a simple, connected graph with 6 vertices and 7 edges

b. a simple, connected graph with 7 vertices and 7 edges, where one vertex has degree 3 and five vertices have degree 2

c. a simple, connected graph with 9 vertices and 8 edges, where one vertex has degree 8.

Technology active

15. By indicating the passages with edges and the intersections and passage endings with vertices, draw a graph to represent the maze shown in the diagram.

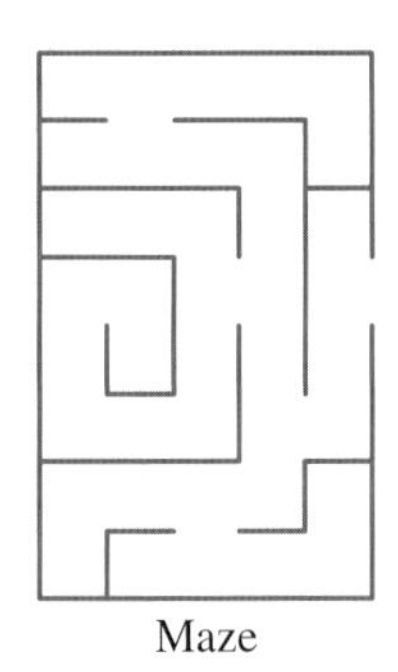
Maze

16. Five teams play a round robin competition.

a. Draw a graph to represent the games played.

b. State what type of graph this is.

c. State what the total number of edges in the graph indicates.

17. The diagram shows the map of some of the main suburbs of Beijing.
   a. Draw a graph to represent the shared boundaries between the suburbs.
   b. State which suburb has the highest degree.
   c. State what type of graph this is.

18. The map shows some of the main highways connecting some of the states on the west coast of the USA.

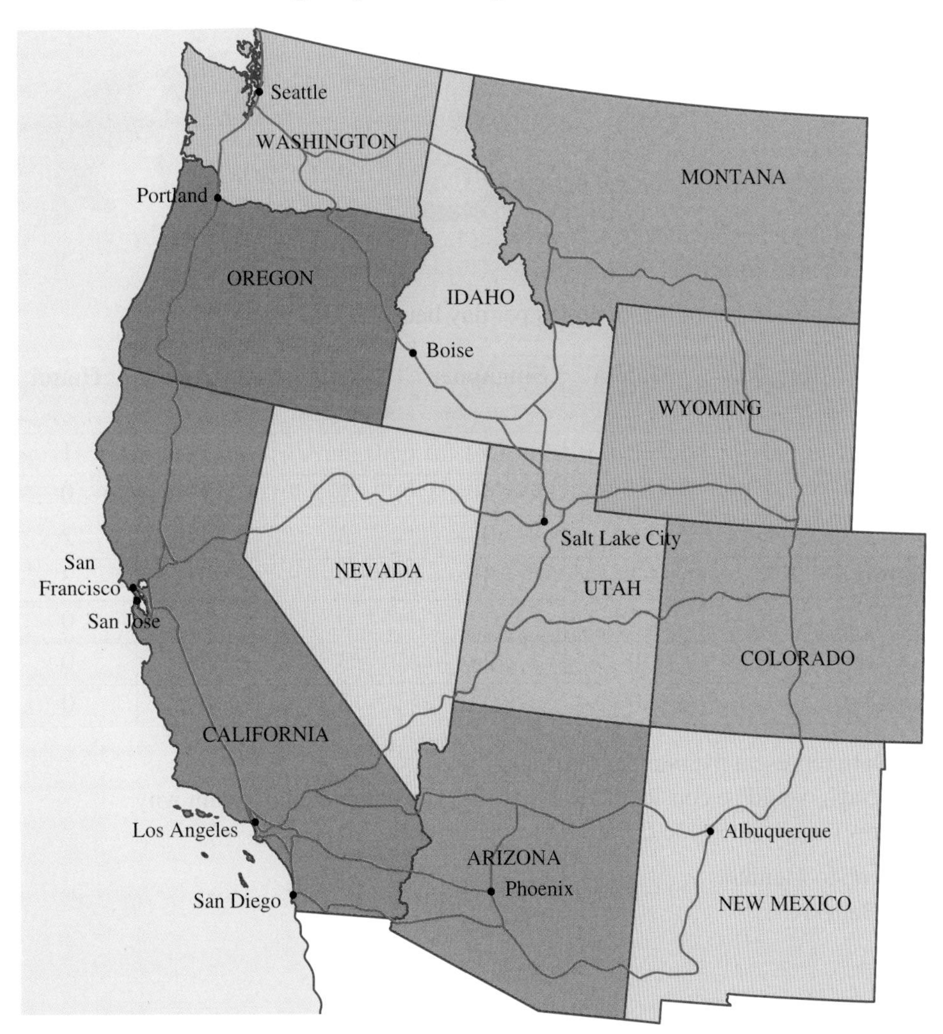

   a. Draw a graph to represent the highways connecting the states shown.
   b. Use your graph to construct an adjacency matrix.
   c. Write down which state has the highest degree.
   d. Write down the state that has a degree of 5.

19. Jetways Airlines operates flights in South East Asia.

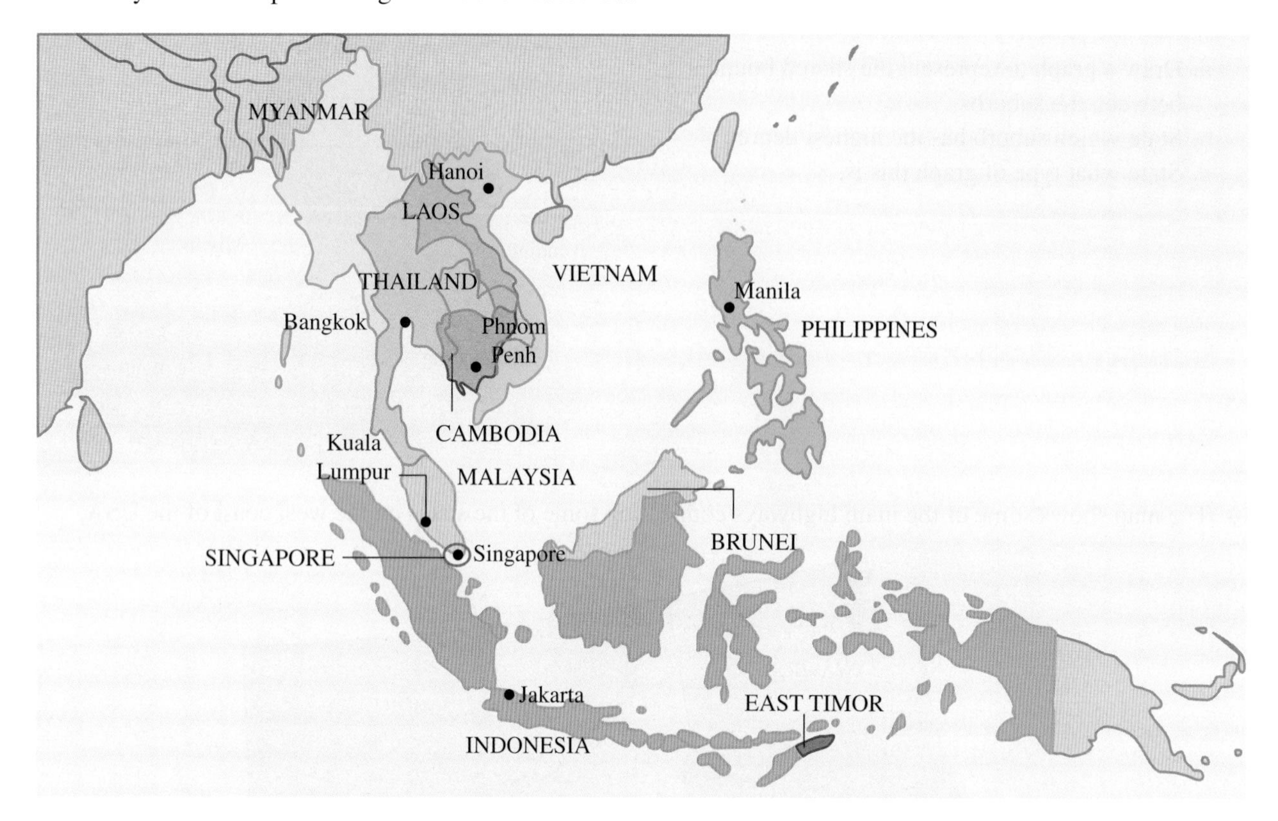

The table indicates the number of direct flights per day between key cities.

From: To:	Bangkok	Manila	Singapore	Kuala Lumpur	Jakarta	Hanoi	Phnom Penh
Bangkok	0	2	5	3	1	1	1
Manila	2	0	4	1	1	0	0
Singapore	5	4	0	3	4	2	3
Kuala Lumpur	3	1	3	0	0	3	3
Jakarta	1	1	4	0	0	0	0
Hanoi	1	0	2	3	0	0	0
Phnom Penh	1	0	3	3	0	0	0

a. Draw a graph to represent the number of direct flights.
b. State if this graph would be considered to be directed or undirected. Explain why.
c. Determine how many routes you can use to travel from:
   i. Phnom Penh to Manila
   ii. Hanoi to Bangkok.

20. Below is a representation of the chemical models of methane (CH_4) and propane (C_3H_8).

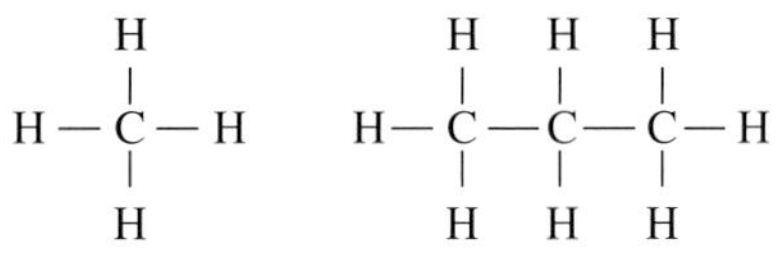

```
     H           H   H   H
     |           |   |   |
 H — C — H   H — C — C — C — H
     |           |   |   |
     H           H   H   H
```

a. When looking at the diagrams, describe what can be said about the vertices that are represented by carbon (C) and hydrogen (H).
b. There are two different chemical molecules with the formula C_4H_{10}. Draw the graphs of these.
c. James and Suet are drawing out different chemical molecules with the formula C_5H_{12}. Two of the diagrams they produce are shown here:

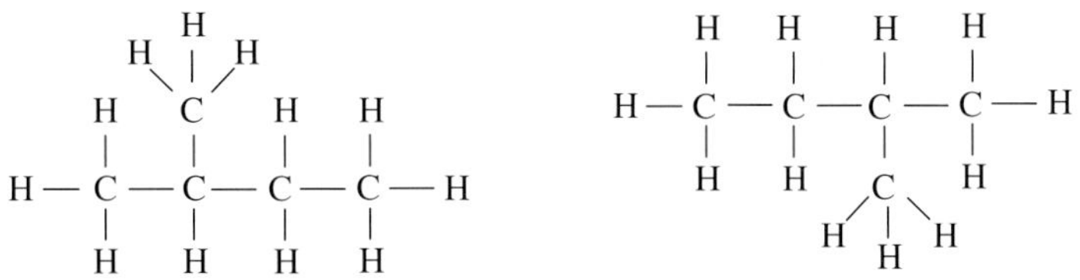

```
         H
       H | H
        \|/                      H   H   H   H
     H   C   H   H               |   |   |   |
     |   |   |   |           H — C — C — C — C — H
 H — C — C — C — C — H           |   |   |   |
     |   |   |   |               H   H   C   H
     H   H   H   H                      /|\
                                       H H H
```

James thinks they are two different molecules and Suet thinks they are the same. Determine who is correct and justify your answer.

6.2 Exam questions

Question 1 (2 marks) TECH-FREE
If two simple graphs are isomorphic, determine if the complement of each graph also isomorphic. Justify your answer.

Question 2 (3 marks) TECH-FREE
A graph G is self-complementary if G is isomorphic to the complement of G. Show that if G is self-complementary, then the number of vertices is equal to $4k$ or $4k + 1$, where k is an integer.

Question 3 (1 mark) TECH-ACTIVE
MC Consider the following adjacency matrix for a graph G.

$$A = \begin{bmatrix} 0 & 1 & 1 & 2 & 0 \\ 1 & 0 & 0 & 0 & 1 \\ 1 & 0 & 0 & 1 & 1 \\ 2 & 0 & 1 & 0 & 0 \\ 0 & 1 & 1 & 0 & 0 \end{bmatrix}$$

The sequence that gives the possible degrees of each vertex in G is
A. (2, 2, 3, 3, 4)
B. (0, 1, 1, 2, 0)
C. (0, 0, 1, 1, 2)
D. (2, 3, 3, 3, 3)
E. (2, 3, 4)

More exam questions are available online.

6.3 Planar graphs and Euler's formula

LEARNING INTENTION

At the end of this subtopic you should be able to:
- understand planarity and be able to determine whether a graph is planar
- use Euler's formula to determine the number of faces, edges or vertices of a planar graph.

6.3.1 Planar graphs

As indicated in subtopic 6.2, graphs can be drawn with intersecting edges. However, in many applications intersections may be undesirable. Consider a graph of an underground railway network. In this case intersecting edges would indicate the need for one rail line to be in a much deeper tunnel, which could add significantly to construction costs.

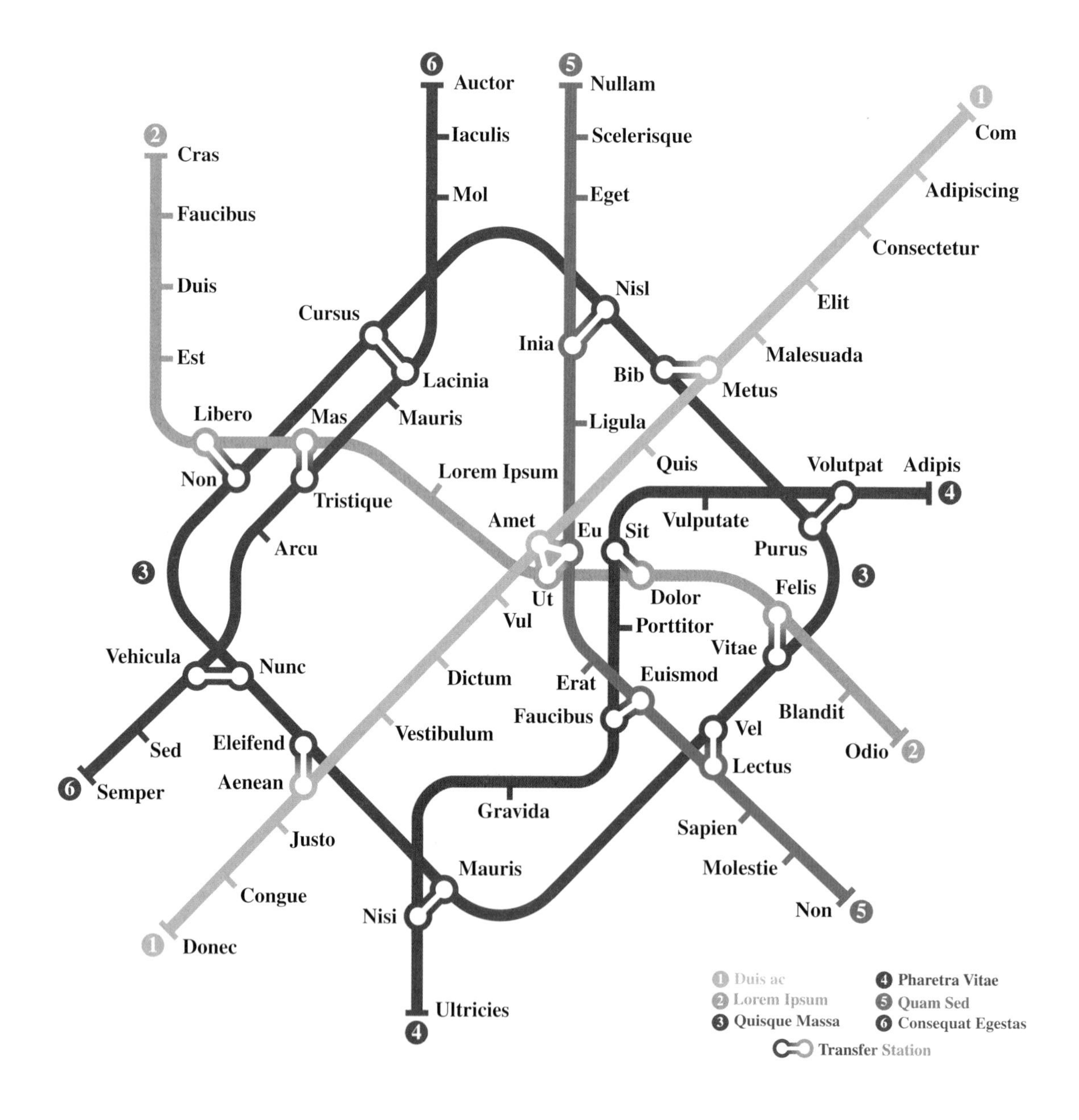

In some cases it is possible to redraw graphs so that they have no intersecting edges. When a graph can be redrawn in this way, it is known as a **planar graph**. For example, in the graph shown below, it is possible to redraw one of the intersecting edges so that it still represents the same information.

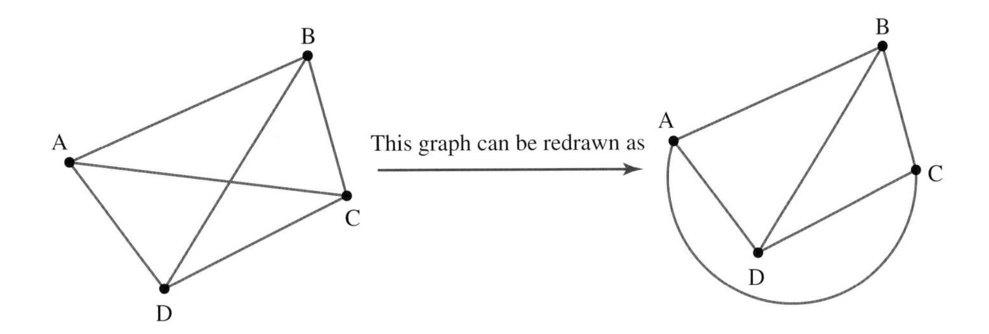

Planar graphs

A planar graph is a graph that can be drawn in the plane (2D) so that no two lines intersect, except at a vertex.

WORKED EXAMPLE 5 Redrawing a graph as a planar graph

Redraw the graph so that it has no intersecting edges.

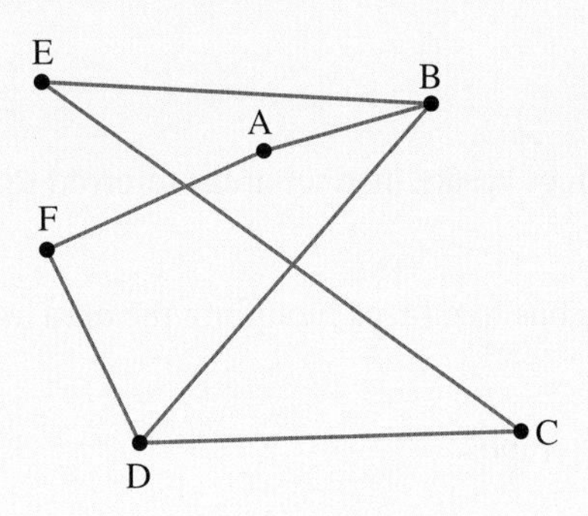

THINK

1. List all connections in the original graph.
2. Draw all vertices and any section(s) of the graph that have no intersecting edges.

WRITE/DRAW

Connections: AB; AF; BD; BE; CD; CE; DF

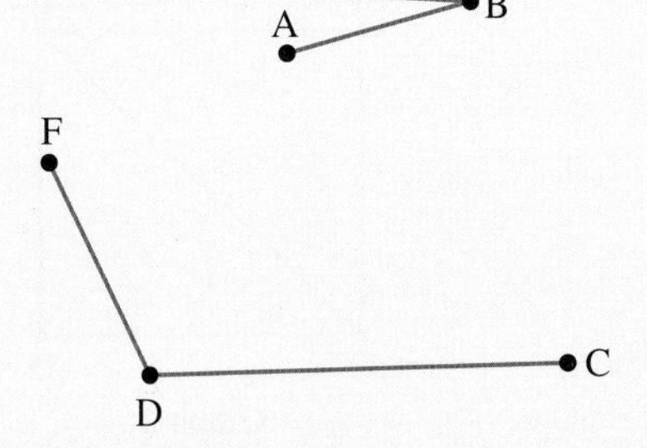

3. Draw any further edges that don't create intersections. Start with edges that have the fewest intersections in the original drawing.

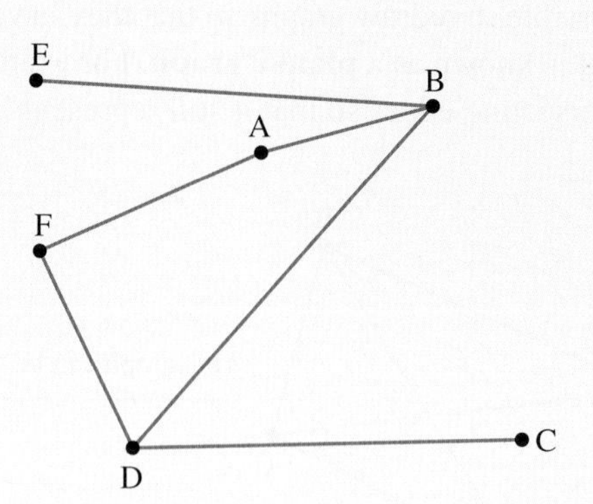

4. Identify any edges yet to be drawn and redraw so that they do not intersect with the other edges.

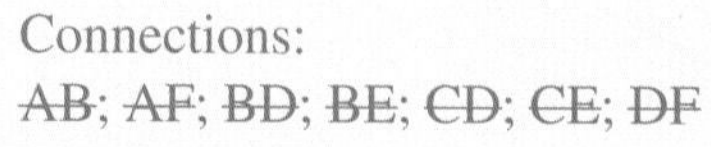

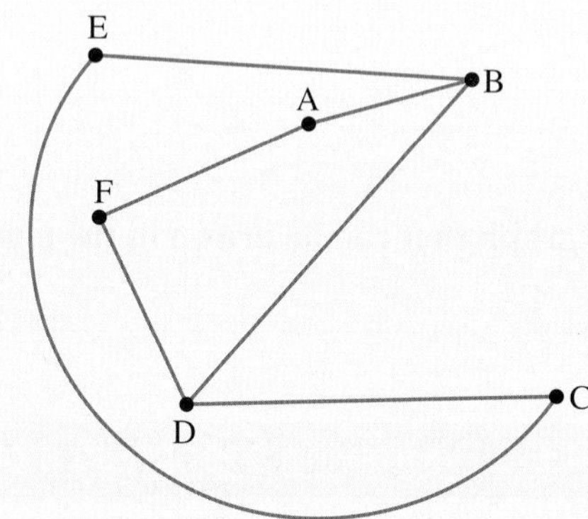

6.3.2 Euler's formula

In all planar graphs, the edges and vertices create distinct areas referred to as **faces**.

The planar graph shown in the diagram has five faces including the area around the outside.

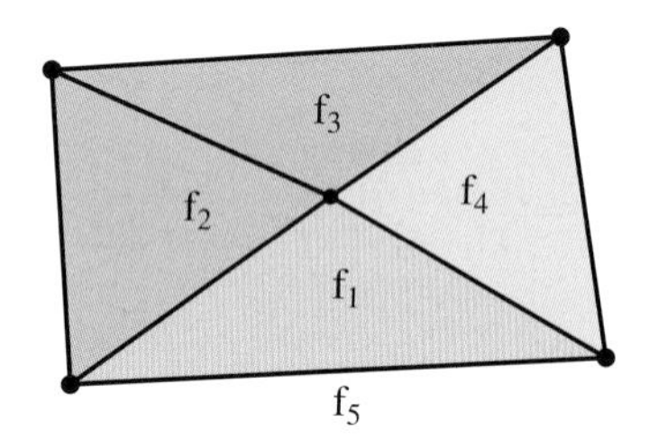

Consider the following group of planar graphs.

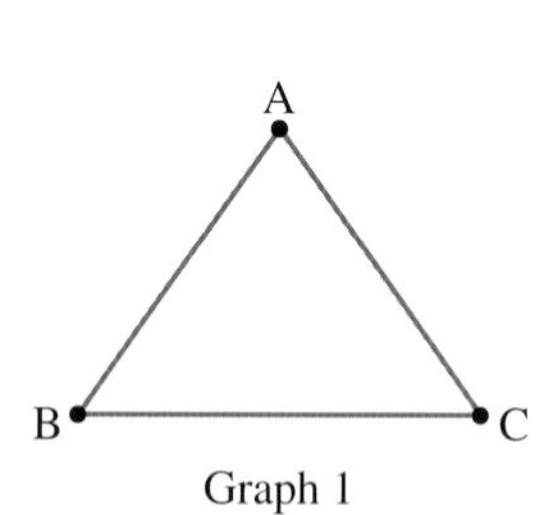

Graph 1

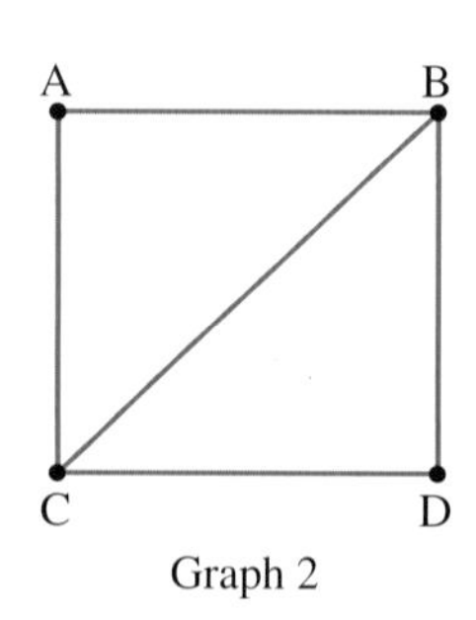

Graph 2

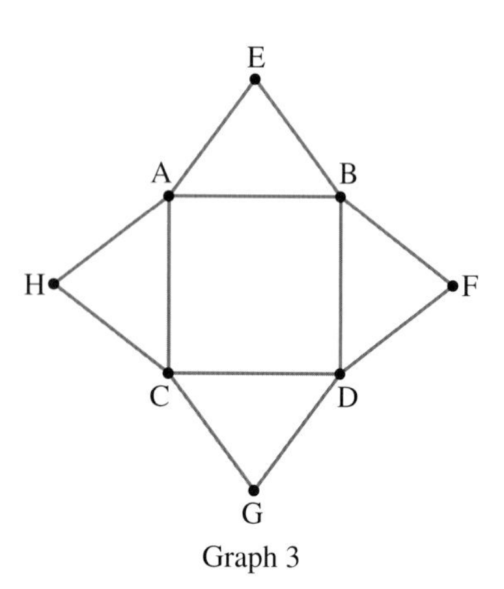

Graph 3

The number of vertices, edges and faces for each graph is summarised in the following table.

Graph	Vertices	Edges	Faces
Graph 1	3	3	2
Graph 2	4	5	3
Graph 3	8	12	6

For each of these graphs, we can obtain a result that is well known for any planar graph: the difference between the vertices and edges added to the number of faces will always equal 2.

Graph 1: $3 - 3 + 2 = 2$

Graph 2: $4 - 5 + 3 = 2$

Graph 3: $8 - 12 + 6 = 2$

This is known as Euler's formula for connected planar graphs.

Euler's formula

$$v - e + f = 2$$

where v is the number of vertices, e is the number of edges and f is the number of faces.

WORKED EXAMPLE 6 Appling Euler's formula

Determine how many faces there will be for a connected planar graph of 7 vertices and 10 edges.

THINK	WRITE
1. Substitute the given values into Euler's formula.	$v - e + f = 2$ $7 - 10 + f = 2$
2. Solve the equation for the unknown value.	$7 - 10 + f = 2$ $f = 2 - 7 + 10$ $f = 5$
3. State the final answer.	There will be 5 faces in a connected planar graph with 7 vertices and 10 edges.

6.3.3 Planarity of different graphs

In subtopic 6.2 we looked at a range of different types of graphs, some of which include complete, regular and Platonic graphs. We can look at these in more detail when determining their planarity.

When considering whether a graph is planar, it is worth noting that if a subgraph is non-planar than the graph itself cannot be planar.

Planarity of complete graphs

Consider the first four complete graphs.

K_1 K_2 K_3 K_4

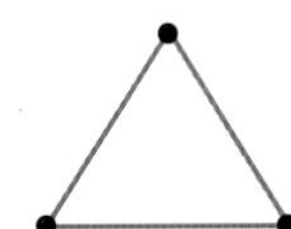

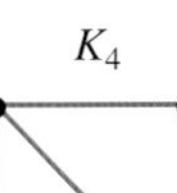

It is clear that these can all be drawn without intersecting lines, and thus are all planar. The complete graph K_5 however cannot be drawn without any intersections. If you try to draw it so that no edges intersect you will be successful up until the final edge. As you can see, there is no way to join the final two vertices without intersecting one of the existing edges. Therefore, K_5 is non-planar.

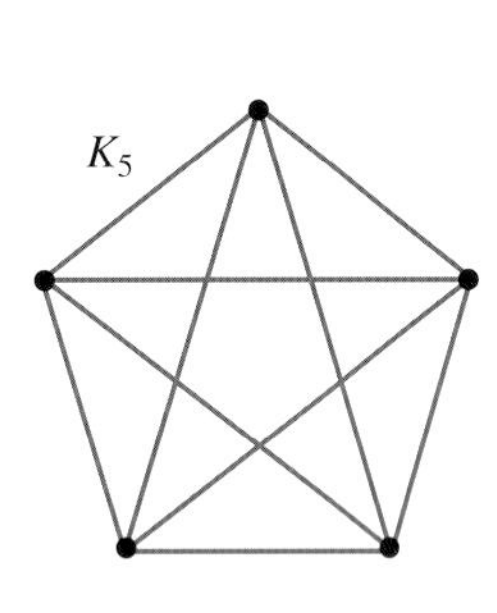

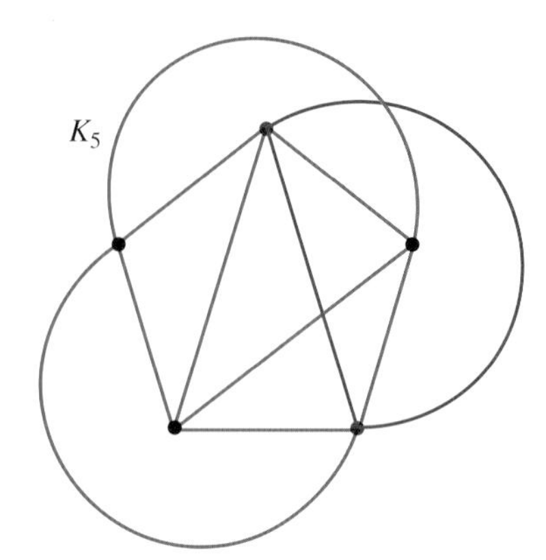

Alternatively, we can use a proof by contradiction to prove that it must be non-planar as follows.

Assume that K_5 is planar.

We know that K_5 is a simple connected graph as each vertex has 1 edge joining to each other vertex. By our assumption, K_5 is therefore a simple planar graph.

Simple planar graphs must follow Euler's formula, $v-e+f=2$.

Since K_5 has 10 edges and 5 vertices it must have 7 faces.

In a simple planar graph, each face must be bounded by at least 3 edges. Therefore:

$$3\times f\leq n\,(\text{boundaries})$$

Each edge is a boundary for exactly 2 faces. Therefore:

$$n\,(\text{boundaries})=2e$$

Combining these gives

$$3f\leq 2e$$

Earlier we determined that $f=7$ and $e=10$ for K_5. This contradicts the above condition:

$$\begin{gathered}3f\leq 2e\\ 3\times 7\leq 2\times 10\\ 21\not\leq 20\end{gathered}$$

Since we have a contradiction, the assumption that K_5 is planar must be false. K_5 is therefore non-planar.

Since K_5 is a subgraph of K_n for $n>5$, K_n is non-planar for $n\geq 5$.

Planarity of regular graphs

There isn't a specific rule that can be used to determine whether a regular graph is planar. For r-regular graphs:

- All are planar for $r=1$ and $r=2$
- All are non-planar for $r\geq 6$
- Planarity must be determined on a case by case basis for $3\leq r\leq 5$

The 5 Platonic regular graphs are all planar.

Planarity

- **Complete graphs K_n are planar for $n \leq 4$ and non-planar for $n \geq 5$.**
- **r-regular graphs are:**
  - **planar for $r = 1$ and $r = 2$.**
  - **non-planar for $r \geq 6$.**
- **The 5 Platonic regular graphs are planar.**
- **If a subgraph is non-planar, then the original graph is also non-planar**

6.3 Exercise

Technology free

1. WE5 Redraw the following graphs so that they have no intersecting edges.

a.

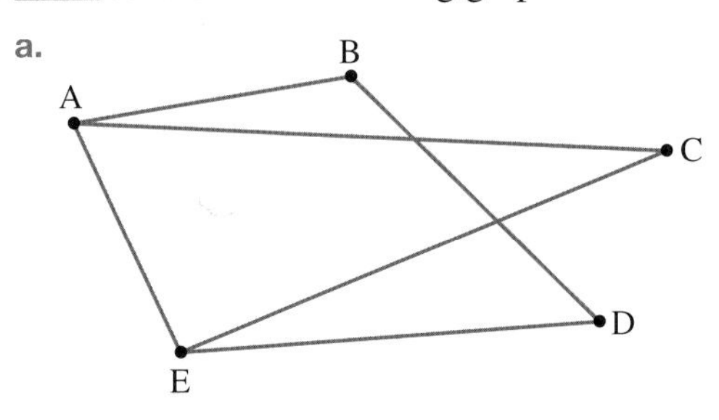

b.

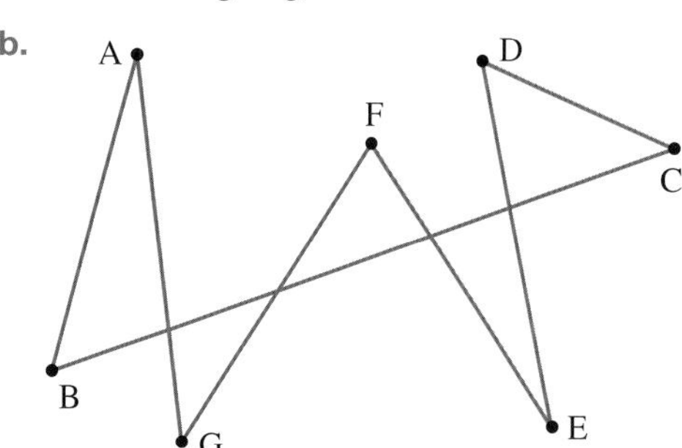

2. Determine which of the following are planar graphs.

a.

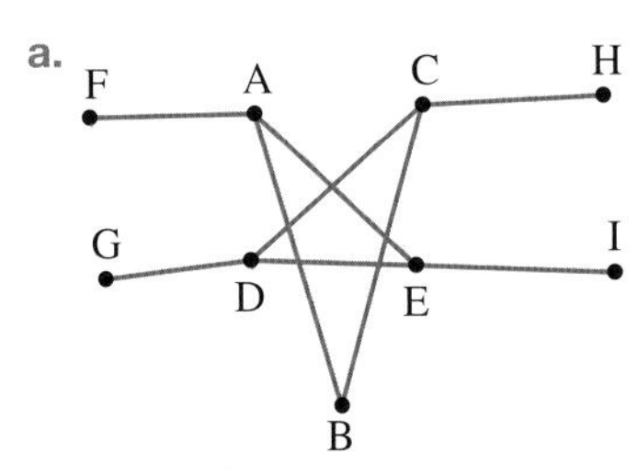

Graph 1

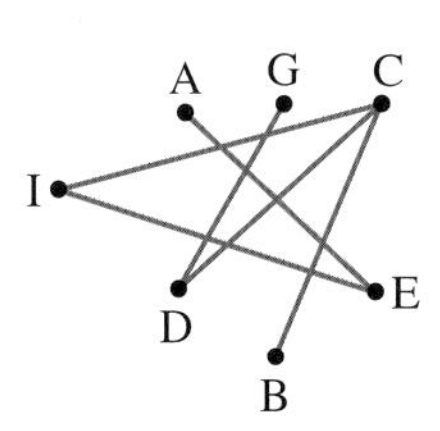

Graph 2

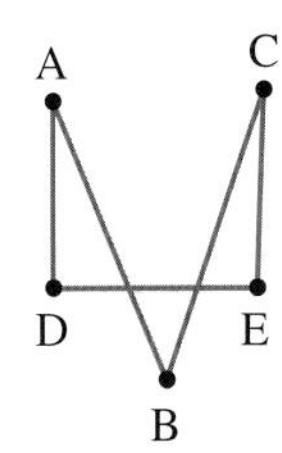

Graph 3

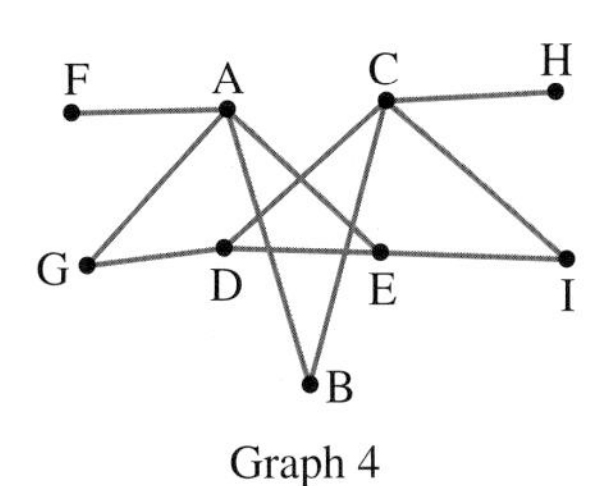

Graph 4

b.

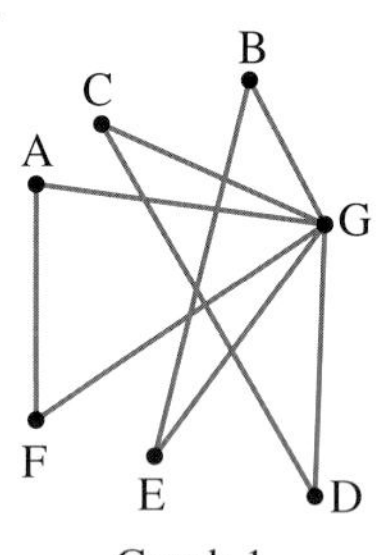

Graph 1

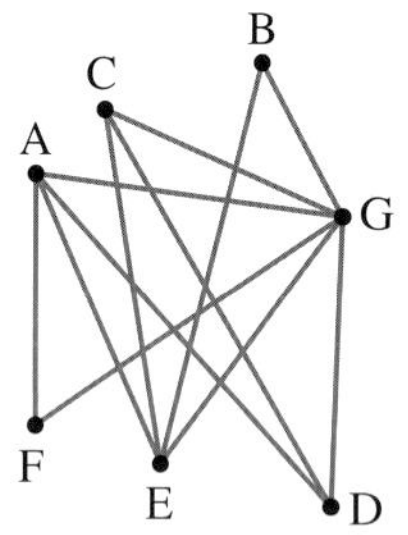

Graph 2

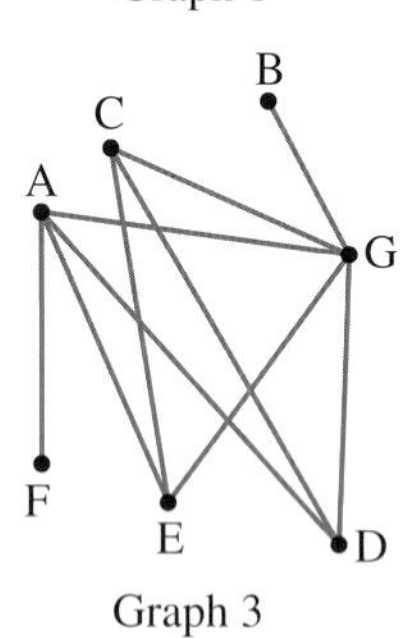

Graph 3

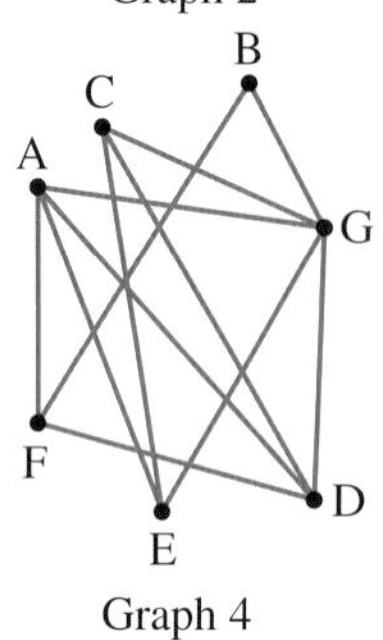

Graph 4

3. Redraw the following graphs to show that they are planar.

a.

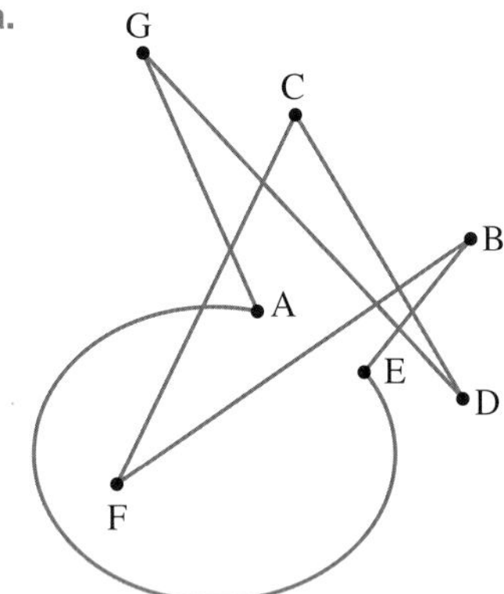

b.

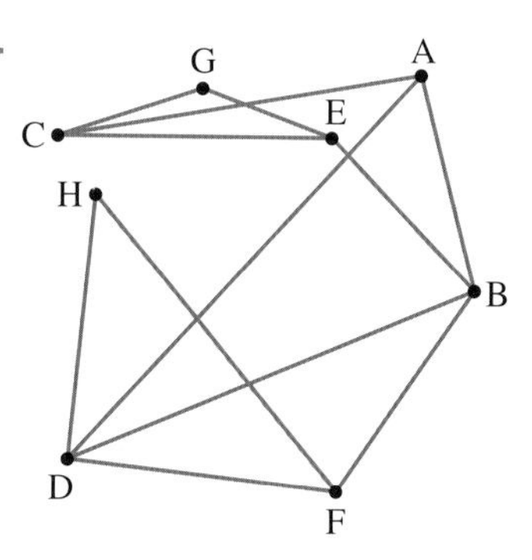

4. Determine which of the following graphs are not planar.

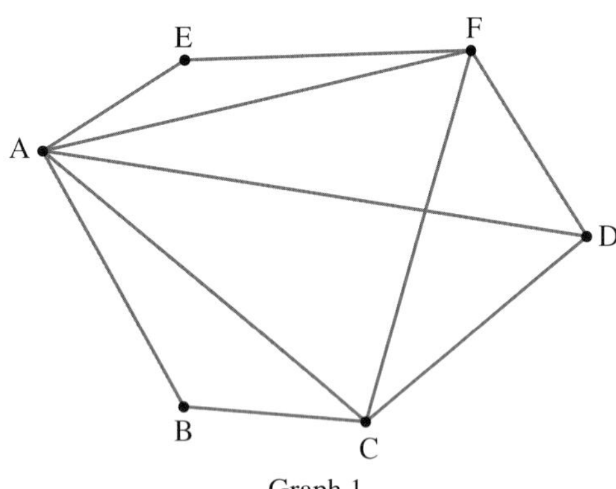

Graph 1

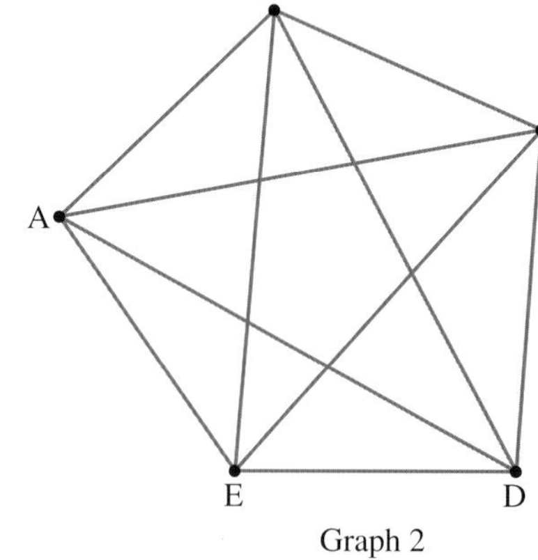

Graph 2

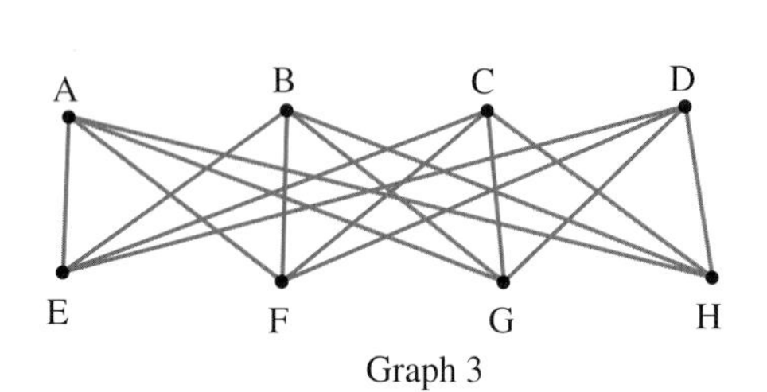

Graph 3

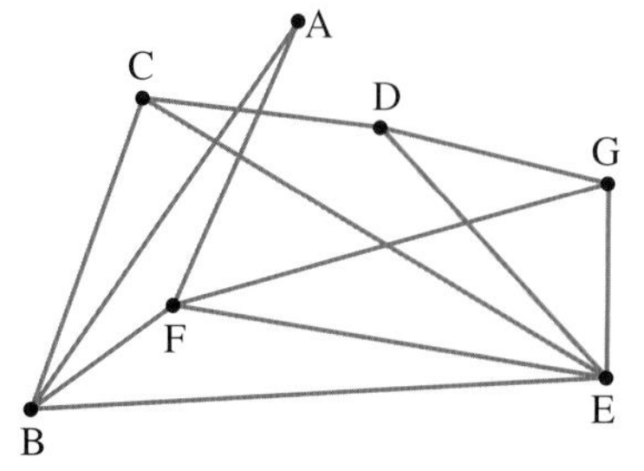

Graph 4

5. WE6 Determine how many faces there will be for a connected planar graph of:

a. 8 vertices and 10 edges

b. 11 vertices and 14 edges.

6. a. For a connected planar graph of 5 vertices and 3 faces, determine how many edges there will be.
   b. For a connected planar graph of 8 edges and 5 faces, determine how many vertices there will be.

7. For each of the following planar graphs, identify the number of faces.

a.
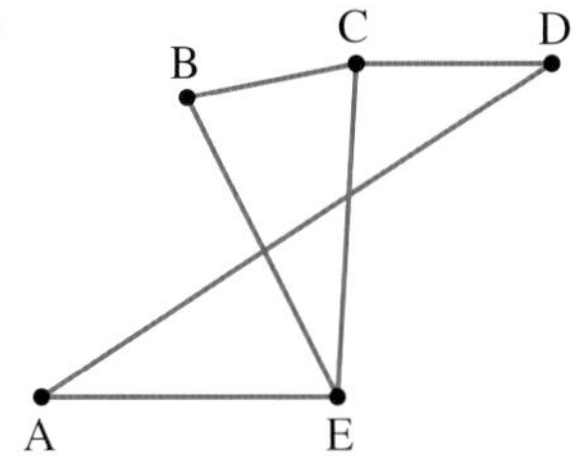

b.
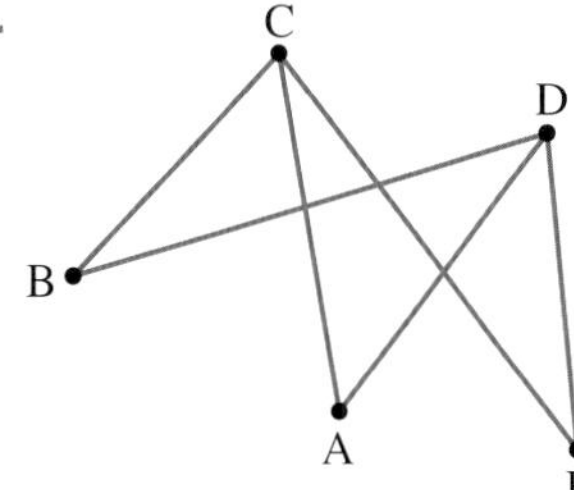

c.
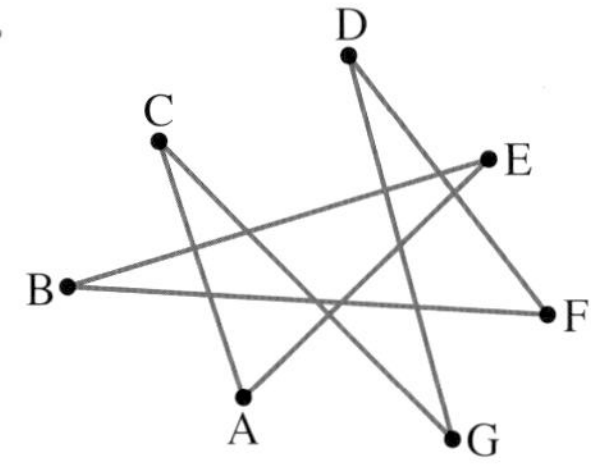

d.
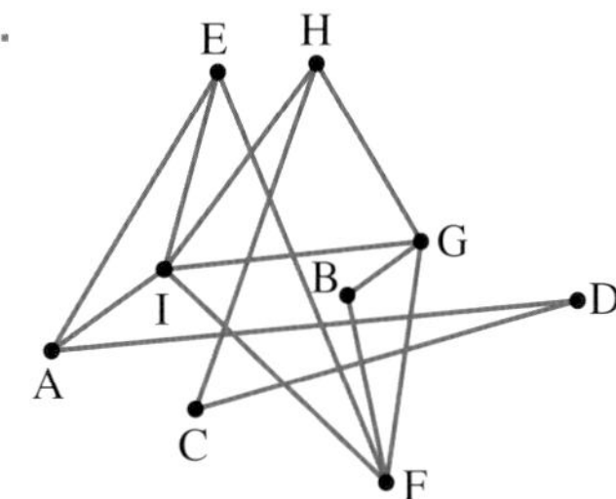

8. Construct a connected planar graph with:

a. 6 vertices and 5 faces

b. 11 edges and 9 faces.

9. Use the following adjacency matrices to draw graphs that have no intersecting edges.

a. $\begin{bmatrix} 0 & 1 & 1 & 1 & 0 \\ 1 & 0 & 1 & 1 & 0 \\ 1 & 1 & 0 & 0 & 1 \\ 1 & 1 & 0 & 0 & 1 \\ 0 & 0 & 1 & 1 & 0 \end{bmatrix}$

b. $\begin{bmatrix} 0 & 0 & 1 & 1 & 0 \\ 0 & 0 & 0 & 1 & 1 \\ 1 & 0 & 0 & 0 & 0 \\ 1 & 1 & 0 & 0 & 1 \\ 0 & 1 & 0 & 1 & 0 \end{bmatrix}$

10. For the graphs in question 9:
   i. identify the number of enclosed faces
   ii. identify the maximum number of additional edges that can be added to maintain a simple planar graph.

11. a. Use the planar graphs shown to copy and complete the table.

Graph	Total edges	Total degrees
Graph 1		
Graph 2		
Graph 3		
Graph 4		

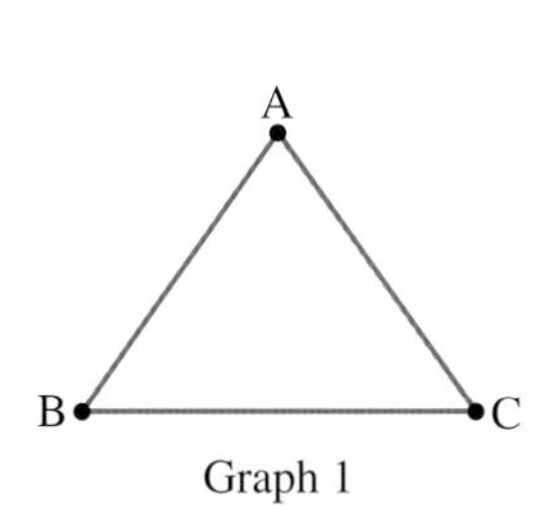

Graph 1

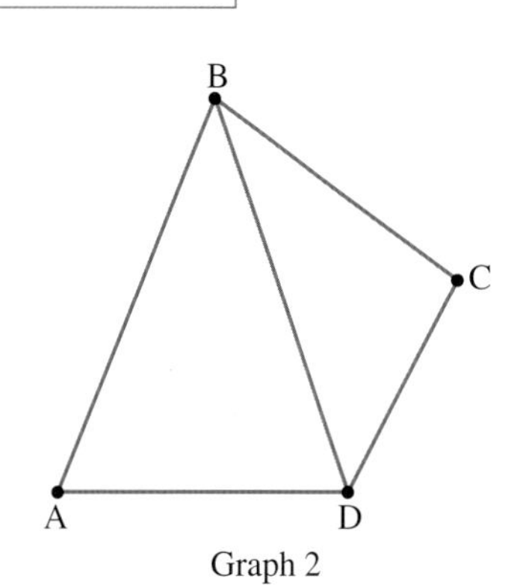

Graph 2

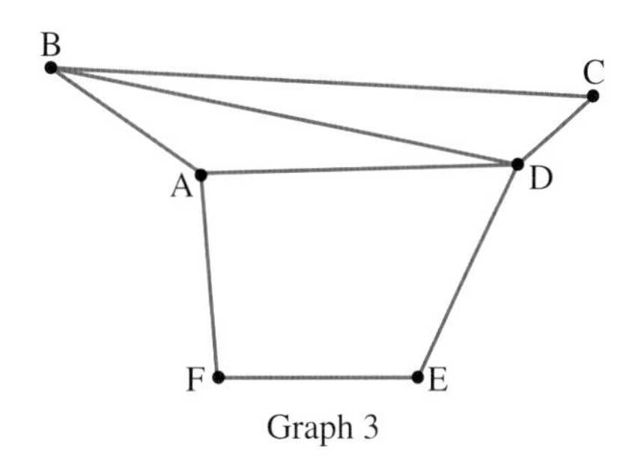

Graph 3

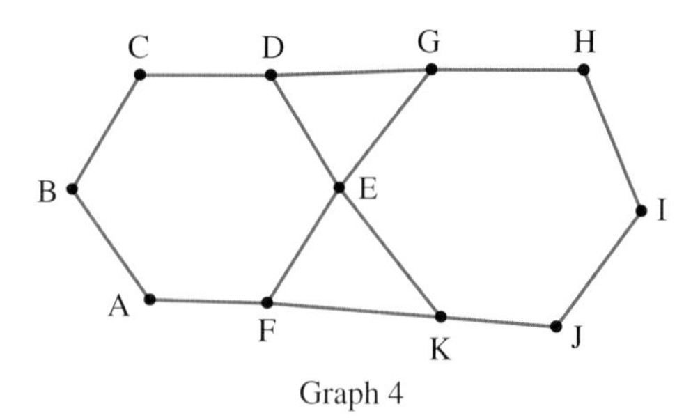

Graph 4

b. Describe the pattern that is evident from the table.

12. a. Use the planar graphs shown to copy and complete the table.

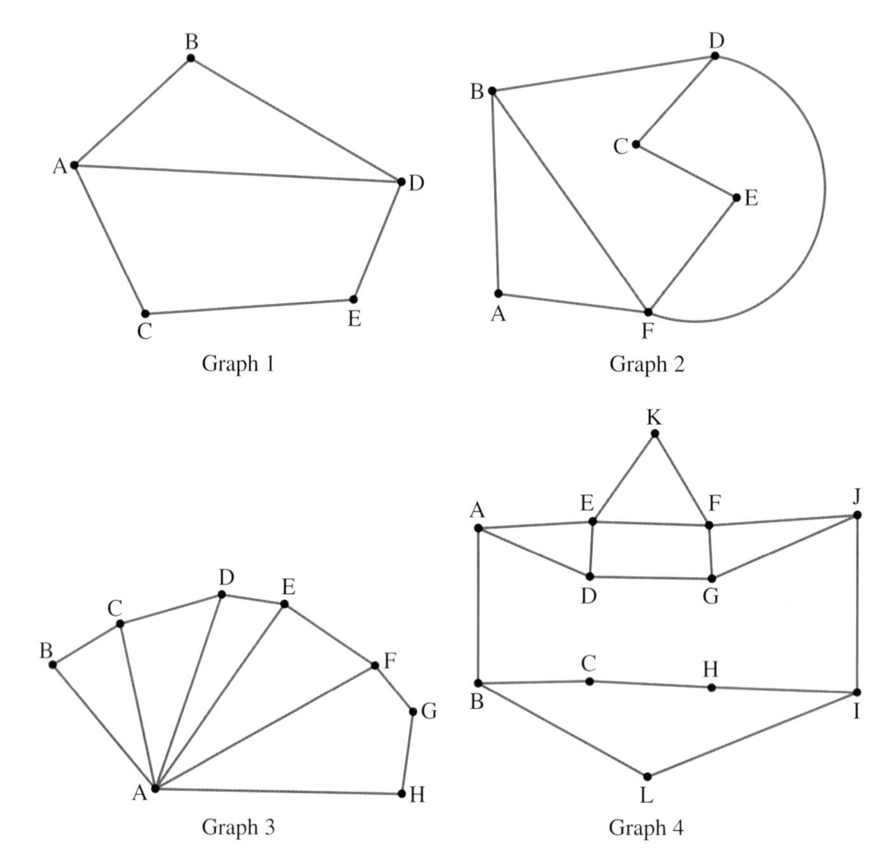

Graph	Total vertices of even degree	Total vertices of odd degree
Graph 1		
Graph 2		
Graph 3		
Graph 4		

b. Describe if there is any pattern evident from this table.

13. Represent the following 3-dimensional shapes as planar graphs.

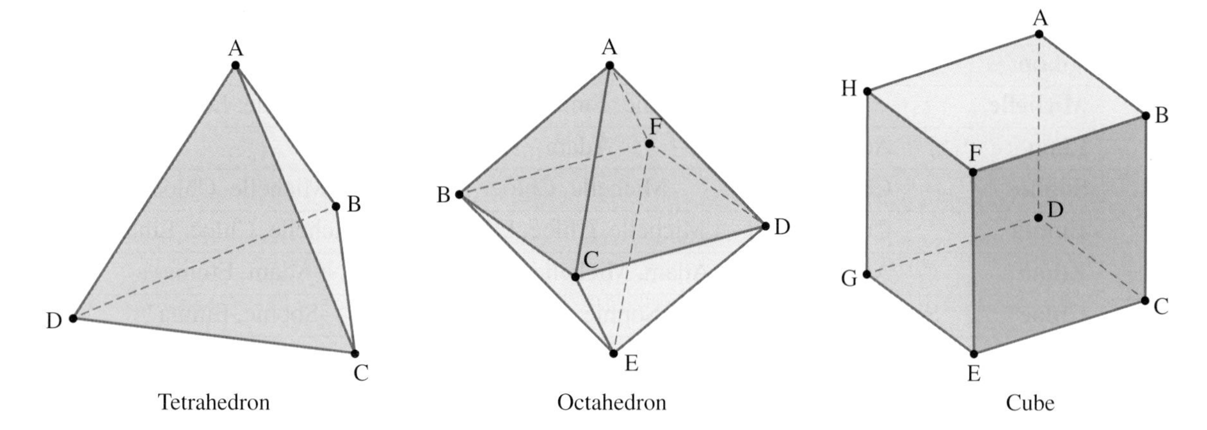

Technology active

14. A section of an electric circuit board is shown in the diagram.
    a. Draw a graph to represent the circuit board, using vertices to represent the labelled parts of the diagram.
    b. State if it is possible to represent the circuit board as a planar graph.

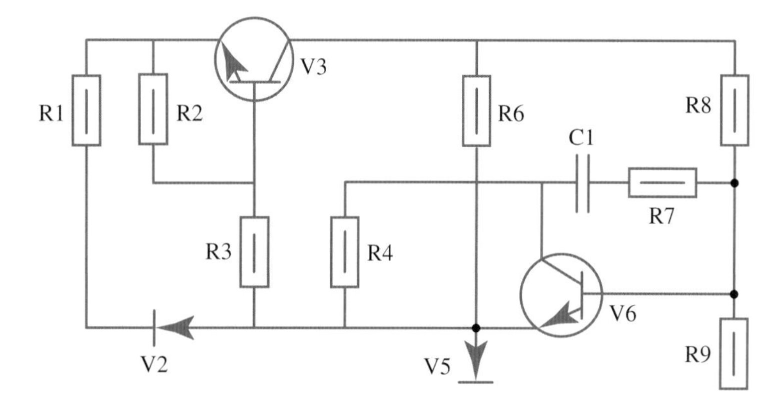

15. The diagram shows a map of a small metro train system.

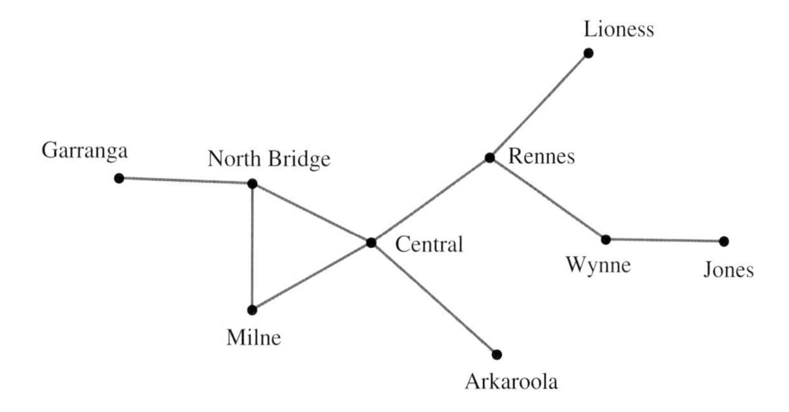

a. Display this information using an adjacency matrix.
b. Describe what the sum of the rows of this adjacency matrix indicate.

16. The table displays the most common methods of communication for a group of people.

	Facebook	Instagram	SMS
Adam	Ethan, Liam	Ethan, Liam	Ethan
Michelle		Sophie, Emma, Ethan	Sophie, Emma
Liam	Adam	Adam	
Sophie	Chloe	Michelle, Chloe	Michelle, Chloe
Emma	Chloe	Michelle, Chloe, Ethan	Michelle, Chloe, Ethan
Ethan	Adam	Adam, Michelle, Emma	Adam, Emma
Chloe	Sophie, Emma	Sophie, Emma	Sophie, Emma

a. Display the information for the entire table in a graph.
b. State who would be the best person to introduce Chloe and Michelle.
c. Display the Instagram information in a separate graph.
d. If Liam and Sophie began communicating through Instagram, determine how many faces the graph from part c would then have.

6.3 Exam questions

Question 1 (1 mark) TECH-FREE
If G is a 3-regular planar graph with 24 vertices, determine how many faces there will be in the planar drawing of G.

Question 2 (2 marks) TECH-FREE
A graph has 6 vertices and 10 edges. Explain whether it is planar, non-planar or either. Justify your answer.

Question 3 (2 marks) TECH-FREE
State whether the following graph is planar. Justify your answer.

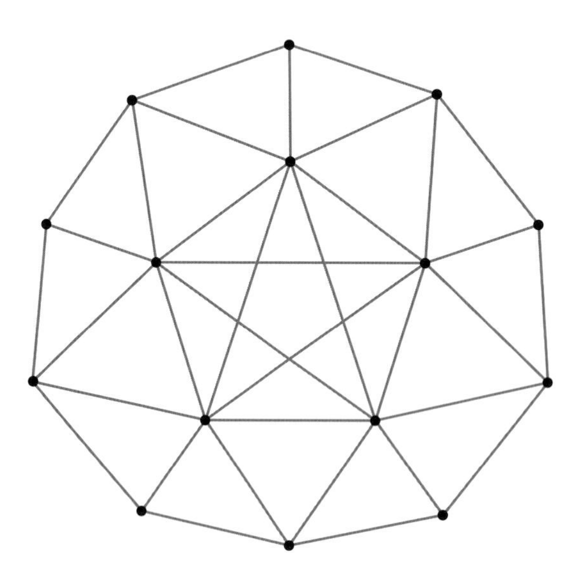

More exam questions are available online.

6.4 Eulerian and Hamiltonian graphs

LEARNING INTENTION

At the end of this subtopic you should be able to:

- define the terms: walk, closed walk, trail, path, cycle and closed trail
- identify Eulerian graphs and semi-Eulerian graphs
- identify Hamiltonian cycles and semi-Hamiltonian cycles.

6.4.1 Traversing connected graphs

Many applications of graphs involve an analysis of movement around a network. These could include fields such as transport, communications or utilities, to name a few. Movement through a simple connected graph is described in terms of starting and finishing at specified vertices by travelling along the edges. This is usually done by listing the labels of the vertices visited in the correct order. In more complex graphs, edges may also have to be indicated, as there may be more than one connection between vertices.

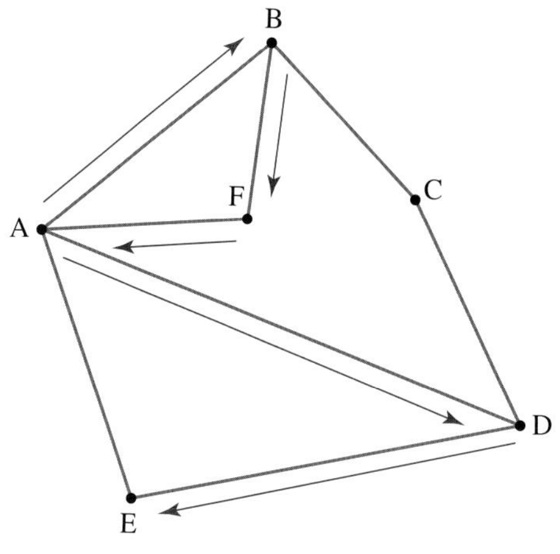

Path: ABFADE

Key terminology for traversing connected graphs

Walk: Any route taken through a graph, including routes that repeat edges and vertices.
Closed walk: A route taken through a graph that starts and ends at the same vertex.
Trail: A walk in which no edges are repeated.
Path: A walk in which no vertices are repeated, except possibly the start and finish.
Cycle: A path beginning and ending at the same vertex.
Simple cycle: A cycle in a graph with no-repeated vertices.
Closed trail: A trail beginning and ending at the same vertex.

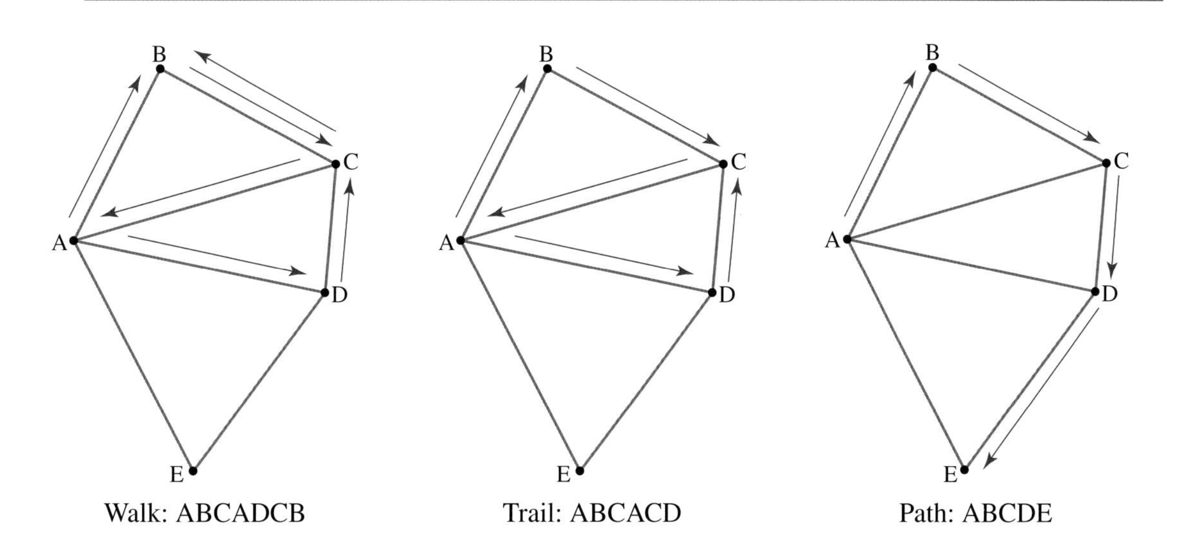

Walk: ABCADCB Trail: ABCACD Path: ABCDE

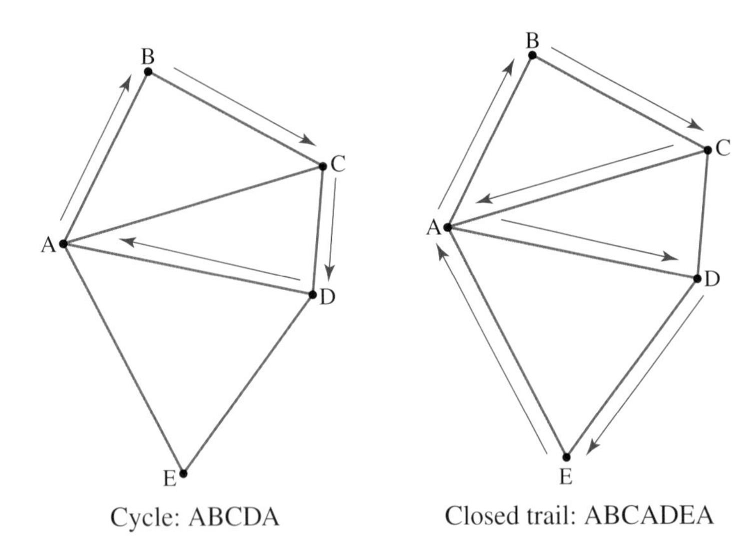

WORKED EXAMPLE 7 Determining routes of a graph

In the following network, identify two different routes: one cycle and one closed trail.

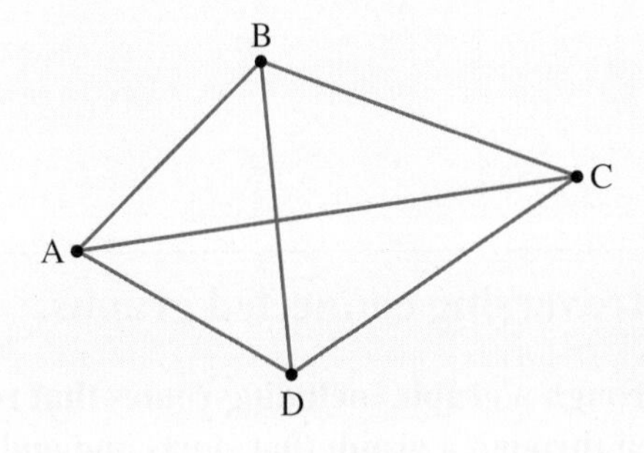

THINK	WRITE
1. For a cycle, identify a route that doesn't repeat a vertex apart from the start/finish.	Cycle: ABDCA
2. For a closed trail, identify a route that doesn't repeat an edge and ends at the starting vertex.	Closed trail: ADBCA

6.4.2 Eulerian graphs and semi-Eulerian graphs

In many practical situations, it is most efficient if a route travels along each edge only once. For example, parcel deliveries and council rubbish collections are most efficient when no road needs to be repeated.

A route that travels along every edge in a connected graph without repeating any edges is called an **Eulerian trail** or a **semi-Eulerian trail**.

Eulerian trails start and end at the same vertex whilst semi-Eulerian trails start and end at different vertices.

Eulerian and semi-Eulerian trails and graphs

- **An Eulerian trail passes along each edge exactly once, starting and ending at the same vertex.**
- **A semi-Eulerian trail passes along each edge exactly once, starting and ending at different vertices.**

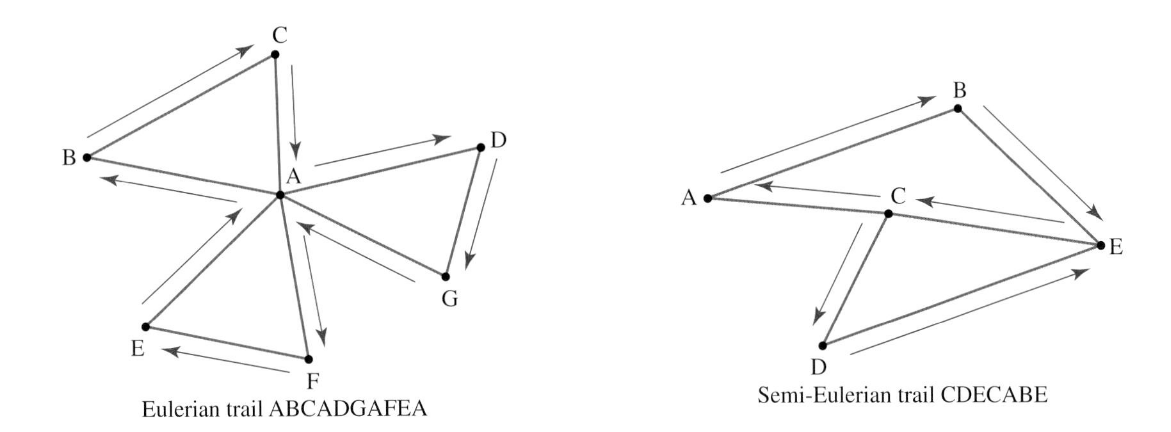

Eulerian trail ABCADGAFEA

Semi-Eulerian trail CDECABE

It is not possible to find an Eulerian or semi-Eulerian trail in all graphs. Thankfully we have a set of conditions that can be used to determine whether a graph contains an Eulerian or semi-Eulerian trail.

If the degrees of all vertices are even, then an Eulerian trail exists. Graphs that contain an Eulerian trail are called **Eulerian graphs**. If the degrees of exactly 2 vertices are odd, then a semi-Eulerian trail exists. Graphs that contain a semi-Eulerian trail are called **semi-Eulerian graphs**.

Conditions for Eulerian and semi-Eulerian graphs

- **If the degrees of all vertices are even, then the graph is Eulerian.**
- **If the degrees of exactly 2 vertices are odd, the graph is semi-Eulerian.**

Königsberg bridge problem

The Königsberg bridge problem is a recreational mathematical puzzle set in the old Prussian city of Königsberg. The problem goes as follows: is it possible to take a walk through the city and cross over each of the seven bridges in the city (as shown in the picture below) exactly once?

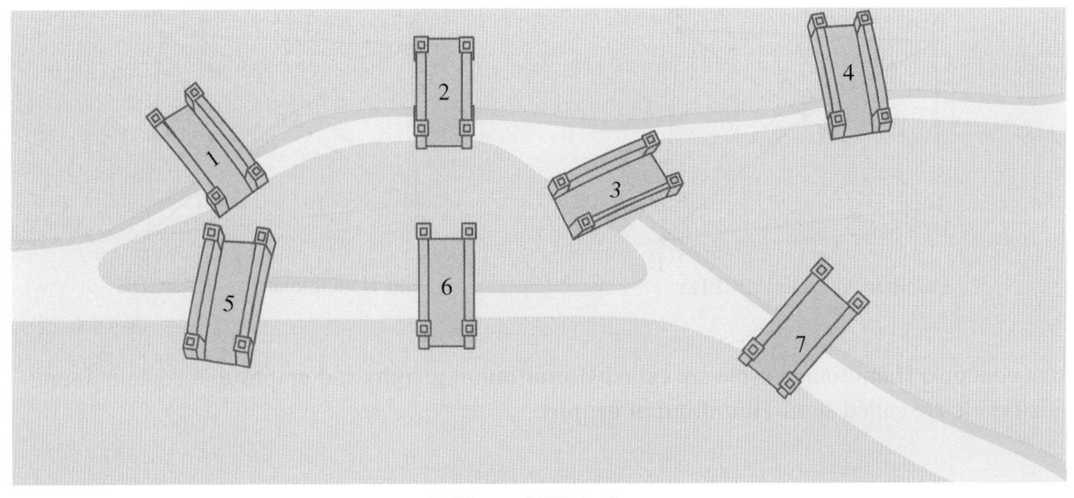

Bridges of Königsberg

In 1735 the Swiss mathematician Leonhard Euler presented a solution to this problem, concluding that such a walk was impossible. Many years after, mathematicians would come to understand that this problem was no different to asking if there is a Eulerian path across the graph shown below with 4 vertices (representing the sections of land) and seven edges (representing the bridges).

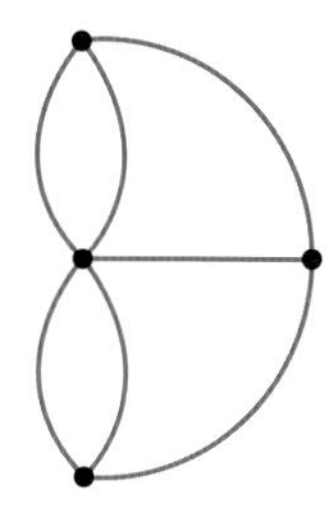

As the degrees of the vertices are all odd (3, 3, 3 and 5) it is not possible for this to be an Eulerian graph or have an Eulerian trail.

6.4.3 Hamiltonian graphs and semi-Hamiltonian graphs

In other situations, it may be more practical if all vertices can be reached without using all of the edges of the graph. For example, if you wanted to visit a selection of the capital cities of Europe, you wouldn't need to use all the available flight routes shown in the diagram.

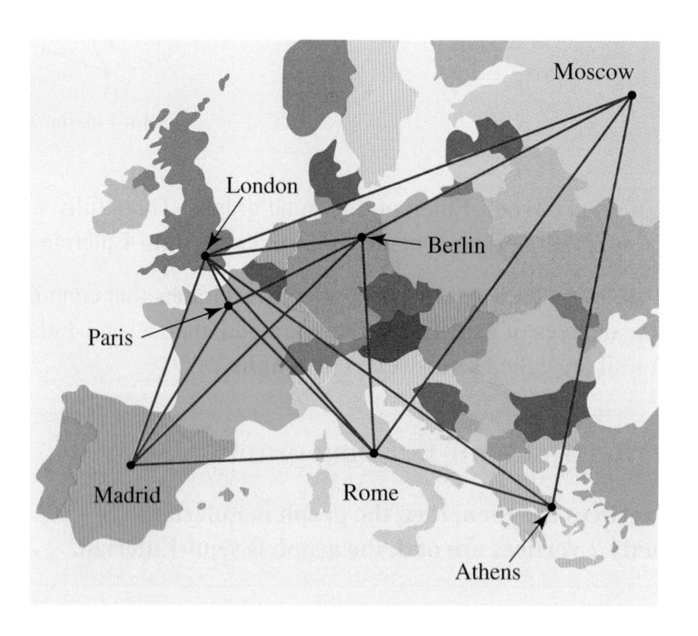

A path that passes through each vertex of a connected graph exactly once is called a **Hamiltonian cycle** or a **semi-Hamiltonian cycle**. Hamiltonian cycles start and end at the same vertex whilst semi-Hamiltonian cycles start and end at different vertices.

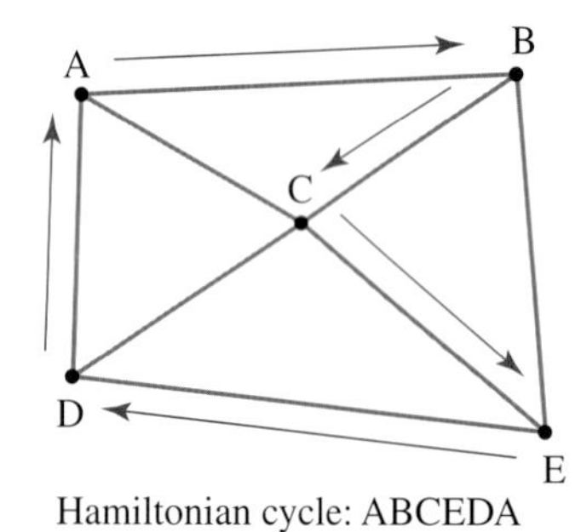

Hamiltonian cycle: ABCEDA

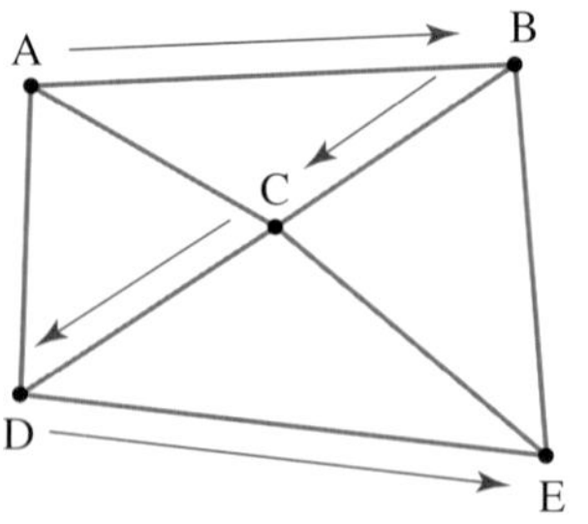

Semi-Hamiltonian cycle: ABCDE

Graphs that contain a Hamiltonian cycle are called **Hamiltonian graphs** and graphs that contain a semi-Hamiltonian cycle are called **semi-Hamiltonian graphs**.

Unlike Eulerian and semi-Eulerian trails there are no specific conditions that can be used to determine whether a graph contains a Hamiltonian or semi-Hamiltonian cycle, so you will need to examine each graph visually to determine whether it contains a Hamiltonian or semi-Hamiltonian cycle.

Hamiltonian and semi-Hamiltonian cycles and graphs

If a graph contains a path that passes through every vertex exactly once, and starts and ends at the same vertex:
- **the path is called a Hamiltonian cycle**
- **the graph is a Hamiltonian graph.**

If a graph contains a path that passes through every vertex exactly once, but starts and ends at different vertices:
- **the path is called a semi-Hamiltonian cycle**
- **the graph is a semi-Hamiltonian graph.**

WORKED EXAMPLE 8 Identifying Eulerian trails and Hamiltonian cycles

Identify a semi-Eulerian trail and a semi-Hamiltonian cycle in the following graph.

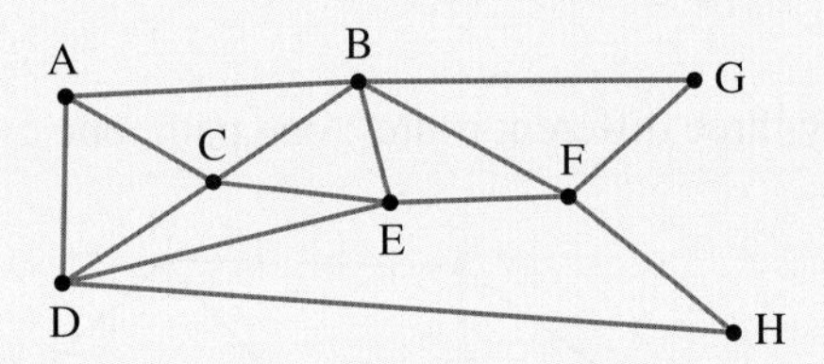

THINK	WRITE/DRAW
1. For a semi-Eulerian trail to exist, there must be exactly 2 vertices with an odd-numbered degree.	deg(A) = 3, deg(B) = 5, deg(C) = 4, deg(D) = 4, deg(E) = 4, deg(F) = 4, deg(G) = 2, deg(H) = 2 As there are only two odd-degree vertices, a semi-Eulerian trail must exist.
2. Identify a route that uses each edge once.	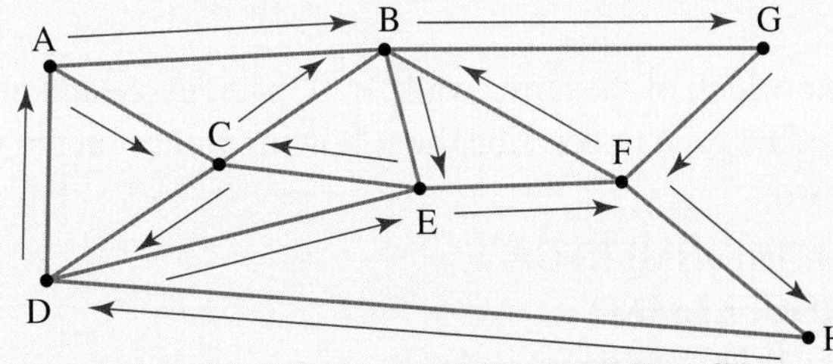 Semi-Eulerian trail: ABGFHDEFBECDACB
3. Identify a route that reaches each vertex once.	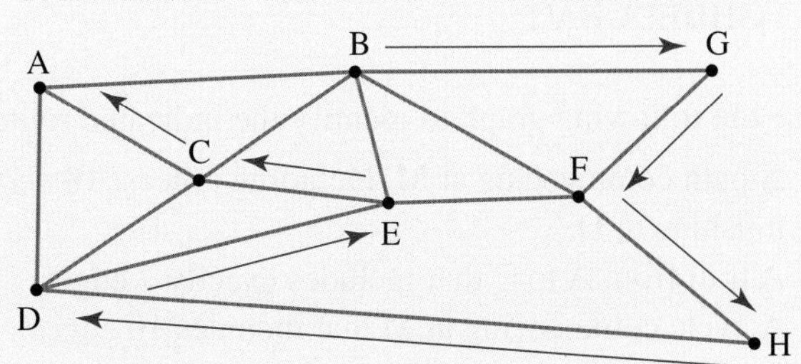
4. State the answer.	Semi-Eulerian cycle: ABGFHDEFBECDACB Semi-Hamiltonian cycle: BGFHDECA

6.4 Exercise

Students, these questions are even better in jacPLUS

Receive immediate feedback and access sample responses

Access additional questions

Track your results and progress

Find all this and MORE in jacPLUS

Technology free

1. WE7 In the following network, identify two different routes: one cycle and one closed trail.

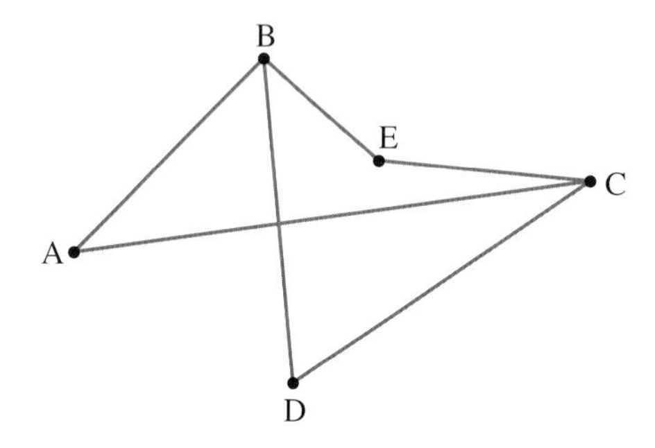

2. In the following network, identify three different routes: one path, one cycle and one closed trail.

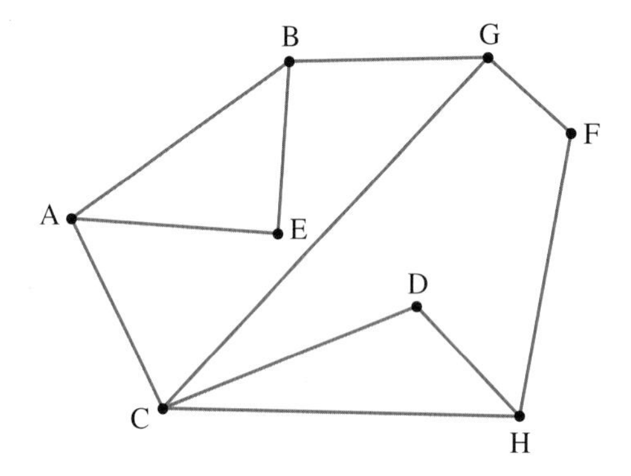

3. State which of the terms walk, trail, path, cycle and closed trail could be used to describe the following routes on the graph shown.
   a. AGHIONMLKFGA
   b. IHGFKLMNO
   c. HIJEDCBAGH
   d. FGHIJEDCBAG

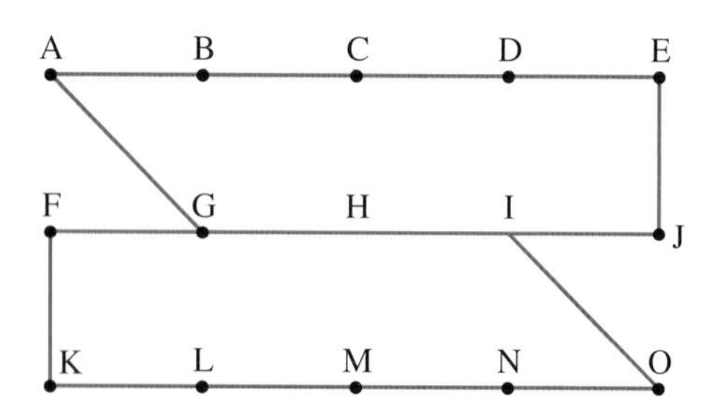

4. Use the following graph to identify the indicated routes.
   a. A path commencing at M, including at least 10 vertices and finishing at D.
   b. A trail from A to C that includes exactly 7 edges.
   c. A cycle commencing at M that includes 10 edges.
   d. A closed trail commencing at F that includes 7 vertices.

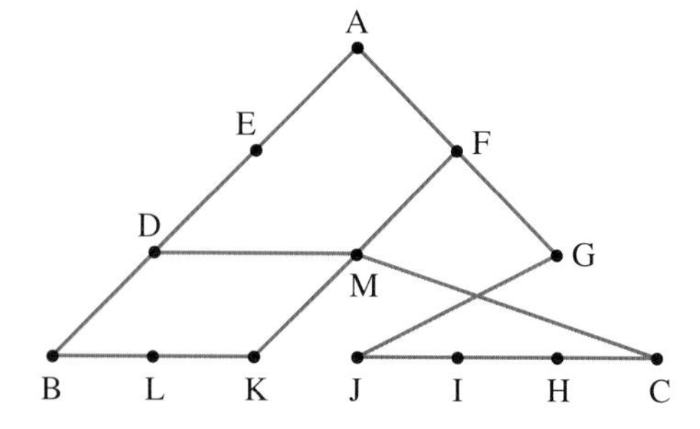

5. WE8 Identify a semi-Eulerian trail and a semi-Hamiltonian cycle in each of the following graphs.

a.

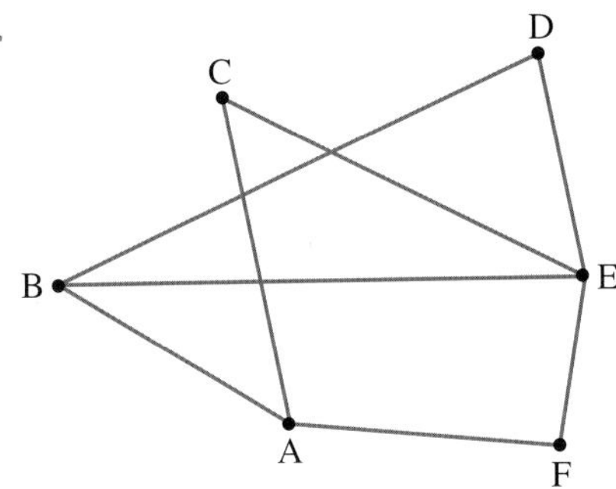

b.

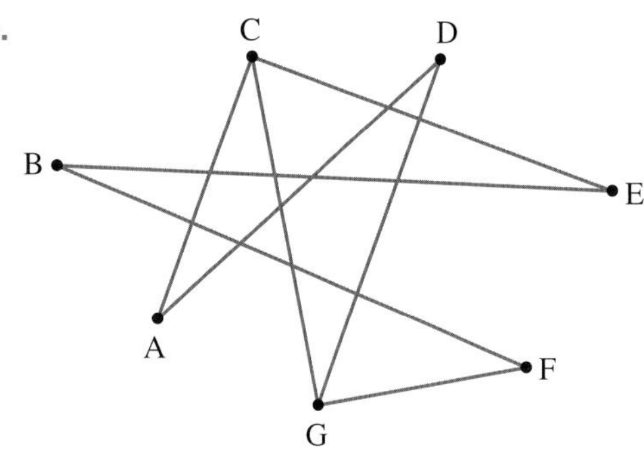

6. Identify an Eulerian trail and a Hamiltonian cycle in each of the following graphs, if they exist.

a.

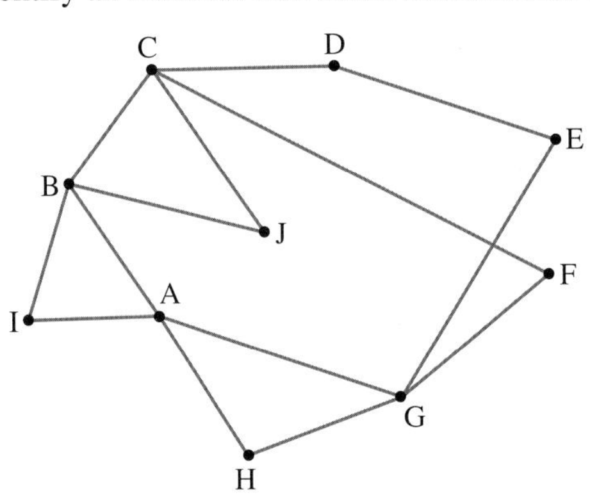

b.

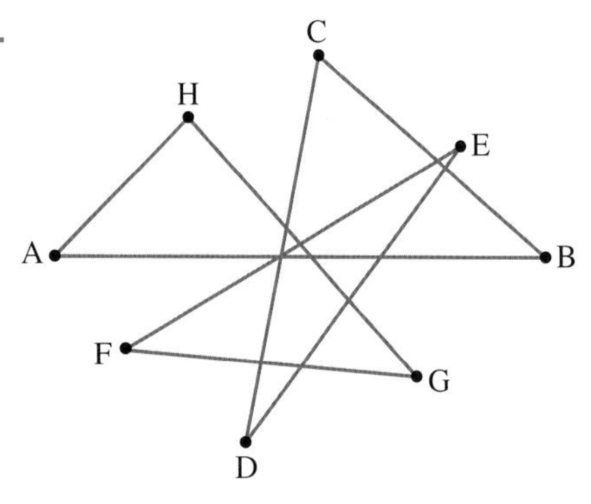

7. a. Identify which of the following graphs have a semi-Eulerian trail.

i.

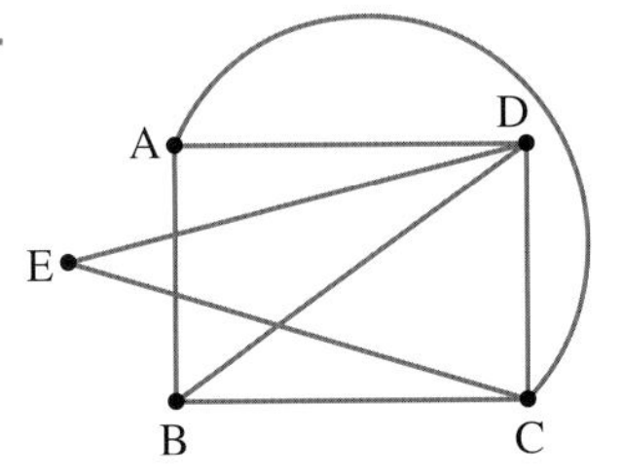

ii.

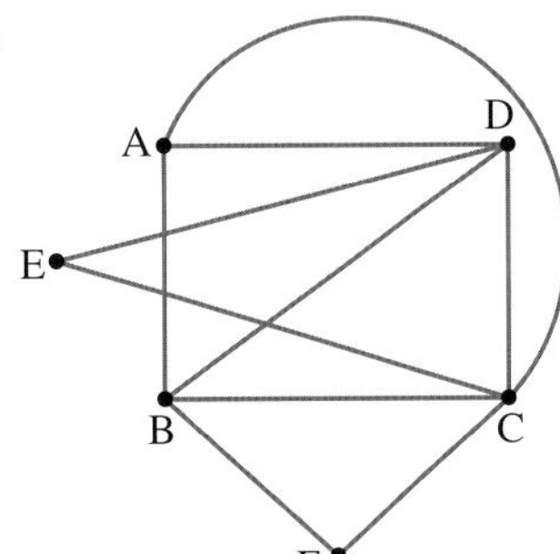

iii.

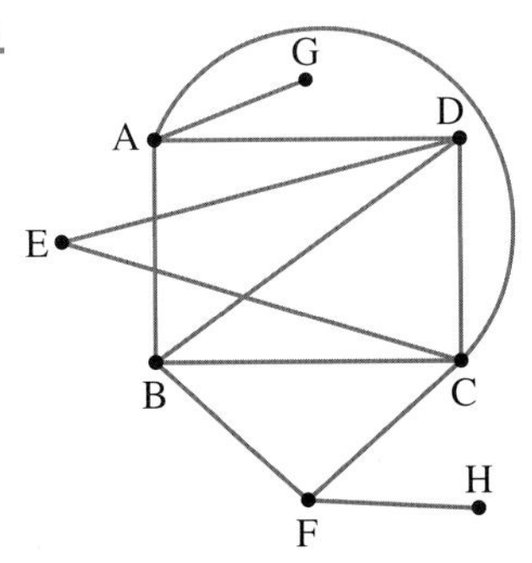

iv.

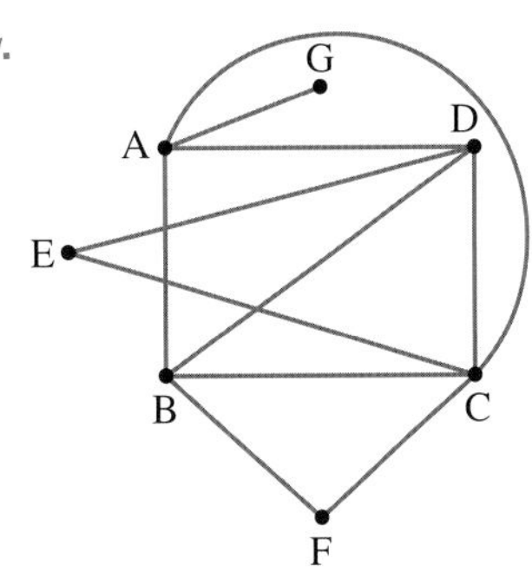

b. Identify the semi-Eulerian trails found.

8. a. Identify which of the graphs from question 7 have a Hamiltonian cycle.
b. Identify the Hamiltonian cycles found.

9. a. Construct adjacency matrices for each of the graphs in question 7.
b. Describe how these might assist with making decisions about the existence of Eulerian trails and semi-Eulerian trails, and Hamiltonian cycles and semi-Hamiltonian cycles?

10. In the following graph, if a semi-Eulerian trail commences at vertex A, determine which vertices it could finish at.

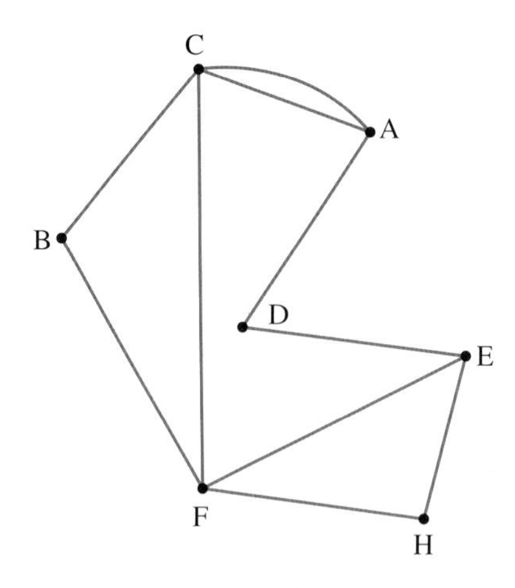

11. In the following graph, determine which vertices a semi-Hamiltonian cycle could finish at if it commences by travelling from:

a. B to E
b. E to A

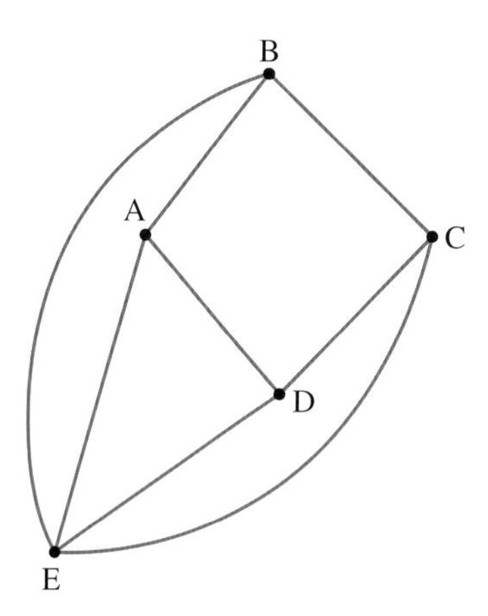

12. In the following graph, other than from G to F, determine between which 2 vertices you must add an edge in order to create a semi-Hamiltonian cycle that commences from vertex:

a. G
b. F

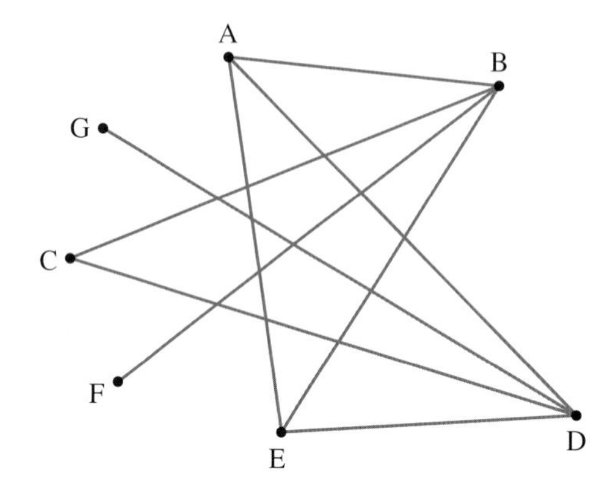

Technology active

13. On the map shown, a school bus route is indicated in yellow. The bus route starts and ends at the school indicated.
   a. Draw a graph to represent the bus route.
   b. Students can catch the bus at stops that are located at the intersections of the roads marked in yellow. State if it is possible for the bus to collect students by driving down each section of the route only once. Explain your answer.
   c. If road works prevent the bus from travelling along the two sections indicated by the Xs, determine if it will be possible for the bus to still collect students on the remainder of the route by travelling each section only once. Explain your answer.

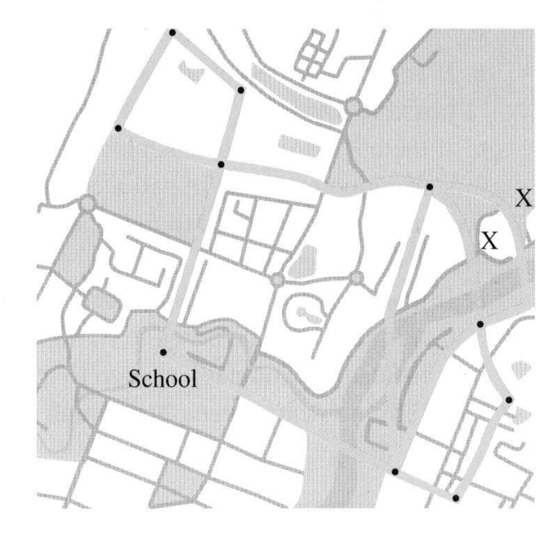

14. The map of an orienteering course is shown. Participants must travel to each of the nine checkpoints along any of the marked paths.

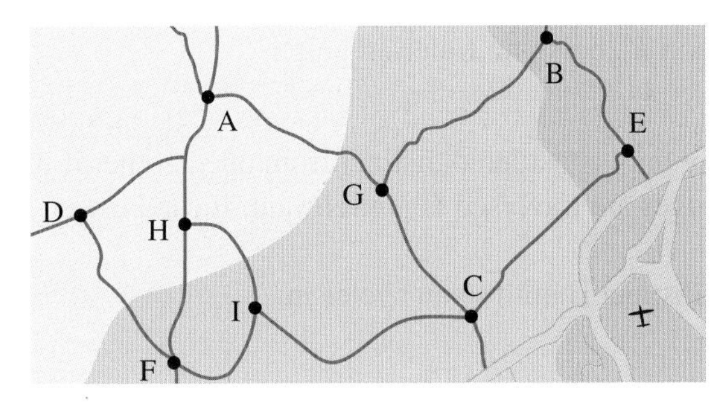

   a. Draw a graph to represent the possible ways of travelling to each checkpoint.
   b. State the degree of checkpoint H.
   c. If participants must start and finish at A and visit every other checkpoint only once, identify two possible routes they could take.
   d. i. If participants can decide to start and finish at any checkpoint, and the paths connecting D and F, H and I, and A and G are no longer accessible, state if it is possible to travel the course by moving along each remaining path only once. Explain your answer.
      ii. Identify the two possible starting points.

15. a. Use the following complete graph to copy and complete the table to identify all of the Hamiltonian cycles commencing at vertex A.

	Hamiltonian cycle
1.	ABCDA
2.	
3.	
4.	
5.	
6.	

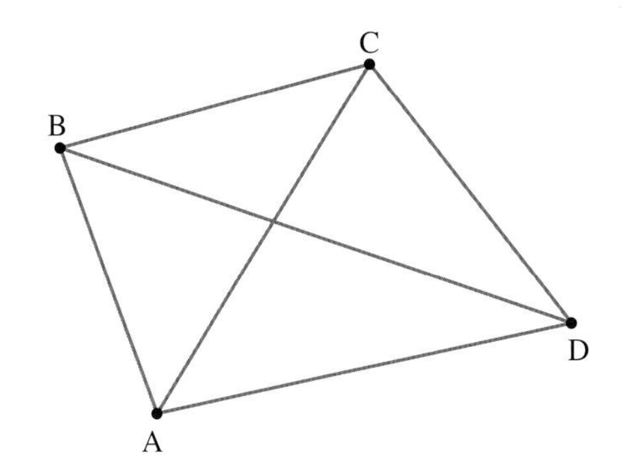

   b. Determine if there are any other Hamiltonian cycles possible.

16. The graph shown outlines the possible ways a tourist bus can travel between eight locations.
    a. If vertex A represents the second location visited, list the possible starting points.
    b. If the bus also visited each location only once, state which of the starting points listed in part **a** could not be correct.
    c. If the bus also needed to finish at vertex D, list the possible paths that could be taken.
    d. If instead the bus company decides to operate a route that travelled to each connection only once, determine what the possible starting and finishing points are.
    e. If instead the company wanted to travel to each connection only once and finish at the starting point, determine which edge of the graph would need to be removed.

6.4 Exam questions

Question 1 (1 mark) TECH-FREE

State what values of n will result in K_n being an Eulerian graph.

Question 2 (4 marks) TECH-FREE

Hamiltonian cycles of a graph are only considered distinct from one another if they use a different set of edges. That is, the order the edges are traversed is not relevant. It can be seen then that K_3 only has 1 Hamiltonian cycle.

a. Determine the number of distinct Hamiltonian cycles in:

  i. K_4 **(1 mark)**

  ii. K_5 **(1 mark)**

b. Determine a rule that connects the number of Hamiltonian cycles with n, the number of vertices in the complete graph K_n. **(2 marks)**

Question 3 (1 mark) TECH-ACTIVE

MC An Eulerian trail for the graph will be possible if only one edge is removed. The number of different ways in which this could this be done is

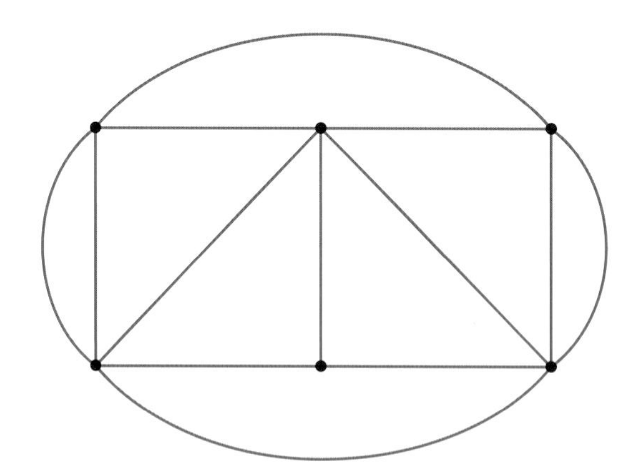

A. 3 **B.** 4 **C.** 5 **D.** 6 **E.** 7

More exam questions are available online.

6.5 Weighted graphs and trees

LEARNING INTENTION

At the end of this subtopic you should be able to:
- recognise weighted graphs and trees
- use Prim's algorithm to find a minimum spanning tree for a weighted graph.

6.5.1 Weighted graphs

In many applications using graphs, it is useful to attach a value to the edges. These values could represent the length of the edge in terms of time or distance, or the costs involved with moving along that section of the path. Such graphs are known as **weighted graphs**.

Weighted graphs can be particularly useful as analysis tools. For example, they can help determine how to travel through a network in the shortest possible time.

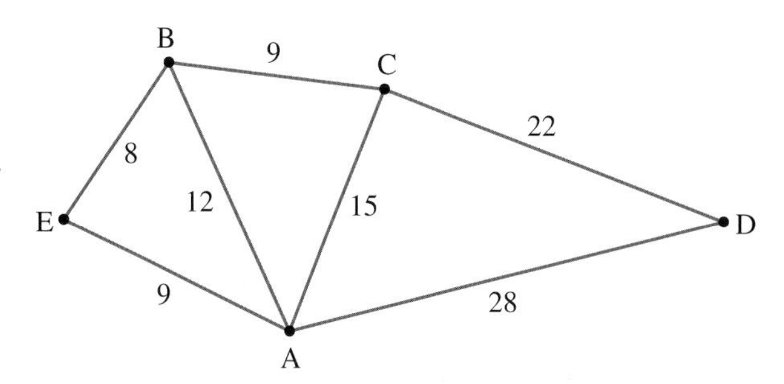

WORKED EXAMPLE 9 Shortest path of a weighted graph

The graph represents the distances in kilometres between eight locations.

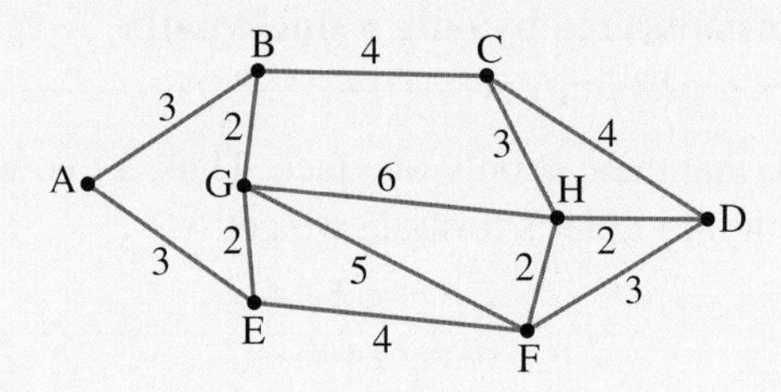

Identify the shortest distance to travel from A to D that goes to all vertices.

THINK	WRITE
1. Identify the Hamiltonian paths that connect the two vertices.	Possible paths: a. ABGEFHCD b. ABCHGEFD c. AEGBCHFD d. AEFGBCHD e. AEFHGBCD
2. Calculate the total distances for each path to find the shortest.	a. $3+2+2+4+2+3+4=20$ b. $3+4+3+6+2+4+3=25$ c. $3+2+2+4+3+2+3=19$ d. $3+4+5+2+4+3+2=23$ e. $3+4+2+6+2+4+4=25$
3. State the final answer.	The shortest distance from A to D that travels to all vertices is 19 km.

6.5.2 Trees

A **tree** is a simple connected graph with no circuits. As such, any pairs of vertices in a tree are connected by a unique path, and the number of edges is always 1 less than the number of vertices.

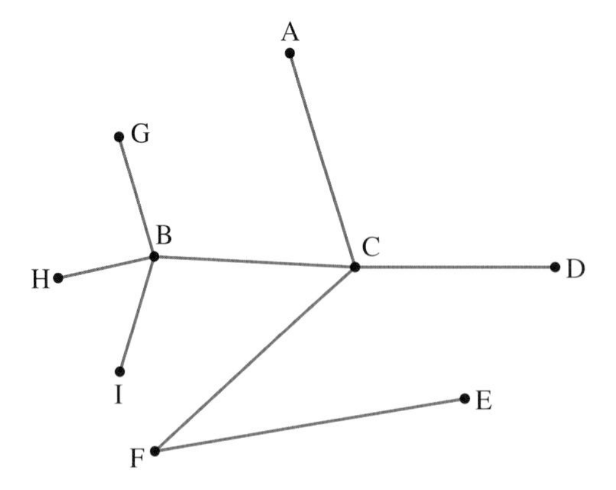

Equivalent conditions for a tree

A simple graph with n vertices is a tree if any of the following conditions are met:

- **it is connected and contains no cycles**
- **it is connected and has $(n - 1)$ edges**
- **it has no cycles and has $(n - 1)$ edges**
- **it is connected but would become disconnected if any edge is removed**
- **it is connected and would form a cycle if any edge is added**
- **any two vertices are connected by only a single path.**

Note: A tree can always be drawn so that there is only one face. Thus, a tree with n vertices will have $(n - 1)$ edges and 1 face, and when this is put into Euler's formula we get:

$$\begin{aligned} v - e + f &= 2 \\ n - (n - 1) + 1 &= 2 \\ n - n + 1 + 1 &= 2 \\ 2 &= 2 \end{aligned}$$

Every tree is therefore a planar graph.

Spanning trees are sub-graphs (graphs that are formed from part of a larger graph) that are trees, and include all of the vertices of the original graph. In practical settings, they can be very useful in analysing network connections. For example, a **minimum spanning tree** for a weighted graph can identify the lowest-cost connections. Spanning trees can be obtained by systematically removing any edges that form a circuit, one at a time.

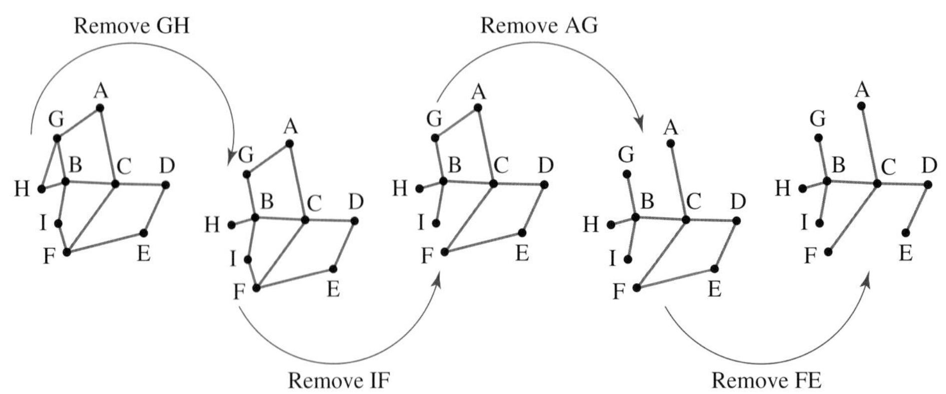

Minimum spanning tree

A minimum spanning tree for a weighted graph is the spanning tree for which the sum of the weighted values on each graph is a minimum.

6.5.3 Prim's algorithm

Prim's algorithm is a set of logical steps that can be used to identify the minimum spanning tree for a weighted connected graph.

Steps for Prim's algorithm

Step 1: Begin at a vertex with low weighted edges.

Step 2: Progressively select edges with the lowest weighting (unless they form a circuit).

Step 3: Continue until all vertices are selected.

WORKED EXAMPLE 10 Applying Prim's algorithm to identify a minimum spanning tree

Use Prim's algorithm to identify the minimum spanning tree of the graph shown.

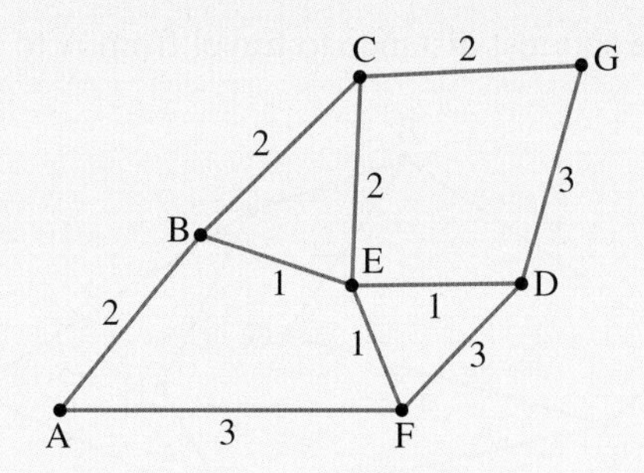

THINK

WRITE/DRAW

1. Draw the vertices of the graph.

C G

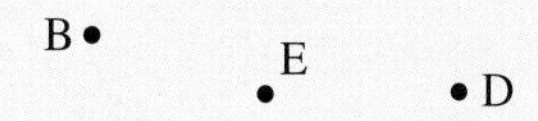

A F

2. Draw in any edges with the lowest weighting that do not complete a circuit.

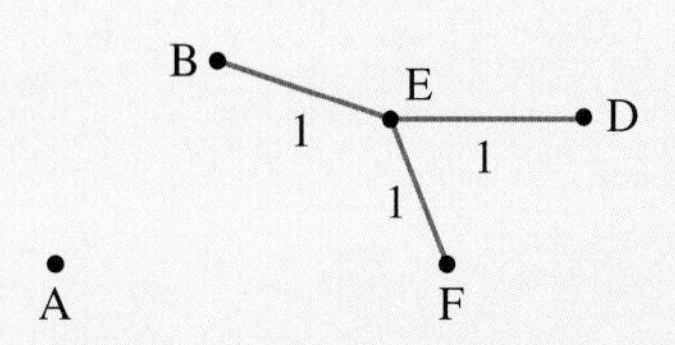

A

3. Draw in any edges with the next lowest weighting that do not complete a circuit. Continue until all vertices are connected.

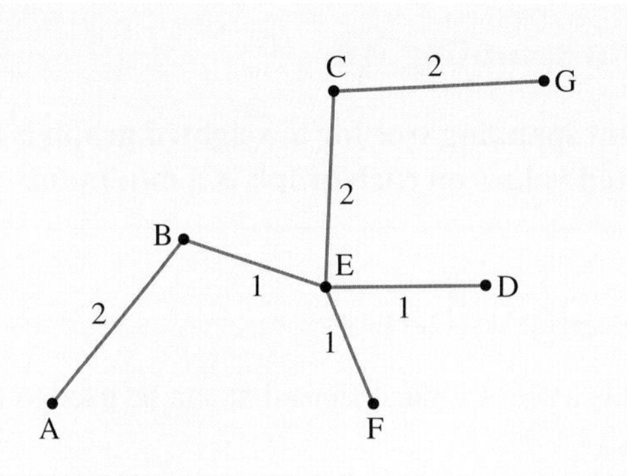

6.5 Exercise

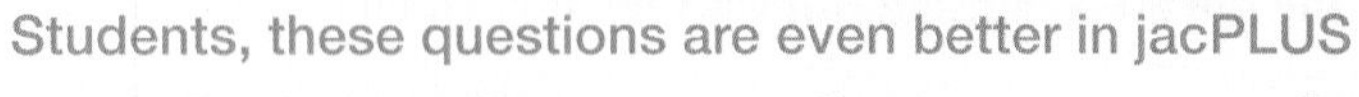

Technology free

1. WE9 Use the graph to identify the shortest distance to travel from A to D that goes to all vertices.

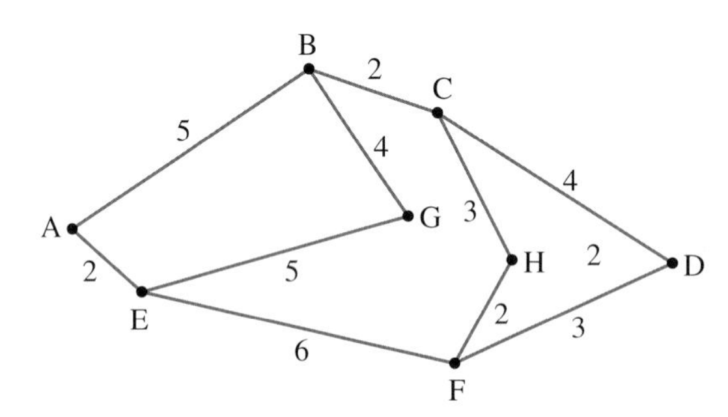

2. Use the graph to identify the shortest distance to travel from A to I that goes to all vertices.

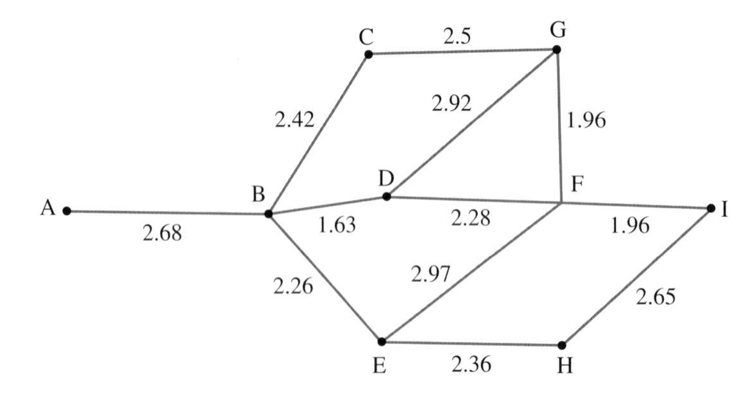

3. Draw three spanning trees for each of the following graphs.

a.

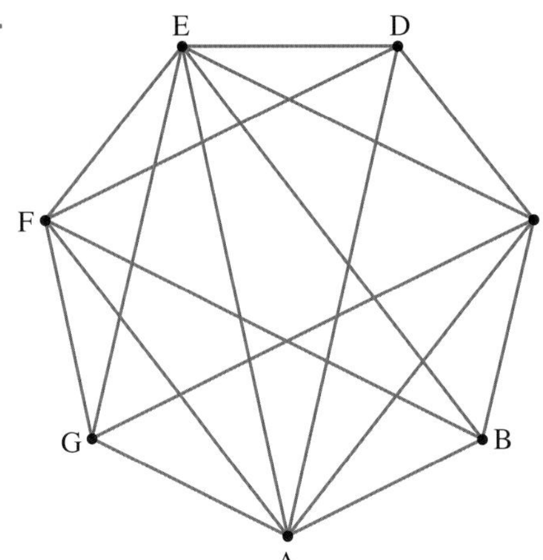

b.

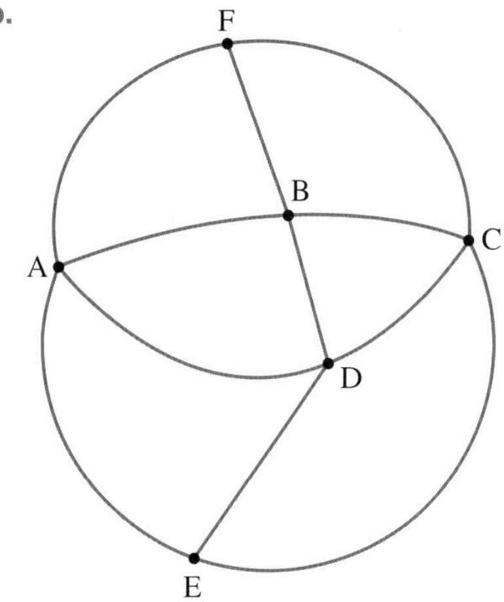

4. A truck starts from the main distribution point at vertex A and makes deliveries at each of the other vertices before returning to A. Determine the shortest route the truck can take.

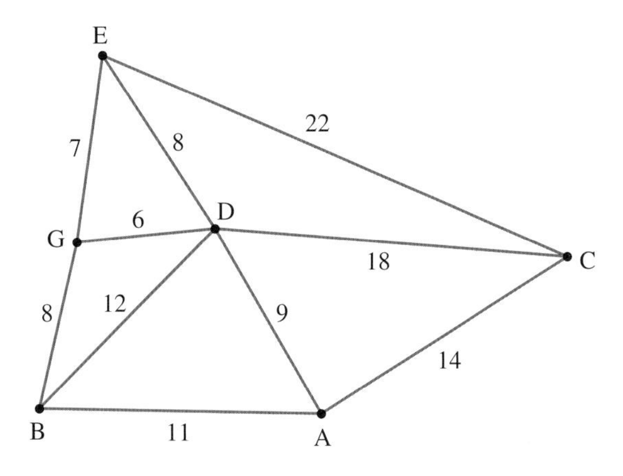

5. For the following trees:

i. add the minimum number of edges to create a semi-Eulerian trail

ii. identify the semi-Eulerian trail created.

a.

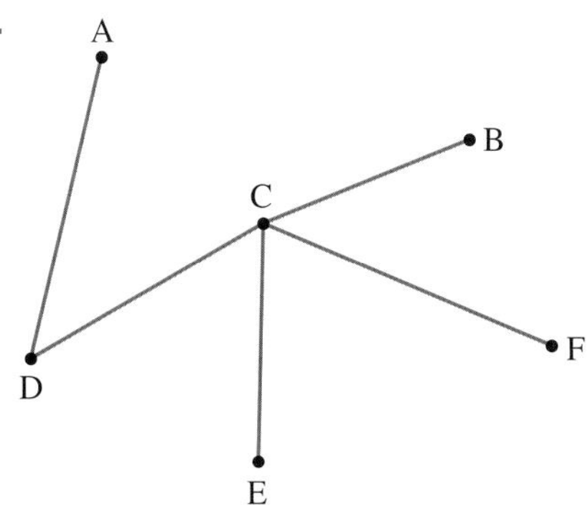

b.

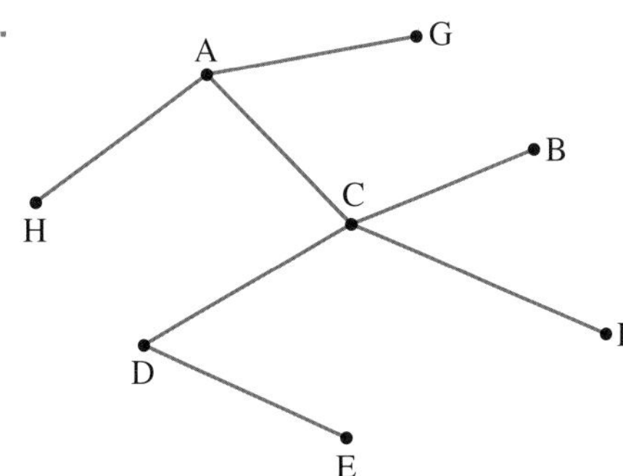

c.

d.

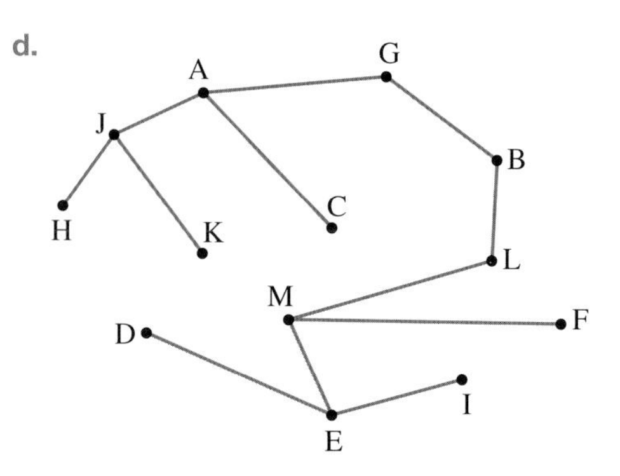

6. **WE10** Use Prim's algorithm to identify the minimum spanning tree of the graph shown.

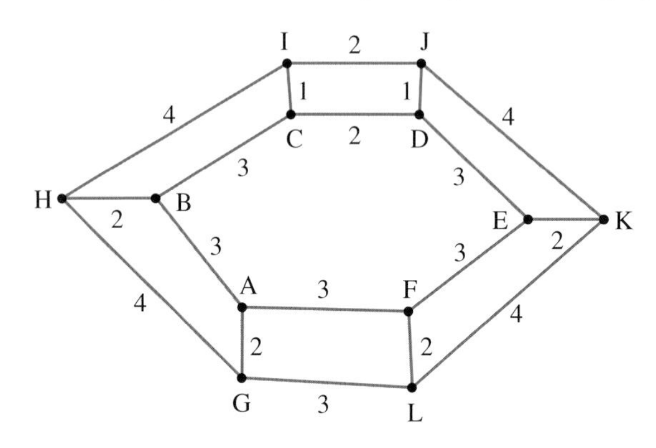

7. Use Prim's algorithm to identify the minimum spanning tree of the graph shown.

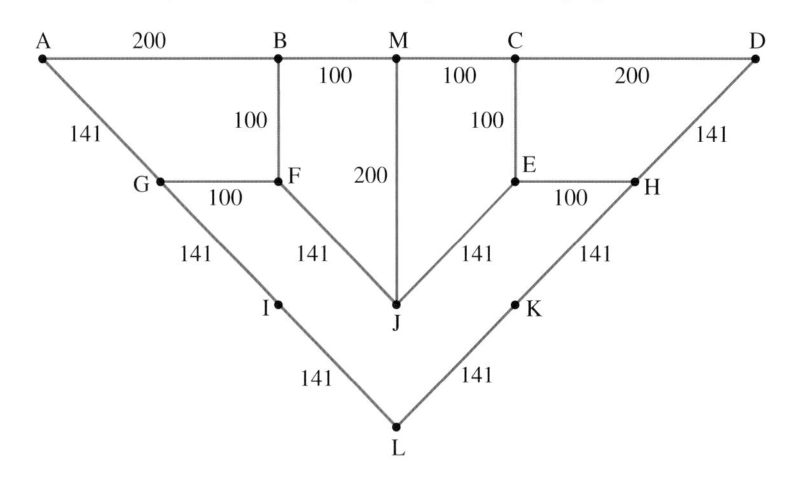

8. Identify the minimum spanning tree for each of the following graphs.

a.

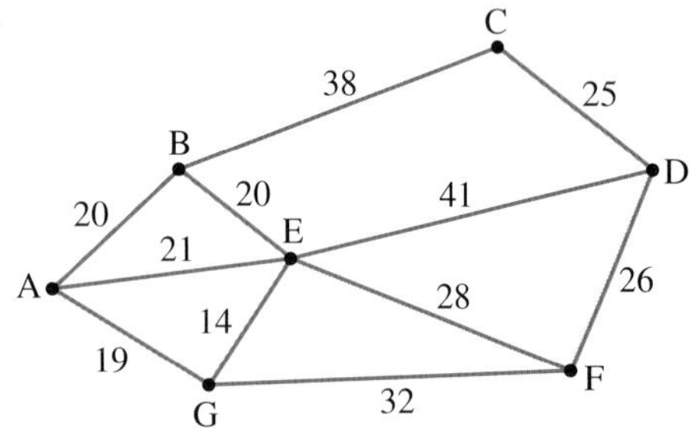

b.

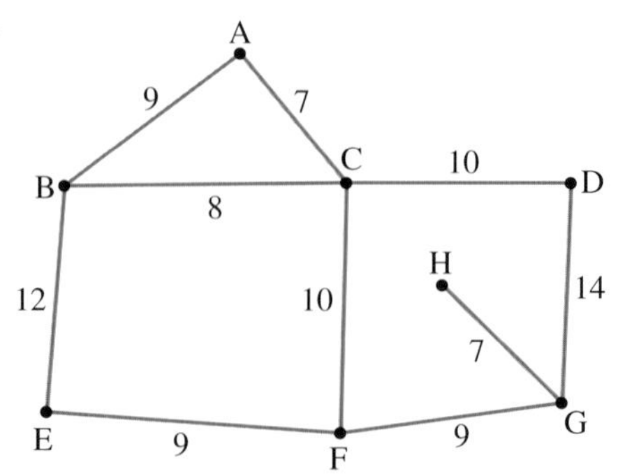

9. Draw diagrams to show the steps you would follow when using Prim's algorithm to identify the minimum spanning tree for the following graph.

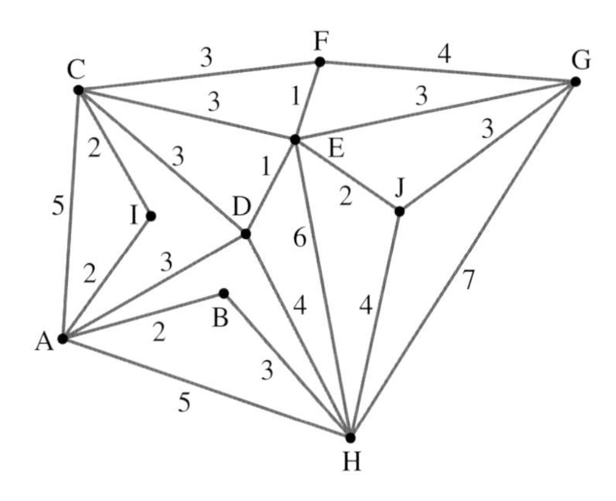

10. Part of the timetable and description for a bus route is shown in the table. Draw a weighted graph to represent the bus route.

Bus stop	Description	Time
Bus depot	The northernmost point on the route	7:00 am
Northsea Shopping Town	Reached by travelling south-east along a highway from the bus depot	7:15 am
Highview Railway Station	Travel directly south along the road from Northsea Shopping Town.	7:35 am
Highview Primary School	Directly east along a road from the railway station	7:40 am
Eastend Medical Centre	Continue east along the road from the railway station.	7:55 am
Eastend Village	South-west along a road from the medical centre	8:05 am
Southpoint Hotel	Directly south along a road from Eastend Village	8:20 am
South Beach	Travel south-west along a road from the hotel.	8:30 am

11. Consider the graph shown.

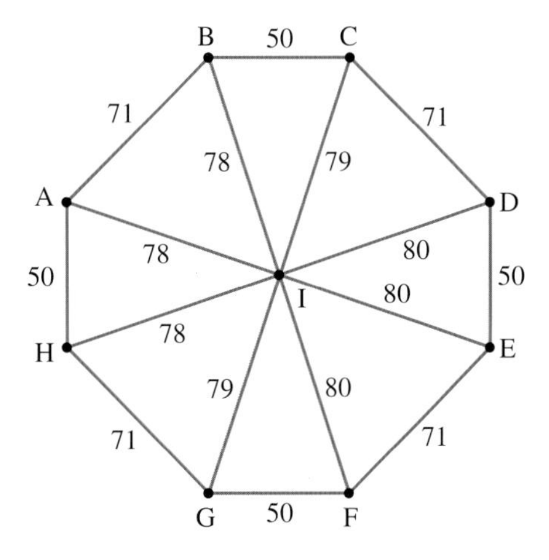

a. Identify the longest and shortest semi-Hamiltonian cycle.
b. Determine the minimum spanning tree for this graph.

12. Consider the graph shown.

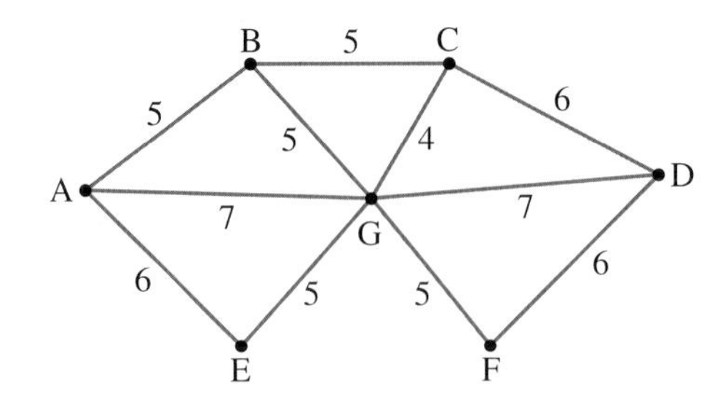

a. If an edge with the highest weighting is removed, identify the shortest semi-Hamiltonian cycle.
b. If the edge with the lowest weighting is removed, identify the shortest semi-Hamiltonian cycle.

13. The weighted graph represents the costs incurred by a salesman when moving between the locations of various businesses.

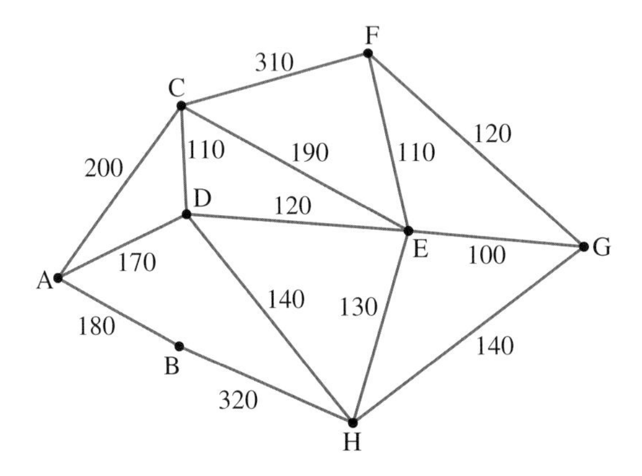

a. Determine the cheapest way of travelling from A to G.
b. Determine the cheapest way of travelling from B to G.
c. If the salesman starts and finishes at E, determine the cheapest way to travel to all vertices.

Technology active

14. The diagrams show two options for the design of a computer network for a small business.
Option 1

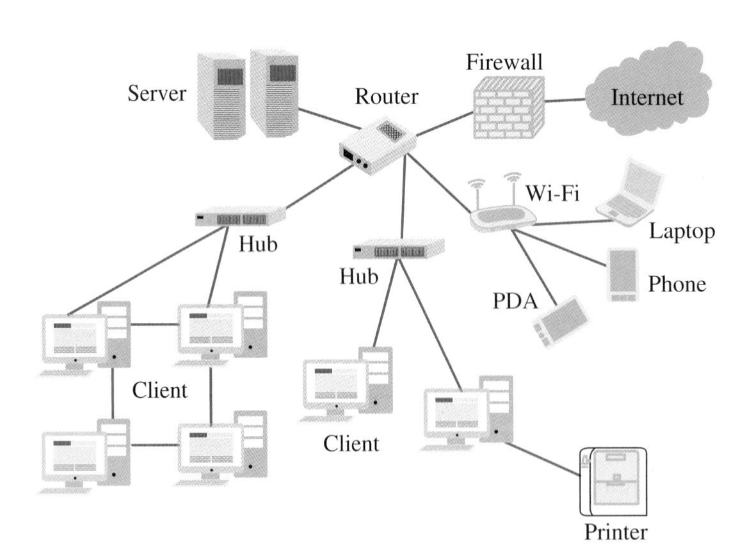

Option 2

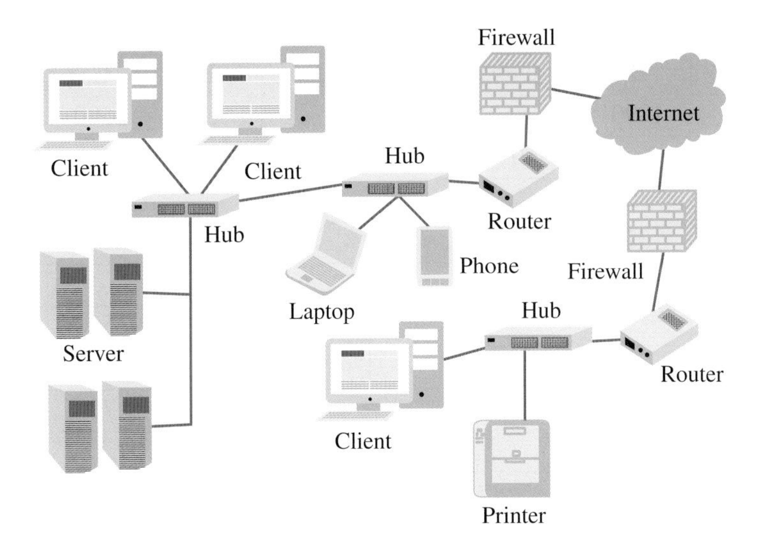

Information relating to the total costs of setting up the network is shown in the following table.

Connected to:	Server	Client	Hub	Router	Firewall	Wi-Fi	Printer
Server			$995	$1050			
Client		$845	$355				$325
Hub			$365	$395			$395
Router	$1050		$395		$395	$395	
Laptop			$295			$325	
Phone			$295			$325	
PDA						$325	
Internet					$855		

a. Use this information to draw a weighted graph for each option.
b. Determine which is the cheaper option.

15. A mining company operates in several locations in Western Australia and the Northern Territory, as shown on the map.

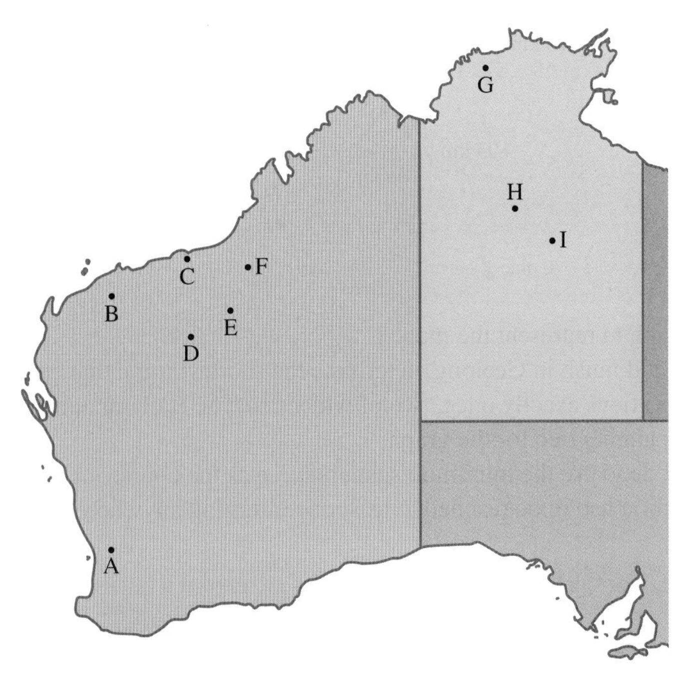

Flights operate between selected locations, and the flight distances (in km) are shown in the following table.

	A	B	C	D	E	F	G	H	I
A		1090		960			2600		2200
B	1090		360	375	435				
C		360							
D	960	375							
E		435							
F							1590	1400	
G	2600					1590		730	
H						1400	730		220
I	2200							220	

a. Show this information as a weighted graph.
b. State if a semi-Hamiltonian cycle exists. Explain your answer.
c. Identify the shortest distance possible for travelling to all sites the minimum number of times if you start and finish at:
 i. A
 ii. G.
d. Draw the minimum spanning tree for the graph.

16. The organisers of the 'Tour de Vic' bicycle race are using the following map to plan the event.

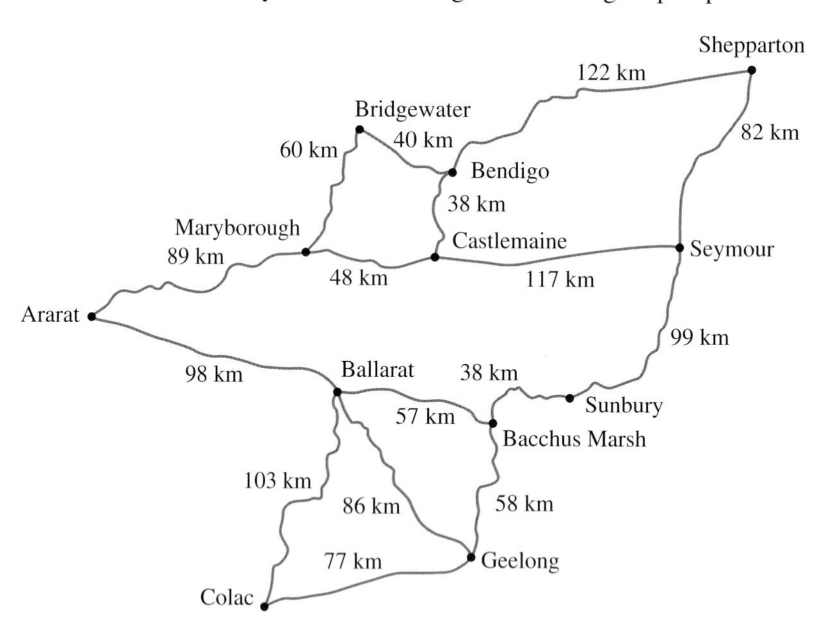

a. Draw a weighted graph to represent the map.
b. If they wish to start and finish in Geelong, determine the shortest route that can be taken that includes a total of nine other locations exactly once, two of which must be Ballarat and Bendigo.
c. Draw the minimal spanning tree for the graph.
d. If the organisers decide to use the minimum spanning tree as the course, calculate the shortest possible distance if each location had to be reached at least once, starting anywhere and finishing anywhere.

6.5 Exam questions

Question 1 (1 mark) TECH-FREE

The following showing the distances (in km) between different camping sites in a national park. Determine the minimum length of cable required to connect all the camps to the same network.

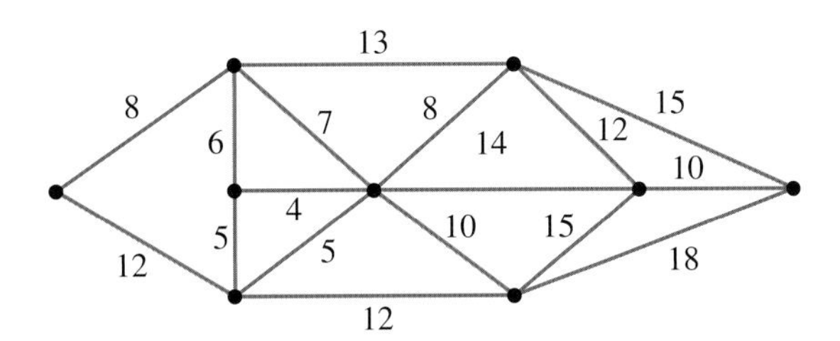

Question 2 (2 marks) TECH-FREE

Let G be a connected graph.

a. Describe what can be said about an edge of G that appears in every spanning tree. **(1 mark)**

b. Describe what can be said about an edge of G that appears in no spanning tree. **(1 mark)**

Question 3 (4 marks) TECH-FREE

Consider the complete graph K_n where the vertices are labelled 1 through to n.

a. Determine the value of the minimum spanning tree if the weight of each edge is the difference of the value of the vertices it connects. **(2 marks)**

b. Determine the value of the minimum spanning tree if the weight of each edge is the sum of the value of the vertices it connects. **(2 marks)**

More exam questions are available online.

6.6 Bipartite graphs and the Hungarian algorithm

LEARNING INTENTION

At the end of this subtopic you should be able to:

- determine if a graph is a bipartite graph
- draw bipartite graphs and complete bipartite graphs
- solve allocation problems, using the Hungarian algorithm where necessary.

6.6.1 Bipartite graphs

Bipartite graphs are graphs in which the vertices can be separated into 2 distinct groups with edges that link vertices from one group to vertices from the other group only. This is stated formally as follows.

Definition of a bipartite graph

The vertices in a bipartite graph can be split into two distinct groups, M and N, so that each edge of the graph joins a vertex in M to a vertex in N.

The groups M and N are often represented using coloured vertices. For example, the bipartite graphs below have vertices coloured pink (representing M) and green (representing N). Note that all edges connect pink and green vertices, never pink and pink or green and green.

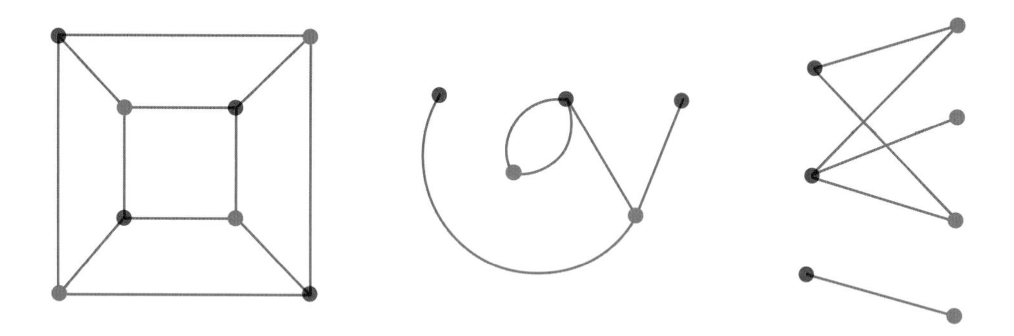

Note also that bipartite graphs may be connected, or disconnected.

Complete bipartite graphs

A **complete bipartite graph** is one where each vertex in M is joined to each vertex in N by one edge. Complete bipartite graphs are denoted by $K_{m,n}$, where m represents the number of vertices in group M and n represents the number of vertices in group N.

Planarity of complete bipartite graphs

All complete bipartite graphs of the form $K_{1,n}$ are planar graphs. It will have $(n+1)$ vertices, n edges and 1 face, thus satisfying Euler's formula: $(n+1)-n+1 = n-n+1+1 = 2$	$K_{1,n}$ n vertices in N
All complete bipartite graphs of the form $K_{2,n}$ are planar graphs. It will have $(n+2)$ vertices, $2n$ edges and n faces, thus satisfying Euler's formula: $(n+2)-2n+n = n-2n+n+2 = 2$	$K_{2,n}$ [ ] n vertices in N
The complete bipartite graph $K_{3,3}$ is non-planar. This means all complete bipartite graphs $K_{m,n}$ are non-planar for $m, n \geq 3$ as they will contain $K_{3,3}$ as a subgraph.	$K_{3,3}$

WORKED EXAMPLE 11 Bipartite graphs

a. **Determine whether the graph shown is bipartite.**

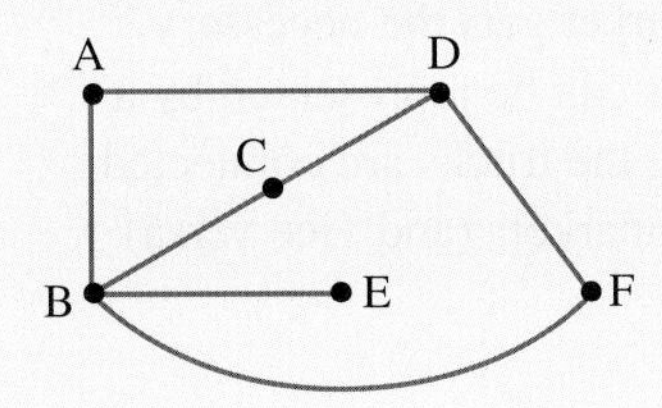

b. **Draw the complete bipartite graph $K_{4,2}$.**

THINK	WRITE/DRAW
a. 1. Start with any vertex and colour it either pink or green. Colour A pink.	a. 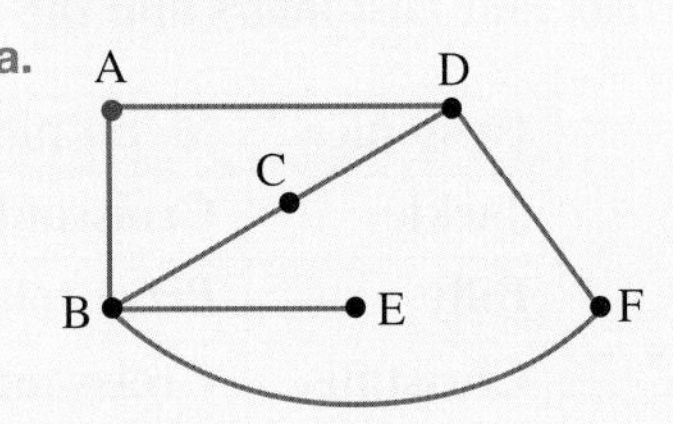
2. Colour the vertices joined to *A* green.	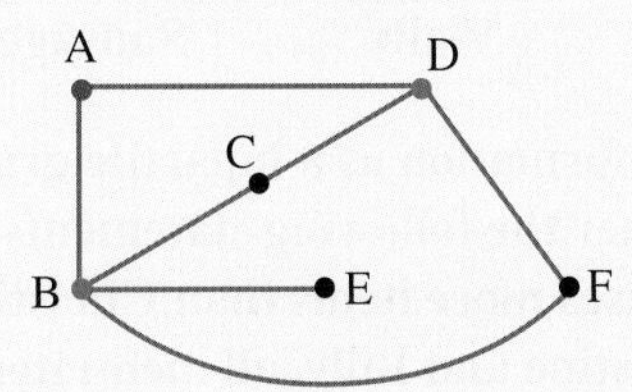
3. Continue colouring vertices in alternating colours until all vertices are coloured. If, at any point, an edge has two vertices of the same colour, stop colouring as the graph is not bipartite.	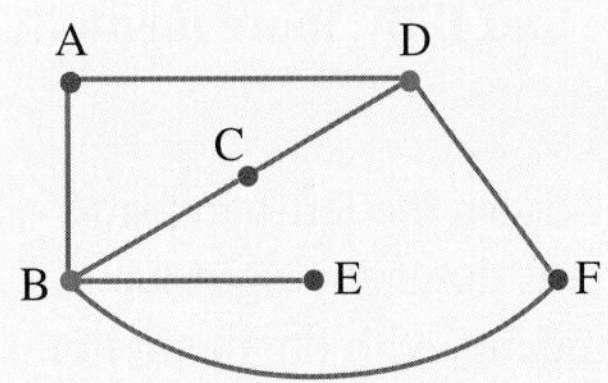
4. If every edge joins a pink vertex to a green vertex then the graph is bipartite.	The graph is bipartite as it can be split into a group of 4 pink vertices connected to a group of 2 green vertices.
b. 1. $K_{4,2}$ is a complete graph. Start by drawing 2 rows (or columns), one with 4 vertices and the other with two vertices.	b.
2. Connect each vertex in one row (or column) to every vertex in the other group.	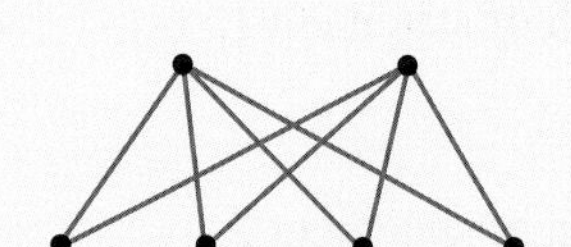

6.6.2 Applications of bipartite graphs

Bipartite graphs have many practical applications. For example, consider a transport company that has three large trucks which deliver to two supermarkets, so that each supermarket gets the necessary supplies, regardless of the truck used. This can be represented by a bipartite graph as the vertices representing the trucks are connected only to the vertices representing the supermarkets (and vice versa).

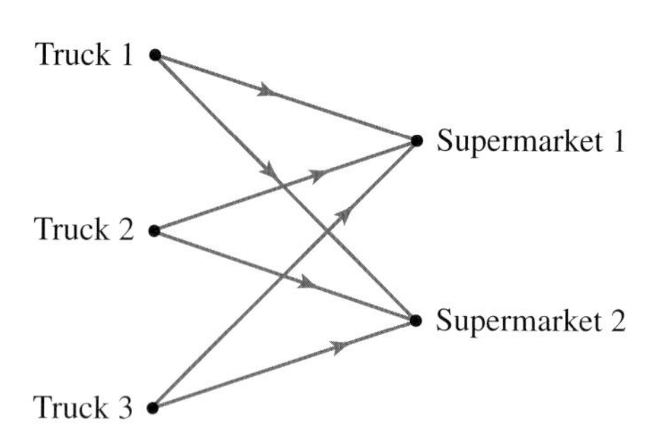

WORKED EXAMPLE 12 Application of bipartite graphs

The following table lists four customers and the menu items that they ordered at a bakery.

Customer	Menu items
Jackie	Croissant, pizza roll
Billy	Pizza roll, donut
Christine	Croissant, pizza roll, sausage roll
Wally	Sausage roll

a. **Represent this information as a bipartite graph.**
b. **Determine whether the following statements are true or false.**
   1. **Jackie purchases more items than Christine.**
   2. **Between Christine and Billy, all menu items are chosen.**
   3. **Between Jackie and Billy, more menu items are chosen than by Christine.**

THINK

a. List the students down the left-hand side and the menu items down the right-hand side. Link the students with the menu items according to the table.

WRITE/DRAW

a.

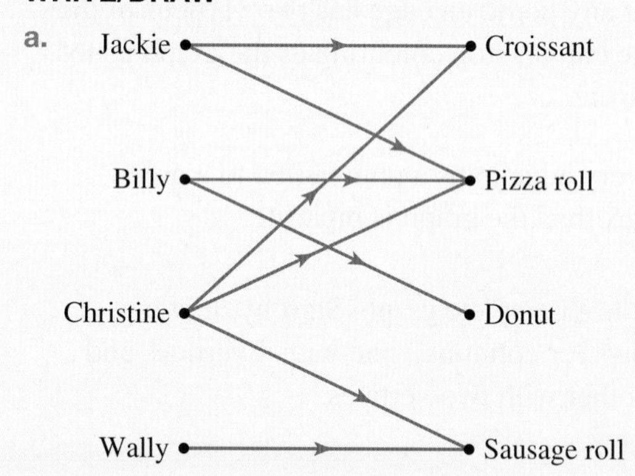

b. 1. Determine the truth of statement 1.

b. Jackie purchases only 2 items, Christine purchases 3.
The statement is false.

2. Determine the truth of statement 2.

Christine purchases a croissant, pizza roll and sausage roll. Billy purchases a pizza roll and a donut. Their choices include all 4 menu items.
The statement is true.

3. Determine the truth of statement 3.

Christine chooses 3 of the menu items and between Jackie and Billy 3 menu items are chosen.
The statement is false.

6.6.3 The assignment or allocation problem

In an office in the city there are four employees and four tasks that need to be completed. Each person can do the task in a different amount of time. What is the best way for their manager to allocate these tasks, one per person, so that the time can be minimised? This is known as optimal allocation.

This case can be modelled by a weighted bipartite graph shown below.

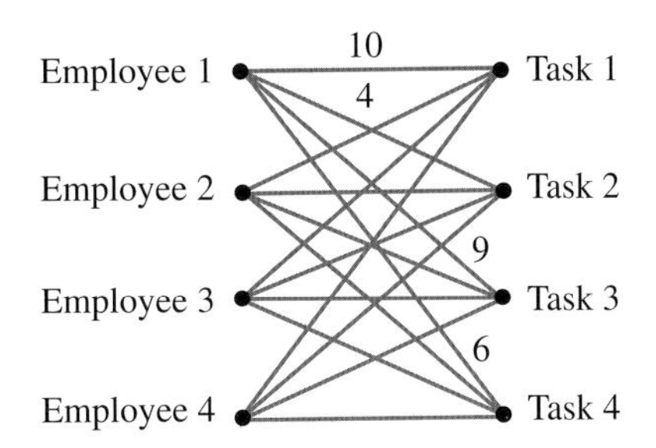

Note that since there are too many overlapping edges it is very difficult to display the weight of all edges on the graph. For this reason, we have only shown the weight of the edges connected to Employee 1.

Since they make it much simpler to read and interpret data, tables are often used to represent weighted bipartite graphs. The table below shows the times (in hours) taken for each of the four employees to complete each of the four tasks.

	Task 1	Task 2	Task 3	Task 4
Employee 1	10	4	9	6
Employee 2	8	11	10	7
Employee 3	6	8	7	9
Employee 4	8	5	3	9

By converting this table into a matrix and performing row reductions we can determine the allocation of employees to tasks that minimises the total time taken to complete all of the tasks. This method is stepped out in the following Worked example.

WORKED EXAMPLE 13 The office allocation problem

In an office in the city there are four employees and four tasks that need to be completed. The times taken by each employee to do the four jobs are given in the following table. Determine the optimal allocation and hence state the minimum time.

	Task A	Task B	Task C	Task D
Employee 1	**10**	**4**	**9**	**6**
Employee 2	**8**	**11**	**10**	**7**
Employee 3	**6**	**8**	**7**	**9**
Employee 4	**8**	**5**	**3**	**9**

THINK

1. Set up the matrix of employees against tasks.

WRITE/DRAW

	A	B	C	D
1	10	4	9	6
2	8	11	10	7
3	6	8	7	9
4	8	5	3	9

2. Perform row reduction by locating the smallest value in each row and subtracting it from all numbers in that row.
The smallest number in Row 1 is 4.
The smallest number in Row 2 is 7.
The smallest number in Row 3 is 6.
The smallest number in Row 4 is 3.

	A	B	C	D
1	6	0	5	2
2	1	4	3	0
3	0	2	1	3
4	5	2	0	6

3. Cover all the zeroes with the smallest number of straight lines; horizontal or vertical but not diagonal.
If the number of lines equals the number of tasks continue to the next step.
If the number of lines does not equal the number of tasks, another method of allocation will need to be used.

	A	B	C	D
~~1~~	~~6~~	~~0~~	~~5~~	~~2~~
~~2~~	~~1~~	~~4~~	~~3~~	~~0~~
~~3~~	~~0~~	~~2~~	~~1~~	~~3~~
~~4~~	~~5~~	~~2~~	~~0~~	~~6~~

There are four lines and four tasks.

4. Draw a bipartite graph, where the zeroes connect the employee to the tasks.

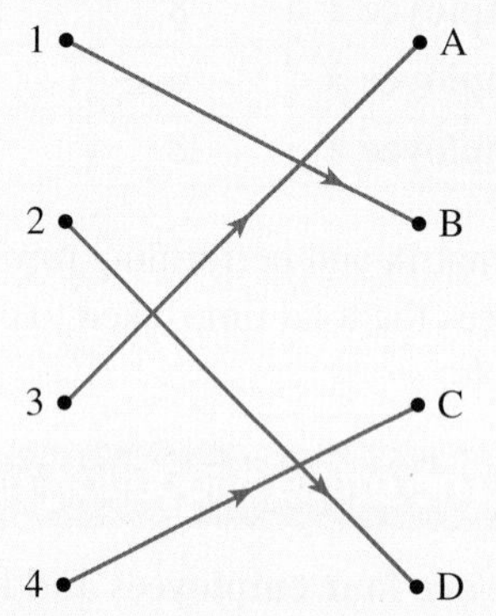

5. Write the possible allocations and determine the minimum number of hours required.

There is only one possible allocation.
$1 \rightarrow B$
$2 \rightarrow D$
$3 \rightarrow A$
$4 \rightarrow C$
Total time: $4 + 7 + 6 + 3 = 20$ hours

6. Write the answer.

The minimum time to complete the 4 tasks is 20 hours.

The Hungarian algorithm

When the number of zeroes does not equal the number of tasks after the first row reduction, the **Hungarian algorithm** needs to be used.

The Hungarian algorithm

Step 1: Subtract the row minimum from each row.

Step 2: Subtract the column minimum from each column.

Step 3: Cover all zeroes with a minimum number of lines.

- **If the number of lines equals the number of tasks, draw a bipartite graph and allocate the tasks.**
- **If the number of lines is less than the number of tasks, continue to step 4.**

Step 4: Find the smallest uncovered number. Subtract this number from all uncovered elements and add it to all elements that are covered twice.

Step 5: Cover all zeros with a minimum number of lines. If the number of lines equals the number of tasks, draw a bipartite graph and allocate tasks; if the number of lines is less than the number of tasks, go back to step 4.

WORKED EXAMPLE 14 Applying the Hungarian algorithm

Four workers need to be allocated to four tasks. The time required for each worker for each task is summarised in the table below. Use the Hungarian algorithm to minimise the time required to complete the tasks by allocating one job to each worker.

	Task 1	Task 2	Task 3	Task 4
Worker 1	80	81	67	90
Worker 2	75	35	47	90
Worker 3	9	67	3	84
Worker 3	6	7	96	21

THINK / **WRITE/DRAW**

1. Set up the matrix of employees against tasks.

	T1	T2	T3	T4
W1	80	81	67	90
W2	75	35	47	90
W3	9	67	3	84
W3	6	7	96	21

2. Perform row reduction by locating the smallest value in each row and subtracting it from all numbers in that row.
The smallest number in Row 1 is 67.
The smallest number in Row 2 is 35.
The smallest number in Row 3 is 3.
The smallest number in Row 4 is 6.
Only 3 lines are required to cover the zeros, so continue to the Hungarian algorithm.

	T1	T2	T3	T4
W1	13	14	0	23
W2	40	0	12	55
W3	6	64	0	81
W4	0	1	90	15

3. Perform a column reduction by subtracting the smallest number in each column from all the numbers in the column. Since the first 3 columns contain a 0, they remain as is, and since the lowest number in the fourth column was 15, subtract 15 from all values in the fourth column.

	T1	T2	T3	T4
W1	13	14	0	8
W2	40	0	12	40
W3	6	64	0	66
W4	0	1	90	0

4. Cover all the zeroes with the smallest number of straight lines; horizontal or vertical but not diagonal. If the number of lines does not equal the number of tasks, continue to step 5.

	T1	T2	T3	T4
W1	13	14	0	8
W2	40	0	12	40
W3	6	64	0	66
W4	0	1	90	0

There are only 3 lines and four tasks.

5. The smallest uncovered number is 6. Subtract 6 from all uncovered elements and add it to all elements that are covered twice.

The smallest uncovered number is 6.

	T1	T2	T3	T4
W1	7	8	0	2
W2	40	0	18	40
W3	0	58	0	60
W4	0	1	96	0

6. Cover all zeroes with a minimum number of lines.

	T1	T2	T3	T4
W1	13	14	0	8
W2	40	0	12	40
W3	6	64	0	66
W4	0	1	90	0

There are four lines and four tasks.

7. Draw a bipartite graph, where the zeroes connect the employee to the tasks.

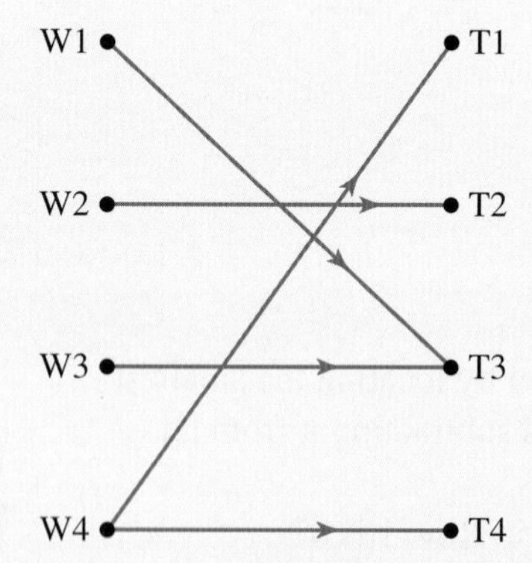

8. Write the possible allocations and determine the minimum number of hours required.	Worker 1 is allocated Task 3. Worker 2 is allocated Task 2. Worker 3 is allocated Task 1 (Task 3 is already allocated). Worker 4 is allocated Task 4 (Task 1 is already allocated). $67 + 35 + 9 + 21 = 132$ hours.
9. Write the answer.	The minimum time required to complete all tasks is 132 hours.

In Worked example 14, the objective was to minimise the time. Some problems require a maximum to be determined. In this process, all elements in the matrix are subtracted from the largest one first. From then on, the procedure is the same as that set out in Worked example 14.

6.6 Exercise

Technology free

1. WE11 Determine which of the following are bipartite graphs.

a.

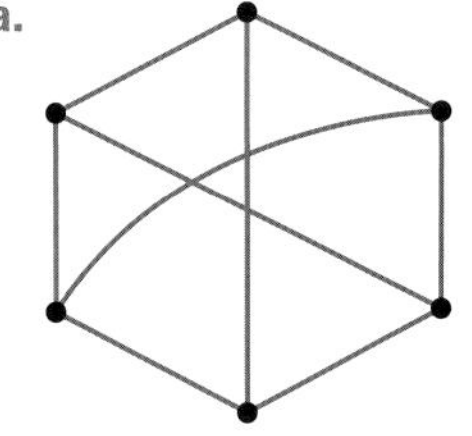

b.

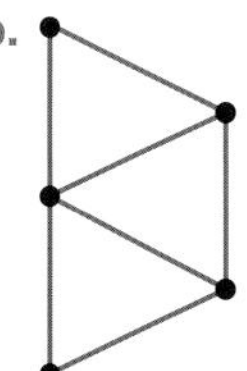

c.

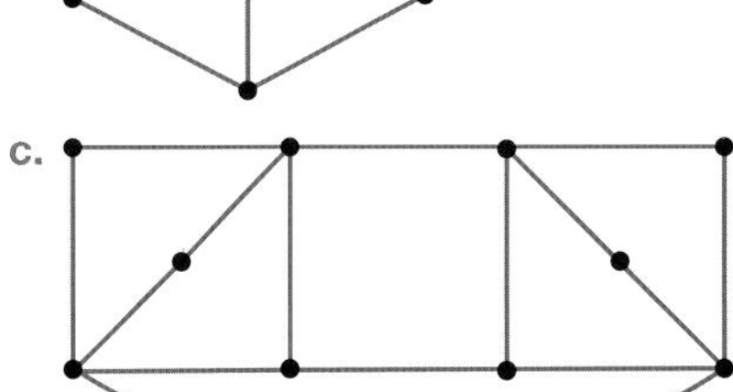

d.

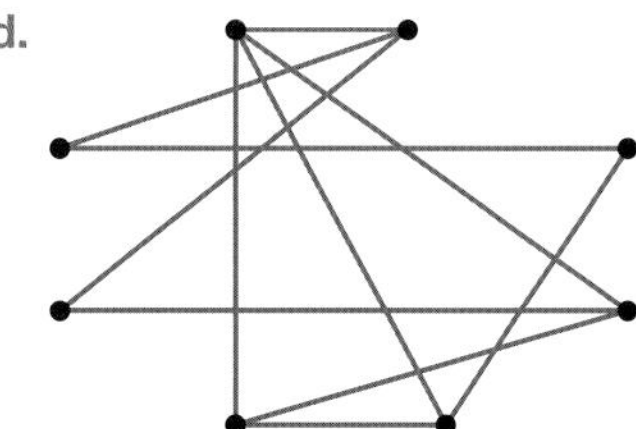

2. For the following complete bipartite graphs:
   i. Draw the graph of each.
   ii. State whether it is planar.

   a. $K_{2,2}$ b. $K_{3,4}$ c. $K_{1,5}$ d. $K_{2,6}$

3. State the number of edges for each of the following:

   a. $K_{3,3}$ b. $K_{5,7}$ c. $K_{m,n}$ d. $K_{x,3x}$

4. **WE12** Five customers (Will, Penny, Fan, Roya and Su Yi) go to the local fish and chip shop for dinner and place the orders as shown in the table.

Diner	Dishes
Will	Fish, chips
Penny	Chips
Fan	Fish, potato cakes, dim sims
Roya	Potato cakes dim sims
Su Yi	Chips, dim sims

a. Represent this information as a bipartite graph.
b. Determine whether the following statements are true or false.

1. Will and Penny between them have more different items than Fan and Roya.
2. Fan and Roya together have tried all the options.
3. Roya and Su Yi between them have more variety than Penny and Fan.
4. Penny and Fan between them have more variety than Roya and Su Yi.

5. Four visitors on a tour to Europe have a choice of five countries to visit. The countries are France, Germany, Italy, Spain and Ireland, they can visit as many of these countries as they wish. Sally decides to visit Italy and Germany, but not the others. Joe decides to spend all his time in France. Mike wants to see Germany, Spain and Ireland. Genevieve is keen to visit all of the countries on this trip.

a. Explain why a bipartite graph is suitable to represent this information.
b. Draw a bipartite graph to represent the information.
c. Determine the degree of the vertex representing Mike.

6. Five teachers can teach a variety of five subjects as indicated by the bipartite graph.
Determine whether the following statements are true or false.

a. Ms Bell and Ms Tran can teach all five subjects between them.
b. Mr Ring and Ms Tran, in total, can teach more subjects than Ms Bell and Ms Jules.
c. Ms Bell and Ms Tran each teach the same number of subjects.
d. Ms Bell and Mr Pho, in total teach fewer subjects than Mr Ring and Ms Tran.
e. Ms Jules teaches fewer subjects than all other teachers.

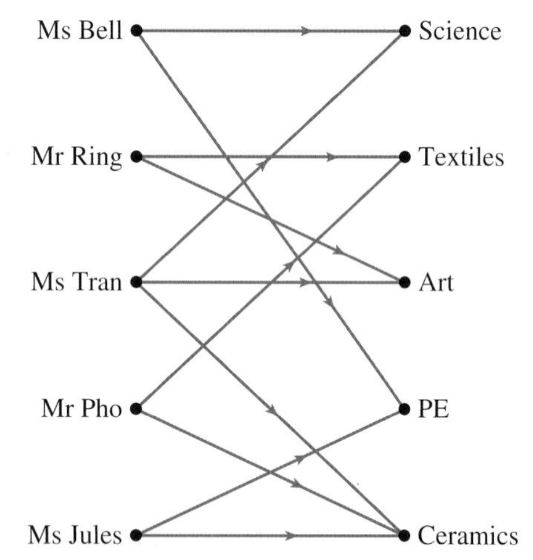

7. **WE13** A shipping company has 4 ships that deliver fuel to four different oil rigs lying off shore. The times taken by each ship to do the four deliveries are given in the following table. Determine the optimal allocation and hence state the minimum time required to complete all four deliveries.

	Rig A	Rig B	Rig C	Rig D
Ship 1	16	14	20	13
Ship 2	15	16	17	16
Ship 3	19	13	13	18
Ship 4	22	26	20	24

8. Perform row reductions on the following matrix, which represents time (in hours) and attempt an optimal allocation for the minimum time. State the minimum time.

$$\begin{bmatrix} 5 & 2 & 6 \\ 1 & 3 & 4 \\ 2 & 4 & 1 \end{bmatrix}$$

9. Perform row reductions on the following matrix, which represents time (in hours) and attempt an optimal allocation for the minimum time. State the minimum time.

$$\begin{bmatrix} 5 & 4 & 8 & 4 \\ 10 & 5 & 7 & 6 \\ 6 & 7 & 8 & 9 \\ 5 & 9 & 4 & 6 \end{bmatrix}$$

10. **WE14** Four workers need to be allocated to four tasks. The time required for each worker for each task is summarised in the table below. Use the Hungarian algorithm to minimise the time required to complete the tasks by allocating one job to each worker.

	Task 1	Task 2	Task 3	Task 4
Worker 1	7	10	10	5
Worker 2	11	10	10	8
Worker 3	5	10	7	4
Worker 4	6	9	9	7

11. Four delivery vans need to deliver to four different supermarkets. The distances of the four drivers from each of the four supermarkets are given in the table below. If the drivers take their loaded vans home in the evening before the delivery day, use the optimal allocation method to minimise the total distance travelled by the vans to reach all four supermarkets.

$$\begin{array}{l} \\ \text{Driver 1} \\ \text{Driver 2} \\ \text{Driver 3} \\ \text{Driver 4} \end{array} \begin{array}{c} \begin{array}{cccc} S1 & S2 & S3 & S4 \end{array} \\ \begin{bmatrix} 7 & 25 & 21 & 6 \\ 13 & 31 & 8 & 16 \\ 23 & 19 & 16 & 15 \\ 22 & 29 & 24 & 10 \end{bmatrix} \end{array}$$

12. A florist wishes to purchase peonies, roses and lilies for three bouquets from three different flower wholesalers. The peonies cost \$40, \$55 and \$60 from the three stores, the roses cost \$65, \$60 and \$70 and the lilies cost \$45, \$50 and \$40. Determine the optimal allocation for the flower order.

Technology active

13. **MC** Consider the following matrix.

$$\begin{bmatrix} 8 & 4 & 8 \\ 4 & 4 & 6 \\ 7 & 6 & 6 \end{bmatrix}$$

The total value of the optimal allocation is

A. 9 B. 11 C. 14 D. 16 E. 19

14. Perform an optimal allocation on the following matrices by conducting row reductions, column reductions and the Hungarian algorithm (where required) until the allocation is complete. State the minimum value.

a. $\begin{bmatrix} 20 & 18 & 13 \\ 21 & 16 & 18 \\ 25 & 28 & 26 \end{bmatrix}$

b. $\begin{bmatrix} 20 & 30 & 40 & 50 \\ 60 & 20 & 30 & 60 \\ 50 & 40 & 50 & 20 \\ 10 & 70 & 40 & 60 \end{bmatrix}$

15. A large holiday park has four maintenance workers and four tasks that need one person to complete each morning.
The time it takes each of the workers to complete the four tasks is summarised in the following table.

	Task 1	Task 2	Task 3	Task 4
Worker 1	50	60	70	80
Worker 2	90	50	60	90
Worker 3	80	70	80	30
Worker 4	30	100	70	90

a. Perform row and column reduction.
b. Perform the Hungarian algorithm.
c. Determine the optimal allocations and display these allocations using a bipartite graph.
d. Determine the total time required to complete all four tasks.

6.6 Exam questions

Question 1 (4 marks) TECH-FREE
Consider the complete bipartite graph $K_{3,4}$.
a. Draw the graph of $K_{3,4}$. **(1 mark)**
b. Determine if the graph has a Eulerian cycle. Justify your answer. **(1 mark)**
c. Determine if the graph has a Hamiltonian cycle. Justify your answer. **(2 marks)**

Question 2 (2 marks) TECH-FREE
a. State which complete bipartite graphs are also Eulerian graphs. **(1 mark)**
b. State which complete bipartite graphs also semi-Eulerian Graphs. **(1 mark)**

Question 3 (3 marks) TECH-FREE
Four swimmers are to compete in a medley relay. During training each swimmer has swum each leg of the relay and their times have been recorded in the following table.

	Leg 1	Leg 2	Leg 3	Leg 4
Mary	2	4	3	5
Jenny	3	5	3	4
Pauline	2	3	4	2
Jacinta	2	4	2	3

Determine the optimal allocation and the minimum time to complete the race.

More exam questions are available online.

6.7 Review

6.7.1 Summary

oc-37049

Hey students! Now that it's time to revise this topic, go online to:

Access the topic summary

Review your results

Watch teacher-led videos

Practise exam questions

Find all this and MORE in jacPLUS

6.7 Exercise

Technology free: short answer

1. a. Identify whether the following graphs are planar or not planar.

i.

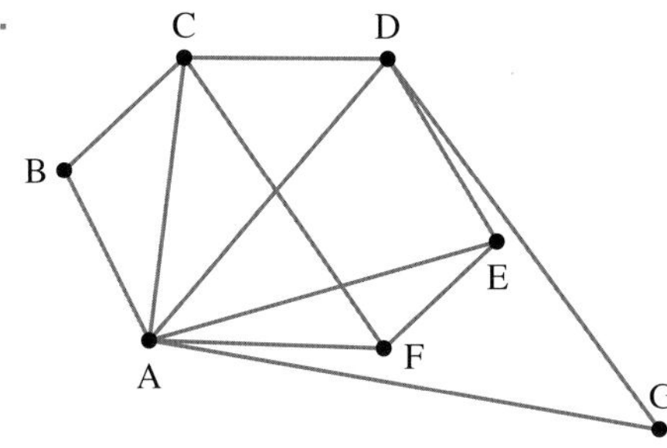

ii.

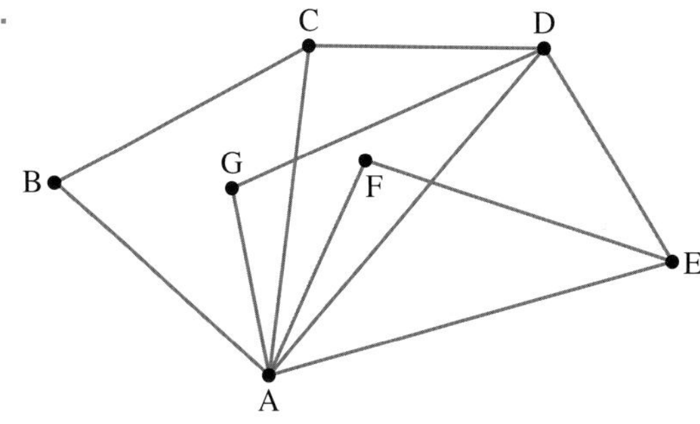

iii.

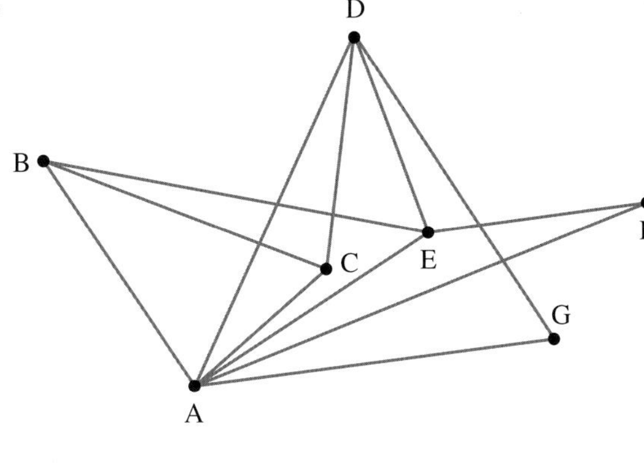

iv.

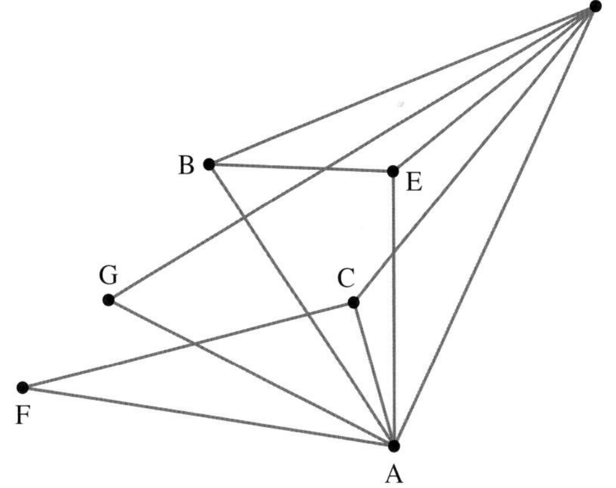

b. Redraw the graphs that are planar without any intersecting edges.

2. Complete the following adjacency matrices.

a. $\begin{bmatrix} 1 & 1 & 0 & 1 \\ & 0 & & 0 \\ & 3 & 1 & \\ & & 1 & 0 \end{bmatrix}$

b. $\begin{bmatrix} 0 & 1 & 2 & 1 & \\ & 0 & & 0 & 1 \\ & 2 & 0 & 2 & \\ & & & 2 & 2 \\ 1 & & 3 & & 0 \end{bmatrix}$

c. $\begin{bmatrix} 0 & & 1 & 3 & 1 & \\ 2 & 0 & & & 1 & \\ & 3 & 0 & 2 & & \\ & 1 & & 2 & 2 & 1 \\ & & 3 & & 3 & 1 \\ 2 & 0 & 1 & & & 0 \end{bmatrix}$

d. $\begin{bmatrix} 0 & & & & 0 & & \\ 2 & 0 & & 1 & & & \\ 1 & 2 & 0 & 1 & 1 & 0 & \\ 3 & & & 0 & & & 1 \\ & 2 & & 0 & 0 & & 3 \\ 1 & 1 & & 2 & 0 & 0 & 2 \\ 0 & 0 & 1 & & & & 0 \end{bmatrix}$

3. Identify which of the following graphs are:

i. simple
ii. complete
iii. planar.

a.

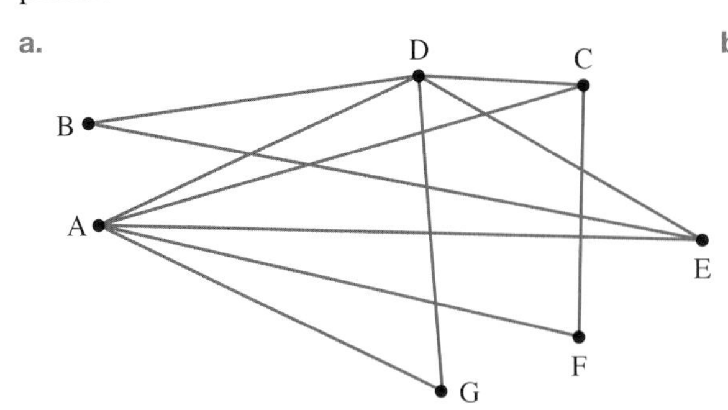

b.

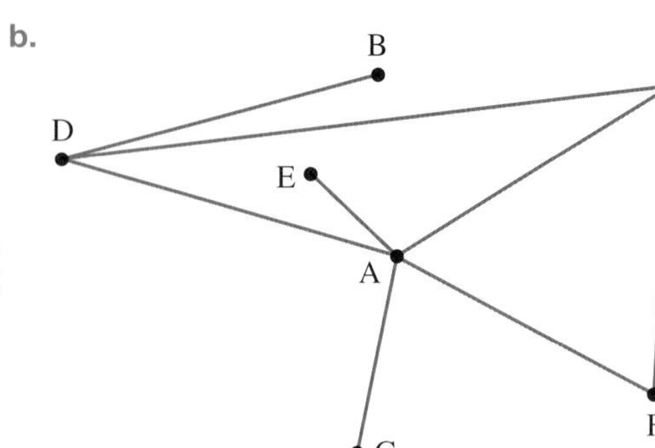

c.

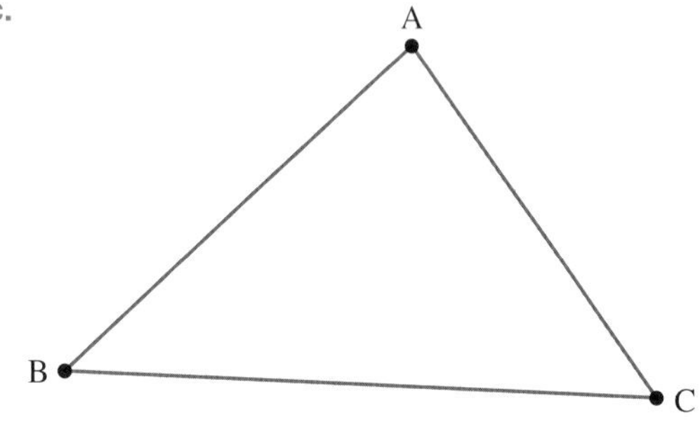

d.

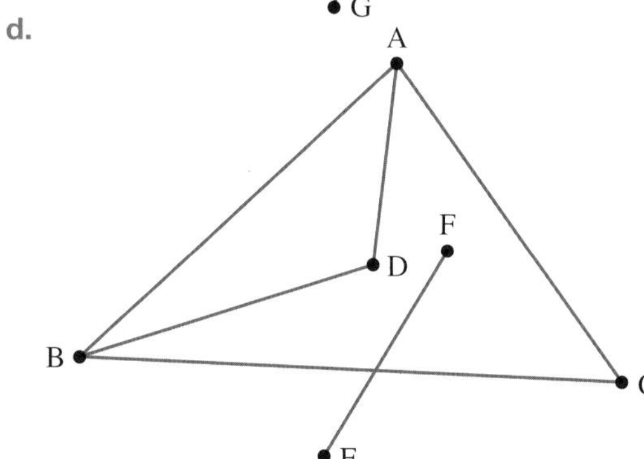

e.

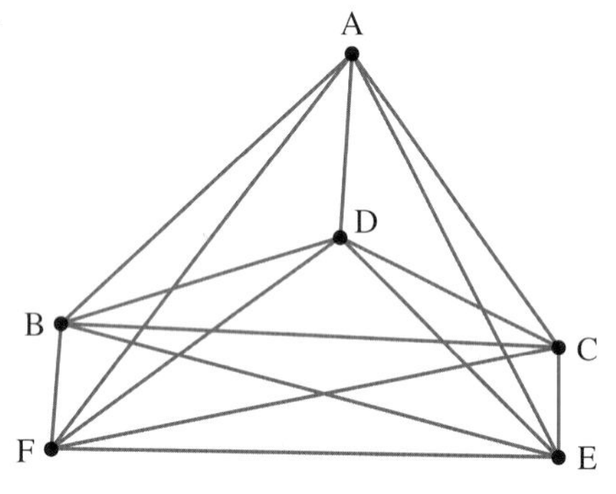

f.

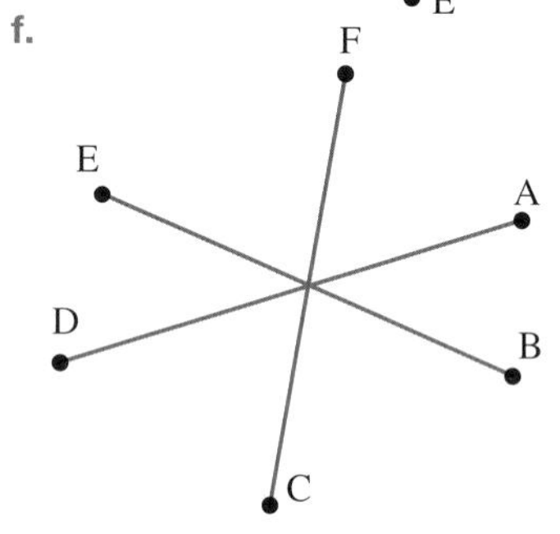

g.

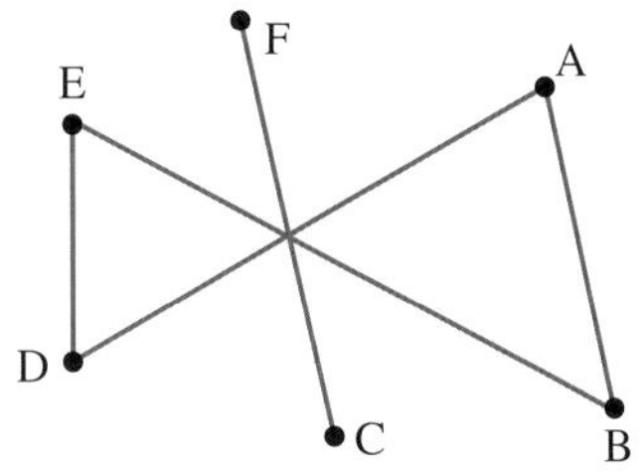

h.

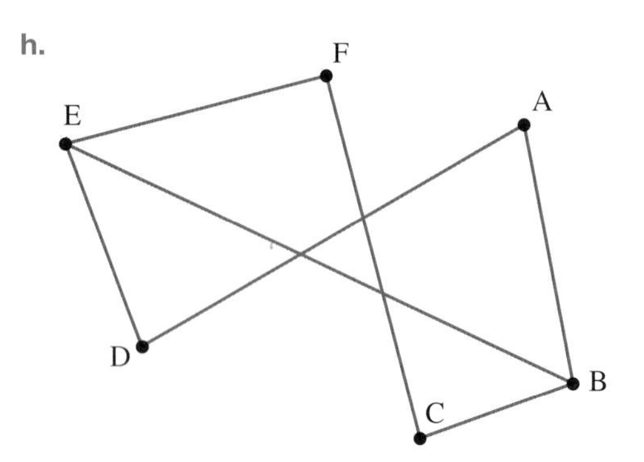

4. **The Bridges of Königsberg problem**
The European city of Königsberg (now called Kaliningrad) is set on the banks of the River Pregel. Seven bridges were arranged as shown to connect the two mainland parts of the city with two large islands.

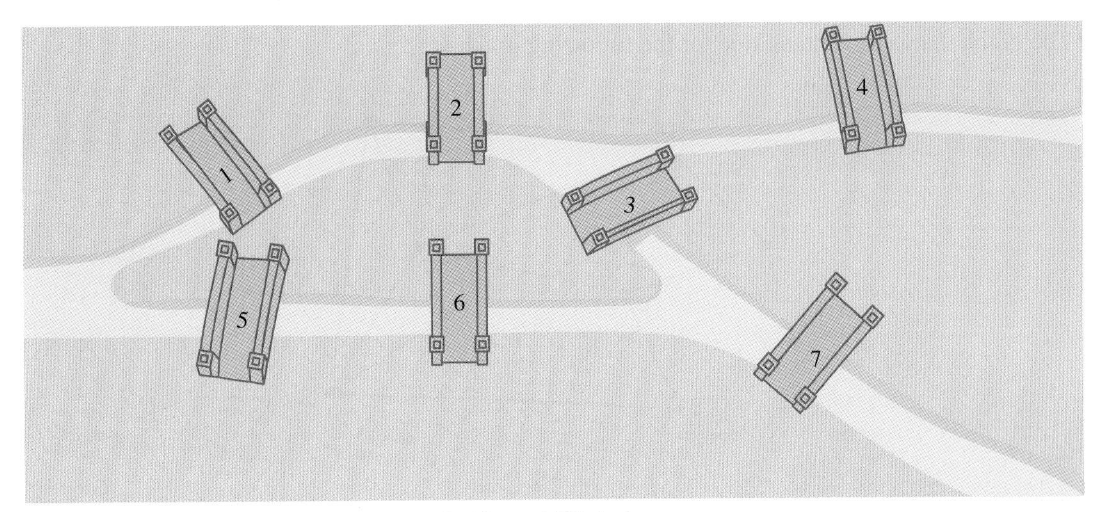

Bridges of Königsberg

For a long time the townspeople wondered it was possible to travel around the city in such a way that all seven bridges would only have to be crossed once each.

a. Determine if there is a way to cross all 7 bridges without crossing any bridge more than once.
b. Explain why or why not.

5. Consider the following graphs
i. Add the minimum number of edges in order to create a semi-Eulerian trail.
ii. State the semi-Eulerian trail created.

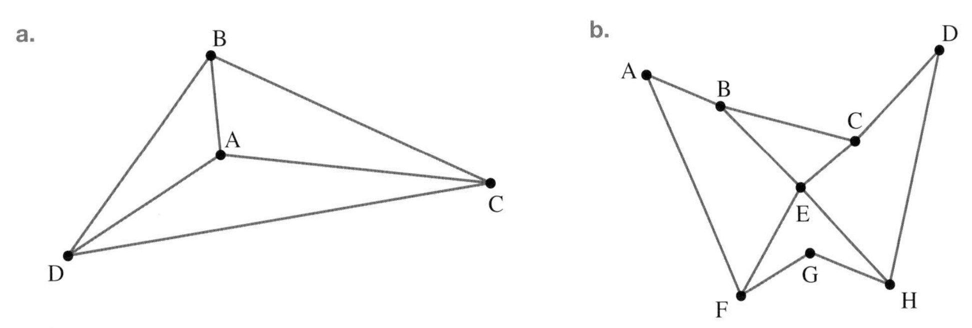

6. a. Determine the shortest distance from start to finish in the following graph.

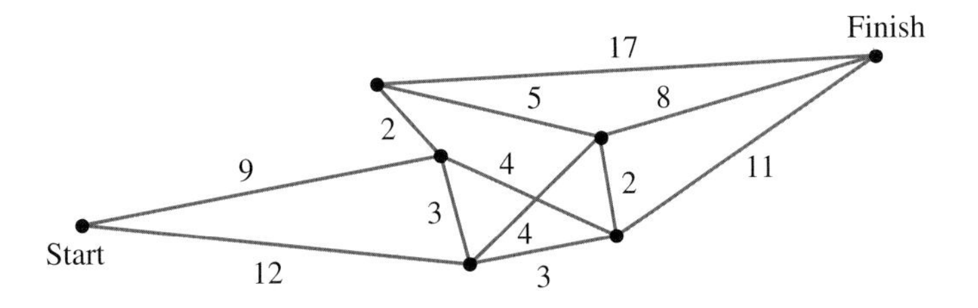

b. Determine the total length of the shortest semi-Hamiltonian cycle from start to finish.
c. Draw the minimum spanning tree for this graph.

Technology active: multiple choice

7. **MC** The minimum number of edges in a connected graph with eight vertices is

A. 4 **B.** 5 **C.** 6 **D.** 7 **E.** 8

8. **MC** The graph that is a spanning tree for the following graph is

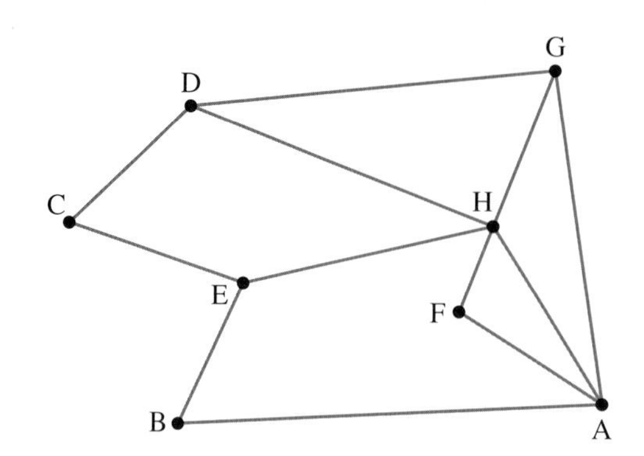

A.

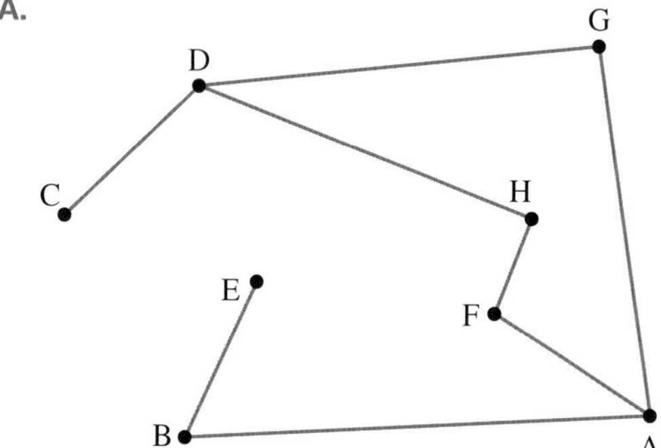

B.

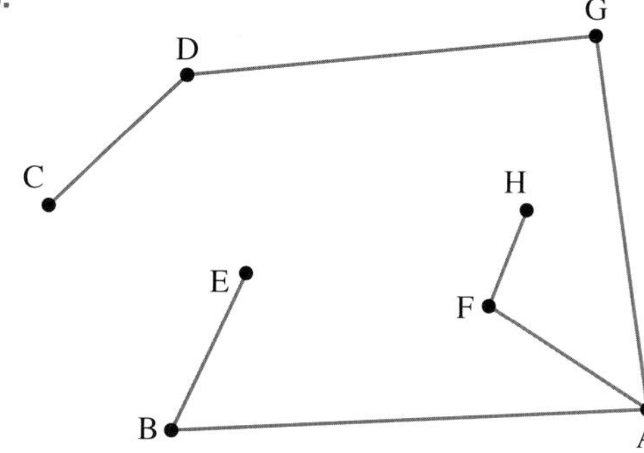

C.

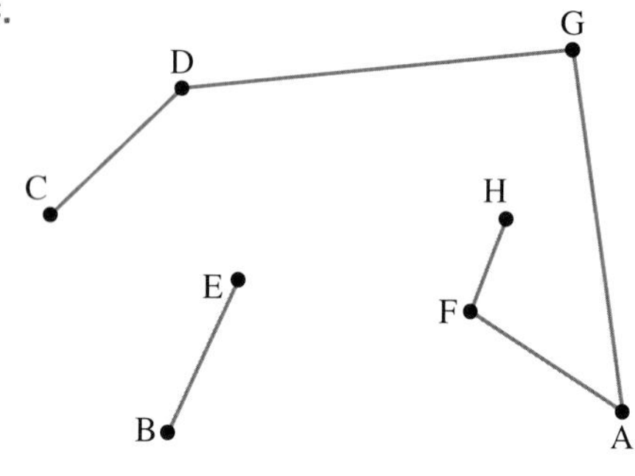

D.

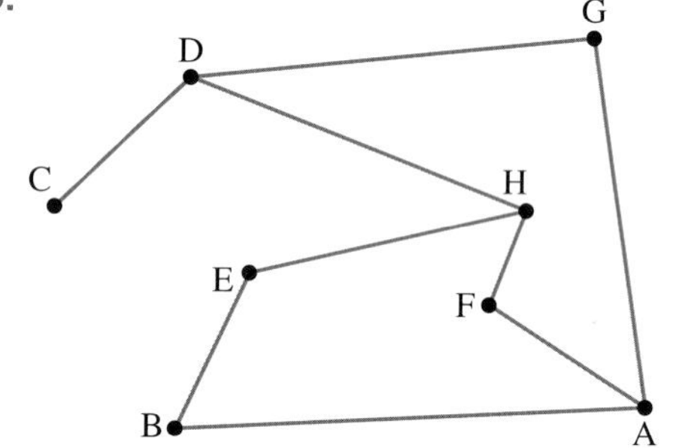

E.

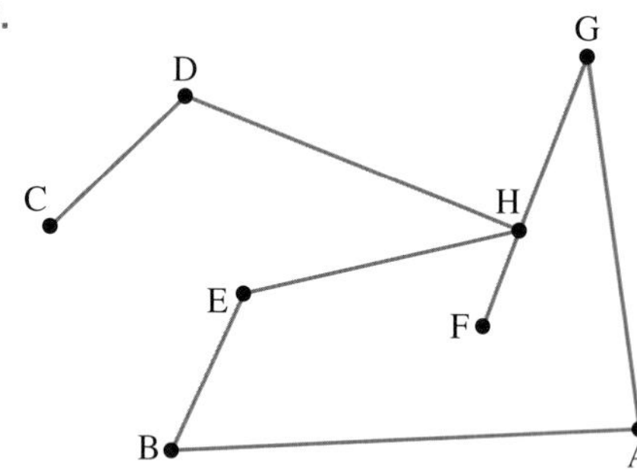

9. **MC** A connected graph with 9 vertices has 10 faces. The number of edges in the graph is

A. 15 **B.** 16 **C.** 17 **D.** 18 **E.** 19

10. **MC** The graph that will not have a semi-Eulerian trail is

A.

B.

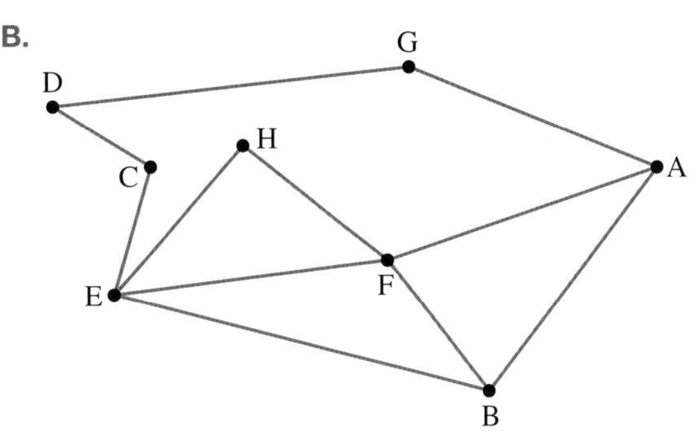

C.

D.

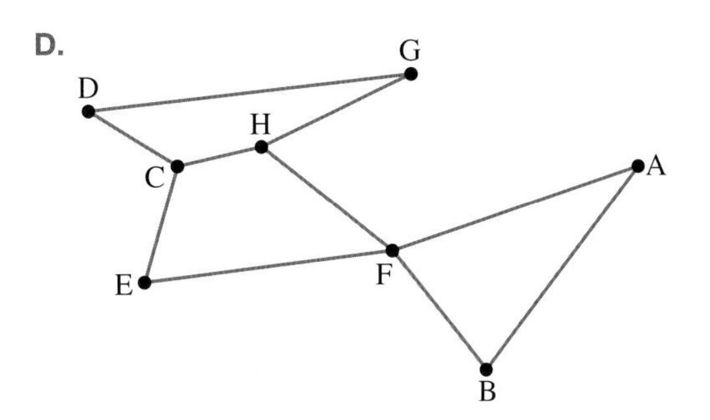

E.

11. **MC** The length of the minimum spanning tree of the following graph is

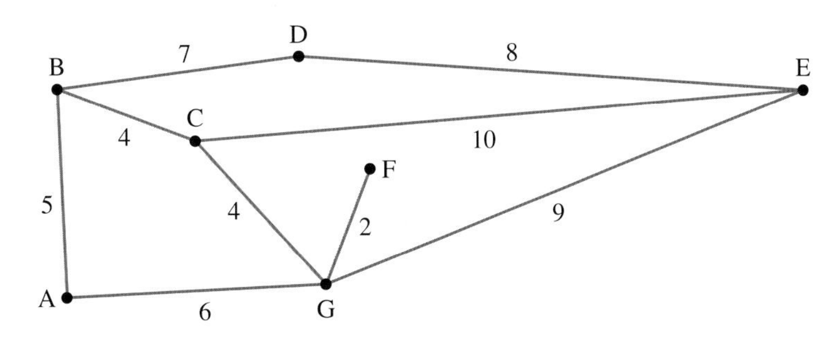

A. 33 **B.** 26 **C.** 34 **D.** 30 **E.** 32

12. **MC** An Eulerian trail can be created in the following graph by adding an edge between the vertices

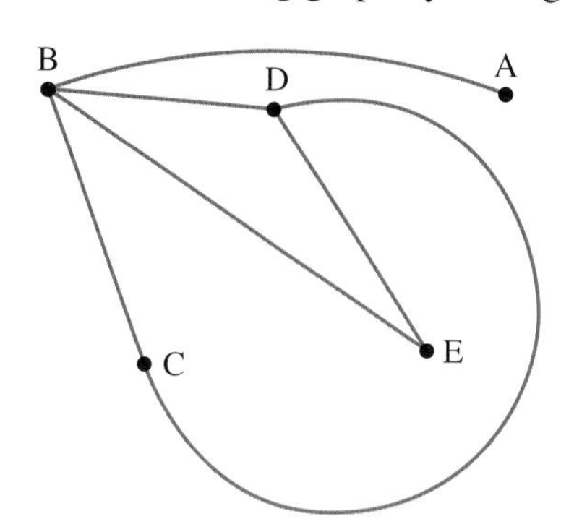

A. A and D B. A and B C. A and C D. B and C E. E and C

13. **MC** The adjacency matrix that represents the following graph is

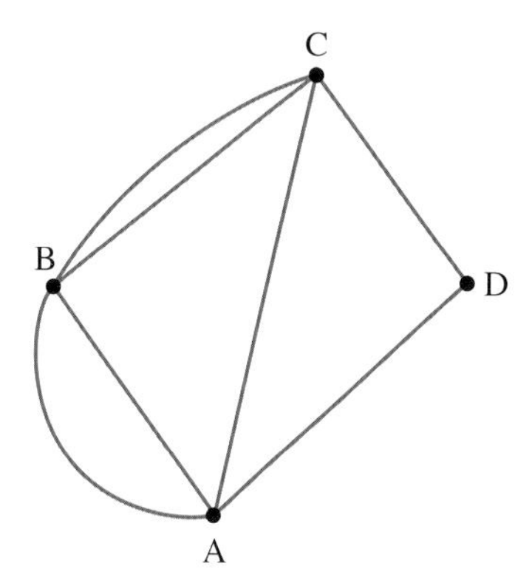

A. $\begin{bmatrix} 0 & 2 & 2 & 2 \\ 2 & 0 & 2 & 0 \\ 2 & 2 & 0 & 1 \\ 2 & 0 & 1 & 0 \end{bmatrix}$

B. $\begin{bmatrix} 0 & 1 & 1 & 0 \\ 1 & 0 & 1 & 0 \\ 1 & 1 & 0 & 1 \\ 1 & 0 & 1 & 0 \end{bmatrix}$

C. $\begin{bmatrix} 0 & 2 & 1 & 1 \\ 2 & 1 & 2 & 0 \\ 1 & 2 & 1 & 1 \\ 1 & 0 & 1 & 1 \end{bmatrix}$

D. $\begin{bmatrix} 1 & 2 & 2 & 1 \\ 2 & 1 & 2 & 0 \\ 2 & 2 & 1 & 1 \\ 1 & 0 & 1 & 1 \end{bmatrix}$

E. $\begin{bmatrix} 0 & 2 & 1 & 1 \\ 2 & 0 & 2 & 0 \\ 1 & 2 & 0 & 1 \\ 1 & 0 & 1 & 0 \end{bmatrix}$

14. **MC** The number of faces in the following planar graph is

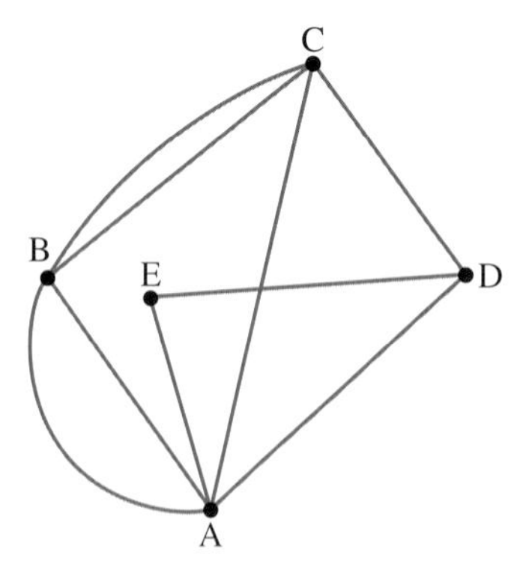

A. 6 B. 7 C. 8 D. 9 E. 10

15. MC A Hamiltonian cycle for the following graph is

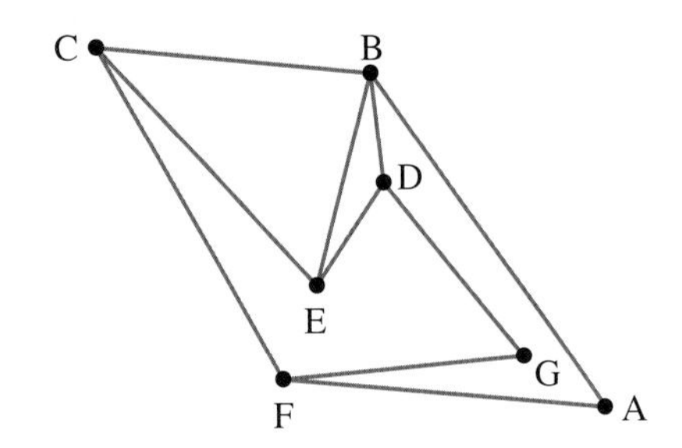

A. ABCEDGFA
B. ABDGFCEA
C. ABDGFCEDEBCFA
D. ABDGFCECFA
E. FCEBDGF

Questions 16 and 17 refer the following matrix.

$$\begin{bmatrix} 15 & 11 & 16 & 6 \\ 13 & 5 & 2 & 12 \\ 4 & 2 & 11 & 14 \\ 12 & 7 & 12 & 10 \end{bmatrix}$$

16. MC The matrix that is the row-reduced matrix for the previous matrix is

A. $\begin{bmatrix} 9 & 5 & 10 & 0 \\ 11 & 3 & 2 & 10 \\ 2 & 0 & 9 & 12 \\ 5 & 0 & 5 & 5 \end{bmatrix}$

B. $\begin{bmatrix} 11 & 6 & 14 & 0 \\ 9 & 3 & 0 & 6 \\ 0 & 0 & 9 & 8 \\ 8 & 5 & 10 & 4 \end{bmatrix}$

C. $\begin{bmatrix} 9 & 5 & 10 & 0 \\ 11 & 3 & 0 & 10 \\ 2 & 0 & 9 & 12 \\ 5 & 0 & 5 & 3 \end{bmatrix}$

D. $\begin{bmatrix} 9 & 5 & 10 & 6 \\ 11 & 3 & 2 & 10 \\ 2 & 2 & 9 & 12 \\ 7 & 5 & 5 & 5 \end{bmatrix}$

E. $\begin{bmatrix} 11 & 9 & 14 & 0 \\ 9 & 3 & 0 & 6 \\ 0 & 0 & 9 & 8 \\ 8 & 5 & 10 & 4 \end{bmatrix}$

17. MC After column reduction is performed to the matrix obtained in question 16, the resultant matrix is

A. $\begin{bmatrix} 7 & 5 & 10 & 0 \\ 9 & 3 & 0 & 10 \\ 0 & 0 & 9 & 12 \\ 3 & 0 & 5 & 3 \end{bmatrix}$

B. $\begin{bmatrix} 7 & 3 & 8 & 1 \\ 9 & 1 & 0 & 5 \\ 0 & 0 & 7 & 7 \\ 5 & 3 & 3 & 0 \end{bmatrix}$

C. $\begin{bmatrix} 7 & 5 & 8 & 0 \\ 9 & 3 & 0 & 10 \\ 0 & 0 & 7 & 12 \\ 3 & 0 & 3 & 5 \end{bmatrix}$

D. $\begin{bmatrix} 11 & 6 & 14 & 0 \\ 9 & 3 & 0 & 6 \\ 0 & 0 & 9 & 8 \\ 8 & 5 & 10 & 4 \end{bmatrix}$

E. $\begin{bmatrix} 11 & 3 & 8 & 0 \\ 9 & 1 & 0 & 6 \\ 0 & 0 & 7 & 8 \\ 8 & 3 & 3 & 4 \end{bmatrix}$

Technology active: extended response

18. The flying distances between the capital cities of Australian mainland states and territories are listed in the following table.

	Adelaide	Brisbane	Canberra	Darwin	Melbourne	Perth	Sydney
Adelaide		2055	1198	3051	732	2716	1415
Brisbane	2055		1246	3429	1671	4289	982
Canberra	1198	1246		4003	658	3741	309
Darwin	3051	3429	4003		3789	4049	4301
Melbourne	732	1671	658	3789		3456	873
Perth	2716	4363	3741	4049	3456		3972
Sydney	1415	982	309	4301	873	3972	

a. Draw a weighted graph to show this information.
b. If technical problems are preventing direct flights from Melbourne to Darwin and from Melbourne to Adelaide, determine the shortest way of flying from Melbourne to Darwin.
c. If no direct flights are available from Brisbane to Perth or from Brisbane to Adelaide, determine the shortest way of getting from Brisbane to Perth.
d. Draw the minimum spanning tree for the graph and state its total distance.

19. The following diagram shows the streets in a suburb of a city with a section of underground tunnels shown in black. Weightings indicate distances in metres. The tunnels are used for utilities such as electricity, gas, water and drainage.

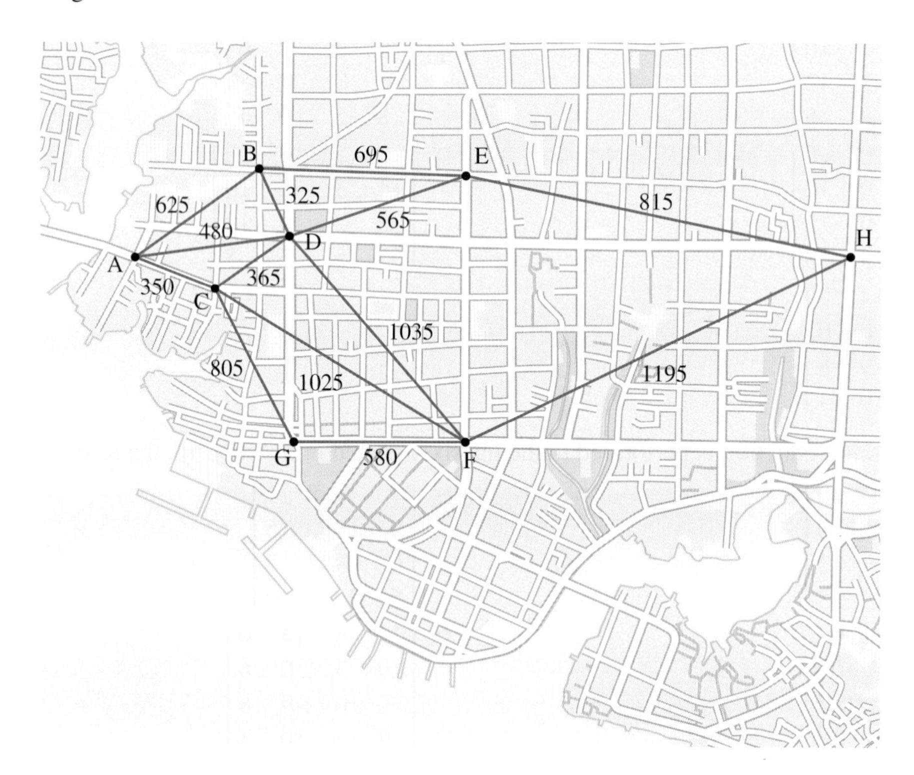

a. i. If the gas company wishes to run a pipeline that minimises its total length but reaches each vertex, determine the total length required.
ii. Draw a graph to show the gas lines.
b. If drainage pipes need to run from H to A, determine the shortest path they can follow and state how long this path will be in total.

c. A single line of cable for a computerised monitoring system needs to be placed so that it starts at D and reaches every vertex once. Determine the minimum length possible, and state the path it must follow.

d. A power line has to run from D so that it reaches every vertex at least once and finishes back at the start. Determine the path it must take to be a minimum.

20. A brochure for a national park includes a map showing the walking trails and available camping sites at the park.

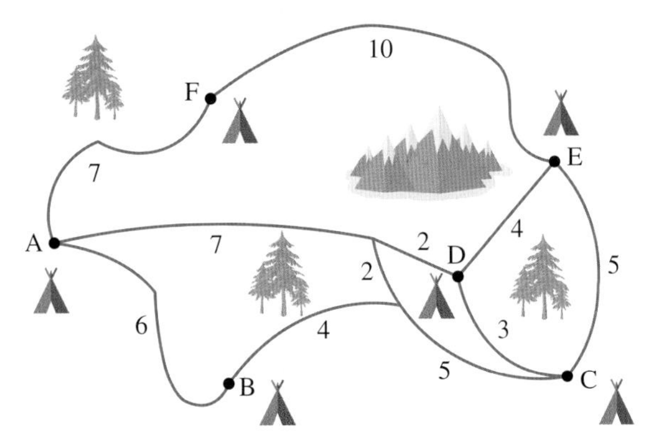

a. Draw a weighted graph to represent all the possible ways of travelling to the camp sites.

b. Draw the adjacency matrix for the graph.

c. Determine if it is possible to walk a route that travels along each edge exactly once. Explain your answer, and indicate the path if it is possible.

d. If the main entrance to the park is situated at A, determine the shortest way to travel to each campsite and return to A.

6.7 Exam questions

Question 1 (2 marks) TECH-FREE

Three houses need to be connected to three utilities: gas water and electricity. Determine if it is possible to connect each house to each utility without the lines intersecting. Justify your answer.

Question 2 (1 mark) TECH-ACTIVE

MC A complete graph with 7 vertices will have a total number of edges of

A. 7 **B.** 8 **C.** 14 **D.** 21

Question 3 (1 mark) TECH-ACTIVE

MC If G is a simple connected 3-regular planar graph where every face is bounded by exactly 3 edges, then the number of edges in G is

A. 4 **B.** 5 **C.** 6 **D.** 7 **E.** 8

Question 4 (5 marks) TECH-ACTIVE

A team of four — Barnie, Ruth, Shelley and Carlos have been selected to play four holes of golf for a social event for their workplace. Each member of the team will play one of the holes. Their scores for their previous rounds of golf for these four holes have been summarised in the following table.

	Hole 1	Hole 2	Hole 3	Hole 4
Barnie	5	7	5	9
Ruth	6	10	10	7
Shelley	7	5	3	8
Carlos	7	8	8	9

a. Perform a row reduction on the matrix formed from this table. **(2 marks)**

b. Perform a column reduction on the matrix from part **a**. **(1 mark)**

c. Apply the Hungarian algorithm if necessary. **(1 mark)**
d. State the optimal team for this event. **(1 mark)**

Question 5 (5 marks) TECH-ACTIVE

A cruise ship takes passengers around Tasmania between the seven locations marked on the map.

The sailing distances between locations are indicated in the table.

	Hobart	Bruny I.	Maria I.	Flinders I.	Devonport	Robbins I.	King I.
Hobart	–	65 km	145 km	595 km	625 km	–	–
Bruny I.	65 km	–	130 km	–	–	715 km	–
Maria I.	145 km	130 km	–	450 km	–	–	–
Flinders I.	595 km	–	450 km	–	330 km	405 km	465 km
Devonport	625 km	–	–	330 km	–	265 km	395 km
Robbins I.	–	715 km	–	405 km	265 km	–	120 km
King I.	–	–	–	465 km	395 km	120 km	–

a. Draw a weighted graph to represent all possible ways of travelling to the locations. **(2 marks)**
b. Determine the shortest route from Hobart to Robbins Island. **(1 mark)**
c. Determine the shortest way of travelling from Hobart to visit each location only once. **(1 mark)**
d. Determine the shortest way of sailing from King Island, visiting each location once and returning to King Island. **(1 mark)**

More exam questions are available online.

Answers

Topic 6 Graph theory

6.2 Introduction to graph theory

6.2 Exercise

1. 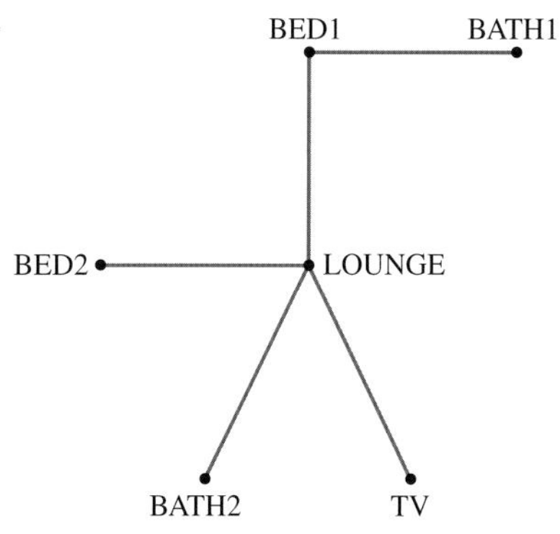

2. 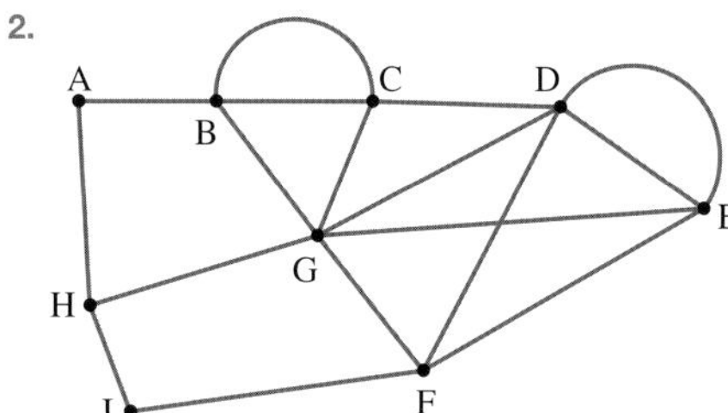

3. a. Edges = 7; Degree sum = 14
   b. Edges = 10; Degree sum = 20
4. a. Edges = 9; Degree sum = 18
   b. Edges = 9; Degree sum = 18
5. a. deg(A) = 5; deg(B) = 3; deg(C) = 4; deg(D) = 1; deg(E) = 1
   b. deg(A) = 0; deg(B) = 2; deg(C) = 2; deg(D) = 3; deg(E) = 3
   c. deg(A) = 4; deg(B) = 2; deg(C) = 2; deg(D) = 2; deg(E) = 4
   d. deg(A) = 1; deg(B) = 2; deg(C) = 1; deg(D) = 1; deg(E) = 3
6. a. Isomorphic
   b. Not isomorphic
   c. Isomorphic
   d. Isomorphic
7. a. $\begin{bmatrix} 0 & 1 & 1 & 1 & 0 \\ 1 & 0 & 1 & 0 & 1 \\ 1 & 1 & 0 & 0 & 0 \\ 1 & 0 & 0 & 0 & 1 \\ 0 & 1 & 0 & 1 & 0 \end{bmatrix}$
   b. $\begin{bmatrix} 0 & 0 & 1 & 1 & 2 & 0 \\ 0 & 0 & 1 & 0 & 0 & 1 \\ 1 & 1 & 1 & 0 & 0 & 0 \\ 1 & 0 & 0 & 0 & 0 & 1 \\ 2 & 0 & 0 & 0 & 0 & 0 \\ 0 & 1 & 0 & 1 & 0 & 1 \end{bmatrix}$
   c. $\begin{bmatrix} 0 & 0 & 1 & 1 & 2 & 0 \\ 0 & 0 & 1 & 0 & 0 & 1 \\ 1 & 1 & 0 & 0 & 0 & 0 \\ 1 & 0 & 0 & 0 & 0 & 3 \\ 2 & 0 & 0 & 0 & 0 & 0 \\ 0 & 1 & 0 & 3 & 0 & 0 \end{bmatrix}$
   d. $\begin{bmatrix} 0 & 1 & 1 & 1 & 0 & 0 \\ 1 & 0 & 1 & 1 & 1 & 1 \\ 1 & 1 & 0 & 0 & 1 & 0 \\ 1 & 1 & 0 & 0 & 0 & 1 \\ 0 & 1 & 1 & 0 & 0 & 4 \\ 0 & 1 & 0 & 1 & 4 & 0 \end{bmatrix}$
8. a. $A \rightarrow (B, C, D)$
   $B \rightarrow (A, C, E)$
   $C \rightarrow (A, B)$
   $D \rightarrow (A, E)$
   $E \rightarrow (B, D)$
   b. $A \rightarrow (C, D, E, E)$
   $B \rightarrow (C, F)$
   $C \rightarrow (A, B, C)$
   $D \rightarrow (A, F)$
   $E \rightarrow (A, A)$
   $F \rightarrow (B, D, F)$
   c. $A \rightarrow (C, D, E, E)$
   $B \rightarrow (C, F)$
   $C \rightarrow (A, B)$
   $D \rightarrow (A, F, F, F)$
   $E \rightarrow (A, A)$
   $F \rightarrow (B, D, D, D)$
   d. $A \rightarrow (B, C, D)$
   $B \rightarrow (A, C, D, E, F)$
   $C \rightarrow (A, B, E)$
   $D \rightarrow (A, B, F)$
   $E \rightarrow (B, C, F, F, F, F)$
   $F \rightarrow (B, D, E, E, E, E)$
9. a. 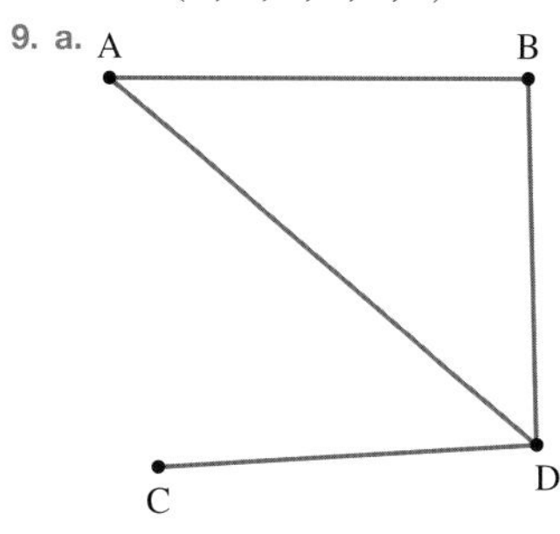

   b. 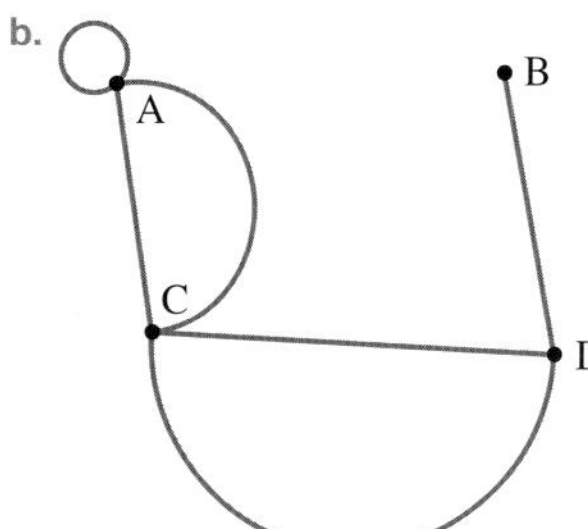

c.

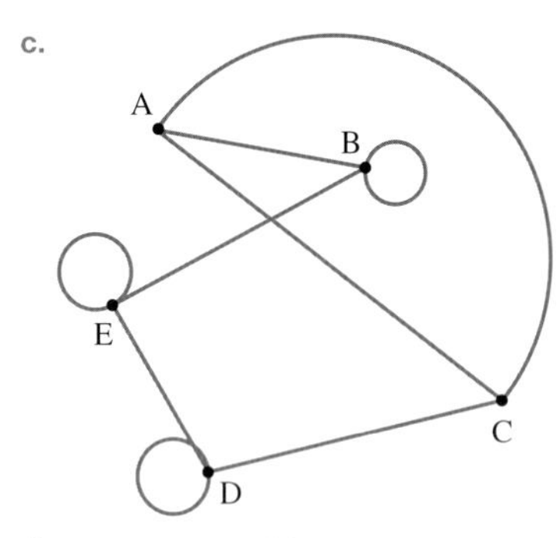

d.

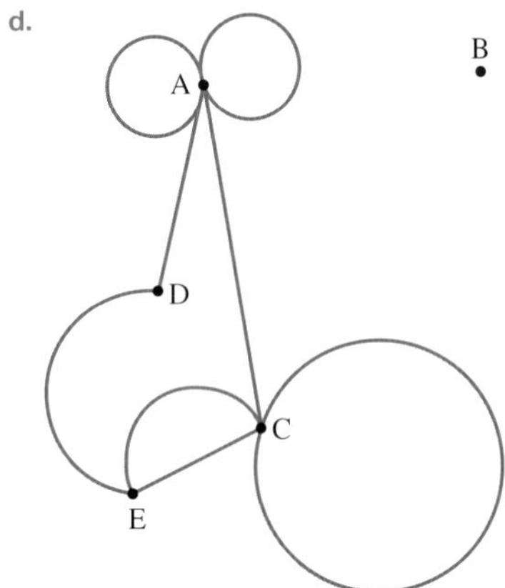

10. a. $\begin{bmatrix} 0 & 0 & 1 \\ 0 & 2 & 2 \\ 1 & 2 & 0 \end{bmatrix}$

b. $\begin{bmatrix} 2 & 1 & 0 & 0 \\ 1 & 0 & 1 & 2 \\ 0 & 1 & 0 & 1 \\ 0 & 2 & 1 & 0 \end{bmatrix}$

c. $\begin{bmatrix} 0 & 0 & 1 & 1 & 0 \\ 0 & 0 & 0 & 0 & 0 \\ 1 & 0 & 0 & 0 & 2 \\ 1 & 0 & 0 & 0 & 1 \\ 0 & 0 & 2 & 1 & 0 \end{bmatrix}$

d. $\begin{bmatrix} 0 & 0 & 0 & 1 & 0 \\ 0 & 0 & 0 & 1 & 0 \\ 0 & 0 & 0 & 0 & 1 \\ 1 & 1 & 0 & 0 & 0 \\ 0 & 0 & 1 & 0 & 1 \end{bmatrix}$

11.

Graph	Simple	Complete	Connected
Graph 1	Yes	No	Yes
Graph 2	Yes	No	Yes
Graph 3	Yes	No	Yes
Graph 4	No	No	Yes
Graph 5	No	Yes	Yes

12. **Graph 1** $\begin{bmatrix} 0 & 1 & 1 & 0 \\ 1 & 0 & 1 & 0 \\ 1 & 1 & 0 & 1 \\ 0 & 0 & 1 & 0 \end{bmatrix}$

Graph 2 $\begin{bmatrix} 0 & 1 & 0 & 0 & 1 & 0 \\ 1 & 0 & 0 & 1 & 0 & 0 \\ 0 & 0 & 0 & 0 & 1 & 1 \\ 0 & 1 & 0 & 0 & 0 & 1 \\ 1 & 0 & 1 & 0 & 0 & 0 \\ 0 & 0 & 1 & 1 & 0 & 0 \end{bmatrix}$

Graph 3 $\begin{bmatrix} 0 & 1 & 0 & 0 & 1 \\ 1 & 0 & 1 & 0 & 0 \\ 0 & 1 & 0 & 1 & 0 \\ 0 & 0 & 1 & 0 & 1 \\ 1 & 0 & 0 & 1 & 0 \end{bmatrix}$

Graph 4 $\begin{bmatrix} 0 & 2 & 1 & 0 & 0 & 0 \\ 2 & 0 & 0 & 0 & 1 & 0 \\ 1 & 0 & 0 & 1 & 0 & 0 \\ 0 & 0 & 1 & 0 & 0 & 0 \\ 0 & 1 & 0 & 0 & 0 & 1 \\ 0 & 0 & 0 & 0 & 1 & 0 \end{bmatrix}$

Graph 5 $\begin{bmatrix} 0 & 1 & 1 & 1 & 1 & 1 \\ 1 & 0 & 1 & 1 & 1 & 1 \\ 1 & 1 & 0 & 1 & 1 & 1 \\ 1 & 1 & 1 & 0 & 1 & 1 \\ 1 & 1 & 1 & 1 & 0 & 1 \\ 1 & 1 & 1 & 1 & 1 & 0 \end{bmatrix}$

13.

Vertices	Edges
2	1
3	3
4	6
5	10
6	15
n	$\dfrac{n(n-1)}{2}$

14. Answers will vary. Possible answers are shown.

a.

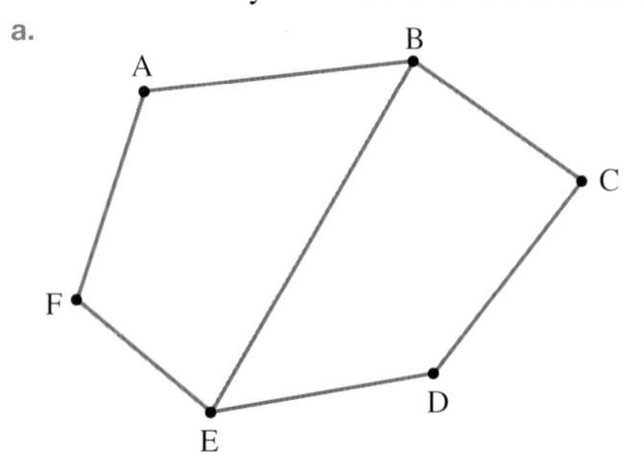

b.

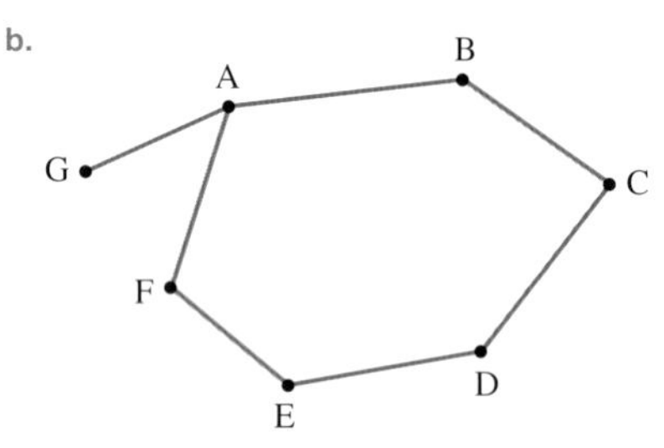

c.

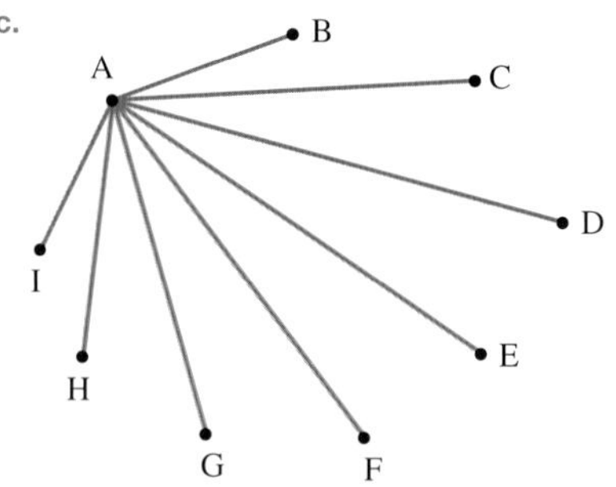

15.

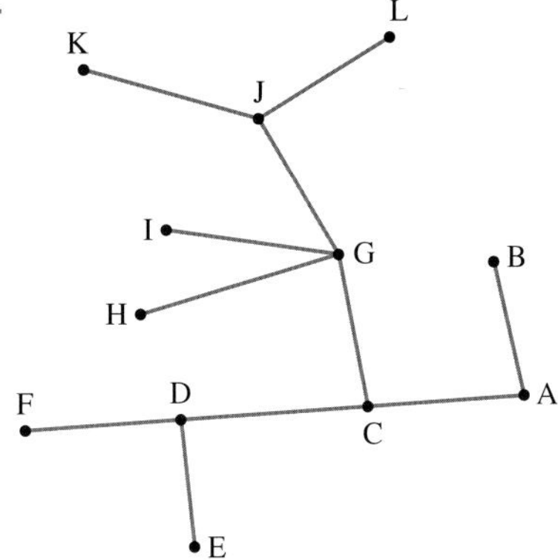

16. a.

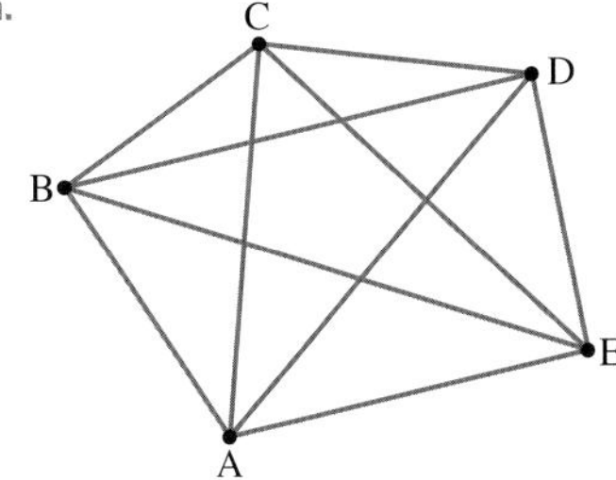

b. Complete graph

c. Total number of games played

17. a.

b. Huairou and Shunyi

c. Simple connected graph

18. a.

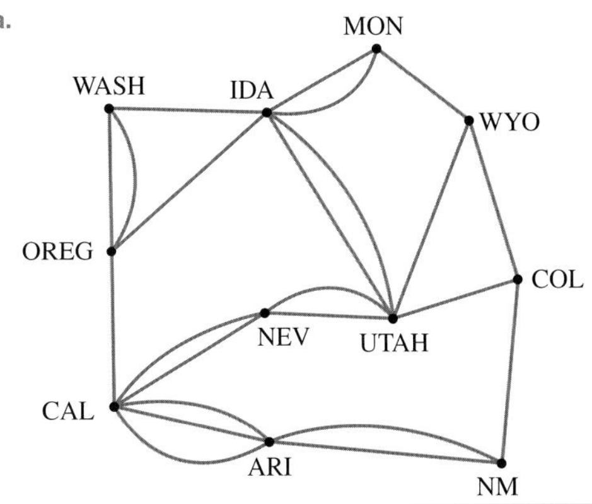

b. See the table at the bottom of the page.*

c. California, Idaho and Utah

d. Arizona

19. a.

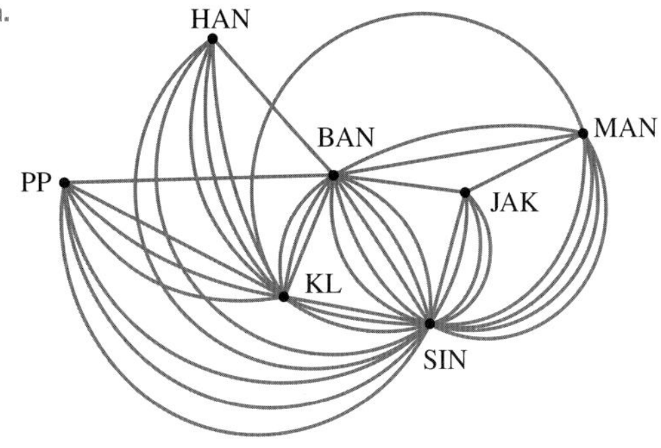

b. Directed, as it would be important to know the direction of the flight.

c. i. 10 ii. 7

20. a. Each carbon atom vertex has degree 4. Each hydrogen atom vertex has degree 1.

b.

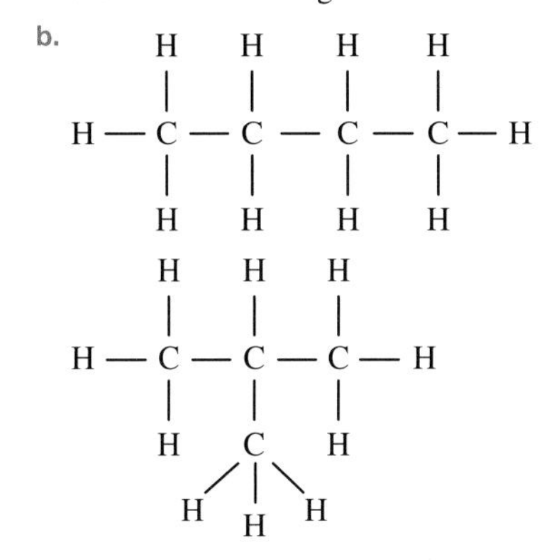

c. Suet is correct as the graphs are isomorphic.

6.2 Exam questions

Note: Mark allocations are available with the fully worked solutions online.

1. Yes, the complements are isomorphic.
2. Sample responses can be found in the worked solutions in the online resources.
3. A

*18. b.

	WASH	OREG	CAL	IDA	NEV	ARI	MON	UTAH	WYO	COL	NM
WASH	0	2	0	1	0	0	0	0	0	0	0
OREG	2	0	1	1	0	0	0	0	0	0	0
CAL	0	1	0	0	2	3	0	0	0	0	0
IDA	1	1	0	0	0	0	1	2	0	0	0
NEV	0	0	2	0	0	0	0	2	0	0	0
ARI	0	0	3	0	0	0	0	0	0	0	2
MON	0	0	0	2	0	0	0	0	1	0	0
UTAH	0	0	0	2	2	0	0	0	1	1	0
WYO	0	0	0	0	0	0	0	1	0	1	0
COL	0	0	0	0	0	0	0	1	1	0	1
NM	0	0	0	0	0	2	0	0	0	1	0

6.3 Planar graphs and Euler's formula

6.3 Exercise

1. a.

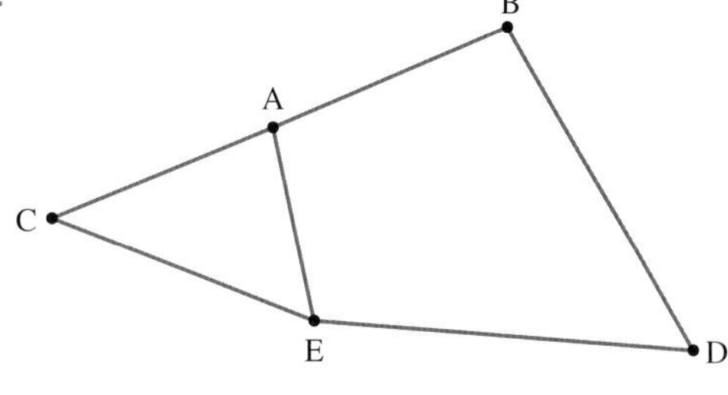

b.

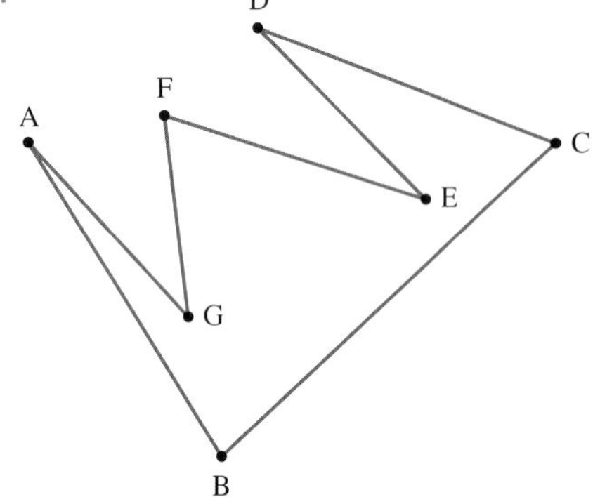

2. a. All of them. b. All of them.

3. a.

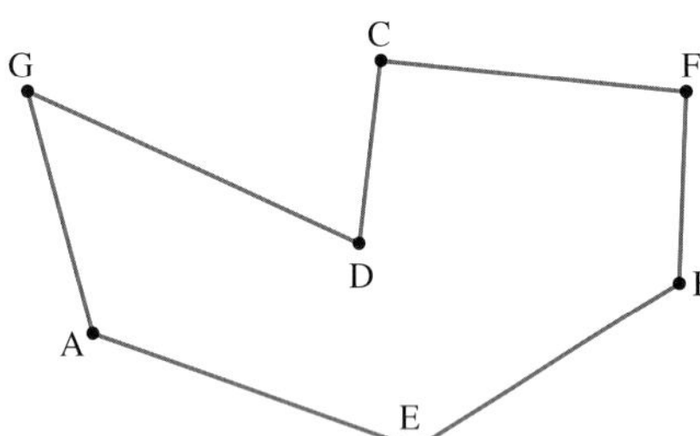

b.

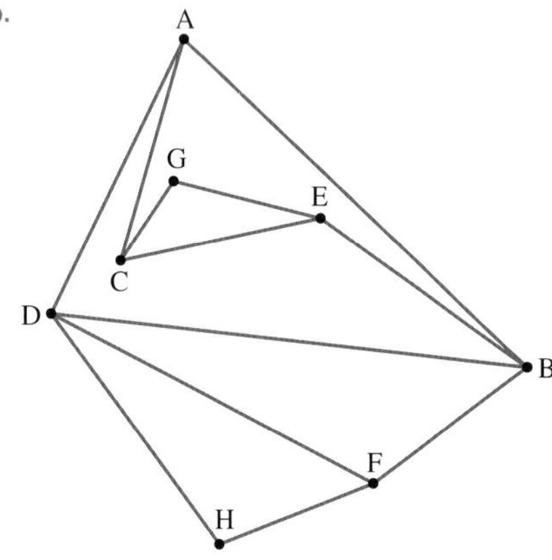

4. Graph 3

5. a. 4 b. 5

6. a. 6 b. 5

7. a. 3 b. 3 c. 2 d. 7

8. a.

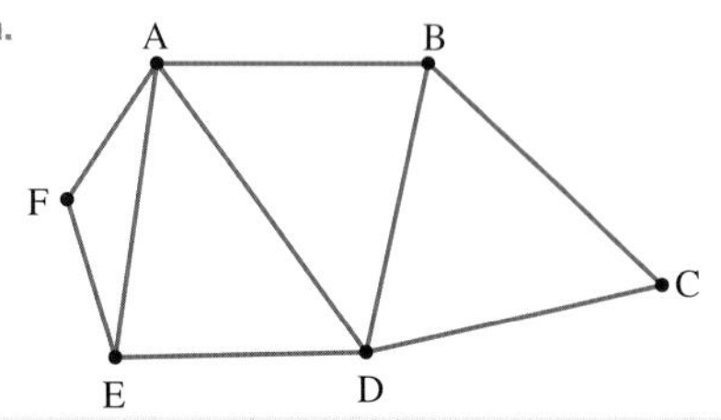

b.

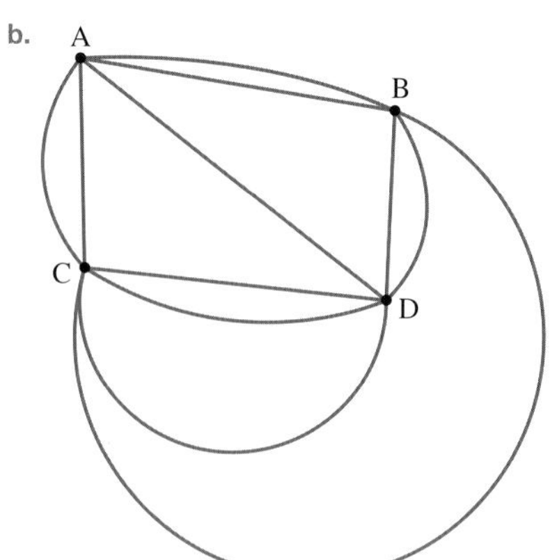

9. a.

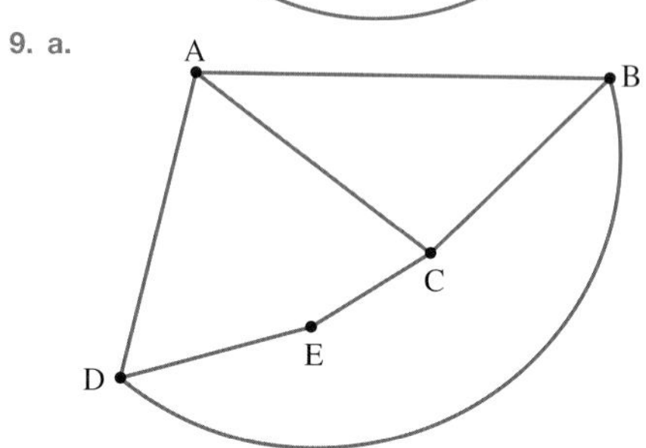

b.

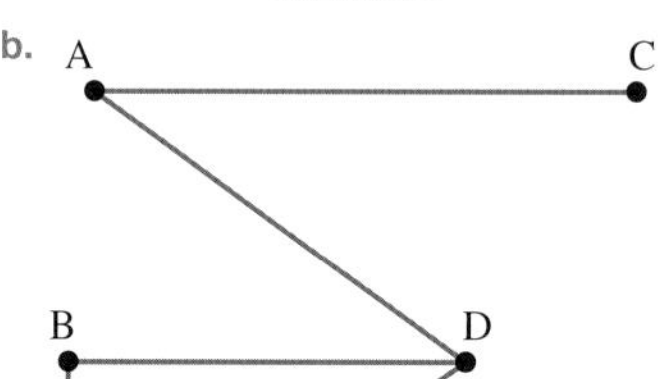

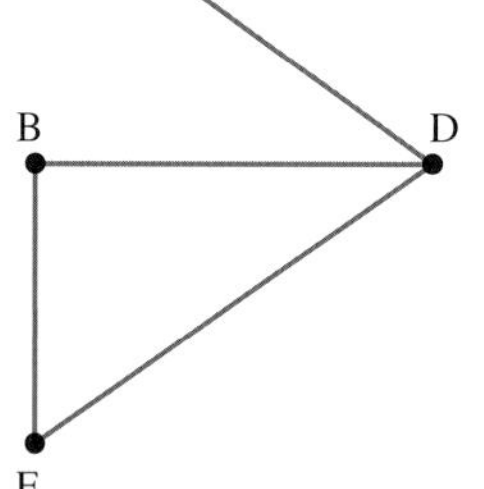

10. a. i. 3 ii. 2

b. i. 1 ii. 4

11. a.

Graph	Total edges	Total degrees
Graph 1	3	6
Graph 2	5	10
Graph 3	8	16
Graph 4	14	28

b. Total degrees $= 2 \times$ total edges

12. a.

Graph	Total vertices of even degree	Total vertices of odd degree
Graph 1	3	2
Graph 2	4	2
Graph 3	4	4
Graph 4	6	6

b. No clear pattern evident.

13. See the figure at the bottom of the page.*

14. a.

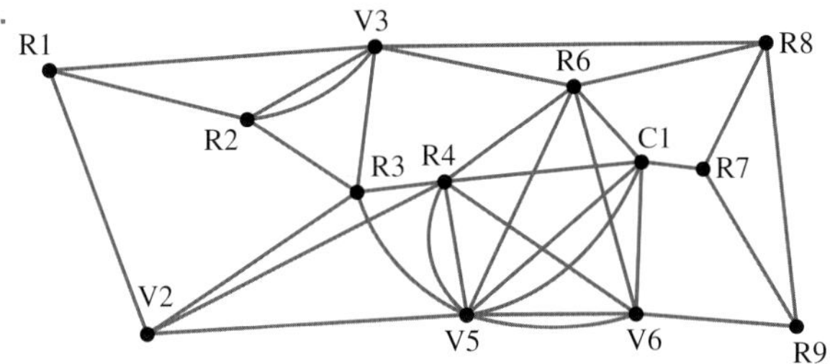

b. No

15. a.

	G	NB	M	C	A	R	L	W	J
G	0	1	0	0	0	0	0	0	0
NB	1	0	1	1	0	0	0	0	0
M	0	1	0	1	0	0	0	0	0
C	0	1	1	0	1	1	0	0	0
A	0	0	0	1	0	0	0	0	0
R	0	0	0	1	0	0	1	1	0
L	0	0	0	0	0	1	0	0	0
W	0	0	0	0	0	1	0	0	1
J	0	0	0	0	0	0	0	1	0

b. The sum of the rows represents the sum of the degree of the vertices, or twice the number of edges (connections).

16. a.

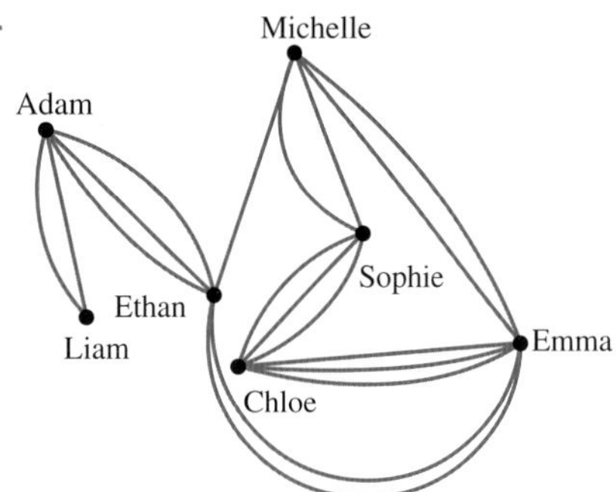

b. Sophie or Emma

c.

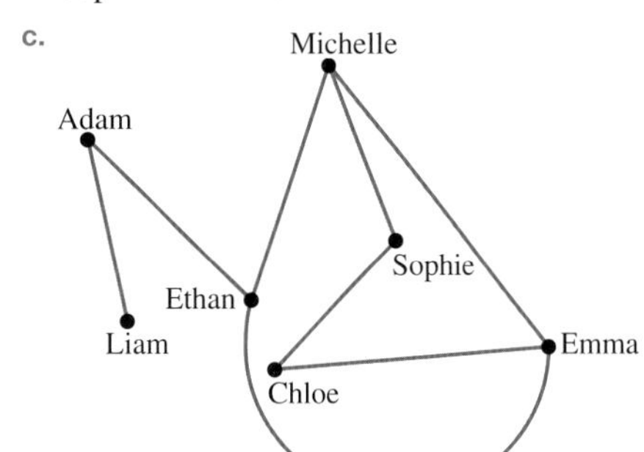

d. 4

*13.

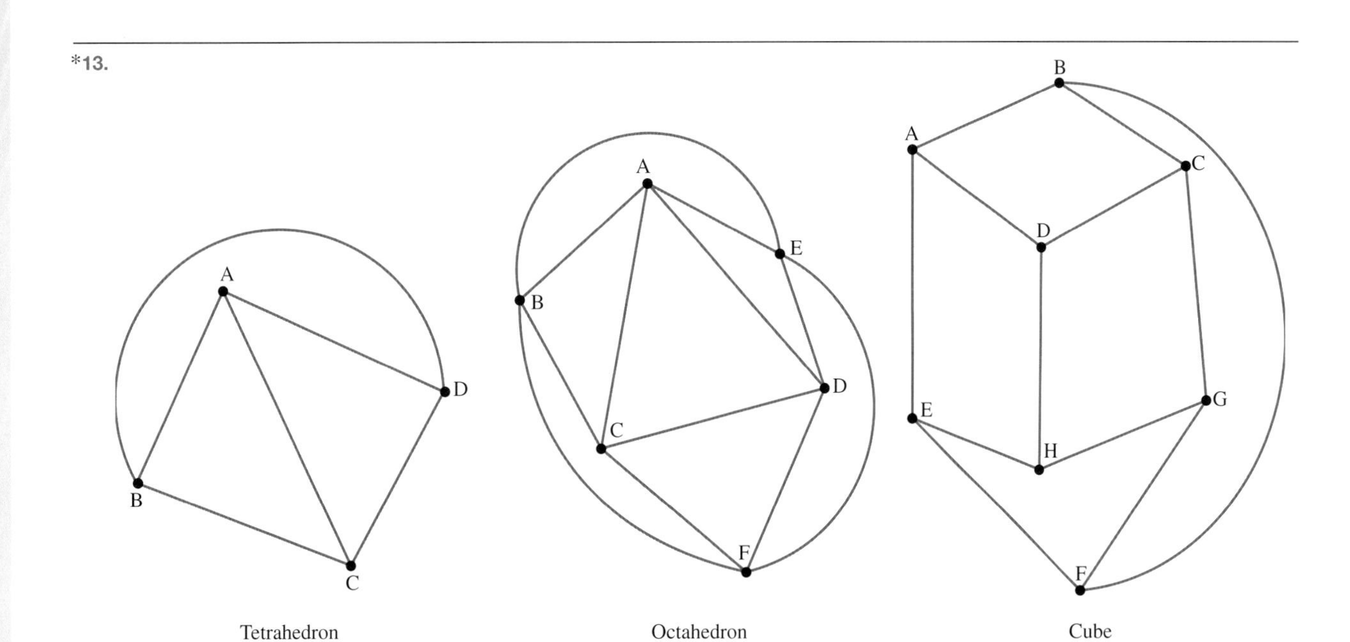

Tetrahedron

Octahedron

Cube

6.3 Exam questions

Note: Mark allocations are available with the fully worked solutions online.

1. 14
2. Either
3. Non-planar, since it contains the subgraph K_5.

6.4 Eulerian and Hamiltonian graphs

6.4 Exercise

1. Cycle: ABECA (others exist)
   Closed trail: BECDB (others exist)
2. Path: ABGFHDC (others exist)
   Cycle: DCGFHD (others exist)
   Closed trail: AEBGFHDCA (others exist)
3. a. Walk
   b. Walk, trail and path
   c. Walk, trail, path, cycle and closed trail
   d. Walk and trail
4. a. MCHIJGFAED
   b. AEDBLKMC
   c. MDEAFGJIHCM
   d. FMCHIJGF
5. a. Semi-Eulerian trail: AFEDBECAB; semi-Hamiltonian cycle: BDECAF
   b. Semi-Eulerian trail: GFBECGDAC; semi-Hamiltonian cycle: BECADGF
6. a. Eulerian trail: AIBAHGFCJBCDEGA; Hamiltonian cycle: none exist
   b. Eulerian trail: ABCDEFGHA (others exist); Hamiltonian cycle: HABCDEFGH (others exist)
7. a. Graphs **i**, **ii** and **iv**
   b. Graph **i**: ACDABDECB (others exist)
   Graph **ii**: CFBCEDBADCA (others exist)
   Graph **iv**: CFBCEDCADBAH (others exist)
8. a. Graphs **i** and **ii**
   b. Graph **i**: CEDABC
   Graph **ii**: CEDABGC
9. a. i. $\begin{bmatrix} 0 & 1 & 1 & 1 & 0 \\ 1 & 0 & 1 & 1 & 0 \\ 1 & 1 & 0 & 1 & 1 \\ 1 & 1 & 1 & 0 & 1 \\ 0 & 0 & 1 & 1 & 0 \end{bmatrix}$

   ii. $\begin{bmatrix} 0 & 1 & 1 & 1 & 0 & 0 \\ 1 & 0 & 1 & 1 & 0 & 1 \\ 1 & 1 & 0 & 1 & 1 & 1 \\ 1 & 1 & 1 & 0 & 1 & 0 \\ 0 & 0 & 1 & 1 & 0 & 0 \\ 0 & 1 & 1 & 0 & 0 & 0 \end{bmatrix}$

   iii. $\begin{bmatrix} 0 & 1 & 1 & 1 & 0 & 0 & 1 & 0 \\ 1 & 0 & 1 & 1 & 0 & 1 & 0 & 0 \\ 1 & 1 & 0 & 1 & 1 & 1 & 0 & 0 \\ 1 & 1 & 1 & 0 & 1 & 0 & 0 & 0 \\ 0 & 0 & 1 & 1 & 0 & 0 & 0 & 0 \\ 0 & 1 & 1 & 0 & 0 & 0 & 0 & 1 \\ 1 & 0 & 0 & 0 & 0 & 0 & 0 & 0 \\ 0 & 0 & 0 & 0 & 0 & 1 & 0 & 0 \end{bmatrix}$

   iv. $\begin{bmatrix} 0 & 1 & 1 & 1 & 0 & 0 & 1 \\ 1 & 0 & 1 & 1 & 0 & 1 & 0 \\ 1 & 1 & 0 & 1 & 1 & 1 & 0 \\ 1 & 1 & 1 & 0 & 1 & 0 & 0 \\ 0 & 0 & 1 & 1 & 0 & 0 & 0 \\ 0 & 1 & 1 & 0 & 0 & 0 & 0 \\ 1 & 0 & 0 & 0 & 0 & 0 & 0 \end{bmatrix}$

   b. The presence of semi-Eulerian trails and Eulerian trails can be identified by using the adjacency matrix to check the degree of the vertices. The presence of Hamiltonian and semi-Hamiltonian cycles can be identified by using the adjacency matrix to check the connections between vertices.
10. E
11. a. A or C b. B or D
12. a. G to C b. F to E
13. a.

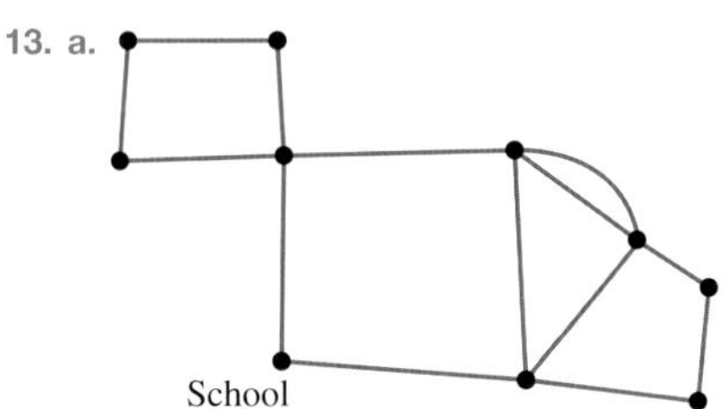

   b. Yes, because the degree of each intersection or corner point is an even number.
   c. Yes, because the degree of each remaining intersection or corner point is still an even number.
14. a.

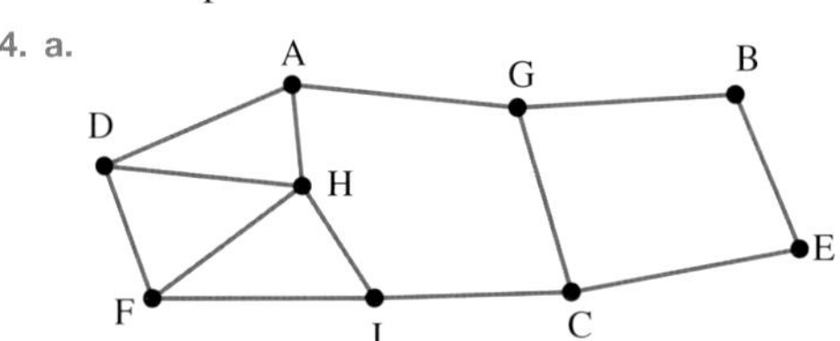

   b. 4
   c. i. ADHFICEBGA
   ii. AHDFICEBGA
   d. i. Two of the checkpoints have odd degrees.
   ii. H and C
15. a.

	Hamiltonian cycle
1.	ABCDA
2.	ABDCA
3.	ACBDA
4.	ACDBA
5.	ADBCA
6.	ADCBA

   b. Yes, commencing on vertices other than A.
16. a. B, C, D, F or G
   b. B or C
   c. None possible
   d. D or E
   e. D to E

6.4 Exam questions

Note: Mark allocations are available with the fully worked solutions online.

1. Odd positive integers
2. a. i. 3
   ii. 12
   b. There are $\frac{(n-1)!}{2}$ Hamiltonian cycles in the complete graph K_n.
3. D

6.5 Weighted graphs and trees

6.5 Exercise

1. 21
2. 20.78
3. a.

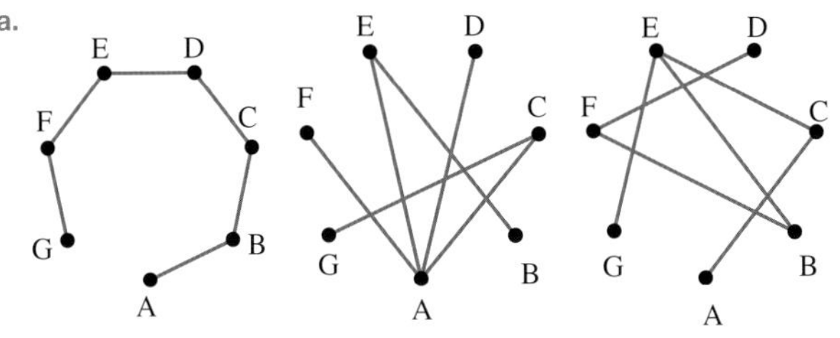

Other possibilities exist.

b.

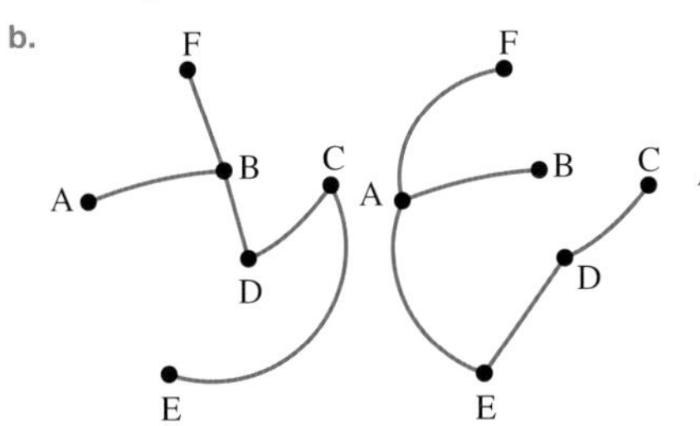

Other possibilities exist.

4. ABGEDCA or ACDEGBA (length 66)
5. a. i.

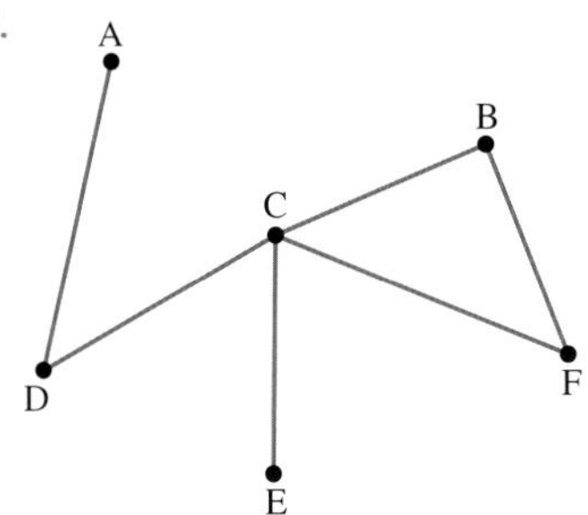

ii. ADCBFCE or ADCFBCE

b. i.

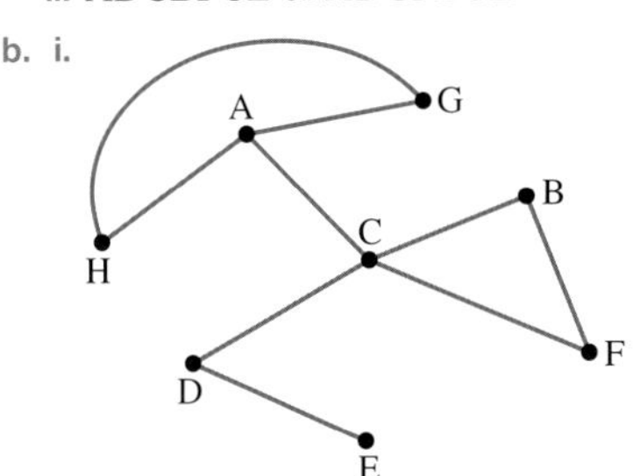

ii. AHGACBFCDE or similar

c. i.

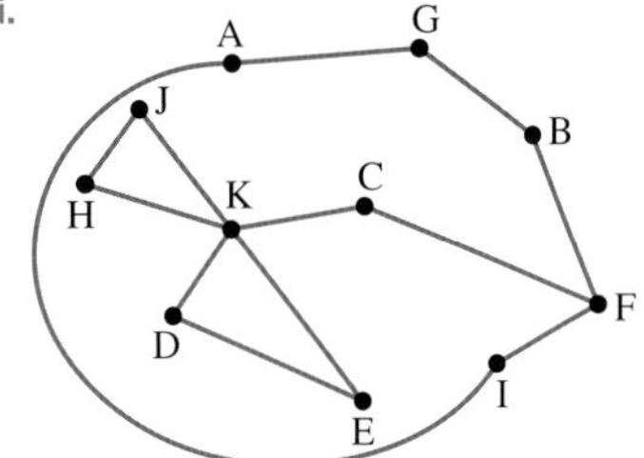

ii. KDEKHJKCFIAGBF or similar

d. i.

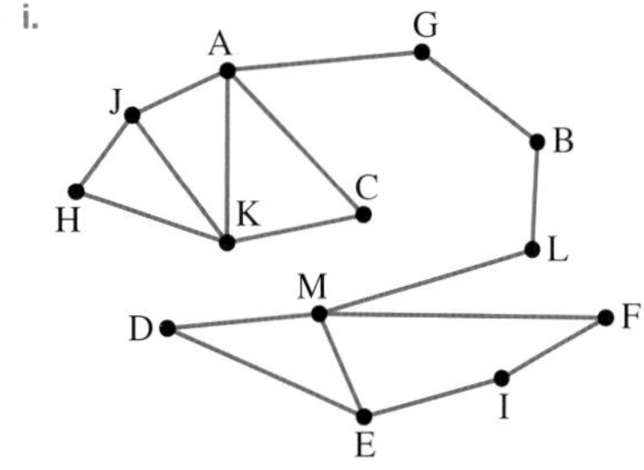

ii. EDMEIFMLBGACKHJKA or similar

6.

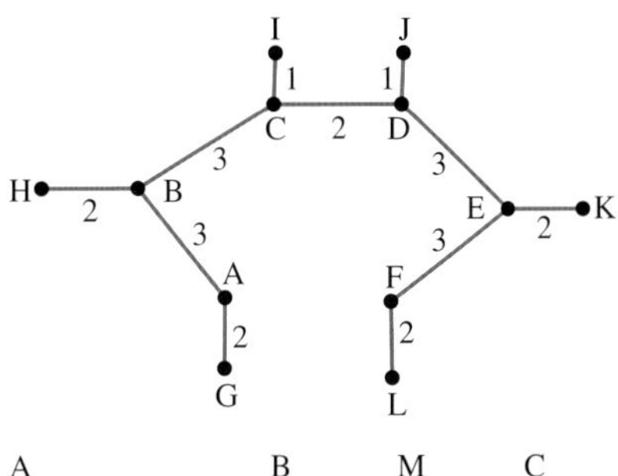

7.

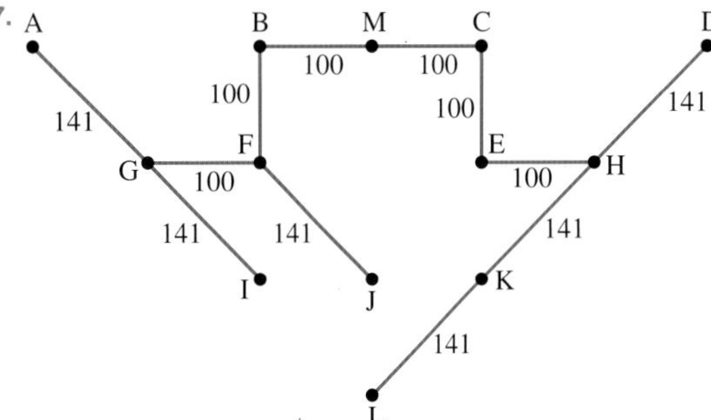

8. a.

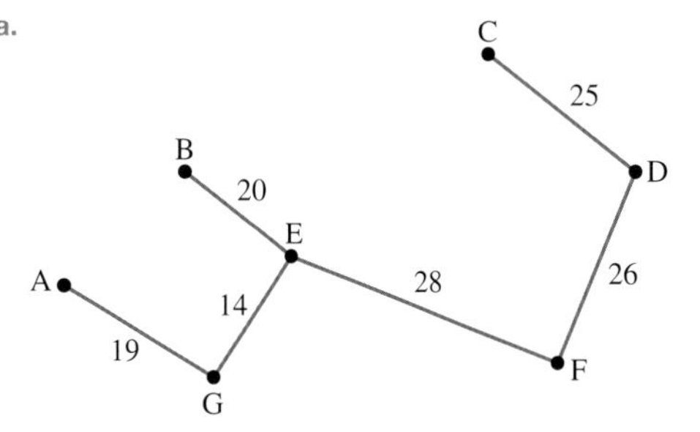

b.

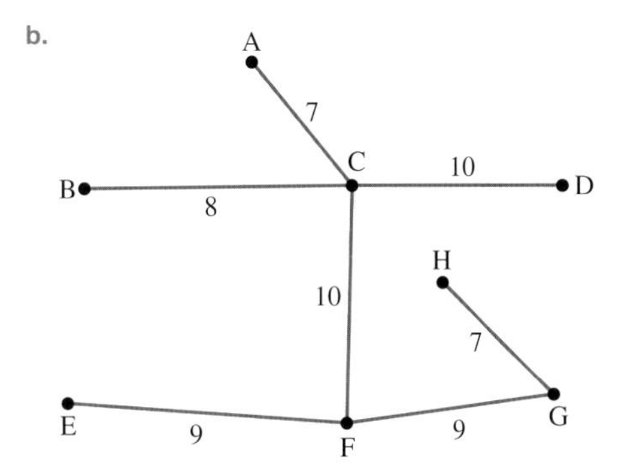

9. Step 1

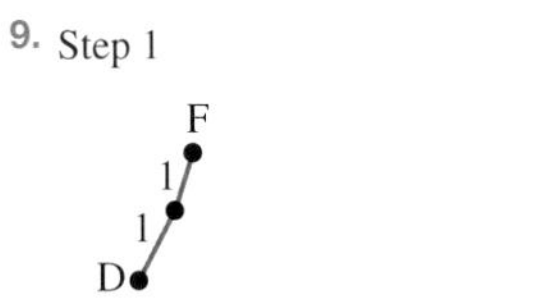

Step 2

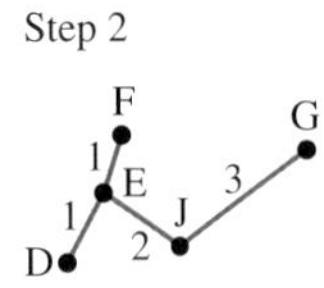

Step 3

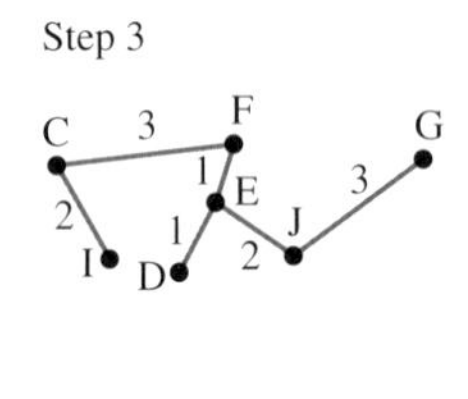

Step 4

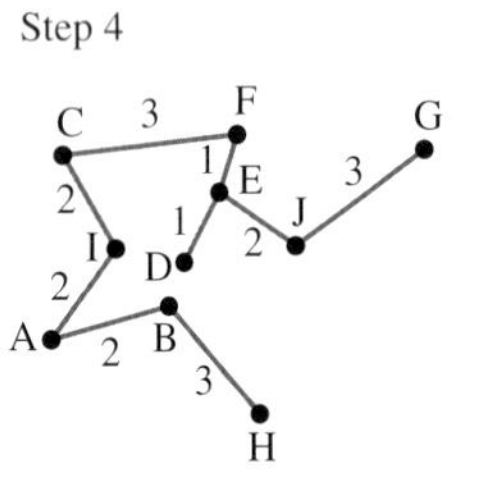

10.

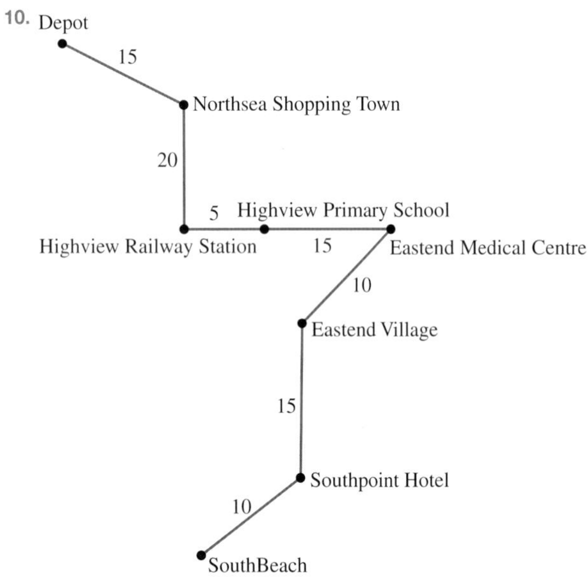

11. a. Longest: IFEDCBAHG (or similar variation of the same values)

Shortest: IAHGFEDCB (or similar variation of the same values)

b.

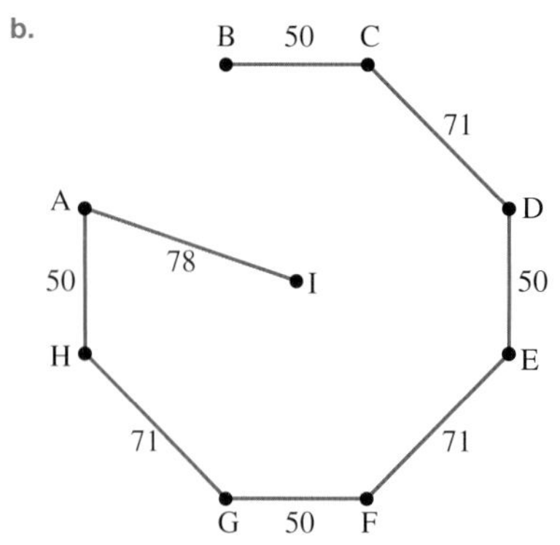

12. a. FDCGBAE (other solutions exist)

b. FDCBAEG (other solutions exist)

13. a. ADEG

b. BHG

c. EGFCDABHE

14. a.

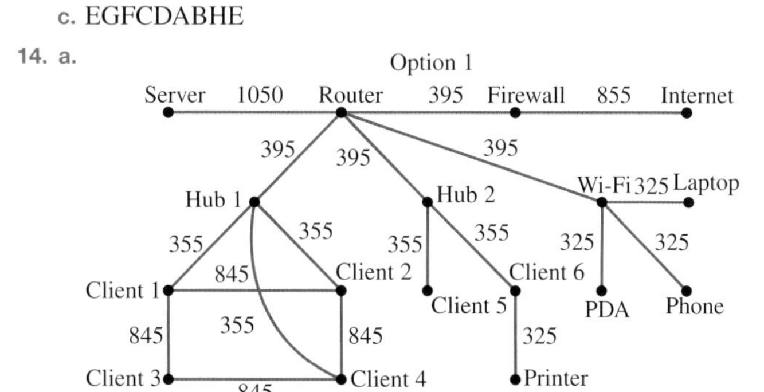

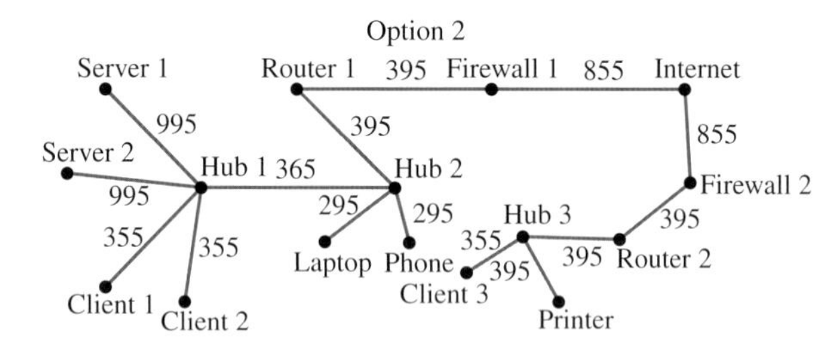

b. Option 2

15. a.

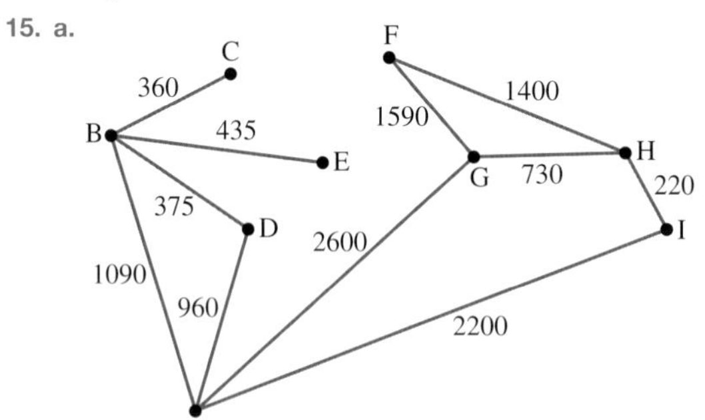

b. No; C and E are both only reachable from B.

c. i. 12 025 **ii.** 12 025

d.

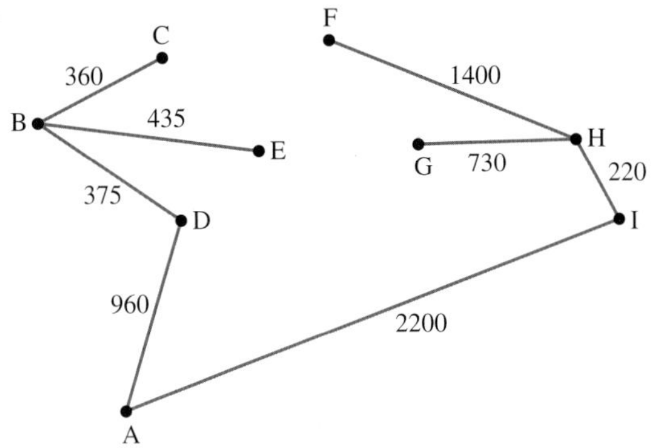

16. a.

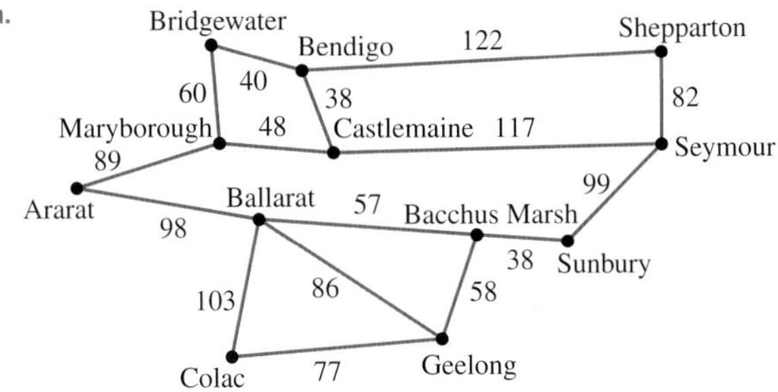

b. 723 km

c.

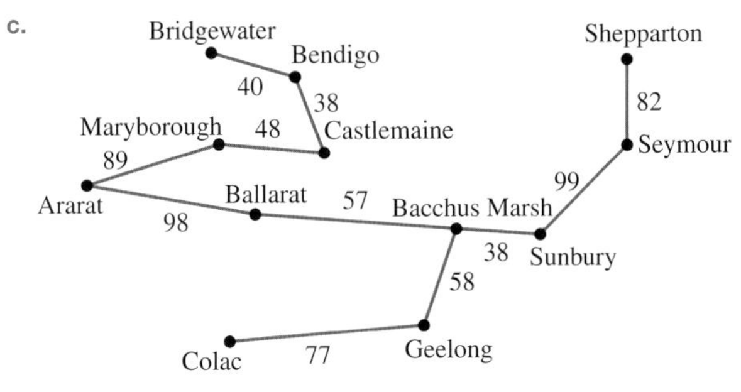

d. 859 km

6.5 Exam questions

Note: Mark allocations are available with the fully worked solutions online.

1. 63 km
2. a. It is a bridge.
   b. It is a loop.
3. a. $(n - 1)$
   b. $3 + 4 + 5 + 6 + \ldots + n + (n + 1)$

6.6 Bipartite graphs and the Hungarian algorithm

6.6 Exercise

1. a. Bipartite
   b. Not bipartite
   c. Bipartite
   d. Not bipartite
2. a. i.

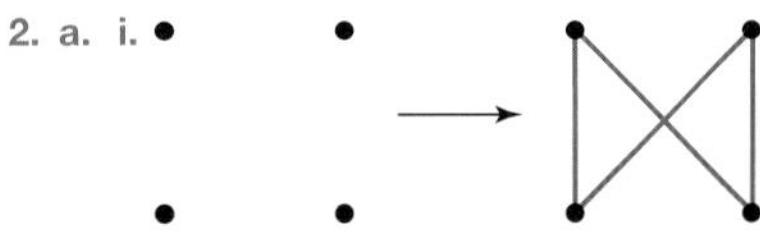

   ii. Planar

   b. i. See the figure at the bottom of the page.*

   ii. Non-planar

   c. i. See the figure at the bottom of the page.*

   ii. Planar

   d. i. See the figure at the bottom of the page.*

   ii. Planar
3. a. 9 b. 35 c. mn d. $3x^2$
4. a.

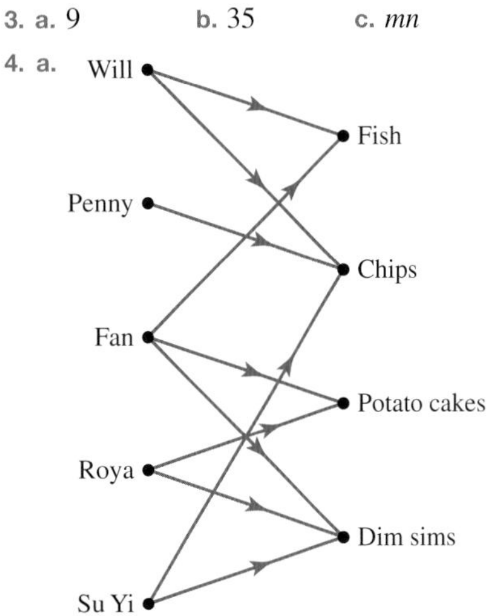

   b. False, False, False, True
5. a. The visitors can be considered the supply vertices and the countries the demand vertices. No two vertices in each of these groups share an edge, so a bipartite graph is suitable.

*2. b. i.

*2. c. i.

*2. d. i.

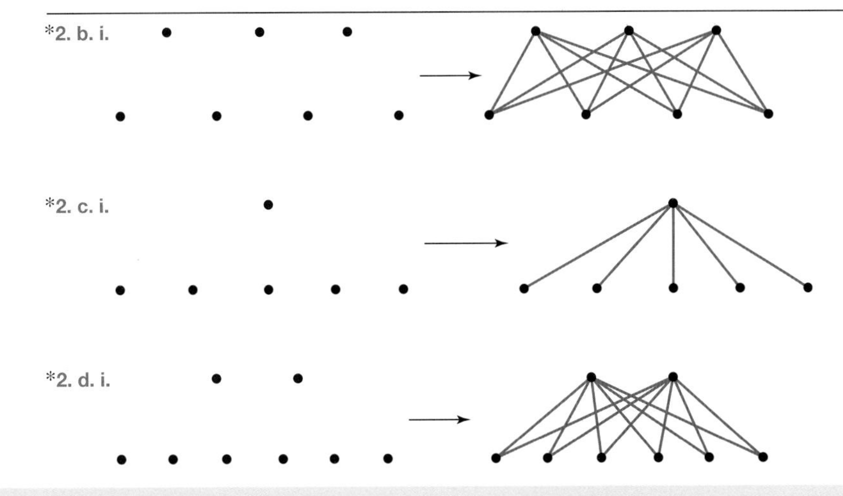

b.

Sally, Joe, Mike, Genevieve — France, Italy, Germany, Spain, Ireland

c. 3

6. a. False
   b. True
   c. False
   d. True
   e. False

7. $S1 \rightarrow D$
   $S2 \rightarrow A$
   $S3 \rightarrow B$
   $S4 \rightarrow C$
   61 hours

8. $R1 \rightarrow C2$
   $R2 \rightarrow C1$
   $R3 \rightarrow C3$
   $2 + 1 + 1 = 4$ hours

9. $R1 \rightarrow C4$
   $R2 \rightarrow C2$
   $R3 \rightarrow C1$
   $R4 \rightarrow C3$
   $4 + 5 + 6 + 4 = 19$ hours

10. 28 hours

11. 44 km

12. $140

13. C

14. a. 54
    b. 90

15. a. $\begin{bmatrix} 0 & 10 & 20 & 30 \\ 40 & 0 & 10 & 40 \\ 50 & 10 & 50 & 0 \\ 0 & 70 & 40 & 60 \end{bmatrix}$

$\begin{bmatrix} 0 & 10 & 10 & 30 \\ 40 & 0 & 0 & 40 \\ 50 & 10 & 40 & 0 \\ 0 & 70 & 30 & 60 \end{bmatrix}$

b. $\begin{bmatrix} 0 & 0 & 0 & 30 \\ 50 & 0 & 0 & 50 \\ 50 & 0 & 30 & 0 \\ 0 & 60 & 20 & 60 \end{bmatrix}$

c. W1, W2, W3, W4 — T1, T2, T3, T4 or W1, W2, W3, W4 — T1, T2, T3, T4

d. 180 hours

6.6 Exam questions

Note: Mark allocations are available with the fully worked solutions online.

1. a.

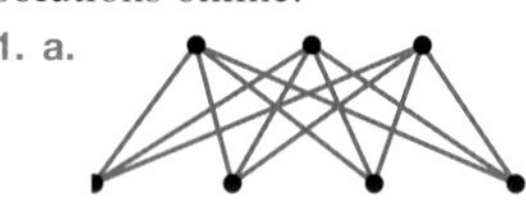

   b. No
   c. No

2. a. $K_{m,n}$, where m, n are even positive integers
   b. $K_{m,n}$, where $m = 2$ and n is an odd positive integer

3. One possible allocation
   $M \rightarrow L1$
   $Je \rightarrow L3$
   $P \rightarrow L2$
   $Ja \rightarrow L4$
   11 minutes

6.7 Review

6.7 Exercise

Technology free: short answer

1. a. i. Planar ii. Planar
   iii. Planar iv. Planar

   b. i.

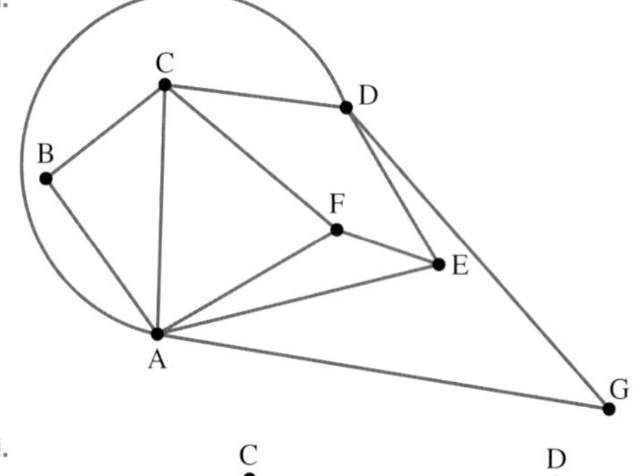

   ii.

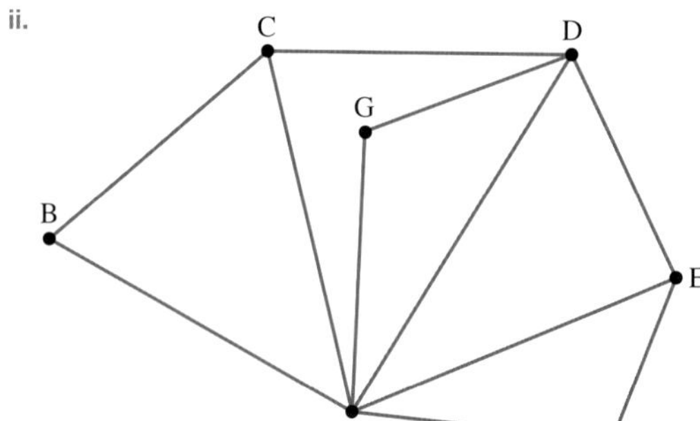

iii.

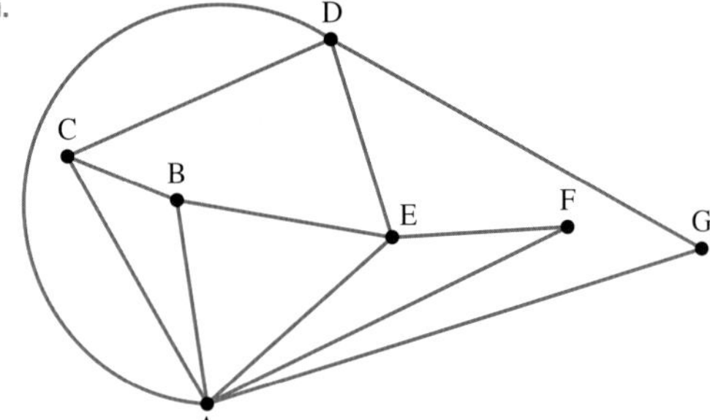

iv.

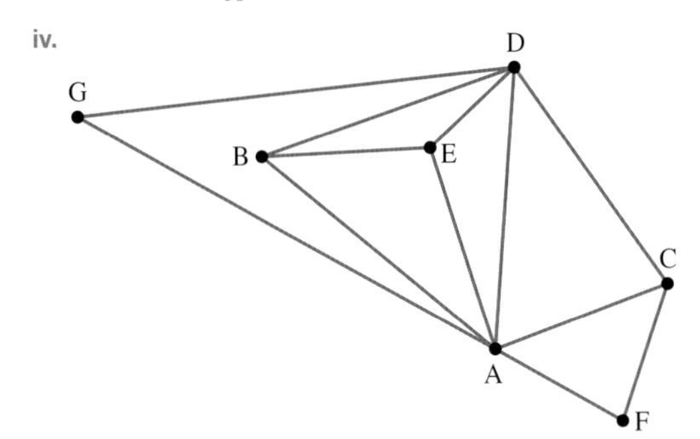

2. a. $\begin{bmatrix} 1 & 1 & 0 & 1 \\ 1 & 0 & 3 & 0 \\ 0 & 3 & 1 & 1 \\ 1 & 0 & 1 & 0 \end{bmatrix}$

b. $\begin{bmatrix} 0 & 1 & 2 & 1 & 1 \\ 1 & 0 & 2 & 0 & 1 \\ 2 & 2 & 0 & 2 & 3 \\ 1 & 0 & 2 & 2 & 2 \\ 1 & 1 & 3 & 2 & 0 \end{bmatrix}$

c. $\begin{bmatrix} 0 & 2 & 1 & 3 & 1 & 2 \\ 2 & 0 & 3 & 1 & 1 & 0 \\ 1 & 3 & 0 & 2 & 3 & 1 \\ 3 & 1 & 2 & 2 & 2 & 1 \\ 1 & 1 & 3 & 2 & 3 & 1 \\ 2 & 0 & 1 & 1 & 1 & 0 \end{bmatrix}$

d. $\begin{bmatrix} 0 & 2 & 1 & 3 & 0 & 1 & 0 \\ 2 & 0 & 2 & 1 & 2 & 1 & 0 \\ 1 & 2 & 0 & 1 & 1 & 0 & 1 \\ 3 & 1 & 1 & 0 & 0 & 2 & 1 \\ 0 & 2 & 1 & 0 & 0 & 0 & 3 \\ 1 & 1 & 0 & 2 & 0 & 0 & 2 \\ 0 & 0 & 1 & 1 & 3 & 2 & 0 \end{bmatrix}$

3. a. Simple, planar
b. Simple, planar
c. Simple, complete, planar
d. Simple, planar
e. Simple, complete
f. Simple, planar
g. Simple, planar
h. Simple, planar

4. a. No
b. In this network, all 4 vertices have an odd degree. For an Eulerian or semi-Eulerian trail to exist 0 or 2 vertices of a connected network must have an odd numbered degree.

5. a. i. 3

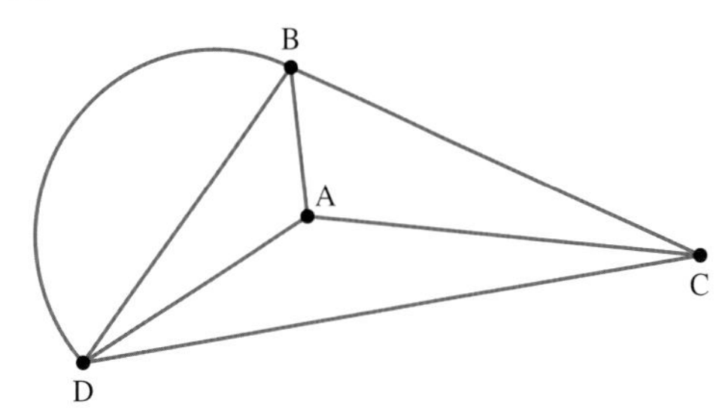

ii. ABDBCADC

b. i.

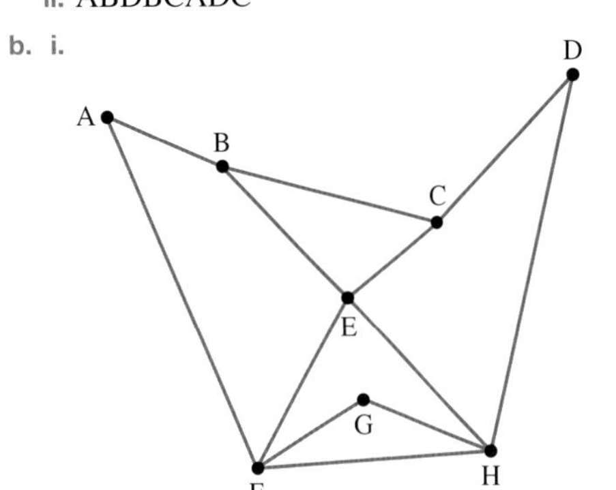

ii. BAFEHGFHDCEBC

6. a. 23
b. 34
c.

Start
9
2
3
3
2
8
Finish

Technology active: multiple choice

7. C
8. B
9. C
10. D
11. D
12. A
13. D
14. A
15. A
16. C
17. A

Technology active: extended response

18. a. See the figure at the bottom of the page.*
    b. Via Canberra (4661 km)
    c. Via Sydney (4954 km)
    d. See the figure at the bottom of the page.*
    The total distance is 9012 km.

19. a. i. 4195 m

ii.

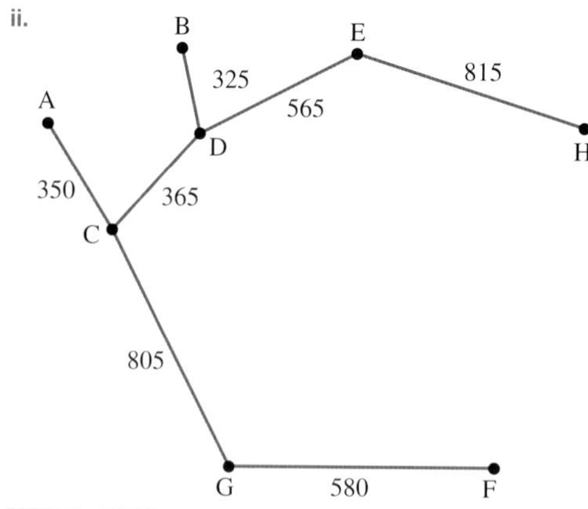

b. HEDA, 1860 m

c. 4905 m, DFGCABEH

d. DEHFGCABD, 5260 m

*18. a.

*18. d.

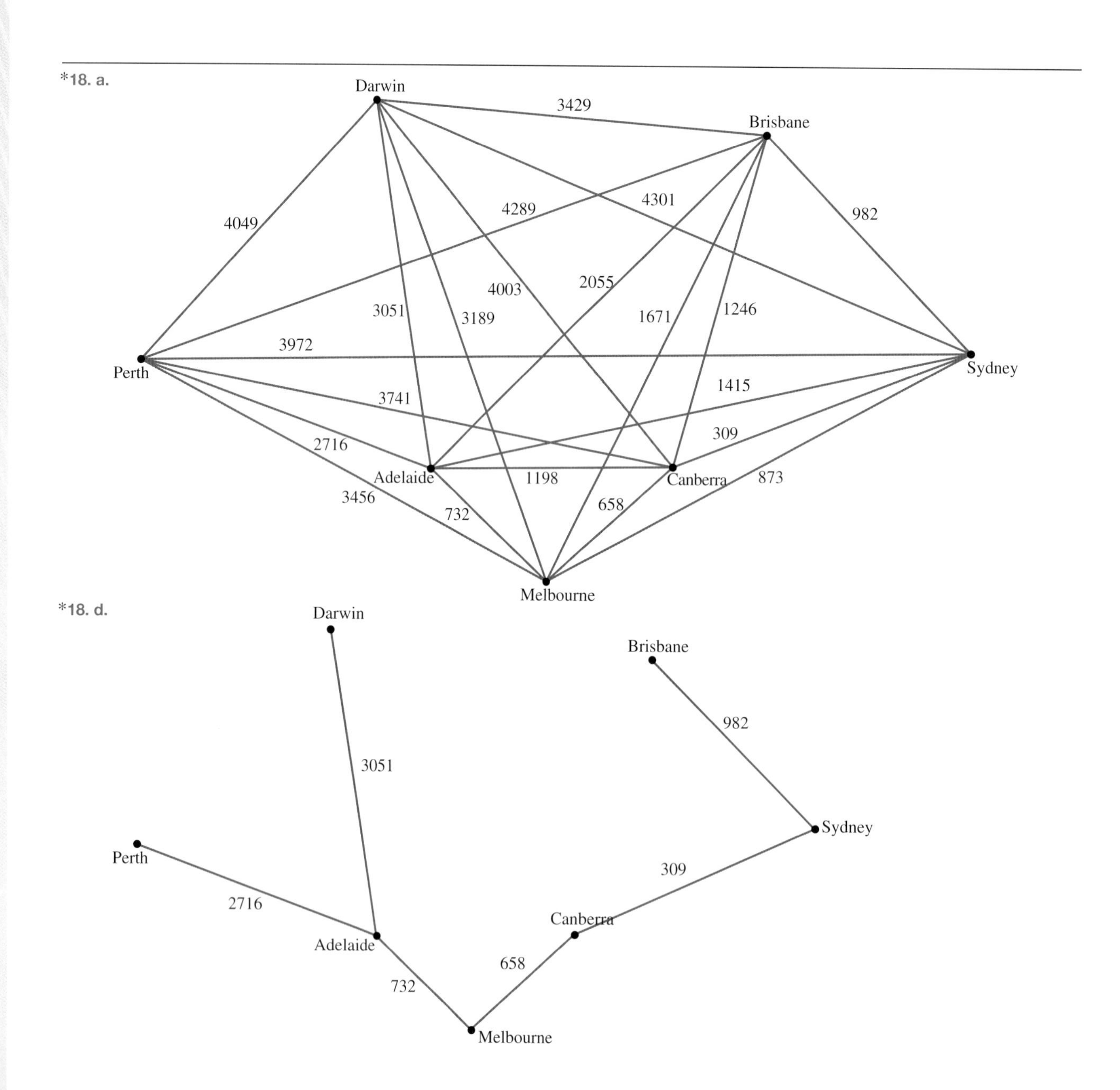

20. a. See the figure at the bottom of the page.*

b. $\begin{bmatrix} 0 & 2 & 1 & 1 & 0 & 0 \\ 2 & 0 & 1 & 1 & 0 & 0 \\ 1 & 1 & 0 & 2 & 1 & 0 \\ 1 & 1 & 2 & 0 & 1 & 0 \\ 0 & 0 & 1 & 1 & 0 & 1 \\ 1 & 0 & 0 & 0 & 1 & 0 \end{bmatrix}$

c. No, as there are more than two vertices of odd degree.

d. AFEDCBA (39)

6.7 Exam questions

Note: Mark allocations are available with the fully worked solutions online.

1. No, since the graph of this situation is $K_{3,3}$, which is non-planar.
2. D
3. C
4. a. $\begin{bmatrix} 0 & 2 & 0 & 4 \\ 0 & 4 & 4 & 1 \\ 4 & 2 & 0 & 5 \\ 0 & 1 & 1 & 2 \end{bmatrix}$

   b. $\begin{bmatrix} 0 & 1 & 0 & 3 \\ 0 & 3 & 4 & 0 \\ 4 & 1 & 0 & 4 \\ 0 & 0 & 1 & 1 \end{bmatrix}$

   c. The minimum number of lines needed to cover all zeros is 4, so all 4 tasks can be allocated. Hungarian algorithm is not necessary.

   d. The minimum time is $5 + 7 + 3 + 8 = 23$
5. a. See the figure at the bottom of the page.*

   b. Hobart–Bruny–Robbins (780 km)

   c. Hobart–Bruny–Robbins–King–Devonport–Flinders–Maria (2075 km)

   d. King–Devonport–Flinders–Maria–Hobart–Bruny–Robbins–King (2220 km)

*20. a.

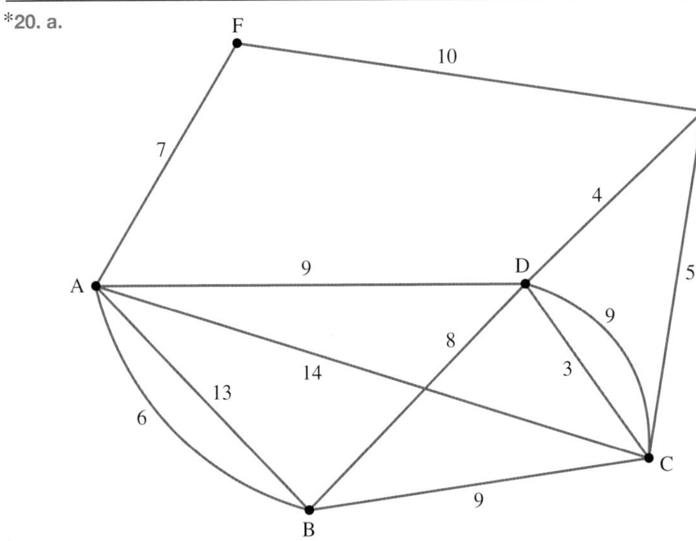

*5. a.

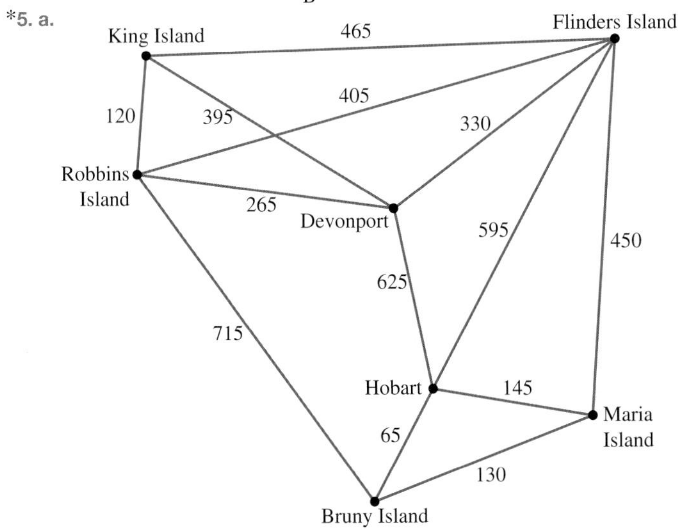

7 Trigonometric ratios and applications

LEARNING SEQUENCE

Fully worked solutions for this chapter are available online.

7.1 Overview

Hey students! Bring these pages to life online

Watch videos

Engage with interactivities

Answer questions and check results

Find all this and MORE in jacPLUS

7.1.1 Introduction

Trigonometry, derived from the Greek words *trigon* (triangle) and *metron* (measurement), is the branch of mathematics that deals with the relationship between the sides and angles of a triangle. Trigonometry was originally devised around 1700 BCE as a tool for astronomers.

Hipparchus, the Greek astronomer and mathematician, is known as the father of trigonometry for compiling the first trigonometric tables around 150 BCE. From the 4th century onwards Indian mathematicians made significant contributions to trigonometry, with Bhaskara the First and Brahmagupta developing formulas for determining sine values. Islamic mathematicians built on the work of the Greeks and Indians and by the 10th century were using all six trigonometric functions. Today, the sine and cosine functions are used in many fields such as building, surveying, navigation and engineering.

KEY CONCEPTS

This topic covers the following key concepts from the VCE Mathematics Study Design:

- radian measure, arc length, sectors and segments
- the sine rule and cosine rule applied to two and three-dimensional situations, including problems involving angles between planes.

7.2 Review of trigonometry

LEARNING INTENTION

At the end of this subtopic you should be able to:
- use trigonometry to solve for unknowns in right-angled triangles
- apply right-angled trigonometry to angles of elevation and depression, and bearings.

7.2.1 Right-angled triangles

In previous years you will have studied the trigonometry of right-angled triangles. We will review this material before considering non–right-angled triangles.

Trigonometric ratios

$$\sin(\theta) = \frac{\text{opposite side}}{\text{hypotenuse}}, \text{ which is abbreviated to } \sin(\theta) = \frac{O}{H}$$

$$\cos(\theta) = \frac{\text{adjacent side}}{\text{hypotenuse}}, \text{ which is abbreviated to } \cos(\theta) = \frac{A}{H}$$

$$\tan(\theta) = \frac{\text{opposite side}}{\text{adjacent side}}, \text{ which is abbreviated to } \tan(\theta) = \frac{O}{A}$$

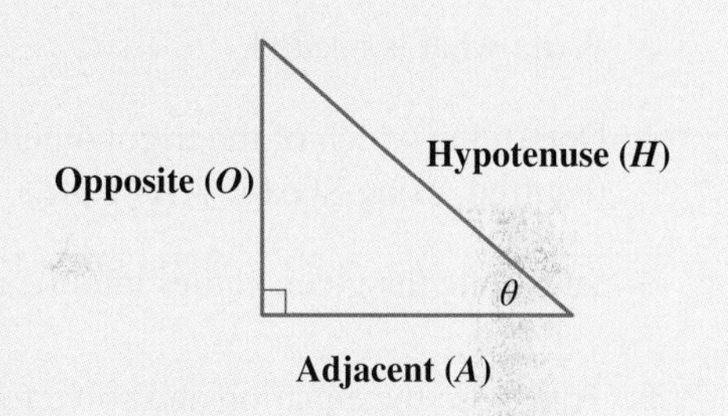

The symbol θ (theta) is one of the many letters of the Greek alphabet used to represent the angle. Other symbols include α (alpha), β (beta) and γ (gamma). Non-Greek letters may also be used.

Writing the mnemonic **SOH–CAH–TOA** each time we perform trigonometric calculations will help us to remember the ratios and solve the problem.

WORKED EXAMPLE 1 Determining unknown side lengths in right-angled triangles

Determine the exact value of the pronumerals.

a.

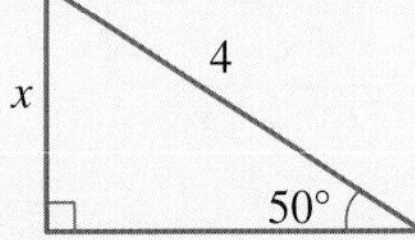

b.

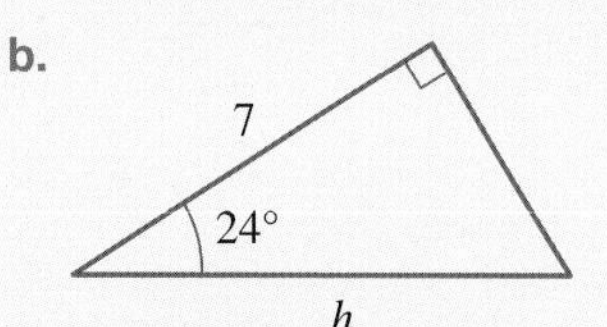

THINK	WRITE/DRAW
a. 1. Label the sides, relative to the marked angles.	a. 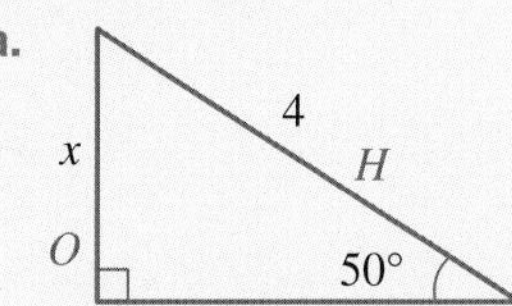
2. Write what is given.	Have: angle and hypotenuse (H)
3. Write what is needed.	Need: opposite (O) side

4. Determine which of the trigonometric ratios is required, using SOH–CAH–TOA. $\sin(\theta) = \dfrac{O}{H}$

5. Substitute the given values into the appropriate ratio. $\sin(50°) = \dfrac{x}{4}$

6. Transpose the equation and solve for x.
$$4 \times \sin(50°) = x$$
$$x = 4\sin(50°)$$

7. ***Note:*** Using a calculator you could determine that $4\sin(50°)$ is ≈ 3.06. ≈ 3.06

b. 1. Label the sides, relative to the marked angles.

b.

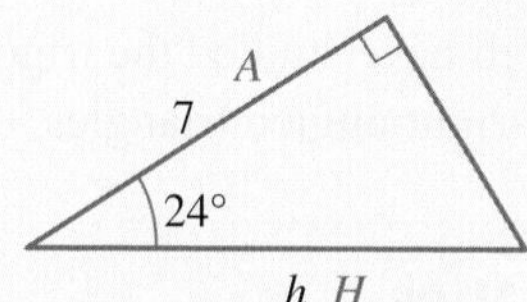

2. Write what is given. Have: angle and adjacent (A) side

3. Write what is needed. Need: hypotenuse (H)

4. Determine which of the trigonometric ratios is required, using SOH–CAH–TOA. $\cos(\theta) = \dfrac{A}{H}$

5. Substitute the given values into the appropriate ratio. $\cos(24°) = \dfrac{7}{h}$

6. Transpose the equation and solve for h. $h = \dfrac{7}{\cos(24°)}$

7. ***Note:*** Using a calculator you could determine that $\dfrac{7}{\cos(24°)}$ is ≈ 7.66. ≈ 7.66

WORKED EXAMPLE 2 Determining unknown angles in right-angled triangles

Determine the angle θ, giving the answer in degrees to one decimal place.

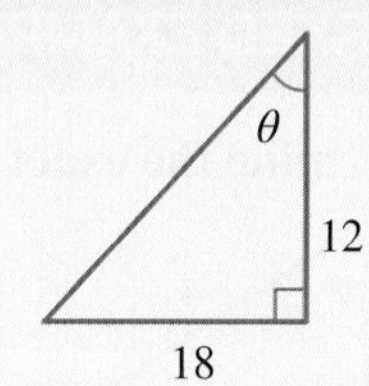

THINK

WRITE/DRAW

1. Label the sides, relative to the marked angles.

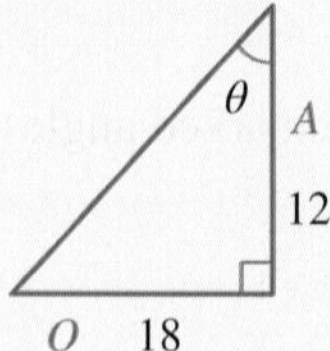

2. Write what is given. Have: opposite (O) and adjacent (A) sides

3. Write what is needed. Need: angle

4. Determine which of the trigonometric ratios is required, using SOH–CAH–TOA. $\tan(\theta) = \dfrac{O}{A}$

5. Substitute the given values into the appropriate ratio. $\tan(\theta) = \frac{18}{12}$

6. Write the answer in terms of inverse tan. $\theta = \tan^{-1}\left(\frac{18}{12}\right)$

7. ***Note:*** Using a calculator you could determine that $\tan^{-1}\left(\frac{3}{2}\right)$ is $\approx 56.3°$. $= \tan^{-1}\left(\frac{3}{2}\right)$
$\approx 56.3°$

7.2.2 Exact values

Most trigonometric values that we will see in this topic will result in values that are fractions involving surds and angles that can only be expressed exactly as inverse trigonometric ratios. However, angles of 30°, 45° and 60° have exact values of sine, cosine and tangent. Consider an equilateral triangle, *ABC*, of side length 2 units.

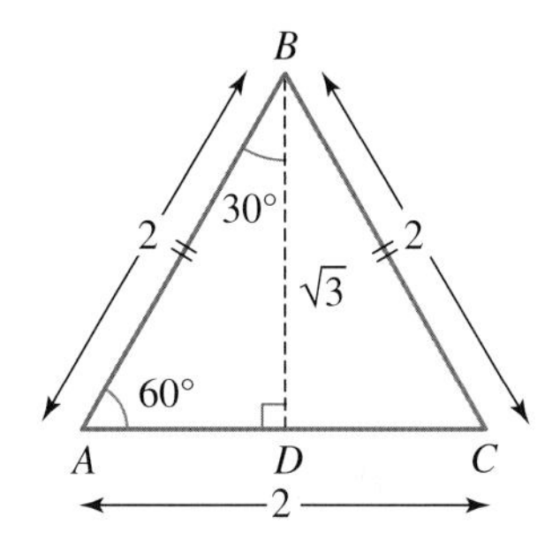

If the triangle is perpendicularly bisected, then two congruent triangles, *ABD* and *CBD*, are obtained. From triangle *ABD* it can be seen that *BD* creates a right-angled triangle with angles of 60° and 30° and base length (*AD*) of 1 unit. The length of *BD* is obtained using Pythagoras' theorem.

Using triangle *ABD* and the three trigonometric ratios, the following exact values are obtained

$$\sin(30°) = \frac{1}{2} \qquad \sin(60°) = \frac{\sqrt{3}}{2}$$
$$\cos(30°) = \frac{\sqrt{3}}{2} \qquad \cos(60°) = \frac{1}{2}$$
$$\tan(30°) = \frac{1}{\sqrt{3}} = \frac{\sqrt{3}}{3} \qquad \tan(60°) = \frac{\sqrt{3}}{1} = \sqrt{3}$$

Consider a right-angled isosceles triangle *EFG* whose equal sides are of 1 unit. The hypotenuse *EG* is obtained by using Pythagoras' theorem.

$$\begin{aligned}(EG)^2 &= (EF)^2 + (FG)^2 \\ &= 1^2 + 1^2 \\ &= 2 \\ EG &= \sqrt{2}\end{aligned}$$

Using triangle *EFG* and the three trigonometric ratios, the following exact values are obtained:

$$\sin(45°) = \frac{1}{\sqrt{2}} = \frac{\sqrt{2}}{2}$$
$$\cos(45°) = \frac{1}{\sqrt{2}} = \frac{\sqrt{2}}{2}$$
$$\tan(45°) = \frac{1}{1} = 1$$

WORKED EXAMPLE 3 Using exact values to determine unknown side lengths

Determine the height of the triangle shown in surd form.

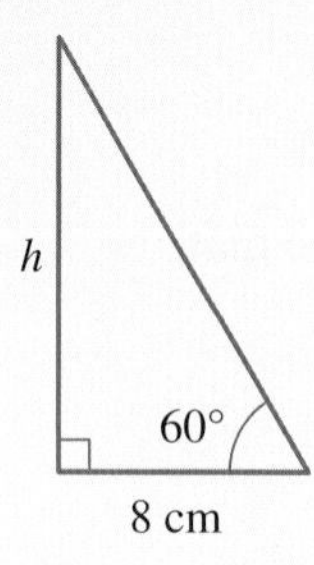

THINK

WRITE/DRAW

1. Label the sides relative to the marked angle.

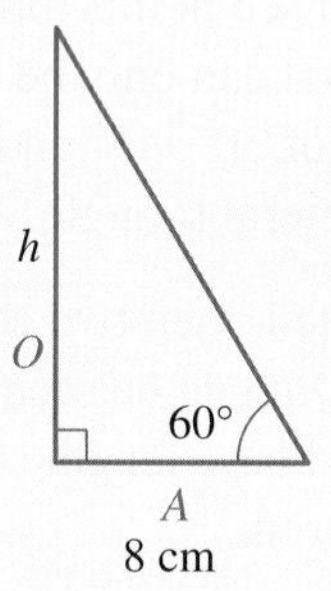

2. Write what is given.
   Have: angle and adjacent (A) side
3. Write what is needed.
   Need: opposite (O) side
4. Determine which of the trigonometric ratios is required, using SOH–CAH–TOA.
   $\tan(\theta) = \frac{O}{A}$
5. Substitute the given values into the appropriate ratio.
   $\tan(60°) = \frac{h}{8}$
6. Substitute exact values where appropriate.
   $\sqrt{3} = \frac{h}{8}$
7. Transpose the equation to determine the required value.
   $h = 8\sqrt{3}$
8. State the answer.
   The triangle's height is $8\sqrt{3}$ cm.

7.2.3 Applications of trigonometry

Trigonometry is especially useful for measuring distances and heights that are difficult or impractical to access. For example, two important applications of right-angled triangles are:

1. Angles of elevation and depression
2. Bearings

Angles of elevation and depression

Angles of elevation and depression are employed when dealing with directions that require us to look up and down respectively.

An **angle of elevation** is the angle between the horizontal and an object that is higher than the observer (for example, the top of a mountain or flagpole).

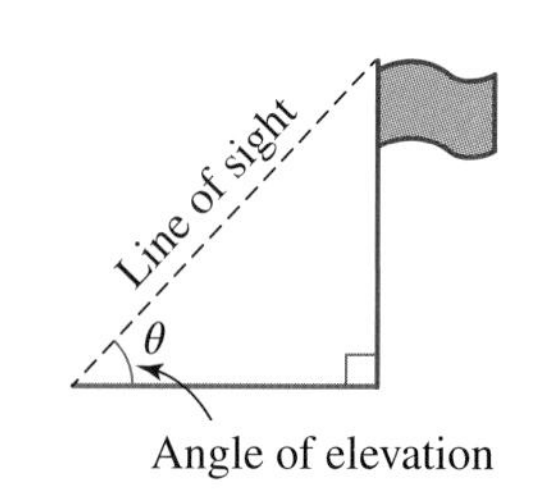

An **angle of depression** is the angle between the horizontal and an object that is lower than the observer (for example, a boat at sea when the observer is on a cliff).

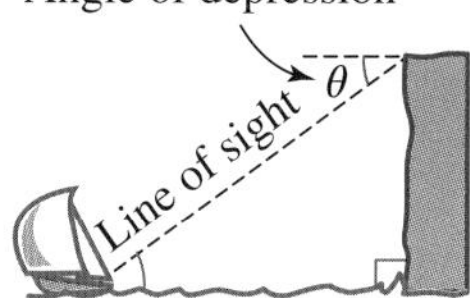

Unless otherwise stated, the angle of elevation or depression is measured and drawn from the horizontal.

Angles of elevation and depression

Angles of elevation and depression are each measured from the horizontal.

When solving problems involving angles of elevation and depression, it is always best to draw a diagram.

The angle of elevation is equal to the angle of depression because they are alternate 'Z' angles.

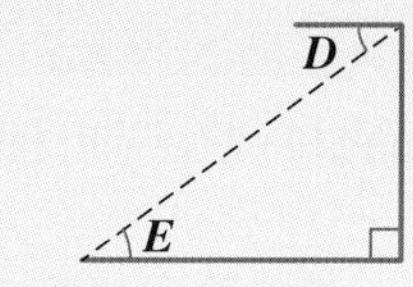

D **and** ***E*** **are alternate angles.**
$\therefore \angle D = \angle E$

WORKED EXAMPLE 4 Applying trigonometric ratios to real-life scenarios (1)

From a cliff 50 metres high, the angle of depression to a boat at sea is 12°. Calculate how far the boat is from the base of the cliff, in metres correct to 2 decimal places.

THINK	WRITE/DRAW
1. Draw a diagram and label all the given information. Include the unknown length, x, and the angle of elevation, 12°.	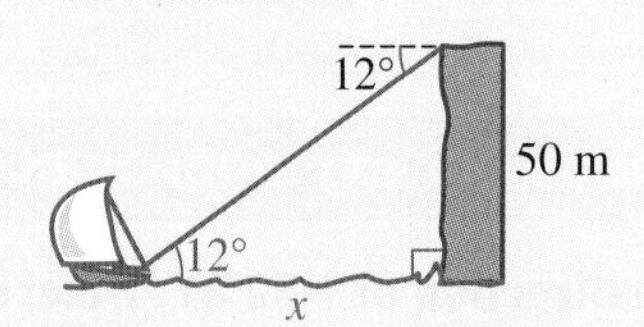
2. Write what is given.	Have: angle and opposite side
3. Write what is needed.	Need: adjacent side
4. Determine which of the trigonometric ratios is required (SOH–CAH–TOA).	$\tan(\theta) = \frac{O}{A}$
5. Substitute the given values into the appropriate ratio.	$\tan(12°) = \frac{50}{x}$
6. Transpose the equation and solve for x.	$x \times \tan(12°) = 50$ $x = \frac{50}{\tan(12°)}$
7. Round the answer to 2 decimal places.	$= 235.23$
8. Answer the question.	The boat is 235.23 m from the base of the cliff.

Bearings

Bearings measure the direction of one object from another. There are two systems used for describing bearings.

True bearings are measured in a clockwise direction, starting from north (0° T).

Conventional or **compass bearings** are measured first, relative to north or south, and second, relative to east or west.

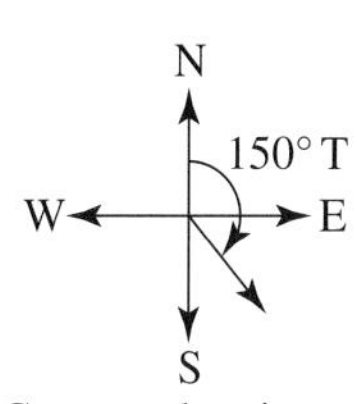

Compass bearing equivalent is S30°E.

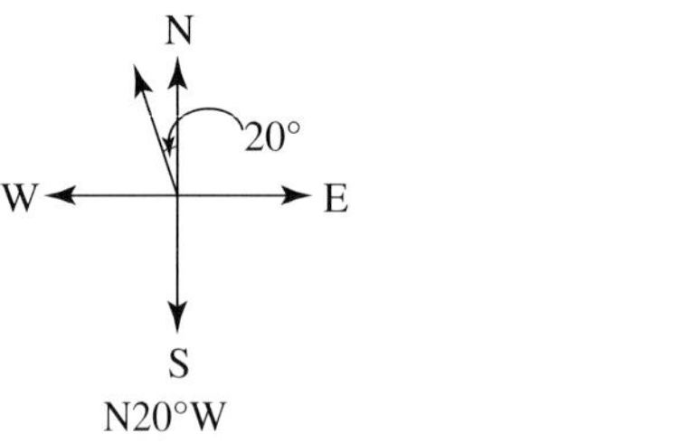

N20°W

True bearing equivalent is 340° T.

S70°E

True bearing equivalent is 110° T.

The two systems are interchangeable. For example, a bearing of 240° T is the same as S60°W.

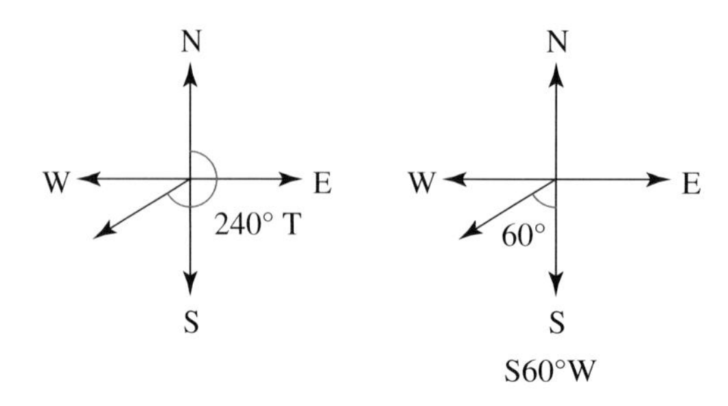

S60°W

When solving questions involving direction, always start with a diagram showing the basic compass points: north, south, east and west.

WORKED EXAMPLE 5 Applying trigonometric ratios to real-life scenarios (2)

A ship sails 40 km in a direction of N52°W. Determine how far west of the starting point it is, in km correct to 2 decimal places.

THINK	WRITE/DRAW
1. Draw a diagram of the situation, labelling each of the compass points and the given information.	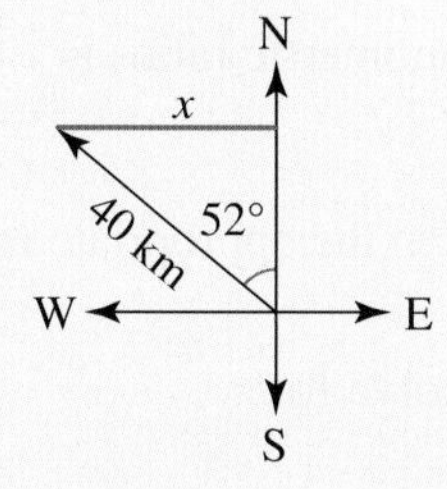
2. Write what is given for the triangle.	Have: angle and hypotenuse
3. Write what is needed for the triangle.	Need: opposite side
4. Determine which of the trigonometric ratios is required (SOH–CAH–TOA).	$\sin(\theta) = \dfrac{O}{H}$
5. Substitute the given values into the appropriate ratio.	$\sin(52°) = \dfrac{x}{40}$
6. Transpose the equation and solve for x.	$40 \times \sin(52°) = x$
7. Round the answer to 2 decimal places.	$x = 31.52$
8. Answer the question.	The ship is 31.52 km west of the starting point.

WORKED EXAMPLE 6 Applying trigonometric ratios to real-life scenarios (3)

A ship sails 10 km east, then 4 km south. Calculate the bearing from its starting point, in degrees correct to 1 decimal place.

THINK	WRITE/DRAW
1. Draw a diagram of the situation, labelling each of the compass points and the given information.	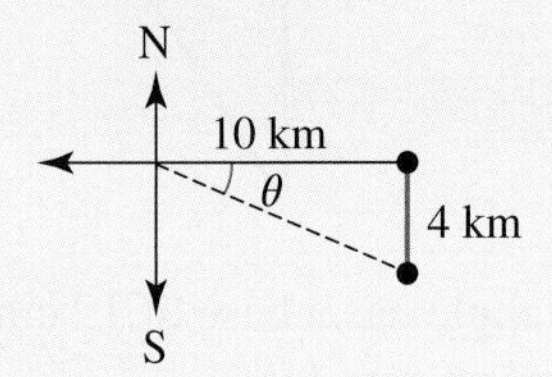
2. Write what is given for the triangle.	Have: adjacent and opposite sides
3. Write what is needed for the triangle.	Need: angle
4. Determine which of the trigonometric ratios is required (SOH–CAH–TOA).	$\tan(\theta)=\frac{O}{A}$
5. Substitute the given values into the appropriate ratio.	$\tan(\theta)=\frac{4}{10}$
6. Transpose the equation and solve for θ, using the inverse tan function.	$\theta=\tan^{-1}\left(\frac{4}{10}\right)$
7. State the value of θ in degrees, rounded to 1 decimal place.	$=21.8°$
8. Express the angle in bearings form. The bearing of the ship was initially 0° T; it has since rotated through an angle of 90° and an additional angle of 21.8°. To obtain the final bearing these values are added.	Bearing $=90°+21.8°$ $=111.8°$ T
9. Answer the question.	The bearing of the ship from its starting point is 111.8° T.

7.2 Exercise

Students, these questions are even better in jacPLUS

Receive immediate feedback and access sample responses

Access additional questions

Track your results and progress

Find all this and MORE in jacPLUS

Technology active

1. WE1 Determine the value of the pronumerals, correct to 2 decimal places.

a.
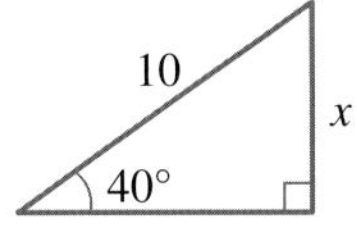

b.
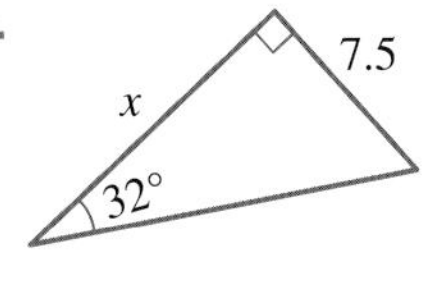

c.
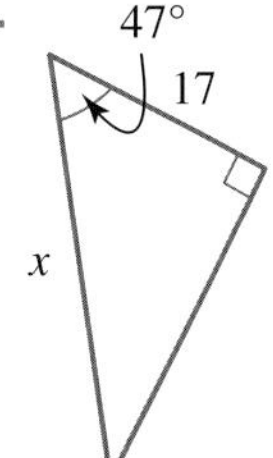

d.
684
62°
x

2. WE2 Calculate the angle θ, giving the answer in degrees correct to 2 decimal places.

a.

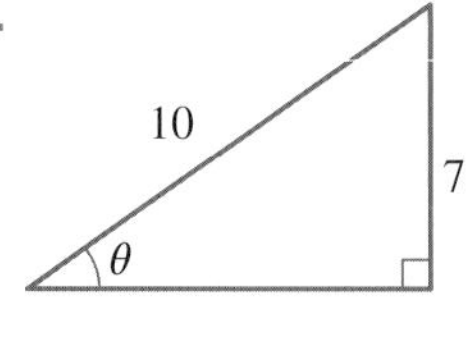

b.

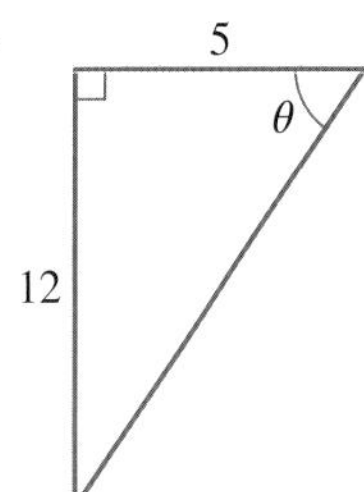

c.

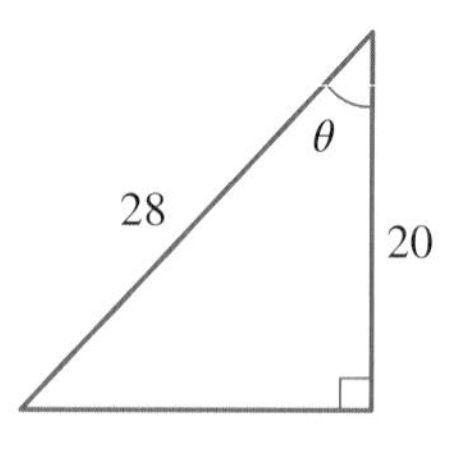

d.

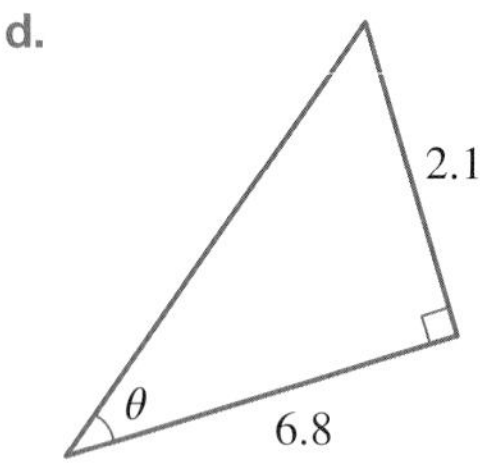

3. WE3 An isosceles triangle has a base of 12 cm and equal angles of 30°. Determine, in the simplest surd form:

   a. the height of the triangle, in cm
   b. the area of the triangle, in cm^2
   c. the perimeter of the triangle, in cm.

4. Calculate the perimeter of the composite shape below, in surd form. The length measurements are in metres.

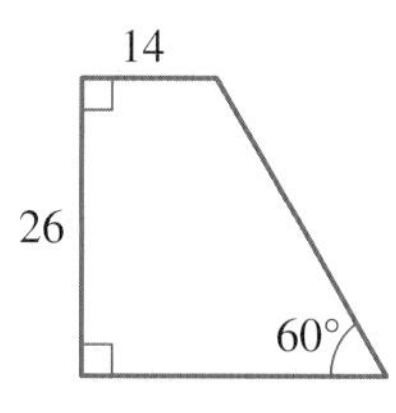

5. Express the following conventional bearings as true bearings, and the true bearings in conventional form.

   a. N35°W
   b. S47°W
   c. N58°E
   d. S17°E
   e. 246° T
   f. 107° T
   g. 321° T
   h. 074° T

6. WE4 From a vertical fire tower 60 m high, the angle of depression to a fire is 6°. Calculate how far away, to the nearest metre, the fire is.

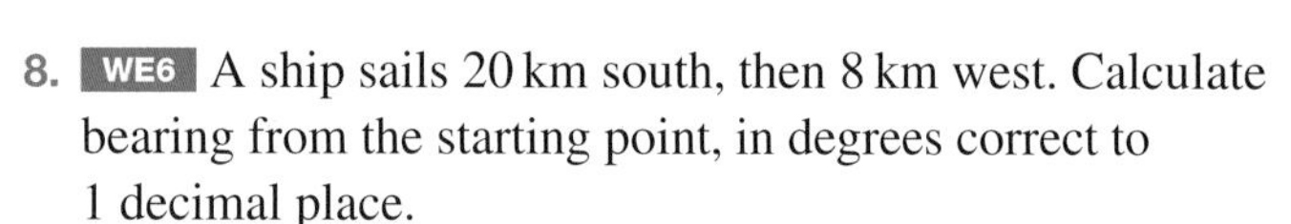

7. WE5 A pair of kayakers paddle 1800 m on a bearing of N20°E. Determine how far north of their starting point they are, to the nearest metre.

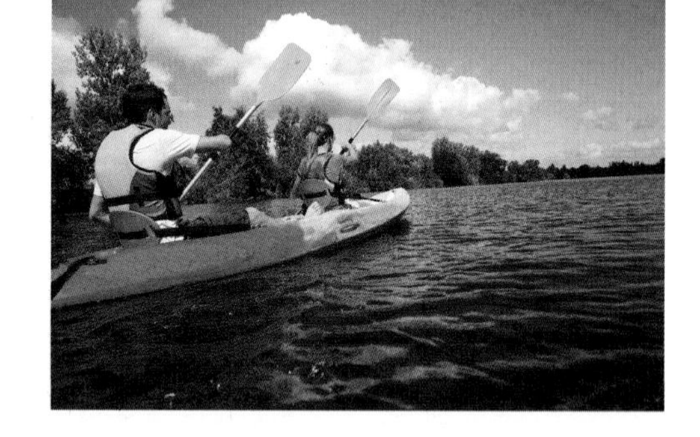

8. WE6 A ship sails 20 km south, then 8 km west. Calculate bearing from the starting point, in degrees correct to 1 decimal place.

9. A cross-country competitor runs 2 km west, then due north for 3 km. Determine the true bearing of the runner from the starting point, in degrees correct to 1 decimal place.

10. A ladder 6.5 m long rests against a vertical wall and makes an angle of 50° to the horizontal ground.

    a. Calculate how high up the wall the ladder reaches, in metres to 2 decimal places.
    b. If the ladder needs to reach 1 m higher, determine the angle it should make with the ground, to the nearest degree.

11. A new skyscraper is proposed for the Melbourne Docklands region. It is to be 500 m tall. Calculate the angle of depression, to the nearest degree, from the top of the building to the island on Albert Park Lake, which is 4.2 km away.

12. An ice-cream cone has a diameter of 6 cm and a sloping edge of 15 cm. Determine the angle at the bottom of the cone, to the nearest degree.

13. A stepladder stands on a floor, with its feet 1.5 m apart. If the angle formed by the legs is 55°, calculate how high above the floor the top of the ladder is, in metres correct to 2 decimal places.

14. In the figure shown, determine the value of the pronumerals, correct to 2 decimal places.

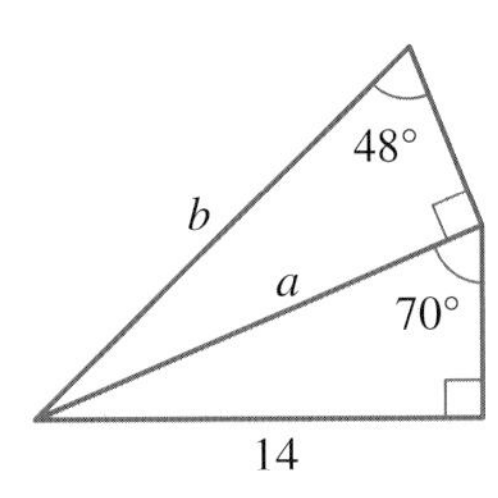

15. From a rescue helicopter 2500 m above the ocean, the angles of depression to two shipwreck survivors are 48° (survivor 1) and 35° (survivor 2).
   a. Draw a labelled diagram that represents the situation.
   b. Calculate how far apart the two survivors are, in metres correct to 2 decimal places.

16. A lookout tower has been erected on top of a mountain. At a distance of 5.8 km, the angle of elevation from the ground to the base of the tower is 15.7°, and the angle of elevation to the observation deck (on the top of the tower) is 15.9°. Determine how high, to the nearest metre, the observation deck is above the top of the mountain.

17. In the figure shown, calculate the value of the pronumeral x, correct to 2 decimal places.

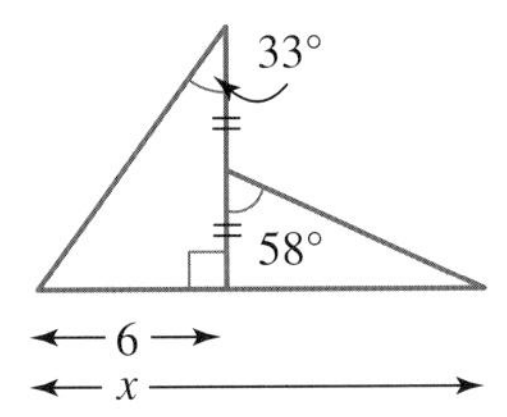

18. A garden bed in the shape of a trapezium is shown. Calculate the volume of garden mulch needed to cover it to a depth of 15 cm. Write your answer in cubic metres correct to 2 decimal places.

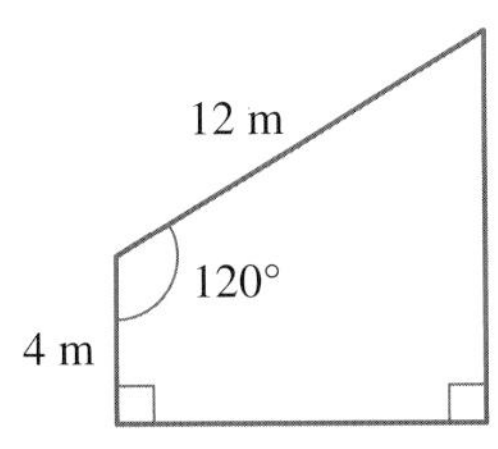

19. Two hikers set out from the same campsite. One walks 7 km in the direction 043° T and the other walks 10 km in the direction 133° T.
   a. Determine the distance between the two hikers, in km to 1 decimal place.
   b. Calculate the bearing of the first hiker from the second, to the nearest degree.

20. A ladder 10 m long rests against a vertical wall at an angle of 55° to the horizontal. It slides down the wall, so that it now makes an angle of 48° with the horizontal.
   a. Calculate the vertical distance that the top of the ladder slid, in metres to 2 decimal places.
   b. Determine if the foot of the ladder moves through the same distance. Justify your answer.

21. A ship sails 30 km on a bearing of 220° T, then 20 km on a bearing of 250° T. Determine:
   a. how far south of the original position it is, in km to 2 decimal places
   b. how far west of the original position it is, in km to 2 decimal places
   c. the true bearing of the ship from its original position, to the nearest degree.

22. The town of Bracknaw is due west of Arley. Chris, in an ultralight plane, starts at a third town, Champton, which is due north of Bracknaw, and flies directly towards Arley at a speed of 40 km/h in a direction of 110° T. She reaches Arley in 3 hours. Determine:
   a. the distance between Arley and Bracknaw, in km to 2 decimal places
   b. the time to complete the journey from Champton to Bracknaw, via Arley, if she increases her speed to 45 km/h between Arley and Bracknaw. Write your answer in hours and minutes, correct to the nearest minute.

23. One of the great pyramids in Egypt has a height of 147 m and stands on a square base of edge length 230 m. Calculate the following angles, stating your answers in degrees rounded to 2 decimal places.
   a. The inclination of the sloping face to the base.
   b. The inclination of the sloping edge to the base.

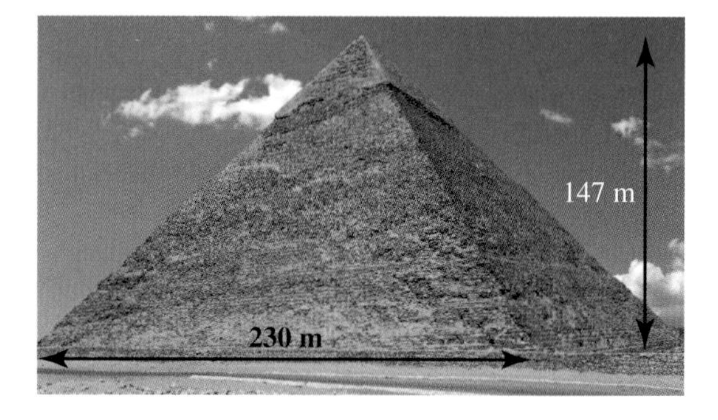

24. A plane is inclined at an angle of 30° to the horizontal. A straight line is drawn on the plane and makes an angle of 60° with the line of greatest slope. Calculate the angle that this line makes with the horizontal, stating your answer in degrees, rounded to 2 decimal places.

25. A and B are two telephone towers. Tower A is 20 metres high and tower B is 30 metres high. Tower B lies directly west of tower A. From a point P due north of tower A, the angles of elevation to the tops of towers A and B are 35° and 25° respectively. Calculate the distance between the towers in metres, rounded to 2 decimal places.

7.2 Exam questions

Question 1 (2 marks) TECH-FREE

The cuboid below has edges 5 cm, 6 cm and 8 cm.

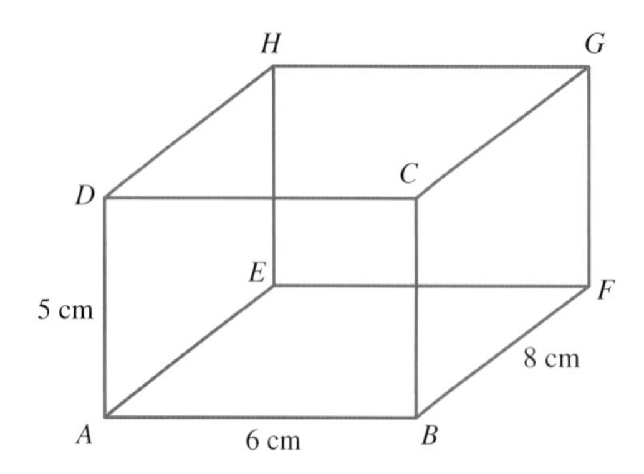

Determine the size of the angle between the diagonal AG and the edge AD. Give your answer in the form $a° - \tan^{-1}\left(\frac{b}{c}\right)$.

Question 2 (1 mark) TECH-ACTIVE

MC From the top of an office building 100 metres high, the angle of depression to a second building, 80 metres away, is 16°. The height of the lower building is

A. 68 m **B.** 71 m **C.** 75 m **D.** 77 m **E.** 79 m

Question 3 (2 marks) TECH-ACTIVE

The gallant knight Sir George, whose eye level is 1.8 metres from the ground, is standing on top of a small mound, 3 metres high. From this vantage point he notices that the angle of elevation of the tall tower, at the top of which the fair damsel Charlotte is being held captive, is 2°. Using his expert judgement he calculates the distance to the foot of the tower as 3 kilometres. Sir George requires a rope twice the height of the tower to aid him to rescue the damsel. Determine what length of rope, rounded to the nearest metre, he should purchase from the conveniently located rope shop.

More exam questions are available online.

7.3 The sine rule

LEARNING INTENTION

At the end of this subtopic you should be able to:

- use the sine rule to calculate unknown side lengths and angles in triangles.

7.3.1 Deriving the sine rule

When working with non–right-angled triangles, it is usual to label the angles A, B and C, and the sides a, b and c, so that side a is the side opposite angle A, side b is opposite angle B and side c is opposite angle C.

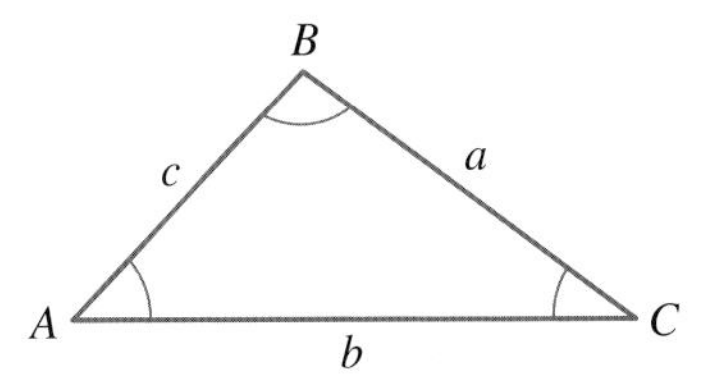

In a non–right-angled triangle, a perpendicular line, h, can be drawn from the angle B to side b.

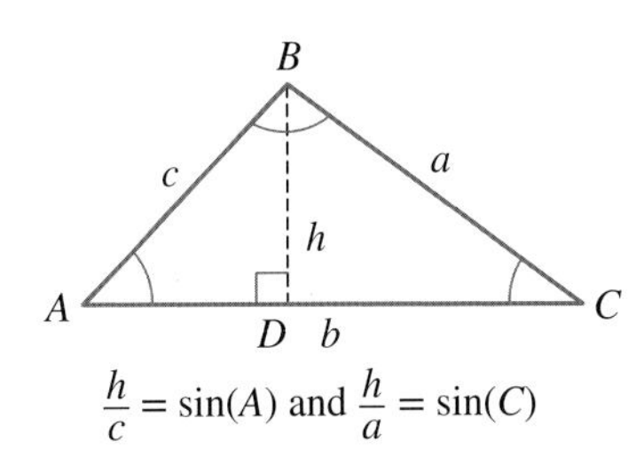

Using triangle ABD, we obtain $\sin(A) = \frac{h}{c}$. Using triangle CBD, we obtain $\sin(C) = \frac{h}{a}$.

Transposing each equation to make h the subject, we obtain $h = c \times \sin(A)$ and $h = a \times \sin(C)$. Equate to get $c \times \sin(A) = a \times \sin(C)$.

Transpose to get

$$\frac{c}{\sin(C)} = \frac{a}{\sin(A)}$$

In a similar way, if a perpendicular line is drawn from angle A to side a, we get

$$\frac{b}{\sin(B)} = \frac{c}{\sin(C)}$$

From this, the sine rule can be stated.

The sine rule

In any triangle ABC: $\dfrac{a}{\sin(A)} = \dfrac{b}{\sin(B)} = \dfrac{c}{\sin(C)}$

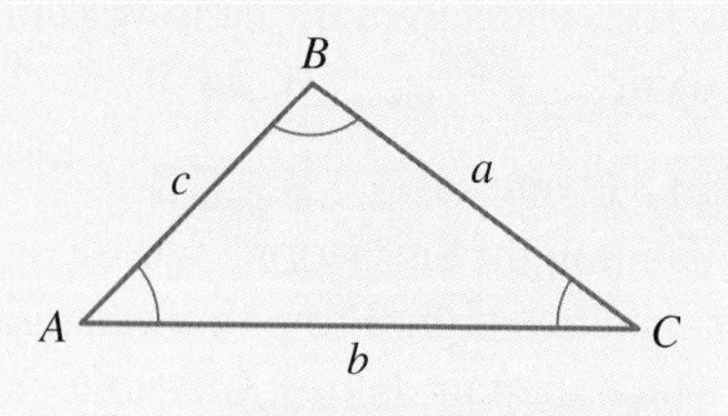

Notes:

1. When using this rule, depending on the values given, any combination of the two equalities may be used to solve a particular triangle.
2. To solve a triangle means to calculate all unknown side lengths and angles.

When to use the sine rule

The sine rule can be used to solve non–right-angled triangles if we are given:

1. **two angles and one side length**
2. **two side lengths and an angle opposite one of these side lengths.**

WORKED EXAMPLE 7 Using the sine rule to determine unknown side lengths and angles

In the triangle ABC, $a = 4$ m, $b = 7$ m and $B = 80°$. Determine the values of A, C and c correct to 2 decimal places.

THINK	WRITE/DRAW
1. Draw a labelled diagram of the triangle ABC and fill in the given information.	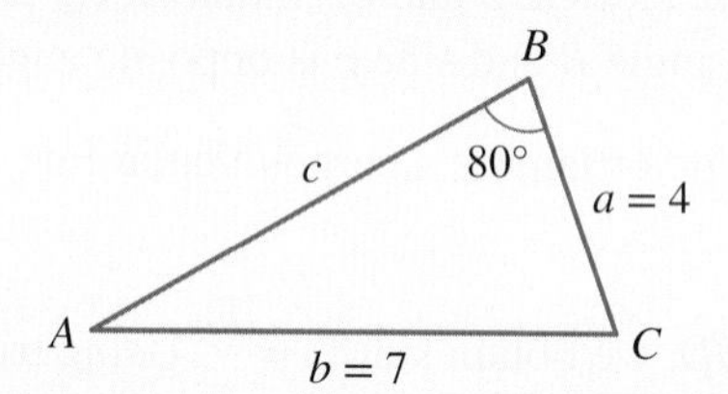
2. Check that one of the criteria for the sine rule has been satisfied.	The sine rule can be used since two side lengths and an angle opposite one of these side lengths have been given.
3. Write the sine rule to calculate A.	To calculate angle A: $\dfrac{a}{\sin(A)} = \dfrac{b}{\sin(B)}$
4. Substitute the known values into the rule.	$\dfrac{4}{\sin(A)} = \dfrac{7}{\sin(80°)}$
5. Transpose the equation to make $\sin(A)$ the subject.	$4 \times \sin(80°) = 7 \times \sin(A)$ $\sin(A) = \dfrac{4 \times \sin(80°)}{7}$
6. Evaluate.	$A = \sin^{-1}\left(\dfrac{4 \times \sin(80°)}{7}\right)$ $= \sin^{-1}(0.5627)$

7. Round the answer to 2 decimal places. $= 34.25°$

8. Determine the value of angle C using the fact that the angle sum of any triangle is 180°.
$C = 180° - (80° + 34.25°)$
$= 65.75°$

9. Write the sine rule to calculate c.
To calculate side length c:
$$\frac{c}{\sin(C)} = \frac{b}{\sin(B)}$$

10. Substitute the known values into the rule.
$$\frac{c}{\sin(65.75°)} = \frac{7}{\sin(80°)}$$

11. Transpose the equation to make c the subject.
$$c = \frac{7 \times \sin(65.75°)}{\sin(80°)}$$

12. Evaluate. Round the answer to 2 decimal places and include the appropriate unit.
$$= \frac{7 \times 0.9118}{0.9848}$$
$$= \frac{6.3823}{0.9848}$$
$$= 6.48 \text{ m}$$

7.3.2 The ambiguous case

When using the sine rule there is one important issue to consider. If we are given two side lengths and an angle opposite one of these side lengths, then sometimes two different triangles can be drawn. For example, if $a = 10$, $c = 6$ and $C = 30°$, two possible triangles could be created.

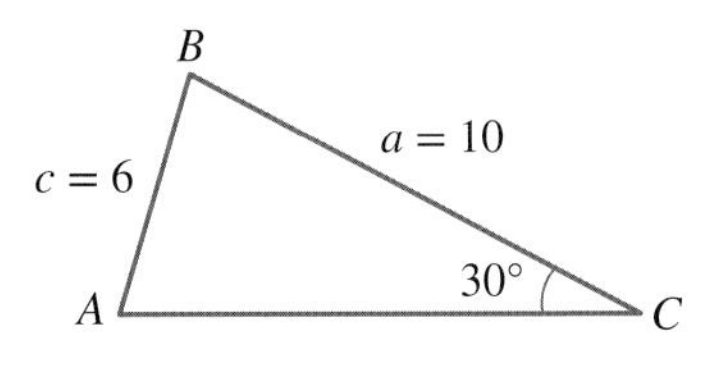

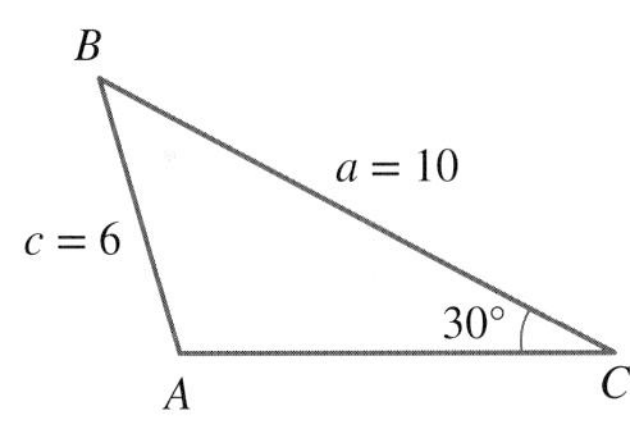

In the first case, angle A is an acute angle, while in the second case, angle A is an obtuse angle. The two values for A will add to 180°.

The ambiguous case does not work for every example. It would be useful to know, before commencing a question, whether or not the ambiguous case exists and, if so, to then determine both sets of solutions.

The ambiguous case of the sine rule

The ambiguous case exists if C is an acute angle and $a > c > a \times \sin(C)$, or any equivalent statement; for example, if B is an acute angle and $a > b > a \times \sin(B)$, and so on.

WORKED EXAMPLE 8 Using the ambiguous case of the sine rule

In the triangle ABC, $a = 10$ m, $c = 6$ m and $C = 30°$.

a. **Show that the ambiguous case exists.**

b. **Determine two possible values of A, and hence two possible values of B and b to 2 decimal places.**

THINK

a. 1. Check that the conditions for an ambiguous case exist, i.e. that C is an acute angle and that $a > c > a \times \sin(C)$.

2. State the answer.

WRITE/DRAW

a. $C = 30°$ so C is an acute angle.
$\sin(C) = \sin(30°) = 0.5$
$a > c > a \times \sin(C)$
$10 > 6 > 10 \sin(30°)$
$10 > 6 > 5$
This is correct.
This is an ambiguous case of the sine rule.

b. Case 1

1. Draw a labelled diagram of the triangle ABC and fill in the given information.

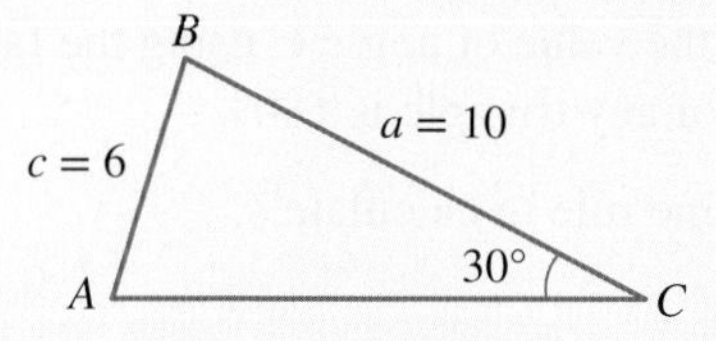

2. Write the sine rule to calculate A.

To calculate angle A:

$$\frac{a}{\sin(A)} = \frac{c}{\sin(C)}$$

3. Substitute the known values into the rule.

$$\frac{10}{\sin(A)} = \frac{6}{\sin(30°)}$$

4. Transpose the equation to make $\sin(A)$ the subject.

$$10 \times \sin(30°) = 6 \times \sin(A)$$

$$\sin(A) = \frac{10 \times \sin(30°)}{6}$$

5. Evaluate angle A, to 2 decimal places.

$$A = \sin^{-1}\left(\frac{10 \times \sin(30°)}{6}\right)$$

$$A = 56.44°$$

6. Determine the value of angle B, using the fact that the angle sum of any triangle is 180°.

$$B = 180° - (30° + 56.44°)$$
$$= 93.56°$$

7. Write the sine rule to calculate b.

To calculate side length b:

$$\frac{b}{\sin(B)} = \frac{c}{\sin(C)}$$

8. Substitute the known values into the rule.

$$\frac{b}{\sin(93.56°)} = \frac{6}{\sin(30°)}$$

9. Transpose the equation to make b the subject and evaluate.

$$b = \frac{6 \times \sin(93.56°)}{\sin(30°)}$$
$$= 11.98\text{ m}$$

Case 2

1. Draw a labelled diagram of the triangle ABC and fill in the given information.

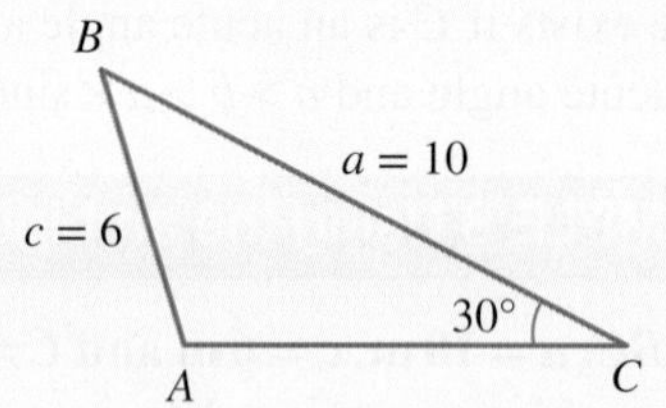

2. Write the alternative value for angle A. Subtract the value obtained for A in Case 1 from 180°.

To calculate the alternative angle A:

If $\sin(A) = 0.8333$, then A could also be:

$$A = 180° - 56.44°$$
$$= 123.56°$$

3. Determine the alternative value of angle B, using the fact that the angle sum of any triangle is 180°.

$$B = 180° - (30° + 123.56°)$$
$$= 26.44°$$

4. Write the sine rule to calculate the alternative b.

To calculate side length b:

$$\frac{b}{\sin(B)} = \frac{c}{\sin(C)}$$

5. Substitute the known values into the rule.

$$\frac{b}{\sin(26.44°)} = \frac{6}{\sin(30°)}$$

6. Transpose the equation to make b the subject and evaluate.

$$b = \frac{6 \times \sin(26.44°)}{\sin(30°)}$$
$$= 5.34 \text{ m}$$

Hence, for Worked example 8 there were two possible solutions as shown by the diagrams below.

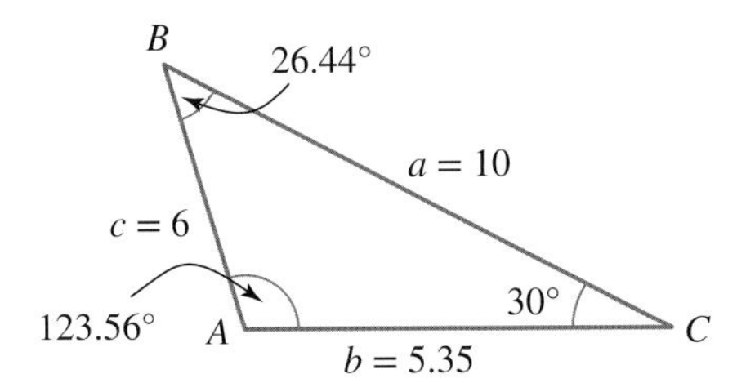

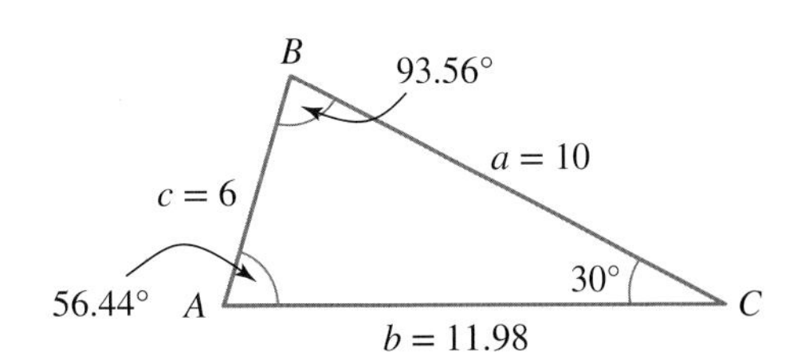

7.3 Exercise

Students, these questions are even better in jacPLUS

Receive immediate feedback and access sample responses

Access additional questions

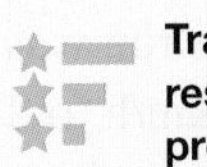

Track your results and progress

Find all this and MORE in jacPLUS

Technology active

1. **WE7** In the triangle ABC, $a = 10$, $b = 12$ and $B = 58°$. Determine the values of A, C and c correct to 1 decimal place.

2. In the triangle ABC, $c = 17.35$, $a = 26.82$ and $A = 101°$. Determine the value of C, B and b correct to 2 decimal places.

3. **WE8** In the triangle ABC, $a = 10$, $c = 8$ and $C = 50°$. Calculate the two possible values of A and hence two possible values of b correct to 2 decimal places.

4. In the triangle ABC, $a = 20$, $b = 12$ and $B = 35°$. Calculate the two possible values for the perimeter of the triangle correct to 2 decimal places.

5. In the triangle ABC, $c = 27$, $C = 42°$ and $A = 105°$. Evaluate B, a and b to 2 decimal places where required.

6. In the triangle ABC, $a = 7$, $c = 5$ and $A = 68°$. Calculate the perimeter of the triangle correct to 2 decimal places.

7. Calculate all unknown sides and angles for the triangle ABC, given $a = 32$, $b = 51$ and $A = 28°$ correct to 2 decimal places.

8. Determine the perimeter of the triangle ABC if $a = 7.8$, $b = 6.2$ and $A = 50°$ to 2 decimal places.

9. **MC** In a triangle ABC, $A = 40°$, $C = 80°$ and $c = 3$. The value of b is closest to:

   A. 2.64 B. 2.86 C. 14 D. 4.38 E. 4.60

10. Calculate all unknown sides and angles for the triangle ABC, given $A = 27°$, $B = 43°$ and $c = 6.4$ correct to 2 decimal places.

11. Calculate all unknown sides and angles for the triangle ABC, given $A = 25°$, $b = 17$ and $a = 13$ correct to 1 decimal place.

12. To calculate the height of a building, Kevin measures the angle of elevation to the top as 48°. He then walks 18 m closer to the building and measures the angle of elevation as 64°. Calculate the height of the building, in metres correct to 2 decimal places.

13. A river has parallel banks that run directly east–west. From one bank Kylie takes a bearing to a tree on the opposite bank. The bearing is 047° T. She then walks 10 m due east and takes a second bearing to the tree. This is 305° T. Determine:
    a. her distance from the second measuring point to the tree, to 2 decimal places
    b. the width of the river, to the nearest metre.

14. A ship sails on a bearing of S20°W for 14 km, then changes direction and sails for 20 km and drops anchor. Its bearing from the starting point is now N65°W.
    a. Calculate how far it is from the starting point, in km to 2 decimal places.
    b. Determine the bearing that the ship sailed the 20 km leg, correct to 1 decimal place.

15. a. A cross-country runner runs at 8 km/h on a bearing of 150° T for 45 minutes, then changes direction to a bearing of 053° T and runs for 80 minutes until he is due east of the starting point.
        i. Calculate how far the second part of the run was, in km to 2 decimal places.
        ii. Calculate his speed for this section, in km/h to 2 decimal places.
        iii. Calculate how far he needs to run to get back to the starting point, in km to 2 decimal places.
    b. From a fire tower, A, a fire is spotted on a bearing of N42°E. From a second tower, B, the fire is on a bearing of N12°W. The two fire towers are 23 km apart, and A is N63°W of B. Determine how far the fire is from each tower, in km to 2 decimal places.

16. A cliff is 37 m high. The rock slopes outward at an angle of 50° to the horizontal, then cuts back at an angle of 25° to the vertical, meeting the ground directly below the top of the cliff.
    Carol wishes to abseil from the top of the cliff to the ground as shown in the diagram. Her climbing rope is 45 m long, and she needs 2 m to secure it to a tree at the top of the cliff. Determine if the rope will be long enough to allow her to reach the ground.

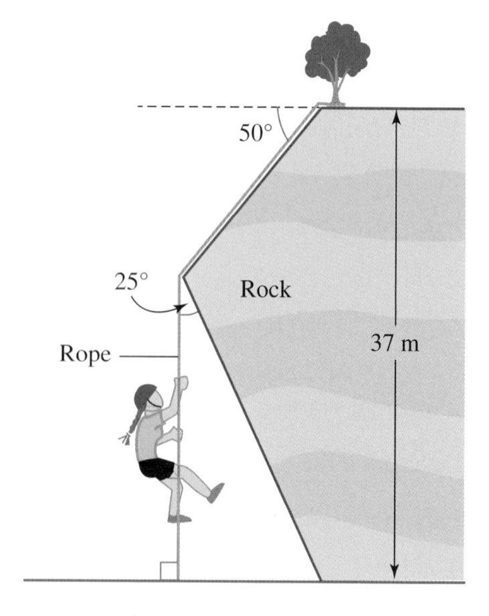

7.3 Exam questions

Question 1 (2 marks) TECH-ACTIVE

In triangle ABC, $AB = 6$ cm, $AC = 8$ cm and angle $ACB = 28°$. Calculate the size of angle ABC correct to 2 decimal places.

Question 2 (2 marks) TECH-ACTIVE

In triangle PQR, PQ is 15 mm and angle RPQ is 40°. If RQ is 12 mm, determine the size of angle PRQ correct to 2 decimal places.

Question 3 (3 marks) TECH-ACTIVE

From point A on a straight road a plane is at an angle of 57° to the vertical. From point B, 2 km away, looking in the same direction, the plane is at an angle of 50° to the vertical. Calculate the altitude of the plane, in km correct to 3 decimal places.

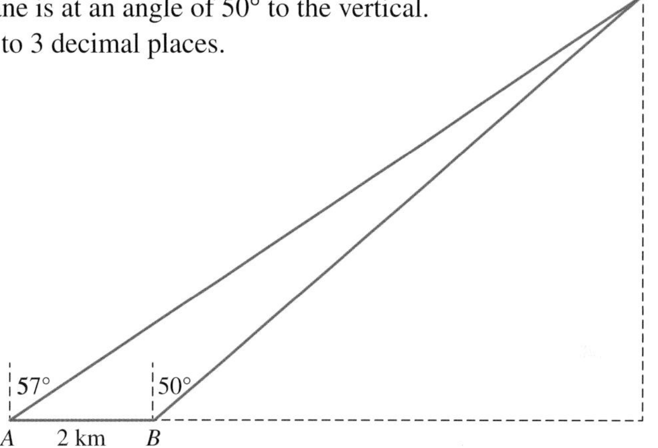

More exam questions are available online.

7.4 The cosine rule

LEARNING INTENTION

At the end of this subtopic you should be able to:

- use the cosine rule to calculate unknown side lengths and angles in triangles.

7.4.1 Deriving the cosine rule

In any non–right-angled triangle ABC, a perpendicular line can be drawn from angle B to side b. Let D be the point where the perpendicular line meets side b, and the length of the perpendicular line be h. Let the length $AD = x$ units. The perpendicular line creates two right-angled triangles, ADB and CDB.

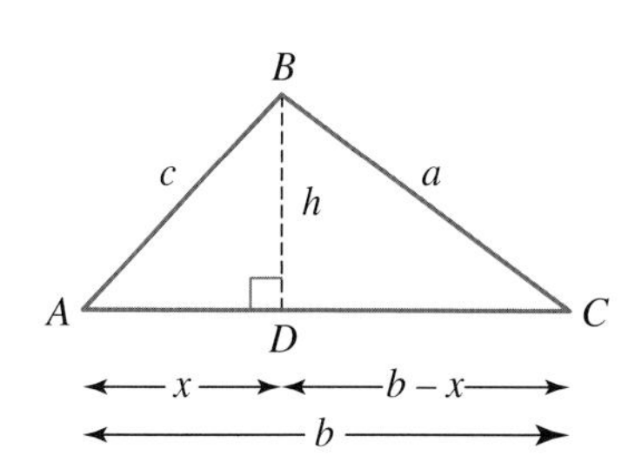

Using triangle ADB and Pythagoras' theorem, we obtain:

$$c^2 = h^2 + x^2 \qquad [1]$$

Using triangle CDB and Pythagoras' theorem, we obtain:

$$a^2 = h^2 + (b - x)^2 \qquad [2]$$

Expanding the brackets in equation [2]:

$$a^2 = h^2 + b^2 - 2bx + x^2$$

Rearranging equation [2] and using $c^2 = h^2 + x^2$ from equation [1]:

$$\begin{aligned} a^2 &= h^2 + x^2 + b^2 - 2bx \\ &= c^2 + b^2 - 2bx \\ &= b^2 + c^2 - 2bx \end{aligned}$$

From triangle ABD, $x = c \times \cos(A)$, therefore $a^2 = b^2 + c^2 - 2bx$ becomes

$$a^2 = b^2 + c^2 - 2bc \times \cos(A)$$

This is called the **cosine rule** and is a generalisation of Pythagoras' theorem.

In a similar way, if the perpendicular line was drawn from angle A to side a or from angle C to side c, the two right-angled triangles would give $c^2 = a^2 + b^2 - 2ab \times \cos(C)$ and $b^2 = a^2 + c^2 - 2ac \times \cos(B)$ respectively. From this, the cosine rule can be stated:

The cosine rule

In any triangle ABC:

$$\mathbf{a^2 = b^2 + c^2 - 2bc\cos(A)}$$
$$\mathbf{b^2 = a^2 + c^2 - 2ac\cos(B)}$$
$$\mathbf{c^2 = a^2 + b^2 - 2ab\cos(C)}$$

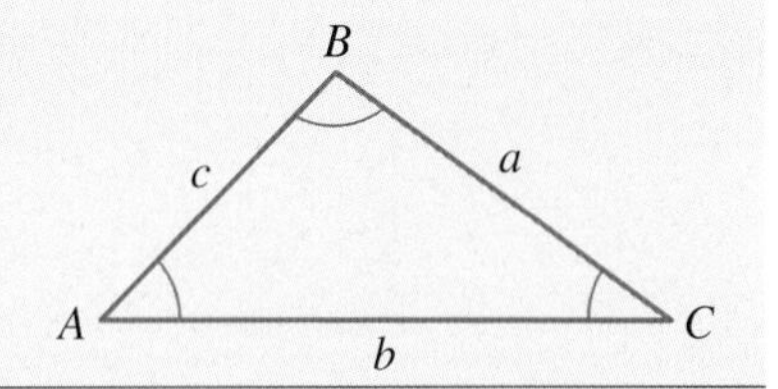

When to use the cosine rule

The cosine rule can be used to solve non–right-angled triangles if we are given:

1. **three sides of the triangle**
2. **two sides of the triangle and the included angle (the angle between the given sides).**

WORKED EXAMPLE 9 Using the cosine rule to determine unknown side lengths

Calculate the length of the third side of triangle ABC correct to 2 decimal places given $a = 6$, $c = 10$ and $B = 76°$.

THINK	WRITE/DRAW
1. Draw a labelled diagram of the triangle ABC and fill in the given information.	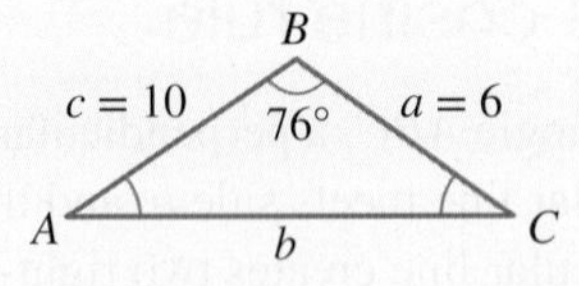
2. Check that one of the criteria for the cosine rule has been satisfied.	Yes, the cosine rule can be used since two side lengths and the included angle have been given.
3. Write the appropriate cosine rule to calculate side b.	To calculate the length of side b: $b^2 = a^2 + c^2 - 2ac\cos(B)$
4. Substitute the given values into the rule.	$= 6^2 + 10^2 - 2 \times 6 \times 10 \times \cos(76°)$
5. Evaluate.	$= 36 + 100 - 120 \times 0.241\,921\,895$ $= 106.969\,372\,5$ $b = \sqrt{106.969\,372\,5}$
6. Round the answer to 2 decimal places.	$= 10.34$ correct to 2 decimal places

Note: Once the third side has been calculated, the sine rule could be used to determine the sizes of the other angles if necessary.

7.4.2 Using the cosine rule to determine unknown angles

If three sides of a triangle are known, an angle could be evaluated by transposing the cosine rule to make cos(A), cos(B) or cos(C) the subject.

Using the cosine rule to determine angles

$$\cos(A) = \frac{b^2 + c^2 - a^2}{2bc}$$

$$\cos(B) = \frac{a^2 + c^2 - b^2}{2ac}$$

$$\cos(C) = \frac{a^2 + b^2 - c^2}{2ab}$$

WORKED EXAMPLE 10 Using the cosine rule to determine unknown angles

Determine the smallest angle in the triangle with sides 4 cm, 7 cm and 9 cm. Give your answer in degrees correct to 1 decimal place.

THINK	WRITE/DRAW
1. Draw a labelled diagram of the triangle, call it ABC and fill in the given information. *Note:* The smallest angle will be opposite the smallest side.	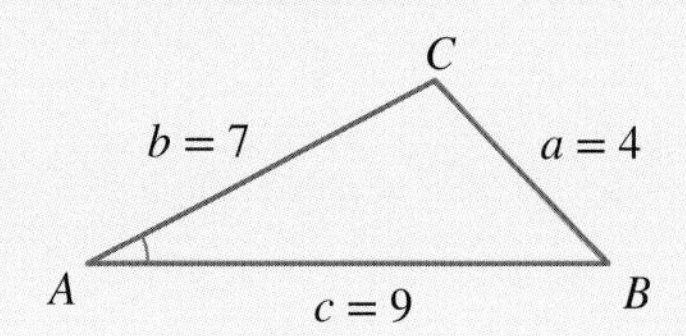 Let $a = 4$, $b = 7$, $c = 9$
2. Check that one of the criteria for the cosine rule has been satisfied.	The cosine rule can be used since three side lengths have been given.
3. Write the appropriate cosine rule to evaluate angle A.	$\cos(A) = \frac{b^2 + c^2 - a^2}{2bc}$
4. Substitute the given values into the rearranged rule.	$= \frac{7^2 + 9^2 - 4^2}{2 \times 7 \times 9}$
5. Evaluate.	$= \frac{49 + 81 - 16}{126}$ $= \frac{114}{126}$
6. Transpose the equation to make A the subject by taking the inverse cos of both sides.	$A = \cos^{-1}\left(\frac{114}{126}\right)$ $= 25.208\,765\,3°$
7. Round the answer to 1 decimal place.	$= 25.2°$

7.4.3 Applying the cosine rule to real-life scenarios

The cosine rule can be used to solve problems involving real-life scenarios. Since it applies to all triangles, not just right-angled triangles there are many situations in which it can be used.

WORKED EXAMPLE 11 Applying the cosine rule to bearings

Two rowers set out from the same point. One rows N70°E for 2000 m and the other rows S15°W for 1800 m. Calculate how far apart the two rowers are, in metres correct to 2 decimal places.

THINK

1. Draw a labelled diagram of the triangle, call it ABC and fill in the given information.

WRITE/DRAW

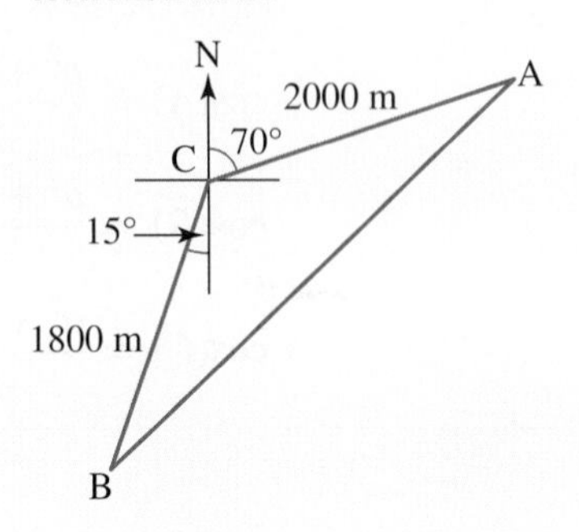

2. Check that one of the criteria for the cosine rule has been satisfied.

The cosine rule can be used since two side lengths and the included angle have been given.

3. Write the appropriate cosine rule to determine the length of side c.

To determine the length of side c:
$c^2 = a^2 + b^2 - 2ab\cos(C)$

4. Substitute the given values into the rule.

$= 2000^2 + 1800^2 - 2 \times 2000 \times 1800 \times \cos(125°)$

5. Evaluate.

$= 40\,000\,000 + 3\,240\,000 - 7\,200\,000 \times (-0.573\,576\,436)$
$= 11\,369\,750.342$
$c = \sqrt{11\,369\,750.342}$
$= 3371.906\,04$

6. Round the answer to 2 decimal places.

$= 3371.91$

7. Answer the question.

The rowers are 3371.91 m apart.

7.4 Exercise

Students, these questions are even better in jacPLUS

Receive immediate feedback and access sample responses

Access additional questions

Track your results and progress

Find all this and MORE in jacPLUS

Technology active

1. WE9 Determine the length of the third side of triangle ABC given $a = 3.4$, $b = 7.8$ and $C = 80°$, to 2 decimal places.

2. In triangle ABC, $b = 64.5$ cm, $c = 38.1$ cm and $A = 58°$. Evaluate the third side, a, in cm correct to 2 decimal places.

3. WE10 Determine the size of the smallest angle in the triangle with sides 6 cm, 4 cm and 8 cm, correct to 1 decimal place.

4. In triangle ABC, $a = 356$, $b = 207$ and $c = 296$. Evaluate the smallest angle, correct to 2 decimal places.

5. WE11 Two rowers set out from the same point. One rows N30°E for 1500 m and the other rows S40°E for 1200 m. Calculate how far apart the two rowers are, correct to the nearest metre.

6. Two rowers set out from the same point. One rows 16.2 km on a bearing of 053° T and the other rows 31.6 km on a bearing of 117° T. Calculate how far apart the two rowers are, correct to the nearest metre.

7. In triangle ABC, $a = 17$, $c = 10$ and $B = 115°$. Determine the size of b to 2 decimal places, and hence determine the size of A and C to 3 decimal places.

8. In triangle ABC, $a = 23.6$, $b = 17.3$ and $c = 26.4$. Determine the size of all the angles, correct to 1 decimal place.

9. In triangle DEF, $d = 3$ cm, $e = 7$ cm and $F = 60°$. Calculate the length of f in exact form.

10. Maria cycles 12 km in a direction of N68°W, then 7 km in a direction of N34°E.
    a. Calculate how far is she from her starting point, in km correct to 2 decimal places.
    b. Calculate the bearing, to the nearest degree, of the starting point from her finishing point.

11. A garden bed is in the shape of a triangle with sides of length 3 m, 4.5 m and 5.2 m.
    a. Calculate the smallest angle, to the nearest degree.
    b. Hence, calculate the area of the garden in square metres correct to 2 decimal places. (*Hint*: Draw a diagram with the longest length as the base of the triangle.)

12. A hockey goal is 3 m wide. When Sophie is 7 m from one post and 5.2 m from the other, she shoots for goal. Calculate within what angle, to the nearest degree, the shot must be made if it is to score a goal.

13. Three circles of radii 5 cm, 6 cm and 8 cm are positioned so that they just touch one another. Their centres form the vertices of a triangle. Calculate the largest angle in the triangle, correct to 1 decimal place.

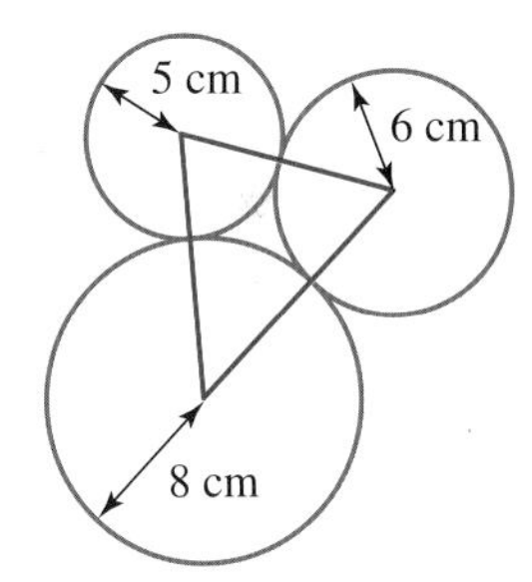

14. An advertising balloon is attached to two ropes 120 m and 100 m long. The ropes are anchored to level ground 35 m apart. Determine how high the balloon can fly, in metres correct to 2 decimal places.

15. A plane flies N70°E for 80 km, then on a bearing of S10°W for 150 km.
    a. Calculate how far the plane is from its starting point, to the nearest km.
    b. Determine the direction of the plane from its starting point, in degrees correct to 1 decimal place.

16. A plane takes off at 10:00 am from an airfield and flies at 120 km/h on a bearing of N35°W. A second plane takes off at 10:05 am from the same airfield and flies on a bearing of S80°E at a speed of 90 km/h. Calculate how far apart the planes are at 10:25 am, in km to 1 decimal place.

17. For the given shape, determine:
    a. the length of the diagonal, in metres to 2 decimal places
    b. the magnitude (size) of angle B, to the nearest degree
    c. the length of x, in metres to 2 decimal places.

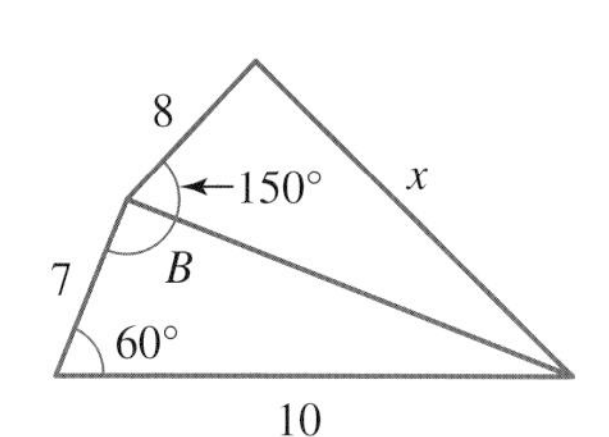

18. From the top of a vertical cliff 68 m high, an observer notices a yacht at sea. The angle of depression to the yacht is 47°. The yacht sails directly away from the cliff, and after 10 minutes the angle of depression is 15°. Determine how fast the yacht sails, in km/h correct to 2 decimal places.

7.4 Exam questions

Question 1 (2 marks) TECH-ACTIVE

Tri looks at a door. One side of the door is 6 metres away while the other side of the door is 7.5 metres away. The angle between his sight lines to either side of the door is 17°. Calculate how wide the door is. Give your answer rounded to the nearest centimetre.

Question 2 (1 mark) TECH-ACTIVE

MC In parallelogram $PQRS$, $\angle QPS = 74°$.

In this parallelogram, $PQ = 18$ cm and $PS = 25$ cm.

The length of the shorter diagonal is closest to

A. 26.5 cm **B.** 30.1 cm **C.** 30.8 cm
D. 26.3 cm **E.** 39.9 cm

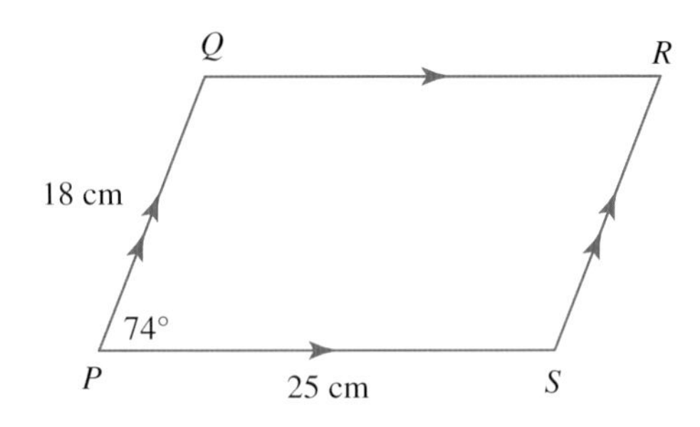

Question 3 (2 marks) TECH-ACTIVE

Determine the value of x in degrees, rounded to 2 decimal places.

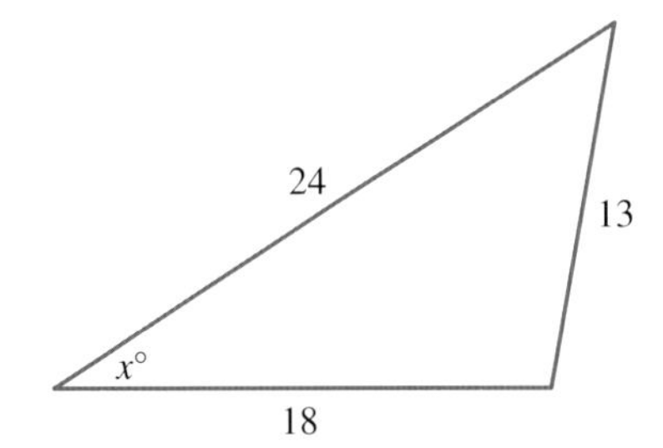

More exam questions are available online.

7.5 Arc length, sectors and segments

LEARNING INTENTION

At the end of this subtopic you should be able to:
- convert angles from radians to degrees and from degrees to radians
- solve problems involving arc length, sectors and segments.

7.5.1 Radian measurement

In all of the trigonometry tasks covered so far, the unit for measuring angles has been the degree. There is another commonly used measurement for angles: the **radian**. This is used in situations involving length and areas associated with circles.

Consider the unit circle, a circle with a radius of 1 unit. OP is the radius.

If OP is rotated $\theta°$ anticlockwise, the point P traces a path along the circumference of the circle to a new point, P_1.

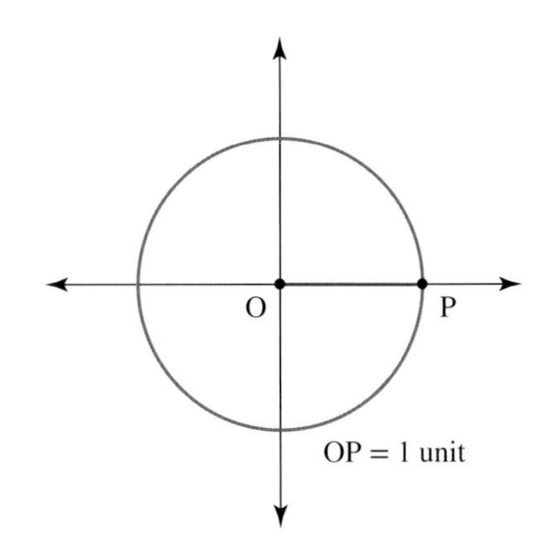

The arc length PP_1 is a radian measurement, symbolised by θ^c.

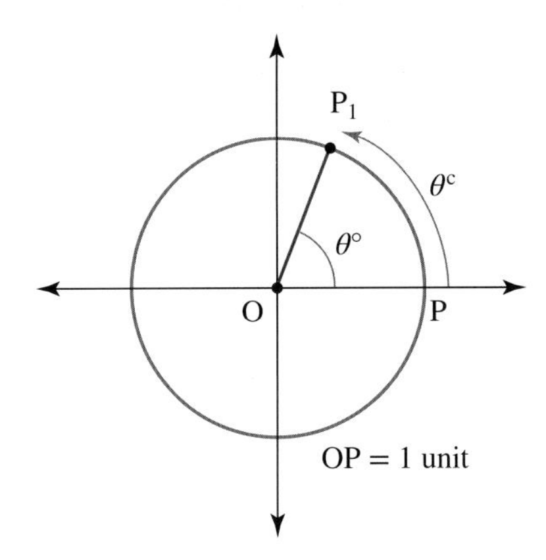

Note: 1^c is equivalent to the angle in degrees formed when the length of PP_1 is 1 unit; in other words, when the arc is the same length as the radius.

If the length OP is rotated 180°, the point P traces out half the circumference. Since the circle has a radius of 1 unit, and $C = 2\pi r$, the arc PP_1 has a length of π.

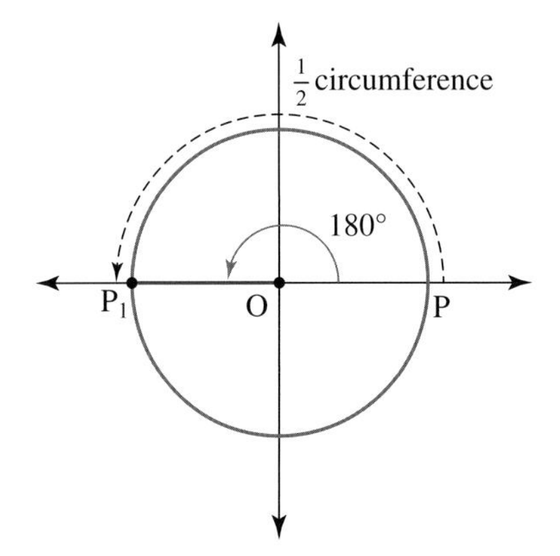

The relationship between degrees and radians is thus established.

$$180° = \pi^c$$

This relationship will be used to convert from one system to another. Rearranging the basic conversion factor gives:

$$\begin{aligned} 180° &= \pi^c \\ 1° &= \frac{\pi^c}{180} \end{aligned}$$

Also, since $\pi^c = 180°$, it follows that $1^c = \dfrac{180°}{\pi}$.

Converting between degrees and radians

To convert an angle in degrees to radians, multiply by $\dfrac{\pi}{180°}$.

To convert an angle in radians to degrees, multiply by $\dfrac{180°}{\pi}$.

Where possible, it is common to have radian values with π in them. It is usual to write radians without any symbol, but degrees must always have a symbol. For example, an angle of 25° must have the degree symbol written, but an angle of $\frac{\pi}{2}$ is understood to be $\frac{\pi}{2}$ radians.

WORKED EXAMPLE 12 Converting between radians and degrees

a. **Convert 135° to radian measure, expressing the answer in terms of π.**

b. **Convert the radian measurement $\frac{4\pi}{5}$ to degrees.**

THINK	WRITE
a. 1. To convert an angle in degrees to radian measure, multiply the angle by $\frac{\pi}{180}$.	**a.** $135° = 135° \times \frac{\pi}{180°}$ $= \frac{135\pi}{180}$
2. Simplify, leaving the answer in terms of π.	$= \frac{3\pi}{4}$
b. 1. To convert radian measure to an angle in degrees, multiply the angle by $\frac{180}{\pi}$. ***Note:*** π cancels out.	**b.** $\frac{4\pi}{5} = \frac{4\pi}{5} \times \frac{180°}{\pi}$ $= \frac{720°}{5}$
2. Simplify.	$= 144°$

7.5.2 Arc length

An arc is a section of the circumference of a circle. The length of an arc is proportional to the angle subtended at the centre.

For example, an angle of 90° will create an arc that is $\frac{1}{4}$ the circumference.

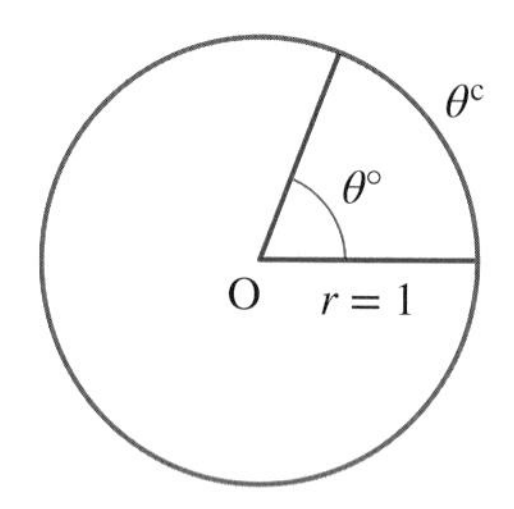

We have already defined an arc length as equivalent to θ radians if the circle has a radius of 1 unit.

Therefore, a simple dilation of the unit circle will enable us to calculate the arc length for any sized circle, as long as the angle is expressed in radians.

If the radius is dilated by a factor of r, the arc length is also dilated by a factor of r.

Therefore, $l = r\theta$, where l represents the arc length, r represents the radius and θ represents an angle measured in radians.

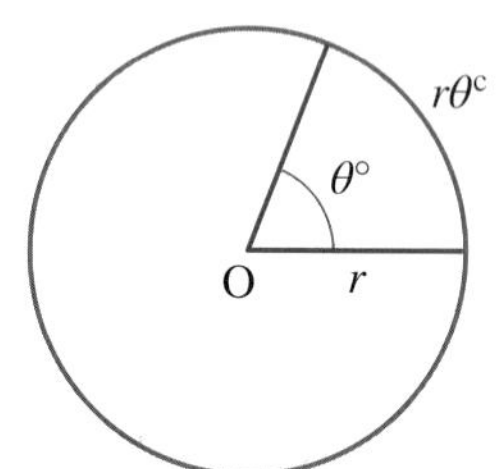

Dilation by factor of r

Arc length of a circle

$$\textbf{arc length } (l) = r\theta$$

Where r represents the radius of the circle and θ represents the angle measured in radians.

WORKED EXAMPLE 13 Calculating the arc length

Determine the length of the arc that subtends an angle of 75° at the centre of a circle with radius 8 cm, in cm to 2 decimal places.

THINK

1. Draw a diagram representing the situation and label it with the given values.

WRITE/DRAW

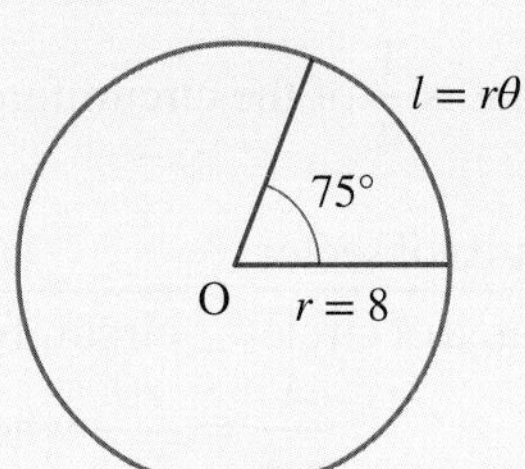

2. Convert the angle from 75° to radian measure by multiplying the angle by $\frac{\pi}{180°}$.

$75° = 75° \times \frac{\pi}{180°}$

$= \frac{75\pi}{180}$

3. Evaluate to 4 decimal places.

$= 1.3090$

4. Write the rule for the length of the arc.

$l = r\theta$

5. Substitute the values into the formula.

$= 8 \times 1.3090$

6. Evaluate to 2 decimal places and include the appropriate unit.

$= 10.4720$

$= 10.47$ cm

Note: In order to use the formula for the length of the arc, the angle must be in radian measure.

WORKED EXAMPLE 14 Determining the angle subtended by an arc

Calculate the angle subtended by a 17 cm arc in a circle of radius 14 cm:

a. in radians, correct to 4 decimal places

b. in degrees, correct to 2 decimal places.

THINK / **WRITE**

a. 1. Write the rule for the length of the arc.

a. $l = r\theta$

2. Substitute the values into the formula.

$17 = 14\theta$

3. Transpose the equation to make θ the subject.

$\theta = \frac{17}{14}$

4. Evaluate to 4 decimal places and include the appropriate unit.

$= 1.214\,285\,714$

$= 1.2143^c$

b. 1. To convert radian measure to an angle in degrees, multiply the angle by $\frac{180°}{\pi}$.

b. $1.2143^c = 1.2143 \times \frac{180°}{\pi}$

2. Evaluate to 2 decimal places.

$= 69.573\,446\,55°$

$= 69.57°$

7.5.3 Area of a sector

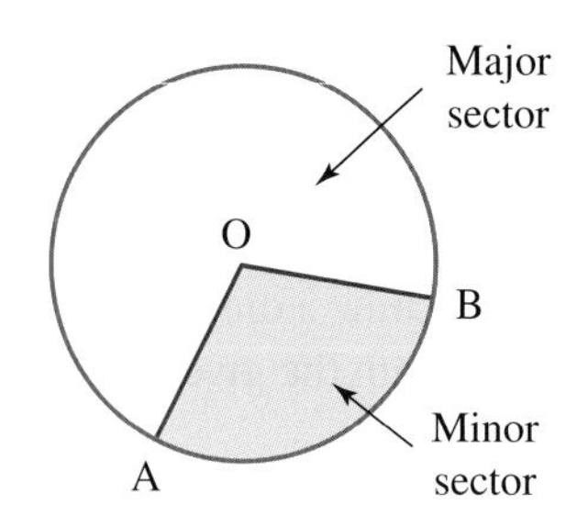

In the diagram, the shaded area is the *minor* sector AOB, and the unshaded area is the major sector AOB.

The area of the sector is proportional to the arc length. For example, an area of $\frac{1}{4}$ of the circle contains an arc that is $\frac{1}{4}$ of the circumference.

$$\text{Thus, in any circle: } \frac{\text{area of sector}}{\text{area of circle}} = \frac{\text{arc length}}{\text{circumference of circle}}$$

$$\frac{A}{\pi r^2} = \frac{r\theta}{2\pi r} \text{ where } \theta \text{ is measured in radians.}$$

$$A = \frac{r\theta \times \pi r^2}{2\pi r}$$

$$= \frac{1}{2} r^2 \theta$$

Area of a sector

$$\textbf{Area of a sector} = \frac{1}{2} r^2 \theta$$

WORKED EXAMPLE 15 Determining the radius of a circle given a sector area and angle

A sector has an area of 157 cm^2 and subtends an angle of 107°. Calculate the radius of the circle, in cm to 2 decimal places.

THINK	WRITE
1. Convert the angle from 107° to radian measure by multiplying the angle by $\frac{\pi}{180°}$.	$107° = 107° \times \frac{\pi}{180°}$ $= \frac{107\pi}{180}$
2. Evaluate to 4 decimal places.	$= 1.8675$
3. Write the rule for the area of a sector.	$A = \frac{1}{2} r^2 \theta$
4. Substitute the values into the formula.	$157 = \frac{1}{2} \times r^2 \times 1.8675$
5. Transpose the equation to make r^2 the subject.	$r^2 = \frac{2 \times 157}{1.8675}$ $r^2 = 168.139\,016\,5$
6. Take the square root of both sides of the equation.	$r = 12.966\,842\,97$
7. Evaluate to 2 decimal places and include the appropriate unit.	$= 12.97$ cm

7.5.4 Area of a segment

A segment is that part of a sector bounded by the arc and the chord.

As can be seen from the diagram:

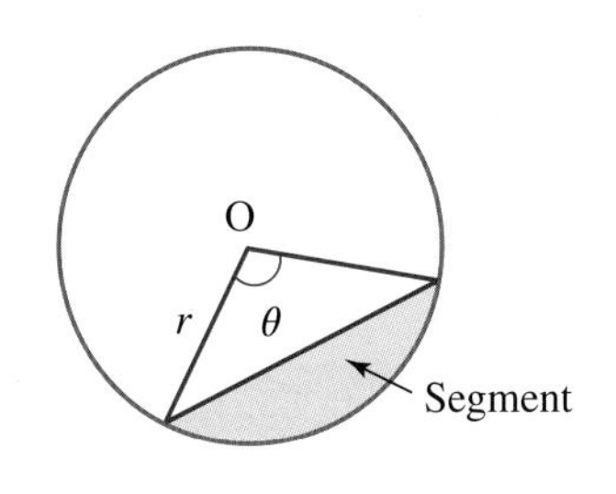

$$\begin{aligned}\text{Area of segment} &= \text{area of sector} - \text{area of triangle}\\ A &= \frac{1}{2}r^2\theta - \frac{1}{2}r^2\sin(\theta°)\\ &= \frac{1}{2}r^2\left(\theta - \sin(\theta°)\right)\end{aligned}$$

Note: θ is in radians and $\theta°$ is in degrees.

Area of a segment

$$\textbf{Area of a segment} = \frac{1}{2}r^2\left(\theta - \sin(\theta°)\right)$$

WORKED EXAMPLE 16 Determining the area of a segment

Determine the area of the segment in a circle of radius 5 cm, subtended by an angle of 40°, in square centimetres correct to 2 decimal places

THINK	WRITE
1. Convert the angle from 40° to radian measure by multiplying the angle by $\frac{\pi}{180°}$.	$40° = 40° \times \frac{\pi}{180°}$ $= \frac{40\pi}{180}$
2. Evaluate to 4 decimal places.	$= 0.6981$
3. Write the rule for the area of a segment.	$A = \frac{1}{2}r^2\left(\theta - \sin(\theta°)\right)$
4. Identify each of the variables.	$r = 5,\ \theta = 0.6981,\ \theta° = 40°$
5. Substitute the values into the formula.	$A = \frac{1}{2} \times 5^2\left(0.6981 - \sin(40°)\right)$
6. Evaluate.	$= \frac{1}{2} \times 25 \times 0.0553$ $= 0.691\,25$
7. Round to 2 decimal places and include the appropriate unit.	$= 0.69\text{ cm}^2$

7.5 Exercise

Students, these questions are even better in jacPLUS

Receive immediate feedback and access sample responses

Access additional questions

Track your results and progress

Find all this and MORE in jacPLUS

Technology free

1. WE 12 Convert the following angles to radian measure, expressing answers in terms of π.

a. 30° b. 60° c. 120° d. 150° e. 225°
f. 270° g. 315° h. 480° i. 72° j. 200°

2. Convert the following radian measurements into degrees.

a. $\frac{\pi}{4}$ b. $\frac{3\pi}{2}$ c. $\frac{7\pi}{6}$ d. $\frac{5\pi}{3}$ e. $\frac{7\pi}{12}$
f. $\frac{17\pi}{6}$ g. $\frac{\pi}{12}$ h. $\frac{13\pi}{10}$ i. $\frac{11\pi}{8}$ j. 8π

Technology active

3. WE13 Determine the length of the arc that subtends an angle of 65° at the centre of a circle of radius 14 cm, in cm correct to 2 decimal places.

4. Calculate the length of the arc that subtends an angle of 153° at the centre of a circle of radius 75 mm, in mm correct to 2 decimal places.

5. WE14 Calculate the angle, correct to 4 decimal places, subtended by a 20 cm arc in a circle of radius 75 cm:

a. in radians
b. in degrees.

6. Calculate the angle subtended by an 8 cm arc in a circle of radius 5 cm:

a. in radians, to 1 decimal place
b. in degrees, to 4 decimal places.

7. WE15 A sector has an area of 825 cm^2 and subtends an angle of 70°. Calculate the radius of the circle, in cm to 2 decimal places.

8. A sector has an area of 309 cm^2 and subtends an angle of 106°. Calculate the radius of the circle, in cm to 2 decimal places.

9. WE16 Determine the area of the segment in a circle of radius 25 cm subtended by an angle of 100°, in square centimetres to 2 decimal places.

10. Calculate the area of the segment of a circle of radius 4.7 m that subtends an angle of 85° at the centre, in square metres to 2 decimal places.

11. Convert the following angles in degrees to radians, giving answers to 4 decimal places.

a. 27° b. 109° c. 243° d. 351° e. 7°
f. 63.7° g. 138.35° h. 274.13° i. 326.88° j. 47.03°

12. Convert the following radian measurements into degrees, giving answers to 2 decimal places.

a. 2.345 b. 0.6103 c. 1 d. 1.61 e. 3.592
f. 7.25 g. 0.182 h. 5.8402 i. 4.073 j. 6.167

13. Determine the length of the arc that subtends an angle of 135° at the centre of a circle of radius 10 cm. Leave the answer in terms of π.

14. An arc of a circle is 27.8 cm long and subtends an angle of 205° at the centre of the circle. Calculate the radius of the circle, in cm to 2 decimal places.

15. An arc of length 8 cm is marked out on the circumference of a circle of radius 13 cm. Determine what angle the arc subtends at the centre of the circle, to the nearest degree.

16. The minute hand of a clock is 35 cm long. Determine how far the tip of the hand travels in 20 minutes, in cm to 2 decimal places.

17. A child's swing is suspended by a rope 3 m long. Calculate the length of the arc it travels if it swings through an angle of 42°, in metres to 2 decimal places.

18. Determine the exact area of the sector of a circle of radius 6 cm with an angle of 100°. Write your answer in terms of π.

19. A garden bed is in the form of a sector of a circle of radius 4 m. The arc of the sector is 5 m long. Determine:
    a. the area of the garden bed, to the nearest square metre
    b. the volume of mulch needed to cover the bed to a depth of 10 cm, to the nearest cubic metre.

20. A sector whose angle is 150° is cut from a circular piece of cardboard whose radius is 12 cm. The two straight edges of the sector are joined so as to form a cone.
    a. Determine the surface area of the cone, in square centimetres to 1 decimal place.
    b. Calculate the radius of the cone, to the nearest centimetre.

21. Two irrigation sprinklers spread water in circular paths with radii of 7 m and 4 m. If the sprinklers are 10 m apart, determine the area of crop that receives water from both sprinklers, in square metres to 2 decimal places.

22. Two circles of radii 3 cm and 4 cm have their centres 5 cm apart. Calculate the area of the intersection of the two circles, in square centimetres to 2 decimal places.

7.5 Exam questions

Question 1 (2 marks) TECH-ACTIVE

A segment of a circle, radius 10 cm, has an angle of 30°. Determine the area of the segment, in square centimetres correct to 3 decimal places.

Question 2 (2 marks) TECH-ACTIVE

The area of a sector of a circle is 500 square centimetres. The central angle is 40°. Calculate the radius, in centimetres correct to 2 decimal places.

Question 3 (2 marks) TECH-FREE

For a particular circle, the area of a sector and the arc length of the same sector are numerically the same. Determine the radius of the circle.

More exam questions are available online.

7.6 Review

doc-37050

7.6.1 Summary

Hey students! Now that it's time to revise this topic, go online to:

 Access the topic summary  Review your results  Watch teacher-led videos  Practise exam questions

Find all this and MORE in jacPLUS

7.6 Exercise

Technology free: short answer

1. A stepladder stands on a floor with its feet 2 m apart. If the angle formed by the legs with the floor is 60°, calculate exactly how high above the floor the top of the ladder is.

2. Two buildings, 15 m and 27 m high, are directly opposite each other across a river. The angle of depression of the top of the smaller building from the top of the taller one is 30°. Determine how wide the river is, in metres to 1 decimal place. *Note:* You may use a scientific calculator to express your answer as a decimal.

3. In the triangle shown, calculate the exact length of side x in cm.

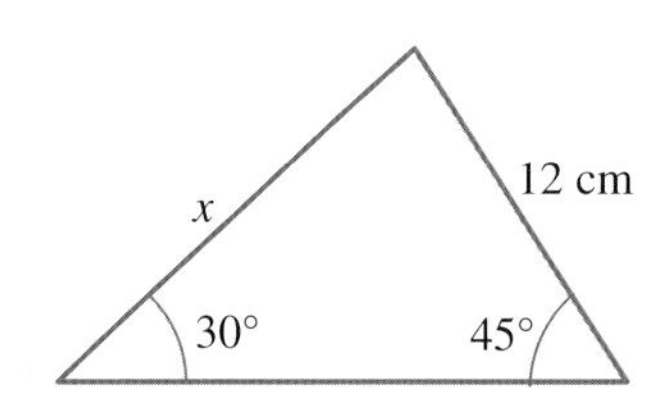

4. A triangle has sides of length 12 m, 15 m and 20 m. If B is the largest angle, evaluate $\cos(B)$, expressing your answer as a simplified fraction.

5. a. Convert the following angles to radian measure, expressing answers in terms of π.
   i. 80° ii. 125° iii. 640°
   b. Convert the following radian measurements into degrees.
   i. $\frac{\pi}{20}$ ii. $\frac{15\pi}{8}$ iii. 7π

6. A paddock is in the shape of a sector with radius of 75 m and an angle of 60°. Using a scientific calculator to help where required, determine:
   a. the amount of fencing needed to enclose the paddock, in metres to 2 decimal places
   b. the area enclosed by the paddock, in square metres to 2 decimal places.

Technology active: multiple choice

7. MC In the triangle shown, the value of θ, to the nearest degree, is

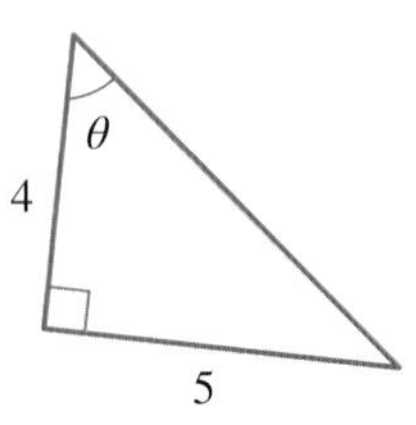

   A. 37°
   B. 39°
   C. 51°
   D. 52°
   E. 53°

8. MC A ladder 4.5 m long rests against a vertical wall, with the foot of the ladder 2 m from the base of the wall. The angle the ladder makes with the wall, to the nearest degree, is
   A. 24° B. 26° C. 35° D. 64° E. 66°

9. **MC** A person stands 18 m from the base of a building and measures the angle of elevation to the top of the building as 62°. If the person is 1.8 m tall, the height of the building, to the nearest metre, is

A. 11 m B. 18 m C. 36 m D. 22 m E. 34 m

10. **MC** A bearing of 310° T is the same as

A. N40°W B. N50°W C. S50°W D. S50°E E. N50°E

11. **MC** In triangle ABC, $a = 10$, $b = 7$ and $B = 40°$. A possible value for C, to the nearest degree, is

A. 37° B. 52° C. 68° D. 73° E. 113°

12. **MC** Two boats start from the same point. One sails due north for 10 km and the other sails south-east for 15 km. Their distance apart is closest to

A. 10.62 km B. 14.83 km C. 17.35 km D. 21.38 km E. 23.18 km

13. **MC** A triangle has sides measuring 5 cm, 8 cm and 10 cm. The largest angle in the triangle, to the nearest degree, is

A. 52° B. 82° C. 98° D. 128° E. 140°

14. **MC** A garden bed is in the shape of a triangle with sides of length 4 m, 5.2 m and 7 m. The volume of topsoil needed to cover the garden to a depth of 250 mm is

A. 2.32 m^3 B. 2.57 m^3 C. 2.81 m^3 D. 3.17 m^3 E. 3.76 m^3

15. **MC** When 75° is converted to radian measure, the value of the angle, expressed in terms of π, is

A. $\frac{12\pi}{5}$ B. $\frac{\pi}{12}$ C. $\frac{5\pi}{24}$ D. $\frac{5\pi}{12}$ E. $\frac{7\pi}{12}$

16. **MC** The area of the shaded region in the figure, to the nearest cm^2, is

A. 800 cm^2
B. 846 cm^2
C. 898 cm^2
D. 952 cm^2
E. 983 cm^2

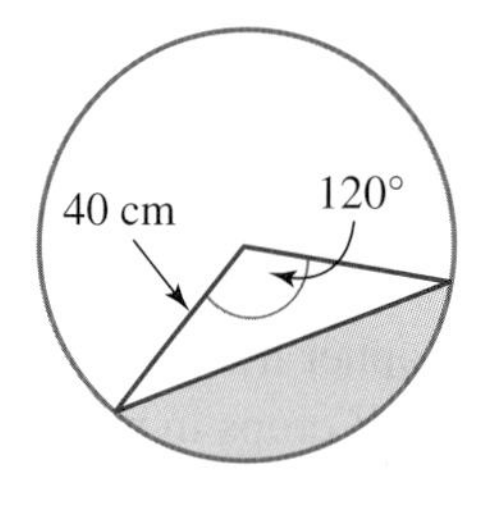

Technology active: extended response

17. Three circles of radii 2 cm, 3 cm and 4 cm are placed so that they just touch each other. A triangle is formed by joining their three centres. Determine:

a. the three angles of the triangle, in degrees correct to 2 decimal places
b. the area of the triangle, in square centimetres correct to 3 decimal places
c. the shaded area, in square centimetres correct to 3 decimal places.

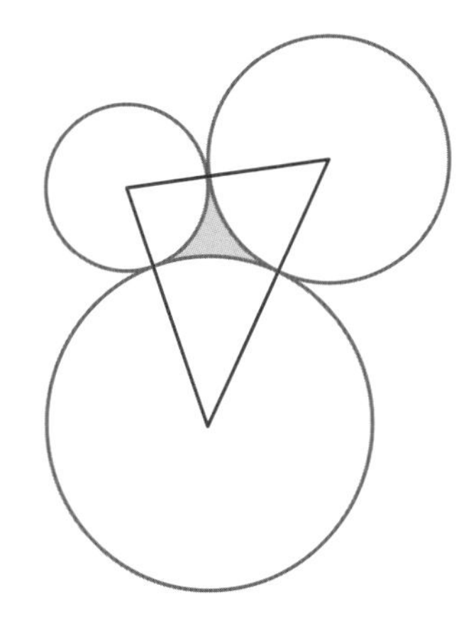

18. A farmer owns a large triangular area of flat land, bounded on one side by an embankment to a river flowing north-east, on a second side by a road that meets the river at a bridge where the angle between river and road is 105°, and on the third side by a long fence.

a. Determine the length of the river frontage, in km correct to 3 decimal places.
The farmer decides to divide the land into two sections of equal area by running a fence from the bridge to a point on the opposite side.

b. Determine on what bearing the fence must be built, correct to 3 decimal places

c. Calculate the length of the fence, in km correct to 3 decimal places.

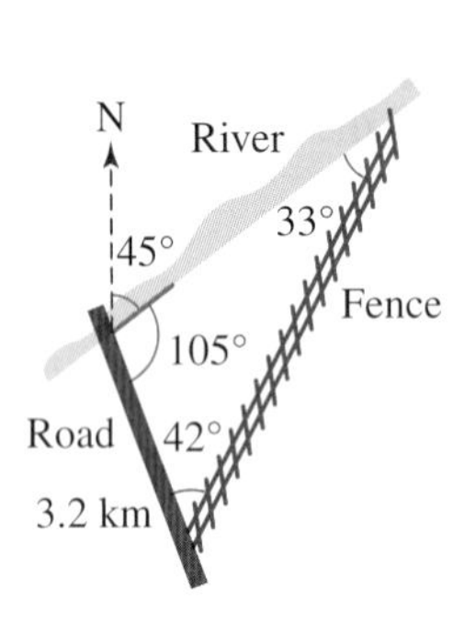

19. a. A four-wheel-drive vehicle leaves a camp site and travels across a flat sandy plain in a direction of S65° E for a distance of 8.2 km. It then heads due south for 6.7 km to reach a waterhole.

i. Calculate how far the waterhole is from the camp site, in km to 2 decimal places.
ii. Determine the bearing of the waterhole from the camp site, in degrees to 2 decimal places.

b. A search plane sets off to find the vehicle. It is on a course that takes it over points A and B, two locations on level ground. At a certain time, from point A, the angle of elevation to the plane is 72°. From point B, the angle of elevation is 47°. If A and B are 3500 m apart, determine the height of the plane off the ground, to the nearest metre.

20. Christopher lives on a farm. He has decided that this year he will plant a variety of crops in his large but unusually shaped vegetable garden. He has divided the vegetable garden into six triangular regions, which he will fence off as shown in the diagram. Christopher needs to calculate the perimeter and area of each region so he can purchase the correct amount of fencing material and seedlings.

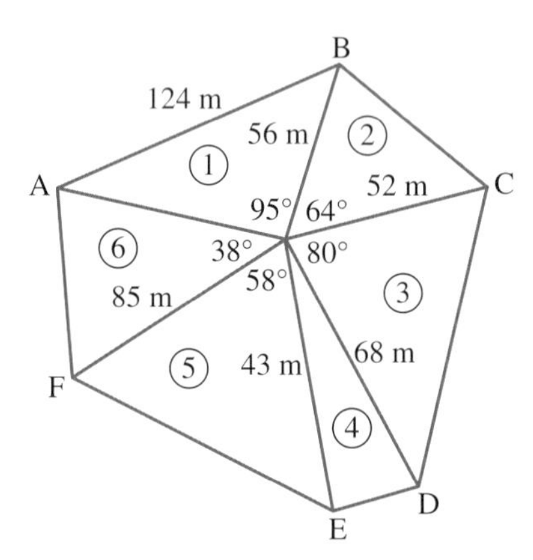

a. Separate each of the regions into single triangles and label each with the information provided.

b. Use the appropriate rules to determine all unknown lengths and relevant angles.

c. Calculate how much fencing material is required to section off the six regions, in metres to 2 decimal places.

d. If fencing material is \$4.50 per metre (and is only sold by the metre), determine the cost to fence the six regions.

7.6 Exam questions

Question 1 (4 marks) TECH-ACTIVE

An orienteering course is as follows:

From S go 7 km to P on a bearing of 050°. Then go 5 km to Q on a bearing of 120°. Calculate how far S is from Q, in km to 2 decimal places, and determine the true bearing of S from Q in degrees to 2 decimal places.

Question 2 (4 marks) TECH-ACTIVE

A yacht sails on a bearing of 045° for 10 km to point P, then sails on a bearing of 250° for 15 km to point Q. Calculate the true bearing of the yacht's starting point from the point Q, in degrees to 2 decimal places.

Question 3 (1 mark) TECH-ACTIVE

MC From a point, A, west of a cliff, the angle of elevation of the top of a cliff from A is 30°. One kilometre due east of A lies B. From B, the angle of elevation of the cliff is 25°. The height of the cliff is

- **A.** 208 m
- **B.** 258 m
- **C.** 299 m
- **D.** 320 m
- **E.** 323 m

Question 4 (1 mark) TECH-ACTIVE

MC Consider the diagram of campsites *A*, *B* and *C*.

Campsite *A* is due west of campsite *B*. Campsite *C* is 5 km away on a bearing of 50°T from campsite *A*. Campsite *C* is on a bearing of 325° from campsite *B*.

The distance between campsite *A* and campsite *B* is

- **A.** 4.11 km
- **B.** 6.08 km
- **C.** 7.78 km
- **D.** 8.71 km
- **E.** 9.07 km

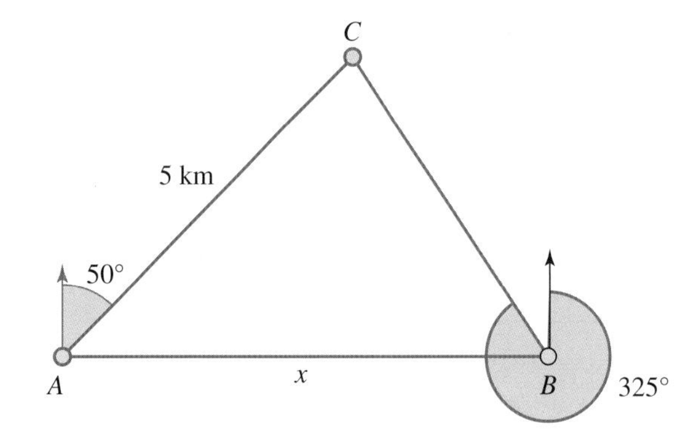

Question 5 (1 mark) TECH-ACTIVE

Calculate the area of the shape in the diagram to the nearest square centimetre.

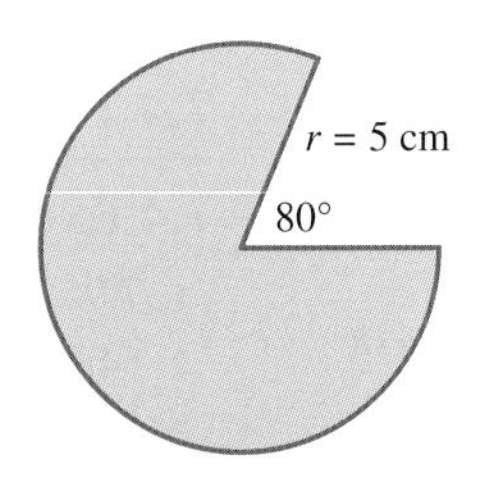

More exam questions are available online.

Answers

Topic 7 Trigonometric ratios and applications

7.2 Review of trigonometry

7.2 Exercise

1. a. 6.43 b. 12.00
   c. 24.93 d. 363.69
2. a. 44.43° b. 67.38°
   c. 44.42° d. 17.16°
3. a. $2\sqrt{3}$ cm b. $12\sqrt{3}$ cm^2
   c. $12 + 8\sqrt{3}$ cm
4. $26\sqrt{3} + 54$ m
5. a. 325° T b. 227° T c. 058° T
   d. 163° T e. S66°W f. S73°E
   g. N39°W h. N74°E
6. 571 m
7. 1691 m
8. S21.8° W or 201.8° T
9. 326.3° T
10. a. 4.98 m b. 67°
11. 7°
12. 23°
13. 1.44 m
14. $a = 14.90$, $b = 20.05$
15. a.

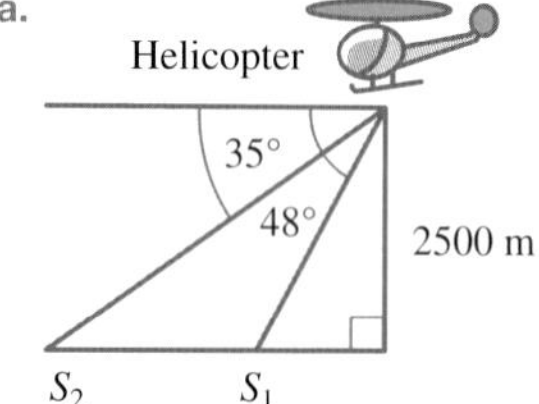

   b. 1319.36 m
16. 22 m
17. $x = 13.39$
18. 10.91 m^3
19. a. 12.2 km b. 348° T or N12°W
20. a. 0.76 m
    b. No, the foot of the ladder moves through a distance of 0.96 m.
21. a. 29.82 km b. 38.08 km c. 232° T
22. a. 112.76 km
    b. 5 hours 30 minutes
23. a 51.96° b 42.11°
24. 14.48°
25. 57.65 m

7.2 Exam questions

Note: Mark allocations are available with the fully worked solutions online.

1. $90° - \tan^{-1}\left(\frac{1}{2}\right)$
2. D
3. 219 m

7.3 The sine rule

7.3 Exercise

1. 45.0°, 77.0°, 13.8
2. 39.42°, 39.58°, 17.41
3. $A = 73.25°$, $b = 8.73$; or $A = 106.75°$, $b = 4.12$
4. 51.90 or 44.86
5. 33°, 38.98, 21.98
6. 19.12
7. $B = 48.44°$, $C = 103.56°$, $c = 66.26$; or $B = 131.56°$, $C = 20.44°$, $c = 23.8$
8. 24.17
9. A
10. $C = 110°$, $a = 3.09$, $b = 4.64$
11. $B = 33.5°$, $C = 121.5°$, $c = 26.2$; or $B = 146.5°$, $C = 8.5°$, $c = 4.6$
12. 43.62 m
13. a. 6.97 m b. 4 m
14. a. 13.11 km b. N20.8°W or 339.2° T
15. a. i. 8.63 km ii. 6.48 km/h iii. 9.90 km
    b. 22.09 km from A and 27.46 km from B
16. Yes, she needs 43 m altogether.

7.3 Exam questions

Note: Mark allocations are available with the fully worked solutions online.

1. 38.75°
2. 126.54°
3. 5.745 km

7.4 The cosine rule

7.4 Exercise

1. 7.95
2. 54.84 cm
3. 29.0°
4. 35.53°
5. 2218 m
6. 28 km
7. 23.08, 41.879°, 23.121°

8. $A = 61.3°$, $B = 40.0°$, $C = 78.8°$
9. $\sqrt{37}$ cm
10. a. 12.57 km b. S35°E or 145°T
11. a. 35° b. 6.73 m^2
12. 23°
13. 70.8°
14. 89.12 m
15. a. 130 km b. S22.2°E or 157.8°T
16. 74.3 km
17. a. 8.89 m b. 77° c. $x = 10.07$ m
18. 1.14 km/h

7.4 Exam questions

Note: Mark allocations are available with the fully worked solutions online.

1. 2.49 m
2. D
3. $x = 32.21°$

7.5 Arc length, sectors and segments

7.5 Exercise

1. a. $\frac{\pi}{6}$ b. $\frac{\pi}{3}$ c. $\frac{2\pi}{3}$
d. $\frac{5\pi}{6}$ e. $\frac{5\pi}{4}$ f. $\frac{3\pi}{2}$
g. $\frac{7\pi}{4}$ h. $\frac{8\pi}{3}$ i. $\frac{2\pi}{5}$
j. $\frac{10\pi}{9}$
2. a. 45° b. 270° c. 210°
d. 300° e. 105° f. 510°
g. 15° h. 234° i. 247.5°
j. 1440°
3. 15.88 cm
4. 200.28 mm
5. a. 0.2667^c b. 15.2789°
6. a. 1.6^c b. 91.6732°
7. 36.75 cm
8. 18.28 cm
9. 237.66 cm^2
10. 5.38 m^2
11. a. 0.4712 b. 1.9024 c. 4.2412
d. 6.1261 e. 0.1222 f. 1.1118
g. 2.4147 h. 4.7845 i. 5.7052
j. 0.8208
12. a. 134.36° b. 34.97° c. 57.30°
d. 92.25° e. 205.81° f. 415.39°
g. 10.43° h. 334.62° i. 233.37°
j. 353.34°
13. $\frac{15\pi}{2}$
14. 7.77 cm
15. 35°
16. 73.30 cm
17. 2.20 m
18. $A = 10\pi$ cm^2
19. a. 10 m^2 b. 1 m^3
20. a. 188.5 cm^2 b. 5 cm
21. 2.95 m^2
22. 6.64 cm^2

7.5 Exam questions

Note: Mark allocations are available with the fully worked solutions online.

1. 1.180 units^2
2. 37.85 cm
3. 2 units

7.6 Review

7.6 Exercise

Technology free: short answer

1. $\sqrt{3}$ m
2. 20.8 m
3. $x = 12\sqrt{2}$ cm
4. $-\frac{31}{360}$
5. a. i. $\frac{4\pi}{9}$ ii. $\frac{25\pi}{36}$ iii. $\frac{32\pi}{9}$
b. i. 9° ii. 337.5° iii. 1260°
6. a. 228.54 m b. 2945.24 m^2

Technology active: multiple choice

7. C
8. B
9. C
10. B
11. D
12. E
13. C
14. B
15. E
16. E

Technology active: extended response

17. a. 44.42°, 57.12°, 78.46°
b. 14.697 cm^2
c. 1.270 cm^2
18. a. 3.931 km
b. N89.887°E or 89.887°T
c. 2.190 km
19. a. i. 12.59 km ii. S36.17° E
b. 2783 m

20. a–b.

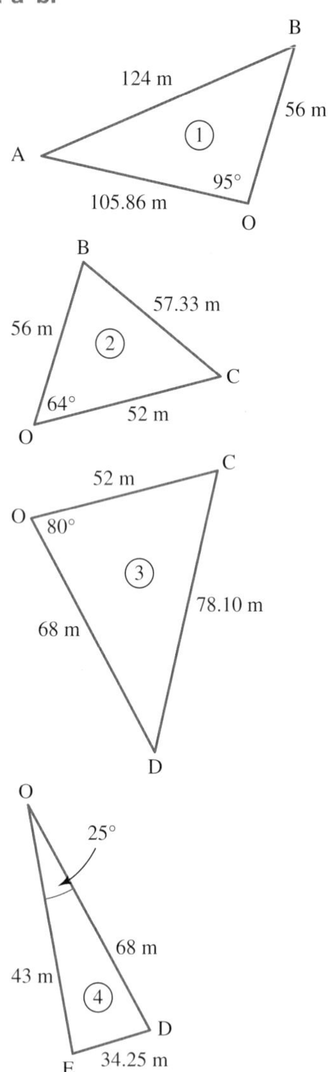

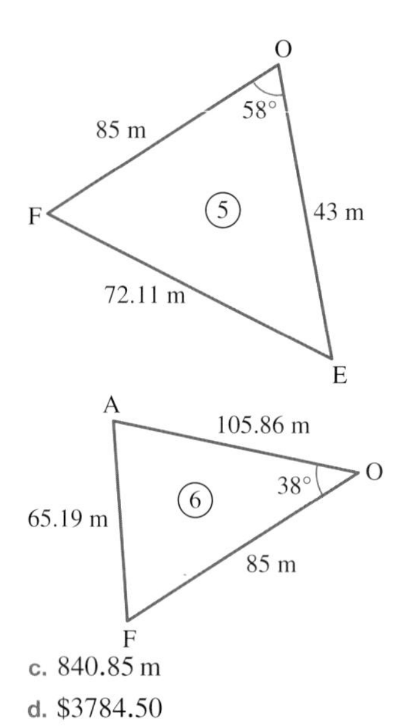

c. 840.85 m

d. $3784.50

7.6 Exam questions

Note: Mark allocations are available with the fully worked solutions online.

1. 9.90 km, 258.36° T
2. 120.11° T
3. B
4. B
5. 61 cm^2

8 Trigonometric identities

LEARNING SEQUENCE

Fully worked solutions for this topic are available online.

8.1 Overview

Hey students! Bring these pages to life online

Watch videos

Engage with interactivities

Answer questions and check results

Find all this and MORE in jacPLUS

8.1.1 Introduction

An identity is a relationship that holds true for all possible values of the variable or variables. This chapter explores trigonometric identities — identities that use functions of one or more angles. You have already used the trigonometric identity $\tan(A) = \dfrac{\sin(A)}{\cos(A)}$. Identities are useful for simplifying expressions involving trigonometric functions, which becomes important in the study of calculus. These trigonometric identities, and calculus more broadly, form the building blocks of architecture and are used across all fields of engineering.

The use of trigonometric identities in architecture is both aesthetic and practical. Many modern buildings, as demonstrated by the images in this chapter, make use of trigonometry, geometry and calculus to create complex, beautiful facades. But this use of mathematics in architecture is more than superficial. Architects must ensure the buildings they design can withhold the loads and forces which act upon these structures. Although these loads and forces are typically defined by vectors, the vector components can be resolved through the use of trigonometric functions relative to the angle the force makes with an axis. Architects also use trigonometry for passive solar design to create comfortable and energy-efficient buildings.

The Russian mathematician Pafnuty Chebyshev (1821–1894) is better known for his work in the fields of probability, statistics, number theory and differential equations, but he also devised recurrence relations for trigonometric multiple angles that we explore later in this chapter.

KEY CONCEPTS

This topic covers the following key concepts from the VCE Mathematics Study Design:

- compound and double angle formulas for sine, cosine and tangent and the identities:
$$\sec^2(x) = 1 + \tan^2(x) \text{ and } \operatorname{cosec}^2(x) = 1 + \cot^2(x)$$
- proof and application of identities between $a\sin(x) + b\cos(x)$ and $r\sin(x \pm \alpha)$ or $r\cos(x \pm \alpha)$ where α is in the first quadrant, the identities for products of sines and cosines expressed as sums and differences, and the identities for addition and differences of sines and cosines expressed as products.

8.2 Pythagorean identities

LEARNING INTENTION

At the end of this subtopic you should be able to:
- use the Pythagorean identities to solve trigonometric unknowns.

8.2.1 Using Pythagorean identities to solve trigonometric unknowns

Recall the Pythagorean theorem $c^2 = a^2 + b^2$ which relates the lengths of the sides of right-angled triangles. Consider the right-angled triangle associated with a point on the unit circle as shown in the diagram.

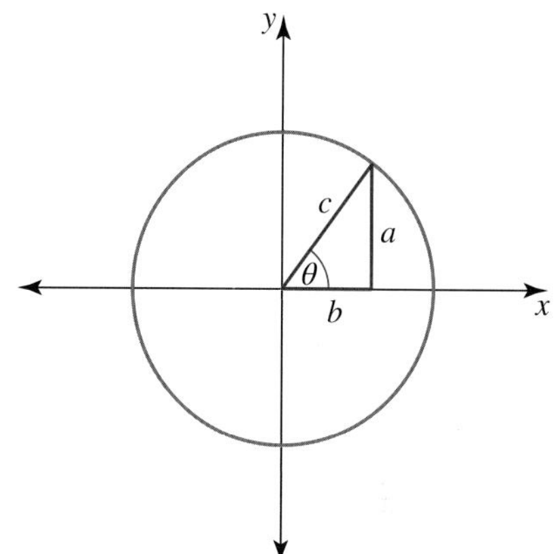

On the unit circle $c = 1$, so $a = \sin(\theta)$ and $b = \cos(\theta)$. By substituting these values into the Pythagorean theorem we get our first Pythagorean identity:

$$c^2 = a^2 + b^2$$

$$1 = \sin^2(\theta) + \cos^2(\theta)$$

A second Pythagorean identity can be derived from $1 = \sin^2(\theta) + \cos^2(\theta)$ by dividing all terms by $\cos^2(\theta)$.

$$\frac{\sin^2(\theta)}{\cos^2(\theta)} + \frac{\cos^2(\theta)}{\cos^2(\theta)} = \frac{1}{\cos^2(\theta)}$$

$$\tan^2(\theta) + 1 = \sec^2(\theta)$$

Similarly, a third Pythagorean identity can be derived from $1 = \sin^2(\theta) + \cos^2(\theta)$ by dividing by all terms by $\sin^2(\theta)$.

$$\frac{\sin^2(\theta)}{\sin^2(\theta)} + \frac{\cos^2(\theta)}{\sin^2(\theta)} = \frac{1}{\sin^2(\theta)}$$

$$1 + \cot^2(\theta) = \mathrm{cosec}^2(\theta)$$

Pythagorean identities

$$\mathbf{1 = \sin^2(\theta) + \cos^2(\theta)}$$

$$\mathbf{\tan^2(\theta) + 1 = \sec^2(\theta)}$$

$$\mathbf{1 + \cot^2(\theta) = \mathrm{cosec}^2(\theta)}$$

We can use these identities to solve a range of problems.

WORKED EXAMPLE 1 Evaluating trigonometric values using trigonometric identities (1)

If $\sin(A) = 0.4$ and $0° < A < 90°$, determine the exact value of $\cos(A)$ correct to 3 decimal places.

THINK	WRITE
1. Use the identity $\sin^2(A) + \cos^2(A) = 1$.	$\sin^2(A) + \cos^2(A) = 1$
2. Substitute 0.4 for $\sin(A)$.	$(0.4)^2 + \cos^2(A) = 1$
3. Solve the equation for $\cos(A)$ correct to 3 decimal places.	$\cos^2(A) = 1 - 0.16$ $= 0.84$ $\cos(A) = \pm\sqrt{0.84}$
4. Retain the positive answer only as cosine is positive in the first quadrant and round the answer as required.	For $0° < A < 90°$, cos is positive so $\cos(A) = \sqrt{0.84} = 0.917$.

WORKED EXAMPLE 2 Evaluating trigonometric values using trigonometric identities (2)

If $\sec(\theta) = -\frac{5}{4}$ and $\theta \in \left[\frac{\pi}{2}, \pi\right]$, determine the value of $\tan(\theta)$.

THINK	WRITE
1. Use the identity $\tan^2(\theta) + 1 = \sec^2(\theta)$ and substitute $\sec(\theta) = -\frac{5}{4}$ into it.	$\tan^2(\theta) + 1 = \sec^2(\theta)$ $\tan^2(\theta) + 1 = \left(-\frac{5}{4}\right)^2$ $\tan^2(\theta) + 1 = \frac{25}{16}$
2. Solve for $\tan^2(\theta)$.	$\tan^2(\theta) = \frac{25}{16} - 1$ $\tan^2(\theta) = \frac{9}{16}$
3. Take the square root of both sides.	$\tan(\theta) = \pm\sqrt{\frac{9}{16}}$ $\tan(\theta) = \pm\frac{3}{4}$
4. Recall that $\theta \in \left[\frac{\pi}{2}, \pi\right]$ and that $\tan(\theta)$ is negative in the second quadrant.	$\tan(\theta) = -\frac{3}{4}$

The Pythagorean identities can be used not only to evaluate trigonometric expressions, but also to simplify them.

WORKED EXAMPLE 3 Using trigonometric identities to simplify expressions (1)

Use a Pythagorean identity to simplify $\cos^2(A)\tan^2(A)$.

THINK	WRITE
1. Use the identity $\tan^2(A)+1=\sec^2(A)$.	Using $\tan^2(A)=\sec^2(A)-1$, $\cos^2(A)\tan^2(A)=\cos^2(A)\left(\sec^2(A)-1\right)$ $=\cos^2(A)\sec^2(A)-\cos^2(A)$
2. As $\sec(A)=\dfrac{1}{\cos(A)}$, $\cos^2(A)\sec^2(A)=1$.	$=1-\cos^2(A)$
3. $\sin^2(A)+\cos^2(A)=1$ so $\sin^2(A)=1-\cos^2(A)$.	$=\sin^2(A)$

Sometimes it is easier to simplify expressions involving $\tan(A)$, $\cot(A)$, $\sec(A)$ and $\text{cosec}(A)$ by rewriting them in terms of $\sin(A)$ and $\cos(A)$.

WORKED EXAMPLE 4 Using trigonometric identities to simplify expressions (2)

Simplify $\dfrac{\tan^2(A)-1}{\tan^2(A)+1}$.

THINK	WRITE
1. Substitute $\tan(A)=\dfrac{\sin(A)}{\cos(A)}$.	$\dfrac{\tan^2(A)-1}{\tan^2(A)+1}=\dfrac{\frac{\sin^2(A)}{\cos^2(A)}-1}{\frac{\sin^2(A)}{\cos^2(A)}+1}$
2. To assist with simplifying the fraction, multiply by $\dfrac{\cos^2(A)}{\cos^2(A)}$.	$=\dfrac{\frac{\sin^2(A)}{\cos^2(A)}-1}{\frac{\sin^2(A)}{\cos^2(A)}+1}\times\dfrac{\cos^2(A)}{\cos^2(A)}$ $=\dfrac{\frac{\sin^2(A)\cancel{\cos^2(A)}}{\cancel{\cos^2(A)}}-\cos^2(A)}{\frac{\sin^2(A)\cancel{\cos^2(A)}}{\cancel{\cos^2(A)}}+\cos^2(A)}$ $=\dfrac{\sin^2(A)-\cos^2(A)}{\sin^2(A)+\cos^2(A)}$
3. Use $\sin^2(A)+\cos^2(A)=1$ to simplify the denominator.	$=\dfrac{\sin^2(A)-\cos^2(A)}{1}$ $=\sin^2(A)-\cos^2(A)$

8.2.2 Quadratic trigonometric equations

In your studies so far, you have factorised quadratic expressions and solved quadratic equations. These skills can also be used to factorise and solve equations when the quadratic is written in terms of a **trigonometric function**.

WORKED EXAMPLE 5 Solving quadratic equations using trigonometric identities (1)

Solve the equation $2\sin^2(A) = \sin(A)$ for A over the domain $0 \le A \le 2\pi$.

THINK	WRITE
1. Write the equation.	$2\sin^2(A) = \sin(A)$
2. Move $\sin(A)$ to the left of the equation to make the equation equal to zero.	$2\sin^2(A) - \sin(A) = 0$
3. This is a quadratic in terms of $\sin(A)$. It can be easier to identify the factors if you rewrite the quadratic in terms of a variable.	Let $a = \sin(A)$. $2a^2 - a = 0$ $a(2a - 1) = 0$
4. Rewrite the equation in terms of $\sin(A)$.	$\sin(A)(2\sin(A) - 1) = 0$
5. Using the Null Factor Law, solve each factor equal to 0.	$\sin(A) = 0$ or $2\sin(A) - 1 = 0$
6. Solve $\sin(A) = 0$.	$\sin(A) = 0 \quad \sin(A) = \frac{1}{2}$ $A = 0, \pi, 2\pi$
7. Solve $\sin(A) = \frac{1}{2}$, remembering that $\sin(A)$ will be positive in quadrants 1 and 2. $\sin\left(\frac{\pi}{6}\right) = \frac{1}{2}$ is a known trigonometric ratio.	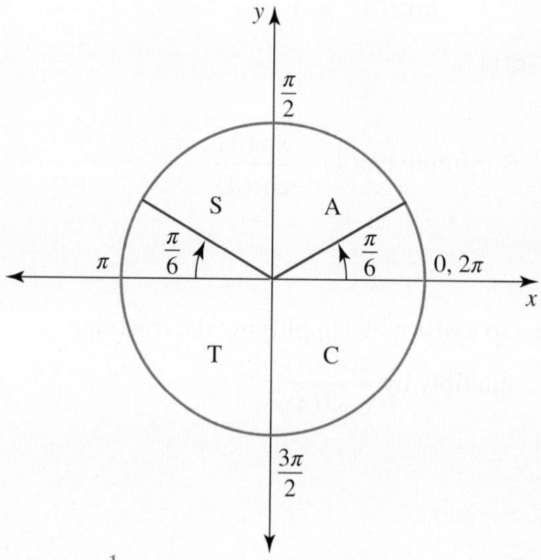 $\sin(A) = \frac{1}{2}$ The quadrant 1 solution becomes: $A = \frac{\pi}{6}$ The quadrant 2 solution becomes: $A = \frac{5\pi}{6}$
8. Combine all 5 solutions to the equation.	$A = 0, \frac{\pi}{6}, \frac{5\pi}{6}, \pi, 2\pi$

Sometimes it is necessary to use the Pythagorean identities to rewrite the expression as a quadratic in terms of one trigonometric function.

WORKED EXAMPLE 6 Factorising quadratic equations using trigonometric identities

Use a Pythagorean identity to factorise $\text{cosec}^2(A) - \cot(A) - 3$.

THINK	WRITE
1. Use the identity $1 + \cot^2(A) = \text{cosec}^2(A)$ and then collect like terms.	$\text{cosec}^2(A) - \cot(A) - 3 = 1 + \cot^2(A) - \cot(A) - 3$ $= \cot^2(A) - \cot(A) - 2$
2. This is a quadratic in terms of cot(A). It can be easier to identify the factors if you rewrite the quadratic in terms of a variable.	Let $a = \cot(A)$. $\cot^2(A) - \cot(A) - 2 = a^2 - a - 2$ $= (a-2)(a+1)$
3. Rewrite the factors in terms of cot(A).	$\text{cosec}^2(A) - \cot(A) - 3 = (\cot(A) - 2)(\cot(A) + 1)$

WORKED EXAMPLE 7 Solving quadratic equations using trigonometric identities (2)

Solve the equation $2\sin^2(A) = \cos(A) + 1$ for A over the domain $0 \le \theta \le 2\pi$.

THINK	WRITE
1. Write the equation.	$2\sin^2(A) = \cos(A) + 1$
2. Make the substitution $\sin^2(A) = 1 - \cos^2(A)$.	$2\left(1 - \cos^2(A)\right) = \cos(A) + 1$
3. Form a quadratic equation by expanding the brackets and then bringing all of the terms to one side. Before we factorise a quadratic, we normally rearrange to write the square term with a positive coefficient. This is a quadratic equation in terms of cos(A).	$2 - 2\cos^2(A) = \cos(A) + 1$ $0 = \cos(A) + 1 + 2\cos^2(A) - 2$ $0 = 2\cos^2(A) + \cos(A) - 1$
4. Factors of quadratic trigonometric equations can be more easily identified if the quadratic is rewritten in terms of a variable.	Let $a = \cos(A)$. $2a^2 + a - 1 = 0$
5. Factorise the equation.	$(2a - 1)(a + 1) = 0$
6. Rewrite the factors in terms of cos(A).	$(2\cos(A) - 1)(\cos(A) + 1) = 0$
7. Using the Null Factor Law, solve each factor equal to 0.	$2\cos(A) - 1 = 0$ or $\cos(A) + 1 = 0$ $\cos(A) = \frac{1}{2}$ $\quad\cos(A) = -1$
8. Solve $\cos(A) = \frac{1}{2}$, remembering that cos(A) will be positive in quadrants 1 and 4. $\cos\left(\frac{\pi}{3}\right) = \frac{1}{2}$ is a known trigonometric ratio.	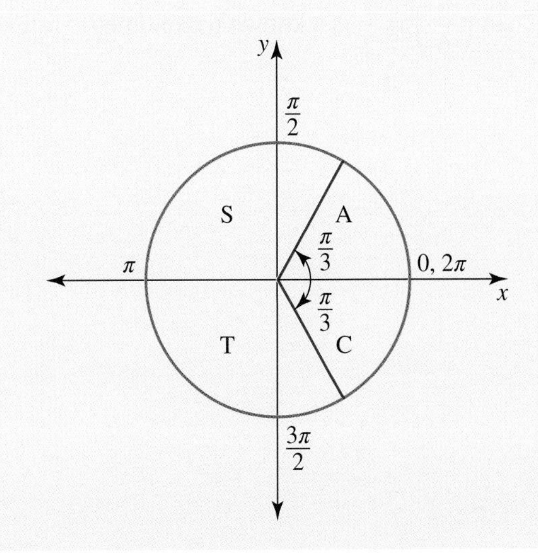

$\cos\left(\frac{\pi}{3}\right) = \frac{1}{2}$

The quadrant 1 solution becomes:

$A = \frac{\pi}{3}$

The quadrant 4 solution becomes:

$A = 2\pi - \frac{\pi}{3}$

$= \frac{5\pi}{3}$

9. Solve $\cos(A) = -1$.
   $\cos(\pi) = -1$ is a known trigonometric ratio.

   $\cos(A) = -1$
   $A = \pi$

10. Combine all solutions.

   $A = \frac{\pi}{3},\ \pi,\ \frac{5\pi}{3}$

WORKED EXAMPLE 8 General solutions to quadratic equations

Determine the general solution to $2\sin^2(2A) - \sin(2A) - 1 = 0$.

THINK / **WRITE**

1. This is a quadratic in terms of $\sin(2A)$. Factors of quadratic trigonometric equations can be more easily identified if the quadratic is rewritten in terms of a variable.

   Let $a = \sin(2A)$.
   $2a^2 - a - 1 = 0$

2. Factorise the equation.

   $(2a + 1)(a - 1) = 0$

3. Rewrite the factors in terms of $\sin(2A)$.

   $(2\sin(2A) + 1)(\sin(2A) - 1) = 0$

4. Using the Null Factor Law, solve each factor equal to 0.

   $2\sin(2A) + 1 = 0$ or $\sin(2A) - 1 = 0$
   $\sin(2A) = -\frac{1}{2}$ $\sin(2A) = 1$

5. Solve $\sin(2A) = -\frac{1}{2}$, remembering that $\sin(2A)$ will be negative in quadrants 3 and 4, and $\sin\left(\frac{\pi}{6}\right) = \frac{1}{2}$ is a known trigonometric ratio.

   $\sin(2A) = -\frac{1}{2}$

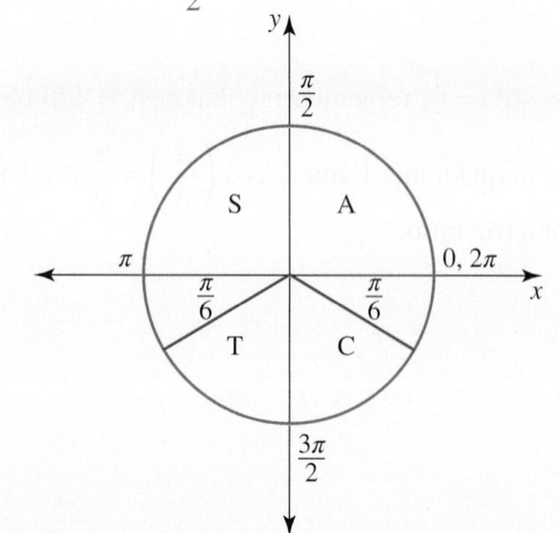

As a general solution is needed, each solution for $2A$ will require $2n\pi$, $n \in Z$.

The quadrant 3 solution becomes:

$$2A = -\pi + \frac{\pi}{6} + 2n\pi$$
$$= -\frac{5\pi}{6} + 2n\pi$$
$$A = -\frac{5\pi}{12} + n\pi$$
$$= \pi\left(-\frac{5}{12} + n\right)$$
$$= \pi\left(\frac{-5 + 12n}{12}\right)$$
$$= \frac{\pi}{12}(12n - 5)$$

The quadrant 4 solution becomes:

$$2A = -\frac{\pi}{6} + 2n\pi$$
$$A = -\frac{\pi}{12} + n\pi$$
$$= \pi\left(-\frac{1}{12} + n\right)$$
$$= \pi\left(\frac{12n - 1}{12}\right)$$
$$= \frac{\pi}{12}(12n - 1)$$

6. Solve $\sin(2A) = 1$ using $\sin\left(\frac{\pi}{2}\right) = 1$.

$\sin(2A) = 1$

The general solution becomes:

$$2A = \frac{\pi}{2} + 2n\pi$$
$$A = \frac{\pi}{4} + n\pi$$
$$= \pi\left(\frac{1}{4} + n\right)$$
$$= \pi\left(\frac{1 + 4n}{4}\right)$$
$$= \frac{\pi}{4}(1 + 4n)$$

7. Combine all solutions.

$$A = \frac{\pi}{12}(12n - 5),\ \frac{\pi}{12}(12n - 1),\ \frac{\pi}{4}(4n + 1),\ n \in Z$$

8.2 Exercise

Students, these questions are even better in jacPLUS

Receive immediate feedback and access sample responses

Access additional questions

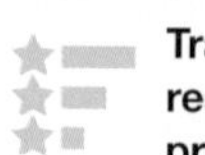
Track your results and progress

Find all this and MORE in jacPLUS

Technology free

1. **WE1** If $\sin(A) = 0.8$ and $0° < A < 90°$, determine the exact values of the following.
   a. $\cos(A)$
   b. $\tan(A)$
2. If $\cos(A) = 0.3$ and $0° < A < 90°$, determine the exact values of the following.
   a. $\sin(A)$
   b. $\tan(A)$
3. Determine all possible values of the following, as exact values.
   a. $\cos(x)$ if $\sin(x) = 0.4$
   b. $\cos(x)$ if $\sin(x) = -0.7$
   c. $\sin(x)$ if $\cos(x) = 0.24$
   d. $\sin(x)$ if $\cos(x) = -0.9$
4. **WE2** If $\sin(\theta) = -\dfrac{3}{8}$ and $\theta \in \left[-\dfrac{\pi}{2}, -\pi\right]$, determine the exact value of $\cos(\theta)$.
5. If $\cot(\theta) = \dfrac{11}{4}$ and $\theta \in \left[0, \dfrac{\pi}{2}\right]$, determine the exact value of $\text{cosec}(\theta)$.
6. Given that $\sin(A) = -\dfrac{\sqrt{5}}{4}$ and $\dfrac{3\pi}{2} < A < 2\pi$, determine:
   a. the exact value of $\cos(A)$
   b. the exact value of $\tan(A)$.
7. **WE3** Use the Pythagorean identity to simplify $\tan^2(A) - \sin^2(A)\tan^2(A)$.
8. **WE4** Simplify $\dfrac{\text{cosec}(A)\cos^2(A)}{1 + \text{cosec}(A)}$.
9. **WE5** Solve each of the following equations over the domain $0 \le x \le 2\pi$.
   a. $\sin^2(x) - \sin(x) = 0$
   b. $\cos^2(x) + \cos(x) = 0$
10. Solve each of the following equations over the domain $0 \le x \le 2\pi$.
    a. $2\sin^2(x) + \sqrt{3}\sin(x) = 0$
    b. $2\cos^2(x) + \cos(x) - 1 = 0$
11. Solve each of the following equations over the domain $0 \le x \le 2\pi$.
    a. $\sin^2(x) + 3\sin(x) - 4 = 0$
    b. $2\sin^2(x) - \sin(x) - 1 = 0$
12. **WE6** Use a Pythagorean identity to factorise each of the following.
    a. $1 + \sin(A) - 2\cos^2(A)$
    b. $\sec^2(x) - \tan(x) - 3$
13. Use a Pythagorean identity to factorise each of the following.
    a. $2\cot^2(\alpha) - \text{cosec}(\alpha) + 1$
    b. $2\cos^2(3x) - 2\sin^2(3x) - 1$
14. Use a Pythagorean identity to factorise each of the following.
    a. $\text{cosec}^2(3A) - 2\text{cosec}(3A) + 2\cot^2(3A) + 1$
    b. $\tan^2(2\beta) - \sec(2\beta) - 1$

15. **WE7** Solve the following equation over the domain $0 \leq A \leq 2\pi$.

$$2\cos^2(A) = 1 + \sin(A)$$

16. Solve each of the following equations over the domain $0 \leq A \leq 2\pi$.

a. $2\sin^2(A) - 1 = 0$

b. $1 + \cos(A) = 2\sin^2(A)$

17. Solve each of the following equations over the domain $0 \leq A \leq 2\pi$.

a. $\sin^2(A) = 1 + \cos(A)$

b. $2\cos^2(A) = 5 + 5\sin(A)$

18. **WE8** Determine the general solution to each of the following equations.

a. $2\sin^2(2A) - 3\sin(2A) + 1 = 0$

b. $2\cos^2(2A) + \cos(2A) - 1 = 0$

19. Determine the general solution to each of the following equations.

a. $2\sin^2(4x) + \sin(4x) = 0$

b. $\cos^2(4x) - \cos(4x) = 0$

20. Determine the general solution to each of the following equations.

a. $2\cos^2(3x) + \sqrt{3}\cos(3x) = 0$

b. $2\sin^2(3x) - \sqrt{3}\sin(3x) = 0$

21. Determine the general solution to each of the following equations.

a. $\tan^2(x) + \left(\sqrt{3} + 1\right)\tan(x) + \sqrt{3} = 0$

b. $\tan^2(x) + \left(\sqrt{3} - 1\right)\tan(x) - \sqrt{3} = 0$

22. Determine the general solution to the following equation.

$$2\sin^3(x) + \sin^2(x) - 2\sin(x) - 1 = 0$$

23. Determine the general solution to the following equation.

$$2\cos^3(x) - \cos^2(x) - 2\cos(x) + 1 = 0$$

24. Determine the general solution to the following equation.

$$\tan^3(x) - \tan^2(x) - \tan(x) + 1 = 0$$

25. Determine the general solution to the following equation.

$$\tan^4(x) - 4\tan^2(x) + 3 = 0$$

26. Solve the following equation for A.

$$\frac{1}{\sin^2(A)} - \frac{1}{\cos^2(A)} - \frac{1}{\tan^2(A)} - \frac{1}{\cot^2(A)} - \frac{1}{\sec^2(A)} - \frac{1}{\operatorname{cosec}^2(A)} = -3,\ A \in [0,\ 2\pi]$$

27. If $\sin^{16}(A) = \frac{1}{9}$, evaluate the following expression.

$$\frac{1}{\cos^2(A)} + \frac{1}{1 + \sin^2(A)} + \frac{2}{1 + \sin^4(A)} + \frac{4}{1 + \sin^8(A)}$$

Technology active

28. Determine the range of $\sin^4(x) + \cos^2(x)$.

8.2 Exam questions

Question 1 (1 mark) TECH-ACTIVE

MC The expression $\frac{1+\tan^2(x)}{\tan^2(x)}$ can be simplified to

A. $\sec^2(x)$ **B.** $\frac{\sin^2(x)}{\cos^2(x)}$ **C.** $\text{cosec}^2(x)$ **D.** $\cot^2(x)$ **E.** 2

Question 2 (1 mark) TECH-ACTIVE

MC Determine which of the following statements is **false**.

A. $\sin\left(\frac{\pi}{6}\right)+\cos\left(\frac{\pi}{6}\right)=\sin\left(\frac{\pi}{3}\right)+\cos\left(\frac{\pi}{3}\right)$

B. $\cos^3(\pi)+\sin^3(\pi)=1$

C. $\cos^2(\pi)+\sin^2(\pi)=1$

D. $\cos(\pi)+\sin(\pi)=-1$

E. $\cos\left(\frac{\pi}{2}\right)+\sin\left(\frac{\pi}{2}\right)=1$

Question 3 (1 mark) TECH-ACTIVE

MC Determine which of the following statements is **false**.

A. $2\sin\left(\frac{\pi}{6}\right)\cos\left(\frac{\pi}{6}\right)=\sin\left(\frac{\pi}{3}\right)$

B. $\cos^2\left(\frac{\pi}{6}\right)-\sin^2\left(\frac{\pi}{6}\right)=\cos\left(\frac{\pi}{6}\right)$

C. $\tan\left(\frac{3\pi}{2}\right)=\frac{\sin\left(\frac{3\pi}{2}\right)}{\cos\left(\frac{3\pi}{2}\right)}$

D. $\frac{1}{\tan\left(\frac{3\pi}{2}\right)}=\frac{\cos\left(\frac{3\pi}{2}\right)}{\sin\left(\frac{3\pi}{2}\right)}$

E. $\sin\left(\frac{\pi}{6}\right)=\sqrt{\frac{1}{2}\left(1-\cos\left(\frac{\pi}{3}\right)\right)}$

More exam questions are available online.

8.3 Compound angle formulas

LEARNING INTENTION

At the end of this subtopic you should be able to:

- use the compound angle formulas to simplify and solve trigonometric equations.

8.3.1 Angle sum and angle difference formulas

Consider a rectangle PQSU with right-angled triangles PQR and PRT as shown. The length of PT is 1.

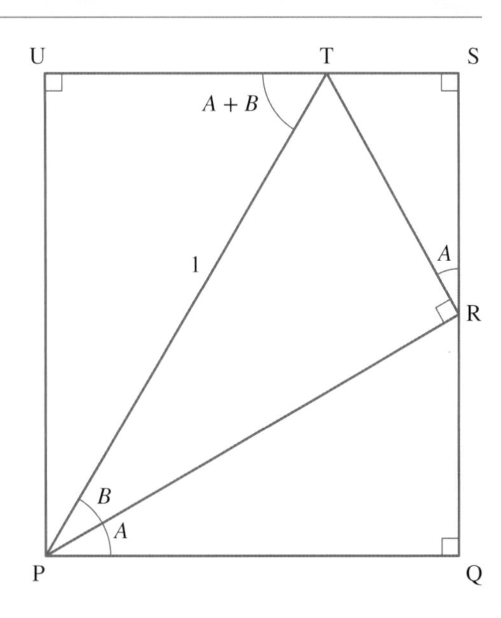

Consider ΔPRT.

$$\cos(B)=\frac{PR}{PT}=\frac{PR}{1}$$
$$PR=\cos(B)$$

$$\sin(B)=\frac{RT}{PT}=\frac{RT}{1}$$
$$RT=\sin(B)$$

Consider ΔPQR.

$$\cos(A) = \frac{PQ}{PR} \qquad\qquad \sin(A) = \frac{RQ}{PR}$$
$$= \frac{PQ}{\cos(B)} \qquad\qquad = \frac{RQ}{\cos(B)}$$
$$PQ = \cos(A)\cos(B) \qquad\qquad RQ = \sin(A)\cos(B)$$

Consider ΔRST.

$$\sin(A) = \frac{ST}{RT} \qquad\qquad \cos(A) = \frac{RS}{RT}$$
$$= \frac{ST}{\sin(B)} \qquad\qquad = \frac{RS}{\sin(B)}$$
$$ST = \sin(A)\sin(B) \qquad\qquad RS = \cos(A)\sin(B)$$

Consider ΔPTU.

As UT $\parallel$ PQ, $\angle UTP = \angle TPQ$ (since they are alternate angles)
$= A + B$

$$\cos(A+B) = \frac{UT}{PT} \qquad\qquad \sin(A+B) = \frac{PU}{PT}$$
$$= \frac{UT}{1} \qquad\qquad = \frac{PU}{1}$$
$$UT = \cos(A+B) \qquad\qquad PU = \sin(A+B)$$

Adding these lengths gives us the following diagram.

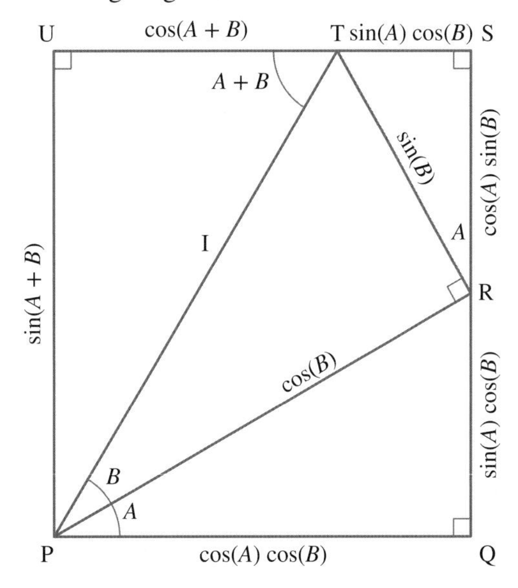

In the rectangle, PU = QS. Therefore, $\sin(A+B) = \sin(A)\cos(B) + \cos(A)\sin(B)$.

As PQ = US, $\cos(A)\cos(B) = \cos(A+B) + \sin(A)\sin(B)$
$\cos(A+B) = \cos(A)\cos(B) - \sin(A)\sin(B)$

If we consider $A + (-B)$ and remember that $\cos(-B) = \cos(B)$ and $\sin(-B) = -\sin(B)$, then the identities become:

$$\sin(A+(-B)) = \sin(A)\cos(-B) + \cos(A)\sin(-B)$$
$$\sin(A-B) = \sin(A)\cos(B) - \cos(A)\sin(B)$$

$$\cos(A + (-B)) = \cos(A)\cos(-B) - \sin(A)\sin(-B)$$
$$\cos(A - B) = \cos(A)\cos(B) + \sin(A)\sin(B)$$

We can also use the identity $\tan(A) = \dfrac{\sin(A)}{\cos(A)}$ to get expressions for $\tan(A + B)$ and $\tan(A - B)$ by using the formulas derived above.

$$\begin{aligned}\tan(A - B) &= \frac{\sin(A - B)}{\cos(A - B)}\\ &= \frac{\sin(A)\cos(B) - \cos(A)\sin(B)}{\cos(A)\cos(B) + \sin(A)\sin(B)}\\ &= \frac{\frac{\sin(A)\cos(B)}{\cos(A)\cos(B)} - \frac{\cos(A)\sin(B)}{\cos(A)\cos(B)}}{1 + \frac{\sin(A)\sin(B)}{\cos(A)\cos(B)}}\\ &= \frac{\tan(A) - \tan(B)}{1 + \tan(A)\tan(B)}\end{aligned}$$

It can similarly be shown that $\tan(A + B) = \dfrac{\tan(A) + \tan(B)}{1 - \tan(A)\tan(B)}$.

These formulas are known collectively as the **compound angle** formulas, or, more specifically as the **angle sum** and **angle difference formulas**.

The angle sum formulas

$$\sin(A + B) = \sin(A)\cos(B) + \cos(A)\sin(B)$$
$$\cos(A + B) = \cos(A)\cos(B) - \sin(A)\sin(B)$$
$$\tan(A + B) = \frac{\tan(A) + \tan(B)}{1 - \tan(A)\tan(B)}$$

The angle difference formulas

$$\sin(A - B) = \sin(A)\cos(B) - \cos(A)\sin(B)$$
$$\cos(A - B) = \cos(A)\cos(B) + \sin(A)\sin(B)$$
$$\tan(A - B) = \frac{\tan(A) - \tan(B)}{1 + \tan(A)\tan(B)}$$

WORKED EXAMPLE 9 Evaluating expressions using the compound angle formulas

Evaluate $\sin(22°)\cos(38°) + \cos(22°)\sin(38°)$, giving your answer as an exact value.

THINK	WRITE
1. The formula $\sin(A + B) = \sin(A)\cos(B) + \cos(A)\sin(B)$ can be used.	Let $A = 22°$ and $B = 38°$. $\sin(A + B) = \sin(A)\cos(B) + \cos(A)\sin(B)$ $= \sin(22°)\cos(38°) + \cos(22°)\sin(38°)$
2. Simplify the expression.	$= \sin(22° + 38°)$ $= \sin(60°)$
3. If the exact value is known, then simplify further.	$= \dfrac{\sqrt{3}}{2}$

The compound angle formulas can be used to simply expressions, particularly when the sum or difference involves an angle with a known trigonometric ratio.

WORKED EXAMPLE 10 Expanding expressions using the compound angle formulas

Expand $2\cos\left(\theta+\frac{\pi}{3}\right)$.

THINK	WRITE
1. The formula $\cos(A+B)=\cos(A)\cos(B)-\sin(A)\sin(B)$ can be used.	Let $A=\theta$ and $B=\frac{\pi}{3}$. $\cos(A+B)=\cos(A)\cos(B)-\sin(A)\sin(B)$ $2\cos\left(\theta+\frac{\pi}{3}\right)=2\left(\cos(\theta)\cos\left(\frac{\pi}{3}\right)-\sin(\theta)\sin\left(\frac{\pi}{3}\right)\right)$
2. Exact values $\cos\left(\frac{\pi}{3}\right)=\frac{1}{2}$ and $\sin\left(\frac{\pi}{3}\right)=\frac{\sqrt{3}}{2}$ are known, so use these to simplify.	$=2\left(\frac{1}{2}\cos(\theta)-\frac{\sqrt{3}}{2}\sin(\theta)\right)$
3. Simplify.	$=\cos(\theta)-\sqrt{3}\sin(\theta)$
4. State the solution.	$2\cos\left(\theta+\frac{\pi}{3}\right)=\cos(\theta)-\sqrt{3}\sin(\theta)$

WORKED EXAMPLE 11 Simplifying expressions using the compound angle formulas

Use one of the compound angle formulas to simplify $\cos\left(\frac{3\pi}{2}-\theta\right)$.

THINK	WRITE
1. The formula $\cos(A-B)=\cos(A)\cos(B)+\sin(A)\sin(B)$ can be used.	Let $A=\frac{3\pi}{2}$ and $B=\theta$. $\cos(A-B)=\cos(A)\cos(B)+\sin(A)\sin(B)$ $\cos\left(\frac{3\pi}{2}-\theta\right)=\cos\left(\frac{3\pi}{2}\right)\cos(\theta)+\sin\left(\frac{3\pi}{2}\right)\sin(\theta)$
2. Simplify using known exact values $\cos\left(\frac{3\pi}{2}\right)=0$ and $\sin\left(\frac{3\pi}{2}\right)=-1$	$=0\times\cos(\theta)-1\times\sin(\theta)$
3. Simplify.	$=-\sin(\theta)$
4. State the solution.	$\cos\left(\frac{3\pi}{2}-\theta\right)=-\sin(\theta)$

8.3.2 Determining exact values

In previous work, we have used the exact values for 30° and 45° $\left(\text{or } \frac{\pi}{6} \text{ and } \frac{\pi}{4}\right)$ and multiples of these values.

As $45° - 30° = 15°$ $\left(\text{or } \frac{\pi}{4} - \frac{\pi}{6} = \frac{\pi}{12}\right)$, it is possible to use the compound angle formulas to evaluate exact values for multiples of 15° $\left(\frac{\pi}{12}\right)$.

WORKED EXAMPLE 12 Calculating exact values using the compound angle formulas (1)

Calculate the exact value of tan $\left(\frac{13\pi}{12}\right)$.

THINK

1. Express $\frac{13\pi}{12}$ in terms of multiples of $\frac{\pi}{4}$ and $\frac{\pi}{6}$.

WRITE

$$\frac{13\pi}{12} = \frac{10\pi}{12} + \frac{3\pi}{12} = \frac{5\pi}{6} + \frac{\pi}{4}$$

2. The compound angle formula $\tan(A+B) = \dfrac{\tan(A) + \tan(B)}{1 - \tan(A)\tan(B)}$ can be used.

Let $A = \frac{5\pi}{6}$ and $B = \frac{\pi}{4}$.

$$\tan(A+B) = \frac{\tan(A) + \tan(B)}{1 - \tan(A)\tan(B)}$$

$$\tan\left(\frac{13\pi}{12}\right) = \tan\left(\frac{5\pi}{6} + \frac{\pi}{4}\right) = \frac{\tan\left(\frac{5\pi}{6}\right) + \tan\left(\frac{\pi}{4}\right)}{1 - \tan\left(\frac{5\pi}{6}\right)\tan\left(\frac{\pi}{4}\right)}$$

3. As $\frac{5\pi}{6}$ is in quadrant 2, tan $\frac{5\pi}{6}$ is negative.

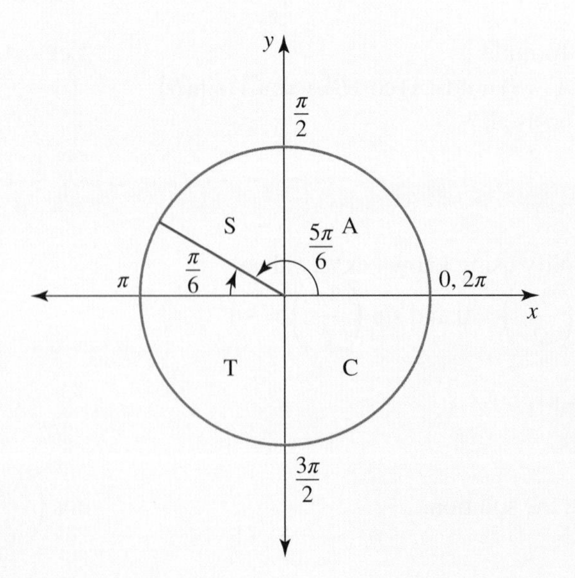

$$\tan\left(\frac{5\pi}{6}\right) = -\tan\left(\frac{\pi}{6}\right)$$
$$= \frac{-\sqrt{3}}{3}$$

4. $\frac{\pi}{4}$ is in quadrant 1, so $\tan\left(\frac{\pi}{4}\right)$ is positive.

$$\tan\left(\frac{\pi}{4}\right) = 1$$

5. Substitute values for $\tan\left(\frac{5\pi}{6}\right)$ and $\tan\left(\frac{\pi}{4}\right)$ and simplify.

$$\tan\left(\frac{13\pi}{12}\right) = \frac{-\frac{\sqrt{3}}{3}+1}{1+\frac{\sqrt{3}}{3}\times 1}$$
$$= \frac{1-\frac{\sqrt{3}}{3}}{1+\frac{\sqrt{3}}{3}}$$
$$= \frac{1-\frac{\sqrt{3}}{3}}{1+\frac{\sqrt{3}}{3}} \times \frac{1-\frac{\sqrt{3}}{3}}{1-\frac{\sqrt{3}}{3}}$$
$$= \frac{\left(1-\frac{\sqrt{3}}{3}\right)^2}{1-\left(\frac{\sqrt{3}}{3}\right)^2}$$
$$= \frac{1-\frac{2\sqrt{3}}{3}+\frac{1}{3}}{1-\frac{1}{3}}$$
$$= \frac{\frac{4-2\sqrt{3}}{3}}{\frac{2}{3}}$$
$$= \frac{4-2\sqrt{3}}{2}$$
$$= 2-\sqrt{3}$$

8.3.3 Using compound angle formulas and Pythagorean identities

Sometimes it is necessary to use the Pythagorean identities to determine the unknown ratios before it is possible to use the compound angle formulas.

WORKED EXAMPLE 13 Calculating exact values using the compound angle formulas (2)

If $\cos(A) = \frac{12}{13}$ and $\sin(B) = \frac{7}{25}$, where $0 < A < \frac{\pi}{2}$ and $\frac{\pi}{2} < B < \pi$, calculate the exact value of $\sin(A-B)$.

THINK

1. To calculate $\sin(A - B)$ we will need to calculate $\sin(A)$ and $\cos(B)$. The Pythagorean identity can be used.

WRITE

$$\sin^2(A) + \cos^2(A) = 1$$

$$\sin^2(A) + \left(\frac{12}{13}\right)^2 = 1$$

$$\sin^2(A) = 1 - \frac{144}{169}$$

$$= \frac{25}{169}$$

$$\sin(A) = \pm\frac{5}{13}$$

As $0 < A < \frac{\pi}{2}$, $\sin(A) > 0$,

$$\sin(A) = \frac{5}{13}$$

$$\sin^2(B) + \cos^2(B) = 1$$

$$\left(\frac{7}{25}\right)^2 + \cos^2(B) = 1$$

$$\cos^2(B) = 1 - \frac{49}{625}$$

$$= \frac{576}{625}$$

$$\cos(B) = \pm\frac{24}{25}$$

As $\frac{\pi}{2} < B < \pi$, $\cos(B) < 0$,

$$\cos(B) = -\frac{24}{25}.$$

THINK

2. Use the compound angle formula $\sin(A - B) = \sin(A)\cos(B) - \cos(A)\sin(B)$ to simplify and solve.

WRITE

$$\sin(A - B) = \sin(A)\cos(B) - \cos(A)\sin(B)$$

$$= \frac{5}{13} \times \left(-\frac{24}{25}\right) - \frac{12}{13} \times \frac{7}{25}$$

$$= -\frac{120}{325} - \frac{84}{325}$$

$$= -\frac{204}{325}$$

TI | THINK / **DISPLAY/WRITE**

1. Ensure that the calculator is in RADIAN mode. On a Calculator page, select MENU, then select:
   3 Algebra
   6 Numerical Solve
   Complete the entry line as:
   $\text{nSolve}\left(\cos(a) = \dfrac{12}{13}, a\right)$
   $\left| 0 < a < \dfrac{\pi}{2}\right.$
   Press CTRL, then press VAR, then type a and press ENTER to store the result as a.

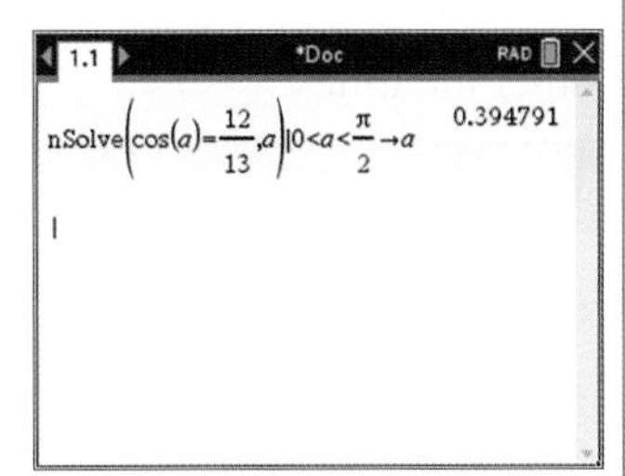

2. Select MENU, then select:
   3 Algebra
   6 Numerical Solve
   Complete the entry line as:
   $\text{nSolve}\left(\sin(b) = \dfrac{7}{25}, b\right)$
   $\left| \dfrac{\pi}{2} < b < \pi\right.$
   Press CTRL, then press VAR, then type b and press ENTER to store the result as b.

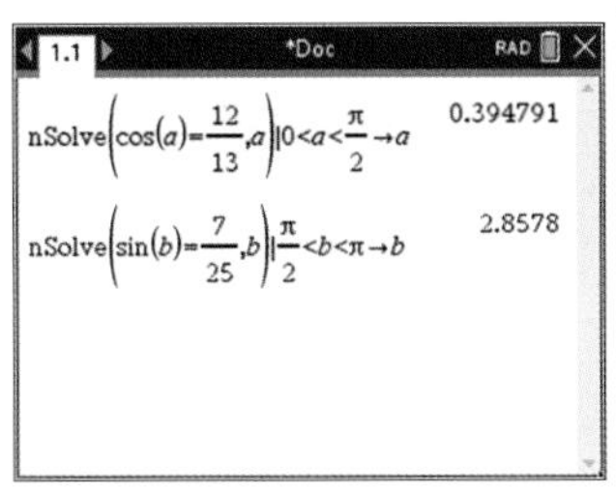

3. Complete the entry line as:
   $\sin(a - b)$
   then press ENTER.

$\text{nSolve}\left(\cos(a)=\frac{12}{13},a\right)|0<a<\frac{\pi}{2} \to a$ 0.394791
$\text{nSolve}\left(\sin(b)=\frac{7}{25},b\right)|\frac{\pi}{2}<b<\pi \to b$ 2.8578
$\sin(a-b)$ -0.627692

4. Press MENU, then select:
   2 Number
   2 Approximate to Fraction
   Then press ENTER.

$\text{nSolve}\left(\sin(b)=\frac{7}{25},b\right)|\frac{\pi}{2}<b<\pi \to b$ 2.8578
$\sin(a-b)$ -0.627692
-0.62769230769232▸approxFraction(5.E-14) $\frac{-204}{325}$

CASIO | THINK / **DISPLAY/WRITE**

1. Ensure the calculator is in **Decimal** and **Rad** mode.
   Select **solve(** from the Keyboard Math3 palette to start the entry line.
   Complete the entry line as:
   $\text{solve}\left(\cos(\text{a}) = \dfrac{12}{13}, \text{a}\right)$
   $\left| 0 < \text{a} < \dfrac{\pi}{2} \Rightarrow \text{A}\right.$
   Then press EXE.
   This will store the result as A.

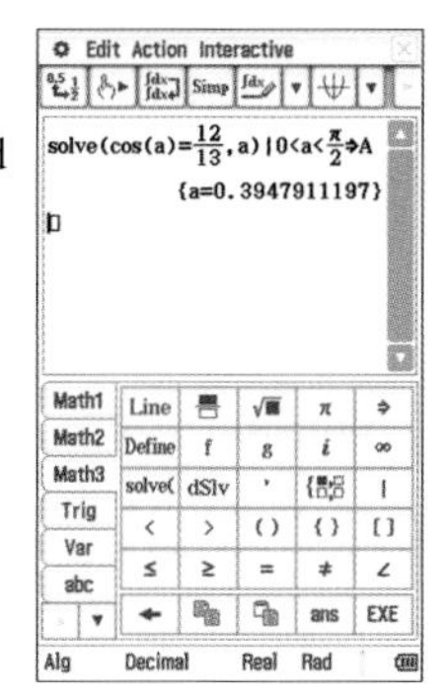

2. Select **solve(** from the Keyboard Math3 palette to start the entry line.
   Complete the entry line as:
   $\text{solve}\left(\sin(\text{b}) = \dfrac{12}{13}, \text{b}\right)$
   $\left| \dfrac{\pi}{2} < \text{b} < \pi \Rightarrow \text{B}\right.$
   Then press EXE.
   This will store the result as B.

solve(cos(a)=$\frac{12}{13}$,a)|0<a<$\frac{\pi}{2}$⇒A
{a=0.3947911197}
solve(sin(b)=$\frac{7}{25}$,b)|$\frac{\pi}{2}$<b<π⇒B
{b=2.857798544}

3. Complete the entry line as:
   $\sin(\text{A} - \text{B})$
   then press EXE.

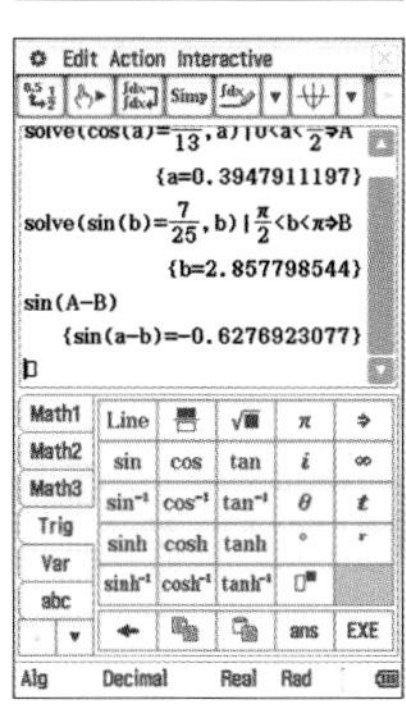

4. Select Interactive > Transformation > Fraction > toFrac
   Then select OK in the wizard window that appears.
   The result appears on the screen.

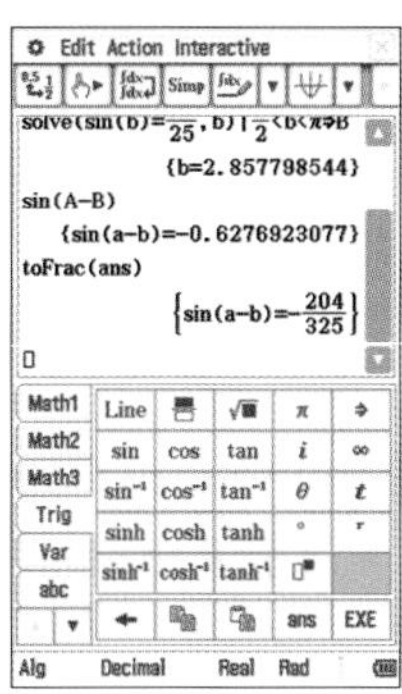

8.3.4 Proofs using the compound angle formulas

The compound angle formulas can be used to prove other identities.

WORKED EXAMPLE 14 Proving equations using the compound angle formulas

Prove that $\sin(A+B)\sin(A-B)=\sin^2(A)-\sin^2(B)$.

THINK	WRITE
1. When trying to decide where to begin, it is often best to begin with the more complicated side (the side with more terms or with the compound angles). In this case, it is the left-hand side.	$\text{LHS} = \sin(A+B)\sin(A-B)$
2. Substitute the compound angle formulas.	$= (\sin(A)\cos(B)+\cos(A)\sin(B)) \times (\sin(A)\cos(B)-\cos(A)\sin(B))$
3. Expand the brackets. Notice in this equation the difference of two squares pattern can be used.	$= \sin^2(A)\cos^2(B)-\cos^2(A)\sin^2(B)$
4. The right-hand side of the identity is written in terms of sine. Use the Pythagorean identity to replace the cosines.	$= \sin^2(A)\left(1-\sin^2(B)\right)-\left(1-\sin^2(A)\right)\sin^2(B)$
5. Simplify. As it has been demonstrated that the left-hand side is equal to the right-hand side, the identity is proven.	$= \sin^2(A)-\sin^2(A)\sin^2(B)-\sin^2(B)+\sin^2(A)\sin^2(B)$ $= \sin^2(A)-\sin^2(B)$ $= \text{RHS}$ $\therefore \sin(A+B)\sin(A-B)=\sin^2(A)-\sin^2(B)$

8.3 Exercise

Technology free

1. WE9 Evaluate the following, giving your answers as exact values.

   a. $\sin(27°)\cos(33°)+\cos(27°)\sin(33°)$
   b. $\cos(47°)\cos(43°)-\sin(47°)\sin(43°)$
   c. $\cos(76°)\cos(16°)+\sin(76°)\sin(16°)$
   d. $\cos(63°)\sin(18°)-\sin(63°)\cos(18°)$

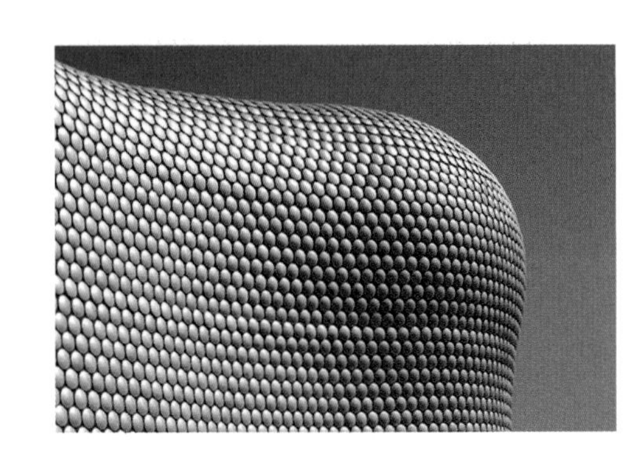

2. WE10 Expand the following.

   a. $\sqrt{2}\sin\left(\theta-\frac{\pi}{4}\right)$
   b. $2\sin\left(\theta+\frac{\pi}{3}\right)$
   c. $2\cos\left(\theta-\frac{\pi}{6}\right)$
   d. $\sqrt{2}\cos\left(\theta+\frac{\pi}{4}\right)$

3. **WE11** Use one of the compound angle formulas to simplify each of the following.

a. $\sin\left(\frac{\pi}{2}-\theta\right)$ b. $\cos\left(\frac{\pi}{2}-\theta\right)$ c. $\sin(\pi+\theta)$ d. $\cos(\pi-\theta)$

4. Use one of the compound angle formulas to simplify each of the following.

a. $\sin\left(\frac{3\pi}{2}-\theta\right)$ b. $\cos\left(\frac{3\pi}{2}+\theta\right)$ c. $\tan(\pi-\theta)$ d. $\tan(\pi+\theta)$

5. Simplify each of the following.

a. $\sin\left(x+\frac{\pi}{3}\right)-\sin\left(x-\frac{\pi}{3}\right)$ b. $\cos\left(\frac{\pi}{3}+x\right)-\cos\left(\frac{\pi}{3}-x\right)$

c. $\cos\left(\frac{\pi}{6}-x\right)-\cos\left(\frac{\pi}{6}+x\right)$ d. $\tan\left(x+\frac{\pi}{4}\right)\tan\left(x-\frac{\pi}{4}\right)$

6. **WE12** Determine the exact values of the following.

a. $\tan\left(\frac{\pi}{12}\right)$ b. $\tan\left(\frac{5\pi}{12}\right)$ c. $\cos\left(\frac{7\pi}{12}\right)$ d. $\sin\left(\frac{11\pi}{12}\right)$

7. **WE13** Given that $\cos(A)=\frac{4}{5}$ and $\sin(B)=\frac{12}{13}$ and that A and B are both acute angles, determine the exact values of:

a. $\cos(A-B)$ b. $\tan(A+B)$

8. Given that $\sin(A)=\frac{5}{13}$ and $\tan(B)=\frac{24}{7}$ and A is obtuse and B is acute, determine the exact values of:

a. $\sin(A+B)$ b. $\cos(A+B)$

9. Given that $\sec(A)=\frac{7}{2}$ and $\text{cosec}(B)=\frac{3}{2}$ and A is acute but B is obtuse, determine the exact values of:

a. $\cos(A+B)$ b. $\sin(A-B)$

10. Given that $\text{cosec}(A)=\frac{1}{a}$ and $\sec(B)=\frac{1}{b}$ and A and B are both acute, evaluate $\tan(A+B)$.

11. Given that $\sin(A)=\frac{a}{a+1}$ and $\cos(B)=\frac{a}{a+2}$ and A and B are both acute, evaluate $\tan(A-B)$.

12. **WE14** Prove that $\cos(A+B)\cos(A-B)=\cos^2(A)-\sin^2(B)$.

13. If ABCD is a cyclic quadrilateral, show that $\cos(A)+\cos(B)+\cos(C)+\cos(D)=0$.

14. If ABC is a triangle, prove that $\tan\left(\frac{A}{2}\right)=\cot\left(\frac{B+C}{2}\right)$.

15. If $\tan(x)-\tan(y)=m$ and $\cot(y)-\cot(x)=n$, prove that $\frac{1}{m}+\frac{1}{n}=\cot(x-y)$.

16. If $\tan(B)=\frac{\sin(A)\cos(A)}{2+\cos^2(A)}$, prove that $3\tan(A-B)=2\tan(A)$.

8.3 Exam questions

Question 1 (1 mark) TECH-ACTIVE

MC When expanded, $\cos(a-b)+\cos(a+b)$ equals

A. 0 **B.** $2\cos(a)\cos(b)$ **C.** $-2\cos(a)\cos(b)$
D. $2\sin(a)\sin(b)$ **E.** $-2\sin(a)\sin(b)$

Question 2 (1 mark) TECH-ACTIVE

MC Determine which of the following is **not** a correct trigonometric identity.

A. $\cos\left(x+\frac{\pi}{4}\right)-\cos\left(x+\frac{\pi}{4}\right)=\sqrt{2}\sin(x)$

B. $\cos\left(x+\frac{\pi}{4}\right)+\cos\left(x+\frac{\pi}{4}\right)=\sqrt{2}\cos(x)$

C. $\sin\left(x+\frac{\pi}{4}\right)-\sin\left(x+\frac{\pi}{4}\right)=\sqrt{2}\cos(x)$

D. $\sin\left(x+\frac{\pi}{4}\right)+\sin\left(x+\frac{\pi}{4}\right)=\sqrt{2}\sin(x)$

E. $\tan\left(x+\frac{\pi}{4}\right)-\tan\left(x+\frac{\pi}{4}\right)=\frac{2\left(\tan^2(x)+1\right)}{1-\tan^2(x)}$

Question 3 (2 marks) TECH-FREE

Simplify $\cos(x+y)\cos(x-y)+\sin(x+y)\sin(x-y)$.

More exam questions are available online.

8.4 Double and half angle formulas

LEARNING INTENTION

At the end of this subtopic you should be able to:

- use the double and half angle formulas to simplify and solve trigonometric equations.

8.4.1 Double angle formulas

If the trigonometric ratio of a particular angle is known, it is possible to determine the trigonometric ratio of multiples of that angle.

Using the compound angle formulas, it is possible to determine formulas for double angles.

Using $\sin(A+B)=\sin(A)\cos(B)+\cos(A)\sin(B)$, we can find an expression for $\sin(2A)$.

$$\begin{aligned}\sin(2A)&=\sin(A+A)\\&=\sin(A)\cos(A)+\cos(A)\sin(A)\\&=2\sin(A)\cos(A)\end{aligned}$$

Similarly, using $\cos(A+B)=\cos(A)\cos(B)-\sin(A)\sin(B)$:

$$\begin{aligned}\cos(2A)&=\cos(A+A)\\&=\cos(A)\cos(A)-\sin(A)\sin(A)\\&=\cos^2(A)-\sin^2(A)\end{aligned}$$

Using $\cos^2(A)=1-\sin^2(A)$: $\cos(2A)=\left(1-\sin^2(A)\right)-\sin^2(A)$

$$=1-2\sin^2(A)$$

Using $\sin^2(A)=1-\cos^2(A)$: $\cos(2A)=\cos^2(A)-\left(1-\cos^2(A)\right)$

$$=2\cos^2(A)-1$$

We can also use the compound angle formula $\tan(A+B) = \dfrac{\tan(A)+\tan(B)}{1-\tan(A)\tan(B)}$ to determine the formula for the double angle $\tan(2A)$.

$$\begin{aligned}\tan(2A) &= \tan(A+A)\\ &= \frac{\tan(A)+\tan(A)}{1-\tan(A)\tan(A)}\\ &= \frac{2\tan(A)}{1-\tan^2(A)}\end{aligned}$$

Double angle formulas

$$\begin{aligned}\sin(2A) &= 2\sin(A)\cos(A)\\ \cos(2A) &= \cos^2(A)-\sin^2(A)\\ &= 2\cos^2(A)-1\\ &= 1-2\sin^2(A)\\ \tan(2A) &= \frac{2\tan(A)}{1-\tan^2(A)}\end{aligned}$$

Using the double angle formulas to determine exact values

The double angle formulas can be used to determine exact values of expressions. The formulas may be used in either direction.

WORKED EXAMPLE 15 Evaluating using double angle formulas (1)

Determine the exact value of $\sin\left(\frac{7\pi}{12}\right)\cos\left(\frac{7\pi}{12}\right)$.

THINK	WRITE
1. As the expression is a multiple of sine and cosine of the same angle, the formula $\sin(2A) = 2\sin(A)\cos(A)$ can be used.	$2\sin(A)\cos(A) = \sin(2A)$ $\sin(A)\cos(A) = \frac{1}{2}\sin(2A)$
2. Substitute $A = \frac{7\pi}{12}$ and simplify.	$\sin\left(\frac{7\pi}{12}\right)\cos\left(\frac{7\pi}{12}\right) = \frac{1}{2}\sin\left(2\times\frac{7\pi}{12}\right)$ $= \frac{1}{2}\sin\left(\frac{7\pi}{6}\right)$

3. $\dfrac{7\pi}{6}$ is in quadrant 3; $\sin\left(\dfrac{\pi}{6}\right) = \dfrac{1}{2}$ is a known trigonometric ratio and is negative in quadrant 3.

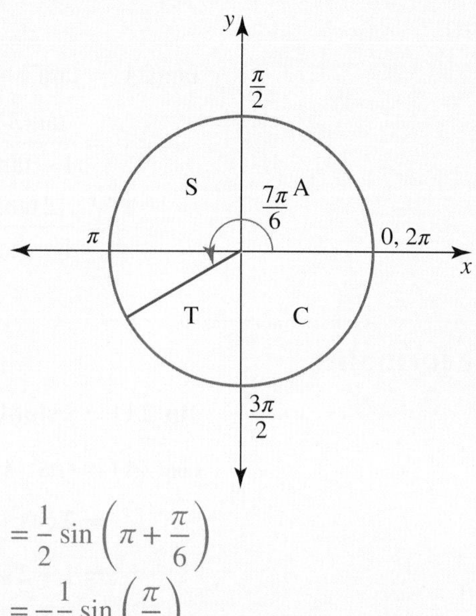

$$= \frac{1}{2}\sin\left(\pi + \frac{\pi}{6}\right)$$
$$= -\frac{1}{2}\sin\left(\frac{\pi}{6}\right)$$
$$= -\frac{1}{2} \times \frac{1}{2}$$
$$= -\frac{1}{4}$$

TI | THINK	DISPLAY/WRITE	CASIO | THINK	DISPLAY/WRITE
Ensure that your calculator is in RADIAN mode. On a Calculator page, complete the entry line as shown.	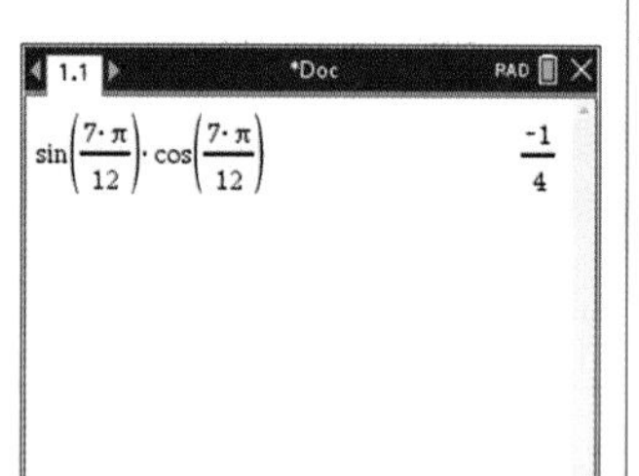	1. Make sure the calculator is in **Standard** and **Rad** mode. Complete the entry line as: $\sin\left(\dfrac{7\pi}{12}\right)\cos\left(\dfrac{7\pi}{12}\right)$	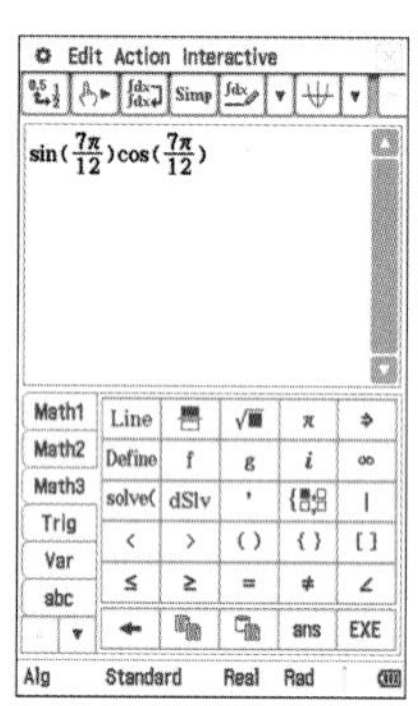
		2. Highlight the entry line. Select Interactive > Transformation > simplify The answer appears on the screen. 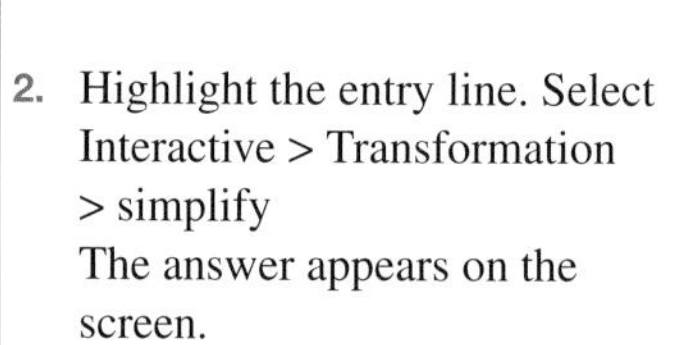	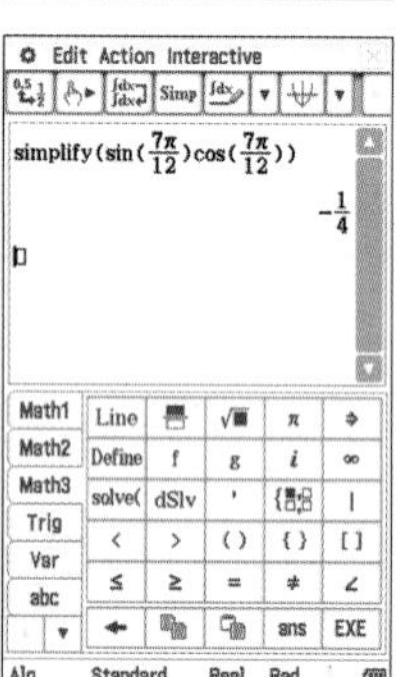

WORKED EXAMPLE 16 Evaluating using double angle formulas (2)

If $0 \le A \le \frac{\pi}{2}$ and $\cos(A) = \frac{1}{4}$, determine the exact values of:

a. **$\cos(2A)$** b. **$\sin(2A)$** c. **$\tan(2A)$**

THINK	WRITE
a. 1. As we know the value for $\cos(A)$, use $\cos(2A) = 2\cos^2(A) - 1$.	a. $\cos(2A) = 2\cos^2(A) - 1$
2. Substitute for $\cos(A)$.	$= 2 \times \left(\frac{1}{4}\right)^2 - 1$ $= \frac{2}{16} - 1$ $= -\frac{14}{16}$ $= -\frac{7}{8}$
b. 1. As $\sin(2A) = 2\sin(A)\cos(A)$, it is necessary to calculate $\sin(A)$. The Pythagorean identity can be used.	b. $\sin^2(A) + \cos^2(A) = 1$ $\sin^2(A) + \left(\frac{1}{4}\right)^2 = 1$ $\sin^2(A) + \frac{1}{16} = 1$ $\sin^2(A) = \frac{15}{16}$ $\sin(A) = \pm\frac{\sqrt{15}}{4}$
2. $0 \le A \le \frac{\pi}{2}$ so $\sin(A) > 0$.	As $0 \le A \le \frac{\pi}{2}$, $\sin(A) = \frac{\sqrt{15}}{4}$.
3. Use the formula to evaluate $\sin(2A)$.	$\sin(2A) = 2\sin(A)\cos(A)$ $= 2 \times \frac{\sqrt{15}}{4} \times \frac{1}{4}$ $= \frac{\sqrt{15}}{8}$
c. 1. As we know the value for $\cos(A)$ and $\sin(A)$ use $\tan(A) = \frac{\sin(A)}{\cos(A)}$.	c. $\tan(A) = \frac{\frac{\sqrt{15}}{4}}{\frac{1}{4}}$ $= \sqrt{15}$
2. Use the formula to evaluate $\tan(2A)$	$\tan(2A) = \frac{2\tan(A)}{1 - \tan^2(A)}$ $= \frac{2\sqrt{15}}{1 - 15}$ $= -\frac{\sqrt{15}}{7}$

8.4.2 Using the double angle formulas to solve trigonometric equations

Previously, the Pythagorean identity was used to rewrite expressions using a single trigonometric ratio. In a similar fashion, the double angle formulas can be used to rewrite the expressions in terms of the same angle.

WORKED EXAMPLE 17 Solving trigonometric equations using double angle formulas

If $\sin(2x) + \sqrt{3}\cos(x) = 0$, solve for $x \in [0, 2\pi]$.

THINK	WRITE
1. The equation involves trigonometric ratios of both x and $2x$. Use the double angle formula to rewrite $\sin(2x)$ in terms of $\sin(x)$ and $\cos(x)$.	$\sin(2x) + \sqrt{3}\cos(x) = 0$ $2\sin(x)\cos(x) + \sqrt{3}\cos(x) = 0$
2. Factorise by identifying the common factor of $\cos(x)$.	$\cos(x)\left(2\sin(x) + \sqrt{3}\right) = 0$
3. Solve each factor equal to 0.	$\cos(x) = 0$ or $2\sin(x) + \sqrt{3} = 0$ $\sin(x) = -\frac{\sqrt{3}}{2}$
4. Solve $\cos(x) = 0$.	$\cos(x) = 0$ $x = \frac{\pi}{2}, \frac{3\pi}{2}$
5. Solve $\sin(x) = -\frac{\sqrt{3}}{2}$, remembering that $\sin(x)$ is negative in quadrants 3 and 4. $\sin\left(\frac{\pi}{3}\right) = \frac{\sqrt{3}}{2}$ is a known trigonometric ratio.	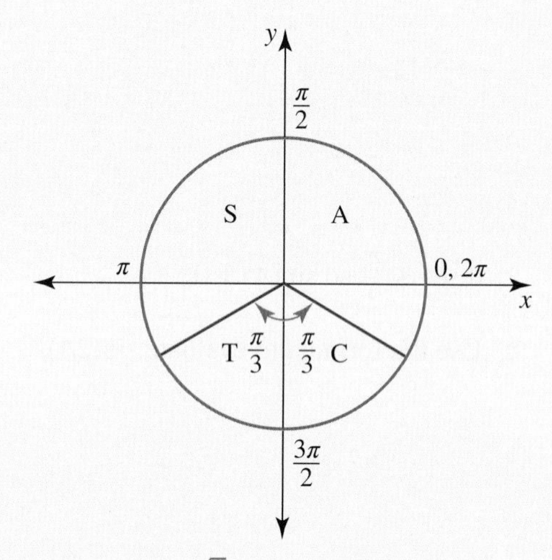 $\sin\left(\frac{\pi}{3}\right) = \frac{\sqrt{3}}{2}$ $x = \pi + \frac{\pi}{3}$ or $x = 2\pi - \frac{\pi}{3}$ $= \frac{4\pi}{3}$ $\quad = \frac{5\pi}{3}$
6. Combine the solutions.	$x = \frac{\pi}{2}, \frac{4\pi}{3}, \frac{3\pi}{2}, \frac{5\pi}{3}$

8.4.3 Proofs using the double angle formulas

The double angle formulas can be combined with the other formulas covered in this chapter and the Pythagorean identities to prove trigonometric identities. When deciding where to start, it is often best to begin with the more complex side and work to simplify it.

WORKED EXAMPLE 18 Proving trigonometric identities using double angle formulas

Prove the identity $\dfrac{\cos(2\alpha)\cos(\alpha)+\sin(2\alpha)\sin(\alpha)}{\sin(3\alpha)\cos(\alpha)-\cos(3\alpha)\sin(\alpha)}=\dfrac{1}{2}\text{cosec}(\alpha)$**.**

THINK	WRITE
1. The left-hand side is the more complex side, so begin with that side.	$\text{LHS}=\dfrac{\cos(2\alpha)\cos(\alpha)+\sin(2\alpha)\sin(\alpha)}{\sin(3\alpha)\cos(\alpha)-\cos(3\alpha)\sin(\alpha)}$
2. Use the angle difference formulas. The numerator can be simplified using $\cos(A-B)=\cos(A)\cos(B)+\sin(A)\sin(B)$, and the denominator can be simplified using $\sin(A-B)=\sin(A)\cos(B)-\cos(A)\sin(B)$.	$=\dfrac{\cos(2\alpha-\alpha)}{\sin(3\alpha-\alpha)}$
3. Simplify.	$=\dfrac{\cos(\alpha)}{\sin(2\alpha)}$
4. Substitute $\sin(2\alpha)=2\sin(\alpha)\cos(\alpha)$ for the denominator.	$=\dfrac{\cos(\alpha)}{2\sin(\alpha)\cos(\alpha)}$
5. Simplify the fraction and use $\text{cosec}(A)=\dfrac{1}{\sin(A)}$.	$=\dfrac{1}{2\sin(\alpha)}$ $=\dfrac{1}{2}\text{cosec}(\alpha)$ $=\text{RHS}$
6. As it has been demonstrated that the left-hand side is equal to the right-hand side, the identity is proven.	$\therefore \dfrac{\cos(2\alpha)\cos(\alpha)+\sin(2\alpha)\sin(\alpha)}{\sin(3\alpha)\cos(\alpha)-\cos(3\alpha)\sin(\alpha)}=\dfrac{1}{2}\text{cosec}(\alpha)$

8.4.4 Proofs using the half angle formulas

The double angle formulas express $\sin(2A)$, $\cos(2A)$ and $\tan(2A)$ in terms of $\sin(A)$, $\cos(A)$ and $\tan(A)$. Using these formulas, it is also possible to express $\sin(A)$, $\cos(A)$ and $\tan(A)$ in terms of $\sin\left(\frac{A}{2}\right)$, $\cos\left(\frac{A}{2}\right)$ and $\tan\left(\frac{A}{2}\right)$, which are the **half angle formulas**.

Half angle formulas

$$\sin(A)=2\sin\left(\frac{A}{2}\right)\cos\left(\frac{A}{2}\right)$$

$$\cos(A)=\cos^2\left(\frac{A}{2}\right)-\sin^2\left(\frac{A}{2}\right)$$

$$=2\cos^2\left(\frac{A}{2}\right)-1$$

$$=1-2\sin^2\left(\frac{A}{2}\right)$$

$$\tan(A)=\frac{2\tan\left(\frac{A}{2}\right)}{1-\tan^2\left(\frac{A}{2}\right)}$$

WORKED EXAMPLE 19 Proving trigonometric identities using half-angle formulas

Prove the identity $\operatorname{cosec}(A) - \cot(A) = \tan\left(\frac{A}{2}\right)$.

THINK	WRITE
1. The left-hand side is the more complex side, so begin with that side.	$\text{LHS} = \operatorname{cosec}(A) - \cot(A)$
2. Remember that $\operatorname{cosec}(A) = \frac{1}{\sin(A)}$ and $\cot A = \frac{\cos(A)}{\sin(A)}$ and make the substitutions.	$= \frac{1}{\sin(A)} - \frac{\cos(A)}{\sin(A)}$
3. As the fractions have a common denominator, simplify the expression by adding the fractions.	$= \frac{1 - \cos(A)}{\sin(A)}$
4. On the right-hand side, the angle is $\frac{A}{2}$, so use the half angle formulas for cos(A) and sin(A). There are three options to use as a substitution for cos(A). In this instance, as $\tan\left(\frac{A}{2}\right) = \frac{\sin\left(\frac{A}{2}\right)}{\cos\left(\frac{A}{2}\right)}$, writing cos(A) in terms of $\sin\left(\frac{A}{2}\right)$ would be the first option to try.	$= \frac{1 - \left(1 - 2\sin^2\left(\frac{A}{2}\right)\right)}{2\sin\left(\frac{A}{2}\right)\cos\left(\frac{A}{2}\right)}$
5. Simplify the fraction.	$= \frac{1 - 1 + 2\sin^2\left(\frac{A}{2}\right)}{2\sin\left(\frac{A}{2}\right)\cos\left(\frac{A}{2}\right)}$ $= \frac{2\sin^2\left(\frac{A}{2}\right)}{2\sin\left(\frac{A}{2}\right)\cos\left(\frac{A}{2}\right)}$ $= \frac{\sin\left(\frac{A}{2}\right)}{\cos\left(\frac{A}{2}\right)}$ $= \tan\left(\frac{A}{2}\right)$ $= \text{RHS}$
6. As it has been demonstrated that the left-hand side is equal to the right-hand side, the identity is proven.	$\therefore \operatorname{cosec}(A) - \cot(A) = \tan\left(\frac{A}{2}\right)$

8.4 Exercise

Students, these questions are even better in jacPLUS

 **Receive immediate feedback and access sample responses**

 **Access additional questions**

 **Track your results and progress**

Find all this and MORE in jacPLUS

Technology free

1. **WE15** Determine the exact values of the following.

   a. $\sin\left(\frac{\pi}{8}\right)\cos\left(\frac{\pi}{8}\right)$ b. $\cos^2\left(\frac{5\pi}{8}\right)-\sin^2\left(\frac{5\pi}{8}\right)$

   c. $2\sin^2\left(\frac{\pi}{8}\right)-1$ d. $\cos^2\left(\frac{\pi}{12}\right)$

2. **WE16** Given that $\sec(A)=\frac{8}{3}$ and $0\le A\le\frac{\pi}{2}$, determine the exact values of the following.

   a. $\cos(2A)$ b. $\sin(2A)$ c. $\tan(2A)$ d. $\cot(2A)$

3. **WE17** Solve each of the following equations for $x\in[0,2\pi]$.

   a. $\sin(2x)-\cos(x)=0$ b. $\sin(x)-\sin(2x)=0$
   c. $\cos(2x)-\cos(x)=0$ d. $\sin(x)-\cos(2x)=0$

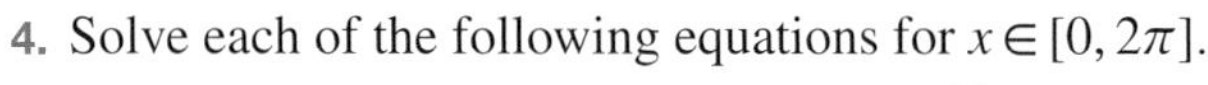

4. Solve each of the following equations for $x\in[0,2\pi]$.

   a. $\tan(x)=\sin(2x)$ b. $\sin(2x)=\sqrt{3}\cos(x)$ c. $\sin(4x)=\sin(2x)$ d. $\cos(2x)=\sin(4x)$

5. **WE18** Prove the following identity.
   $$\frac{\sin(2A)\cos(A)-\cos(2A)\sin(A)}{\cos(2A)\cos(A)+\sin(2A)\sin(A)}=\tan(A)$$

6. Prove the following identity.
   $$\frac{\cos(2A)\cos(A)+\sin(2A)\sin(A)}{\sin(2A)\cos(A)-\cos(2A)\sin(A)}=\cot(A)$$

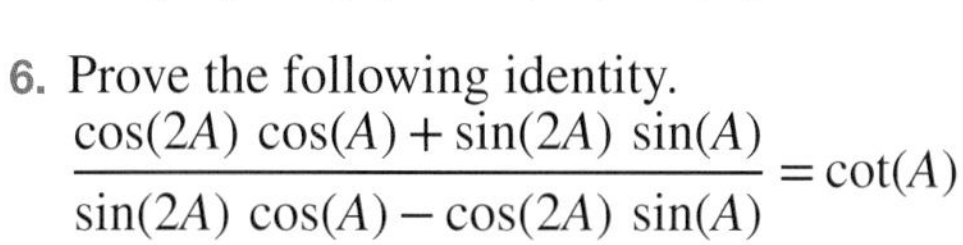

7. **WE19** Prove the following identity.
   $$\frac{\sin(A)}{1-\cos(A)}=\cot\left(\frac{A}{2}\right)$$

8. Prove the following identity.
   $$\frac{\sin(A)}{1+\cos(A)}=\tan\left(\frac{A}{2}\right)$$

9. Prove the following identity.
   $$\frac{1-\cos(A)+\sin(A)}{1+\cos(A)+\sin(A)}=\tan\left(\frac{A}{2}\right)$$

10. Prove the following identity.
    $$\frac{\sin\left(\frac{A}{2}\right)+\sin(A)}{1+\cos(A)+\cos\left(\frac{A}{2}\right)}=\tan\left(\frac{A}{2}\right)$$

11. Prove the following identity.

$$\sin(2A) = \frac{2\tan(A)}{1+\tan^2(A)}$$

12. Prove the following identity.

$$\cos(2A) = \frac{1-\tan^2(A)}{1+\tan^2(A)}$$

13. Prove the following identity.

$$\frac{\cos^3(A)-\sin^3(A)}{\cos(A)-\sin(A)} = 1+\frac{1}{2}\sin(2A)$$

14. Prove the following identity.

$$\frac{\cos^3(A)+\sin^3(A)}{\cos(A)+\sin(A)} = 1-\frac{1}{2}\sin(2A)$$

15. In a triangle ABC with side lengths a, b and c, where C is a right angle, c is the hypotenuse and a is opposite A, show that:

a. $\sin(2A) = \dfrac{2ab}{c^2}$

b. $\cos(2A) = \dfrac{b^2-a^2}{c^2}$

c. $\tan(2A) = \dfrac{2ab}{b^2-a^2}$

16. In a triangle ABC with side lengths a, b and c, where C is a right angle, c is the hypotenuse and a is opposite A, show that:

a. $\sin\left(\dfrac{A}{2}\right) = \sqrt{\dfrac{c-b}{2c}}$

b. $\cos\left(\dfrac{A}{2}\right) = \sqrt{\dfrac{c+b}{2c}}$

c. $\tan\left(\dfrac{A}{2}\right) = \sqrt{\dfrac{c-b}{c+b}}$

17. Chebyshev, the Russian mathematician mentioned in the overview, developed a number of recurrence relationships for trigonometric multiple angles.

a. Prove the relationship $\cos(nx) = 2\cos(x)\cos[(n-1)x] - \cos[(n-2)x]$.

b. Use this relationship and the formula for $\cos(2x)$ to demonstrate that $\cos(5x) = 16\cos^5(x) - 20\cos^3(x) + 5\cos(x)$.

18. Chebyshev's recurrence formula for multiple angles of the sine function is given by $\sin nx = 2\cos(x)\sin[(n-1)x] - \sin[(n-2)x]$.

a. Prove Chebyshev's recurrence formula for multiple angles of the sine function.

b. Use this formula and the formula for $\sin(2x)$ to demonstrate that $\sin(6x) = \cos(x)\left(32\sin^5(x) - 32\sin^3(x) + 6\sin(x)\right)$.

8.4 Exam questions

Question 1 (3 marks) TECH-FREE

Simplify $\cos(4\theta)$.

Question 2 (2 marks) TECH-FREE

a. Simplify $\sin(2\theta)\cos(\theta) - \sin(\theta)\cos(2\theta)$.

b. Simplify $\sin(6\theta)\cos(2\theta) - \sin(2\theta)\cos(6\theta)$.

Question 3 (1 mark) TECH-ACTIVE

MC Determine which of the following is **not** equal to $\cot\left(\frac{8\pi}{3}\right)$.

A. $\tan\left(-\frac{\pi}{6}\right)$ **B.** $\frac{\cos\left(\frac{8\pi}{3}\right)}{\sin\left(\frac{8\pi}{3}\right)}$ **C.** $\frac{1}{\tan\left(\frac{8\pi}{3}\right)}$ **D.** $\frac{1-\tan^2\left(\frac{2\pi}{3}\right)}{2\tan\left(\frac{2\pi}{3}\right)}$ **E.** $\frac{1-\tan^2\left(\frac{2\pi}{6}\right)}{2\tan\left(\frac{2\pi}{6}\right)}$

More exam questions are available online.

8.5 Converting $a\cos(x) + b\sin(x)$ to $r\cos(x \pm \alpha)$ or $r\sin(x \pm \alpha)$

LEARNING INTENTION

At the end of this subtopic you will be able to:

- prove and apply the identities between $a\sin(x) + b\cos(x)$ and $r\sin(x \pm \alpha)$ or $r\cos(x \pm \alpha)$.

8.5.1 Expressing $a\cos(x) + b\sin(x)$ in the form $r\cos(x \pm \alpha)$ or $r\sin(x \pm \alpha)$

In Topic 7, you learned how to graph trigonometric functions. Using the compound angle formulas, it is possible to rewrite expressions of the form $a\cos(x) + b\sin(x)$ as either $r\cos(x \pm \alpha)$ or $r\sin(x \pm \alpha)$. This makes it easier to visualise what the graph might look like and to identify the maximum and minimum values.

WORKED EXAMPLE 20 Converting $a\cos(x) + b\sin(x)$ to $r\cos(x - \alpha)$

Express $\sqrt{3}\cos(x) + \sin(x)$ in the form $r\cos(x-\alpha)$.

THINK	WRITE
1. Let $\sqrt{3}\cos(x) + \sin(x)$ equal $r\cos(x-\alpha)$ and use the compound angle formulas to expand $r\cos(x-\alpha)$.	Let $\sqrt{3}\cos(x) + \sin(x) = r\cos(x-\alpha)$. $\sqrt{3}\cos(x) + \sin(x) = r\cos(x-\alpha)$ $= r(\cos(x)\cos(\alpha) + \sin(x)\sin(\alpha))$ $= r\cos(x)\cos(\alpha) + r\sin(x)\sin(\alpha)$
2. Equate the coefficients of $\cos(x)$ and $\sin(x)$.	Equating the coefficients of $\cos(x)$ and $\sin(x)$: $\sqrt{3} = r\cos(\alpha)$ [1] $1 = r\sin(\alpha)$ [2]
3. As r is a factor of both equations, it can be eliminated by calculating $\frac{[2]}{[1]}$. This will also allow us to use $\tan(\alpha) = \frac{\sin(\alpha)}{\cos(\alpha)}$.	$\frac{[2]}{[1]}: \frac{1}{\sqrt{3}} = \frac{r\sin(\alpha)}{r\cos(\alpha)}$ $= \frac{\sin(\alpha)}{\cos(\alpha)}$ $= \tan(\alpha)$
4. From equations [1] and [2], both $\sin(\alpha)$ and $\cos(\alpha)$ are positive, so α is in quadrant 1. Remember $\tan\left(\frac{\pi}{6}\right) = \frac{1}{\sqrt{3}}$. Solve for α.	$\alpha = \frac{\pi}{6}$

5. Use either equation [1] or [2] to solve for r.

$$[1]: \sqrt{3} = r\cos\left(\frac{\pi}{6}\right)$$
$$= r \times \frac{\sqrt{3}}{2}$$
$$r = \sqrt{3} \times \frac{2}{\sqrt{3}}$$
$$= 2$$

6. State the final result.

$$\sqrt{3}\cos(x) + \sin(x) = 2\cos\left(x - \frac{\pi}{6}\right)$$

8.5.2 General transformations and applications

To make it easier to visualise the graph and to interpret the function, r is normally a positive number and α is generally in the first quadrant. Note that $-r\cos(x) = r\cos(x+\pi)$ and $-r\sin(x) = r\sin(x+\pi)$. Additionally, $r\cos(x) = r\sin\left(x + \frac{\pi}{2}\right)$ and $r\sin(x) = r\cos\left(x - \frac{\pi}{2}\right)$. This means that it is possible to write $a\cos(x) + b\sin(x)$ as either a sine or a cosine function, and the value of α will vary depending on the function chosen.

For r to be positive and α to be in quadrant 1, the following transformations are used:

$$a\sin(x) \pm b\cos(x) \text{ in the form } r\sin(x \pm \alpha)$$
$$a\cos(x) \pm b\sin(x) \text{ in the form } r\cos(x \mp \alpha)$$

WORKED EXAMPLE 21 Using $r\sin(x \pm \alpha)$ and $r\cos(x \pm \alpha)$ to sketch functions

Given the function $f(x) = 2\sin(x) - 2\sqrt{3}\cos(x)$, $x \in [0, 2\pi]$:

a. **express $f(x)$ in the form $r\sin(x - \alpha)$**
b. **determine the coordinates of the maximum and minimum of $f(x)$**
c. **sketch $f(x)$**
d. **solve $f(x) = 2$.**

THINK

a. 1. Let $2\sin(x) - 2\sqrt{3}\cos(x)$ equal $r\sin(x-\alpha)$ and use the compound angle formulas to expand $r\sin(x-\alpha)$.

WRITE

a. Let $2\sin(x) - 2\sqrt{3}\cos(x) = r\sin(x-\alpha)$.

$$2\sin(x) - 2\sqrt{3}\cos(x) = r\sin(x-\alpha)$$
$$= r(\sin(x)\cos(\alpha) - \cos(x)\sin(\alpha))$$
$$= r\sin(x)\cos(\alpha) - r\cos(x)\sin(\alpha)$$

2. Equate the coefficients of $\cos(x)$ and $\sin(x)$.

Equating the coefficients of $\cos(x)$ and $\sin(x)$:

$$2 = r\cos(\alpha) \quad [1]$$
$$2\sqrt{3} = r\sin(\alpha) \quad [2]$$

3. As r is a factor of both equations, it can be eliminated by calculating $\frac{[2]}{[1]}$. This will also allow us to use $\tan(\alpha) = \frac{\sin(\alpha)}{\cos(\alpha)}$.

$$\frac{[2]}{[1]}: \frac{2\sqrt{3}}{2} = \frac{r\sin(\alpha)}{r\cos(\alpha)}$$
$$\sqrt{3} = \frac{\sin(\alpha)}{\cos(\alpha)}$$
$$= \tan(\alpha)$$

4. From equations [1] and [2], both $\sin(\alpha)$ and $\cos(\alpha)$ are positive, so α is in quadrant 1. Remember $\tan\left(\frac{\pi}{3}\right) = \sqrt{3}$. Solve for α.

$$\alpha = \frac{\pi}{3}$$

5. Use either equation [1] or [2] to solve for r.

$$\begin{aligned}[1]: 2 &= r\cos\left(\frac{\pi}{3}\right)\\ &= r \times \frac{1}{2}\\ r &= 2 \times 2\\ &= 4\end{aligned}$$

6. State the final result.

$$2\sin(x) - 2\sqrt{3}\cos(x) = 4\sin\left(x - \frac{\pi}{3}\right)$$

b. 1. r will be the amplitude of the graph. There is no vertical shift.

b. $r = 4$; therefore, the amplitude of the graph is 4. This means that the maximum is 4 and the minimum is −4.

2. Identify where the maximum occurs. (The largest possible value of a sine function is 1 and $\sin\left(\frac{\pi}{2}\right) = 1$.) Note we are looking for $x \in [0, 2\pi]$.

The maximum will occur when

$$\begin{aligned}\sin\left(x - \frac{\pi}{3}\right) &= 1\\ x - \frac{\pi}{3} &= \frac{\pi}{2}\\ x &= \frac{\pi}{2} + \frac{\pi}{3}\\ &= \frac{3\pi}{6} + \frac{2\pi}{6}\\ &= \frac{5\pi}{6}\end{aligned}$$

The maximum point is $\left(\frac{5\pi}{6}, 4\right)$.

3. Identify where the minimum occurs. (The smallest possible value of a sine function is -1 and $\sin\left(\frac{3\pi}{2}\right) = 1$.) Note we are looking for $x \in [0, 2\pi]$.

The minimum will occur when

$$\begin{aligned}\sin\left(x - \frac{\pi}{3}\right) &= -1\\ x - \frac{\pi}{3} &= \frac{3\pi}{2}\\ x &= \frac{3\pi}{2} + \frac{\pi}{3}\\ &= \frac{9\pi}{6} + \frac{2\pi}{6}\\ &= \frac{11\pi}{6}\end{aligned}$$

The minimum point is $\left(\frac{11\pi}{6}, -4\right)$.

c. Use the maximum and minimum points to sketch the function over the required domain.

c. $f(x) = 2\sin(x) - 2\sqrt{3}\cos(x),\ x \in [0, 2\pi]$

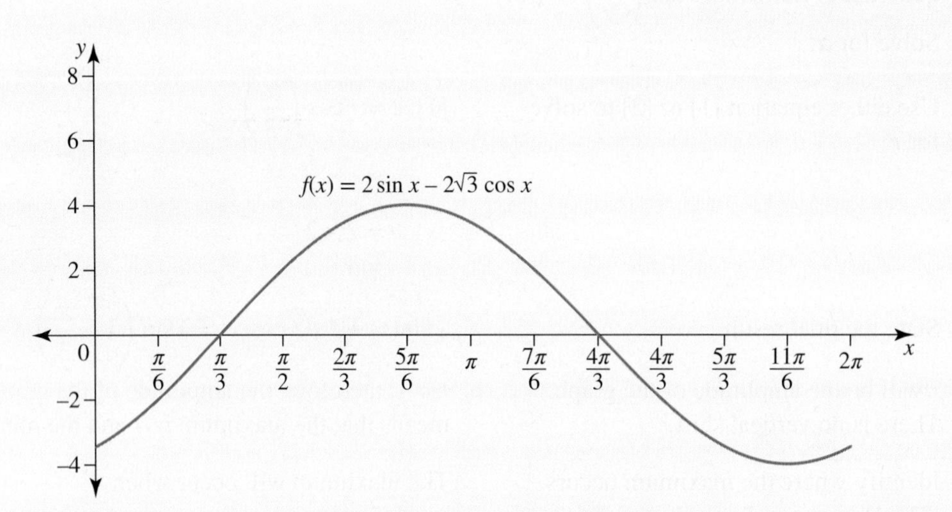

d. 1. To solve $f(x)$, use the form $r\sin(x - \alpha)$.

d. $2\sin(x) - 2\sqrt{3}\cos(x) = 2$

$$4\sin\left(x - \frac{\pi}{3}\right) = 2$$

2. Solve for x. Remember that $\sin\left(\frac{\pi}{6}\right) = \frac{1}{2}$ and sine is positive in quadrants 1 and 2.

$$\sin\left(x - \frac{\pi}{3}\right) = \frac{1}{2}$$

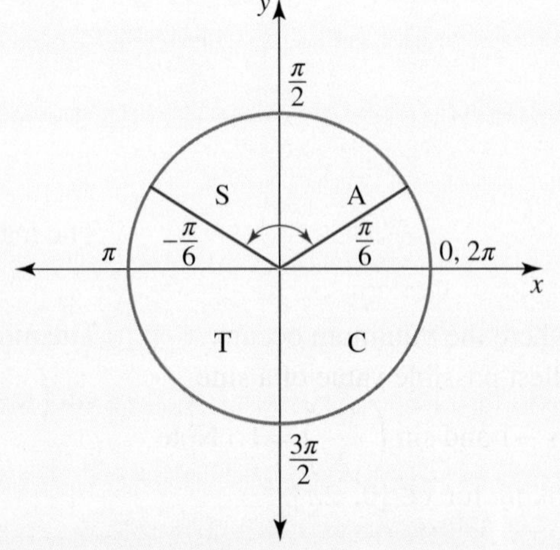

$$\sin\left(\frac{\pi}{6}\right) = \frac{1}{2}$$

Therefore : $x - \frac{\pi}{3} = \frac{\pi}{6}$ or $x - \frac{\pi}{3} = \pi - \frac{\pi}{6}$

$$x = \frac{\pi}{6} + \frac{2\pi}{6} \qquad x = \pi - \frac{\pi}{6} + \frac{\pi}{3}$$

$$= \frac{\pi}{6} + \frac{2\pi}{6} \qquad = \frac{6\pi}{6} - \frac{\pi}{6} + \frac{2\pi}{6}$$

$$= \frac{\pi}{2} \qquad = \frac{7\pi}{6}$$

3. Combine the solutions.

$$x = \frac{\pi}{2} \text{ or } x = \frac{7\pi}{6}$$

8.5 Exercise

Students, these questions are even better in jacPLUS

Receive immediate feedback and access sample responses

Access additional questions

Track your results and progress

Find all this and MORE in jacPLUS

Technology free

1. WE20 Express $\cos(x) + \sin(x)$ in the form $r\cos(x - \alpha)$.

2. Express $\sqrt{3}\cos(x) - \sin(x)$ in the form $r\cos(x + \alpha)$.
3. Express the following in the form $r\sin(x + \alpha)$.
   a. $\sin(x) + \cos(x)$
   b. $\sqrt{3}\sin(x) + \cos(x)$
4. Express the following in the form $r\sin(x - \alpha)$.
   a. $\sin(x) - \cos(x)$
   b. $\sqrt{3}\sin(x) - \cos(x)$
5. Express the following in the form $r\cos(x - \alpha)$. (*Note:* You may use a scientific calculator in part b.)
   a. $4\sqrt{2}\cos(x) + 4\sqrt{2}\sin(x)$
   b. $7\cos(x) + 24\sin(x)$
6. Express the following in the form $r\cos(x + \alpha)$.
   a. $\cos(x) - \sin(x)$
   b. $\cos(x) - \sqrt{3}\sin(x)$
7. WE21 Given the function $f(x) = \sqrt{2}\sin(x) - \sqrt{2}\cos(x),\ x \in [0,\ 2\pi]$:
   a. express $f(x)$ in the form $r\sin(x - \alpha)$
   b. determine the coordinates of the maximum and minimum of $f(x)$
   c. sketch $f(x)$
   d. solve $f(x) = 1$.

Technology active

8. Consider the function $f(x) = 5\sin(x) - 12\cos(x), x \in [0,\ 2\pi]$. In your answers, use 3 decimal places where necessary.
   a. Express $f(x)$ in the form $r\cos(x + \alpha)$.
   b. Determine the coordinates of the maximum and minimum of $f(x)$.
   c. Sketch $f(x)$.
   d. Solve $f(x) = 6.5$.
9. Consider the function $f(x) = 3\sin(x) - 4\cos(x), x \in [0°,\ 360°]$. In your answers, use 2 decimal places where necessary.
   a. Express $f(x)$ in the form $r\sin(x - \alpha)$.
   b. Determine the coordinates of the maximum and minimum of $f(x)$.
   c. Sketch $f(x)$.
   d. Solve $f(x) = 2.5$.

10. Consider the function $f(x) = 12\cos(x) + 5\sin(x)$, $x \in [0, 2\pi]$. In your answers, use 2 decimal places where necessary.

  a. Express $f(x)$ in the form $r\cos(x - \alpha)$.
  b. Determine the coordinates of the maximum and minimum of $f(x)$.
  c. Sketch $f(x)$.
  d. Solve $f(x) = 6.5$ and hence determine the general solution of $12\cos(x) + 5\sin(x) = 6.5$.

11. a. Express $24\cos(y) - 7\sin(y)$ in the form $r\cos(y + \alpha)$, giving your answer to 3 decimal places as necessary.
  b. A particle moves in a straight line. Its displacement, x, at time t seconds is given by $x = 30 + 24\cos(3t) - 7\sin(3t)$ for $t \geq 0$. Calculate the maximum and minimum displacement from the origin, O, and the first time these occur, to the nearest hundredth of a second.

12. a. Express $2\sin(x) + 2\sqrt{3}\cos(x)$ in the form $r\sin(x + \alpha)$.
  b. The depth of water near a pier changes with the tides according to the rule $h(t) = 5 + 2\sin\left(\frac{\pi t}{12}\right) + 2\sqrt{3}\cos\left(\frac{\pi t}{12}\right)$ where t hours is the time, for $t \geq 0$, and h is depth in metres. Determine the maximum and minimum depth and the first times these occur, to the nearest hour.
  c. Identify the first two times when the depth of the water is 7 metres.
  d. Sketch the graph of the depth of the water, showing one cycle.

13. Calculate the maximum and minimum values of each of the following.

  a. $\dfrac{4}{5 + 2.4\cos(x) - 3.2\sin(x)}$
  b. $\dfrac{\sqrt{3}}{2 - 4\cos(x) - 4\sqrt{3}\sin(x)}$

14. a. At time t, the current I in a circuit is given by $I = 40\cos(300t) + 9\sin(300t)$. Express the current in the form $r\cos(nt - \alpha)$ where α is in decimal degrees, correct to 2 decimal places.
  b. At a time t, the current I in a circuit is given by $I = 11\cos(800t) - 60\sin(800t)$. Express the current in the form $r\cos(nt + \alpha)$ where α is in decimal degrees, correct to 2 decimal places.

8.5 Exam questions

Question 1 (3 marks) TECH-FREE

Solve the trigonometric equation $\sqrt{3}\cos(x) + \sin(x) = 1$ for $0 \leq x \leq 2\pi$ by converting the LHS of the equation into the form $r\cos(x - \alpha)$.

Question 2 (4 marks) TECH-FREE

Sketch the graph of $y = \cos(x) + \sqrt{3}\sin(x)$ over the domain $x \in [0, 2\pi]$.

Question 3 (1 mark) TECH-ACTIVE

MC $\cos(x) - \sin(x)$ can be rewritten as:

A. $\sqrt{2}\sin\left(x + \frac{\pi}{4}\right)$
B. $\sqrt{2}\sin\left(x + \frac{3\pi}{4}\right)$
C. $\sqrt{2}\sin\left(x + \frac{5\pi}{4}\right)$
D. $\sqrt{2}\sin\left(x + \frac{7\pi}{4}\right)$
E. $\sqrt{2}\sin\left(x + \frac{\pi}{2}\right)$

More exam questions are available online.

8.6 Review

8.6.1 Summary

c-37051

Hey students! Now that it's time to revise this topic, go online to:

Access the topic summary

Review your results

Watch teacher-led videos

Practise exam questions

Find all this and MORE in jacPLUS

8.6 Exercise

Technology free: short answer

1. Use the diagram shown to determine the exact values of:

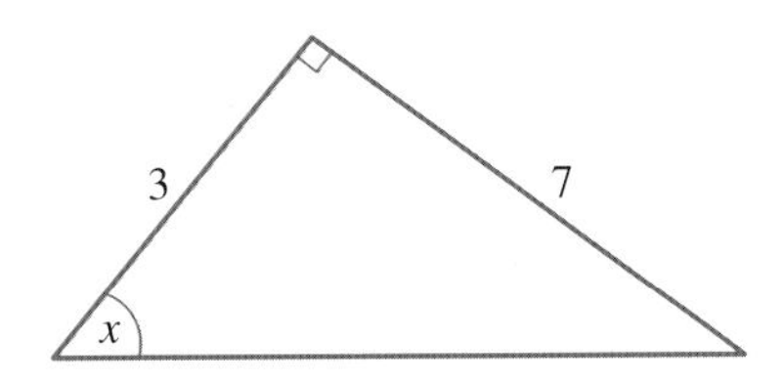

   a. $\cos(x)$
   b. $\sin(x)$

2. Simplify $\left(2\tan\left(\frac{x}{2}\right)\right)^2 - \left(2\sec\left(\frac{x}{2}\right)\right)^2$.

3. If $\sec(x) = a$, $a > 1$, $0 < x < \frac{\pi}{2}$, determine expressions for:

   a. $\cot(x)$
   b. $\operatorname{cosec}(x)$

4. Determine the values of $\cos(x)$ and $\sin(x)$ if $\cos(2x) = \frac{7}{25}$.

5. If $\sin(A) = \frac{1}{9}$, $\frac{\pi}{2} < A < \pi$, calculate the exact value of $\sin(2A)$.

6. If $\cos(A) = \frac{1}{3}$ and $\cos(B) = \frac{1}{4}$, where A and B are both acute angles, calculate $\cos(A + B)$ as exact values.

7. Use multiple angle formulas to simplify $\frac{\sin(3A)}{\sin(A)} + \frac{\cos(3A)}{\cos(A)}$.

8. Determine the exact value of $\tan\left(\frac{13\pi}{12}\right)$.

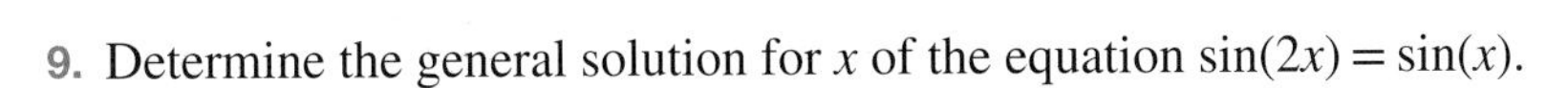

9. Determine the general solution for x of the equation $\sin(2x) = \sin(x)$.

10. Use appropriate formulas to simplify $\frac{\sin(A) + \sin(B)}{\cos(A) + \cos(B)}$.

Technology active: multiple choice

11. **MC** Determine which of the following statements is true.

A. $\left(2\sin\left(\frac{x}{2}\right)\right)^2 + \left(2\cos\left(\frac{x}{2}\right)\right)^2 = 1$

B. $\left(2\sin\left(\frac{x^2}{2}\right)\right)^2 + \left(2\cos\left(\frac{x^2}{2}\right)\right)^2 = 1$

C. $\left(2\sin^2\left(\frac{x}{2}\right)\right)^2 + \left(2\cos^2\left(\frac{x}{2}\right)\right)^2 = 1$

D. $\left(\sin\left(\frac{x}{2}\right)\right)^2 + \left(\cos\left(\frac{x}{2}\right)\right)^2 = 1$

E. $\sqrt{\sin^2\left(\frac{x^2}{2}\right)} + \sqrt{\cos^2\left(\frac{x^2}{2}\right)} = 1$

12. **MC** $\frac{\sin(A)}{1+\cos(A)} + \frac{1+\cos(A)}{\sin(A)}$ is equal to

A. $2\sin(A)$

B. $2\cos(A)$

C. 2

D. $\frac{2}{\sin(A)}$

E. $\frac{2}{\cos(A)}$

13. **MC** Determine which of the following is **not** a correct trigonometric identity.

A. $\sin(2x) = \frac{2\tan(x)}{1+\tan^2(x)}$

B. $\cos(2x) = \frac{1-\tan^2(x)}{1+\tan^2(x)}$

C. $\frac{1-\cos(2x)}{\sin(2x)} = \tan(x)$

D. $\frac{\sin(2x)}{1-\cos(2x)} = \cot(x)$

E. $\frac{\sin(3x)}{\sin(x)} - \frac{\cos(3x)}{\cos(x)} = 2$

14. **MC** Determine which of the following is **not** true.

A. $\left(\sec(x)+\tan(x)\right)\left(\sec(x)-\tan(x)\right) = 1$

B. $\left(\text{cosec}(x)+\cot(x)\right)\left(\text{cosec}(x)-\cot(x)\right) = 1$

C. $\frac{1+\sin(x)}{\cos(x)} = \sec(x)+\tan(x)$

D. $\frac{1-\cos(x)}{\sin(x)} = \text{cosec}(x)-\cot(x)$

E. $\frac{\sin(x)}{\text{cosec}(x)} + \frac{\cos(x)}{\sec(x)} = 2$

15. **MC** Determine which of the following is **not** a correct trigonometric identity.

A. $\sec^2(x)+\text{cosec}^2(x) = \sec^2(x)\text{cosec}^2(x)$

B. $\left(\sec^2(x)-1\right)\left(\text{cosec}^2(x)-1\right) = 1$

C. $\cot(x)+\tan(x) = \sec(x)\text{cosec}(x)$

D. $\frac{1}{1-\sin(x)} - \frac{1}{1+\sin(x)} = 2\text{cosec}^2(x)$

E. $\sec^2(x)-\text{cosec}^2(x) = \tan^2(x)-\cot^2(x)$

Technology active: extended response

16. a. Prove that $\tan(\theta) = \cot(\theta) - 2\cot(2\theta)$.

b. Hence, demonstrate that $\tan(\theta) + 2\tan(2\theta) = \cot(\theta) - 4\cot(4\theta)$.

c. Use appropriate formulas to simplify $\tan(\theta) + 2\tan(2\theta) + 4\tan(4\theta)$.

17. At a time t, the current in a circuit, I_1, is given by $I_1 = 40\cos(800t) + 9\sin(800t)$ and the current in another circuit, I_2, is given by $I_2 = 9\cos(800t) - 40\sin(800t)$.
    a. Express I_1 in the form $r\cos(nt - \alpha)$ and I_2 in the form $r\cos(nt + \alpha)$, where α is in degrees correct to 2 decimal places.
    b. Express the total current, $I_T = I_1 + I_2$, as a single cosine function.
18. If $\alpha + \beta = \theta$ and $\tan(\alpha) : \tan(\beta) = a : b$, prove that $\sin(\alpha - \beta) : \sin(\theta) = (a - b) : (a + b)$.
19. If $\sin^4(\theta) + \cos^4(\theta) = 1$, prove that θ is a multiple of $\frac{\pi}{2}$.
20. Prove that if $\sin(\alpha - \beta) = \sin^2(\alpha) - \sin^2(\beta)$, then either $\alpha - \beta = n\pi$ or $\alpha + \beta = 2n\pi + \frac{\pi}{2}$, $n \in Z$.

8.6 Exam questions

Question 1 (3 marks)

Show that $\dfrac{\sin(2\theta)\cos(\theta) - \sin(\theta)\cos(2\theta)}{\sin(6\theta)\cos(2\theta) - \sin(2\theta)\cos(6\theta)} = \dfrac{1}{4}\sec(\theta)\sec(2\theta)$.

Question 2 (2 marks)

If $\frac{1}{4}\sec(\theta)\sec(2\theta) = -\frac{1}{4}$, calculate θ, $0 \le \theta \le 2\pi$.

Question 3 (1 mark)

MC Determine which of the following statements is **false**.

A. $2\sin\left(\frac{\pi}{4}\right)\cos\left(\frac{\pi}{4}\right) = 1$

B. $\cos^2\left(\frac{\pi}{4}\right) - \sin^2\left(\frac{\pi}{4}\right) = 0$

C. $\tan\left(\frac{\pi}{2}\right) - \tan\left(\frac{5\pi}{2}\right) = 0$

D. $\tan\left(\frac{\pi}{3}\right) = \dfrac{2\tan\left(\frac{\pi}{6}\right)}{1 - \tan^2\left(\frac{\pi}{3}\right)}$

E. $\cos\left(\frac{\pi}{6}\right) = \sqrt{\frac{1}{2}\left(1 + \cos\left(\frac{\pi}{3}\right)\right)}$

Question 4 (2 marks)

If $\cot^2(2x) = \frac{4\sqrt{2}}{7}$, and $\frac{\pi}{2} \le x \le \pi$, determine an exact value for $\sin(2x)$.

Question 5 (2 marks)

If $\cos(2\theta) = \frac{1-a}{b}$, determine $\cos(\theta)$ in terms of a and b.

More exam questions are available online.

Answers

Topic 8 Trigonometric identities

8.2 Pythagorean identities

8.2 Exercise

1. a. 0.6 b. $\frac{4}{3}$
2. a. $\sqrt{0.91}$ b. 3.18
3. a. $\pm\sqrt{0.84}$ b. $\pm\sqrt{0.51}$
   c. $\pm\sqrt{0.9424}$ d. $\pm\sqrt{0.19}$
4. $\cos(\theta) = -\frac{\sqrt{55}}{8}$
5. $\text{cosec}(\theta) = \frac{\sqrt{137}}{4}$
6. a. $\frac{\sqrt{11}}{4}$ b. $-\sqrt{\frac{5}{11}}$
7. $\sin^2(A)$
8. $1 - \sin(A)$
9. a. $0, \frac{\pi}{2}, \pi, 2\pi$ b. $\frac{\pi}{2}, \pi, \frac{3\pi}{2}$
10. a. $0, \pi, \frac{4\pi}{3}, \frac{5\pi}{3}, 2\pi$ b. $\frac{\pi}{3}, \pi, \frac{5\pi}{3}$
11. a. $\frac{\pi}{2}$ b. $\frac{\pi}{2}, \frac{7\pi}{6}, \frac{11\pi}{6}$
12. a. $(2\sin(A) - 1)(\sin(A) + 1)$
    b. $(\tan(x) - 2)(\tan(x) + 1)$
13. a. $(2\,\text{cosec}(\alpha) + 1)(\text{cosec}(\alpha) - 1)$
    b. $(1 - 2\sin(3x))(1 + 2\sin(3x))$ or $(2\cos(3x) + 1)(2\cos(3x) - 1)$
14. a. $(\text{cosec}(3A) - 1)(3\,\text{cosec}(3A) + 1)$
    b. $(\sec(2\beta) + 1)(\sec(2\beta) - 2)$
15. $\frac{\pi}{6}, \frac{5\pi}{6}, \frac{3\pi}{2}$
16. a. $\frac{\pi}{4}, \frac{3\pi}{4}, \frac{5\pi}{4}, \frac{7\pi}{4}$ b. $\frac{\pi}{3}, \pi, \frac{5\pi}{3}$
17. a. $\frac{\pi}{2}, \pi, \frac{3\pi}{2}$ b. $\frac{3\pi}{2}$
18. a. $\frac{\pi}{12}(12n + 1), \frac{\pi}{12}(12n + 5), \frac{\pi}{4}(4n + 1), n \in Z$
    b. $\frac{\pi}{3}(6n \pm 1), \frac{\pi}{2}(2n + 1), n \in Z$
19. a. $\frac{\pi}{24}(12n - 5), \frac{\pi}{24}(12n - 1), \frac{n\pi}{4}, n \in Z$
    b. $\frac{n\pi}{2}, \frac{\pi}{8}(2n + 1), n \in Z$
20. a. $\frac{\pi}{18}(12n + 5), \frac{\pi}{18}(12n - 5), \frac{\pi}{6}(2n + 1), n \in Z$
    b. $\frac{\pi}{9}(6n + 1), \frac{2\pi}{9}(3n + 1), \frac{n\pi}{3}, n \in Z$
21. a. $\frac{\pi}{3}(3n - 1), \frac{\pi}{4}(4n - 1), n \in Z$
    b. $\frac{\pi}{3}(3n - 1), \frac{\pi}{4}(4n + 1), n \in Z$
22. $\frac{\pi}{6}(12n + 1), \frac{\pi}{6}(12n + 5), \frac{\pi}{2}(2n + 1), n \in Z$
23. $\frac{\pi}{3}(6n \pm 1), n\pi, n \in Z$
24. $\frac{\pi}{4}(2n + 1), n \in Z$
25. $\frac{\pi}{4}(2n + 1), \frac{\pi}{3}(3n \pm 1), n \in Z$
26. $\frac{\pi}{4}, \frac{3\pi}{4}, \frac{5\pi}{4}, \frac{7\pi}{4}$
27. 9
28. $\left[\frac{3}{4}, 1\right]$

8.2 Exam questions

Note: Mark allocations are available with the fully worked solutions online.

1. C
2. B
3. C

8.3 Compound angle formulas

8.3 Exercise

1. a. $\frac{\sqrt{3}}{2}$ b. 0 c. $\frac{1}{2}$ d. $-\frac{\sqrt{2}}{2}$
2. a. $\sin(\theta) - \cos(\theta)$ b. $\sin(\theta) + \sqrt{3}\cos(\theta)$
   c. $\sqrt{3}\cos(\theta) + \sin(\theta)$ d. $\cos(\theta) - \sin(\theta)$
3. a. $\cos(\theta)$ b. $\sin(\theta)$
   c. $-\sin(\theta)$ d. $-\cos(\theta)$
4. a. $-\cos(\theta)$ b. $\sin(\theta)$
   c. $-\tan(\theta)$ d. $\tan(\theta)$
5. a. $\sqrt{3}\cos(x)$ b. $-\sqrt{3}\sin(x)$
   c. $\sin(x)$ d. -1
6. a. $2 - \sqrt{3}$ b. $2 + \sqrt{3}$
   c. $\frac{\sqrt{2} - \sqrt{6}}{4}$ d. $\frac{\sqrt{6} - \sqrt{2}}{4}$
7. a. $\frac{56}{65}$ b. $-\frac{63}{16}$
8. a. $-\frac{253}{325}$ b. $-\frac{204}{325}$
9. a. $-\frac{8\sqrt{5}}{21}$ b. $-\frac{19}{21}$
10. $\frac{ab + \sqrt{1 - a^2}\sqrt{1 - b^2}}{b\sqrt{1 - a^2} - a\sqrt{1 - b^2}}$

11. $\dfrac{a^2 - 2\sqrt{a+1}\sqrt{2a+1}}{a\left(\sqrt{2a+1} + 2\sqrt{a+1}\right)}$

12-16. Sample responses can be found in the worked solutions in the online resources.

8.3 Exam questions

Note: Mark allocations are available with the fully worked solutions online.

1. B
2. A
3. $\cos(2y)$

8.4 Double and half angle formulas

8.4 Exercise

1. a. $\dfrac{\sqrt{2}}{4}$ b. $-\dfrac{\sqrt{2}}{2}$
   c. $-\dfrac{\sqrt{2}}{2}$ d. $\dfrac{\sqrt{3}+2}{4}$
2. a. $-\dfrac{23}{32}$ b. $\dfrac{3\sqrt{55}}{32}$
   c. $-\dfrac{3\sqrt{55}}{23}$ d. $-\dfrac{23\sqrt{55}}{165}$
3. a. $\dfrac{\pi}{6}, \dfrac{\pi}{2}, \dfrac{5\pi}{6}, \dfrac{3\pi}{2}$
   b. $0, \dfrac{\pi}{3}, \pi, \dfrac{5\pi}{3}, 2\pi$
   c. $0, \dfrac{2\pi}{3}, \dfrac{4\pi}{3}, 2\pi$
   d. $\dfrac{\pi}{6}, \dfrac{5\pi}{6}, \dfrac{3\pi}{2}$
4. a. $0, \dfrac{\pi}{4}, \dfrac{3\pi}{4}, \pi, \dfrac{5\pi}{4}, \dfrac{7\pi}{4}, 2\pi$
   b. $\dfrac{\pi}{3}, \dfrac{\pi}{2}, \dfrac{2\pi}{3}, \dfrac{3\pi}{2}$
   c. $0, \dfrac{\pi}{6}, \dfrac{\pi}{2}, \dfrac{5\pi}{6}, \pi, \dfrac{7\pi}{6}, \dfrac{3\pi}{2}, \dfrac{11\pi}{6}, 2\pi$
   d. $\dfrac{\pi}{12}, \dfrac{\pi}{4}, \dfrac{5\pi}{12}, \dfrac{3\pi}{4}, \dfrac{13\pi}{12}, \dfrac{5\pi}{4}, \dfrac{17\pi}{12}, \dfrac{7\pi}{4}$

5–18. Sample responses can be found in the worked solutions in the online resources.

8.4 Exam questions

Note: Mark allocations are available with the fully worked solutions online.

1. $8\cos^4(\theta) - 8\cos^2(\theta) + 1$
2. a. $\sin(\theta)$ b. $\sin(4\theta)$
3. D

8.5 Converting $a\cos(x) + b\sin(x)$ to $r\cos(x \pm \alpha)$ or $r\sin(x \pm \alpha)$

8.5 Exercise

1. $\sqrt{2}\cos\left(x - \dfrac{\pi}{4}\right)$
2. $2\cos\left(x + \dfrac{\pi}{6}\right)$
3. a. $\sqrt{2}\sin\left(x + \dfrac{\pi}{4}\right)$ b. $2\sin\left(x + \dfrac{\pi}{6}\right)$
4. a. $\sqrt{2}\sin\left(x - \dfrac{\pi}{4}\right)$ b. $2\sin\left(x - \dfrac{\pi}{6}\right)$
5. a. $8\cos\left(x - \dfrac{\pi}{4}\right)$ b. $25\cos(x - 1.287)$
6. a. $\sqrt{2}\cos\left(x + \dfrac{\pi}{4}\right)$ b. $2\cos\left(x + \dfrac{\pi}{3}\right)$
7. a. $f(x) = 2\sin\left(x - \dfrac{\pi}{4}\right)$
   b. Maximum $\left(\dfrac{3\pi}{4}, 2\right)$, minimum $\left(\dfrac{7\pi}{4}, -2\right)$
   c.

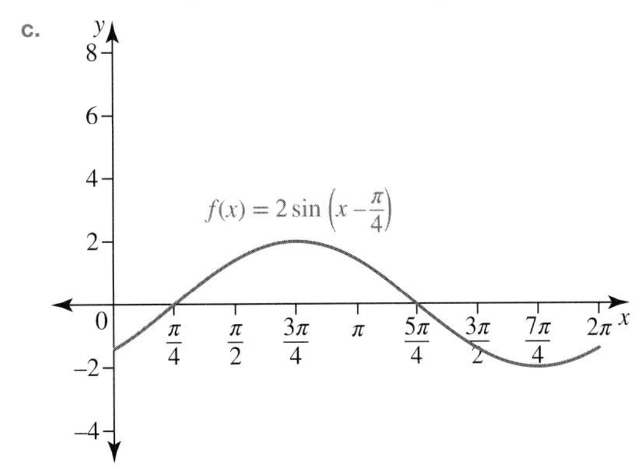

   d. $\dfrac{5\pi}{12}, \dfrac{13\pi}{12}$
8. a. $f(x) = 13\cos(x - 2.747)$
   b. Maximum $(2.747, 13)$, minimum $(5.888, -13)$
   c.

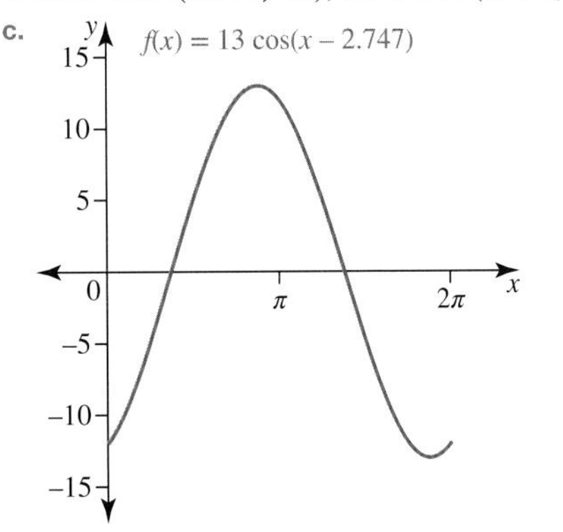

   d. 1.7, 3.794

9. a. $f(x) = 5\sin(x° - 53.13°)$

b. Maximum $(143.13°, 5)$, minimum $(323.13°, -5)$

c. See the image at the bottom of the page.*

d. $83.13°, 203.13°$

10. a. $f(x) = 13\cos(x - 0.395)$

b. Maximum $(0.395, 13)$, minimum $(3.537, -13)$

c.

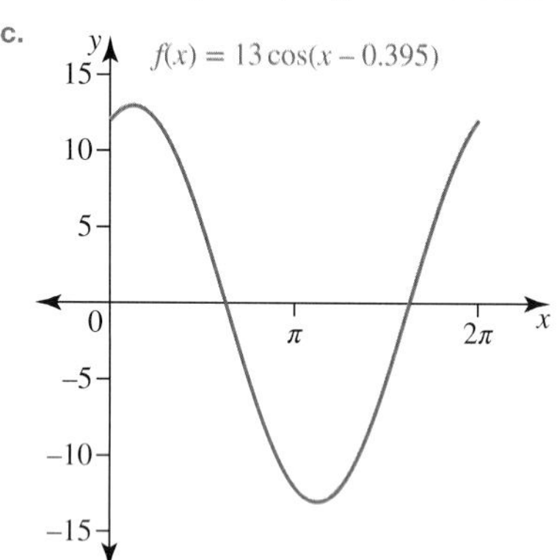

d. $x = 1.442,\ 5.631$; general solution:
$x = 1.442 + 2n\pi,\ 5.631 + 2n\pi,\ n \in Z$

11. a. $f(x) = 25\cos(x + 0.284)$

b. Maximum of 55 after 2.00 seconds, minimum of 5 after 0.95 seconds

12. a. $2\sin(x) + 2\sqrt{3}\cos(x) = 4\sin\left(x + \frac{\pi}{3}\right)$

b. Maximum of 9 after 2 hours, minimum of 1 after 14 hours

c. 6, 22 hours

d. See the image at the bottom of the page.*

13. a. Maximum 4, minimum $\frac{4}{9}$

b. Maximum $\frac{\sqrt{3}}{10}$, minimum $-\frac{\sqrt{3}}{6}$

14. a. $I = 41\cos(300t - 12.68°)$

b. $I = 61\cos(800t - 79.61°)$

8.5 Exam questions

Note: Mark allocations are available with the fully worked solutions online.

1. $x = \frac{\pi}{2}, \frac{11\pi}{6}$

2.

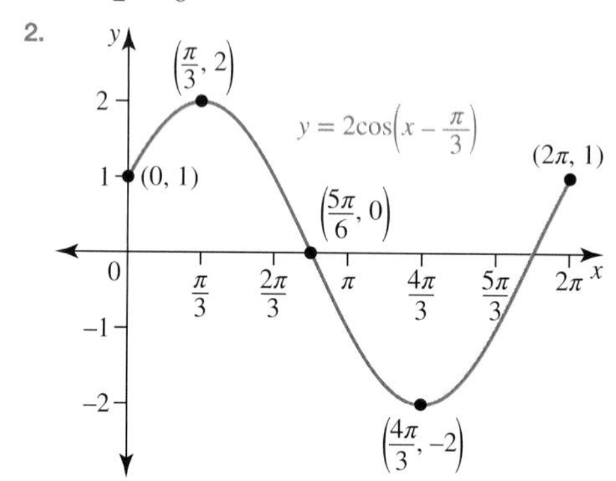

3. B

*9. c.

*12. d.

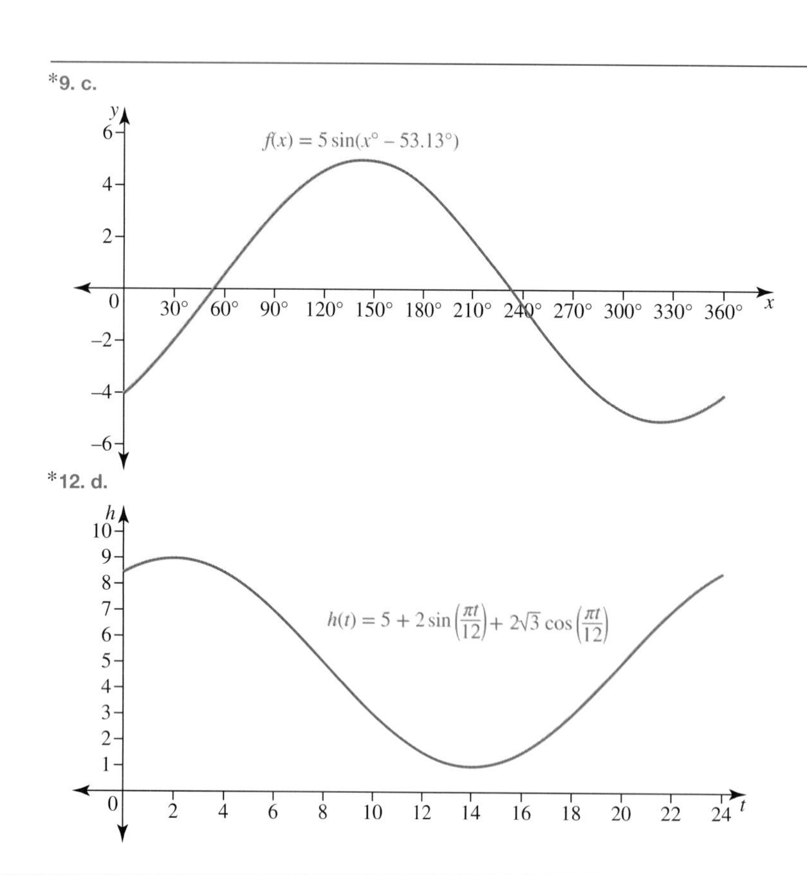

8.6 Review

8.6 Exercise

Technology free: short answer

1. a. $\frac{3\sqrt{58}}{58}$ b. $\frac{7\sqrt{58}}{58}$
2. -4
3. a. $\frac{\sqrt{a^2-1}}{a^2-1}$ b. $\frac{a\sqrt{a^2-1}}{a^2-1}$
4. $\cos(x)=\pm\frac{4}{5}$, $\sin(x)=\pm\frac{3}{5}$
5. $-\frac{8\sqrt{5}}{81}$
6. $\frac{1-2\sqrt{30}}{12}$
7. $4\cos(2A)$
8. $2-\sqrt{3}$
9. $\frac{\pi}{3}(6n\pm 1),\ n\pi,\ n\in Z$
10. $\tan\left(\frac{A+B}{2}\right)$

Technology active: multiple choice

11. D
12. D
13. D
14. E
15. D

Technology active: extended response

16. a. Sample responses can be found in the worked solutions in the online resources.
    b. Sample responses can be found in the worked solutions in the online resources.
    c. $\cot(\theta)-8\cot(8\theta)$
17. a. $I_1=41\cos(800t-12.68°)$, $I_2=41\cos(800t+77.32°)$
    b. $I_T=41\sqrt{2}\cos(800t+32.32°)$

18–20. Sample responses can be found in the worked solutions in the online resources.

8.6 Exam questions

Note: Mark allocations are available with the fully worked solutions online.

1. Sample responses can be found in the worked solutions in the online resources.
2. There are no solutions.
3. C
4. $\sin(2x)=-\frac{7}{9}$
5. $\cos(\theta)=\sqrt{\frac{1-a+b}{2b}}$

9 Vectors in the plane

LEARNING SEQUENCE

Fully worked solutions for this topic are available online.

9.1 Overview

Hey students! Bring these pages to life online

Watch videos

Engage with interactivities

Answer questions and check results

Find all this and MORE in jacPLUS

9.1.1 Introduction

A vector is a directed line segment that is described using both magnitude and direction. The length of the line segment is the magnitude of the vector and the direction is determined by a reference to the angle from a fixed line. Wind is a vector as it is described using magnitude and direction, for example a north-easterly wind of 20 km/h.

Consider the two yachts shown in the photograph. Could they be in danger of a collision? One yacht is travelling at a speed of 15 km/h and the other at a speed of 12 km/h. Their speeds give no indication of their direction. Their respective directions are as important as their speed for a correct analysis of their path.

Vectors are used in physics to describe and analyse any quantity that has both magnitude and direction, such as motion and force. Vectors can be used by coaches in many sporting areas to illustrate the importance of the angle of contact with the ball or the optimum position to kick a goal. For example, by watching film clips with vectors superimposed over their golf swings, professional golfers can determine the perfect angle to hold their club to ensure the best outcome.

KEY CONCEPTS

This topic covers the following key concepts from the VCE Mathematics Study Design:

- the representation of plane vectors as directed lines segments, magnitude and direction of a plane vector, and unit vectors
- geometric representation of addition, subtraction (triangle and/or parallelogram rules), scalar multiple of a vector and linear combination of plane vectors
- the representation of a plane vector as an ordered pair (a, b) in the form $a\underset{\sim}{i} + b\underset{\sim}{j} = a\begin{pmatrix}1\\0\end{pmatrix} + b\begin{pmatrix}0\\1\end{pmatrix} = \begin{pmatrix}a\\b\end{pmatrix}$
- simple vector algebra (addition, subtraction, multiplication by a scalar, linear combination) using these forms
- scalar (dot) product of two plane vectors, perpendicular and parallel vectors, projection of one vector onto another vector, and angle between two vectors
- geometric proofs with vectors
- application of vectors to displacement, velocity, resultant velocity, relative velocity, statics and motion under a constant force.

9.2 Vectors and scalars

LEARNING INTENTION

At the end of this subtopic you should be able to:
- use vector notation
- calculate addition and subtraction with vectors
- draw vectors multiplied by scalars
- use vectors to solve real-world problems.

9.2.1 Scalar and vector quantities

In mathematics, one of the important distinctions is between **scalar** quantities and **vector** quantities. Scalar quantities have **magnitude** only; vector quantities have *direction as well as magnitude*. Most of the quantities that we use are scalar, and include such measurements as time (for example 1.2 s; 15 min), mass (3.4 kg; 200 t) and area (3 cm^2; 400 ha). For some quantities it is very important that we know both magnitude and direction.

Consider a yacht travelling at 12 km/h in a north-westerly direction and a wind of 15 km/h blowing the yacht in an easterly direction as shown.

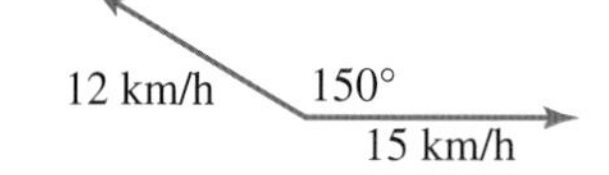

In what direction will the yacht move and at what speed? That is, what is the resultant velocity? The resultant velocity depends not only on the speed of the yacht and the wind but the direction of both.

Vectors and scalars

A vector is a quantity that has magnitude and direction.

A scalar is a quantity measured by a real number only.

9.2.2 Vector notation

A vector is shown graphically as a line, with a *tail* (start) and *head* (end). The length of the line indicates the magnitude and the orientation of the line indicates its direction.

In the figure, the head of the vector is at point B (indicated with an arrow), and the tail is at point A.

When writing this vector, we can use the points A and B to indicate the start and end points with a special arrow to indicate that it is a vector: $\overrightarrow{AB}$. Some textbooks use a single letter, in bold, such as ***w***, but this is difficult to write using pen and paper, so $\underset{\sim}{w}$ can also be used. The symbol (~) is called a *tilde*.

9.2.3 Equality of vectors

Vector equality

Two vectors are equal if *both* their magnitude and direction are equal.

In the figure shown, the following statements can be made:

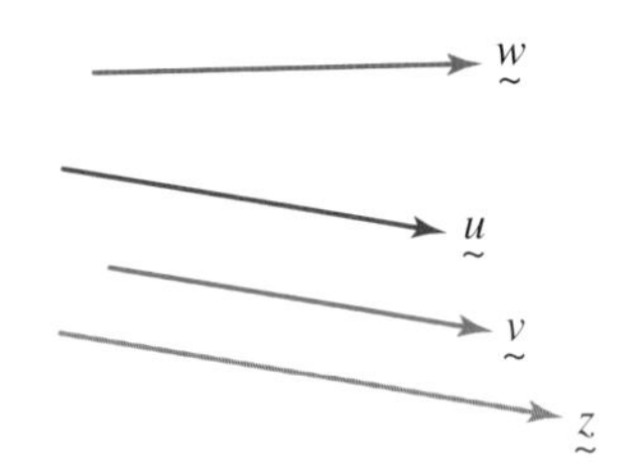

$\underset{\sim}{u} = \underset{\sim}{v}$
$\underset{\sim}{u} \neq \underset{\sim}{w}$ (directions are not equal)
$\underset{\sim}{u} \neq \underset{\sim}{z}$ (magnitudes are not equal).

9.2.4 Addition of vectors — the triangle rule

Consider a vector, $\underset{\sim}{u}$, that measures the travel from A to B and another vector, $\underset{\sim}{v}$, that measures the subsequent travel from B to C. The net result is as if the person travelled directly from A to C (vector $\underset{\sim}{w}$). Therefore, we can say that $\underset{\sim}{w} = \underset{\sim}{u} + \underset{\sim}{v}$.

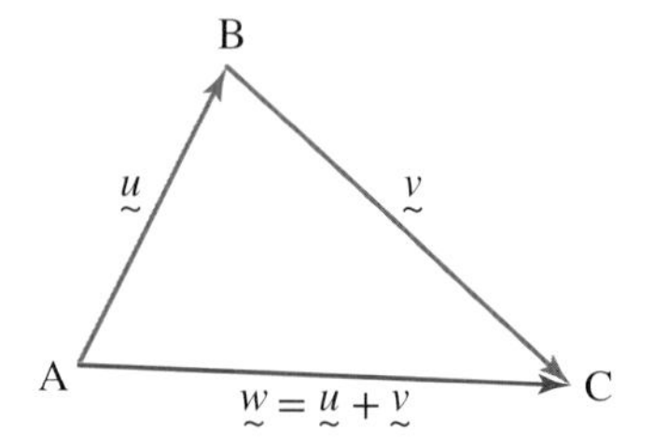

Adding vectors

To add two vectors, take the tail of one vector and join it to the head of another. The result of this addition is the vector from the *tail of the first vector* to the *head of the second vector*.

Returning to the yacht sailing north-east at 12 km/h with an easterly wind of 15 km/h, we can see that the velocities of both can be represented as the sum of two vectors.

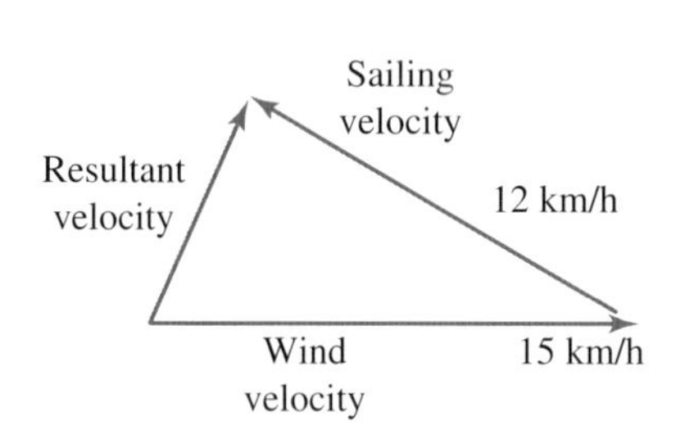

From this figure we are able to get a rough idea of the magnitude and direction of the resultant velocity. In the following sections, we will learn techniques for calculating the resultant magnitude and direction accurately.

9.2.5 The negative of a vector

If $\underset{\sim}{u}$ is the vector from A to B, then $-\underset{\sim}{u}$ is the vector from B to A.

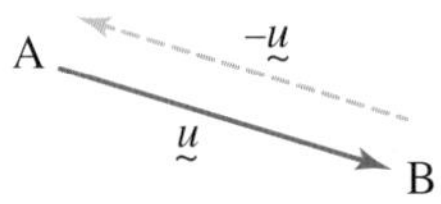

Subtracting vectors

We can subtract vectors by adding the *negative* of the second vector to the first vector.

WORKED EXAMPLE 1 Drawing vector sums and differences

Using the vectors shown, draw the results of:

a. $\underset{\sim}{u} + \underset{\sim}{v}$　　b. $-\underset{\sim}{u}$　　c. $\underset{\sim}{u} - \underset{\sim}{v}$　　d. $\underset{\sim}{v} - \underset{\sim}{u}$

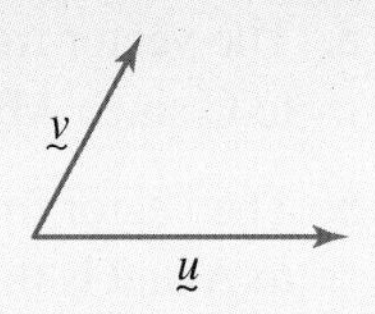

THINK	WRITE
a. 1. Move $\underset{\sim}{v}$ so that its tail is at the head of $\underset{\sim}{u}$.	a.  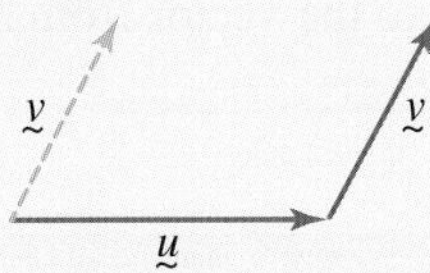
2. Join the tail of $\underset{\sim}{u}$ to the head of $\underset{\sim}{v}$ to obtain $\underset{\sim}{u} + \underset{\sim}{v}$.	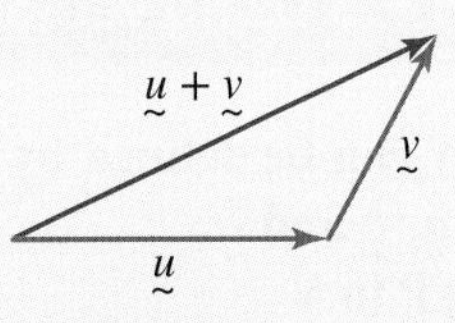
b. Reverse the arrow on $\underset{\sim}{u}$ to obtain $-\underset{\sim}{u}$.	b.
c. 1. Reverse $\underset{\sim}{v}$ to get $-\underset{\sim}{v}$.	c. 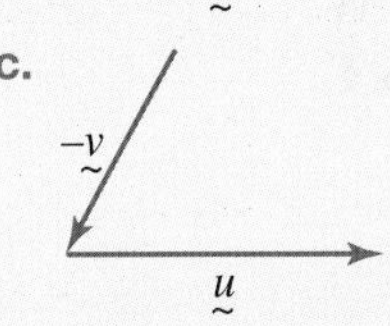
2. Join the tail of $-\underset{\sim}{v}$ to the head of $\underset{\sim}{u}$ to get $-\underset{\sim}{v} + \underset{\sim}{u}$, which is the same as $\underset{\sim}{u} - \underset{\sim}{v}$ or $\underset{\sim}{u} + (-\underset{\sim}{v})$.	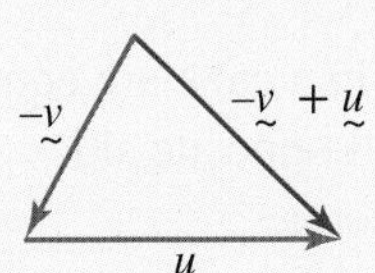
d. 1. Reverse $\underset{\sim}{u}$ to get $-\underset{\sim}{u}$. The vectors are now 'aligned properly' with the head of $-\underset{\sim}{u}$ joining the tail of $\underset{\sim}{v}$.	d.
2. Join the tail of $-\underset{\sim}{u}$ to the head of $\underset{\sim}{v}$ to get $\underset{\sim}{v} - \underset{\sim}{u}$. Note that this is the same as $(-\underset{\sim}{u} + \underset{\sim}{v})$	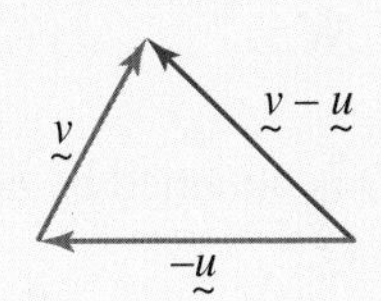

WORKED EXAMPLE 2 Using vector notation in 2D shapes

The parallelogram ABCD can be defined by the two vectors $\overrightarrow{AB} = \underset{\sim}{b}$ and $\overrightarrow{BC} = \underset{\sim}{c}$.
In terms of these vectors, determine:

a. **the vector from A to D**
b. **the vector from C to D**
c. **the vector from D to B.**

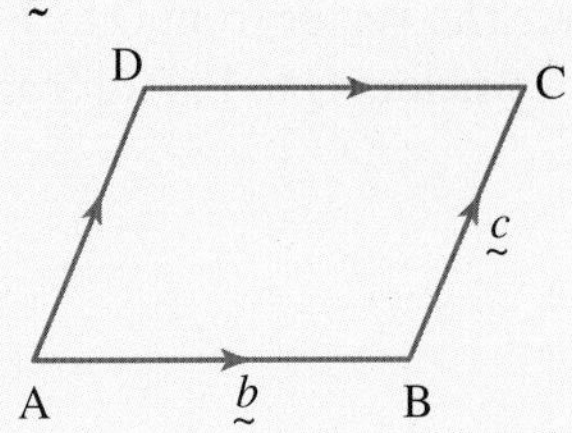

THINK

a. The vector from A to D is equal to the vector from B to C since ABCD is a parallelogram.

b. The vector from C to D is equal to the vector from B to A and is the reverse of B to A, which is $\underset{\sim}{b}$.

c. The vector from D to B is obtained by adding the vector from D to A to the vector from A to B.

WRITE

a. $\overrightarrow{AD} = \underset{\sim}{c}$

b. $\overrightarrow{CD} = -\underset{\sim}{b}$

c. $\overrightarrow{DB} = -\underset{\sim}{c} + \underset{\sim}{b}$
$= \underset{\sim}{b} - \underset{\sim}{c}$

WORKED EXAMPLE 3 Using vector notation in 3D shapes

A cube PQRSTUVW can be defined by the three vectors $\overrightarrow{PQ} = \underset{\sim}{a}$, $\overrightarrow{QV} = \underset{\sim}{b}$ and $\overrightarrow{PS} = \underset{\sim}{c}$ as shown. Express in terms of $\underset{\sim}{a}$, $\underset{\sim}{b}$ and $\underset{\sim}{c}$:

a. **the vector joining P to V**
b. **the vector joining P to W**
c. **the vector joining U to Q**
d. **the vector joining S to W**
e. **the vector joining Q to T.**

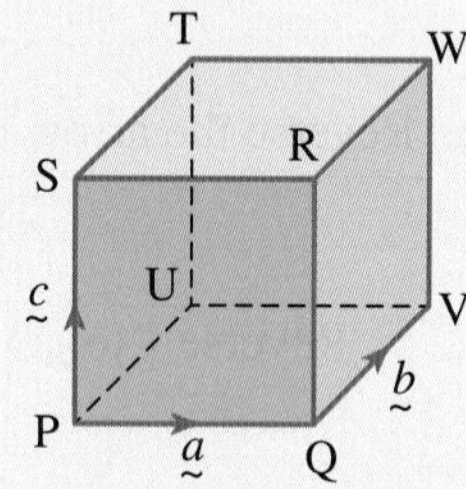

THINK

All of the opposite sides in a cube are equal in length and parallel. Therefore all opposite sides can be expressed as the same vector.

a. The vector from P to V is obtained by adding the vector from P to Q to the vector from Q to V.

b. The vector from P to W is obtained by adding the vectors P to V and V to W.

c. The vector from U to Q is obtained by adding the vectors U to P and P to Q.

d. The vector from S to W is obtained by adding the vectors S to R and R to W.

e. The vector from Q to T is obtained by adding the vectors Q to P, P to S and S to T.

WRITE

a. $\overrightarrow{PV} = \overrightarrow{PQ} + \overrightarrow{QV}$
$\overrightarrow{PV} = \underset{\sim}{a} + \underset{\sim}{b}$

b. $\overrightarrow{PW} = \overrightarrow{PV} + \overrightarrow{VW}$
$\overrightarrow{PW} = \underset{\sim}{a} + \underset{\sim}{b} + \underset{\sim}{c}$

c. $\overrightarrow{UQ} = \overrightarrow{UP} + \overrightarrow{PQ}$
$\overrightarrow{UQ} = -\underset{\sim}{b} + \underset{\sim}{a}$
$= \underset{\sim}{a} - \underset{\sim}{b}$

d. $\overrightarrow{SW} = \overrightarrow{SR} + \overrightarrow{RW}$
$\overrightarrow{SW} = \underset{\sim}{a} + \underset{\sim}{b}$

e. $\overrightarrow{QT} = \overrightarrow{QP} + \overrightarrow{PS} + \overrightarrow{ST}$
$\overrightarrow{QT} = -\underset{\sim}{a} + \underset{\sim}{c} + \underset{\sim}{b}$
$= \underset{\sim}{b} + \underset{\sim}{c} - \underset{\sim}{a}$

9.2.6 Multiplying a vector by a scalar

Multiplication of a vector by a *positive* number (scalar) affects only the *magnitude* of the vector, not the direction. For example, if a vector $\underset{\sim}{u}$ has a direction of north and a magnitude of 10, then the vector $3\underset{\sim}{u}$ has a direction of north and magnitude of 30.

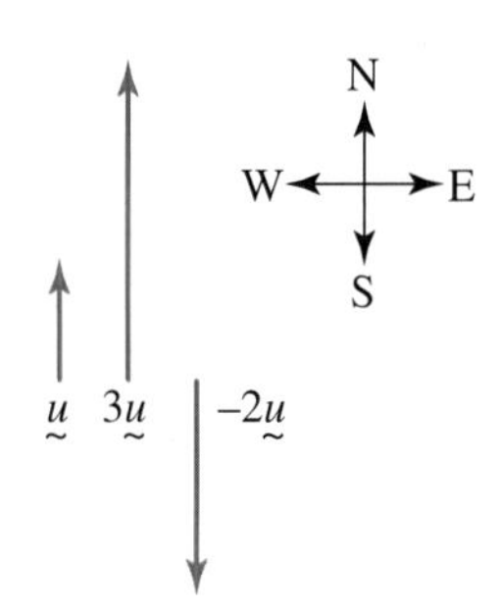

If the scalar is *negative*, then the direction is reversed. Therefore, $-2\underset{\sim}{u}$ has a direction of south and a magnitude of 20.

Scalar multiples of a vector are all parallel as multiplication by a positive scalar only affects the magnitude of a vector, and multiplication by a negative scalar only affects the vector's magnitude and reverses its direction.

WORKED EXAMPLE 4 Drawing vector multiplied by scalars

Use the vectors shown to draw the result of:

a. $2\underset{\sim}{r} + 3\underset{\sim}{s}$

b. $2\underset{\sim}{s} - 4\underset{\sim}{r}$.

THINK | **WRITE**

a. 1. Increase the magnitude of $\underset{\sim}{r}$ by a factor of 2 and $\underset{\sim}{s}$ by a factor of 3.

a.

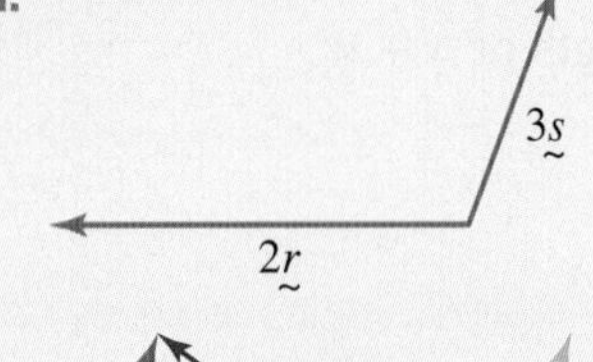

2. Move the tail of $3\underset{\sim}{s}$ to the head of $2\underset{\sim}{r}$. Then join the tail of $2\underset{\sim}{r}$ to the head of $3\underset{\sim}{s}$ to get $2\underset{\sim}{r} + 3\underset{\sim}{s}$.

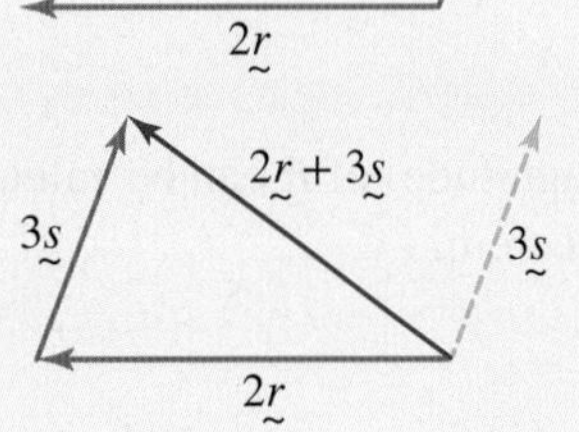

b. 1. Increase the magnitude of $\underset{\sim}{s}$ by a factor of 2 and $\underset{\sim}{r}$ by a factor of 4.

b.

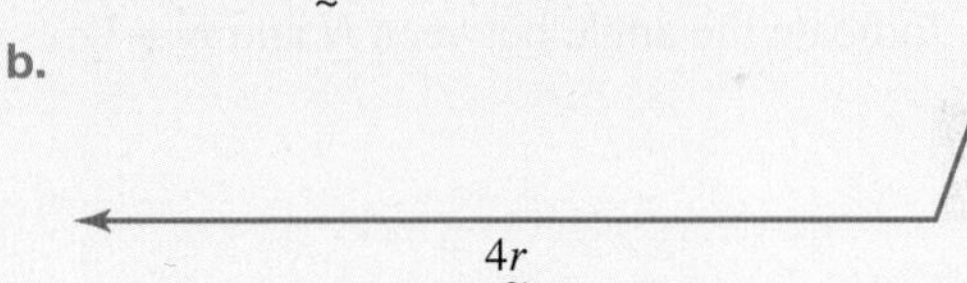

2. Reverse the arrow on $4\underset{\sim}{r}$ to get $-4\underset{\sim}{r}$.

3. Join the tail of $-4\underset{\sim}{r}$ to the head of $2\underset{\sim}{s}$.

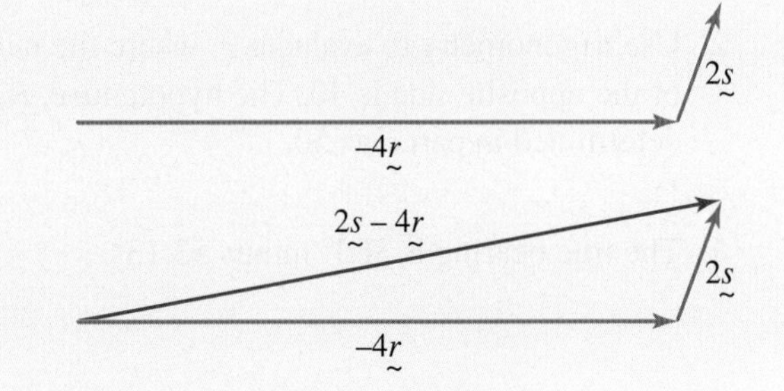

Vectors can be used to solve real-world problems involving movement and direction. Direction is typically referenced using a compass and expressed as the degrees from north. This is known as a true bearing. We can then use Pythagoras' theorem and trigonometry to solve displacement and bearing problems through vector addition.

WORKED EXAMPLE 5 Applications of vectors

A boat travels 30 km north and then 40 km west.

a. Make a vector drawing of the path of the boat.

b. Draw the vector that represents the net displacement of the boat.

c. Determine the magnitude of the net displacement.

d. Calculate the bearing (from true north) of this net displacement vector.

THINK	WRITE
a. 1. Set up vectors (tail to head), one pointing north, the other west.	**a.** 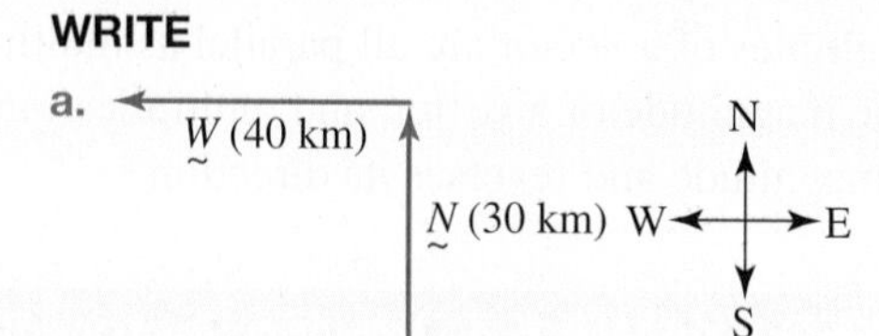
2. Indicate the distances as 30 km and 40 km respectively.	
b. Join the tail of the $\underset{\sim}{N}$ vector with the head of the $\underset{\sim}{W}$ vector.	**b.** 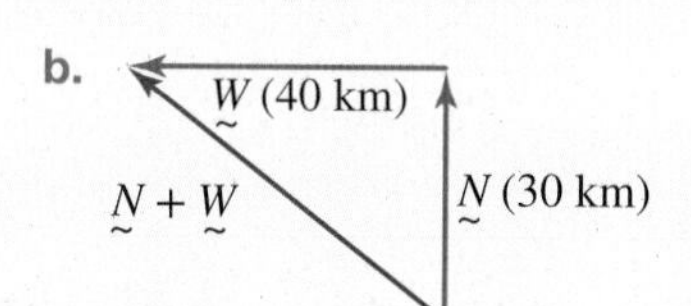
c. 1. Let R km = length of $\underset{\sim}{N}+\underset{\sim}{W}$.	**c.** 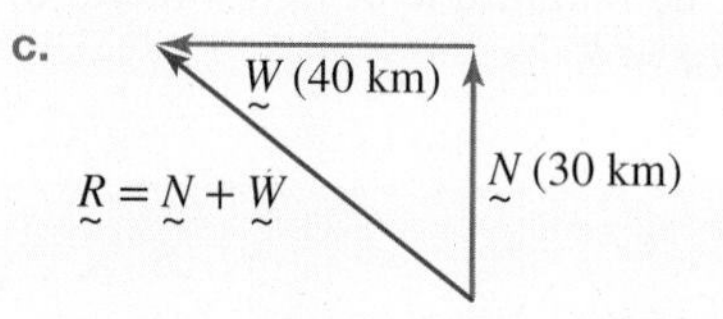
2. The length (magnitude) of $\underset{\sim}{R}$ can be calculated using Pythagoras' theorem.	$R=\sqrt{30^2+40^2}$ $=\sqrt{900+1600}$ $=50$ km
d. 1. Indicate the angle between $\underset{\sim}{N}$ and $\underset{\sim}{N}+\underset{\sim}{W}$ as θ.	**d.** $\underset{\sim}{W}$ (40 km), $\underset{\sim}{N}+\underset{\sim}{W}$, $\underset{\sim}{N}$ (30 km), θ
2. Use trigonometry to evaluate θ, where the magnitude of the opposite side is 40. The hypotenuse, $\underset{\sim}{R}$, was determined in part **c** as 50.	$\sin\theta=\frac{40}{50}$ $=0.8$ $\theta=53.13°$
3. The true bearing is 360° minus 53.13°.	Therefore, the true bearing is: $360°-53.13°=306.87°$T or N53.13°W

9.2 Exercise

Students, these questions are even better in jacPLUS

Receive immediate feedback and access sample responses

Access additional questions

Track your results and progress

Find all this and MORE in jacPLUS

Technology free

1. WE1 Using the vectors shown, draw the results of:

a. $\underset{\sim}{r}+\underset{\sim}{s}$ b. $\underset{\sim}{r}-\underset{\sim}{s}$ c. $\underset{\sim}{s}-\underset{\sim}{r}$

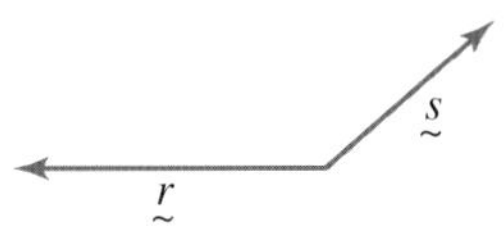

2. WE2 The pentagon ABCDE shown can be defined by the four vectors, $\underset{\sim}{s}$, $\underset{\sim}{t}$, $\underset{\sim}{u}$ and $\underset{\sim}{v}$. Determine, in terms of these four vectors:

a. the vector from A to D
b. the vector from A to B
c. the vector from D to A
d. the vector from B to E
e. the vector from C to A.

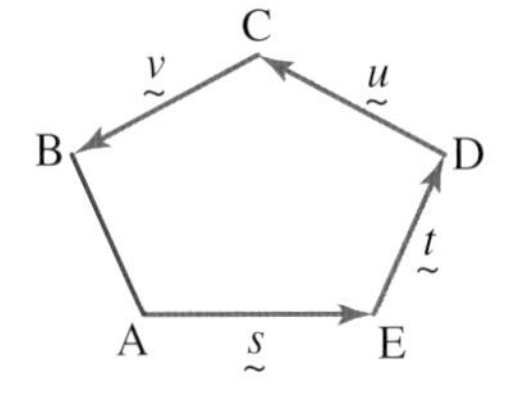

3. In the rectangle ABCD, the vector joining A to B is denoted by $\underset{\sim}{u}$ and the vector joining B to C is $\underset{\sim}{v}$. Determine which pairs of points are joined by:

a. $\underset{\sim}{u}+\underset{\sim}{v}$
b. $\underset{\sim}{u}-\underset{\sim}{v}$
c. $\underset{\sim}{v}-\underset{\sim}{u}$
d. $3\underset{\sim}{u}+2\underset{\sim}{v}-2\underset{\sim}{u}-\underset{\sim}{v}$

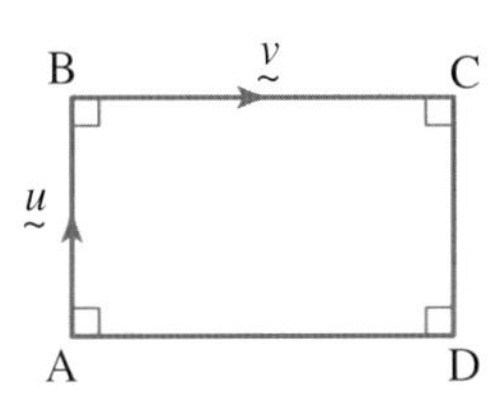

4. WE3 A rectangular prism (box) CDEFGHIJ can be defined by three vectors $\underset{\sim}{r}$, $\underset{\sim}{s}$ and $\underset{\sim}{t}$ as shown.
Express in terms of $\underset{\sim}{r}$, $\underset{\sim}{s}$ and $\underset{\sim}{t}$:

a. the vector joining C to H
b. the vector joining C to J
c. the vector joining G to D
d. the vector joining F to I
e. the vector joining H to E
f. the vector joining D to J
g. the vector joining C to I
h. the vector joining J to C.

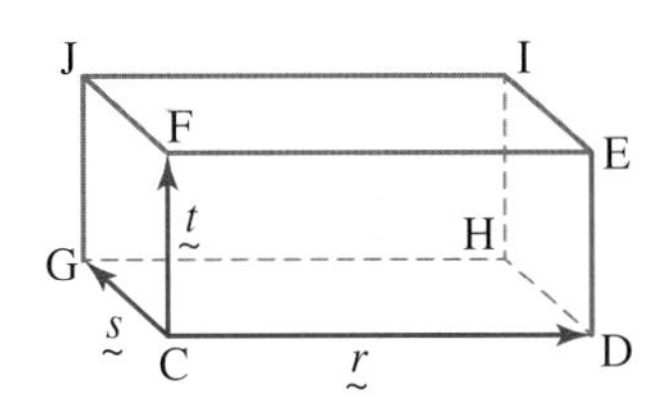

5. In terms of vectors $\underset{\sim}{a}$ and $\underset{\sim}{b}$ in the figure, define the vector joining O to D.

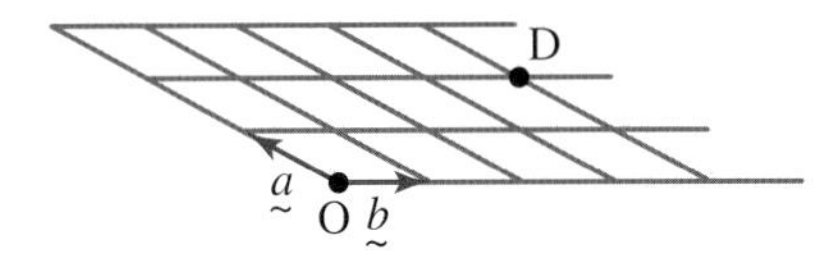

6. In terms of vectors $\underset{\sim}{a}$ and $\underset{\sim}{b}$, define the vector joining E to O.

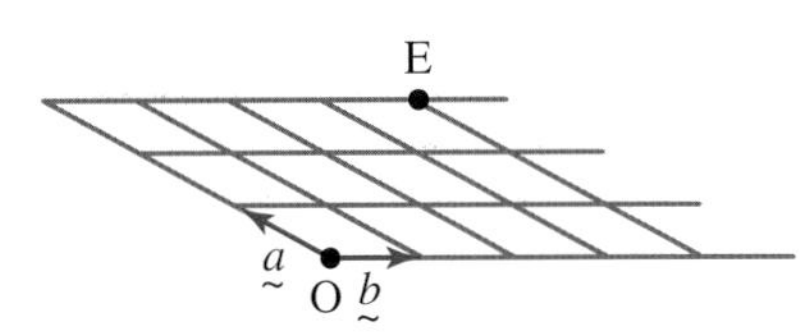

7. WE4 Using the vectors shown, draw the results of:

a. $2\underset{\sim}{r}+2\underset{\sim}{s}$ b. $2\underset{\sim}{r}-2\underset{\sim}{s}$ c. $3\underset{\sim}{s}-4\underset{\sim}{r}$

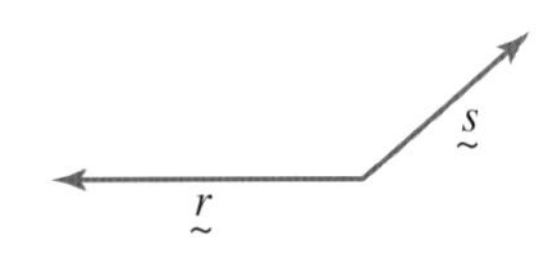

8. For speed, velocity, displacement, force, volume and angle; determine which are vector quantities. Justify your answer.

9. For speed, time, acceleration, velocity, length and displacement; determine which are scalar quantities. Justify your answer.

10. A 2-dimensional vector can be determined by its length and its angle with respect to (say) true north. State what quantities could be used to represent a 3-dimensional vector.

Technology active

11. A pilot plans to fly 300 km north then 400 km east.
    a. Make a vector drawing of her flight plan.
    b. Show the resulting net displacement vector.
    c. Calculate the length (magnitude) of this net displacement vector.
    d. Calculate the bearing (from true north) of this net displacement vector, in degrees to 1 decimal place.

12. Another pilot plans to travel 300 km east, then 300 km north-east. Show that the resultant bearing is 67.5 degrees and determine how far east of its starting point the plane has travelled, to 1 decimal place.

13. An aeroplane travels 400 km west, then 600 km north. Calculate how far the aeroplane is from its starting point and determine the bearing of the resultant displacement. Give your answers to 1 decimal place.

14. Using technology or a piece of graph paper, draw a vector, $\underset{\sim}{a}$, that is 3 units east and 5 units north of the origin. Draw another vector, $\underset{\sim}{b}$, that is 5 units east and 3 units north of the origin. On the same graph, draw the following vectors.
    a. $\underset{\sim}{a}+\underset{\sim}{b}$
    b. $\underset{\sim}{a}+3\underset{\sim}{b}$
    c. $\underset{\sim}{a}-\underset{\sim}{b}$
    d. $\underset{\sim}{b}-\underset{\sim}{a}$
    e. $3\underset{\sim}{b}-4\underset{\sim}{a}$

15. Determine the direction, in degrees to 1 decimal place, and magnitude, to 2 decimal places, of a vector joining point A to point B, where B is 10 m east and 4 m north of A.

16. Consider a parallelogram defined by the vectors $\underset{\sim}{a}$ and $\underset{\sim}{b}$, and its associated diagonals, as shown.

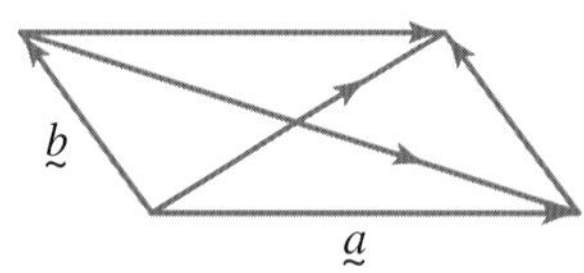

Show that the vector sum of the diagonal vectors is $2\underset{\sim}{a}$.

17. Show, *by construction*, that for any vectors $\underset{\sim}{u}$ and $\underset{\sim}{v}$:

$$3(\underset{\sim}{u}+\underset{\sim}{v})=3\underset{\sim}{u}+3\underset{\sim}{v}$$

(This is called the *Distributive Law*.)

18. Show, *by construction*, that for any three vectors $\underset{\sim}{a}$, $\underset{\sim}{b}$ and $\underset{\sim}{c}$:

$$(\underset{\sim}{a}+\underset{\sim}{b})+\underset{\sim}{c}=\underset{\sim}{a}+(\underset{\sim}{b}+\underset{\sim}{c})$$

(This is called the *Associative Law*.)

19. Show, *by construction*, that for any two vectors $\underset{\sim}{r}$ and $\underset{\sim}{s}$:

$$3\underset{\sim}{r} - \underset{\sim}{s} = -(\underset{\sim}{s} - 3\underset{\sim}{r})$$

20. A girl walks the following route: 400 m north, 300 m east, 200 m north, 500 m west, 600 m south, 200 m east.
Make a vector drawing of these six paths. Determine the net displacement vector.

9.2 Exam questions

Question 1 (1 mark) TECH-ACTIVE

MC OABC is a parallelogram with $\overrightarrow{OA}$ = and $\overrightarrow{OC} = \underset{\sim}{c}$.

The sum of the diagonal vectors $\overrightarrow{OB}$ and $\overrightarrow{AC}$ is

A. $2\underset{\sim}{a} + 2\underset{\sim}{c}$ **B.** $2\underset{\sim}{a} - 2\underset{\sim}{c}$ **C.** $2\underset{\sim}{a}$ **D.** $2\underset{\sim}{c}$ **E.** $\underset{\sim}{a} - \underset{\sim}{c}$

Question 2 (2 marks) TECH-ACTIVE

Clemantine, an intrepid hiker, walks 10 km east through rugged terrain, then 6 km due north. After a suitable break, she then walks a further 3 km west and 3 km south, then camps for the night. Determine how far she is from her starting point as an exact value, and determine the bearing of her camp from her starting point, in degrees rounded to 2 decimal places.

Question 3 (2 marks) TECH-ACTIVE

A river current flows parallel to the river banks at 5 km/h. A boat heads directly across the river from one river bank to the other side at 12 km/h. Determine the actual velocity and direction of the boat, in degrees to 2 decimal places.

More exam questions are available online.

9.3 Position vectors in the plane

LEARNING INTENTION

At the end of this subtopic you should be able to:
- determine the magnitude and direction of a vector
- express vectors in polar and Cartesian form
- determine unit vectors
- calculate vectors between two points
- apply vectors to real-world situations.

9.3.1 Cartesian form of a vector

As a vector has both magnitude and direction, it can be represented in 2-dimensional planes.

In the figure, the vector $\underset{\sim}{u}$ joins the point A to point B.

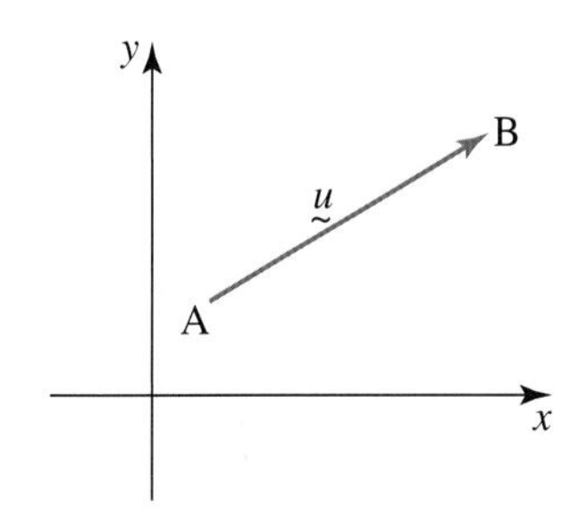

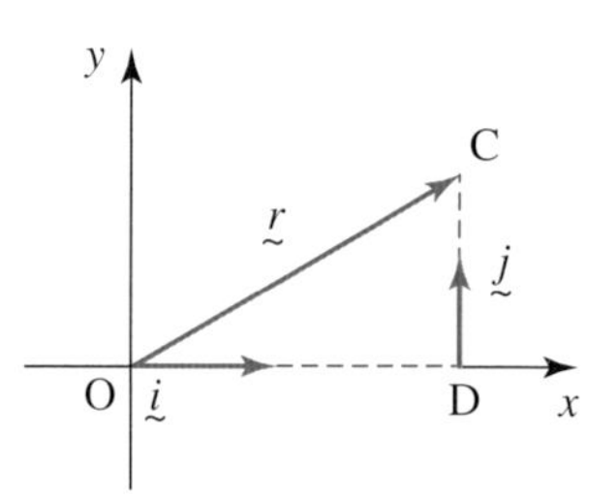

Using the Cartesian plane, an **identical vector**, $\underset{\sim}{r}$, can be considered to join the origin with the point C.

It is easy to see that $\underset{\sim}{r}$ is made up of two components: one along the x-axis and one parallel to the y-axis. Let $\underset{\sim}{i}$ be a vector along the x-axis with magnitude 1. Similarly, let $\underset{\sim}{j}$ be a vector along the y-axis with magnitude 1. Vectors $\underset{\sim}{i}$ and $\underset{\sim}{j}$ are known as **unit vectors**, and are discussed in section 9.3.5.

The vector $\underset{\sim}{r}$ is the **position vector** of point C relative to the origin; that is, $\overrightarrow{OC} = \underset{\sim}{r}$.

Note: Unit vectors $\underset{\sim}{i}$ and $\underset{\sim}{j}$ may also be shown as $\hat{\mathbf{i}}$ and $\hat{\mathbf{j}}$ respectively.

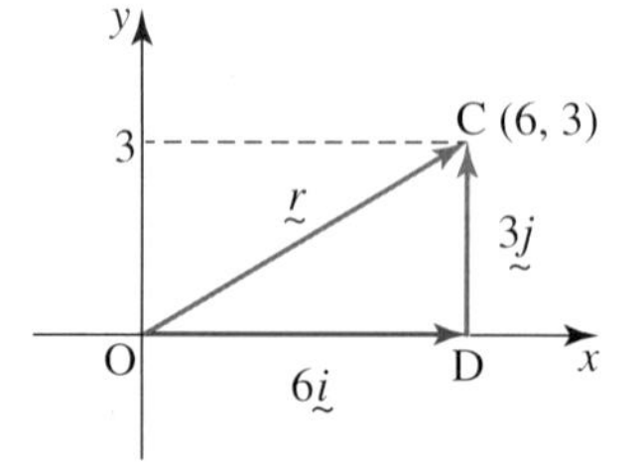

With vectors, it is equivalent to travel along $\underset{\sim}{r}$ from the origin directly to C, or to travel first along the x-axis to D and then parallel to the y-axis to C. In either case we started at the origin and ended up at C. Clearly, then, $\underset{\sim}{r}$ is made up of some multiple of $\underset{\sim}{i}$ in the x-direction and some multiple of $\underset{\sim}{j}$ in the y-direction.

For example, if the point C has coordinates (6, 3) then $\overrightarrow{OC} = \underset{\sim}{r} = 6\underset{\sim}{i} + 3\underset{\sim}{j}$. This is the **Cartesian form of a vector**.

Cartesian form of a vector

The Cartesian form of a position vector from the origin to the point (x, y) is given by:

$$\underset{\sim}{r} = x\underset{\sim}{i} + y\underset{\sim}{j}$$

9.3.2 Ordered pair notation and column vector notation

The vector $\underset{\sim}{r} = 6\underset{\sim}{i} + 3\underset{\sim}{j}$ can be expressed as an ordered pair (6, 3) or in column vector notation as $\begin{bmatrix} 6 \\ 3 \end{bmatrix}$.

Similarly, the vector $\underset{\sim}{r} = x\underset{\sim}{i} + y\underset{\sim}{j}$ can be expressed as an ordered pair (x, y) or in column vector notation as 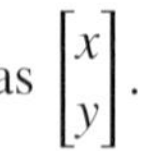 $\begin{bmatrix} x \\ y \end{bmatrix}$.

9.3.3 The magnitude of a vector

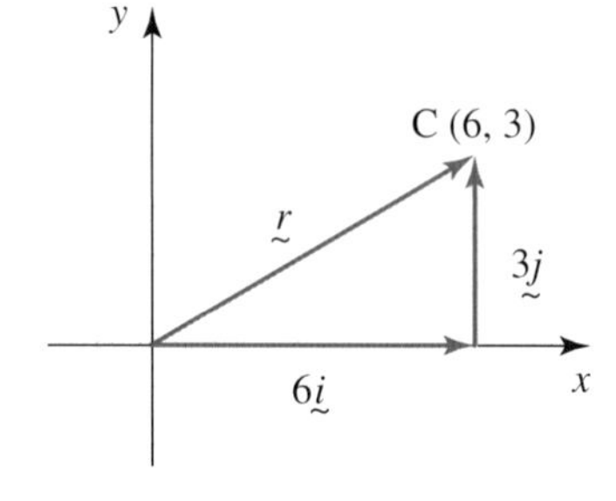

By using Pythagoras' theorem on a position vector, we can determine its length, or *magnitude*.

Consider the vector $\underset{\sim}{r}$ shown.

The magnitude of $\underset{\sim}{r}$, denoted as $|\underset{\sim}{r}|$ or r, is given by:

$$\begin{aligned} |\underset{\sim}{r}| &= \sqrt{6^2 + 3^2} \\ &= \sqrt{45} \\ &= 3\sqrt{5} \end{aligned}$$

Vector magnitude

The magnitude of a position vector, $\underset{\sim}{r} = x\underset{\sim}{i} + y\underset{\sim}{j}$, is given by:

$$r = |\underset{\sim}{r}| = \sqrt{x^2 + y^2}$$

The zero vector

A vector that has a magnitude of zero is called the **zero vector**. The zero vector has no magnitude, so has no effect when added to or subtracted from any other vector.

9.3.4 The polar form of a vector

Using prior knowledge of trigonometry, the angle (θ) that $\underset{\sim}{r}$ makes with the positive x-axis (anticlockwise from the positive x-axis) gives the direction of $\underset{\sim}{r}$.

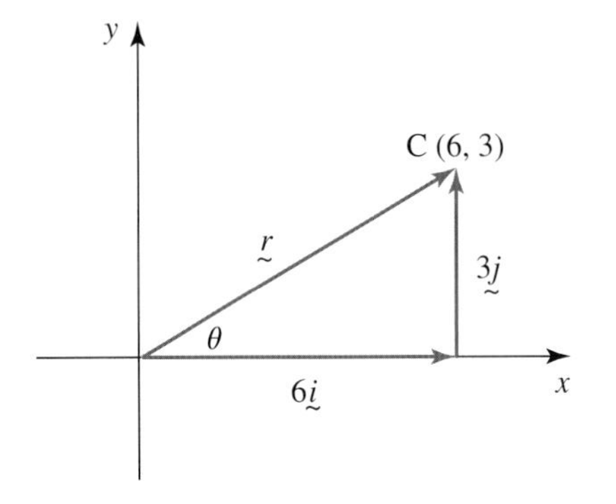

The angle θ can be calculated as:

$$\begin{aligned}\theta &= \tan^{-1}\left(\frac{3}{6}\right)\\ &= \tan^{-1}(0.5)\\ &= 0.464 \text{ radians}\\ &= 26.6°\end{aligned}$$

> **Direction of a vector**
>
> **The direction, θ, of the vector $\underset{\sim}{r} = x\underset{\sim}{i} + y\underset{\sim}{j}$ is defined as the angle it makes with the positive direction of the x-axis. It can be calculated as:**
>
> $$\theta = \tan^{-1}\left(\frac{y}{x}\right)$$

The result obtained by this method needs to be adjusted if the angle is in the 2nd, 3rd or 4th quadrants.

This allows us to express the vector in **polar form** as $\left[3\sqrt{5}, 26.6°\right]$.

> **Polar form of a vector**
>
> **The polar form of a vector $\underset{\sim}{r}$ with magnitude r and direction θ is:**
>
> $$\underset{\sim}{r} = [r, \theta]$$

WORKED EXAMPLE 6 Determining the magnitude and bearing of a vector

Using the vector shown, determine:

a. **the magnitude of $\underset{\sim}{r}$ (state your answer as an exact value)**
b. **the direction of $\underset{\sim}{r}$ (state your answer in degrees rounded to the nearest degree)**
c. **the true bearing of $\underset{\sim}{r}$**
d. **the expression of $\underset{\sim}{r}$ in polar form.**

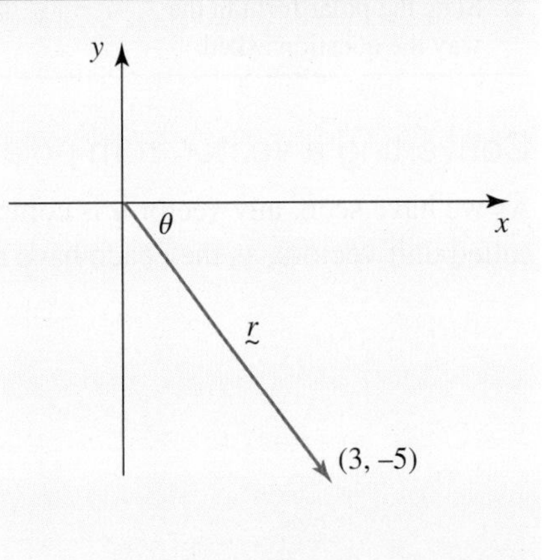

THINK	WRITE
a. 1. Use Pythagoras' theorem or the rule for magnitude of a vector with the x- and y-components 3 and -5 respectively.	**a.** $\lvert \underset{\sim}{r} \rvert = \sqrt{3^2 + (-5)^2}$
2. Simplify the surd.	$\lvert \underset{\sim}{r} \rvert = \sqrt{9+25}$ $= \sqrt{34}$ (= 5.831 to 3 decimal places)
b. 1. The angle is in the 4th quadrant since $x = 3$ and $y = -5$. Use trigonometry to calculate the angle θ from the x- and y-component values, recalling $\tan\theta = \dfrac{\text{opp}}{\text{adj}}$.	**b.** $\theta = \tan^{-1}\left(\dfrac{-5}{3}\right)$
2. Use a calculator to simplify.	$\theta = -59°$
c. The negative sign implies that the direction is 59° clockwise from the x-axis. The true bearing from north is the angle measurement from the positive y-axis to the vector $\underset{\sim}{r}$.	**c.** True bearing $= 90° + 59°$ $= 149°$
d. 1. Recall the polar form of a vector is $[r, \theta]$.	**d.** $[r, \theta]$
2. Write $\underset{\sim}{r}$ in polar form.	$\underset{\sim}{r}$ is a vector of magnitude $\sqrt{34}$ in a direction of $-59°$ from the positive direction of the x-axis. In polar form, $\underset{\sim}{r} = \left[\sqrt{34}, -59°\right]$.

TI | THINK	DISPLAY/WRITE
1. Ensure that your calculator is in DEGREE mode. On a Calculator page, complete the entry line as: $[3 \;\; -5]$ Press MENU, then select: 7 Matrix and Vector C Vector 4 Convert to Polar then press enter. Do the same thing again, but this time press ctrl enter to get the approximate value of the direction.	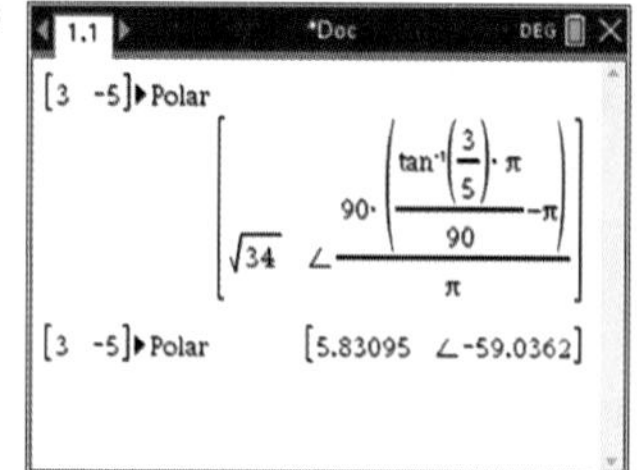
2. State the polar form in the way the question asked.	$\underset{\sim}{r} = \left[\sqrt{34}, -59°\right]$

CASIO | THINK	DISPLAY/WRITE
1. Ensure that your calculator is in Deg mode. On a Main screen page, complete the entry line as: toPol $([3 \;\; -5])$ Then change the mode from Standard to Decimal and repeat to get the approximate value of the direction.	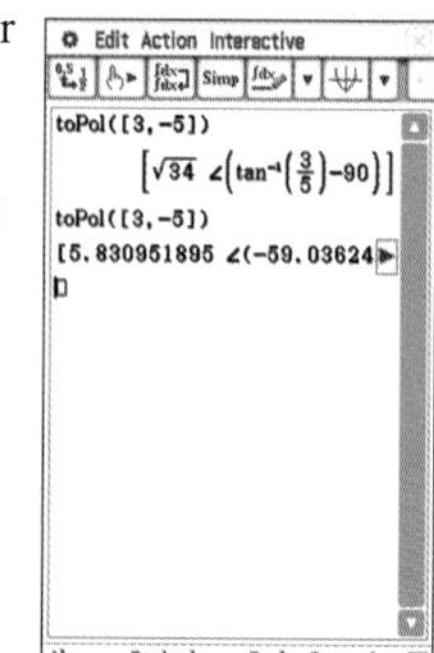
2. State the polar form in the way the question asked.	$\underset{\sim}{r} = \left[\sqrt{34}, -59°\right]$

Converting a vector from polar form into x and y components

As we have seen, any vector $\underset{\sim}{u}$ is composed of x- and y- components denoted by $x\underset{\sim}{i}$, $y\underset{\sim}{j}$. The vectors $\underset{\sim}{i}$ and $\underset{\sim}{j}$ are called unit vectors, as they each have a magnitude of 1. This allows us to resolve a vector into its components.

x and y vector components

If a 2-dimensional vector $\underset{\sim}{u}$ makes an angle of θ with the positive x-axis and it has a magnitude of r, then we can calculate its x- and y-components using the formulas:

$$x = |r| \cos(\theta)$$
$$y = |r| \sin(\theta)$$

WORKED EXAMPLE 7 Expressing a vector in $\underset{\sim}{i}$ and $\underset{\sim}{j}$ notation

Consider the vector $\underset{\sim}{u}$, whose magnitude is 30 and whose bearing (from N) is 310°. Determine its x- and y-components and write $\underset{\sim}{u}$ in terms of $\underset{\sim}{i}$ and $\underset{\sim}{j}$, that is, in Cartesian form. (State the coordinates rounded to 2 decimal places.)

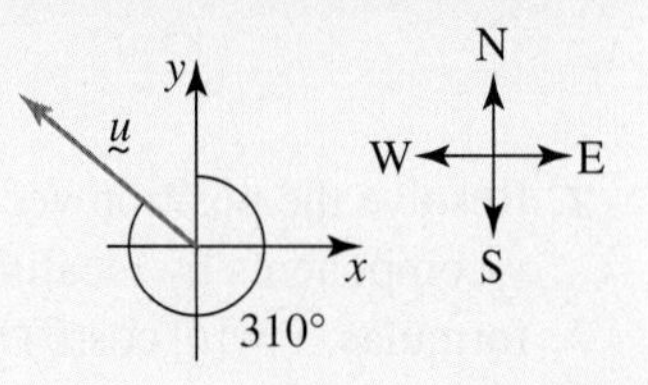

THINK / **WRITE**

1. Change the bearing into an angle with respect to the positive x-axis (θ).

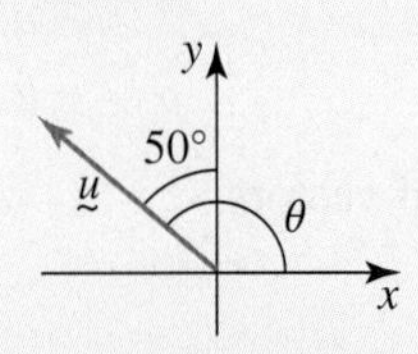

2. The angle between $\underset{\sim}{u}$ and the positive y-axis is $360° - 310° = 50°$.

3. Calculate θ.

$$\theta = 90° + 50° \\ = 140°$$

4. Determine the x- and y-components using trigonometry.

$$x = |\underset{\sim}{u}| \cos(\theta) \quad y = |\underset{\sim}{u}| \sin(\theta)$$
$$= 30\cos(140) \quad = 30\sin(140)$$
$$= -22.98 \quad = 19.28$$

5. Express $\underset{\sim}{u}$ as a vector in Cartesian form.

$$\underset{\sim}{u} = -22.98\underset{\sim}{i} + 19.28\underset{\sim}{j}$$

TI | THINK / **DISPLAY/WRITE**

1. Determine the angle that $\underset{\sim}{u}$ makes with the positive x-axis.

$\theta = 50° + 90° = 140°$

2. Ensure that your calculator is in DEGREE mode.
On a Calculator page, complete the entry line as:
(30∠140)
Press MENU, then select:
7 Matrix and Vector
C Vector
5 Convert to Rectangular
then press ctrl enter.

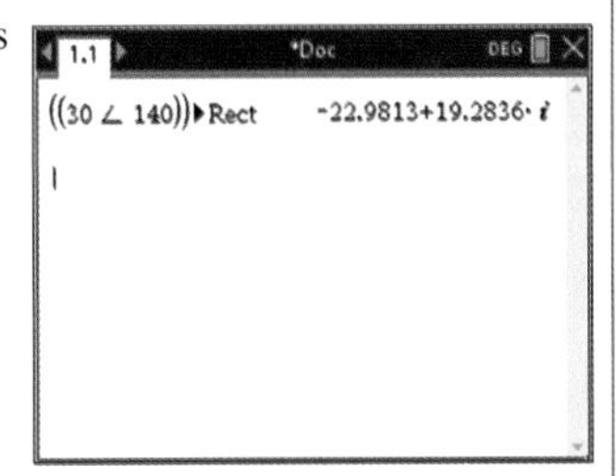

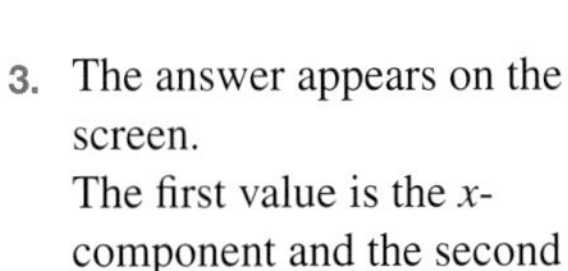

3. The answer appears on the screen.
The first value is the x-component and the second value is the y-component.

$\underset{\sim}{u} = -22.98\underset{\sim}{i} + 19.28\underset{\sim}{j}$

CASIO | THINK / **DISPLAY/WRITE**

1. Determine the angle that $\underset{\sim}{u}$ makes with the positive x-axis.

$\theta = 50° + 90° = 140°$

2. 2. Ensure that your calculator is in Deg and Decimal mode.
On a Main screen using the Action > Vector menu, complete the entry line as:
toRect ([30∠(140)])

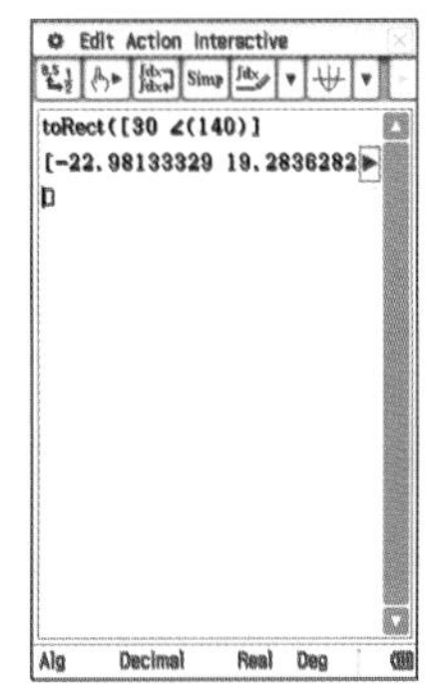

3. The answer appears on the screen. The first value is the x-component and the second value is the y-component.

$\underset{\sim}{u} = -22.98\underset{\sim}{i} + 19.28\underset{\sim}{j}$

WORKED EXAMPLE 8 Applications of vectors

A bushwalker walks 16 km in a direction of bearing 050°, then walks 12 km in a direction of bearing 210°. Determine the resulting position of the hiker giving magnitude and direction from the starting point. State the resultant vector in polar form.

THINK

1. Draw a clear diagram to represent the situation.

WRITE

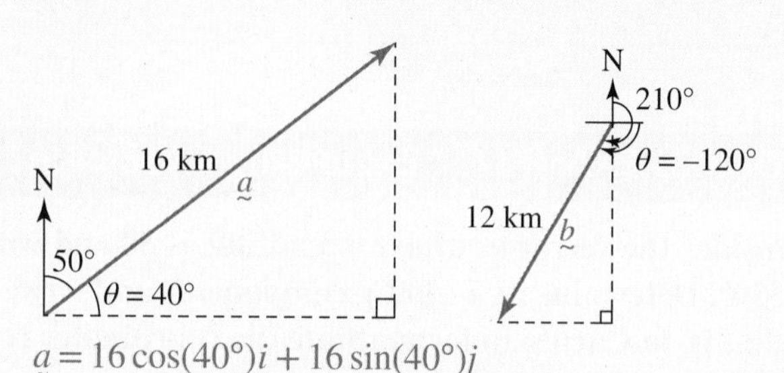

2. Resolve the position vectors into their x- and y-components by recalling the component formulas, $x = |\underset{\sim}{u}|\cos(\theta)$ and $y = |\underset{\sim}{u}|\sin(\theta)$, where θ is the angle from the x-axis.

$\underset{\sim}{a} = 16\cos(40°)\underset{\sim}{i} + 16\sin(40°)\underset{\sim}{j}$

$\underset{\sim}{b} = 12\cos(-120°)\underset{\sim}{i} + 12\sin(-120°)\underset{\sim}{j}$

3. Simplify position vectors.

$\underset{\sim}{a} = 12.2567\underset{\sim}{i} + 10.2846\underset{\sim}{j}$

$\underset{\sim}{b} = -6\underset{\sim}{i} - 10.3923\underset{\sim}{j}$

4. Use the triangle rule of addition of vectors.

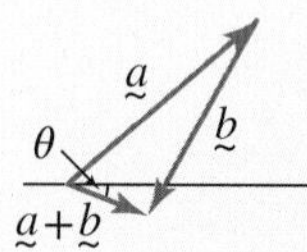

$$\begin{aligned}\underset{\sim}{a} + \underset{\sim}{b} &= (12.2567 - 6)\underset{\sim}{i} + (10.2846 - 10.3923)\underset{\sim}{j}\\ &= 6.2567\underset{\sim}{i} - 0.1077\underset{\sim}{j}\end{aligned}$$

5. Calculate the angle θ by recalling the direction formula, $\theta = \tan^{-1}\left(\dfrac{y}{x}\right)$.

$$\begin{aligned}\theta &= \tan^{-1}\left(\frac{-0.1077}{6.2567}\right)\\ &= -0.986°\end{aligned}$$

6. Determine the magnitude.

$$\begin{aligned}|\underset{\sim}{a} + \underset{\sim}{b}| &= \sqrt{x^2 + y^2}\\ &= \sqrt{6.26^2 + (-0.11)^2}\\ &= \sqrt{39.2}\\ &= 6.26\end{aligned}$$

7. State the resultant vector polar form.

The bushwalker's final position is 6.26 km at an angle of −0.986° from the starting point: [6.26, −0.986°].

9.3.5 Unit vectors

A unit vector is a vector with a magnitude of 1. Unit vectors are denoted by a 'hat' on top of the vector, $\underset{\sim}{\hat{u}}$. We often want to determine a unit vector in the direction of another vector. To do this, we can simply divide the vector by its magnitude.

Unit vectors

The unit vector in the direction of $\underset{\sim}{u}$ is:

$$\underset{\sim}{\hat{u}} = \frac{\underset{\sim}{u}}{|\underset{\sim}{u}|}$$

WORKED EXAMPLE 9 Determining a unit vector

Determine the unit vector in the direction of $\underset{\sim}{u}$. Confirm the unit vector has a magnitude of 1.

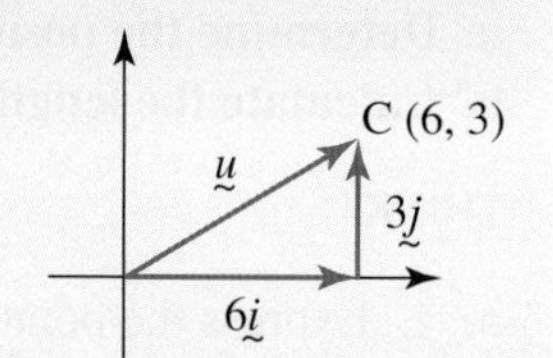

THINK	WRITE
1. Express the vector in component form.	$\underset{\sim}{u} = 6\underset{\sim}{i} + 3\underset{\sim}{j}$
2. Calculate the magnitude of the vector $\underset{\sim}{u}$ using Pythagoras' theorem.	$r = \sqrt{6^2 + 3^2}$ $= \sqrt{45}$ $= 3\sqrt{5}$
3. Divide the original vector by its magnitude to get $\hat{\underset{\sim}{u}}$.	$\hat{\underset{\sim}{u}} = \dfrac{1}{3\sqrt{5}}(6\underset{\sim}{i} + 3\underset{\sim}{j})$ $= \dfrac{2\sqrt{5}}{5}\underset{\sim}{i} + \dfrac{\sqrt{5}}{5}\underset{\sim}{j}$
4. Confirm that $\hat{\underset{\sim}{u}}$ has a magnitude of 1.	$\lvert\hat{\underset{\sim}{u}}\rvert = \sqrt{x^2 + y^2}$ $= \sqrt{\dfrac{20}{25} + \dfrac{5}{25}}$ $= \sqrt{\dfrac{25}{25}}$ $= 1$

TI | THINK	DISPLAY/WRITE	CASIO | THINK	DISPLAY/WRITE
On a Calculator page, press MENU, then select: 7 Matrix and Vector C Vector 1 Unit Vector Complete the entry line as: unitV ([6 3]) then press enter.	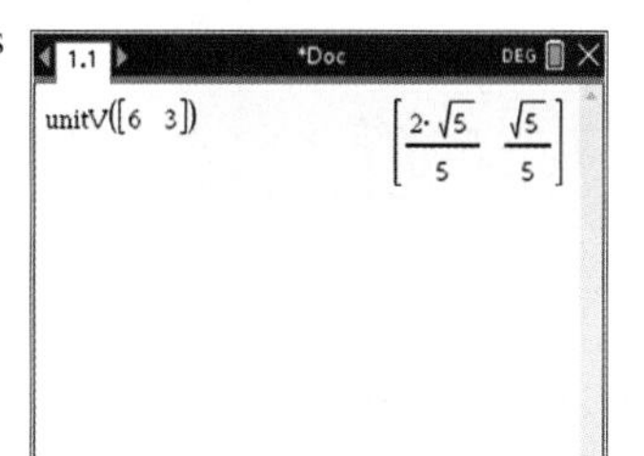	On a Main screen, using the Action > Vector menu, complete the entry line as: unitV ([6 3]) then press enter.	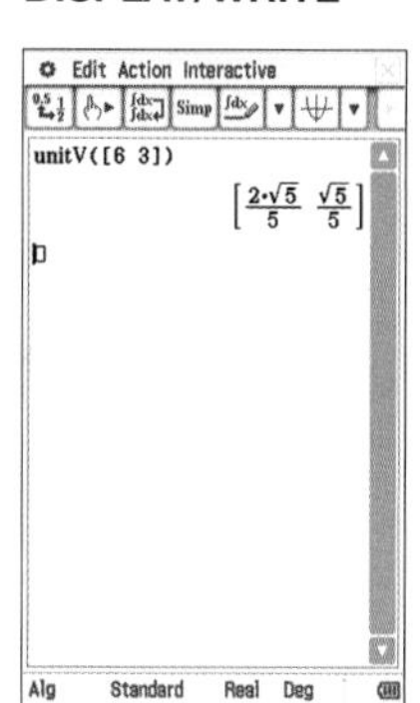

9.3.6 A vector between two points

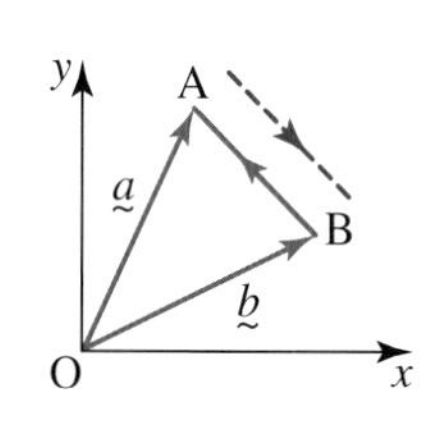

In the figure shown, $\underset{\sim}{a}$ is the position vector of point A, $\overrightarrow{OA}$, and $\underset{\sim}{b}$ is the position vector of point B, $\overrightarrow{OB}$, relative to the origin.

The vector describing the location of A relative to B, $\overrightarrow{BA}$, is easily found using vector addition as $-\underset{\sim}{b} + \underset{\sim}{a}$ or $\underset{\sim}{a} - \underset{\sim}{b}$.

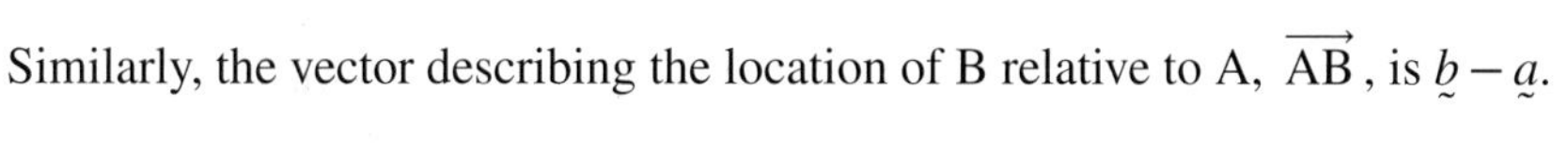

Similarly, the vector describing the location of B relative to A, $\overrightarrow{AB}$, is $\underset{\sim}{b} - \underset{\sim}{a}$.

Vectors between two points

If A and B are points defined by position vectors $\underset{\sim}{a}$ and $\underset{\sim}{b}$ respectively, then:

$$\overrightarrow{AB} = \underset{\sim}{b} - \underset{\sim}{a}$$

WORKED EXAMPLE 10 Calculate the vector between two points

a. Determine the position vector locating point B (3, − 3) from point A (2, 5).
b. Calculate the length of this vector.

THINK	WRITE
a. 1. Express the point A as a position vector $\underset{\sim}{a}$.	a. Let $\overrightarrow{OA} = \underset{\sim}{a} = 2\underset{\sim}{i} + 5\underset{\sim}{j}$
2. Express the point B as a position vector $\underset{\sim}{b}$.	Let $\overrightarrow{OB} = \underset{\sim}{b} = 3\underset{\sim}{i} - 3\underset{\sim}{j}$
3. The location of B relative to A, ($\overrightarrow{AB}$), is $\underset{\sim}{b} - \underset{\sim}{a}$.	$\overrightarrow{AB} = \underset{\sim}{b} - \underset{\sim}{a}$ $= 3\underset{\sim}{i} - 3\underset{\sim}{j} - (2\underset{\sim}{i} + 5\underset{\sim}{j})$ $= \underset{\sim}{i} - 8\underset{\sim}{j}$
b. The length of ($\overrightarrow{AB}$) is $\lvert \underset{\sim}{b} - \underset{\sim}{a} \rvert$.	b. $\lvert \overrightarrow{AB} \rvert = \lvert \underset{\sim}{b} - \underset{\sim}{a} \rvert$ $= \sqrt{1^2 + (-8)^2}$ $= \sqrt{65}$ (or 8.06)

We are now in a position to resolve the problem of calculating, accurately, the resultant velocity of the yacht experiencing wind from another direction.

First redraw the diagram to show the addition of vectors.

Taking the direction of the wind as the $\underset{\sim}{i}$ direction:

Wind velocity, $\underset{\sim}{a} = 15\underset{\sim}{i} + 0\underset{\sim}{j}$

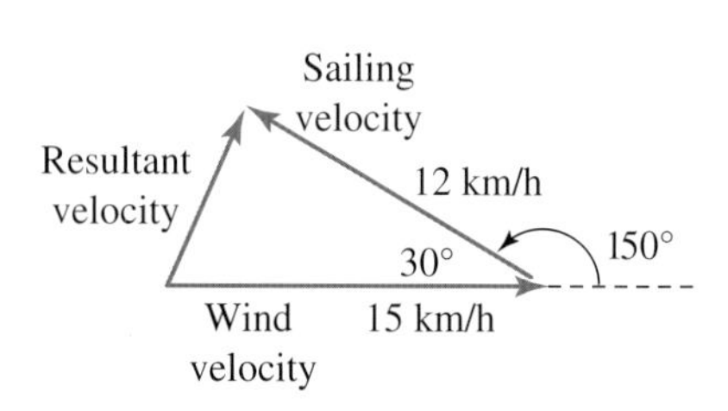

$$\begin{aligned}\text{Yacht velocity, } \underset{\sim}{b} &= 12\cos(150°)\underset{\sim}{i} + 12\sin(150°)\underset{\sim}{j}\\ &= -10.3923\underset{\sim}{i} + 6\underset{\sim}{j}\end{aligned}$$

$$\begin{aligned}\text{Resultant velocity, } \underset{\sim}{a} + \underset{\sim}{b} &= (15 - 10.3923)\underset{\sim}{i} + 6\underset{\sim}{j}\\ &= 4.6077\underset{\sim}{i} + 6\underset{\sim}{j}\end{aligned}$$

$$\begin{aligned}\text{Magnitude} &= \lvert \underset{\sim}{a} + \underset{\sim}{b} \rvert\\ &= 7.6 \text{ km/h}\end{aligned}$$

$$\begin{aligned}\text{Direction, } \theta &= \tan^{-1}\left(\frac{6}{4.6077}\right)\\ &= 52.48°\end{aligned}$$

9.3 Exercise

Technology active

1. State the x, y components of the following vectors.

   a. $3\underset{\sim}{i}+4\underset{\sim}{j}$ b. $6\underset{\sim}{i}-3\underset{\sim}{j}$ c. $3.4\underset{\sim}{i}+\sqrt{2}\underset{\sim}{j}$

2. WE6a, b For each of the following, determine:

   i. the magnitude of the vector, in exact surd form for parts **a** and **b** and correct to 2 decimal places for parts **c** and **d**.

   ii. the direction of each vector, in degrees to 1 decimal place where appropriate. (Express the direction with respect to the positive x-axis.)

   a.  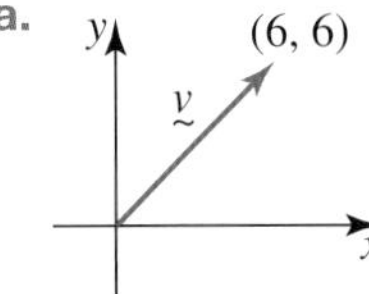

   b. 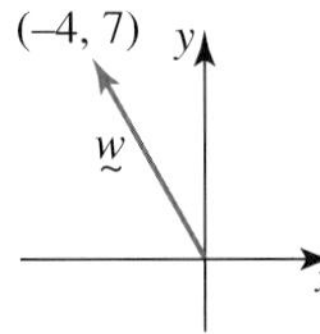

   c. 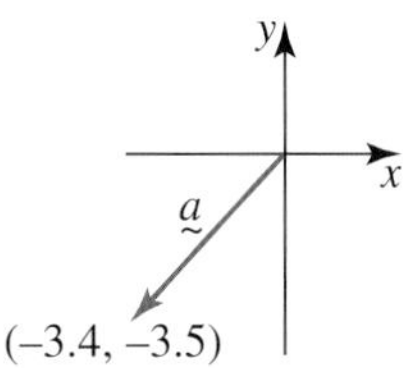

   d. 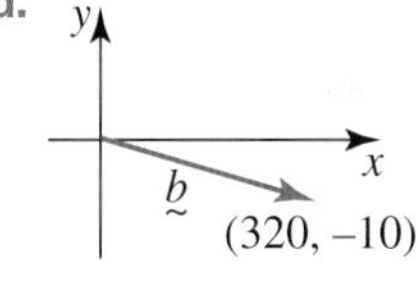

3. i. WE6c, d Determine the true bearing of each vector in question **2**.

   ii. Express each vector in question **2** in polar form.

4. WE7 Consider the vector $\underset{\sim}{w}$ shown. Its magnitude is 100 and its bearing is 210° True.

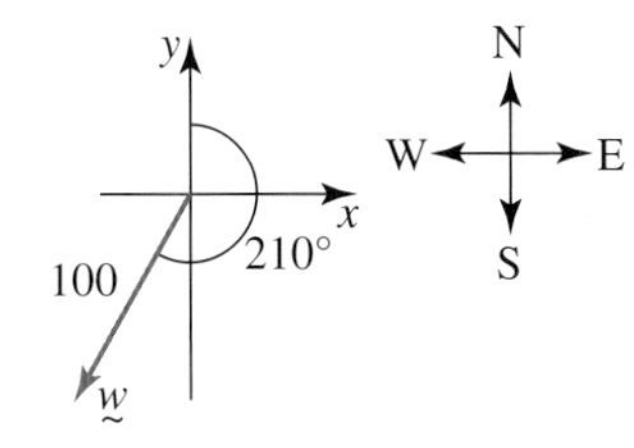

   Determine the x- and y-components of $\underset{\sim}{w}$, and express them as *exact* values (surds). State the answer in the form $\underset{\sim}{w}=x\underset{\sim}{i}+y\underset{\sim}{j}$.

5. An aeroplane travels on a bearing of 147° for 457 km. Express its position as a vector in terms of $\underset{\sim}{i}$ and $\underset{\sim}{j}$ (to 1 decimal place).

6. A ship travels on a bearing of 331° for 125 km. Express its position as a vector in terms of $\underset{\sim}{i}$ and $\underset{\sim}{j}$ (to 1 decimal place).

7. WE8 A pilot flies 420 km in a direction 45° south of east and then 200 km in a direction 60° south of east. Calculate the resultant displacement from the starting position giving both magnitude and direction. State the final vector in polar form to 1 decimal place.

8. The instructions to Black-eye the Pirate's hidden treasure say: 'Take 20 steps in a north-easterly direction and then 30 steps in a south-easterly direction.' However, a rockfall blocks the first part of the route in the north-easterly direction. State how you could head directly to the treasure. Write your answer correct to 1 decimal place.

9. Two scouts are in contact with home base. Scout A is 15 km from home base in a direction 30° north of east. Scout B is 12 km from home base in a direction 40° west of north. Determine how far is scout B from scout A, in km to 1 decimal place.

10. **WE9** Determine unit vectors in the direction of the given vector for the following:

a. 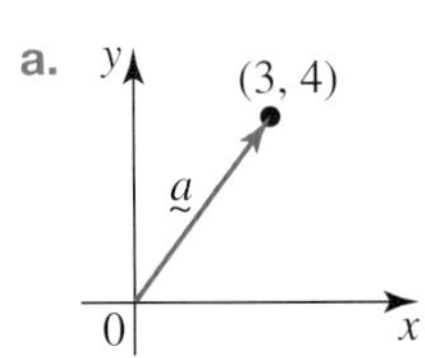

b. 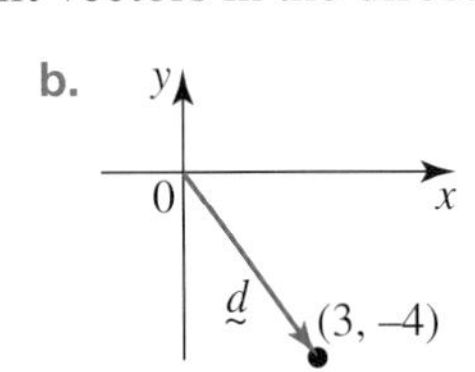

c. $\underset{\sim}{b} = 4\underset{\sim}{i} + 3\underset{\sim}{j}$

d. $\underset{\sim}{e} = -4\underset{\sim}{i} + 3\underset{\sim}{j}$

e. $\underset{\sim}{c} = \underset{\sim}{i} + \sqrt{2}\underset{\sim}{j}$

11. Not all unit vectors are smaller than the original vectors. Consider the vector $\underset{\sim}{v} = 0.3\underset{\sim}{i} + 0.4\underset{\sim}{j}$. Show that the unit vector in the direction of $\underset{\sim}{v}$ is twice as long as $\underset{\sim}{v}$.

12. Determine the unit vector in the direction of $\underset{\sim}{w} = -0.1\underset{\sim}{i} - 0.02\underset{\sim}{j}$, correct to 2 decimal places.

13. Consider the points A (0, 1) and B (4, 5) in the figure. A vector joining A to B can be drawn.

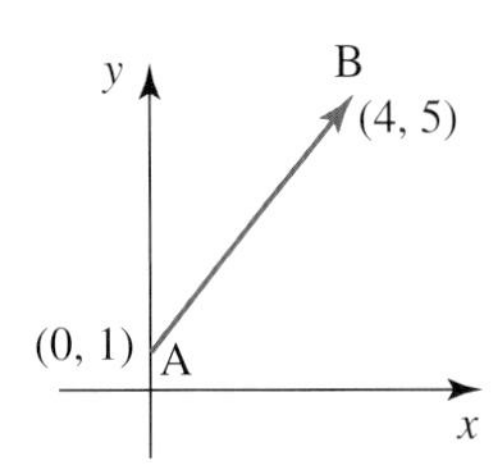

a. Show that an equivalent position vector is given by: $4\underset{\sim}{i} + 4\underset{\sim}{j}$.

b. Similarly, show that an equivalent position vector joining B to A is given by: $-4\underset{\sim}{i} - 4\underset{\sim}{j}$.

14. **WE10** For each of the following pairs of points determine:

i. the position vectors locating the second point from the first point

ii. the length of this vector as an exact value.

a. (0, 2), (4, −5)
b. (2, 3), (5, 4)
c. (4, −5), (0, 2)
d. (5, 4), (2, 3)
e. (3, 7), (5, 7)
f. (7, −3), (3, −3)

15. Determine the position vectors from question 14, by going from the second point to the first.

16. Determine unit vectors in the direction of the position vectors for each of the vectors of question 14.

17. Let $u = 5\underset{\sim}{i} - 2\underset{\sim}{j}$ and $\underset{\sim}{e} = -2\underset{\sim}{i} + 3\underset{\sim}{j}$.

a. Determine:

i. $|\underset{\sim}{u}|$
ii. $|\underset{\sim}{e}|$
iii. $\hat{\underset{\sim}{u}}$
iv. $\hat{\underset{\sim}{e}}$
v. $\underset{\sim}{u} + \underset{\sim}{e}$
vi. $|\underset{\sim}{u} + \underset{\sim}{e}|$

b. Confirm or reject the statement that $|\underset{\sim}{u}| + |\underset{\sim}{e}| = |\underset{\sim}{u} + \underset{\sim}{e}|$.

18. Let $\underset{\sim}{u} = -3\underset{\sim}{i} + 4\underset{\sim}{j}$ and $e = 5\underset{\sim}{i} - \underset{\sim}{j}$.

a. Determine:

i. $|\underset{\sim}{u}|$
ii. $|\underset{\sim}{e}|$
iii. $\hat{\underset{\sim}{u}}$
iv. $\hat{\underset{\sim}{e}}$
v. $\underset{\sim}{u} + \underset{\sim}{e}$
vi. $|\underset{\sim}{u} + \underset{\sim}{e}|$

b. Confirm or reject the statement that $|\underset{\sim}{u}| + |\underset{\sim}{e}| = |\underset{\sim}{u} + \underset{\sim}{e}|$.

19. To calculate the *distance* between two vectors, $\underset{\sim}{a}$ and $\underset{\sim}{b}$, simply calculate $|\underset{\sim}{a}-\underset{\sim}{b}|$. Calculate the distance between these pairs of vectors:

    a. $3\underset{\sim}{i}+2\underset{\sim}{j}$ and $2\underset{\sim}{i}+3\underset{\sim}{j}$

    b. 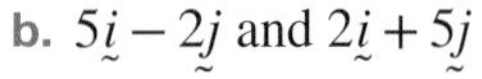 $5\underset{\sim}{i}-2\underset{\sim}{j}$ and $2\underset{\sim}{i}+5\underset{\sim}{j}$

20. A river flows through the jungle from west to east at a speed of 3 km/h. An explorer wishes to cross the river by boat, and attempts this by travelling at 5 km/h due north.
    Determine:

    a. the vector representing the velocity of the river
    b. the vector representing the velocity of the boat
    c. the resultant (net) vector of the boat's journey
    d. the true bearing of the boat's journey, in degrees correct to 1 decimal place
    e. the magnitude of the net vector
    f. what bearing the boat should travel so that it arrives at the opposite bank of the river due north of the starting position, in degrees to 1 decimal place.

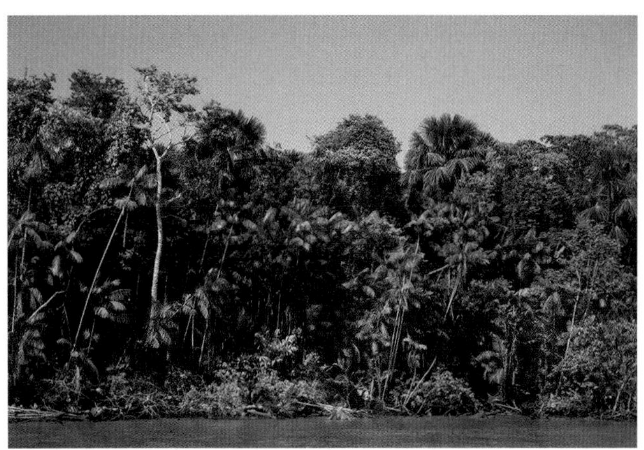

21. A boat travels east at 20 km/h, while another boat travels south at 15 km/h. Determine:

    a. a vector representing each boat and the difference between the boats
    b. the magnitude of the difference vector
    c. the true bearing of the difference vector, in degrees to 1 decimal place.

22. Consider the vector $\underset{\sim}{u}=3\underset{\sim}{i}+4\underset{\sim}{j}$ and the vector $\underset{\sim}{v}=4\underset{\sim}{i}-3\underset{\sim}{j}$. Determine the angles of each of these vectors with respect to the x-axis. Show that these two vectors are perpendicular to each other. Also show that the products of each vector's corresponding x- and y-components add up to 0. Confirm if this is a pattern for all perpendicular vectors.

23. A river has a current of 4 km/h westward. A boat which is capable of travelling at 12 km/h is attempting to cross the river by travelling due north. Determine:

    a. the vector representing the net velocity of the boat
    b. the bearing of the actual motion of the boat, in degrees to 1 decimal place
    c. how long it takes to cross the river, if the river is 500 m wide (from north to south), in minutes to 1 decimal place. (*Hint:* The maximum 'speed' of the boat is still 12 km/h.)

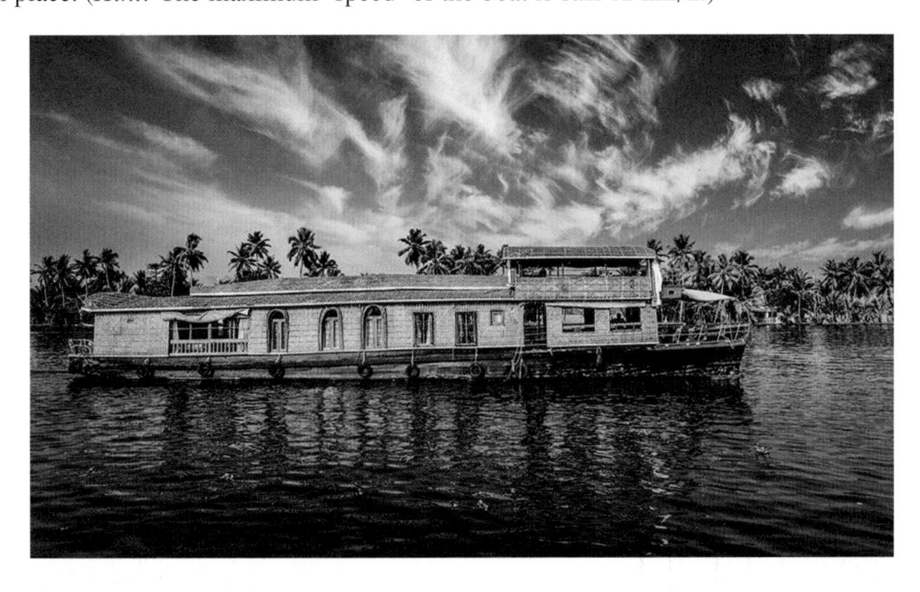

9.3 Exam questions

Question 1 (1 mark) TECH-ACTIVE

MC The magnitude of the vector $\underset{\sim}{a} = 3\underset{\sim}{i} + 4\underset{\sim}{j}$ is equal to

A. 7 **B.** 5 **C.** $\sqrt{7}$ **D.** $\sqrt{5}$ **E.** 25

Question 2 (1 mark) TECH-ACTIVE

MC The vector with a magnitude of $20\sqrt{5}$ is

A. $2\underset{\sim}{i} - 15\underset{\sim}{j}$ **B.** $7\underset{\sim}{i} + 5\underset{\sim}{j}$ **C.** $20\underset{\sim}{i} + 40\underset{\sim}{j}$ **D.** $6\underset{\sim}{i} + 5\underset{\sim}{j}$ **E.** $80\underset{\sim}{i} + 5\underset{\sim}{j}$

Question 3 (1 mark) TECH-FREE

If $a = i + 3j$ and $\underset{\sim}{b} = -i - 2j$, determine $|\underset{\sim}{a} - \underset{\sim}{b}|$.

More exam questions are available online.

9.4 Scalar multiplication of vectors

LEARNING INTENTION

At the end of this subtopic you should be able to:

- use scalar multiplication on vectors
- determine midpoints
- solve vector equations.

9.4.1 Scalar multiplication and midpoints

When a vector in $\underset{\sim}{i}$ and $\underset{\sim}{j}$ form is multiplied by a scalar, each coefficient is multiplied by the scalar.

Scalar multiplication

If $\underset{\sim}{a} = x\underset{\sim}{i} + y\underset{\sim}{j}$, then:

$$k\underset{\sim}{a} = kx\underset{\sim}{i} + ky\underset{\sim}{j}$$

For example, if $\underset{\sim}{a} = \underset{\sim}{i} - 2\underset{\sim}{j}$, then $2\underset{\sim}{a} = 2\underset{\sim}{i} - 4\underset{\sim}{j}$ and $-\underset{\sim}{a} = -\underset{\sim}{i} + 2\underset{\sim}{j}$.

Given the vectors $\overrightarrow{OA} = \underset{\sim}{a} = x_1\underset{\sim}{i} + y_1\underset{\sim}{j}$ and $\overrightarrow{OB} = \underset{\sim}{b} = x_2\underset{\sim}{i} + y_2\underset{\sim}{j}$, the two vectors are equal, $\underset{\sim}{a} = \underset{\sim}{b}$, if and only if $x_1 = x_2$ and $y_1 = y_2$.

WORKED EXAMPLE 11 Scalar multiplication of vectors

If $\underset{\sim}{a} = x\underset{\sim}{i} - 3\underset{\sim}{j}$ and $\underset{\sim}{b} = 4\underset{\sim}{i} - 5\underset{\sim}{j}$, determine the value of x if the vector $\underset{\sim}{c} = 2\underset{\sim}{a} - 3\underset{\sim}{b}$ is parallel to the y-axis.

THINK	WRITE
1. Substitute the vectors $\underset{\sim}{a}$ and $\underset{\sim}{b}$ into $\underset{\sim}{c} = 2\underset{\sim}{a} - 3\underset{\sim}{b}$.	$\underset{\sim}{c} = 2\underset{\sim}{a} - 3\underset{\sim}{b}$ $= 2(x\underset{\sim}{i} - 3\underset{\sim}{j}) - 3(4\underset{\sim}{i} - 5\underset{\sim}{j})$
2. Multiply the vectors $\underset{\sim}{a}$ and $\underset{\sim}{b}$ by the scalars 2 and −3.	$\underset{\sim}{c} = (2x\underset{\sim}{i} - 6\underset{\sim}{j}) - (12\underset{\sim}{i} - 15\underset{\sim}{j})$

3. Simplify	$\underset{\sim}{c} = (2x - 12)\underset{\sim}{i} + (-6 + 15)\underset{\sim}{j}$ $\underset{\sim}{c} = (2x - 12)\underset{\sim}{i} + 9\underset{\sim}{j}$
4. As vector $\underset{\sim}{c}$ is parallel to the y-axis, its $\underset{\sim}{i}$ component must be zero. Equate the $\underset{\sim}{i}$ component to zero.	$2x - 12 = 0$
5. Solve for x.	$2x = 12$ $x = 6$

WORKED EXAMPLE 12 Determining midpoints

Determine the vector representing the midpoint of the line segment AB if A = (1, 2) and B = (4, 3).

THINK	WRITE
1. Draw a diagram to represent this situation.	Let M be the midpoint of AB.

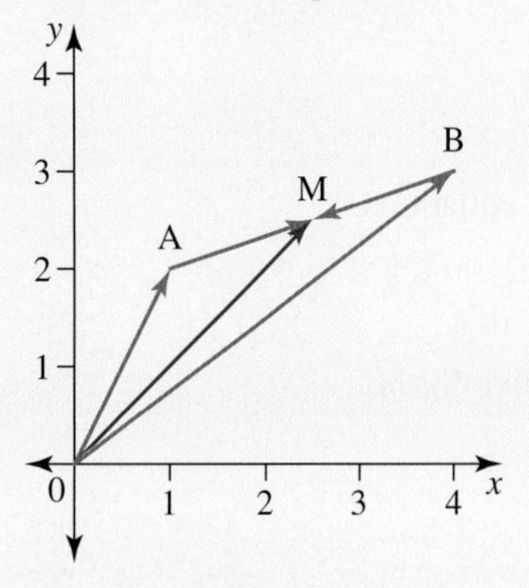

2. Write $\overrightarrow{OA}$ and $\overrightarrow{OB}$ in component form.	$\overrightarrow{OA} = \underset{\sim}{i} + 2\underset{\sim}{j}$ $\overrightarrow{OB} = 4\underset{\sim}{i} + 3\underset{\sim}{j}$
3. Write $\overrightarrow{AB}$ in component form.	$\overrightarrow{AB} = \overrightarrow{AO} + \overrightarrow{OB}$ $= \overrightarrow{OB} - \overrightarrow{OA}$ $= 4\underset{\sim}{i} + 3\underset{\sim}{j} - \underset{\sim}{i} - 2\underset{\sim}{j}$ $= 3\underset{\sim}{i} + \underset{\sim}{j}$
4. Write $\overrightarrow{OM}$ in component form.	$\overrightarrow{OM} = \overrightarrow{OA} + \frac{1}{2}\overrightarrow{AB}$ $= \underset{\sim}{i} + 2\underset{\sim}{j} + \frac{1}{2}(3\underset{\sim}{i} + \underset{\sim}{j})$ $= \frac{5}{2}\underset{\sim}{i} + \frac{5}{2}\underset{\sim}{j}$

Further consideration of Worked example 12 demonstrates that the vector representing the midpoint, M, of the line segment AB can be determined using the formula:

Midpoint of a line segment AB

$$\overrightarrow{OM} = \frac{\overrightarrow{OA} + \overrightarrow{OB}}{2}$$

9.4.2 Solving vector problems

When solving vector problems, we may often need to solve simultaneous, linear or even non-linear equations to solve for the unknown in each case.

WORKED EXAMPLE 13 Solving vector problems simultaneously

Given the vectors $\underset{\sim}{a} = 2\underset{\sim}{i} - 4\underset{\sim}{j}$, $\underset{\sim}{b} = 3\underset{\sim}{i} - 8\underset{\sim}{j}$ and $\underset{\sim}{c} = \underset{\sim}{i} + 4\underset{\sim}{j}$, determine the values of the scalars m and n if $\underset{\sim}{c} = m\underset{\sim}{a} + n\underset{\sim}{b}$.

THINK	WRITE
1. Substitute the vectors $\underset{\sim}{a}$ and $\underset{\sim}{b}$ into $\underset{\sim}{c} = m\underset{\sim}{a} + n\underset{\sim}{b}$.	$\underset{\sim}{c} = m\underset{\sim}{a} + n\underset{\sim}{b}$ $\underset{\sim}{i} + 4\underset{\sim}{j} = m(2\underset{\sim}{i} - 4\underset{\sim}{j}) + n(3\underset{\sim}{i} - 8\underset{\sim}{j})$
2. Multiply the vectors $\underset{\sim}{a}$ and $\underset{\sim}{b}$ by the scalars m and n.	$\underset{\sim}{i} + 4\underset{\sim}{j} = (2m\underset{\sim}{i} - 4m\underset{\sim}{j}) + (3n\underset{\sim}{i} - 8n\underset{\sim}{j})$
3. Simplify.	$\underset{\sim}{i} + 4\underset{\sim}{j} = (2m + 3n)\underset{\sim}{i} - (4m + 8n)\underset{\sim}{j}$
4. Equate the components.	Since the $\underset{\sim}{i}$ components are equal: $1 = 2m + 3n$ [1] Since the $\underset{\sim}{j}$ components are equal: $4 = -4m - 8n$ [2]
5. Solve the simultaneous equations.	$2 \times [1]$ $2 = 4m + 6n$ $4 = -4m - 8n$ [2] Adding gives $6 = -2n \Rightarrow n = -3$
6. Substitute into [1] to solve for m.	$2m = 1 - 3n$ $2m = 1 + 9 = 10$ $m = 5$
7. Write the answer.	$\underset{\sim}{c} = 5\underset{\sim}{a} - 3\underset{\sim}{b}$

9.4.3 Parallel vectors

Two vectors $\underset{\sim}{a}$ and $\underset{\sim}{b}$ are parallel if $\underset{\sim}{a} = k\underset{\sim}{b}$, where $k \in R$. Conversely, if $\underset{\sim}{a} = k\underset{\sim}{b}$, then $\underset{\sim}{a}$ is parallel to $\underset{\sim}{b}$. It is important to note that vectors that have opposite directions will be parallel.

WORKED EXAMPLE 14 Solving vector equations

Given the vectors $\underset{\sim}{r} = x\underset{\sim}{i} + 3\underset{\sim}{j}$ and $\underset{\sim}{s} = 5\underset{\sim}{i} - 6\underset{\sim}{j}$, calculate the value of x in each case if:

a. the length of the vector $\underset{\sim}{r}$ is 7

b. the vector $\underset{\sim}{r}$ is parallel to the vector $\underset{\sim}{s}$.

THINK	WRITE
a. 1. Determine the magnitude of the vector in terms of the unknown value.	a. $\underset{\sim}{r} = x\underset{\sim}{i} + 3\underset{\sim}{j}$ $\lvert\underset{\sim}{r}\rvert = \sqrt{x^2 + 3^2}$ $= \sqrt{x^2 + 9}$
2. Equate the length of the vector to the given value.	Since $\lvert\underset{\sim}{r}\rvert = 7$: $\sqrt{x^2 + 9} = 7$

3. Square both sides and solve for the unknown value. Both answers are acceptable values.

$$x^2 + 9 = 49$$
$$x^2 = 40$$
$$x = \pm\sqrt{40}$$
$$= \pm\sqrt{4 \times 10}$$
$$= \pm 2\sqrt{10}$$

b. 1. If two vectors are parallel, then one is a scalar multiple of the other. Substitute for the given vectors and expand.

b. $\underset{\sim}{r} = k\underset{\sim}{s}$
$$x\underset{\sim}{i} + 3\underset{\sim}{j} = k(5\underset{\sim}{i} - 6\underset{\sim}{j})$$
$$= 5k\underset{\sim}{i} - 6k\underset{\sim}{j}$$

2. For the two vectors to be equal, both components must both be equal.

From the $\underset{\sim}{j}$ component, $3 = -6k$, so $k = -\frac{1}{2}$.
From the $\underset{\sim}{i}$ component, $x = 5k$.

3. Solve for the unknown value in this case.

$x = 5k$ but $k = -\frac{1}{2}$, so $x = -\frac{5}{2}$.

9.4 Exercise

Students, these questions are even better in jacPLUS

Receive immediate feedback and access sample responses

Access additional questions

Track your results and progress

Find all this and MORE in jacPLUS

Technology free

1. **WE11** If $\underset{\sim}{a} = 4\underset{\sim}{i} - 5\underset{\sim}{j}$ and $\underset{\sim}{b} = 3\underset{\sim}{i} + y\underset{\sim}{j}$, determine the value of y if the vector $\underset{\sim}{c} = 3\underset{\sim}{a} + 2\underset{\sim}{b}$ is parallel to the x-axis.

2. The vector $\overrightarrow{CD} = 7\underset{\sim}{i} - 5\underset{\sim}{j}$, the coordinates of point C are $(x, -3)$ and the coordinates of point D are $(4, y)$. Determine the values of x and y.

3. **WE12** Determine the vector representing the midpoint of the line segment AB if A = (3, 4) and B = (7, 8).

4. a. Determine the value of x if the vector $x\underset{\sim}{i} + 3\underset{\sim}{j}$ is parallel to the vector $5\underset{\sim}{i} - 6\underset{\sim}{j}$.
   b. Determine the value of y if the vector $-4\underset{\sim}{i} + 5\underset{\sim}{j}$ is parallel to the vector $6\underset{\sim}{i} + y\underset{\sim}{j}$.

5. a. For the vectors $\underset{\sim}{a} = -2\underset{\sim}{i} + 3\underset{\sim}{j}$ and $\underset{\sim}{b} = 4\underset{\sim}{i} - 2\underset{\sim}{j}$, show that the vector $2\underset{\sim}{a} + 3\underset{\sim}{b}$ is parallel to the x-axis.
   b. For the vectors $\underset{\sim}{c} = 3\underset{\sim}{i} - 5\underset{\sim}{j}$ and $\underset{\sim}{d} = 4\underset{\sim}{i} - 3\underset{\sim}{j}$, show that the vector $4\underset{\sim}{c} - 3\underset{\sim}{d}$ is parallel to the y-axis.
   c. Given the vectors $\underset{\sim}{a} = x\underset{\sim}{i} - 5\underset{\sim}{j}$ and $\underset{\sim}{b} = 5\underset{\sim}{i} + 3\underset{\sim}{j}$, determine the value of x if the vector $3\underset{\sim}{a} + 4\underset{\sim}{b}$ is parallel to the y-axis.
   d. If $\underset{\sim}{c} = 5\underset{\sim}{i} - 3\underset{\sim}{j}$ and $\underset{\sim}{d} = 4\underset{\sim}{i} + y\underset{\sim}{j}$, determine the value of y if the vector $5\underset{\sim}{c} + 7\underset{\sim}{d}$ is parallel to the x-axis.

6. a. Given the points A (3, −2), B (4, −5) and C (1, 4), determine the vectors $\overrightarrow{AB}$ and $\overrightarrow{BC}$. What can be said about the points A, B and C?
   b. Given the points D (5, −3), E (2, 1) and F (8, −7), determine the vectors $\overrightarrow{DE}$ and $\overrightarrow{EF}$. What can be said about the points D, E and F?

7. a. Given the points A $(5, -2)$ and B $(-1, 3)$, determine:
   i. the vector $\overrightarrow{AB}$
   ii. the distance between the points A and B
   iii. a unit vector parallel to $\overrightarrow{AB}$.

   b. Two points C and D are given by $(-4, 3), (2, -5)$ respectively. Determine:
   i. the vector $\overrightarrow{CD}$
   ii. the distance between the points C and D
   iii. a unit vector parallel to $\overrightarrow{DC}$.

8. **WE13** Given the vectors $\underset{\sim}{a} = 5\underset{\sim}{i} + 2\underset{\sim}{j}$, $\underset{\sim}{b} = -3\underset{\sim}{i} - 4\underset{\sim}{j}$ and $\underset{\sim}{c} = \underset{\sim}{i} - 8\underset{\sim}{j}$, determine the values of the scalars m and n if $\underset{\sim}{c} = m\underset{\sim}{a} + n\underset{\sim}{b}$.

9. Given the vectors $\underset{\sim}{r} = x\underset{\sim}{i} + y\underset{\sim}{j}$ and $\underset{\sim}{s} = 4\underset{\sim}{i} - 3\underset{\sim}{j}$, determine the values of x and y if $2\underset{\sim}{r} + 3\underset{\sim}{s} = 8\underset{\sim}{i} + \underset{\sim}{j}$.

10. **WE14** Given the vectors $\underset{\sim}{r} = x\underset{\sim}{i} - 4\underset{\sim}{j}$ and $\underset{\sim}{s} = 3\underset{\sim}{i} + 5\underset{\sim}{j}$, calculate the value of x if:
    a. the length of the vector $\underset{\sim}{r}$ is 6
    b. the vector $\underset{\sim}{r}$ is parallel to the vector $\underset{\sim}{s}$.

11. Calculate the value of y for the vectors $\underset{\sim}{a} = 4\underset{\sim}{i} + y\underset{\sim}{j}$ and $\underset{\sim}{b} = 2\underset{\sim}{i} - 5\underset{\sim}{j}$ if:
    a. the vectors $\underset{\sim}{a}$ and $\underset{\sim}{b}$ are equal in length
    b. the vector $\underset{\sim}{a}$ is parallel to the vector $\underset{\sim}{b}$.

12. a. Given the vectors $\underset{\sim}{a} = x\underset{\sim}{i} + 3\underset{\sim}{j}$ and $\underset{\sim}{b} = -2\underset{\sim}{i} + y\underset{\sim}{j}$, and also given that $3\underset{\sim}{a} + 4\underset{\sim}{b} = 4\underset{\sim}{i} + \underset{\sim}{j}$, determine the values of x and y.
    b. For the vectors $\underset{\sim}{a} = 4\underset{\sim}{i} - 5\underset{\sim}{j}$, $\underset{\sim}{b} = -7\underset{\sim}{i} + 3\underset{\sim}{j}$ and $m\underset{\sim}{a} + n\underset{\sim}{b} = 8\underset{\sim}{i} + 13\underset{\sim}{j}$, determine the values of m and n.
    c. If $\underset{\sim}{a} = 5\underset{\sim}{i} - 6\underset{\sim}{j}$, $\underset{\sim}{b} = -2\underset{\sim}{i} + 4\underset{\sim}{j}$ and $m\underset{\sim}{a} + n\underset{\sim}{b} = 2\underset{\sim}{j}$, determine the values of m and n.

9.4 Exam questions

Question 1 (1 mark) TECH-ACTIVE

MC Consider the following relationships between vectors $\underset{\sim}{p}$, $\underset{\sim}{q}$ and $\underset{\sim}{r}$.

$\underset{\sim}{p} = 2\underset{\sim}{q} - \underset{\sim}{r}$ and $\underset{\sim}{r} = \underset{\sim}{q} + \underset{\sim}{p}$

Out of the following, the statement which is true is

A. $\underset{\sim}{p} = \dfrac{\underset{\sim}{q}}{2}$
B. $\underset{\sim}{p} = 2\underset{\sim}{q}$
C. $\underset{\sim}{q} = 0, \underset{\sim}{p} = \underset{\sim}{r}$
D. $\underset{\sim}{q} = 0, \underset{\sim}{p} = -\underset{\sim}{r}$
E. $\underset{\sim}{p} = \underset{\sim}{q} + \underset{\sim}{r}$

Question 2 (1 mark) TECH-ACTIVE

MC $\underset{\sim}{a} = \left(\frac{3}{7}\underset{\sim}{i} - \frac{8}{5}\underset{\sim}{j}\right)$, if $k\underset{\sim}{a} = 15\underset{\sim}{i} + m\underset{\sim}{j}$, the values of k and m are

A. $k = 35$, $m = 56$
B. $k = 35$, $m = -56$
C. $k = 7$, $m = 56$
D. $k = 7$, $m = -56$
E. $k = -7$, $m = 35$

Question 3 (2 marks) TECH-FREE

$\underset{\sim}{r} = x\underset{\sim}{i} + 4\underset{\sim}{j}$ and $s = -5\underset{\sim}{i} + 6\underset{\sim}{j}$. If $3\underset{\sim}{r} + \underset{\sim}{s} = 2\underset{\sim}{i} + 18\underset{\sim}{j}$, determine the value of x.

More exam questions are available online.

9.5. The scalar (dot) product

LEARNING INTENTION

At the end of this subtopic you should be able to:
- determine dot products
- calculate the angle between two vectors
- determine parallel and perpendicular vectors.

9.5.1 Calculating the dot product

In a previous section, we studied the result of multiplying a vector by a scalar. In this section, we study the multiplication of two vectors.

The scalar or *dot* product of two vectors, $\underset{\sim}{a}$ and $\underset{\sim}{b}$, is denoted by $\underset{\sim}{a} \cdot \underset{\sim}{b}$.

There are two ways of calculating the **dot product**. The first method follows from its definition. (The second method is shown later.) Consider the two vectors $\underset{\sim}{a}$ and $\underset{\sim}{b}$.

Definition of the dot product

The dot product $\underset{\sim}{a} \cdot \underset{\sim}{b}$ is defined by:

$$\underset{\sim}{a} \cdot \underset{\sim}{b} = |\underset{\sim}{a}||\underset{\sim}{b}|\cos(\theta)$$

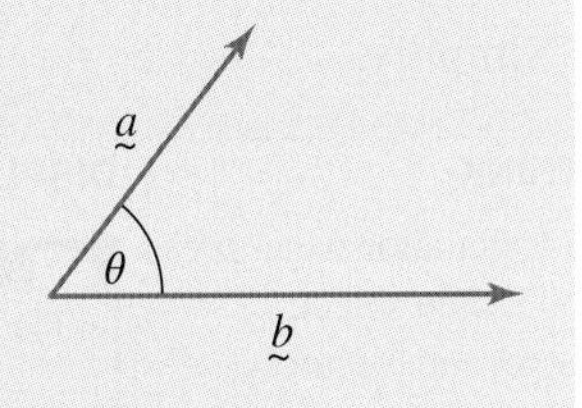

where θ is the angle between (the positive directions of) $\underset{\sim}{a}$ and $\underset{\sim}{b}$.

Note: The vectors are not aligned as for addition or subtraction, but their two tails are joined.

Properties of the dot product

- The dot product is a scalar. It is the result of multiplying three scalar quantities: the magnitudes of the two vectors and the cosine of the angle between them.
- The order of multiplication is unimportant (*commutative* property); thus,

$$\underset{\sim}{a} \cdot \underset{\sim}{b} = \underset{\sim}{b} \cdot \underset{\sim}{a}$$

- The dot product is *distributive*; thus,

$$\underset{\sim}{c} \cdot (\underset{\sim}{a} + \underset{\sim}{b}) = \underset{\sim}{c} \cdot \underset{\sim}{a} + \underset{\sim}{c} \cdot \underset{\sim}{b}$$

- Since the angle between $\underset{\sim}{a}$ and itself is 0°

$$\underset{\sim}{a} \cdot \underset{\sim}{a} = |\underset{\sim}{a}|^2$$

WORKED EXAMPLE 15 Determining dot products

Let $\underset{\sim}{a} = 3\underset{\sim}{i} + 4\underset{\sim}{j}$ and $\underset{\sim}{b} = 3\underset{\sim}{i}$. Determine $\underset{\sim}{a} \cdot \underset{\sim}{b}$.

THINK	WRITE
1. Calculate the magnitudes of $\underset{\sim}{a}$ and $\underset{\sim}{b}$.	$\lvert\underset{\sim}{a}\rvert = \sqrt{3^2 + 4^2}$ $= 5$ $\lvert\underset{\sim}{b}\rvert = \sqrt{3^2}$ $= 3$
2. Draw a right-angled triangle showing the angle that $\underset{\sim}{a}$ makes with the positive x-axis since $\underset{\sim}{b}$ is along the x-axis.	
3. Calculate $\cos(\theta)$, knowing that $a = 5$ and the x-component of $\underset{\sim}{a}$ is 3.	$\cos(\theta) = \frac{3}{5}$
4. Determine the value of $\underset{\sim}{a} \cdot \underset{\sim}{b}$ using the equation with the angle between 2 vectors.	$\underset{\sim}{a} \cdot \underset{\sim}{b} = \lvert\underset{\sim}{a}\rvert \times \lvert\underset{\sim}{b}\rvert \times \cos(\theta)$ $= 5 \times 3 \times \frac{3}{5}$
5. Simplify.	$= 9$

TI | THINK

On a Calculator page, press MENU, then select:
7 Matrix and Vector
C Vector
3 Dot Product
Complete the entry line as:
dotP ([3 4], [3 0])
then press enter.

DISPLAY/WRITE

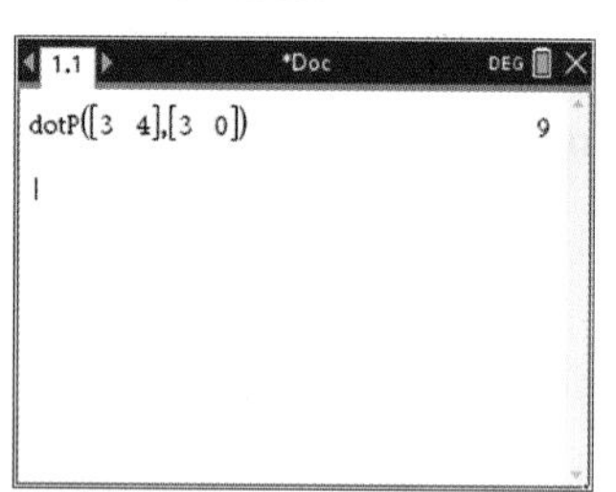

CASIO | THINK

On a Main screen, using the Action > Vector menu, complete the entry line as:
dotP ([3 4], [3 0])
then press enter.

DISPLAY/WRITE

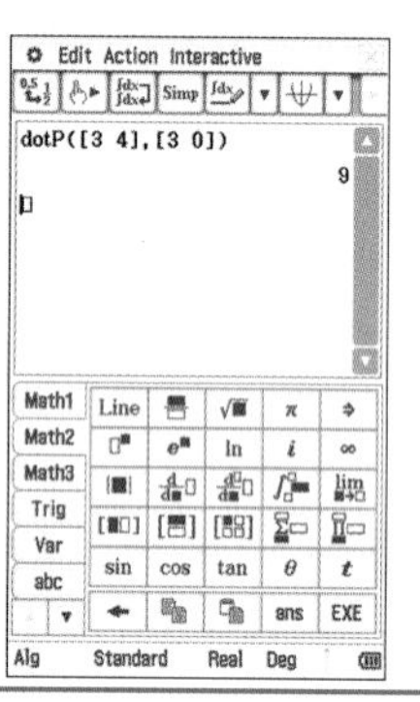

9.5.2 The scalar product of vectors expressed in component form

Consider the dot product of the unit vectors $\underset{\sim}{i}$ and $\underset{\sim}{j}$. Firstly, consider $\underset{\sim}{i} \cdot \underset{\sim}{i}$ in detail. By definition, $\lvert\underset{\sim}{i}\rvert = 1$ and, since the angle between them is 0°, $\cos(\theta) = 1$; thus, $\underset{\sim}{i} \cdot \underset{\sim}{i} = 1$. To summarise these results:

$$\underset{\sim}{i} \cdot \underset{\sim}{i} = 1 \text{ (since } \theta = 0°)$$

$$\underset{\sim}{j} \cdot \underset{\sim}{j} = 1 \text{ (since } \theta = 0°)$$

$$\underset{\sim}{i} \cdot \underset{\sim}{j} = 0 \text{ (since } \theta = 90°)$$

Using this information, we can develop another way to calculate the dot product of any vector.

Let $\underset{\sim}{a} = x_1\underset{\sim}{i} + y_1\underset{\sim}{j}$ and $\underset{\sim}{b} = x_2\underset{\sim}{i} + y_2\underset{\sim}{j}$ where x_1, y_1, x_2, y_2, are constants. Then we can write $\underset{\sim}{a} \cdot \underset{\sim}{b}$ as:

$$\begin{aligned}\underset{\sim}{a} \cdot \underset{\sim}{b} &= (x_1\underset{\sim}{i} + y_1\underset{\sim}{j}) \cdot (x_2\underset{\sim}{i} + y_2\underset{\sim}{j}) \\ &= x_1x_2(\underset{\sim}{i} \cdot \underset{\sim}{i}) + x_1y_2(\underset{\sim}{i} \cdot \underset{\sim}{j}) + y_1x_2(\underset{\sim}{j} \cdot \underset{\sim}{i}) + y_1y_2(\underset{\sim}{j} \cdot \underset{\sim}{j})\end{aligned}$$

Considering the various unit vector dot products (in brackets), the 'like' products ($\underset{\sim}{i} \cdot \underset{\sim}{i}$ and $\underset{\sim}{j} \cdot \underset{\sim}{j}$) are 1; the rest are 0. Therefore:

The dot product using components

For vectors $\underset{\sim}{a} = x_1\underset{\sim}{i} + y_1\underset{\sim}{j}$ and $\underset{\sim}{b} = x_2\underset{\sim}{i} + y_2\underset{\sim}{j}$:

$$\underset{\sim}{a} \cdot \underset{\sim}{b} = x_1x_2 + y_1y_2$$

This is a very important result.

We only need to multiply the corresponding x- and y-components of two vectors to determine their dot product.

WORKED EXAMPLE 16 Determining the dot product using components

Let $\underset{\sim}{a} = 3\underset{\sim}{i} + 4\underset{\sim}{j}$ and $\underset{\sim}{b} = 6\underset{\sim}{i} - 4\underset{\sim}{j}$. Determine $\underset{\sim}{a} \cdot \underset{\sim}{b}$.

THINK	WRITE
1. Write down $\underset{\sim}{a} \cdot \underset{\sim}{b}$ using the equation $\underset{\sim}{a} \cdot \underset{\sim}{b} = x_1x_2 + y_1y_2$.	$\underset{\sim}{a} \cdot \underset{\sim}{b} = (3\underset{\sim}{i} + 4\underset{\sim}{j}) \cdot (6\underset{\sim}{i} - 4\underset{\sim}{j})$
2. Multiply the corresponding components.	$\underset{\sim}{a} \cdot \underset{\sim}{b} = 3 \times 6 + 4 \times -4$
3. Simplify.	$= 18 - 16$ $= 2$

9.5.3 Calculating the angle between two vectors

We now have two formulas for calculating the dot product:

$$\underset{\sim}{a} \cdot \underset{\sim}{b} = |\underset{\sim}{a}|\,|\underset{\sim}{b}| \cos(\theta)$$

$$\underset{\sim}{a} \cdot \underset{\sim}{b} = x_1x_2 + y_1y_2$$

Combining these two formulas allows us to calculate the angle between the vectors.

Rearranging the two equations, we obtain the result:

Angle between two vectors

$$\cos(\theta) = \frac{x_1x_2 + y_1y_2}{|\underset{\sim}{a}||\underset{\sim}{b}|}$$

WORKED EXAMPLE 17 Calculating the angle between two vectors

Let $\underset{\sim}{a} = 4\underset{\sim}{i} + 3\underset{\sim}{j}$ and $\underset{\sim}{b} = 2\underset{\sim}{i} - 3\underset{\sim}{j}$. Calculate the angle between $\underset{\sim}{a}$ and $\underset{\sim}{b}$ to the nearest degree.

THINK	WRITE
1. Determine the dot product using the equation $\underset{\sim}{a} \cdot \underset{\sim}{b} = x_1x_2 + y_1y_2$.	$\underset{\sim}{a} \cdot \underset{\sim}{b} = (4\underset{\sim}{i} + 3\underset{\sim}{j}) \cdot (2\underset{\sim}{i} - 3\underset{\sim}{j})$ $= 4 \times 2 + 3 \times (-3)$
2. Simplify.	$= -1$
3. Calculate the magnitude of each vector.	$\lvert\underset{\sim}{a}\rvert = \sqrt{4^2 + 3^2}$ $= \sqrt{25}$ $= 5$ $\lvert\underset{\sim}{b}\rvert = \sqrt{2^2 + (-3)^2}$ $= \sqrt{13}$
4. Substitute results into the equation $\cos(\theta) = \dfrac{x_1x_2 + y_1y_2}{\lvert\underset{\sim}{a}\rvert \lvert\underset{\sim}{b}\rvert}$.	$\cos(\theta) = \dfrac{-1}{5\sqrt{13}}$
5. Simplify the result for $\cos(\theta)$.	$= -0.05547$
6. Take $\cos^{-1}$ of both sides to obtain θ and round the answer to the nearest degree.	$\theta = \cos^{-1}(-0.05547)$ $= 93°$

TI | THINK	DISPLAY/WRITE	CASIO | THINK	DISPLAY/WRITE
On a Calculator page, complete the entry line as shown, then press ctrl enter.	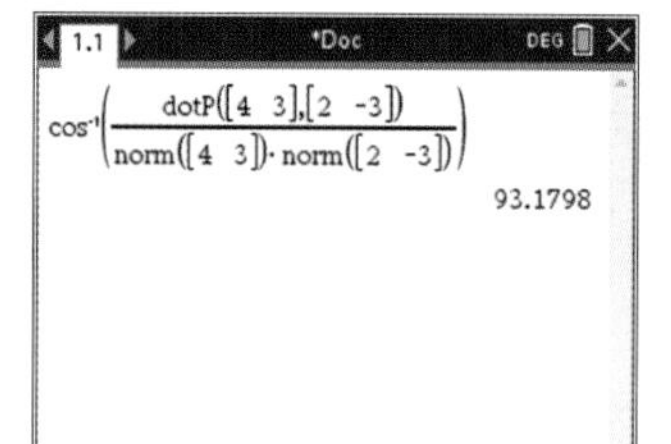	On a Main screen, complete the entry line as shown.	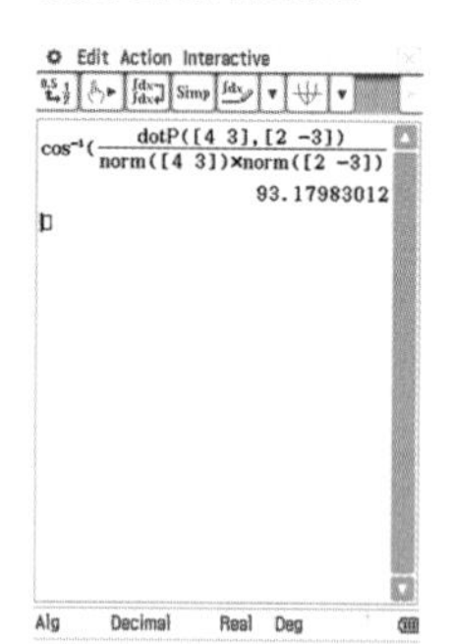

9.5.4 Special results of the dot product

Perpendicular vectors

If two vectors are perpendicular then the angle between them is 90° and the equation $\underset{\sim}{a} \cdot \underset{\sim}{b} = \lvert\underset{\sim}{a}\rvert \lvert\underset{\sim}{b}\rvert \cos(\theta)$ becomes:

$$\begin{aligned}\underset{\sim}{a} \cdot \underset{\sim}{b} &= \lvert\underset{\sim}{a}\rvert \lvert\underset{\sim}{b}\rvert \cos(90°) \\ &= \lvert\underset{\sim}{a}\rvert \lvert\underset{\sim}{b}\rvert \times 0 \qquad (\text{since } \cos(90°) = 0) \\ &= 0\end{aligned}$$

Perpendicular vectors

If $\underset{\sim}{a} \cdot \underset{\sim}{b} = 0$, then $\underset{\sim}{a}$ and $\underset{\sim}{b}$ are perpendicular.

WORKED EXAMPLE 18 Determining constants for perpendicular vectors

Determine the constant m if the vectors $\underset{\sim}{a} = 4\underset{\sim}{i} + 3\underset{\sim}{j}$ and $\underset{\sim}{b} = -3\underset{\sim}{i} + m\underset{\sim}{j}$ are perpendicular.

THINK	WRITE
1. Calculate the dot product using the equation $\underset{\sim}{a} \cdot \underset{\sim}{b} = x_1x_2 + y_1y_2$.	$\underset{\sim}{a} \cdot \underset{\sim}{b} = (4\underset{\sim}{i} + 3\underset{\sim}{j}) \cdot (-3\underset{\sim}{i} + m\underset{\sim}{j})$
2. Simplify.	$= -12 + 3m$
3. Set $\underset{\sim}{a} \cdot \underset{\sim}{b}$ equal to zero since $\underset{\sim}{a}$ and $\underset{\sim}{b}$ are perpendicular.	$\underset{\sim}{a} \cdot \underset{\sim}{b} = -12 + 3m$ $= 0$
4. Solve the equation for m.	$m = 4$

Parallel vectors

If vector $\underset{\sim}{a}$ is parallel to vector $\underset{\sim}{b}$, then $\underset{\sim}{a} = k\underset{\sim}{b}$ where $k \in R$.

Note: When applying the dot product to parallel vectors, θ (the angle between them) may be either $0°$ or $180°$ depending on whether the vectors are in the same or opposite directions.

WORKED EXAMPLE 19

Let $\underset{\sim}{a} = 5\underset{\sim}{i} + 2\underset{\sim}{j}$. Determine a vector parallel to $\underset{\sim}{a}$ such that the dot product is 87.

THINK	WRITE
1. Let the required vector $\underset{\sim}{b} = k\underset{\sim}{a}$.	Let $\underset{\sim}{b} = k(5\underset{\sim}{i} + 2\underset{\sim}{j})$ $= 5k\underset{\sim}{i} + 2k\underset{\sim}{j}$
2. Determine the dot product of $\underset{\sim}{a} \cdot \underset{\sim}{b}$.	$\underset{\sim}{a} \cdot \underset{\sim}{b} = (5\underset{\sim}{i} + 2\underset{\sim}{j}) \cdot (5k\underset{\sim}{i} + 2k\underset{\sim}{j})$
3. Simplify.	$= 25k + 4k$ $= 29k$
4. Equate the result to the given dot product 87.	$29k = 87$
5. Solve for k.	$k = 3$
6. Substitute $k = 3$ into vector $\underset{\sim}{b}$.	$\underset{\sim}{b} = 15\underset{\sim}{i} + 6\underset{\sim}{j}$

9.5.5 Geometric proofs using vectors

Vectors can also be used to prove various geometric properties, such as the diagonals of a square are perpendicular. This property will be proved using vectors in Worked example 20. The following properties of vectors will be useful in proofs.

- $\underset{\sim}{a} \cdot \underset{\sim}{b} = \underset{\sim}{0}$ if the two vectors $\underset{\sim}{a}$ and $\underset{\sim}{b}$ are perpendicular
- $\underset{\sim}{a} \cdot \underset{\sim}{b} = \underset{\sim}{b} \cdot \underset{\sim}{a}$
- $\underset{\sim}{a} \cdot \underset{\sim}{a} = |\underset{\sim}{a}|^2$
- $(\underset{\sim}{a} + \underset{\sim}{b}) \cdot \underset{\sim}{c} = \underset{\sim}{a} \cdot \underset{\sim}{c} + \underset{\sim}{b} \cdot \underset{\sim}{c}$
- $(\underset{\sim}{a} + \underset{\sim}{b}) \cdot (\underset{\sim}{c} + \underset{\sim}{d}) = \underset{\sim}{a} \cdot \underset{\sim}{c} + \underset{\sim}{a} \cdot \underset{\sim}{d} + \underset{\sim}{b} \cdot \underset{\sim}{c} + \underset{\sim}{b} \cdot \underset{\sim}{d}$

WORKED EXAMPLE 20 Geometric proofs using vectors

Prove that the diagonals of a square are perpendicular.

THINK	WRITE
1. Let OABC be a square.	
2. State the properties of the square: opposites sides are equal and parallel, and all sides are equal in length.	Let $\overrightarrow{OA} = \underset{\sim}{a}$ and $\overrightarrow{OC} = \underset{\sim}{c}$. Since OABC is a square, $\underset{\sim}{a} = \overrightarrow{OA} = \overrightarrow{CB}$ and $\underset{\sim}{c} = \overrightarrow{OC} = \overrightarrow{AB}$ and $\lvert\underset{\sim}{a}\rvert = \lvert\underset{\sim}{c}\rvert$.
3. Determine a vector expression for the diagonal OB.	$\overrightarrow{OB} = \overrightarrow{OA} + \overrightarrow{AB}$ $= \underset{\sim}{a} + \underset{\sim}{c}$
4. Determine a vector expression for the diagonal AC. Remember direction is important.	$\overrightarrow{AC} = \overrightarrow{AO} + \overrightarrow{OC}$ $\overrightarrow{AC} = -\overrightarrow{OA} + \overrightarrow{OC}$ $= -\underset{\sim}{a} + \underset{\sim}{c}$ $= \underset{\sim}{c} - \underset{\sim}{a}$
5. Evaluate the dot product of the diagonals of the square by expanding and simplifying.	$\overrightarrow{OB} \cdot \overrightarrow{AC} = (\underset{\sim}{a} + \underset{\sim}{c}) \cdot (\underset{\sim}{c} - \underset{\sim}{a})$ $= \underset{\sim}{a} \cdot \underset{\sim}{c} - \underset{\sim}{a} \cdot \underset{\sim}{a} + \underset{\sim}{c} \cdot \underset{\sim}{c} - \underset{\sim}{c} \cdot \underset{\sim}{a}$ $= -\lvert\underset{\sim}{a}\rvert^2 + \lvert\underset{\sim}{c}\rvert^2$
6. Since OABC is a square, $\underset{\sim}{a}$ and $\underset{\sim}{c}$ are equal in length.	$= 0$
7. State the conclusion.	The diagonals of a square are perpendicular.

When a proof involves midpoints, it may be easier to let the unknown vectors be $2\underset{\sim}{a}$ and $2\underset{\sim}{b}$. This is illustrated in Worked example 21.

WORKED EXAMPLE 21 Geometric proof involving midpoints

Prove that the diagonals of a parallelogram bisect each other.

THINK	WRITE
1. Let OABC be a parallelogram.	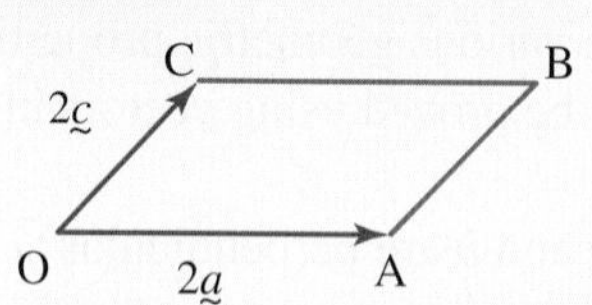
2. State the properties of the parallelogram: opposites sides are equal and parallel.	Let $\overrightarrow{OA} = 2\underset{\sim}{a}$ and $\overrightarrow{OC} = 2\underset{\sim}{c}$. Since OABC is a parallelogram, $2\underset{\sim}{a} = \overrightarrow{OA} = \overrightarrow{CB}$ and $2\underset{\sim}{c} = \overrightarrow{OC} = \overrightarrow{AB}$.
3. Determine a vector expression for the diagonal OB.	$\overrightarrow{OB} = \overrightarrow{OA} + \overrightarrow{AB}$ $= 2\underset{\sim}{a} + 2\underset{\sim}{c}$

4. Let M be the midpoint of OB. Write a vector expression for OM.

Let M be the midpoint of OB.

$$\overrightarrow{OM} = \frac{1}{2}\overrightarrow{OB}$$
$$= \frac{1}{2}(2\underset{\sim}{a} + 2\underset{\sim}{c})$$
$$= \underset{\sim}{a} + \underset{\sim}{c}$$

5. Determine a vector expression for the diagonal AC. Remember direction is important.

$$\overrightarrow{AC} = \overrightarrow{AO} + \overrightarrow{OC}$$
$$\overrightarrow{AC} = -\overrightarrow{OA} + \overrightarrow{OC}$$
$$= -2\underset{\sim}{a} + 2\underset{\sim}{c}$$
$$= 2\underset{\sim}{c} - 2\underset{\sim}{a}$$

6. Let N be the midpoint of AC. Write a vector expression for AN.

Let N be the midpoint of AC.

$$\overrightarrow{AN} = \frac{1}{2}\overrightarrow{AC}$$
$$= \frac{1}{2}(2\underset{\sim}{c} - 2\underset{\sim}{a})$$
$$= \underset{\sim}{c} - \underset{\sim}{a}$$

7. Write a vector expression for ON and simplify.

$$\overrightarrow{ON} = \overrightarrow{OA} + \overrightarrow{AN}$$
$$= 2\underset{\sim}{a} + (\underset{\sim}{c} - \underset{\sim}{a})$$
$$= \underset{\sim}{a} + \underset{\sim}{c}$$
$$= \overrightarrow{OM}$$

8. Since the two vectors start at the point O and are in the same direction, the points N and M must be the same point. State your conclusion.

Since the two vectors start at the point O and are in the same direction, the points N and M must be the same point.
The diagonals of a parallelogram bisect each other.

9.5 Exercise

Students, these questions are even better in jacPLUS

Receive immediate feedback and access sample responses

Access additional questions

Track your results and progress

Find all this and MORE in jacPLUS

Technology active

1. **WE15** Determine the dot product of the vectors $3\underset{\sim}{i} + 3\underset{\sim}{j}$ and $6\underset{\sim}{i} + 2\underset{\sim}{j}$ using the equation $\underset{\sim}{a} \cdot \underset{\sim}{b} = |\underset{\sim}{a}|\,|\underset{\sim}{b}|\cos(\theta)$.

2. Compare the result from question 1 with that obtained by calculating the dot product using the equation $\underset{\sim}{a} \cdot \underset{\sim}{b} = x_1x_2 + y_1y_2$.

3. **WE16** Determine $\underset{\sim}{a} \cdot \underset{\sim}{b}$ in each of the following cases.
   a. $\underset{\sim}{a} = 2\underset{\sim}{i} + 3\underset{\sim}{j},\ \underset{\sim}{b} = 3\underset{\sim}{i} + 3\underset{\sim}{j}$
   b. $\underset{\sim}{a} = 4\underset{\sim}{i} - 2\underset{\sim}{j},\ \underset{\sim}{b} = 5\underset{\sim}{i} + \underset{\sim}{j}$

4. Determine $\underset{\sim}{a} \cdot \underset{\sim}{b}$ in each of the following cases.

   a. $\underset{\sim}{a} = -\underset{\sim}{i} + 4\underset{\sim}{j},\ \underset{\sim}{b} = 3\underset{\sim}{i} - 7\underset{\sim}{j}$

   b. $\underset{\sim}{a} = 5\underset{\sim}{i} + 9\underset{\sim}{j},\ \underset{\sim}{b} = 2\underset{\sim}{i} - 4\underset{\sim}{j}$

5. Determine $\underset{\sim}{a} \cdot \underset{\sim}{b}$ in each of the following cases.

   a. $\underset{\sim}{a} = -3\underset{\sim}{i} + \underset{\sim}{j},\ \underset{\sim}{b} = \underset{\sim}{i} + 4\underset{\sim}{k}$

   b. $\underset{\sim}{a} = 10\underset{\sim}{i},\ \underset{\sim}{b} = -2\underset{\sim}{i}$

6. Determine $\underset{\sim}{a} \cdot \underset{\sim}{b}$ in each of the following cases.

   a. $\underset{\sim}{a} = 3\underset{\sim}{i} + 5\underset{\sim}{j},\ \underset{\sim}{b} = \underset{\sim}{i}$

   b. $\underset{\sim}{a} = 6\underset{\sim}{i} - 2\underset{\sim}{j},\ \underset{\sim}{b} = -\underset{\sim}{i} - 4\underset{\sim}{j}$

7. Let $\underset{\sim}{a} = x\underset{\sim}{i} + y\underset{\sim}{j}$. Show that $\underset{\sim}{a} \cdot \underset{\sim}{a} = x^2 + y^2$.

8. Let $\underset{\sim}{a} = 2\underset{\sim}{i} - 5\underset{\sim}{j}$ and let $\underset{\sim}{b} = -\underset{\sim}{i} - 2\underset{\sim}{j}$. Determine their dot product.

9. Let $\underset{\sim}{a} = 3\underset{\sim}{i} + 2\underset{\sim}{j},\ \underset{\sim}{b} = \underset{\sim}{i} - 2\underset{\sim}{j}$ and $\underset{\sim}{c} = 5\underset{\sim}{i} - 2\underset{\sim}{j}$. Demonstrate, using these vectors, the property:

$$\underset{\sim}{c} \cdot (\underset{\sim}{a} - \underset{\sim}{b}) = \underset{\sim}{c} \cdot \underset{\sim}{a} - \underset{\sim}{c} \cdot \underset{\sim}{b}$$

   Formally, this means that vectors are *distributive* over subtraction.

10. Repeat question 9 for the property:

$$\underset{\sim}{c} \cdot (\underset{\sim}{a} + \underset{\sim}{b}) = \underset{\sim}{c} \cdot \underset{\sim}{a} + \underset{\sim}{c} \cdot \underset{\sim}{b}$$

   Formally, this means that vectors are *distributive* over addition.

11. Consider the vectors $\underset{\sim}{a}$ and $\underset{\sim}{b}$.

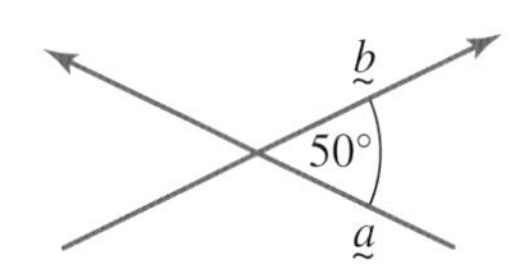

   Their magnitudes are 7 and 8 respectively. Calculate $\underset{\sim}{a} \cdot \underset{\sim}{b}$, to the nearest whole number.

12. Determine the dot product of the following pairs of vectors.

   a. $4\underset{\sim}{i} - 3\underset{\sim}{j}$ and $7\underset{\sim}{i} + 4\underset{\sim}{j}$

   b. $\underset{\sim}{i} + 2\underset{\sim}{j}$ and $-9\underset{\sim}{i} + 4\underset{\sim}{j}$

13. Determine the dot product of the following pairs of vectors.

   a. $8\underset{\sim}{i} + 3\underset{\sim}{j}$ and $-2\underset{\sim}{i} - 3\underset{\sim}{j}$

   b. $5\underset{\sim}{i} - 5\underset{\sim}{j}$ and $5\underset{\sim}{i} + 5\underset{\sim}{j}$

14. **WE17** Calculate the angle between each pair of vectors in questions 12 and 13 to the nearest degree.

15. Determine the angle between the vectors $2\underset{\sim}{i} - 3\underset{\sim}{j}$ and $-4\underset{\sim}{i} + 6\underset{\sim}{j}$, stating your answer rounded to the nearest degree.

16. **WE18** Determine the constant m, if the vectors $\underset{\sim}{b} = m\underset{\sim}{i} + 3\underset{\sim}{j}$ and $\underset{\sim}{a} = 6\underset{\sim}{i} - 2\underset{\sim}{j}$ are perpendicular.

17. Determine the constant m, such that $\underset{\sim}{b} = m\underset{\sim}{i} - 2\underset{\sim}{j}$ is perpendicular to $\underset{\sim}{a} = 4\underset{\sim}{i} - 3\underset{\sim}{j}$.

18. **WE19** Let $\underset{\sim}{a} = 2\underset{\sim}{i} + 4\underset{\sim}{j}$. Determine a vector parallel to $\underset{\sim}{a}$ such that their dot product is 40.

19. Let $\underset{\sim}{a} = 4\underset{\sim}{i} - 3\underset{\sim}{j}$. Determine a vector parallel to $\underset{\sim}{a}$ such that their dot product is 80.

20. **WE20** Prove that the diagonals of a rhombus are perpendicular.

21. The diagram shows a semicircle, centre O and diameter AB. Let $\overrightarrow{OA} = \tilde{a}$ and $\overrightarrow{OC} = \tilde{c}$.

   a. Write a vector expression for:

      i. $\overrightarrow{OB}$

      ii. $\overrightarrow{BC}$

      iii. $\overrightarrow{AC}$

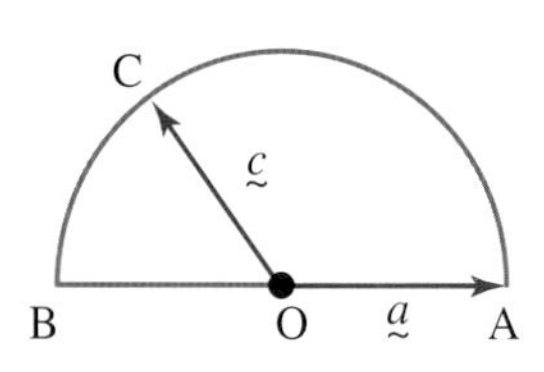

   b. Hence, prove that the angle in a semicircle, $\angle ACB$, is $90°$.

22. **WE21** Prove that the diagonals of a rhombus bisect each other.

23. OAB is an isosceles triangle with sides OA and OB equal in length. Prove that the line from the vertex O to the midpoint of the base AB is perpendicular to the base.

24. Prove that the lengths of the diagonals of a rectangle are equal.

9.5 Exam questions

Question 1 (1 mark) TECH-ACTIVE

MC If $\underset{\sim}{a} = i + 3j$ and $\underset{\sim}{b} = -i - 2j$, then $\underset{\sim}{a} \cdot \underset{\sim}{b}$, the scalar product, is equal to

A. 7 **B.** -7 **C.** 5 **D.** $-5\sqrt{2}$ **E.** $\sqrt{7}$

Question 2 (1 mark) TECH-ACTIVE

If $\underset{\sim}{a} = i + 3j$ and $\underset{\sim}{b} = -i - 2j$, determine the value of θ, the angle between $\underset{\sim}{a}$ and $\underset{\sim}{b}$ in degrees correct to 1 decimal place.

Question 3 (2 marks) TECH-FREE

$\underset{\sim}{r} = ai + (a-1)j$, $\underset{\sim}{s} = 3ai + (a+1)j$. If r and s are perpendicular, determine the value(s) of a.

More exam questions are available online.

9.6 Projections of vectors — scalar and vector resolutes

LEARNING INTENTION

At the end of this subtopic you should be able to:

- calculate vector and scalar resolutes.

9.6.1 Introduction

Previously we have resolved a vector parallel and perpendicular to the x- and y-axes. In this section, we consider a generalisation of this process where we resolve one vector parallel and perpendicular to another vector.

Consider the two vectors, $\underset{\sim}{a}$ and $\underset{\sim}{b}$, shown. The angle between them, as for a dot product, is given by θ. It can be shown that $\underset{\sim}{b}$ is made up of a **projection** acting in the direction of $\underset{\sim}{a}$ and another projection acting perpendicular to $\underset{\sim}{a}$.

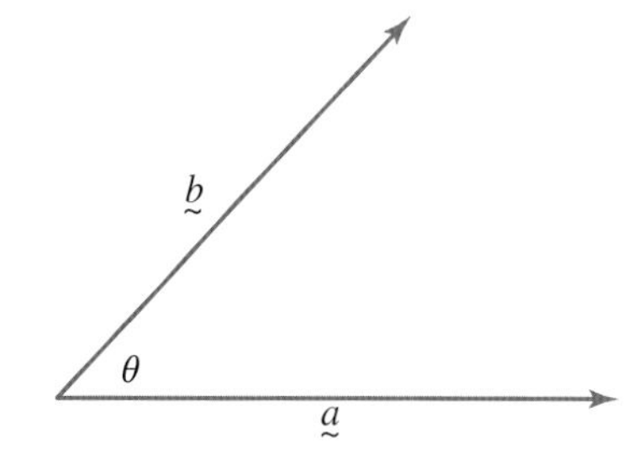

Firstly we wish to determine the projection in the direction of $\underset{\sim}{a}$.

9.6.2 The scalar resolute

To obtain the projection of $\underset{\sim}{b}$ in the direction of $\underset{\sim}{a}$, we perform the following construction:

1. Drop a perpendicular line from the head of $\underset{\sim}{b}$ to $\underset{\sim}{a}$ (this is perpendicular to $\underset{\sim}{a}$). This line joins $\underset{\sim}{a}$ at point A.
2. We wish to determine the length of the line OA.

This construction is shown.

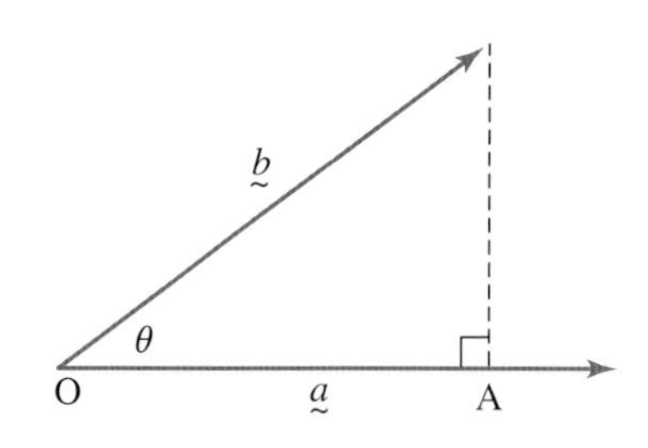

Let the length of $\underset{\sim}{b}$ (its magnitude) be denoted by $|\underset{\sim}{b}|$. Then, from trigonometry:

$$\text{OA} = |\underset{\sim}{b}| \cos(\theta)$$

But from the definition of the dot product:

$$\begin{aligned} \underset{\sim}{a} \cdot \underset{\sim}{b} &= |\underset{\sim}{a}|\,|\underset{\sim}{b}| \cos(\theta) \\ \underset{\sim}{a} \cdot \underset{\sim}{b} &= |\underset{\sim}{a}|\,\text{OA} \end{aligned}$$

Therefore, solving for OA:

$$\begin{aligned} \text{OA} &= \frac{\underset{\sim}{a} \cdot \underset{\sim}{b}}{|\underset{\sim}{a}|} \\ &= \left(\frac{\underset{\sim}{a}}{|\underset{\sim}{a}|}\right) \underset{\sim}{b} \end{aligned}$$

But we know that $\frac{\underset{\sim}{a}}{|\underset{\sim}{a}|} = \hat{\underset{\sim}{a}}$, the unit vector in the direction of $\underset{\sim}{a}$, and therefore

$$\text{OA} = \hat{\underset{\sim}{a}} \cdot \underset{\sim}{b}$$

This quantity, the length OA, is called the **scalar resolute** of $\underset{\sim}{b}$ on $\underset{\sim}{a}$. It effectively indicates 'how much' of $\underset{\sim}{b}$ is in the direction of $\underset{\sim}{a}$.

Scalar resolute

The scalar resolute of $\underset{\sim}{b}$ on $\underset{\sim}{a}$ is given by:

$$\underset{\sim}{b} \cdot \hat{\underset{\sim}{a}} = \frac{\underset{\sim}{b} \cdot \underset{\sim}{a}}{|\underset{\sim}{a}|}$$

WORKED EXAMPLE 22 Calculating scalar resolutes

Let $\underset{\sim}{a} = 3\underset{\sim}{i} + 4\underset{\sim}{j}$ and $\underset{\sim}{b} = 6\underset{\sim}{i} - 2\underset{\sim}{j}$. Calculate:

a. the scalar resolute of $\underset{\sim}{b}$ on $\underset{\sim}{a}$

b. the scalar resolute of $\underset{\sim}{a}$ on $\underset{\sim}{b}$.

THINK	WRITE
a. 1. Calculate the magnitude of $\underset{\sim}{a}$.	a. $|\underset{\sim}{a}| = \sqrt{3^2 + 4^2}$ $= 5$
2. Determine the dot product of $\underset{\sim}{a}$ and $\underset{\sim}{b}$.	$\underset{\sim}{a} \cdot \underset{\sim}{b} = 3 \times 6 + 4 \times (-2)$ $= 18 - 8$ $= 10$

3. Calculate the scalar resolute of $\underset{\sim}{b}$ on $\underset{\sim}{a}$.

$$\underset{\sim}{b} \cdot \hat{\underset{\sim}{a}} = \frac{\underset{\sim}{b} \cdot \underset{\sim}{a}}{\underset{\sim}{a}} = \frac{10}{5} = 2$$

b. 1. Calculate the magnitude of $\underset{\sim}{b}$.

b. $$|\underset{\sim}{b}| = \sqrt{6^2 + (-2)^2} = \sqrt{40}$$

2. Recall the dot product of $\underset{\sim}{a}$ and $\underset{\sim}{b}$ as calculated in part **a**.

$$\underset{\sim}{a} \cdot \underset{\sim}{b} = 10$$

3. Calculate the scalar resolute of $\underset{\sim}{a}$ on $\underset{\sim}{b}$.

$$\underset{\sim}{a} \cdot \hat{\underset{\sim}{b}} = \frac{\underset{\sim}{a} \cdot \underset{\sim}{b}}{|\underset{\sim}{b}|} = \frac{10}{\sqrt{40}} = \frac{\sqrt{10}}{2}$$

Notes:

- The two scalar resolutes are not equal.
- The scalar resolute of $\underset{\sim}{b}$ on $\underset{\sim}{a}$ can easily be evaluated as $\dfrac{\underset{\sim}{b} \cdot \underset{\sim}{a}}{|\underset{\sim}{a}|}$.

9.6.3 Vector resolutes

Consider, now, the vector joining O to A as shown. Its magnitude is just the scalar resolute $(\hat{\underset{\sim}{a}} \cdot \underset{\sim}{b})$, while its direction is the same as $\underset{\sim}{a}$, that is $\hat{\underset{\sim}{a}}$. This quantity is called the **vector resolute** of $\underset{\sim}{b}$ parallel to $\underset{\sim}{a}$ and is denoted by the symbol $\underset{\sim}{b}_{||}$.

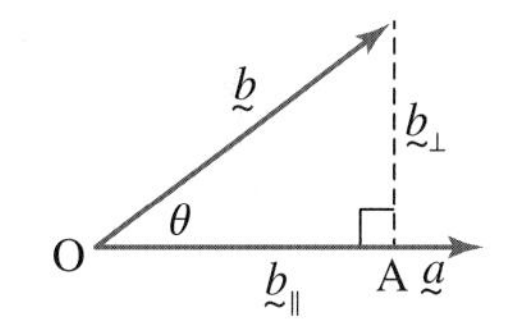

Parallel vector resolutes

The vector resolute of $\underset{\sim}{b}$ parallel to $\underset{\sim}{a}$ is given by:

$$\underset{\sim}{b}_{||} = (\hat{\underset{\sim}{a}} \cdot \underset{\sim}{b})\, \hat{\underset{\sim}{a}}$$

Consider the geometry of the above figure. The original vector can be seen to be the sum of two other vectors, namely $\underset{\sim}{b}_{||}$ and $\underset{\sim}{b}_{\perp}$. This second vector is called the vector resolute of $\underset{\sim}{b}$ perpendicular to $\underset{\sim}{a}$ and can be computed simply as follows:

$\underset{\sim}{b} = \underset{\sim}{b}_{||} + \underset{\sim}{b}_{\perp}$ (by addition of vectors)

$\underset{\sim}{b}_{\perp} = \underset{\sim}{b} - \underset{\sim}{b}_{||}$ (by rearranging the vector equation)

By substitution for $\underset{\sim}{b}_{||}$ from the equation $b_{||} = (\hat{\underset{\sim}{a}} \cdot \underset{\sim}{b})\,\hat{\underset{\sim}{a}}$, we can determine the vector resolute of $\underset{\sim}{b}$ perpendicular to $\underset{\sim}{a}$:

Perpendicular vector resolutes

The vector resolute of $\underset{\sim}{b}$ perpendicular to $\underset{\sim}{a}$ is given by:

$$\underset{\sim}{b}_{\perp} = \underset{\sim}{b} - (\hat{\underset{\sim}{a}} \cdot \underset{\sim}{b})\,\hat{\underset{\sim}{a}}$$

Once $\underset{\sim}{b}_{||}$ has been calculated, simply subtract it from $\underset{\sim}{b}$ to get $\underset{\sim}{b}_{\perp}$.

WORKED EXAMPLE 23 Calculating vector resolutes

Let $\underset{\sim}{a} = -2\underset{\sim}{i} + 3\underset{\sim}{j}$ and $\underset{\sim}{b} = 4\underset{\sim}{i} + 2\underset{\sim}{j}$. Calculate:

a. **the scalar resolute of $\underset{\sim}{b}$ on $\underset{\sim}{a}$**
b. **the vector resolute of $\underset{\sim}{b}$ parallel to $\underset{\sim}{a}$, namely $\underset{\sim}{b}_{||}$**
c. **the vector resolute of $\underset{\sim}{b}$ perpendicular to $\underset{\sim}{a}$, namely $\underset{\sim}{b}_{\perp}$.**

THINK	WRITE
a. 1. Calculate the magnitude of $\underset{\sim}{a}$.	a. $\lvert\underset{\sim}{a}\rvert = \sqrt{(-2)^2 + 3^2}$ $= \sqrt{13}$
2. Determine $\hat{\underset{\sim}{a}}$.	$\hat{\underset{\sim}{a}} = \dfrac{\underset{\sim}{a}}{\lvert\underset{\sim}{a}\rvert}$ $= \dfrac{1}{\sqrt{13}}(-2\underset{\sim}{i} + 3\underset{\sim}{j})$
3. Determine the scalar resolute using $\hat{\underset{\sim}{a}} \cdot \underset{\sim}{b}$.	$\hat{\underset{\sim}{a}} \cdot \underset{\sim}{b} = \dfrac{1}{\sqrt{13}}(-2\underset{\sim}{i} + 3\underset{\sim}{j}) \cdot (4\underset{\sim}{i} + 2\underset{\sim}{j})$
4. Simplify.	$= \dfrac{1}{\sqrt{13}}(-8 + 6)$ $= -\dfrac{2}{\sqrt{13}}$
b. 1. Determine $\underset{\sim}{b}_{||}$ using equation $b_{||} = (\hat{\underset{\sim}{a}} \cdot \underset{\sim}{b})\,\hat{\underset{\sim}{a}}$	b. $\underset{\sim}{b}_{||} = (\hat{\underset{\sim}{a}} \cdot \underset{\sim}{b})\,\hat{\underset{\sim}{a}}$
2. Simplify.	$= \left(-\dfrac{2}{\sqrt{13}}\right)\left[\dfrac{1}{\sqrt{13}}(-2\underset{\sim}{i} + 3\underset{\sim}{j})\right]$ $= -\dfrac{2}{13}(-2\underset{\sim}{i} + 3\underset{\sim}{j})$ $= \dfrac{4}{13}\underset{\sim}{i} - \dfrac{6}{13}\underset{\sim}{j}$
c. 1. Determine $\underset{\sim}{b}_{\perp}$ by subtraction of $\underset{\sim}{b}_{||}$ from $\underset{\sim}{b}$ as in equation $\underset{\sim}{b}_{\perp} = \underset{\sim}{b} - \underset{\sim}{b}_{||}$.	c. $\underset{\sim}{b}_{\perp} = \underset{\sim}{b} - \underset{\sim}{b}_{||}$ $= 4\underset{\sim}{i} + 2\underset{\sim}{j} - \left(\dfrac{4}{13}\underset{\sim}{i} - \dfrac{6}{13}\underset{\sim}{j}\right)$
2. Simplify by subtracting $\underset{\sim}{i}$, $\underset{\sim}{j}$ and $\underset{\sim}{k}$ components.	$\underset{\sim}{b}_{\perp} = \dfrac{48}{13}\underset{\sim}{i} - \dfrac{20}{13}\underset{\sim}{j}$

9.6 Exercise

Students, these questions are even better in jacPLUS

Receive immediate feedback and access sample responses

Access additional questions

Track your results and progress

Find all this and MORE in jacPLUS

Technology free

1. WE22 For each of the following pairs of vectors, determine:
   i. the scalar resolute of $\underset{\sim}{a}$ on $\underset{\sim}{u}$
   ii. the scalar resolute of $\underset{\sim}{u}$ on $\underset{\sim}{a}$.
   a. $\underset{\sim}{u} = 2\underset{\sim}{i} + 3\underset{\sim}{j}$ and $\underset{\sim}{a} = 4\underset{\sim}{i} + 5\underset{\sim}{j}$
   b. $\underset{\sim}{u} = 5\underset{\sim}{i} - 2\underset{\sim}{j}$ and $\underset{\sim}{a} = 3\underset{\sim}{i} - \underset{\sim}{j}$
   c. $\underset{\sim}{u} = -2\underset{\sim}{i} + 6\underset{\sim}{j}$ and $\underset{\sim}{a} = \underset{\sim}{i} - 4\underset{\sim}{j}$
   d. $\underset{\sim}{u} = 3\underset{\sim}{i} - 2\underset{\sim}{j}$ and $\underset{\sim}{a} = -4\underset{\sim}{i} - 3\underset{\sim}{j}$
   e. $\underset{\sim}{u} = 8\underset{\sim}{i} - 6\underset{\sim}{j}$ and $\underset{\sim}{a} = -5\underset{\sim}{i} + \underset{\sim}{j}$

2. WE23 For each pair of vectors $\underset{\sim}{a}$ and $\underset{\sim}{b}$, determine:
   i. the scalar resolute of $\underset{\sim}{b}$ on $\underset{\sim}{a}$
   ii. the vector resolute of $\underset{\sim}{b}$, parallel to $\underset{\sim}{a}$, namely $\underset{\sim}{b}_{\parallel}$
   iii. the vector resolute of $\underset{\sim}{b}$, perpendicular to $\underset{\sim}{a}$, namely $\underset{\sim}{b}_{\perp}$.
   a. $\underset{\sim}{a} = 3\underset{\sim}{i} - \underset{\sim}{j}$; $\underset{\sim}{b} = 2\underset{\sim}{i} + 5\underset{\sim}{j}$
   b. $\underset{\sim}{a} = 4\underset{\sim}{i} + 5\underset{\sim}{j}$; $\underset{\sim}{b} = 8\underset{\sim}{i} + 10\underset{\sim}{j}$
   c. $\underset{\sim}{a} = 4\underset{\sim}{i} + 3\underset{\sim}{j}$; $\underset{\sim}{b} = -3\underset{\sim}{i} + 4\underset{\sim}{j}$
   d. $\underset{\sim}{a} = \underset{\sim}{i} + \underset{\sim}{j}$; $\underset{\sim}{b} = 2\underset{\sim}{i} + \underset{\sim}{j}$
   e. $\underset{\sim}{a} = 2\underset{\sim}{i} + 3\underset{\sim}{j}$; $\underset{\sim}{b} = 2\underset{\sim}{i} - 3\underset{\sim}{j}$
   f. $\underset{\sim}{a} = 3\underset{\sim}{i} + \underset{\sim}{j}$; $\underset{\sim}{b} = 2\underset{\sim}{j}$

Technology active

3. An injured bushwalker is located at a position relative to a camp given by the vector $2\underset{\sim}{i} + 3\underset{\sim}{j}$.
   A searcher heads off from the camp in a direction parallel to the vector $3\underset{\sim}{i} + 4\underset{\sim}{j}$. All measurements are in kilometres.
   a. Calculate how far the searcher is from the camp when closest to the bushwalker, in km to 1 decimal place.
   b. Determine the minimum distance between the searcher and the bushwalker, correct to the nearest 100 m.

4. A distressed yacht is located at a position given by the vector $5\underset{\sim}{i} - 2\underset{\sim}{j}$ relative to a cruiser.
   A rescue boat is sent off from the cruiser and travels in a direction parallel to the vector $3\underset{\sim}{i} - \underset{\sim}{j}$. If all measurements are in kilometres calculate, to the nearest metre, how close the rescue boat gets to the yacht.

5. Given the two vectors $\underset{\sim}{a} = 4\underset{\sim}{i} - 3\underset{\sim}{j}$ and $\underset{\sim}{b} = \underset{\sim}{i} - 2\underset{\sim}{j}$, determine:
   a. the scalar resolute of $\underset{\sim}{a}$ in the direction of $\underset{\sim}{b}$
   b. the vector resolute of $\underset{\sim}{a}$ in the direction of $\underset{\sim}{b}$
   c. the vector resolute of $\underset{\sim}{a}$ perpendicular $\underset{\sim}{b}$.

6. Given the two vectors $\underset{\sim}{r} = 2\underset{\sim}{i} - 3\underset{\sim}{j}$ and $\underset{\sim}{s} = 3\underset{\sim}{i} - 4\underset{\sim}{j}$, determine:
   a. the vector component of $\underset{\sim}{r}$ parallel to $\underset{\sim}{s}$
   b. the vector component of $\underset{\sim}{r}$ perpendicular to $\underset{\sim}{s}$.

7. Two points A and B are given by (5, 2) and (4, −3) respectively. Let E be the point on $\overrightarrow{OB}$ such that $\overrightarrow{OE}$ is the vector resolute of $\overrightarrow{OA}$ onto $\overrightarrow{OB}$.
   a. Determine $\left|\overrightarrow{OE}\right|$.
   b. Show that the vector $\overrightarrow{OC} = -3\underset{\sim}{i} - 4\underset{\sim}{j}$ is perpendicular and equal in length to the vector $\overrightarrow{OB}$.
   c. Point D is placed so that OCDE is a rectangle. Show that the area of this rectangle is equal to $\overrightarrow{OA} \cdot \overrightarrow{OB}$.

8. Two points A and B are given by $(-1, 5)$ and $(3, -2)$ respectively. Let C be the point on $\overrightarrow{OB}$ such that $\overrightarrow{OC}$ is the vector resolute of $\overrightarrow{OA}$ onto $\overrightarrow{OB}$.
   a. Determine the scalar resolute of $\overrightarrow{OA}$ onto $\overrightarrow{OB}$.
   b. Show that the vector $\overrightarrow{OE} = -2\underset{\sim}{i} - 3\underset{\sim}{j}$ is perpendicular and equal in length to the vector $\overrightarrow{OB}$.
   c. Point D is placed so that OCDE is a rectangle. Show that the area of this rectangle is equal to $\overrightarrow{OA} \cdot \overrightarrow{OB}$.
9. Given the two vectors $\underset{\sim}{a} = 3\underset{\sim}{i} - \underset{\sim}{j}$ and $\underset{\sim}{b} = \underset{\sim}{i} + \underset{\sim}{j}$, determine:
   a. a unit vector parallel to $\underset{\sim}{b}$
   b. the scalar resolute of $\underset{\sim}{a}$ in the direction of $\underset{\sim}{b}$
   c. the vector resolute of $\underset{\sim}{a}$ in the direction of $\underset{\sim}{b}$
   d. the vector resolute of $\underset{\sim}{a}$ perpendicular to $\underset{\sim}{b}$.
10. For the two vectors $\underset{\sim}{a} = 3\underset{\sim}{i} + 4\underset{\sim}{j}$ and $\underset{\sim}{b} = 2\underset{\sim}{i} - \underset{\sim}{j}$, determine the vector resolute of $\underset{\sim}{a}$ in the direction of $\underset{\sim}{b}$.
11. If $\underset{\sim}{r} = 4\underset{\sim}{i} + \underset{\sim}{j}$ and $\underset{\sim}{s} = -3\underset{\sim}{i} + 4\underset{\sim}{j}$, determine the vector resolute of $\underset{\sim}{r}$ perpendicular to $\underset{\sim}{s}$.
12. a. If $|\underset{\sim}{a}| = 4$, $|\underset{\sim}{b}| = 3$ and $\underset{\sim}{a} \cdot \underset{\sim}{b} = 0$, determine $|\underset{\sim}{a} + \underset{\sim}{b}|$ and $|\underset{\sim}{a} - \underset{\sim}{b}|$.
    b. If $|\underset{\sim}{a}| = 5$, $|\underset{\sim}{b}| = 12$ and $\underset{\sim}{a} \cdot \underset{\sim}{b} = 0$, determine $|\underset{\sim}{a} + \underset{\sim}{b}|$ and $|\underset{\sim}{a} - \underset{\sim}{b}|$.
    c. If $|\underset{\sim}{r}| = 7$, $|\underset{\sim}{s}| = 24$ and $\underset{\sim}{r} \cdot \underset{\sim}{s} = 0$, determine $|\underset{\sim}{r} + \underset{\sim}{s}|$ and $|\underset{\sim}{r} - \underset{\sim}{s}|$.
    d. Given that $|\underset{\sim}{a}| = a$, $|\underset{\sim}{b}| = b$ and $\underset{\sim}{a} \cdot \underset{\sim}{b} = 0$, deduce $|\underset{\sim}{a} + \underset{\sim}{b}|$ and $|\underset{\sim}{a} - \underset{\sim}{b}|$.
13. If $|\underset{\sim}{a}| = 4$, $|\underset{\sim}{b}| = 3$ and $\underset{\sim}{a} \cdot \underset{\sim}{b} = 2$, calculate:
    a. $(\underset{\sim}{a} + \underset{\sim}{b}) \cdot (\underset{\sim}{a} - \underset{\sim}{b})$
    b. $(\underset{\sim}{a} + 2\underset{\sim}{b}) \cdot (\underset{\sim}{a} - \underset{\sim}{b})$
    c. $(\underset{\sim}{a} + \underset{\sim}{b}) \cdot (\underset{\sim}{a} - 2\underset{\sim}{b})$
14. If $|\underset{\sim}{a}| = 6$, $|\underset{\sim}{b}| = 7$ and $\underset{\sim}{a} \cdot \underset{\sim}{b} = -4$, calculate:
    a. $|\underset{\sim}{a} + \underset{\sim}{b}|$
    b. $|\underset{\sim}{a} - \underset{\sim}{b}|$
    c. $|3\underset{\sim}{a} - 2\underset{\sim}{b}|$
15. a. If the angle between the vectors $2\underset{\sim}{i} + \underset{\sim}{j}$ and $\underset{\sim}{i} + y\underset{\sim}{j}$ is 45°, determine the value of y.
    b. If the angle between the vectors $x\underset{\sim}{i} - 2\sqrt{3}\underset{\sim}{j}$ and $-3\underset{\sim}{i} + \sqrt{3}\underset{\sim}{j}$ is 150°, determine the value of x.
16. a. Determine a unit vector in the xy plane perpendicular to $3\underset{\sim}{i} - 4\underset{\sim}{j}$.
    b. Determine a unit vector in the xy plane perpendicular to $-5\underset{\sim}{i} + 12\underset{\sim}{j}$.
    c. Determine a unit vector in the xy plane perpendicular to $7\underset{\sim}{i} + 24\underset{\sim}{j}$.
    d. Determine a unit vector in the xy plane perpendicular to $a\underset{\sim}{i} + b\underset{\sim}{j}$.

9.6 Exam questions

Question 1 (1 mark) TECH-ACTIVE

MC If $\underset{\sim}{a} = 4\underset{\sim}{i} + 3\underset{\sim}{j}$ and $\underset{\sim}{b} = -5\underset{\sim}{i} + 2\underset{\sim}{j}$, the scalar resolute of $\underset{\sim}{b}$ on $\underset{\sim}{a}$ is equal to

A. $\frac{-14}{5}$ B. $\frac{14}{5}$ C. $\frac{14}{\sqrt{29}}$ D. $\frac{-14}{\sqrt{29}}$ E. $\frac{14}{\sqrt{5}}$

Question 2 (1 mark) TECH-ACTIVE

MC If $\underset{\sim}{u} = 6\underset{\sim}{i} + 7\underset{\sim}{j}$ and $\underset{\sim}{v} = -3\underset{\sim}{i} + 4\underset{\sim}{j}$, the vector resolute of $\underset{\sim}{u}$ parallel to $\underset{\sim}{v}$ is equal to

A. $\frac{2}{5}\underset{\sim}{v}$ B. $-\frac{2}{5}\underset{\sim}{v}$ C. $-6\underset{\sim}{i} + 8\underset{\sim}{j}$ D. $6\underset{\sim}{i} + 8\underset{\sim}{j}$ E. $6\underset{\sim}{i} - 8\underset{\sim}{j}$

Question 3 (3 marks) TECH-FREE

If $\underset{\sim}{r} = \sqrt{5}(14\underset{\sim}{i} + 18\underset{\sim}{j})$ and $\underset{\sim}{s} = 2\underset{\sim}{i} + \underset{\sim}{j}$, calculate the vector resolute of $\underset{\sim}{r}$ perpendicular to $\underset{\sim}{s}$.

More exam questions are available online.

9.7 Review

9.7.1 Summary

doc-37052

Hey students! Now that it's time to revise this topic, go online to:

Access the topic summary  Review your results  Watch teacher-led videos  Practise exam questions

Find all this and MORE in jacPLUS

9.7 Exercise

Technology free: short answer

1. If $\underset{\sim}{a} = 4\underset{\sim}{i} - 3\underset{\sim}{j}$ and $\underset{\sim}{b} = 2\underset{\sim}{i} - 4\underset{\sim}{j}$, evaluate $4\underset{\sim}{a} - 2.5\underset{\sim}{b}$.
2. A boat sails 5 km due east from H, turns northward at a bearing of 45° (N45°E) for a distance of 10 km and then travels due north for a further 5 km to point X. Determine the position vector from H to X.
3. Let $\underset{\sim}{a} = 3\underset{\sim}{i} - 5\underset{\sim}{j}$ and $\underset{\sim}{b} = -4\underset{\sim}{i} + \underset{\sim}{j}$. Determine:
   a. $\underset{\sim}{a} + \underset{\sim}{b}$
   b. $\underset{\sim}{a} - \underset{\sim}{b}$
   c. $\underset{\sim}{a} \cdot \underset{\sim}{b}$
   d. $\hat{\underset{\sim}{a}}$
4. Determine the value(s) of p such that $p\underset{\sim}{i} + 2(1 - 3p)\underset{\sim}{j}$ is perpendicular to $2p\underset{\sim}{i} + 3\underset{\sim}{j}$.
5. Using the vectors $\underset{\sim}{a} = \underset{\sim}{i} - 2\underset{\sim}{j}$ and $\underset{\sim}{b} = 2\underset{\sim}{i} + 3\underset{\sim}{j}$, determine:
   a. the scalar resolute of $\underset{\sim}{a}$ on $\underset{\sim}{b}$
   b. the vector resolute of $\underset{\sim}{b}$ parallel to $\underset{\sim}{a}$.
6. a. Given the points A(3, −4), B(x, y) and C(7, 8), calculate the values of x and y if $\overrightarrow{AB} = \overrightarrow{BC}$.
   b. Given the points A(x, 3), B(2, −1), C(7, −2) and D(2, −4), calculate the value of x if $\overrightarrow{AB}$ is parallel to $\overrightarrow{DC}$.
   c. Given the points A(3, 5), B (−1, y), C (2, 7) and D(−6, 3), calculate the value of y if $\overrightarrow{AB}$ is parallel to $\overrightarrow{DC}$.
7. Given the vector $\underset{\sim}{a} = x\underset{\sim}{i} + y\underset{\sim}{j}$ and $\underset{\sim}{b} = 2\underset{\sim}{i} + 3\underset{\sim}{j}$, determine the value of x and y if the length of the vector $\underset{\sim}{a}$ is $\sqrt{34}$ and $\underset{\sim}{a} \cdot \underset{\sim}{b} = 9$.

Technology active: multiple choice

8. **MC** A girl travels 4 km north and then 2 km south. The net displacement vector is
   A. 6 km north B. 6 km south C. 2 km north D. −2 km north E. 2 km
9. **MC** Consider the following relationships between vectors $\underset{\sim}{u}$, $\underset{\sim}{v}$ and $\underset{\sim}{w}$.
   $\underset{\sim}{u} = 2\underset{\sim}{v} + \underset{\sim}{w}$
   $\underset{\sim}{w} = \underset{\sim}{v} - \underset{\sim}{u}$
   Out of the following, the statement that is true is
   A. $\underset{\sim}{u} = \underset{\sim}{w}$ B. $\underset{\sim}{u} = \underset{\sim}{v}$ C. $\underset{\sim}{u} = \frac{2}{3}\underset{\sim}{v}$ D. $\underset{\sim}{u} = 3\underset{\sim}{v}$ E. $\underset{\sim}{u} = \frac{3}{2}\underset{\sim}{v}$

10. **MC** A vector with a bearing of 60 degrees from N and a magnitude of 10 has

A. x-component $= \frac{\sqrt{3}}{2}$, y-component $= \frac{1}{2}$

B. x-component $= \frac{1}{2}$, y-component $= \frac{\sqrt{3}}{2}$

C. x-component $= 5\sqrt{3}$, y-component $= 5$

D. x-component $= 5$, y-component $= 5\sqrt{3}$

E. x-component $= -\frac{1}{2}$, y-component $= -\frac{\sqrt{3}}{2}$

11. **MC** A unit vector in the direction of $3\underset{\sim}{i} - 4\underset{\sim}{j}$ is

A. $\frac{3}{5}\underset{\sim}{i} + \frac{4}{5}\underset{\sim}{j}$
B. $\frac{3}{5}\underset{\sim}{i} - \frac{4}{5}\underset{\sim}{j}$
C. $\underset{\sim}{i} - \underset{\sim}{j}$
D. $\frac{3}{25}\underset{\sim}{i} - \frac{4}{25}\underset{\sim}{j}$
E. $\underset{\sim}{i} + \underset{\sim}{j}$

12. **MC** The dot product of $\underset{\sim}{a} = 3\underset{\sim}{i} - 3\underset{\sim}{j}$ and $\underset{\sim}{b} = \underset{\sim}{i} - 2\underset{\sim}{j}$ is

A. 3
B. 12
C. 21
D. 15
E. 9

13. **MC** If $\underset{\sim}{a} = 5i + 4j$, then a vector that is perpendicular to $\underset{\sim}{a}$ is

A. $-5\underset{\sim}{i} - 4\underset{\sim}{j}$
B. $3\underset{\sim}{i} + 4\underset{\sim}{j}$
C. $-5\underset{\sim}{i}$
D. $-4\underset{\sim}{i} + 5\underset{\sim}{j}$
E. $-4\underset{\sim}{j}$

14. **MC** If $(\underset{\sim}{a} - \underset{\sim}{b}) \cdot (\underset{\sim}{a} + \underset{\sim}{b}) = 0$, then

A. $\underset{\sim}{a}$ is parallel to $\underset{\sim}{b}$

B. $\underset{\sim}{a}$ and $\underset{\sim}{b}$ have equal magnitudes

C. $\underset{\sim}{a}$ is perpendicular to $\underset{\sim}{b}$

D. $\underset{\sim}{a}$ is a multiple of $\underset{\sim}{b}$

E. $\underset{\sim}{a} = \underset{\sim}{b}$

15. **MC** If $(\underset{\sim}{a} - \underset{\sim}{b}) \cdot (\underset{\sim}{a} + \underset{\sim}{b}) = |\underset{\sim}{b}|^2$, then

A. $\underset{\sim}{a} = \underset{\sim}{b}$

B. $\underset{\sim}{a}$ must be equal to the zero vector, $\underset{\sim}{0}$

C. $\underset{\sim}{a}$ is perpendicular to $\underset{\sim}{b}$

D. $|\underset{\sim}{a}|$ must be equal to $\sqrt{2}\,|\underset{\sim}{b}|$

E. $\underset{\sim}{a}$ is parallel to $\underset{\sim}{b}$

16. **MC** Consider the two vectors shown.
Their dot product is

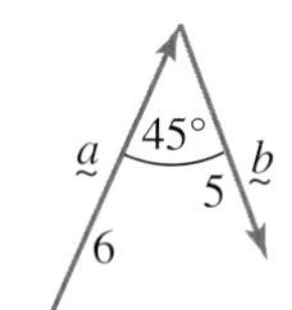

A. 30
B. 21.2
C. −21.2
D. −30
E. There is insufficient data to determine the dot product.

17. **MC** The angle between the vectors $2\underset{\sim}{i} + 3\underset{\sim}{j}$ and $2\underset{\sim}{i} - 3\underset{\sim}{j}$ is closest to

A. 0°
B. 67°
C. 90°
D. 113°
E. 122°

Technology active: extended response

18. The parallelogram OXYZ has O at the origin. The vector joining O to Z is given by $5\underset{\sim}{i}$ while the vector joining O to X is given by $2\underset{\sim}{i} + 7\underset{\sim}{j}$. You may wish to use technology to answer the following.

a. Sketch the parallelogram, labelling all vertices.
b. State the vectors joining Z to Y and Y to X.
c. State the vectors which represent the diagonals of the parallelogram.
d. Determine the cosine of the angle between the diagonals. Express your answer in simplest surd form.
e. Calculate the angle that OX makes with the x-axis, in degrees to 1 decimal place.
f. State the vector resolute of the vector joining O to X in the direction of OZ.
g. Let P be a point on the extended line of XY, such that the vector joining P to Z is perpendicular to OY. Determine the coordinates of P.
h. Calculate the area of the parallelogram.

19. Given the two vectors $\underset{\sim}{a} = 4\underset{\sim}{i} - 5\underset{\sim}{j}$ and $\underset{\sim}{b} = 2\underset{\sim}{i} - \underset{\sim}{j}$, determine:

a. a unit vector parallel to $\underset{\sim}{b}$

b. the scalar resolute of $\underset{\sim}{a}$ in the direction of $\underset{\sim}{b}$

c. the angle between the vectors $\underset{\sim}{a}$ and $\underset{\sim}{b}$, in degrees to 3 decimal places

d. the vector resolute of $\underset{\sim}{a}$ in the direction of $\underset{\sim}{b}$

e. the vector resolute of $\underset{\sim}{a}$ perpendicular $\underset{\sim}{b}$.

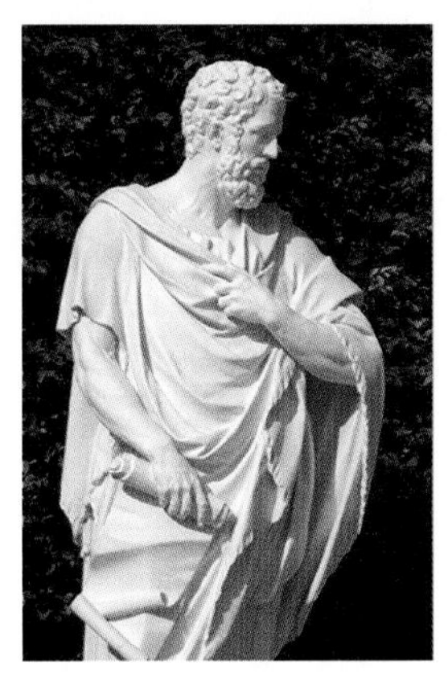

20. a. OAB is a triangle in which M is the midpoint of AB. Let $\overrightarrow{OA} = \underset{\sim}{a}$ and $\overrightarrow{OB} = \underset{\sim}{b}$. Express OM and AM in terms of $\underset{\sim}{a}$ and $\underset{\sim}{b}$ and hence show that $\left|\overrightarrow{OA}\right|^2 + \left|\overrightarrow{OB}\right|^2 = 2\left|\overrightarrow{AM}\right|^2 + 2\left|\overrightarrow{OM}\right|^2$.

This result is known as Apollonius' theorem. Apollonius of Perga (262–190 BCE) was a Greek astronomer. He is also noted for naming the conic sections. There is a crater on the moon named after him.

b. If $\underset{\sim}{a} = \cos(A)\underset{\sim}{i} + \sin(A)\underset{\sim}{j}$ and $\underset{\sim}{b} = \cos(B)\underset{\sim}{i} + \sin(B)\underset{\sim}{j}$, determine $\underset{\sim}{a} \cdot \underset{\sim}{b}$ and hence show that $\cos(A - B) = \cos(A)\cos(B) + \sin(A)\sin(B)$.

9.7 Exam questions

Question 1 (1 mark) TECH-ACTIVE

MC Out of the following, the vector which is a unit vector is

A. $\underset{\sim}{i} + \underset{\sim}{j}$

B. $\sqrt{2}(2\underset{\sim}{i} - \underset{\sim}{j})$

C. $\frac{1}{11}(-3\underset{\sim}{i} + 2\underset{\sim}{j})$

D. $\frac{1}{\sqrt{21}}\left(4\underset{\sim}{i} + \sqrt{5}\underset{\sim}{j}\right)$

E. $\frac{1}{2}(\underset{\sim}{i} - \underset{\sim}{j})$

Question 2 (2 marks) TECH-ACTIVE

An airship is headed east at a speed of 50 km/h. A wind blows from the southwest at 10 km/h. Determine the velocity of the airship relative to the ground, rounded to one decimal place.

Question 3 (1 mark) TECH-FREE

Calculate the magnitude of the position vector $\begin{bmatrix} -7 \\ 1 \end{bmatrix}$.

Question 4 (2 marks) TECH-ACTIVE

$\underset{\sim}{a} \cdot \underset{\sim}{b} = -\frac{\sqrt{2}\,|\underset{\sim}{a}|\,|\underset{\sim}{b}|}{2}$. Determine the angle between $\underset{\sim}{a}$ and $\underset{\sim}{b}$.

Question 5 (3 marks) TECH-ACTIVE

OAB is a triangle, with O the origin. $\overrightarrow{OA} = \underset{\sim}{a}$, $\overrightarrow{OB} = \underset{\sim}{b}$, M is the midpoint of $\overrightarrow{OB}$, N is the midpoint of $\overrightarrow{OA}$ and Q is the midpoint of $\overrightarrow{AB}$. Show the ONQM is a parallelogram.

More exam questions are available online.

Answers

Topic 9 Vectors in the plane

9.2 Vectors and scalars

9.2 Exercise

1. a.

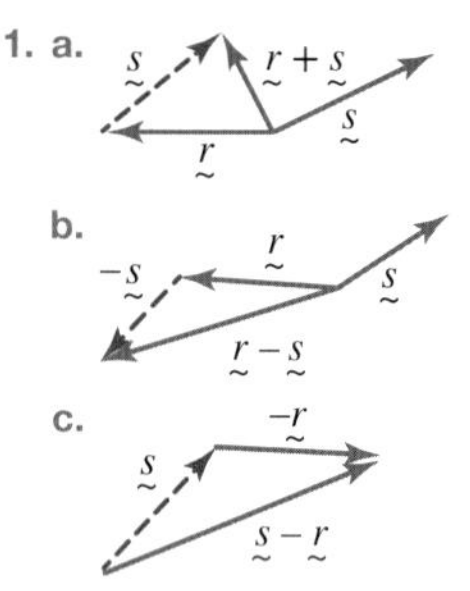

2. a. $\underset{\sim}{s}+\underset{\sim}{t}$ b. $\underset{\sim}{s}+\underset{\sim}{t}+\underset{\sim}{u}+\underset{\sim}{v}$ c. $-\underset{\sim}{s}-\underset{\sim}{t}$
   d. $-\underset{\sim}{u}-\underset{\sim}{v}-\underset{\sim}{t}$ e. $-\underset{\sim}{u}-\underset{\sim}{t}-\underset{\sim}{s}$

3. a. A to C b. D to B
   c. B to D d. A to C

4. a. $\underset{\sim}{r}+\underset{\sim}{s}$ b. $\underset{\sim}{s}+\underset{\sim}{t}$ c. $\underset{\sim}{r}-\underset{\sim}{s}$
   d. $\underset{\sim}{r}+\underset{\sim}{s}$ e. $\underset{\sim}{t}-\underset{\sim}{s}$ f. $\underset{\sim}{s}+\underset{\sim}{t}-\underset{\sim}{r}$
   g. $\underset{\sim}{r}+\underset{\sim}{s}+\underset{\sim}{t}$ h. $-\underset{\sim}{s}-\underset{\sim}{t}$

5. $2\underset{\sim}{a}+4\underset{\sim}{b}$

6. $-3\underset{\sim}{a}-4\underset{\sim}{b}$

7. a. Same as **1 a** except scaled by a factor of 2.
   b. Same as **1 b** except scaled by a factor of 2.
   c.

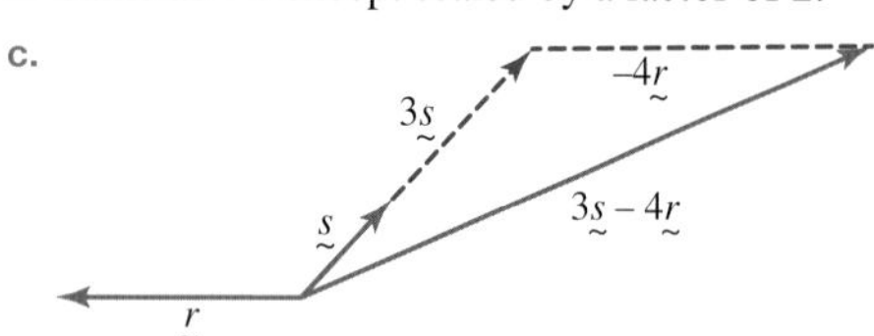

8. Displacement, velocity, force

9. Speed, time, length

10. 1 magnitude and 2 angles

11. a, b.

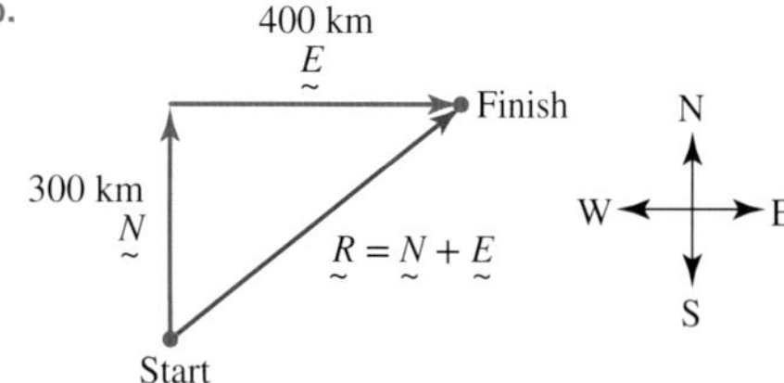

   c. 500 km
   d. 53.1° clockwise from N

12. 512.1 km

13. 721.1 km, 326.3° (clockwise from N)

14. Each part of the answer has coordinates as shown in the diagram *a*, *b*, ... *j*. The original vectors $\underset{\sim}{a}$ and $\underset{\sim}{b}$ are also drawn.

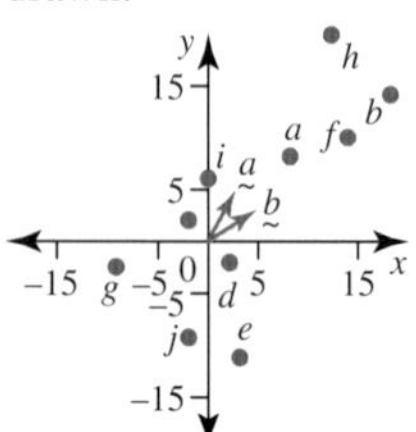

15. Magnitude = 10.77, direction 068.2° T.

16–19. Sample responses can be found in the worked solutions in the online resources.

20. $\underset{\sim}{0}$

9.2 Exam questions

Note: Mark allocations are available with the fully worked solutions online.

1. D
2. $\sqrt{58}$, 23.20°
3. 13 km/h
   22.62°

9.3 Position vectors in the plane

9.3 Exercise

1. a. 3, 4 b. 6, −3 c. 3.4, $\sqrt{2}$

2. a. i. $6\sqrt{2}$ ii. 45°
   b. i. $\sqrt{65}$ ii. 119.7°
   c. i. 4.88 ii. 225.8°
   d. i. 320.16 ii. 358.2°

3. i. a. 045° b. 330.3°
   c. 224.2° d. 091.8°
   ii. a. $\left[6\sqrt{2}, 45°\right]$ b. $\left[\sqrt{65}, 119.7°\right]$
   c. $[4.88, 225.8°]$ d. $[320.16, 358.2°]$

4. $\underset{\sim}{w}=-50\underset{\sim}{i}-50\sqrt{3}\underset{\sim}{j}$

5. $248.9\underset{\sim}{i}-383.3\underset{\sim}{j}$

6. $-60.6\underset{\sim}{i}+109.3\underset{\sim}{j}$

7. $[615.4, -49.8]$

8. 36 steps 11.3° south of east

9. 20.8 km

10. a. $\frac{3}{5}\underset{\sim}{i}+\frac{4}{5}\underset{\sim}{j}$ b. $\frac{3}{5}\underset{\sim}{i}-\frac{4}{5}\underset{\sim}{j}$ c. $\frac{4}{5}\underset{\sim}{i}+\frac{3}{5}\underset{\sim}{j}$
    d. $-\frac{4}{5}\underset{\sim}{i}+\frac{3}{5}\underset{\sim}{j}$ e. $\frac{1}{\sqrt{3}}\underset{\sim}{i}+\frac{\sqrt{2}}{\sqrt{3}}\underset{\sim}{j}$

11. Sample responses can be found in the worked solutions in the online resources.

12. $-0.98\underset{\sim}{i}-0.20\underset{\sim}{j}$

13. Sample responses can be found in the worked solutions in the online resources.

14. a. i. $4\underset{\sim}{i}-7\underset{\sim}{j}$ ii. $\sqrt{65}$
    b. i. $3\underset{\sim}{i}+\underset{\sim}{j}$ ii. $\sqrt{10}$
    c. i. $-4\underset{\sim}{i}+7\underset{\sim}{j}$ ii. $\sqrt{65}$
    d. i. $-3\underset{\sim}{i}-\underset{\sim}{j}$ ii. $\sqrt{10}$
    e. i. $2\underset{\sim}{i}$ ii. 2
    f. i. $-4\underset{\sim}{i}$ ii. 4

15. a. $-4\underset{\sim}{i}+7\underset{\sim}{j}$ b. $-3\underset{\sim}{i}-\underset{\sim}{j}$ c. $4\underset{\sim}{i}-7\underset{\sim}{j}$
    d. $3\underset{\sim}{i}+\underset{\sim}{j}$ e. $-2\underset{\sim}{i}$ f. $4\underset{\sim}{i}$

16. a. $\frac{4}{\sqrt{65}}\underset{\sim}{i} - \frac{7}{\sqrt{65}}\underset{\sim}{j}$ b. $\frac{3}{\sqrt{10}}\underset{\sim}{i} + \frac{1}{\sqrt{10}}\underset{\sim}{j}$

c. $-\frac{4}{\sqrt{65}}\underset{\sim}{i} + \frac{7}{\sqrt{65}}\underset{\sim}{j}$ d. $-\frac{3}{\sqrt{10}}\underset{\sim}{i} - \frac{1}{\sqrt{10}}\underset{\sim}{j}$

e. $\underset{\sim}{i}$ f. $-\underset{\sim}{i}$

17. a. i. $\sqrt{29}$ ii. $\sqrt{13}$

iii. $\frac{5}{\sqrt{29}}\underset{\sim}{i} - \frac{2}{\sqrt{29}}\underset{\sim}{j}$ iv. $-\frac{2}{\sqrt{13}}\underset{\sim}{i} + \frac{3}{\sqrt{13}}\underset{\sim}{j}$

v. $3\underset{\sim}{i} + \underset{\sim}{j}$ vi. $\sqrt{10}$

b. Reject, because magnitude is different.

18. a. i. 5 ii. $\sqrt{26}$

iii. $-\frac{3}{5}\underset{\sim}{i} + \frac{4}{5}\underset{\sim}{j}$ iv. $\frac{5}{\sqrt{26}}\underset{\sim}{i} - \frac{1}{\sqrt{26}}\underset{\sim}{j}$

v. $2\underset{\sim}{i} + 3\underset{\sim}{j}$ vi. $\sqrt{13}$

b. Reject, because the magnitude is different.

19. a. $\sqrt{2}$ b. $\sqrt{58}$

20. a. $3\underset{\sim}{i}$ b. $5\underset{\sim}{j}$ c. $3\underset{\sim}{i} + 5\underset{\sim}{j}$

d. 031.0°T e. $\sqrt{34}$ km/h f. 329.0°

21. a. $\underset{\sim}{a} = 20\underset{\sim}{i}, \underset{\sim}{b} = -15\underset{\sim}{j}, 20\underset{\sim}{j} + 15\underset{\sim}{j}$

b. 25

c. 053.1°

22. 53.1°, −36.9°. Difference = 90°; sample responses can be found in the worked solutions in the online resources.

23. a. $-4\underset{\sim}{i} + 12\underset{\sim}{j}$

b. 341.6°

c. 2.5 minutes

9.3 Exam questions

Note: Mark allocations are available with the fully worked solutions online.

1. B
2. C
3. $|\underset{\sim}{a} - \underset{\sim}{b}| = \sqrt{29}$

9.4 Scalar multiplication of vectors

9.4 Exercise

1. $\frac{15}{2}$

2. $x = -3, y = -8$

3. $\overrightarrow{OM} = 5\underset{\sim}{i} + 6\underset{\sim}{j}$

4. a. $-\frac{5}{2}$ b. $-\frac{15}{2}$

5. a, b. Sample responses can be found in the worked solutions in the online resources.

c. $-\frac{20}{3}$ d. $\frac{15}{7}$

6. a. $\overrightarrow{AB} = \underset{\sim}{i} - 3\underset{\sim}{j}$

$\overrightarrow{BC} = -3(\underset{\sim}{i} - 3\underset{\sim}{j})$

As $\overrightarrow{AB}$ and $\overrightarrow{BC}$ are parallel and share a point B, the points A, B and C must lie on the same line.

b. $\overrightarrow{DE} = -3\underset{\sim}{i} + 4\underset{\sim}{j}$

$\overrightarrow{EF} = -2(-3\underset{\sim}{i} + 4\underset{\sim}{j})$

As $\overrightarrow{DE}$ and $\overrightarrow{EF}$ are parallel and share a point E, the points D, E and F must lie on the same line.

7. a. i. $-6\underset{\sim}{i} + 5\underset{\sim}{j}$ ii. $\sqrt{61}$ iii. $\frac{1}{\sqrt{61}}(-6\underset{\sim}{i} + 5\underset{\sim}{j})$

b. i. $6\underset{\sim}{i} - 8\underset{\sim}{j}$ ii. 10 iii. $\frac{1}{5}(-3\underset{\sim}{i} + 4\underset{\sim}{j})$

8. $m = 2, n = 3$

9. $x = -2, y = 5$

10. a. $\pm 2\sqrt{5}$ b. $-\frac{12}{5}$

11. a. $\pm\sqrt{13}$ b. −10

12. a. $x = 4, y = -2$

b. $m = -5, n = -4$

c. $m = \frac{1}{2}, n = \frac{5}{4}$

9.4 Exam questions

Note: Mark allocations are available with the fully worked solutions online.

1. A
2. B
3. $x = \frac{7}{3}$

9.5 The scalar (dot) product

9.5 Exercise

1. 23.99
2. Dot product = 24; more accurate, since no angle needed.
3. a. 15 b. 18
4. a. −31 b. −26
5. a. 1 b. −20
6. a. 3 b. 2
7. Sample responses can be found in the worked solutions in the online resources.
8. 8
9. Sample responses can be found in the worked solutions in the online resources.
10. Sample responses can be found in the worked solutions in the online resources.
11. −36
12. a. 16 b. −1
13. a. −25 b. 0
14. a. 67° b. 93° c. 144° d. 90°
15. 180°
16. $m = 1$
17. $m = \frac{3}{2}$
18. $4\underset{\sim}{i} + 8\underset{\sim}{j}$

19. $\frac{64}{5}\underset{\sim}{i} - \frac{48}{5}\underset{\sim}{j}$

20-24. Sample responses can be found in the worked solutions in the online resources.

9.5 Exam questions

Note: Mark allocations are available with the fully worked solutions online.

1. B
2. $\theta = 171.9°$
3. $a = \pm\frac{1}{2}$

9.6 Projections of vectors — scalar and vector resolutes

9.6 Exercise

1. a. i. $\frac{23\sqrt{13}}{13}$ ii. $\frac{23\sqrt{41}}{41}$

b. i. $\frac{17\sqrt{29}}{29}$ ii. $\frac{17\sqrt{10}}{10}$

c. i. $-\frac{13\sqrt{10}}{10}$ ii. $-\frac{26\sqrt{17}}{17}$

d. i. $-\frac{6\sqrt{13}}{13}$ ii. $-\frac{6}{5}$

e. i. $-\frac{23}{5}$ ii. $-\frac{23\sqrt{26}}{13}$

2. a. i. $\frac{1}{\sqrt{10}}$

ii. $\underset{\sim}{b}_{||} = \frac{3}{10}\underset{\sim}{i} - \frac{1}{10}\underset{\sim}{j}$

iii. $\underset{\sim}{b}_{\perp} = \frac{17}{10}\underset{\sim}{i} + \frac{51}{10}\underset{\sim}{j}$

b. i. $2\sqrt{41}$

ii. $\underset{\sim}{b}_{||} = 8\underset{\sim}{i} + 10\underset{\sim}{j}$

iii. $\underset{\sim}{b}_{\perp} = \underset{\sim}{0}$

c. i. 0

ii. $\underset{\sim}{b}_{\perp} = \underset{\sim}{0}$

iii. $\underset{\sim}{b}_{\perp} = -3\underset{\sim}{i} + 4\underset{\sim}{j}$

d. i. $\frac{3}{\sqrt{2}}$

ii. $\underset{\sim}{b}_{||} = \frac{3}{2}\underset{\sim}{i} + \frac{3}{2}\underset{\sim}{j}$

iii. $\underset{\sim}{b}_{\perp} = \frac{1}{2}\underset{\sim}{i} - \frac{1}{2}\underset{\sim}{j}$

e. i. $-\frac{5}{\sqrt{13}}$

ii. $\underset{\sim}{b}_{||} = \frac{-10}{13}\underset{\sim}{i} - \frac{15}{13}\underset{\sim}{j}$

iii. $\underset{\sim}{b}_{\perp} = \frac{36}{13}\underset{\sim}{i} - \frac{24}{13}\underset{\sim}{j}$

f. i. $\frac{2}{\sqrt{10}}$

ii. $\underset{\sim}{b}_{||} = \frac{3}{5}\underset{\sim}{i} + \frac{1}{5}\underset{\sim}{j}$

iii. $\underset{\sim}{b}_{\perp} = -\frac{3}{5}\underset{\sim}{i} + \frac{9}{5}\underset{\sim}{j}$

3. a. 3.6 km

b. 0.2 km or 200 metres

4. 316 metres

5. a. $\frac{10}{\sqrt{5}}$ b. $2(\underset{\sim}{i} - 2\underset{\sim}{j})$ c. $2\underset{\sim}{i} + \underset{\sim}{j}$

6. a. $\frac{18}{25}(3\underset{\sim}{i} - 4\underset{\sim}{j})$ b. $-\frac{1}{25}(4\underset{\sim}{i} + 3\underset{\sim}{j})$

7. a. 2.8

b. Sample responses can be found in the worked solutions in the online resources.

c. Sample responses can be found in the worked solutions in the online resources.

8. a. $-\sqrt{13}$

b. Sample responses can be found in the worked solutions in the online resources.

c. Sample responses can be found in the worked solutions in the online resources.

9. a. $\frac{1}{\sqrt{2}}(\underset{\sim}{i} + \underset{\sim}{j})$ b. $\sqrt{2}$

c. $\underset{\sim}{i} + \underset{\sim}{j}$ d. $2\underset{\sim}{i} - 2\underset{\sim}{j}$

10. $\frac{2}{5}(2\underset{\sim}{i} - \underset{\sim}{j})$

11. $\frac{1}{25}(76\underset{\sim}{i} + 57\underset{\sim}{j})$

12. a. 5, 5 b. 13, 13

c. 25, 25 d. $\sqrt{a^2 + b^2}$, $\sqrt{a^2 + b^2}$

13. a. 7 b. 0 c. −4

14. a. $\sqrt{77}$ b. $\sqrt{93}$ c. $2\sqrt{142}$

15. a. $3, -\frac{1}{3}$ b. 2

16. a. $\pm\frac{1}{5}(4\underset{\sim}{i} + 3\underset{\sim}{j})$ b. $\pm\frac{1}{13}(12\underset{\sim}{i} + 5\underset{\sim}{j})$

c. $\pm\frac{1}{25}(24\underset{\sim}{i} - 7\underset{\sim}{j})$ d. $\pm\frac{1}{\sqrt{a^2 + b^2}}(b\underset{\sim}{i} - a\underset{\sim}{j})$

9.6 Exam questions

Note: Mark allocations are available with the fully worked solutions online.

1. A
2. A
3. $\frac{-22}{\sqrt{5}}\underset{\sim}{i} + \frac{44}{\sqrt{5}}\underset{\sim}{j}$

9.7 Review

9.7 Exercise

Technology free: short answer

1. $11\underset{\sim}{i} - 2\underset{\sim}{j}$
2. $(5 + 5\sqrt{2})\underset{\sim}{i} + (5 + 5\sqrt{2})\underset{\sim}{j}$
3. a. $-\underset{\sim}{i} - 4\underset{\sim}{j}$ b. $7\underset{\sim}{i} - 6\underset{\sim}{j}$
   c. -17 d. $\frac{3}{\sqrt{34}}\underset{\sim}{i} - \frac{5}{\sqrt{34}}\underset{\sim}{j}$
4. $\frac{9 \pm \sqrt{69}}{2}$
5. a. $-\frac{4}{\sqrt{3}}$ b. $-\frac{4}{5}(i - 2j)$
6. a. $x = 5, y = 2$ b. $x = 12$ c. $y = 3$
7. $x = -3, y = 5$ or $x = \frac{75}{13}, y = -\frac{11}{13}$

Technology active: multiple choice

8. C
9. E
10. C
11. B
12. E
13. D
14. B
15. D
16. C
17. D

Technology active: extended response

18. a.

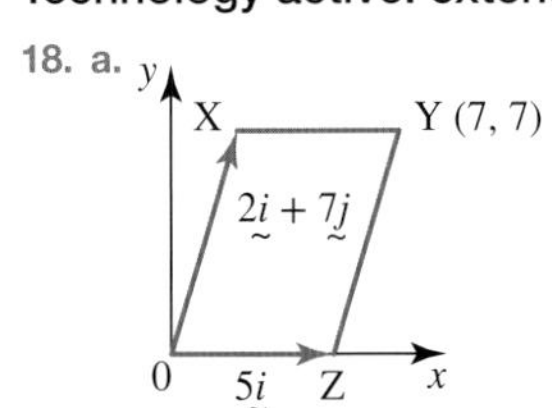

b. $2\underset{\sim}{i} + 7\underset{\sim}{j}, -5\underset{\sim}{i}$
c. $7\underset{\sim}{i} + 7\underset{\sim}{j}, -3\underset{\sim}{i} + 7\underset{\sim}{j}$
d. $\frac{2\sqrt{29}}{29}$
e. $74.1°$
f. $2\underset{\sim}{i}$
g. $(-2, 7)$
h. 35 square units

19. a. $\frac{1}{\sqrt{5}}(2\underset{\sim}{i} - \underset{\sim}{j})$ b. $\frac{13}{\sqrt{5}}$ c. $24.775°$
   d. $\frac{13}{5}(2\underset{\sim}{i} - \underset{\sim}{j})$ e. $-\frac{6}{5}(\underset{\sim}{i} + 2\underset{\sim}{j})$
20. Sample responses can be found in the worked solutions in the online resources.

9.7 Exam questions

Note: Mark allocations are available with the fully worked solutions online.

1. D
2. 57.5 km/h
3. $5\sqrt{2}$
4. $135°$
5. Sample responses can be found in the worked solutions in the online resources.

10 Complex numbers

LEARNING SEQUENCE

Fully worked solutions for this topic are available online.

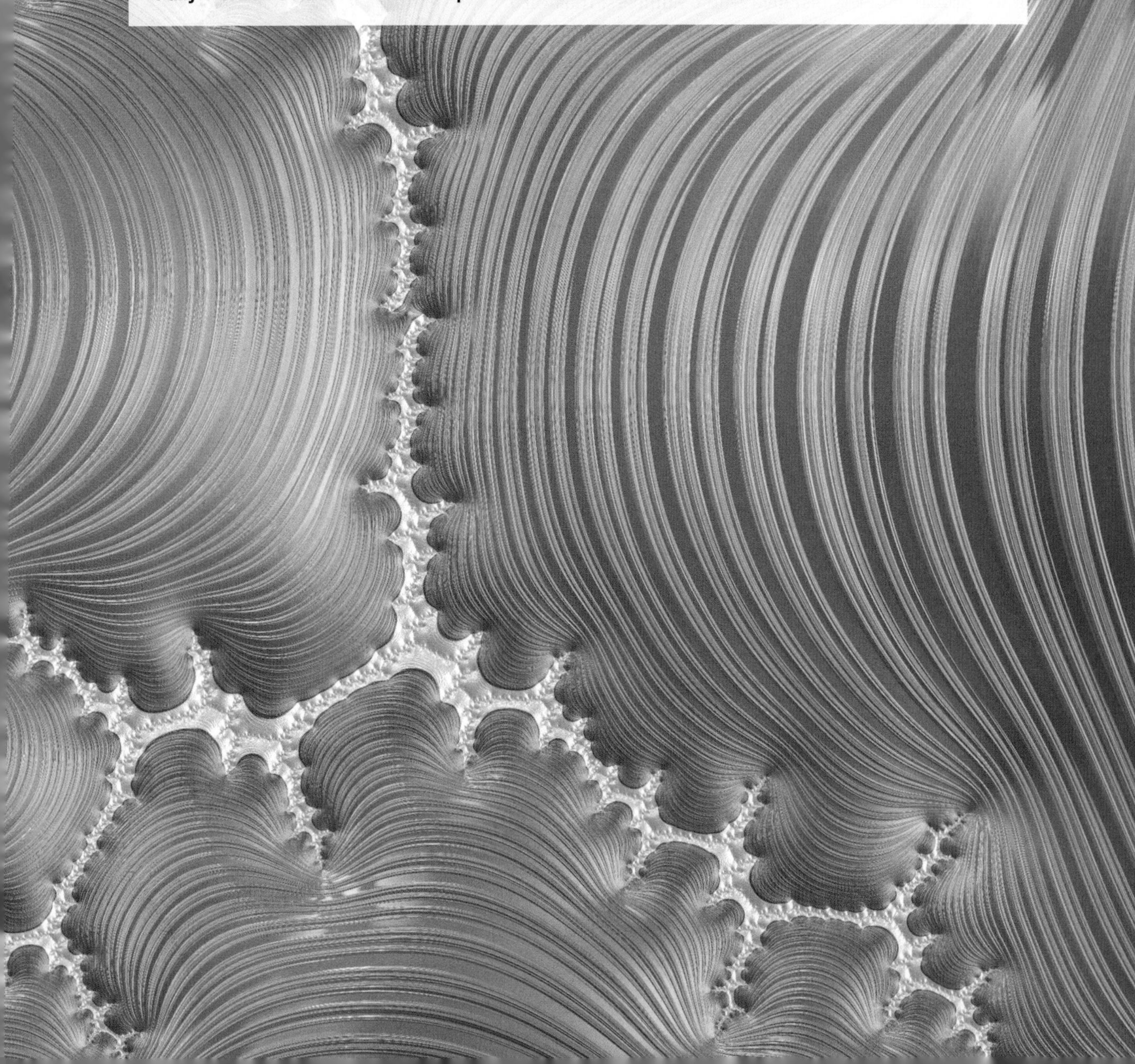

10.1 Overview

Hey students! Bring these pages to life online

Watch videos

Engage with interactivities

Answer questions and check results

Find all this and MORE in jacPLUS

10.1.1 Introduction

In 1545, the Italian mathematician Girolamo Cardano proposed what was then a startling mathematical expression:

$$40 = \left(5 + \sqrt{-15}\right)\left(5 - \sqrt{-15}\right)$$

This was a valid expression, yet it included the square root of a negative number, which seemed 'impossible.

The definition of real numbers included whole numbers, fractions, decimals, irrational and rational numbers as subsets of the real number set. Whenever the square root of a negative number was encountered, it did not fit into any of these subsets and hence could not be classified as a real number. However, the solutions to quadratic equations sometimes force us to consider the square root of a negative number. Consider the two equations $x^2 + 2x + 26 = 0$ and $x^2 - 4x + 29 = 0$. How do the solutions of these equations relate to properties of the associated parabolas?

Up to this point, you have encountered solutions of quadratic equations that were classified as having rational, irrational or no real solutions. The study of complex numbers provides solutions to these previously unsolvable equations.

Why did the square roots of negative numbers become central to the study of a new set of numbers called complex numbers? It was partly curiosity and partly because people such as the Ancient Greek mathematician Diophantus and the 17th-century German mathematician Leibniz had found that real numbers could not solve all equations. Eventually it was shown that complex numbers could solve previously unsolvable problems. They are now used extensively in the fields of physics and engineering in areas such as electrical circuits and electromagnetic waves. Combined with calculus, complex numbers form an important part of the field of mathematics known as complex analysis.

KEY CONCEPTS

This topic covers the following key concepts from the VCE Mathematics Study Design:

- definition and properties of the complex numbers, C, arithmetic, modulus of a complex number, and the representation of complex numbers on an Argand diagram
- general solution of quadratic equations (with real coefficients) of a single variable over C, and conjugate roots
- lines, rays, circles and ellipses
- regions defined in the complex plane using combinations of the above
- use of the modulus of a complex number and the argument of a non-zero complex number to prove basic identities
- conversion between Cartesian and polar form of complex numbers
- multiplication, division, and powers of complex numbers in polar form and their geometric interpretation.

10.2 Introduction to complex numbers

LEARNING INTENTION

At the end of this subtopic you should be able to:
- express square roots of negative numbers using the imaginary number i
- determine the real and imaginary components of complex numbers.

10.2.1 Square root of a negative number

The quadratic equation $x^2 + 1 = 0$ has no solutions for x in the Real Number System R because the equation yields $x = \pm\sqrt{-1}$ and there is no real number which, when squared, gives -1 as the result.

If, however, we define an **imaginary number** denoted by i such that:

$$i^2 = -1$$

Then the answer to the quadratic $x^2 + 1 = 0$, $x = \pm\sqrt{-1}$ becomes $x = \pm\sqrt{i^2} = \pm i$.

For the general case $x^2 + a^2 = 0$, with $a \in R$, we can write:

$$\begin{aligned} x &= \pm\sqrt{-a^2} \\ &= \pm\sqrt{-1 \times a^2} \\ &= \pm\sqrt{i^2 \times a^2} \\ &= \pm ai \end{aligned}$$

Powers of i will produce $\pm i$ or ± 1. As $i^2 = -1$, if follows:

$$\begin{aligned} i^3 &= i^2 \times i \\ &= -1 \times i \\ &= -i \end{aligned} \qquad \begin{aligned} i^4 &= i^2 \times i^2 \\ &= -1 \times -1 \\ &= 1 \end{aligned} \qquad \begin{aligned} i^5 &= i^2 \times i^3 \\ &= -1 \times -i \\ &= i \end{aligned} \qquad \begin{aligned} i^6 &= \left(i^2\right)^3 \\ &= (-1)^3 \\ &= -1 \end{aligned}$$

The pattern is, quite obviously, even powers of i result in 1 or -1 and odd powers of i result in i or $-i$.

10.2.2 Definition of a complex number

A **complex number** (generally denoted by the letter z) is defined as a quantity consisting of a real number added to a multiple of the imaginary unit i. For real numbers a and b, $a + bi$ is a complex number. This is referred to as the standard or **Cartesian form**.

$C = \{z: z = a + bi \text{ where } a, b \in R\}$ defines the set of complex numbers.

The real part of z is a and is written as $\text{Re}(z)$. That is, $\text{Re}(z) = a$.

The imaginary part of z is b and is written as $\text{Im}(z)$. That is, $\text{Im}(z) = b$.

Every real number x can be written as $a + 0i$, so the set of real numbers is a subset of the set of complex numbers. That is, $R \subset C$.

Definition of a complex number

A complex number is of the form:

$$z = a + bi$$

where $a, b \in R$.

WORKED EXAMPLE 1 Simplifying in terms of i

Using the imaginary number i, write a simplified expression for each of the following.

a. $\sqrt{-16}$ **b.** $\sqrt{-5}$.

THINK	WRITE
a. 1. Express the square root of -16 as the product of the square root of 16 and the square root of -1.	**a.** $\sqrt{-16} = \sqrt{16} \times \sqrt{-1}$
2. Substitute i^2 for -1.	$= \sqrt{16} \times \sqrt{i^2}$
3. Take the square root of 16 and i^2.	$= 4i$
b. 1. Express the square root of -5 as the product of the square root of 5 and the square root of -1.	**b.** $\sqrt{-5} = \sqrt{5} \times \sqrt{-1}$
2. Substitute i^2 for -1.	$= \sqrt{5} \times \sqrt{i^2}$
3. Simplify.	$= i\sqrt{5}$

TI | THINK

a. On a Calculator page, complete the entry as shown.
Note: Make sure that the 'Real or Complex' setting is set to 'Rectangular'.

DISPLAY/WRITE

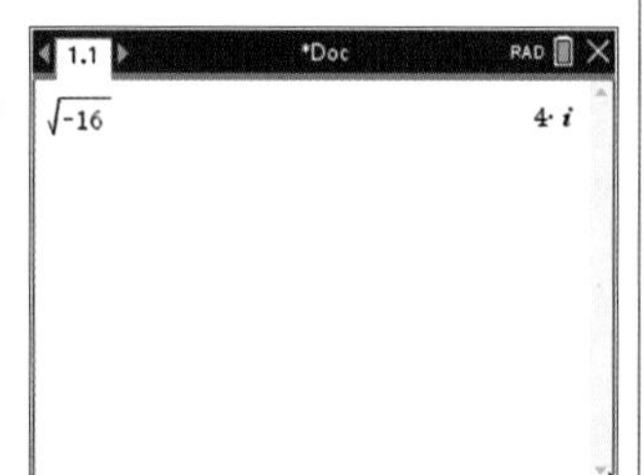

CASIO | THINK

a. On a Main screen, complete the entry as shown.
Note: Make sure that the 'Real or Cplx' setting is set to 'Cplx'.

DISPLAY/WRITE

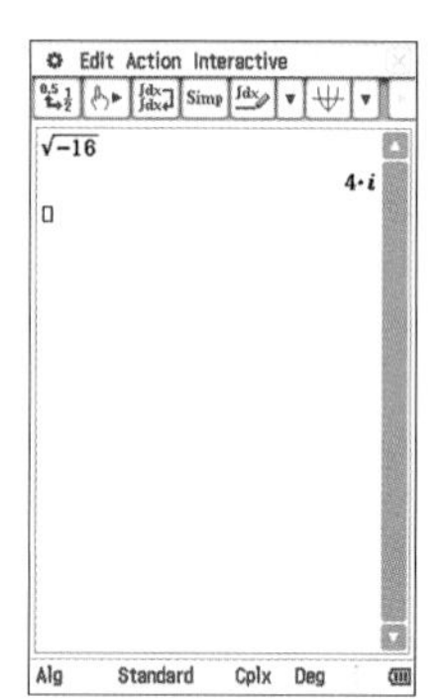

WORKED EXAMPLE 2 Decomposing complex numbers into real and imaginary parts

Write down the real and imaginary parts of the following complex numbers, z.

a. $z = -3 + 2i$ **b.** $z = -\frac{1}{2}i$

THINK	WRITE
a. 1. The real part is the 'non-i' term.	**a.** $\text{Re}(z) = -3$
2. The imaginary part is the coefficient of the i term.	$\text{Im}(z) = 2$
b. 1. The real part is the 'non-i' term.	**b.** $\text{Re}(z) = 0$
2. The imaginary part is the coefficient of the i term.	$\text{Im}(z) = -\frac{1}{2}$

WORKED EXAMPLE 3 Evaluating powers of i

Write $i^8 + i^5$ in the form $a + bi$ where a and b are real numbers.

THINK	WRITE
1. Simplify both i^8 and i^5 using the lowest possible power of i, recalling the index laws.	$i^8 = \left(i^2\right)^4 = (-1)^4 = 1$ $i^5 = i^4 \times i = \left(i^2\right)^2 \times i = (-1)^2 \times i = 1 \times i = i$
2. Add the two answers.	$i^8 + i^5 = 1 + i$

WORKED EXAMPLE 4 Simplifying complex expressions

Simplify $z = i^4 - 2i^2 + 1$ and $w = i^6 - 3i^4 + 3i^2 - 1$ and show that $z + w = -4$.

THINK	WRITE
1. Replace terms with the lowest possible powers of i (remember $i^2 = -1$).	$z = i^4 - 2i^2 + 1$ $= \left(i^2\right)^2 - 2 \times (-1) + 1$ $= (-1)^2 + 2 + 1$ $= 4$ $w = i^6 - 3i^4 + 3i^2 - 1$ $= \left(i^2\right)^3 - 3\left(i^2\right)^2 + 3 \times (-1) - 1$ $= (-1)^3 - 3(-1)^2 - 3 - 1$ $= -1 - 3 - 3 - 1$ $= -8$
2. Add the two answers.	$z + w = i^4 - 2i^2 + 1 + i^6 - 3i^4 + 3i^2 - 1$ $= 4 - 8$ $= -4$

WORKED EXAMPLE 5 Evaluating real and imaginary parts

Evaluate each of the following.

a. $\textbf{Re}(7 + 6i)$

b. $\textbf{Im}(10)$

c. $\textbf{Re}\left(2 + i - 3i^3\right)$

d. $\textbf{Im}\left(\dfrac{1 - 3i - i^2 - i^3}{2}\right)$

THINK	WRITE
a. Recall the real part of the complex number $a + bi$ is a, so the real part of $7 + 6i$ is 7.	a. $\text{Re}(7 + 6i) = 7$
b. The number 10 can be expressed in complex form as $10 + 0i$ and so the imaginary part is 0.	b. $\text{Im}(10) = \text{Im}(10 + 0i)$ $= 0$
c. 1. Simplify $2 + i - 3i^3$ to the lowest possible power of i (remember $i^2 = -1$).	c. $\text{Re}(2 + i - 3i^3) = \text{Re}(2 + i - 3i \times i^2)$ $= \text{Re}(2 + i + 3i)$ $= \text{Re}(2 + 4i)$
2. The real part is 2.	$= 2$

d. 1. Simplify the numerator of $\left(\frac{1-3i-i^2-i^3}{2}\right)$ to the lowest possible value of i.

d. $\text{Im}\left(\frac{1-3i-i^2-i^3}{2}\right) = \text{Im}\left(\frac{1-3i+1+i}{2}\right)$

$= \text{Im}\left(\frac{2-2i}{2}\right)$

2. Simplify by dividing the numerator by 2.

$= \text{Im}\frac{(\not{2}(1-i))}{\not{2}}$

$= \text{Im}(1-i)$

3. The imaginary part is -1.

$= -1$

TI | THINK	DISPLAY/WRITE	CASIO | THINK	DISPLAY/WRITE
c. On a Calculator page, press MENU, then select: 2 Number 9 Complex Number Tools 2 Real Part Complete the entry line as shown. *Note:* The symbol i can be found by pressing the button.	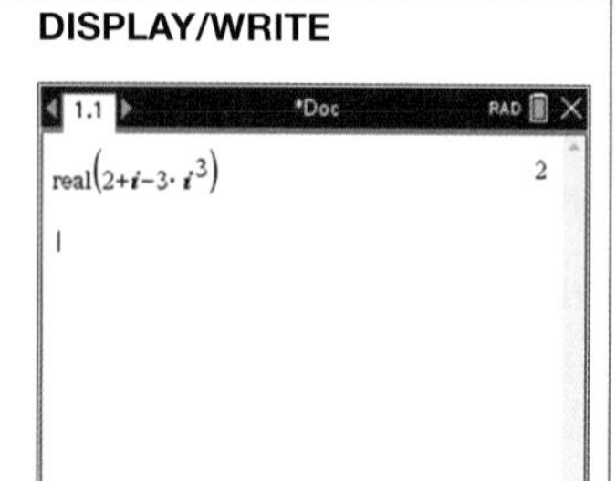	**c.** On a Main screen, select: Action Complex re Complete the entry as shown.	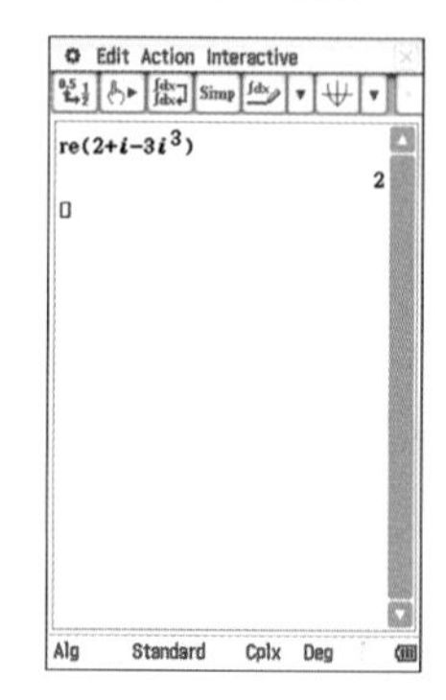
d. On a Calculator page, press MENU, then select: 2 Number 9 Complex Number Tools 3 Imaginary Part Complete the entry line as shown.		**d.** On a Main screen, select: Action Complex im Complete the entry as shown.	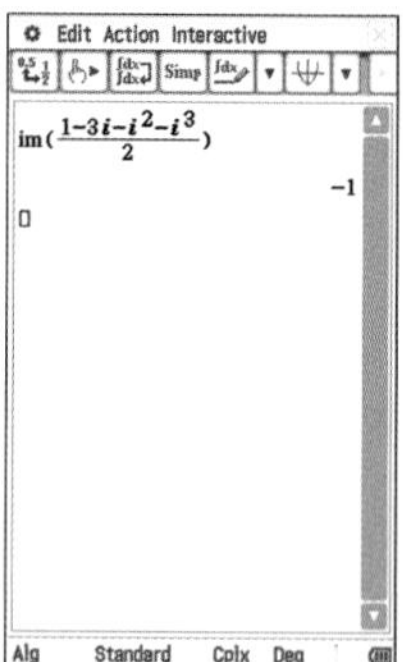

10.2 Exercise

Students, these questions are even better in jacPLUS

Receive immediate feedback and access sample responses

Access additional questions

Track your results and progress

Find all this and MORE in jacPLUS

Technology free

1. **WE1** Using the imaginary number i, write down expressions for:

a. $\sqrt{-9}$ **b.** $\sqrt{-25}$ **c.** $\sqrt{-49}$ **d.** $\sqrt{-3}$

2. Using the imaginary number i, write down expressions for:

a. $\sqrt{-11}$ **b.** $\sqrt{-7}$ **c.** $\sqrt{-\frac{4}{9}}$ **d.** $\sqrt{-\frac{36}{25}}$

3. **WE2** Write down the real and imaginary parts, respectively, of the following complex numbers, z.

a. $9+5i$
b. $5-4i$
c. $-3-8i$
d. $11i-6$

4. Write down the real and imaginary parts, respectively, of the following complex numbers, z.

a. 27
b. $2i$
c. $-5+i$
d. $-17i$

5. **WE3** Write each of the following in the form $a+bi$, where a and b are real numbers.

a. i^9+i^{10}
b. i^9-i^{10}
c. $i^{12}+i^{15}$
d. i^7-i^{11}

6. Write each of the following in the form $a+bi$, where a and b are real numbers.

a. $i^5+i^6-i^7$
b. $i\left(i^{13}+i^{16}\right)$
c. $2i-i^2+2i^3$
d. $3i+i^4-5i^5$

7. **WE4** Simplify $z=i^6+3i^7-2i^{10}-3$ and $w=4i^8-3i^{11}+3$ and show that $z+w=5$.

8. **WE5** Evaluate each of the following.

a. $\text{Re}(-5+4i)$
b. $\text{Re}(15-8i)$
c. $\text{Re}(12i)$
d. $\text{Im}(1-6i)$

9. Evaluate each of the following.

a. $\text{Im}(3+2i)$
b. $\text{Im}(8)$
c. $\text{Re}(i^5-3i^4+6i^6)$
d. $\text{Im}\left(\dfrac{4i^9-5i^{14}-2i^7}{3}\right)$

10. Write $3-\dfrac{i^3-i+2}{i^2-i^4}$ in the form $a+bi$, where a and b are real numbers.

11. Simplify each of the following.

a. $\dfrac{6i^3}{\sqrt{-9}}$
b. $\dfrac{20i^4}{\sqrt{-100}}$
c. $\dfrac{10i^5}{\sqrt{-50}}$
d. $\dfrac{8i^6}{\sqrt{-16}}$

12. Evaluate the following.

a. $\text{Re}(3(4-6i)+i^8)$
b. $\text{Im}(2(2-5i)+i^7)$
c. $\text{Re}(3(2-5i)-4i^{10})$
d. $\text{Im}(4(2+3i)-5i^9)$

13. If $f(n)=1+i+i^2+i^3+i^4+...+i^n$, evaluate the following.

a. $f(6)$
b. $f(7)$
c. $f(2014)$
d. $f(2015)$

14. If $g(n)=1-i+i^2-i^3+i^4+...+(-i)^n$, evaluate the following.

a. $g(6)$
b. $g(7)$
c. $g(2014)$
d. $g(2015)$

10.2 Exam questions

Question 1 (2 marks) TECH-FREE

a. Simplify each of the following. $\sqrt{-4}$, $\sqrt{-9}$, $\sqrt{-4}\times\sqrt{-9}$ and $\sqrt{-4\times-9}$. **(1 mark)**

b. Consider the following: $-1=i^2=i\times i=\sqrt{-1}\times\sqrt{-1}=\sqrt{-1\times-1}=\sqrt{1}=1$
Determine where the logic breaks down. **(1 mark)**

Question 2 (3 marks) TECH-FREE

a. If n is a natural number evaluate i^{4n}. **(1 mark)**

b. If n is a natural number evaluate i^{4n+3}. **(1 mark)**

c. If n is an even natural number show that $(-1)^{\frac{n}{2}} = i^n$. **(1 mark)**

Question 3 (2 marks) TECH-FREE

Let $f(n) = i + 2i^2 + 3i^3 + \ldots + ni^n$. Determine:

a. $\text{Re}(f(10))$ **(1 mark)**

b. $\text{Im}(f(11))$ **(1 mark)**

More exam questions are available online.

10.3 Basic operations on complex numbers

LEARNING INTENTION

At the end of this subtopic you should be able to:

- add and subtract complex numbers
- multiply complex numbers by a scalar
- multiply complex numbers in Cartesian form.

10.3.1 Complex number arithmetic

Complex numbers can be added, subtracted, multiplied and divided. In general, the solutions obtained when performing these operations are presented in the standard form $z = a + bi$.

Addition of complex numbers

Addition is performed by adding the real and imaginary parts separately.

You can think of the real parts of the numbers being like terms and the imaginary parts of the numbers being like terms.

Addition of complex numbers

If $z = m + ni$ and $w = p + qi$, then:

$$z + w = (m + p) + (n + q)i$$

Subtraction of complex numbers

If we write $z - w$ as $z + -w$, we can use the rule for addition of complex numbers to obtain:

$$\begin{aligned} z + -w &= (m + ni) + -(p + qi) \\ &= m + ni - p - qi \\ &= (m - p) + (n - q)i \end{aligned}$$

Subtraction of complex numbers

If $z = m + ni$ and $w = p + qi$, then:

$$z - w = (m - p) + (n - q)i$$

WORKED EXAMPLE 6 Addition and subtraction of complex numbers

For $z = 8 + 7i$, $w = -12 + 5i$ and $u = 1 + 2i$, calculate the following.

a. $z + w$ b. $w - z$ c. $u - w + z$

THINK	WRITE
a. Use the addition rule for complex numbers.	a. $z + w = (8 + 7i) + (-12 + 5i)$ $= (8 - 12) + (7 + 5)i$ $= -4 + 12i$
b. Use the subtraction rule for complex numbers.	b. $w - z = (-12 + 5i) - (8 + 7i)$ $= (-12 - 8) + (5 - 7)i$ $= -20 - 2i$
c. Use both the addition rule and the subtraction rule.	c. $u - w + z = (1 + 2i) - (-12 + 5i) + (8 + 7i)$ $= (1 + 12 + 8) + (2 - 5 + 7)i$ $= 21 + 4i$

Multiplication by a constant (or scalar)

Consider part a in Worked example 6, where $z + w = -4 + 12i$. The real and imaginary parts share a common factor of 4. Hence, the equation could be rewritten as:

$$\begin{aligned} z + w &= -4 + 12i \\ &= 4(-1 + 3i) \end{aligned}$$

This is the equivalent of the complex number $-1 + 3i$ multiplied by a constant (or scalar), 4.

If we consider the general form $z = a + bi$ and $k \in R$, then:

Multiplying a complex number by a constant

For a complex number $z = a + bi$ multiplied by a constant $k \in R$:

$$\begin{aligned} kz &= k(a + bi) \\ &= ka + kbi \end{aligned}$$

WORKED EXAMPLE 7 Scalar multiplication of complex numbers

If $z = 3 + 5i$, $w = 4 - 2i$ and $v = 6 + 10i$, evaluate the following.

a. $3z + w$ b. $2z - v$ c. $4z - 3w + 2v$

THINK	WRITE
a. 1. Calculate $3z + w$ by substituting values for z and w.	a. $3z + w = 3(3 + 5i) + (4 - 2i)$ $= (9 + 15i) + (4 - 2i)$
2. Use the rule for adding complex numbers.	$= (9 + 4) + (15 - 2)i$ $= 13 + 13i$ (or $13(1 + i)$)
b. 1. Calculate $2z - v$ by substituting values for z and v.	b. $2z - v = 2(3 + 5i) - (6 + 10i)$
2. Use the rule for subtraction of complex numbers.	$= 6 + 10i - 6 - 10i$ $= 0 + 0i$ $= 0$

c. 1. Calculate $4z - 3w + 2v$ by substituting values for z, w and v.

2. Use the addition rule and the subtraction rule to simplify.

c. $4z - 3w + 2v = 4(3 + 5i) - 3(4 - 2i) + 2(6 + 10i)$
$= 12 + 20i - 12 + 6i + 12 + 20i$
$= 12 + 46i$

TI | THINK

c. On a Calculator page, define z, w and v as shown.
Then complete the entry line as $4z - 3w + 2v$.

DISPLAY/WRITE

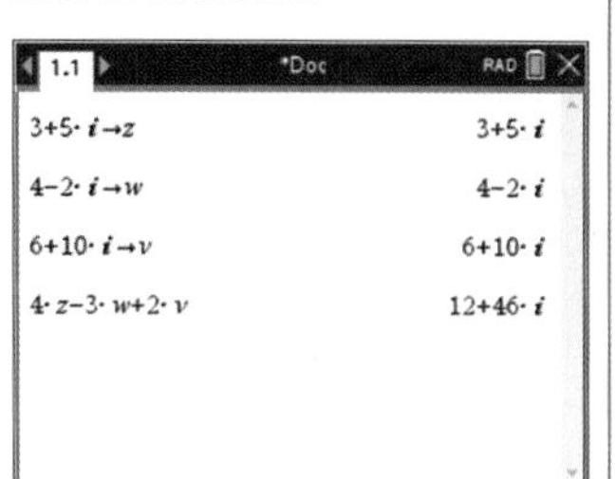

CASIO | THINK

c. On a Main screen, define z, w and v as shown.
Then complete the entry line as $4z - 3w + 2v$.

DISPLAY/WRITE

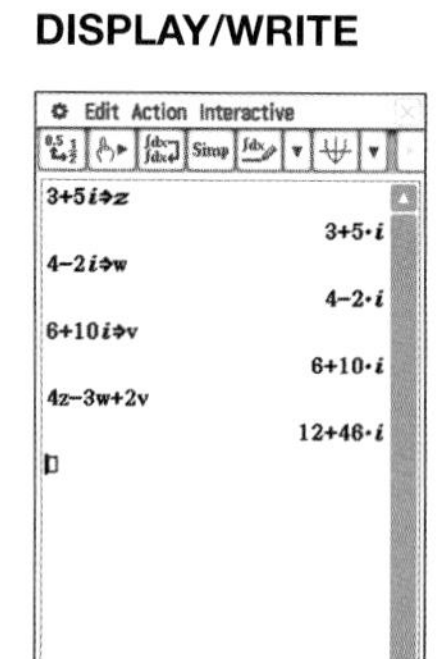

Multiplication of two complex numbers

Multiplication of two complex numbers, can also result in a complex number.

To multiply complex numbers, simply treat them like algebraic expressions and expand the brackets.

$$\text{If } z = m + ni \text{ and } w = p + qi,$$
$$\begin{aligned} \text{then } z \times w &= (m + ni)(p + qi) \\ &= mp + mqi + npi + nqi^2 \\ &= (mp - nq) + (mq + np)i \text{ (since } i^2 = -1). \end{aligned}$$

Multiplication of complex numbers

If $z = m + ni$ and $w = p + qi$, then

$$\mathbf{z \times w = (mp - nq) + (mq + np)i.}$$

WORKED EXAMPLE 8 Multiplication of complex numbers (1)

If $z = 6 - 2i$ and $w = 3 + 4i$, express zw in standard form.

THINK

1. Expand the brackets.

2. Express in the form $a + bi$ by substituting -1 for i^2 and simplifying the expression using the addition and subtraction rules.

WRITE

$zw = (6 - 2i)(3 + 4i)$
$= 18 + 24i - 6i - 8i^2$

$= 18 + 24i - 6i + 8$
$= 26 + 18i$

WORKED EXAMPLE 9 Multiplication of complex numbers (2)

Simplify $(2-3i)(2+3i)$.

THINK	WRITE
1. Expand the brackets.	$(2-3i)(2+3i)=4+6i-6i-9i^2$
2. Substitute -1 for i^2 and simplify the expression.	$=4-9\times(-1)$ $=13$

WORKED EXAMPLE 10 Determining real and imaginary components of complex products

Determine $\text{Re}\left(z^2w\right)+\text{Im}\left(zw^2\right)$ for $z=4+i$ and $w=3-i$.

THINK	WRITE
1. Express z^2w in the form $a+bi$.	$z^2w=(4+i)^2(3-i)$ $=(16+8i+i^2)(3-i)$ $=(16+8i-1)(3-i)$ $=(15+8i)(3-i)$ $=45-15i+24i-8i^2$ $=53+9i$
2. The real part, $\text{Re}\left(z^2w\right)$, is 53.	$\text{Re}\left(z^2w\right)=53$
3. Express zw^2 in the form $a+bi$.	$zw^2=(4+i)(3-i)^2$ $=(4+i)(9-6i+i^2)$ $=(4+i)(8-6i)$ $=32-24i+8i-6i^2$ $=38-16i$
4. The imaginary part, $\text{Im}\left(zw^2\right)$, is -16.	$\text{Im}\left(zw^2\right)=-16$
5. Calculate the value of $\text{Re}\left(z^2w\right)+\text{Im}\left(zw^2\right)$.	$\text{Re}\left(z^2w\right)+\text{Im}\left(zw^2\right)=53-16$ $=37$

10.3.2 Equality of two complex numbers

Two complex numbers are equal if their real parts are equal and their imaginary parts are also equal. That is, if $z=m+ni$ and $w=p+qi$, then $z=w$ if and only if $m=p$ and $n=q$.

The condition 'if and only if (sometimes written in short form as **iff**) means that both of the following situations must apply.

- If $z=w$, then $m=p$ and $n=q$.
- If $m=p$, and $n=q$ then $z=w$.

WORKED EXAMPLE 11 Solving equations using equality of complex numbers

Determine the values of x and y that satisfy $(3+4i)(x+yi)=29+22i$.

THINK	WRITE
1. Write the left-hand side of the equation.	$\text{LHS}=(3+4i)(x+yi)$
2. Expand the left-hand side of the equation.	$=3x+3yi+4xi+4yi^2$
3. Express the left-hand side in the form $a+bi$.	$=(3x-4y)+(4x+3y)i$
4. Equate the real parts and imaginary parts of both sides of the equation to create a pair of simultaneous equations.	$3x-4y=29$ [1] $4x+3y=22$ [2]
5. Simultaneously solve [1] and [2] for x and y.	$9x-12y=87$ [3]
Multiply equation [1] by 3 and equation [2] by 4 so that y can be eliminated.	$16x-12y=88$ [4]
6. Add the two new equations and solve for x.	Adding equations [3] and [4]: $25x=175$ $x=7$
7. Substitute $x=7$ into equation [1] and solve for y.	Substituting $x=7$ into equation [1]: $3(7)-4y=29$ $21-4y=29$ $-4y=8$ $y=-2$
8. State the solution.	Therefore, $x=7$ and $y=-2$.
9. Check the solution by substituting these values into equation [2].	Check: $4\times7+3\times-2=22$.

10.3 Exercise

Technology free

1. **WE6** For $z=5+3i$, $w=-1-4i$, $u=6-11i$ and $v=2i-3$, calculate the following.

   a. $z+w$ b. $u-z$ c. $w+v$

2. For $z=5+3i$, $w=-1-4i$, $u=6-11i$ and $v=2i-3$, calculate the following.

   a. $u-v$ b. $w-z-u$ c. $v+w-z$

3. **WE7** If $z=-3+2i$, $w=-4+i$ and $u=-8-5i$, evaluate the following.

   a. $3w$ b. $2u+z$ c. $4z-3u$

4. If $z=-3+2i$, $w=-4+i$ and $u=-8-5i$, evaluate the following.

   a. $3z+u+2w$ b. $2z-7w+9u$ c. $3(z+2u)-4w$

5. **WE8** For $z = 5 + 3i$, $w = -1 - 4i$, $u = 6 - 11i$ and $v = 2i - 3$, calculate each of the following in the standard form $a + bi$.

a. zw
b. uv
c. wu

6. For $z = 5 + 3i$, $w = -1 - 4i$, $u = 6 - 11i$ and $v = 2i - 3$, calculate each of the following in the standard form $a + bi$.

a. zu
b. u^2
c. $u(wv)$

7. **WE9** Simplify the following.

a. $(10 + 7i)(9 - 3i)$
b. $(3 - 4i)(5 + 4i)$
c. $(8 - 2i)(4 - 5i)$

8. Simplify the following.

a. $(5 + 6i)(5 - 6i)$
b. $(2i - 7)(2i + 7)$
c. $(9 - 7i)^2$

9. For $z = -1 - 3i$ and $w = 2 - 5i$, calculate $z^2 w$.

10. **WE10** Determine $\text{Re}\left(z^2\right) - \text{Im}(zw)$ for $z = 1 + i$ and $w = 4 - i$.

11. For $z = 3 + 5i$, $w = 2 - 3i$ and $u = 1 - 4i$, determine:

a. $\text{Im}\left(u^2\right)$
b. $\text{Re}\left(w^2\right)$
c. $\text{Re}(uw) + \text{Im}(zw)$

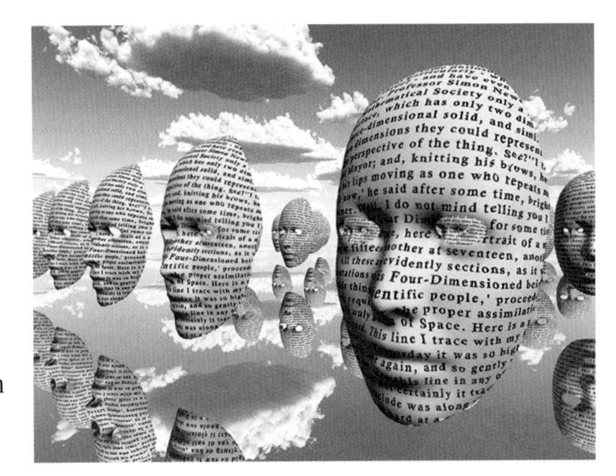

12. For $z = 3 + 5i$, $w = 2 - 3i$ and $u = 1 - 4i$, determine:

a. $\text{Re}(zu) - \text{Im}\left(w^2\right)$
b. $\text{Re}\left(z^2\right) - \text{Re}(zw) - \text{Im}(uz)$
c. $\text{Re}\left(u^2 w\right) + \text{Im}\left(zw^2\right)$

13. **WE11** Determine the values of a and b that satisfy each of the following.

a. $(2 + 3i)(a + bi) = 16 + 11i$
b. $(5 - 4i)(a + bi) = 1 - 4i$
c. $(3i - 8)(a + bi) = -23 - 37i$
d. $(7 + 6i)(a + bi) = 4 - 33i$

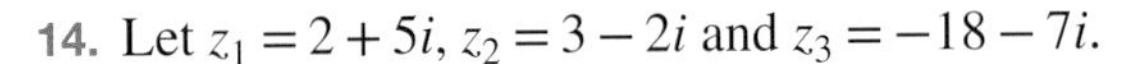

14. Let $z_1 = 2 + 5i$, $z_2 = 3 - 2i$ and $z_3 = -18 - 7i$.

a. Verify the associative law, that is, $z_1 + (z_2 + z_3) = (z_1 + z_2) + z_3$.
b. Evaluate $5z_1 - 2z_2 + 3z_3$.
c. Evaluate $\text{Re}(2z_1 - 3z_2 + 4z_3) + \text{Im}(2z_1 - 3z_2 + 4z_3)$.
d. Evaluate the real numbers x and y if $xz_1 + yz_2 = z_3$.

15. Evaluate the values of the real numbers x and y if:

a. $x(3 - 2i) + y(4 + 5i) = 23$
b. $x(3 - 2i) - y(4 - 7i) = 13i$
c. $x(4 - 3i) + y(3 - 4i) = 14 - 7i$
d. $x(3 - 5i) + y(4 + 7i) + 7 + 43i = 0$

10.3 Exam questions

Question 1 (5 marks) TECH-FREE

Let $u = 3 - 4i$ and $v = 4 + 5i$. Evaluate each of the following.

a. $u + v$ **(1 mark)**

b. $2u - 3v$ **(1 mark)**

c. uv **(1 mark)**

d. $\text{Re}\left(u^2 - 2uv + v^2\right)$ **(1 mark)**

e. $\text{Im}\left(u^2 + 2uv + v^2\right)$ **(1 mark)**

Question 2 (4 marks) TECH-FREE

Let $u = 3 - 4i$ and $v = 4 + 5i$. Determine the values of the real numbers x and y if:

a. $xu + yv = 17 - 2i$ **(2 marks)**

b. $xu^2 + yv^2 + 5 + 88i = 0$ **(2 marks)**

Question 3 (4 marks) TECH-FREE

Let $u = 2 - 3i$ and $v = 3 + 4i$. Determine the values of the real numbers x and y if:

a. $(x + yi)\,u = v$ **(2 marks)**

b. $(x + yi)\,v = u$ **(2 marks)**

More exam questions are available online.

10.4 Complex conjugates and division of complex numbers

LEARNING INTENTION

At the end of this subtopic you should be able to:

- determine the conjugate of a complex number
- divide complex numbers.

10.4.1 The conjugate of a complex number

The **conjugate of a complex number** is obtained by changing the sign of the imaginary component.

If $z = a + bi$, the conjugate $\bar{z}$ of z is defined as $\bar{z} = a - bi$.

Conjugates are useful because the multiplication (or addition) of a complex number and its conjugate results in a real number.

The conjugate of a complex number

For a complex number $z = a + bi$, the conjugate is:

$$\bar{z} = a - bi$$

Multiplication of a complex number and its conjugate results in a real number:

$$\begin{aligned} z\bar{z} &= (a + bi)(a - bi) \\ &= a^2 - abi + abi - b^2i^2 \\ &= a^2 + b^2 \end{aligned}$$

You will use this result when dividing complex numbers.

Note: Compare this expression with the formula for the difference of two squares:

$$(m-n)(m+n)=m^2-n^2$$

Addition of a complex number and its conjugate results in two times the real component of the complex number:

$$\begin{aligned} z+\bar{z} &= a+bi+a-bi \\ &= 2a \end{aligned}$$

WORKED EXAMPLE 12 Determining complex conjugates

Write the conjugate of each of the following complex numbers.

a. $\mathbf{8+5i}$ b. $\mathbf{-2-3i}$ c. $\mathbf{4+i\sqrt{5}}$

THINK	WRITE
a. Change the sign of the imaginary component.	a. $8-5i$
b. Change the sign of the imaginary component.	b. $-2+3i$
c. Change the sign of the imaginary component.	c. $4-i\sqrt{5}$

TI | THINK

a. On a Calculator page, complete the entry line as conj$(8+5i)$.

DISPLAY/WRITE

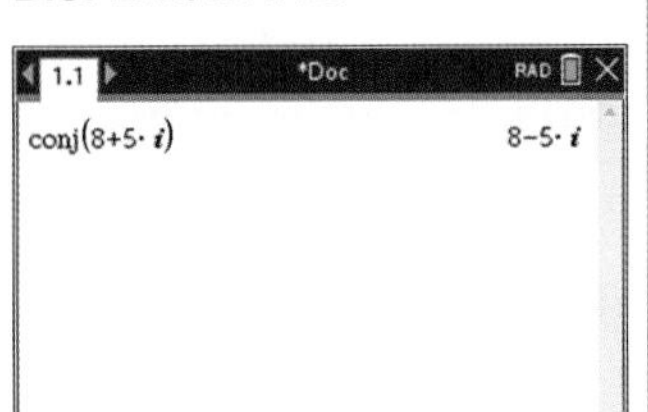

CASIO | THINK

a. On a Main screen, complete the entry line as conj$(8+5i)$.

DISPLAY/WRITE

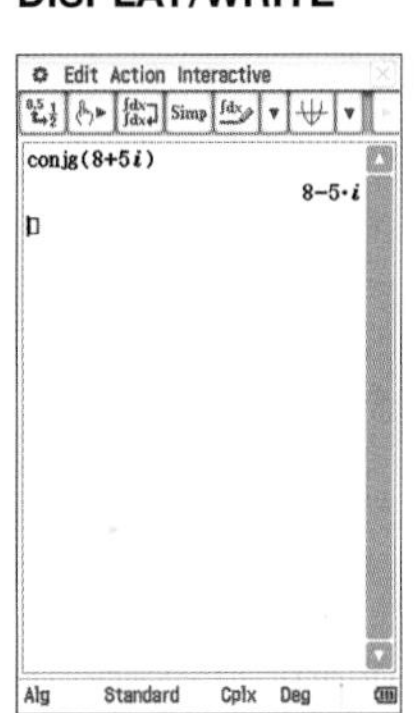

WORKED EXAMPLE 13 Verifying the sum of conjugates is equal to the conjugate of a sum

If $z=5-2i$ and $w=7-i$, show that $\overline{z+w}=\bar{z}+\bar{w}$.

THINK	WRITE
1. Add the conjugates $\bar{z}$ and $\bar{w}$.	$\bar{z}+\bar{w}=(5+2i)+(7+i)=12+3i$
2. Add z to w.	$z+w=(5-2i)+(7-i)=12-3i$
3. Write down the conjugate of $z+w$.	$\overline{z+w}=12+3i$
4. The conjugate of $z+w$ equals the sum of the individual conjugates.	$\overline{z+w}=\bar{z}+\bar{w}$

10.4.2 Division of complex numbers

The application of conjugates to division of two complex numbers will now be investigated. Consider the complex numbers $z=a+bi$ and $w=c+di$. To calculate $\frac{z}{w}$ in the form $x+yi$ we must multiply both the numerator and denominator by the conjugate of w to make the denominator a real number only.

$$\frac{z}{w} = \frac{a+bi}{c+di}$$

$$= \frac{a+bi}{c+di} \times \frac{c-di}{c-di}$$ Multiply by the conjugate of $c+di$.

$$= \frac{(ac+bd)+(bc-ad)\,i}{c^2+d^2}$$ Simplify the expressions in the numerator and in the denominator.

$$= \frac{ac+bd}{c^2+d^2} + \frac{(bc-ad)\,i}{c^2+d^2}$$ Express in the form $x+yi$.

Dividing complex numbers

To express the quotient of complex numbers $\frac{z}{w}$ in the form $a+bi$ multiply the numerator and denominator by the conjugate of the denominator.

$$\frac{z}{w} = \frac{z}{w} \times \frac{\overline{w}}{\overline{w}}$$

WORKED EXAMPLE 14 Division of complex numbers

Express $\frac{2+i}{2-i}$ in the form $a+bi$.

THINK	WRITE
1. Multiply both the numerator and denominator by the conjugate of $2-i$ to make the denominator real.	$\frac{2+i}{2-i} = \frac{(2+i)}{(2-i)} \times \frac{(2+i)}{(2+i)}$
2. Expand the expressions obtained in the numerator and denominator.	$= \frac{4+4i+i^2}{4-i^2}$
3. Substitute -1 for i^2 and simplify the expression.	$= \frac{4+4i-1}{4+1}$ $= \frac{3+4i}{5}$ $= \frac{3}{5} + \frac{4i}{5}$

TI | THINK	DISPLAY/WRITE	CASIO | THINK	DISPLAY/WRITE
On a Calculator page, complete the entry line as $\frac{2+i}{2-i}$.		On a Main screen, complete the entry line as $\frac{2+i}{2-i}$.	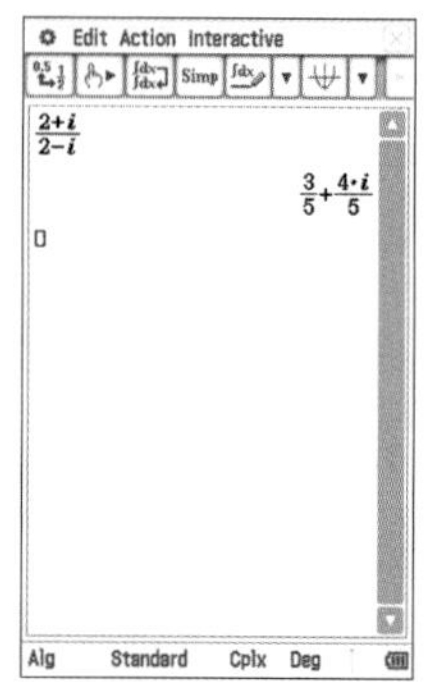

10.4.3 Multiplicative inverse of a complex number

Given a non-zero complex number z, there exists a complex number w such that $zw = 1$, with w being the **multiplicative inverse** of z denoted by $w = z^{-1} = \frac{1}{z}$. Hence, $zz^{-1} = 1$.

WORKED EXAMPLE 15 Determining the multiplicative inverse of a complex number

If $z = 3 + 4i$, determine z^{-1}.

THINK	WRITE
1. Write z^{-1} as a rational expression: $z^{-1} = \frac{1}{z}$.	$z^{-1} = \frac{1}{z} = \frac{1}{3+4i}$
2. Multiply both the numerator and denominator by the conjugate of $3 + 4i$.	$= \frac{1}{(3+4i)} \times \frac{(3-4i)}{(3-4i)}$ $= \frac{3-4i}{25}$
3. Write the expression in the form $a + bi$.	$= \frac{3}{25} - \frac{4i}{25}$

In general terms, the multiplicative inverse of a complex number can be written in standard (Cartesian) form, $a + bi$.

Multiplicative inverse of a complex number

If $z = a + bi$, then:

$$z^{-1} = \frac{a - bi}{a^2 + b^2}$$

WORKED EXAMPLE 16 Simplifying complex quotients

If $z = 3 + i$ and $w = \frac{2}{4-i}$, determine $\text{Im}(4z - w)$.

THINK	WRITE
1. Substitute for z and w in $4z - w$.	$4z - w = 4(3+i) - \frac{2}{4-i}$
2. Express $4z - w$ with a common denominator.	$= \frac{4(3+i)(4-i) - 2}{4-i}$ $= \frac{4(13+i) - 2}{4-i}$ $= \frac{50+4i}{4-i}$
3. Remove i from the denominator by multiplying the numerator and denominator by the conjugate of $4 - i$.	$= \frac{(50+4i)}{(4-i)} \times \frac{(4+i)}{(4+i)}$ $= \frac{196+66i}{17}$

4. Simplify the expression so that it is in the form $a+bi$.	$=\frac{196}{17}+\frac{66i}{17}$
5. State the imaginary component of $4z-w$.	$\text{Im}(4z-w)=\frac{66}{17}$

WORKED EXAMPLE 17 Proofs with complex numbers

Prove that $\overline{z_1 z_2}=\bar{z}_1\bar{z}_2$.

THINK	WRITE
1. When asked to 'prove, you should not use actual values for the pronumerals. Use general values of $z_1, z_2, \bar{z}_1$ and $\bar{z}_2$.	Let $z_1=a+bi$. $\bar{z}_1=a-bi$. Let $z_2=c+di$. $\bar{z}_2=c-di$.
2. Generally, in a proof, do not work both sides of the equation at once. Calculate the LHS first.	$\text{LHS}=\overline{(a+bi)\times(c+di)}$ $=\overline{ac+adi+bci+bdi^2}$ $=\overline{(ac-bd)+(ad+bc)i}$ $=(ac-bd)-(ad+bc)i$
3. Calculate the RHS and show that it equals the LHS.	$\text{RHS}=(a-bi)(c-di)$ $=ac-adi-bci+bdi^2$ $=(ac-bd)-(ad+bc)i$ $=\text{LHS}$ Hence, $\overline{z_1 z_2}=\bar{z}_1\bar{z}_2$.

10.4 Exercise

Students, these questions are even better in jacPLUS

Receive immediate feedback and access sample responses

Access additional questions

Track your results and progress

Find all this and MORE in jacPLUS

Technology free

1. WE12 Write down the conjugate of each of the following complex numbers.

a. $7+10i$ b. $5-9i$ c. $3+12i$ d. $\sqrt{7}-3i$ e. $2i+5$ f. $-6-\sqrt{11}i$

2. WE13 If $z=6+3i$ and $w=3-4i$, show that $\overline{z-w}=\bar{z}-\bar{w}$.

3. **WE14** Express $\dfrac{2+i}{3-i}$ in the form $a+bi$.

4. Express each of the following in the form $a+bi$.

a. $\dfrac{1-i}{1+i}$ b. $\dfrac{3-2i}{2+3i}$ c. $\dfrac{2+5i}{4-3i}$

5. Express each of the following in the form $a+bi$.

a. $\dfrac{4-3i}{5+2i}$ b. $\dfrac{4-5i}{2-7i}$ c. $\dfrac{2+\sqrt{3}i}{\sqrt{5}-\sqrt{2}i}$

6. **WE15** Determine z^{-1} if z is equal to:

a. $2-i$ b. $3+i$ c. $4-3i$

7. Determine z^{-1} if z is equal to:

a. $5+4i$ b. $2i-3$ c. $\sqrt{3}-i\sqrt{2}$

8. If $676z = 10-24i$, express z^{-1} in the form $a+bi$.

9. **WE16** If $z = 2-i$ and $w = \dfrac{1}{3+i}$, determine each of the following.

a. $\text{Re}(z+w)$ b. $\text{Im}(w-z)$ c. $\text{Re}(z^{-1}+w^{-1})$ d. $\text{Im}(3z+2w)$ e. $\text{Re}(4w-2z)$

10. Write $\dfrac{2+i}{1+i}+\dfrac{9-2i}{2-i}+\dfrac{7+i}{1-i}$ in the form $a+bi$.

11. Simplify $\dfrac{(2+5i)^2(5i-2)}{3(4+7i)-2(5+8i)}$.

12. Determine the conjugate of $(5-6i)(3-8i)$.

13. Let $z = 6+8i$ and $w = 10-3i$.
Show that $\overline{zw} = \overline{z}\times\overline{w}$.

14. Let $z = 4+i$ and $w = 1+3i$.
Show that $\overline{\left(\dfrac{z}{w}\right)} = \dfrac{\overline{z}}{\overline{w}}$.

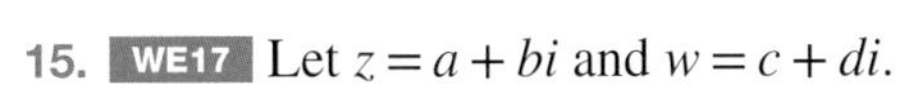

15. **WE17** Let $z = a+bi$ and $w = c+di$.

a. Show that $\overline{zw} = \overline{z}\times\overline{w}$.

b. Show that $\overline{\left(\dfrac{z}{w}\right)} = \dfrac{\overline{z}}{\overline{w}}$.

16. If $z = -5-4i$ and $w = 2i$, calculate $\text{Re}(z\overline{w}+\overline{z}w)$.

17. If $z_1 = 2+3i$, $z_2 = -4-i$ and $z_3 = 5-i$, calculate:

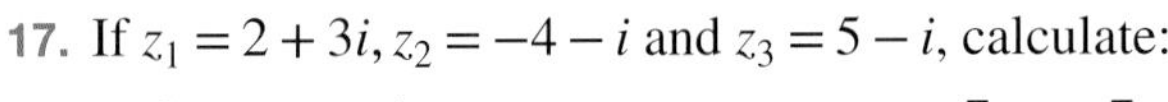

a. $2z_1 - z_2 - 4z_3$ b. $z_1\overline{z}_2 + z_2\overline{z}_3$ c. $\overline{z_1 z_2 z_3} - \overline{z_1} z_2 z_3$

18. If $z_1 = a+bi$ and $z_2 = c+di$, show that $(z_1z_2)^{-1} = {z_1}^{-1}{z_2}^{-1}$.

19. a. If $z = 1+i$, evaluate z^4, z^8 and z^{12}.
b. Deduce from your results in **a** that $z^{4n} = (2i)^{2n},\ n \in N$.

20. If $z = a + bi$, determine the values of a and b such that $\frac{z-1}{z+1} = z + 2$.

21. Determine values for a and b so that $z = a + bi$ satisfies $\frac{z+i}{z+2} = i$.

22. If $z = 2 - 3i$ and $w = 1 - 2i$:

a. calculate:

i. $z\bar{z}$

ii. $w\bar{w}$

b. show that:

i. $\overline{z+w} = \bar{z} + \bar{w}$

ii. $\overline{zw} = \bar{z} \times \bar{w}$

iii. $\overline{\left(\frac{z}{w}\right)} = \frac{\bar{z}}{\bar{w}}$

c. calculate:

i. $\overline{\left(\frac{1}{z}\right)}$

ii. $\overline{\left(\frac{1}{w}\right)}$

d. calculate $\overline{z^2 + w^2}$

e. calculate $\overline{z + zw}$

f. calculate $z^{-1}w^{-1}$.

10.4 Exam questions

Question 1 (5 marks) TECH-FREE

Let $u = 3 - 4i$ and $v = 4 + 5i$, evaluate each of the following.

a. $\frac{1}{u}$ **(1 mark)**

b. $\frac{1}{v}$ **(1 mark)**

c. $\frac{u}{v}$ **(1 mark)**

d. $\frac{v}{u}$ **(1 mark)**

e. $\frac{1}{uv}$ **(1 mark)**

Question 2 (2 marks) TECH-FREE

Let $u = a + bi$ and $v = c + di$. Simplify $(u + v)^{-1}$.

Question 3 (2 marks) TECH-FREE

Determine the value of the complex number z in Cartesian form if $\frac{z-i}{z+i} = -2i$.

More exam questions are available online.

10.5 The complex plane (the Argand plane)

LEARNING INTENTION

At the end of this subtopic you should be able to:
- plot complex numbers on the Argand plane
- demonstrate how addition and subtraction of complex numbers can be thought of as adding and subtracting vectors in two-dimensional vectors
- show the result of multiplication by i on a complex number in the Argand plane.

10.5.1 Plotting numbers in the complex plane

Thus far we have represented complex numbers in the standard form $z = a + bi$. This is referred to as the Cartesian form. We know from previous studies that an ordered pair of real numbers (x, y) can be represented on the Cartesian plane. Similarly, we can consider the complex number $z = a + bi$ as $z = x + yi$, consisting of the ordered pair of real numbers (x, y), which can be plotted as a point (x, y) on the complex number plane.

This is known as the **Argand plane** or an **Argand diagram** in recognition of the work done in this area by the Swiss mathematician Jean-Robert Argand.

The horizontal axis is referred to as the real axis and the vertical axis is referred to as the imaginary axis.

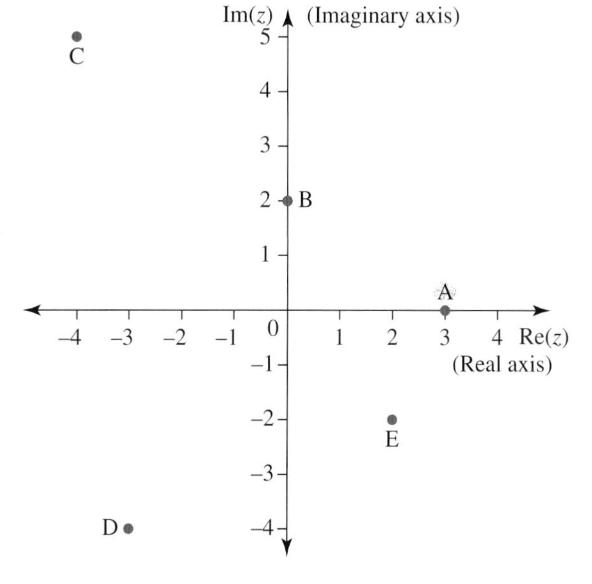

For example, the complex numbers $3 + 0i$ (A), $0 + 2i$ (B), $-4 + 5i$ (C), $-3 - 4i$ (D) and $2 - 2i$ (E) are shown on the Argand diagram.

Note: The imaginary axis is labelled 1, 2, 3 ... (not $i, 2i, 3i$ ...).

10.5.2 Geometrically multiplying a complex number by a scalar

WORKED EXAMPLE 18 Representing complex numbers on an Argand plane

Given the complex number $z = 1 + 2i$, represent the complex numbers z and $2z$ on the same Argand plane. Comment on their relative positions.

THINK	WRITE
1. Draw an Argand plane.	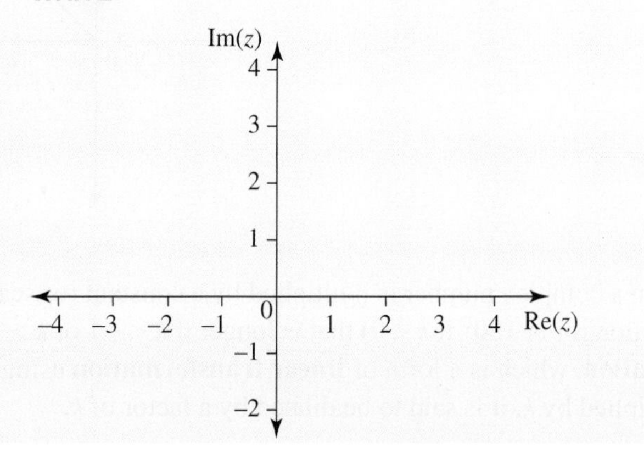

2. The complex number $z = 1 + 2i$ is of the form $z = x + yi$.
It can be plotted on the Argand plane at $\text{Re}(z) = 1$ and $\text{Im}(z) = 2$.

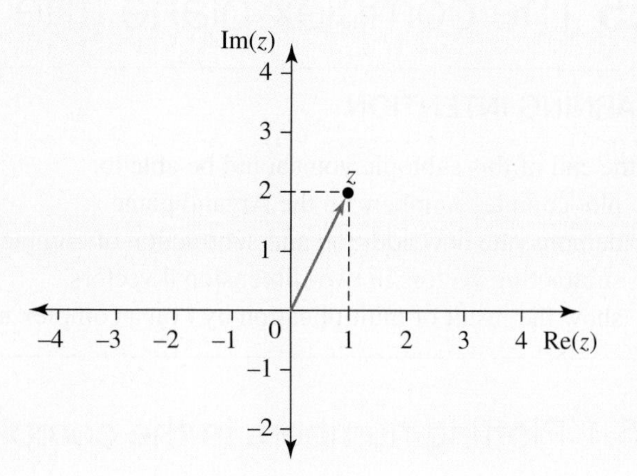

3. The complex number z has been multiplied by a scalar of 2. Calculate the complex number $2z$ from $z = 1 + 2i$.

$z = 1 + 2i$
$2z = 2 + 4i$

4. The complex number $2z = 2 + 4i$ is of the form $z = x + yi$. It can be plotted on the Argand plane at $\text{Re}(z) = 2$ and $\text{Im}(z) = 4$.
Note: The complex numbers are represented by the points only; the arrows and lines are only drawn only for convenience.

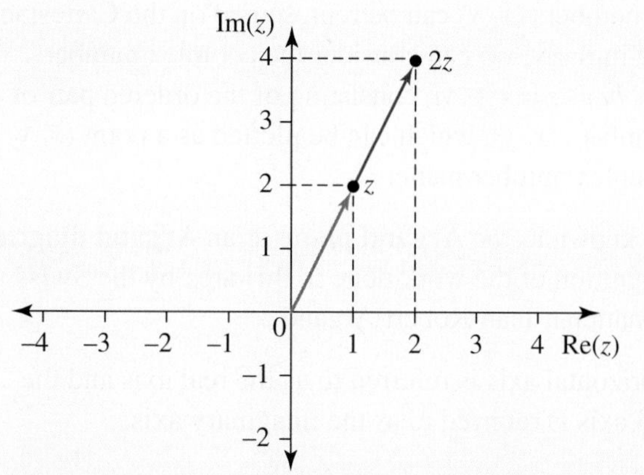

We can see that $2z$ is a complex number with twice the length of z.

If $z = x + yi$, then $kz = kx + kyi$, where $k \in R$. The following diagram shows the situation for $x > 0$, $y > 0$ and $k > 1$.

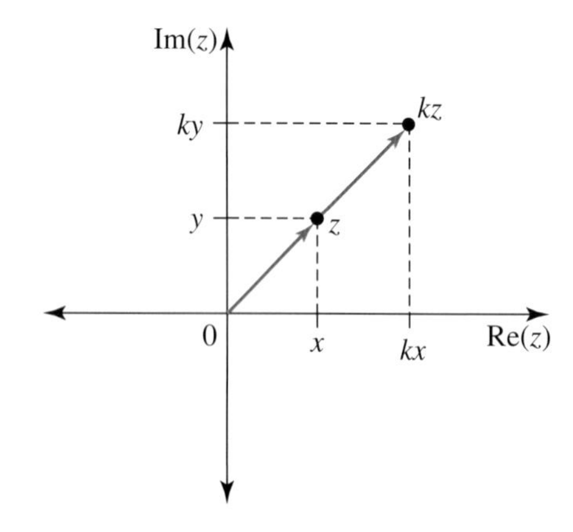

When a complex number is multiplied by a constant (or scalar), k, this produces a line segment in the same direction (or at 180° if $k < 0$) that is longer if $k < -1$ or $k > 1$, or shorter if $-1 < k < 0$ or $0 < k < 1$. This is called a **dilation**, which is a form of **linear transformation** using multiplication. When the complex number z is multiplied by k, it is said to be dilated by a factor of k.

10.5.3 Addition of complex numbers in the complex plane

If $z_1 = x_1 + y_1 i$ and $z_2 = x_2 + y_2 i$, then:

$$z_1 + z_2 = (x_1 + x_2) + (y_1 + y_2)\,i$$

If we now draw line segments from the origin (the point $0 + 0i$) to the points z_1 and z_2, then the addition of two complex numbers can be achieved using the same procedure as adding two vectors.

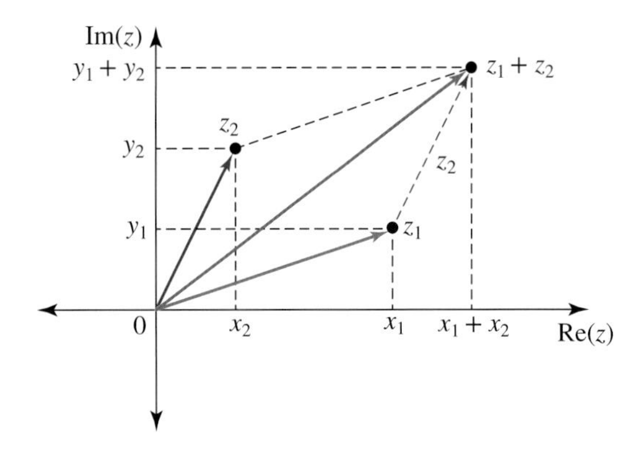

10.5.4 Subtraction of complex numbers in the complex plane

If $z_1 = x_1 + y_1 i$ and $z_2 = x_2 + y_2 i$, then:

$$\begin{aligned} z_1 - z_2 &= z_1 + (-z_2) \\ &= (x_1 + y_1 i) - (x_2 + y_2 i) \\ z_1 - z_2 &= (x_1 - x_2) + (y_1 - y_2)\,i \end{aligned}$$

The subtraction of two complex numbers can be achieved using the same procedure as subtracting two vectors.

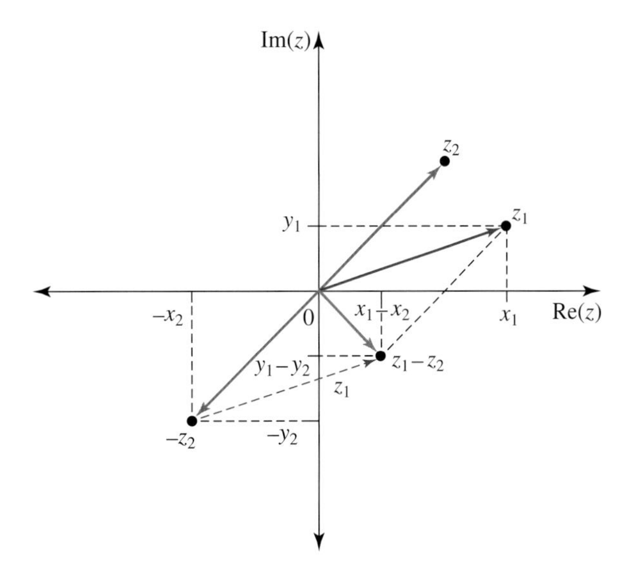

WORKED EXAMPLE 19 Representing sums and products on an Argand plane

Given the complex numbers $u = 1 + i$ and $v = 2 - 3i$, represent the following complex numbers on separate Argand planes.

a. $u + v$

b. $u - v$

THINK	WRITE
a. 1. The complex numbers $u = 1 + i$ and $v = 2 - 3i$ are of the form $z = x + yi$. They can be plotted on an Argand plane at: u: $\text{Re}(z) = 1$ and $\text{Im}(z) = 1$ v: $\text{Re}(z) = 2$ and $\text{Im}(z) = -3$	a. $u = 1 + i$ $v = 2 - 3i$ 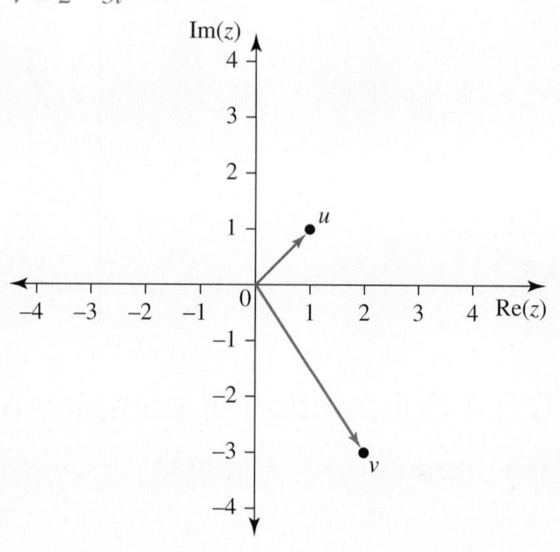
2. Evaluate the complex number $u + v$ and plot this complex number on the same Argand plane. $u + v$: $\text{Re}(z) = -1$ and $\text{Im}(z) = -2$	$u + v = 1 + i + 2 - 3i$ $= 3 - 2i$ 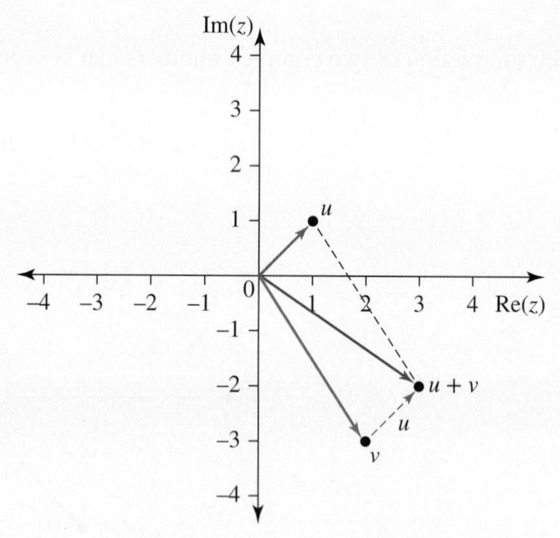

b. Plot the complex numbers u and v on an Argand plane.
Evaluate the complex number $u - v$ and plot this complex number on the same Argand plane.
$u - v$: $\text{Re}(z) = -1$ and $\text{Im}(z) = 4$

b.

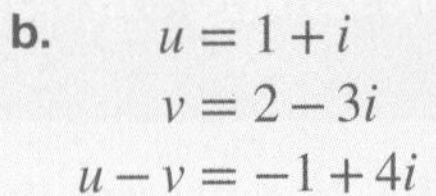

$u = 1 + i$
$v = 2 - 3i$
$u - v = -1 + 4i$

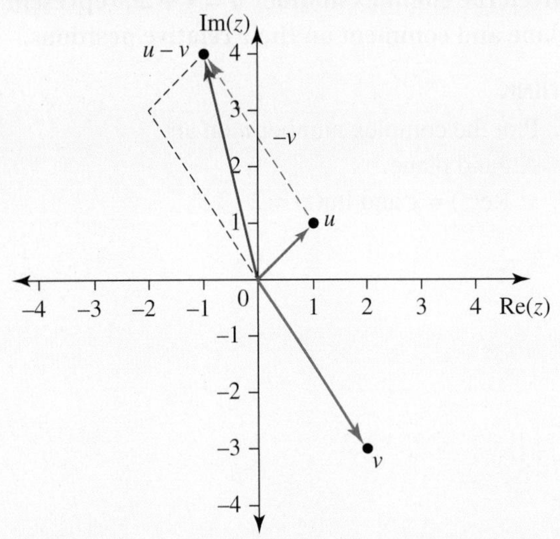

10.5.5 Geometrical representation of a conjugate of a complex number

If $z = x + yi$, then the conjugate of z is given by $\bar{z} = x - yi$.

It can be seen that $\bar{z}$ is the **reflection** of the complex number in the real axis. Reflections are another form of linear transformation, which will be studied in more detail in Topic 11.

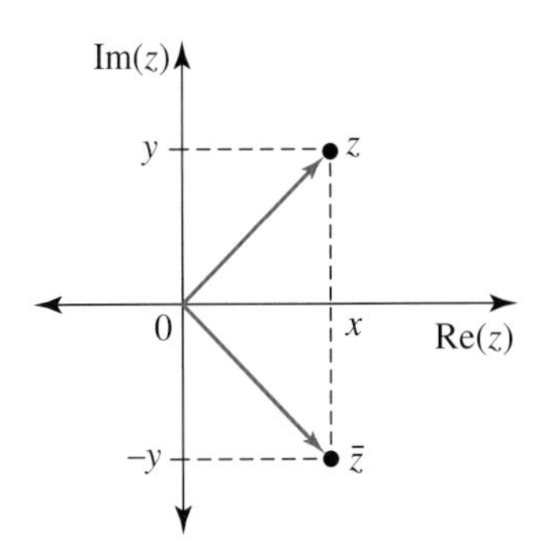

10.5.6 Multiplication by i

If $z = x + yi$, then iz is given by:

$$
\begin{aligned}
iz &= i(x + yi) \\
&= ix + i^2y \\
&= -y + xi
\end{aligned}
$$

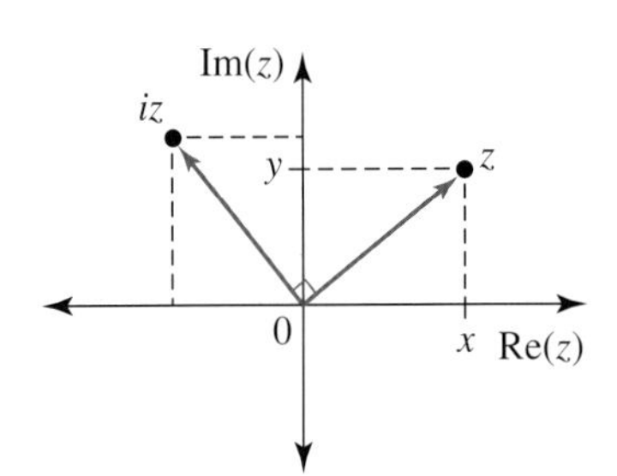

It is clear from the diagram that iz is a **rotation** of 90° anticlockwise from z. Rotation is another form of transformation, using multiplication. The complex number z is said to be rotated 90° anticlockwise from z.

WORKED EXAMPLE 20 Representing the conjugate and multiplication by i on an Argand plane

Given the complex number $u = 1 + 2i$, represent the complex numbers $\overline{u}$ and iu on the same Argand plane and comment on their relative positions.

THINK	WRITE
1. Plot the complex number u on an Argand plane. u: $\text{Re}(z) = 1$ and $\text{Im}(z) = 2$	$u = 1 + 2i$
2. Evaluate the complex numbers $\overline{u}$ and iu, and plot these complex numbers on the same Argand diagram. $\overline{u}$: $\text{Re}(z) = 1$ and $\text{Im}(z) = -2$ iu: $\text{Re}(z) = -2$ and $\text{Im}(z) = 1$	$\overline{u} = 1 - 2i$ $iu = i(1 + 2i)$ $= i + 2i^2$ $= -2 + i$

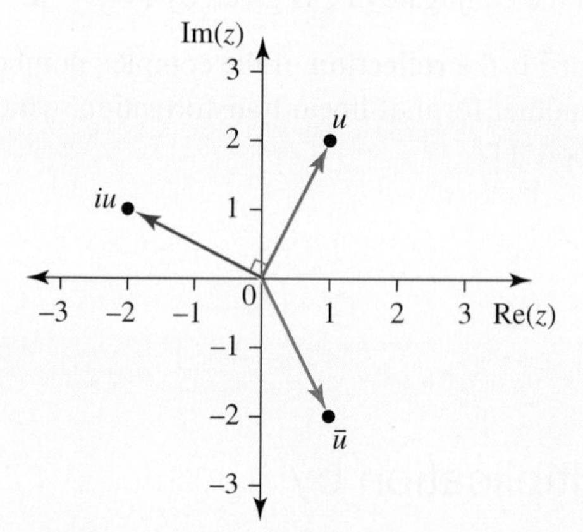

THINK	WRITE
3. Comment on their relative positions.	$\overline{u}$ is the reflection of the complex number in the real axis. iu is a rotation of 90° anticlockwise from u.

10.5 Exercise

Technology free

1. WE18 Given the complex number $z = 2 + i$, represent the complex numbers z and $3z$ on the same Argand diagram. Comment on their relative positions.

2. Given the complex number $z = 4 - 2i$, represent the complex numbers z and $-\frac{z}{2}$ on the same Argand diagram. Comment on their relative positions.

3. WE19 Given the complex numbers $u = 3 - 2i$ and $v = 1 + 2i$, represent the following complex numbers on separate Argand diagrams.

   a. $u + v$ b. $u - v$

4. Given the complex numbers $u = 3 - 2i$ and $v = 1 + 2i$, evaluate and draw the complex numbers $u + v$ and $u - v$.

5. WE20 Given the complex number $u = -3 - 2i$, represent the complex numbers $\overline{u}$ and iu on the same Argand diagram and comment on their relative positions.

6. Given the complex number $v = -2 + 3i$, represent the complex numbers $\overline{v}$ and $i^2 v$ on the same Argand diagram and comment on their relative positions.

7. a. Given $z = 2 - 3i$, evaluate and plot each of the following on the same Argand diagram. Comment on their relative positions.

   i. $2z$ ii. $\overline{z}$ iii. iz

   b. Given $z = -4 - 4i$, evaluate and plot each of the following on the same Argand diagram. Comment on their relative positions.

   i. $-\frac{z}{2}$ ii. $\overline{z}$ iii. iz

8. a. Given the complex numbers $u = -1 - 2i$ and $v = 2 + 3i$, evaluate and plot each of the following on separate Argand diagrams.

   i. $u + v$ ii. $u - v$

   b. Given the complex numbers $u = 2 - 3i$ and $v = 1 + 4i$, evaluate and plot each of the following on separate Argand diagrams.

   i. $u + v$ ii. $u - v$

9. a. Given $z = 1 - i$, plot each of the following on one Argand diagram.

   i. z ii. $i^2 z$ iii. $i^3 z$

   b. Given $z = -2 + 3i$, plot each of the following on the same Argand diagram.

   i. z ii. $i^2 z$ iii. $i^3 z$

10. a. If $u = 1 + i$, evaluate and plot each of the following on one Argand diagram.

   i. u ii. $\frac{1}{u}$

   b. If $u = -\sqrt{3} - i$, evaluate and plot each of the following on one Argand diagram.

   i. u ii. $\frac{1}{u}$

11. a. Given that $u = 2 + 2i$ and $v = -1 - i$, evaluate and plot each of the following on one Argand diagram.

i. uv

ii. $\frac{u}{v}$

b. Given that $u = 2i$ and $v = 1 - i$, evaluate and plot each of the following on one Argand diagram.

i. uv

ii. $\frac{u}{v}$

12. a. Given $z = 1 - i$, plot each of the following on one Argand diagram.

i. z

ii. z^2

iii. z^3

b. Given $z = -1 + i$, plot each of the following on one Argand diagram.

i. z

ii. z^2

iii. z^3

13. a. If $\underset{\sim}{u} = 3\underset{\sim}{i} - 4\underset{\sim}{j}$ and $\underset{\sim}{v} = 4\underset{\sim}{i} + 3\underset{\sim}{j}$, show that the vectors $\underset{\sim}{u}$ and $\underset{\sim}{v}$ are perpendicular.

b. Evaluate the image of the point $(3, -4)$ when it is rotated 90° anticlockwise.

c. Consider the vectors $\underset{\sim}{u} = a\underset{\sim}{i} + b\underset{\sim}{j}$ and $\underset{\sim}{v} = -b\underset{\sim}{i} + a\underset{\sim}{j}$ where $a, b \in R$.

i. Show that $\underset{\sim}{u}$ and $\underset{\sim}{v}$ are perpendicular.

ii. Explain how this relates to the complex numbers $z = a + bi$ and iz.

iii. Complex numbers and vectors are said to be isomorphic. What does this mean?

10.5 Exam questions

Question 1 (4 marks) TECH-FREE

Let $u = 3 - 4i$. Plot $\bar{u}$, iu, i^2u and i^3u on an Argand diagram and comment on their relative positions.

Question 2 (2 marks) TECH-FREE

A triangle is formed when the complex number $u = a + bi$ and $\bar{u}$ are plotted on an Argand diagram with the origin O. Determine the area of this triangle.

Question 3 (2 marks) TECH-FREE

A triangle is formed when the complex number $u = a + bi$ and iu are plotted on an Argand diagram with the origin O. Determine the area of this triangle.

More exam questions are available online.

10.6 Complex numbers in polar form

LEARNING INTENTION

At the end of this subtopic you should be able to:

- express complex numbers in polar form
- convert between Cartesian and polar forms of complex numbers.

10.6.1 The modulus of z

The **magnitude** (or **modulus** or **absolute value**) of the complex number $z = x + yi$ is the length of the line segment joining the origin to the point z. It is denoted by:

$$|z|, \ |x + yi| \text{ or mod } z.$$

The modulus of z is calculated using Pythagoras' theorem.

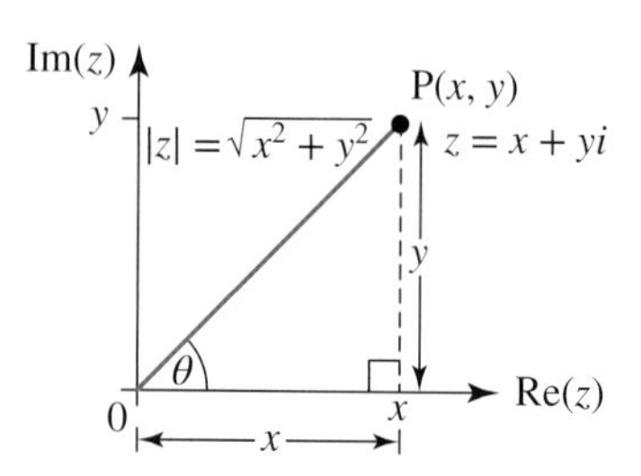

The magnitude of a complex number

The magnitude of a complex number $z = x + yi$ is:

$$|z| = \sqrt{x^2 + y^2}$$

Recall that $z\bar{z} = x^2 + y^2$ and note that this is the square of the magnitude. Therefore, $|z|^2 = z\bar{z} = x^2 + y^2$.

WORKED EXAMPLE 21 Calculating the modulus of a complex number

Calculate the modulus of the complex number $z = 8 - 6i$.

THINK	WRITE
Calculate the modulus by rule. Because it forms the hypotenuse of a right-angled triangle, the modulus is always greater than or equal to Re(z) or Im(z).	$\|z\| = \sqrt{8^2 + (-6)^2}$ $= \sqrt{100}$ $= 10$

TI | THINK	DISPLAY/WRITE	CASIO | THINK	DISPLAY/WRITE
On a Calculator page, complete the entry line as $\|8 - 6i\|$.	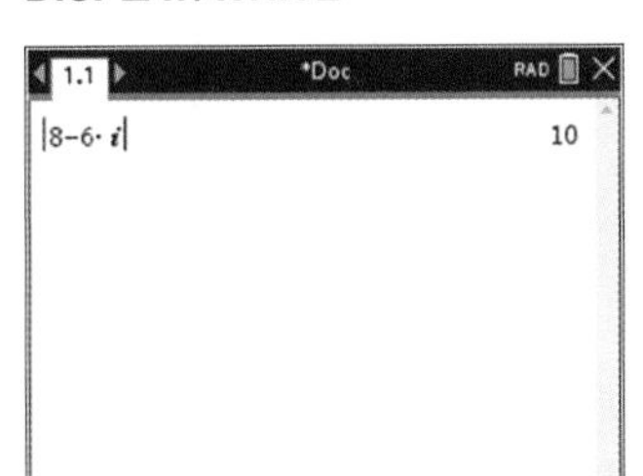	On a Main screen, complete the entry line as $\|8 - 6i\|$.	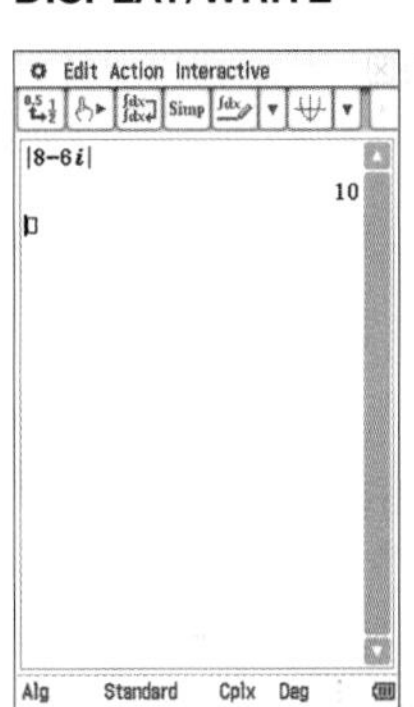

WORKED EXAMPLE 22 Calculating the modulus of the difference of complex numbers

If $z = 4 + 2i$ and $w = 7 + 6i$, represent the position of $w - z$ on an Argand diagram and calculate $|w - z|$.

THINK	WRITE
1. Calculate $w - z$.	$w - z = 7 + 6i - (4 + 2i)$ $= 3 + 4i$
2. Represent it on an Argand diagram as a directed line segment OP.	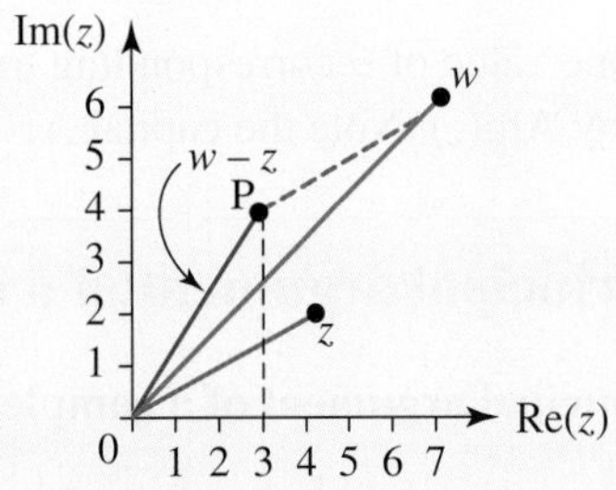
3. Use Pythagoras theorem to determine the length of OP.	$\|OP\|^2 = 3^2 + 4^2 = 25$ $\|OP\| = 5$ So $\|w - z\| = 5$.

WORKED EXAMPLE 23 Applications to areas represented by complex numbers

Represent $z_1 = 2 + 3i$, $z_2 = 5 - 2i$ and $z_3 = -4 - 2i$ on the complex number plane and calculate the area of the shape formed when the three points are connected by straight line segments.

THINK	WRITE
1. Show the connected points on the complex number plane.	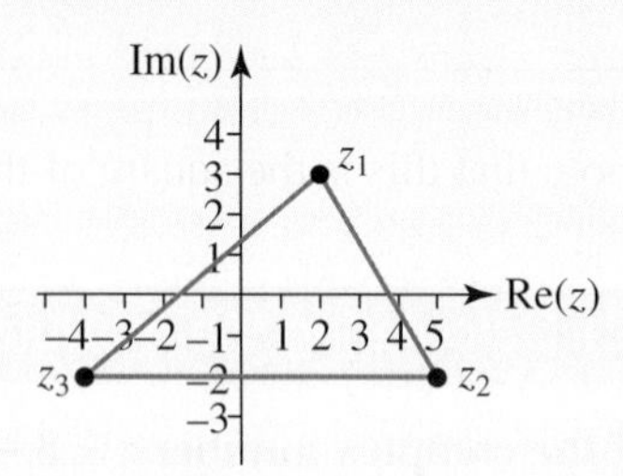
2. Calculate the area of the triangle obtained. The length of the base and height can be found by inspection (base = 9, height = 5).	$\text{Area of triangle} = \frac{1}{2} \times 9 \times 5$ $= 22.5$ square units.

10.6.2 The argument of z

The **argument** of z, $\arg(z)$, is the angle measurement anticlockwise from the positive real axis.

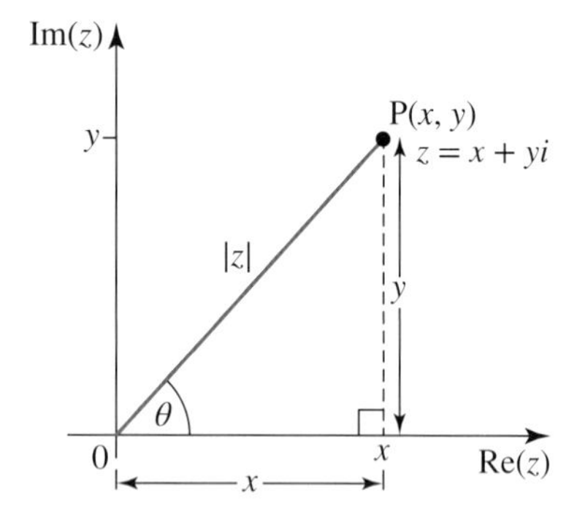

In the figure, $\arg(z) = \theta$, where

$\sin(\theta) = \frac{y}{|z|}$ and $\cos(\theta) = \frac{x}{|z|}$ or $\tan(\theta) = \frac{y}{x}$.

Using this we can determine the argument for any complex number as $\theta = \tan^{-1}\left(\frac{y}{x}\right)$.

Note that there are an infinite number of arguments for any non-zero complex number z since adding or subtracting multiples of 2 radians (or 360°) does not change the position on the Argand plane.

The argument of a complex number

The argument of a complex number $z = x + yi$ is:

$$\arg(z) = \theta = \tan^{-1}\left(\frac{y}{x}\right),\ x \neq 0$$

To ensure that there is only one value of θ corresponding to z, we refer to the *principal value* or principal argument of θ and denote it by $\text{Arg}(z)$. Note the capital A.

The principal argument of a complex number

The principal argument of a complex number $z = x + yi$ is:

$$\text{Arg}(z) = \theta \in (-\pi,\ \pi]$$

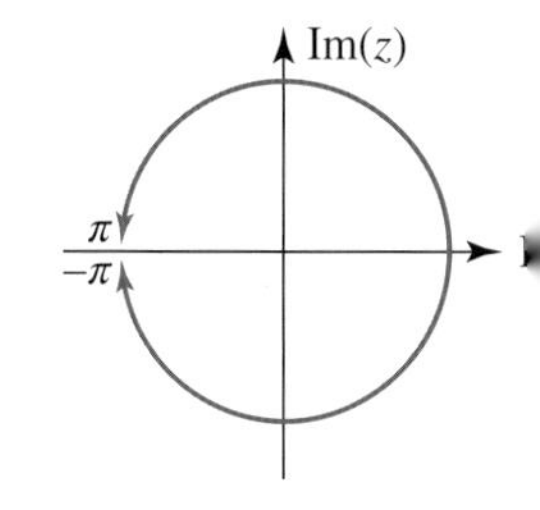

Exact values

It is useful to remember the exact values of the trigonometric ratios for the commonly used angles $\frac{\pi}{6}$, $\frac{\pi}{4}$ and $\frac{\pi}{3}$. To remember these we can draw a couple of triangles that can be analysed to determine whichever exact value you require. The triangle on the left is an isosceles right-angled triangle with two sides of length 1. The triangle on the right is an equilateral triangle with sides of length 2. The vertical line in the middle creates right-angled triangles with angles $\frac{\pi}{3}$ and $\frac{\pi}{6}$. Using the definitions of the sine, cosine and tangent ratios, you will be able to determine the exact values for the angles $\frac{\pi}{6}$, $\frac{\pi}{4}$ and $\frac{\pi}{3}$.

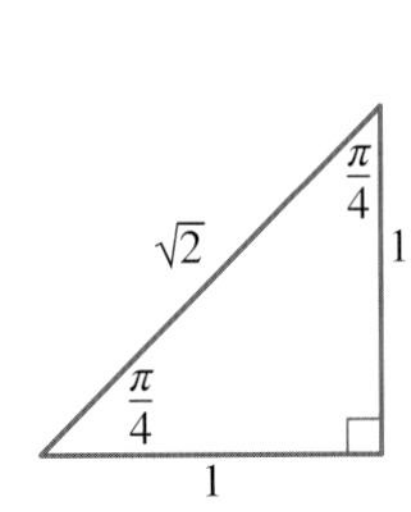

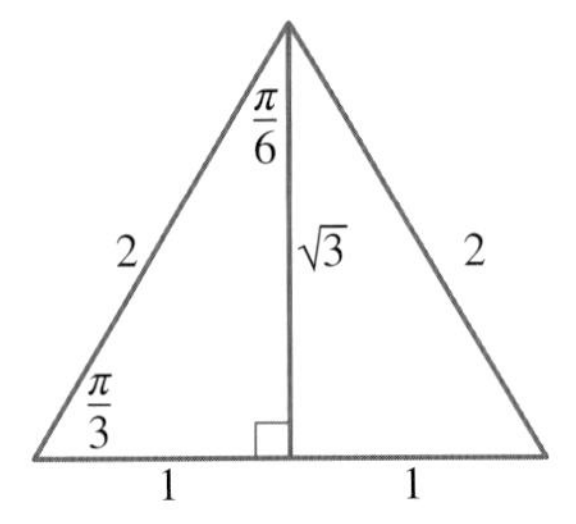

These exact values are summarised in the following table.

Angle	$\sin(\theta)$	$\cos(\theta)$	$\tan(\theta)$
$\frac{\pi}{6}$	$\frac{1}{2}$	$\frac{\sqrt{3}}{2}$	$\frac{1}{\sqrt{3}}$
$\frac{\pi}{4}$	$\frac{1}{\sqrt{2}}$	$\frac{1}{\sqrt{2}}$	1
$\frac{\pi}{3}$	$\frac{\sqrt{3}}{2}$	$\frac{1}{2}$	$\sqrt{3}$

WORKED EXAMPLE 24 Calculating the arguments of complex numbers

Calculate the argument of z for each of the following in the interval $(-\pi, \pi]$.

a. $z = 4 + 4i$ **b.** $z = 1 - \sqrt{3}i$

THINK

a. 1. Plot z.

2. Sketch the triangle that has sides in this 1 : 1 ratio.

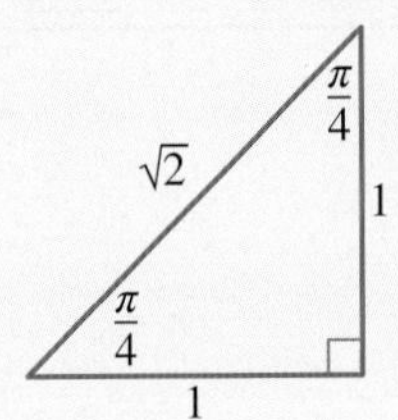

WRITE

a.

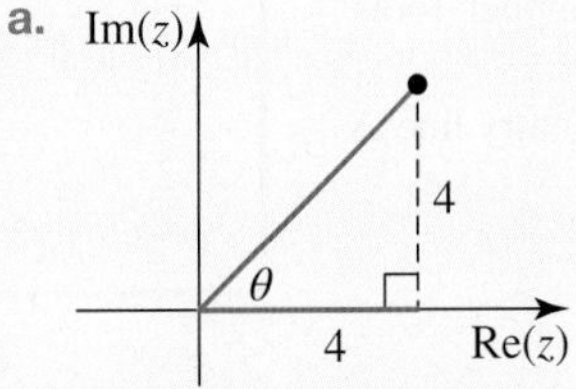

From the diagram:

$\theta = \frac{\pi}{4}$

$\therefore \text{Arg}(z) = \frac{\pi}{4}$

3. This result can be verified using an inverse trigonometric ratio, $\theta = \tan^{-1}\left(\frac{y}{x}\right)$.

Check:

$$\theta = \tan^{-1}\left(\frac{4}{4}\right)$$
$$= \frac{\pi}{4}$$

b. 1. Plot z.

b.

Im(z) 1 θ Re(z) $\sqrt{3}$

2. Sketch the triangle that has sides in this ratio.

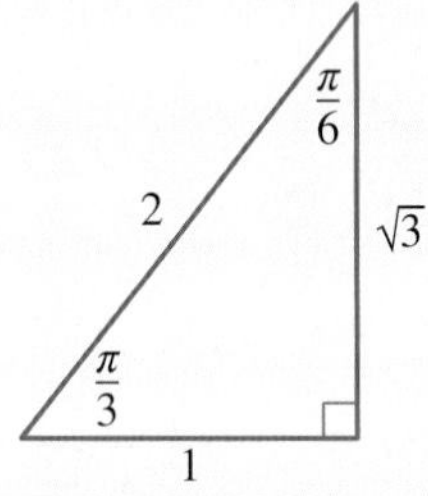

Remember: $-\pi < \theta \leq \pi$.

From the diagram:

$$\theta = -\frac{\pi}{3}$$
$$\therefore \text{Arg}(z) = -\frac{\pi}{3}$$

3. This result can be verified using an inverse trigonometric ratio, $\theta = \tan^{-1}\left(\frac{y}{x}\right)$.

Check:

$$\theta = \tan^{-1}\left(\frac{-\sqrt{3}}{1}\right)$$
$$= -\frac{\pi}{3}$$

TI | THINK	DISPLAY/WRITE	CASIO | THINK	DISPLAY/WRITE
a. On a Calculator page, press MENU, then select: 2 Number 9 Complex Number Tools 4 Polar Angle Complete the entry line as shown.	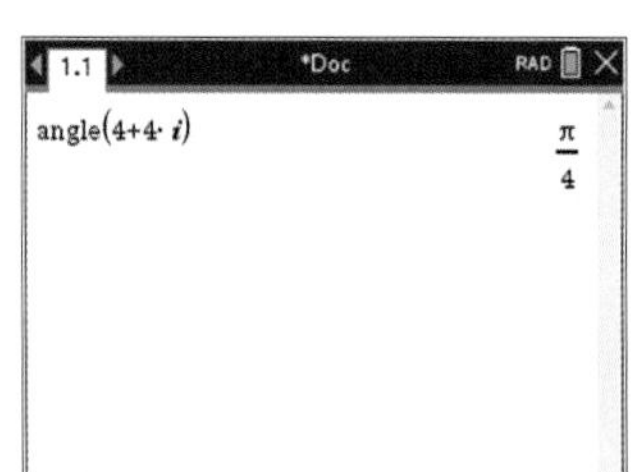	a. On a Main screen, complete the entry line as arg(4 + 4i).	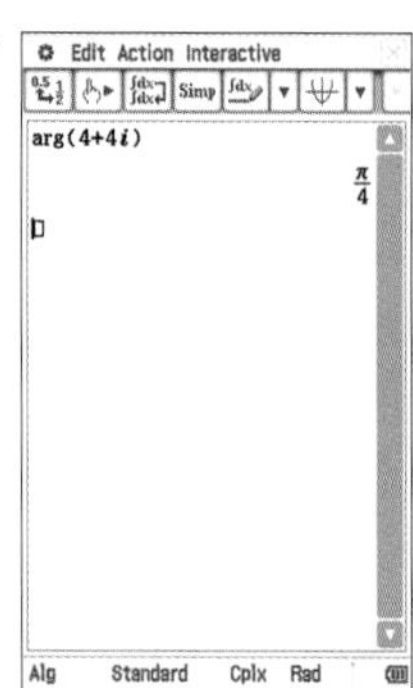

WORKED EXAMPLE 25 Converting arguments into principal arguments

Convert each of the following into principal arguments.

a. $\dfrac{7\pi}{4}$ b. $-\dfrac{5\pi}{2}$

THINK	WRITE
a. 1. Sketch the angle.	a.  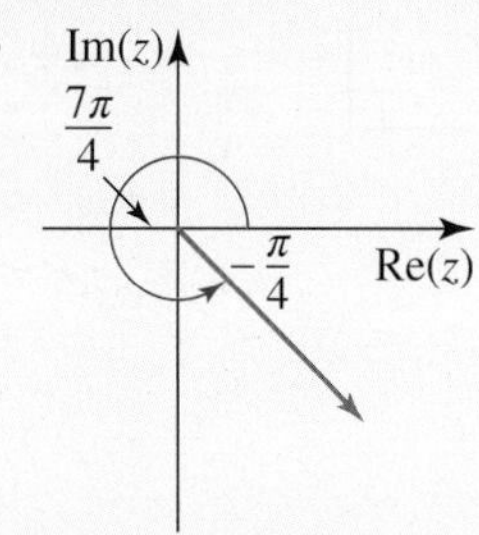
2. Since the given angle is positive, subtract multiples of 2π until it lies in the range $(-\pi, \pi]$.	$\text{Arg}(z) = \dfrac{7\pi}{4} - 2\pi$ $= -\dfrac{\pi}{4}$
b. 1. Sketch the angle.	b. 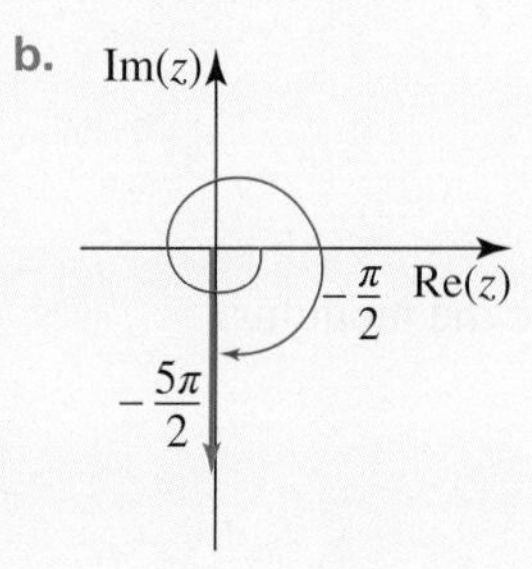
2. Since the given angle is negative, add multiples of 2π until it lies in the range $(-\pi, \pi]$.	$\text{Arg}(z) = -\dfrac{5\pi}{2} + 2\pi$ $= -\dfrac{\pi}{2}$

WORKED EXAMPLE 26 Determining the modulus and argument of complex numbers

Determine the modulus and principal argument for each of the following complex numbers.

a. $-\sqrt{3} + i$ b. $-\sqrt{2} - \sqrt{2}i$

THINK	WRITE
a. 1. Plot z.	a. 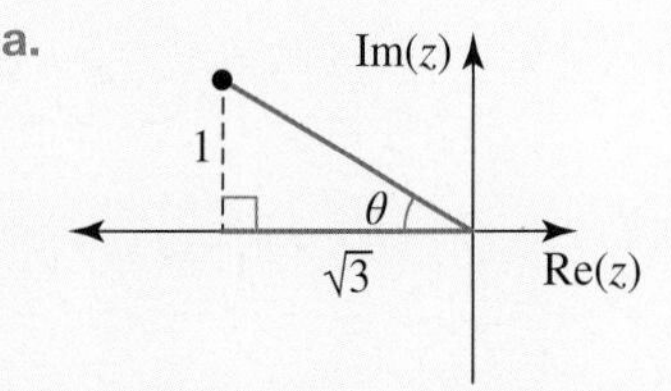

2. This triangle has sides in the same ratio as

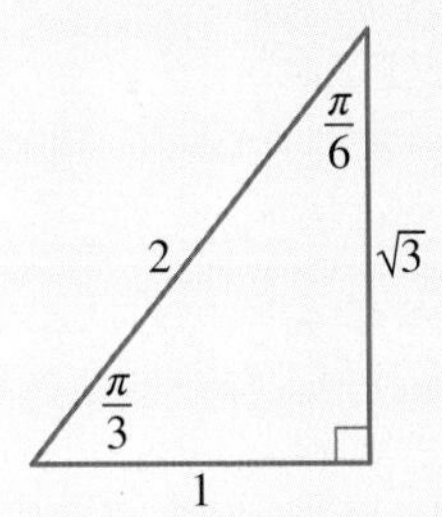

$$|z| = \sqrt{\left(\sqrt{3}\right)^2 + (1)^2}$$
$$= 2$$
$$\theta = \tan^{-1}\left(\frac{1}{\sqrt{3}}\right)$$
$$= \frac{\pi}{6}$$
$$\text{Arg}(z) = \pi - \frac{\pi}{6}$$
$$= \frac{5\pi}{6}$$

b. 1. Plot z.

b.

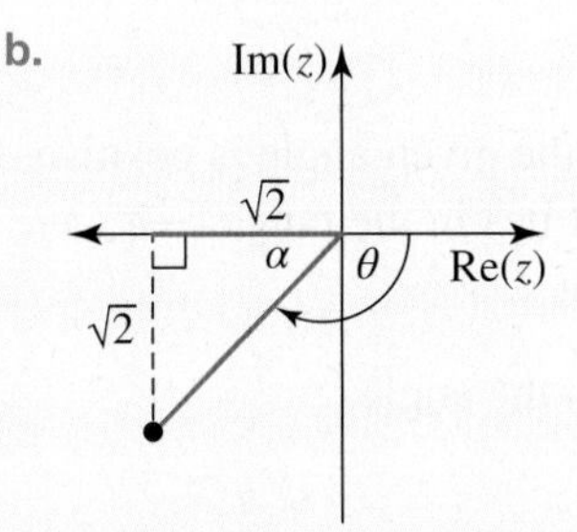

2. Determine the modulus.

$$|z| = \sqrt{\left(-\sqrt{2}\right)^2 + \left(-\sqrt{2}\right)^2}$$
$$= \sqrt{2+2}$$
$$= \sqrt{4}$$
$$= 2$$

3. The triangle in the third quadrant will be used to determine α but the answer will be finally expressed as θ and $\text{Arg}(z)$.

$$\alpha = \tan^{-1}\left(\frac{y}{x}\right)$$
$$= \tan^{-1}\left(\frac{-\sqrt{2}}{-\sqrt{2}}\right)$$
$$= \frac{\pi}{4}$$
$$\text{Arg}(z) = \theta = -\pi + \frac{\pi}{4}$$
$$= -\frac{3\pi}{4}$$

10.6.3 Expressing complex numbers in polar form

Now that we know how to calculate the modulus and arguments of complex numbers, we can express them in a new form. The **polar form** of a complex number can be used to express any complex number in terms of its modulus and argument.

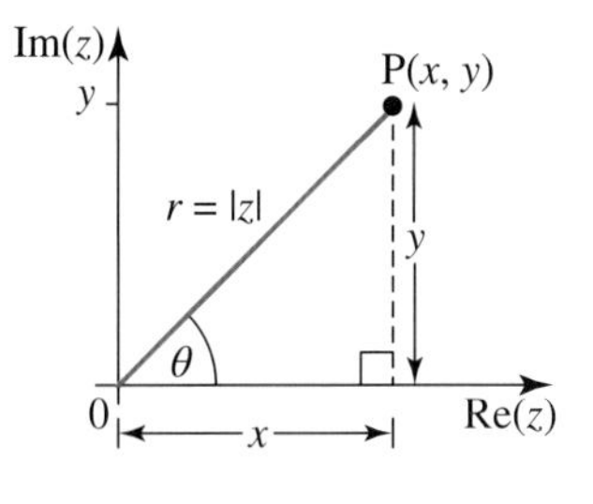

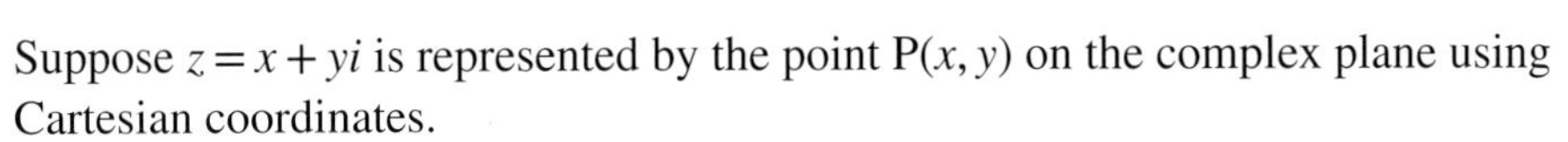

Suppose $z = x + yi$ is represented by the point P(x, y) on the complex plane using Cartesian coordinates.

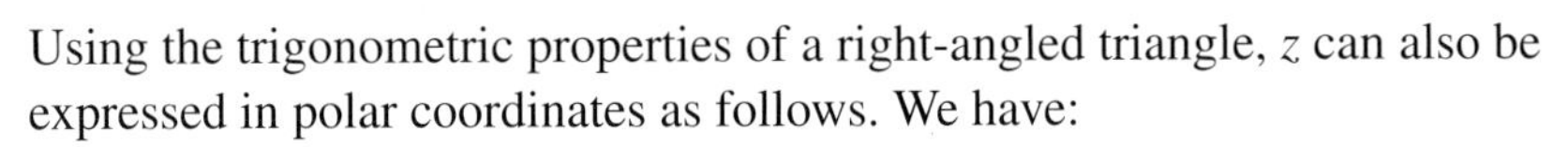

Using the trigonometric properties of a right-angled triangle, z can also be expressed in polar coordinates as follows. We have:

$$\cos(\theta) = \frac{x}{r} \quad \text{or} \quad x = r\cos(\theta)$$

$$\sin(\theta) = \frac{y}{r} \quad \text{or} \quad y = r\sin(\theta)$$

where $|z| = r = \sqrt{x^2 + y^2}$ and $\theta = \text{Arg}(z)$.

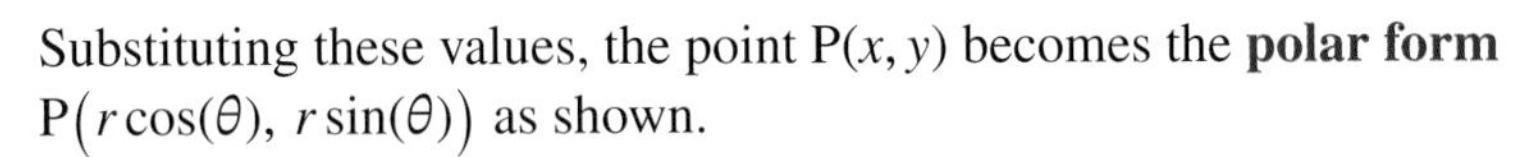

Substituting these values, the point P(x, y) becomes the **polar form** P$\big(r\cos(\theta),\ r\sin(\theta)\big)$ as shown.

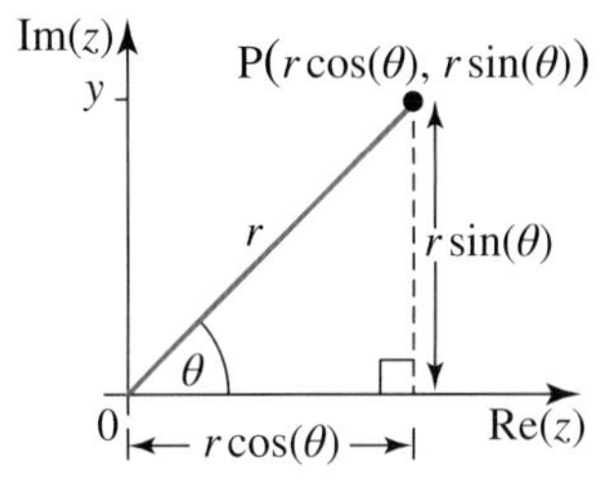

$z = x + yi$ in Cartesian form becomes:

$z = r\cos(\theta) + r\sin(\theta)i$

$= r\big(\cos(\theta) + i\sin(\theta)\big)$

$= r\,\text{cis}(\theta)$, where $\text{cis}(\theta)$ is the abbreviated form of $\cos(\theta) + i\sin(\theta)$.

(***Note:*** The acronym 'cis' is pronounced 'sis'.)

Polar form of complex numbers

The polar form of a complex number is:

$$z = r\,\text{cis}(\theta)$$

To convert from Cartesian form to polar form, use the following formulas:

$$r = \sqrt{x^2 + y^2}$$
$$\theta = \text{Arg}(z)$$

You may find it useful to sketch the complex number on an Argand plane to help determine the principal argument.

WORKED EXAMPLE 27 Expressing complex numbers in polar form

Express each of the following in polar form, using the principal argument.

a. $z = 1 + i$ **b.** $z = 1 - \sqrt{3}i$

THINK	WRITE
a. 1. Sketch z.	**a.** Im(z), $z = 1 + i$, r, 1, θ, 1, Re(z)
2. Calculate the value of r using $r = \lvert z \rvert = \sqrt{x^2 + y^2}$. Determine θ from $\tan(\theta) = \left(\frac{y}{x}\right)$.	$r = \sqrt{1^2 + 1^2}$ $= \sqrt{2}$ $\tan(\theta) = \frac{1}{1}$ $\theta = \tan^{-1}(1)$
The angle θ is in the range $(-\pi, \pi]$, which is required.	$\theta = \frac{\pi}{4}$
3. Substitute the values of r and θ in $z = r\,\text{cis}(\theta)$.	$z = \sqrt{2}\,\text{cis}\left(\frac{\pi}{4}\right)$
b. 1. Sketch z.	**b.** 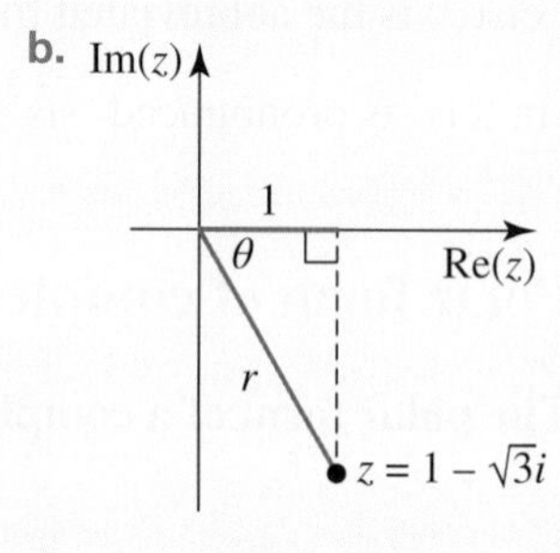
2. Calculate the value of r and θ.	$r = \sqrt{1 + \left(\sqrt{3}\right)^2}$ $= 2$ $\tan(\theta) = -\frac{\sqrt{3}}{1}$ $\theta = \tan^{-1}\left(-\sqrt{3}\right)$ $\theta = -\frac{\pi}{3}$
3. Substitute the values of r and θ into $r\,\text{cis}(\theta)$.	$z = 2\,\text{cis}\left(-\frac{\pi}{3}\right)$

TI | THINK	DISPLAY/WRITE	CASIO | THINK	DISPLAY/WRITE
b. 1. On a Calculator page, enter $1 - \sqrt{3}i$, then press MENU and select: 2 Number 9 Complex Number Tools 6 Convert to Polar	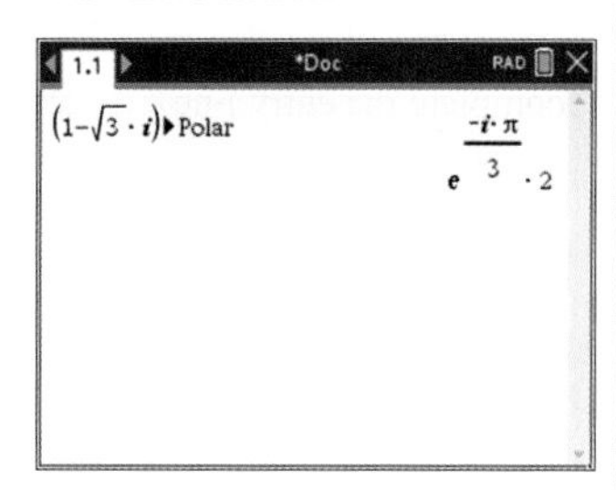	**b. 1.** On a Main screen, select: Action Complex compToPol and complete the entry line as shown.	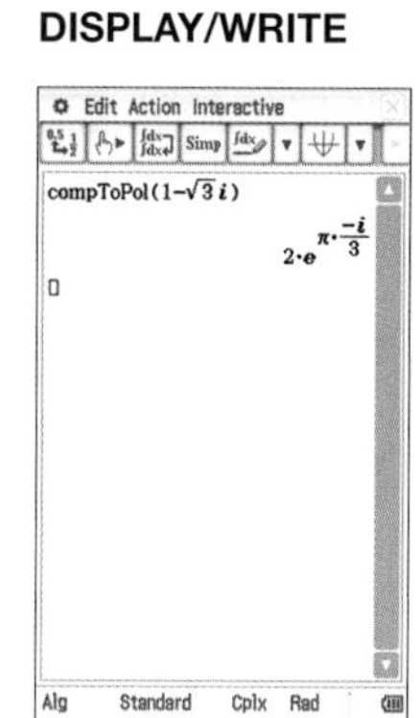
2. Note that the calculator shows the polar form as $r \times e^{i\theta}$. From this form you will need to identify r and θ to express it in the form $z = r\,\text{cis}(\theta)$.	$r = 2$ $\theta = -\dfrac{\pi}{3}$ $z = 2\,\text{cis}\left(-\dfrac{\pi}{3}\right)$	**2.** Note that the calculator shows the polar form as $r \times e^{i\theta}$. From this form you will need to identify r and θ to express it in the form $z = r\,\text{cis}(\theta)$.	$r = 2$ $\theta = -\dfrac{\pi}{3}$ $z = 2\,\text{cis}\left(-\dfrac{\pi}{3}\right)$

10.6.4 Converting from polar form to Cartesian form

To convert from polar form to Cartesian form, use the following formulas:

$$x = r\cos(\theta)$$
$$y = r\sin(\theta)$$

WORKED EXAMPLE 28 Converting from polar form to Cartesian form

Express $3\,\text{cis}\left(\dfrac{\pi}{4}\right)$ in Cartesian (or standard $a + bi$) form.

THINK	WRITE
1. Sketch z.	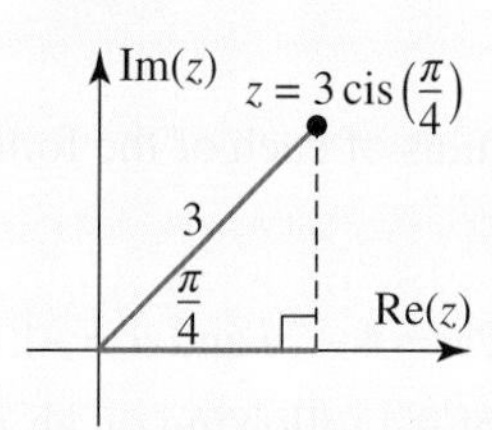
2. Express $3\,\text{cis}\left(\dfrac{\pi}{4}\right)$ in Cartesian form.	$3\,\text{cis}\left(\dfrac{\pi}{4}\right) = 3\cos\left(\dfrac{\pi}{4}\right) + 3\sin\left(\dfrac{\pi}{4}\right)i$
3. Simplify using exact values from the following triangle: 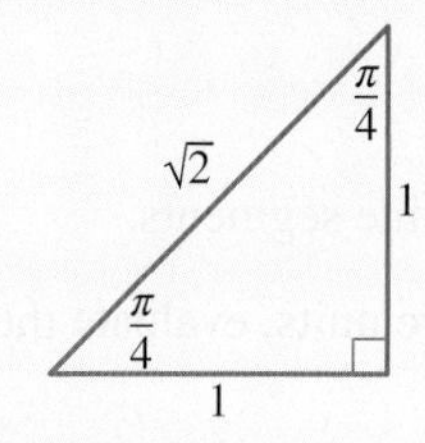	$= 3 \times \dfrac{1}{\sqrt{2}} + 3 \times \dfrac{1}{\sqrt{2}}i$ $= \dfrac{3}{\sqrt{2}} + \left(\dfrac{3}{\sqrt{2}}\right)i$

TI | THINK

On a Calculator page, enter $3\angle\frac{\pi}{4}$, then press MENU and select:
2 Number
9 Complex Number Tools
7 Convert to Rectangular.

DISPLAY/WRITE

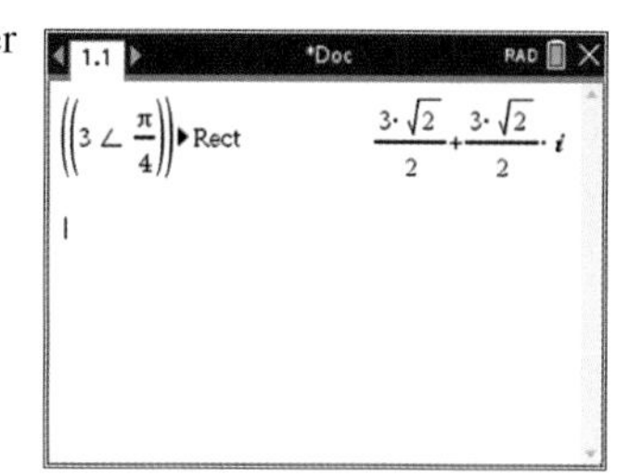

In Cartesian form this is $z = \frac{3\sqrt{2}}{2} + \frac{3\sqrt{2}}{2}i$.

CASIO | THINK

On a Main screen, complete the entry line as shown.
The result shows the x and y values in $z = x + yi$.

DISPLAY/WRITE

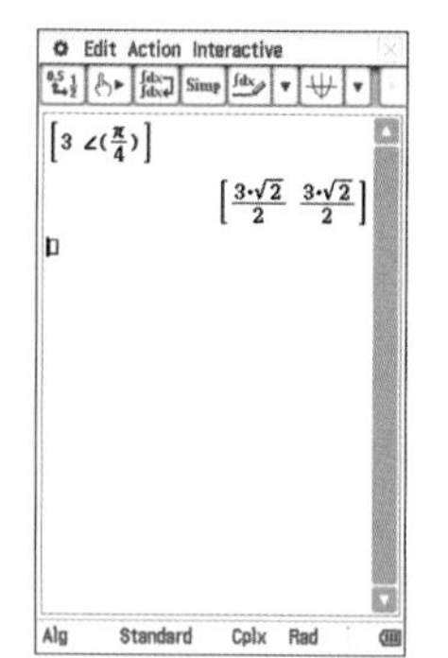

In Cartesian form this is $z = \frac{3\sqrt{2}}{2} + \frac{3\sqrt{2}}{2}i$.

10.6 Exercise

Students, these questions are even better in jacPLUS

Receive immediate feedback and access sample responses

Access additional questions

Track your results and progress

Find all this and MORE in jacPLUS

Technology free

1. WE21 Calculate the modulus of each of the following complex numbers.
 a. $z = 5 + 12i$ b. $z = \sqrt{5} - 2i$
2. Calculate the modulus of each of the following complex numbers.
 a. $z = -4 + 7i$ b. $z = -3 - 6i$
3. Calculate the modulus of each of the following complex numbers.
 a. $z = \sqrt{3} + \sqrt{2}i$ b. $z = (2 + i)^2$
4. WE22 If $z = 3 + i, w = 4 - 3i$ and $u = -2 + 5i$:
 i. represent each of the following on an Argand plane
 ii. calculate the magnitude in each case.
 a. $z - w$ b. $u + z$ c. $w - u$ d. $w + z$ e. $z + w - u$ f. z^2
5. a. WE23 Represent the points $z_1 = -3 + 0i, z_2 = 2 + 5i, z_3 = 7 + 5i$ and $z_4 = 9 + 0i$ on the complex number plane.
 b. Calculate the area of the shape formed when the four points are connected by straight line segments in the order z_1 to z_2 to z_3 to z_4.
6. a. Show the points $z = -1 + 3i, u = 3$ and $w = 3 + 12i$ on the complex number plane.
 b. Calculate the area of the triangle produced by joining the three points with straight line segments.
7. a. If the complex numbers $u = 3 - 4i, \overline{u}, v$ and $\overline{v}$ form a square with an area of 64 square units, evaluate the complex number v.
 b. If the complex numbers $u = -2 + 5i, \overline{u}, v$ and $\overline{v}$ form a rectangle with an area of 60 square units, evaluate the complex number v.

c. If $a, b, c \in R^+$, evaluate the area of the rectangle formed by the complex numbers $u = a + bi$, $\overline{u}$, $v = -c - bi$ and $\overline{v}$.

8. a. Evaluate the area of the triangle that is formed by the complex numbers $u = 4 + 3i$ and iu and the origin O.
b. Evaluate the area of the triangle that is formed by the complex numbers $u = 12 + 5i$ and iu and the origin O.
c. If $a, b \in R^+$, evaluate the area of the triangle that is formed by the complex numbers $u = a + bi$, iu and the origin O.

9. a. Evaluate the area of the square that is formed by the complex numbers $u = 4 - 3i$, iu, $u + iu$ and the origin O.
b. If the area of the square formed by the complex numbers $u = 6 + bi$, iu, $u + iu$ and the origin O is equal to 50 square units, evaluate the value of b.
c. If $a, b \in R^+$, evaluate the area of the square that is formed by the complex numbers $u = a + bi$, iu, $u + iu$ and the origin O.

10. a. Evaluate the equation and area of the circle that passes through the complex numbers $u = 3 - 4i$, iu and the origin O.
b. Evaluate the equation and area of the circle that passes through the complex numbers $u = 5 + 12i$, iu and the origin O.

11. If $a, b \in R^+$, evaluate the equation and area of the circle that passes through the complex numbers $u = a + bi$, iu and the origin O.

12. **WE24** Determine the principal argument of z for each of the following in the interval $(-\pi, \pi]$.

a. $z = \sqrt{3} + i$
b. $z = 5 - 5i$
c. $z = -2 - 2\sqrt{3}i$

13. Determine the principal argument of z for each of the following in the interval $(-\pi, \pi]$.

a. $z = 3i$
b. $z = -\sqrt{7}$
c. $z = -6i$
d. $z = 55$

14. **WE25** Convert each of the following into principal arguments.

a. $\frac{3\pi}{2}$ b. $-\frac{11\pi}{6}$ c. $\frac{15\pi}{8}$ d. $-\frac{5\pi}{4}$

15. Convert each of the following into principal arguments.

a. $\frac{19\pi}{6}$ b. $\frac{20\pi}{7}$ c. $-\frac{18\pi}{5}$ d. $-\frac{13\pi}{12}$

16. **WE26** Determine the modulus and principal argument of each of the following complex numbers.

a. $3 - 3i$ b. $-5 + 5i$ c. $-1 - \sqrt{3}i$ d. $4\sqrt{3} + 4i$

17. **WE27** Express each of the following in polar form, using the principal argument.

a. $z = -1 + i$ b. $z = \sqrt{6} + \sqrt{2}i$ c. $z = -\sqrt{5} - \sqrt{5}i$

18. Express each of the following in polar form, using the principal argument.

a. $z = \sqrt{5} - \sqrt{15}i$ b. $z = -\frac{1}{2} - \frac{\sqrt{3}}{2}i$ c. $z = -\frac{1}{4} + \frac{1}{4}i$

19. WE28 Express each of the following complex numbers in Cartesian form.

a. $2\text{cis}\left(\frac{2\pi}{3}\right)$ b. $3\text{cis}\left(\frac{\pi}{4}\right)$ c. $\sqrt{5}\text{cis}\left(\frac{5\pi}{6}\right)$

20. Express each of the following complex numbers in Cartesian form.

a. $4\text{cis}\left(-\frac{\pi}{3}\right)$ b. $\sqrt{7}\text{cis}\left(-\frac{7\pi}{4}\right)$ c. $8\text{cis}\left(\frac{\pi}{2}\right)$

10.6 Exam questions

Question 1 (2 marks) TECH-FREE

a. Let $u = \sqrt{3} + i$. Express u, $\overline{u}$, $\frac{1}{u}$ and $-u$ in polar form. **(1 mark)**

b. Let $w = 5\sqrt{2}\text{cis}\left(-\frac{3\pi}{4}\right)$. Express w in Cartesian form. **(1 mark)**

Question 2 (1 mark) TECH-FREE

Let $u = 1 - i$ and $v = -1 - i$. Express $u,\ v,\ u + v$ and $\frac{u}{v}$ in polar form.

Question 3 (3 marks) TECH-FREE

Let $u = a - \sqrt{6}i$. Determine the value of the real number a if:

a. $|u| = 9$ **(1 mark)**

b. $\text{Arg}(u) = -\frac{\pi}{3}$ **(1 mark)**

c. $u^2 = 8\text{cis}\left(-\frac{2\pi}{3}\right)$ **(1 mark)**

More exam questions are available online.

10.7 Basic operations on complex numbers in polar form

LEARNING INTENTION

At the end of this subtopic you should be able to:

- multiply and divide complex numbers in polar form using de Moivre's theorem
- evaluate powers of complex numbers in polar form using de Moivre's theorem.

10.7.1 Addition and subtraction in polar form

In general there is no simple way to add or subtract complex numbers given in the polar form $r\,\text{cis}(\theta)$. For addition or subtraction, the complex numbers need to be expressed in Cartesian form first.

10.7.2 Multiplication, division and powers in polar form

De Moivre's theorem is very useful for multiplying and dividing complex numbers. It applies to complex numbers in polar form and states that the product of two complex numbers can be calculated by multiplying their moduli and adding their arguments. It also applies to division, which is similar to multiplication but uses the inverse operations, division of moduli and subtraction of arguments.

> **De Moivre's theorem**
>
> **If $z_1 = r_1\text{cis}(\theta_1)$ and $z_2 = r_2\text{cis}(\theta_2)$, then**
>
> $$z_1 \times z_2 = r_1 \times r_2\text{cis}(\theta_1 + \theta_2)$$
>
> **and**
>
> $$\frac{z_1}{z_2} = \frac{r_1}{r_2}\text{cis}(\theta_1 - \theta_2).$$

The diagrams below illustrate this geometrically.

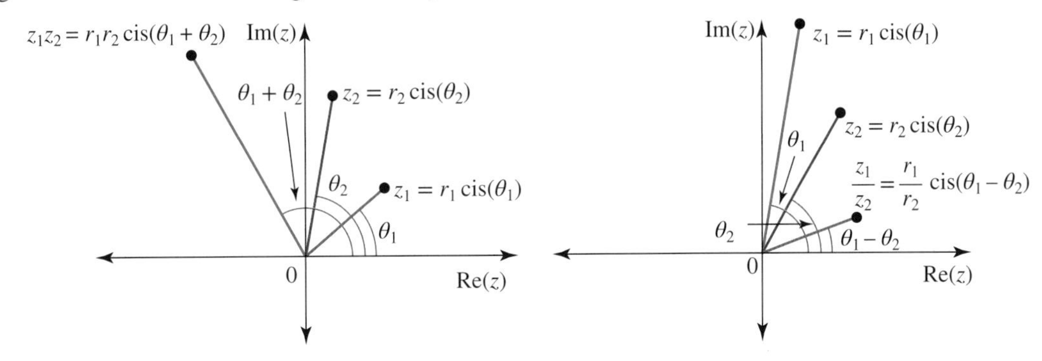

WORKED EXAMPLE 29 Calculating products of complex numbers in polar form (1)

Express $5\,\text{cis}\left(\frac{\pi}{4}\right) \times 2\,\text{cis}\left(\frac{5\pi}{6}\right)$ in the form $r\,\text{cis}(\theta)$ where $\theta \in (-\pi, \pi]$.

THINK	WRITE
1. Simplify by recalling the multiplication version of de Moivre's theorem: $z_1 \times z_2 = r_1 \times r_2\text{cis}(\theta_1 + \theta_2)$.	$5\,\text{cis}\left(\frac{\pi}{4}\right) \times 2\,\text{cis}\left(\frac{5\pi}{6}\right) = (5 \times 2)\,\text{cis}\left(\frac{\pi}{4} + \frac{5\pi}{6}\right)$ $= 10\,\text{cis}\left(\frac{13\pi}{12}\right)$
2. Sketch this number.	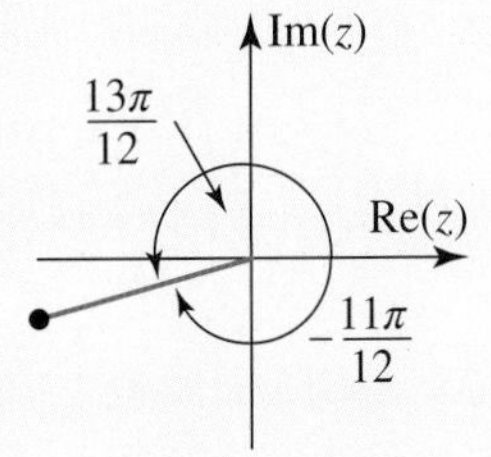
3. Subtract 2π from θ to express the answer in the required form where $\theta \in (-\pi, \pi]$.	$10\,\text{cis}\left(\frac{13\pi}{12}\right) = 10\,\text{cis}\left(-\frac{11\pi}{12}\right)$

WORKED EXAMPLE 30 Calculating products of complex numbers in polar form (2)

Express z_1z_2 in Cartesian form if $z_1 = \sqrt{2}\,\text{cis}\left(\frac{5}{6}\right)$ and $z_2 = \sqrt{6}\,\text{cis}\left(-\frac{\pi}{3}\right)$.

THINK	WRITE
1. Recall the formula $z_1 \times z_2 = r_1 \times r_2\,\text{cis}(\theta_1 + \theta_2)$.	$z_1 \times z_2 = \sqrt{2}\,\text{cis}\left(\frac{5\pi}{6}\right) \times \sqrt{6}\,\text{cis}\left(-\frac{\pi}{3}\right)$ $= \left(\sqrt{2} \times \sqrt{6}\right)\text{cis}\left(\frac{5\pi}{6} - \frac{\pi}{3}\right)$ $= 2\sqrt{3}\,\text{cis}\left(\frac{\pi}{2}\right)$
2. Simplify and write the result in Cartesian form.	$= 2\sqrt{3}\cos\left(\frac{\pi}{2}\right) + 2\sqrt{3}\sin\left(\frac{\pi}{2}\right)i$ $= 2\sqrt{3} \times 0 + 2\sqrt{3} \times 1i$ $= 2\sqrt{3}i$

WORKED EXAMPLE 31 Converting to polar form and using de Moivre's theorem

If $z = 5\sqrt{3} + 5i$ and $w = 3 + 3\sqrt{3}i$, express the product zw in polar form.

THINK	WRITE
1. Convert z to polar form.	$z = r_1\,\text{cis}(\theta_1)$ $r_1 = \lvert z \rvert$ $= \sqrt{\left(5\sqrt{3}\right)^2 + (5)^2}$ $= \sqrt{100}$ $= 10$ $\theta_1 = \tan^{-1}\left(\frac{5}{5\sqrt{3}}\right)$ $= \tan^{-1}\left(\frac{1}{\sqrt{3}}\right)$ $= \frac{\pi}{6}$ $z = 10\,\text{cis}\left(\frac{\pi}{6}\right)$

2. Convert w to polar form.

$$w = r_2 \operatorname{cis}(\theta_2)$$
$$r_2 = |w|$$
$$= \sqrt{(3)^2 + \left(3\sqrt{3}\right)^2}$$
$$= \sqrt{36}$$
$$= 6$$
$$\theta_2 = \tan^{-1}\left(\frac{3\sqrt{3}}{3}\right)$$
$$= \tan^{-1}\left(\sqrt{3}\right)$$
$$= \frac{\pi}{3}$$
$$w = 6 \operatorname{cis}\left(\frac{\pi}{3}\right)$$

3. Determine zw using de Moivre's theorem: $z_1 \times z_2 = r_1 \times r_2 \operatorname{cis}(\theta_1 + \theta_2)$.

$$zw = 10 \operatorname{cis}\left(\frac{\pi}{6}\right) \times 6 \operatorname{cis}\left(\frac{\pi}{3}\right)$$
$$= 60 \operatorname{cis}\left(\frac{\pi}{6} + \frac{\pi}{3}\right)$$
$$= 60 \operatorname{cis}\left(\frac{\pi}{2}\right)$$

WORKED EXAMPLE 32 Calculating quotients of complex numbers in polar form

Express $10 \operatorname{cis}\left(-\frac{\pi}{3}\right) \div 5 \operatorname{cis}\left(\frac{5\pi}{6}\right)$ in the form $r \operatorname{cis}(\theta)$ where $\theta \in (-\pi, \pi]$.

THINK | **WRITE**

1. Simplify by recalling the division version of de Moivre's theorem: $\frac{z_1}{z_2} = \frac{r_1}{r_2} \operatorname{cis}(\theta_1 - \theta_2)$.

$$10 \operatorname{cis}\left(-\frac{\pi}{3}\right) \div 5 \operatorname{cis}\left(\frac{5\pi}{6}\right) = 2 \operatorname{cis}\left(-\frac{\pi}{3} - \frac{5\pi}{6}\right)$$
$$= 2 \operatorname{cis}\left(-\frac{7\pi}{6}\right)$$

2. Sketch this number.

Im(z)
2
$\frac{\pi}{6}$
$\frac{5\pi}{6}$
$-\frac{7\pi}{6}$
Re(z)

3. State θ, the principal argument.

$$\operatorname{Arg}(z) = \frac{5\pi}{6}$$

4. State the result in polar form.

$$z = 2 \operatorname{cis}\left(\frac{5\pi}{6}\right)$$

TI | THINK	DISPLAY/WRITE
1. On a Calculator page, complete the entry as shown.	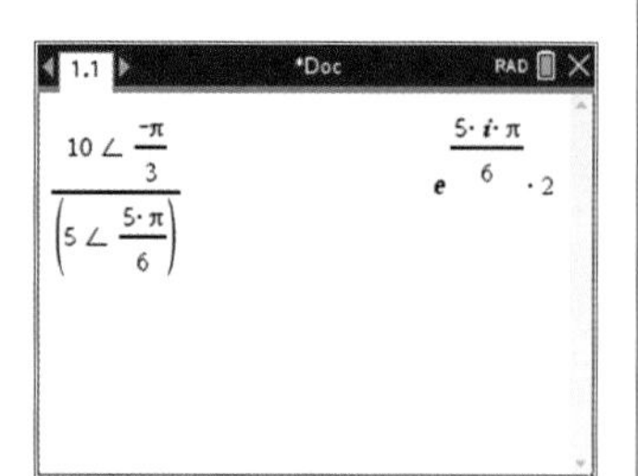
2. Note that the calculator shows the polar form as $r \times e^{i\theta}$. From this form you will need to identify r and θ to express it in the form $z = r\,\text{cis}(\theta)$.	$r = 2$ $\theta = \frac{5\pi}{6}$ $z = 2\,\text{cis}\left(\frac{5\pi}{6}\right)$

CASIO | THINK	DISPLAY/WRITE
1. On a Main screen, write the division using the polar form $r \times e^{\theta i}$ as shown. Then select: Action Complex compToPol	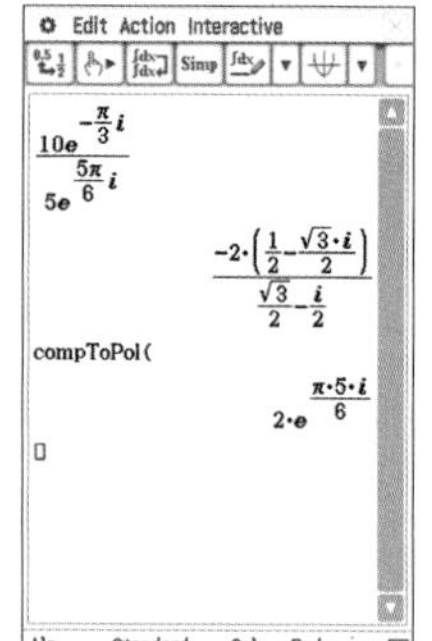
2. Note that the calculator shows the polar form as $r \times e^{i\theta}$. From this form you will need to identify r and θ to express it in the form $z = r\,\text{cis}(\theta)$.	$r = 2$ $\theta = \frac{5\pi}{6}$ $z = 2\,\text{cis}\left(\frac{5\pi}{6}\right)$

Index powers of z

A logical consequence of the multiplication version of de Moivre's theorem involves the calculation of a power of z.

$z^n = z \times z \times ... \times z$ (n times)

> **Powers of z**
>
> **If $z = r\,\text{cis}(\theta)$, then $z^n = r^n\,\text{cis}(n\theta)$.**
>
> **In other words,**
>
> $$(r\,\text{cis}(\theta))^n = r^n\,\text{cis}(n\theta)$$

The proofs required to establish these rules are outside the scope of this course and are not included here.

WORKED EXAMPLE 33 Calculating powers of complex numbers in polar form

If $z = 2\,\text{cis}\left(\frac{5\pi}{6}\right)$, calculate z^3.

THINK	WRITE
1. Recall de Moivre's theorem for index powers.	$z^n = r^n\,\text{cis}(n\theta)$
2. Substitute the values into de Moivre's theorem and simplify.	$z^3 = 2^3\,\text{cis}\left(3 \times \frac{5\pi}{6}\right)$ $= 8\,\text{cis}\left(\frac{5\pi}{2}\right)$
3. Remember $-\pi < \text{Arg}(z) \le \pi$.	$\text{Arg}\left(\frac{5\pi}{2}\right) = \frac{\pi}{2}$
4. Write the answer.	$z^3 = 8\,\text{cis}\left(\frac{\pi}{2}\right)$

Negative powers of z

Your earlier studies have shown that $z^{-1} = \frac{1}{z}$. Similarly, $z^{-3} = \frac{1}{z^3}$.

WORKED EXAMPLE 34 Calculating negative powers of complex numbers in polar form

Evaluate $(1 - i)^{-4}$.

THINK	WRITE
1. Convert $(1 - i)$ to polar form.	$\|1 - i\| = \sqrt{1^2 + (-1)^2}$ $= \sqrt{2}$ $\text{Arg}(1 - i) = \tan^{-1}(-1)$ $= -\frac{\pi}{4}$ $1 - i = \sqrt{2}\,\text{cis}\left(-\frac{\pi}{4}\right)$
2. Write de Moivre's theorem and substitute in the values.	$z^n = r^n\,\text{cis}(n\theta)$ $(1 - i)^{-4} = \left(\sqrt{2}\right)^{-4}\text{cis}\left(4 \times \frac{\pi}{4}\right)$
3. Simplify the expression, recalling the exact values of $\cos(\pi)$ and $\sin(\pi)$.	$= \frac{1}{4}\,\text{cis}(\pi)$ $= \frac{1}{4}\left(\cos(\pi) + i\sin(\pi)\right)$
4. Write the final expression.	$(1 - i)^{-4} = -\frac{1}{4}$

10.7.3 Powers of complex numbers in Cartesian form

Whole powers of z

As with real numbers, powers of complex numbers can be written as:

$$z^n = z \times z \times z \times z \times \ldots \times z \text{ to } n \text{ factors.}$$

Since $z = a + bi$ is a binomial (containing two terms), we can express z^n using Pascal's triangle to generate the coefficients of each term.

```
                          1
                       1     1
                    1     2     1
                 1     3     3     1
              1     4     6     4     1
5th row  →  1     5    10    10     5     1     and so on.
```

$(a+bi)^5$ can therefore be expanded using the elements of the fifth row of Pascal's triangle:

$$\begin{aligned}(a+bi)^5 &= 1a^5+5a^4(bi)^1+10a^3(bi)^2+10a^2(bi)^3+5a(bi)^4+(bi)^5\\ &= 1a^5+5a^4bi+10a^3b^2i^2+10a^2b^3i^3+5ab^4i^4+b^5i^5\\ &= 1a^5+5a^4bi-10a^3b^2-10a^2b^3i+5ab^4+b^5i\\ &= 1a^5-10a^3b^2+5ab^4+5a^4bi-10a^2b^3i+b^5i\\ &= 1a^5-10a^3b^2+5ab^4+(5a^4b-10a^2b^3+b^5)\,i \text{ grouped into standard form.}\end{aligned}$$

$\text{Re}[(a+bi)^5] = 1a^5-10a^3b^2+5ab^4$

$\text{Im}[(a+bi)^5] = 5a^4b-10a^2b^3+b^5$

WORKED EXAMPLE 35 Expanding powers of complex numbers in Cartesian form

Use Pascal's triangle to expand $(2-3i)^3$.

THINK	WRITE
1. Use the third row of Pascal's triangle to expand (1 3 3 1). Use brackets to keep the negative sign of the second term.	$(2-3i)^3 = 1\left(2^3\right)+3(2)^2(-3i)+3(2)(-3i)^2+(-3i)^3$
2. Simplify the expression.	$= 8-36i+54i^2-27i^3$ $= 8-36i-54+27i$ $= -46-9i$

10.7.4 Trigonometric proofs with complex numbers

In Topic 8 you explored trigonometric identities and the different ways to prove whether statements that are written as functions of one or more angles are true. It is possible to combine de Moivre's theorem and binomial expansion of complex numbers to prove some of these multi-angle identities.

If $z = r\,\text{cis}(\theta)$, then de Moivre's theorem tells us that:

$$\begin{aligned}z^n &= r^n\,\text{cis}(n\theta)\\ &= r^n\left(\cos(n\theta)+i\sin(n\theta)\right) \quad [1]\end{aligned}$$

We can also write:

$$z^n = r^n\left(\cos(\theta)+i\sin(\theta)\right)^n \quad [2]$$

Equating [1] and [2] gives us:

$$\cos(n\theta)+i\sin(n\theta) = \left(\cos(\theta)+i\sin(\theta)\right)^n$$

Equating the real and imaginary parts of equations [1] and [2] gives us the following identities.

Identities for $\sin(n\theta)$ and $\cos(n\theta)$

$$\mathbf{\sin(n\theta) = Re\left[\left(\cos(\theta)+i\sin(\theta)\right)^n\right]}$$
$$\mathbf{\cos(n\theta) = Im\left[\left(\cos(\theta)+i\sin(\theta)\right)^n\right]}$$

It may be necessary to use the Pythagorean identity $\sin^2(\theta)+\cos^2(\theta)=1$ as part of your proof.

WORKED EXAMPLE 36 Using de Moivre's theorem to prove trigonometric identities

Use de Moivre's theorem to prove:

a. $\cos(3\theta)=4\cos^3(\theta)-3\cos(\theta)$

b. $\sin(3\theta)=3\sin(\theta)-4\sin^3(\theta)$

THINK	WRITE
1. Both proofs involve 3θ, so it will be necessary to determine $(\cos(\theta)+i\sin(\theta))^3$.	$(\cos(\theta)+i\sin(\theta))^3$
2. Determine the expansion using the third row (1 3 3 1) of Pascal's triangle.	$(\cos(\theta)+i\sin(\theta))^3 = 1(\cos(\theta))^3+3(\cos(\theta))^2(i\sin(\theta))$ $+3(\cos(\theta))(i\sin(\theta))^2+1(i\sin(\theta))^3$ $=\cos^3(\theta)+3i\cos^2(\theta)\sin(\theta)$ $+3i^2\cos(\theta)\sin^2(\theta)+i^3\sin^3(\theta)$
3. Simplify the expression, remembering that $i^2=-1$ and $i^3=-i$.	$=\cos^3(\theta)+3i\cos^2(\theta)\sin(\theta)$ $-3\cos(\theta)\sin^2(\theta)-i\sin^3(\theta)$
4. Group the real and imaginary terms together.	$=\cos^3(\theta)-3\cos(\theta)\sin^2(\theta)$ $+3i\cos^2(\theta)\sin(\theta)-i\sin^3(\theta)$ $(\cos(\theta)+i\sin(\theta))^3=\cos^3(\theta)-3\cos(\theta)\sin^2(\theta)$ $+i(3\cos^2(\theta)\sin(\theta)-\sin^3(\theta))$
a. 1. State the rule for $\cos(n\theta)$.	$\cos(n\theta)=\text{Re}\left[(\cos(\theta)+i\sin(\theta))^n\right]$
2. Use the expansion of $(\cos(\theta)+i\sin(\theta))^3$ to determine $\cos(3\theta)$.	$\cos(3\theta)=\text{Re}\left[(\cos(\theta)+i\sin(\theta))^3\right]$ $=\cos^3(\theta)-3\cos(\theta)\sin^2(\theta)$
3. We want to prove that $\cos(3\theta)=4\cos^3(\theta)-3\cos(\theta)$, so use the Pythagorean identity $\sin^2(\theta)=1-\cos^2(\theta)$ to replace $\sin^2\theta$.	$\cos(3\theta)=\cos^3(\theta)-3\cos(\theta)(1-\cos^2(\theta))$ $=\cos^3(\theta)-3\cos(\theta)+3\cos^3(\theta)$ $\therefore\cos(3\theta)=4\cos^3(\theta)-3\cos(\theta)$
b. 1. State the rule for $\sin(n\theta)$.	$\sin(n\theta)=\text{Im}\left[(\cos(\theta)+i\sin(\theta))^n\right]$
2. Use the expansion of $(\cos(\theta)+i\sin(\theta))^3$ to determine $\sin(3\theta)$.	$\sin(3\theta)=\text{Im}\left[(\cos(\theta)+i\sin(\theta))^3\right]$ $=3\cos^2(\theta)\sin(\theta)-\sin^3(\theta)$
3. We want to prove that $\sin(3\theta)=3\sin(\theta)-4\sin^3(\theta)$, so use the Pythagorean identity $\cos^2(\theta)=1-\sin^2(\theta)$ to replace $\sin^2(\theta)$.	$\sin(3\theta)=3(1-\sin^2(\theta))\sin(\theta)-\sin^3(\theta)$ $=3\sin(\theta)-3\sin^3(\theta)-\sin^3(\theta)$ $\therefore\sin(3\theta)=3\sin(\theta)-4\sin^3(\theta)$

10.7 Exercise

Students, these questions are even better in jacPLUS

Receive immediate feedback and access sample responses

Access additional questions

Track your results and progress

Find all this and MORE in jacPLUS

Technology free

1. **WE29** Express each of the following in the form $r\,\text{cis}(\theta)$ where $\theta \in (-\pi, \pi]$.

   a. $2\,\text{cis}\left(\frac{\pi}{4}\right) \times 3\,\text{cis}\left(\frac{\pi}{2}\right)$

   b. $5\,\text{cis}\left(\frac{2\pi}{3}\right) \times 4\,\text{cis}\left(-\frac{\pi}{3}\right)$

   c. $6\,\text{cis}\left(\frac{3\pi}{4}\right) \times \sqrt{5}\,\text{cis}(\pi)$

   d. $\sqrt{3}\,\text{cis}\left(-\frac{5\pi}{6}\right) \times \sqrt{2}\,\text{cis}\left(-\frac{\pi}{2}\right)$

   e. $\sqrt{7}\,\text{cis}\left(-\frac{7\pi}{12}\right) \times 2\,\text{cis}\left(\frac{5\pi}{12}\right)$

2. **WE30** Express the resultant complex numbers in question 1 in Cartesian form.

3. **WE31** Express the following products in polar form.

   a. $(2+2i)\left(\sqrt{3}+i\right)$

   b. $\left(\sqrt{3}-3i\right)\left(2\sqrt{3}-2i\right)$

   c. $\left(-4+4\sqrt{3}i\right)(-1-i)$

4. **WE32** Express $12\,\text{cis}\left(\frac{5\pi}{6}\right) \div 4\,\text{cis}\left(\frac{\pi}{3}\right)$ in the form $r\,\text{cis}(\theta)$ where $\theta \in (-\pi, \pi]$.

5. Express each of the following in the form $r\,\text{cis}(\theta)$ where $\theta \in (-\pi, \pi]$.

   a. $36\,\text{cis}\left(\frac{3\pi}{4}\right) \div 9\,\text{cis}\left(-\frac{\pi}{6}\right)$

   b. $\sqrt{20}\,\text{cis}\left(-\frac{\pi}{2}\right) \div \sqrt{5}\,\text{cis}\left(-\frac{\pi}{5}\right)$

   c. $4\sqrt{3}\,\text{cis}\left(\frac{4\pi}{7}\right) \div \sqrt{6}\,\text{cis}\left(\frac{11\pi}{14}\right)$

   d. $3\sqrt{5}\,\text{cis}\left(-\frac{7\pi}{12}\right) \div 2\sqrt{10}\,\text{cis}\left(\frac{5\pi}{6}\right)$

6. **WE33** If $z = \sqrt{3}\,\text{cis}\left(\frac{3\pi}{4}\right)$, calculate z^3, giving your answer in polar form.

7. If $w = 2\,\text{cis}\left(-\frac{\pi}{4}\right)$, express each of the following in polar form.

   a. w^4

   b. w^5

8. **WE34** Evaluate $(1-i)^{-3}$.

9. If $z = 1 - i$ and $w = -\sqrt{3} + i$, write the following in standard form.

   a. z^{-4}

   b. w^{-3}

   c. w^{-5}

   d. $\frac{z^3}{w^4}$

   e. $z^2 w^3$

10. Determine $(2+2i)^2(1-\sqrt{3}i)^4$ in standard form.

11. Write $\dfrac{\left(\sqrt{3}-i\right)^6}{\left(2-2\sqrt{3}i\right)^3}$ in the form $x+yi$.

12. WE35 Use Pascal's triangle to expand $(4+5i)^4$.

Technology active

13. If $z=\sqrt{2}\,\text{cis}\left(\frac{3\pi}{4}\right)$ and $w=\sqrt{3}\,\text{cis}\left(\frac{\pi}{6}\right)$, determine the modulus and the argument of $\frac{z^6}{w^4}$.

14. If $z=4+i$ and $w=-3-2i$, determine $(z+w)^9$.

15. Evaluate z^6+w^4, if $z=\sqrt{2}-\sqrt{2}i$ and $w=2-2i$.

16. If $z_1=\sqrt{5}\,\text{cis}\left(-\frac{2\pi}{5}\right)$, $z_2=2\,\text{cis}\left(\frac{3\pi}{8}\right)$ and $z_3=\sqrt{10}\,\text{cis}\left(\frac{\pi}{12}\right)$, determine the modulus and the argument of $\frac{{z_1}^2\times{z_2}^3}{{z_3}^4}$.

17. WE36 Use de Moivre's theorem to prove:

a. $\sin(2\theta)=2\sin(\theta)\cos(\theta)$

b. $\cos(2\theta)=\cos^2(\theta)-\sin^2(\theta)$
$=2\cos^2(\theta)-1$

18. Use de Moivre's theorem to prove:

a. $\sin(4\theta)=4\sin(\theta)\cos^3(\theta)-4\cos(\theta)\sin^3(\theta)$

b. $\cos(4\theta)=\cos^4(\theta)-6\cos^2(\theta)\sin^2(\theta)+\sin^4(\theta)$
$=8\cos^4(\theta)-8\cos^2(\theta)+1$

10.7 Exam questions

Question 1 (4 marks) TECH-FREE

a. Let $u=-1-\sqrt{3}i$. Express u^{12} in Cartesian form. **(2 marks)**

b. Let $w=-1-i$. Express $\frac{1}{w^5}$ in Cartesian form. **(2 marks)**

Question 2 (4 marks) TECH-FREE

Simplify $\dfrac{(1-i)^{10}}{\left(\sqrt{3}-i\right)^6}$, stating your answer in Cartesian form.

Question 3 (6 marks) TECH-FREE

If $u=r\,\text{cis}\left(\frac{\pi}{3}\right)$ and $v=4\,\text{cis}(\theta)$, determine the values of r and θ if:

a. $uv=-12$ **(2 marks)**

b. $\frac{u}{v}=-12i$ **(2 marks)**

c. $u^2v^2=32$ **(2 marks)**

More exam questions are available online.

10.8 Solving quadratic equations with complex roots

LEARNING INTENTION

At the end of this subtopic you should be able to:
- solve quadratic equations over the complex numbers
- solve quartic equations which are reducible to quadratics over the complex numbers.

10.8.1 Linear factors of real quadratic polynomials

The **roots of an equation** are also known as the solutions of an equation. Geometrically, the roots of a function are the x-intercepts of the function.

Up to this point, linear factors of real quadratic polynomials have been limited to those over the set of real numbers. Now, linear factors can be determined over the complex number field.

WORKED EXAMPLE 37 Determining the linear factors of quadratic polynomials

Determine the linear factors over C for $z^2 - 4z + 29$.

THINK	WRITE
1. Complete the square. *Note:* The sum of two squares has no real factors.	$z^2 - 4z + 29$ $= z^2 - 4z + 4 + (29 - 4)$ $= (z-2)^2 + 25$
2. Write the expression as the difference of two squares.	$= (z-2)^2 - 25i^2$ (using $i^2 = -1$) $= (z-2)^2 - (5i)^2$
3. Write the expression as a product of linear factors.	$= (z - 2 + 5i)(z - 2 - 5i)$
4. Write the answer.	The linear factors of $z^2 - 4z + 29$ are $(z - 2 + 5i)(z - 2 - 5i)$.

10.8.2 The general solution of real quadratic equations

Consider the quadratic equation $az^2 + bz + c = 0$, where the coefficients a, b and c are real. Recall that the roots depend upon the discriminant, $\Delta = b^2 - 4ac$.
- If $\Delta > 0$, the equation has two distinct real roots.
- If $\Delta = 0$, the equation has one real repeated root.
- If $\Delta < 0$, the equation has no real roots.

With the introduction of complex numbers, it can now be stated that if $\Delta 0$, then the equation has one pair of **complex conjugate roots** of the form $a \pm bi$.

WORKED EXAMPLE 38 Solving quadratic equations

Solve for z, given $z^2 - 6z + 25 = 0$.

THINK	WRITE
Method 1: Using completing the square	$z^2 - 6z + 25 = 0$
1. Complete the square.	$z^2 - 6x + 9(25 - 9) = 0$
	$(z-3)^2 + 16 = 0$
	$(z-3)^2 = -16$
2. Substitute -1 with i^2.	$(z-3)^2 = 16i^2$
3. Solve for z.	$z - 3 = \pm 4i$
4. State the two solutions for z.	$z = 3 \pm 4i$
Method 2: Using the general quadratic formula	$z = \frac{-b \pm \sqrt{b^2 - 4ac}}{2a}$
1. Write the general quadratic formula.	
2. State the values for a, b and c.	$z^2 - 6z + 25 = 0$ $a = 1,\ b = -6,\ c = 25$
3. Substitute the values into the formula and simplify.	$z = \frac{6 \pm \sqrt{(-6)^2 - 4(1)(25)}}{2}$ $= \frac{6 \pm \sqrt{-64}}{2}$ $= \frac{6 \pm \sqrt{64i^2}}{2}$ $= \frac{6 \pm 8i}{2}$ $= 3 \pm 4i$
4. State the two solutions.	$z = 3 \pm 4i$

10.8.3 The relationship between roots and coefficients

For a quadratic with real coefficients, we have seen that if the discriminant is negative, then the roots occur in complex conjugate pairs. Here we determine a relationship between the roots and the coefficients.

Given a quadratic $az^2 + bz + c = 0$ where $a \neq 0$, then:

$$z^2 + \frac{b}{a}z + \frac{c}{a} = 0$$

Let the roots of the quadratic equation be α and β, so the factors are $(z - \alpha)(z - \beta)$. Expanding the brackets, the expression becomes:

$$z^2 - (\alpha + \beta)z + \alpha\beta = 0 \text{ or}$$

$$z^2 - (\text{sum of roots})\, z + \text{product of roots} = 0$$

$$\text{so that } \alpha + \beta = -\frac{b}{a} \text{ and } \alpha\beta = \frac{c}{a}.$$

This gives us a relationship between the roots of a quadratic equation and its coefficients. Rather than formulating a problem as solving a quadratic equation, we now consider the reverse problem. That is forming a quadratic equation with real coefficients, given one of the roots.

WORKED EXAMPLE 39 Determining a quadratic given one of its roots

Determine the equation of the quadratic $P(z)$, with real coefficients given that $P(4-3i)=0$.

THINK	WRITE
1. State the value of the second root, recalling that complex conjugate roots are of the form $a \pm bi$.	$P(4-3i)=0$ $\Rightarrow 4-3i$ is a root of $P(z)$. $\Rightarrow 4+3i$ is also a root of $P(z)$. Let $\alpha = 4-3i$ and let $\beta = 4+3i$.
2. Determine the sum of the roots.	$\alpha + \beta = 8$
3. Determine the product of the roots.	$\alpha\beta = 16 - 9i^2 = 25$
4. State the quadratic equation.	$P(z) = (z-4+3i)(z-4-3i)$ $= z^2 - 8z + 25$

10.8 Exercise

Students, these questions are even better in jacPLUS

Receive immediate feedback and access sample responses

Access additional questions

Track your results and progress

Find all this and MORE in jacPLUS

Technology free

1. WE37 Determine the linear factors over C for $z^2 - 6z + 25$.
2. Determine the linear factors over C for $z^2 + 4z + 7$.
3. WE38 Solve for z, given $z^2 - 4z + 29 = 0$.
4. Solve for z, given $z^2 + 2z + 26 = 0$.
5. Calculate the roots of each of the following.
   a. $z^2 + 2z + 17 = 0$
   b. $z^2 - 4z + 20 = 0$
   c. $z^2 - 6z + 13 = 0$
   d. $z^2 + 10z + 41 = 0$
6. WE39 Determine the quadratic $P(z)$ with real coefficients given that $P(-5-2i)$.
7. Determine the quadratic $P(z)$ with real coefficients given that $P(5i)$.
8. Form a quadratic with integer coefficients for each of the following cases.
   a. -2 and $\frac{1}{3}$ are the roots.
   b. $2-6i$ is a root.
   c. $-2+3i$ is a root.
   d. $-4-5i$ is a root.

9. Form a quadratic with integer coefficients for each of the following cases.
   a. $\frac{1}{2}$ and $-\frac{1}{5}$ are the roots.
   b. $4-\sqrt{3}i$ is a root.
   c. $-5-\sqrt{7}i$ is a root.
   d. $-3-\sqrt{8}i$ is a root.

10. a. Solve for z if $z^2+(2-i)z+(3-i)=0$.
    b. Determine the two numbers such that their sum is 6 and their product is 10.
    c. i. Show that $(3-2i)^2=5-12i$.
       ii. Solve for z if $z^2+(4-6i)z=10$.

11. a. Solve for z if $z^2+(4-i)z+(5-2i)=0$.
    b. Determine the two numbers such that their sum is 8 and their product is 25.
    c. i. Show that $(4-3i)^2=7-24i$.
       ii. Solve for z if $z^2+\sqrt{11}z+(1+6i)=0$.

12. Solve for z in each of the following.
    a. $z(4-z)=8$
    b. $z(6-z)=10$
    c. $z(2-z)=26$
    d. $z(8-z)=41$

13. Solve for z in each of the following.
    a. $(z+2)(z-6)+25=0$
    b. $(z-2)(z+8)+30=0$
    c. $(z+2)(z-4)+12=0$
    d. $(z+6)(z+4)+7=0$

14. Determine the roots of each of the following.
    a. $4z^2+12z+10=0$
    b. $4z^2+20z+29=0$
    c. $4z^2-12z+13=0$
    d. $9z^2-42z+53=0$

10.8 Exam questions

Question 1 (2 marks) TECH-FREE

Solve for z in each of the following.

a. $z^2+12z+52=0$ **(1 mark)**

b. $2z^2-5z+6=0$ **(1 mark)**

Question 2 (2 marks) TECH-FREE

a. Form a quadratic with integer coefficients that has $2-\sqrt{3}i$ as one of its roots. **(1 mark)**

b. The quadratic $z^2+bz+41=0$ has $3+\sqrt{c}i$ as one of its roots, determine the values of the real numbers b and c. **(1 mark)**

Question 3 (3 marks) TECH-FREE

Determine the real numbers a and b if $(a+bi)^2=7-24i$.

More exam questions are available online.

10.9 Lines, rays, circles, ellipses and regions in the complex plane

LEARNING INTENTION

At the end of this subtopic you should be able to:
- identify and sketch lines, rays, circles and ellipses in the complex plane
- sketch regions defined in the complex plane.

So far in this topic, complex numbers have been used to represent points on the Argand plane. If we consider z as a complex variable, we can sketch relations or regions of the Argand plane.

10.9.1 Lines

If $z = x + yi$, then $\text{Re}(z) = x$ and $\text{Im}(z) = y$. The equation $a\,\text{Re}(z) + b\,\text{Im}(z) = c$ where a, b and $c \in R$ represents the line $ax + by = c$.

WORKED EXAMPLE 40 Sketching lines in the complex plane (1)

Determine the Cartesian equation and sketch the graph defined by $\{z: 2\,\text{Re}(z) - 3\,\text{Im}(z) = 6\}$.

THINK	WRITE/DRAW
1. Consider the equation.	$2\,\text{Re}(z) - 3\,\text{Im}(z) = 6$ As $z = x + yi$, then $\text{Re}(z) = x$ and $\text{Im}(z) = y$. This is a straight line with the Cartesian equation $2x - 3y = 6$.
2. Evaluate the axial intercepts.	When $y = 0$, $2x = 6 \Rightarrow x = 3$. $(3, 0)$ is the intercept with the real axis. When $x = 0$, $-3y = 6 \Rightarrow y = -2$. $(0, -2)$ is the intercept with the imaginary axis.
3. Identify and sketch the equation.	The equation represents the line $2x - 3y = 6$.

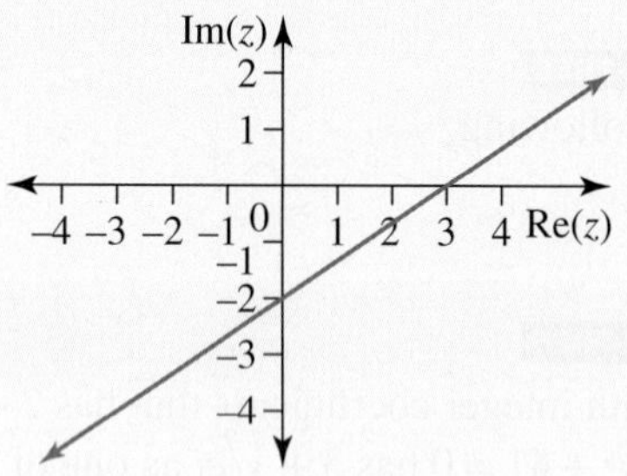

Im(z)
2
1
–4 –3 –2 –1 0 1 2 3 4 Re(z)
–1
–2
–3
–4

Lines in the complex plane can also be represented as a set of points that are equidistant from two other fixed points. The equations of a line in the complex plane can thus have multiple representations.

WORKED EXAMPLE 41 Sketching lines in the complex plane (2)

Determine the Cartesian equation and sketch the graph defined by $\{z: |z-2i| = |z+2|\}$.

THINK	WRITE/DRAW
1. Consider the equation as a set of points.	$\lvert z-2i\rvert = \lvert z+2\rvert$ Substitute $z=x+yi$: $\lvert x+yi-2i\rvert = \lvert x+yi+2\rvert$
2. Group the real and imaginary parts together.	$\lvert x+(y-2)i\rvert = \lvert (x+2)+yi\rvert$
3. Use the definition of the modulus.	$\sqrt{x^2+(y-2)^2} = \sqrt{(x+2)^2+y^2}$
4. Square both sides, expand, and cancel like terms.	$x^2+y^2-4y+4 = x^2+4x+4+y^2$ $-4y = 4x$
5. Identify the required line.	$y=-x$
6. Identify the line geometrically.	The line is the set of points that is equidistant from the two points $(0, 2)$ and $(-2, 0)$.
7. Sketch the required line.	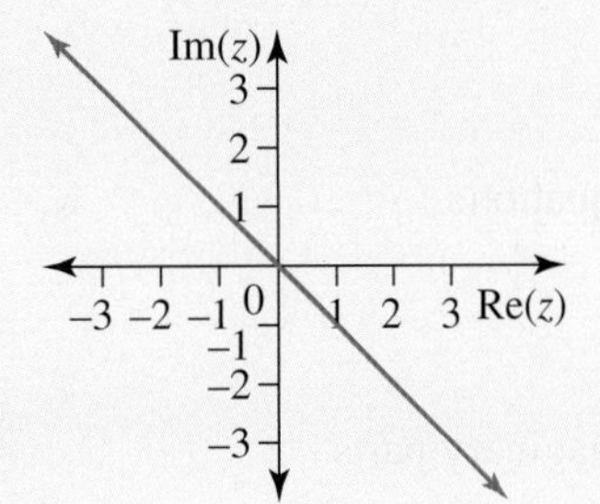

10.9.2 Rays

A ray is a half-line and is usually defined in terms of the argument of the complex number. When drawing rays, remember that $\text{Arg}(z) = \theta \in (-\pi, \pi]$. The argument of a complex number $z = x + yi$ is $\theta = \tan^{-1}\left(\dfrac{y}{x}\right)$, $x > 0$, where θ is measured anticlockwise from the positive real axis.

$\text{Arg}(z) = \theta$ represents the set of all points on the half-line or ray that has one end at the origin and makes an angle of θ with the positive real axis. Note that the end point, in this case the origin, is not included in the set since $x > 0$ and $y > 0$. We indicate this by placing a small open circle at this point.

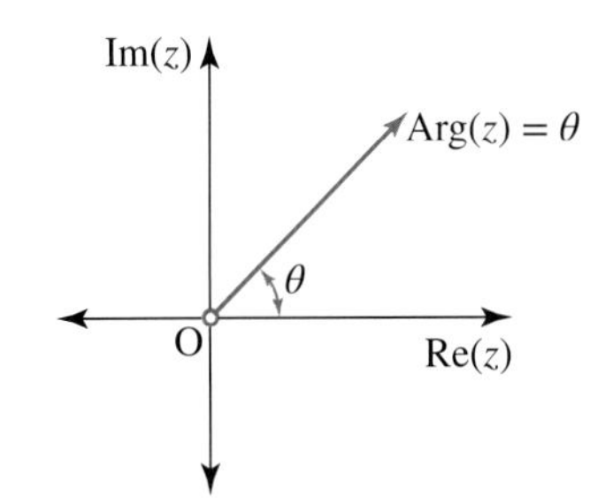

WORKED EXAMPLE 42 Sketching rays in the complex plane

a. **Describe and sketch the graph defined by $\left\{z: \text{Arg}(z) = \dfrac{-\pi}{4}\right\}$.**

b. **Determine the Cartesian equation and sketch the graph defined by $\left\{z: \text{Arg}(z-1+i) = \dfrac{-\pi}{4}\right\}$.**

THINK / **WRITE/DRAW**

a. 1. Consider the given equation.

a. $\text{Arg}(z) = \frac{-\pi}{4}$

2. Recognise the equation as a ray.

The equation is of the form $\text{Arg}(z) = \theta$.

3. Identify the point from which the ray starts.

The ray starts from the point $(0, 0)$, not including the point.

4. Determine the angle the ray makes.

The ray makes an angle of –45° with the positive real axis.

5. Describe the ray.

The ray starts from $(0, 0)$, making an angle of –45° with the positive real axis.

6. Sketch the required ray.

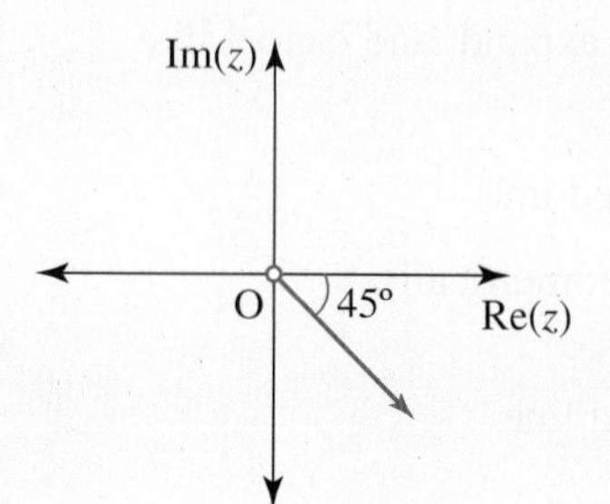

b. 1. Consider the given equation.

b.

$$\text{Arg}(z - 1 + i) = \frac{-\pi}{4}$$

2. Substitute $z = x + yi$.

$$\text{Arg}(x + yi - 1 + i) = \frac{-\pi}{4}$$

3. Group the real and imaginary parts.

$$\text{Arg}((x - 1) + (y + 1)i) = \frac{-\pi}{4}$$

4. Use the definition of the argument.

$$\tan^{-1}\left(\frac{y+1}{x-1}\right) = \frac{-\pi}{4} \text{ for } x > 1$$

5. Simplify.

$$\frac{y+1}{x-1} = \tan\left(\frac{-\pi}{4}\right) \text{ for } x > 1$$

$$\frac{y+1}{x-1} = -1 \text{ for } x > 1$$

6. State the Cartesian equation of the ray.

$$y + 1 = -(x - 1) \text{ for } x > 1$$

$$y = -x \text{ for } x > 1$$

7. Identify the point from which the ray starts.

The ray starts from the point $(1, -1)$, not including the point

8. Determine the angle the ray makes with the positive real axis.

The ray makes an angle of −45° with the positive real axis.

9. Sketch the ray.

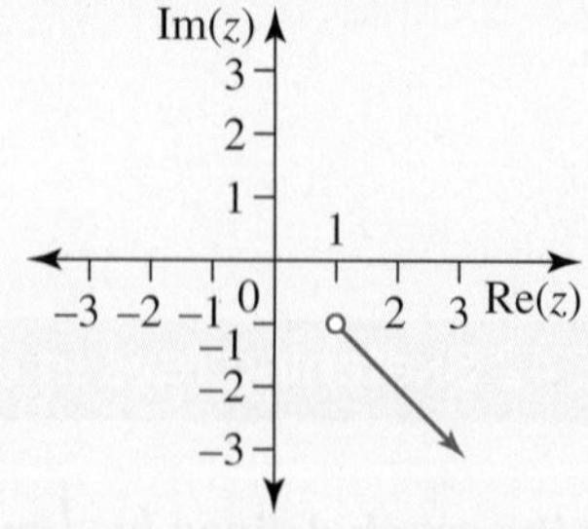

10. Alternatively, use translations from the answer of part **a**.

The ray from the origin making an angle of –45° with the positive real axis has been translated 1 unit to the right parallel to the real axis and 1 unit down parallel to the imaginary axis.

TI | THINK	DISPLAY/WRITE
1. On a Calculator page define z as $x + yi$. Type in the equation using the keyword 'angle' instead of 'arg'.	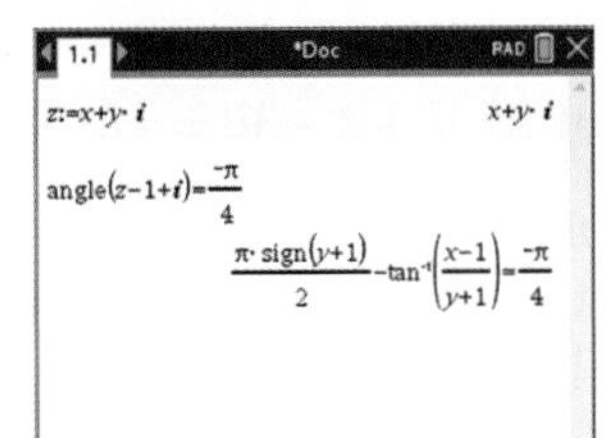
2. Type in 'solve' followed by the result of the above equation with 'y' at the end.	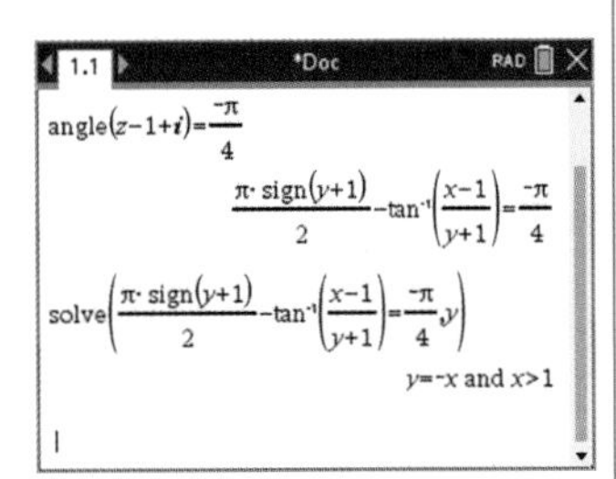
3. On a Graphs page, type in the equation as follows, using the 'given that' symbol, | to restrict the domain.	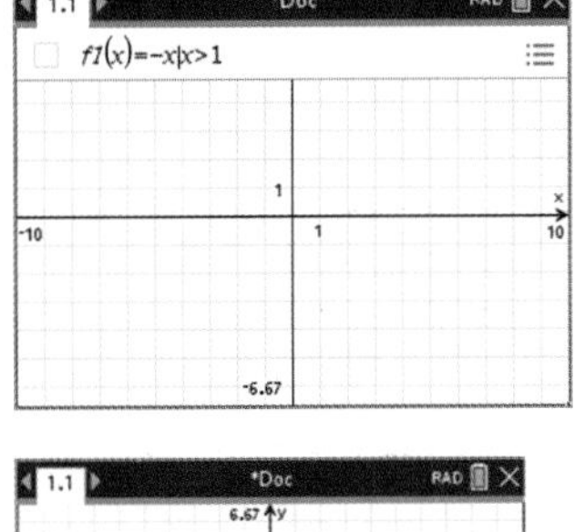
4. The ray will appear, though the end point will not be shown with an open circle as it should be.	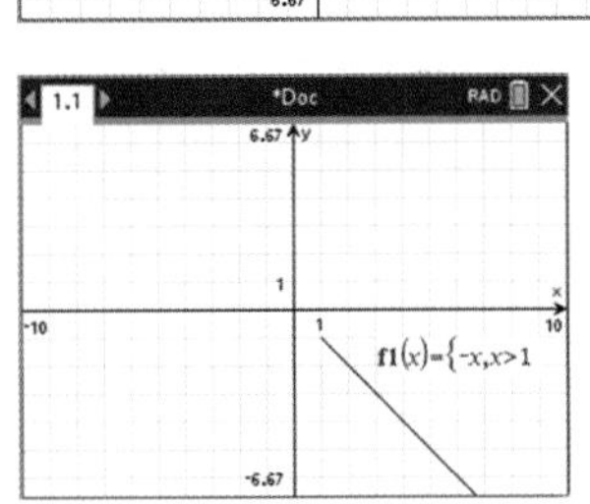

CASIO | THINK	DISPLAY/WRITE
1. On a Main screen, complete the entry as shown using the arg and solve Actions.	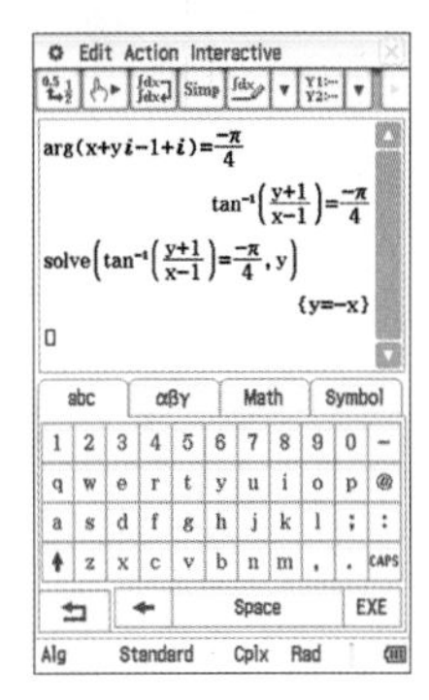
2. On a Graphs screen, type in the equation as follows, using the 'given that' symbol, | to restrict the domain. The ray will appear. Even though the end point will not be shown with an open circle as it should be, trying to trace to the point will show it is undefined.	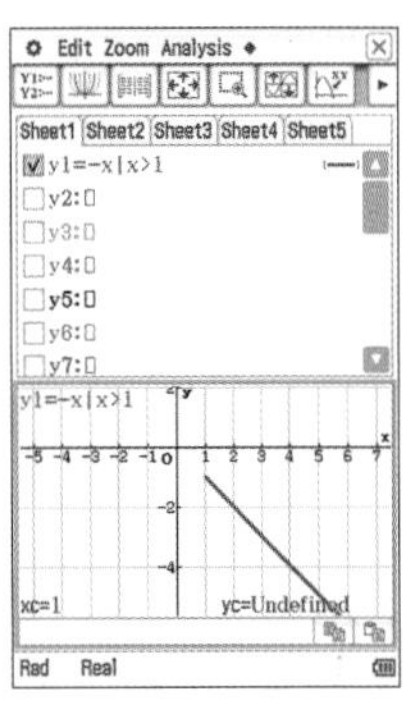

10.9.3 Circles

The equation $|z| = r$ where $z = x + yi$ is given by $|z| = \sqrt{x^2 + y^2} = r$. Expanding this produces $x^2 + y^2 = r^2$. This represents a circle with centre at the origin and radius r. Geometrically, $|z| = r$ represents the set of points, or what is called the locus of points, in the Argand plane that are at r units from the origin.

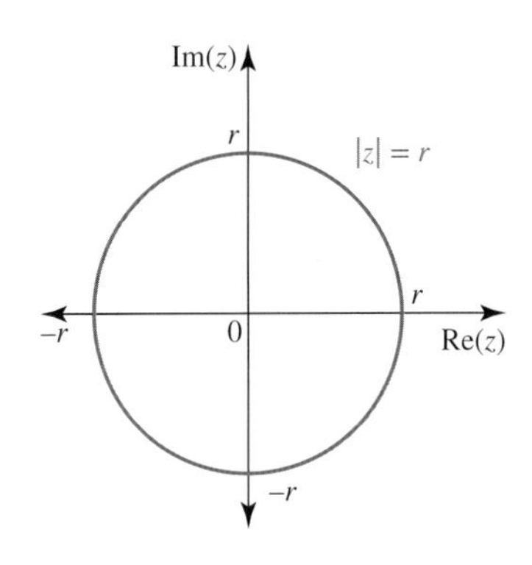

WORKED EXAMPLE 43 Sketching circles in the complex plane

Determine the Cartesian equation and sketch the graph of $\{z: |z+2-3i| = 4\}$.

THINK	WRITE/DRAW
1. Consider the equation.	$|z+2-3i| = 4$ Substitute $z = x + yi$: $|x+yi+2-3i| = 4$
2. Group the real and imaginary parts.	$|(x+2)+i(y-3)| = 4$
3. Use the definition of the modulus.	$\sqrt{(x+2)^2+(y-3)^2} = 4$
4. Square both sides.	$(x+2)^2+(y-3)^2 = 16$
5. Sketch and identify the graph of the Argand plane.	The equation represents a circle with centre at $(-2, 3)$ and radius 4.

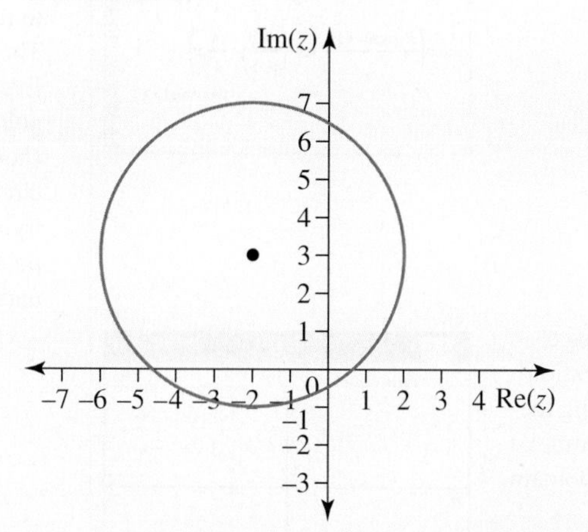

TI | THINK	DISPLAY/WRITE
1. On a Calculator page, define z as $x + yi$. Type in the equation $|z+2-3i| = 4$.	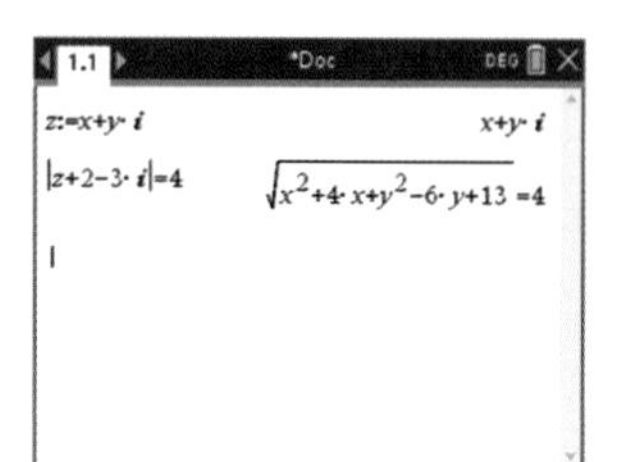
2. Square both sides of the resulting equation.	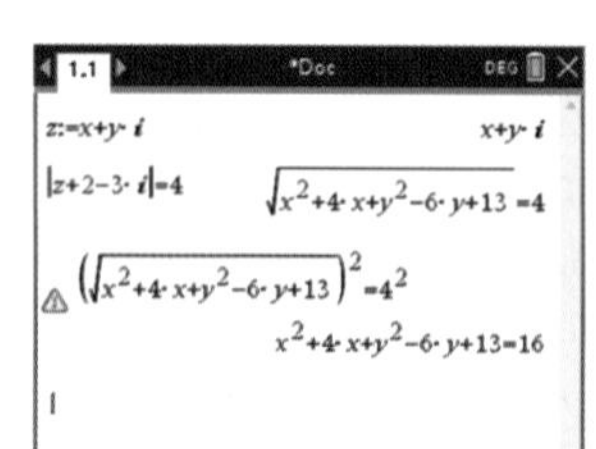

CASIO | THINK	DISPLAY/WRITE
1. On a Main screen, define z as $x + yi$.	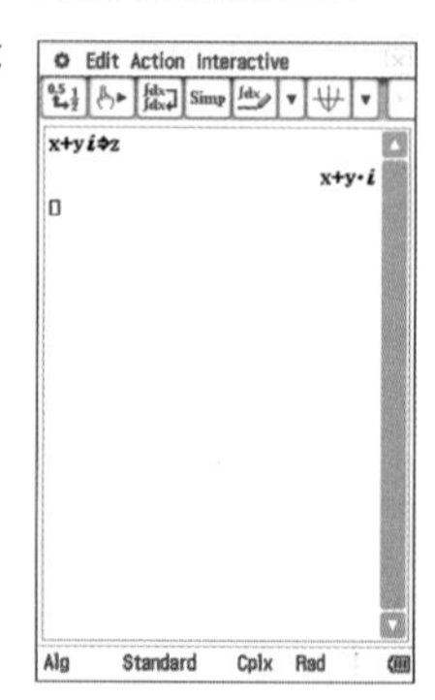
2. Type in 'cExpand' followed by the equation as shown. The equation of the circle is therefore $(x+2)^2+(y-3)^2 = 16$.	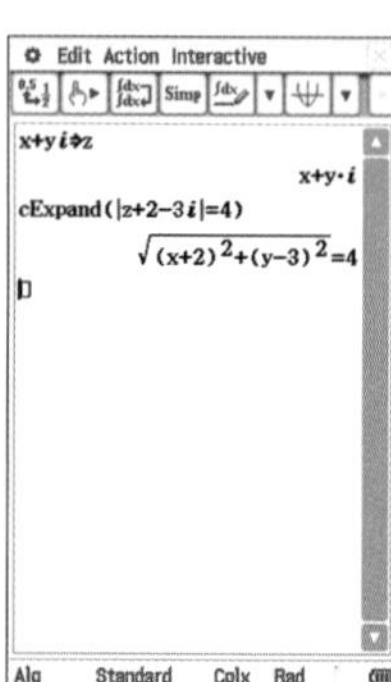

3. Press Menu, then select:
3 Algebra
5 Complete the square
Complete the entry line with the equation from the previous step, with 'x, y' at the end.

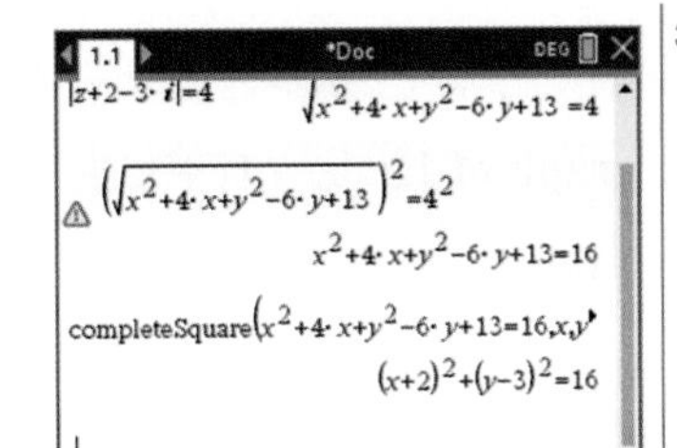

4. On a Graphs page, delete $f1(x)$ and select
6 Relation. Type the equation of the circle in as shown.

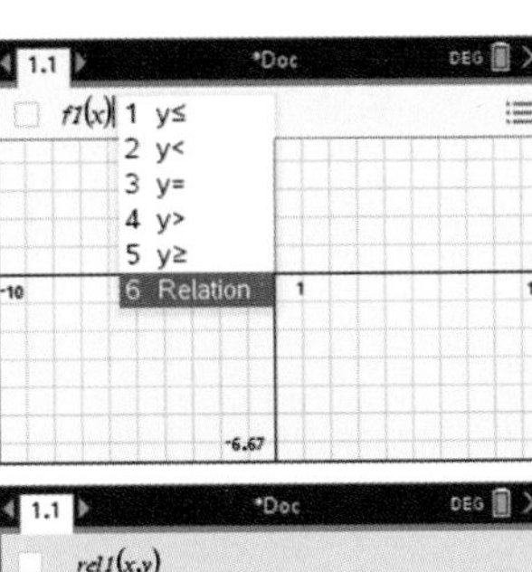

5. The circle will appear. You may need to edit the window setting to see the entire graph using:
Menu 4:
Window/Zoom

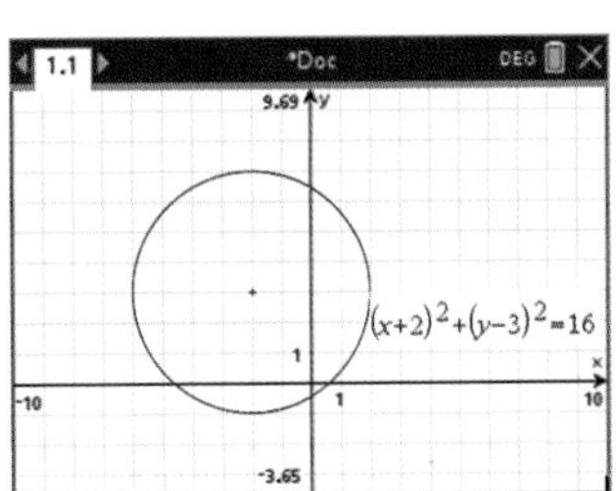

3. To sketch relations go to a Conics screen, type in the equation and click on the top left button, below the settings button.

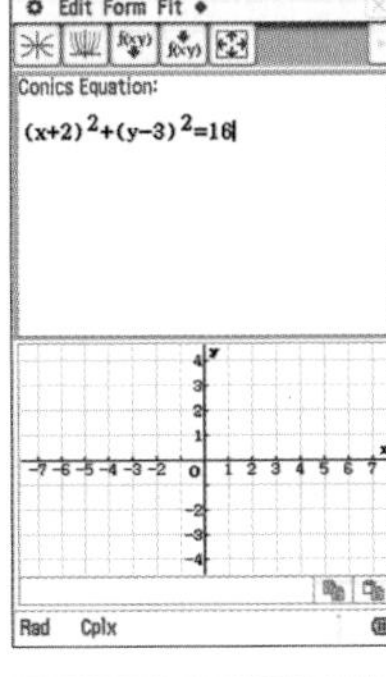

4. The graph will appear on the screen as shown.

10.9.4 Ellipses

The general form of an ellipse is $\dfrac{x^2}{a^2} + \dfrac{y^2}{b^2} = 1$.

The definition of an ellipse is that for all points on the curve the sum of the distances to two fixed points is a constant.

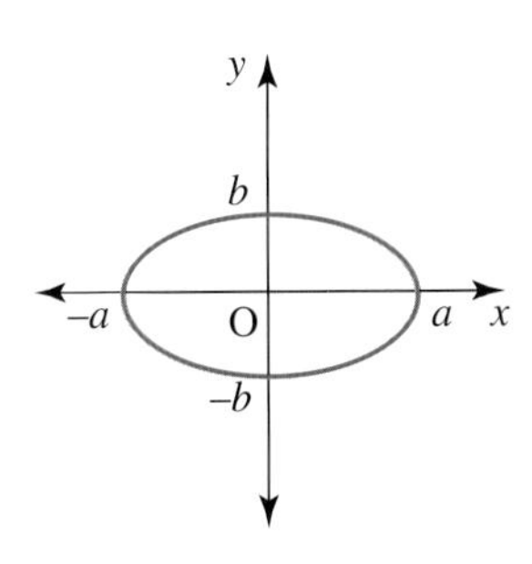

For example, the equation $|z-1| + |z+1| = 4$ would represent an ellipse where the sum of the distances from the variable point z to the points $(1, 0)$ and $(-1, 0)$ is equal to 4.

This is illustrated in the following worked example.

WORKED EXAMPLE 44 Sketching ellipses in the complex plane

Determine the Cartesian equation and sketch the graph of $\{z: |z-1|+|z+1|=4\}$.

THINK	WRITE/DRAW
1. Consider the equation.	$\lvert z-1\rvert+\lvert z+1\rvert=4$
2. Substitute $z=x+yi$.	$\lvert x+yi-1\rvert+\lvert x+yi+1\rvert=4$
3. Group the real and imaginary parts.	$\lvert (x-1)+yi\rvert+\lvert (x+1)+yi\rvert=4$
4. Use the definition of the modulus.	$\sqrt{(x-1)^2+y^2}+\sqrt{(x+1)^2+y^2}=4$
5. Rearrange the equation.	$\sqrt{(x-1)^2+y^2}=4-\sqrt{(x+1)^2+y^2}$
6. Square both sides.	$(x-1)^2+y^2=16-8\sqrt{(x+1)^2+y^2}+(x+1)^2+y^2$
7. Expand and simplify.	$x^2-2x+1+y^2=x^2+2x+1+y^2+16-8\sqrt{(x+1)^2+y^2}$ $8\sqrt{(x+1)^2+y^2}=16+4x$ $2\sqrt{(x+1)^2+y^2}=4+x$
8. Square both sides.	$4\left((x+1)^2+y^2\right)=16+8x+x^2$
9. Expand and simplify.	$4x^2+8x+4+4y^2=16+8x+x^2$ $3x^2+4y^2=12$ $\dfrac{x^2}{4}+\dfrac{y^2}{3}=1$
10. Identify and sketch the graph.	The equation represents an ellipse with centre at $(0,0)$, x-intercepts at $(2,0)$ and $(-2,0)$, and y-intercepts at $\left(0,\sqrt{3}\right)$ and $\left(0,-\sqrt{3}\right)$.

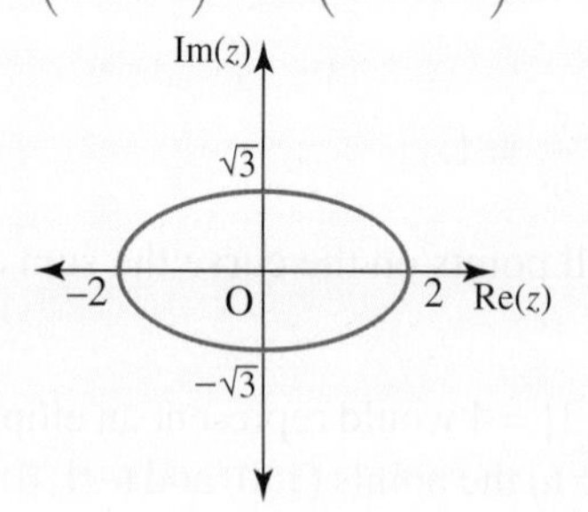

10.9.5 Regions in the complex plane

When graphing a linear inequality on a Cartesian plane, the solution is a region on one side of the line, called a half plane. Similarly, sets of points in the complex plane can be represented by regions. For example, $\{z: \text{Re}(z)\geq 4\}$ is the set of points on or to the right of the vertical line $x=4$. The boundary line is a solid line if it is included or dotted if it is not included. Special care needs to be taken when the inequality involves the argument of the complex number. This is illustrated in the following worked example.

WORKED EXAMPLE 45 Sketching regions in the complex plane (1)

Sketch the region defined by $\left\{z: \text{Arg}(z) \leq \frac{\pi}{4}\right\}$.

THINK	WRITE/DRAW
1. Consider the given set.	$\text{Arg}(z) \leq \frac{\pi}{4}$
2. Determine the boundary line.	$\text{Arg}(z) = \frac{\pi}{4}$ is a ray from but not including the point $(0, 0)$ making an angle of 45° with the positive real axis.
3. State the restrictions on θ for Arg(z).	$\text{Arg}(z) = \theta \in (-\pi, \pi]$
4. Apply to the given inequality.	$\text{Arg}(z) \leq \frac{\pi}{4}$ $\therefore -\pi < \theta \leq \frac{\pi}{4}$
5. Sketch the region, dotting the negative real axis as it is not included. Show $(0, 0)$ as an open circle as it is also not included in the region.	

Regions may involve the intersection of two sets of points in the complex plane. This is illustrated in the following worked example.

WORKED EXAMPLE 46 Sketching regions in the complex plane (2)

Sketch the region defined by $\{z: 0 \leq \text{Re}(z) \leq 4\} \cap \{z: -1 \leq \text{Im}(z) < 3\}$.

THINK	WRITE/DRAW
1. Consider the first set, where $z = x + yi$.	$0 \leq \text{Re}(z) \leq 4$ $\therefore 0 \leq x \leq 4$
2. Determine the boundary lines and state the region.	The boundary lines are $x = 0$ and $x = 4$ The region is on and between the lines.
3. Sketch the region.	

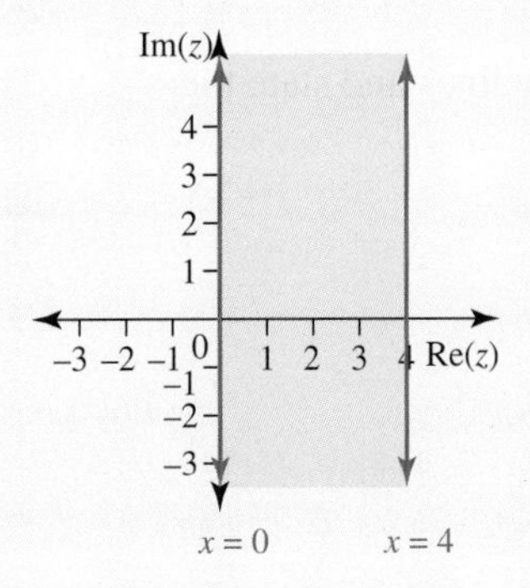

4. Consider the second set. $-1 \leq \text{Im}(z) < 3$

$\therefore -1 \leq y < 3$

5. Determine the boundary lines and state the region. The boundary lines are $y = -1$ and $y = 3$. The region includes $y = -1$ up to a dotted $y = 3$.

6. Sketch the region.

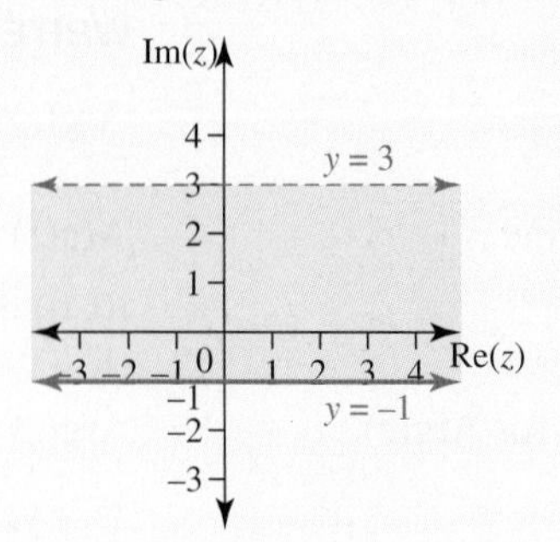

7. Shade the intersection, or overlap, of the two regions. Dot the boundary line $y = 3$ as it is not included. The points of intersection (4, 3) and (0, 3) are open circles.

The required region is:

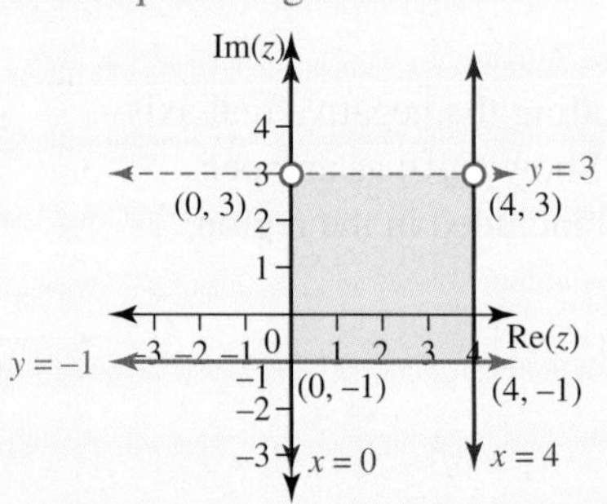

WORKED EXAMPLE 47 Sketching regions in the complex plane (3)

a. Sketch the region defined by $\{z: 2 < |z| \leq 4\}$.

b. Sketch the region defined by $\left\{z: \frac{\pi}{3} \leq \text{Arg}(z) \leq \frac{2\pi}{3}\right\}$.

c. Hence, sketch the region defined by $\{z: 2 < |z| \leq 4\} \cap \left\{z: \frac{\pi}{3} \leq \text{Arg}(z) \leq \frac{2\pi}{3}\right\}$.

THINK	WRITE/DRAW
a. 1. Consider the given set, where $z = x + yi$.	a. $2 < \lvert z \rvert \leq 4$
2. Square each side.	$2 < \sqrt{x^2 + y^2} \leq 4$ $4 < x^2 + y^2 \leq 16$
3. Determine the boundary lines and state the region.	The boundary lines are two circles, centre (0, 0). The circle with radius 2 is dotted. The circle with radius 4 is solid.

4. Identify and sketch the region.

The region lies between the two concentric circles of radius 2 and 4, not including the circumference of the circle with radius 2.

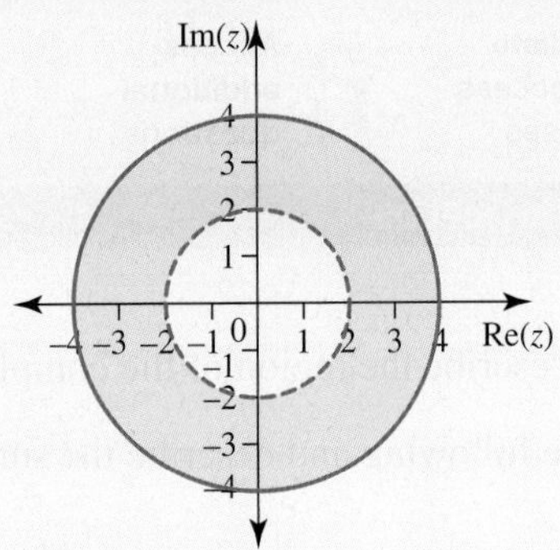

b. 1. Consider the given set.

b. $\frac{\pi}{3} \le \text{Arg}(z) \le \frac{2\pi}{3}$

2. Determine the boundary lines.

$\text{Arg}(z) = \frac{\pi}{3}$ and $\text{Arg}(z) = \frac{2\pi}{3}$

The boundary lines are rays from but not including $(0, 0)$ at angles of 60° and 120° from the positive real axis.

3. State the restrictions on θ for $\text{Arg}(z)$.

$\text{Arg}(z) = \theta \in (-\pi, \pi]$

4. Apply to the given inequality.

$\frac{\pi}{3} \le \text{Arg}(z) \le \frac{2\pi}{3}$

$\therefore \frac{\pi}{3} \le \theta \le \frac{2\pi}{3}$

5. Identify and sketch the region, shading between the two rays, with an open circle at $(0, 0)$.

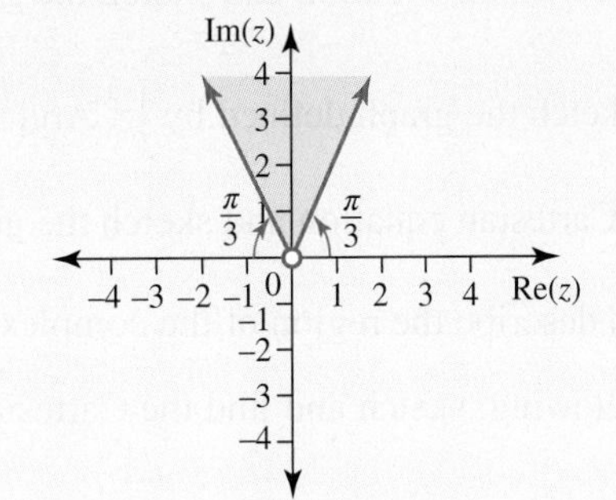

c. 1. On one Argand diagram, sketch both regions from parts **a** and **b**.

2. Identify and sketch the required region, which is the common region. Open circles are needed where the rays intersect with the circle of radius 2 units.

c.

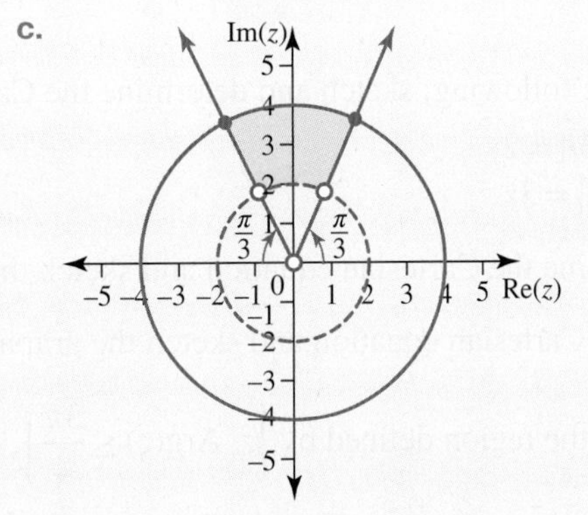

10.9 Exercise

Students, these questions are even better in jacPLUS

Receive immediate feedback and access sample responses

Access additional questions

Track your results and progress

Find all this and MORE in jacPLUS

1. **WE40** Sketch and describe the region of the complex plane defined by $\{z: 4\,\text{Re}(z) + 3\,\text{Im}(z) = 12\}$.
2. Illustrate each of the following and describe the subset of the complex plane.
   a. $\{z: \text{Im}(z) = 2\}$
   b. $\{z: \text{Re}(z) + 2\,\text{Im}(z) = 4\}$
3. **WE41** Sketch and describe the region of the complex plane defined by $\{z: |z + 3i| = |z - 3|\}$.
4. Sketch and describe the region of the complex plane defined by $\{z: |z - i| = |z + 3i|\}$.
5. Determine the Cartesian equation and sketch each of the following sets.
   a. $\{z: |z + 4| = |z - 2i|\}$
   b. $\{z: |z + 2 - 3i| = |z - 2 + 3i|\}$
6. **WE42**
   a. Describe and sketch the graph defined by $\left\{z: \text{Arg}(z) = \frac{\pi}{6}\right\}$.
   b. Determine the Cartesian equation and sketch the graph defined by $\left\{z: \text{Arg}(z - 1) = \frac{\pi}{6}\right\}$.
7. a. Describe and sketch the graph defined by $\left\{z: \text{Arg}(z) = \frac{\pi}{4}\right\}$.
   b. Determine the Cartesian equation and sketch the graph defined by $\left\{z: \text{Arg}(z + 2) = \frac{\pi}{4}\right\}$.
8. **WE43** Sketch and describe the region of the complex plane defined by $\{z: |z - 3 + 2i| = 4\}$.
9. For each of the following, sketch and find the Cartesian equation of the set, and describe the region.
   a. $\{z: |z| = 3\}$
   b. $\{z: |z| = 2\}$
10. For each of the following, sketch and determine the Cartesian equation of the set.
    a. $\{z: |z + 2 - 3i| = 2\}$
    b. $\{z: |z - 3 + i| = 3\}$
11. **WE44** Determine the Cartesian equation and sketch the graph of $\{z: |z - 3| + |z + 3| = 12\}$.
12. Determine the Cartesian equation and sketch the graph of $\{z: |z + 2| + |z - 2| = 6\}$.
13. **WE45** Sketch the region defined by $\left\{z: \text{Arg}(z) \le \frac{3\pi}{4}\right\}$.
14. Sketch the region defined by $\left\{z: \text{Arg}(z) \ge \frac{\pi}{3}\right\}$.
15. **WE46** Sketch the region defined by $\{z: -2 \le \text{Re}(z) \le 2\} \cap \{z: 3 \le \text{Im}(z) < 5\}$.
16. Sketch the region defined by $\{z: 0 < \text{Re}(z) < 6\} \cap \{z: 0 \le \text{Im}(z) \le 4\}$.

17. Describe and sketch the region defined by $\{z: 1 \leq |z-2| \leq 2\}$.

18. Describe and sketch the region defined by $\{z: |z| \leq 3\} \cap \{z: |z-3| \leq 3\}$.

19. WE47

a. Sketch the region defined by $\{z: 3 < |z| \leq 5\}$.

b. Sketch the region defined by $\left\{z: \frac{\pi}{4} \leq \text{Arg}(z) \leq \frac{3\pi}{4}\right\}$.

c. Hence, sketch the region defined by $\{z: 3 < |z| \leq 5\} \cap \left\{z: \frac{\pi}{4} \leq \text{Arg}(z) \leq \frac{3\pi}{4}\right\}$.

20. a. Sketch the region defined by $\{z: 3 \leq |z| \leq 6\}$.
b. Sketch the region defined by $\{z: \text{Re}(z) + \text{Im}(z) \geq 3\}$.
c. Hence, sketch the region defined by $\{z: 3 \leq |z| \leq 6\} \cap \{z: \text{Re}(z) + \text{Im}(z) \geq 3\}$.

10.9 Exam questions

Question 1 (4 marks) TECH-FREE

Sketch and describe each of the following sets, clearly indicating which boundaries are included.

a. $\{z: |z-2| = |z-4|\}$ **(2 marks)**

b. $\{z: |z+4i| = |z-4|\}$ **(2 marks)**

Question 2 (2 marks) TECH-FREE

Determine the Cartesian equation and sketch the graph defined by $\left\{z: \text{Arg}(z-2) = \frac{\pi}{6}\right\}$.

Question 3 (2 marks) TECH-FREE

Describe and sketch the region defined by $\{z: 1 \leq |z| \leq 2\}$.

More exam questions are available online.

10.10 Review

10.10.1 Summary

doc-37053

Hey students! Now that it's time to revise this topic, go online to:

 Access the topic summary

 Review your results

 Watch teacher-led videos

 Practise exam questions

Find all this and MORE in jacPLUS

10.10 Exercise

Technology free: short answer

1. Simplify $i^6 - i^3(i^2 - 1)$.
2. Let $u = 5 - i$ and $v = 4 + 3i$. Evaluate the expression $2u - v$.
3. If $z = 3 - 8i$, determine the value of:
   a. $\text{Im}\left(z^2\right)$
   b. a and b if $z^3 = a + bi$.
4. If $z = 6 - 2i$ and $w = 5 + 3i$, express $\dfrac{z}{w}$ in the form $a + bi,\ a,\ b \in R$.
5. If $z = -7 - 7i$, express z in polar form.
6. Determine the roots of the equation $z^2 + 2z + 5 = 0$.
7. Determine the Cartesian equation and sketch the graph of each of the following.
   a. $\{z: |z - 4| = 2\,|z - 1|\}$
   b. $\{z: |z - 4| = |z - 1|\}$

Technology active: multiple choice

8. **MC** The value of $\text{Im}\left[i\left(2i^4 - 3i^2 + 5i\right)\right]$ is
   A. 0 B. −5 C. 5 D. 10 E. 18
9. **MC** If $z = 5 - 12i$, $\text{Re}\left(z^{-1}\right)$ is
   A. 5 B. 12 C. $\dfrac{12}{169}$ D. $\dfrac{16}{169}$ E. $\dfrac{5}{169}$
10. **MC** If $z = 8 - 7i$ and $w = 3 + 4i$, then $\text{Im}\left(w^2\right) + \text{Re}\left(z^2\right)$ is equal to
    A. 76 B. 39 C. 105 D. 56 E. 41
11. **MC** If $z = 3 - 50i$ and $w = 5 + 65i$, the value of $|z + w|$ is
    A. 64 B. 15 C. 17 D. 225 E. 45
12. **MC** The perimeter of the triangle formed by the line segments connecting the points $2 - 4i$, $14 - 4i$ and $2 + i$ is
    A. 13 B. 30 C. 10 D. 17 E. 22

13. **MC** The Argand diagram that correctly represents $z = 2\sqrt{5} - 4i$ is

A. 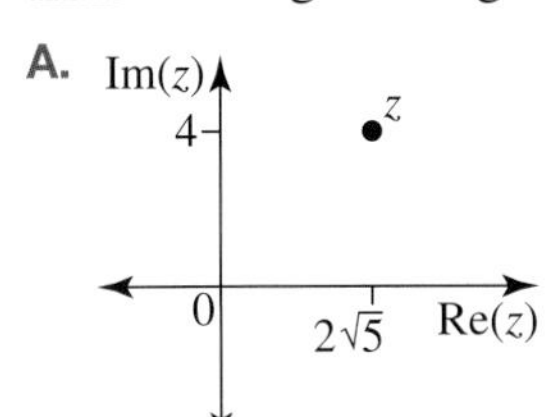

B. 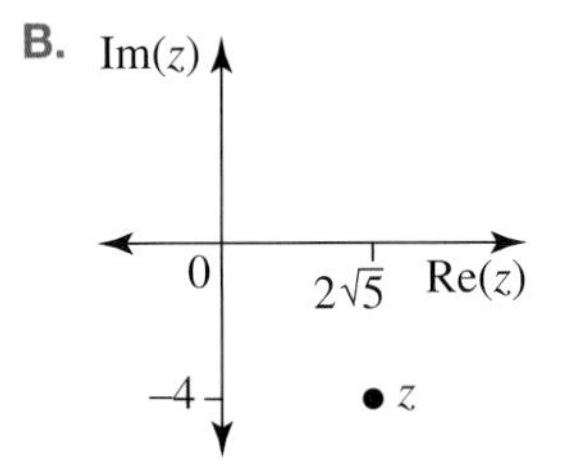

C. 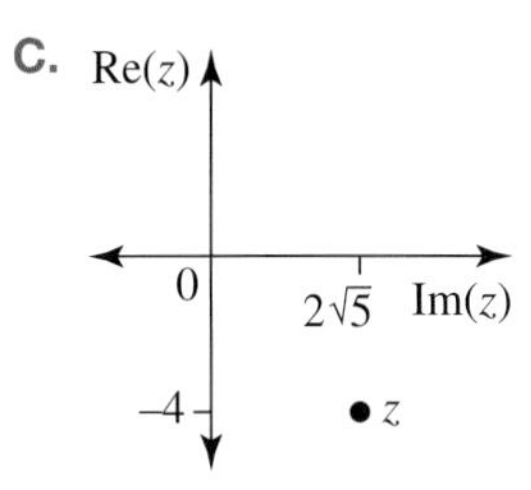

D. 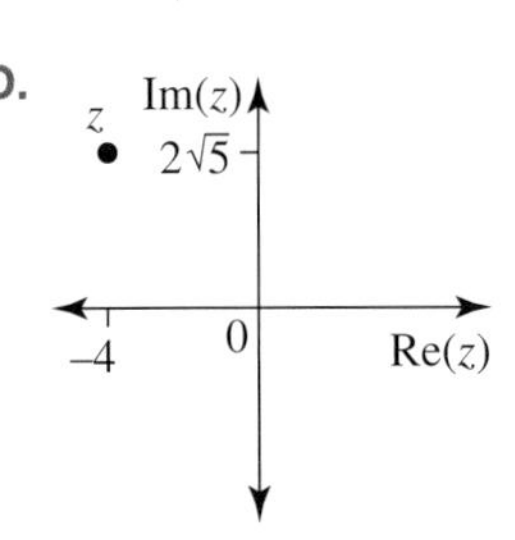

E. None of these.

14. **MC** The principal argument of $4\sqrt{3} - 4i$ is

A. $\frac{\pi}{6}$ B. $\frac{\pi}{3}$ C. $\frac{5\pi}{6}$ D. $-\frac{\pi}{6}$ E. $-\frac{\pi}{3}$

15. **MC** In polar form, $5i$ is

A. $\text{cis}(5\pi)$ B. $\text{cis}\left(\frac{5\pi}{2}\right)$ C. $5\,\text{cis}(5\pi)$ D. $5\,\text{cis}\left(\frac{\pi}{4}\right)$ E. $5\,\text{cis}\left(\frac{\pi}{2}\right)$

16. **MC** The Cartesian form of $\sqrt{3}\,\text{cis}\left(-\frac{7\pi}{6}\right)$ is

A. $\frac{1}{2} + \frac{\sqrt{3}}{2}i$ B. $-\frac{1}{2} + \frac{\sqrt{3}}{2}i$ C. $-\frac{\sqrt{3}}{2} + \frac{1}{2}i$ D. $-\frac{3}{2} + \frac{\sqrt{3}}{2}i$ E. $-\frac{1}{2} - \frac{\sqrt{3}}{2}i$

17. **MC** $\sqrt{5}\,\text{cis}\left(-\frac{\pi}{3}\right) \times \sqrt{8}\,\text{cis}\left(-\frac{\pi}{6}\right)$ is equal to

A. $6\sqrt{2i}$ B. $-2\sqrt{10i}$ C. $-6\sqrt{3}$ D. $-6i$ E. $4\sqrt{5}$

Technology active: extended response

18. Consider the complex number z such that $z = 3 + 2i$.
    a. Determine the value for iz, i^2z, i^3z and i^4z. (Give answers in standard $x + yi$ form.)
    b. Comment on the value of i^4z.
    c. Plot each number from part **a** on the same Argand diagram.
    d. Use a pair of compasses to draw a circle whose centre is at the origin and which passes through each point on the diagram.
    e. Determine the radius of the circle, giving your answer in exact (surd) form.
    f. Carefully study the five points on your diagram. What transformation is required to transform:
        i. Determine the transformation required to transform point z into point iz.
        ii. Determine the transformation required to transform point iz into point i^2z.
        iii. Determine the transformation required to transform point i^2z into point i^3z.
    g. On the Argand diagram, determine what transformation takes place when a complex number is multiplied by i.
    h. For a complex number z such that $z = x + yi$, describe the curve that all points representing numbers of the form zi^n (that is, z, zi, zi^2, zi^3 and so on) would lie on in the Argand plane.

19. a. For any complex number $z = x + yi$ where both x and y are real, describe the transformation required to obtain $\frac{1}{z}$.

b. For any complex number $z = x + yi$ where both x and y are real, describe the transformation required to obtain iz.

20. Sketch the following regions.

a. $\{z: 1 < |z| \le 2\}$

b. $\left\{z: \frac{\pi}{6} \le \text{Arg}(z) \le \frac{5\pi}{6}\right\}$

c. Hence, sketch the region defined by $\{z: 1 < |z| \le 2\} \cap \left\{z: \frac{\pi}{6} \le \text{Arg}(z) \le \frac{5\pi}{6}\right\}$.

10.10 Exam questions

Question 1 (3 marks) TECH-FREE

Simplify the expression $\frac{2i}{1+i} - \frac{3}{2-i}$, giving your answer in the form $a + bi$.

Question 2 (1 mark) TECH-ACTIVE

MC If $f(i) = \frac{1 + i + i^2 + ... + i^{11}}{4}$, then the statement below that is true is

A. $f(i) = 2 + i$
B. $\text{Re}\left[f(i)\right] = 5$
C. $\text{Im}\left[f(i)\right] - \frac{1}{4}$
D. $f(i) = 0$
E. $\text{Re}\left[f(i)\right] = \frac{1}{4}$

Question 3 (1 mark) TECH-ACTIVE

MC If $z = 8 - 7i$ and $w = 3 + 4i$, then $3z - 2w$ is equal to

A. $30 - 13i$
B. $30 - 29i$
C. $18 - 29i$
D. 18
E. $18 + 29i$

Question 4 (1 mark) TECH-ACTIVE

MC If $z = -1 - \sqrt{3}i$ and $w = 2 + 2i$, then $\frac{w^4}{z^3}$ is equal to

A. $-4 + 4i$
B. $2\sqrt{3}$
C. $-4i$
D. -8
E. $\frac{1}{2i}$

Question 5 (5 marks) TECH-FREE

a. Evaluate the complex number z in Cartesian form if $\frac{z+1}{z-1} = \frac{1}{2}(3 + i)$. **(2 marks)**

b. Determine the real numbers a and b if $(a + bi)^2 = 5 - 12i$. **(3 marks)**

More exam questions are available online.

Answers

Topic 10 Complex numbers

10.2 Introduction to complex numbers

10.2 Exercise

1. a. $3i$ b. $5i$ c. $7i$ d. $\sqrt{3}\,i$
2. a. $\sqrt{11}\,i$ b. $\sqrt{7}\,i$ c. $\frac{2}{3}i$ d. $\frac{6}{5}i$
3. a. $9, 5$ b. $5, -4$ c. $-3, -8$ d. $-6, 11$
4. a. $27, 0$ b. $0, 2$ c. $-5, 1$ d. $0, -17$
5. a. $-1+i$ b. $1+i$ c. $1-i$ d. $0+0i$
6. a. $-1+2i$ b. $-1+i$ c. $1+0i$ d. $1-2i$
7. $z=-2-3i$, $w=7+3i$; sample responses can be found in the worked solutions in the online resources.
8. a. -5 b. 15 c. 0 d. -6
9. a. 2 b. 0 c. -9 d. 2
10. $4-i$
11. a. -2 b. $-2i$ c. $\sqrt{2}$ d. $2i$
12. a. 13 b. -11 c. 10 d. 7
13. a. i b. 0 c. i d. 0
14. a. $-i$ b. 0 c. $-i$ d. 0

10.2 Exam questions

Note: Mark allocations are available with the fully worked solutions online.

1. a.
$$\sqrt{-4}=2i$$
$$\sqrt{-9}=3i$$
$$\sqrt{-4}\times\sqrt{-9}=-6$$
$$\sqrt{-4\times-9}=6$$
   b. $\sqrt{a\times b}=\sqrt{a}\times\sqrt{b}$ only when $a, b\geq 0$.
2. a. $i^{4n}=1$
   b. $i^{4n+3}=-i$
   c. Sample responses can be found in the worked solutions in the online resources.
3. a. $\text{Re}\left(f(10)\right)=-6$
   b. $\text{Im}\left(f(11)\right)=-6$

10.3 Basic operations on complex numbers

10.3 Exercise

1. a. $4-i$ b. $1-14i$ c. $-4-2i$
2. a. $9-13i$ b. $-12+4i$ c. $-9-5i$
3. a. $-12+3i$ b. $-19-8i$ c. $12+23i$
4. a. $-25+3i$ b. $-50-48i$ c. $-41-28i$
5. a. $7-23i$ b. $4+45i$ c. $-50-13i$
6. a. $63-37i$ b. $-85-132i$ c. $176-61i$
7. a. $111+33i$ b. $31-8i$ c. $22-48i$
8. a. 61 b. -53 c. $32-126i$
9. $14+52i$
10. -3
11. a. -8 b. -5 c. -9
12. a. 35 b. -30 c. -115
13. a. $a=5, b=-2$ b. $a=\frac{21}{41}$, $b=-\frac{16}{41}$
   c. $a=1, b=5$ d. $a=-2$, $b=-3$
14. a. $-13-4i$ b. $-50+8i$
   c. -89 d. $x=-3, y=-4$
15. a. $x=5, y=2$ b. $x=4, y=3$
   c. $x=5, y=-2$ d. $x=3, y=-4$

10.3 Exam questions

Note: Mark allocations are available with the fully worked solutions online.

1. a. $7+i$ b. $-6-23i$ c. $32-i$
   d. -80 e. 14
2. a. $x=3$, $y=2$ b. $x=2$, $y=-1$
3. a. $x=-\frac{6}{13}$, $y=\frac{17}{13}$
   b. $x=-\frac{6}{25}$, $y=-\frac{17}{25}$

10.4 Complex conjugates and division of complex numbers

10.4 Exercise

1. a. $7-10i$ b. $5+9i$ c. $3-12i$
   d. $\sqrt{7}+3i$ e. $5-2i$ f. $-6+\sqrt{11}\,i$
2. Sample responses can be found in the worked solutions in the online resources.
3. $\frac{1}{2}+\frac{1}{2}i$
4. a. $0-i$
   b. $0-i$
   c. $-\frac{7}{25}+\frac{26}{25}i$
5. a. $\frac{14}{29}-\frac{23}{29}i$
   b. $\frac{43}{53}+\frac{18}{53}i$
   c. $\frac{2\sqrt{5}-\sqrt{6}}{7}+\frac{2\sqrt{2}+\sqrt{15}}{7}i$
6. a. $\frac{2}{5}+\frac{1}{5}i$ b. $\frac{3}{10}-\frac{1}{10}i$ c. $\frac{4}{25}+\frac{3}{25}i$
7. a. $\frac{5}{41}-\frac{4}{41}i$ b. $-\frac{3}{13}-\frac{2}{13}i$ c. $\frac{\sqrt{3}}{5}+\frac{\sqrt{2}}{5}i$
8. $10+24i$
9. a. $\frac{23}{10}$ b. $\frac{9}{10}$ c. $\frac{17}{5}$
   d. $-\frac{16}{5}$ e. $-\frac{14}{5}$
10. $\frac{17}{2}+\frac{9}{2}i$
11. -29

12. $-33 + 58i$

13–15. Sample responses can be found in the worked solutions in the online resources.

16. -16

17. a. $-12 + 11i$ b. $-30 - 19i$ c. 0

18. Sample responses can be found in the worked solutions in the online resources.

19. a. $-4, 16, -64$

b. Sample responses can be found in the worked solutions in the online resources.

20. $a = -1, b = \pm\sqrt{2}$

21. $a = -\frac{1}{2}, b = \frac{1}{2}$

22. a. i. 13 ii. 5

b. Sample responses can be found in the worked solutions in the online resources.

c. i. $\frac{2}{13} - \frac{3}{13}i$

ii. $\frac{1}{5} - \frac{2}{5}i$

d. $-8 + 16i$

e. $-2 + 10i$

f. $-\frac{4}{65} + \frac{7}{65}i$

10.4 Exam questions

Note: Mark allocations are available with the fully worked solutions online.

1. a. $\frac{3}{25} + \frac{4}{25}i$ b. $\frac{4}{41} - \frac{5}{41}i$

c. $-\frac{8}{41} - \frac{3}{41}i$ d. $-\frac{8}{25} + \frac{31}{25}i$

e. $\frac{32}{1025} + \frac{1}{1025}i$

2. $\frac{a + c - (b + d)i}{(a + c)^2 + (b + d)^2}$

3. $z = \frac{4}{5} - \frac{3}{5}i$

10.5 The complex plane (the Argand plane)

10.5 Exercise

1. $3z = 3 + 6i$, $3z$ has a length of 3 times that of z.

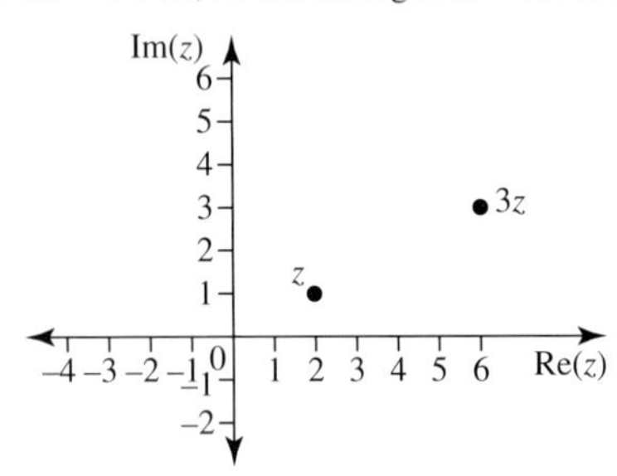

2. $-\frac{z}{2} = -2 + i$; $-\frac{z}{2}$ is in the opposite direction to z and half its length.

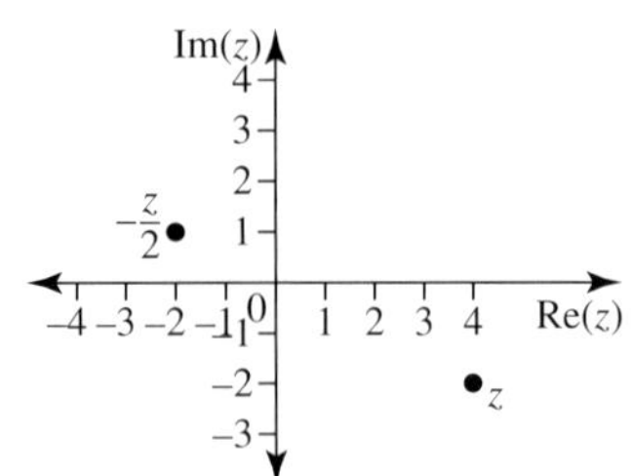

3. a. $u + v = 4$

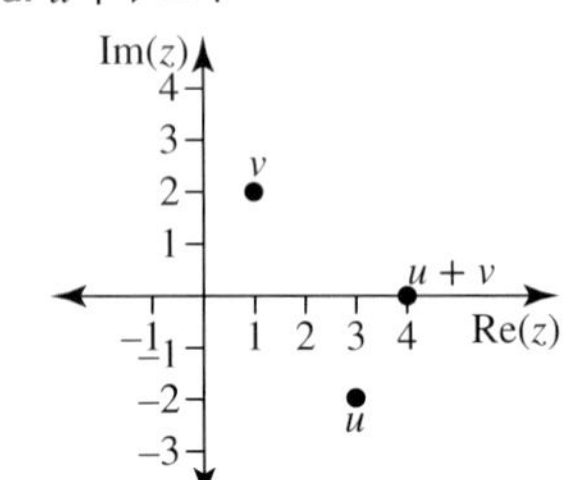

b. $u - v = 2 - 4i$

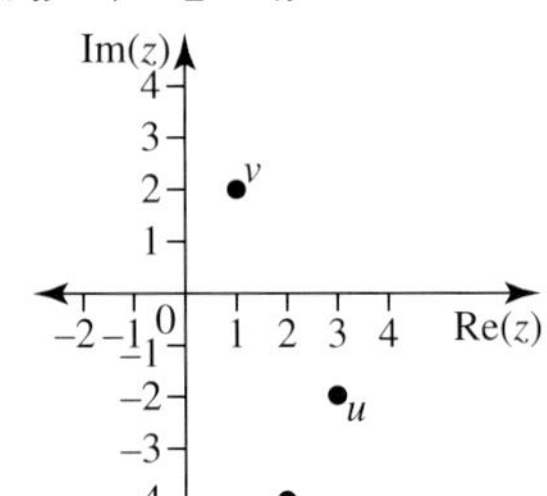

4.

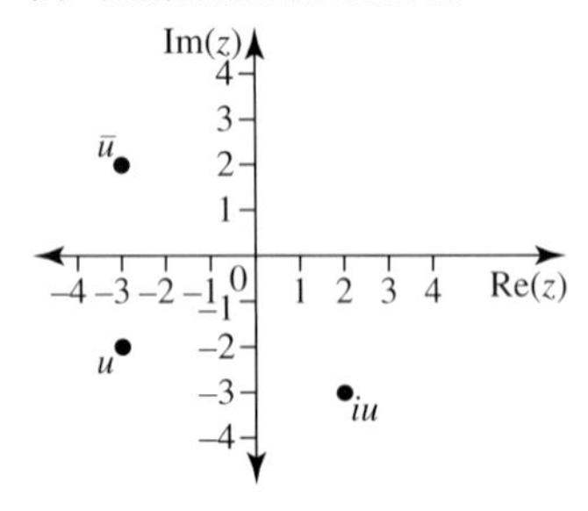

5. $\overline{u} = -3 + 2i$, $iu = 2 - 3i$, $\overline{u}$ is the reflection of the complex number in the real axis and iu is a rotation of 90° anticlockwise from u.

6. $\overline{v} = -2 - 3i$, $i^2 v = 2 - 3i$, $\overline{v}$ is the reflection of the complex number in the real axis and $i^2 v$ is a rotation of 180° anti-clockwise from v.

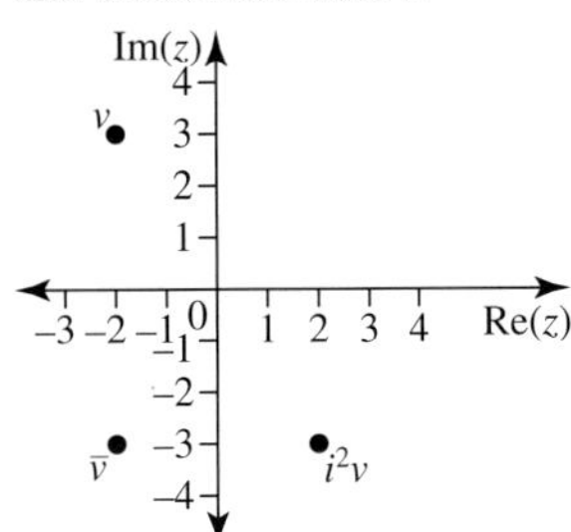

7. a. $2z = 4 - 6i$, $\overline{z} = 2 + 3i$, $iz = 3 + 2i$

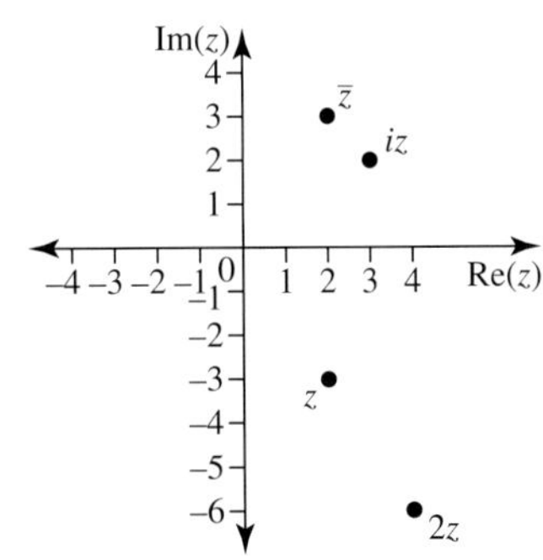

b. $-\dfrac{z}{2} = 2 + 2i$, $\overline{z} = -4 + 4i$, $iz = 4 - 4i$

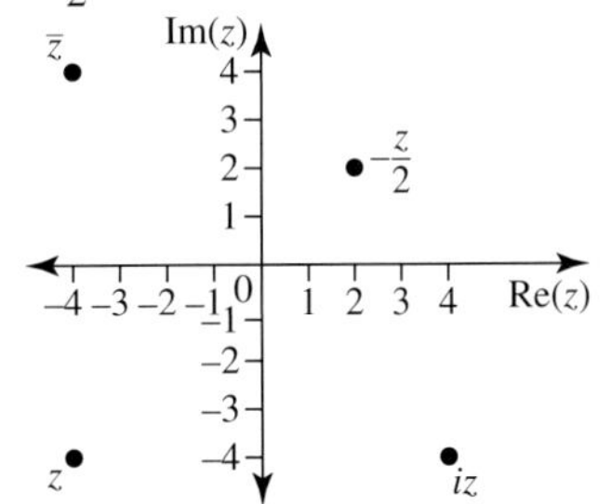

8. a. i. $u + v = 1 + i$

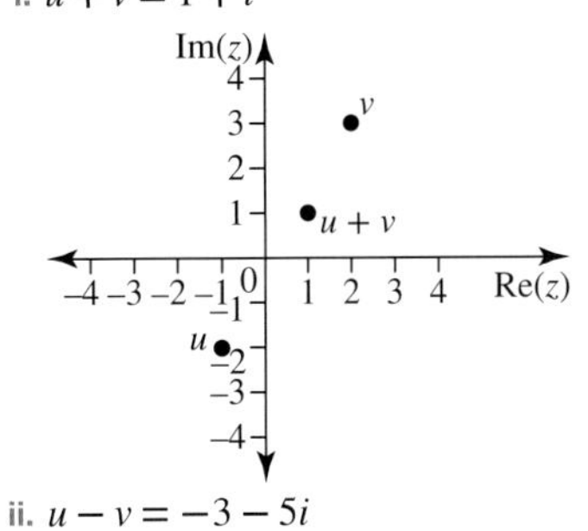

ii. $u - v = -3 - 5i$

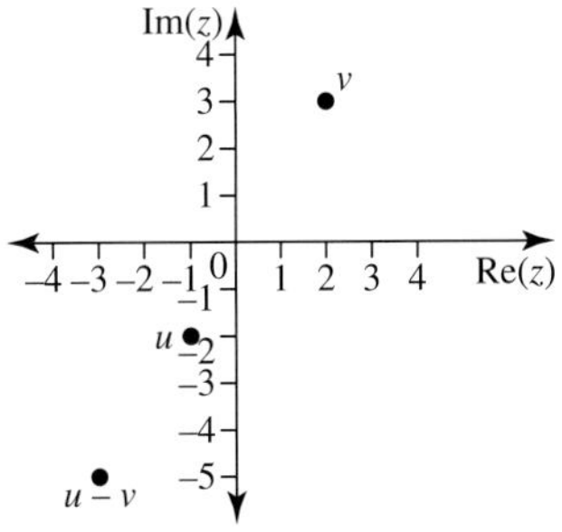

b. i. $u + v = 3 + i$

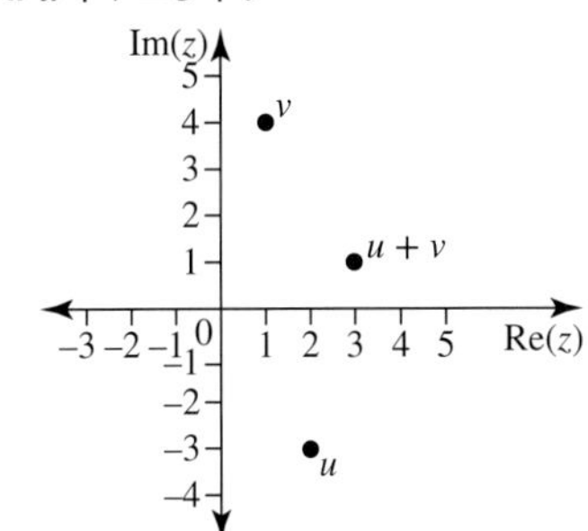

ii. $u - v = 1 - 7i$

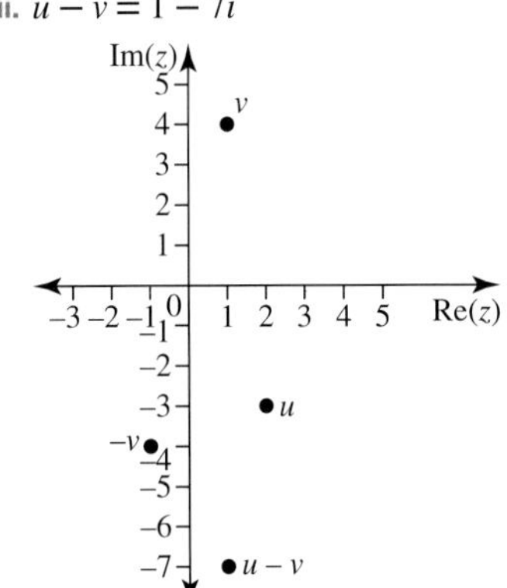

9. a. $z = 1 - i$, $i^2 z = -1 + i$, $i^3 z = -1 - i$

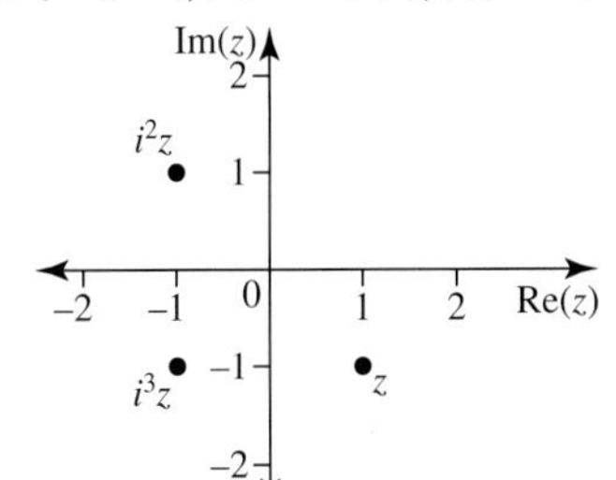

b. $z = -2 + 3i$, $i^2 z = 2 - 3i$, $i^3 z = 3 + 2i$

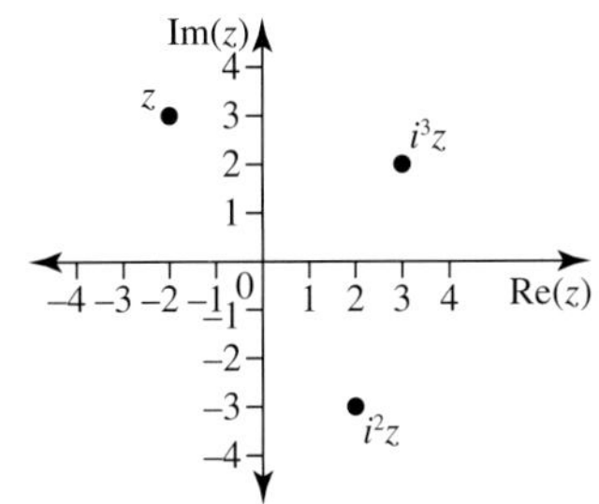

10. a. $u = 1 + i$, $\dfrac{1}{u} = \dfrac{1}{2} - \dfrac{1}{2}i$

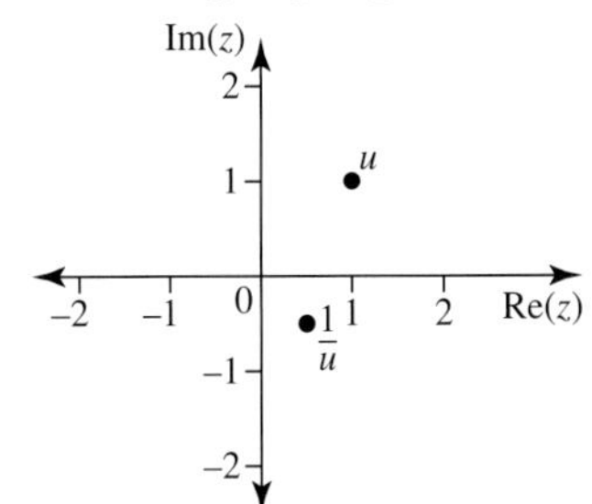

b. $u = -\sqrt{3} - i, \ \dfrac{1}{u} = -\dfrac{\sqrt{3}}{4} + \dfrac{1}{4}i$

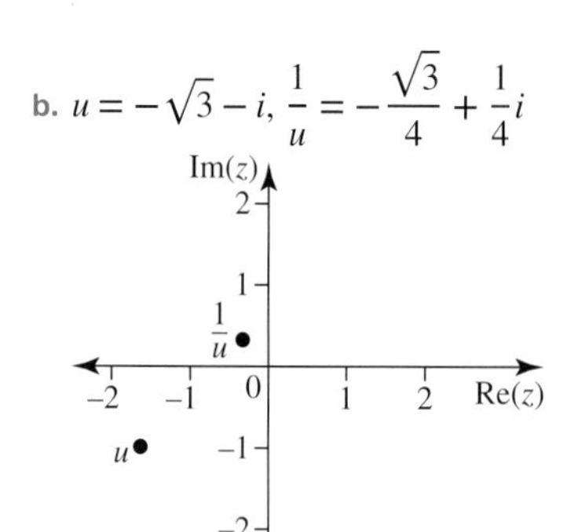

11. a. $uv = -4i, \ \dfrac{u}{v} = -2$

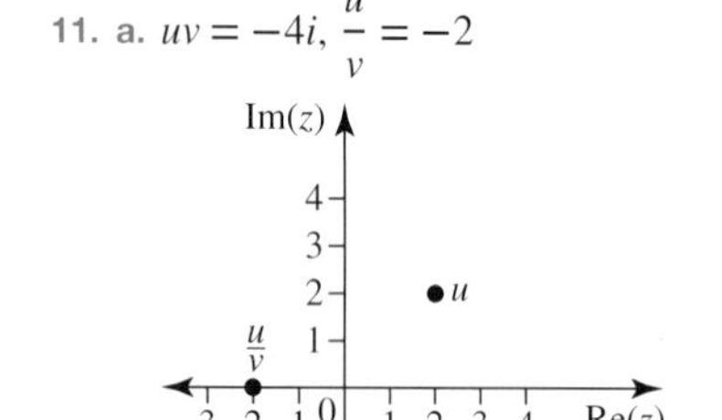

b. $uv = 2 + 2i, \ \dfrac{u}{v} = -1 + i$

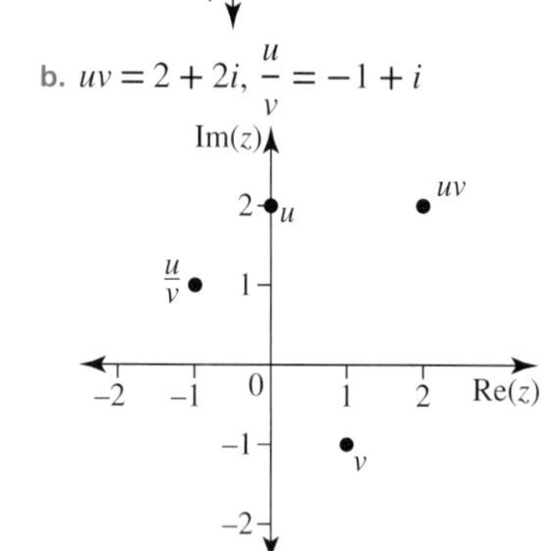

12. a. $z^2 = -2i, z^3 = -2 - 2i$

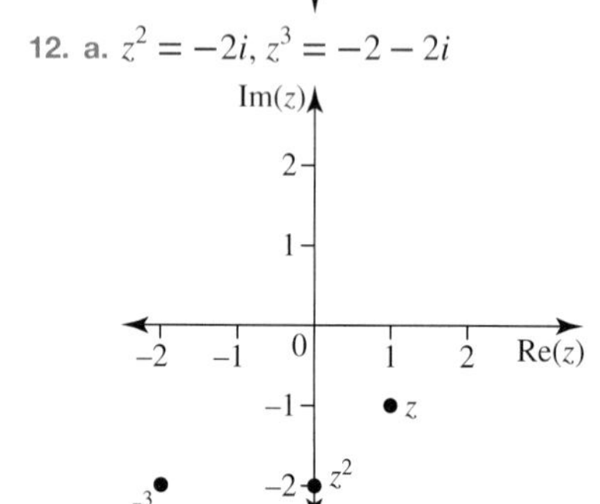

b. $z^2 = -2i, z^3 = 2 + 2i$

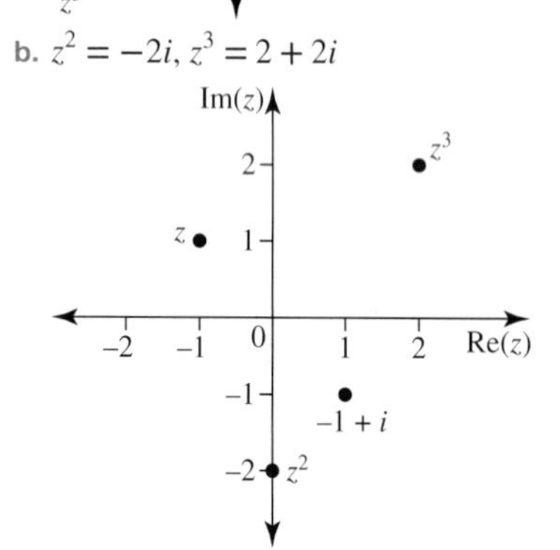

13. a. Sample responses can be found in the worked solutions in the online resources.

b. (4, 3)

c. i. Sample responses can be found in the worked solutions in the online resources.

ii. Vector $\underset{\sim}{v}$ is a rotation of vector $\underset{\sim}{u}$ through an angle of 90°, which is similar to the complex numbers z and iz.

iii. Vectors and complex numbers have a similar structure. The term 'isomorphic' means having the same form and is used in many branches of mathematics to identify mathematical objects that have the same structural properties.

10.5 Exam questions

Note: Mark allocations are available with the fully worked solutions online.

1.

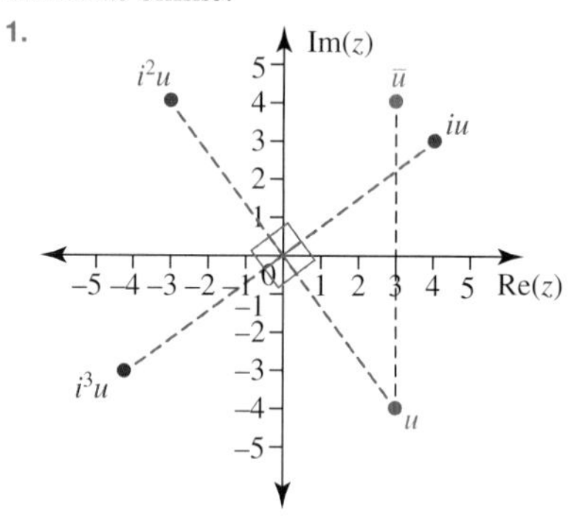

2. $|ab|$

3. $\dfrac{1}{2}\left(a^2 + b^2\right)$

10.6 Complex numbers in polar form

10.6 Exercise

1. a. 13 b. 3

2. a. $\sqrt{65}$ b. $3\sqrt{5}$

3. a. $\sqrt{5}$ b. 5

4. a. i.

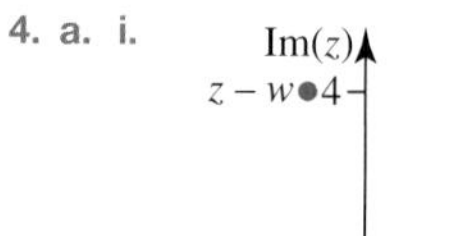

ii. $\sqrt{17}$

b. i.

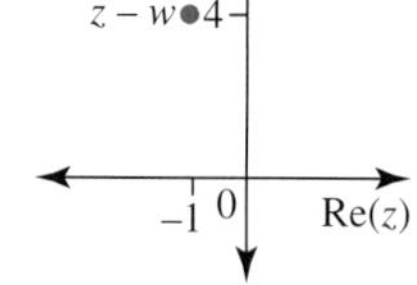

ii. $\sqrt{37}$

c. i.

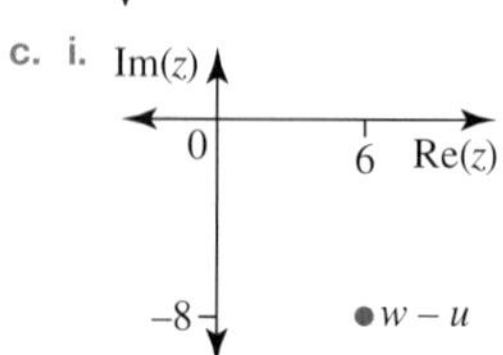

ii. 10

d. i.

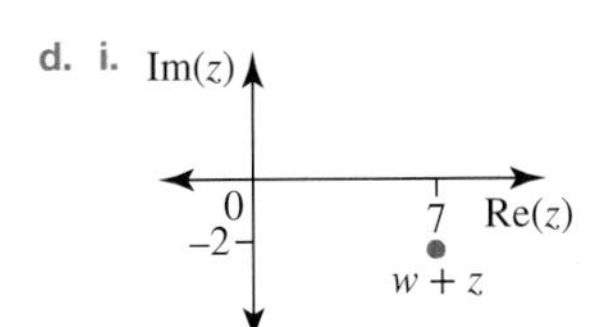

ii. $\sqrt{53}$

e. i.

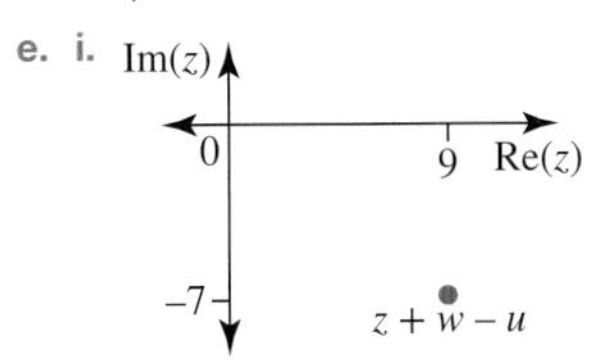

ii. $\sqrt{130}$

f. i.

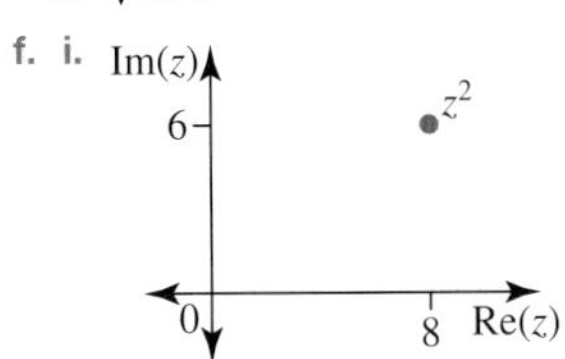

ii. 10

5. a.

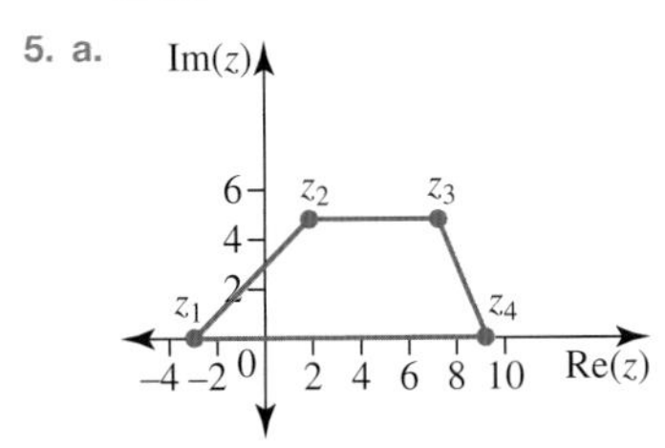

b. 42.5 square units

6. a.

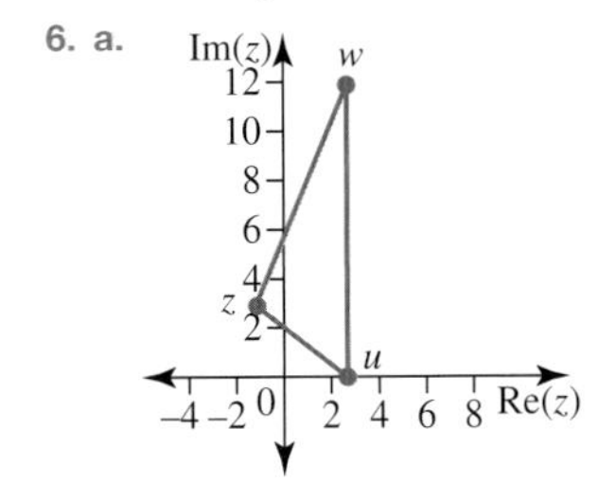

b. 24 square units

7. a. $v = -5 + 4i$ or $11 + 4i$

b. $v = 4 - 5i$ or $8 - 5i$

c. $2b(a + c)$

8. a. $\frac{25}{2}$ b. $\frac{169}{2}$ c. $\frac{1}{2}\left(a^2 + b^2\right)$

9. a. 25 b. $b = \pm\sqrt{14}$ c. $a^2 + b^2$

10. a. $x^2 - 7x + y^2 + y = 0,\ \frac{25\pi}{2}$

b. $x^2 + 7x + y^2 - 17y = 0,\ \frac{169\pi}{2}$

11. $x^2 - (a - b)x + y^2 - (a + b)y = 0,\ \frac{\pi}{2}\left(a^2 + b^2\right)$

12. a. $\frac{\pi}{6}$ b. $-\frac{\pi}{4}$ c. $-\frac{2\pi}{3}$

13. a. $\frac{\pi}{2}$ b. π c. $-\frac{\pi}{2}$ d. 0

14. a. $-\frac{\pi}{2}$ b. $\frac{\pi}{6}$ c. $-\frac{\pi}{8}$ d. $\frac{3\pi}{4}$

15. a. $-\frac{5\pi}{6}$ b. $\frac{6\pi}{7}$ c. $\frac{2\pi}{5}$ d. $\frac{11\pi}{12}$

16. a. $3\sqrt{2}, -\frac{\pi}{4}$ b. $5\sqrt{2}, \frac{3\pi}{4}$

c. $2, -\frac{2\pi}{3}$ d. $8, \frac{\pi}{6}$

17. a. $\sqrt{2}\,\text{cis}\left(\frac{3\pi}{4}\right)$ b. $2\sqrt{2}\,\text{cis}\left(\frac{\pi}{6}\right)$

c. $\sqrt{10}\,\text{cis}\left(-\frac{3\pi}{4}\right)$

18. a. $2\sqrt{5}\,\text{cis}\left(-\frac{\pi}{3}\right)$ b. $\text{cis}\left(-\frac{2\pi}{3}\right)$

c. $\frac{\sqrt{2}}{4}\,\text{cis}\left(\frac{3\pi}{4}\right)$

19. a. $-1 + \sqrt{3}\,i$ b. $\frac{3\sqrt{2}}{2} + \frac{3\sqrt{2}}{2}i$

c. $-\frac{\sqrt{15}}{2} + \frac{\sqrt{5}}{2}i$

20. a. $2 - 2\sqrt{3}i$ b. $\frac{\sqrt{14}}{2} + \frac{\sqrt{14}}{2}i$

c. $0 + 8i$

10.6 Exam questions

Note: Mark allocations are available with the fully worked solutions online.

1. a. $u = 2\text{cis}\left(\frac{\pi}{6}\right)$

$\bar{u} = 2\text{cis}\left(-\frac{\pi}{6}\right)$

$\frac{1}{u} = \frac{1}{2}\text{cis}\left(-\frac{\pi}{6}\right)$

$-u = 2\text{cis}\left(-\frac{5\pi}{6}\right)$

b. $w = -5 - 5i$

2. $u = \sqrt{2}\text{cis}\left(-\frac{\pi}{4}\right)$

$v = \sqrt{2}\text{cis}\left(-\frac{3\pi}{4}\right)$

$u + v = 2\text{cis}\left(-\frac{\pi}{2}\right)$

$\frac{u}{v} = \text{cis}\left(\frac{\pi}{2}\right)$

3. a. $a = \pm 5\sqrt{5}$

b. $a = \sqrt{2}$

c. $a = \sqrt{2}$

10.7 Basic operations on complex numbers in polar form

10.7 Exercise

1. a. $6\operatorname{cis}\left(\frac{3\pi}{4}\right)$ b. $20\operatorname{cis}\left(\frac{\pi}{3}\right)$
   c. $6\sqrt{5}\operatorname{cis}\left(-\frac{\pi}{4}\right)$ d. $\sqrt{6}\operatorname{cis}\left(\frac{2\pi}{3}\right)$
   e. $2\sqrt{7}\operatorname{cis}\left(-\frac{\pi}{6}\right)$
2. a. $-3\sqrt{2}+3\sqrt{2}i$ b. $10+10\sqrt{3}i$
   c. $3\sqrt{10}-3\sqrt{10}i$ d. $-\frac{\sqrt{6}}{2}+\frac{3\sqrt{2}}{2}i$
   e. $\sqrt{21}-\sqrt{7}i$
3. a. $4\sqrt{2}\operatorname{cis}\left(\frac{5\pi}{12}\right)$
   b. $8\sqrt{3}\operatorname{cis}\left(-\frac{\pi}{2}\right)$
   c. $8\sqrt{2}\operatorname{cis}\left(-\frac{\pi}{12}\right)$
4. $3\operatorname{cis}\left(\frac{\pi}{2}\right)$
5. a. $4\operatorname{cis}\left(\frac{11\pi}{12}\right)$ b. $2\operatorname{cis}\left(-\frac{3\pi}{10}\right)$
   c. $2\sqrt{2}\operatorname{cis}\left(-\frac{3\pi}{14}\right)$ d. $\frac{3\sqrt{2}}{4}\operatorname{cis}\left(\frac{7\pi}{12}\right)$
6. $3\sqrt{3}\operatorname{cis}\left(\frac{\pi}{4}\right)$
7. a. $16\operatorname{cis}(\pi)$ b. $32\operatorname{cis}\left(\frac{3\pi}{4}\right)$
8. $-\frac{1}{4}+\frac{1}{4}i$
9. a. $-\frac{1}{4}+0i$
   b. $0-\frac{1}{8}i$
   c. $\frac{\sqrt{3}}{64}-\frac{1}{64}i$
   d. $\frac{\sqrt{2}}{8}\cos\left(\frac{\pi}{12}\right)-\frac{\sqrt{2}}{8}\sin\left(\frac{\pi}{12}\right)i$
   e. $16+0i$
10. $-64\sqrt{3}-64i$
11. $1+0i$
12. $-1519-720i$
13. $\frac{8}{9}, -\frac{\pi}{6}$
14. $16-16i$
15. $-64+64i$
16. $\frac{2}{5}, -\frac{\pi}{120}$
17. Sample responses can be found in the worked solutions in the online resources.
18. Sample responses can be found in the worked solutions in the online resources.

10.7 Exam questions

Note: Mark allocations are available with the fully worked solutions online.

1. a. $u^{12}=4096$
   b. $\frac{1}{w^5}=\frac{1}{8}-\frac{1}{8}i$
2. $\frac{1}{2}i$
3. a. $r=3,\ \theta=\frac{2\pi}{3}$
   b. $r=48,\ \theta=\frac{5\pi}{6}$
   c. $r=\sqrt{2},\ \theta=-\frac{\pi}{3}$

10.8 Solving quadratic equations with complex roots

10.8 Exercise

1. $(z-3+4i)(z-3-4i)$
2. $\left(z+2+\sqrt{3}i\right)\left(z+2-\sqrt{3}i\right)$
3. $2\pm 5i$
4. $-1\pm 5i$
5. a. $-1\pm 4i$ b. $2\pm 4i$
   c. $3\pm 2i$ d. $-5\pm 4i$
6. $z^2+10z+29$
7. z^2+25
8. a. $3z^2+5z-2$ b. $z^2-4z+40$
   c. $z^2+4z+13$ d. $z^2+8z+41$
9. a. $10z^2-3z-1$ b. $z^2-8z+19$
   c. $z^2+10z+32$ d. $z^2+6z+17$
10. a. $-1+2i,\ -1-i$
    b. $3\pm i$
    c. i. Sample responses can be found in the worked solutions in the online resources.
       ii. $-5+5i,\ 1+i$
11. a. $-2+\left(\frac{\sqrt{5}+1}{2}\right)i, -2+\left(\frac{1-\sqrt{5}}{2}\right)i$
    b. $4\pm 3i$
    c. i. Sample responses can be found in the worked solutions in the online resources.
       ii. $-\left(\frac{\sqrt{11}+4}{2}\right)+\frac{3}{2}i, -\left(\frac{\sqrt{11}-4}{2}\right)-\frac{3}{2}i$
12. a. $2\pm 2i$ b. $3\pm i$
    c. $1\pm 5i$ d. $4\pm 5i$
13. a. $2\pm 3i$ b. $-3\pm\sqrt{5}i$
    c. $1\pm\sqrt{3}i$ d. $-5\pm\sqrt{6}i$

14. a. $-\frac{3}{2} \pm \frac{1}{2}i$ b. $-\frac{5}{2} \pm i$ c. $\frac{3}{2} \pm i$ d. $\frac{7}{3} \pm \frac{2}{3}i$

10.8 Exam questions

Note: Mark allocations are available with the fully worked solutions online.

1. a. $z = -6 \pm 4i$ b. $z = \frac{5 \pm \sqrt{23}i}{4}$

2. a. $z^2 - 4z + 7$ b. $b = -6,\ c = 32$

3. $a = \pm 4,\ b = \mp 3$

10.9 Lines, rays, circles, ellipses and regions in the complex plane

10.9 Exercise

1. The line $4x + 3y = 12$

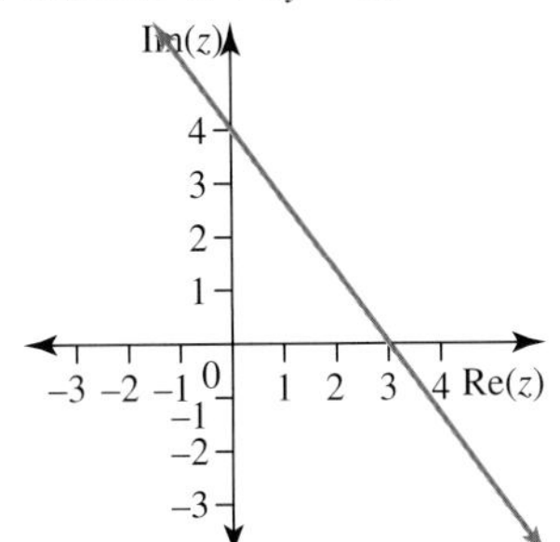

2. a. $y = 2$; line

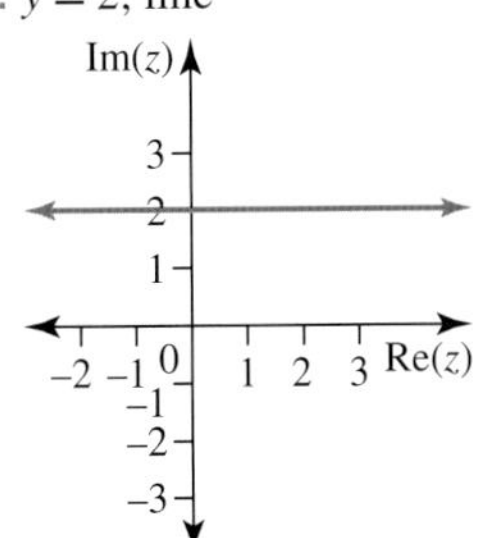

b. $x + 2y = 4$; line

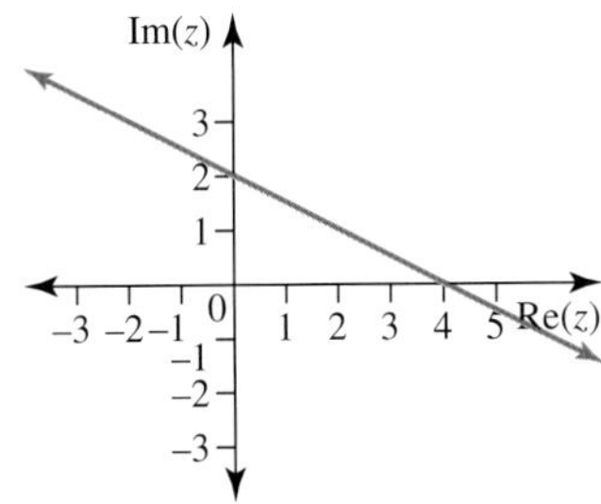

3. Line $y = -x$; the set of points equidistant from $(0, -3)$ and $(3, 0)$

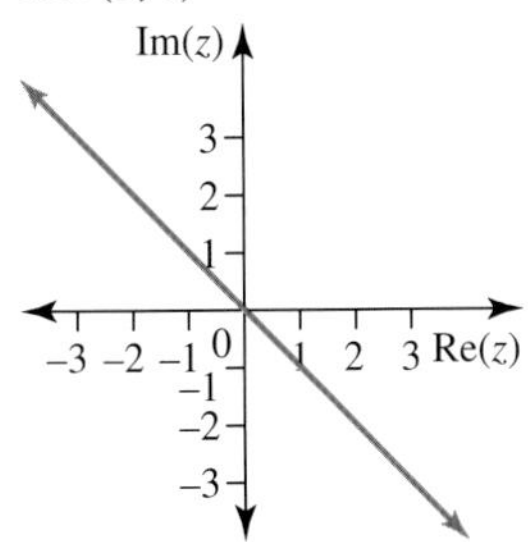

4. Line $y = -1$; the set of points equidistant from $(0, 1)$ and $(0, -3)$

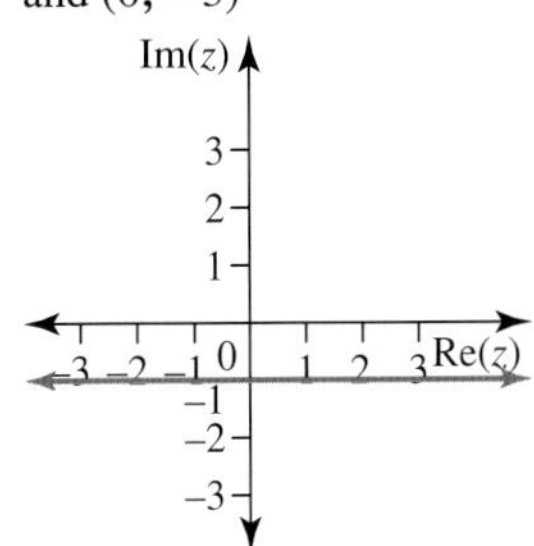

5. a. $y = -2x - 3$

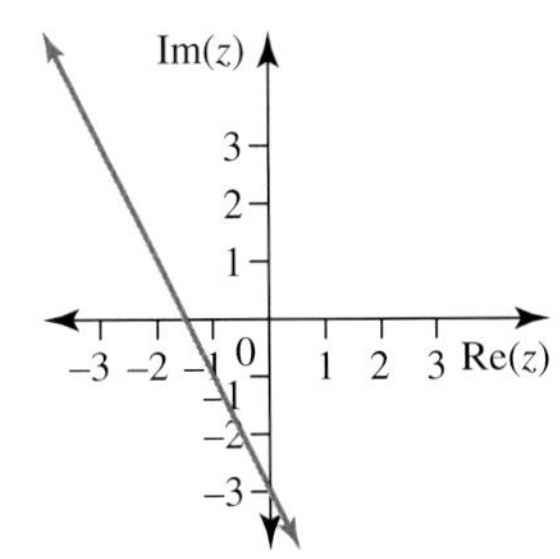

b. $y = \frac{2x}{3}$

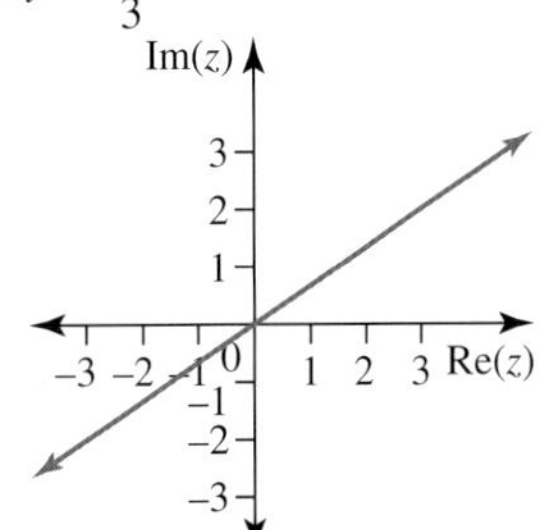

6. a. The ray makes an angle of 30° with the positive real axis, starting from but not including $(0, 0)$.

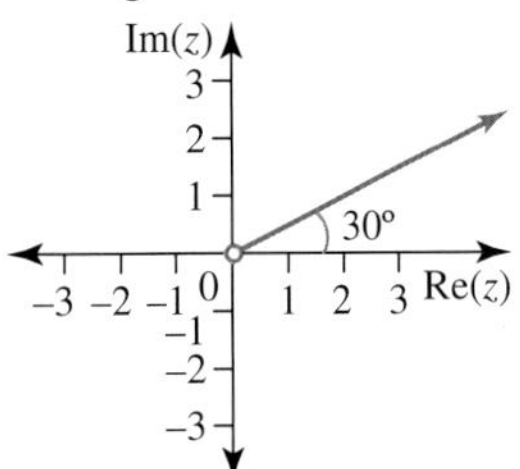

b. The ray starts from the point $(1, 0)$, not including the point, making an angle of 30° with the positive real axis.

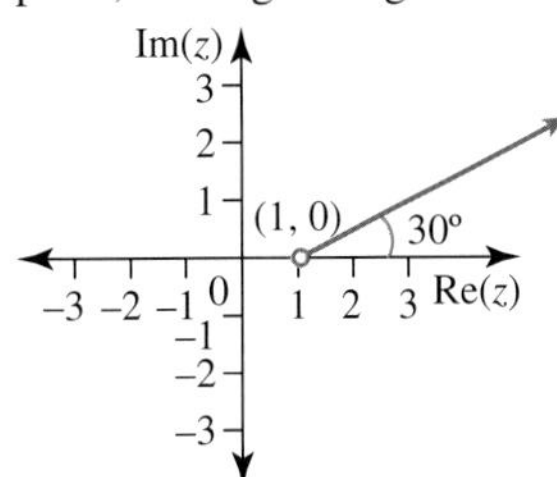

7. a. The ray makes an angle of 45° with the positive real axis, starting from but not including $(0, 0)$.

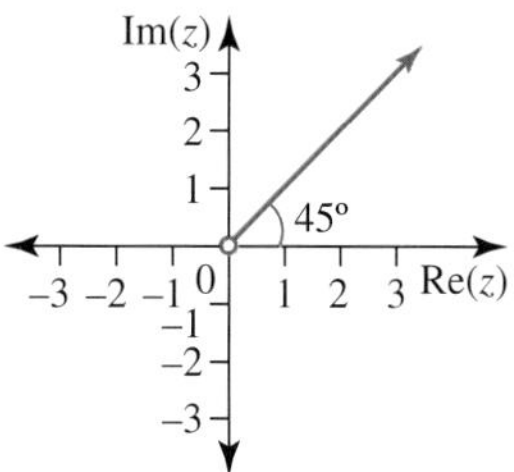

b. The ray starts from the point $(-2, 0)$, not including the point, making an angle of 45° with the positive real axis.

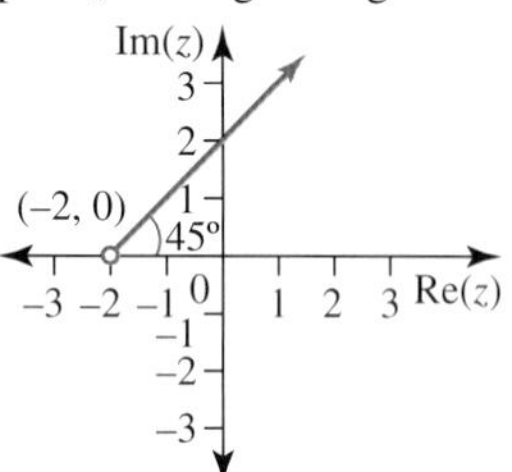

8. $(x-3)^2+(y+2)^2=16$; circle with centre $(3, -2)$ and radius 4.

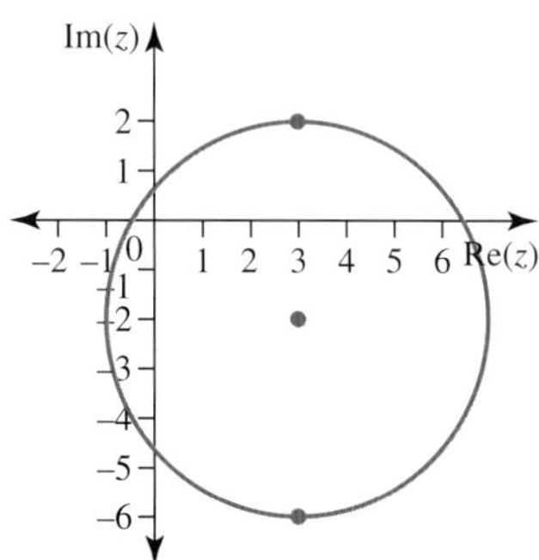

9. a. $x^2+y^2=9$; circle with centre $(0, 0)$ and radius 3.

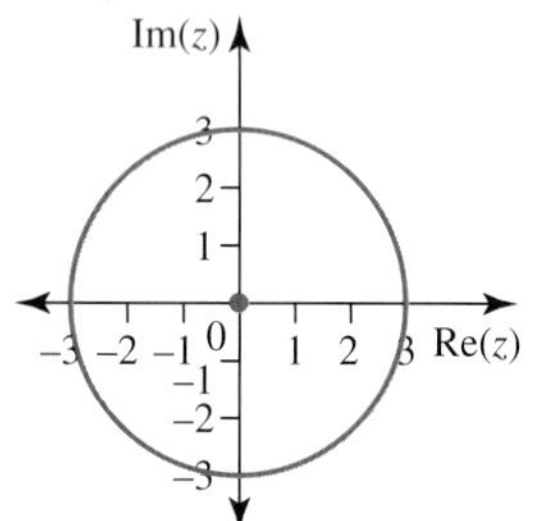

b. $x^2+y^2=4$; circle with centre $(0, 0)$ and radius 2.

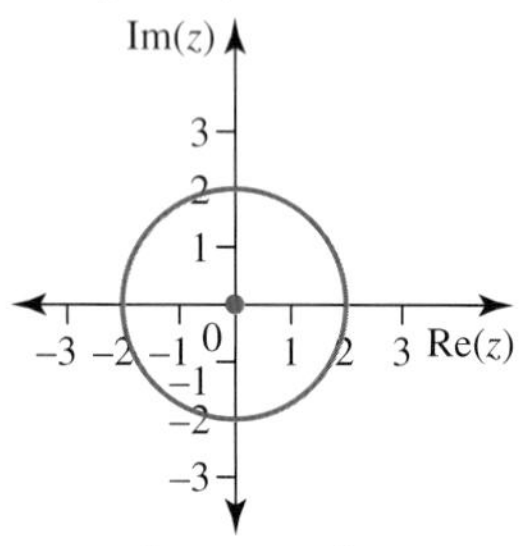

10. a. $(x+2)^2+(y-3)^2=4$; circle with centre $(-2, 3)$ and radius 2.

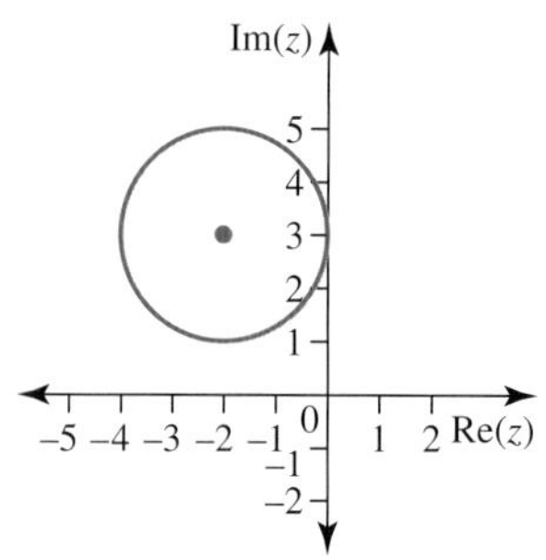

b. $(x-3)^2+(y+1)^2=9$; circle with centre $(3, -1)$ and radius 3.

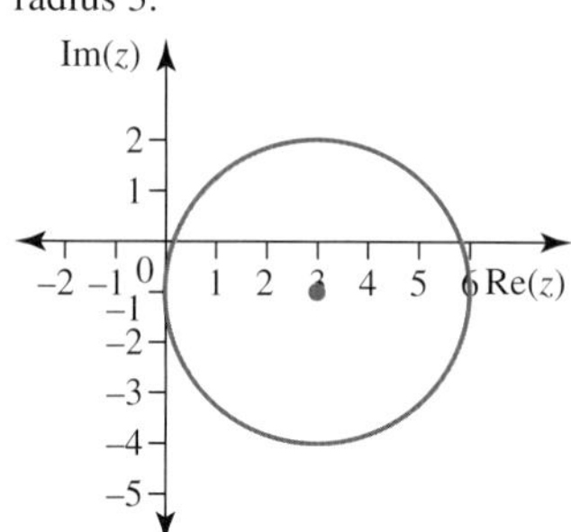

11. $\dfrac{x^2}{48}+\dfrac{y^2}{36}=1$

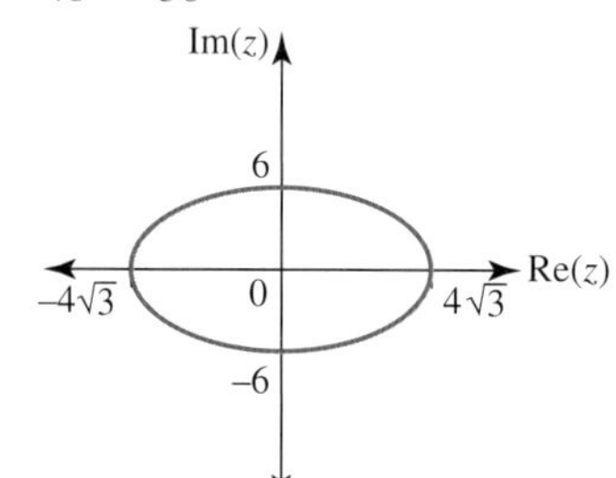

12. $\dfrac{x^2}{9}+\dfrac{y^2}{5}=1$

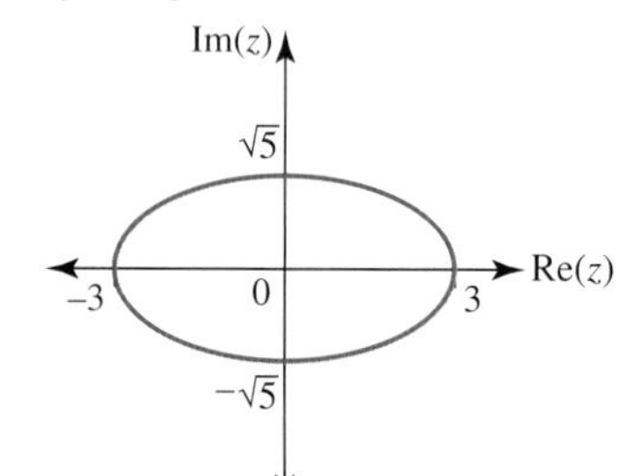

13.

Im(z)

$\dfrac{3\pi}{4}$

Re(z)

14.

Im(z)

60°

Re(z)

15.

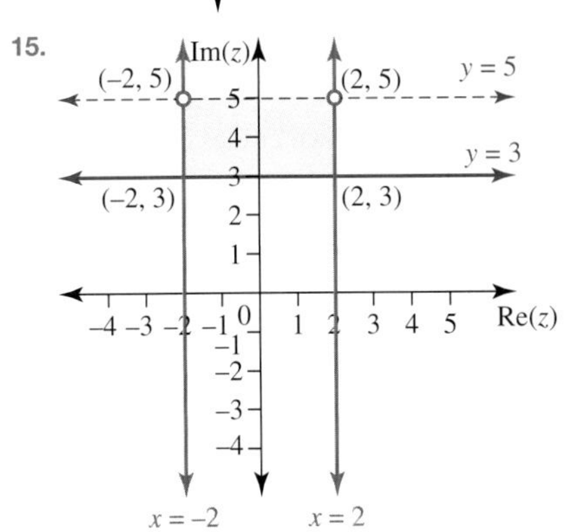

16.

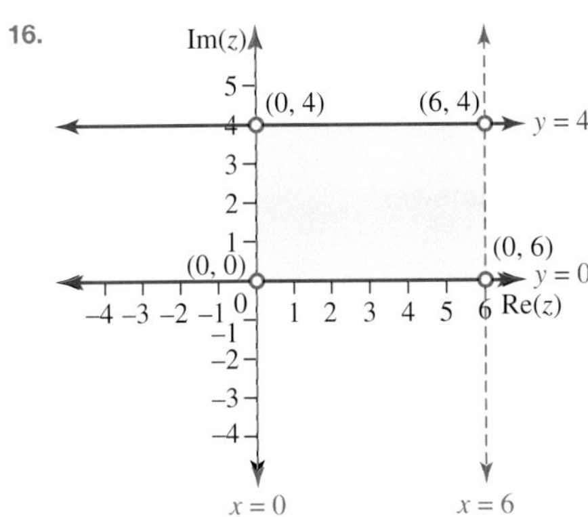

17. The region lies on and between the two circles centred at $(2, 0)$ with radii of 1 and 2 units.

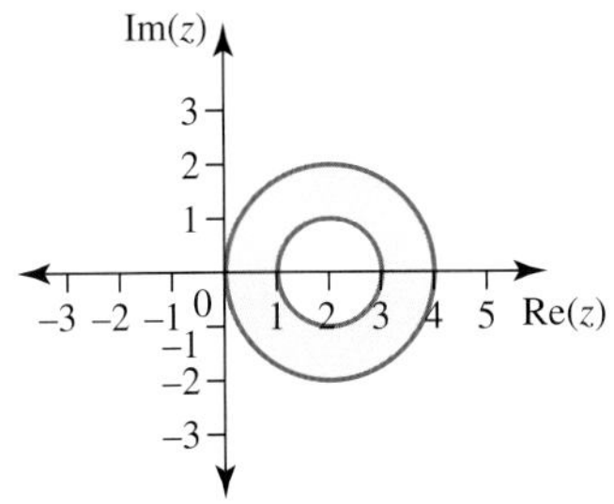

18. The required region is the overlap of these two circles, including points of intersection of the circumferences.

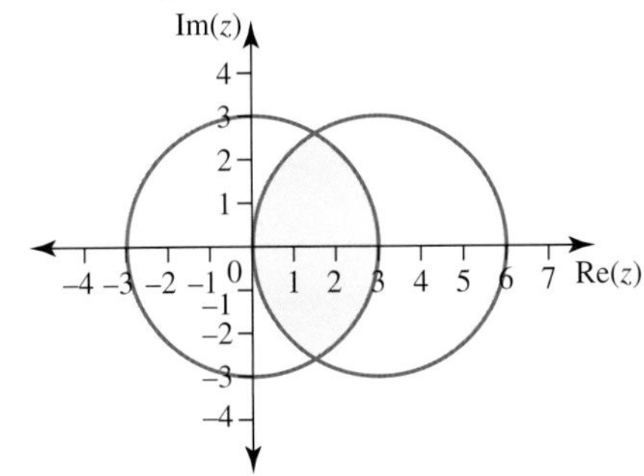

19. a.

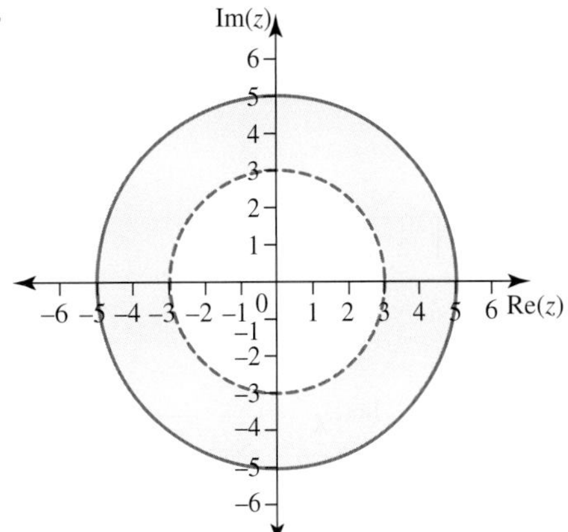

b.

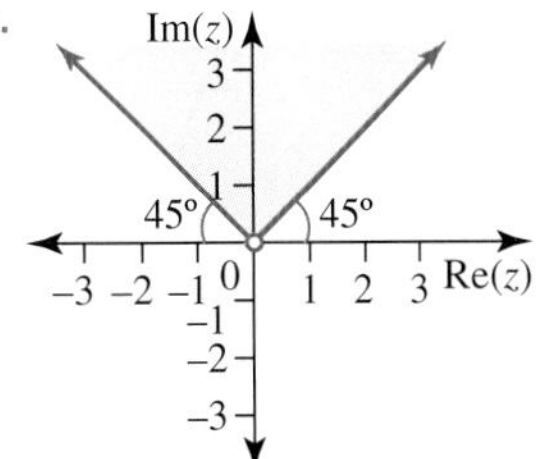

c.

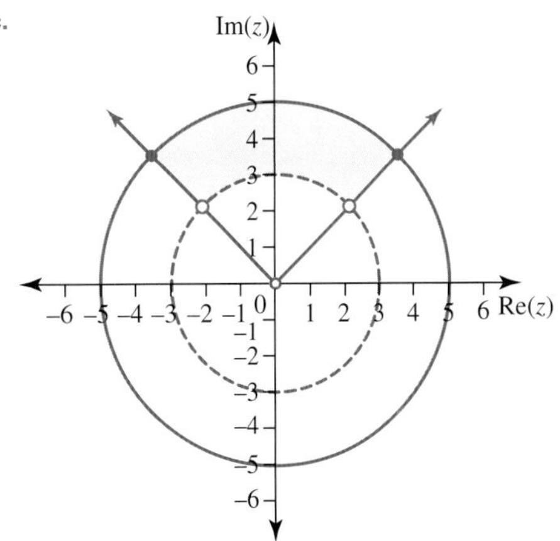

20. a.

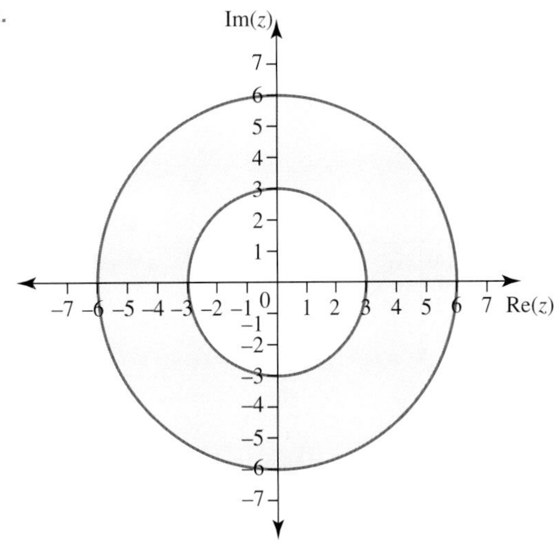

b.

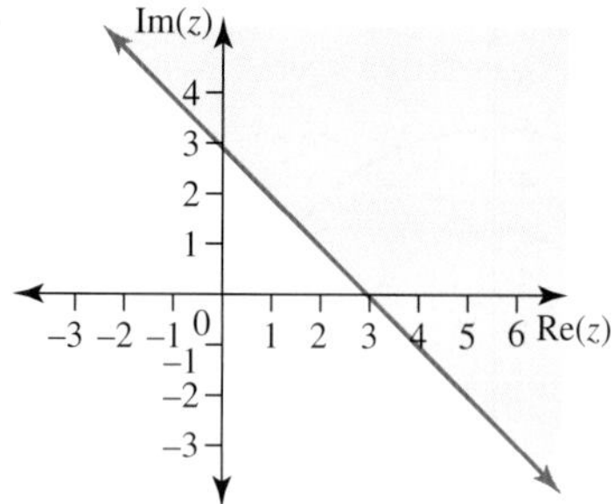

c.

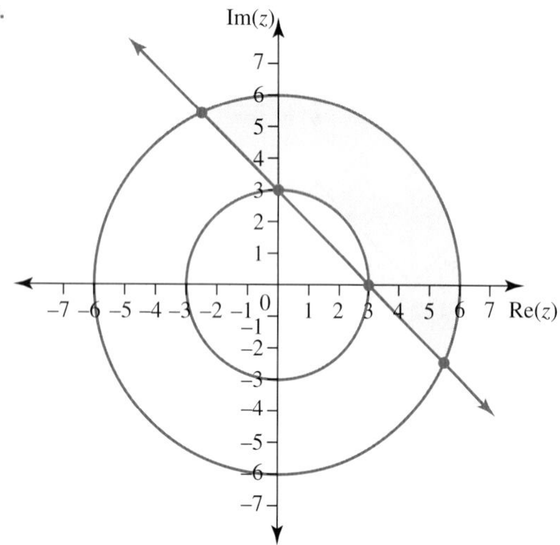

10.9 Exam questions

Note: Mark allocations are available with the fully worked solutions online.

1. a. $x = 3$; line

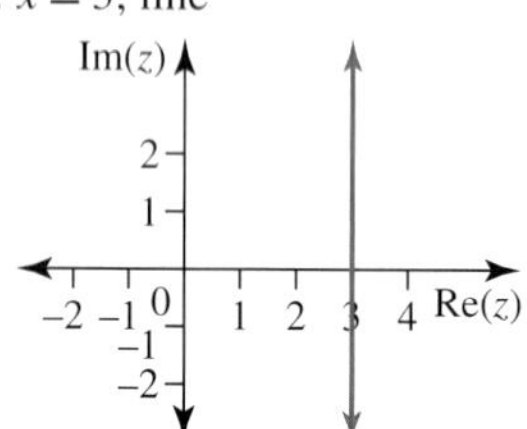

b. $y = -x$

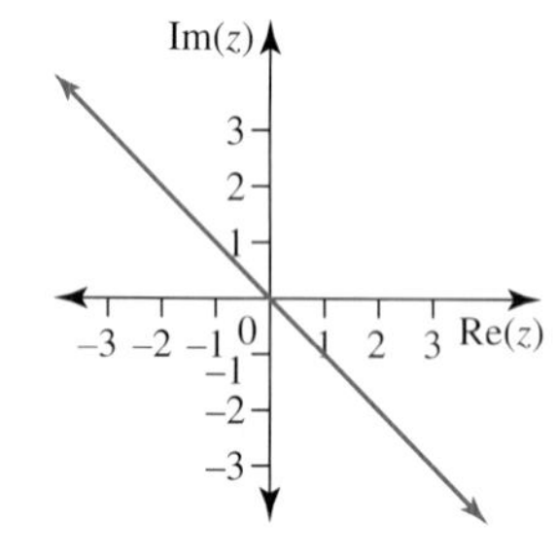

2. A ray from $(2, 0)$ making an angle of $\frac{\pi}{6}$ or 30° with the real axis.

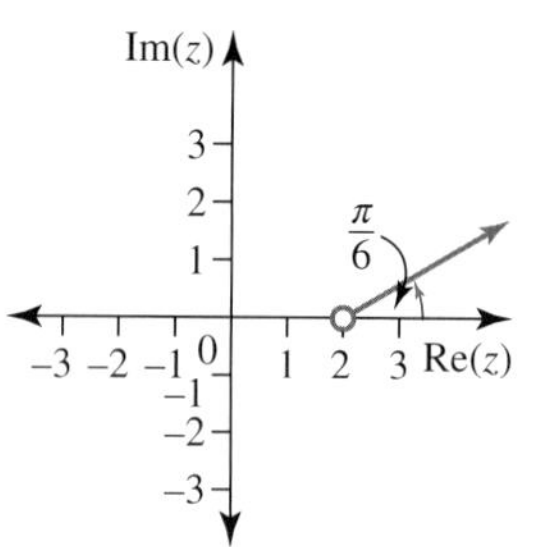

3. The region lies on and between the two circles centred at $(0, 0)$ of radius 1 and 2 units.

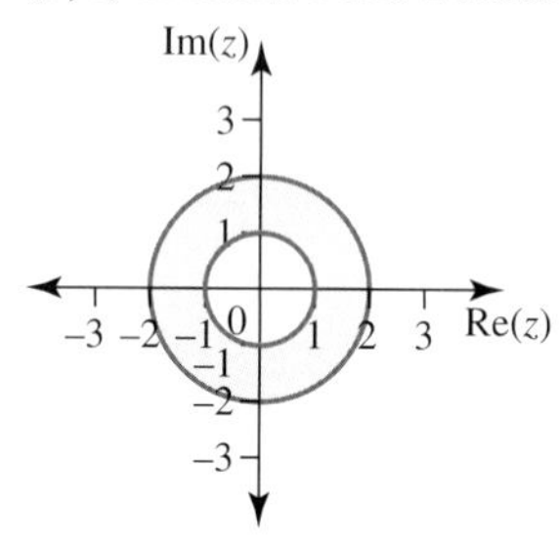

10.10 Review

10.10 Exercise

Technology free: short answer

1. $-1 - 2i$
2. $6 - 5i$
3. a. -48
   b. $a = -549$, $b = 296$
4. $\frac{12}{17} - \frac{14}{17}i$
5. $7\sqrt{2}\,\text{cis}\left(-\frac{3\pi}{4}\right)$
6. $-1 \pm 2i$
7. a. Circle $x^2 + y^2 = 4$

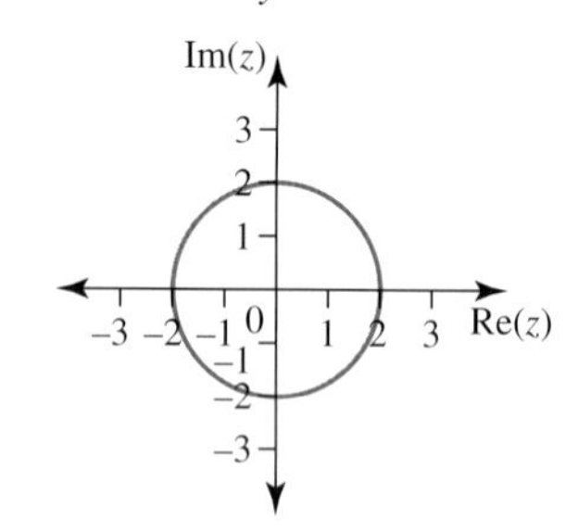

b. Line $x = \frac{5}{2}$

$x = \frac{5}{2}$

Technology active: multiple choice

8. C
9. E
10. B
11. C
12. B
13. B
14. D
15. E
16. D
17. B

Technology active: extended response

18. a. $iz = -2 + 3i, i^2z = -3 - 2i, i^3z = 2 - 3i, i^4z = 3 + 2i$

b. $i^4z = z$

c.

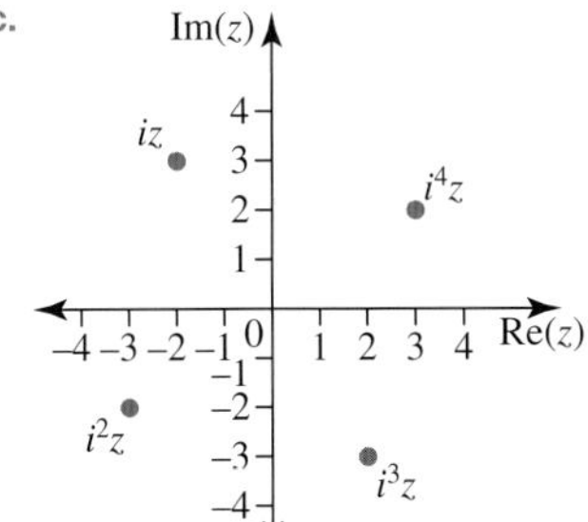

d.

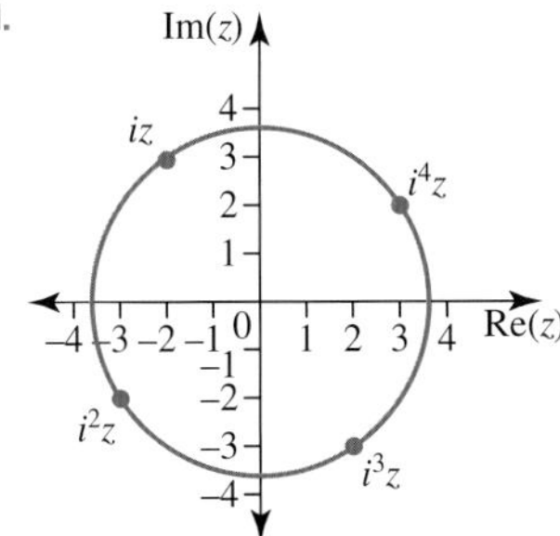

e. $\sqrt{13}$

f. ii.–iii. Rotation of $\frac{\pi}{2}$ (90°) anticlockwise about the origin

g. Rotation of $\frac{\pi}{2}$ (90°) anticlockwise about the origin

h. All points will lie on a circle with the origin as the centre and radius $\sqrt{x^2 + y^2}$

19. a. Reflecting z in the real axis followed by a dilation of $\frac{1}{|z|}$

b. Rotating z through $\frac{\pi}{2}$ (90°) in an anticlockwise direction

20. a.

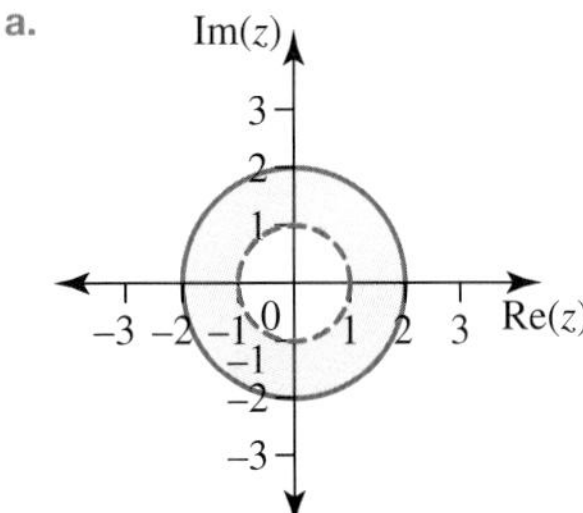

b.

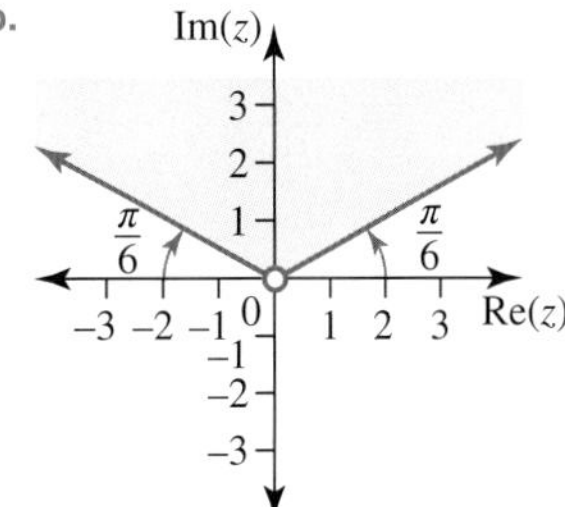

c.

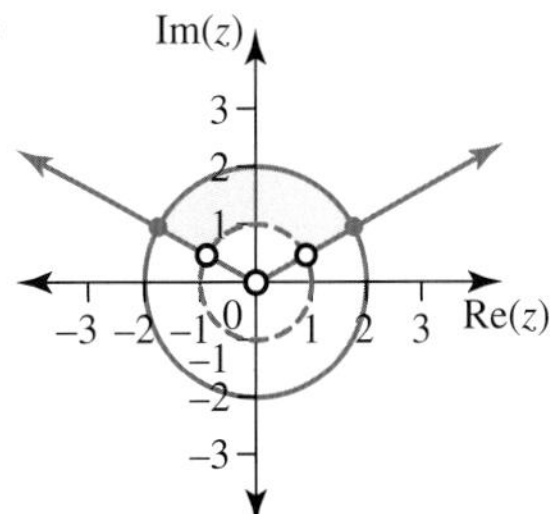

10.10 Exam questions

Note: Mark allocations are available with the fully worked solutions online.

1. $-\frac{1}{5} + \frac{2}{5}i$
2. D
3. C
4. D
5. a. $z = 3 - 2i$

b. $a = \pm 3,\ b = \mp 2$

11 Transformations

LEARNING SEQUENCE

Fully worked solutions for this topic are available online.

11.1 Overview

Hey students! Bring these pages to life online

Watch videos

Engage with interactivities

Answer questions and check results

Find all this and MORE in jacPLUS

11.1.1 Introduction

Following on from the introduction to matrices in Topic 5, we will now investigate geometric transformations with matrices. A transformation is a movement or change in a geometric shape through translation, reflection, rotation and dilation. A translation can be thought of as a 'slide', a reflection as a 'mirror image', a rotation as a 'spin' about a point and a dilation as an enlargement or reduction of a shape or figure.

One of the most visually arresting examples of the manipulation of images through transformations is in the fascinating art of Maurits Cornelis Escher. Although this Dutch artist was not formally mathematically trained, he drew great inspiration from the mathematical ideas he read about.

Early animations, such as cartoons, were originally drawn by hand — a very time-consuming process. Today the transformation of points, and hence images, through matrix transformations is the basis of computer based animation. In this topic you will see how matrix transformations of reflection, dilation and rotation can be combined into a single new matrix. This combination of the movements of several individual matrices into one operation allows movement of every vertex in three-dimensional space. When matrix transformations are combined with the imagination of animators, designers and storytellers, images are created that have the power to shock, awe and inspire.

KEY CONCEPTS

This topic covers the following key concepts from the VCE Mathematics Study Design:

- points in the plane, coordinates and their representation as 2×1 matrices (column vectors)
- translations of the plane $T(x,\ y) \rightarrow (x + a,\ y + b)$
- linear transformations of the plane $T(x, y) \rightarrow (ax + by,\ cx + dy)$ and $T\begin{pmatrix} x \\ y \end{pmatrix} = \begin{pmatrix} a & b \\ c & d \end{pmatrix}\begin{pmatrix} x \\ y \end{pmatrix}$ as a map of the plane onto itself, rotations about the origin and reflection in a line through the origin
- the effect of translation, linear transformations and their inverse transformations, and compositions of these transformations on subsets of the plane such as points, lines, shapes and graphs of functions and relations
- invariance of properties under transformation, and the relationship between the determinant of a transformation matrix and the effect of the linear transformation on the area of a bounded region of the plane.

11.2 Translations

LEARNING INTENTION

At the end of this subtopic you should be able to:
- determine matrix translations of points, lines, curves and objects.

11.2.1 Matrix transformations

A **transformation** is a function which maps the points of a set X, called the **pre-image** or original point onto a set of points Y, called the **image**.

A transformation is a change of position of points, lines, curves or shapes in a plane, or a change in shape due to an enlargement or reduction by a scale factor.

Each point of the plane is transformed or mapped onto another point.

Matrix transformations

The transformation, *T*, is written as:

$$T: \begin{bmatrix} x \\ y \end{bmatrix} \rightarrow \begin{bmatrix} x' \\ y' \end{bmatrix}$$

The matrix $\begin{bmatrix} x \\ y \end{bmatrix}$ is the column matrix representing the coordinates of the point $P(x, y)$, the pre-image or original point. The matrix $\begin{bmatrix} x' \\ y' \end{bmatrix}$ represents the coordinates of the point $P'(x', y')$, the image of $P(x, y)$ after a transformation.

Any transformation that can be represented by a 2×2 matrix, $\begin{bmatrix} a & b \\ c & d \end{bmatrix}$, is called a **linear transformation**. The origin never moves under a linear transformation.

An **invariant point** or **fixed point** is a point of the domain of the function which is mapped onto itself after a transformation, that is, the pre-image point is the same as the image point and is unchanged by the transformation.

$$\begin{bmatrix} x \\ y \end{bmatrix} = \begin{bmatrix} x' \\ y' \end{bmatrix} \Rightarrow x' = x \text{ and } y' = y$$

For example, a reflection in the line $y = x$ leaves every point on the line $y = x$ unchanged.

The transformations which will be studied in this topic are:
- translations
- reflections
- rotations
- dilations.

11.2.2 Translations

A **translation** is a transformation where each point in the plane is moved a given distance in a horizontal and/or vertical direction.

Consider a marching band marching in perfect formation. As the leader of the marching band moves from a position P(x, y) to a new position P′(x', y'), which is a steps across and b steps up, all members of the band will also move to a new position P′($x+a$, $y+b$). Their new position could be defined as P′(x', y') = P′($x+a$, $y+b$) where a represents the horizontal translation and b represents the vertical translation and P′ is the image of P.

Translation matrix equation

The matrix equation for a translation can be given as:

$$\underset{\mathbf{P'}}{\begin{bmatrix} x' \\ y' \end{bmatrix}} = \underset{\mathbf{P}}{\begin{bmatrix} x \\ y \end{bmatrix}} + \underset{\mathbf{T}}{\begin{bmatrix} a \\ b \end{bmatrix}}$$

where a represents the horizontal translation and b the vertical translation.

The matrix $\begin{bmatrix} a \\ b \end{bmatrix}$ is called the **translation matrix** or column vector and is denoted by T. It represents the horizontal and vertical displacement, as shown in the diagram:

$$x' = x + a$$
$$y' = y + b$$

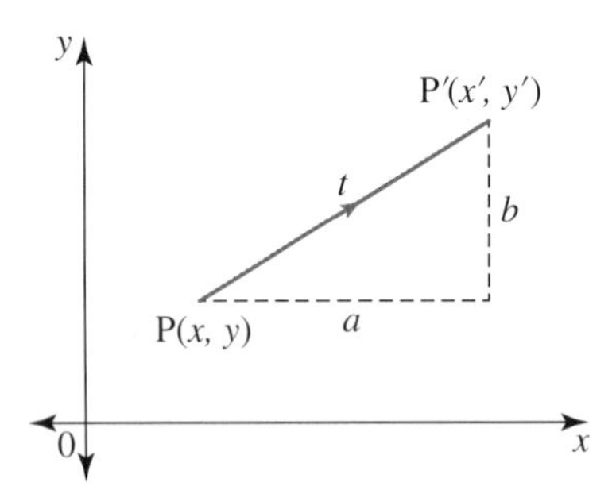

Each x-coordinate is moved a units parallel to the x-axis and each y-coordinate in moved b units parallel to the y-axis.

Note t (lower case) denotes the translation itself and T (upper case) denotes the matrix of the translation.

WORKED EXAMPLE 1 Determining a matrix translation of a point

A cyclist in a race needs to move from the front position at (0, 0) across 2 positions, to the left, so that the other cyclists can pass. Write the translation matrix and determine the cyclist's new position.

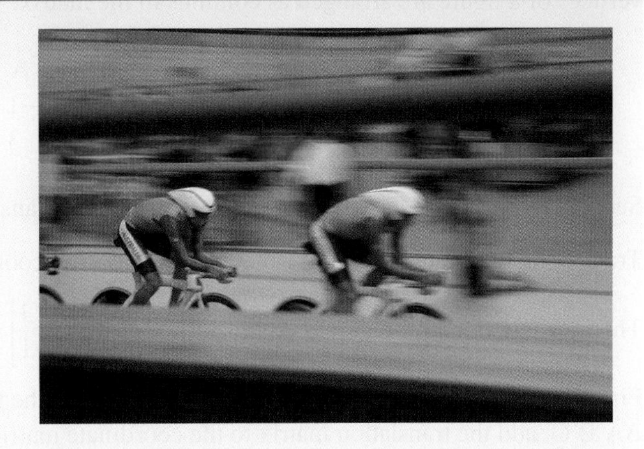

THINK	WRITE
1. Write down the translation matrix, T, using the information given.	The cyclist moves across to the left by 2 units. Translating 2 units to the left means each x-coordinate decreases by 2. $T=\begin{bmatrix}-2\\0\end{bmatrix}$
2. Recall and apply the matrix transformation for a translation equation.	$\overset{P'}{\begin{bmatrix}x'\\y'\end{bmatrix}}=\overset{P}{\begin{bmatrix}x\\y\end{bmatrix}}+\overset{T}{\begin{bmatrix}-2\\0\end{bmatrix}}=\begin{bmatrix}x-2\\y+0\end{bmatrix}$
3. Substitute the pre-image point into the matrix equation.	The pre-image point is $(0,0)$. $\overset{P'}{\begin{bmatrix}x'\\y'\end{bmatrix}}=\begin{bmatrix}0-2\\0+0\end{bmatrix}$
4. State the cyclist's new position by calculating the coordinates of the image point from the matrix equation. *Note:* The y-coordinate has not changed; that is, the cyclist is still in front, but to the left.	$\overset{P'}{\begin{bmatrix}x'\\y'\end{bmatrix}}=\overset{P}{\begin{bmatrix}0\\0\end{bmatrix}}+\overset{T}{\begin{bmatrix}-2\\0\end{bmatrix}}=\begin{bmatrix}-2\\0\end{bmatrix}$ The new position is $(-2,0)$.

11.2.3 Translations of an object

Matrix addition can be used to determine the Cartesian coordinates of a translated object when an object is moved or translated from one location to another on the coordinate plane without changing its size or orientation.

Consider the triangle ABC with coordinates $A(-1,3), B(0,2)$ and $C(-2,1)$. It is to be moved 3 units to the right and 1 unit down. To calculate the coordinates of the vertices of the translated $\Delta A'B'C'$, we can use matrix addition.

First, the coordinates of the triangle ΔABC can be written as a **coordinate matrix**. The coordinates of the vertices of a figure are arranged as columns in the matrix.

$$\Delta ABC = \begin{array}{c} \begin{array}{ccc} A & B & C \end{array} \\ \begin{bmatrix} -1 & 0 & -2 \\ 3 & 2 & 1 \end{bmatrix} \end{array}$$

Secondly, translating the triangle 3 units to the right means each x-coordinate increases by 3.

Translating the triangle 1 unit down means that each y-coordinate decreases by 1.

The translation matrix that will do this is $\begin{bmatrix} 3 & 3 & 3 \\ -1 & -1 & -1 \end{bmatrix}$.

Finally, to determine the coordinates of the vertices of the translated triangle $\Delta A'B'C'$ add the translation matrix to the coordinate matrix.

$$\begin{array}{c} \begin{array}{ccc} A & B & C \end{array} \\ \begin{bmatrix} -1 & 0 & -2 \\ 3 & 2 & 1 \end{bmatrix} \end{array} + \begin{bmatrix} 3 & 3 & 3 \\ -1 & -1 & -1 \end{bmatrix} = \begin{array}{c} \begin{array}{ccc} A' & B' & C' \end{array} \\ \begin{bmatrix} 2 & 3 & 1 \\ 2 & 1 & 0 \end{bmatrix} \end{array}$$

The coordinates of the vertices of the translated triangle $\Delta A'B'C' = \begin{bmatrix} 2 & 3 & 1 \\ 2 & 1 & 0 \end{bmatrix}$ are $A'(2, 2)$, $B'(3, 1)$ and $C'(1, 0)$.

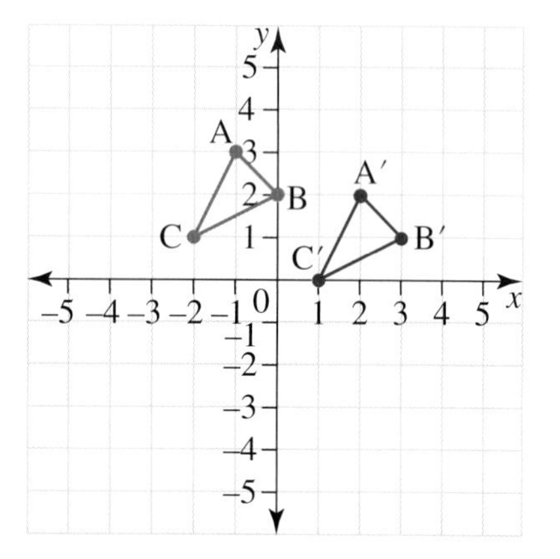

WORKED EXAMPLE 2 Determining a matrix translation of a shape

Determine the translation matrix if ΔABC with coordinates A(−1, 3), B(0, 2) and C(−2, 1) is translated to ΔA′B′C′ with coordinates A′(2, 4), B′(3, 3) and C′(1, 2).

THINK

1. Write the coordinates of ΔABC as a coordinate matrix.

WRITE

The coordinates of the vertices of a figure are arranged as columns in the matrix.

$$\Delta ABC = \begin{array}{c} \begin{array}{ccc} A & B & C \end{array} \\ \begin{bmatrix} -1 & 0 & -2 \\ 3 & 2 & 1 \end{bmatrix} \end{array}$$

2. Write the coordinates of the vertices of the translated triangle $\Delta A'B'C'$ as a coordinate matrix.

$$\Delta A'B'C' = \begin{array}{c} \begin{array}{ccc} A' & B' & C' \end{array} \\ \begin{bmatrix} 2 & 3 & 1 \\ 4 & 3 & 2 \end{bmatrix} \end{array}$$

3. Calculate the translation matrix by recalling the matrix equation: $P' = P + T$. Rearrange the equation to make T the subject and apply the rules of matrix subtraction.

$$\overset{P'}{\begin{bmatrix} 2 & 3 & 1 \\ 4 & 3 & 2 \end{bmatrix}} = \overset{P}{\begin{bmatrix} -1 & 0 & -2 \\ 3 & 2 & 1 \end{bmatrix}} + T$$

$$T = \overset{P'}{\begin{bmatrix} 2 & 3 & 1 \\ 4 & 3 & 2 \end{bmatrix}} - \overset{P}{\begin{bmatrix} -1 & 0 & -2 \\ 3 & 2 & 1 \end{bmatrix}}$$

4. Translating the triangle 3 units to the right means that each x-coordinate increases by 3. Translating the triangle 1 unit up means that each y-coordinate increases by 1.

The translation matrix is:

$$T = \begin{bmatrix} 3 & 3 & 3 \\ 1 & 1 & 1 \end{bmatrix}$$

11.2.4 Invariant properties of translations

Properties which are unchanged by a transformation are called **invariants**. For translations, properties such as length, shape and area are invariant since they are unchanged when a line, curve or shape undergoes a translation.

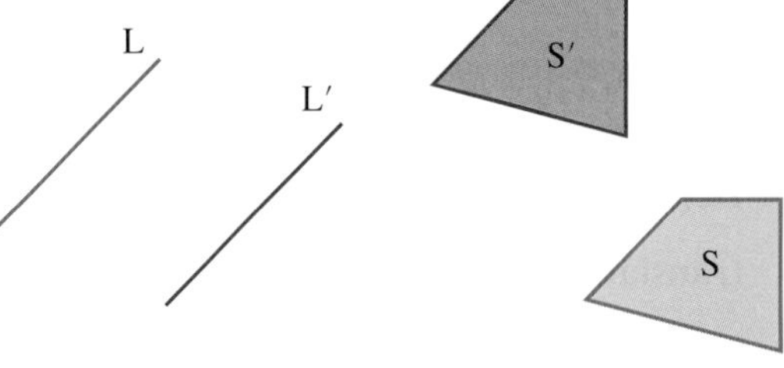

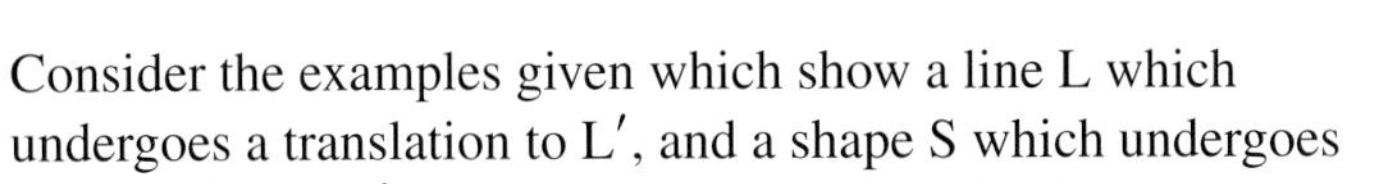

Consider the examples given which show a line L which undergoes a translation to L′, and a shape S which undergoes a translation to S′. Notice that the length of the line L is unchanged under the translation, and the length of sides, shape and area of S is unchanged under translation.

11.2.5 Translations of a curve

A translation of a curve maps every original point (x, y) of the curve onto a new unique and distinct image point (x', y').

Consider the parabola with the equation $y = x^2$.

If the parabola is translated 3 units in the positive direction of the x-axis, what is the image equation and what happens to the coordinates?

As seen from the table of values, each coordinate (x, y) has a new coordinate pair or image point $(x + 3, y)$.

x	y	(x, y)	$x' = x + 3$	$y' = y$	(x', y')
−3	9	(−3, 9)	−3 + 3	9	(0, 9)
−2	4	(−2, 4)	−2 + 3	4	(1, 4)
−1	1	(−1, 1)	−1 + 3	1	(2, 1)
0	0	(0, 0)	0 + 3	0	(3, 0)
1	1	(1, 1)	1 + 3	1	(4, 1)
2	4	(2, 4)	2 + 3	4	(5, 4)
3	9	(3, 9)	3 + 3	9	(6, 9)

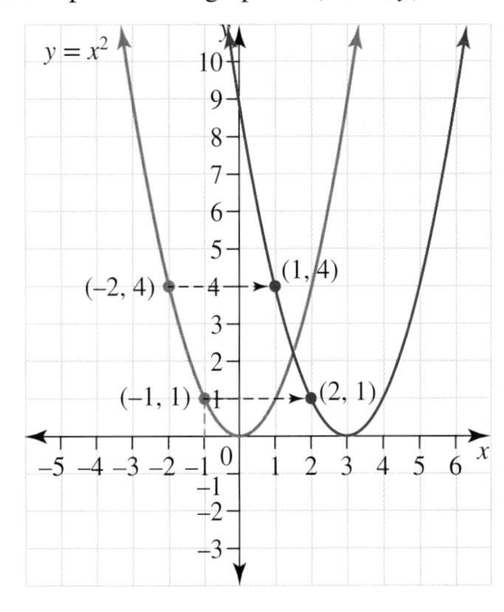

The matrix equation for the translation of any point on the curve $y = x^2$ can be written as:

$$\begin{bmatrix} x' \\ y' \end{bmatrix} = \begin{bmatrix} x \\ y \end{bmatrix} + \begin{bmatrix} 3 \\ 0 \end{bmatrix}$$

The image equations for the two coordinates are $x' = x + 3$ and $y' = y$.

Rearranging the image equations to make the pre-image coordinates the subject, we get

$$x' = x + 3 \leftrightarrow x = x' - 3 \text{ and } y = y'.$$

To determine the image equation, substitute the image expressions into the pre-image equation.

$$y = x^2$$

$$y = y' \qquad x = x' - 3$$

$$\therefore y' = (x' - 3)^2$$

The equation of the parabola, $y = x^2$, after a translation of 3 units in the positive direction of the x-axis is $y = (x - 3)^2$.

WORKED EXAMPLE 3 Determining a matrix translation of a line

Determine the equation of the image of the line with equation $y = x + 1$ after it is transformed by the translation matrix $T = \begin{bmatrix} 2 \\ 1 \end{bmatrix}$.

THINK	WRITE
1. Recall the matrix equation for the transformation and apply it to the given transformation.	$\begin{bmatrix} x' \\ y' \end{bmatrix} = \begin{bmatrix} x \\ y \end{bmatrix} + \begin{bmatrix} 2 \\ 1 \end{bmatrix}$
2. State the image equations for the two coordinates.	$x' = x + 2$ and $y' = y + 1$
3. Rearrange the equations to make the pre-image coordinates x and y the subjects.	$x = x' - 2$ and $y = y' - 1$
4. Substitute the image equations into the pre-image equation to determine the image equation.	$y = x + 1$ $y' - 1 = (x' - 2) + 1$ $y' = x'$ The image equation is $y = x$.
5. Graph the image and pre-image equation to verify the translation.	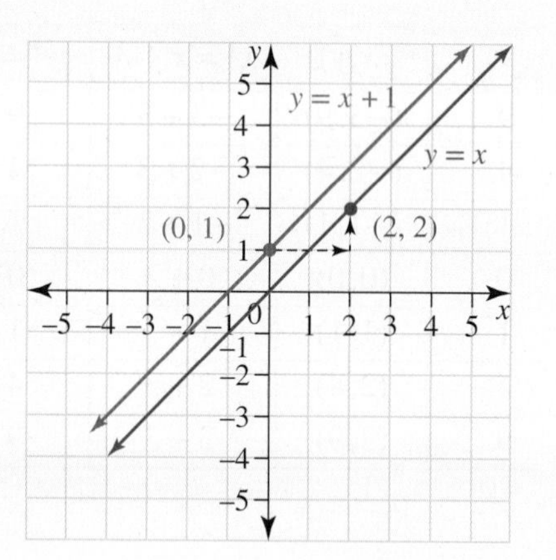

WORKED EXAMPLE 4 Determining a matrix translation of a parabola

Determine the equation of the image of the parabola with equation $y = x^2$ after it is transformed by the translation matrix $T = \begin{bmatrix} -3 \\ 1 \end{bmatrix}$.

THINK | **WRITE**

1. Recall the matrix equation for the transformation and apply it to the given translation.

$$\begin{bmatrix} x' \\ y' \end{bmatrix} = \begin{bmatrix} x \\ y \end{bmatrix} + \begin{bmatrix} -3 \\ 1 \end{bmatrix}$$

2. State the image equations for the two coordinates.

$x' = x - 3$ and $y' = y + 1$

3. Rearrange the equations to make the pre-image coordinates x and y the subjects.

$x = x' + 3$ and $y = y' - 1$

4. Substitute the image expressions into the pre-image equation to determine the image equation.

$$y = x^2$$
$$y' - 1 = (x' + 3)^2$$
$$y' = (x' + 3)^2 + 1$$

The image equation is $y = (x + 3)^2 + 1$.

5. Graph the image and pre-image equation to verify the translation.

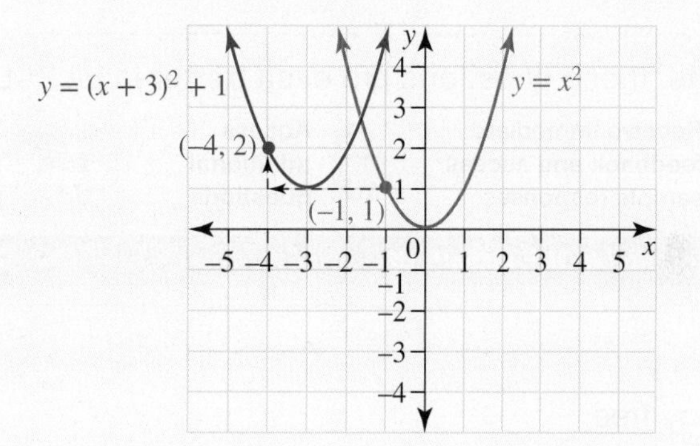

WORKED EXAMPLE 5 Determining the translation that maps one line onto another

Determine a translation matrix that maps the line with equation $y = x$ onto the line with equation $y = x - 7$.

THINK | **WRITE**

1. Write a matrix equation for a translation.

$$\begin{bmatrix} x' \\ y' \end{bmatrix} = \begin{bmatrix} x \\ y \end{bmatrix} + \begin{bmatrix} a \\ b \end{bmatrix} = \begin{bmatrix} x + a \\ y + b \end{bmatrix}$$

2. State the separate equations for the x- and y-coordinates of the image and then rearrange them to make x and y the subjects.

$x' = x + a \quad y' = y + b$

$x = x' - a \quad y = y' - b$

3. Substitute these into the equation of the line $y = x$ and rearrange to make y' the subject.

$$y = x$$
$$y' - b = x' - a$$
$$y' = x' - a + b$$

4. Recall that the equation of the image is $y' = x' - 7$. Determine values of a and b that satisfy this equation.

$y' = x' - 7$
$\therefore -a + b = -7$
$a = 7,\ b = 0$ satisfy this equation.

5. State the translation that maps the line $y = x$ onto the line $y = x - 7$.
Note: There are infinitely many translations that map the line $y = x$ onto the line $y = x - 7$. Any values of a and b for which $b - a = -7$ will be valid answers.

$T = \begin{bmatrix} 7 \\ 0 \end{bmatrix}$

11.2 Exercise

Students, these questions are even better in jacPLUS

Receive immediate feedback and access sample responses

Access additional questions

Track your results and progress

Find all this and MORE in jacPLUS

Technology free

1. WE1 A chess player moves his knight 1 square to the right and 2 squares up from position $(2, 5)$. Write the translation matrix and determine the new position of the knight.

2. Determine the image of the point $(-1, 0)$ using the matrix equation for the translation $\begin{bmatrix} x' \\ y' \end{bmatrix} = \begin{bmatrix} x \\ y \end{bmatrix} + \begin{bmatrix} -5 \\ 2 \end{bmatrix}$.
3. Determine the image of the point $(1, 2)$ using the matrix equation for translation $\begin{bmatrix} x' \\ y' \end{bmatrix} = \begin{bmatrix} x \\ y \end{bmatrix} + \begin{bmatrix} 3 \\ -2 \end{bmatrix}$.
4. Determine the image of the point $(3, -4)$ using the matrix equation for translation $\begin{bmatrix} x' \\ y' \end{bmatrix} = \begin{bmatrix} x \\ y \end{bmatrix} + \begin{bmatrix} -1 \\ 2 \end{bmatrix}$.
5. WE2 Determine the translation matrix if ΔABC with coordinates $A(0, 0), B(2, 3)$ and $C(-3, 4)$ is translated to $\Delta A'B'C'$ with coordinates $A'(1, -2), B'(3, 1)$ and $C'(-2, 2)$.
6. Determine the translation matrix if ΔABC with coordinates $A(3, 0), B(2, 4)$ and $C(-2, -5)$ is translated to $\Delta A'B'C'$ with coordinates $A'(4, 2), B'(3, 6)$ and $C'(-1, -3)$.
7. The image equations are given by $x' = x + 2$ and $y' = y + 1$. Express the translation in matrix equation form.

8. a. On a Cartesian plane, draw $\Delta ABC = \begin{bmatrix} 0 & 1 & -2 \\ 0 & 3 & 1 \end{bmatrix}$ and $\Delta A'B'C' = \begin{bmatrix} 2 & 3 & 0 \\ -1 & 2 & 0 \end{bmatrix}$.

b. Calculate the translation matrix if $\Delta ABC = \begin{bmatrix} 0 & 1 & -2 \\ 0 & 3 & 1 \end{bmatrix}$ is translated to $\Delta A'B'C' = \begin{bmatrix} 2 & 3 & 0 \\ -1 & 2 & 0 \end{bmatrix}$.

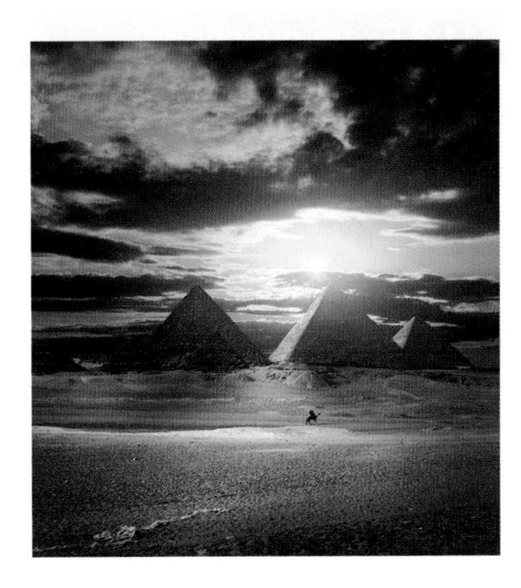

9. **WE3** Determine the equation of the image of the line with equation $y = x - 3$ after it is transformed by the translation matrix $T = \begin{bmatrix} -1 \\ 3 \end{bmatrix}$.

10. Determine the equation of the image of the line with equation $y = x - 1$ after it is transformed by the translation matrix $T = \begin{bmatrix} 3 \\ -2 \end{bmatrix}$.

11. Determine the equation of the image of the line with equation $y = x + 3$ after it is transformed by the translation matrix $T = \begin{bmatrix} -2 \\ 1 \end{bmatrix}$.

12. **WE4** Determine the equation of the image of the parabola with equation $y = x^2$ after it is transformed by the translation matrix $T = \begin{bmatrix} 2 \\ -1 \end{bmatrix}$.

13. Determine the equation of the image of the parabola with equation $y = x^2 + 1$ after it is transformed by the translation matrix $T = \begin{bmatrix} -3 \\ 0 \end{bmatrix}$.

14. Determine the equation of the image of the parabola with equation $y = x^2 - 2$ after it is transformed by the translation matrix $T = \begin{bmatrix} 7 \\ -4 \end{bmatrix}$.

15. **WE5** Determine a translation matrix that maps the line with equation $y = x$ onto the line with equation $y = x + 2$.

16. Determine a translation matrix that maps the parabola with equation $y = x^2$ onto the parabola with equation $y = (x - 7)^2 + 3$.

17. Determine the translation equation that maps the parabola with equation $y = x^2$ onto the parabola with equation $y = x^2 - 4x + 10$.

18. Determine the translation equation that maps the circle with equation $x^2 + y^2 = 9$ onto the circle with equation $(x - 1)^2 + y^2 = 9$.

19. Determine the translation equation that maps the parabola with equation $y = x^2$ onto the parabola with equation $y = (x - a)^2 + b$.

20. Determine the translation equation that maps the circle with equation $x^2 + y^2 = r^2$ onto the circle with equation $(x - a)^2 + y^2 = r^2$.

11.2 Exam questions

Question 1 (1 mark) TECH-ACTIVE

MC The coordinates of the image of $(-3, 2)$ under the transformation given by $\begin{bmatrix} 5 \\ -3 \end{bmatrix}$ are

A. $(-2, 1)$ **B.** $(2, -1)$ **C.** $(-6, 7)$ **D.** $(6, -7)$ **E.** $(6, 7)$

Question 2 (1 mark) TECH-ACTIVE

MC The image equation when the line with equation $y = x - 5$ is transformed by the translation matrix $T = \begin{bmatrix} 3 \\ -1 \end{bmatrix}$ is

A. $y' = 3x' - 1$ **B.** $y' = -3x' - 1$ **C.** $y' = x' - 9$ **D.** $y' = x' - 7$ **E.** $y' = x' - 1$

Question 3 (3 marks) TECH-FREE

Determine the image equation when the equation $y = x^2 + 1$ is transformed by the translation matrix $T = \begin{bmatrix} 2 \\ -1 \end{bmatrix}$.

More exam questions are available online.

11.3 Reflections and rotations

LEARNING INTENTION

At the end of this subtopic you should be able to:

- determine reflections and rotations of points, lines, curves and objects using matrices.

11.3.1 Reflections

A **reflection** is a transformation in a line. This line is called a line of reflection, and the image point is a mirror image of the pre-image point.

Line of reflection

Pre-image Image

In a reflection:

- the image, P′, and the pre-image, P, are equidistant from the line of reflection.
- the line of reflection is perpendicular to the line joining the image, P′, and the pre-image, P.

The following reflections will be considered:

- reflection in the x-axis $(y = 0)$
- reflection in the y-axis $(x = 0)$
- reflections in lines that pass through the origin.
- reflection in the line $y = x \tan(\theta)$.

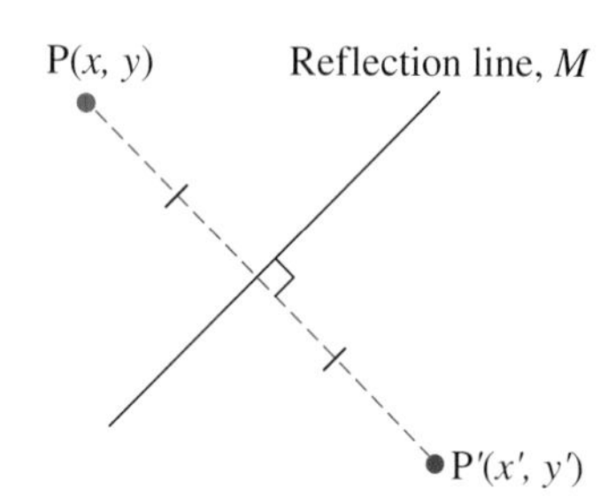

11.3.2 Reflection in the x-axis $(y=0)$

The reflection in the x-axis maps the point $P(x, y)$ onto the point $P'(x', y')$, giving the image point $(x', y') = (x, -y)$.

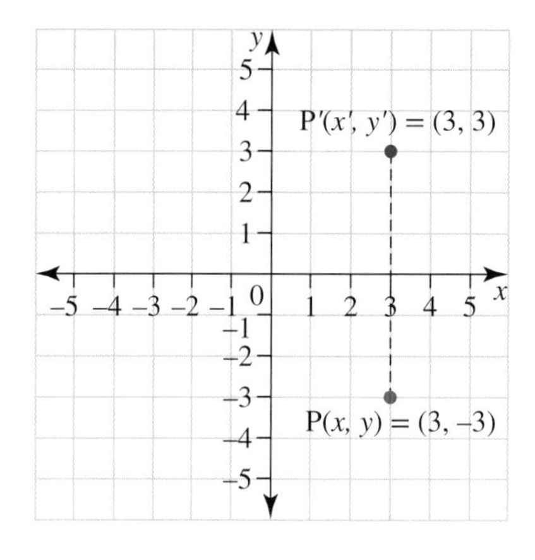

The matrix for a reflection in the x-axis is:

$$M_x = \begin{bmatrix} 1 & 0 \\ 0 & -1 \end{bmatrix}$$

In matrix form, the reflection for any point in the x-axis is

$$\overset{P'}{\begin{bmatrix} x' \\ y' \end{bmatrix}} = \overset{T}{\begin{bmatrix} 1 & 0 \\ 0 & -1 \end{bmatrix}} \overset{P}{\begin{bmatrix} x \\ y \end{bmatrix}}$$

11.3.3 Reflection in the y-axis $(x=0)$

The reflection in the y-axis maps the point $P(x, y)$ onto the point $P'(x', y')$, giving the image point $(x', y') = (-x, y)$.

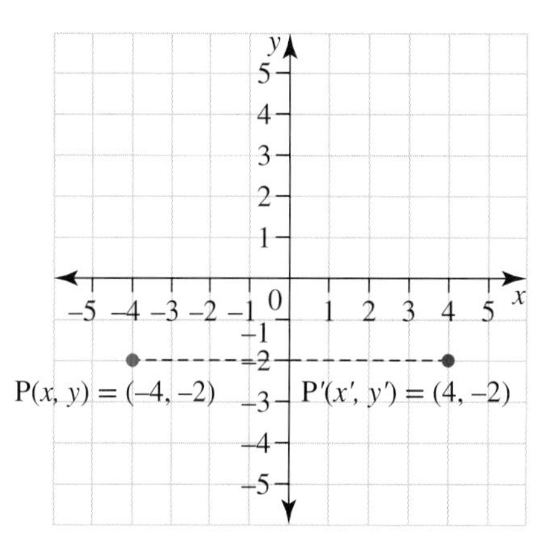

The matrix for a reflection in the y-axis is:

$$M_y = \begin{bmatrix} -1 & 0 \\ 0 & 1 \end{bmatrix}$$

In matrix form, the reflection for any point in the y-axis is:

$$\overset{P'}{\begin{bmatrix} x' \\ y' \end{bmatrix}} = \overset{T}{\begin{bmatrix} -1 & 0 \\ 0 & 1 \end{bmatrix}} \overset{P}{\begin{bmatrix} x \\ y \end{bmatrix}}$$

WORKED EXAMPLE 6 Determining the reflection of a point using matrices

Determine the image of the point $(-2, 3)$:

a. after a reflection in the x-axis

b. after it is transformed by the matrix $\begin{bmatrix} -1 & 0 \\ 0 & 1 \end{bmatrix}$. Comment on your answer.

THINK	WRITE
a. 1. Recall the reflection matrix for reflection in the x-axis.	a. $M_x = \begin{bmatrix} 1 & 0 \\ 0 & -1 \end{bmatrix}$
2. Apply the matrix equation for reflection in the x-axis matrix, $P' = TP$.	$\begin{bmatrix} x' \\ y' \end{bmatrix} = \begin{bmatrix} 1 & 0 \\ 0 & -1 \end{bmatrix} \begin{bmatrix} x \\ y \end{bmatrix}$
3. Substitute the pre-image point into the matrix equation.	The pre-image point is $(-2, 3)$. $\begin{bmatrix} x' \\ y' \end{bmatrix} = \begin{bmatrix} 1 & 0 \\ 0 & -1 \end{bmatrix} \begin{bmatrix} -2 \\ 3 \end{bmatrix}$

4. Calculate the coordinates of the image point.

$$\begin{bmatrix} x' \\ y' \end{bmatrix} = \begin{bmatrix} 1 & 0 \\ 0 & -1 \end{bmatrix}\begin{bmatrix} -2 \\ 3 \end{bmatrix} = \begin{bmatrix} -2 \\ -3 \end{bmatrix}$$

The image point is $(-2, -3)$.

b. 1. Recall the matrix equation, $P' = TP$.

b. $$\begin{bmatrix} x' \\ y' \end{bmatrix} = \begin{bmatrix} -1 & 0 \\ 0 & 1 \end{bmatrix}\begin{bmatrix} x \\ y \end{bmatrix}$$

2. Substitute the pre-image point into the matrix equation.

The pre-image point is $(-2, 3)$.

$$\begin{bmatrix} x' \\ y' \end{bmatrix} = \begin{bmatrix} -1 & 0 \\ 0 & 1 \end{bmatrix}\begin{bmatrix} -2 \\ 3 \end{bmatrix}$$

3. Calculate the coordinates of the image point.

$$\begin{bmatrix} x' \\ y' \end{bmatrix} = \begin{bmatrix} -1 & 0 \\ 0 & 1 \end{bmatrix}\begin{bmatrix} -2 \\ 3 \end{bmatrix} = \begin{bmatrix} 2 \\ 3 \end{bmatrix}$$

The image point is $(2, 3)$. It is a reflection in the y-axis from the pre-image point.

TI | THINK

a. On a Calculator page, complete the entry line as:

$$\begin{bmatrix} 1 & 0 \\ 0 & -1 \end{bmatrix} \times \begin{bmatrix} -2 \\ 3 \end{bmatrix}$$

then press ENTER.

Note: Matrix templates can be found by pressing the templates button.

DISPLAY/WRITE

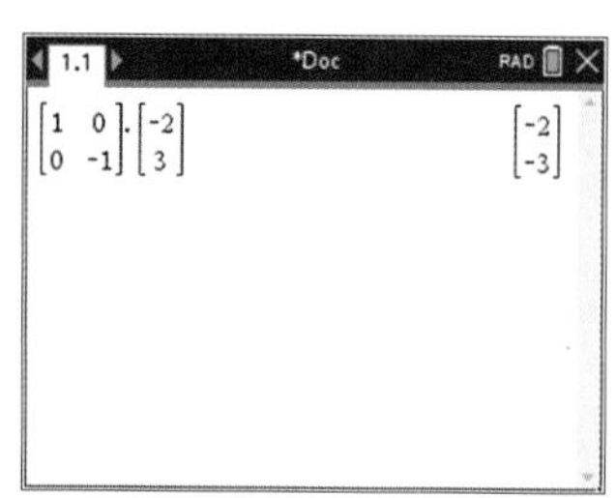

The image point is $(-2, -3)$.

CASIO | THINK

a. On a Main screen, complete the entry line as:

$$\begin{bmatrix} 1 & 0 \\ 0 & -1 \end{bmatrix} \times \begin{bmatrix} -2 \\ 3 \end{bmatrix}$$

then press EXE.

DISPLAY/WRITE

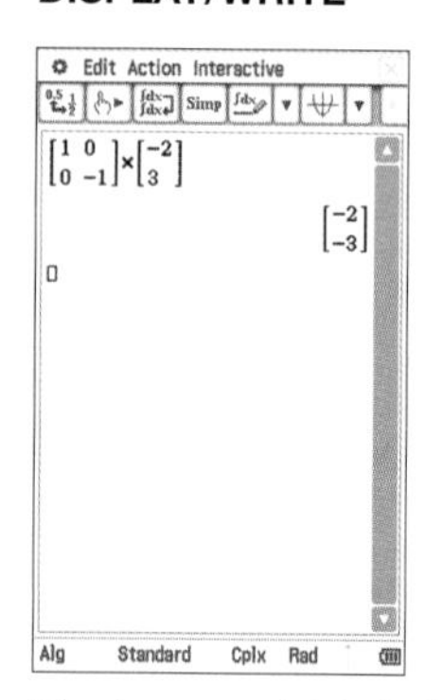

The image point is $(-2, -3)$.

WORKED EXAMPLE 7 Determining a reflection in the y-axis using matrices

Determine the equation of the image of the graph of $y = (x + 1)^2$ after it is reflected in the y-axis.

THINK

1. Recall the reflection matrix for reflection in the y-axis and apply it to the matrix equation.

2. Determine the image coordinates.

WRITE

$$M_y = \begin{bmatrix} -1 & 0 \\ 0 & 1 \end{bmatrix}$$

$$\begin{bmatrix} x' \\ y' \end{bmatrix} = \begin{bmatrix} -1 & 0 \\ 0 & 1 \end{bmatrix}\begin{bmatrix} x \\ y \end{bmatrix}$$

$x' = -x$

$y' = y$

3. Rearrange the equations to make the pre-image coordinates x and y the subjects.

$x = -x'$

$y = y'$

4. Substitute the image expressions into the pre-image equation $y = (x + 1)^2$ to determine the equation of the image.

$$y = (x + 1)^2$$
$$y' = (-x' + 1)^2$$
$$= [-(x' - 1)]^2$$
$$y' = (x' - 1)^2$$

The equation of the image is $y = (x - 1)^2$.

5. Graph the image and the pre-image equation to verify the reflection.

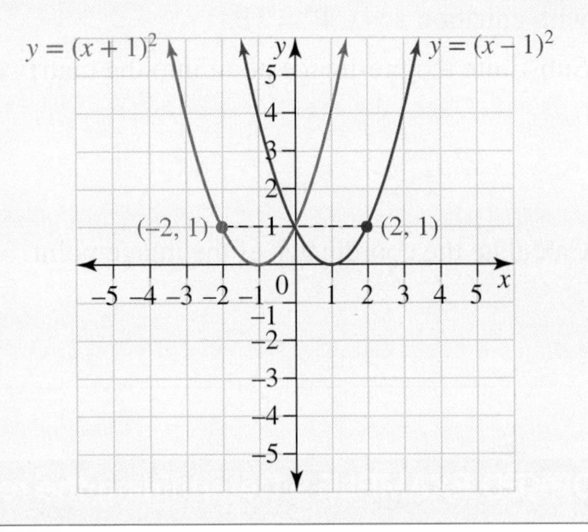

11.3.4 Reflection in a line that passes through the origin (0, 0)

Reflection in the line with equation $y = x$

A reflection in the line $y = x$ maps the point $P(x, y)$ onto the point $P'(x', y')$, giving the image point $(x', y') = (y, x)$.

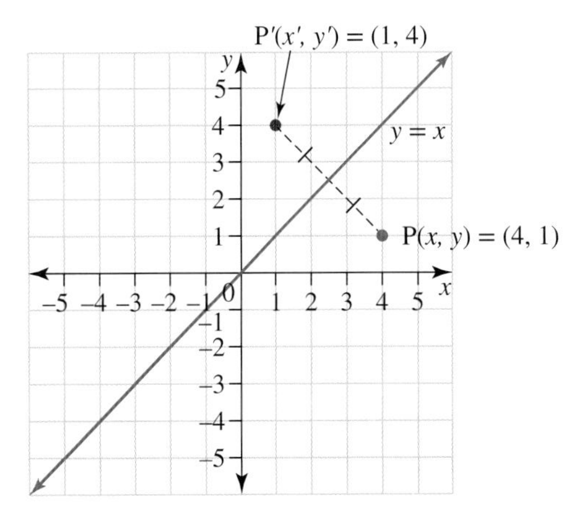

The matrix for a reflection in the line $y = x$ is $M_{y=x} = \begin{bmatrix} 0 & 1 \\ 1 & 0 \end{bmatrix}$.

In matrix form, a reflection in the line $y = x$ is

$$\begin{bmatrix} x' \\ y' \end{bmatrix} = \begin{bmatrix} 0 & 1 \\ 1 & 0 \end{bmatrix} \begin{bmatrix} x \\ y \end{bmatrix}.$$

A reflection in the line $y = -x$ maps the point $P(x, y)$ onto the point $P'(x', y')$, giving the image point $(x', y') = (-y, -x)$.

The matrix for a reflection in the line $y = -x$ is

$$M_{y=-x} = \begin{bmatrix} 0 & -1 \\ -1 & 0 \end{bmatrix}$$

Similarly, in matrix form, a reflection in the line $y = -x$ is

$$\begin{bmatrix} x' \\ y' \end{bmatrix} = \begin{bmatrix} 0 & -1 \\ -1 & 0 \end{bmatrix} \begin{bmatrix} x \\ y \end{bmatrix}$$

WORKED EXAMPLE 8 Determining the reflection in the line $y = x$ using matrices

Determine the coordinates of the image of the point (3, 1) after a reflection in the line with equation $y = x$.

THINK	WRITE
1. Recall the reflection matrix for reflection in the line $y = x$.	$M_{y=x} = \begin{bmatrix} 0 & 1 \\ 1 & 0 \end{bmatrix}$
2. Use the matrix equation for a reflection about the line with equation $y = x$: $P' = TP$.	$\begin{bmatrix} x' \\ y' \end{bmatrix} = \begin{bmatrix} 0 & 1 \\ 1 & 0 \end{bmatrix} \begin{bmatrix} x \\ y \end{bmatrix}$
3. Substitute the pre-image point into the matrix equation.	The pre-image point is (3, 1). $\begin{bmatrix} x' \\ y' \end{bmatrix} = \begin{bmatrix} 0 & 1 \\ 1 & 0 \end{bmatrix} \begin{bmatrix} 3 \\ 1 \end{bmatrix}$
4. Calculate the coordinates of the image point.	$\begin{bmatrix} x' \\ y' \end{bmatrix} = \begin{bmatrix} 0 & 1 \\ 1 & 0 \end{bmatrix} \begin{bmatrix} 3 \\ 1 \end{bmatrix} = \begin{bmatrix} 1 \\ 3 \end{bmatrix}$ The image point is (1, 3).

WORKED EXAMPLE 9 Determining the reflection in the line $y = -x$ using matrices

Determine the equation of the image of $y = x^2$ after a reflection in the line $y = -x$.

THINK	WRITE
1. Recall the matrix for a reflection in the line $y = -x$.	$M_{y=-x} = \begin{bmatrix} 0 & -1 \\ -1 & 0 \end{bmatrix}$
2. Write the matrix equation for a reflection in the line $y = -x$; $P' = TP$.	$\begin{bmatrix} x' \\ y' \end{bmatrix} = \begin{bmatrix} 0 & -1 \\ -1 & 0 \end{bmatrix} \begin{bmatrix} x \\ y \end{bmatrix}$
3. Multiply the matrices.	$\begin{bmatrix} x' \\ y' \end{bmatrix} = \begin{bmatrix} -y \\ -x \end{bmatrix}$
4. Determine the image equations.	$x' = -y$ and $y' = -x$
5. Rearrange the equations to make x and y the subjects.	$y = -x'$ and $x = -y'$
6. Substitute the image expressions into the pre-image equation $y = x^2$ and make y the subject.	$y = x^2$ $-x' = (-y')^2$ $-x' = y'^2$ $\therefore y' = \pm\sqrt{-x'}$
7. Answer the question.	The equation of the image is $y = \pm\sqrt{-x}$.

8. Graph the image and the pre-image to verify the reflection.

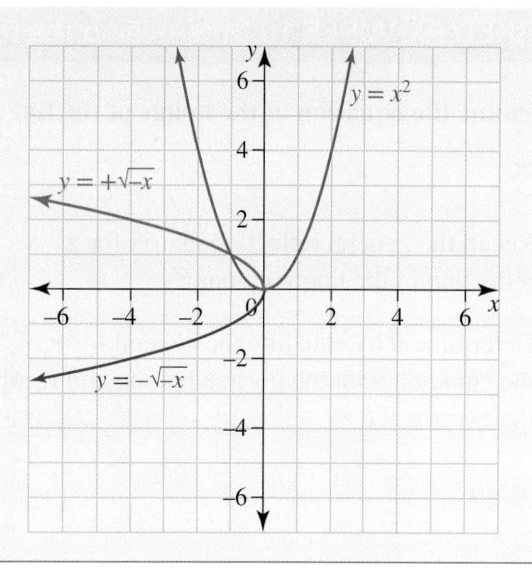

11.3.5 Reflection in the line $y = x\tan(\theta)$

The line $y = x\ \tan(\theta)$ might be more easily recognised as $y = mx$, where m is the gradient of the line which passes through the origin.

Remember that the gradient $m = \dfrac{y_2 - y_1}{x_2 - x_1}$ and tangent ratio $= \dfrac{\text{rise}}{\text{run}}$.

Therefore the tangent and gradient ratios provide the same information: $\dfrac{\text{rise}}{\text{run}} = \dfrac{y_2 - y_1}{x_2 - x_1}$.

Carefully examine these diagrams that illustrate reflection of the points (1, 0) and (0, 1) in the line $y = x\tan(\theta)$.

Note the following from these diagrams.

For the point A(1, 0):

1. Point A is reflected to a point equidistant from and perpendicular to, the line $y = x\tan(\theta)$.
2. The angle from the x-axis to A′ is 2θ.
3. The x-coordinate of the right-angled triangle is $\cos(2\theta)$.
4. The y-coordinate of this triangle is $\sin(2\theta)$.
5. Hence, point $(1, 0) \rightarrow \left(\cos(2\theta), \sin(2\theta)\right)$.

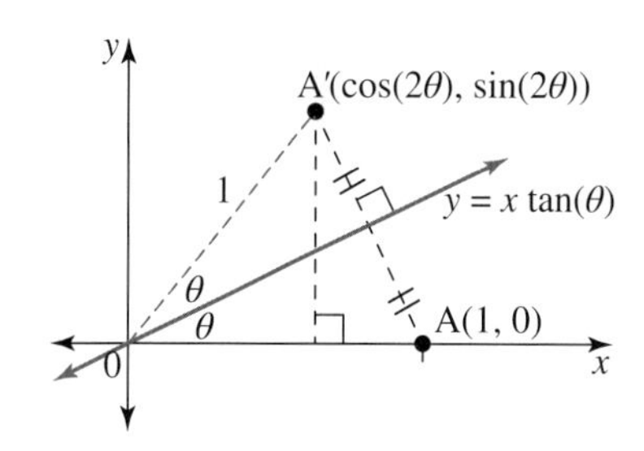

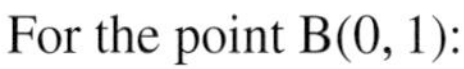

For the point B(0, 1):

1. Point B is reflected to a point equidistant from, and perpendicular to, the line $y = x\ \tan(\theta)$.
2. $\angle\text{MOB} = 90° - \theta$; therefore, $\angle\text{MOB}' = 90° - \theta$.
3. Therefore, $\angle\text{XOB}' = (90° - \theta) - \theta = 90° - 2\theta$.
4. The x-coordinate $= \cos(90° - 2\theta)$.
5. The y-coordinate $= -\sin(90° - 2\theta)$ because the angle is in the fourth quadrant.
6. Hence, point $(0, 1) \rightarrow [\cos(90° - 2\theta), -\sin(90° - 2\theta)]$.
7. Using trigonometric ratios, this simplifies to yield $(\sin(2\theta), -\cos(2\theta))$. (Remember that $\sin(30°) = \cos(60°)$, etc.) Hence, the general reflection matrix in the line $y = x\ \tan(\theta)$ is:

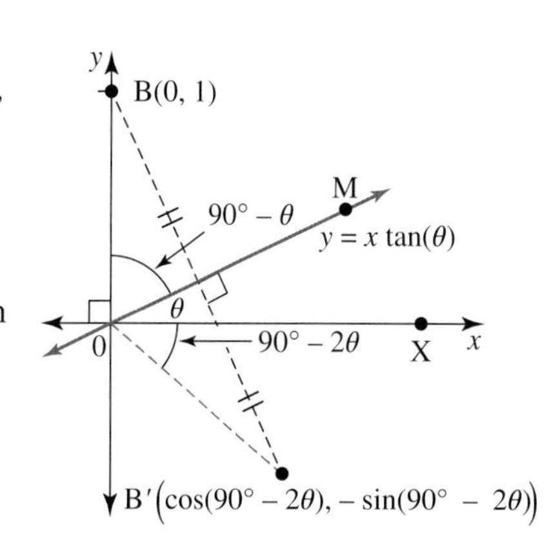

$$M_{y = x\tan(\theta)} = \begin{bmatrix} \cos(2\theta) & \sin(2\theta) \\ \sin(2\theta) & -\cos(2\theta) \end{bmatrix}$$

WORKED EXAMPLE 10 Determining the reflection in any line using matrices

Determine the equation of the image of the line $y = -x - 1$ after a reflection in the line $y = \sqrt{3}x$.

THINK	WRITE
1. Recall the general reflection matrix for a reflection in the line $y = x\tan(\theta)$.	$M_{y=x\tan(\theta)} = \begin{bmatrix} \cos(2\theta) & \sin(2\theta) \\ \sin(2\theta) & -\cos(2\theta) \end{bmatrix}$
2. Determine θ by equating the general form of the equation with the given equation and recall the exact value.	$y = \sqrt{3}x$ $y = x\tan(\theta)$ $\therefore \tan(\theta) = \sqrt{3}$ $\theta = \tan^{-1}\left(\sqrt{3}\right)$ $\therefore \theta = \frac{\pi}{3}$
3. Substitute $\frac{\pi}{3}$ for θ into the general reflection matrix for a reflection in the line $y = x\tan(\theta)$.	$M_{y=x\tan(\theta)} = \begin{bmatrix} \cos(2\theta) & \sin(2\theta) \\ \sin(2\theta) & -\cos(2\theta) \end{bmatrix}$ $M_{y=\sqrt{3}x} = \begin{bmatrix} \cos\left(\frac{2\pi}{3}\right) & \sin\left(\frac{2\pi}{3}\right) \\ \sin\left(\frac{2\pi}{3}\right) & -\cos\left(\frac{2\pi}{3}\right) \end{bmatrix}$
4. Simplify the reflection matrix.	$= \begin{bmatrix} -\frac{1}{2} & \frac{\sqrt{3}}{2} \\ \frac{\sqrt{3}}{2} & \frac{1}{2} \end{bmatrix}$
5. State the matrix equation for a reflection in the line $y = \sqrt{3}x$.	$\begin{bmatrix} x' \\ y' \end{bmatrix} = M_{y=\sqrt{3}x} \begin{bmatrix} x \\ y \end{bmatrix}$ $\begin{bmatrix} x' \\ y' \end{bmatrix} = \begin{bmatrix} -\frac{1}{2} & \frac{\sqrt{3}}{2} \\ \frac{\sqrt{3}}{2} & \frac{1}{2} \end{bmatrix} \begin{bmatrix} x \\ y \end{bmatrix}$
6. Rearrange the matrix equation to make the pre-image point (x, y) the subject.	$\begin{bmatrix} x \\ y \end{bmatrix} = \begin{bmatrix} -\frac{1}{2} & \frac{\sqrt{3}}{2} \\ \frac{\sqrt{3}}{2} & \frac{1}{2} \end{bmatrix}^{-1} \begin{bmatrix} x' \\ y' \end{bmatrix}$

7. Determine the inverse of $\begin{bmatrix} \frac{1}{2} & \frac{\sqrt{3}}{2} \\ \frac{\sqrt{3}}{2} & -\frac{1}{2} \end{bmatrix}$ and simplify the matrix equation.

$$\begin{bmatrix} x \\ y \end{bmatrix} = \frac{1}{-\frac{1}{4}-\frac{3}{4}} \begin{bmatrix} -\frac{1}{2} & -\frac{\sqrt{3}}{2} \\ -\frac{\sqrt{3}}{2} & \frac{1}{2} \end{bmatrix} \begin{bmatrix} x' \\ y' \end{bmatrix}$$

$$= \begin{bmatrix} -\frac{1}{2} & \frac{\sqrt{3}}{2} \\ \frac{\sqrt{3}}{2} & \frac{1}{2} \end{bmatrix} \begin{bmatrix} x' \\ y' \end{bmatrix}$$

8. Multiply the matrices to determine the pre-image expressions.

$$x = -\frac{1}{2}x' + \frac{\sqrt{3}}{2}y'$$

$$y = \frac{\sqrt{3}}{2}x' + \frac{1}{2}y'$$

9. Substitute the image expressions into the pre-image equation $y = -x - 1$ to determine the equation of the image.

$y = -x - 1$ becomes

$$\frac{\sqrt{3}}{2}x' + \frac{1}{2}y' = \frac{1}{2}x' - \frac{\sqrt{3}}{2}y' - 1$$

$$\frac{1}{2}y' + \frac{\sqrt{3}}{2}y' = \frac{1}{2}x' - \frac{\sqrt{3}}{2}x' - 1$$

10. Make y' the subject and simplify the equations by rationalising the denominator.

$$\frac{1+\sqrt{3}}{2}y' = \frac{1-\sqrt{3}}{2}x' - 1$$

$$y' = \left(\sqrt{3}-2\right)x' + 1 - \sqrt{3}$$

11. Answer the question.

The equation of the image of the line $y = -x - 1$ is $y = (\sqrt{3}-2)x + 1 - \sqrt{3}$.

12. Sketch the image and the pre-image graphs to verify the reflection.

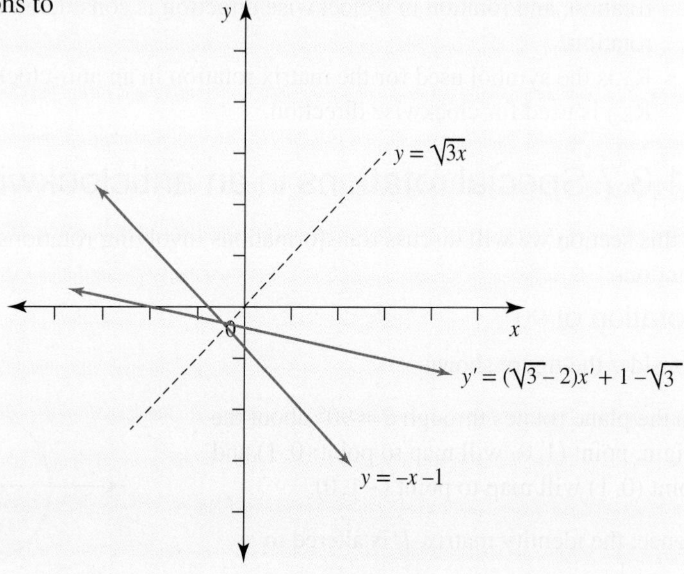

Matrix reflections

A summary of the matrices for reflections are shown in the following table.

Reflection in	Matrix	Matrix equation
x-axis	$M_x = \begin{bmatrix} 1 & 0 \\ 0 & -1 \end{bmatrix}$	$\begin{bmatrix} x' \\ y' \end{bmatrix} = \begin{bmatrix} 1 & 0 \\ 0 & -1 \end{bmatrix}\begin{bmatrix} x \\ y \end{bmatrix}$
y-axis	$M_y = \begin{bmatrix} -1 & 0 \\ 0 & 1 \end{bmatrix}$	$\begin{bmatrix} x' \\ y' \end{bmatrix} = \begin{bmatrix} -1 & 0 \\ 0 & 1 \end{bmatrix}\begin{bmatrix} x \\ y \end{bmatrix}$
line $y = x$	$M_{y=x} = \begin{bmatrix} 0 & 1 \\ 1 & 0 \end{bmatrix}$	$\begin{bmatrix} x' \\ y' \end{bmatrix} = \begin{bmatrix} 0 & 1 \\ 1 & 0 \end{bmatrix}\begin{bmatrix} x \\ y \end{bmatrix}$
line $y = -x$	$M_{y=-x} = \begin{bmatrix} 0 & -1 \\ -1 & 0 \end{bmatrix}$	$\begin{bmatrix} x' \\ y' \end{bmatrix} = \begin{bmatrix} 0 & -1 \\ -1 & 0 \end{bmatrix}\begin{bmatrix} x \\ y \end{bmatrix}$
line $y = x\tan(\theta)$	$M_{y=x\tan(\theta)} = \begin{bmatrix} \cos(2\theta) & \sin(2\theta) \\ \sin(2\theta) & -\cos(2\theta) \end{bmatrix}$	$\begin{bmatrix} x' \\ y' \end{bmatrix} = \begin{bmatrix} \cos(2\theta) & \sin(2\theta) \\ \sin(2\theta) & -\cos(2\theta) \end{bmatrix}\begin{bmatrix} x \\ y \end{bmatrix}$

11.3.6 Rotations

A **rotation** is a linear transformation in which the point P(x, y) is rotated about a fixed point called the centre or **point of rotation**. This point is usually taken as the origin. The pre-image point, P(x, y), can be rotated through an angle, θ, in an anti-clockwise or clockwise direction to get the image point, P$'(x', y')$.

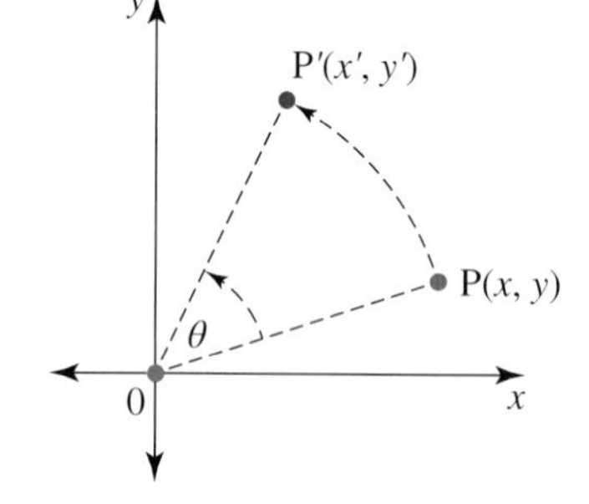

- In a rotation, each point rotates through the same angle of rotation, θ.
- The pre-image, P, and the image, P$'$, are equidistant from the origin. That is, $OP = OP'$.
- Rotation in an anti-clockwise direction is considered to be a positive rotation, and rotation in a clockwise direction is considered to be a negative rotation.
- R_θ is the symbol used for the matrix rotation in an anti-clockwise direction; $R_{-\theta}$ is used for clockwise direction.

11.3.7 Special rotations in an anticlockwise direction

In this section we will discuss transformations involving rotations of $90°$, $180°$, $270°$ and $360°$, as well as general rotations.

Rotation of 90°

Consider the figure shown.

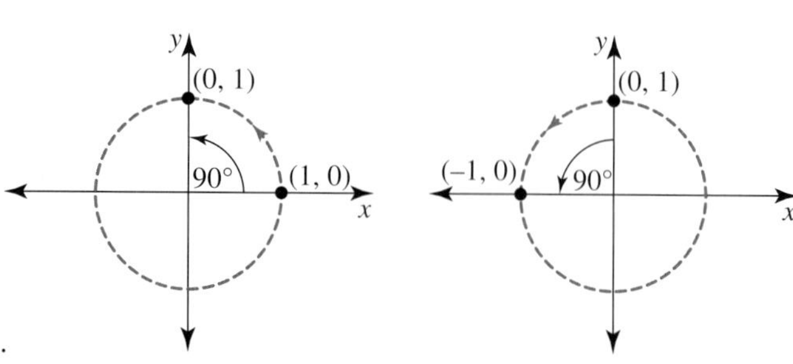

As the plane rotates through $\theta = 90°$ about the origin, point $(1, 0)$ will map to point $(0, 1)$ and point $(0, 1)$ will map to point $(-1, 0)$.

Hence, the identity matrix, I, is altered to

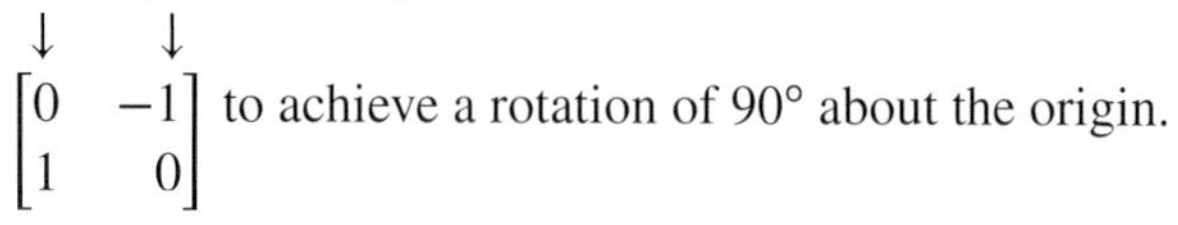

$\begin{matrix} \downarrow & \downarrow \end{matrix}$
$\begin{bmatrix} 0 & -1 \\ 1 & 0 \end{bmatrix}$ to achieve a rotation of $90°$ about the origin.

It is most important that you recognise the pattern that is displayed by the columns in the matrix and the coordinates of the image points. This concept forms the basis of the next section of work and totally eliminates 'remembering' formulas so that you will be able to understand what is happening to the points.

Hence, $R_{90^\circ} = \begin{bmatrix} 0 & -1 \\ 1 & 0 \end{bmatrix}$ and is the matrix of rotation.

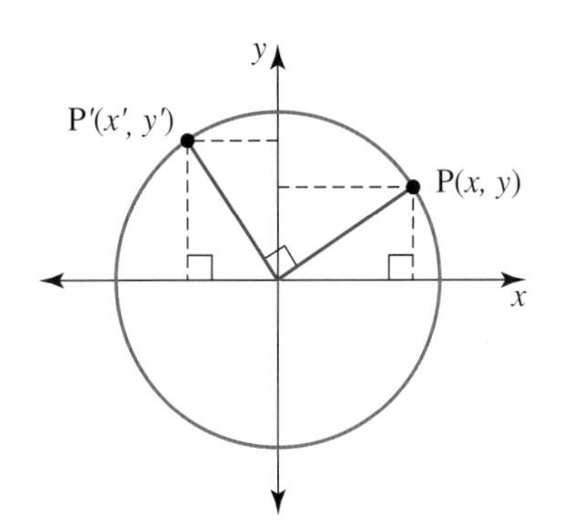

In general terms

$$(x, y) \rightarrow (-y, x)$$

$$\begin{bmatrix} x' \\ y' \end{bmatrix} = \begin{bmatrix} 0 & -1 \\ 1 & 0 \end{bmatrix} \begin{bmatrix} x \\ y \end{bmatrix}$$

$$x' = -y$$

$$y' = x$$

As mentioned earlier, these rotation matrices should not be learned. They are quite similar and can be too readily confused. Sketch the original $(1, 0)$ and $(0, 1)$ points and then use their images to build the rotation matrices.

Rotation of 180°

In the diagrams below, notice that point $(1, 0)$ is mapped onto point $(-1, 0)$ and point $(0, 1)$ is mapped onto $(0, -1)$.

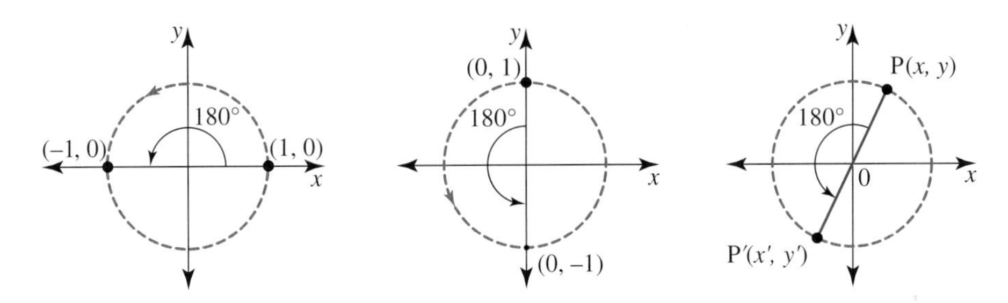

Therefore, $R_{180^\circ} = \begin{bmatrix} -1 & 0 \\ 0 & -1 \end{bmatrix}$ where $(x, y) \rightarrow (-x, -y)$.

Rotation of 270°

In the diagrams below, notice that point $(1, 0)$ is mapped onto point $(0, -1)$ and point $(0, 1)$ is mapped onto point $(1, 0)$.

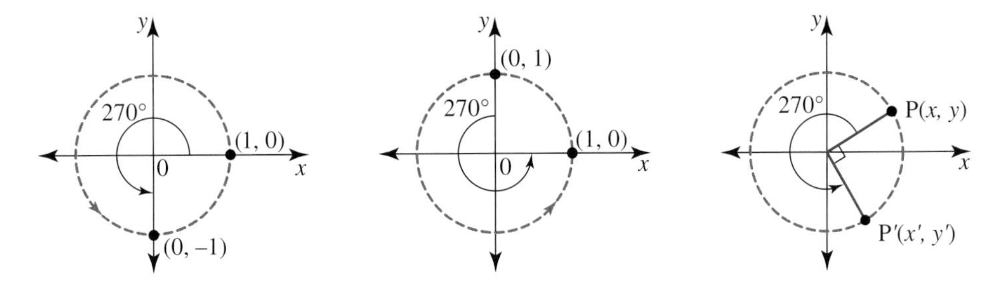

Therefore, $R_{270^\circ} = \begin{bmatrix} 0 & 1 \\ -1 & 0 \end{bmatrix}$ where $(x, y) \rightarrow (y, -x)$.

Rotation of 360°

$R_{360^\circ} = \begin{bmatrix} 0 & 1 \\ 1 & 0 \end{bmatrix}$ because R_{360° essentially leaves the original unchanged (or mapped onto itself).

General rotation of θ

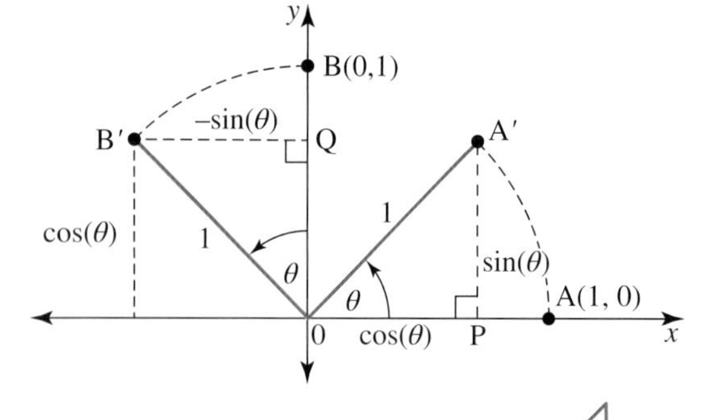

Consider the points A(1, 0) and B(0, 1) that are rotated through angle θ about the origin in an anti-clockwise direction.

Careful examination of the diagram shows that point A(1, 0) is mapped onto point $A'(\cos(\theta), \sin(\theta))$ and point B(0, 1) is mapped onto point $B'(-\sin(\theta), \cos(\theta))$, where

$$\cos(\theta) = x \text{ (horizontal)}$$
$$\text{and } \sin(\theta) = y \text{ (vertical)}$$

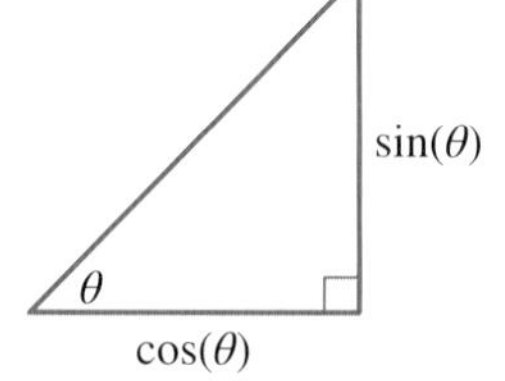

R_θ is the matrix rotation for an anticlockwise rotation through θ about the origin.

$$R_\theta = \begin{bmatrix} \cos(\theta) & -\sin(\theta) \\ \sin(\theta) & \cos(\theta) \end{bmatrix}$$

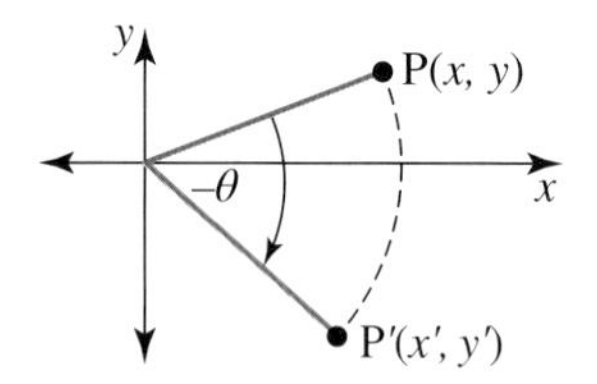

$R_{-\theta}$ is the matrix rotation when θ is taken in a clockwise, negative rotation about the origin as shown.

$$R_{-\theta} = \begin{bmatrix} \cos(-\theta) & -\sin(-\theta) \\ \sin(-\theta) & \cos(-\theta) \end{bmatrix}$$
$$= \begin{bmatrix} \cos(\theta) & \sin(\theta) \\ -\sin(\theta) & \cos(\theta) \end{bmatrix}$$

since $\cos(-\theta) = \cos(\theta)$ and $\sin(-\theta) = -\sin(\theta)$.

Both R_θ and $R_{-\theta}$ can be used to confirm the specific cases of R_{90}, R_{180}, and R_{270}.

Matrix rotations

In summary:

- **The matrix rotation for an anti-clockwise rotation through θ about the origin is:**

$$\mathbf{R}_\theta = \begin{bmatrix} \cos(\theta) & -\sin(\theta) \\ \sin(\theta) & \cos(\theta) \end{bmatrix}$$

- **The matrix rotation for a clockwise rotation through θ about the origin is:**

$$\mathbf{R}_{-\theta} = \begin{bmatrix} \cos(\theta) & \sin(\theta) \\ -\sin(\theta) & \cos(\theta) \end{bmatrix}$$

- **The matrix equation for an anti-clockwise through θ about the origin is:**

$$\begin{bmatrix} x' \\ y' \end{bmatrix} = \begin{bmatrix} \cos(\theta) & -\sin(\theta) \\ \sin(\theta) & \cos(\theta) \end{bmatrix} \begin{bmatrix} x \\ y \end{bmatrix}$$

- **The matrix equation for a clockwise rotation through θ about the origin is:**

$$\begin{bmatrix} x' \\ y' \end{bmatrix} = \begin{bmatrix} \cos(\theta) & \sin(\theta) \\ -\sin(\theta) & \cos(\theta) \end{bmatrix} \begin{bmatrix} x \\ y \end{bmatrix}$$

WORKED EXAMPLE 11 Determining a rotation of a point using matrices

The point (2, −2) is rotated $\frac{\pi}{4}$ about the origin in an anti-clockwise direction. Determine the coordinates of the image of the point after this transformation.

THINK	WRITE
1. Sketch the point (2, −2) and the angle $\frac{\pi}{4}$.	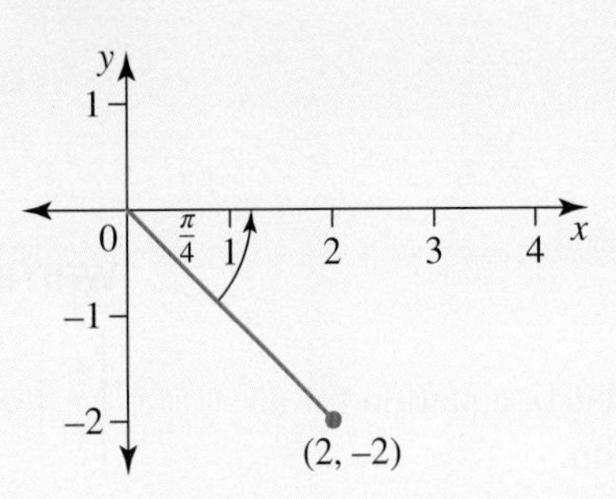
2. Recall the rotation matrix equation for an anti-clockwise rotation.	$\begin{bmatrix} x' \\ y' \end{bmatrix} = \begin{bmatrix} \cos(\theta) & -\sin(\theta) \\ \sin(\theta) & \cos(\theta) \end{bmatrix} \begin{bmatrix} x \\ y \end{bmatrix}$
3. Substitute the pre-image point, (2, −2), and $\frac{\pi}{4}$ for θ and evaluate.	$\begin{bmatrix} x' \\ y' \end{bmatrix} = \begin{bmatrix} \cos\left(\frac{\pi}{4}\right) & -\sin\left(\frac{\pi}{4}\right) \\ \sin\left(\frac{\pi}{4}\right) & \cos\left(\frac{\pi}{4}\right) \end{bmatrix} \begin{bmatrix} 2 \\ -2 \end{bmatrix}$ $\begin{bmatrix} x' \\ y' \end{bmatrix} = \begin{bmatrix} \frac{1}{\sqrt{2}} & -\frac{1}{\sqrt{2}} \\ \frac{1}{\sqrt{2}} & \frac{1}{\sqrt{2}} \end{bmatrix} \begin{bmatrix} 2 \\ -2 \end{bmatrix}$ $= \begin{bmatrix} \frac{2}{\sqrt{2}} + \frac{2}{\sqrt{2}} \\ \frac{2}{\sqrt{2}} - \frac{2}{\sqrt{2}} \end{bmatrix}$ $= \begin{bmatrix} \frac{4}{\sqrt{2}} \\ 0 \end{bmatrix}$
4. Rationalise the denominator and simplify.	$= \begin{bmatrix} \frac{4}{\sqrt{2}} \times \frac{\sqrt{2}}{\sqrt{2}} \\ 0 \end{bmatrix}$ $= \begin{bmatrix} 2\sqrt{2} \\ 0 \end{bmatrix}$
5. State the coordinates of the image point.	The image point is $(2\sqrt{2}, 0)$.

WORKED EXAMPLE 12 Determining a rotation of a line using matrices

Determine the equation of the image of the line $y = -x + 4$ under a rotation of 30° about the origin in an anti-clockwise direction.

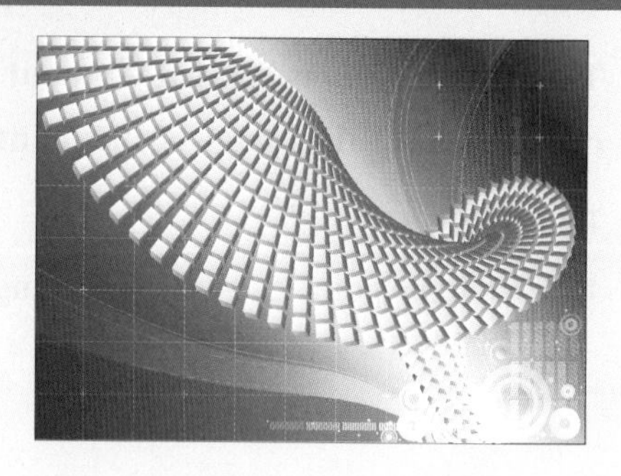

THINK

WRITE

1. Recall the rotation matrix equation for an anti-clockwise rotation.

$$\begin{bmatrix} x' \\ y' \end{bmatrix} = \begin{bmatrix} \cos(\theta) & -\sin(\theta) \\ \sin(\theta) & \cos(\theta) \end{bmatrix} \begin{bmatrix} x \\ y \end{bmatrix}$$

2. Substitute 30° for θ and evaluate.

$$\begin{bmatrix} x' \\ y' \end{bmatrix} = \begin{bmatrix} \cos(30°) & -\sin(30°) \\ \sin(30°) & \cos(30°) \end{bmatrix} \begin{bmatrix} x \\ y \end{bmatrix}$$

$$\begin{bmatrix} x' \\ y' \end{bmatrix} = \begin{bmatrix} \frac{\sqrt{3}}{2} & -\frac{1}{2} \\ \frac{1}{2} & \frac{\sqrt{3}}{2} \end{bmatrix} \begin{bmatrix} x \\ y \end{bmatrix}$$

3. Rearrange the matrix equation to make the pre-image point (x, y) the subject.

$$\begin{bmatrix} x \\ y \end{bmatrix} = \begin{bmatrix} \frac{\sqrt{3}}{2} & -\frac{1}{2} \\ \frac{1}{2} & \frac{\sqrt{3}}{2} \end{bmatrix}^{-1} \begin{bmatrix} x' \\ y' \end{bmatrix}$$

4. Determine the inverse of $\begin{bmatrix} \frac{\sqrt{3}}{2} & \frac{1}{2} \\ -\frac{1}{2} & \frac{\sqrt{3}}{2} \end{bmatrix}$ and simplify the matrix equation.

$$\begin{bmatrix} x \\ y \end{bmatrix} = \frac{1}{\left(\frac{\sqrt{3}}{2} \times \frac{\sqrt{3}}{2}\right) - \left(-\frac{1}{2} \times \frac{1}{2}\right)} \begin{bmatrix} \frac{\sqrt{3}}{2} & \frac{1}{2} \\ -\frac{1}{2} & \frac{\sqrt{3}}{2} \end{bmatrix} \begin{bmatrix} x' \\ y' \end{bmatrix}$$

$$\begin{bmatrix} x \\ y \end{bmatrix} = \frac{1}{\frac{3}{4} + \frac{1}{4}} \begin{bmatrix} \frac{\sqrt{3}}{2} & \frac{1}{2} \\ -\frac{1}{2} & \frac{\sqrt{3}}{2} \end{bmatrix} \begin{bmatrix} x' \\ y' \end{bmatrix}$$

$$\begin{bmatrix} x \\ y \end{bmatrix} = \begin{bmatrix} \frac{\sqrt{3}}{2} & \frac{1}{2} \\ -\frac{1}{2} & \frac{\sqrt{3}}{2} \end{bmatrix} \begin{bmatrix} x' \\ y' \end{bmatrix}$$

5. Multiply the matrices to determine the pre-image expressions.

$$x = \frac{\sqrt{3}}{2}x' + \frac{1}{2}y'$$

$$y = -\frac{1}{2}x' + \frac{\sqrt{3}}{2}y'$$

6. Substitute the image expressions into the pre-image equation $y = -x + 4$ to determine the equation of the image.

$$y = -x + 4$$

$$-\frac{1}{2}x' + \frac{\sqrt{3}}{2}y' = -\left(\frac{\sqrt{3}}{2}x' + \frac{1}{2}y'\right) + 4$$

$$-\frac{1}{2}x' + \frac{\sqrt{3}}{2}y' = -\frac{\sqrt{3}}{2}x' - \frac{1}{2}y' + 4$$

7. Make y' the subject and simplify the equations by rationalising the denominator.

$$y' = \frac{\left(1-\sqrt{3}\right)x'}{\sqrt{3}+1} + \frac{8}{\sqrt{3}+1}$$

$$= \left(\sqrt{3}-2\right)x' + 4\left(\sqrt{3}-1\right)$$

8. Answer the question.

The equation of the image of the line $y = -x + 4$ is $y' = \left(\sqrt{3}-2\right)x' + 4\left(\sqrt{3}-1\right)$.

9. Use technology to sketch the image and the pre-image graphs to verify the rotation.

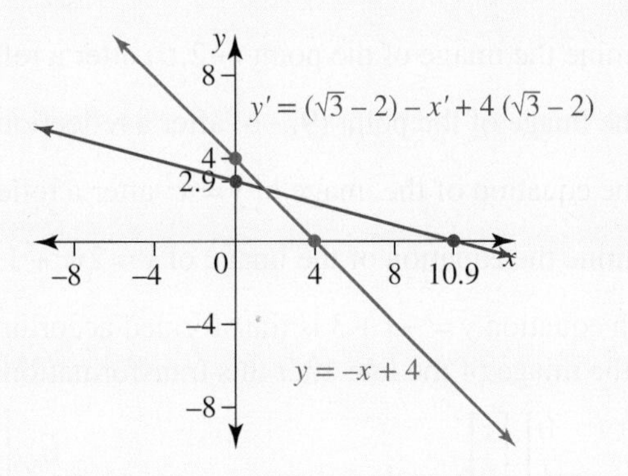

11.3.8 Invariant properties of reflections and rotations

Recall that properties which are unchanged by a transformation are called invariants. For reflections and rotations, invariant properties include length, area and angles between corresponding sides.

Invariant points are points which are unchanged by a transformation. For reflections, any point on the line of reflection is an invariant point, and for rotations, the centre of rotation is the only invariant point (unless the rotation is a multiple of 360°, in which case all points are invariant points).

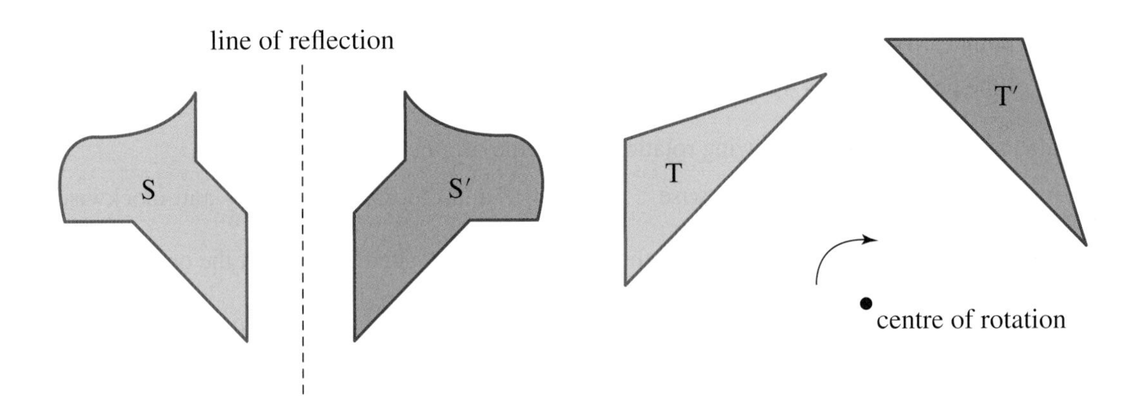

11.3 Exercise

Students, these questions are even better in jacPLUS

Receive immediate feedback and access sample responses

Access additional questions

Track your results and progress

Find all this and MORE in jacPLUS

Technology free

1. WE6a Determine the image of the point $(-3, -1)$ after a reflection in the x-axis.
2. WE6b Determine the image of the point $(5, -2)$ after it is transformed by the matrix $\begin{bmatrix} -1 & 0 \\ 0 & 1 \end{bmatrix}$.
3. WE7 Determine the equation of the image of the graph of $y = (x-2)^2$ after it is reflected in the y-axis.
4. Determine the equation of the image of the graph of $y = x^2 + 1$ after it is reflected in the x-axis.
5. WE8 Determine the image of the point $(-2, 5)$ after a reflection in the line with equation $y = x$.
6. Determine the image of the point $(9, -6)$ after a reflection in the line with equation $y = -x$.
7. Determine the equation of the image of $y = x^2$ after a reflection in the line $y = x$.
8. WE9 Determine the equation of the image of $y = 2x^2 + 1$ after a reflection in the line $y = -x$.
9. The line with equation $y = -x + 3$ is transformed according to the matrix equations given. Determine the equation of the image of the line after this transformation.
   a. $\begin{bmatrix} x' \\ y' \end{bmatrix} = \begin{bmatrix} 1 & 0 \\ 0 & -1 \end{bmatrix} \begin{bmatrix} x \\ y \end{bmatrix}$
   b. $\begin{bmatrix} x' \\ y' \end{bmatrix} = \begin{bmatrix} -1 & 0 \\ 0 & 1 \end{bmatrix} \begin{bmatrix} x \\ y \end{bmatrix}$
10. The parabola with equation $y = x^2 + 2x + 1$ is transformed according to the matrix equations given. Determine the equation of the image of the parabola after this transformation.
   a. $\begin{bmatrix} x' \\ y' \end{bmatrix} = \begin{bmatrix} 1 & 0 \\ 0 & -1 \end{bmatrix} \begin{bmatrix} x \\ y \end{bmatrix}$
   b. $\begin{bmatrix} x' \\ y' \end{bmatrix} = \begin{bmatrix} -1 & 0 \\ 0 & 1 \end{bmatrix} \begin{bmatrix} x \\ y \end{bmatrix}$

Technology active

11. WE10 Determine the equation of the image of the line $y = x + 1$ after a reflection in the line $y = \frac{\sqrt{3}}{3}x$.
12. WE11 The point $(5, 4)$ is rotated $\frac{\pi}{3}$ about the origin in a clockwise direction. Determine the coordinates of the image of the point after this transformation.
13. Determine the matrices for the following rotations about the origin.
   a. $90°$ clockwise
   b. $180°$ clockwise
   c. $45°$ anti-clockwise
   d. $\frac{\pi}{6}$ anti-clockwise
14. a. Determine the coordinates of the image of the point A $(7, -6)$ rotated $270°$ about the origin in a clockwise direction.
   b. Determine the coordinates of the image of the point A $(7, -6)$ rotated $90°$ about the origin in an anti-clockwise direction.
   c. Show that a clockwise rotation of $270°$ is the same as an anti-clockwise rotation of $90°$ about the origin.
15. WE12 Determine the equation of the image of the line $y = -3x + 1$ under a rotation of $45°$ about the origin in an anti-clockwise direction.

16. Determine the equation of the image of the line $y = 2x + 1$ under a rotation of 90° about the origin in a clockwise direction.

11.3 Exam questions

Question 1 (1 mark) TECH-ACTIVE

MC The coordinates of the point $(-5, -2)$ after a reflection in the y-axis are

A. $(-2, 5)$ **B.** $(5, -2)$ **C.** $(-5, 2)$ **D.** $(5, 2)$ **E.** $(2, 5)$

Question 2 (1 mark) TECH-ACTIVE

MC An anticlockwise rotation about the origin of 60° is best represented by

A. $\frac{1}{2}\begin{bmatrix} 1 & \sqrt{3} \\ -\sqrt{3} & 1 \end{bmatrix}$ **B.** $\frac{1}{2}\begin{bmatrix} 2 & \sqrt{3} \\ -\sqrt{3} & 2 \end{bmatrix}$ **C.** $\frac{1}{2}\begin{bmatrix} 1 & -\sqrt{3} \\ \sqrt{3} & 1 \end{bmatrix}$

D. $\frac{1}{2}\begin{bmatrix} 2 & -\sqrt{3} \\ \sqrt{3} & 2 \end{bmatrix}$ **E.** $\frac{1}{2}\begin{bmatrix} 1 & \sqrt{3} \\ \sqrt{3} & 1 \end{bmatrix}$

Question 3 (3 marks) TECH-FREE

Determine the equation of the image of the line $2y - 3x = 1$ under a rotation of 90° about the origin in a clockwise direction.

More exam questions are available online.

11.4 Dilations

LEARNING INTENTION

At the end of this subtopic you should be able to:

- determine dilations of points, lines, curves and objects using matrices.

11.4.1 Dilations from the x- and y-axes

A **dilation** is a linear transformation that changes the size of a figure. The figure is enlarged or reduced parallel to either axis or both axes. A dilation requires a centre point and a scale factor.

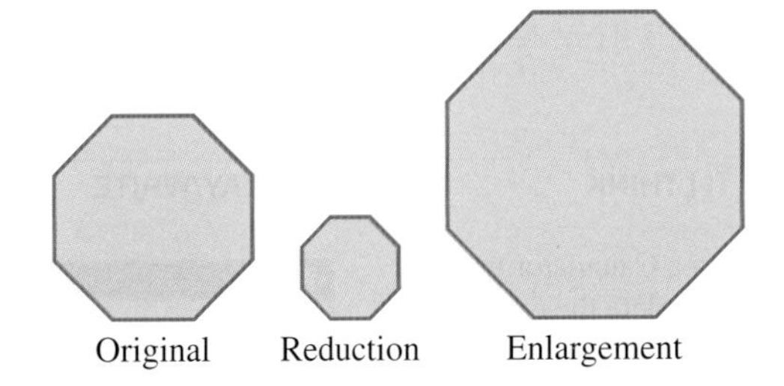

A dilation is defined by a scale factor, denoted in general terms by λ.

If $\lambda > 1$, the figure is enlarged.

If $0 < \lambda < 1$, the figure is reduced.

Dilation from the y-axis or parallel to the x-axis

A dilation (by a factor of a) from the y-axis or parallel to the x-axis is represented by the following matrix equation.

$$\begin{bmatrix} x' \\ y' \end{bmatrix} = \begin{bmatrix} a & 0 \\ 0 & 1 \end{bmatrix}\begin{bmatrix} x \\ y \end{bmatrix} = \begin{bmatrix} ax \\ y \end{bmatrix}$$

The points (x, y) are transformed onto points with the same y-coordinate.

The point moves away from the y-axis in the direction of the x-axis by a factor of a. This determines the horizontal enlargement of the figure if $a > 1$ or the horizontal compression if $0 < a < 1$.

Dilations from the x-axis or parallel to the y-axis

A dilation (by a factor of b) from the x-axis or parallel to the y-axis is represented by the following matrix equation.

$$\begin{bmatrix} x' \\ y' \end{bmatrix} = \begin{bmatrix} 1 & 0 \\ 0 & b \end{bmatrix} \begin{bmatrix} x \\ y \end{bmatrix} = \begin{bmatrix} x \\ by \end{bmatrix}$$

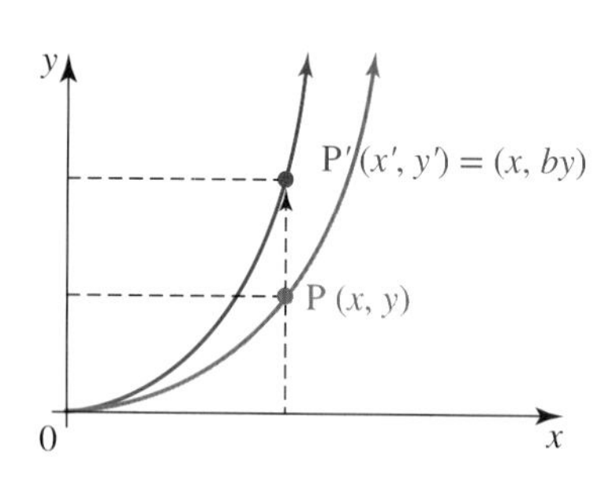

The point moves away from the x-axis in the direction of the y-axis by a factor of b. This determines the vertical enlargement of the figure if $b > 1$ or if $0 < b < 1$, the vertical compression.

WORKED EXAMPLE 13 Determining a dilation of a point using matrices

Determine the coordinates of the image of the point (3, −4) under a dilation of factor $\frac{1}{2}$ from the x-axis.

THINK	WRITE
1. Recall the dilation matrix for dilations from the x-axis.	$\begin{bmatrix} 1 & 0 \\ 0 & b \end{bmatrix}$
2. Identify the dilation factor and use the matrix equation for dilation by substituting the value of b.	The dilation factor is $b = \frac{1}{2}$. $\begin{bmatrix} x' \\ y' \end{bmatrix} = \begin{bmatrix} 1 & 0 \\ 0 & \frac{1}{2} \end{bmatrix} \begin{bmatrix} x \\ y \end{bmatrix}$
3. Substitute the pre-image point into the matrix equation.	The pre-image point is (3, −4). $\begin{bmatrix} x' \\ y' \end{bmatrix} = \begin{bmatrix} 1 & 0 \\ 0 & \frac{1}{2} \end{bmatrix} \begin{bmatrix} 3 \\ -4 \end{bmatrix}$
4. Calculate the coordinates of the image point.	$\begin{bmatrix} x' \\ y' \end{bmatrix} = \begin{bmatrix} 1 & 0 \\ 0 & \frac{1}{2} \end{bmatrix} \begin{bmatrix} 3 \\ -4 \end{bmatrix} = \begin{bmatrix} 3 \\ -2 \end{bmatrix}$ The image point is (3, −2).

TI | THINK

On a Calculator page, complete the entry line as:

$\begin{bmatrix} 1 & 0 \\ 0 & \frac{1}{2} \end{bmatrix} \times \begin{bmatrix} 3 \\ -4 \end{bmatrix}$

then press ENTER.

Note: Matrix templates can be found by pressing the templates button.

DISPLAY/WRITE

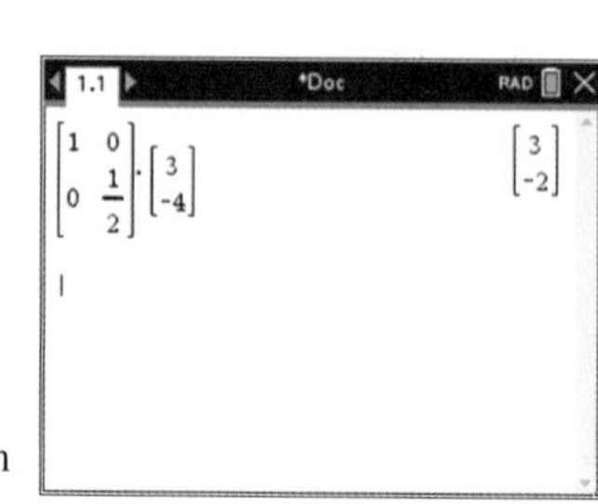

The image point is (3, −2).

CASIO | THINK

On the Main screen, complete the entry line as:

$\begin{bmatrix} 1 & 0 \\ 0 & \frac{1}{2} \end{bmatrix} \times \begin{bmatrix} 3 \\ -4 \end{bmatrix}$

then press EXE.

DISPLAY/WRITE

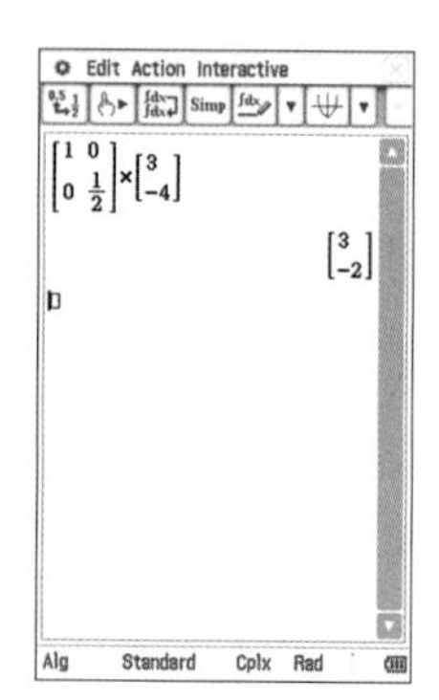

The image point is (3, −2).

WORKED EXAMPLE 14 Determining a dilation of a curve using matrices

Determine the equation of the image of the parabola with equation $y = x^2$ after it is dilated by a factor of 2 from the y-axis.

THINK	WRITE
1. Recall the matrix equation for dilation.	$\begin{bmatrix} x' \\ y' \end{bmatrix} = \begin{bmatrix} 2 & 0 \\ 0 & 1 \end{bmatrix} \begin{bmatrix} x \\ y \end{bmatrix} = \begin{bmatrix} 2x \\ y \end{bmatrix}$
2. Determine the expressions of the image coordinates in terms of the pre-image coordinates.	$x' = 2x$ and $y' = y$
3. Rearrange the equations to make the pre-image coordinates x and y the subjects.	$x = \frac{x'}{2}$ and $y = y'$
4. Substitute the image values into the pre-image equation to determine the image equation.	$y = x^2$ $y' = \left(\frac{x'}{2}\right)^2$ $= \frac{(x')^2}{4}$ The equation of the image is $y = \frac{x^2}{4}$.
5. Graph the image and the pre-image equations to verify the translation.	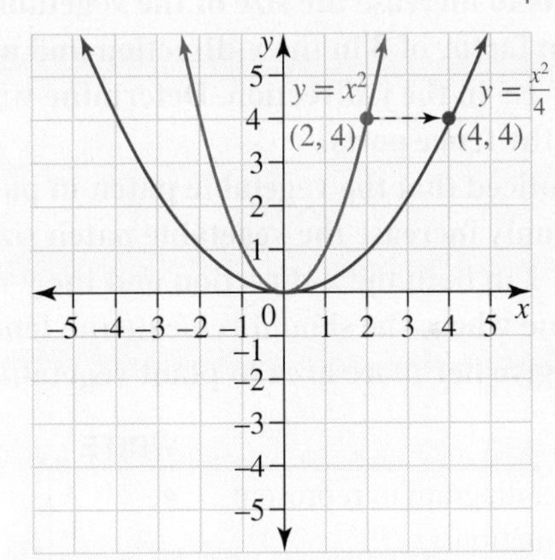

11.4.2 Dilation from both x- and y-axes

Dilation from both axes

A dilation parallel to both the x-axis and y-axis can be represented by the matrix equation:

$$\begin{bmatrix} x' \\ y' \end{bmatrix} = \begin{bmatrix} a & 0 \\ 0 & b \end{bmatrix} \begin{bmatrix} x \\ y \end{bmatrix} = \begin{bmatrix} ax \\ by \end{bmatrix}$$

where a and b are the dilation factors in the x-axis and y-axis directions respectively.

- When $a \neq b$ the object is skewed.
- When $a = b = \lambda$, the size of the object is enlarged or reduced by the same factor, and the matrix equation is:

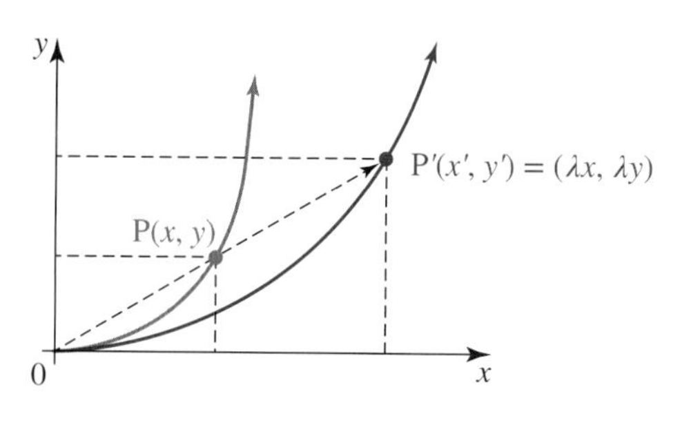

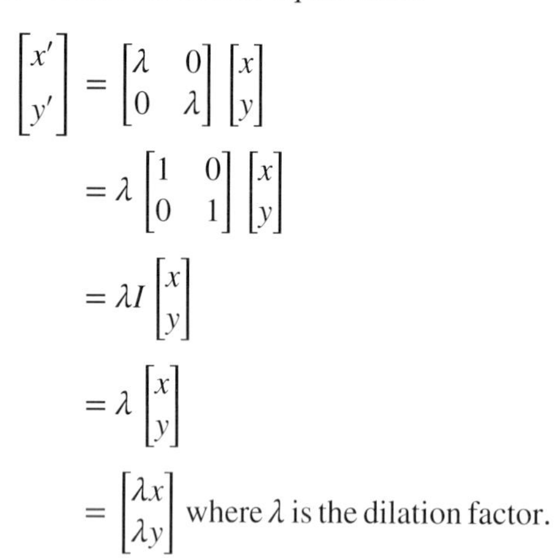

$$\begin{bmatrix} x' \\ y' \end{bmatrix} = \begin{bmatrix} \lambda & 0 \\ 0 & \lambda \end{bmatrix} \begin{bmatrix} x \\ y \end{bmatrix}$$
$$= \lambda \begin{bmatrix} 1 & 0 \\ 0 & 1 \end{bmatrix} \begin{bmatrix} x \\ y \end{bmatrix}$$
$$= \lambda I \begin{bmatrix} x \\ y \end{bmatrix}$$
$$= \lambda \begin{bmatrix} x \\ y \end{bmatrix}$$
$$= \begin{bmatrix} \lambda x \\ \lambda y \end{bmatrix} \text{ where } \lambda \text{ is the dilation factor.}$$

WORKED EXAMPLE 15 Performing dilations in real-world situations using matrices

Jo has fenced a rectangular vegetable patch with fence posts at A(0, 0), B(3, 0), C(3, 4) and D (0, 4).

a. She wants to increase the size of the vegetable patch by a dilation factor of 3 in the x-direction and a dilation factor of 1.5 in the y-direction. Determine where Jo should relocate the fence posts.

b. Jo has noticed that the vegetable patch in part a is too long and can only increase the vegetable patch size by a dilation factor of 2 in both the x-direction and the y-direction. Determine where she should relocate the fence posts and if this will give her more area to plant vegetables. Explain.

THINK	WRITE
a. 1. Draw a diagram to represent this situation.	a. 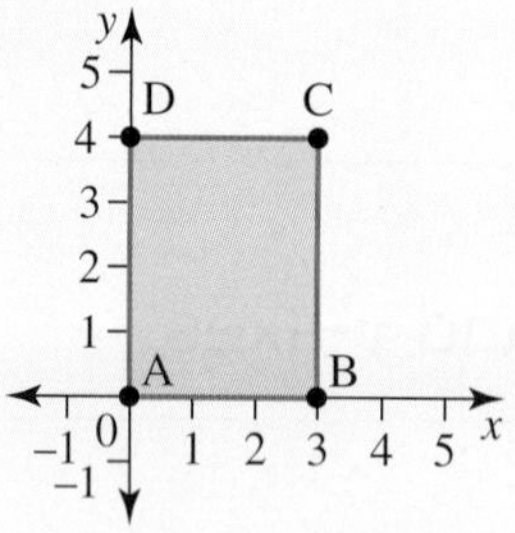 (diagram: rectangle ABCD with A(0, 0), B(3, 0), C(3, 4), D(0, 4) on x–y axes)
2. State the coordinates of the vegetable patch as a coordinate matrix.	The coordinates of the vegetable patch ABCD can be written as a coordinate matrix. $\begin{bmatrix} 0 & 3 & 3 & 0 \\ 0 & 0 & 4 & 4 \end{bmatrix}$
3. State the dilation matrix.	$\begin{bmatrix} 3 & 0 \\ 0 & 1.5 \end{bmatrix}$

4. Multiply the dilation matrix by the coordinate matrix to calculate the new fence post coordinates.

$$\begin{bmatrix} 3 & 0 \\ 0 & 1.5 \end{bmatrix}\begin{bmatrix} 0 & 3 & 3 & 0 \\ 0 & 0 & 4 & 4 \end{bmatrix} = \begin{bmatrix} 0 & 9 & 9 & 0 \\ 0 & 0 & 6 & 6 \end{bmatrix}$$

The new fence posts are located at $A'(0,0), B'(9,0), C'(9,6)$ and $D'(0,6)$.

b. 1. State the coordinates of the vegetable patch as a coordinate matrix.

b. The coordinates of the vegetable patch $2 \times 2 \times 12 = 48$ unites^2 can be written as a coordinate matrix:

$$\begin{bmatrix} 0 & 3 & 3 & 0 \\ 0 & 0 & 4 & 4 \end{bmatrix}$$

2. State the dilation matrix.

The dilation matrix is $\begin{bmatrix} 2 & 0 \\ 0 & 2 \end{bmatrix}$.

3. Calculate the new fence post coordinates $A'B'C'D'$ by multiplying the dilation matrix by the coordinate matrix.

$$\begin{bmatrix} 2 & 0 \\ 0 & 2 \end{bmatrix}\begin{bmatrix} 0 & 3 & 3 & 0 \\ 0 & 0 & 4 & 4 \end{bmatrix} = 2\begin{bmatrix} 1 & 0 \\ 0 & 1 \end{bmatrix}\begin{bmatrix} 0 & 3 & 3 & 0 \\ 0 & 0 & 4 & 4 \end{bmatrix}$$

$$= \begin{bmatrix} 0 & 6 & 6 & 0 \\ 0 & 0 & 8 & 8 \end{bmatrix}$$

The new fence posts are located at $A'(0,0), B'(6,0), C'(6,8)$ and $D'(0,8)$.

4. Draw a diagram of the original vegetable patch, and the two transformed vegetable patches on the same Cartesian plane.

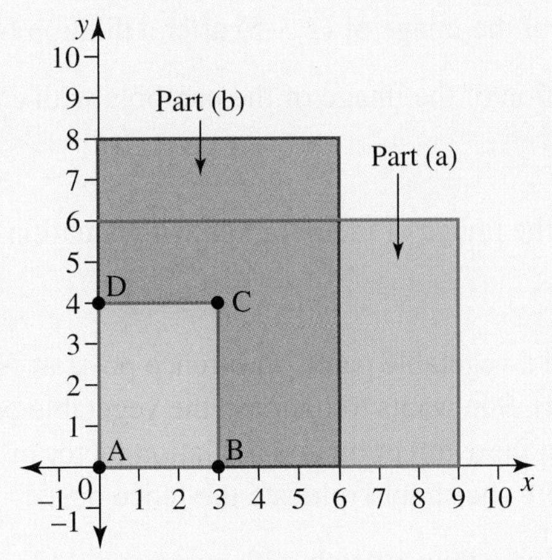

5. Determine the area for each vegetable patch.

The vegetable patch size when dilated by a factor of 3 in the x-direction and a dilation factor of 1.5 in the y-direction gives an area of 54 units^2.

When dilated by a factor of 2 in both the x-direction and the y-direction, the vegetable patch has an area of 48 units^2. The farmer will have less area to plant vegetables in the second option.

Note that you could use simply multiply the original area by the determinant of the dilation matrices.

a. $3 \times 1.5 \times 12 = 54$ units^2

b. $2 \times 2 \times 12 = 48$ units^2

11.4.3 Invariant properties of dilations

Unlike translations, reflections and rotations, dilations generally have few, if any, invariant properties. Lengths of sides, angles and areas are usually all affected by a dilation. For dilations from the x-axis (parallel to the y-axis), points on the x-axis are invariant points, and for dilations from the y-axis (parallel to the x-axis), points on the y-axis are invariant points.

11.4 Exercise

Students, these questions are even better in jacPLUS

Receive immediate feedback and access sample responses

Access additional questions

Track your results and progress

Find all this and MORE in jacPLUS

Technology free

1. WE13 Determine the coordinates of the image of the point $(2, -1)$ after a dilation of factor 3 from the x-axis.
2. Determine the coordinates of the image of $(-1, 4)$ after a dilation factor of 2 parallel to the y-axis.
3. A man standing in front of a carnival mirror looks like he has been dilated 3 times wider. Write a matrix equation for this situation.
4. Determine the coordinates of the image of the point $(-1, 4)$ after a dilation by $\begin{bmatrix} 2 & 0 \\ 0 & 1 \end{bmatrix}$ from the y-axis.
5. Determine the coordinates of the image of $(2, -5)$ after a dilation of 3 parallel to the x-axis.
6. WE14 Determine the equation of the image of the parabola with equation $y = x^2$ after it is dilated by a factor of 3 from the y-axis.
7. Determine the equation of the image of the parabola with equation $y = x^2$ after it is dilated by a factor of $\frac{1}{2}$ from the x-axis.
8. WE15 A farmer has fenced a vegetable patch with fence posts at A(0, 0), B(3, 0), C(3, 4) and D(0, 4). She wants to increase the vegetable patch size by a dilation factor of 1.5 in the x-direction and a dilation factor of 3 in the y-direction. Determine where she should relocate the fence posts.

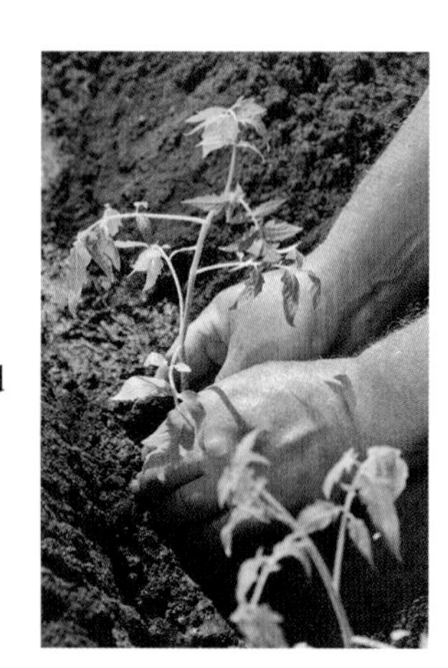

9. Jack wants to plant flowers on a flower patch with corners at A(2, 1), B(4, 1), C(3, 2) and D(1, 2). He wants to increase the flower patch size by a dilation factor of 2 in both the x-direction and the y-direction. Determine where he should relocate the new corners of the flower patch.
10. 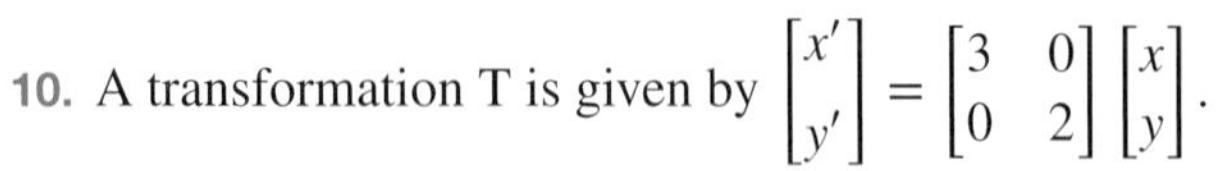
A transformation T is given by $\begin{bmatrix} x' \\ y' \end{bmatrix} = \begin{bmatrix} 3 & 0 \\ 0 & 2 \end{bmatrix} \begin{bmatrix} x \\ y \end{bmatrix}$.
    a. Determine the coordinates of image of the point A $(-1, 3)$.
    b. Describe the transformation represented by T.
11. 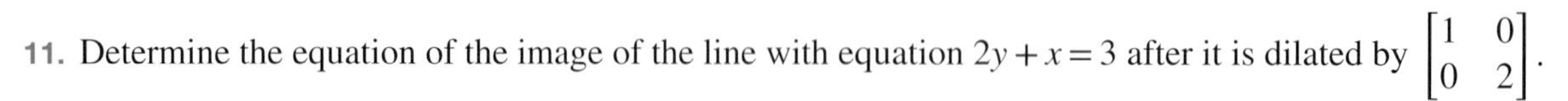
Determine the equation of the image of the line with equation $2y + x = 3$ after it is dilated by $\begin{bmatrix} 1 & 0 \\ 0 & 2 \end{bmatrix}$.
12. Determine the equation of the image of the parabola with equation $y = x^2 - 1$ after it is dilated by $\begin{bmatrix} 3 & 0 \\ 0 & 1 \end{bmatrix}$.
13. Determine the equation of the image of the hyperbola with equation $y = \frac{1}{x+1}$ after it is dilated by a factor of 2 from the y-axis.

14. The equation $y = 2\sqrt{x}$ is transformed according to $\begin{bmatrix} x' \\ y' \end{bmatrix} = \begin{bmatrix} 2 & 0 \\ 0 & 3 \end{bmatrix} \begin{bmatrix} x \\ y \end{bmatrix}$.

a. Determine the transformations represented by this matrix equation.
b. Determine the equation of the image of $y = 2\sqrt{x}$ after this transformation.

15. Determine the equation of the image of the circle with equation $x^2 + y^2 = 4$ after it is transformed according to $\begin{bmatrix} x' \\ y' \end{bmatrix} = \begin{bmatrix} 2 & 0 \\ 0 & 1 \end{bmatrix} \begin{bmatrix} x \\ y \end{bmatrix}$.

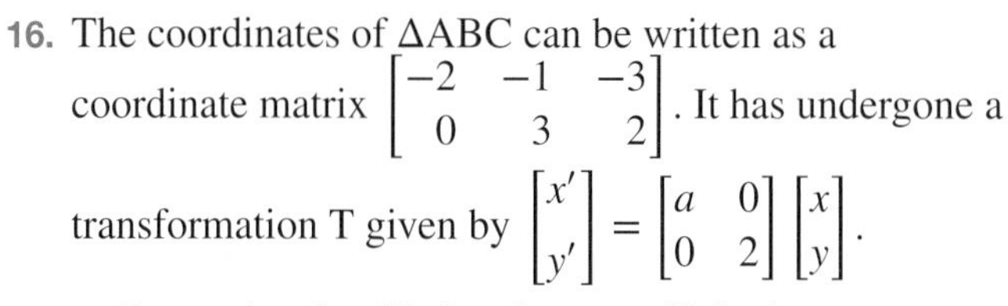

16. The coordinates of ΔABC can be written as a coordinate matrix $\begin{bmatrix} -2 & -1 & -3 \\ 0 & 3 & 2 \end{bmatrix}$. It has undergone a transformation T given by $\begin{bmatrix} x' \\ y' \end{bmatrix} = \begin{bmatrix} a & 0 \\ 0 & 2 \end{bmatrix} \begin{bmatrix} x \\ y \end{bmatrix}$.

a. Determine the dilation factor, a, if the image coordinate point A$'$ is $(-3, 0)$.
b. Calculate the coordinates of the vertices of ΔA$'$B$'$C$'$.

Technology active

17. Calculate the dilation factor from the y-axis when the graph of $y = \frac{1}{x^2}$ maps on to the graph of $y = \frac{1}{3x^2}$.

18. a. Determine the equation of the image of $x + 2y = 2$ under a dilation by a factor of 3 parallel to the x-axis.
b. Determine if there is an invariant point.

11.4 Exam questions

Question 1 (3 marks) TECH-FREE

A transformation T is given by

$$\begin{bmatrix} x' \\ y' \end{bmatrix} = \begin{bmatrix} 4 & 0 \\ 0 & 3 \end{bmatrix} \begin{bmatrix} x \\ y \end{bmatrix}$$

a. Determine the image of point A $(-3, 1)$. **(1 mark)**
b. Describe the transformation represented by T. **(2 marks)**

Question 2 (3 marks) TECH-FREE

Determine the image equation when the hyperbola $y = \frac{1}{x-2}$ is dilated by a factor of 3 from the x-axis.

Question 3 (1 mark) TECH-ACTIVE

MC The image of the point $(-6, 2)$ after a dilation of factor 2 from the x-axis is

A. $(-12, 2)$ **B.** $(-6, 4)$ **C.** $(-3, 2)$ **D.** $(-3, 1)$ **E.** $(-6, 2)$

More exam questions are available online.

11.5 Combinations of transformations

LEARNING INTENTION

At the end of this subtopic you should be able to:

- apply a combination of transformations using matrices
- determine the inverse of a matrix
- using the determinant to calculate the area of an image.

11.5.1 Double transformation matrices

A **combined transformation** is made up of two or more transformations.

If a linear transformation T_1 of a plane is followed by a second linear transformation T_2, then the results may be represented by a single transformation matrix T.

When transformation T_1 is applied to the point $P(x, y)$ it results in $P'(x', y')$.

When transformation T_2 is then applied to $P'(x', y')$ it results in $P''(x'', y'')$.

Summarising in matrix form:

$$\begin{bmatrix} x' \\ y' \end{bmatrix} = T_1 \begin{bmatrix} x \\ y \end{bmatrix}$$

$$\begin{bmatrix} x'' \\ y'' \end{bmatrix} = T_2 \begin{bmatrix} x' \\ y' \end{bmatrix}$$

Substituting $T_1 \begin{bmatrix} x \\ y \end{bmatrix}$ for $\begin{bmatrix} x' \\ y' \end{bmatrix}$ into $\begin{bmatrix} x'' \\ y'' \end{bmatrix} = T_2 \begin{bmatrix} x' \\ y' \end{bmatrix}$ results in $\begin{bmatrix} x' \\ y' \end{bmatrix} = T_2 T_1 \begin{bmatrix} x \\ y \end{bmatrix}$.

Combined transformations

To form the single transformation matrix T, the first transformation matrix T_1 must be *pre-multiplied* by the second transformation matrix T_2. The order of multiplication is important.

$$\mathbf{T = T_2 T_1}$$

Common transformation matrices used for combinations of transformations	
$M_x = \begin{bmatrix} 1 & 0 \\ 0 & -1 \end{bmatrix}$	Reflection in the x-axis
$M_y = \begin{bmatrix} -1 & 0 \\ 0 & 1 \end{bmatrix}$	Reflection in the y-axis
$M_{y=x} = \begin{bmatrix} 0 & 1 \\ 1 & 0 \end{bmatrix}$	Reflection in the line $y = x$

Common transformation matrices used for combinations of transformations	
$R_\theta = \begin{bmatrix} \cos(\theta) & -\sin(\theta) \\ \sin(\theta) & \cos(\theta) \end{bmatrix}$	Anti-clockwise (positive) rotation about origin
$D_{a,1} = \begin{bmatrix} a & 0 \\ 0 & 1 \end{bmatrix}$	Dilation in one direction parallel to the x-axis or from the y-axis
$D_{1,b} = \begin{bmatrix} 1 & 0 \\ 0 & b \end{bmatrix}$	Dilation in one direction parallel to the y-axis or from the x-axis
$D_{a,b} = \begin{bmatrix} a & 0 \\ 0 & b \end{bmatrix}$	Dilation parallel to both the x- and y-axes (a and b are the dilation factors)

Note: Translations are not linear transformations. The combined effect of two translations $\begin{bmatrix} a \\ b \end{bmatrix}$ and $\begin{bmatrix} c \\ d \end{bmatrix}$ is found by addition, $\begin{bmatrix} a+c \\ b+d \end{bmatrix}$.

WORKED EXAMPLE 16 Determining a single transformation matrix

Determine the single transformation matrix T that describes a reflection in the x-axis followed by a dilation of factor 3 from the y-axis.

THINK	WRITE
1. Determine the transformation matrices being used.	T_1 = reflection in the x-axis $T_1: M_x = \begin{bmatrix} 1 & 0 \\ 0 & -1 \end{bmatrix}$ T_2 = dilation of factor 3 from the y-axis $T_2: D_{3,1} = \begin{bmatrix} 3 & 0 \\ 0 & 1 \end{bmatrix}$
2. State the combination of transformations matrix and simplify.	$T = T_2T_1$ $T = D_{3,1}M_x$ $T = \begin{bmatrix} 3 & 0 \\ 0 & 1 \end{bmatrix}\begin{bmatrix} 1 & 0 \\ 0 & -1 \end{bmatrix}$ $= \begin{bmatrix} 3 & 0 \\ 0 & -1 \end{bmatrix}$
3. State the single transformation matrix.	The single transformation matrix is: $T = \begin{bmatrix} 3 & 0 \\ 0 & -1 \end{bmatrix}$

WORKED EXAMPLE 17 Determining the effect of a combination of transformations

Calculate the coordinates of the image of the point P (2, 3) under a reflection in the y-axis followed by a rotation of 90° about the origin in an anti-clockwise direction.

THINK	WRITE
1. State the transformation matrices, T_1 and T_2.	T_1 = reflection in the y-axis T_2 = rotation of 90° anti-clockwise $T_1 = M_y = \begin{bmatrix} -1 & 0 \\ 0 & 1 \end{bmatrix}$ $T_2 = R_{90^\circ} = \begin{bmatrix} \cos(90^\circ) & -\sin(90^\circ) \\ \sin(90^\circ) & \cos(90^\circ) \end{bmatrix}$
2. Determine the single transformation matrix and simplify.	$T = T_2T_1$ $T = R_{90^\circ}M_y$ $T = \begin{bmatrix} \cos(90^\circ) & -\sin(90^\circ) \\ \sin(90^\circ) & \cos(90^\circ) \end{bmatrix} \begin{bmatrix} -1 & 0 \\ 0 & 1 \end{bmatrix}$ $= \begin{bmatrix} 0 & -1 \\ 1 & 0 \end{bmatrix} \begin{bmatrix} -1 & 0 \\ 0 & 1 \end{bmatrix}$ $= \begin{bmatrix} 0 & -1 \\ -1 & 0 \end{bmatrix}$ The single transformation matrix is $\begin{bmatrix} 0 & -1 \\ -1 & 0 \end{bmatrix}$.
3. State the single transformation matrix equation.	$\begin{bmatrix} x' \\ y' \end{bmatrix} = \begin{bmatrix} 0 & -1 \\ -1 & 0 \end{bmatrix} \begin{bmatrix} x \\ y \end{bmatrix}$
4. Substitute the pre-image (2, 3) into the matrix equation.	$\begin{bmatrix} x' \\ y' \end{bmatrix} = \begin{bmatrix} 0 & -1 \\ -1 & 0 \end{bmatrix} \begin{bmatrix} 2 \\ 3 \end{bmatrix}$
5. Calculate the coordinates of the image point.	$\begin{bmatrix} x' \\ y' \end{bmatrix} = \begin{bmatrix} 0 & -1 \\ -1 & 0 \end{bmatrix} \begin{bmatrix} 2 \\ 3 \end{bmatrix}$ $\begin{bmatrix} x' \\ y' \end{bmatrix} = \begin{bmatrix} -3 \\ -2 \end{bmatrix}$
6. State the answer.	The coordinates of the image point are (−3, −2).

TI | THINK	DISPLAY/WRITE
Ensure the calculator is in Degree mode. On a Calculator page, complete the entry line as: $\begin{bmatrix}\cos(90) & -\sin(90)\\ \sin(90) & \cos(90)\end{bmatrix} \times \begin{bmatrix}-1 & 0\\ 0 & 1\end{bmatrix} \times \begin{bmatrix}2\\3\end{bmatrix}$ then press ENTER. *Note:* Matrix templates can be found by pressing the templates button.	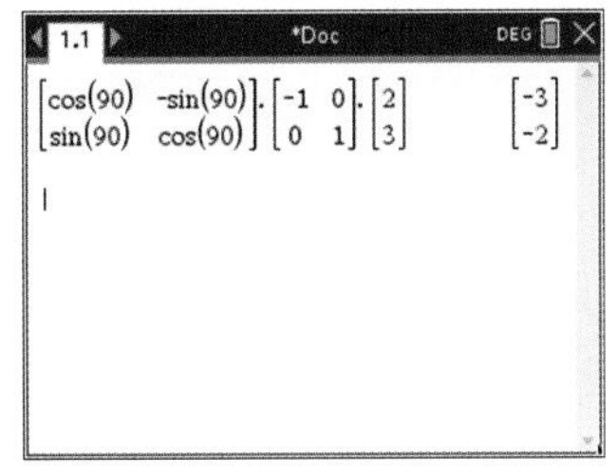 The image point is $(-3, -2)$.

CASIO | THINK	DISPLAY/WRITE
Ensure the calculator is in Degree mode. On a Main screen, complete the entry line as: $\begin{bmatrix}\cos(90) & -\sin(90)\\ \sin(90) & \cos(90)\end{bmatrix} \times \begin{bmatrix}-1 & 0\\ 0 & 1\end{bmatrix} \times \begin{bmatrix}2\\3\end{bmatrix}$ then press EXE.	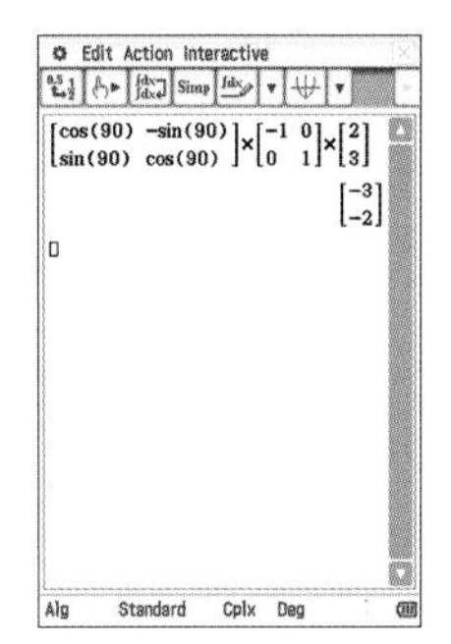 The image point is $(-3, -2)$.

11.5.2 Inverse transformation matrices

The inverse of a transformation matrix will transform the image of a point or shape back to its original position.

Note: $\begin{bmatrix}x'\\y'\end{bmatrix} = \text{T}\begin{bmatrix}x\\y\end{bmatrix}$

Inverse transformation matrices

By determining the inverse of a matrix, we get:

$$\begin{bmatrix}x\\y\end{bmatrix} = \text{T}^{-1}\begin{bmatrix}x'\\y'\end{bmatrix}$$

WORKED EXAMPLE 18 Determining the inverse transformation of a point

Determine the coordinates of the pre-image point, $P(x, y)$, under a reflection in the x-axis followed by a rotation of 90° about the origin in a clockwise direction of the image point $P'(-3, -2)$.

THINK	WRITE
1. State the transformation matrices, T_1 and T_2.	T_1 = reflection in the x-axis T_2 = rotation of 90° clockwise $T_1 = M_x = \begin{bmatrix}1 & 0\\ 0 & -1\end{bmatrix}$ $T_2 = R_{-90^\circ} = \begin{bmatrix}\cos(90^\circ) & \sin(90^\circ)\\ -\sin(90^\circ) & \cos(90^\circ)\end{bmatrix}$

2. Determine the single transformation matrix and simplify.

$T = T_2 T_1$

$T = R_{-90^\circ} M_x$

$T = \begin{bmatrix} \cos(90^\circ) & \sin(90^\circ) \\ -\sin(90^\circ) & \cos(90^\circ \end{bmatrix} \begin{bmatrix} 1 & 0 \\ 0 & -1 \end{bmatrix}$

$= \begin{bmatrix} 0 & 1 \\ -1 & 0 \end{bmatrix} \begin{bmatrix} 1 & 0 \\ 0 & -1 \end{bmatrix}$

$= \begin{bmatrix} 0 & -1 \\ -1 & 0 \end{bmatrix}$

The single transformation matrix is $\begin{bmatrix} 0 & -1 \\ -1 & 0 \end{bmatrix}$.

3. State the single transformation matrix equation.

$\begin{bmatrix} x' \\ y' \end{bmatrix} = \begin{bmatrix} 0 & -1 \\ -1 & 0 \end{bmatrix} \begin{bmatrix} x \\ y \end{bmatrix}$

4. Substitute the image point P′(−3, −2) into the matrix equation.

$\begin{bmatrix} -3 \\ -2 \end{bmatrix} = \begin{bmatrix} 0 & -1 \\ -1 & 0 \end{bmatrix} \begin{bmatrix} x \\ y \end{bmatrix}$

5. Using $\begin{bmatrix} x \\ y \end{bmatrix} = T^{-1} \begin{bmatrix} x' \\ y' \end{bmatrix}$, rearrange the equation to make the image point (x, y) the subject.

$\begin{bmatrix} x \\ y \end{bmatrix} = \begin{bmatrix} 0 & -1 \\ -1 & 0 \end{bmatrix}^{-1} \begin{bmatrix} -3 \\ -2 \end{bmatrix}$

6. Determine the inverse matrix for the single transformation matrix.

$T^{-1} = \begin{bmatrix} 0 & -1 \\ -1 & 0 \end{bmatrix}$

7. Simplify the matrix equation.

$\begin{bmatrix} x \\ y \end{bmatrix} = \begin{bmatrix} 0 & -1 \\ -1 & 0 \end{bmatrix}^{-1} \begin{bmatrix} -3 \\ -2 \end{bmatrix}$

$= \begin{bmatrix} 0 & -1 \\ -1 & 0 \end{bmatrix} \begin{bmatrix} -3 \\ -2 \end{bmatrix}$

$\therefore \begin{bmatrix} x \\ y \end{bmatrix} = \begin{bmatrix} 2 \\ 3 \end{bmatrix}$

8. State the answer.

The coordinates of the pre-image point are (2, 3).

WORKED EXAMPLE 19 Determining the inverse transformation of an object

A triangle ABC is transformed under the transformation matrix $T = \begin{bmatrix} 1 & 2 \\ -3 & 4 \end{bmatrix}$ to give vertices at A′(−1, −7), B′(4, 18) and C′(11, 7). Determine the vertices A, B and C.

THINK | **WRITE**

1. Write the image vertices, A′(−1, −7), B′(4, 18) and C′(11, 7), of as a coordinate matrix.

$\begin{bmatrix} -1 & 4 & 11 \\ -7 & 18 & 7 \end{bmatrix}$

2. Write the pre-image vertices, A, B and C, of as a coordinate matrix.	Let $A=(a,b), B=(c,d), C=(e,f)$. $\begin{bmatrix} a & c & e \\ b & d & f \end{bmatrix}$
3. Recall the matrix equation and substitute known values.	$\begin{bmatrix} -1 & 4 & 11 \\ -7 & 18 & 7 \end{bmatrix} = \begin{bmatrix} 1 & 2 \\ -3 & 4 \end{bmatrix} \begin{bmatrix} a & c & e \\ b & d & f \end{bmatrix}$
4. Rearrange the equation to make the image points the subject.	$\begin{bmatrix} a & c & e \\ b & d & f \end{bmatrix} = \begin{bmatrix} 1 & 2 \\ -3 & 4 \end{bmatrix}^{-1} \begin{bmatrix} -1 & 4 & 11 \\ -7 & 18 & 7 \end{bmatrix}$
5. Determine the inverse of the transformation matrix.	$T^{-1} = \frac{1}{10}\begin{bmatrix} 4 & -2 \\ 3 & 1 \end{bmatrix}$
6. Simplify the matrix equation.	$\begin{bmatrix} a & c & e \\ b & d & f \end{bmatrix} = \begin{bmatrix} 1 & 2 \\ -3 & 4 \end{bmatrix}^{-1} \begin{bmatrix} -1 & 4 & 11 \\ -7 & 18 & 7 \end{bmatrix}$ $\begin{bmatrix} a & c & e \\ b & d & f \end{bmatrix} = \frac{1}{10}\begin{bmatrix} 4 & -2 \\ 3 & 1 \end{bmatrix} \begin{bmatrix} -1 & 4 & 11 \\ -7 & 18 & 7 \end{bmatrix}$ $= \frac{1}{10}\begin{bmatrix} 10 & -20 & 30 \\ -10 & 30 & 40 \end{bmatrix}$ $= \begin{bmatrix} 1 & -2 & 3 \\ -1 & 3 & 4 \end{bmatrix}$
7. Answer the question.	The vertices of triangle ABC are A(1, −1), B(−2, 3) and C(3, 4).

TI | THINK	DISPLAY/WRITE	CASIO | THINK	DISPLAY/WRITE
On a Calculator page, complete the entry line as: $\begin{bmatrix} 1 & 2 \\ -3 & 4 \end{bmatrix}^{-1} \times \begin{bmatrix} -1 & 4 & 11 \\ -7 & 18 & 7 \end{bmatrix}$ then press ENTER. *Note:* Matrix templates can be found by pressing the templates button.	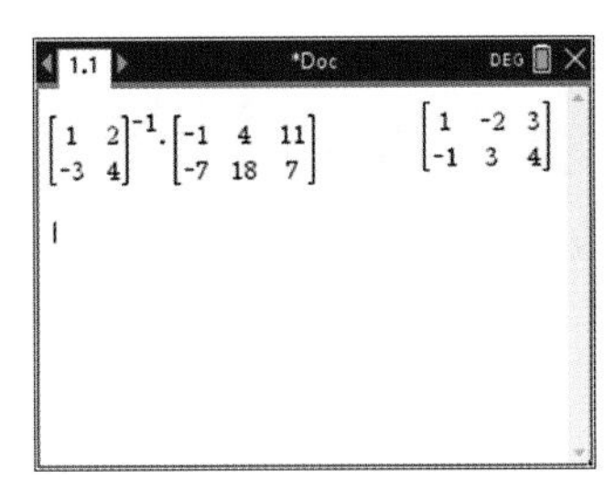 The vertices of triangle ABC are $A(1,-1)$, $B(-2,3)$ and $C(3,4)$.	On a Main screen, complete the entry line as: $\begin{bmatrix} 1 & 2 \\ -3 & 4 \end{bmatrix}^{-1} \times \begin{bmatrix} -1 & 4 & 11 \\ -7 & 18 & 7 \end{bmatrix}$ then press EXE.	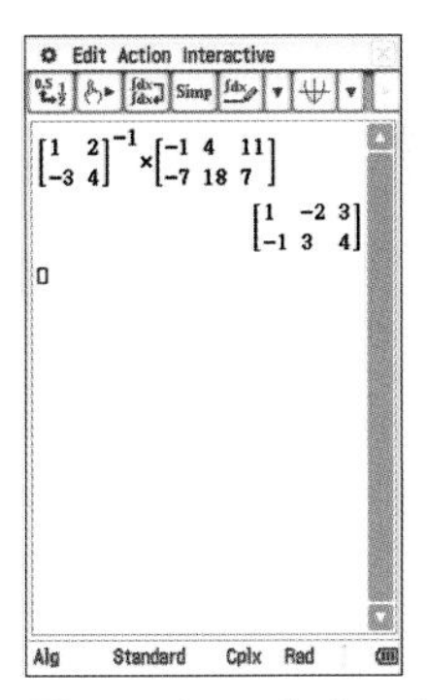 The vertices of triangle ABC are $A(1,-1)$, $B(-2,3)$ and $C(3,4)$.

11.5.3 Interpreting the determinant of the transformation matrix

A single transformation matrix is represented by T. When a shape is transformed by this transformation matrix, the magnitude of the determinant of matrix T gives the ratio of the image area to the original area.

Determinant and area of an image

Area of image = |det(T)| × area of object

where |det(T)| represents the area scale factor for the transformation.

If det(T) is negative, then the transformation will have involved some reflection.

Recall that the determinant of a 2×2 matrix, $\text{T} = \begin{bmatrix} a & b \\ c & d \end{bmatrix}$, is $\det(\text{T}) = ad - bc$.

WORKED EXAMPLE 20 Using determinants to determine areas of transformed images

The triangle ABC is mapped by the transformation represented by $\text{T} = \begin{bmatrix} 1 & -\sqrt{3} \\ \sqrt{3} & 1 \end{bmatrix}$ onto the triangle A′B′C′. Given that the area of ΔABC is 8 units², calculate the area of A′B′C′.

THINK	WRITE
1 Recall the determinant formula and calculate for the given matrix.	$\text{T} = \begin{bmatrix} 1 & -\sqrt{3} \\ \sqrt{3} & 1 \end{bmatrix}$ $\det(\text{T}) = 1 + 3 = 4$
2 Analyse the determinant.	The determinant is 4, which means that the area of the image is 4 times the area of the original object.
3 Calculate the area of the image, ΔA′B′C′.	Area of image = |det(T)| × area of object Area of $\Delta \text{A}'\text{B}'\text{C}' = 4 \times 8$ $= 32$
4 Answer the question.	The area of ΔA′B′C′ is 32 units².

TI | THINK | **DISPLAY/WRITE**

1. On a Calculator page, press MENU, then select:
7 Matrix & Vector
3 Determinant
complete the entry line as:
$\det\left(\begin{bmatrix} 1 & -\sqrt{3} \\ \sqrt{3} & 1 \end{bmatrix}\right)$
then press ENTER.
Note: Matrix templates can be found by pressing the templates button.

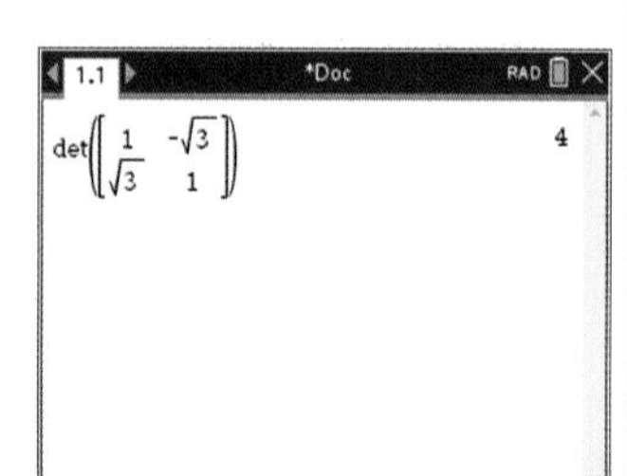

2. Complete the next entry line as: ans×8 then press ENTER.

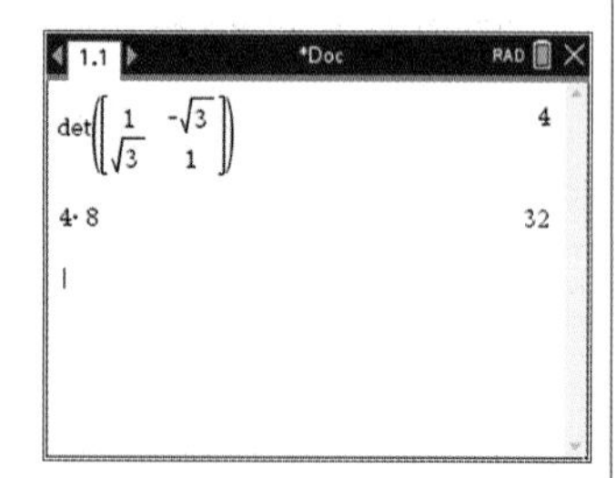

CASIO | THINK | **DISPLAY/WRITE**

1. On a Main screen, complete the entry line as:
$\det\left(\begin{bmatrix} 1 & -\sqrt{3} \\ \sqrt{3} & 1 \end{bmatrix}\right)$
then press EXE.

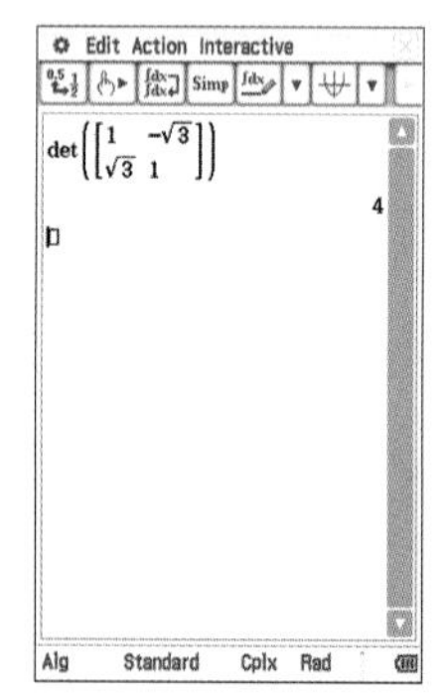

2. Complete the next entry line as: ans×8 then press EXE.

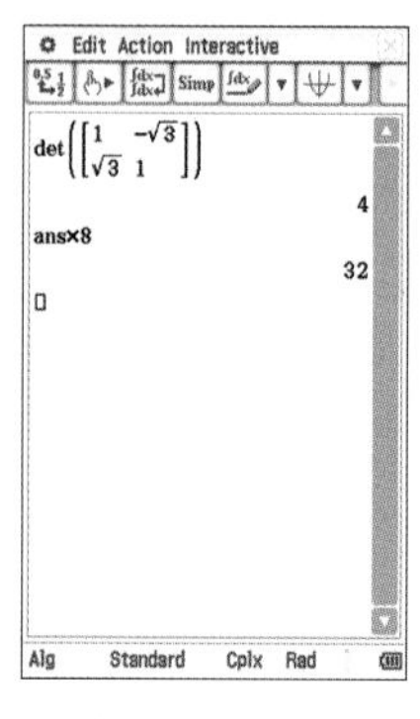

11.5 Exercise

Students, these questions are even better in jacPLUS

Receive immediate feedback and access sample responses

Access additional questions

Track your results and progress

Find all this and MORE in jacPLUS

Technology free

1. WE16 Determine the single transformation matrix T that describes a reflection in the y-axis followed by a dilation factor of 3 from the x-axis.

2. Determine the single transformation matrix T that describes a reflection in the line $y = x$ followed by a dilation of factor 2 from both the x- and y-axes.

3. WE17 Calculate the coordinates of the image of the point P$(1, -3)$ under a reflection in the x-axis followed by a rotation of 180° about the origin in an anti-clockwise direction.

4. Calculate the coordinates of the image of the point P$(-2, 2($ under a reflection in the line $y = x$ followed by a rotation of 45° about the origin in an anti-clockwise direction.

5. Describe fully a sequence of two geometrical transformations represented by $T = \begin{bmatrix} 0 & -1 \\ 1 & 0 \end{bmatrix} \begin{bmatrix} 2 & 0 \\ 0 & 2 \end{bmatrix}$.

6. Determine the equation of the image of $y = x^2$ under a double transformation: a reflection in the x-axis followed by a dilation factor of 2 parallel to both the x- and y-axes.

7. Determine the equation of the image of $y = \sqrt{x}$ under a double transformation: a reflection in the y-axis followed by a dilation of 3 parallel to the x-axis.

8. a. State the transformations that have undergone $T\left(\begin{bmatrix} x' \\ y' \end{bmatrix}\right) = \begin{bmatrix} 1 & 0 \\ 0 & -1 \end{bmatrix} \begin{bmatrix} 0 & 1 \\ 1 & 0 \end{bmatrix} \begin{bmatrix} x \\ y \end{bmatrix}$.
   b. Determine the image of the curve with equation $2x - 3y = 12$.

9. a. State the transformations that have undergone $T\left(\begin{bmatrix} x' \\ y' \end{bmatrix}\right) = \begin{bmatrix} 2 & 0 \\ 0 & -1 \end{bmatrix} \begin{bmatrix} x \\ y \end{bmatrix} + \begin{bmatrix} 1 \\ 2 \end{bmatrix}$.
   b. Determine the image of the curve with equation $y = 2x^2 - 1$.

10. State the image of P(x, y) for a translation of $\begin{bmatrix} 2 \\ -1 \end{bmatrix}$ followed by a reflection in the x-axis.

Technology active

11. WE18 Determine the coordinates of the pre-image point, P(x, y), under a reflection in the x-axis followed by a rotation of 90° about the origin in an anti-clockwise direction of the image point P$'(1, -2)$.

12. a. Calculate the coordinates of the image of point P$'(x', y')$ when the point P(x, y) undergoes a double transformation: a reflection in the y-axis followed by a translation of 4 units in the positive direction of the x-axis.
    b. Reverse the order of the pair of transformations in part **a** and determine if the image is different.

13. Determine the coordinates of the pre-image point, P(x, y), under a reflection in the y-axis followed by a dilation of factor 3 from the y-axis of the image point P$'(-3, 6)$.

14. **WE19** A triangle ABC is transformed under the transformation matrix $T = \begin{bmatrix} 3 & 2 \\ 5 & 8 \end{bmatrix}$ to give vertices at $A'(0, 0), B'(4, 18)$ and $C'(9, 15)$. Determine the vertices A, B and C.

15. **WE20** The triangle ABC is mapped by the transformation represented by $T = \begin{bmatrix} 3 & 1 \\ 1 & 2 \end{bmatrix}$ onto the triangle $A'B'C'$. Given that the area of ABC is 4 units^2, calculate the area of $A'B'C'$.

16. A rectangle ABCD is transformed under the transformation matrix $T = \begin{bmatrix} 3 & 2 \\ 5 & 8 \end{bmatrix}$, to give vertices at $A'(0, 0), B'(3, 0), C'(3, 2)$ and $D'(0, 2)$.
   a. Determine the vertices of the square ABCD.
   b. Calculate the area of the figure ABCD.

17. The triangle ΔABC is mapped by the transformation represented by $T = \begin{bmatrix} 3 & -1 \\ 1 & 2 \end{bmatrix}$ onto the triangle $\Delta A'B'C'$. Given that the area of ΔABC is 10 units^2, calculate the area of $\Delta A'B'C'$.

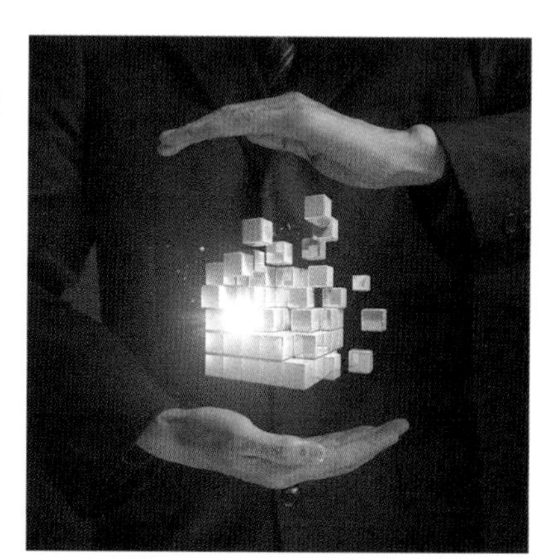

18. A rectangle ABCD with vertices at $A(0,\ 0), B(2, 0), C(2, 3)$ and $D(0, 3)$ is transformed under the transformation matrix $T = \begin{bmatrix} 2 & -1 \\ 1 & 2 \end{bmatrix}$. Calculate the new area of the transformed rectangle.

19. If D_λ denotes a dilation factor of λ parallel to both axes, determine what single dilation would be equivalent to D_λ^2.

20. Check whether the transformation 'a reflection in the y-axis followed by a reflection in the line $y = x$' is the same as 'a reflection in the line $y = x$ followed by a reflection in the y-axis'.

11.5 Exam questions

Question 1 (1 mark) TECH-ACTIVE

MC The transformation $\begin{bmatrix} 2 & 0 \\ 0 & 2 \end{bmatrix}$ is

A. a translation of 2 units in both the x- and y-directions.
B. a dilation of factor 2 parallel to the y-axis.
C. a dilation of factor 2 from the y-axis.
D. a dilation of factor 2 parallel to both the x- and y-axes
E. a dilation of factor 2 from the x-axis.

Question 2 (3 marks) TECH-FREE

Determine the image of the point $(-1,\ 3)$ transformed by the following matrices in order:

Dilation $\begin{bmatrix} 2 & 0 \\ 0 & 1 \end{bmatrix}$, reflection $\begin{bmatrix} -1 & 0 \\ 0 & 1 \end{bmatrix}$, translation $\begin{bmatrix} -4 \\ 5 \end{bmatrix}$

Question 3 (3 marks) TECH-FREE

A rectangle ABCD with vertices $A(0,\ 0)$, $B(0,\ 2)$, $C(3,\ 2)$ and $D(3,\ 0)$ is transformed under the transformation matrix $T = \begin{bmatrix} -1 & 0 \\ 0 & -1 \end{bmatrix}$.

Describe the effect of the transformation on the rectangle's position and area.

More exam questions are available online.

11.6 Review

11.6.1 Summary

doc-37054

Hey students! Now that it's time to revise this topic, go online to:

 Access the topic summary

 Review your results

 Watch teacher-led videos

 Practise exam questions

Find all this and MORE in jacPLUS

11.6 Exercise

Technology free: short answer

1. Determine the single transformation matrix for a dilation of factor 7 from the y-axis followed by a reflection in the line $y = x$.
2. Determine the transformation matrix for a dilation of factor λ from the x-axis followed by a reflection in the line $y = x$, then a reflection in the x-axis.
3. State the coordinates of the image of the point $P(x, y)$ under a translation defined by $\begin{bmatrix} 2 \\ -1 \end{bmatrix}$ followed by a reflection in the x-axis.

4. Determine the matrix equation that represents a dilation of factor 3 from the y-axis followed by a translation of 2 units in the negative direction of the x-axis and 5 units in the positive direction of the y-axis.
5. Determine the coordinates of the pre-image point $D(a, b)$ under a transformation defined by $T = \begin{bmatrix} 0 & 2 \\ -1 & 0 \end{bmatrix}$ of the image point $D'(10, 3)$.
6. Determine the equation of the image of the graph of $y = \sqrt{x}$ after a dilation of factor 3 from the y -axis followed by a translation of 2 units in the negative direction of the x-axis and 5 units in the positive direction of the y-axis.
7. A triangle ABC with vertices $A(2, -1)$, $B(-4, 0)$ and $C(5, 2)$ is rotated 45° anti-clockwise about the origin.
   a. Calculate the coordinates of vertices A', B' and C' of the rotated triangle.
   b. Compare the area of the triangle $A'B'C'$ to that of triangle ABC.

Technology active: multiple choice

8. **MC** The translation matrix that maps point $P(5, -3)$ onto point $P'(x', y')$ giving the image point $(-4, 7)$ is

   A. $\begin{bmatrix} 9 \\ 10 \end{bmatrix}$
   B. $\begin{bmatrix} -9 \\ 10 \end{bmatrix}$
   C. $\begin{bmatrix} -10 \\ 9 \end{bmatrix}$
   D. $\begin{bmatrix} -9 \\ -10 \end{bmatrix}$
   E. $\begin{bmatrix} 9 \\ -10 \end{bmatrix}$

9. **MC** The equation of the image of the parabola $y = x^2 + 2x + 1$ when it undergoes a reflection in the x-axis is

   A. $y = -x^2 - 2x - 1$
   B. $y = x^2 - 2x + 1$
   C. $y = -x^2 + 2x + 1$
   D. $y = x^2 - 2x - 1$
   E. $y = x^2 + 2x - 1$

10. MC The rotation about the origin that the matrix $\frac{\sqrt{2}}{2}\begin{bmatrix} 1 & 1 \\ -1 & 1 \end{bmatrix}$ represents is

A. Anticlockwise by 30°
B. Clockwise by 30°
C. Anticlockwise by 45°
D. Clockwise by 45°
E. Anticlockwise by 60°

11. MC The point $(a, -b)$ is rotated anticlockwise through 45° about the origin. The coordinates of the image point are

A. $\left(\frac{a+b}{\sqrt{2}}, \frac{a-b}{\sqrt{2}}\right)$
B. $\left(\frac{a+b}{\sqrt{2}}, \frac{-a+b}{\sqrt{2}}\right)$
C. $\left(\frac{a-b}{\sqrt{2}}, \frac{a+b}{\sqrt{2}}\right)$
D. $\left(\frac{a+b}{\sqrt{2}}, \frac{-(a+b)}{\sqrt{2}}\right)$
E. $\left(\frac{a-b}{\sqrt{2}}, \frac{a-b}{\sqrt{2}}\right)$

12. MC The image of the point $(-3, -2)$ after a dilation of factor 3 parallel to the y-axis is

A. $(-3, -6)$
B. $(-9, -2)$
C. $(-9, -6)$
D. $(9, 6)$
E. $(3, 6)$

13. MC The coordinates of the image of $(-1, 3)$ under the transformation $\begin{bmatrix} 3 & 0 \\ 0 & 1 \end{bmatrix}$ are

A. $(-3, 3)$
B. $(3, -3)$
C. $(-3, 9)$
D. $(3, -9)$
E. $(3, 3)$

14. MC The coordinates of the image of $(2, 5)$ under the transformation $\begin{bmatrix} -2 & 0 \\ 0 & 1 \end{bmatrix}$ are

A. $(0, 5)$
B. $(-4, 5)$
C. $(-4, -10)$
D. $(2, -10)$
E. $(4, -10)$

15. MC The transformation $\begin{bmatrix} 1 & 0 \\ 0 & -1 \end{bmatrix}$ is

A. a vertical translation of 1 unit down
B. a horizontal translation of 1 unit to the left
C. a reflection in the line $y = x$.
D. a reflection in the y-axis.
E. a reflection in the x-axis.

16. MC The transformation $T\left(\begin{bmatrix} x' \\ y' \end{bmatrix}\right) = \begin{bmatrix} 0 & 1 \\ 1 & 0 \end{bmatrix}\begin{bmatrix} 1 & 0 \\ 0 & -1 \end{bmatrix}\begin{bmatrix} x \\ y \end{bmatrix}$ can be described as a reflection in

A. the line $y = x$, then the y-axis
B. the line $y = x$, then the x-axis
C. the x-axis, then the line $y = x$
D. the y-axis, then the line $y = x$
E. none of these

17. MC The transformation $T : R^2 \to R^2$ with rule $T\left(\begin{bmatrix} x \\ y \end{bmatrix}\right) = \begin{bmatrix} 2 & 0 \\ 0 & -3 \end{bmatrix}\begin{bmatrix} x \\ y \end{bmatrix} + \begin{bmatrix} -1 \\ 4 \end{bmatrix}$ maps the curve with equation $y = x^2$ to the curve with equation

A. $y = 4 - \frac{3(x-1)^2}{4}$
B. $y = 4 + \frac{3(x+1)^2}{4}$
C. $y = -4 + \frac{3(x+1)^2}{4}$
D. $y = 4 + \frac{3(x-1)^2}{4}$
E. $y = 4 - \frac{3(x+1)^2}{4}$

Technology active: extended response

18. Avril has plotted all of the locations for a cartoon on the Cartesian plane. The vertices of the outline of a character's house have coordinates $(-3, 5), (-3, 7), (5, 7)$ and $(5, 5)$. She wants to move the points so that they are reflected in the y-axis.
To add more animation to her cartoon, she wants not only to reflect the outline of the house in the y-axis, but also to combine it with a dilation factor of 2 in the x-axis and then a translation of 2 units across in the negative x-direction and 5 units down in the negative y-direction.
    a. Plot the points on a set of axes.
    b. State the transformation matrix for a reflection in the y-axis.

c. State the transformation matrix equation.
d. Determine the new coordinates of the house.
e. Plot the new coordinates of the house on the same set of axes as the original coordinates.
f. Write the transformation matrix for each of the transformations.
g. State the single transformation matrix required to animate the house.
h. State the transformation matrix equation.
i. Determine the new coordinates of the house after the combinations of transformations.

19. A computer designer wants to animate a marshmallow being squashed. The marshmallow is modelled by the unit circle equation $x^2 + y^2 = 1$. To make it look squashed, it has to undergo the following transformations for the animation:

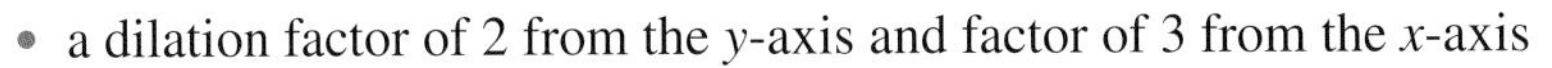

- a dilation factor of 2 from the y-axis and factor of 3 from the x-axis
- a translation of 1 unit across the x-axis and 2 units up the y-axis.

a. State the transformation matrix for each of the transformations.
b. State the transformation matrix equation required to squash the marshmallow.
c. Determine the equation of the image of the squashed marshmallow. State the shape of the squashed marshmallow.
d. On the same set of axes, draw both the pre-image and image equations representing the marshmallow.
e. Determine the area of the squashed marshmallow using the determinant.

20. Mark wants to build a deck for entertaining outside. To plan out his deck, he uses a coordinate grid where each square represents 1 metre. The coordinates for the vertices of his deck are $(2, 2), (8, 2), (2, 6)$ and $(8, 6)$.

a. Calculate the area of the deck.
b. Mark decides to increase the deck by dilating it by a factor of 1.5 from the x-axis. State the transformation matrix required to transform the deck.
c. Determine the new coordinates of the deck.
d. Calculate the new area of the deck.
e. Mark ordered enough decking wood to cover an area of 40 m^2. Determine if he has enough to build his new deck.
f. On the blueprint of Mark's house, the coordinates of the corners of the garage are $(1, 5), (1, 25), (31, 5)$ and $(31, 25)$. On the blueprint, 1 unit represents 1 cm.
   i. If the scale of the blueprint is $\frac{1}{30}$ of the actual structure, determine the coordinates of the garage when it is built.
   ii. Calculate the area of Mark's garage.
   iii. Mark has two cars. Each car needs a parking area of 15 m^2 in the garage. Determine if he has enough space to park both cars in his garage.
g. Mark decides to transform a garden bed in the form of a triangle with vertices at $A(-2, -2), B(0, 2)$ and $C(2, -2)$ using the transformation equations $x' = x + y$ and $y' = x - y$.
   i. Determine the general transformation matrix.
   ii. Determine the new coordinates of Mark's garden bed.
   iii. Calculate the scale factor by which the area has been increased.

11.6 Exam questions

Question 1 (1 mark) TECH-ACTIVE

MC The image of the point $(-a, b)$ after using the matrix equation for translation $\begin{bmatrix} x' \\ y' \end{bmatrix} = \begin{bmatrix} x \\ y \end{bmatrix} + \begin{bmatrix} -5 \\ 4 \end{bmatrix}$ is

A. $\left(-(5+a), -(b-4)\right)$ **B.** $\left(-(5+a), (b+4)\right)$ **C.** $\left((5+a), -(b-4)\right)$

D. $\left((5+a), (b-4)\right)$ **E.** $\left((5+a), (b+4)\right)$

Question 2 (1 mark) TECH-ACTIVE

MC The matrix equation for the translation that maps the circle $(x-p)^2 + (y+q)^2 = 1$ onto the circle $x^2 + y^2 = 1$ is

A. $\begin{bmatrix} x' \\ y' \end{bmatrix} = \begin{bmatrix} x \\ y \end{bmatrix} + \begin{bmatrix} -p \\ q \end{bmatrix}$ **B.** $\begin{bmatrix} x' \\ y' \end{bmatrix} = \begin{bmatrix} x \\ y \end{bmatrix} + \begin{bmatrix} p \\ -q \end{bmatrix}$ **C.** $\begin{bmatrix} x' \\ y' \end{bmatrix} = \begin{bmatrix} x \\ y \end{bmatrix} - \begin{bmatrix} p \\ -q \end{bmatrix}$

D. $\begin{bmatrix} x' \\ y' \end{bmatrix} = \begin{bmatrix} x \\ y \end{bmatrix} - \begin{bmatrix} -p \\ -q \end{bmatrix}$ **E.** $\begin{bmatrix} x' \\ y' \end{bmatrix} = \begin{bmatrix} x \\ y \end{bmatrix} + \begin{bmatrix} -p \\ -q \end{bmatrix}$

Question 3 (1 mark) TECH-ACTIVE

MC The single transformation matrix T that describes a reflection in the line $y = x$ followed by a dilation of factor 2 from both the x- and y-axes is

A. $T = \begin{bmatrix} 2 & 0 \\ 0 & 2 \end{bmatrix} \begin{bmatrix} 0 & 1 \\ 1 & 0 \end{bmatrix}$ **B.** $T = \begin{bmatrix} 0 & 1 \\ 1 & 0 \end{bmatrix} \begin{bmatrix} 2 & 0 \\ 0 & 2 \end{bmatrix}$ **C.** $T = \begin{bmatrix} 0 & 2 \\ 2 & 0 \end{bmatrix}$

D. $T = \begin{bmatrix} 2 & 0 \\ 0 & 2 \end{bmatrix}$ **E.** $T = \begin{bmatrix} 2 & 1 \\ 1 & 2 \end{bmatrix}$

Question 4 (1 mark) TECH-ACTIVE

MC The point $P(a, b)$ comes under a reflection in the y-axis followed by a dilation of factor 3 from the y-axis. The coordinates of the image point $P'(x, y)$ are

A. $(-a, -3b)$ **B.** $(a, 3b)$ **C.** $(a, -3b)$ **D.** $(-3a, b)$ **E.** $(-3b, a)$

Question 5 (1 mark) TECH-ACTIVE

MC The equation of the image of $y = x^2$ under a double transformation of a reflection in the line $y = -x$ followed by a clockwise rotation of 90° about the origin is

A. $y = \sqrt{x}$ **B.** $y = -\sqrt{x}$ **C.** $y = -x^2$ **D.** $y = x^2$ **E.** none of these

More exam questions are available online.

Answers

Topic 11 Transformations

11.2 Translations

11.2 Exercise

1. $\begin{bmatrix} x' \\ y' \end{bmatrix} = \begin{bmatrix} 3 \\ 7 \end{bmatrix}$, (3, 7)
2. $\begin{bmatrix} x' \\ y' \end{bmatrix} = \begin{bmatrix} -6 \\ 2 \end{bmatrix}$, (−6, 2)
3. $\begin{bmatrix} x' \\ y' \end{bmatrix} = \begin{bmatrix} 4 \\ 0 \end{bmatrix}$, (4, 0)
4. $\begin{bmatrix} x' \\ y' \end{bmatrix} = \begin{bmatrix} 2 \\ -2 \end{bmatrix}$, (2, −2)
5. $T = \begin{bmatrix} 1 & 1 & 1 \\ -2 & -2 & -2 \end{bmatrix}$
6. $T = \begin{bmatrix} 1 & 1 & 1 \\ 2 & 2 & 2 \end{bmatrix}$
7. $\begin{bmatrix} x' \\ y' \end{bmatrix} = \begin{bmatrix} x \\ y \end{bmatrix} + \begin{bmatrix} 2 \\ 1 \end{bmatrix}$
8. a.

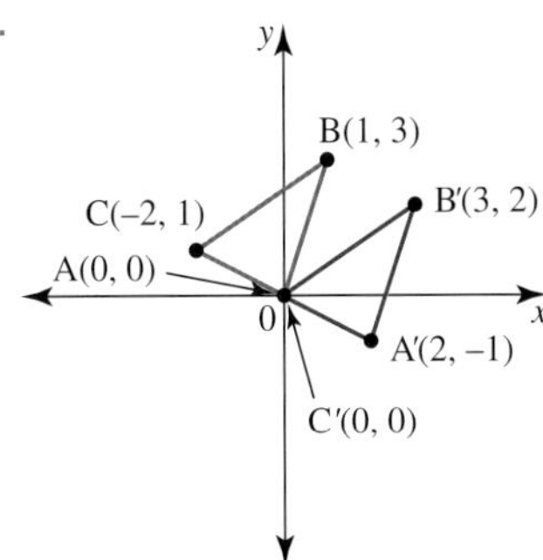

   b. $T = \begin{bmatrix} 2 & 2 & 2 \\ -1 & -1 & -1 \end{bmatrix}$
9. $y = x + 1$
10. $y = x - 6$
11. $y = x + 6$
12. $y = (x - 2)^2 - 1$
13. $y = (x + 3)^2 + 1$
14. $y = (x - 7)^2 - 6$
15. $T = \begin{bmatrix} 0 \\ 2 \end{bmatrix}$
16. $T = \begin{bmatrix} 7 \\ 3 \end{bmatrix}$
17. $\begin{bmatrix} x' \\ y' \end{bmatrix} = \begin{bmatrix} x \\ y \end{bmatrix} + \begin{bmatrix} 2 \\ 6 \end{bmatrix}$
18. $\begin{bmatrix} x' \\ y' \end{bmatrix} = \begin{bmatrix} x \\ y \end{bmatrix} + \begin{bmatrix} 1 \\ 0 \end{bmatrix}$
19. $\begin{bmatrix} x' \\ y' \end{bmatrix} = \begin{bmatrix} x \\ y \end{bmatrix} + \begin{bmatrix} a \\ b \end{bmatrix}$
20. $\begin{bmatrix} x' \\ y' \end{bmatrix} = \begin{bmatrix} x \\ y \end{bmatrix} + \begin{bmatrix} a \\ 0 \end{bmatrix}$

11.2 Exam questions

Note: Mark allocations are available with the fully worked solutions online.

1. B
2. C
3. $y = (x - 2)^2$

11.3 Reflections and rotations

11.3 Exercise

1. $\begin{bmatrix} x' \\ y' \end{bmatrix} = \begin{bmatrix} -3 \\ 1 \end{bmatrix}$, (−3, 1)
2. $\begin{bmatrix} x' \\ y' \end{bmatrix} = \begin{bmatrix} -5 \\ -2 \end{bmatrix}$, (−5, −2)
3. $y = (x + 2)^2$
4. $y = -x^2 - 1$
5. $\begin{bmatrix} x' \\ y' \end{bmatrix} = \begin{bmatrix} 5 \\ -2 \end{bmatrix}$, (5, −2)
6. $\begin{bmatrix} x' \\ y' \end{bmatrix} = \begin{bmatrix} 6 \\ -9 \end{bmatrix}$, (6, −9)
7. $y = \pm\sqrt{x}$
8. $y = \pm\dfrac{\sqrt{-2(x+1)}}{2}$
9. a. $y = x - 3$ b. $y = x + 3$
10. a. $y = -\left(x^2 + 2x + 1\right)$ b. $y = x^2 - 2x + 1$
11. $y = -\left(\sqrt{3} - 2\right)x + 1 - \sqrt{3}$
12. $\left(\dfrac{5}{2} + 2\sqrt{3}, -\dfrac{5\sqrt{3}}{2} + 2\right)$
13. a. $\begin{bmatrix} 0 & 1 \\ -1 & 0 \end{bmatrix}$
    b. $\begin{bmatrix} -1 & 0 \\ 0 & -1 \end{bmatrix}$
    c. $\dfrac{1}{\sqrt{2}}\begin{bmatrix} 1 & -1 \\ 1 & 1 \end{bmatrix}$
    d. $\dfrac{1}{2}\begin{bmatrix} \sqrt{3} & -1 \\ 1 & \sqrt{3} \end{bmatrix}$
14. a. (6, 7)
    b. (6, 7)
    c. The points are the same.
15. $y = -\dfrac{x}{2} + \dfrac{\sqrt{2}}{4}$
16. $y = \dfrac{-x + 1}{2}$

11.3 Exam questions

Note: Mark allocations are available with the fully worked solutions online.

1. B
2. C
3. $2x + 3y = 1$

11.4 Dilations

11.4 Exercise

1. $\begin{bmatrix} x' \\ y' \end{bmatrix} = \begin{bmatrix} 2 \\ -3 \end{bmatrix}, (2, -3)$

2. $\begin{bmatrix} x' \\ y' \end{bmatrix} = \begin{bmatrix} -1 \\ 8 \end{bmatrix}, (-1, 8)$

3. $\begin{bmatrix} x' \\ y' \end{bmatrix} = \begin{bmatrix} 3 & 0 \\ 0 & 1 \end{bmatrix} \begin{bmatrix} x \\ y \end{bmatrix}$

4. $\begin{bmatrix} x' \\ y' \end{bmatrix} = \begin{bmatrix} -2 \\ 4 \end{bmatrix}, (-2, 4)$

5. $\begin{bmatrix} x' \\ y' \end{bmatrix} = \begin{bmatrix} 6 \\ -5 \end{bmatrix}, (6, -5)$

6. $y = \left(\frac{x}{3}\right)^2 = \frac{x^2}{9}$

7. $y = \frac{x^2}{2}$

8. $(0, 0), (4.5, 0), (4.5, 12), (0, 12)$

9. $(4, 2), (8, 2), (6, 4), (2, 4)$

10. a. $\begin{bmatrix} x' \\ y' \end{bmatrix} = \begin{bmatrix} -3 \\ 6 \end{bmatrix}, (-3, 6)$

    b. A dilation of 3 parallel to the x-axis and a dilation of 2 parallel to the y-axis.

11. $x + y = 3$

12. $y = \left(\frac{x}{3}\right)^2 - 1 = \frac{x^2}{9} - 1$

13. $y = \frac{2}{x + 2}$

14. a. A dilation of 2 parallel to the x-axis and a dilation of 3 parallel to the y-axis.

    b. $y = 3\sqrt{2x}$

15. $\frac{x^2}{4} + y^2 = 4$

16. a. $a = \frac{3}{2}$

    b. $\begin{bmatrix} -3 & -\frac{3}{2} & -\frac{9}{2} \\ 0 & 6 & 4 \end{bmatrix}$;

    A′ $(-3, 0)$, B′ $\left(-\frac{3}{2}, 6\right)$, C′ $\left(-\frac{9}{2}, 4\right)$

17. $a = \pm\frac{1}{\sqrt{3}}$ or $\pm\frac{\sqrt{3}}{3}$

18. a. $y = -\frac{x}{6} + 1$

    b. Invariant point is $(0, 1)$.

11.4 Exam questions

Note: Mark allocations are available with the fully worked solutions online.

1. a. $(-3, 1) \to (-12, 3)$

   b. A dilation of factor 4 parallel to the x-axis and a dilation of factor 3 parallel to the y-axis.

2. $y = \frac{3}{x - 2}$

3. B

11.5 Combinations of transformations

11.5 Exercise

1. $T = \begin{bmatrix} -1 & 0 \\ 0 & 3 \end{bmatrix}$

2. $T = \begin{bmatrix} 0 & 2 \\ 2 & 0 \end{bmatrix}$

3. $(-1, -3)$

4. $(2\sqrt{2}, 0)$

5. Dilation of factor 2 parallel to both axes followed by an anticlockwise rotation of 90° about the origin.

6. $y = -\frac{x^2}{2}$

7. $y = \sqrt{-\frac{x}{3}}$

8. a. Reflection in line $y = x$ followed by a reflection in x-axis.

   b. $3x + 2y = -12$ or $y = -\frac{3}{2}x - 6$

9. a. Dilation of factor 2 parallel to the x-axis followed by a reflection in the x-axis and translation of 1 unit in the positive x-direction and 2 units in the positive y-direction.

   b. $y = -\frac{1}{2}(x - 1)^2 + 3$

10. $(x + 2, -y + 1)$

11. $(-2, 1)$

12. a. $(-x + 4,\ y)$

    b. Yes: $(-x - 4,\ y)$

13. $(1, 6)$

14. A$(0, 0)$, B$\left(\frac{-2}{17}, \frac{17}{7}\right)$, C$(3, 0)$

15. 20 units^2

16. a. A$(0, 0)$, B$\left(\frac{12}{7}, \frac{-15}{14}\right)$, C$\left(\frac{10}{7}, \frac{-9}{14}\right)$, D$\left(\frac{-2}{7}, \frac{3}{7}\right)$

    b. $\frac{3}{7}\text{units}^2$

17. 70 units^2

18. 30 units^2

19. D_λ^2 gives a dilation factor of λ^2 parallel to both axes.

20. Not the same.

11.5 Exam questions

Note: Mark allocations are available with the fully worked solutions online.

1. D

2. $(-2, 3)$

3. Original rectangle is reflected in both the axes. The area is uncharged.

11.6 Review

11.6 Exercise

Technology free: short answer

1. $\begin{bmatrix} 0 & 1 \\ 7 & 0 \end{bmatrix}$

2. $\begin{bmatrix} 1 & 0 \\ 0 & -1 \end{bmatrix}\begin{bmatrix} 0 & 1 \\ 1 & 0 \end{bmatrix}\begin{bmatrix} 1 & 0 \\ 0 & \lambda \end{bmatrix}$

3. $(x+2, -y+1)$ or $(x+2, -(y-1))$

4. $\begin{bmatrix} x' \\ y' \end{bmatrix} = \begin{bmatrix} 3 & 0 \\ 0 & 1 \end{bmatrix}\begin{bmatrix} x \\ y \end{bmatrix} + \begin{bmatrix} -2 \\ 5 \end{bmatrix}$

5. $(-3, 5)$

6. $y = \dfrac{\sqrt{3(x+2)}}{3} + 5$

7. a. $A'\left(\dfrac{1}{\sqrt{2}}, -\dfrac{3}{\sqrt{2}}\right)$, $B'\left(-2\sqrt{2}, 2\sqrt{2}\right)$, $C'\left(\dfrac{7}{\sqrt{2}}, -\dfrac{3}{\sqrt{2}}\right)$

   b. The areas are the same.

Technology active: multiple choice

8. B
9. A
10. D
11. C
12. A
13. A
14. B
15. E
16. B
17. E

Technology active: extended response

18. a.

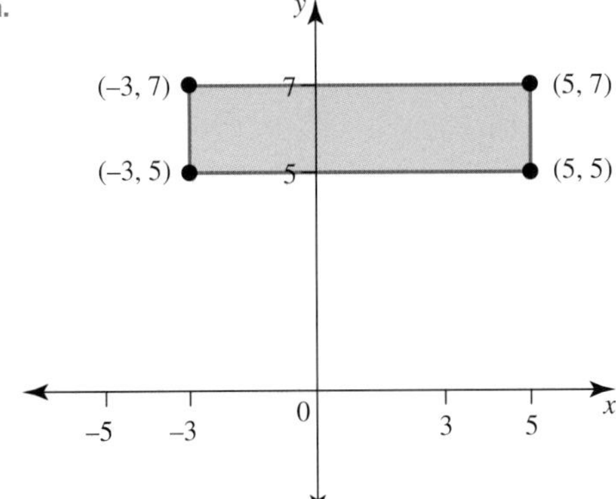

b. $\begin{bmatrix} -1 & 0 \\ 0 & 1 \end{bmatrix}$

c. $\begin{bmatrix} x' \\ y' \end{bmatrix} = \begin{bmatrix} -1 & 0 \\ 0 & 0 \end{bmatrix}\begin{bmatrix} x \\ y \end{bmatrix}$

d. $\begin{bmatrix} x' \\ y' \end{bmatrix} = \begin{bmatrix} -1 & 0 \\ 0 & 1 \end{bmatrix}\begin{bmatrix} -3 & -3 & 5 & 5 \\ 5 & 7 & 7 & 5 \end{bmatrix}$

$\begin{bmatrix} x' \\ y' \end{bmatrix} = \begin{bmatrix} 3 & 3 & -5 & -5 \\ 5 & 7 & 7 & 5 \end{bmatrix}$

$(3, 5), (3, 7), (-5, 7), (-5, 5)$

e.

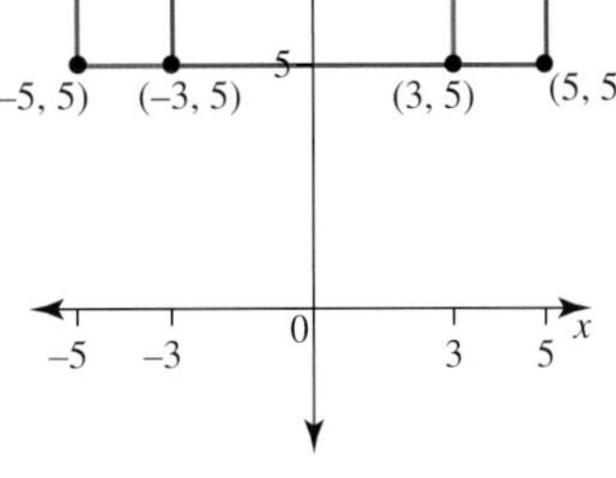

f. $\begin{bmatrix} -1 & 0 \\ 0 & 1 \end{bmatrix}, \begin{bmatrix} 2 & 0 \\ 0 & 1 \end{bmatrix}, \begin{bmatrix} -2 \\ -5 \end{bmatrix}$

g. $\begin{bmatrix} -2 & 0 \\ 0 & 1 \end{bmatrix}$

h. $\begin{bmatrix} x' \\ y' \end{bmatrix} = \begin{bmatrix} -2 & 0 \\ 0 & 1 \end{bmatrix}\begin{bmatrix} x \\ y \end{bmatrix} + \begin{bmatrix} -2 \\ -5 \end{bmatrix}$

i. $\begin{bmatrix} x' \\ y' \end{bmatrix} = \begin{bmatrix} -2 & 0 \\ 0 & 1 \end{bmatrix}\begin{bmatrix} -3 & -3 & 5 & 5 \\ 5 & 7 & 7 & 5 \end{bmatrix}$

$= \begin{bmatrix} 4 & 4 & -12 & -12 \\ 0 & 2 & 2 & 0 \end{bmatrix}$

$(4, 0), (4, 2), (-12, 2), (-12, 0)$

19. a. $\begin{bmatrix} 2 & 0 \\ 0 & 3 \end{bmatrix}, \begin{bmatrix} 1 \\ 2 \end{bmatrix}$

b. $\begin{bmatrix} x' \\ y' \end{bmatrix} = \begin{bmatrix} 2 & 0 \\ 0 & 3 \end{bmatrix}\begin{bmatrix} x \\ y \end{bmatrix} + \begin{bmatrix} 1 \\ 2 \end{bmatrix}$

c. $\dfrac{(x-1)^2}{4} + \dfrac{(y-2)}{9} = 1$; oval shape

d.

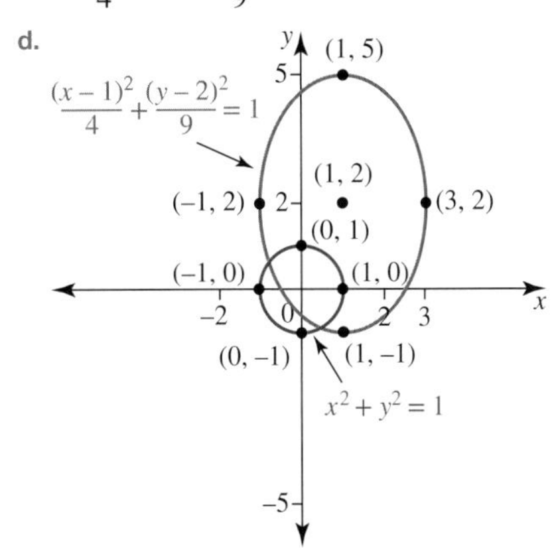

e. 6π units^2

20. a.

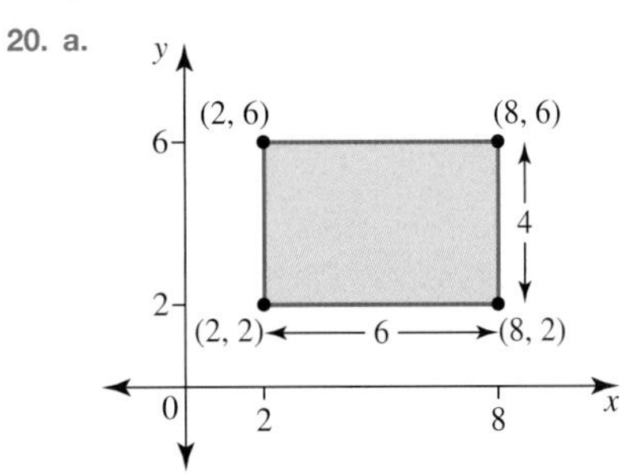

Area $= 6 \times 4 = 24\,\text{m}^2$

b. $\begin{bmatrix} 1 & 0 \\ 0 & 1.5 \end{bmatrix}$

c. $(2,3), (8,3), (2,9), (8,9)$

d. Area $= 6 \times 6 = 36$ m^2

e. Yes, he only needs 36 m^2 and has 4 m^2 to spare.

f. i. (30, 150), (30, 750), (930, 150), (930, 750)

ii. 54 m^2

iii. He needs 30 m^2 for the car; he has enough space.

g. i. $\begin{bmatrix} 1 & 1 \\ 1 & -1 \end{bmatrix}$

ii. $A'(-4,0)$, $B'(2,-2)$, $C'(0,4)$

iii. Area factor $= 2$

11.6 Exam questions

Note: Mark allocations are available with the fully worked solutions online.

1. B
2. B
3. C
4. D
5. D

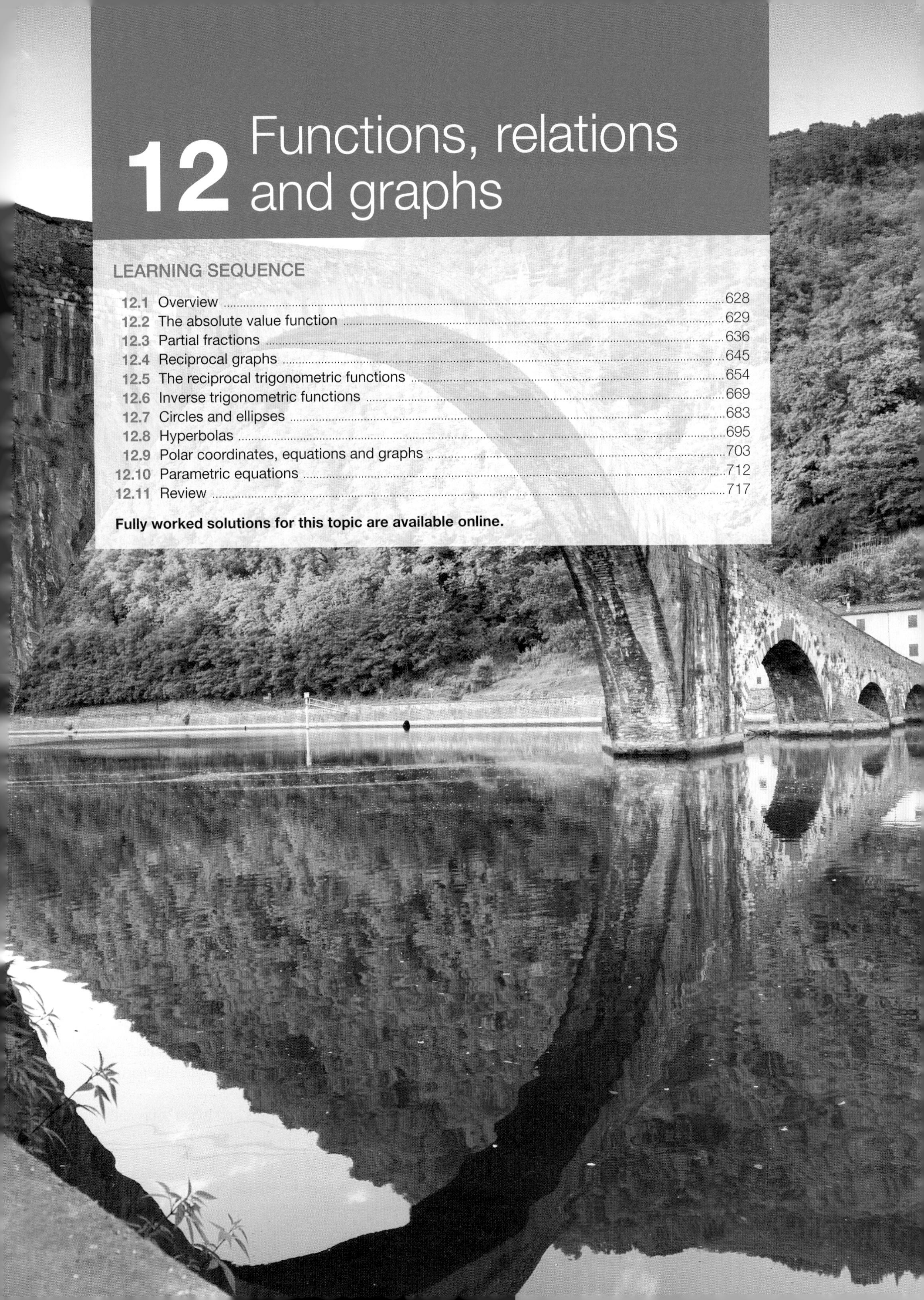

12 Functions, relations and graphs

LEARNING SEQUENCE

Fully worked solutions for this topic are available online.

12.1 Overview

Hey students! Bring these pages to life online

Watch videos

Engage with interactivities

Answer questions and check results

Find all this and MORE in jacPLUS

12.1.1 Introduction

Functions define special relationships between two variables. Specifically, relationships where for every input value (x) there is one output value (y). Functions were first defined in the late 17th century by Leibnitz as any quantity varying along a curve that is described by an equation, in 1734 Euler introduced the familiar $y = f(x)$ notation.

Functions are used to model relationships across a variety of disciplines, from forensic science and computer programs to climate science. For example, climate scientists use data about sea level, temperature, carbon dioxide and Arctic ice extent to define functions and allow them to create the complex models required to predict the extent of global climate change.

If we understand the relationship between two variables, it is possible to model their relationship. Knowing what a function looks like helps us to understand why functions behave the way they do. In this section we will explore the graphing of the modulus function, reciprocal functions, reciprocal trigonometric functions and the inverse trigonometric functions.

Sometimes variables are related implicitly, that is not by a mathematical function, we can still graph these relationships for example conic sections include the graphs of circles, ellipses and hyperbolae.

Finally, in this section we will also graph using polar coordinates and graph parametric equations, where the two variables x and y are defined in terms of a third variable called a parameter. Polar coordinates are very useful when dealing with three dimensions, and is used extensively in mathematics, engineering and astronomy to accurately describe physical phenomena such as the movement of planets and stars.

KEY CONCEPTS

This topic covers the following key concepts from the VCE Mathematics Study Design:

- identities from equating coefficients of polynomials, rational functions and their decompositions into partial fractions with denominators expressed as products of linear and irreducible quadratic terms
- graphs of simple reciprocal functions, including graphs of the reciprocal circular functions cosecant, secant and cotangent, and simple transformations of these
- graphs of the restricted circular functions of sine, cosine and tangent over principal domains and their respective inverse functions $\sin^{-1}$, $\cos^{-1}$, and $\tan^{-1}$ (students should be familiar with alternative notations, arcsin, arccos and arctan), and simple transformations of these graphs
- locus definition and construction in the plane of lines, parabolas, circles, ellipses and hyperbolas and their Cartesian, polar and parametric forms and graphs
- the absolute value function, its graph and simple transformations of its graph.

12.2 The absolute value function

LEARNING INTENTION

At the end of this subtopic you should be able to:
- sketch graphs and solve equations involving the absolute value function.

12.2.1 Definition of the absolute value functions

The absolute value or modulus of a real number, x, is denoted by $|x|$. It is a function which flips the sign of negative numbers to make them positive.

The absolute value of a real number

The absolute value of a real number, x, is defined by:

$$|x| = \begin{cases} x & \text{if } x \geq 0 \\ -x & \text{if } x < 0 \end{cases}$$

For example, $|3| = 3$ and $|-3| = -(-3) = 3$.

12.2.2 The graph of the modulus function

To see what the absolute value or modulus looks like visually consider the graph of $y = |x|$. By the definition of the absolute value, the graph of $y = |x|$ is defined as $y = x$ for $x \geq 0$ and $y = -x$ for $x < 0$. This gives the graph a characteristic V-shape as the section of the graph of $y = x$ which is negative is reflected to be positive in the graph of $y = |x|$.

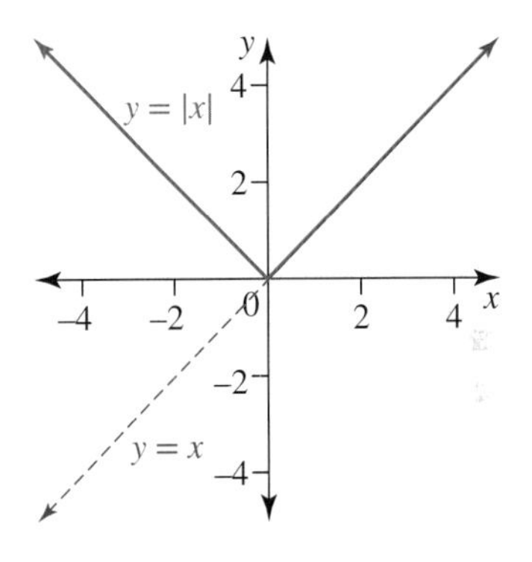

Note that the graph is continuous and is an example of a hybrid or piece-wise graph, which is a function defined by different rules on different domains $y = |x| = \begin{cases} x & \text{if } x \geq 0 \\ -x & \text{if } x < 0 \end{cases}$. The point at which the graph sharply changes direction is called a vertex and it occurs when the function inside the modulus changes sign (crosses the x-axis). The domain of is R and the range is $\{y : y \geq 0\}$ or $R^+ \cup \{0\}$.

12.2.3 Graphs of transformed modulus functions

The graphs of simple modulus functions of the form $y = |x \pm a| \pm b$ can be sketched by finding the coordinates of the vertex and the coordinates of the axis intercepts. The vertex occurs when the function inside the modulus changes sign (crosses the x-axis). If the function is expressed as a piece-wise function, the vertex occurs at the x-value where the equations change.

Note also that $|a - b| = |-(b - a)| = |-1|\,|b - a| = |b - a|$ since $|-1| = 1$.

Important result regarding the absolute value

For any two real numbers a, b:

$$|a - b| = |b - a|$$

WORKED EXAMPLE 1 Sketching simple transformations of the modulus function (1)

Sketch the graphs of:

a. $y = |x - 2|$ b. $y = |x| - 2$

THINK	WRITE
a. 1. State the equation as a piece-wise function.	a. $y = \lvert x-2 \rvert = \begin{cases} x-2 & \text{if } x \ge 2 \\ 2-x & \text{if } x < 2 \end{cases}$
2. Determine the coordinates of the vertex and the axis intercepts.	x-intercept: $(y = 0)$ $\lvert x-2 \rvert = 0$ $x = 2$ The x-intercept is a vertex and has coordinates $(2, 0)$. y-intercept: $(x = 0)$ $y = \lvert 0-2 \rvert$ $= \lvert -2 \rvert$ $= 2$ The coordinates of the y-intercept are $(0, 2)$.
3. Sketch the graph showing the coordinates of the vertex and axis intercepts. 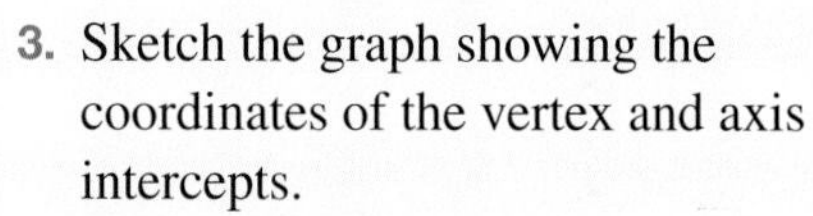 ***Note:*** the graph is the graph of $y = \lvert x \rvert$ translated 2 units to the right.	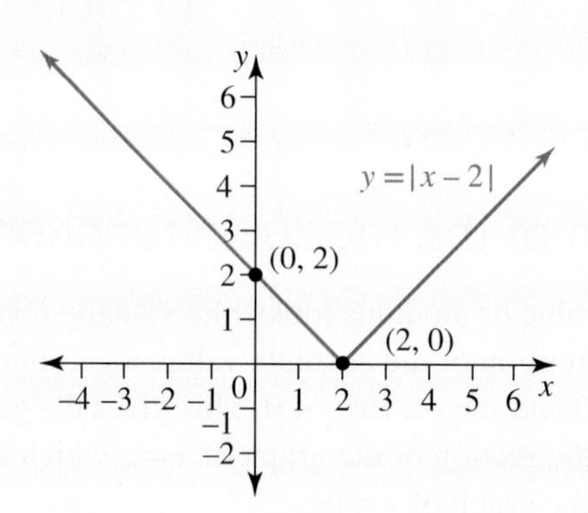
b. 1. State the equation as a piece-wise function.	b. $y = \lvert x \rvert - 2 = \begin{cases} x-2 & \text{if } x \ge 0 \\ -x-2 & \text{if } x < 0 \end{cases}$
2. Determine the coordinates of the vertex and the axis intercepts.	x-intercept: $(y = 0)$ $\lvert x \rvert - 2 = 0$ $\lvert x \rvert = 2$ $x = -2, 2$ The coordinates of the x-intercept are $(-2, 0)$ and $(2, 0)$. y-intercept: $(x = 0)$ $y = \lvert 0 \rvert - 2$ $= -2$ The y-intercept is a vertex and has coordinates $(0, -2)$.
3. Sketch the graph showing the coordinates of the vertex and axis intercepts. 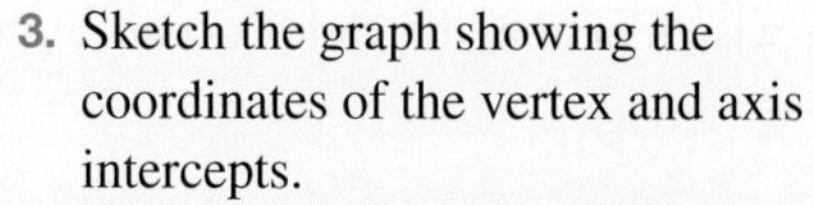 ***Note:*** the graph is the graph of $y = \lvert x \rvert$ translated 2 units down.	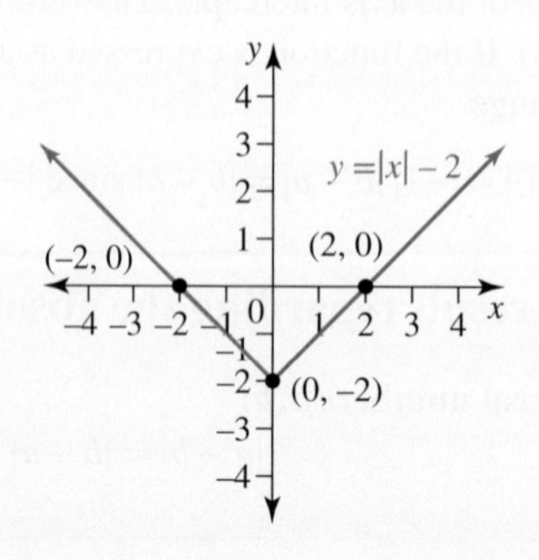

WORKED EXAMPLE 2 Sketching transformations of the modulus functions (2)

Sketch the graph of $y = |2x - 1| - 3$, stating its range.

THINK	WRITE
1. Express $\|2x - 1\|$ in a piece-wise manner.	$\|2x-1\| = \begin{cases} 2x-1 & \text{if } x \geq \frac{1}{2} \\ 1-2x & \text{if } x < \frac{1}{2} \end{cases}$
2. State the equation $y = \|2x - 1\| - 3$ as a piece-wise function.	$y = \|2x-1\| - 3 = \begin{cases} 2x-4 & \text{if } x \geq \frac{1}{2} \\ -2-2x & \text{if } x < \frac{1}{2} \end{cases}$
3. Determine the y-intercept.	y-intercept: $(x = 0)$ $y = \|2(0) - 1\| - 3$ $= 1 - 3$ $= -2$ The coordinates of the y-intercept are $(0, -2)$.
4. Determine the x-intercepts.	x-intercept: $(y = 0)$ $0 = \|2x - 1\| - 3$ $\|2x - 1\| = 3$ $2x - 1 = \pm 3$ $2x = -2$ or $2x = 4$ $x = -1$ or $x = 2$ The coordinates of the x-intercepts are $(2, 0)$ and $(-1, 0)$.
5. Determine the coordinates of the vertex.	The vertex occurs when $x = \frac{1}{2}$. The coordinates of the vertex are $\left(\frac{1}{2}, -3\right)$.
6. Sketch the graph and state the range.	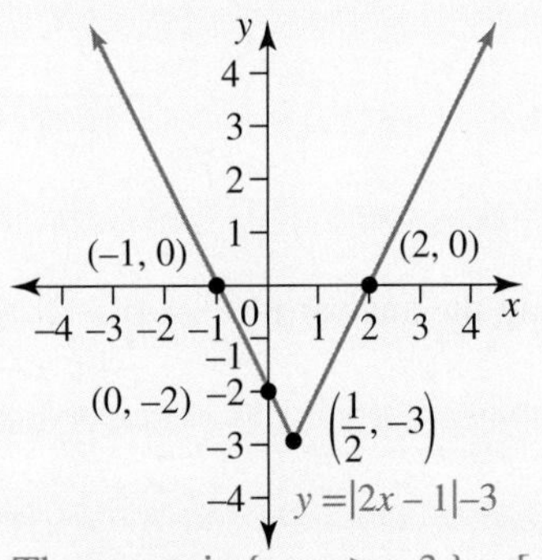 The range is $\{y : y \geq -3\} = [-3, \infty)$

12.2.4 Solving equations and inequations involving the modulus function

Note that if $|x| = 3$ then $x = \pm 3$. To solve $|x| < 3$ then this is equivalent to $-3 < x < 3$ or in interval notation $(-3, 3)$. We can also observe this graphically as the graph of $y = |x|$ is below the line $y = 3$ for values of x between $-3 < x < 3$, or in interval notation, $(-3, 3)$.

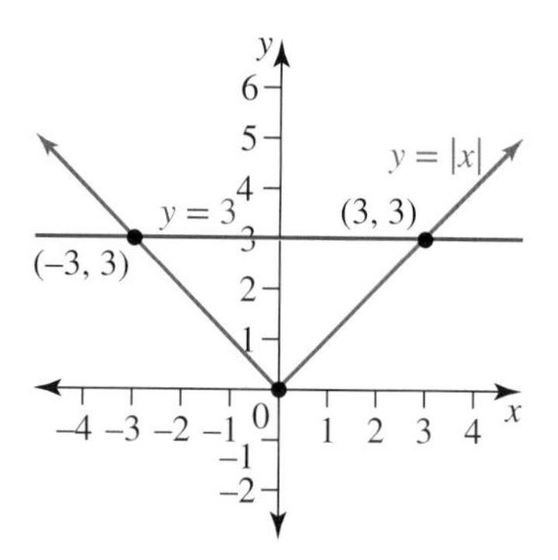

$|x| \le 3$ is equivalent to values of x between $-3 \le x \le 3$ or in interval notation $[-3, 3]$, where the endpoints are included.

Also, if $|x| > 3$ then this is equivalent to $x < -3$ and $x > 3$ or in interval notation $(-\infty, -3) \cup (3, \infty)$. We can also observe this graphically as when the graph of $y = |x|$ is above the line $y = 3$.

WORKED EXAMPLE 3 Solving equations and inequations involving the modulus function

Solve the following for x.

a. $|x-2| = 4$ 　　 b. $|x-2| \le 4$

THINK	WRITE
a. 1. Use the definition of the modulus to set up two equations to solve.	a. $|x-2| = 4$ $x-2 = -4$ or $x-2 = 4$
2. Solve the equations and state the solutions.	$x-2 = 4$ $x = 6$ $x-2 = -4$ $x = -2$ The solutions are $x = -2, 6$.
b. 1. Use a graphical approach, sketch the graphs of $y = |x-2|$ and $y = 4$.	b.
2. State the result, giving the answer in interval notation.	$|x-2| \le 4$, $-4 \le x-2 \le 4$ $-2 \le x \le 6 = [-2, 6]$

12.2.5 Sketching and solving quadratic functions involving modulus function

When graphing a function $y = |f(x)|$ note that

$$y = |f(x)| = \begin{cases} f(x) & \text{if } f(x) \geq 0 \\ -f(x) & \text{if } f(x) < 0 \end{cases}$$

This results in a reflection in the x-axis at the x-intercepts for the sections when the graph of $y = f(x)$ is below the x-axis. Note also that $|ax^2 + bx + c| = |-(ax^2 + bx + c)|$ since $|-1| = 1$.

WORKED EXAMPLE 4 Solving quadratic functions involving the modulus function

Sketch the graph of $y = |x^2 + 4x - 5|$ and solve $|x^2 + 4x - 5| < 16$.

THINK	WRITE
1. First sketch the graph of $y = x^2 + 4x - 5$ by factorising to determine the x-intercepts.	$y = x^2 + 4x - 5$ $y = (x + 5)(x - 1)$ This graph has x-intercepts $x = -5$ and at $x = 1$ and crosses the y-axis at $y = -5$.
2. Determine the turning point.	The turning point occurs midway between the x-intercepts at $x = -2$. $y = (-2)^2 + 4(-2) - 5$ $= 4 - 8 - 5$ $= -9$ The turning point is at $(-2, -9)$.
3. Graph the function $y = x^2 + 4x - 5$ and identify where the graph is above and below the x-axis.	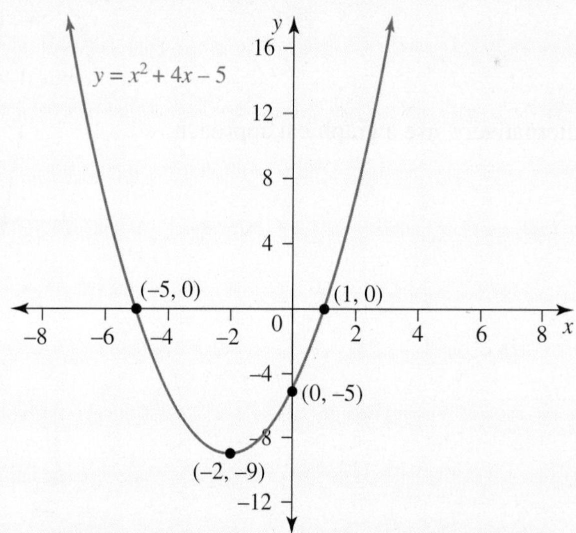 The graph is above or on the x-axis for $x \leq -5$ and $x \geq 1$. The graph is below the x-axis for $-5 < x < 1$.

4. Sketch the graph of $y = |x^2 + 4x - 5|$ by reflecting the section of the graph that is below the x-axis, $x \in (-5, 1)$, in the x-axis.

The minimum turning point will now become a maximum at the point $(-2, 9)$.

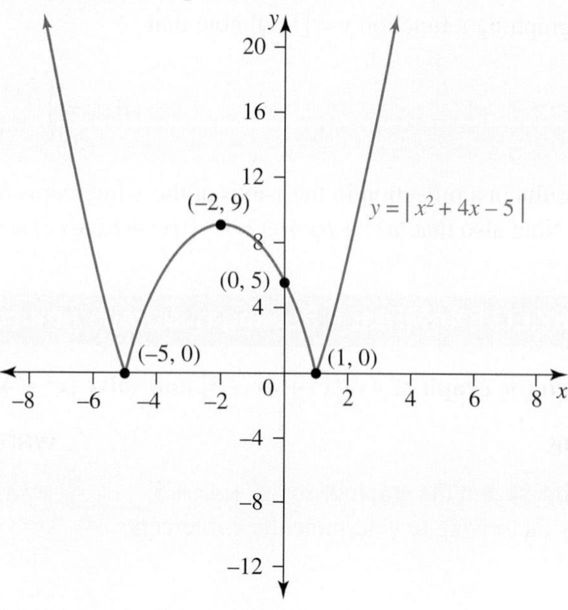

5. First solve the equality.

$|x^2 + 4x - 5| = 16$ means $x^2 + 4x - 5 = \pm 16$

$$x^2 + 4x - 5 = 16$$
$$x^2 + 4x - 21 = 0$$
$$(x + 7)(x - 3) = 0$$
$$x = -7, 3$$

or $x^2 + 4x - 5 = -16$

$$x^2 + 4x + 11 = 0$$
$$\Delta = 16 - 4 \times 11 < 0$$

So this equation has no solutions.

6. Alternatively, use a graphical approach.

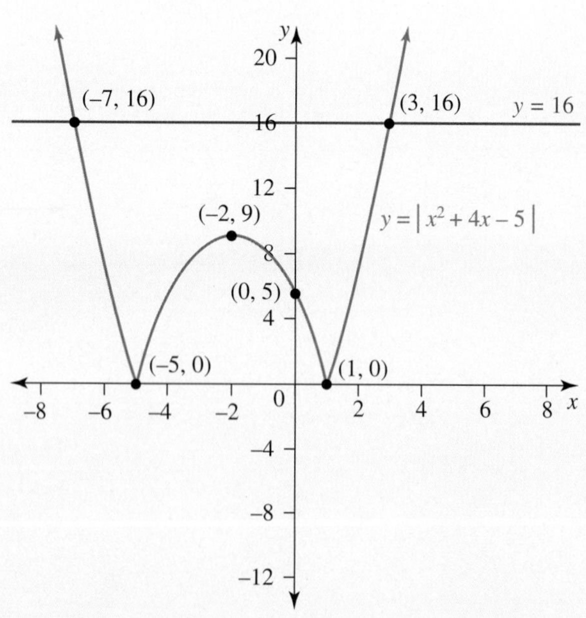

7. State the result, giving the answer in interval notation.

The graph of $y = |x^2 + 4x - 5|$ is under the graph of the line $y = 16$ for $-7 < x < 3$, or in interval notation, $x \in (-7, 3)$.

12.2 Exercise

Students, these questions are even better in jacPLUS

Receive immediate feedback and access sample responses

Access additional questions

Track your results and progress

Find all this and MORE in jacPLUS

Technology free

1. WE1 Sketch the graphs of each of the following.

   a. $y = |x - 4|$ b. $y = |x| - 4$

2. Sketch the graphs of each of the following.

   a. $y = |x + 3|$ b. $y = |x| + 3$

3. WE2 Sketch the graph of $y = |3x - 2| - 1$, stating the range.
4. Sketch the graph of $y = |4x + 3| + 2$, stating the range.
5. Sketch the graph of $y = 5 - |2x + 3|$, stating the range.
6. Sketch the graph of $y = 3 - |1 - 2x|$, stating the range.
7. WE3 For each of the following, solve for x.

   a. $|x - 4| = 6$ b. $|x - 4| > 6$

8. For each of the following, solve for x.

   a. $|x + 3| = 5$ b. $|x + 3| \leq 5$

9. For each of the following, solve for x.

   a. $|3x + 2| = 5$ b. $|3x + 2| \geq 5$

10. For each of the following, solve for x.

    a. $|7 - 3x| = 2$ b. $|7 - 3x| \leq 2$

11. For each of the following, solve for x.

    a. $|2x + 5| = x$ b. $2x + 5 = |x|$ c. $|2x + 5| = |x|$

12. Determine all values of x for which

    a. $|x - 3| = \frac{x}{2} + 5$ b. $|2 - 3x| < \frac{x}{2} + 6$

13. Determine all values of x for which

    a. $|3x - 2| = |x + 4|$ b. $|4x - 3| > |x + 5|$

14. Determine all values of x for which

    a. $|5 - 7x| = |3 - 2x|$ b. $|7 - 2x| \leq |5 - 3x|$

15. WE4 Sketch the graph of $y = |x^2 - 6x + 5|$ and solve $|x^2 - 6x + 5| < 5$
16. a. Sketch the graph of $y = |9 - x^2|$ b. Solve $|x^2 - 9| > 3$
17. a. Sketch the graph of $y = |x^2 - 3x - 4|$ b. Solve $|4 + 3x - x^2| \geq 6$
18. Sketch the graph of $y = |8 - 2x - x^2|$ and solve $|8 - 2x - x^2| \leq 16$

Technology active

19. Solve for x

a. $|8+2x|=x^2$

b. $|4-3x|=x^2$

20. Solve for x

a. $|3x^2-2|=x^2$

b. $|3-4x^2|=x^2$

21. Solve for x

a. $|4x-3|=|x^2+5|$

b. $|3x^2-2|=|x+4|$

22. Solve for x

a. $|3-2x|=\sqrt{x}$

b. $|5-4x|=\sqrt{x}$

23. Sketch and define the function $y=|3x+2|+|3x-2|$ and solve $|3x+2|+|3x-2|\leq 6$.

24. Sketch and define the function $y=|2x+3|-|2x-3|$ and solve $|2x+3|-|2x-3|=2$.

12.2 Exam questions

Question 1 (5 marks) TECH-FREE

Given the function $f(x)=4-|3-2x|$

a. Sketch the graph of $y=f(x)$ stating the range. **(3 marks)**

b. Determine $\{x: f(x)<3\}$. **(2 marks)**

Question 2 (2 marks) TECH-FREE

Solve $|5-2x|\geq x$.

Question 3 (3 marks) TECH-FREE

Sketch the graph of $y=|x^2+2x-3|$ and solve $|3-2x-x^2|<5$.

More exam questions are available online.

12.3 Partial fractions

LEARNING INTENTION

At the end of this subtopic you should be able to:

- decompose a rational function into its partial fractions, using equating coefficients for cases when the denominator can be expressed as products of linear factors and irreducible quadratic terms.

12.3.1 Rational functions

A rational function is the ratio of two polynomial functions where the denominator function is not equal to zero. It can be written as $R(x)=\dfrac{P(x)}{Q(x)}$, where $P(x)$ and $Q(x)$ are polynomials and $Q(x)\neq 0$.

For example: $f(x)=\dfrac{x}{x-3}$, $x\in R/\{3\}$ and $g(x)=\dfrac{3x^2}{x^2-16}$, $x\in R/\{-4,4\}$ are rational functions.

Equating coefficients

Consider a rational function, $R(x)=\dfrac{mx+k}{ax^2+bx+c}$, $a\neq 0$ where the numerator is a linear function, and the denominator is a quadratic function. If the discriminant $\Delta=b^2-4ac>0$, then the denominator can be expressed as a pair of linear factors, $ax^2 + bx + c = (ax + \alpha)(x + \beta)$ and the rational function can be expressed as the sum of two partial fractions.

This known as partial fraction decomposition.

Partial fraction decomposition for linear factors

$$\frac{mx+k}{ax^2+bx+c} = \frac{mx+k}{(ax+\alpha)(x+\beta)} = \frac{A}{(ax+\alpha)} + \frac{B}{(x+\beta)}, \text{ where } A \text{ and } B \text{ are constants.}$$

The methods of determining A and B are illustrated in the following worked examples.

WORKED EXAMPLE 5 Decomposing rational fractions into partial fractions (1)

Express the following as partial fractions.

a. $\dfrac{12}{9-x^2}$

b. $\dfrac{24x}{9x^2-16}$

THINK

a. 1. Factorise the denominator into linear factors, using the difference of two squares.

2. Write the expression as partial fractions, where A and B are constants to be found.

3. Add the fractions, by forming a common denominator.

4. Expand the numerator, collect like terms, factorise and expand the denominator.

5. Since the denominators are equal, the numerators are also equal. Equate coefficients, of x and the term independent of x. This gives two simultaneous equations, for the two unknowns, A and B.

6. Solve the simultaneous equations.

WRITE/DRAW

a. $\dfrac{12}{9-x^2} = \dfrac{12}{(3+x)(3-x)}$

$= \dfrac{12}{(3+x)(3-x)}$

$= \dfrac{A}{3+x} + \dfrac{B}{3-x}$

$= \dfrac{A(3-x)+B(3+x)}{(3+x)(3-x)}$

$= \dfrac{3A - Ax + 3B + Bx}{(3+x)(3-x)}$

$= \dfrac{3A + 3B + Bx - Ax}{(3+x)(3-x)}$

$= \dfrac{x(B-A)+3(A+B)}{9-x^2}$

$\dfrac{12}{9-x^2} = \dfrac{x(B-A)+3(A+B)}{9-x^2}$

$12 = x(B-A) + 3(A+B)$

$3x(B-A) + 3(A+B) = 12$

[1] $B - A = 0$

[2] $3(A+B) = 12$

[1] $\Rightarrow A = B$ substitute into [2]

$6A = 12 \ \Rightarrow A = B = 2$

7. An **alternative method** can be used to determine the unknowns, A and B. Equating the numerators from working above.

$12 = A(3 - x) + B(3 + x)$

8. Substitute an appropriate value of x. The value of $x = 3$ eliminates A.

Substitute $x = 3$
$12 = 6B$
$B = 2$

9. Substitute an appropriate value of x. The value of $x = -3$ eliminates B.

Substitute $x = -3$
$12 = 6A$
$A = 2$

10. Express the rational function as its partial fraction decomposition.

$$\frac{12}{9-x^2} = \frac{2}{3+x} + \frac{2}{3-x}$$

b. 1. Factorise the denominator into linear factors, using the difference of two squares.

b. $$\frac{24x}{9x^2-16} = \frac{24x}{(3x+4)(3x-4)}$$

2. Write the expression as partial fractions, where A and B are constants to be found.

$$= \frac{24x}{(3x+4)(3x-4)}$$
$$= \frac{A}{3x+4} + \frac{B}{3x-4}$$

3. Add the fractions, by forming a common denominator.

$$= \frac{A(3x-4)+B(3x+4)}{(3x+4)(3x-4)}$$

4. Expand the numerator, collect like terms, factorise and expand the denominator.

$$= \frac{3Ax-4A+3Bx+4B}{(3x+4)(3x-4)}$$
$$= \frac{3Ax+3Bx-4A+4B}{(3x+4)(3x-4)}$$
$$= \frac{3x(A+B)+4(B-A)}{9x^2-24}$$

5. Since the denominators are equal, the numerators are also equal. Equate coefficients, of x and the term independent of x. This gives two simultaneous equations, for the two unknowns, A and B.

$3x(A+B) + 3(B-A) = 24x$
[1] $3(A+B)x = 24x$
[1] $A + B = 8$
[2] $3(B-A) = 0$

6. Solve the simultaneous equations

[2] $\Rightarrow A = B$ substitute into [1]
$2A = 8 \Rightarrow A = B = 4$

7. Express the rational function as its partial fraction decomposition.

$$\frac{24x}{9x^2-16} = \frac{4}{3x+4} + \frac{4}{3x-4}$$

WORKED EXAMPLE 6 Decomposing rational fractions into partial fractions (2)

Express the following as partial fractions.

a. $\dfrac{12x+4}{x^2-2x-3}$

b. $\dfrac{2x-5}{3-2x^2-5x}$

THINK	WRITE/DRAW
a. 1. Factorise the denominator into linear factors.	a. $\dfrac{12x+4}{x^2-2x-3}=\dfrac{12x+4}{(x-3)(x+1)}$
2. Write the expression as partial fractions, where A and B are constants to be found.	$=\dfrac{12x+4}{(x-3)(x+1)}$ $=\dfrac{A}{x-3}+\dfrac{B}{x+1}$
3. Add the fractions, by forming a common denominator.	$=\dfrac{A(x+1)+B(x-3)}{(x-3)(x+1)}$
4. Expand the numerator, collect like terms and expand the denominator.	$=\dfrac{Ax+A+Bx-3B}{(x-3)(x+1)}$ $=\dfrac{x(A+B)+(A-3B)}{x^2-2x-3}$
5. Since the denominators are equal, the numerators are also equal. Equate coefficients, of x and the term independent of x, this gives two simultaneous equations, for the two unknowns, A and B.	$x(A+B)+(A-3B)=12x+4$ [1] $A+B=12$ [2] $A-3B=4$
6. Solve the simultaneous equations by elimination. Add the two equations to eliminate B, and solve for A.	$3\times$[1] $3A+3B=36$ [2] $A-3B=4$ $4A=40$ $A=10$
7. Substitute back into [1] to evaluate B.	$B=12-A=12-10$ $B=2$
8. An **alternative method**, can be used to determine the unknowns, A and B, equating the numerators from working above.	$A(x+1)+B(x-3)=12x+4$
9. Let $x=3$, to eliminate B.	$4A=40$ $A=10$
10. Let $x=-1$, to eliminate A.	$-4B=-8$ $B=2$
11. Express the rational function as its partial fraction decomposition.	$\dfrac{12x+4}{x^2-2x-3}=\dfrac{10}{x-3}+\dfrac{2}{x+1}$
b. 1. Multiply both numerator and denominator by -1 to express the denominator with a positive coefficient of x^2.	b. $\dfrac{2x-5}{3-2x^2-5x}=\dfrac{5-2x}{2x^2+5x-3}$
2. Factorise the denominator into linear factors.	$\dfrac{5-2x}{2x^2+5x-3}=\dfrac{5-2x}{(2x-1)(x+3)}$

3. Write the expression as partial fractions, where A and B are constants to be found.

$$\frac{5-2x}{(2x-1)(x+3)} = \frac{A}{2x-1} + \frac{B}{x+3}$$

4. Add the fractions, by forming a common denominator.

$$= \frac{A}{2x-1} + \frac{B}{x+3}$$
$$= \frac{A(x+3)+B(2x-1)}{(2x-1)(x+3)}$$

5. Expand the numerator, collect like terms, factorise and expand the denominator.

$$= \frac{Ax+3A+2Bx-B}{(2x-1)(x+3)}$$
$$= \frac{x(A+2B)+(3A-B)}{2x^2+5x-3}$$

6. Since the denominators are equal, the numerators are also equal. Equate coefficients, of x and the term independent of x. This gives two simultaneous equations, for the two unknowns, A and B.

$x(A+2B)+(3A-B) = 5-2x$
[1] $A+2B=-2$
[2] $3A-B=5$

7. Solve the simultaneous equations by eliminating B.

[1] $A+2B=-2$
$2\times$[2] $6A-2B=10$
$7A=8 \Rightarrow A=\frac{8}{7}$

8. Determine the value of B.

$B=3A-5=\frac{24}{7}-5=-\frac{11}{7}$

9. Express the rational function as its partial fraction decomposition.

$$\frac{2x-5}{3-2x^2-5x} = \frac{8}{7(2x-1)} - \frac{11}{7(x+3)}$$

Perfect squares

For a rational expression of the form $\frac{mx+k}{ax^2+bx+c}$ when $a \neq 0$ and $\Delta = b^2-4ac=0$, the quadratic function in the denominator can be expressed as a perfect square.

Partial fraction decomposition for a perfect square

$$\frac{mx+k}{ax^2+bx+c} = \frac{mx+k}{(px+\alpha)^2} = \frac{A}{(px+\alpha)} + \frac{B}{(px+\alpha)^2},$$ **where A and B are constants.**

The methods of determining A and B are illustrated in the following worked examples.

WORKED EXAMPLE 7 Decomposing rational fractions into partial fractions (3)

Express the following as partial fractions.

a. $\frac{3x+2}{x^2-4x+4}$

b. $\frac{4x}{4x^2-20x+25}$

THINK	WRITE/DRAW
a. 1. Factorise the denominator as a perfect square.	**a.** $\dfrac{3x+2}{x^2-4x+4}=\dfrac{3x+2}{(x-2)^2}$
2. Write the expression as partial fractions, where A and B are constants to be found.	$=\dfrac{A}{x-2}+\dfrac{B}{(x-2)^2}$
3. Add the fractions, by forming the lowest common denominator and expanding the numerator.	$=\dfrac{A(x-2)+B}{(x-2)^2}$ $=\dfrac{Ax+(B-2A)}{(x-2)^2}$
4. Since the denominators are equal, the numerators are also equal. Equate coefficients, of x, this gives the value of A, and equate the term independent of x, this gives an equation which can be solved for B.	$Ax+(B-2A)=3x+2$ [1] $A=3$ [2] $B-2A=2$ $B=2A+2=6+2=8$
5. Express the rational function as its partial fraction decomposition.	$\dfrac{3x+2}{x^2-4x+4}=\dfrac{3}{x-2}+\dfrac{8}{(x-2)^2}$
b. 1. Factorise the denominator as a perfect square.	**b.** $\dfrac{4x}{4x^2-20x+25}=\dfrac{4x}{(2x-5)^2}$
2. Write the expression as partial fractions, where A and B are constants to be found.	$=\dfrac{A}{2x-5}+\dfrac{B}{(2x-5)^2}$
3. Add the fractions, by forming the lowest common denominator and expanding the numerator.	$=\dfrac{A(2x-5)+B}{(2x-5)^2}$ $=\dfrac{2Ax+(B-5A)}{(2x-5)^2}$
4. Since the denominators are equal, the numerators are also equal. Equate the coefficients of x, this gives the value of A. Equate the term independent of x, this gives an equation which can be solved for B.	[1] $2A=4$ $A=2$ [2] $B-5A=0$ $B=5A=10$
5. Express the rational function as its partial fraction decomposition.	$\dfrac{4x}{4x^2-20x+25}=\dfrac{2}{2x-5}+\dfrac{10}{(2x-5)^2}$

Rational functions involving non-linear factors

If the denominator of the rational function includes a quadratic expression of the form ax^2+bx+c, where $b^2-4ac<0$ then that quadratic factor is said to be irreducible.

Partial fraction decomposition for an irreducible quadratic

$$\frac{px^2+qx+r}{x\left(x^2+a^2\right)}=\frac{A}{x}+\frac{Bx+C}{x^2+a^2}$$

where: p, q, r, a, A, B and C are constants

The methods of determining A, B and C are illustrated in the following worked examples.

WORKED EXAMPLE 8 Decomposing rational fractions into partial fractions (4)

Express the following as partial fractions.

a. $\dfrac{9}{x^2+9x}$

b. $\dfrac{2x+1}{x^3+6x}$

THINK	WRITE/DRAW
a. 1. Factorise the denominator by taking out the common factor of x. Note that x^2+9 is the sum of two squares and has no real factors.	a. $\dfrac{9}{x^3+9x}=\dfrac{9}{x(x^2+9)}$
2. Write the expression as partial fractions, where A, B and C are constants to be found.	$\dfrac{9}{x(x^2+9)}=\dfrac{A}{x}+\dfrac{Bx+C}{x^2+9}$
3. Add the fractions, by forming the lowest common denominator.	$=\dfrac{A(x^2+9)+(Bx+C)x}{x(x^2+9)}$
4. Expand the numerator and collect like terms.	$=\dfrac{A(x^2+9)+(Bx+C)x}{x(x^2+9)}$ $=\dfrac{Ax^2+9A+Bx^2+Cx}{x(x^2+9)}$ $=\dfrac{x^2(A+B)+Cx+9A}{x(x^2+9)}$
5. Since the denominators are equal, the numerators are also equal. Equate the terms independent of x, the coefficients of x, and the coefficients of x^2. Solve for A, B and C.	$x^2(A+B)+Cx+9A=9$ [1] $9A=9,\ A=1$ [2] $C=0$ [3] $A+B=0,\ B=-A=-1$
6. Express the rational function as its partial fraction decomposition.	$\dfrac{9}{x^3+9x}=\dfrac{1}{x}-\dfrac{x}{x^2+9}$
b. 1. Factorise the denominator by taking out the common factor of x. Note that x^2+6 is the sum of two squares and has no real factors.	b. $\dfrac{2x+1}{x^3+6x}=\dfrac{2x+1}{x(x^2+6)}$
2. Write the expression as partial fractions, where A, B and C are constants to be found.	$\dfrac{2x+1}{x(x^2+6)}=\dfrac{A}{x}+\dfrac{Bx+C}{x^2+6}$

3. Add the fractions, by forming the lowest common denominator.

$$= \frac{A(x^2+6)+(Bx+C)x}{x(x^2+6)}$$

4. Expand the numerator and collect like terms.

$$= \frac{A(x^2+6)+(Bx+C)x}{x(x^2+6)}$$

$$= \frac{Ax^2+6A+Bx^2+Cx}{x(x^2+6)}$$

$$= \frac{x^2(A+B)+Cx+6A}{x(x^2+6)}$$

5. Since the denominators are equal, the numerators are also equal. Equate the term independent of x, the coefficients of x, and the coefficients of x^2.
Solve for A, B and C.

$x^2(A+B)+Cx+6A=2x+1$

[1] $6A=1 \quad A=\frac{1}{6}$

[2] $C=2$

[3] $A+B=0, \quad B=-A=-\frac{1}{6}$

6. Express the rational function as its partial fraction decomposition.

$$\frac{2x+1}{x^3+6x}=\frac{1}{6x}-\frac{x}{6(x^2+6)}+\frac{2}{x^2+6}$$

TI | THINK

DISPLAY/WRITE

b. On a Calculator page, complete the entry as shown.

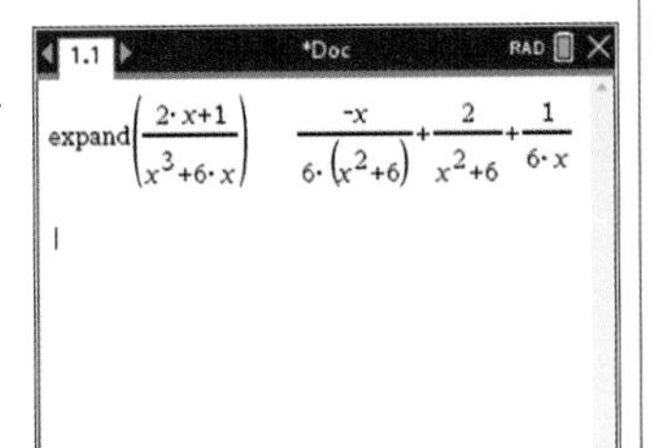

CASIO | THINK

DISPLAY/WRITE

b. On a Main screen, complete the entry as shown.

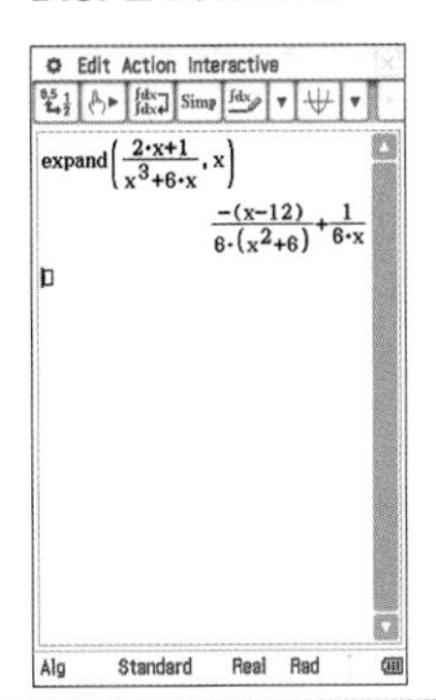

12.3 Exercise

Students, these questions are even better in jacPLUS

Receive immediate feedback and access sample responses

Access additional questions

Track your results and progress

Find all this and MORE in jacPLUS

Technology free

1. **WE5** Express the following as partial fractions.

a. $\frac{8}{16-x^2}$

b. $\frac{8x}{4x^2-9}$

2. Express the following as partial fractions.

a. $\frac{10x}{x^2-25}$

b. $\frac{5}{25-4x^2}$

3. Express the following as partial fractions.

a. $\frac{5x-3}{x^2-49}$

b. $\frac{3x-5}{4-9x^2}$

4. **WE6** Express the following as partial fractions.

a. $\frac{8x-14}{x^2-2x-8}$

b. $\frac{7-14x}{6-x-2x^2}$

5. Express the following as partial fractions.

a. $\frac{5x-5}{x^2-3x-4}$

b. $\frac{2x+18}{x^2-3x-10}$

6. Express the following as partial fractions.

a. $\frac{6x}{x^2+4x-5}$

b. $\frac{18}{10-x-2x^2}$

7. **WE7** Express the following as partial fractions.

a. $\frac{3x+2}{x^2+8x+16}$

b. $\frac{3x+5}{9x^2-12x+4}$

8. Express the following as partial fractions.

a. $\frac{x-3}{x^2-10x+25}$

b. $\frac{5x+4}{25x^2-20x+4}$

9. **WE8** Express the following as partial fractions.

a. $\frac{5x+4}{x^3+4x}$

b. $\frac{5x-3}{x^3+2x}$

10. Express the following as partial fractions.

a. $\frac{3-5x}{x^3+3x}$

b. $\frac{2x+5}{x^3+5x}$

12.3 Exam questions

Question 1 (1 mark) TECH-ACTIVE

MC The partial fraction decomposition of $\frac{x}{(x+2)^3}$ can be expressed A, B, C, where, and are non-zero real numbers, as

A. $\frac{A}{(x+2)}+\frac{B}{(x+2)^3}$

B. $\frac{A}{(x+2)^2}+\frac{B}{(x+2)^3}$

C. $\frac{Ax}{(x+2)^2}+\frac{B}{(x+2)^3}$

D. $\frac{A}{(x+2)}+\frac{B}{(x+2)^2}+\frac{C}{(x+2)^3}$

E. $\frac{A}{(x+2)}+\frac{B}{(x+2)^2}+\frac{Cx}{(x+2)^3}$

Question 2 (1 mark) TECH-ACTIVE

MC The partial fractions decomposition of $\frac{a}{a^2 - x^2}$ is given by

A. $\frac{a}{a-x} + \frac{a}{a+x}$

B. $\frac{1}{a-x} + \frac{1}{a+x}$

C. $\frac{1}{a-x} - \frac{1}{a+x}$

D. $\frac{1}{2(a-x)} + \frac{1}{2(a+x)}$

E. $\frac{1}{2(a+x)} - \frac{1}{2(a-x)}$

Question 3 (1 mark) TECH-ACTIVE

MC The partial fractions decomposition of $\frac{x}{x^2\left(x^2+a^2\right)}$ is given by

A. $\frac{A}{x} + \frac{B}{x^2+a^2}$

B. $\frac{A}{x} + \frac{Bx}{x^2+a^2}$

C. $\frac{A}{x} + \frac{B}{x^2} + \frac{Cx+D}{x^2+a^2}$

D. $\frac{A}{x} + \frac{B}{x^2} + \frac{Cx}{x^2+a^2}$

E. $\frac{A}{x} + \frac{B}{x^2} + \frac{C}{x^2+a^2}$

More exam questions are available online.

12.4 Reciprocal graphs

LEARNING INTENTION

At the end of this subtopic you should be able to:

- sketch graphs of reciprocal functions.

12.4.1 Sketching reciprocal graphs

This technique involves sketching the graph of $y = \frac{1}{f(x)}$ from the graph of $y = f(x)$.

1. When $f(x) \to 0$, $y = \frac{1}{f(x)} \to \infty$, so the graph of $y = \frac{1}{f(x)}$ approaches the vertical asymptote(s).
2. Therefore, the graph of $y = \frac{1}{f(x)}$ will have vertical asymptotes at the x-intercepts of $y = f(x)$.
3. When $f(x) \to \infty$, $\frac{1}{f(x)} \to 0$, so the graph of $y = \frac{1}{f(x)}$ approaches the horizontal asymptote (the x-axis in this case).
4. These graphs also have common points:
   (a) When $f(x) = \pm 1$, $\frac{1}{f(x)} = \pm 1$. The graphs are in the same quadrant.
   (b) When $f(x) < 0$, $\frac{1}{f(x)} < 0$.
   (c) When $f(x) > 0$, $\frac{1}{f(x)} > 0$.
5. The x-intercepts of $f(x)$ determine the equations of the vertical asymptotes for the reciprocal of the functions.
6. The minimum turning point of $f(x)$ gives the maximum turning point of the reciprocal function.

7. The maximum turning point of $f(x)$ gives the minimum turning point of the reciprocal function.

Note: If $y = \dfrac{-1}{f(x)}$ then:

- for $f(x) = 1$, $y = -1$ and for $f(x) = -1$, $y = 1$
- for $f(x) < 0$, $y > 0$ and for $f(x) > 0$, $y < 0$.

WORKED EXAMPLE 9 Sketching a reciprocal graph (1)

Sketch the graph of the function $y = \dfrac{1}{x^2 - 9}$, $x \neq \pm 3$.

THINK | **WRITE/DRAW**

1. Sketch the graph of the function $y = x^2 - 9$.

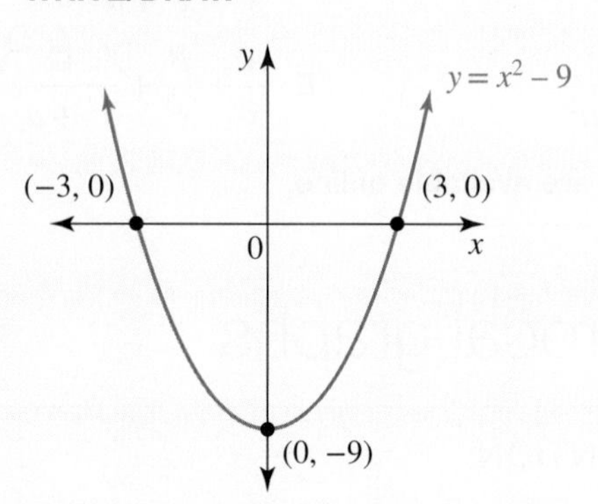

2. Work out the asymptotes for $y = \dfrac{1}{x^2 - 9}$.

The x-intercepts of $y = x^2 - 9$ are $x = \pm 3$.
The vertical asymptotes for $y = \dfrac{1}{x^2 - 9}$ are $x = \pm 3$.
The horizontal asymptote is $y = 0$.

3. For the y-intercept, $x = 0$.

The y-intercept is $\left(0, -\dfrac{1}{9}\right)$.

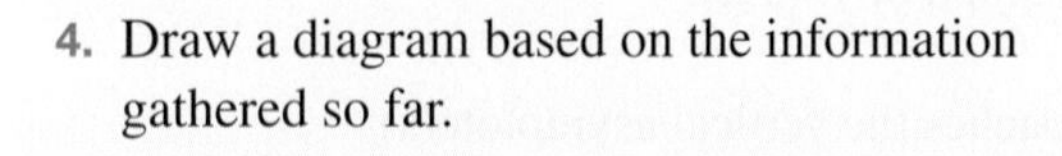

4. Draw a diagram based on the information gathered so far.

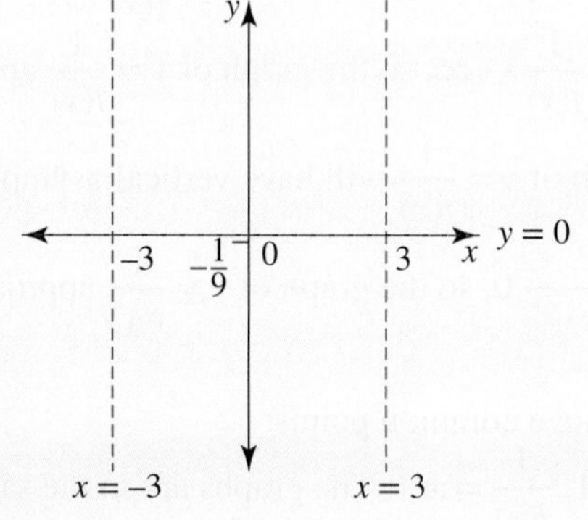

5. As $x \to -\infty$, y approaches 0 from the positive direction ($y \to 0^+$).
As x approaches −3 from the negative direction ($x \to -3^-$), $y \to \infty$.

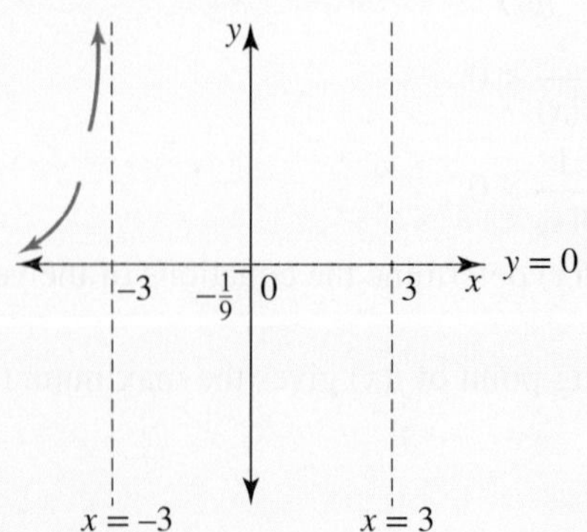

6. As $x \to -3^+$, $y \to -\infty$.
As $x \to +3^-$, $y \to -\infty$.

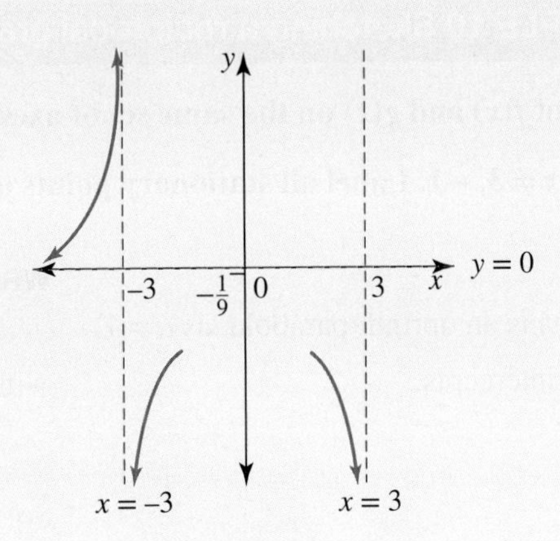

7. As $x \to +3^+$, $y \to \infty$.
As $x \to \infty$, $y \to 0^+$.

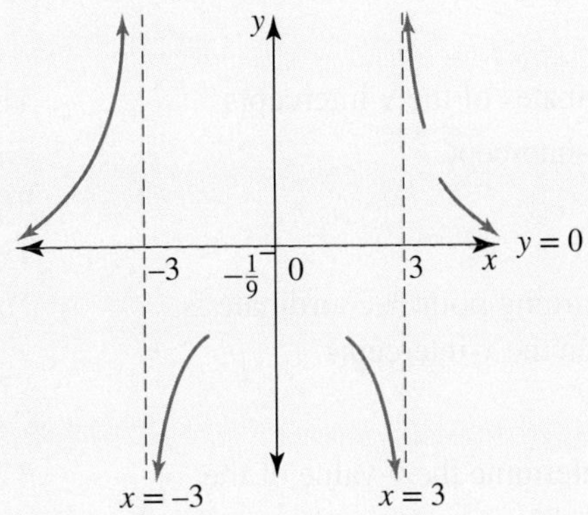

8. To determine the shape of the graph near the y-intercept, evaluate the value of y when x is ± 1 and ± 2.

9. Sketch the graph of $y = \dfrac{1}{x^2 - 9}$.

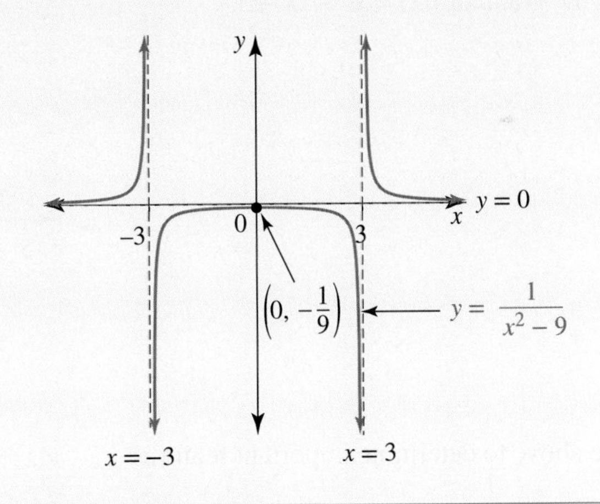

The following examples show a different approach to sketching reciprocal functions.

WORKED EXAMPLE 10 Sketching a reciprocal graph (2)

Sketch the graphs of $f(x)$ and $g(x)$ on the same set of axes where: $f(x) = x^2 - 2x - 3$ and $g(x) = \dfrac{1}{x^2 - 2x - 3}$, $x \neq 3, -1$. Label all stationary points and asymptotes.

THINK	WRITE/DRAW
1. The graph of $f(x)$ is an upright parabola, as $a = 1$.	
2. Calculate the x-intercepts.	x-intercepts: $x^2 - 2x - 3 = 0$ $(x-3)(x+1) = 0$ So: $x - 3 = 0$ or $x + 1 = 0$ $x = 3$ or $x = -1$
3. State the coordinates of the x-intercepts.	The x-intercepts are $(3, 0)$ and $(-1, 0)$.
4. Calculate the y-intercept.	y-intercept: $f(0) = -3$ The y-intercept is $(0, -3)$.
5. The vertex or turning point x-coordinate is halfway between the x-intercepts.	Turning point: $x = \dfrac{-1+3}{2} = 1$
6. Substitute to determine the y-value of the turning point.	$f(1) = (1)^2 - 2(1) - 3$ $= 1 - 2 - 3$ $= -4$ The turning point is $(1, -4)$.
7. Sketch the graph of $f(x) = x^2 - 2x - 3$.	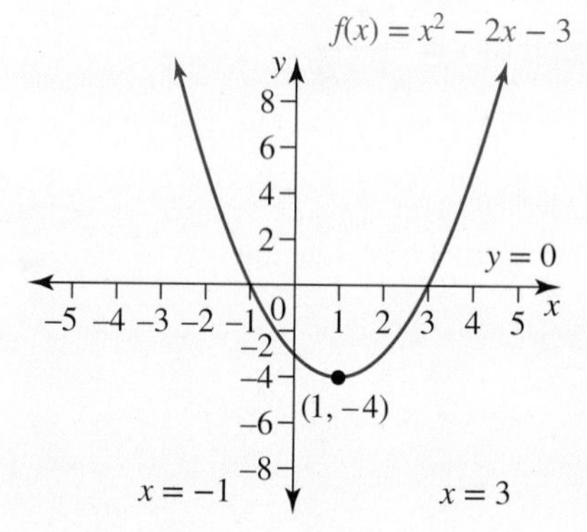
8. Use the above to determine important features for $g(x) = \dfrac{1}{x^2 - 2x - 3}$.	$g(x) = \dfrac{1}{(x^2 - 2x - 3)}$ $= \dfrac{1}{(x-3)(x+1)}$
9. Vertical asymptotes occur where $f(x)$ has its x-intercepts.	Vertical asymptotes: $x = 3$ and $x = -1$ As $x \to \pm\infty$, $\left\lvert \dfrac{1}{x^2 - 2x - 3} \right\rvert \to 0^+$, and so $g(x) \to 0^+$.
10. Determine the horizontal asymptotes.	The horizontal asymptote is $y = 0$.

11. The reciprocal of the turning point for $f(x)$ is a turning point for $g(x)$.	The reciprocal of the turning point $(1, -4)$ is $\left(1, -\frac{1}{4}\right)$.
12. $f(x)$ and $g(x)$ intercept at $y = 1$ and $y = -1$.	
13. As $g(x) = \frac{1}{f(x)}$, the graphs of $f(x)$ and $g(x)$ are in the same quadrants.	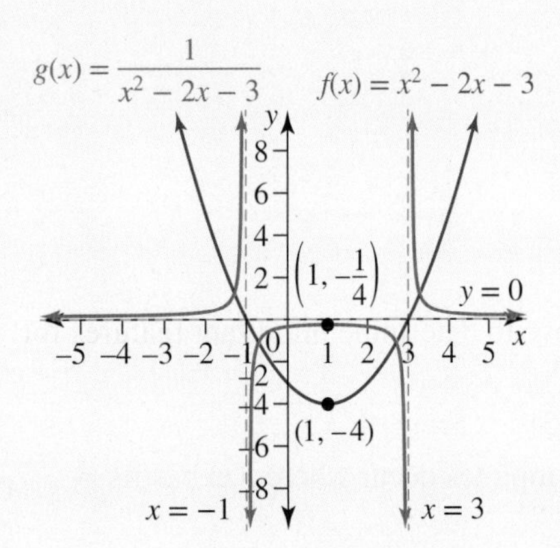
14. Sketch the graph of $g(x)$ on the same axes as $f(x)$.	

WORKED EXAMPLE 11 Sketching a reciprocal graph (3)

Sketch the graphs of $f(x)$ and $g(x)$ on the same set of axes where: $f(x) = -(x+3)^2$ and $g(x) = \frac{-1}{(x+3)^2}, x \neq -3$.

THINK	WRITE/DRAW
1. Work out important features for $f(x) = -(x+3)^2$. This is a parabola reflected in the x-axis, as $a = -1$. Calculate the x-intercept(s).	x-intercepts: $-(x+3)^2 = 0$ $(x+3)^2 = 0$ $x + 3 = 0$ $x = -3$
2. State the coordinates of the x-intercept.	The x-intercept is $(-3, 0)$.
3. Calculate the y-intercept.	The y-intercept: $f(0) = -(0+3)^2$ $= -(9)$ $= -9$ The y-intercept is $(0, -9)$.
4. As the graph touches the x-axis at $(-3, 0)$, it must also turn at this point. Hence, $(-3, 0)$ is the turning point.	The turning point is $(-3, 0)$.

5. Sketch the graph of $f(x)$.

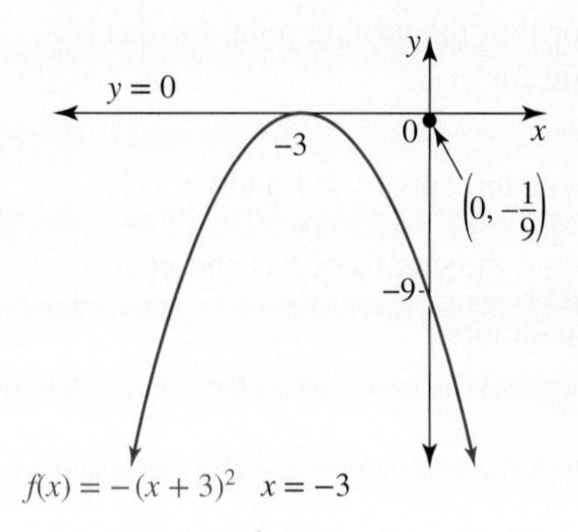

6. Use the above to determine important features for $g(x) = \dfrac{-1}{(x+3)^2}$.

For $g(x) = \dfrac{-1}{(x+3)^2}$:

7. Vertical asymptotes occur where $f(x)$ has its x-intercepts.

Vertical asymptote: $x = -3$

8. Determine the horizontal asymptotes.

As $x \to \pm\infty$, $\left|\dfrac{-1}{(x+3)^2}\right| \to 0^-$, and so $g(x) \to 0^-$.

9. We cannot take the reciprocal of the turning point for $f(x)$ as the reciprocal of 0 is not defined. It was determined in step 7 above that this was the vertical asymptote.

The horizontal asymptote is $y = 0$.

10. The y-intercept of $g(x)$ is the reciprocal of the y-intercept of $f(x)$.

The y-intercept is $\left(0, -\dfrac{1}{9}\right)$.

11. Since $g(x) = \dfrac{1}{f(x)}$, then $g(x) = 1$ or -1 when $f(x) = 1$ or -1.

12. As $g(x) = \dfrac{1}{f(x)}$, the graphs of $f(x)$ and $g(x)$ are in the same quadrants.

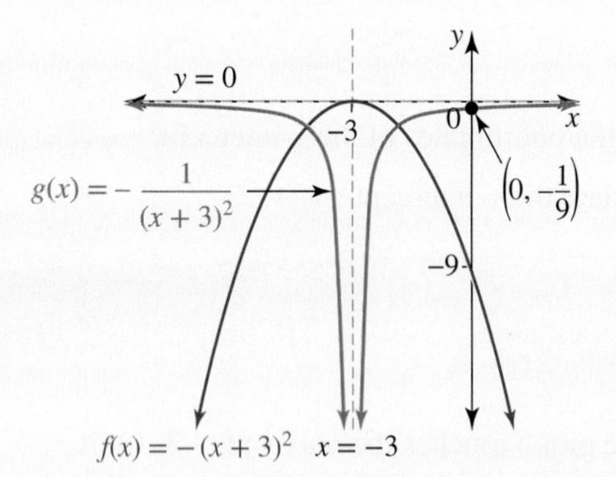

13. Sketch the graph of $g(x)$ on the same axes as $f(x)$.

WORKED EXAMPLE 12 Sketching a reciprocal graph (4)

Sketch the graphs of $f(x)$ and $g(x)$ on the same set of axes where:
$f(x) = x^2 + 4x + 5$ and $g(x) = \dfrac{1}{x^2 + 4x + 5}$.

THINK	WRITE/DRAW
Work out important features for $f(x) = x^2 + 4x + 5$.	
1. This is an upright parabola, as $a = 1$.	
2. Determine the x-value of the turning point by completing the square.	$f(x) = x^2 + 4x + 5$ $= (x^2 + 4x + 4) + 1$ $= (x+2)^2 + 1$ The turning point is at $x = -2$.
3. Evaluate $f(x)$ when $x = -2$.	$f(-2) = (-2)^2 + 4(-2) + 5$ $= 4 - 8 + 5$ $= 1$ The turning point is $(-2, 1)$.
4. As the parabola is upright and turns at $(-2, 1)$ it is completely above the x-axis and hence there is no x-intercept.	There is no x-intercept.
5. Calculate the y-intercept.	y-intercept: $f(0) = 5$ The y-intercept is $(0, 5)$. By symmetry $(-4, 5)$ is also on the curve.
6. Sketch the graph of $f(x)$.	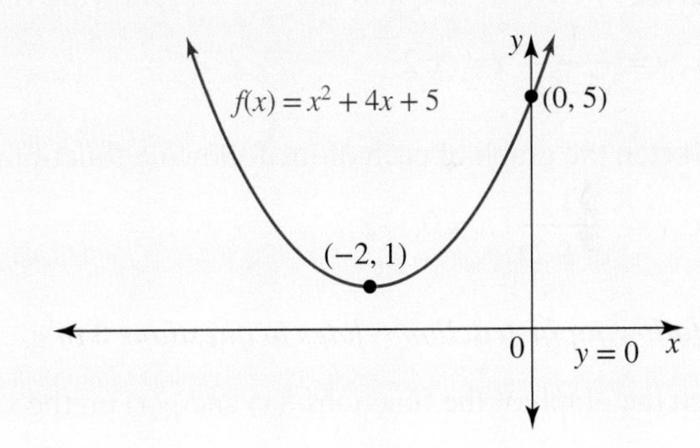
7. Use the above to determine important features for $g(x) = \dfrac{1}{x^2 + 4x + 5}$.	
8. Since there are no x-intercepts for $f(x)$, $g(x)$ has no vertical asymptotes.	There are no vertical asymptotes.
9. Determine the horizontal asymptote.	As $x \to \pm\infty$, $\left\lvert \dfrac{1}{x^2 + 4x + 5} \right\rvert \to 0^+$, and so $g(x) \to 0^+$. The horizontal asymptote is $y = 0$.
10. The vertex of $g(x)$ is the reciprocal of the vertex of $f(x)$.	The vertex is $(-2, 1)$.

11. The y-intercept for $g(x)$ is the reciprocal of the y-intercept of $f(x)$.

The y-intercept is $\left(0, \frac{1}{5}\right)$.

12. Since $g(x) = \frac{1}{f(x)}$ the graphs of $f(x)$ and $g(x)$ are in the same quadrants.

13. Sketch the graph of $g(x)$ on the same axes as $f(x)$.

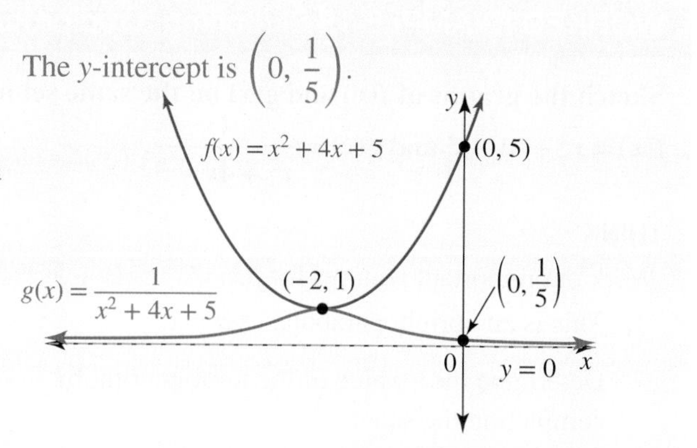

12.4 Exercise

Students, these questions are even better in jacPLUS

Receive immediate feedback and access sample responses

Access additional questions

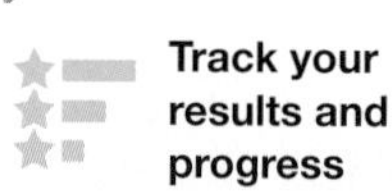

Track your results and progress

Find all this and MORE in jacPLUS

Technology free

1. **WE9** Sketch the graph of each of the following functions

a. $y = \frac{1}{x^2 - 4}, x \neq \pm 2$

b. $y = \frac{1}{x + 2}, x \neq -2$

2. Sketch the graph of each of the following functions.

a. $y = \frac{1}{x^2 + 2x}, x \neq 0, -2$

b. $y = \frac{1}{x^2 - 4x + 3}, x \neq 3, 1$

The following instruction relates to questions 3 to 8.

Sketch the graph of the functions $f(x)$ and $g(x)$ on the same set of axes. Label all turning points and asymptotes.

3. **WE10** $f(x) = x - 4$, $g(x) = \frac{1}{x - 4}, x \neq 4$

4. $f(x) = x^2 - 4x$, $g(x) = \frac{1}{x^2 - 4x}, x \neq 0, 4$

5. **WE11** $f(x) = (x - 4)^2$, $g(x) = \frac{1}{(x - 4)^2}, x \neq 4$

6. $f(x) = (x + 3)^2$, $g(x) = \frac{1}{(x + 3)^2}, x \neq -3$

7. **WE12** $f(x) = x^2 + 2$, $g(x) = \frac{1}{x^2 + 2}$

8. $f(x) = x^2 + 2x + 4$, $g(x) = \frac{1}{x^2 + 2x + 4}$

9. Sketch the graph of each of the following functions.

a. $y = \dfrac{1}{-2x+5}, x \neq \dfrac{5}{2}$

b. $y = \dfrac{1}{-x^2+2x}, x \neq 0, 2$

10. Sketch the graph of each of the following functions.

a. $y = \dfrac{1}{4x^2-9}, x \neq \pm\dfrac{3}{2}$

b. $y = \dfrac{1}{-2x^2+5x-3}, x \neq \dfrac{3}{2}, 1$

The following instruction relates to questions 11 to 16.

Sketch the graphs of $f(x)$ and $g(x)$ on the same set of axes. Label all turning points and asymptotes.

11. $f(x) = 3 - x$, $g(x) = \dfrac{1}{3-x}, x \neq 3$

12. $f(x) = x^2 + 3x + 2$, $g(x) = \dfrac{1}{x^2+3x+2}, x \neq -1, -2$

13. $f(x) = 3x + x^2$, $g(x) = \dfrac{1}{3x+x^2}, x \neq -3, 0$

14. $f(x) = 3x^2 - 8x - 3$, $g(x) = \dfrac{1}{3x^2-8x-3}, x \neq \dfrac{-1}{3}, 3$

15. $f(x) = -x^2 + 4x - 4$, $g(x) = \dfrac{1}{-x^2+4x-4}, x \neq 2$

16. $f(x) = x^2 + x + \dfrac{1}{4}$, $g(x) = \dfrac{1}{x^2+x+\frac{1}{4}}, x \neq \dfrac{-1}{2}$

12.4 Exam questions

Question 1 (3 marks) TECH-FREE

Sketch the graph of $y = \dfrac{1}{x^2+2x-3}$.

Question 2 (4 marks) TECH-FREE

Sketch the graphs of each of the following

a. $y = \dfrac{1}{x^2+9}$ **(2 marks)**

b. $y = \dfrac{1}{x^2+6x+9}$ **(2 marks)**

Question 3 (3 marks) TECH-FREE

Given the function $y = \dfrac{1}{x^2+kx+16}$, determine the values of k, for which the function has:

a. two straight line vertical asymptotes **(1 mark)**

b. one straight line vertical asymptote **(1 mark)**

c. no straight line vertical asymptotes. **(1 mark)**

More exam questions are available online.

12.5 The reciprocal trigonometric functions

LEARNING INTENTION

At the end of this subtopic you should be able to:
- evaluate exact values of reciprocal trigonometric ratios
- sketch graphs of reciprocal trigonometric functions.

12.5.1 Naming the reciprocal trigonometric functions

The reciprocal of the sine function is called the **cosecant function**; it may be abbreviated to cosec or csc. It is defined as $\text{cosec}(x) = \dfrac{1}{\sin(x)}$, provided that $\sin(x) \neq 0$.

The reciprocal of the cosine function is called the **secant function**, often abbreviated to sec. It is defined as $\sec(x) = \dfrac{1}{\cos(x)}$, provided that $\cos(x) \neq 0$.

The reciprocal of the tangent function is called the **cotangent function**, often abbreviated to cot. It is defined as $\cot(x) = \dfrac{1}{\tan(x)} = \dfrac{\cos(x)}{\sin(x)}$, provided that $\sin(x) \neq 0$.

Note that the reciprocal functions are not the same as the inverse trigonometric functions.

These functions can also be defined in terms of right-angled triangles.

$$\text{cosec}(\theta) = \frac{1}{\sin(\theta)} = \frac{1}{\left(\frac{\text{opposite}}{\text{hypotenuse}}\right)} = \frac{\text{hypotenuse}}{\text{opposite}} = \frac{c}{b}$$

$$\sec(\theta) = \frac{1}{\cos(\theta)} = \frac{1}{\left(\frac{\text{adjacent}}{\text{hypotenuse}}\right)} = \frac{\text{hypotenuse}}{\text{adjacent}} = \frac{c}{a}$$

$$\cot(\theta) = \frac{1}{\tan(\theta)} = \frac{1}{\left(\frac{\text{opposite}}{\text{adjacent}}\right)} = \frac{\text{adjacent}}{\text{opposite}} = \frac{a}{b}$$

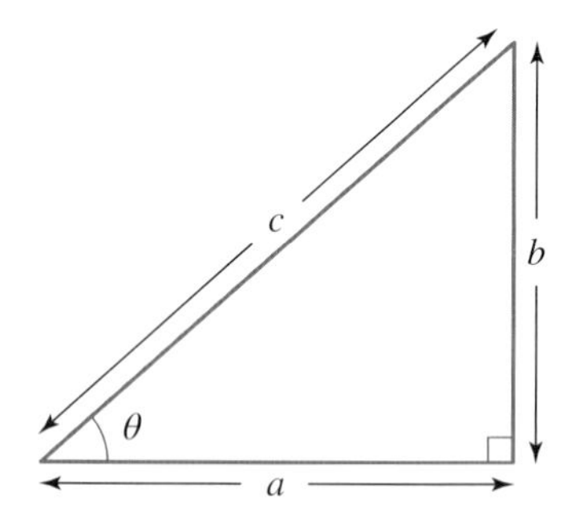

12.5.2 Exact values

The exact values for the reciprocal trigonometric functions for angles that are multiples of $\dfrac{\pi}{6}$ and $\dfrac{\pi}{4}$ (or 30° and 45°) can be found from the corresponding trigonometric values by finding the reciprocal. Note that we may need to simplify the resulting expression or rationalise the denominator.

WORKED EXAMPLE 13 Determining exact values of reciprocal trigonometric functions (1)

Determine the exact value of cosec $\left(\dfrac{5\pi}{4}\right)$.

THINK	WRITE
1. Rewrite cosec $\left(\dfrac{5\pi}{4}\right)$ by recalling $\text{cosec}(\theta) = \dfrac{1}{\sin(\theta)}$.	$\text{cosec}\left(\dfrac{5\pi}{4}\right) = \dfrac{1}{\sin\left(\frac{5\pi}{4}\right)}$

2. $\frac{5\pi}{4}$ is in quadrant 3. Sine is negative in that quadrant.

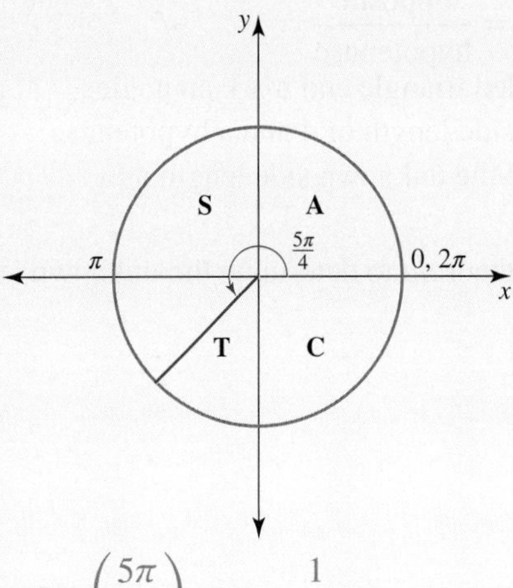

$$\text{cosec}\left(\frac{5\pi}{4}\right) = \frac{1}{\sin\left(\pi + \frac{\pi}{4}\right)}$$
$$= \frac{1}{-\sin\left(\frac{\pi}{4}\right)}$$

3. Use $\sin\left(\frac{\pi}{4}\right) = \frac{1}{\sqrt{2}}$ and simplify the fraction.

$$\text{cosec}\left(\frac{5\pi}{4}\right) = \frac{1}{-\frac{1}{\sqrt{2}}}$$
$$= 1 \times \frac{-\sqrt{2}}{1}$$
$$= -\sqrt{2}$$

In a circle of radius 1, for an acute angle θ, it is possible to locate lengths that are $\sin(\theta)$, $\cos(\theta)$, $\tan(\theta)$, $\sec(\theta)$, $\text{cosec}(\theta)$ and $\cot(\theta)$.

Through the use of right-angled triangles and the unit circle, if one trigonometric ratio for an angle is known, it is possible to calculate all of the ratios.

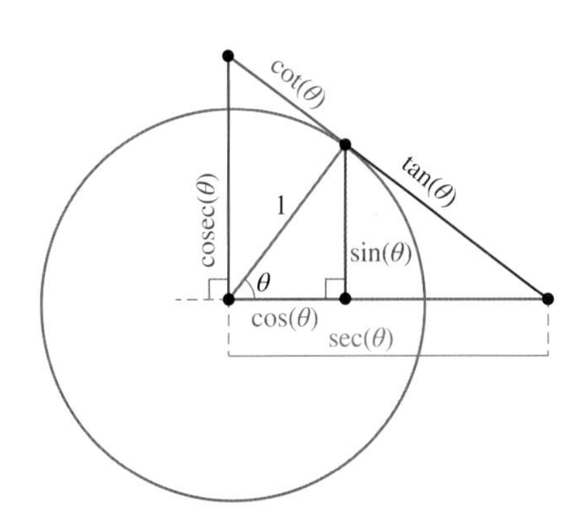

WORKED EXAMPLE 14 Determining exact values of reciprocal trigonometric functions (2)

If $\text{cosec}(\theta) = \frac{7}{4}$ and $\frac{\pi}{2} \le \theta \le \pi$, determine the exact value of $\cot(\theta)$.

THINK

1. Recall the definition $\text{cosec}(\theta) = \frac{1}{\sin(\theta)}$ and solve for $\sin(\theta)$.

WRITE

$$\text{cosec}(\theta) = \frac{1}{\sin(\theta)}$$
$$\frac{7}{4} = \frac{1}{\sin(\theta)}$$
$$\sin(\theta) = \frac{4}{7}$$

2. Recall that $\sin(\theta) = \dfrac{\text{opposite}}{\text{hypotenuse}}$.
Draw a right-angled triangle and mark an angle with an opposite side length of 4 and a hypotenuse of length 7. Label the unknown side length as x.

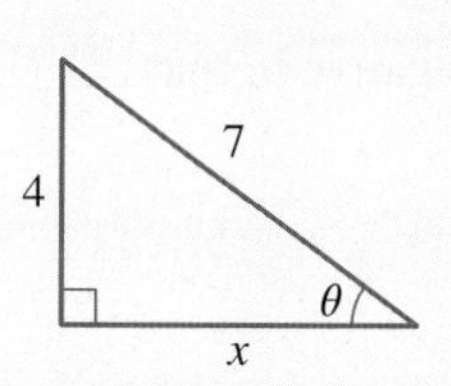

3. Use Pythagoras' theorem to determine the unknown side length.

$$\begin{aligned} x &= \sqrt{7^2 - 4^2} \\ &= \sqrt{49 - 16} \\ &= \sqrt{33} \end{aligned}$$

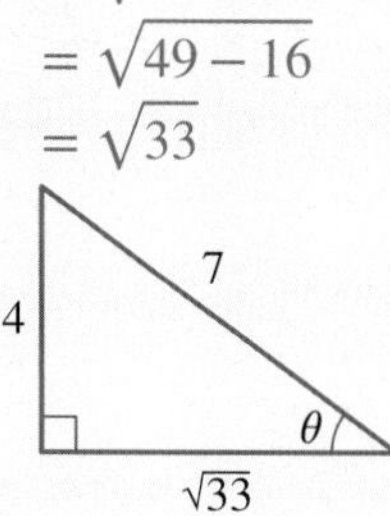

4. Given that $\dfrac{\pi}{2} < \theta < \pi$, θ is in the second quadrant, and although $\sin(\theta)$ is positive, $\tan(\theta)$ is negative.
Determine $\tan(\theta)$ using the value of x from step 3.

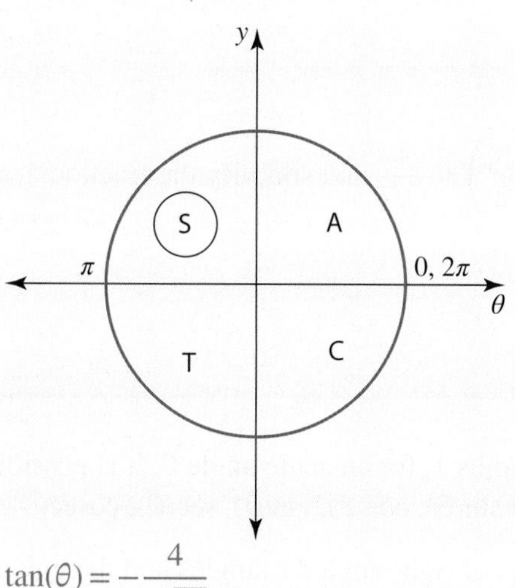

$$\tan(\theta) = -\frac{4}{\sqrt{33}}$$

5. Use $\tan(\theta)$ to calculate $\cot(\theta)$.

$$\begin{aligned} \cot(\theta) &= \frac{1}{\tan(\theta)} \\ &= \frac{1}{-\frac{4}{\sqrt{33}}} \\ &= -\frac{\sqrt{33}}{4} \end{aligned}$$

$$\therefore \cot(\theta) = -\frac{\sqrt{33}}{4}$$

12.5.3 Sketching the reciprocal trigonometric functions

We have looked at how the graph of $f(x)$ can be used to sketch the graph of $\frac{1}{f(x)}$. This method will be used to graph $\sec(x) = \frac{1}{\cos(x)}$, $\operatorname{cosec}(x) = \frac{1}{\sin(x)}$ and $\cot(x) = \frac{1}{\tan(x)}$.

The graph of $y = \sec(x)$

Consider the graph of $y = \cos(x)$.

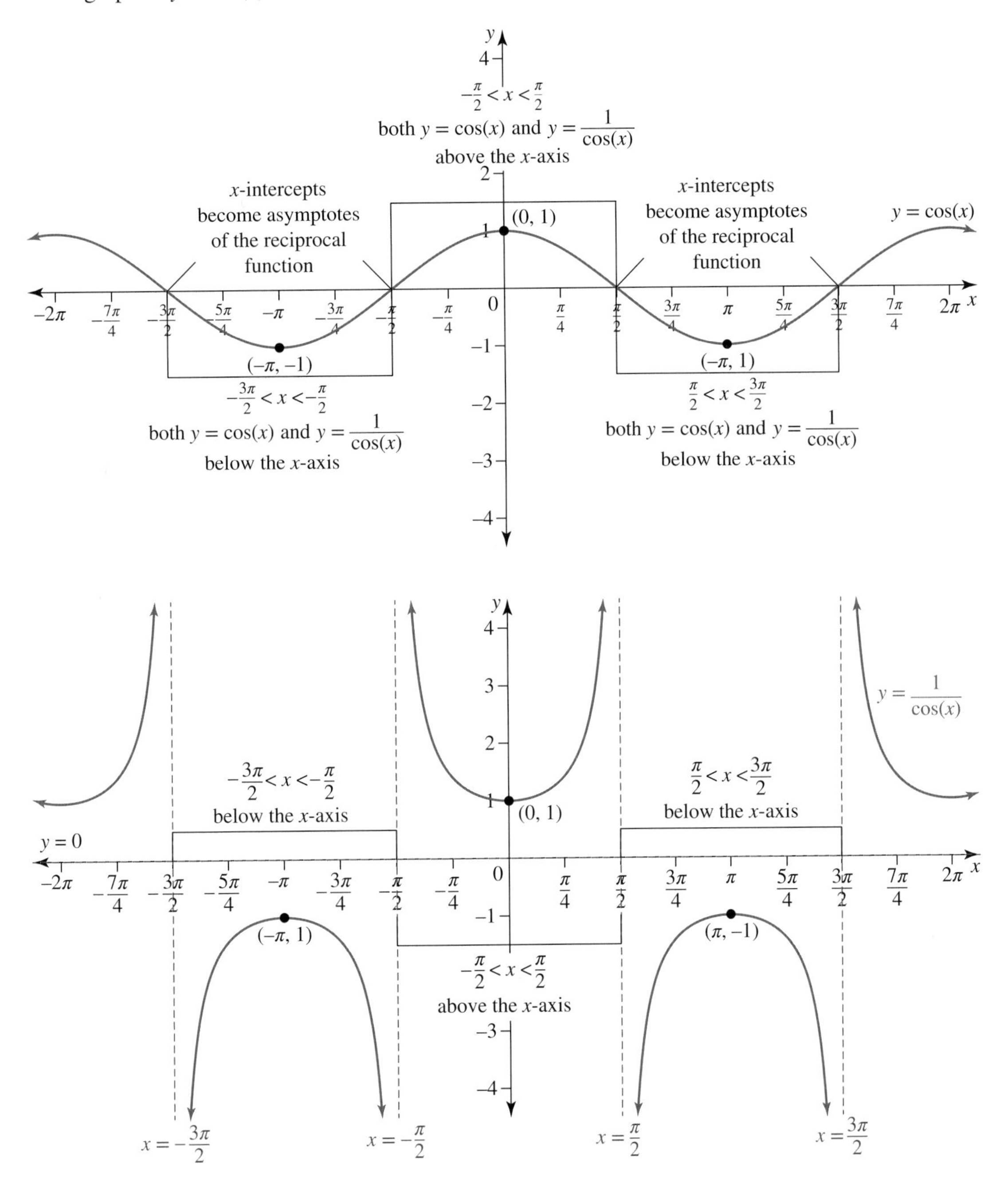

The graph of $y = \operatorname{cosec}(x)$

In a similar fashion, the graph of $y = \sin(x)$ can be used to sketch the graph of $y = \frac{1}{\sin(x)}$ (also known as $y = \operatorname{cosec}(x)$).

Consider the graph of $y = \sin(x)$.

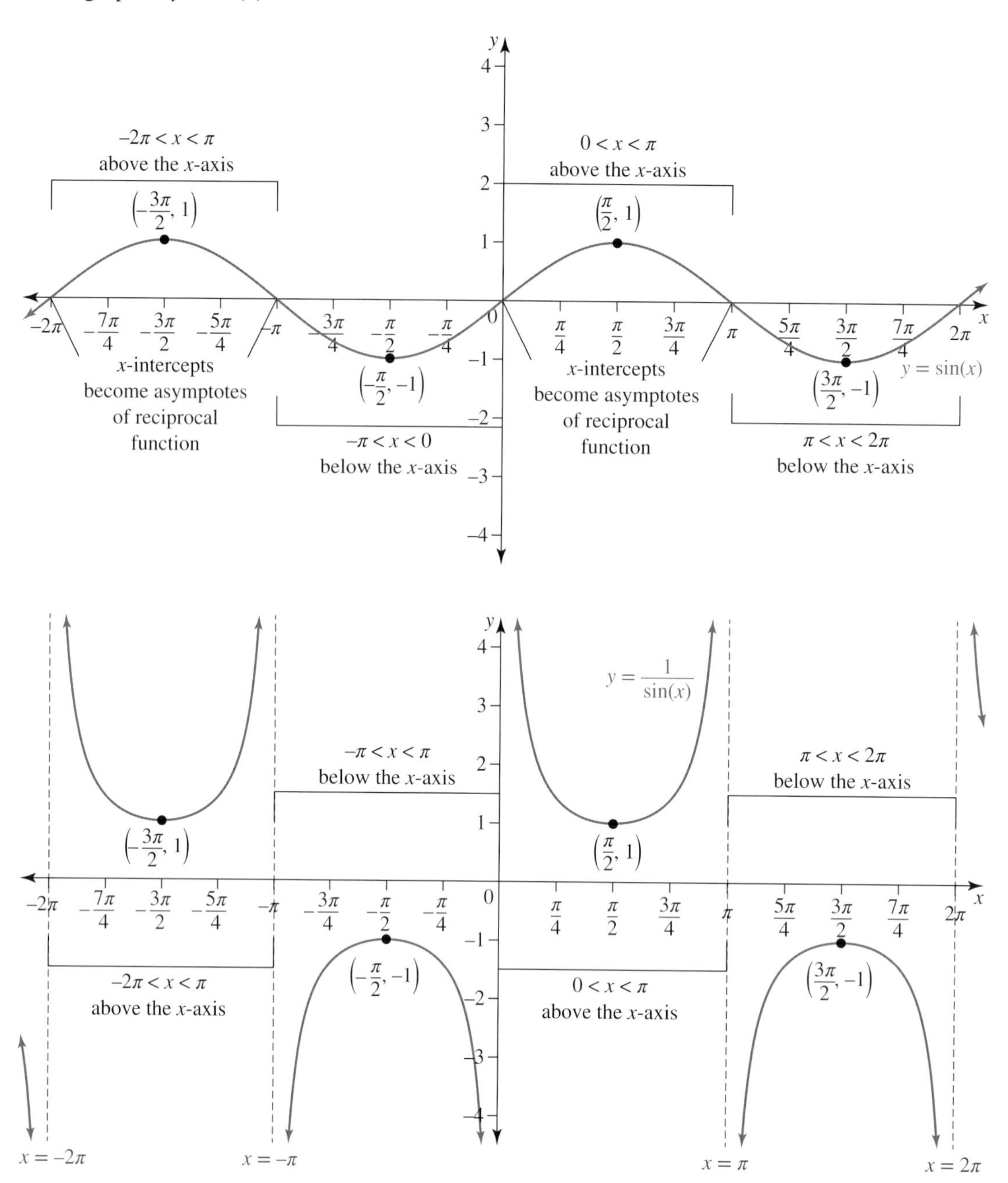

WORKED EXAMPLE 15 Sketching graphs of sec(x)

Use the graph of $y = 2\cos(x)$ to sketch $y = \frac{1}{2}\sec(x)$ over the domain $-2\pi \le x \le 2\pi$.

THINK	WRITE
1. Rewrite $y = \frac{1}{2}\sec(x)$ in terms of $\cos(x)$.	$y = \frac{1}{2}\sec(x)$ $= \frac{1}{2} \times \frac{1}{\cos(x)}$ $= \frac{1}{2\cos(x)}$

2. Sketch $y = 2\cos(x)$ by identifying the amplitude, period, horizontal shift and vertical shift. Identify the end points at $x = -2\pi$ and $x = 2\pi$.

$y = 2\cos(x)$
Amplitude: 2 Period: 2π
Horizontal shift: 0 Vertical shift: 0
$2\cos(-2\pi) = 2\cos(2\pi) = 2$

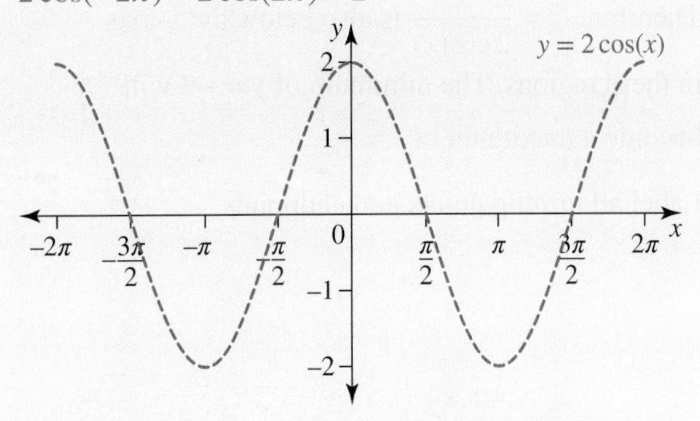

3. Determine the x-intercepts and hence the vertical asymptotes for the reciprocal graph.

The x-intercepts occur at $x = -\frac{3\pi}{2}$, $x = -\frac{\pi}{2}$, $x = \frac{\pi}{2}$ and $x = \frac{3\pi}{2}$. These will be the vertical asymptotes for the reciprocal function.

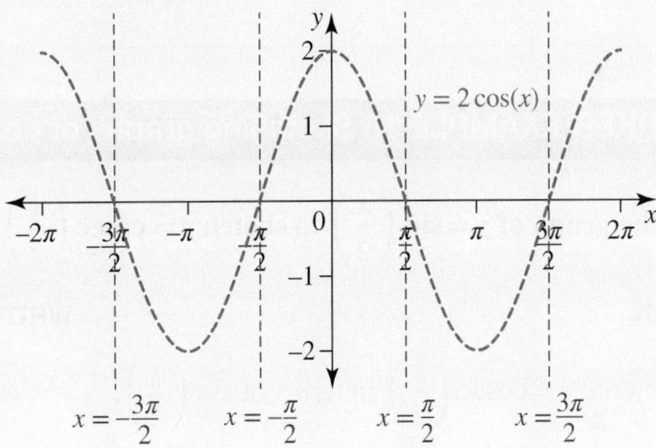

4. The graph of $y = 2\cos(x)$ is above the x-axis in the regions $-2\pi \le x < -\frac{-3\pi}{2}$, $-\frac{\pi}{2} < x < \frac{\pi}{2}$ and $\frac{3\pi}{2} < x \le 2\pi$. The graph of $y = \frac{1}{2\cos(x)}$ will also be above the x-axis in these regions.
A maximum value of $y = 2$ is reached in the original graph, meaning that a minimum of $y = \frac{1}{2}$ will be reached in the reciprocal function.

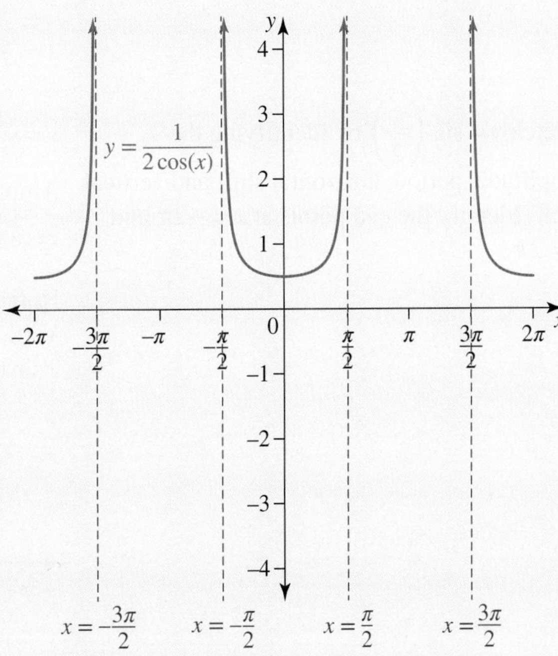

5. The graph of $y = 2\cos(x)$ is below the x-axis in the regions $-\frac{3\pi}{3} < x < -\frac{\pi}{2}$ and $\frac{\pi}{2} < x < \frac{3\pi}{2}$.
Therefore, $y = \frac{1}{2\cos(x)}$ is also below the x-axis in these regions. The minimum of $y = -2$ will become a maximum of $y = -\frac{1}{2}$.
6. Label all turning points and endpoints.

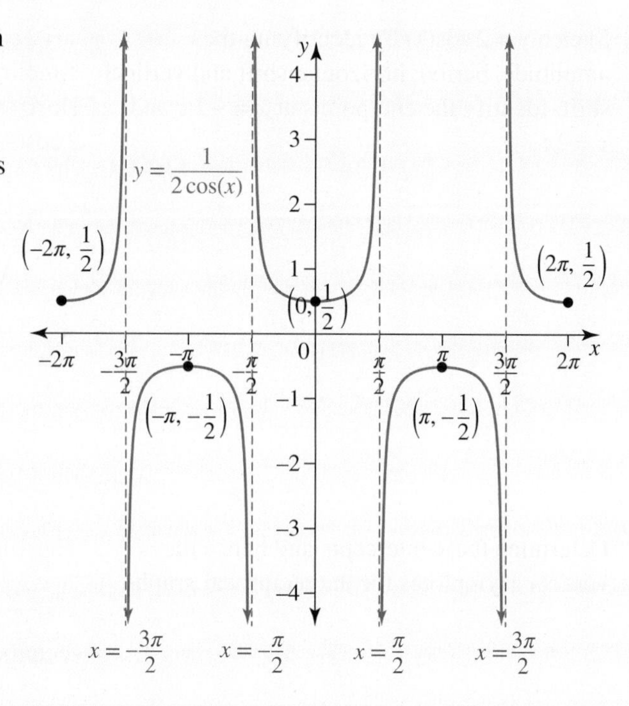

WORKED EXAMPLE 16 Sketching graphs of cosec(x)

Use the graph of $y = \sin\left(\frac{x}{2}\right)$ to sketch $y = \text{cosec}\left(\frac{x}{2}\right)$ over the domain $-2\pi \le x \le 2\pi$.

THINK

1. Rewrite $y = \text{cosec}\left(\frac{x}{2}\right)$ in terms of $\sin\left(\frac{x}{2}\right)$.

WRITE

$$y = \text{cosec}\left(\frac{x}{2}\right)$$
$$= \frac{1}{\sin\left(\frac{x}{2}\right)}$$

2. Sketch $y = \sin\left(\frac{x}{2}\right)$ by identifying the amplitude, period, horizontal shift and vertical shift. Identify the end points at $x = -2\pi$ and $x = 2\pi$.

$y = \sin\left(\frac{x}{2}\right)$
Amplitude: 1
Period: $\frac{2\pi}{\frac{1}{2}} = 4\pi$
Horizontal shift: 0
Vertical shift: 0
$2\sin(-2\pi) = 2\sin(2\pi) = 0$

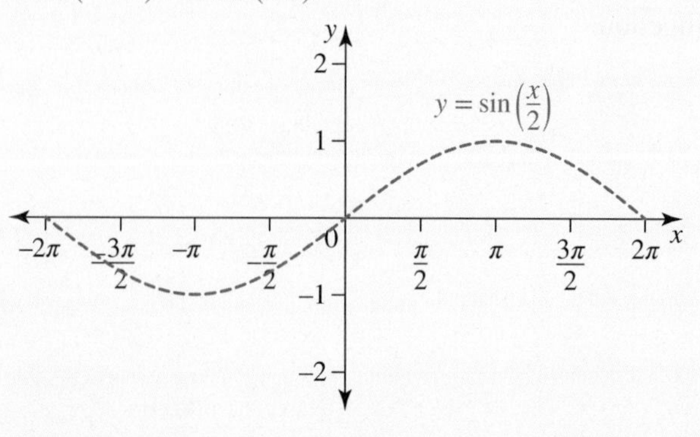

3. Determine the x-intercepts and hence the vertical asymptotes for the reciprocal graph.

The x-intercepts occur at $x = -2\pi$, $x = 0$ and $x = 2\pi$. These will be the vertical asymptotes for the reciprocal function.

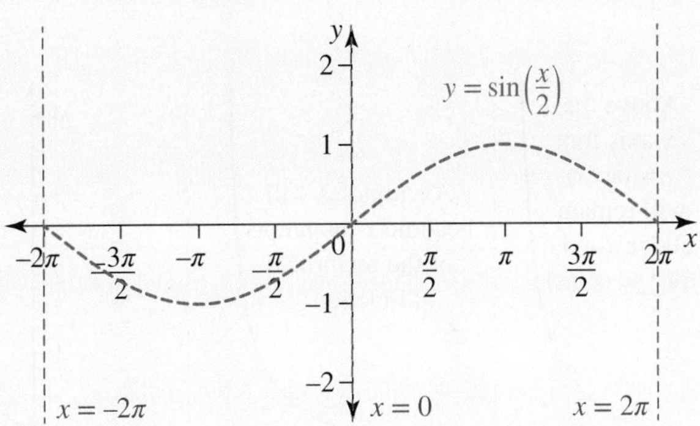

4. The graph of $y = \sin\left(\frac{x}{2}\right)$ is above the x-axis in the region $0 < x \leq 2\pi$. The graph of $y = \frac{1}{\sin\left(\frac{x}{2}\right)}$ will also be above the x-axis in these regions.
A maximum value of $y = 1$ is reached in the original graph, meaning that a minimum of $y = 1$ will be reached in the reciprocal function.

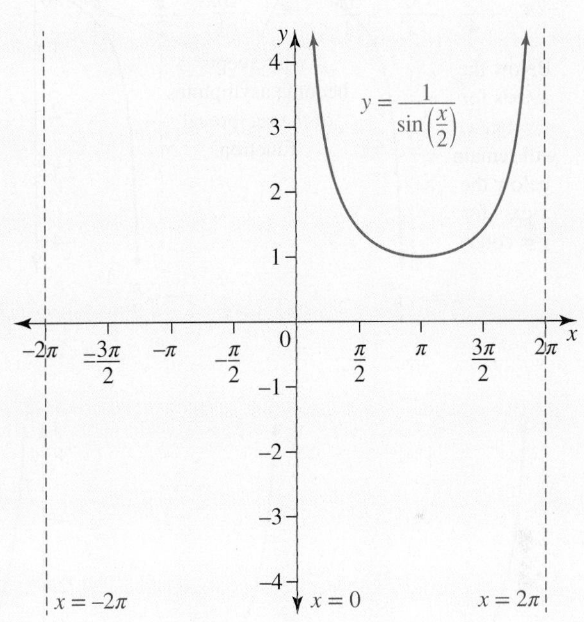

5. The graph of $y = \sin\left(\frac{x}{2}\right)$ is below the x-axis in the region $-2\pi \leq x < 0$. Therefore, $y = \frac{1}{\sin\left(\frac{x}{2}\right)}$ is also below the x-axis in these regions. The minimum of $y = -1$ will become a maximum of $y = -1$.
6. Label all turning points.

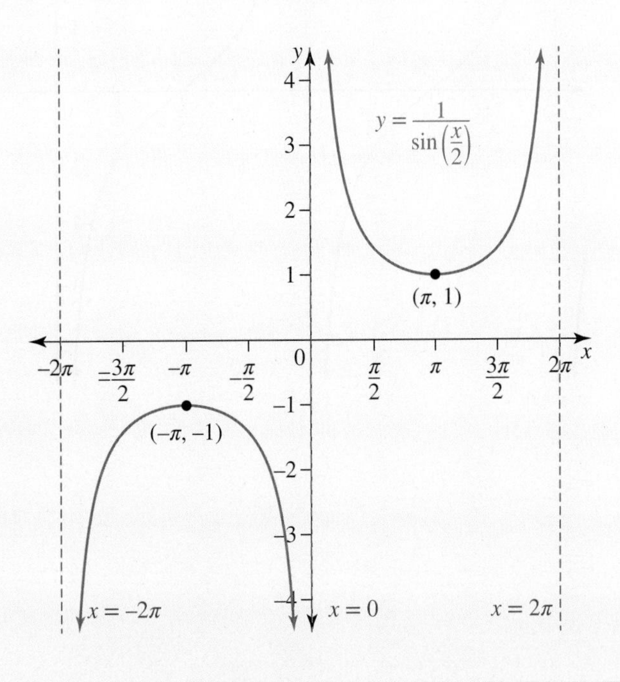

The graph of $y = \cot(x)$

The graph of $y = \tan(x)$ can also be used to sketch the graph of $y = \dfrac{1}{\tan(x)}$ (also known as $y = \cot(x)$).

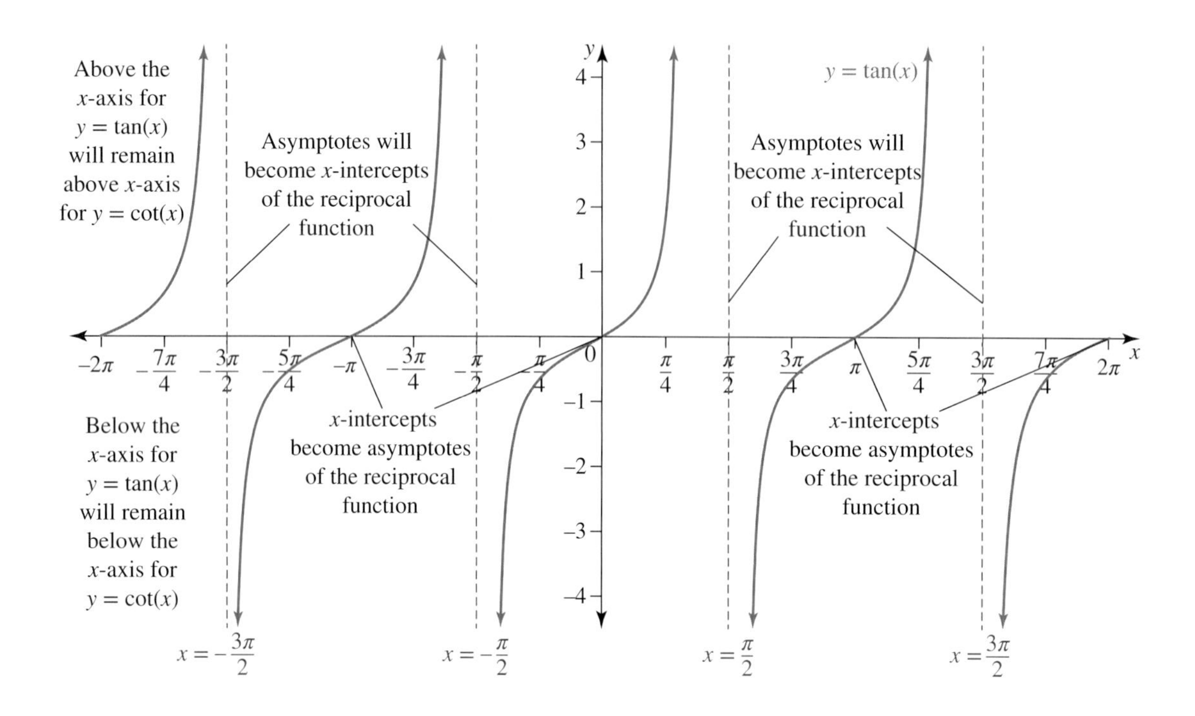

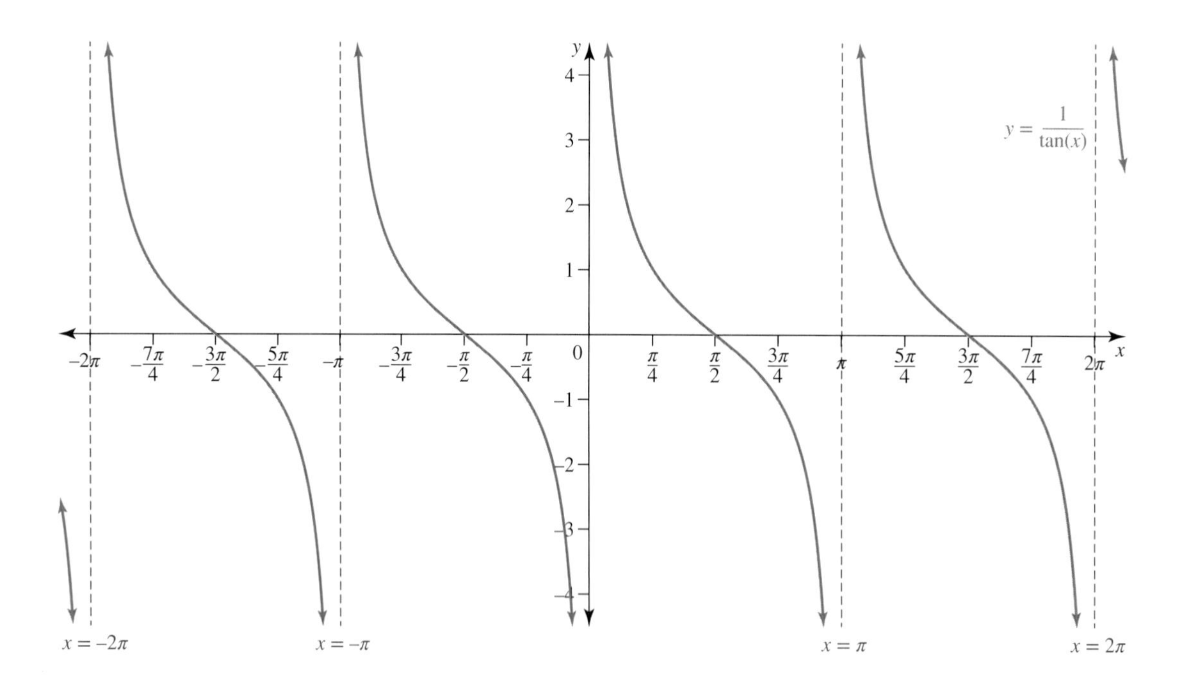

WORKED EXAMPLE 17 Sketching graphs of cot(x)

Use the graph of $y = \tan\left(\frac{x}{2}\right)$ to sketch $y = \cot\left(\frac{x}{2}\right)$ over the domain $-2\pi \le x \le 2\pi$.

THINK	WRITE
1. Rewrite $y = \cot\left(\frac{x}{2}\right)$ in terms of $\tan\left(\frac{x}{2}\right)$.	$y = \cot\left(\frac{x}{2}\right)$ $= \dfrac{1}{\tan\left(\frac{x}{2}\right)}$
2. Sketch $y = \tan\left(\frac{x}{2}\right)$ by identifying the period, horizontal shift and vertical shift. Identify the end points at $x = -2\pi$ and $x = 2\pi$.	$y = \tan\left(\frac{x}{2}\right)$ Period: $\dfrac{\pi}{\frac{1}{2}} = 2\pi$ Horizontal shift: 0 Vertical shift: 0 $\tan(-2\pi) = \tan(2\pi) = 0$ 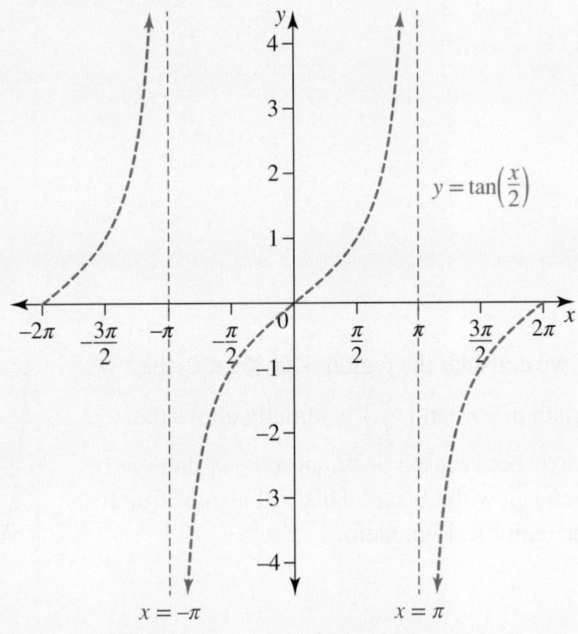
3. The asymptotes of $y = \tan\left(\frac{x}{2}\right)$ will become the x-intercepts of $y = \cot\left(\frac{x}{2}\right)$.	The x-intercepts of $y = \cot\left(\frac{x}{2}\right)$ will be $x = \pm\pi$.

4. Determine the x-intercepts and hence the vertical asymptotes for the reciprocal graph.

The x-intercepts of $y = \tan\left(\frac{x}{2}\right)$ are $x = -2\pi$, $x = 0$ and $x = 2\pi$. These will become the vertical asymptotes of $y = \cot\left(\frac{x}{2}\right)$.

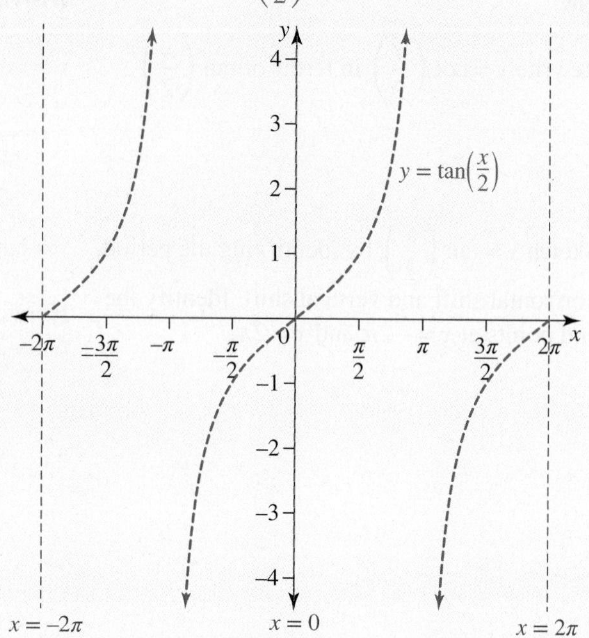

5. If we consider the region $-2\pi \le x < 0$, the graph of $y = \tan\left(\frac{x}{2}\right)$ is initially above the x-axis between $x = -2\pi$ and $x = -\pi$ and is then below the x-axis. This will also be true for the reciprocal function.

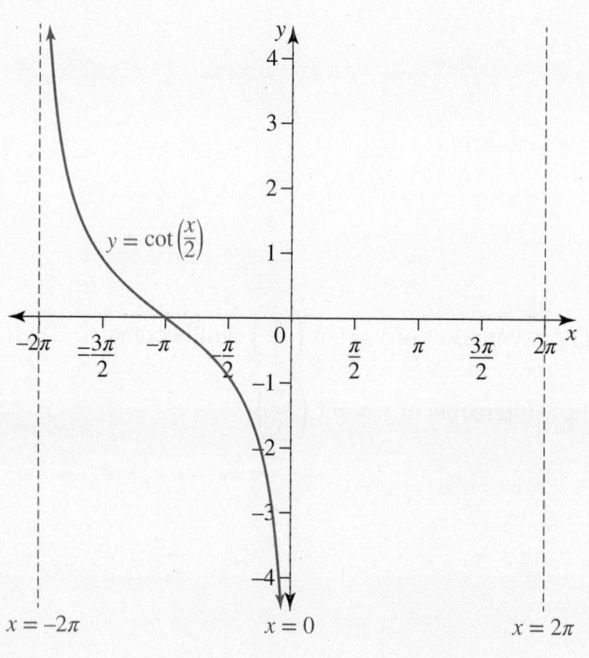

6. In a similar fashion, the graph for $x = 0$ to $x = 2\pi$ can be obtained.

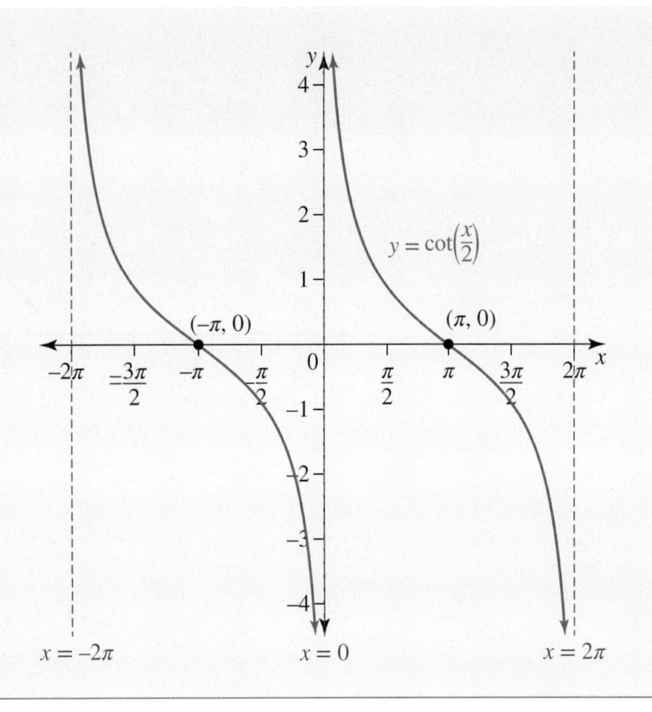

Transformations of the reciprocal functions

Transformations can also be applied to the reciprocal functions.

WORKED EXAMPLE 18 Transformations of inverse trigonometric functions

Sketch the graph of $y = \sec\left(x + \frac{\pi}{4}\right) + 1$ over the domain $[-\pi, 2\pi]$.

THINK	WRITE
1. Sketch the graph of $y = \cos\left(x + \frac{\pi}{4}\right)$ to sketch the graph of $y = \sec\left(x + \frac{\pi}{4}\right)$.	$y = \cos\left(x + \frac{\pi}{4}\right)$ Amplitude: 1 Period: 2π Horizontal shift: $\frac{\pi}{4}$ left Vertical shift: 0 End Points: $\cos\left(-\pi + \frac{\pi}{4}\right) = \cos\left(-\frac{3\pi}{4}\right) = -\frac{\sqrt{2}}{2}$ $\cos\left(2\pi + \frac{\pi}{4}\right) = \cos\left(\frac{\pi}{4}\right) = \frac{\sqrt{2}}{2}$

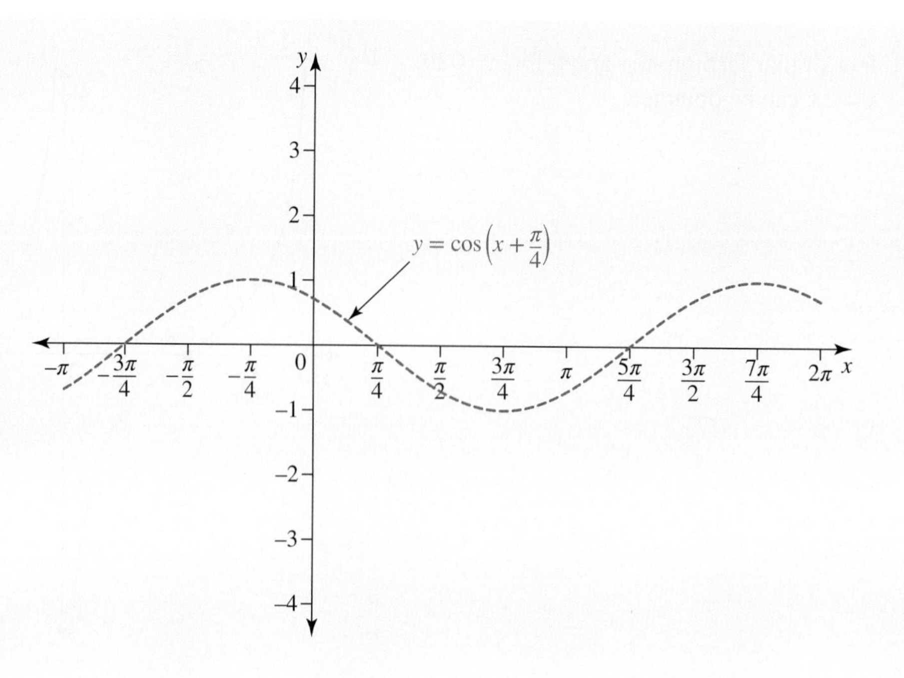

2. Now consider $y = \sec\left(x + \frac{\pi}{4}\right)$. The x-intercepts of $y = \cos\left(x + \frac{\pi}{4}\right)$ will become the asymptotes. Sections of the graph above the x-axis will remain above the x-axis, and sections below the x-axis will remain below the x-axis.

$y = \sec\left(x + \frac{\pi}{4}\right)$

The asymptotes will be $x = -\frac{\pi}{4}$, $x = \frac{3\pi}{4}$ and $x = \frac{7\pi}{4}$.

The graph of $y = \cos\left(x + \frac{\pi}{4}\right)$ and $y = \sec\left(x + \frac{\pi}{4}\right)$:

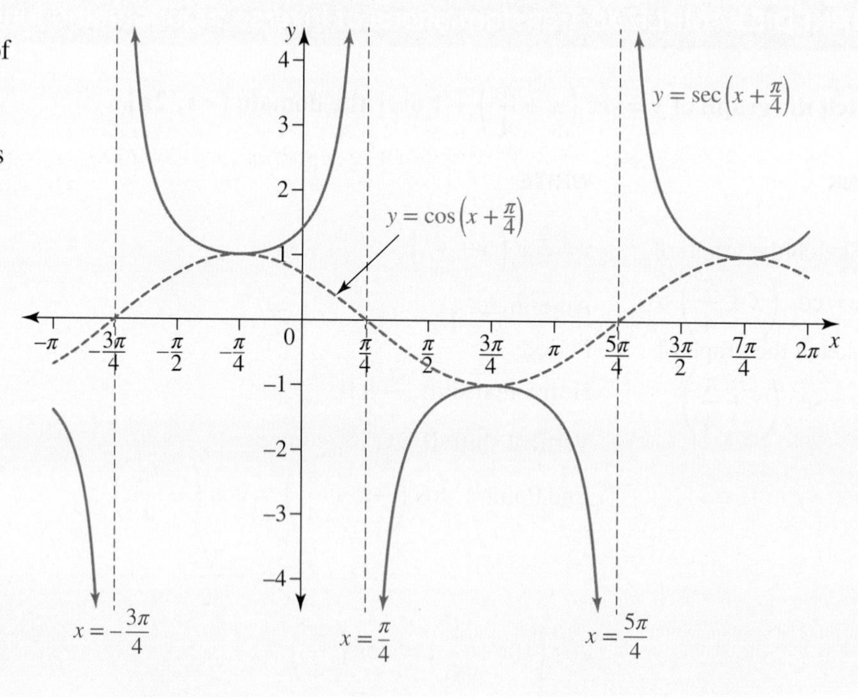

3. To graph $y = \sec\left(x + \frac{\pi}{4}\right) + 1$, move $y = \sec\left(x + \frac{\pi}{4}\right)$ up 1.

4. Label all turning points, axis intercepts and endpoints.

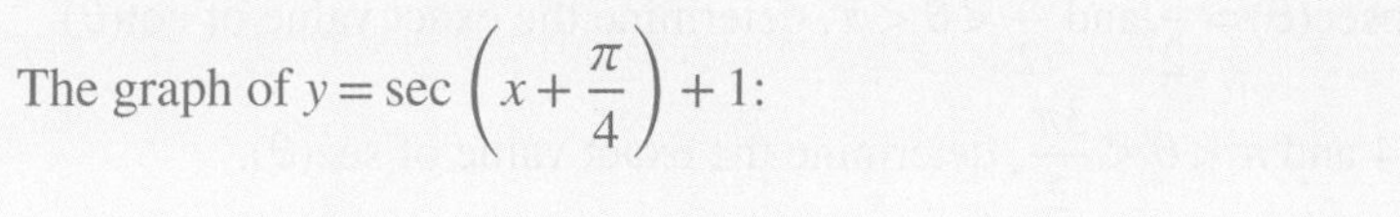

The graph of $y = \sec\left(x + \frac{\pi}{4}\right) + 1$:

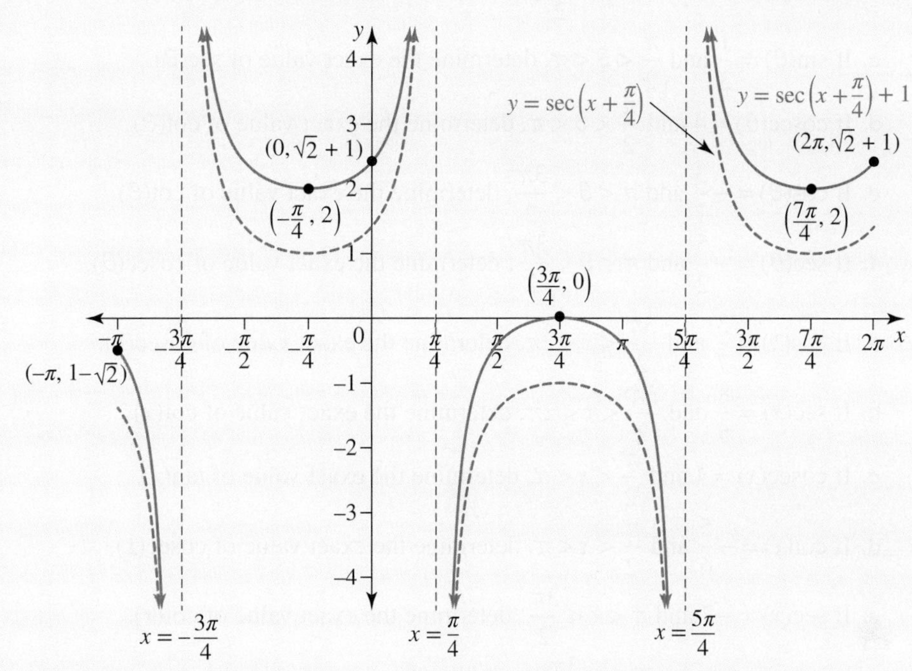

12.5 Exercise

Students, these questions are even better in jacPLUS

Receive immediate feedback and access sample responses

Access additional questions

Track your results and progress

Find all this and MORE in jacPLUS

Technology free

1. **WE13** Determine the exact value of each of the following.

a. $\text{cosec}\left(\frac{2\pi}{3}\right)$ b. $\text{cosec}\left(\frac{\pi}{3}\right)$ c. $\text{cosec}\left(\frac{5\pi}{6}\right)$ d. $\text{cosec}\left(\frac{7\pi}{4}\right)$

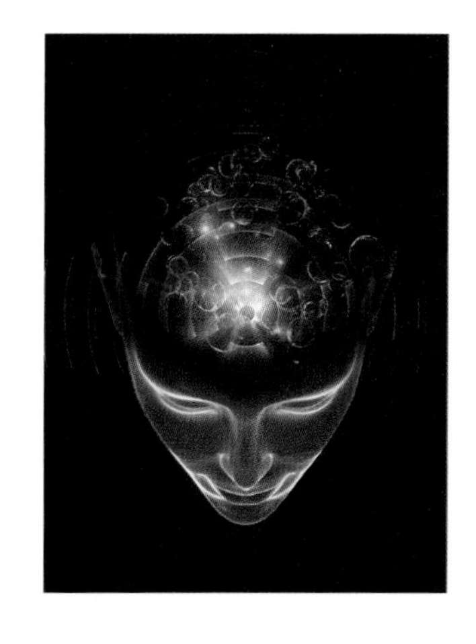

2. Determine the exact value of each of the following.

a. $\sec\left(\frac{\pi}{6}\right)$ b. $\sec\left(-\frac{7\pi}{6}\right)$ c. $\sec\left(\frac{4\pi}{3}\right)$ d. $\sec\left(-\frac{7\pi}{4}\right)$

3. Determine the exact value of each of the following.

a. $\cot\left(\frac{\pi}{6}\right)$ b. $\cot\left(\frac{2\pi}{3}\right)$ c. $\cot\left(\frac{7\pi}{4}\right)$ d. $\cot\left(\frac{11\pi}{6}\right)$

4. a. **WE14** If $\operatorname{cosec}(\theta) = \frac{5}{2}$ and $\frac{\pi}{2} < \theta < \pi$, determine the exact value of $\cot(\theta)$.

b. If $\cot(\theta) = 4$ and $\pi < \theta < \frac{3\pi}{2}$, determine the exact value of $\sec(\theta)$.

c. If $\sin(\theta) = \frac{1}{3}$ and $\frac{\pi}{2} < \theta < \pi$, determine the exact value of $\sec(\theta)$.

d. If $\operatorname{cosec}(\theta) = 4$ and $\frac{\pi}{2} < \theta < \pi$, determine the exact value of $\cot(\theta)$.

e. If $\cos(\theta) = -\frac{3}{7}$ and $\pi < \theta < \frac{3\pi}{2}$, determine the exact value of $\cot(\theta)$.

f. If $\sec(\theta) = -\frac{5}{2}$ and $\pi < \theta < \frac{3\pi}{2}$, determine the exact value of $\operatorname{cosec}(\theta)$.

5. a. If $\cos(x) = \frac{3}{7}$ and $\frac{3\pi}{2} < x < 2\pi$, determine the exact value of $\operatorname{cosec}(x)$.

b. If $\sec(x) = \frac{8}{5}$ and $\frac{3\pi}{2} < x < 2\pi$, determine the exact value of $\cot(x)$.

c. If $\operatorname{cosec}(x) = 4$ and $\frac{\pi}{2} < x < \pi$, determine the exact value of $\tan(x)$.

d. If $\cot(x) = -\frac{5}{6}$ and $\frac{\pi}{2} < x < \pi$, determine the exact value of $\operatorname{cosec}(x)$.

e. If $\sec(x) = -7$ and $\pi < x < \frac{3\pi}{2}$, determine the exact value of $\cot(x)$.

f. If $\cot(x) = 4$ and $\pi < x < \frac{3\pi}{2}$, determine the exact value of $\operatorname{cosec}(x)$.

6. a. **WE15** Use the graph of $y = 4\cos(x)$ to sketch $y = \frac{1}{4}\sec(x)$ over the domain $-2\pi \le x \le 2\pi$.

b. Use the graph of $y = 2\sin(x)$ to sketch $y = \frac{1}{2}\operatorname{cosec}(x)$ over the domain $-2\pi \le x \le 2\pi$.

7. Sketch $y = \frac{1}{4}\operatorname{cosec}(x)$ over the domain $[-\pi, \pi]$.

8. a. **WE16** Use the graph of $y = \sin(2x)$ to sketch $y = \operatorname{cosec}(2x)$ over the domain $-2\pi \le x \le 2\pi$.
b. Use the graph of $y = \cos(2x)$ to sketch $y = \sec(2x)$ over the domain $-2\pi \le x \le 2\pi$.

9. Sketch $y = \sec\left(\frac{x}{2}\right)$ over the domain $[-\pi, \pi]$.

10. a. **WE17** Use the graph of $y = \tan(2x)$ to sketch $y = \cot(2x)$ over the domain $-2\pi \le x \le 2\pi$.
b. Use the graph of $y = \tan(3x)$ to sketch $y = \cot(3x)$ over the domain $-2\pi \le x \le 2\pi$.

11. Sketch $y = \cot\left(\frac{x}{3}\right)$ over the domain $[-3\pi, 3\pi]$.

Technology active

12. a. **WE18** Sketch the graph of $y = \frac{1}{2}\sec\left(x + \frac{\pi}{4}\right) - 1$ over the domain $[-\pi, 2\pi]$.

b. Sketch the graph of $y = \cot\left(x + \frac{\pi}{4}\right) + 1$ over the domain $[-\pi, 2\pi]$.

13. Sketch the following over the domain of $[-\pi, \pi]$.

a. $y = 2\sec(x) - 1$

b. $y = \dfrac{2}{\sin\left(x + \frac{\pi}{4}\right)}$

c. $y = 0.25\,\text{cosec}\left(x - \dfrac{\pi}{4}\right)$

d. $y = 3\sec\left(2x + \dfrac{\pi}{2}\right) - 2$

14. a. If $\text{cosec}(x) = \dfrac{p}{q}$ where $p, q \in R^+$ and $\dfrac{\pi}{2} < x < \pi$, evaluate $\sec(x) - \cot(x)$ in terms of p and q.

b. If $\sec(x) = \dfrac{a}{b}$ where $a, b \in R^+$ and $\dfrac{3\pi}{2} < x < 2\pi$, evaluate $\cot(x) - \text{cosec}(x)$ in terms of p and q.

15. Use the graph of $y = \sin(x) + 2$ to sketch $y = \dfrac{1}{\sin(x) + 2}$ over the domain $\left[-\dfrac{5\pi}{2}, \dfrac{5\pi}{2}\right]$.

16. a. Use the graph of $y = \cos^2(x)$ to sketch $y = \dfrac{1}{\cos^2(x)}$ over the domain $\left[-\dfrac{3\pi}{2}, \dfrac{3\pi}{2}\right]$. Sketch both graphs on the same set of axes.

b. Hence, determine the graph of $y = \tan^2(x)$ for the same domain.

12.5 Exam questions

Question 1 (3 marks) TECH-FREE

Sketch the graph of the function $f: [-3\pi, 3\pi] \to R,\ f(x) = 2\cot\left(\dfrac{x}{2}\right)$.

Question 2 (3 marks) TECH-FREE

Sketch the graph of the function $f: (-3\pi, 3\pi) \to R,\ f(x) = 3\sec\left(\dfrac{x}{2}\right)$.

Question 3 (3 marks) TECH-FREE

Sketch the graph of the function $f: [-3\pi, 3\pi] \to R,\ f(x) = 2\,\text{cosec}\left(\dfrac{x}{2}\right)$.

More exam questions are available online.

12.6 Inverse trigonometric functions

LEARNING INTENTION

At the end of this subtopic you should be able to:
- sketch inverse trigonometric functions
- evaluate angles using the inverse trigonometric functions.

12.6.1 Inverse trigonometric functions

All circular functions are periodic and are many-to-one functions. The inverses of these functions therefore cannot not be functions. However, by restricting the domain so that the circular functions are one-to-one functions, then the inverses are functions.

The inverse sine function

The sine function, $y = \sin(x)$ is a many-to-one function, so its inverse does not exist as a function.

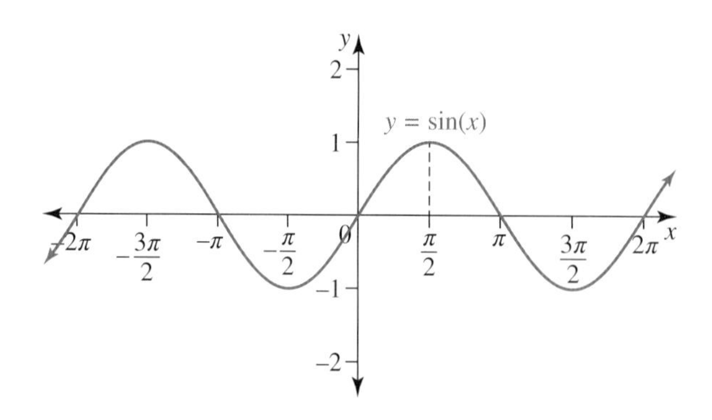

However there are many restrictions of the domain which make the sine function one-to-one, such as $\left[-\frac{3\pi}{2}, -\frac{\pi}{2}\right]$, $\left[-\frac{\pi}{2}, \frac{\pi}{2}\right]$ or $\left[\frac{\pi}{2}, \frac{3\pi}{2}\right]$. By convention, we agree that $\left[-\frac{\pi}{2}, \frac{\pi}{2}\right]$ will be the domain of the restricted sine function, and $[-1, 1]$ is the range.

$$f: \left[-\frac{\pi}{2}, \frac{\pi}{2}\right] \to [-1, 1],\ f(x) = \sin(x)$$

And the inverse function will be $f: [-1, 1] \to \left[-\frac{\pi}{2}, \frac{\pi}{2}\right]$, $f(x) = \sin^{-1}(x)$

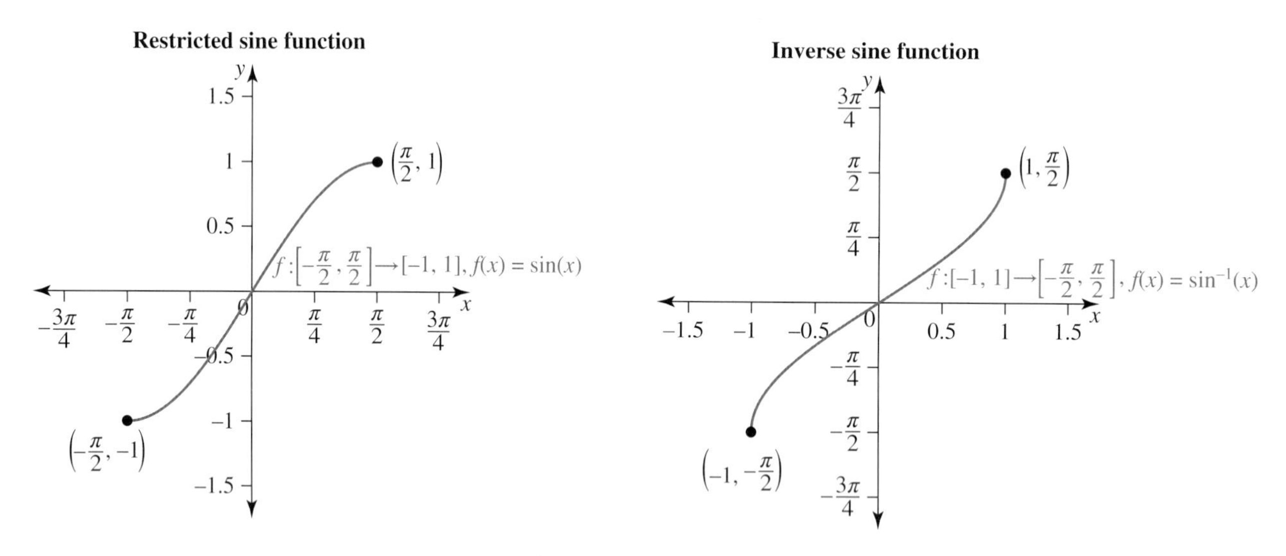

The inverse function can be written as $y = \sin^{-1}(x)$ or $y = \arcsin(x)$. Be careful not to confuse $y = \sin^{-1}(x)$ and $y = \frac{1}{\sin(x)}$.

Although there are an infinite number of solutions to $\sin(x) = \frac{1}{2}$, there is only one solution to $x = \sin^{-1}\left(\frac{1}{2}\right)$ because $x \in \left[-\frac{\pi}{2}, \frac{\pi}{2}\right]$, so $\sin^{-1}\left(\frac{1}{2}\right) = \frac{\pi}{6}$. Note this is also the solution that your calculator returns.

Properties of the inverse sine function

For the inverse sine function $f: [-1, 1] \rightarrow \left[-\frac{\pi}{2}, \frac{\pi}{2}\right], f(x) = \sin^{-1}(x)$,

$$\sin\left(\sin^{-1}(x)\right) = x, \quad x \in [-1, 1]$$
$$\sin^{-1}\left(\sin(x)\right) = x, \quad x \in \left[-\frac{\pi}{2}, \frac{\pi}{2}\right]$$

WORKED EXAMPLE 19 Evaluating inverse sine expressions

Evaluate each of the following.

a. $\sin^{-1}(2)$

b. $\sin^{-1}\left(\sin\left(\frac{5\pi}{6}\right)\right)$

c. $\sin\left(\sin^{-1}(0.5)\right)$

THINK	WRITE
a. 1. Write an equivalent statement.	a. Let $x = \sin^{-1}(2)$ $\sin(x) = 2$
2. The range of $\sin(x)$ is $[-1, 1]$ so there is no solution to $\sin(x) = 2$.	As $2 \notin [-1, 1]$ there is no solution to $\sin(x) = 2$.
b. 1. Write an equivalent statement.	b. Let $x = \sin^{-1}\left(\frac{5\pi}{6}\right)$ Since $\sin\left(\frac{5\pi}{6}\right) = \frac{1}{2}$, $x = \sin^{-1}\left(\frac{1}{2}\right)$ $\sin(x) = \frac{1}{2}$
2. $\sin(x) = \frac{1}{2}$ and $x \in \left[-\frac{\pi}{2}, \frac{\pi}{2}\right]$.	The only solution is $x = \frac{\pi}{6}$ $\sin^{-1}\left(\sin\left(\frac{5\pi}{6}\right)\right) = \frac{\pi}{6}$
c. Recall that $\sin\left(\sin^{-1}(x)\right) = x$ if $x \in [-1, 1]$.	c. $\sin\left(\sin^{-1}(0.5)\right) = 0.5$

The inverse cosine function

The cosine function, $y = \cos(x)$ is a many-to-one function, so its inverse does not exist as a function.

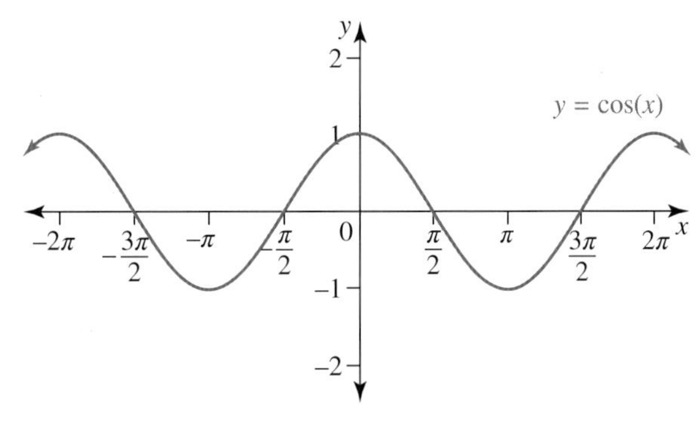

However there are many restrictions of the domain which make the cosine function one-to-one, such as $[-\pi, 0]$, $[0, \pi]$ or $[\pi, 2\pi]$. By convention, we agree that $[0, \pi]$ will be the domain of the restricted cosine function, and $[-1, 1]$ is the range.

$$f: [0, \pi] \to [-1, 1],\ f(x) = \cos(x)$$

And the inverse function will be $f: [-1, 1] \to [0, \pi],\ f(x) = \cos^{-1}(x)$.

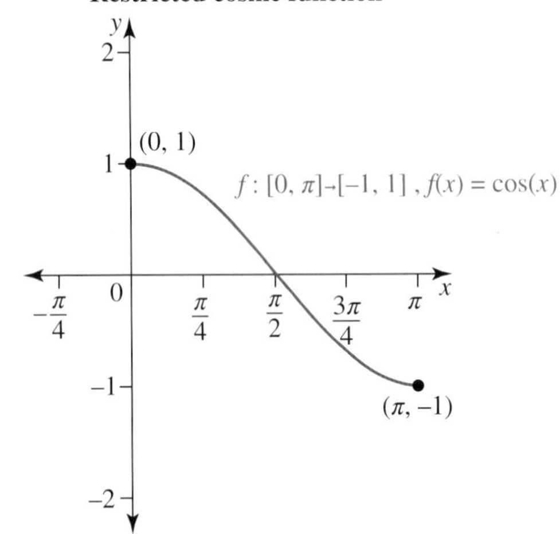

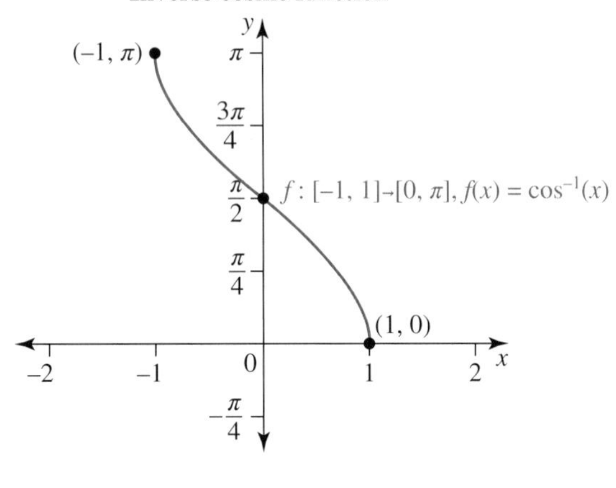

The inverse function can be written as $y = \cos^{-1}(x)$ or $y = \arccos(x)$. Be careful not to confuse $y = \cos^{-1}(x)$ and $y = \dfrac{1}{\cos(x)}$.

Although there are an infinite number of solutions to $\cos(x) = \dfrac{\sqrt{2}}{2}$, there is only one solution to $x = \cos^{-1}\left(\dfrac{\sqrt{2}}{2}\right)$ because $x \in [0, \pi]$, so $\cos^{-1}\left(\dfrac{\sqrt{2}}{2}\right) = \dfrac{\pi}{4}$. Note this is also the solution that your calculator returns.

Properties of the inverse cosine function

For the inverse cosine function $f: [-1, 1] \to [0, \pi],\ f(x) = \cos^{-1}(x)$,

$$\cos\left(\cos^{-1}(x)\right) = x, \quad x \in [-1, 1]$$

$$\cos^{-1}\left(\cos(x)\right) = x, \quad x \in [0, \pi]$$

WORKED EXAMPLE 20 Evaluating inverse cosine expressions

Evaluate each of the following.

a. $\cos^{-1}\left(\dfrac{3}{2}\right)$

b. $\cos^{-1}\left(\cos\left(\dfrac{5\pi}{4}\right)\right)$

c. $\cos\left(\cos^{-1}\left(\dfrac{\pi}{12}\right)\right)$

THINK	WRITE
a. 1. Write an equivalent statement.	**a.** Let $x = \cos^{-1}\left(\frac{3}{2}\right)$ $\cos(x) = \frac{3}{2}$
2. The range of $\cos(x)$ is $[-1, 1]$ so there is no solution to $\cos^{-1}\left(\frac{3}{2}\right)$.	As $\frac{3}{2} \notin [-1, 1]$ there is no solution to $\cos(x) = \frac{3}{2}$.
b. 1. Write an equivalent statement.	**b.** Let $x = \cos^{-1}\left(\cos\left(\frac{5\pi}{4}\right)\right)$ Since $\cos\left(\frac{5\pi}{4}\right) = -\frac{\sqrt{2}}{2}$ $x = \cos^{-1}\left(\cos\left(\frac{5\pi}{4}\right)\right)$ $= \cos^{-1}\left(-\frac{\sqrt{2}}{2}\right)$ $\cos(x) = -\frac{\sqrt{2}}{2}$
2. $\cos(x) = -\frac{\sqrt{2}}{2}$ and $x \in [0, \pi]$.	The only solution is $x = \frac{3\pi}{4}$ $\cos^{-1}\left(\cos\left(\frac{5\pi}{4}\right)\right) = \frac{3\pi}{4}$
c. Recall that $\cos\left(\cos^{-1}(x)\right) = x$ if $x \in [0, \pi]$.	**c.** $\cos\left(\cos^{-1}\left(\frac{\pi}{12}\right)\right) = \frac{\pi}{12}$

WORKED EXAMPLE 21 Evaluating inverse trigonometric expressions (1)

Determine the exact value of $\cos\left(\sin^{-1}\left(\frac{1}{3}\right)\right)$.

THINK	WRITE
1. The inverse trigonometric functions are angles.	Let $\theta = \sin^{-1}\left(\frac{1}{3}\right)$. $\sin(\theta) = \frac{1}{3}$
2. Draw a right-angled triangle and label the side lengths using the definition of the trigonometric ratios.	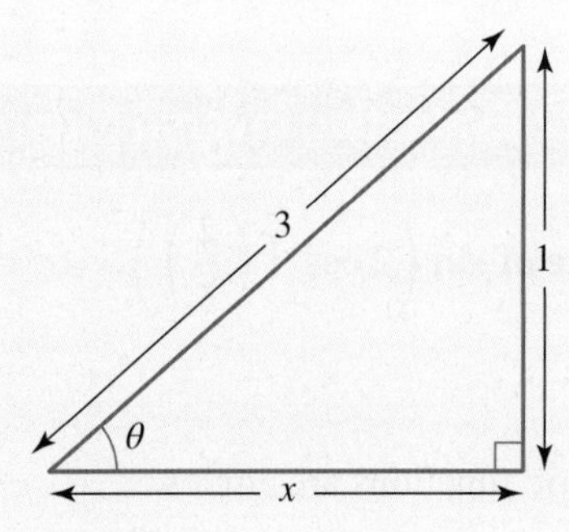

3. Calculate the value of the third side using Pythagoras' theorem.

$$x^2 + 1^2 = 3^2$$
$$x^2 + 1 = 9$$
$$x^2 = 9 - 1$$
$$x^2 = 8$$
$$x = 2\sqrt{2}$$

4. Evaluate $\cos\left(\sin^{-1}\left(\frac{1}{3}\right)\right)$.

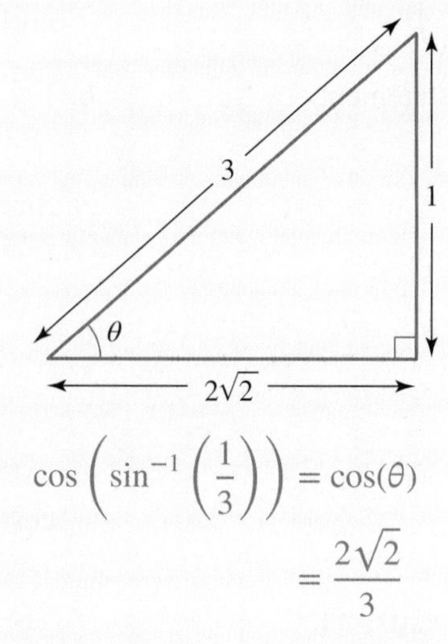

$$\cos\left(\sin^{-1}\left(\frac{1}{3}\right)\right) = \cos(\theta)$$
$$= \frac{2\sqrt{2}}{3}$$

Double angle formulae

We may need to use the double angle formulae. Recall the formulas as stated below.

Double angle formulae

$$\sin(2A) = 2\sin(A)\cos(A)$$
$$\cos(2A) = \cos^2(A) - \sin^2(A)$$
$$= 2\cos^2(A) - 1$$
$$= 1 - 2\sin^2(A)$$
$$\tan(2A) = \frac{2\tan(A)}{1 - \tan^2(A)}$$

WORKED EXAMPLE 22 Evaluating inverse trigonometric expressions (2)

Determine the exact value of $\sin\left(2\cos^{-1}\left(\frac{2}{5}\right)\right)$.

THINK

1. The inverse trigonometric functions are angles.

WRITE

Let $\theta = \cos^{-1}\left(\frac{2}{5}\right)$.

$$\cos(\theta) = \frac{2}{5}$$

2. Draw a right angle triangle and label the side lengths using the definition of the trigonometric ratios.

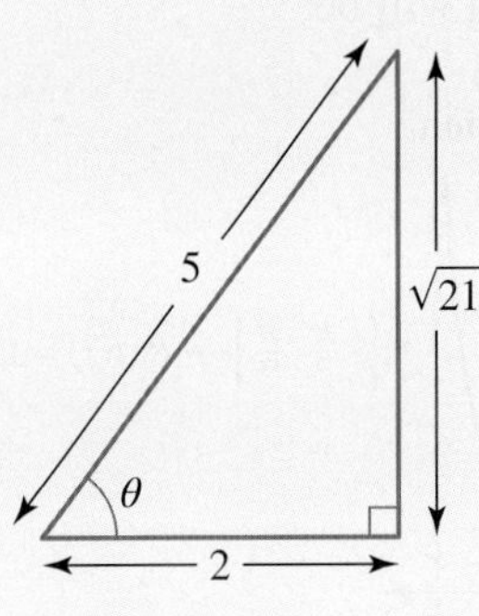

3. Calculate the value of the third side using Pythagoras' theorem.

$$x^2 + 2^2 = 5^2$$
$$x^2 + 4 = 25$$
$$x^2 = 25 - 4$$
$$x^2 = 21$$
$$x = \sqrt{21}$$

4. Use an appropriate double angle formula to evaluate $\sin\left(2\cos^{-1}\left(\frac{2}{5}\right)\right)$.

$$\begin{aligned}\sin\left(2\cos^{-1}\left(\frac{2}{5}\right)\right) &= \sin(2\theta)\\ &= 2\sin(\theta)\cos(\theta)\\ &= 2\times\frac{\sqrt{21}}{5}\times\frac{2}{5}\\ &= \frac{4\sqrt{21}}{5}\end{aligned}$$

The inverse tangent function

The tangent function, $y = \tan(x)$ is a many-to-one function, so its inverse does not exist as a function.

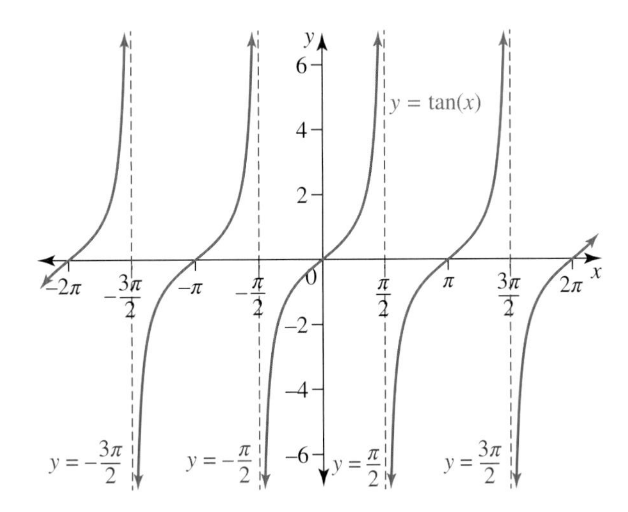

However there are many restrictions of the domain which make the tangent function one-to-one, such as $\left(-\frac{3\pi}{2}, -\frac{\pi}{2}\right), \left(-\frac{\pi}{2}, \frac{\pi}{2}\right)$ or $\left(\frac{\pi}{2}, \frac{3\pi}{2}\right)$. By convention, we agree that $\left(-\frac{\pi}{2}, \frac{\pi}{2}\right)$ will be the domain of the restricted tangent function, and R is the range. Note that we must have an open interval, since the function is not defined at $x = \pm\frac{\pi}{2}$, at these points we have vertical asymptotes.

$$f: \left(-\frac{\pi}{2}, \frac{\pi}{2}\right) \to R, f(x) = \tan(x)$$

And the inverse function will be:

Restricted tangent function

$f:\left[-\frac{\pi}{2}, \frac{\pi}{2}\right] \rightarrow R, f(x) = \tan(x)$

$x = -\frac{\pi}{2}$ $x = \frac{\pi}{2}$

Inverse tangent function

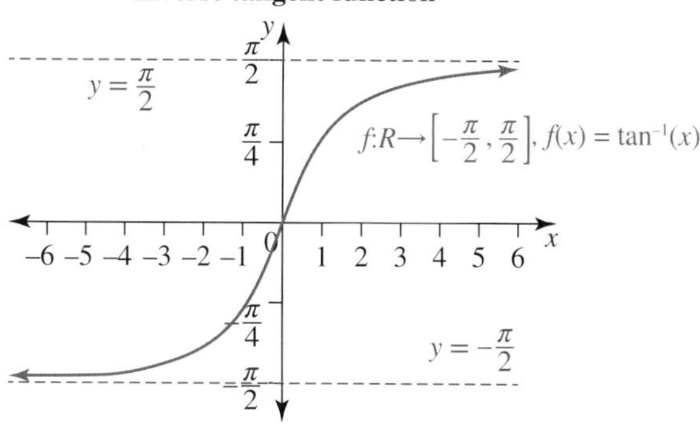

Notice that the inverse tangent function has horizontal asymptotes at $y = \pm\frac{\pi}{2}$.

The inverse function can be written as $y = \tan^{-1}(x)$ or $y = \arctan(x)$. Be careful not to confuse $y = \tan^{-1}(x)$ and $y = \frac{1}{\tan(x)}$.

Although there are an infinite number of solutions to $\tan(x) = 1$, there is only one solution to $x = \tan^{-1}(1)$ because $x \in \left(-\frac{\pi}{2}, \frac{\pi}{2}\right)$, so $\tan^{-1}(1) = \frac{\pi}{4}$. Note this is also the solution that your calculator returns.

Properties of the inverse tangent function

For the inverse tangent function $f: R \rightarrow \left(-\frac{\pi}{2}, \frac{\pi}{2}\right)$, $f(x) = \tan^{-1}(x)$,

$$\tan\left(\tan^{-1}(x)\right) = x, \quad x \in R$$

$$\tan^{-1}\left(\tan(x)\right) = x, \quad x \in \left(-\frac{\pi}{2}, \frac{\pi}{2}\right)$$

WORKED EXAMPLE 23 Evaluating inverse tangent expressions (1)

Evaluate each of the following.

a. $\tan\left(\tan^{-1}(2)\right)$

b. $\tan^{-1}\left(\tan\left(\frac{11\pi}{6}\right)\right)$

THINK	WRITE
a. Recall that $\tan\left(\tan^{-1}(x)\right) = x$ if $x \in R$.	a. $\tan\left(\tan^{-1}(2)\right) = 2$

b. 1. Recall the symmetry of the tangent function.	**b.** $\tan\left(\frac{11\pi}{6}\right) = \tan\left(-\frac{\pi}{6}\right)$
	$\tan^{-1}\left(\tan\left(-\frac{11\pi}{6}\right)\right) = \tan^{-1}\left(\tan\left(-\frac{\pi}{6}\right)\right)$
2. Recall that $\tan^{-1}(\tan(x)) = x$ if $x \in \left(-\frac{\pi}{2}, \frac{\pi}{2}\right)$.	$\tan^{-1}\left(\tan\left(-\frac{\pi}{6}\right)\right) = -\frac{\pi}{6}$

In some cases we may need to use the double angle formula for tan, $\tan(2A) = \frac{2\tan(A)}{1 - \tan^2(A)}$.

WORKED EXAMPLE 24 Evaluating inverse tangent expressions (2)

Determine the exact value of $\tan\left(2\tan^{-1}\left(\frac{1}{2}\right)\right)$.

THINK	WRITE
1. The inverse trigonometric functions are angles.	Let $\theta = \tan^{-1}\left(\frac{1}{2}\right)$ so that $\tan(\theta) = \frac{1}{2}$
2. Use the double angle formula to evaluate $\tan\left(2\tan^{-1}\left(\frac{1}{2}\right)\right)$.	$\tan\left(2\tan^{-1}\left(\frac{1}{2}\right)\right) = \tan(2\theta)$ $= \frac{2\tan(\theta)}{1-\tan^2(\theta)}$ $= \frac{2 \times \frac{1}{2}}{1 - \left(\frac{1}{2}\right)^2}$ $= \frac{1}{1 - \frac{1}{4}}$ $= \frac{1}{\left(\frac{3}{4}\right)}$ $= \frac{4}{3}$

12.6.2 Graphs of transformations of inverse trigonometric functions

To sketch graphs involving inverse trigonometric functions recall the domain and range of each of the functions.

Domain and range of inverse trigonometric functions

For $y = \sin^{-1}(x)$ the domain is $[-1, 1]$ and the range is $\left[-\frac{\pi}{2}, \frac{\pi}{2}\right]$.

For $y = \cos^{-1}(x)$ the domain is $[-1, 1]$ and the range is $[0, \pi]$.

For $y = \tan^{-1}(x)$ the domain is R and the range is $\left(-\frac{\pi}{2}, \frac{\pi}{2}\right)$.

For inverse trigonometric functions, that have been dilated or translated, we can apply these dilations and translations to determine the domain and range of the transformed function and sketch the graphs.

WORKED EXAMPLE 25 Sketching graphs of transformed inverse trigonometric functions

Sketch the graphs of each of the following, stating coordinates of the endpoints and axial intercepts.

a. $y = 4\sin^{-1}(x+2)$ b. $y = 6\cos^{-1}\left(\frac{3-x}{6}\right)$ c. $y = 4\tan^{-1}\left(\frac{x}{4}\right)$

THINK	WRITE
a. 1. $y=\sin^{-1}(x)$ has a domain of $[-1, 1]$.	a. $-1 \le x+2 \le 1$
2. Solve the inequality.	$-3 \le x \le -1$
3. State the domain.	$y=4\sin^{-1}(x+2)$ has a maximal domain of $-3 \le x \le -1$ or $[-3, -1]$.
4. Determine the range and the endpoints.	When $x=-3$: $y = 4\sin^{-1}(-1)$ $= 4 \times \left(-\frac{\pi}{2}\right)$ $= -2\pi$ When $x=-1$: $y = 4\sin^{-1}(1)$ $= 4 \times \frac{\pi}{2}$ $= 2\pi$ The range is therefore $[-2\pi, 2\pi]$ and the endpoints have coordinates $(-3, -2\pi)$ and $(-1, 2\pi)$.
5. Determine the axial intercepts.	The graph crosses the x-axis when $y=0$ at $(-2, 0)$. The graph does not cross the y-axis since when $x=0$, $y=4\sin^{-1}(2)$ and this does not exist.
6. Sketch the graph.	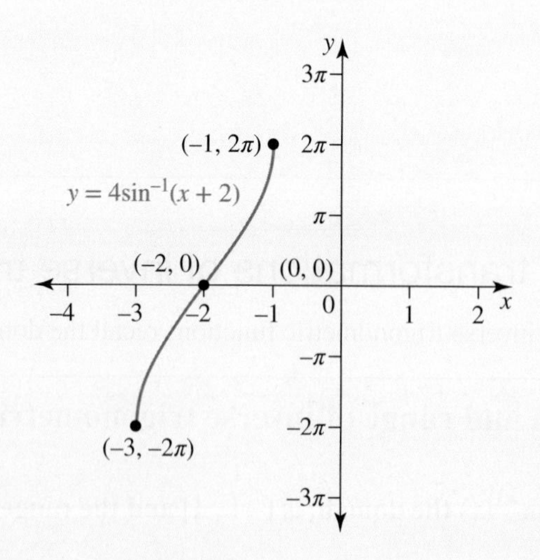
b. 1. $y=\cos^{-1}(x)$ has a domain of $[-1, 1]$.	b. $-1 \le \frac{3-x}{6} \le 1$

2. Solve the inequality.	$-6 \leq 3-x \leq 6$ $-9 \leq -x \leq 3$ $-3 \leq x \leq 9$
3. State the domain.	$y=6\cos^{-1}\left(\frac{3-x}{6}\right)$ has a maximal domain of $-3 \leq x \leq 9$ or $[-3, 9]$.
4. Determine the range and the endpoints.	When $x=-3$: $y=6\cos^{-1}(1)$ $=6\times 0$ $=0$ When $x=9$: $y=6\cos^{-1}(-1)$ $=6\times\pi$ $=6\pi$ The range is therefore $[0, 6\pi]$ and the endpoints have coordinates $(-3, 0)$ and $(9, 6\pi)$.
5. Determine the axial intercepts.	The graph crosses the x-axis when $y=0$ at $(-3,\ 0)$. The graph crosses the y-axis when $x=0$: $y=6\cos^{-1}\left(\frac{1}{2}\right)$ $=6\times\frac{\pi}{3}$ $=2\pi$
6. Sketch the graph.	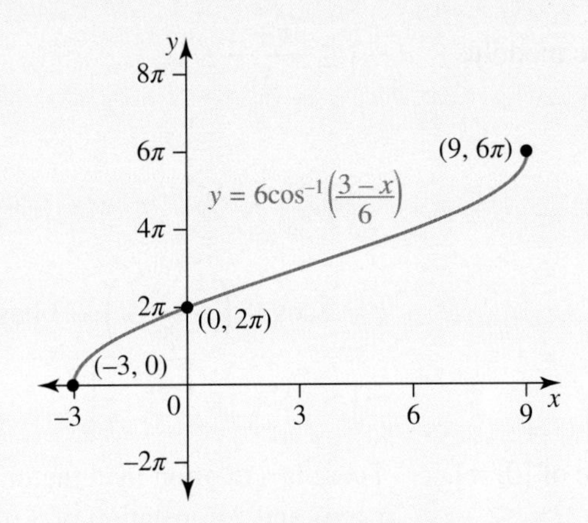

c. 1. $y=\tan^{-1}(x)$ has a domain of R.	**c.** The domain of $y=4\tan^{-1}\left(\frac{x}{4}\right)$ is R.
2. Determine the range.	The range is $4\left(-\frac{\pi}{2}, \frac{\pi}{2}\right)=(-2\pi,\ 2\pi)$,
3. Determine the equations of the asymptotes.	The graph has horizontal asymptotes at $y=\pm 2\pi$
4. Determine the axial intercept.	The graph passes through the origin.

5. Sketch the graph.

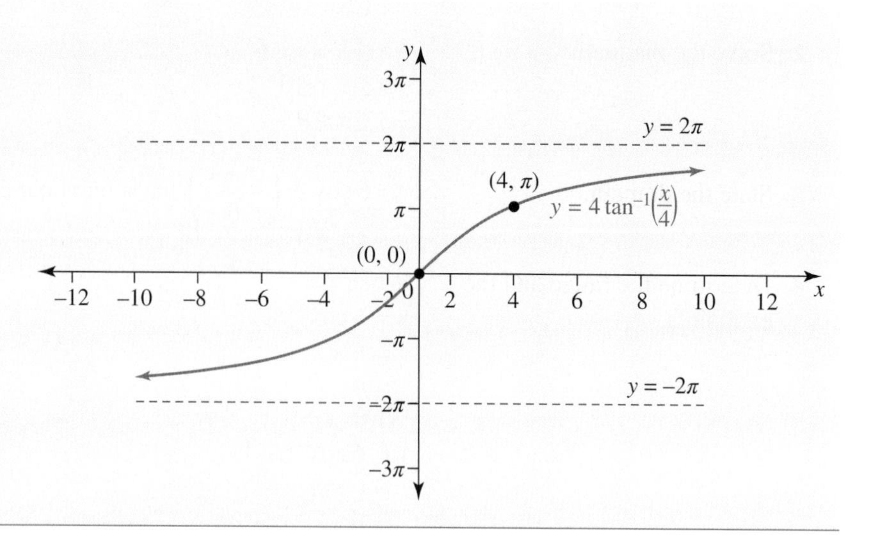

WORKED EXAMPLE 26 Stating the domain and range

State the implied domain and range of each of the following.

a. $y = 2\cos^{-1}\left(\dfrac{3x-2}{5}\right) - 3$ b. $y = 4\tan^{-1}\left(\dfrac{2x-7}{6}\right) + 1$

THINK	WRITE
a. 1. $y = \cos^{-1}(x)$ has a domain of $[-1, 1]$.	a. $\left\lvert\dfrac{3x-2}{5}\right\rvert \le 1$
2. Use the definition of the modulus function.	$-1 \le \dfrac{3x-2}{5} \le 1$
3. Solve the inequality.	$-5 \le 3x - 2 \le 5$ $-3 \le 3x \le 7$
4. State the domain.	$y = 2\cos^{-1}\left(\dfrac{3x-2}{5}\right) - 3$ has a maximal domain of $-1 \le x \le \dfrac{7}{3}$ or $\left[-1, \dfrac{7}{3}\right]$.
5. $y = \cos^{-1}(x)$ has a range of $[0, \pi]$.	There is a dilation by a factor of 2 parallel to the y-axis and a translation of 3 units down parallel to the y-axis. The range is from $2 \times 0 - 3$ to $2 \times \pi - 3$.
6. State the range.	$y = 2\cos^{-1}\left(\dfrac{4x-3}{5}\right) - 3$ has a range of $[-3, 2\pi - 3]$.
b. 1. $y = \tan^{-1}(x)$ has a domain of R.	b. $y = 4\tan^{-1}\left(\dfrac{2x-7}{6}\right) + 1$ has a domain of R.
2. $y = \tan^{-1}(x)$ has a range of $\left(-\dfrac{\pi}{2}, \dfrac{\pi}{2}\right)$.	There is a dilation by a factor of 4 parallel to the y-axis and a translation of 1 unit up parallel to the y-axis. The range is from $4 \times \left(\dfrac{-\pi}{2}\right) + 1$ to $4 \times \dfrac{\pi}{2} + 1$, not including the end-points.

3. State the range.	$y = 4\tan^{-1}\left(\frac{2x-7}{6}\right) + 1$ has a range of $(-2\pi + 1, 2\pi + 1)$.

12.6 Exercise

Students, these questions are even better in jacPLUS

Receive immediate feedback and access sample responses

Access additional questions

Track your results and progress

Find all this and MORE in jacPLUS

Technology free

1. **WE19** Evaluate each of the following.

   a. $\sin^{-1}(1.1)$
   b. $\sin^{-1}\left(\sin\left(\frac{5\pi}{3}\right)\right)$
   c. $\sin\left(\sin^{-1}(0.9)\right)$

2. **WE20** Evaluate each of the following.

   a. $\cos^{-1}(1.2)$
   b. $\cos^{-1}\left(\cos\left(\frac{7\pi}{6}\right)\right)$
   c. $\cos\left(\cos^{-1}\left(\frac{\pi}{6}\right)\right)$

3. **WE21** Determine the exact value of $\cos\left(\sin^{-1}\left(\frac{1}{5}\right)\right)$.

4. Determine the exact value of $\sin\left(\cos^{-1}\left(\frac{3}{7}\right)\right)$.

5. **WE22** Determine the exact value of $\sin\left(2\cos^{-1}\left(\frac{4}{7}\right)\right)$.

6. Determine the exact value of $\cos\left(2\sin^{-1}\left(\frac{3}{8}\right)\right)$.

7. **WE23** Evaluate:

   a. $\tan^{-1}\left(\tan\left(\frac{7\pi}{6}\right)\right)$
   b. $\tan\left(\tan^{-1}(1.1)\right)$.

8. Evaluate:

   a. $\tan^{-1}\left(\tan\left(\frac{5\pi}{3}\right)\right)$
   b. $\tan\left(\tan^{-1}\left(\frac{5}{4}\right)\right)$.

9. **WE24** Determine the exact value of $\tan\left(2\tan^{-1}\left(\frac{1}{3}\right)\right)$.

10. Determine the exact value of $\cot\left(2\tan^{-1}\left(\frac{1}{4}\right)\right)$.

11. Evaluate each of the following.

a. $\sin\left(\cos^{-1}\left(\frac{2}{9}\right)\right)$ b. $\tan\left(\cos^{-1}\left(-\frac{2}{3}\right)\right)$

12. Evaluate each of the following.

a. $\tan\left(\sin^{-1}\left(-\frac{5}{6}\right)\right)$ b. $\sin\left(\tan^{-1}\left(\frac{5}{8}\right)\right)$

13. Evaluate each of the following.

a. $\cos\left(\sin^{-1}\left(\frac{2}{5}\right)\right)$ b. $\cos\left(\tan^{-1}\left(-\frac{7}{4}\right)\right)$

14. **WE25** Sketch the graphs of each of the following, stating coordinates of the endpoints and axial intercepts.

a. $y=3\sin^{-1}(2-x)$ b. $y=2\cos^{-1}(3-x)$ c. $y=\tan^{-1}(2x)$

15. Sketch the graphs of each of the following, stating coordinates of the endpoints and axial intercepts.

a. $y=\sin^{-1}\left(\frac{3x-1}{2}\right)$ b. $y=\cos^{-1}\left(\frac{4x+1}{2}\right)$ c. $y=4\tan^{-1}\left(\frac{x}{2}\right)$

16. Sketch the graphs of each of the following, stating coordinates of the endpoints and axial intercepts.

a. $y=\sin^{-1}\left(\frac{2-5x}{4}\right)$ b. $y=\cos^{-1}\left(\frac{5-2x}{10}\right)$ c. $y=3\tan^{-1}\left(\frac{x}{3}\right)$

17. **WE26** State the implied domain and range of each of the following.

a. $y=\frac{4}{\pi}\cos^{-1}(3x+5)-3$ b. $y=\frac{8}{\pi}\tan^{-1}(10x)+3$ c. $y=2\sin^{-1}(3x-1)+\pi$

18. State the implied domain and range of each of the following.

a. $y=3\cos^{-1}(2x-5)-\pi$ b. $y=3\sin^{-1}\left(\frac{2x-5}{4}\right)-2\pi$ c. $y=\frac{6}{\pi}\tan^{-1}\left(\frac{3x-5}{4}\right)+2$

12.6 Exam questions

Question 1 (1 mark) TECH-ACTIVE

MC Consider the function, f, defined by $f(x)=\dfrac{1}{\cos^{-1}\left(\frac{3-4x}{5}\right)}$. The maximal domain of f is

A. $x\in R\backslash\left\{\frac{3}{4}\right\}$ **B.** $\left(-\frac{1}{2}, 2\right)$ **C.** $\left[-\frac{1}{2}, 2\right]$ **D.** $\left[-\frac{1}{2}, 2\right)$ **E.** $\left(-\frac{1}{2}, 2\right]$

Question 2 (6 marks) TECH-FREE

Sketch the graph of the following, stating coordinates of any axial intercepts and the coordinates of any endpoints.

a. $y=\sin^{-1}\left(\frac{3x-2}{4}\right)$ **(3 marks)**

b. $y=\cos^{-1}\left(\frac{3-2x}{6}\right)$ **(3 marks)**

Question 3 (3 marks) TECH-FREE

Sketch the graph of $y=\frac{4}{\pi}\tan^{-1}(3x)$ stating the equations of any asymptotes.

More exam questions are available online.

12.7 Circles and ellipses

LEARNING INTENTION

At the end of this subtopic you should be able to:
- sketch graphs of circles and ellipses.

12.7.1 The circle

The circle belongs to the family of conics. That is, a circle is a curve produced by the intersection of a plane with a cone.

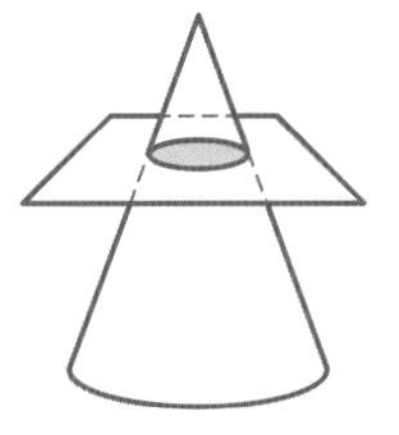

A circle is the path or locus traced out by a point at a constant distance (the radius) from a fixed point (the centre).

Consider the circles shown below. The first circle has its centre at the origin and radius r.

Let P(x, y) be a point on the circle. By Pythagoras' theorem, $x^2 + y^2 = r^2$.

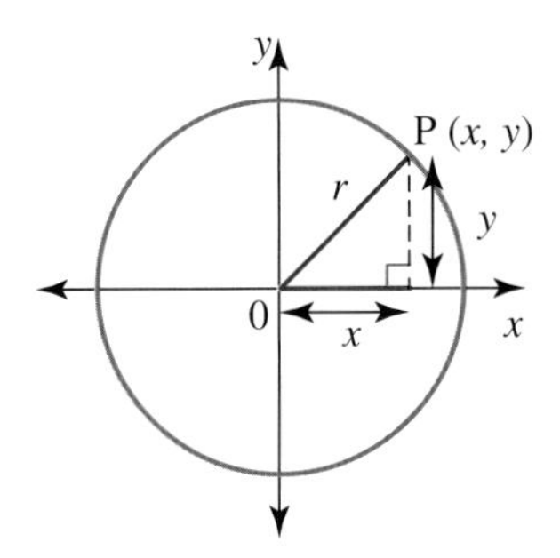

Equation of a circle centred at the origin

The equation of a circle with centre (0, 0) and radius r is:

$$x^2 + y^2 = r^2$$

If the circle is translated h units to the right, parallel to the x-axis, and k units upwards, parallel to the y-axis, then the equation becomes $(x - h)^2 + (y - k)^2 = r^2$.

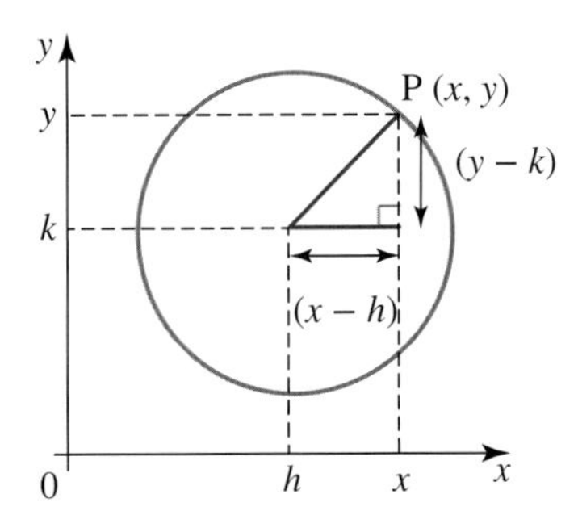

Domain: $[h - r, h + r]$

Range: $[k - r, k + r]$

Equation of a circle

The equation of a circle with centre (h, k) and radius r is:

$$(x-h)^2+(y-k)^2=r^2$$

WORKED EXAMPLE 27 Sketching the graph of a circle (1)

Sketch the graph $4x^2+4y^2=25$. Stating the centre, radius, domain and range.

THINK	WRITE/DRAW
1. Express the equation in standard form by dividing both sides by 4.	$x^2+y^2=r^2$ $4x^2+4y^2=25$ $x^2+y^2=\frac{25}{4}$ $x^2+y^2=\left(\frac{5}{2}\right)^2$
2. State the coordinates of the centre.	Centre (0, 0)
3. Determine the length of the radius by taking the square root of both sides.	$r^2=\left(\frac{5}{2}\right)^2$ $r=\left(\frac{5}{2}\right)$ Radius = 2.5 units
4. Sketch the graph.	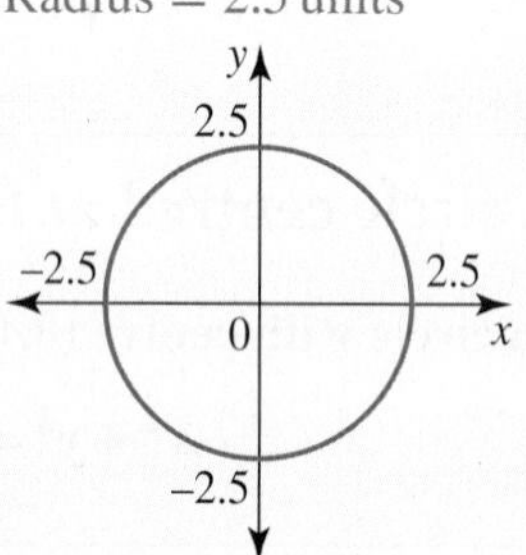
5. State the domain and range.	Domain [−2.5, 2.5] Range [−2.5, 2.5]

WORKED EXAMPLE 28 Sketching the graph of a circle (2)

Sketch the graph of $(x-2)^2+(y+3)^2=16$, clearly showing the centre, radius, domain and range.

THINK	WRITE/DRAW
1. Express the equation in standard form by expressing 16 as 4^2.	$(x-h)^2+(y-k)^2=r^2$ $(x-2)^2+(y+3)^2=4^2$
2. State the coordinates of the centre.	Centre (2, −3)

3. State the length of the radius.

$$r^2 = 4^2$$
$$r = 4$$
Radius = 4 units

4. Sketch the graph.

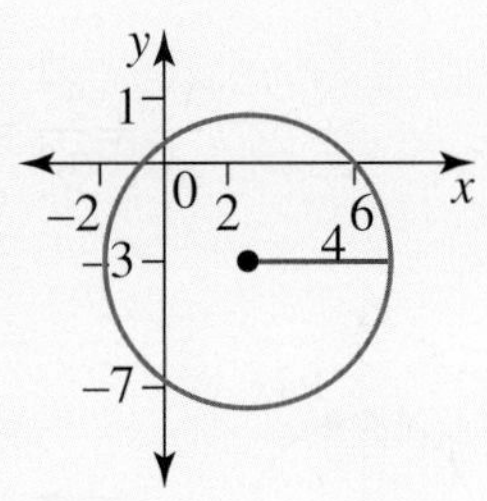

5. State the domain and range.

Domain [−2, 6]
Range [−7, 1]

WORKED EXAMPLE 29 Sketching the graph of a circle (3)

Sketch the graph of the circle $x^2 + 2x + y^2 - 6y + 6 = 0$, stating the domain and range.

THINK | **WRITE/DRAW**

1. Express the equation in standard form using the 'completing the square' method twice.

$$(x-h)^2 + (y-k)^2 = r^2$$
$$x^2 + 2x + y^2 - 6y + 6 = 0$$
$$(x^2 + 2x + 1) - 1 + (y^2 - 6y + 9) - 9 + 6 = 0$$
$$(x+1)^2 + (y-3)^2 - 4 = 0$$
$$(x+1)^2 + (y-3)^2 = 4$$
$$(x+1)^2 + (y-3)^2 = 2^2$$

2. State the coordinates of the centre.

Centre (−1, 3)

3. State the length of the radius.

$$r^2 = 2^2$$
$$r = 2$$
Radius = 2 units

4. Sketch the graph.

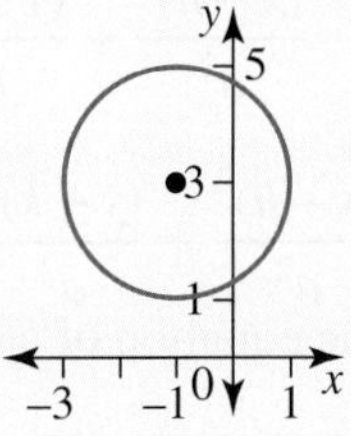

5. State the domain and range.

Domain [−3, 1]
Range [1, 5]

12.7.2 Graphs of ellipses

If a circle with Cartesian equation $x^2 + y^2 = 1$ is dilated by a factor a from the y-axis and by a factor b from the x-axis then all points P(x, y) on the circle become the points P′(ax, by) as shown. The basic equation of an ellipse is:

$$\frac{x^2}{a^2} + \frac{y^2}{b^2} = 1$$

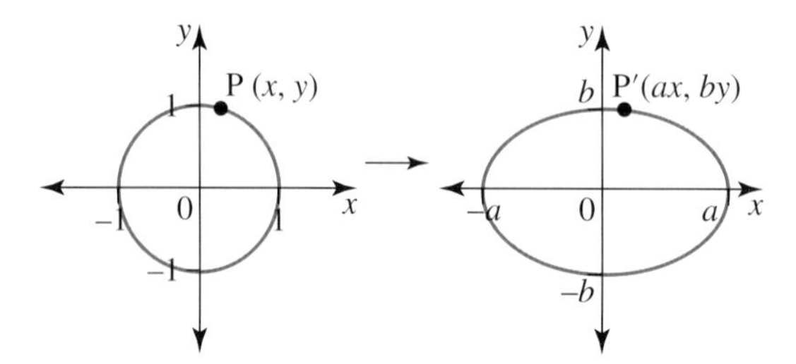

Its graph is shaped like an elongated circle.

This ellipse:

1. is centred at (0, 0)
2. has vertices at $(-a, 0)$, $(a, 0)$ (found by letting $y = 0$ and solving), and $(0, -b)$ and $(0, b)$ (found by letting $x = 0$ and solving).

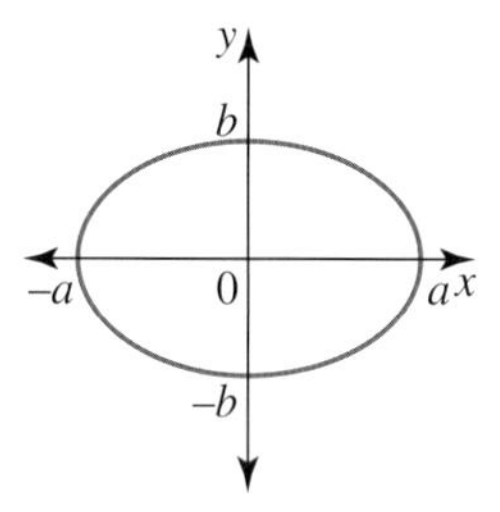

If this curve were shifted h units to the right and k units up, then the centre would move to (h, k) and its equation would become:

$$\frac{(x-h)^2}{a^2} + \frac{(y-k)^2}{b^2} = 1$$

Note: If $a = b$ then the equation becomes $\frac{(x-h)^2}{a^2} + \frac{(y-k)^2}{a^2} = 1$ and can be rearranged to $(x-h)^2 + (y-k)^2 = a^2$ (by multiplying both sides by a^2). This is the equation of a circle.

Ellipses

For an ellipse in the form $\frac{(x-h)^2}{a^2} + \frac{(y-k)^2}{b^2} = 1$ we can deduce the following, which will help us to sketch the ellipse:

1. **(h, k) are the coordinates of the centre of the ellipse.**
2. **The vertices are $(-a+h, k), (a+h, k), (h, -b+k), (h, b+k)$.**
3. **The domain is $[-a+h, a+h]$**
4. **The range is $[-b+k, b+k]$**

Notes:

1. a and b are lengths and so are positive values, $a > 0, b > 0$.
2. a is half the length of the **major axis**, the length of the semi-major axis (axis of symmetry parallel to the x-axis if $a > b$ or parallel to the y-axis if $a < b$).
3. b is half the length of the **minor axis**, the length of the semi-minor axis (axis of symmetry parallel to the y-axis if $a > b$ or parallel to the x-axis if $a < b$).

WORKED EXAMPLE 30 Sketching the graph of an ellipse (1)

Sketch the graph of $\frac{(x-1)^2}{25} + \frac{(y-2)^2}{9} = 1$, showing x- and y-intercepts and stating the domain and range.

THINK	WRITE/DRAW
1. Compare $\frac{(x-1)^2}{25} + \frac{(y-2)^2}{9} = 1$ with $\frac{(x-h)^2}{a^2} + \frac{(y-k)^2}{b^2} = 1$.	$h = 1, k = 2$ and so the centre is $(1, 2)$. $a^2 = 25 \quad b^2 = 9$ $a = 5 \quad b = 3$
2. The major axis is parallel to the x-axis as $a > b$.	
3. The extreme points (vertices) parallel to the x-axis for the ellipse are: $(-a+h, k) \quad (a+h, k)$	Vertices are: $(-5+1, 2) = (-4, 2) \quad (5+1, 2) = (6, 2)$ Domain is $[-4, 6]$
4. The extreme points (vertices) parallel to the y-axis for the ellipse are: $(h, -b+k) \quad (h, b+k)$	Vertices are: $(1, -3+2) = (1, -1) \quad (1, 3+2) = (1, 5)$ Range is $[-1, 5]$
5. Determine the x- and y-intercepts.	x-intercepts: $y = 0 \Rightarrow \frac{(x-1)^2}{25} + \frac{4}{9} = 1$ $\Rightarrow 9x^2 - 18x - 116 = 0$ $x = \frac{3 \pm 5\sqrt{5}}{3}$ y-intercepts: $x = 0 \Rightarrow \frac{1}{25} + \frac{(y-2)^2}{9} = 1$ $\Rightarrow 25y^2 - 100y - 116 = 0$ $y = \frac{10 \pm 6\sqrt{6}}{5}$
6. Sketch the graph of the ellipse.	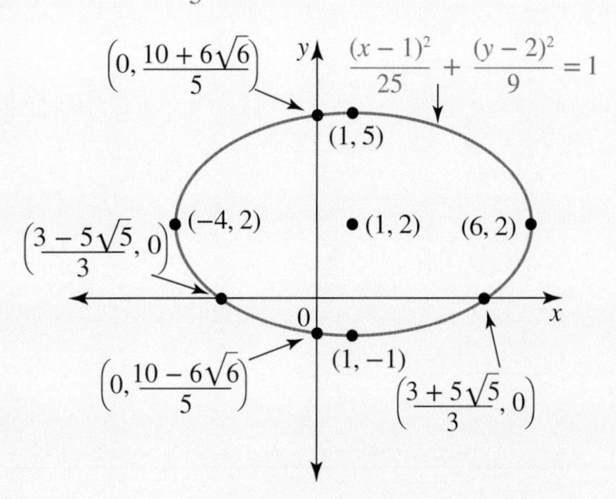

WORKED EXAMPLE 31 Sketching the graph of an ellipse (2)

Sketch the graph of $\frac{(x-2)^2}{9}+\frac{(y+4)^2}{16}=1$, showing x- and y-intercepts and stating the domain and range.

THINK	WRITE/DRAW
1. Compare $\frac{(x-2)^2}{9}+\frac{(y+4)^2}{16}=1$ with $\frac{(x-h)^2}{a^2}+\frac{(y-k)^2}{b^2}=1$.	$h=2$, $k=-4$. So the centre is $(2, -4)$. $a^2=9 \quad b^2=16$ $a=3 \quad b=4$
2. The major axis is parallel to the y-axis as $b>a$.	
3. The extreme points (vertices) parallel to the x-axis for the ellipse are: $(-a+h, k) \quad (a+h, k)$	Vertices are: $(-3+2, -4) = (-1, -4)$ $\quad (3+2, -4) = (5, -4)$ Domain is $[-1, 5]$
4. The extreme points (vertices) parallel to the y-axis for the ellipse are: $(h, -b+k) \quad (h, b+k)$	Vertices are: $(2, -4-4) = (2, -8)$ $\quad (2, 4-4) = (2, 0)$ Range is $[-8, 0]$
5. Determine the x- and y-intercepts.	x-intercept: $y=0 \Rightarrow \frac{(x-2)^2}{9}+\frac{16}{16}=1$ $\Rightarrow (x-2)^2=0$ $x=2$ y-intercepts: $x=0 \Rightarrow \frac{4}{9}+\frac{(y+4)^2}{16}=1$ $\Rightarrow 9y^2+72y+64=0$ $y=\frac{-12\pm4\sqrt{5}}{3}$
6. Sketch the graph of the ellipse.	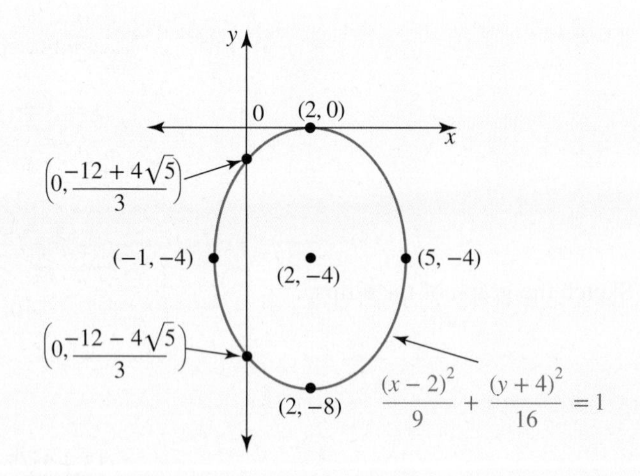

WORKED EXAMPLE 32 Sketching the graph of an ellipse (3)

Sketch the graph of $5x^2 + 9(y-2)^2 = 45$.

THINK	WRITE/DRAW
1. Rearrange and simplify by dividing both sides by 45 to make the RHS = 1.	$\frac{5x^2}{45} + \frac{9(y-2)^2}{45} = \frac{45}{45}$
2. Simplify by cancelling.	$\frac{x^2}{9} + \frac{(y-2)^2}{5} = 1$
3. Compare $\frac{x^2}{9} + \frac{(y-2)^2}{5} = 1$ with $\frac{(x-h)^2}{a^2} + \frac{(y-k)^2}{b^2} = 1$.	$h = 0, k = 2$ and so the centre is $(0, 2)$. $a^2 = 9 \quad b^2 = 5$ as $a, b > 0$ $a = 3 \quad b = \sqrt{5}$
4. Major axis is parallel to the x-axis as $a > b$.	
5. The extreme points (vertices) parallel to the x-axis for the ellipse are: $(-a+h, k) \quad (a+h, k)$	Vertices are: $(-3+0, 2) \quad (3+0, 2)$ $= (-3, 2) \quad = (3, 2)$ Domain is $[-3, 3]$
6. The extreme points (vertices) parallel to the y-axis for the ellipse are: $(h, -b+k) \quad (h, b+k)$	Vertices are: $(0, -\sqrt{5}+2) \quad (0, \sqrt{5}+2)$ or $(0, 2-\sqrt{5}) \quad (0, 2+\sqrt{5})$ $\approx (0, -0.24) \quad \approx (0, 4.24)$ Range is$[2-\sqrt{5}, 2+\sqrt{5}]$
7. Determine the x- and y-intercepts.	x-intercepts: $y = 0 \Rightarrow 5x^2 + 36 = 45$ $x = \frac{\pm 3\sqrt{5}}{5}$ y-intercepts: $x = 0 \Rightarrow 9(y-2)^2 = 45$ $y = 2 \pm \sqrt{5}$
8. Sketch the graph of the ellipse.	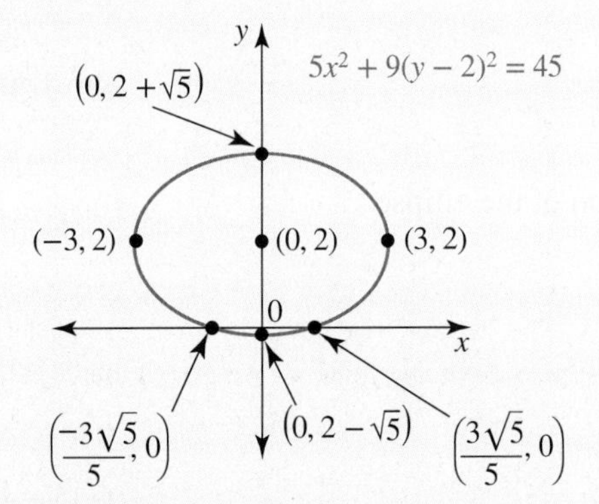

WORKED EXAMPLE 33 Sketching the graph of an ellipse (4)

Sketch the graph of the relation described by the rule: $25x^2 + 150x + 4y^2 - 8y + 129 = 0$.

THINK

1. Factorise, complete the square and simplify. Then divide both sides by 100 to make the RHS = 1.

WRITE/DRAW

$$25x^2 + 150x + 4y^2 - 8y + 129 = 0$$
$$25(x^2 + 6x + 9) - 225 + 4(y^2 - 2y + 1) - 4 + 129$$
$$= 25(x+3)^2 + 4(y-1)^2 = 100$$
$$\frac{(x+3)^2}{4} + \frac{(y-1)^2}{25} = 1\text{; ellipse}$$

2. Compare $\frac{(x+3)^2}{4} + \frac{(y-1)^2}{25} = 1$ with $\frac{(x-h)^2}{a^2} + \frac{(y-k)^2}{b^2} = 1$.

$h = -3$, $k = 1$

So the centre is $(-3, 1)$.

$a^2 = 4 \qquad b^2 = 25$

$a = 2 \qquad b = 5$ as $a, b > 0$

3. The major axis is parallel to the y-axis as $b > a$.

4. The extreme points (vertices) parallel to the x-axis for the ellipse are: $(-a + h, k)$ $\quad (a + h, k)$

Vertices are: $(-2 - 3, 1) = (-5, 1)$ $\quad (2 - 3, 1) = (-1, 1)$

Domain is $[-5, -1]$

5. The extreme points (vertices) parallel to the y-axis for the ellipse are: $(h, -b + k)$ $\quad (h, b + k)$

Vertices are: $(-3, -5 + 1) = (-3, -4)$ $\quad (-3, 5 + 1) = (-3, 6)$

Range is $[-4, 6]$

6. Determine the x- and y-intercepts.

x-intercepts:

$y = 0 \Rightarrow 25x^2 + 125x + 129 = 0$

$$x = \frac{-125 \pm \sqrt{125^2 - 12\,900}}{50}$$
$$= \frac{-15 \pm 4\sqrt{6}}{5}$$

y-intercepts:

$x = 0 \Rightarrow 4y^2 - 8y + 129 = 0$

$\Rightarrow \Delta < 0$

No y-intercepts.

7. Sketch the graph of the ellipse.

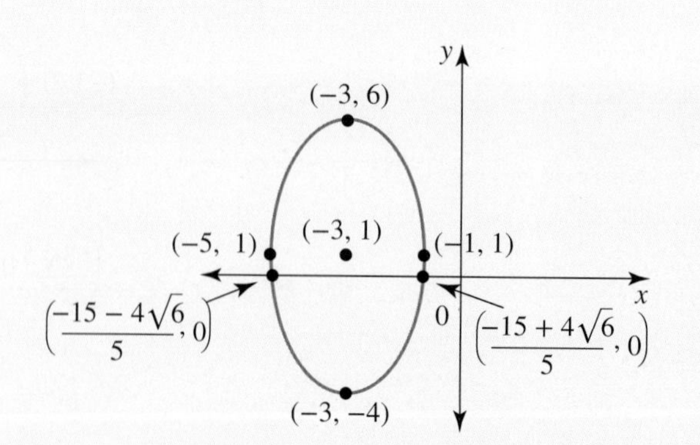

12.7.3 Loci

A **locus** (plural *loci*) is the path traced by a point that moves according to a condition or rule.

A locus can be defined by a description, for example, '6 units from the x-axis', or by an equation, for example, '$y = x^2 + 4x + 1$'.

WORKED EXAMPLE 34 Determining the equation of a locus (1)

State the equation(s) of the locus of a point $P(x, y)$ such that:

a. P is 3 units from the y-axis

b. P is 4 units from the x-axis.

THINK

WRITE/DRAW

a. 1. Since the point P has to be 3 units from the y-axis, this could be three units in the positive or negative direction.

a.

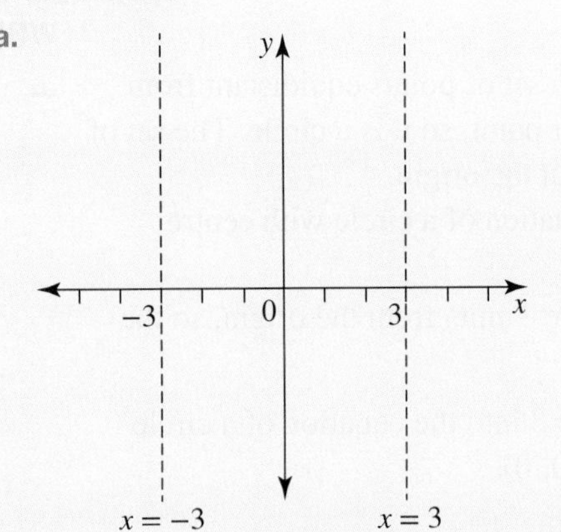

2. Draw a diagram with points 3 units from the y-axis. These points create two vertical lines.

3. Since the lines are vertical, going through 3 and −3, the equations of the loci are $x = -3$ and $x = 3$.

The equations of the loci are:
$x = -3$ and $x = 3$.

b. 1. Since the point P has to be 4 units from the x-axis, this could be 4 units in the positive or negative direction.

b.

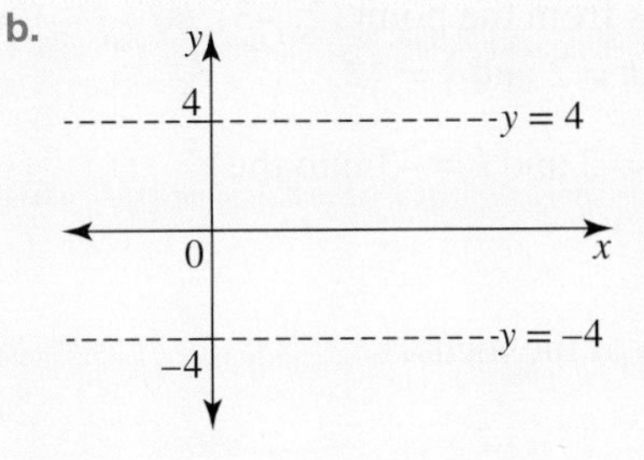

2. Draw a diagram with points 4 units from the x-axis. These points create two horizontal lines.

3. Since the lines are horizontal, going through 4 and −4, the equations of the loci are $y = -4$ and $y = 4$.

The equations of the loci are:
$y = -4$ and $y = 4$.

Locus from a point

A set of points **equidistant** (the same distance) from a point is represented by a circle.

The equation for a locus that describes a circle with centre (0, 0) and radius r is $x^2 + y^2 = r^2$.

The equation for a locus that describes a circle with centre (h, k) and radius r is $(x-h)^2 + (y-k)^2 = r^2$.

WORKED EXAMPLE 35 Determining the equation of a locus (2)

State the equation of each of the following loci.

a. **A set of points 3 units from the origin**

b. **A set of points 2 units from the point (2, –3)**

THINK	WRITE
a. 1. The locus is a set of points equidistant from one particular point, so it is a circle. The set of points is about the origin. Write the equation of a circle with centre (0, 0).	a. $x^2 + y^2 = r^2$
2. The points are 3 units from the origin, so the radius is 3.	$r = 3$
3. Substitute $r = 3$ into the equation of a circle with centre (0, 0).	$x^2 + y^2 = 3^2$ $x^2 + y^2 = 9$
4. Write the equation of the locus.	The equation of the locus that is a set of points 3 units from the origin is $x^2 + y^2 = 9$.
b. 1. The locus is a set of points equidistant from one point, so it is a circle. The locus is about (2, –3), that is, not the origin. Use the formula: $(x-h)^2 + (y-k)^2 = r^2$.	b. $(x-h)^2 + (y-k)^2 = r^2$
2. The locus is 2 units from the point (2, –3), so the radius (r) is 2, $h = 2$ and $k = -3$.	$r = 2$, $h = 2$, $k = -3$
3. Substitute $r = 2$, $h = 2$ and $k = -3$ into the formula.	$(x-2)^2 + (y+3)^2 = 2^2$ $(x-2)^2 + (y+3)^2 = 4$
4. Write the equation of the locus.	The equation of the locus that is a set of points 2 units from the point (2, –3) is $(x-2)^2 + (y+3)^2 = 4$.

Distance between two points

The distance between two points (x_1, y_1) and (x_2, y_2) is given by the equation:

$$d = \sqrt{(x_2 - x_1)^2 + (y_2 - y_1)^2}$$

WORKED EXAMPLE 36 Determining the equation of a locus (3)

Determine the equation of the locus of $P(x, y)$ given that P is equidistant from $R(1, 2)$ and $S(3, -2)$.

THINK	WRITE/DRAW
1. Draw a diagram to show points S and R and the moving point P.	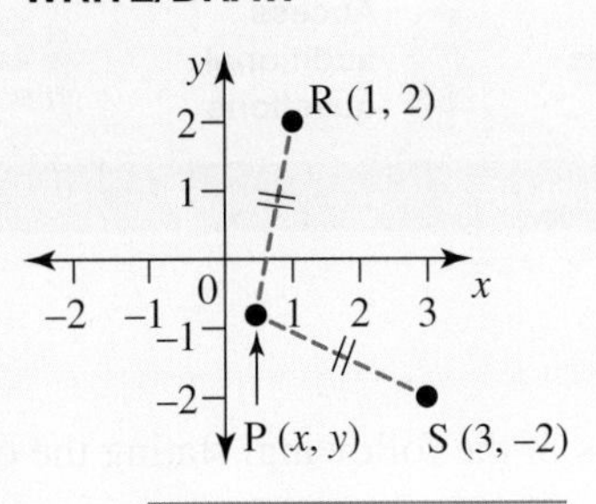
2. Since point P is equidistant from points R and S(PR = PS), the distance formula can be used to derive the equation. Write the formula for the distance between two points.	$d = \sqrt{(x_2 - x_1)^2 + (y_2 - y_1)^2}$
3. Identify (x_1, y_1) and (x_2, y_2) for the points S and P.	For S and P: $(x_1, y_1) = (3, -2)$ $(x_2, y_2) = (x, y)$
4. Substitute the known information into the formula.	$d_{PS} = \sqrt{(x-3)^2 + (y+2)^2}$
5. Repeat steps 3 and 4 for the points R and P.	For R and P: $(x_1, y_1) = (1, 2)$ $(x_2, y_2) = (x, y)$ $d_{PR} = \sqrt{(x-1)^2 + (y-2)^2}$
6. As PS = PR, equate the two expressions.	$\sqrt{(x-3)^2 + (y+2)^2} = \sqrt{(x-1)^2 + (y-2)^2}$
7. Simplify by first squaring both sides.	$(x-3)^2 + (y+2)^2 = (x-1)^2 + (y-2)^2$
8. Expand both sides.	$x^2 - 6x + 9 + y^2 + 4y + 4$ $= x^2 - 2x + 1 + y^2 - 4y + 4$
9. Move all terms to the left and simplify to determine the equation of the locus of $P(x, y)$.	$x^2 - 6x + 9 + y^2 + 4y + 4 - x^2$ $+ 2x - 1 - y^2 + 4y - 4 = 0$ $8y - 4x + 8 = 0$ $2y - x + 2 = 0$ (or $2y = x - 2$, so $y = \frac{1}{2}x - 1$)
10. Write the equation of the locus.	The equation of the locus is $y = \frac{1}{2}x - 1$.

12.7 Exercise

Students, these questions are even better in jacPLUS

Receive immediate feedback and access sample responses

Access additional questions

Track your results and progress

Find all this and MORE in jacPLUS

Technology free

1. WE27 Sketch the graphs of the following, stating the centre and radius of each.

 a. $x^2 + y^2 = 49$
 b. $x^2 + y^2 = 4^2$

2. Sketch the graphs of the following, stating the centre and radius of each.

 a. $x^2 + y^2 = 36$
 b. $x^2 + y^2 = 81$

3. WE28 Sketch the graphs of the following, clearly showing the centre and the radius.

 a. $(x-1)^2 + (y-2)^2 = 5^2$
 b. $(x+2)^2 + (y+3)^2 = 62$

4. Sketch the graphs of the following, clearly showing the centre and the radius.

 a. $(x+3)^2 + (y-1)^2 = 49$
 b. $(x-4)^2 + (y+5)^2 = 64$
 c. $x^2 + (y+3)^2 = 4$
 d. $(x-5)^2 + y^2 = 100$

5. WE29 Sketch the graphs of the following circles.

 a. $x^2 + 4x + y^2 + 8y + 16 = 0$
 b. $x^2 - 10x + y^2 - 2y + 10 = 0$

6. Sketch the graphs of the following circles.

 a. $x^2 - 14x + y^2 + 6y + 9 = 0$
 b. $x^2 + 8x + y^2 - 12y - 12 = 0$
 c. $x^2 + y^2 - 18y - 19 = 0$
 d. $2x^2 - 4x + 2y^2 + 8y - 8 = 0$

7. WE30 Sketch the following ellipses.

 a. $\dfrac{(x-1)^2}{9} + \dfrac{(y+2)^2}{4} = 1$
 b. $\dfrac{(x+5)^2}{25} + \dfrac{(y-2)^2}{16} = 1$

8. Sketch the following ellipses.

 a. $\dfrac{(x+5)^2}{49} + \dfrac{(y+1)^2}{25} = 1$
 b. $\dfrac{(x-2)^2}{169} + \dfrac{(y-3)^2}{25} = 1$
 c. $\dfrac{(x-5)^2}{36} + y^2 = 1$
 d. $x^2 + 9(y+2)^2 = 9$

9. WE31 Sketch the graph of $\dfrac{(x-2)^2}{4} + \dfrac{(y+3)^2}{9} = 1$.

10. Sketch the graph of $\dfrac{x^2}{9} + \dfrac{(y+3)^2}{16} = 1$.

11. WE32 Sketch the graph of $9(x-5)^2 + 16(y+1)^2 = 144$.

12. Sketch the graph of $16x^2 + 25y^2 = 400$.

13. WE33 Sketch the graph of $9x^2 - 72x + y^2 - 4y + 112 = 0$.

14. Sketch the graph of $9x^2 + 16y^2 + 32y - 128 = 0$.

15. WE34 State the equation of the locus of a point $P(x, y)$ such that:

a. P is 3 units from the y-axis

b. P is 5 units from the x-axis.

16. State the equation of the locus of a point $P(x, y)$ such that:

a. P is 2 units from $x = 2$

b. P is 4 units from $y = -1$.

17. WE35 State the equation of each of the following loci.

a. A set of points 2 units from (0, 0)

b. A set of points 5 units from (0, 0)

18. State the equation of each of the following loci.

a. A set of points 2 units from (2, 3)

b. A set of points 5 units from (2, –1)

19. WE36 Determine the equation of the locus of $P(x, y)$ given that P is equidistant from R and S in each of the following.

a. R(0, 2) and S(4, 2)

b. R(1, 5) and S(1, –1)

20. Determine the equation of the locus of $P(x, y)$ given that P is equidistant from R and S in each of the following.

a. R(–1, –1) and S(3, –1)

b. R(0, 3) and S(0, 5)

12.7 Exam questions

Question 1 (4 marks) TECH-FREE

a. Determine the equation of the circle which has a domain of $[1, 7]$ and range of $[-2, 4]$. **(2 marks)**

b. Determine the equation of the ellipse which has a domain of $[-4, 6]$ and range of $[-2, 6]$. **(2 marks)**

Question 2 (3 marks) TECH-FREE

Show that $x^2 - 4x + y^2 + 6y - 12 = 0$ represents a circle, sketch its graph, stating the coordinates of the centre and the domain and range.

Question 3 (3 marks) TECH-FREE

Show that $49x^2 + 294x + 25y^2 - 200y - 384 = 0$ represents an ellipse, sketch its graph, stating the coordinates of the centre and the domain and range.

More exam questions are available online.

12.8 Hyperbolas

LEARNING INTENTION

At the end of this subtopic you should be able to:

- sketch graphs of hyperbolas.

12.8.1 Horizontal hyperbolas

Horizontal hyperbolas have the following important characteristics.

1. The equation of a horizontal hyperbola centred at (0, 0) is $\frac{x^2}{a^2} - \frac{y^2}{b^2} = 1$.
2. If this curve were shifted h units to the right and k units up, then the centre would move to (h, k) and its equation would become $\frac{(x-h)^2}{a^2} - \frac{(y-k)^2}{b^2} = 1$.

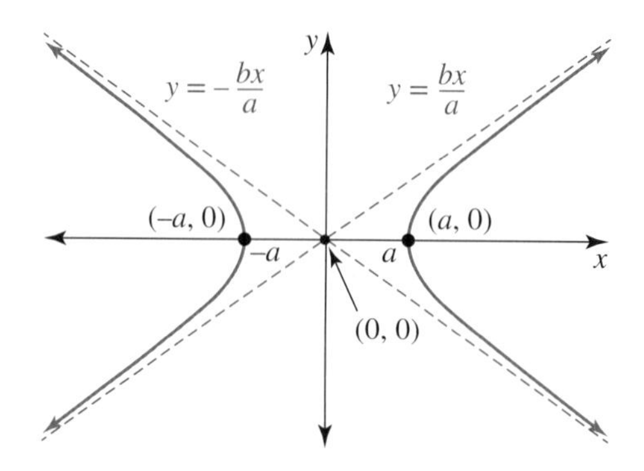

3. The basic form of a horizontal hyperbola centred at (0, 0) is shown.
4. The vertices for this curve are at $(-a, 0)$ and $(a, 0)$ and the two asymptotes are given by $y = -\frac{bx}{a}$ and $y = \frac{bx}{a}$.

Horizontal hyperbolas that are not centred at the origin

When the hyperbola is not centred at (0,0):

1. **For the curve of the function $\frac{(x-h)^2}{a^2} - \frac{(y-k)^2}{b^2} = 1$, the points on $\frac{x^2}{a^2} - \frac{y^2}{b^2} = 1$ are moved h units to the right and k units up (or x has been replaced with $(x-h)$, and y replaced with $(y-k)$).**
2. **Therefore, the vertices are $(-a+h, k)$ and $(a+h, k)$, and the centre is at (h, k).**
3. **The asymptotes are at $y - k = \frac{b}{a}(x-h)$ and $y - k = \frac{-b}{a}(x-h)$ or $y = \frac{b}{a}(x-h) + k$ and $y = \frac{-b}{a}(x-h) + k$.**

12.8.2 Vertical hyperbolas

Vertical hyperbolas have the following characteristics.

1. The equation of a vertical hyperbola centred at (0, 0) is $\frac{y^2}{b^2} - \frac{x^2}{a^2} = 1$.
2. If this graph is shifted h units to the right and k units up, then its centre would move to (h, k) and its equation would become $\frac{(y-k)^2}{b^2} - \frac{(x-h)^2}{a^2} = 1$
3. The basic form of a vertical asymptote centred at (0, 0) is shown.
4. The vertices are on the y-axis at $(0, -b)$ and $(0, b)$ and the two asymptotes are given by $y = \frac{-bx}{a}$ and $y = \frac{bx}{a}$

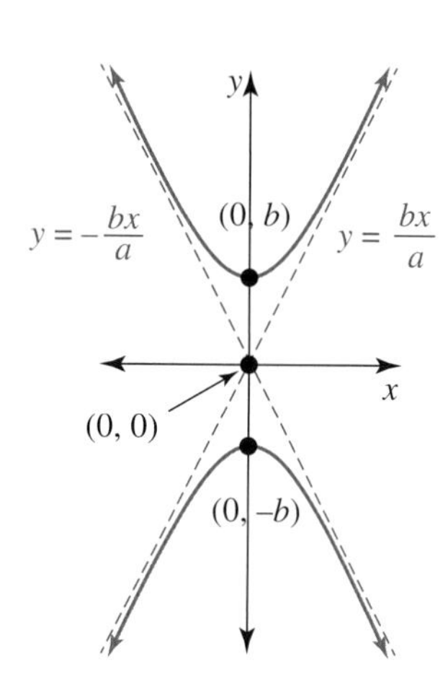

Vertical hyperbolas that are not centred at the origin.

When the hyperbola is not centred at the (0, 0):

1. **For the curve** $\frac{(y-k)^2}{b^2} - \frac{(x-h)^2}{a^2} = 1$, **the points on** $\frac{y^2}{b^2} - \frac{x^2}{a^2} = 1$ **are moved** h **units to the right and** k **units up (or** x **has been replaced with** $(x-a)$ **and** y **replaced with** $(y-k)$**).**
2. **Therefore, the vertices are** $(h, -b+k)$ **and** $(h, b+k)$**, and the centre is at** (h, k)**.**
3. **The asymptotes are at** $y-k = \frac{b}{a}(x-h)$ **and** $y-k = \frac{-b}{a}(x-h)$ **or** $y = \frac{b}{a}(x-h)+k$ **and** $y = \frac{-b}{a}(x-h)+k$

To sketch hyperbolic relations:

1. Rearrange the equation into the appropriate general form and determine the values of a and b.
2. Write down the coordinates of the centre.
3. State the coordinates of the vertices.
4. Write down the equations of the asymptotes.
5. Sketch a hyperbolic graph that fits the above information.

WORKED EXAMPLE 37 Sketching graphs of hyperbolas (1)

Sketch the graphs of the hyperbolas with the following equations:

a. $\frac{x^2}{9} - \frac{y^2}{25} = 1$

b. $\frac{y^2}{16} - \frac{x^2}{9} = 1$

THINK	WRITE/DRAW
a. 1. The equation is in the correct form, so read off the values of a, b, h and k.	a. As $h=0$, $k=0$, there are no translations. $a^2 = 9$ $b^2 = 25$ $a = 3$ $b = 25$
2. Write the coordinates of the centre.	The centre is at (0, 0).
3. Write the coordinates of the vertices.	The vertices are (−3, 0) and (3, 0).
4. Write the equations of the asymptotes.	The asymptotes are $y = \frac{-5x}{3}$ and $y = \frac{5x}{3}$.
5. Draw the asymptotes, plot the vertices and centre, and then sketch the hyperbola.	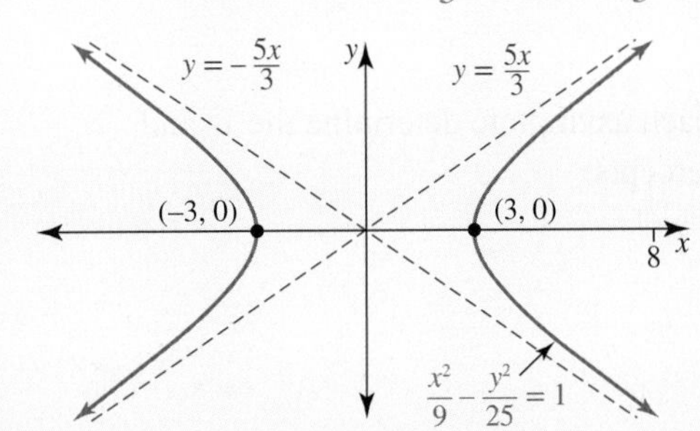
b. 1. The equation is in the correct form, so read off the values of a and b.	b. As there are no translations. $b^2 = 16$ $a^2 = 9$ $b = 4$ $a = 3$
2. Write the coordinates of the centre.	The centre is at (0, 0)

3. Write the coordinates of the vertices. — The vertices are $(0, -4)$ and $(0, 4)$.

4. Write the equations of the asymptotes. — The asymptotes are $y = \frac{-4x}{3}$ and $y = \frac{4x}{3}$.

5. Draw the asymptotes, plot the vertices, and then sketch the hyperbola.

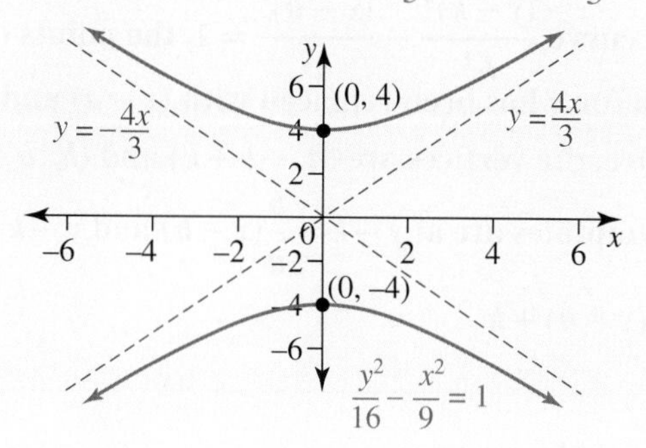

WORKED EXAMPLE 38 Sketching graphs of hyperbolas (2)

Sketch the graphs of the hyperbolas with the following equations:

a. $\frac{(x-3)^2}{16} - \frac{(y-2)^2}{9} = 1$

b. $\frac{(y+2)^2}{25} - \frac{(x-1)^2}{16} = 1$

THINK	WRITE/DRAW
a. 1. The equation is in the correct form, so read off the values of a, b, h and k.	a. $h = 3, k = 2$ $a^2 = 16 \quad b^2 = 9$ $a = 4 \quad b = 3$
2. Write the coordinates of the centre.	The centre is (3, 2).
3. Write the coordinates of the vertices.	The vertices are $(-4+3, 2)$ and $(4+3, 2)$ or $(-1, 2)$ and $(7, 2)$.
4. Write the equations of the asymptotes.	The asymptotes: $y - 2 = \frac{-3}{4}(x-3) \qquad y - 2 = \frac{3}{4}(x-3)$ $4(y-2) = -3(x-3) \qquad 4(y-2) = 3(x-3)$ $4y - 8 = -3x + 9 \qquad 4y - 3x = -9$ $4y + 3x = 17 \qquad 4y - 3x = -1$
5. For each asymptote determine the x- and y-intercepts.	For $4y + 3x = 17$: $x = 0, 4y = 17$, $y = \frac{17}{4}$, $\left(0, \frac{17}{4}\right)$; $y = 0, 3x = 17$ For $4y - 3x = -1$: $x = 0, 4y = -1$, $y = -\frac{1}{4}$, $\left(0, -\frac{1}{4}\right)$; $y = 0, -3x = -1$
6. The x- and y-intercepts for $4y - 3x = -1$ are too close to each other so use one of these points, say $\left(\frac{1}{3}, 0\right)$, and the centre to sketch this line — as both asymptotes intersect here.	$x = \frac{17}{3} \qquad x = \frac{1}{3}$ $\left(\frac{17}{3}, 0\right) \qquad \left(\frac{1}{3}, 0\right)$

7. Determine the x-intercepts of the graph.

$y=0$: $\dfrac{(x-3)^2}{16}-\dfrac{(-2)^2}{9}=1$

$x=\dfrac{9\pm 4\sqrt{13}}{3}$

8. Plot the vertices and centre, and then sketch the hyperbola.

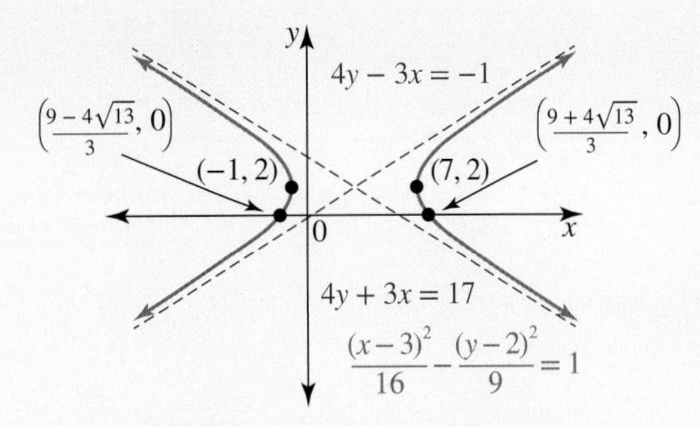

b. 1. The equation is in the correct form, so read off the values of a, b, h and b.

b. $h=1,\ k=-2$

$b^2=25 \quad a^2=16$

$b=5 \quad a=4$

2. Write the coordinates of the centre.

The centre is at $(1,-2)$

3. Write the coordinates of the vertices.

The vertices are $(1,-5-2)$ and $(1,5-2)$ or $(1,-7)$ and $(1,3)$

4. Write the equations of the asymptotes.

The asymptotes:

$y+2=\dfrac{-5}{4}(x-1) \qquad y+2=\dfrac{5}{4}(x-1)$

$4(y+2)=-5(x-1) \qquad 4(y+2)=5(x-1)$

$4y+8=-5x+5 \qquad 4y+8=5x-5$

$4y+5x=-3 \qquad 4y-5x=-13$

5. For each asymptote determine the x- and y-intercepts. Note that both asymptotes intersect at the centre $(1,-7)$.

For $4y+5x=-3$ For $4y-5x=-13$

$x=0,\ 4y=-3 \qquad x=0,\ 4y=-13$

$y=-\dfrac{3}{4} \qquad y=-\dfrac{13}{4}$

$\left(0,\ -\dfrac{3}{4}\right) \qquad \left(0,\ -\dfrac{13}{4}\right)$

$y=0,\ 5x=-3 \qquad y=0,\ -5x=-13$

$x=-\dfrac{3}{5} \qquad x=\dfrac{13}{5}$

$\left(-\dfrac{3}{5},0\right) \qquad \left(\dfrac{13}{5},0\right)$

6. Determine the y-intercepts of the graph:

$x=0$: $\dfrac{(y+2)^2}{25}-\dfrac{(-1)^2}{16}=1$

$y=\dfrac{-2\pm\sqrt{17}}{4}$

7. Plot the vertices and the centre then sketch the hyperbola.

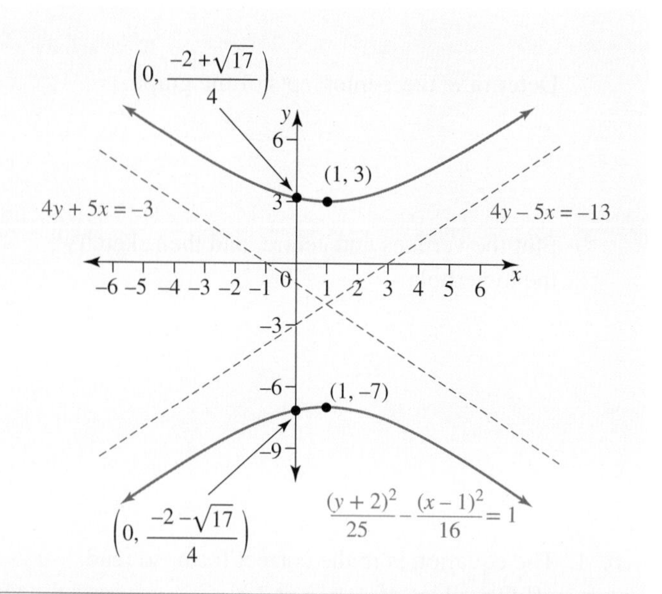

WORKED EXAMPLE 39 Sketching the graph of a hyperbola (3)

Sketch the graph of the hyperbola with equation $6x^2 - 9(y-2)^2 = 54$.

THINK	WRITE/DRAW
1. Rearrange the equation by dividing both sides by 54 to make the RHS = 1.	$6x^2 - 9(y-2)^2 = 54$ $\frac{6x^2}{54} - \frac{9(y-2)^2}{54} = \frac{54}{54}$
2. Simplify by cancelling.	$\frac{x^2}{9} - \frac{(y-2)^2}{6} = 1$
3. Read off the values of h and k. Work out values of a and b.	$h = 0$, $k = 2$, translation of 2 units up $a^2 = 9 \quad b^2 = 6$ $a = 3 \quad b = \sqrt{6}$ as a and $b > 0$
4. Write the coordinates of the centre.	The centre is at (0, 2).
5. Write the coordinates of the vertices.	The vertices are: $(-3+0, 2)$ and $(3+0, 2)$ or $(-3, 2)$ and $(3, 2)$.
6. Write the equations of the asymptotes.	The asymptotes are: $y - 2 = \frac{-\sqrt{6}}{3}(x - 0)$ and $y - 2 = \frac{\sqrt{6}}{3}x$ $3(y-2) = -\sqrt{6}x \quad 3(y-2) = \sqrt{6}x$ $3y - 6 = -\sqrt{6}x \quad 3y - 6 = \sqrt{6}x$ $3y + \sqrt{6}x = 6 \quad 3y - \sqrt{6}x = 6$
7. Write the x- and y-intercepts for the asymptotes.	Intercepts for $3y + \sqrt{6}x = 6$ are $(\sqrt{6}, 0)$ and $(0, 2)$. Intercepts for $3y - \sqrt{6}x = 6$ are $(-\sqrt{6}, 0)$ and $(0, 2)$.

8. Draw the asymptotes, plot the vertices and centre, and then sketch the hyperbola.

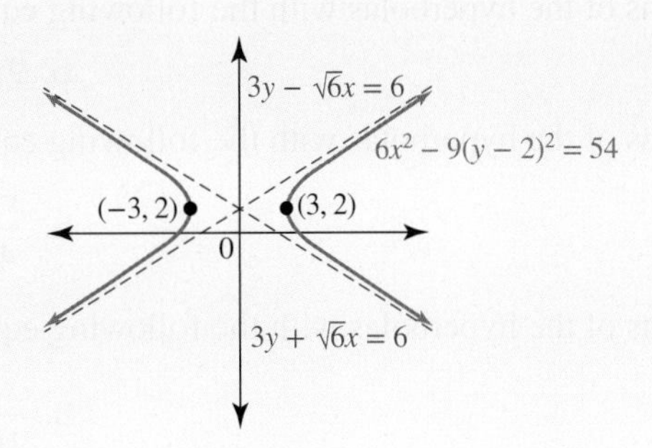

12.8 Exercise

Students, these questions are even better in jacPLUS

Receive immediate feedback and access sample responses

Access additional questions

Track your results and progress

Find all this and MORE in jacPLUS

Technology free

1. WE37 Sketch the following hyperbolas, stating the coordinates of the centre, vertices and asymptotes.
   a. $\frac{x^2}{16} - \frac{y^2}{9} = 1$
   b. $\frac{y^2}{25} - \frac{x^2}{16} = 1$
2. Sketch the following hyperbolas, stating the coordinates of the centre, vertices and asymptotes.
   a. $\frac{x^2}{64} - \frac{y^2}{36} = 1$
   b. $\frac{x^2}{144} - \frac{y^2}{25} = 1$
3. WE38 Sketch the following hyperbolas.
   a. $\frac{(x+1)^2}{64} - \frac{(y-2)^2}{36} = 1$
   b. $\frac{(y+1)^2}{9} - \frac{(x-1)^2}{25} = 1$
4. Sketch the following hyperbolas.
   a. $\frac{(x-3)^2}{25} - \frac{(y-3)^2}{4} = 1$
   b. $\frac{(y-3)^2}{4} - \frac{(x-3)^2}{25} = 1$
5. Sketch the following hyperbolas.
   a. $\frac{(x-1)^2}{16} - \frac{y^2}{9} = 1$
   b. $\frac{(x+3)^2}{144} - \frac{y^2}{25} = 1$
6. Sketch the following hyperbolas with equations.
   a. $\frac{x^2}{9} - \frac{(y+2)^2}{9} = 1$
   b. $x^2 - (y-3)^2 = 4$
7. WE39 Sketch the following hyperbolas with equations.
   a. $4x^2 - 9y^2 = 36$
   b. $4(x-5)^2 - 9(y+3)^2 = 36$
8. Sketch the graphs of the hyperbolas with the following equations.
   a. $25x^2 - 6y^2 = 400$
   b. $9x^2 - 16y^2 = 144.$

9. Sketch the graphs of the hyperbolas with the following equations.

a. $x^2 - y^2 = 25$

b. $9x^2 - 25y^2 = 225$.

10. Sketch the graphs of the hyperbolas with the following equations.

a. $\frac{x^2}{9} - \frac{y^2}{4} = 1$

b. $\frac{x^2}{4} - \frac{y^2}{9} = 1$

11. Sketch the graphs of the hyperbolas with the following equations.

a. $\frac{x^2}{9} - \frac{y^2}{16} = 1$

b. $\frac{x^2}{16} - \frac{y^2}{4} = 1$

12. Sketch the graphs of the hyperbolas with the following equations.

a. $\frac{(x-2)^2}{9} - \frac{(y-3)^2}{4} = 1$

b. $\frac{(x-2)^2}{25} - \frac{(y-3)^2}{4} = 1$

13. Sketch the graphs of the hyperbolas with the following equations.

a. $\frac{x^2}{9} - \frac{(y-1)^2}{25} = 1$

b. $\frac{(x-1)^2}{25} - \frac{(y+1)^2}{9} = 1$

14. Sketch the graph of the hyperbola with the equation:

a. $16(x-2)^2 - 9(y+5)^2 = 144$

b. $25(x-3)^2 - 9(y+2)^2 = 225$

15. Sketch the graph of the hyperbola with the equation:

a. $36(x-4)^2 - 4(y-2)^2 = 144$

b. $9x^2 - 16(y+1)^2 = 144$

16. A boomerang manufacturer's specifications for a particular model of boomerang can be seen in the diagram. Determine an equation for the dashed curve drawn through the boomerangs if the equation of one asymptote is $y = \frac{\sqrt{5}}{2}x$.

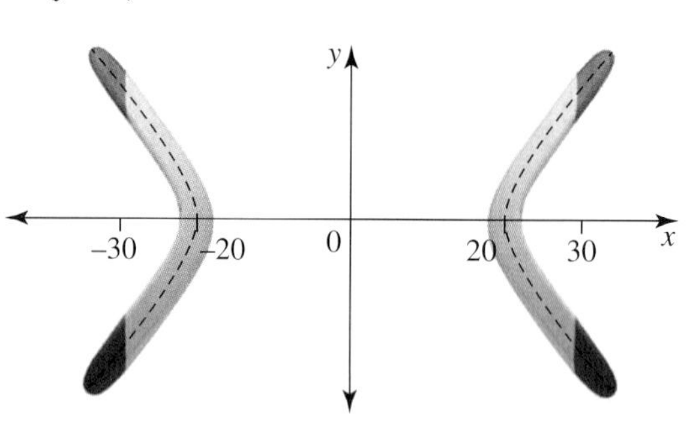

12.8 Exam questions

Question 1 (3 marks) TECH-FREE

Determine the equation of the hyperbola that has asymptotes given by $y = \pm\frac{x}{2}$ and vertex at the points $(\pm 4, 0)$.

Question 2 (4 marks) TECH-FREE

A hyperbola has the equation $\frac{(x-h)^2}{a^2} - \frac{(y-k)^2}{b^2} = 1$, asymptotes given by $y = \frac{3x}{2}$ and $y = -\frac{3x}{2} - 6$ and a vertex at the point $(0, -3)$.

Determine the values of h, k, a and b.

Question 3 (4 marks) TECH-FREE

Show that $9x^2 - 54x - 16y^2 - 64y - 127 = 0$ represents a hyperbola, sketch its graph, stating the coordinates of the centre, the domain and range and the equations of any asymptotes.

More exam questions are available online.

12.9 Polar coordinates, equations and graphs

LEARNING INTENTION

At the end of this subtopic you should be able to:
- convert between polar and Cartesian coordinates.

12.9.1 Polar coordinates

In the Cartesian coordinate system, a point, P, is located using (x, y) coordinates. The same point can be located by stating the distance of the point from the origin, the radius, r, and the angle, θ, it makes with the positive x-direction. These are known as **polar coordinates**. We write the polar coordinates of point P as $[r, \theta]$.

Note: θ may be given in degrees or radians.

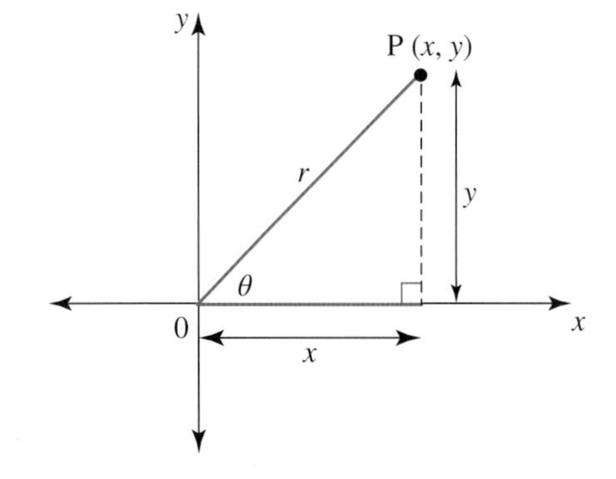

WORKED EXAMPLE 40 Plotting polar coordinates

Plot the following polar coordinates.

a. $[2, 60°]$ b. $\left[-3, \dfrac{2\pi}{3}\right]$

THINK

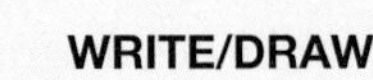
WRITE/DRAW

a. 1. Draw the positive x-direction.
2. Rotate 60° anticlockwise.
3. Extend the line 2 units.

a.

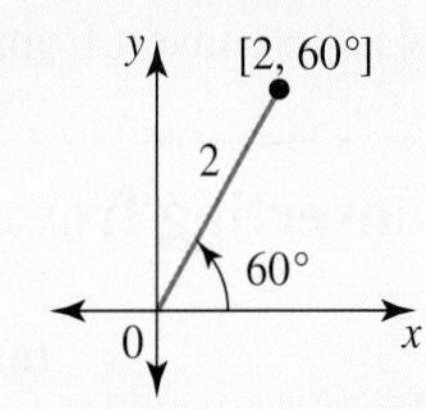

b. 1. Draw the positive x-direction.
2. Rotate $\dfrac{2\pi}{3}$ anticlockwise.
3. Extend the line 3 units in the opposite direction.
Note: $\left[-3, \dfrac{2\pi}{3}\right]$ is the same as $\left[3, \dfrac{5\pi}{3}\right]$. Why? Can you write another set of coordinates for the same point?

b.

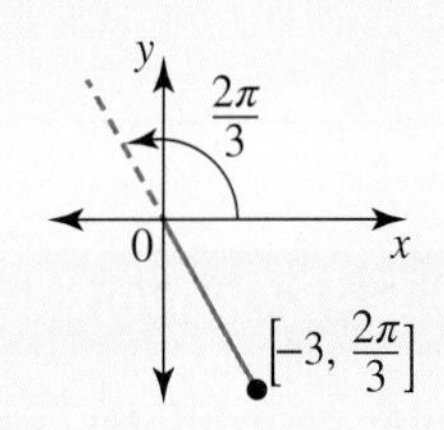

Converting from polar coordinates to Cartesian coordinates

Looking at the initial polar coordinates diagram and using trigonometry we can obtain the following equivalences which allow us to convert from polar coordinates to Cartesian coordinates.

Converting from polar to Cartesian coordinates

$$x = r\cos(\theta)$$
$$y = r\sin(\theta)$$

WORKED EXAMPLE 41 Converting polar coordinates into Cartesian coordinates

Convert $\left[2, \frac{2\pi}{3}\right]$ to Cartesian coordinates.

THINK

1. Determine the x-coordinate.

WRITE

$$x = r\cos(\theta)$$
$$= 2\cos\left(\frac{2\pi}{3}\right)$$
$$= 2 \times \left(-\frac{1}{2}\right)$$
$$x = -1$$

2. Determine the y-coordinate.

$$y = r\sin(\theta)$$
$$= 2\sin\left(\frac{2\pi}{3}\right)$$
$$= 2 \times \frac{\sqrt{3}}{2}$$
$$y = \sqrt{3}$$

3. State the Cartesian coordinates.

Hence, the Cartesian coordinates are $(-1, \sqrt{3})$.

Converting from Cartesian coordinates to polar coordinates

Looking at the initial polar coordinates diagram and using trigonometry we can obtain the following equivalences which allow us to convert from Cartesian coordinates to polar coordinates.

Converting from Cartesian to polar coordinates

$$\tan(\theta) = \frac{y}{x}, \quad \theta = \tan^{-1}\left(\frac{y}{x}\right)$$
$$r = \sqrt{x^2 + y^2}$$

WORKED EXAMPLE 42 Converting Cartesian coordinates into polar coordinates

Convert (3, −4) to polar coordinates, writing the angle in degrees correct to 1 decimal place.

THINK

1. Evaluate r.

WRITE/DRAW

$$r = \sqrt{x^2 + y^2}$$
$$= \sqrt{3^2 + (-4)^2}$$
$$= \sqrt{9 + 16}$$
$$= \sqrt{25}$$
$$r = 5$$

2. Evaluate θ.
Note that, as $\tan(\theta)$ is negative, θ is in the fourth quadrant.

$$\tan(\theta) = \frac{y}{x}$$
$$= -\frac{4}{3} (\theta \text{ in 4th quadrant})$$

3. State the polar coordinates (choose $\theta = -53.1°$ in this case).

so $\theta = \tan^{-1}\left(-\frac{4}{3}\right)$
$$\theta = -53.1° \text{ or } 360 - 53.1°$$
$$\theta = -53.1° \text{ or } 306.9°$$
Hence, the polar coordinates are [5, −53.1°].

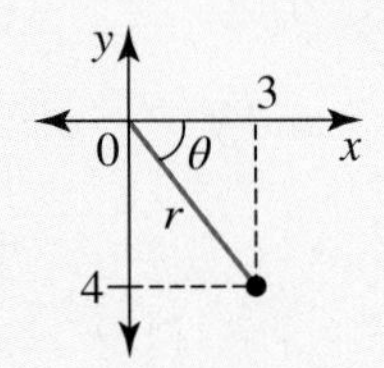

Polar equations

A polar equation is an equation written in terms of r and/or θ.

Using the conversions for $x = r\cos(\theta)$ and $y = r\sin(\theta)$ from polar coordinates, we can convert Cartesian equations into polar equations.

WORKED EXAMPLE 43 Converting Cartesian equations into polar equations

Convert the following Cartesian equations into polar equations.

a. $x^2 + y^2 = 25$
b. $y = 2x$
c. $2x - 3y = 5$
d. $x^2 + y^2 + 6x - 8y = 0$
e. $\frac{x^2}{16} + \frac{y^2}{9} - 1$

THINK

a. 1. Substitute the polar expressions for x and y.

WRITE/DRAW

a. $x^2 + y^2 = 25$
Since $x = r\cos(\theta)$
and $y = r\sin(\theta)$
$$(r\cos(\theta))^2 + (r\sin(\theta))^2 = 25$$

2. Expand and simplify.
(Use the identity $\cos^2(\theta) + \sin^2(\theta) = 1$.)

$$r^2\cos^2(\theta) + r^2\sin^2(\theta) = 25$$
$$r^2\left(\cos^2(\theta) + \sin^2(\theta)\right) = 25$$
$$r^2 = 25$$
$$r = 5$$

3. Alternatively, because $x^2 + y^2 = 25$ represents a circle of radius 5 units, the polar equation must be $r = 5$.

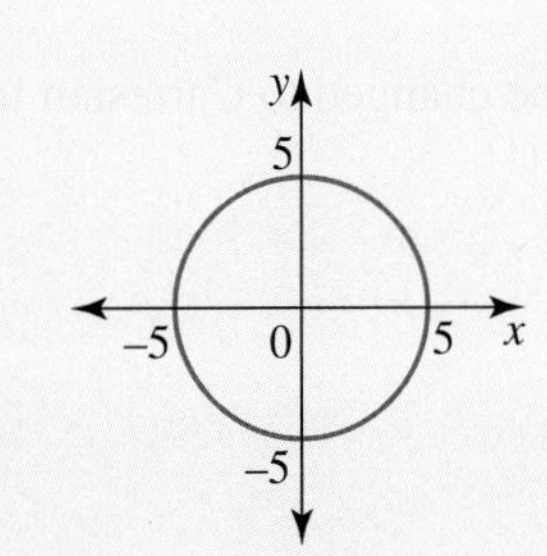

b. 1. Substitute the polar expressions for x and y.

b. $y = 2x$

Since $x = r\cos(\theta)$ and $y = r\sin(\theta)$

Then $r\sin(\theta) = 2r\cos(\theta)$

2. Divide both sides by $r\cos(\theta)$ and recall the identity $\dfrac{\sin(\theta)}{\cos(\theta)} = \tan(\theta)$.

$\dfrac{\sin(\theta)}{\cos(\theta)} = 2$

$\tan(\theta) = 2$

Since $\dfrac{\sin(\theta)}{\cos(\theta)} = \tan(\theta)$

3. Evaluate θ.

$\theta = \tan^{-1}(2)$

$\theta = 63.4°$ or 1.107^c

c. 1. Substitute the polar expressions for x and y.

c. $2x - 3y = 5$

Since $x = r\cos(\theta)$ and $y = r\sin(\theta)$

Then $2r\cos(\theta) - 3r\sin(\theta) = 5$

2. Simplify.

$r\left(2\cos(\theta) - 3\sin(\theta)\right) = 5$

$r = \dfrac{5}{2\cos(\theta) - 3\sin(\theta)}$

d. 1. Substitute the polar expressions for x and y.

d. $x^2 + y^2 + 6x - 8y = 0$

Since $x = r\cos(\theta)$ and $y = r\sin(\theta)$

$r^2\cos^2(\theta) + r^2\sin^2(\theta) + 6r\cos(\theta) - 8r\sin(\theta) = 0$

$r^2\left(\cos^2(\theta) + \sin^2(\theta)\right) + r(6\cos(\theta) - 8\sin(\theta)) = 0$

2. Note that $\cos^2(\theta) + \sin^2(\theta) = 1$.

$r^2 + r\left(6\cos(\theta) - 8\sin(\theta)\right) = 0$

3. Divide both sides by r, $r \neq 0$.

$r + 6\cos(\theta) - 8\sin(\theta) = 0$

Hence, $r = 8\sin(\theta) - 6\cos(\theta)$.

e. 1. Substitute the polar expressions for x and y.

e. $\dfrac{x^2}{16} + \dfrac{y^2}{9} = 1$

Since $x = r\cos(\theta)$ and $y = r\sin(\theta)$

$\dfrac{\left(r\cos(\theta)\right)^2}{16} + \dfrac{\left(r\sin(\theta)\right)^2}{9} = 1$

2. Simplify.

$\dfrac{r^2\cos^2(\theta)}{16} + \dfrac{r^2\sin^2(\theta)}{9} = 1$

$\dfrac{9r^2\cos^2(\theta) + 16r^2\sin^2(\theta)}{144} = 1$

$r^2\left(9\cos^2(\theta) + 16\sin^2(\theta)\right) = 144$

$r^2 = \dfrac{144}{9\cos^2(\theta) + 16\sin^2(\theta)}$

Similarly, polar equations can be changed to Cartesian form.

WORKED EXAMPLE 44 Converting polar equations into Cartesian equations

Convert the following polar equations into Cartesian equations.

a. $r = 4\cos(\theta)$ b. $\tan(\theta) = 2$ c. $r = \dfrac{2}{1+\sin(\theta)}$

THINK	WRITE
a. 1. Express the equation in terms of r^2 by multiplying both sides of the equation by r.	a. $r = 4\cos(\theta)$ $r^2 = 4r\cos(\theta)$
2. Substitute the Cartesian expressions for r and θ.	Since $r^2 = x^2 + y^2$ and $x = r\cos(\theta)$ Then $x^2 + y^2 = 4x$
3. Simplify by 'completing the square'. *Note:* This is the equation of a circle of radius 2 units and centre (2, 0).	$x^2 - 4x + y^2 = 0$ $x^2 - 4x + 4 - 4 + y^2 = 0$ $(x-2)^2 + y^2 = 4$
b. 1. Substitute $\tan(\theta) = \dfrac{y}{x}$.	b. $\tan(\theta) = 2$ As $\tan(\theta) = \dfrac{y}{x}$, $\dfrac{y}{x} = 2$
2. Simplify by making y the subject.	$y = 2x$
c. 1. Simplify the equation by multiplying both sides of the equation by $(1+\sin(\theta))$.	c. $r = \dfrac{2}{1+\sin(\theta)}$ $r(1+\sin(\theta)) = 2$ $r + r\sin(\theta) = 2$
2. Substitute the Cartesian expressions for r and θ.	Since $y = r\sin(\theta)$ $r + y = 2$
3. Make r the subject.	$r = 2 - y$
4. Express the equation in terms of r^2 by squaring both sides.	$r^2 = (2-y)^2$
5. Substitute for r^2.	Since $r^2 = x^2 + y^2$ $x^2 + y^2 = (2-y)^2$
6. Expand and simplify. *Note:* This is the equation of a parabola.	$= 4 - 4y + y^2$ $x^2 = 4 - 4y$ $4y = 4 - x^2$ $y = 1 - \dfrac{x^2}{4}$

Polar graphs

Polar equations can be graphed using polar coordinates. This is often a better alternative than converting polar equations to the sometimes more complicated Cartesian equation form.

When using polar equations, θ is assumed to be measured in radians.

WORKED EXAMPLE 45 Plotting polar graphs (1)

Sketch the graph of $r = \theta$ for $0 < \theta < 4\pi$ using a CAS calculator.

THINK

The graph looks like the spiral shown.

The process used to sketch the graph on a CAS calculator is stepped out below.

WRITE

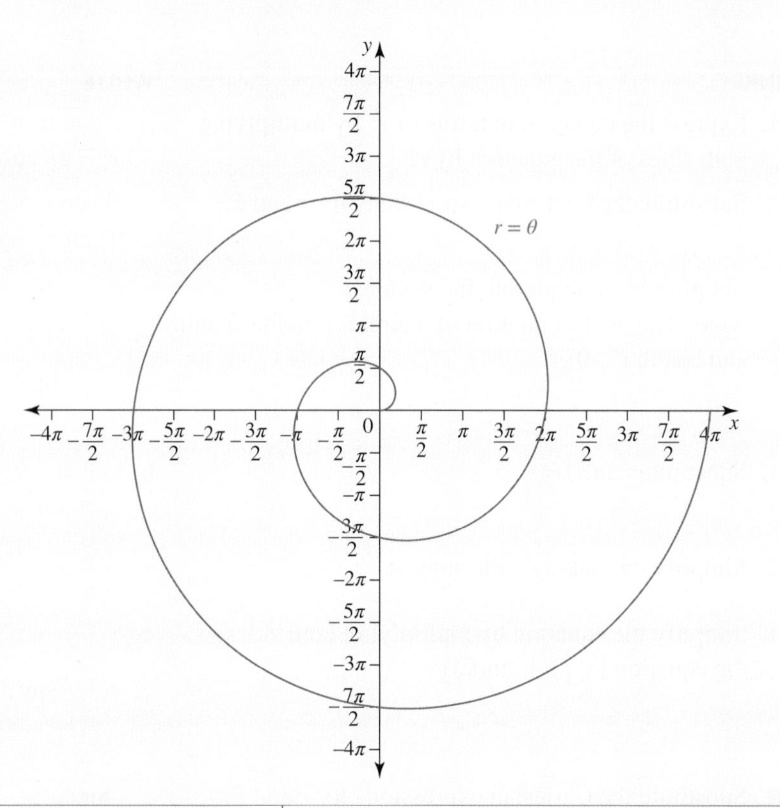

TI | THINK

1. On a Graphs page polar equations can be graphed by selecting:
MENU
3 Graph Entry/Edit
5 Polar

DISPLAY/WRITE

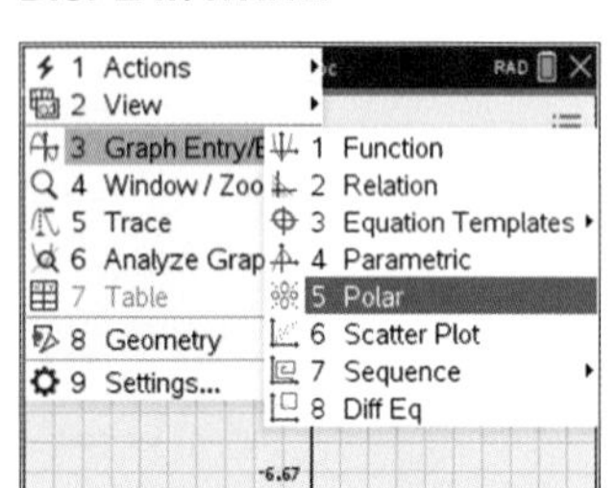

2. Complete the entry line as shown.
$r1(\theta) = \theta$
$0 \le \theta \le 4\pi \;\; \theta step = \dfrac{\pi}{12}$

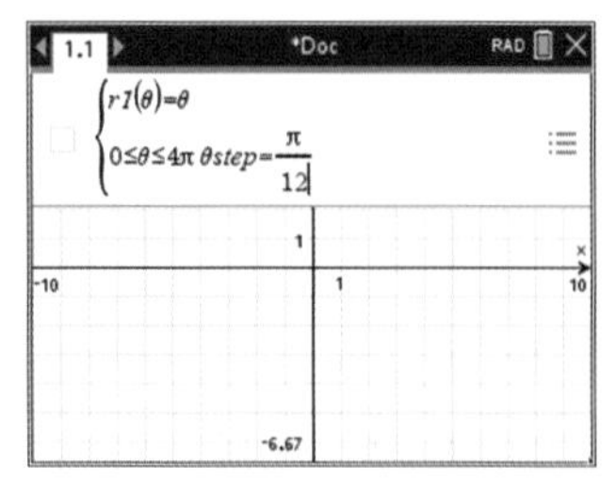

CASIO | THINK

1. In the Graph Editor app, select the polar graphical format.
2. Complete the entry line for $r1 = \theta$ as shown.

DISPLAY/WRITE

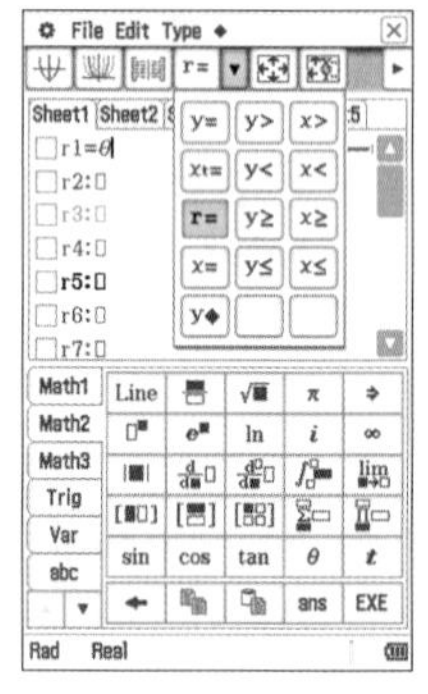

2. In the Window settings screen, set $t\theta$ min to 0 and max to 4π to create a scale that will sketch the polar function.

3. Press the ENTER button. The graph will appear on the screen.
Adjust the zoom and axis scale settings to make the entire graph visible.

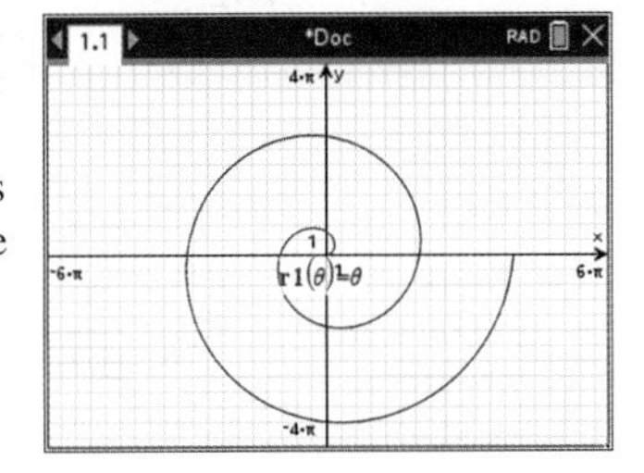

3. Select the graphing icon in the toolbar to create the sketch on the screen.
The Zoom > Square option will allow the design to be drawn using a graphing ratio of 1 : 1.

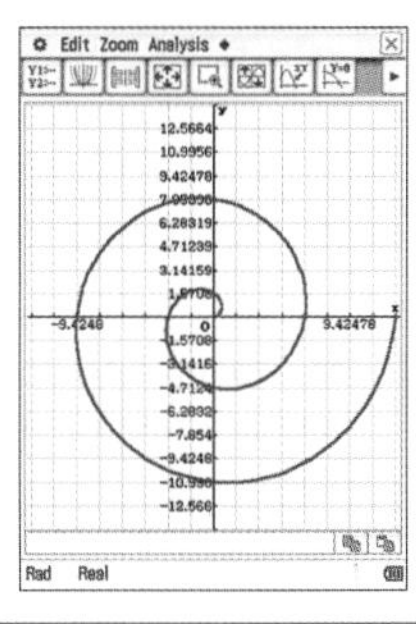

WORKED EXAMPLE 46 Plotting polar graphs (2)

Sketch the graph of $r = 8$ for $0 \leq \theta \leq 2\pi$.

THINK

1. Construct a table of values for $0 \leq \theta \leq 2\pi$ and determine the corresponding r values.

WRITE/DRAW

θ	0	$\frac{\pi}{6}$	$\frac{\pi}{3}$	$\frac{\pi}{2}$	$\frac{2\pi}{3}$	$\frac{5\pi}{6}$
r	8	8	8	8	8	8

θ	π	$\frac{7\pi}{6}$	$\frac{4\pi}{3}$	$\frac{3\pi}{2}$	$\frac{5\pi}{3}$	$\frac{11\pi}{6}$	2π
r	8	8	8	8	8	8	8

2. Sketch the graph, using a protractor and ruler to plot each of the points from the table. Remember r is the distance from the centre (the origin).

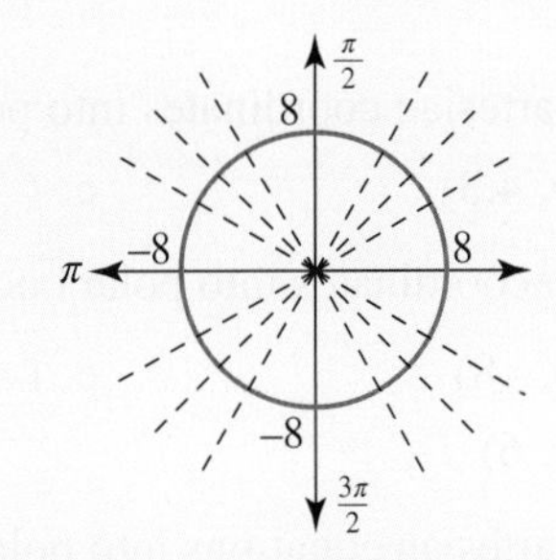

WORKED EXAMPLE 47 Plotting polar graphs (3)

Sketch the graph of $r = 2\sin(\theta)$ for $0 \leq \theta \leq 2\pi$ using a calculator.

THINK

Using the method described in Worked example 45, sketch the graph on a CAS calculator.

WRITE/DRAW

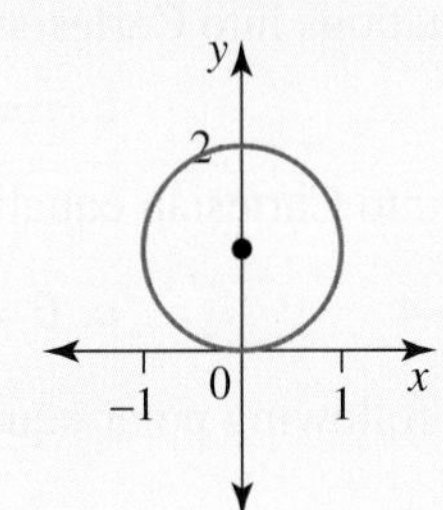

12.9 Exercise

Students, these questions are even better in jacPLUS

Receive immediate feedback and access sample responses

Access additional questions

Track your results and progress

Find all this and MORE in jacPLUS

Technology free

1. **WE40** Plot the following polar coordinates on the same graph.

   a. [2, 0°] b. [5, 180°] c. [0.5, 270°] d. [3, 90°]

2. Plot the following polar coordinates on the same graph.

   a. $\left[1, \frac{\pi}{3}\right]$ b. $\left[1, -\frac{\pi}{3}\right]$ c. $\left[1, \frac{2\pi}{3}\right]$ d. $\left[1, \frac{5\pi}{3}\right]$

3. **WE41** Convert the following to Cartesian coordinates.

   a. [2, 45°] b. [5, 30°] c. [3, 60°] d. [2.7, 90°]
   e. [1.5, 120°] f. [12, 210°]

Technology active

4. Convert the following to Cartesian coordinates.

   a. $\left[2.6, \frac{\pi}{2}\right]$ b. $[7.8, \pi]$ c. $\left[10, \frac{\pi}{3}\right]$ d. $\left[9.1, \frac{5\pi}{3}\right]$

5. **WE42** Convert the following Cartesian coordinates into polar coordinates, expressing θ in degrees.

   a. (5, 0) b. (0, 4.3) c. (–30, 0) d. (0, –9)

6. Convert the following Cartesian coordinates into polar coordinates, expressing θ in radians.

   a. (–5, –12) b. (6, –8) c. $(-1, \sqrt{3})$ d. (–2, –2)
   e. $(2\sqrt{3}, -2)$ f. (5, 6)

7. **WE43** Convert the following Cartesian equations into polar equations.

   a. $x = 3$ b. $y = 2$ c. $x^2 + y^2 = 9$ d. $x^2 + y^2 = 36$

8. Convert the following Cartesian equations into polar equations. Write your answer to part **a** in degrees correct to 1 decimal place.

   a. $y = 5x$ b. $y = x$ c. $3x–4y = 1$ d. $5x + y = 7$

9. **WE44** Convert the following polar equations into Cartesian equations.

   a. $r = 2$ b. $r = 5$ c. $r = 6\sin(\theta)$ d. $r = 2\cos(\theta)$

10. Convert the following polar equations into Cartesian equations.

    a. $\tan(\theta) = 3$ b. $\tan(\theta) = -4$ c. $\theta = \frac{\pi}{4}$ d. $\theta = \frac{3\pi}{4}$

11. **WE45** Sketch the graph of each of the following polar equations for $0 \le \theta \le 4\pi$.

    a. $r = \theta$ b. $r = -\theta$

12. Sketch the graph of each of the following polar equations for $0 \le \theta \le 4\pi$.

    a. $r = 2\theta$ b. $r = -\frac{1}{2}\theta$

13. WE46 Sketch the graph of each of the following polar equations for $0 \le \theta \le 2\pi$.

a. $r = 2$ b. $r = 4$

14. Sketch the graph of each of the following polar equations for $0 \le \theta \le 2\pi$.

a. $r = 1.5$ b. $r = -2$

15. WE47 Sketch the graph of each of the following polar equations for $0 \le \theta \le 2\pi$.

a. $r = \sin(\theta)$ b. $r = 1.5 \sin(\theta)$

16. Sketch the graph of each of the following polar equations for $0 \le \theta \le 2\pi$.

a. $r = -3 \sin(\theta)$ b. $r = 4 \sin(\theta)$

17. Locate each of the following points on the same graph.

a. $[3, 45°]$ b. $[4, 100°]$ c. $[1, 300°]$ d. $[2.5, -30°]$

18. Convert the following Cartesian equations into polar equations.

a. $x^2 + y^2 - 10x + 6y = 0$ b. $x^2 + y^2 + 6x + 8y = 0$ c. $x^2 + y^2 - 12y = 0$

d. $x^2 + y^2 - 2x = 0$ e. $\frac{x^2}{9} + \frac{y^2}{4} = 1$ f. $\frac{x^2}{4} + \frac{y^2}{25} = 1$

19. Convert the following polar equations into Cartesian equations.

a. $r \cos(\theta) = 4$ b. $r \sin(\theta) = -1$ c. $r = 4 \sin(\theta) - 2 \cos(\theta)$

d. $r = 6 \sin(\theta) + 8 \cos(\theta)$ e. $r = \frac{3}{1 + \sin(\theta)}$ f. $r = \frac{4}{1 - \cos(\theta)}$

20. Sketch the following polar equations for $0 \le \theta \le 2\pi$.

a. $r = \cos(\theta)$ b. $r = 2 \cos(\theta)$

c. $r = 3 \cos(\theta)$ d. $r = -4 \cos(\theta)$

21. Sketch $r = 3 \sin(\theta) + 4 \cos(\theta)$ for $0 \le \theta \le 2\pi$.

a. Comment on the shape of the curve.

b. State:

i. the y-intercept(s)
ii. the x-intercept(s)
iii. the length of the diameter
iv. the length of the radius
v. the coordinates of the centre
vi. the Cartesian equation of the curve.

22. Sketch $r = 5 \sin(\theta) + 12 \cos(\theta)$ for $0 \le \theta \le 2\pi$.

a. Comment on the shape of the curve.

b. State:

i. the y-intercept(s)
ii. the x-intercept(s)
iii. the length of the diameter
iv. the length of the radius
v. the coordinates of the centre
vi. the Cartesian equation of the curve.

12.9 Exam questions

Question 1 (4 marks) TECH-FREE

a. Convert each of the following to polar coordinates, expressing θ in radians.

i. $\left(-1, \sqrt{3}\right)$ **(1 mark)**

ii. $\left(-\sqrt{3}, -1\right)$ **(1 mark)**

b. Convert each of the following to Cartesian coordinates.

i. $\left[4\sqrt{2}, -\frac{3\pi}{4}\right]$ **(1 mark)**

ii. $\left[8, \frac{2\pi}{3}\right]$ **(1 mark)**

Question 2 (7 marks) TECH-FREE

Convert each of the following Cartesian equations into polar equations.

a. $x^2 + y^2 = 49$ **(2 marks)**

b. $5x - 12y = 13$ **(2 marks)**

c. $\frac{x^2}{49} + \frac{y^2}{25} = 1$ **(3 marks)**

Question 3 (7 marks) TECH-FREE

Convert each of the following polar equations into Cartesian equations and sketch the graphs.

a. $r = 7\sin(2\theta)$ **(2 marks)**

b. $\tan(\theta) + 1 = 0$ **(2 marks)**

c. $r = \frac{3}{1 + \cos(\theta)}$ **(3 marks)**

More exam questions are available online.

12.10 Parametric equations

LEARNING INTENTION

At the end of this subtopic you should be able to:

- determine the Cartesian equation of a curve expressed by parametric equations.

Parametric equations of a curve express the coordinates of the point of a curve as functions of a variable called a parameter.

For example, the graph of an ellipse in the Cartesian plane cannot be expressed as a function $y = f(x)$, since some values of x are associated with more than one value of y (a relation).

By viewing the curve as the path traced out by the point $P[x(t), y(t)]$ as the independent variable, t, ranges through some interval, $x(t)$ and $y(t)$ are expressed as functions of the single variable, t.

Simple algebraic techniques can be used to convert a set of parametric equations into Cartesian form.

The following worked example shows how to eliminate the parameter and form a single Cartesian equation involving only x and y.

WORKED EXAMPLE 48 Determining the Cartesian equation from parametric equations (1)

Determine the Cartesian equation of the curve whose parametric equations are given by:

a. $x = t, y = t + 2$ **b.** $x = t + 1, y = t^2$.

THINK	WRITE
a. 1. Write the parametric equations and number them.	**a.** $x = t$ [1] $y = t + 2$ [2]
2. Substitute equation [1] into equation [2]. That is, replace t with x in the second equation. This produces a Cartesian equation of a straight line.	Substituting [1] into [2] gives $y = x + 2$.
b. 1. Write the parametric equations and number them.	**b.** $x = t + 1$ [1] $y = t^2$ [2]
2. Rearrange equation [1] to write t in terms of x.	Equation [1] becomes $t = x - 1$
3. Substitute this expression for t into equation [2]. This produces a Cartesian equation of a parabola.	Substituting $t = x - 1$ into [2] gives $y = (x - 1)^2$.

Let us now look at some examples that involve another parameter. When parametric equations involve trigonometric functions, we often need to make use of standard trigonometric identities to eliminate the parameter. In this section the following trigonometric identities may be used.

The Pythagorean identities:

$$\sin^2(\theta) + \cos^2(\theta) = 1$$

$$\sec^2(\theta) = 1 + \tan^2(\theta)$$

$$\text{cosec}^2(\theta) = 1 + \cot^2(\theta)$$

The double-angle identities:

$$\sin(2\theta) = 2\sin(\theta)\cos(\theta)$$

$$\cos(2\theta) = \cos^2(\theta) - \sin^2(\theta)$$

$$\cos(2\theta) = 2\cos^2(\theta) - 1$$

$$\cos(2\theta) = 1 - 2\sin^2(\theta)$$

$$\tan(2\theta) = \frac{2\tan(\theta)}{1 - \tan^2(\theta)}$$

WORKED EXAMPLE 49 Determining the Cartesian equation from parametric equations (2)

Determine the Cartesian equation of the curve whose parametric equations are given by:

a. $x = 3\cos(\theta), y = 2\sin(\theta)$ **b.** $x = \sin(\theta), y = \cos(2\theta)$.

THINK	WRITE
a. 1. Write the parametric equations and number them.	**a.** $x = 3\cos(\theta)$ [1] $y = 2\sin(\theta)$ [2]
2. Square equation [1] and then rearrange it to isolate the trigonometric function. Number the equation.	Squaring [1] gives $x^2 = 9\cos^2(\theta)$ or $\frac{x^2}{9} = \cos^2(\theta)$ [3]

3. Square equation [2] and then rearrange it to isolate the trigonometric function. Number the equation.

Squaring [2] gives $y^2 = 4\sin^2(\theta)$

or $\dfrac{y^2}{4} = \sin^2(\theta)$ [4]

4. Add equations [3] and [4].

Adding [3] and [4] gives

$$\frac{x^2}{9} + \frac{y^2}{4} = \cos^2(\theta) + \sin^2(\theta)$$

5. Use the Pythagorean identity $\cos^2(\theta) + \sin^2(\theta) = 1$ to simplify. This produces a Cartesian equation of an ellipse.

$$\frac{x^2}{9} + \frac{y^2}{4} = 1$$

b. 1. Write the parametric equations and number them.

b. $x = \sin(\theta)$ [1]

$y = \cos(2\theta)$ [2]

2. Use the double-angle identity $\cos(2\theta) = 1 - 2\sin^2(\theta)$ to rewrite equation [2]. Number the equation.

Since $\cos(2\theta) = 1 - \sin^2(\theta)$,
[2] becomes $y = 1 - 2\sin^2(\theta)$ [3]

3. Substitute equation [1] into equation [3]. This produces a Cartesian equation of a section of a parabola with domain and range $[-1, 1]$.

Substituting [1] into [3] gives $y = 1 - 2x^2$.

TI | THINK

DISPLAY/WRITE

a. 1. On a Graphs page parametric equations can be graphed by selecting:
MENU
3 Graph Entry/Edit
4 Parametric

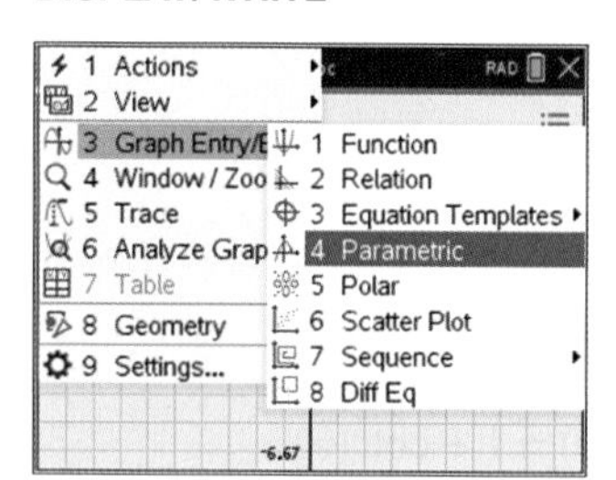

2. Complete the entry line as shown.
$x1(t) = 3\cos(t)$
$y1(t) = 2\sin(t)$

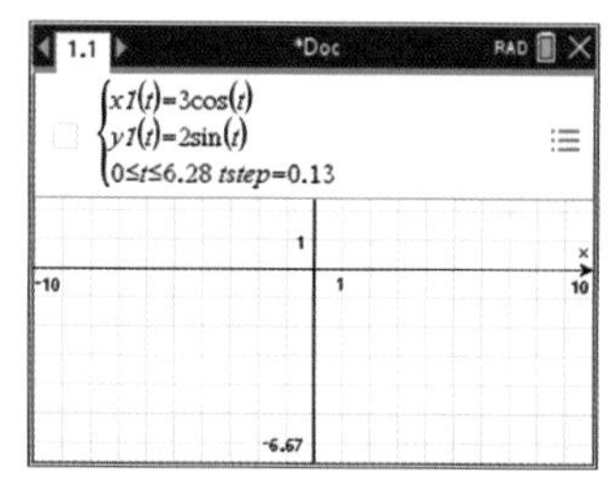

3. Press the ENTER button. The graph will appear on the screen. Adjust the zoom and axis scale settings to make the entire graph visible.

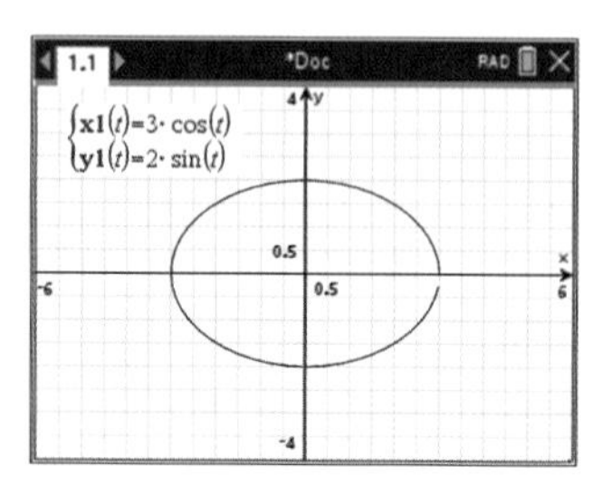

CASIO | THINK

DISPLAY/WRITE

a. 1. In the Graph Editor app, select the parametric graphical format.

2. Complete the entry line as shown:
xt1 = 3 cos(t)
yt1 = 2 sin(t)
Select the graphing icon in the toolbar to create the sketch on the screen. The axes scales can be adjusted accordingly.

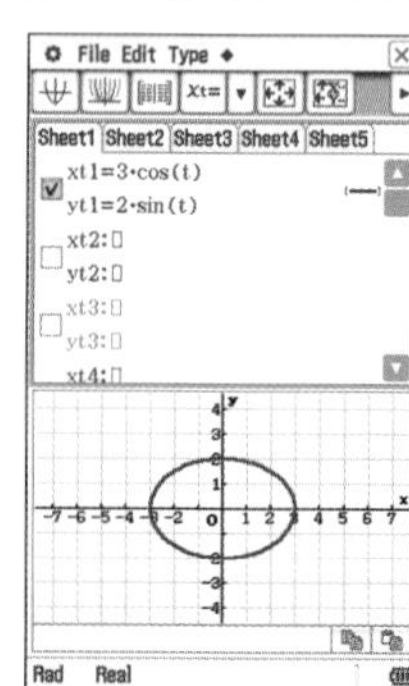

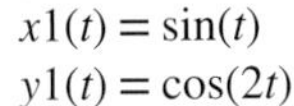

b. Repeat on a new graphs page for
$x1(t) = \sin(t)$
$y1(t) = \cos(2t)$

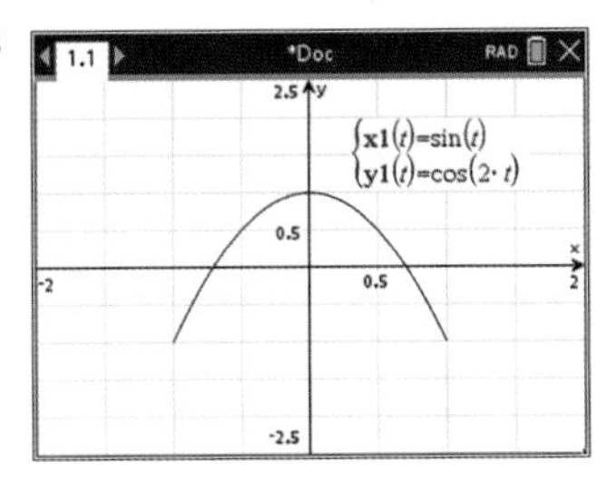

b. Repeat the process for the new graph of:
$xt2 = \sin(t)$
$yt2 = \cos(2t)$
The axes scales can be adjusted accordingly.

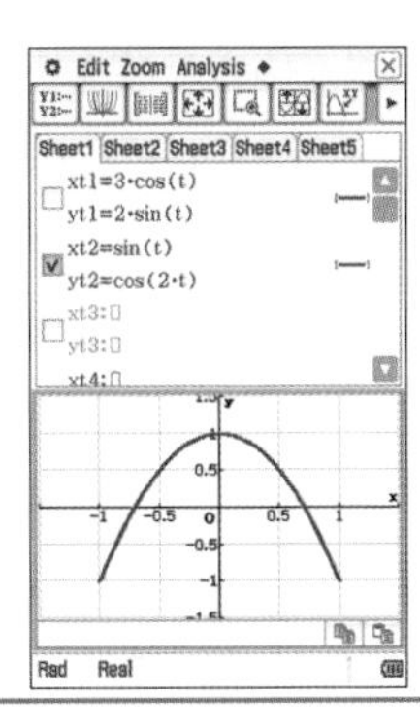

12.10 Exercise

Students, these questions are even better in jacPLUS

Receive immediate feedback and access sample responses

Access additional questions

Track your results and progress

Find all this and MORE in jacPLUS

Technology free

1. WE48 Determine the Cartesian equation of the curve whose parametric equations are given by:

   a. $x = t,\ y = t - 7$
   b. $x = t + 1,\ y = t^2 + 4.$

2. Determine the Cartesian equation of the curve whose parametric equations are given by:

   a. $x = t,\ y = \dfrac{1}{t}$
   b. $x = t^3,\ y = t^2.$

3. WE49 Determine the Cartesian equations and sketch the graphs of the curves whose parametric equations are given by:

   a. $x = \sin(\theta),\ y = \cos(\theta)$
   b. $x = 4\cos(\theta),\ y = 2\sin(\theta).$

4. Determine the Cartesian equation of the curve whose parametric equations are given by:

   a. $x = \cos(\theta),\ y = \sin(\theta)$
   b. $x = 2\sec(\theta),\ y = \tan(\theta).$

The following instruction relates to questions 5 to 14.

Determine the Cartesian equation of the curve with the given parametric equation.

5. $x = 2t + 3,\ y = 4t^2 - 9$
6. $x = t^2 + t,\ y = t^2 - t$
7. $x = \dfrac{2t}{1 + t^2},\ y = \dfrac{1 - t^2}{1 + t^2}.$
8. $x = -\sec(\theta),\ y = \tan(\theta)$
9. $x = \cos(2\theta),\ y = \sin(\theta)$
10. $x = 1 - \sin(t),\ y = 3 - 2\cos(t)$
11. $x = \cos(\theta),\ y = \sin(2\theta)$
12. $x = \sec^2(\theta) - 1,\ y = \tan(\theta)$

13. $x = \cos(2\theta),\ y = \sin(2\theta)$

14. $x = 2\cos^3(\theta),\ y = 2\sin^3(\theta)$.

Technology active

15. a. Determine the Cartesian equation of the curve whose parametric equations are given by:

$$x = t\cos(t),\ y = t\sin(t),\ t > 0.$$

b. The curve described by the parametric equations in part a is called a spiral. Using technology, sketch the graph of this spiral for $0 < t < 4\pi$.

16. Lissajous figures are sometimes used in graphic design as logos or as screen savers. The general parametric equations to describe a Lissajous figure are:

$$x = A\sin(at + c),\ y = B\sin(bt).$$

Using technology, sketch the graph of a Lissajous figure for $A = 4,\ B = 4,\ a = 1,\ b = 3$ and $c = \frac{\pi}{2}$.

Do you recognise this shape as a well-known logo?

Try some different values for A, B, a, b and c.

12.10 Exam questions

Question 1 (4 marks) TECH-FREE

Given the parametric equations $x = \cos(t)$ and $y = \cos(2t)$ for $0 \le t \le 2\pi$, determine the Cartesian equation and sketch the graph, stating the domain and range.

Question 2 (4 marks) TECH-FREE

Given the parametric equations $x = 4\cos(t)$ and $y = 3\sin(t)$ for $0 \le t \le 2\pi$, determine the Cartesian equation and sketch the graph, stating the domain and range.

Question 3 (4 marks) TECH-FREE

Given the parametric equations $x = 3\sec(t)$ and $y = 4\tan(t)$ for $0 \le t \le 2\pi$, determine the Cartesian equation and sketch the graph, stating the domain and range.

More exam questions are available online.

12.11 Review

12.11.1 Summary

doc-37931

Hey students! Now that it's time to revise this topic, go online to:

Access the topic summary

Review your results

Watch teacher-led videos

Practise exam questions

Find all this and MORE in jacPLUS

12.11 Exercise

Technology free: short answer

1. Sketch the graph of each of the following stating the range.
   a. $y = 2 - |3x + 2|$
   b. $y = \dfrac{1}{x^2 + 3x + 2}$
   c. $y = |x^2 + 3x + 2|$ and solve $|x^2 + 3x + 2| \leq 1$
2. Sketch the graph of each of the following functions.
   a. $f(x) = 2\sin^{-1}\left(\dfrac{3x+2}{4}\right)$
   b. $f: [-4, 4] \to R,\ f(x) = 2\sec\left(\dfrac{\pi x}{2}\right)$
   c. $f: (-4, 4) \to R,\ f(x) = 2\cot\left(\dfrac{\pi x}{2}\right)$
3. Sketch graphs of the following, clearly showing the key features.
   a. $x^2 + y^2 = 9$
   b. $\dfrac{(x+3)^2}{9} + \dfrac{(y+1)^2}{4} = 1$
   c. $\dfrac{x^2}{64} - \dfrac{y^2}{36} = 1$
4. a. For the equation $\dfrac{x^2}{25} + \dfrac{y^2}{16} = 1$, determine:
      i. the length of the major axis
      ii. the length of the minor axis
      iii. the coordinates of the x- and y-intercepts.
   b. Hence, sketch the graph.
5. Transform the following polar equations into Cartesian equations.
   a. $r = 2$
   b. $\theta = -\dfrac{\pi}{4}$
   c. $r\cos(\theta) = 6$
   d. $r = \dfrac{3}{1 + \sin(\theta)}$
6. Determine the Cartesian equation of the curve with the following parametric equations.
   a. $x = \dfrac{-5}{t},\ y = -5t$
   b. $x = 2^t,\ y = 4^t$
   c. $x = 2\sin(t),\ y = 3t$

Technology active: multiple choice

7. **MC** Given the function $f: [0, 2] \to R$, $f(x) = |2 - x| - 2$, the range of the function is

A. R **B.** $[0, \infty)$ **C.** $(-\infty, -2]$ **D.** $[0, 2]$ **E.** $[-2, 0]$

8. **MC** The graph of $y = \dfrac{1}{x^2 + 2x + 3}$, has

A. a maximum turning point at $\left(-1, \dfrac{1}{2}\right)$

B. a maximum turning point at $\left(1, \dfrac{1}{3}\right)$

C. a minimum turning point at $\left(-1, \dfrac{1}{2}\right)$

D. a minimum turning point at $\left(1, \dfrac{1}{3}\right)$

E. no turning points and no vertical asymptotes

9. **MC** The domain of the function $y = \dfrac{1}{\cos^{-1}(3 - x)}$ is

A. R **B.** $(2, 4]$ **C.** $[2, 4)$ **D.** $[2, 4]$ **E.** $[2, 3) \cup (3, 4]$

10. **MC** The graph of the function $f: \left[-\dfrac{\pi}{2}, \dfrac{\pi}{2}\right] \to R$, $f(x) = 3\cot(3x)$, has

A. vertical asymptotes at $x = 0$ and $x = \pm\dfrac{\pi}{3}$ and crosses the x-axis at $x = \pm\dfrac{\pi}{6}$ and $x = \pm\dfrac{\pi}{2}$

B. vertical asymptotes at $x = \pm\dfrac{\pi}{6}$ and $x = \pm\dfrac{\pi}{2}$ and crosses the x-axis at $x = 0$ and $x = \pm\dfrac{\pi}{3}$

C. vertical asymptotes at $x = \pm\dfrac{\pi}{6}$ and $x = \pm\dfrac{\pi}{2}$ and does not crosses the x-axis

D. crosses the x-axis at $x = \pm\dfrac{\pi}{6}$ and $x = \pm\dfrac{\pi}{2}$ and has no vertical asymptotes

E. does not crosses the x-axis and has no vertical asymptotes

11. **MC** The circle with the equation $x^2 + 8x + y^2 - 6y + 1 = 0$ has its centre and radius respectively as

A. $(4, -3)$, 24 **B.** $(-4, 3)$, $6\sqrt{2}$ **C.** $(-4, 3)$, 24 **D.** $(3, -4)$, $2\sqrt{6}$ **E.** $(-4, 3)$, $2\sqrt{6}$

12. **MC** The polar coordinates of $(-3, 0)$ are

A. $\left[3, \dfrac{3\pi}{2}\right]$ **B.** $[-3, \pi]$ **C.** $[3, \pi]$ **D.** $[-3, 2\pi]$ **E.** $[3, 0]$

13. **MC** The Cartesian equation $x + y = 0$ can be expressed in polar form as

A. $\theta = \dfrac{3\pi}{4}$ **B.** $\theta = \dfrac{\pi}{4}$ **C.** $\theta = -\dfrac{\pi}{4}$ **D.** $r = 1$ **E.** $r = \theta$

14. **MC** The graph of the polar equation $r = 3\cos(4\theta)$ is a 'rose'. The number of leaves is

A. 12 **B.** 8 **C.** 4 **D.** 3 **E.** 2

15. **MC** The graph of the polar equation $r = \dfrac{2}{3 - \cos(\theta)}$ is a

A. straight line **B.** parabola **C.** circle **D.** ellipse **E.** hyperbola

16. **MC** The graph of the parametric equation $x = 5\cos(t)$, $y = 12\sin(t)$ is a

A. straight line **B.** parabola **C.** circle **D.** ellipse **E.** hyperbola

Technology active: extended response

17. For the hyperbola shown:

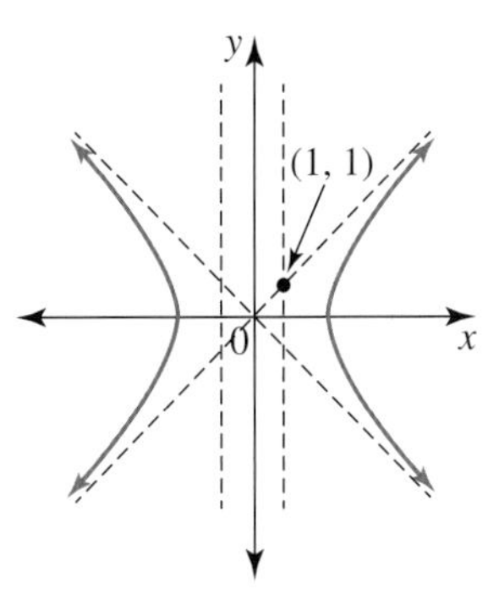

   a. state the gradient of each asymptote
   b. determine the equations of each asymptote
   c. given that the equation of the hyperbola is $\frac{x^2}{a^2} - \frac{y^2}{b^2} = 1$, state the relationship between a and b
   d. determine the equation of the hyperbola, given that the graph has a vertex at $\left(\sqrt{2}, 0\right)$.

18. A rectangle is bounded by the straight lines with equations $x = 2, x = 2a, y = b$, and $y = -b$, where $a > b + 1$.
   a. Sketch the rectangle on the Cartesian plane.
   b. Determine the equation of the largest ellipse that can be enclosed within the rectangle.

19. Plot the graph of the polar equation $r = 3\ \sin(\theta)$ where $0 \le \theta \le 2\pi$.
   a. Explain why the graph is the same when $0 \le \theta \le \pi$.
   b. By experimenting with different values for θ step (the amount by which θ is increased), develop and justify an argument that polar graphs with the general equation $r = a\ \sin(\theta)$ are not circular.
   c. Determine the relationship between θ step and the shape of the graph obtained.
   d. Given that polar graphs with the general equation $r = a\ \sin(\theta)$ are in fact circular, what assumptions are made when an equation is determined only by inspection of its graph?

20. a. Show that the graph defined in polar form $r = \frac{3}{2 - \cos(\theta)}$ and the graph of the parametric equations $x = 1 + 2\cos(t),\ y = \sqrt{3}\sin(t)$ both give the ellipse $3x^2 - 6x + 4y^2 - 9 = 0$
   b. Show that the graph defined in polar form $r = \frac{6}{1 - 2\cos(\theta)}$ and the graph of the parametric equations $x = -4 + 2\sec(t),\ y = 2\sqrt{3}\tan(t)$ both give the hyperbola $3x^2 + 24x - 4y^2 + 36 = 0$

12.11 Exam questions

Question 1 (4 marks) TECH-FREE

Express $r = \frac{1}{1 + \cos(\theta)}$ in Cartesian form.

Question 2 (2 marks) TECH-FREE

Determine the Cartesian equation with the parametric form $x = \sqrt{t + 2},\ y = t^3$.

Question 3 (1 mark) TECH-ACTIVE

MC The equation $p\left(x^2 + 2x\right) + y^2 = 1$ will represent an ellipse if

A. $p > 0$
B. $p > 1$
C. $-1 < p < 0$
D. $p = 1$
E. $p = \pm 1$

Question 4 (1 mark) TECH-ACTIVE

MC The equation for the hyperbola shown is

A. $\dfrac{(y+1)^2}{9} - x^2 = 1$

B. $\dfrac{(y-1)^2}{9} - x^2 = 1$

C. $\dfrac{(y-1)^2}{9} - \dfrac{x^2}{36} = 1$

D. $\dfrac{x^2}{36} - \dfrac{(y-1)^2}{9} = 1$

E. $\dfrac{(y-1)^2}{9} - \dfrac{x^2}{6} = 1$

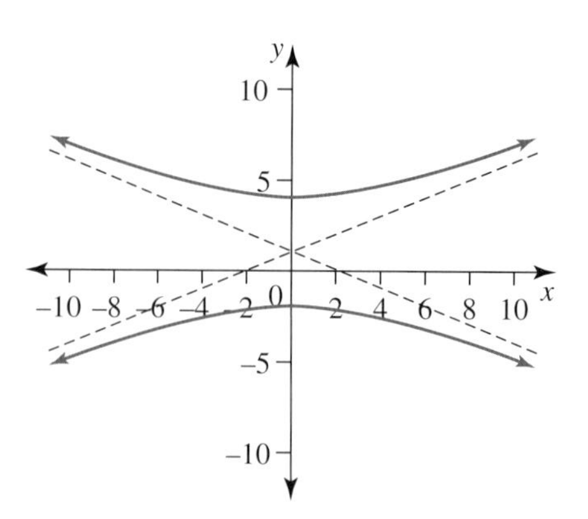

Question 5 (4 marks) TECH-ACTIVE

a. Show that the graph defined in polar form $r = 2$ and the graph of the parametric equations $x = 2\cos(t),\ y = 2\sin(t)$ both give the circle $x^2 + y^2 = 4$. **(2 marks)**

b. Show that the graph defined in polar form $r = \dfrac{2}{1-\cos(\theta)}$ and the graph of the parametric equations $x = \cos(2t),\ y = 2\sqrt{2}\cos(t)$ both give the parabola $y^2 = 4(x+1)$. **(2 marks)**

More exam questions are available online.

Answers

Topic 12 Functions, relations and graphs

12.2 The absolute value function

12.2 Exercise

1. a.

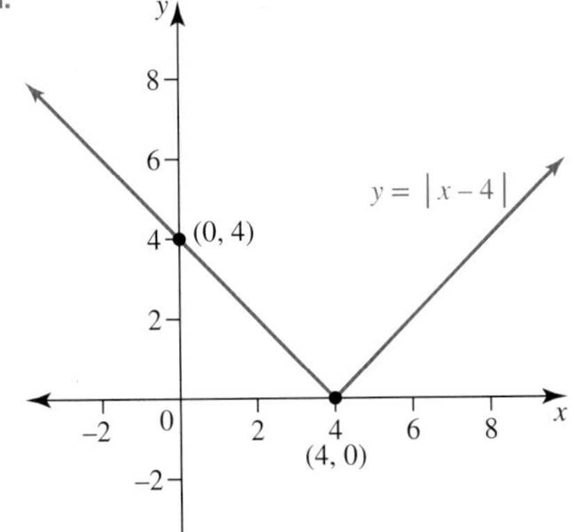

b.

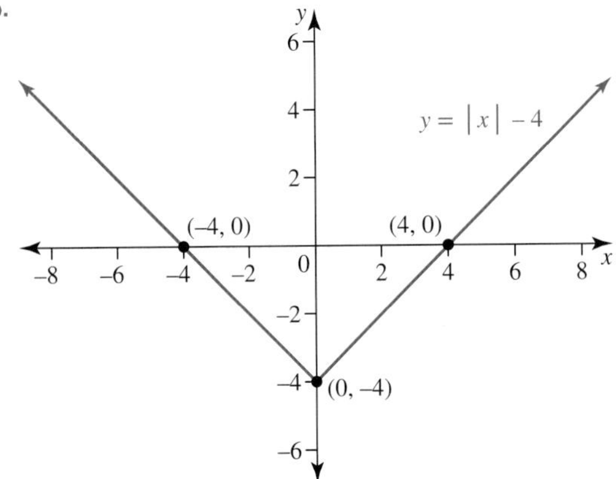

2. a.

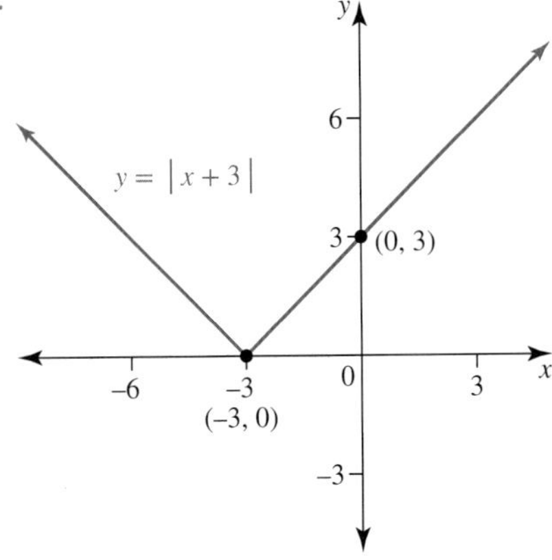

b.

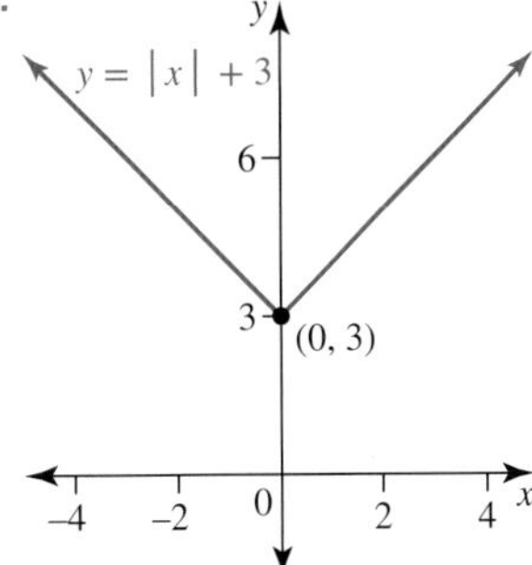

3.

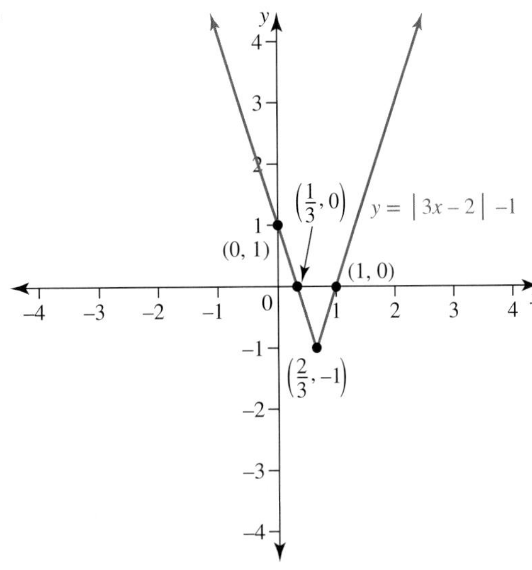

Range $[-1, \infty)$

4.

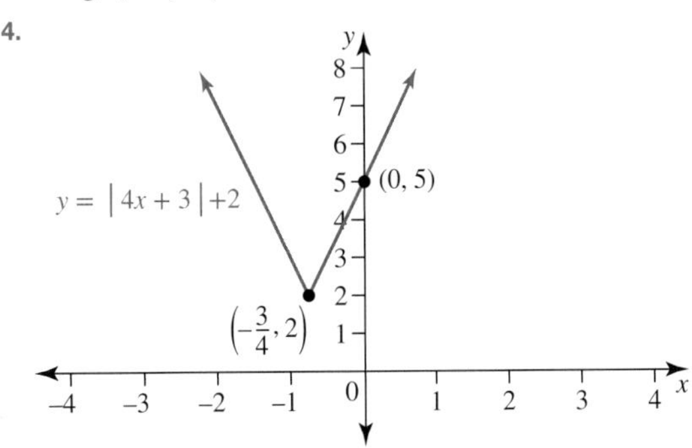

Range $[2, \infty)$

5.

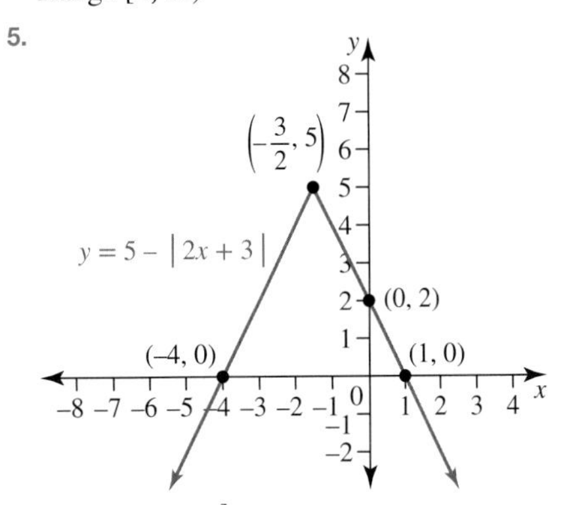

Range $(-\infty, 5]$

6.

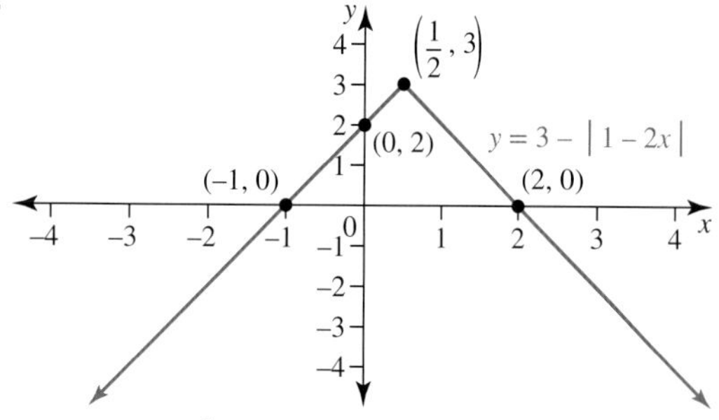

Range $(-\infty, 3]$

7. a. $-2, 10$ b. $(-\infty, -2) \cup (10, \infty)$

8. a. $-8, 2$ b. $[-8, 2]$

9. a. $-\frac{7}{3}, 1$ b. $\left(-\infty, -\frac{7}{3}\right] \cup [1, \infty)$

10. a. $\frac{5}{3}, 3$ b. $\left[\frac{5}{3}, 3\right]$

11. a. No solution

b. $-\frac{5}{3}$

c. $-5, -\frac{5}{3}$

12. a. $-\frac{4}{3}, 16$ b. $\left(-\frac{8}{7}, \frac{16}{5}\right)$

13. a. $-\frac{1}{2}, 3$ b. $\left(-\infty, -\frac{2}{5}\right) \cup \left(\frac{8}{3}, \infty\right)$

14. a. $\frac{2}{5}, \frac{8}{9}$ b. $(-\infty, -2] \cup \left[\frac{12}{5}, \infty\right)$

15.

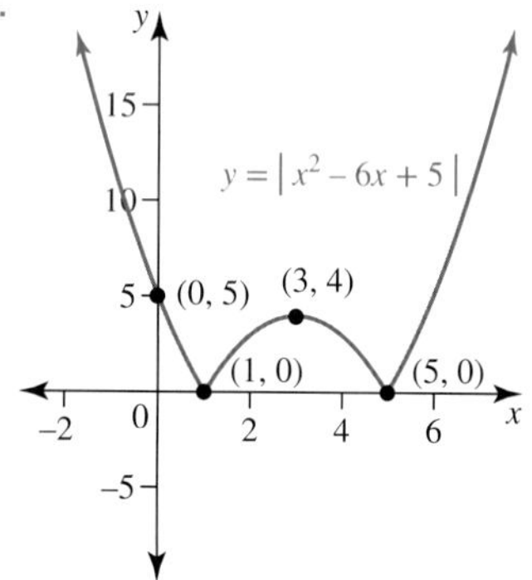

$x \in (0, 6)$

16. a.

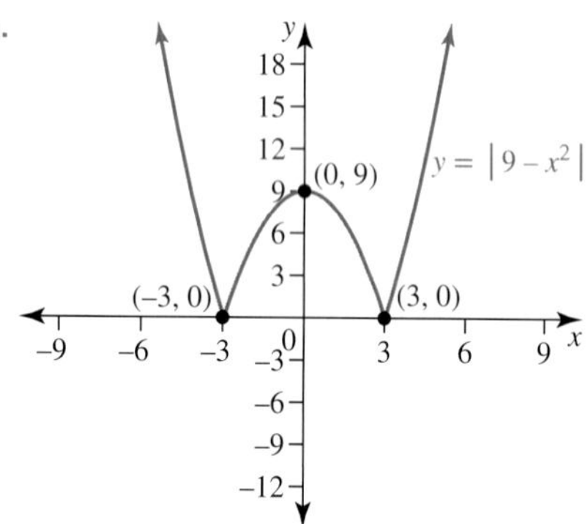

b. $x \in \left(-\sqrt{6}, \sqrt{6}\right) \cup \left(-\infty, -2\sqrt{3}\right) \cup \left(2\sqrt{3}, \infty\right)$

17. a.

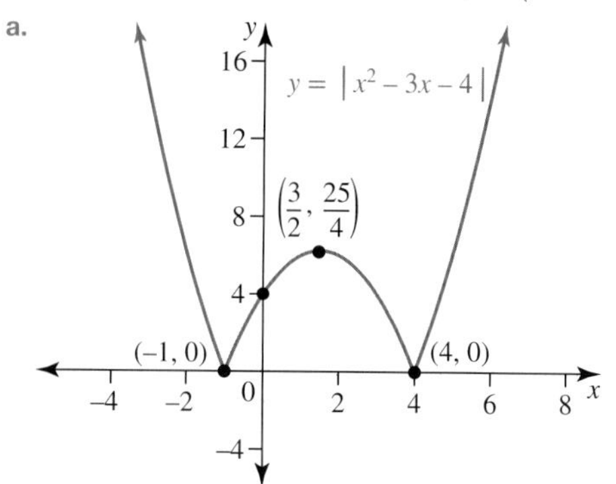

b. $x \in [1, 2] \cup (-\infty, -2] \cup [5, \infty)$

18.

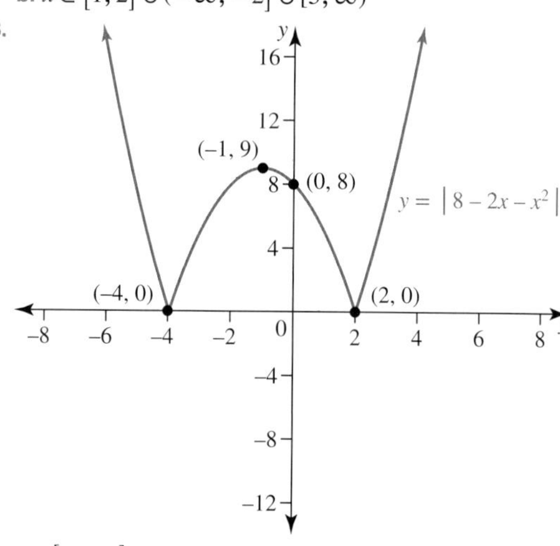

$x \in [-6, 4]$

19. a. $-2, 4$ b. $-4, 1$

20. a. $\pm 1, \ \pm \frac{\sqrt{2}}{2}$

b. $\pm 1, \ \pm \frac{\sqrt{15}}{3}$

21. a. $-2 \pm \sqrt{2}$

b. $\frac{1 \pm \sqrt{73}}{6}$

22. a. $1, \frac{9}{4}$

b. $1, \frac{25}{16}$

23. $y = \begin{cases} 6x \text{ if } x \geq \frac{2}{3} \\ 4 \quad \text{if } -\frac{2}{3} \leq x \leq \frac{2}{3} \\ -6x \text{ if } x \leq -\frac{2}{3} \end{cases}$

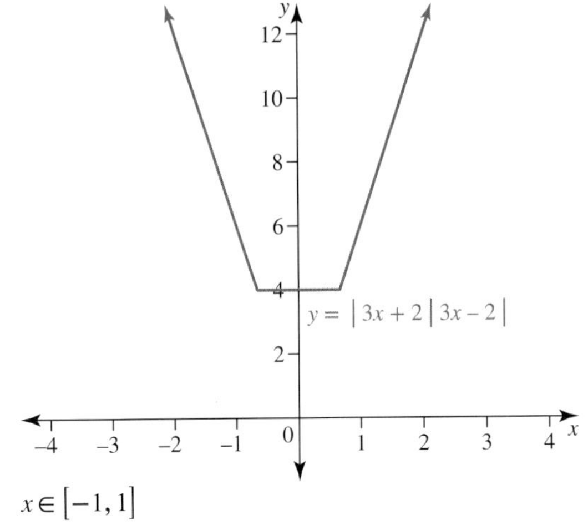

$x \in [-1, 1]$

24. $y = \begin{cases} 6 \text{ if } x \geq \frac{3}{2} \\ 4x \text{ if } -\frac{3}{2} \leq x \leq \frac{3}{2} \\ -6 \text{ if } x \leq -\frac{3}{2} \end{cases}$

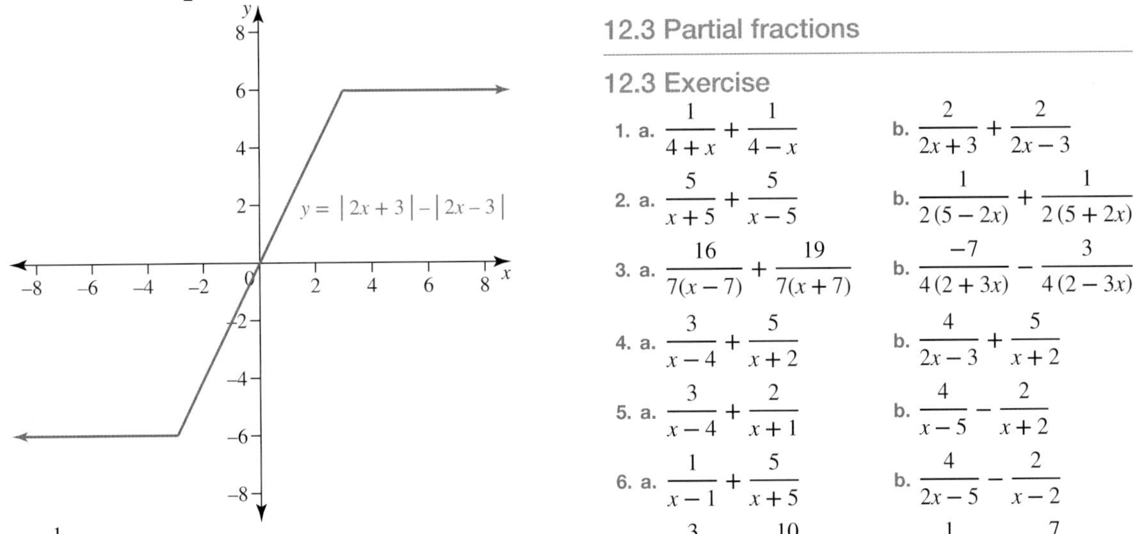

$x = \frac{1}{2}$

12.2 Exam questions

Note: Mark allocations are available with the fully worked solutions online.

1. a. See the image at the bottom of the page*
   Range $(-\infty, 4]$
   b. $x \in (-\infty, 1) \cup (2, \infty)$
2. $\left(-\infty, \frac{5}{3}\right] \cup [5, \infty)$
3.

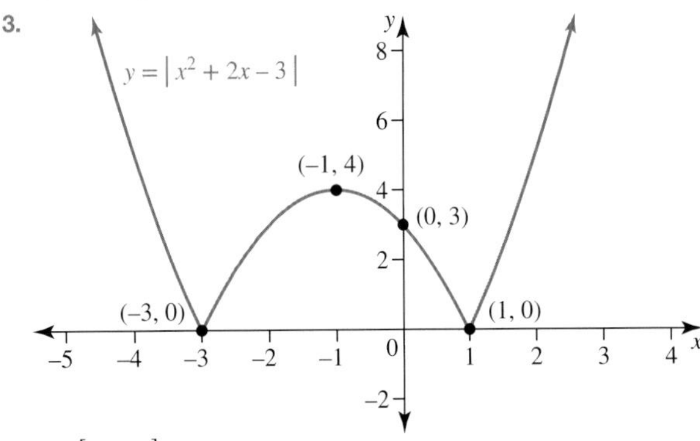

$x \in [-4, 2]$

12.3 Partial fractions

12.3 Exercise

1. a. $\frac{1}{4+x} + \frac{1}{4-x}$ b. $\frac{2}{2x+3} + \frac{2}{2x-3}$
2. a. $\frac{5}{x+5} + \frac{5}{x-5}$ b. $\frac{1}{2(5-2x)} + \frac{1}{2(5+2x)}$
3. a. $\frac{16}{7(x-7)} + \frac{19}{7(x+7)}$ b. $\frac{-7}{4(2+3x)} - \frac{3}{4(2-3x)}$
4. a. $\frac{3}{x-4} + \frac{5}{x+2}$ b. $\frac{4}{2x-3} + \frac{5}{x+2}$
5. a. $\frac{3}{x-4} + \frac{2}{x+1}$ b. $\frac{4}{x-5} - \frac{2}{x+2}$
6. a. $\frac{1}{x-1} + \frac{5}{x+5}$ b. $\frac{4}{2x-5} - \frac{2}{x-2}$
7. a. $\frac{3}{x+4} - \frac{10}{(x+4)^2}$ b. $\frac{1}{3x-2} + \frac{7}{(3x-2)^2}$
8. a. $\frac{1}{x-5} + \frac{2}{(x-5)^2}$ b. $\frac{1}{5x-2} + \frac{6}{(5x-2)^2}$
9. a. $\frac{1}{x} - \frac{x}{x^2+4} + \frac{5}{x^2+4}$ b. $\frac{-3}{2x} + \frac{3x}{2(x^2+2)} + \frac{5}{x^2+2}$

---

*1. a.

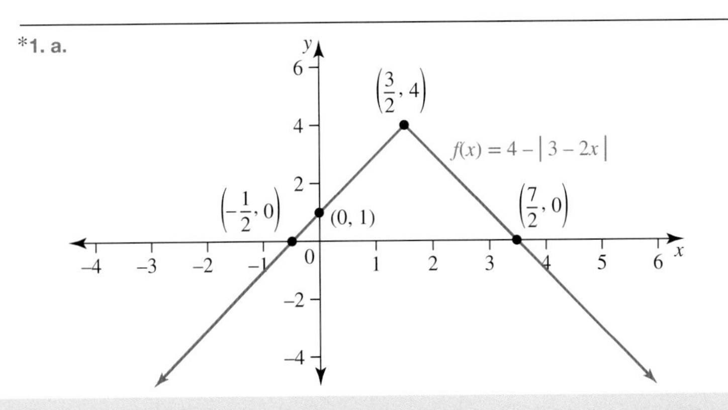

10. a. $\frac{1}{x} - \frac{x}{x^2+3} - \frac{5}{x^2+3}$ b. $\frac{1}{x} - \frac{x}{x^2+5} + \frac{2}{x^2+5}$

12.3 Exam questions

Note: Mark allocations are available with the fully worked solutions online.

1. D
2. E
3. B

12.4 Reciprocal graphs

12.4 Exercise

1. a.

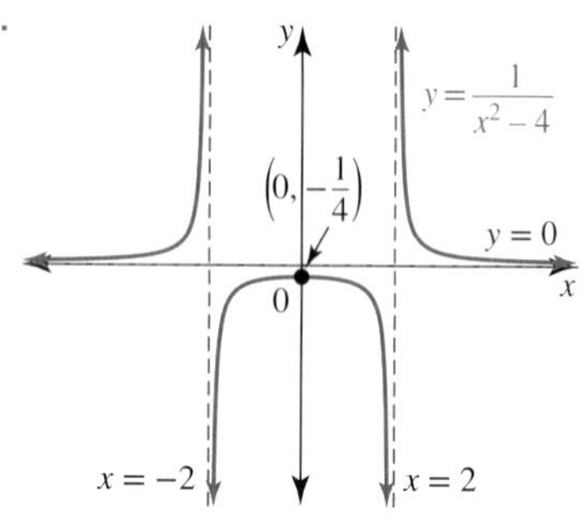

Turning point $\left(0, -\frac{1}{4}\right)$

b.

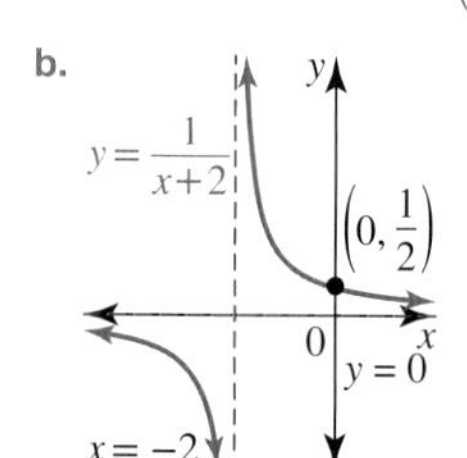

2. a.

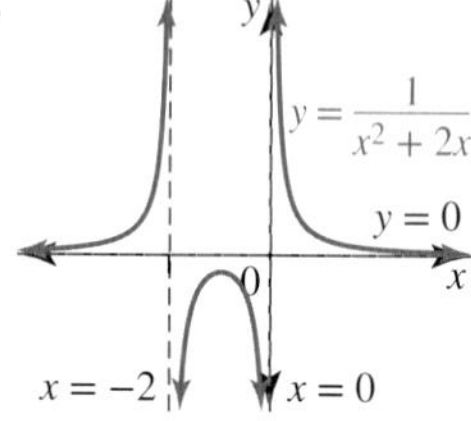

Turning point (−1, −1)

b.

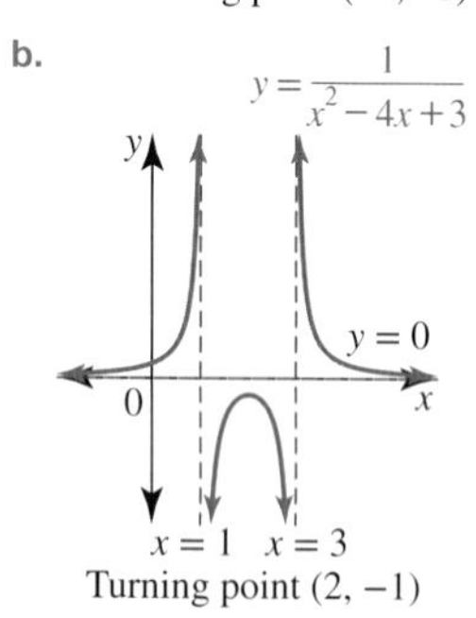

3.

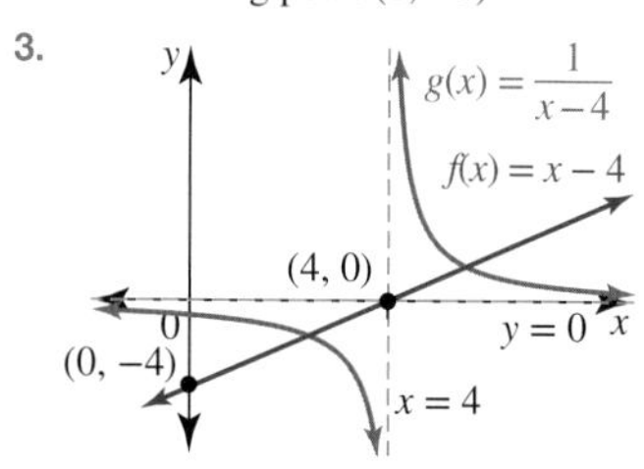

4.

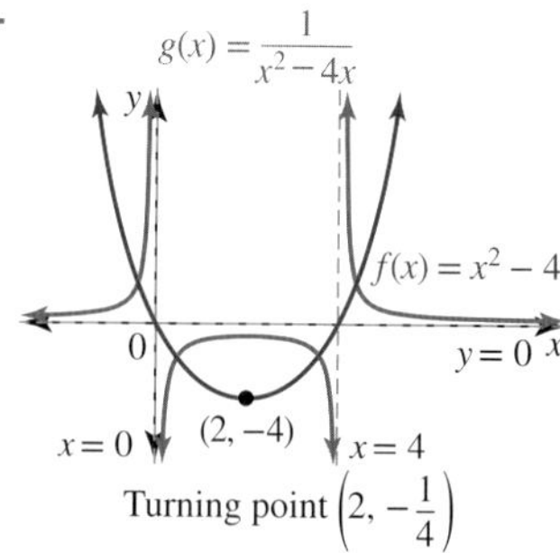

5.

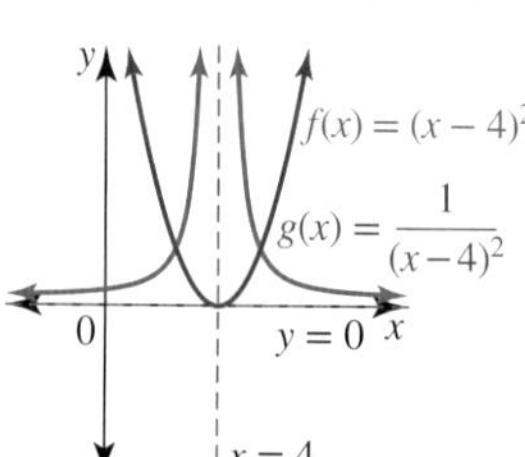

6.

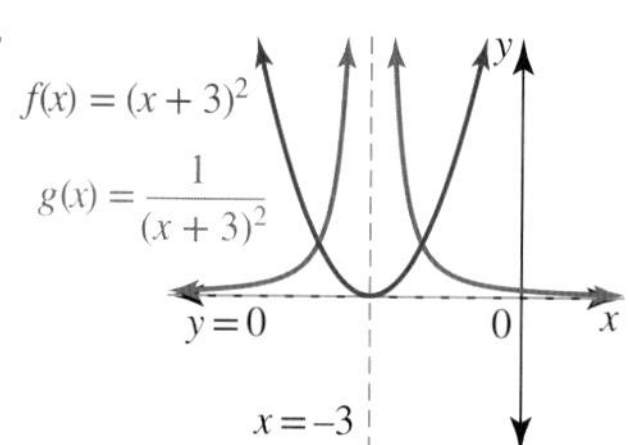

7.

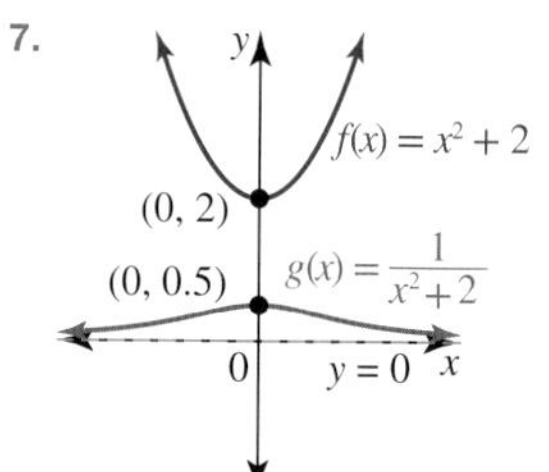

Turning point $\left(0, \frac{1}{2}\right)$

8.

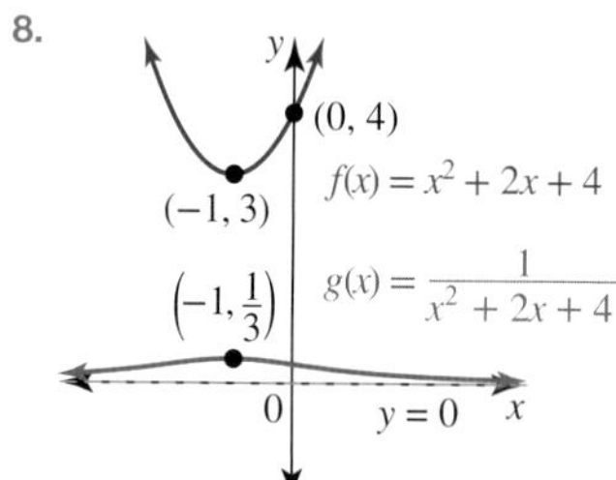

Turning point $\left(-1, \frac{1}{3}\right)$

9. a.

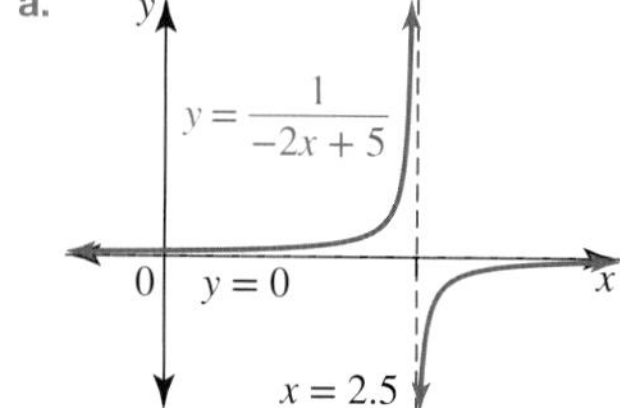

b.

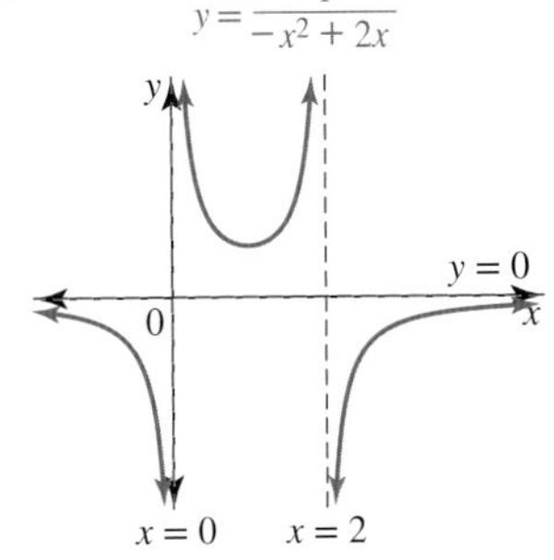

Turning point (1, 1)

10. a.

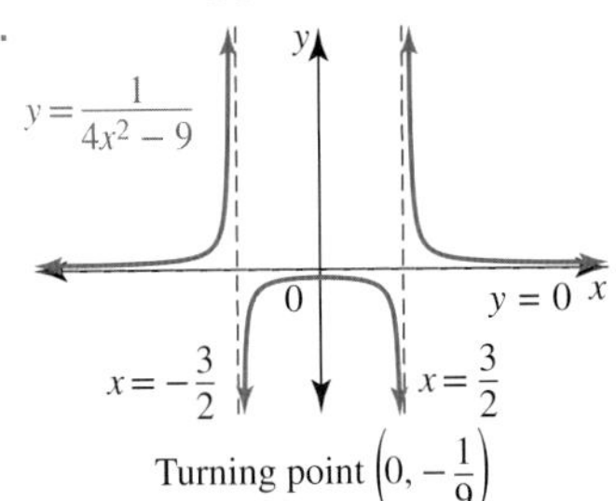

Turning point $\left(0, -\frac{1}{9}\right)$

b.

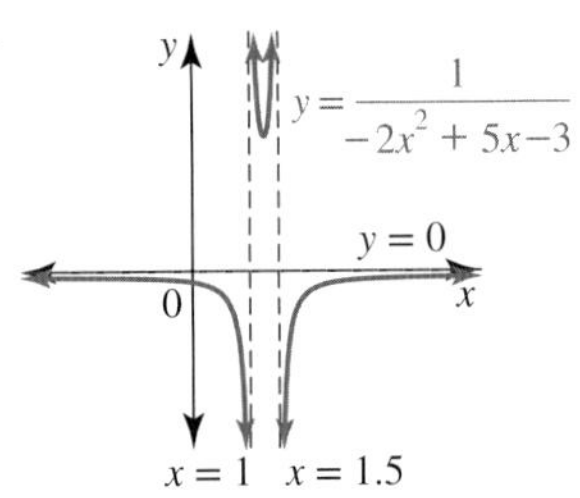

Turning point $\left(\frac{5}{4}, 8\right)$

11.

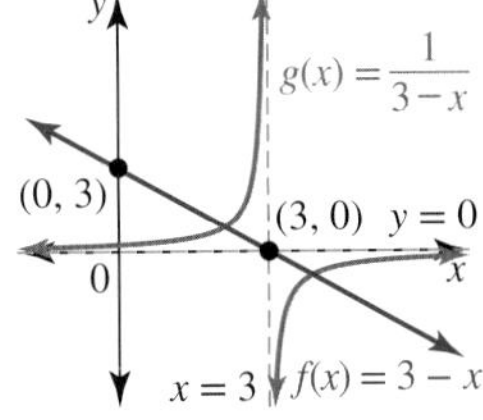

12.

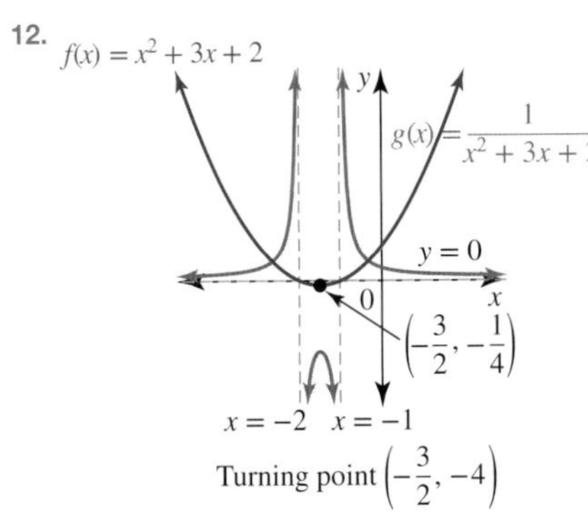

Turning point $\left(-\frac{3}{2}, -4\right)$

13.

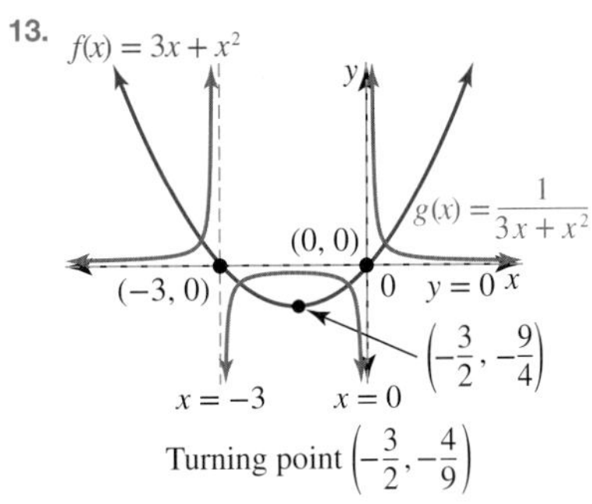

Turning point $\left(-\frac{3}{2}, -\frac{4}{9}\right)$

14.

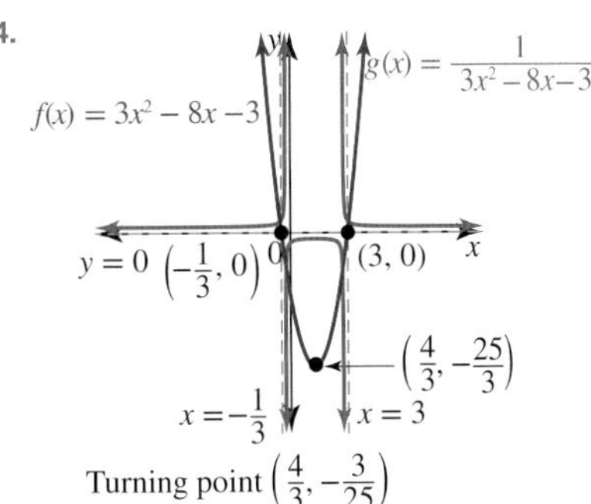

Turning point $\left(\frac{4}{3}, -\frac{3}{25}\right)$

15.

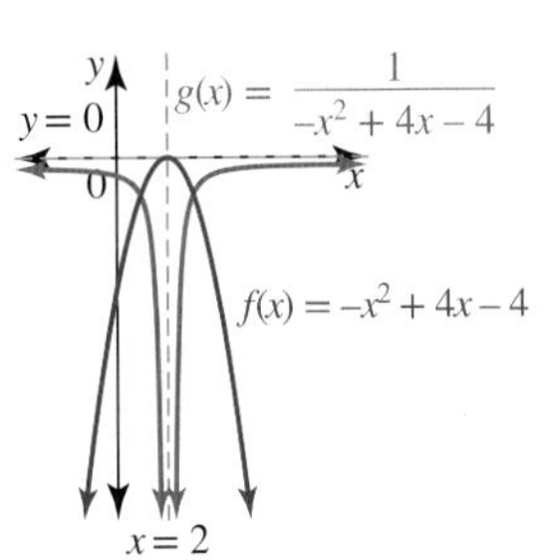

16.

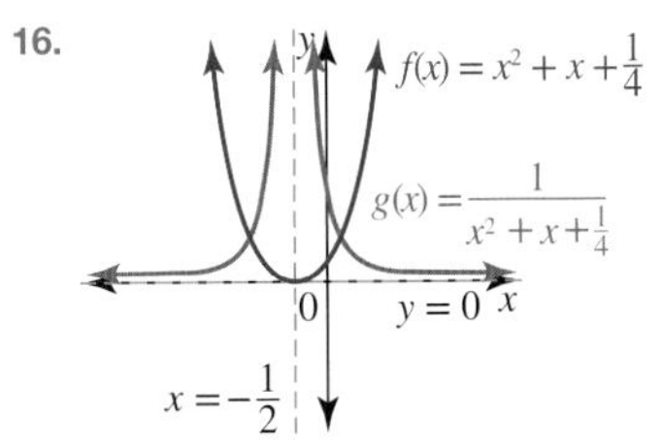

12.4 Exam questions

Note: Mark allocations are available with the fully worked solutions online.

1.

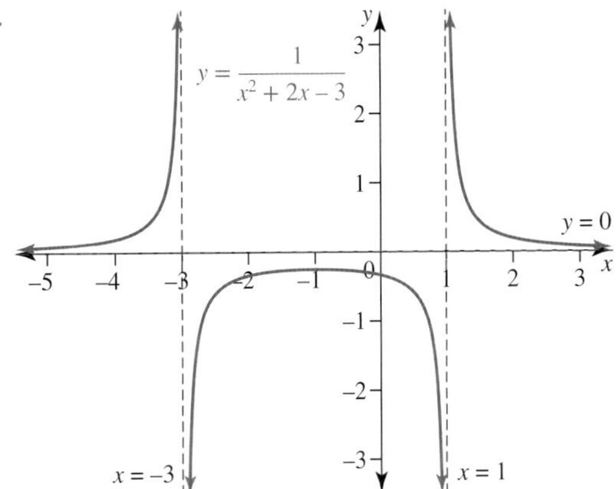

2. a.

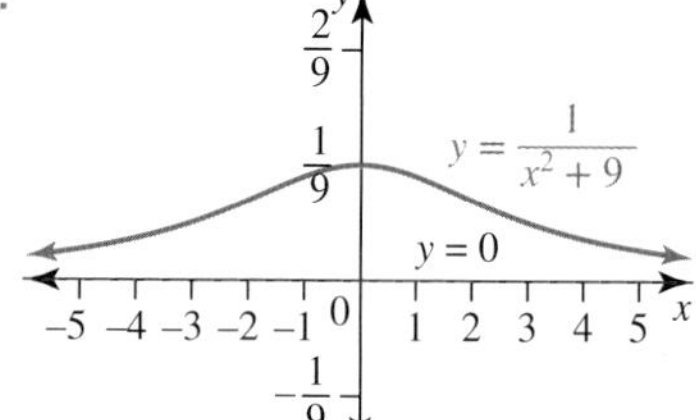

b.

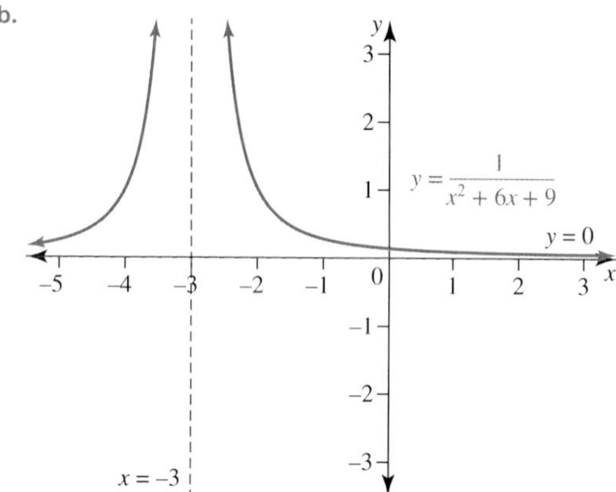

3. a. $(-\infty, -8) \cup (8, \infty)$ b. ± 8
c. $(-8, 8)$

12.5 The reciprocal trigonometric functions

12.5 Exercise

1. a. $\frac{2\sqrt{3}}{3}$ b. $\frac{2\sqrt{3}}{3}$ c. 2 d. $-\sqrt{2}$

2. a. $\frac{2\sqrt{3}}{3}$ b. $-\frac{2\sqrt{3}}{3}$ c. -2 d. $\sqrt{2}$

3. a. $\sqrt{3}$ b. $-\frac{\sqrt{3}}{3}$ c. -1 d. $-\sqrt{3}$

4. a. $-\frac{\sqrt{21}}{2}$ b. $-\frac{\sqrt{17}}{4}$ c. $-\frac{3\sqrt{8}}{8}$ d. $-\sqrt{15}$
e. $\frac{3\sqrt{40}}{40}$ f. $-\frac{5\sqrt{21}}{21}$

5. a. $-\frac{7\sqrt{40}}{40}$ b. $-\frac{5\sqrt{39}}{39}$ c. $-\frac{\sqrt{15}}{15}$
d. $\frac{\sqrt{61}}{6}$ e. $\frac{\sqrt{3}}{12}$ f. $-\sqrt{17}$

6. a. See the image at the bottom of the page*
b. See the image at the bottom of the page**

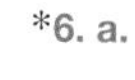
*6. a.

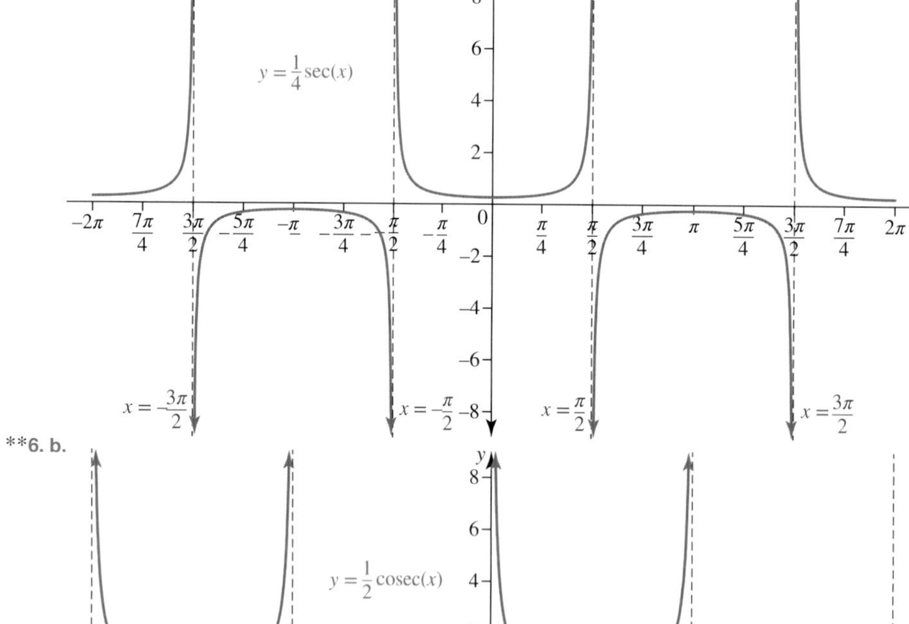

**6. b.

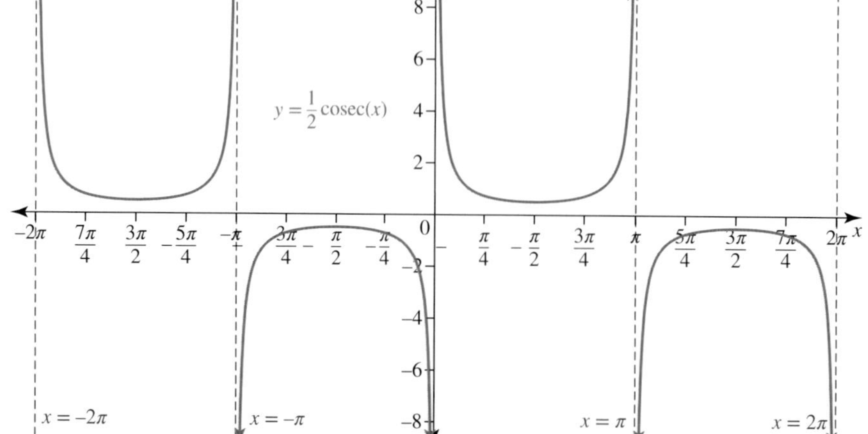

7.

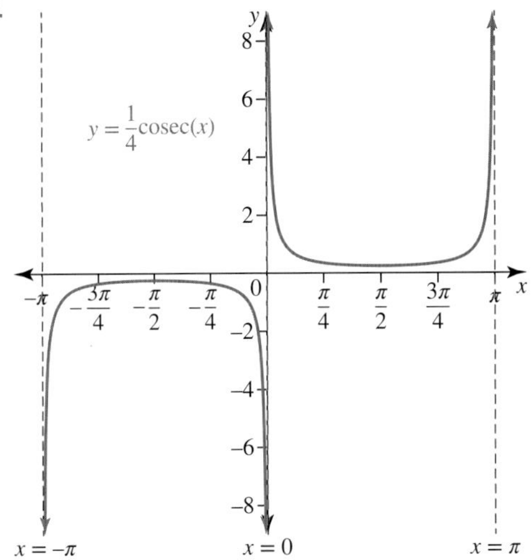

9.

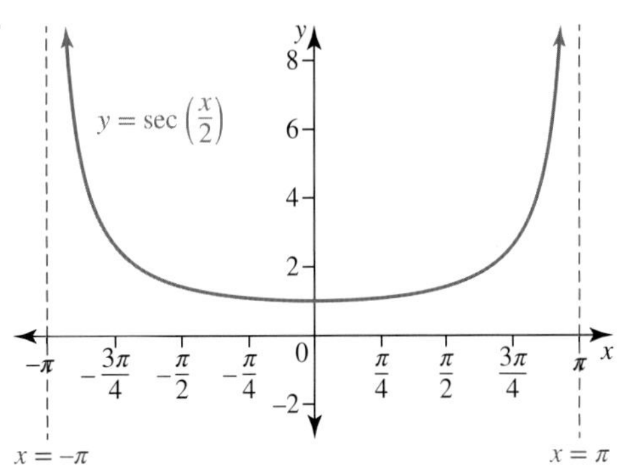

8. a. See the image at the bottom of the page*

b. See the image at the bottom of the page**

---

*8. a.

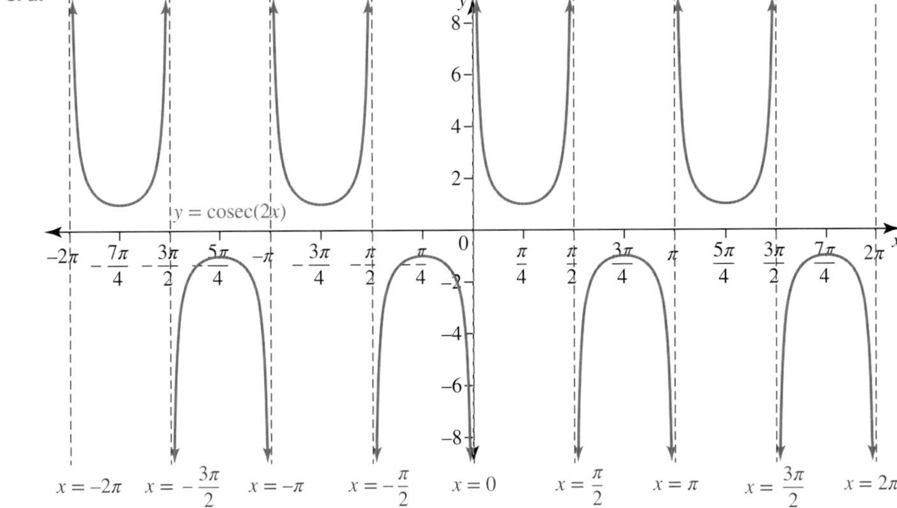

**8. b.

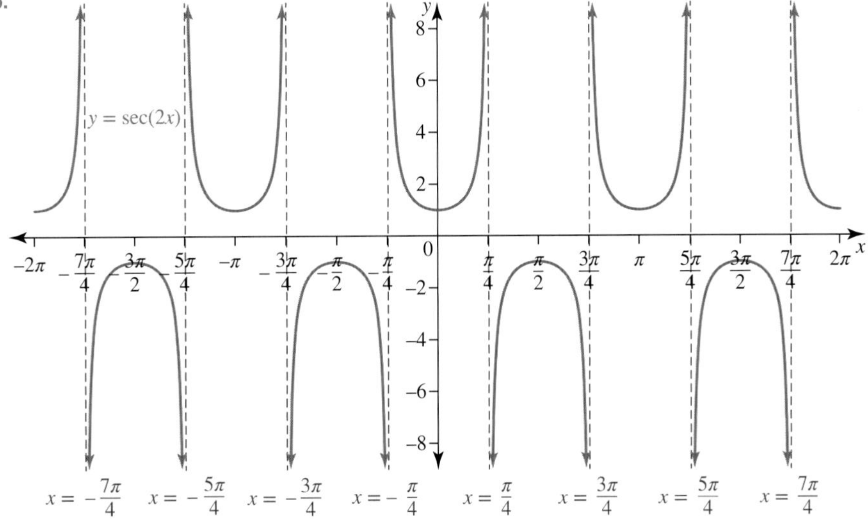

10. a.

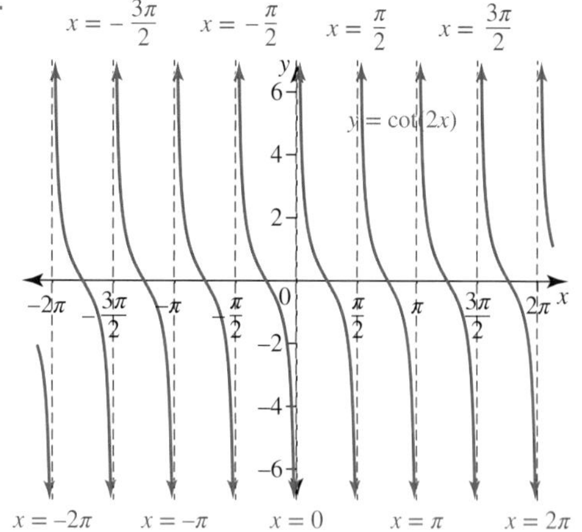

b.

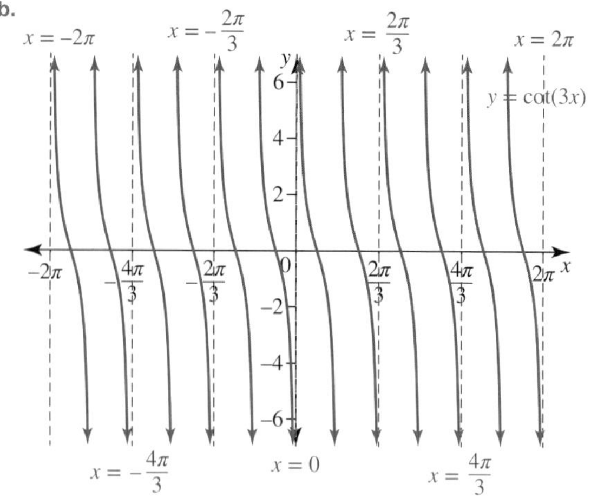

11. See the image at the bottom of the page*

12. a. See the image at the bottom of the page*

***11.**

***12. a.**

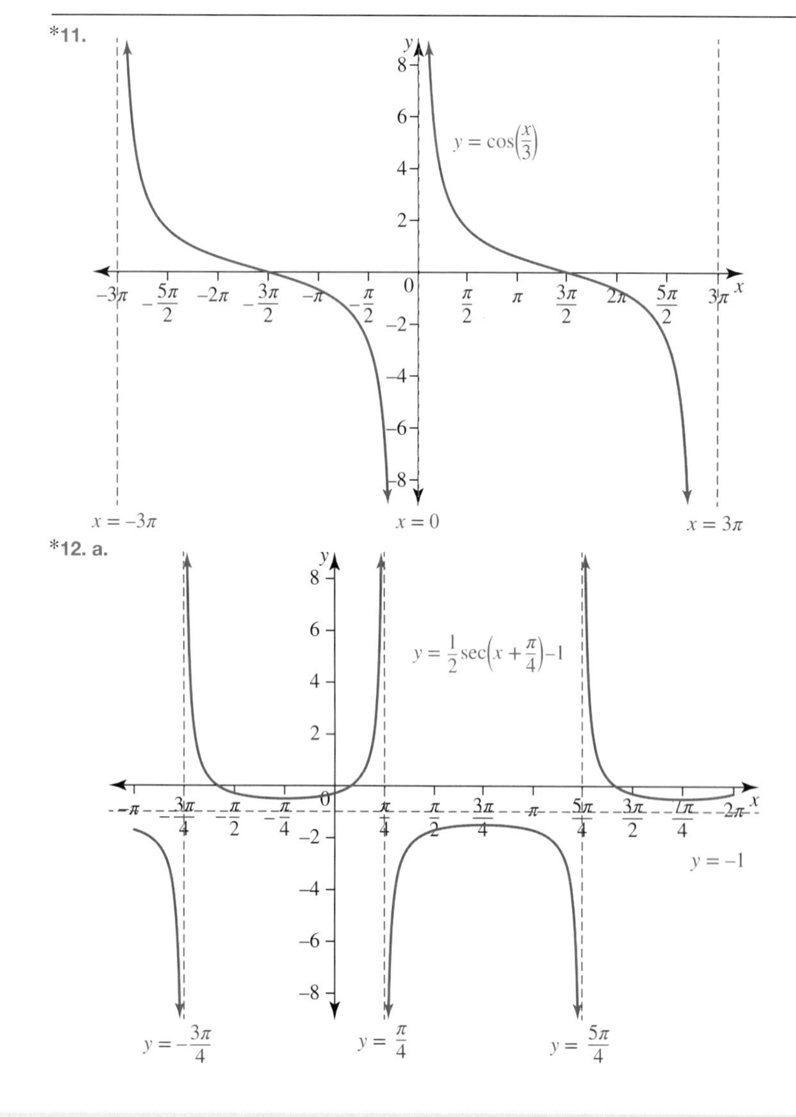

b. See the image at the bottom of the page**

13. a.

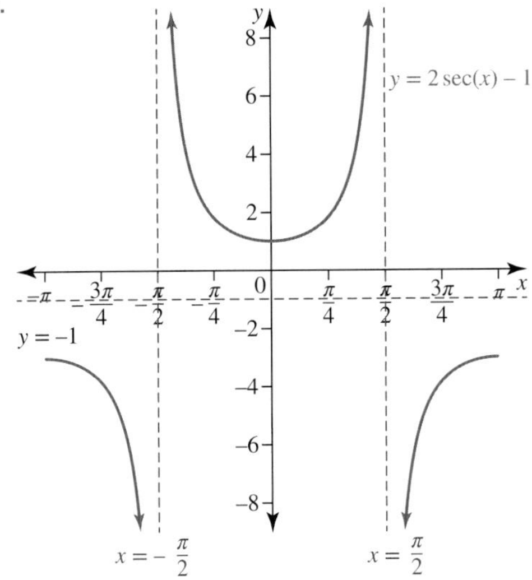

b.

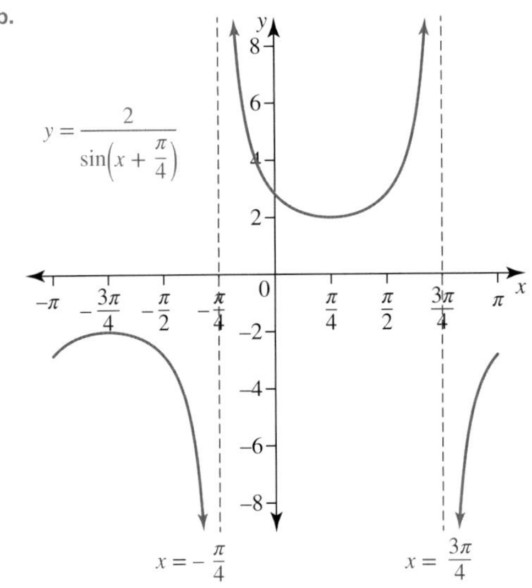

c.

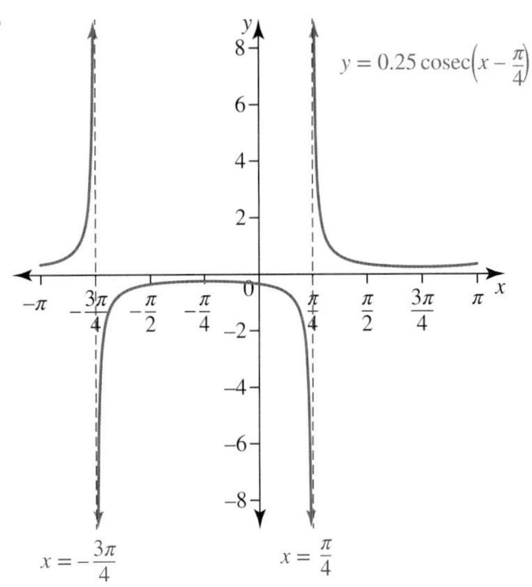

d.

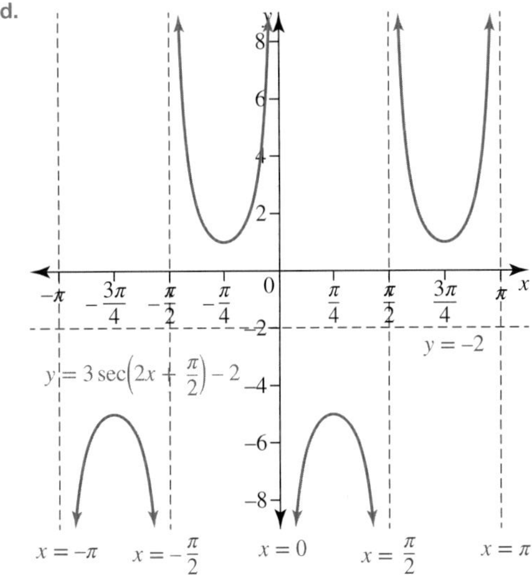

****12. b.**

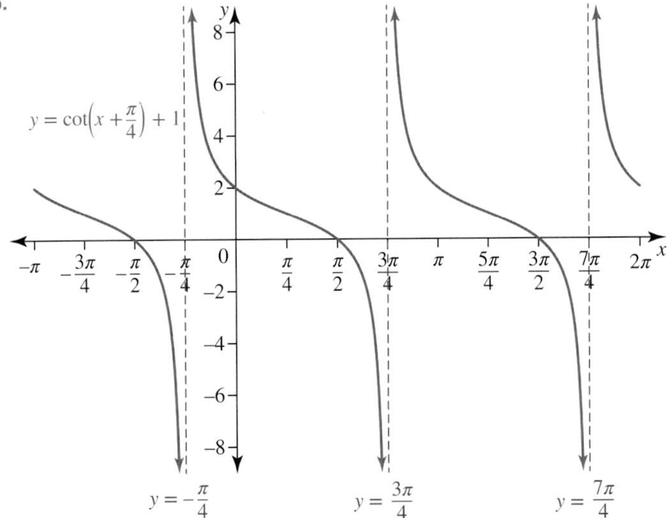

14. a. $\dfrac{p^2 - q^2 - pq}{q\sqrt{p^2 - q^2}}$ b. $\dfrac{\sqrt{a^2 - b^2}}{a + b}$

15.

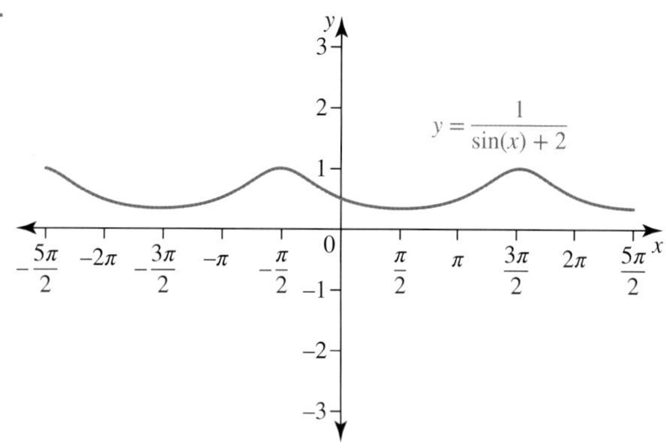

16. a. See the image at the bottom of the page*

b. See the image at the bottom of the page**

12.5 Exam questions

Note: Mark allocations are available with the fully worked solutions online.

1.

$f: [-3\pi, 3\pi] \rightarrow R, f(x) = 2\cot\left(\frac{x}{2}\right)$

(−3π, 0)

(3π, 0)

x = −2π

x = 0

x = 2π

---

*16. a.

**16. b.

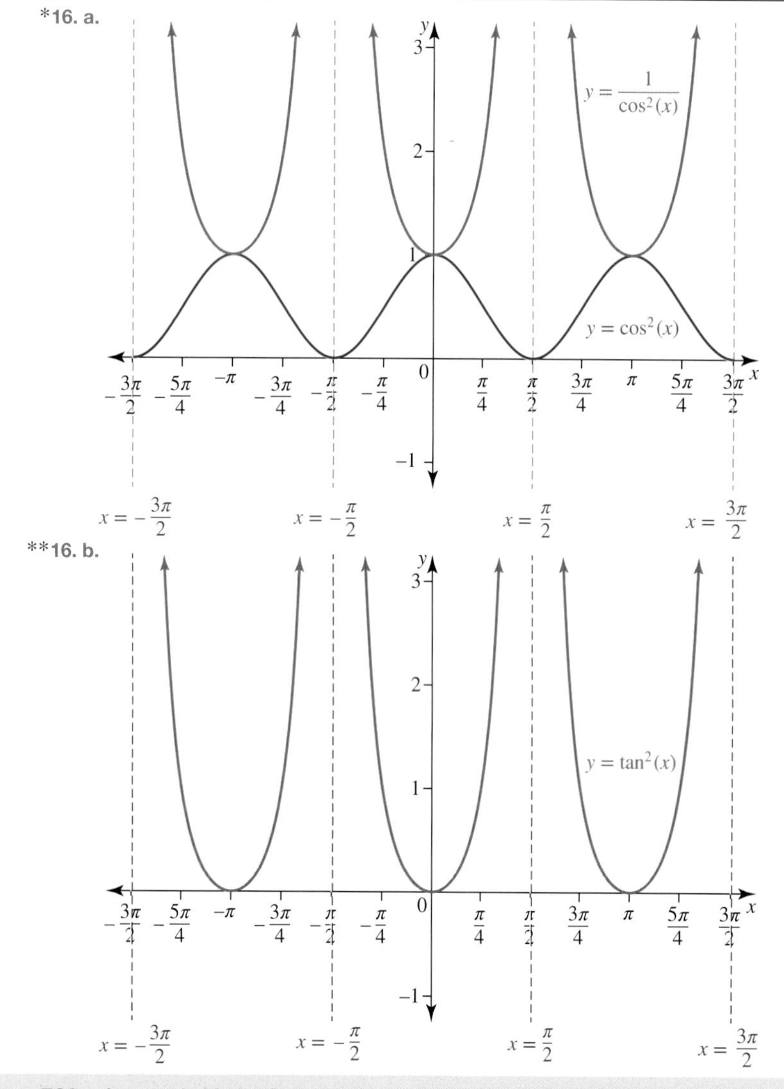

2.

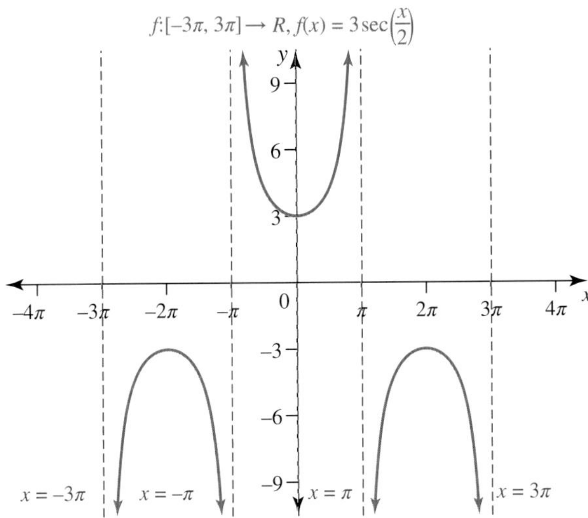

3.

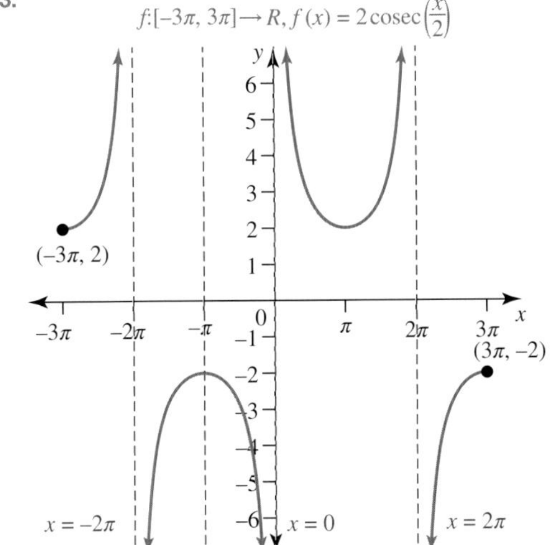

12.6 Inverse trigonometric functions

12.6 Exercise

1. a. Does not exist b. $-\frac{\pi}{3}$ c. 0.9

2. a. Does not exist b. $\frac{5\pi}{6}$ c. $\frac{\pi}{6}$

3. $\frac{2\sqrt{6}}{5}$

4. $\frac{2\sqrt{10}}{7}$

5. $\frac{8\sqrt{33}}{49}$

6. $\frac{23}{32}$

7. a. $\frac{\pi}{6}$ b. 1.1

8. a. $-\frac{\pi}{3}$ b. $\frac{5}{4}$

9. $\frac{3}{4}$

10. $\frac{15}{8}$

11. a. $\frac{\sqrt{77}}{9}$ b. $-\frac{\sqrt{5}}{2}$

12. a. $-\frac{5\sqrt{11}}{11}$ b. $\frac{5\sqrt{89}}{89}$

13. a. $\frac{\sqrt{21}}{5}$ b. $\frac{4\sqrt{65}}{65}$

14. a.

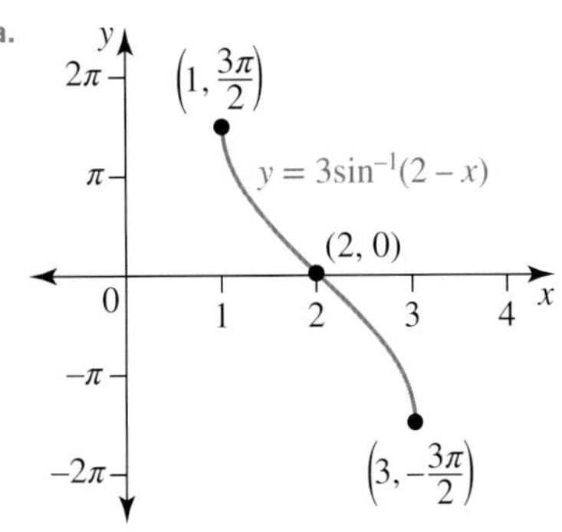

b.

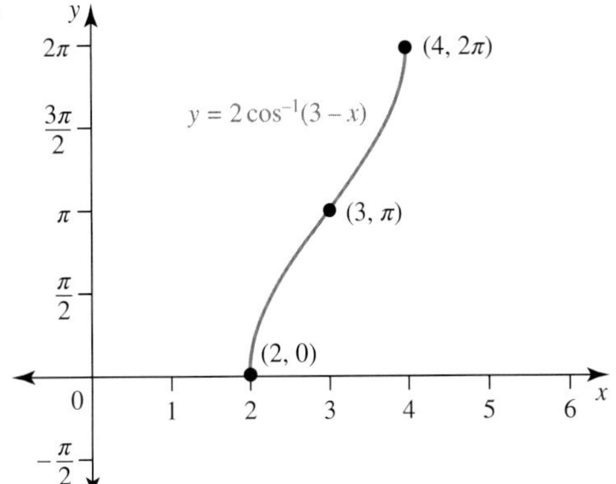

c.

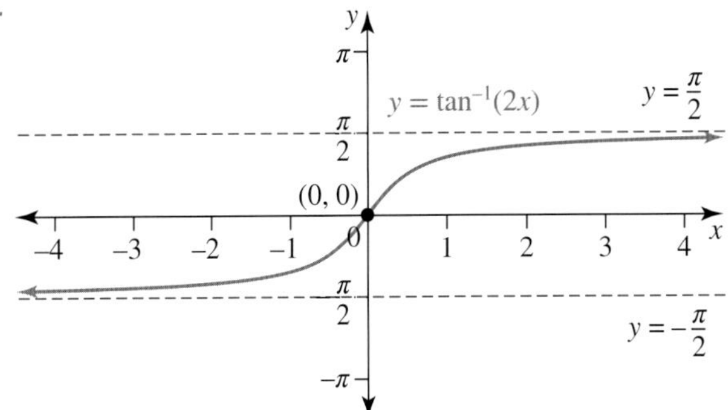

15. a.

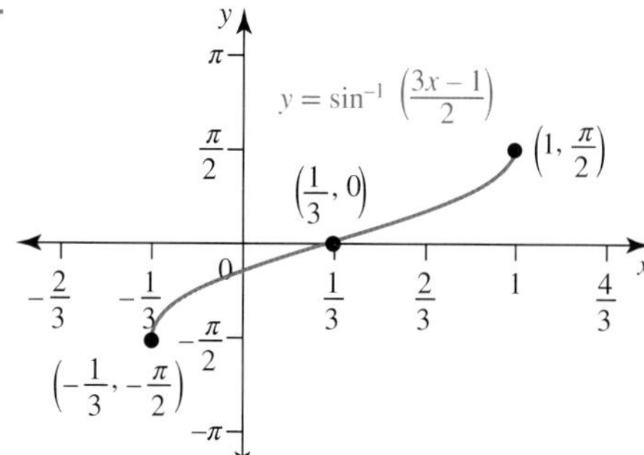

b.

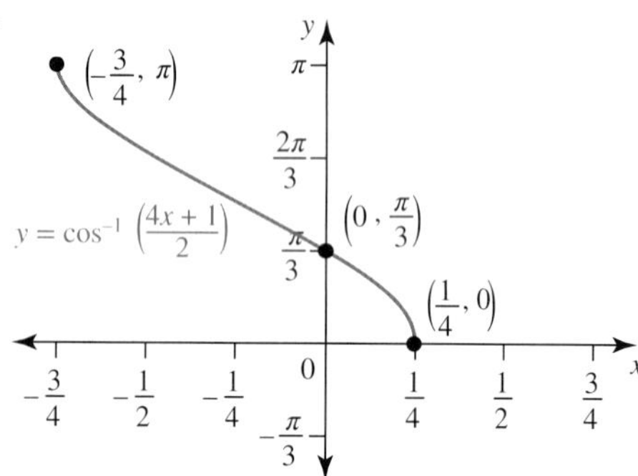

c.

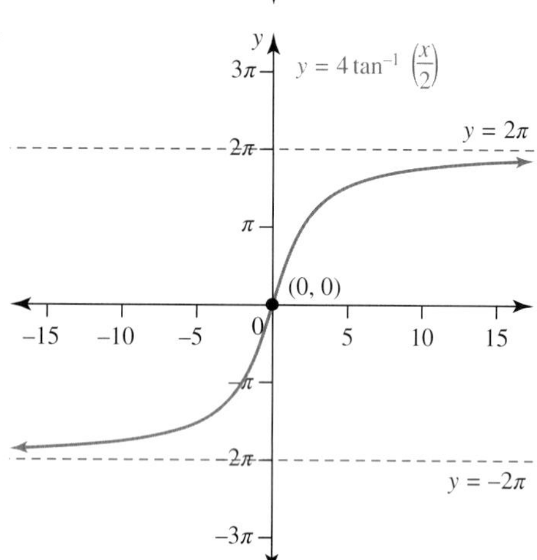

16. a.

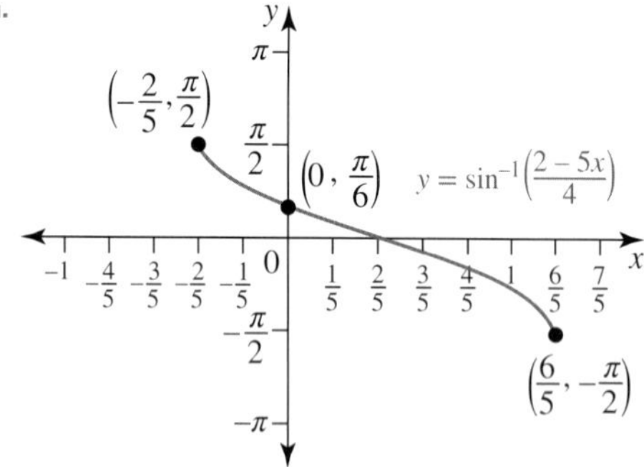

b.

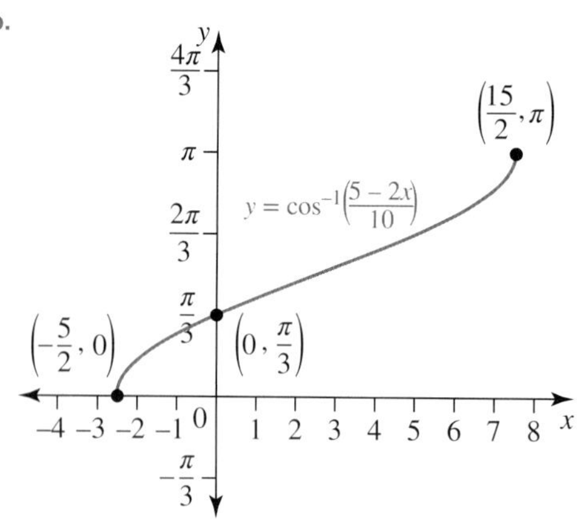

c.

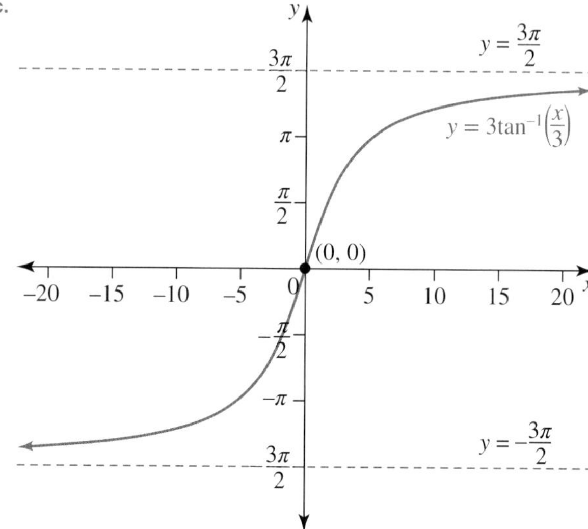

17. a. Domain $\left[-2, -\frac{4}{3}\right]$, range $[-3, 1]$

b. Domain R, range $(-1, 7)$

c. Domain $\left[0, \frac{2}{3}\right]$, range $[0, 2\pi]$

18. a. Domain $[2, 3]$, range $[-\pi, 2\pi]$

b. Domain $\left[\frac{1}{2}, \frac{9}{2}\right]$, range $\left[-\frac{7\pi}{2}, -\frac{\pi}{2}\right]$

c. Domain R, range $(-1, 5)$

12.6 Exam questions

Note: Mark allocations are available with the fully worked solutions online.

1. E

2. a. Endpoints $\left(-\frac{2}{3}, 2\right), \left(2, \frac{\pi}{2}\right)$

   Intercepts $\left(\frac{2}{3}, 0\right), \left(0, -\frac{\pi}{6}\right)$

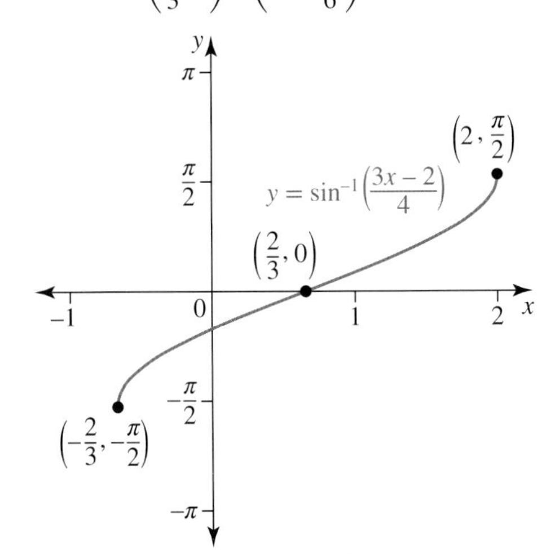

   b. Endpoints $\left(-\frac{3}{2}, 0\right), \left(\frac{9}{2}, \pi\right)$

   Intercepts $\left(-\frac{3}{2}, 0\right), \left(0, \frac{\pi}{3}\right)$

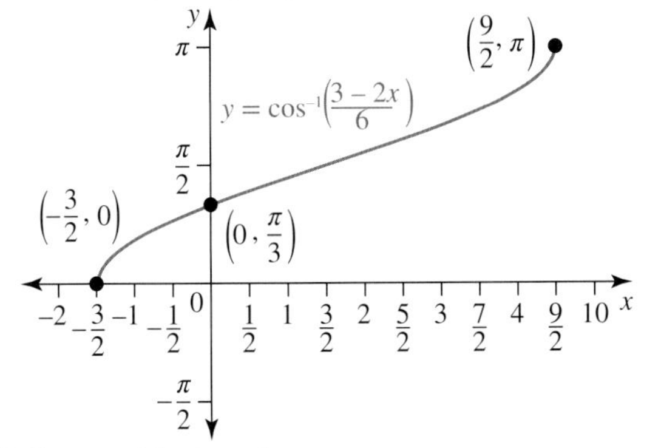

3. $(0, 0)$, $y = \pm 2$ horizontal asymptotes.

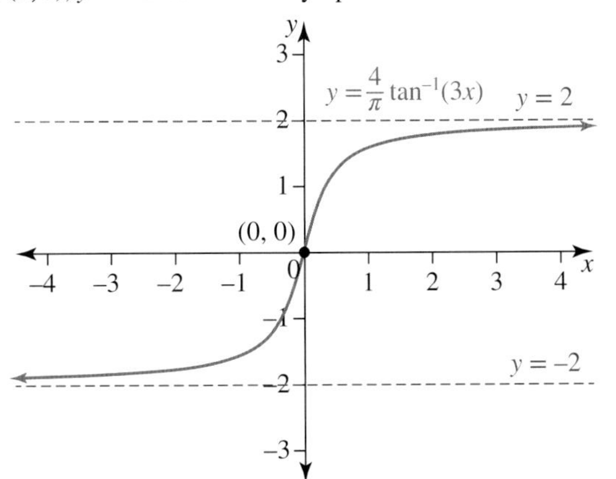

12.7 Circles and ellipses

12.7 Exercise

1. a.

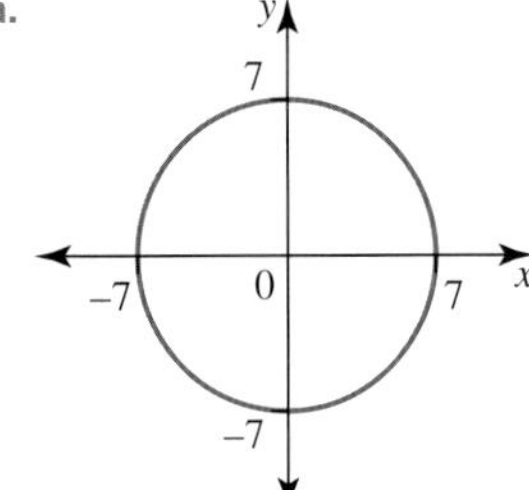

   Centre (0, 0); radius 7

   b.

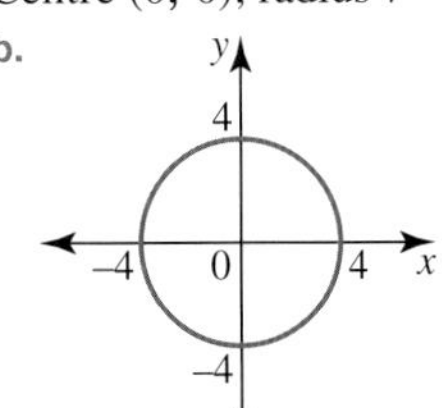

   Centre (0, 0); radius 4

2. a.

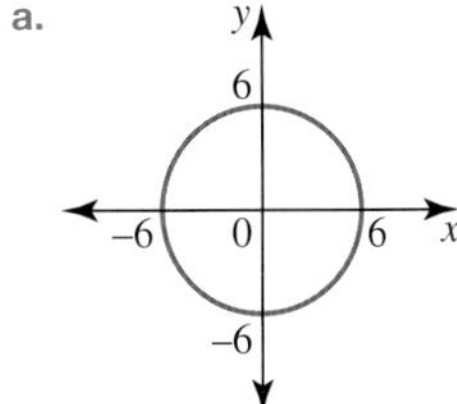

   Centre (0, 0); radius 6

   b.

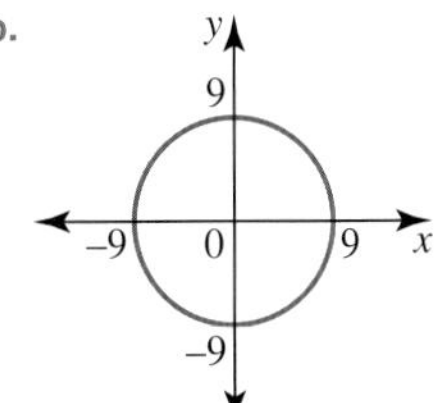

   Centre (0, 0); radius 9

3. a.

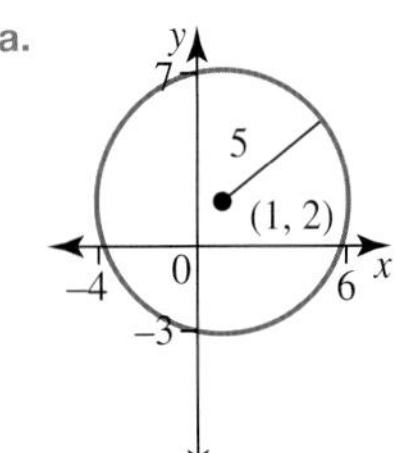

   b.

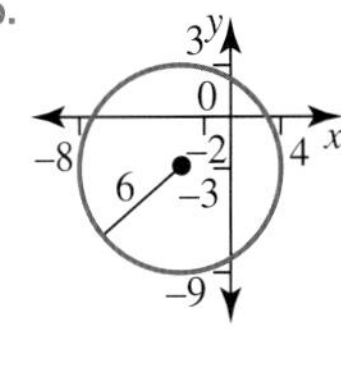

4. a.

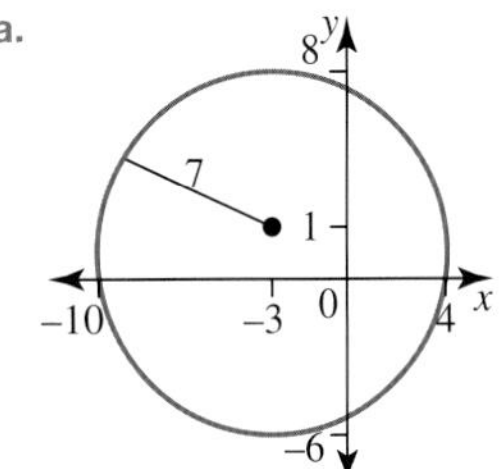

b.

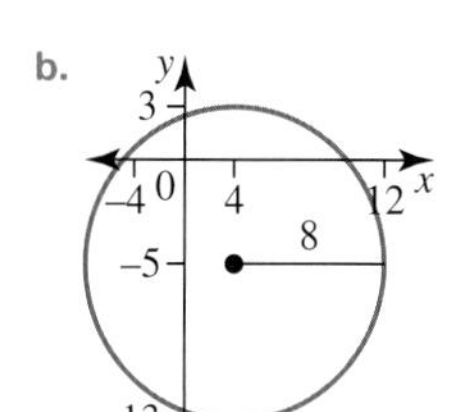

c.

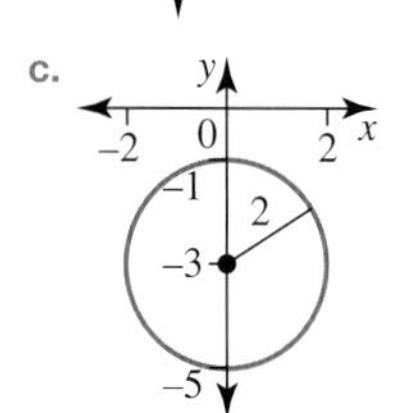

d.

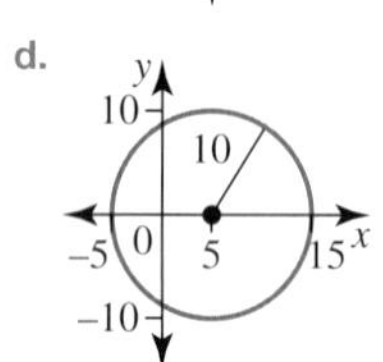

5. a. $(x + 2)^2 + (y + 4)^2 = 2^2$

b. $(x - 5)^2 + (y - 1)^2 = 4^2$

6. a. $(x - 7)^2 + (y + 3)^2 = 7^2$

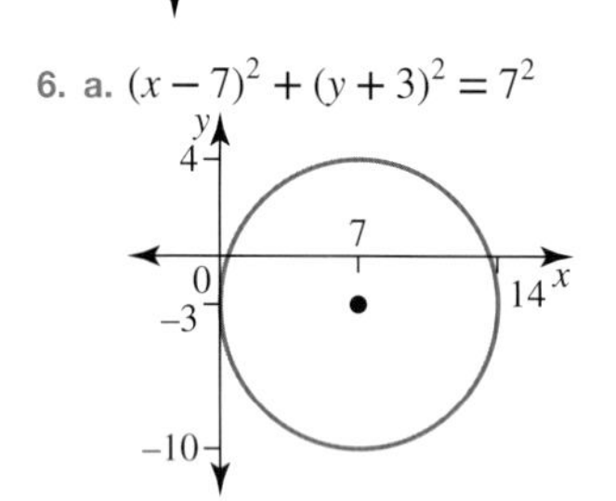

b. $(x + 4)^2 + (y - 6)^2 = 8^2$

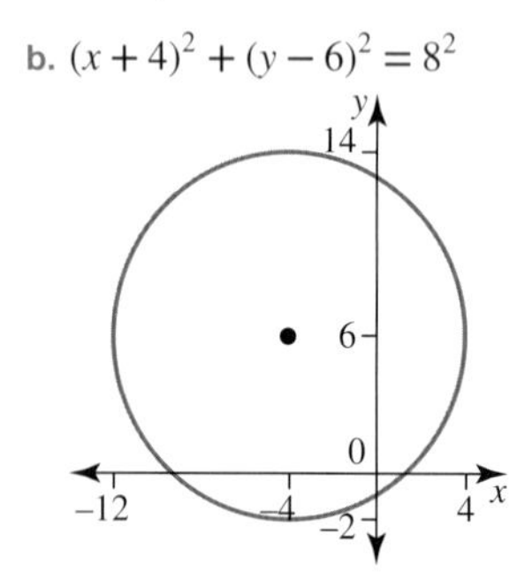

c. $x^2 + (y - 9)^2 = 10^2$

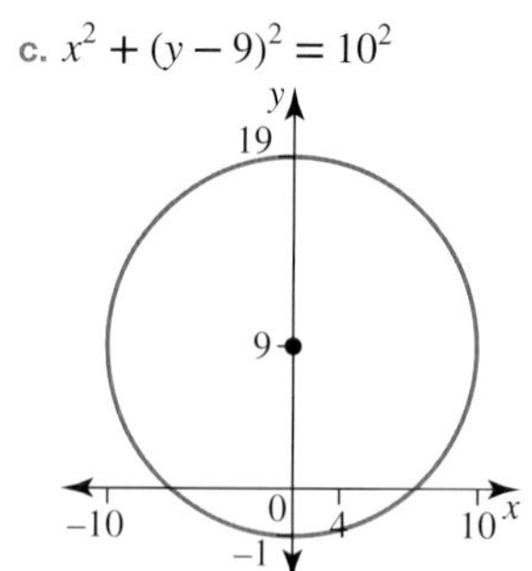

d. $(x - 1)^2 + (y + 2)^2 = 3^2$

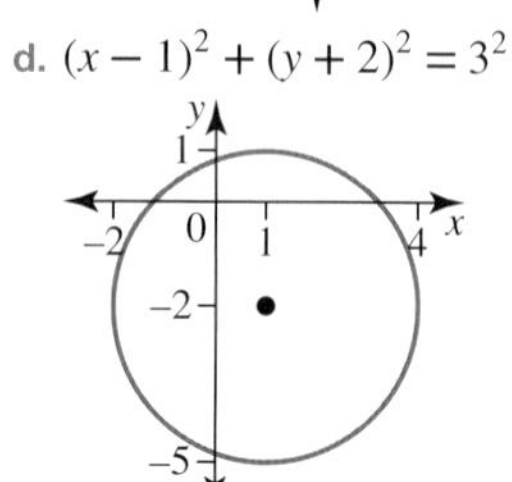

7. a.

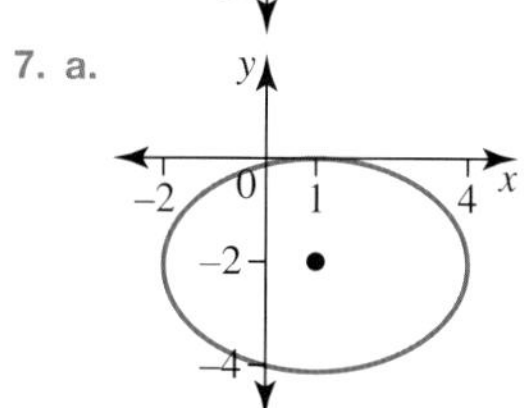

b.

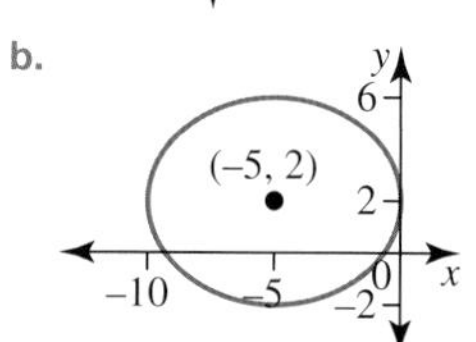

8. a.

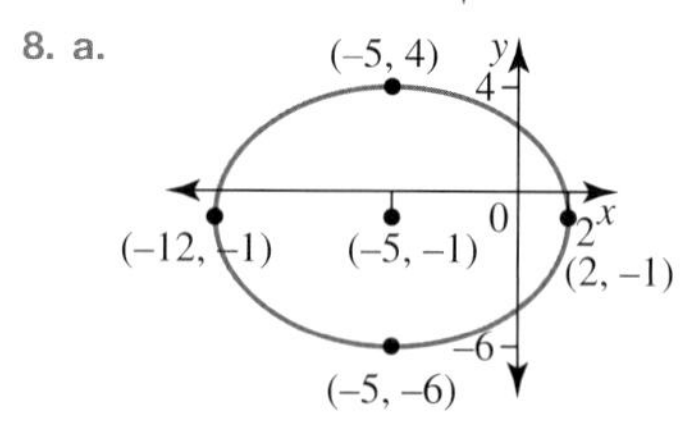

b.

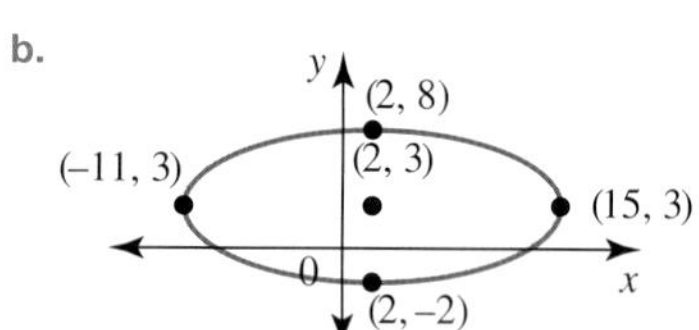

c.

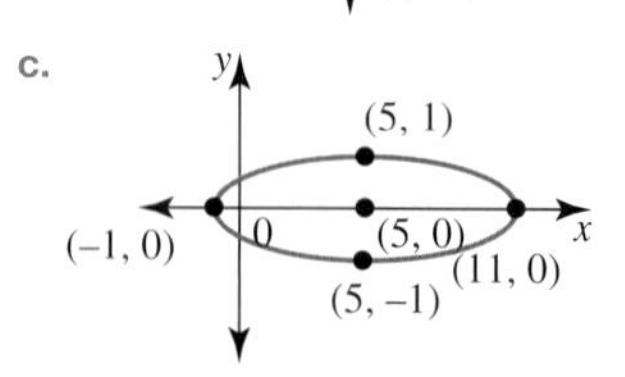

d.

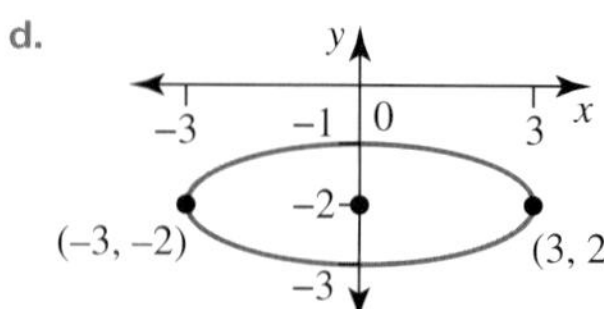

9.

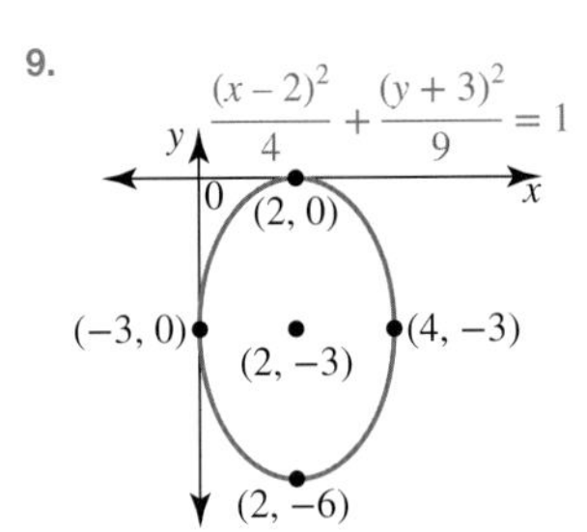

10.

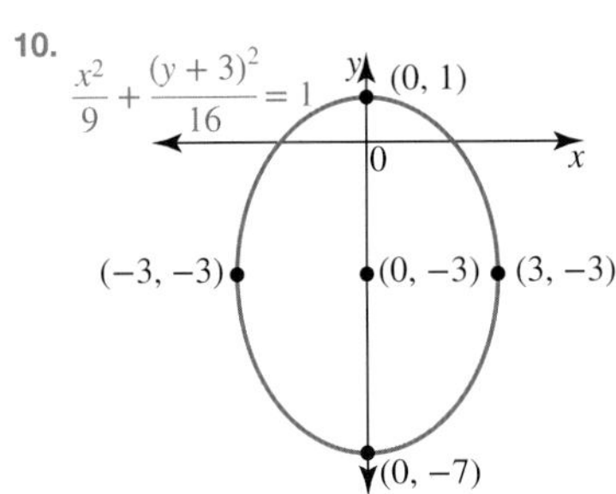

11.

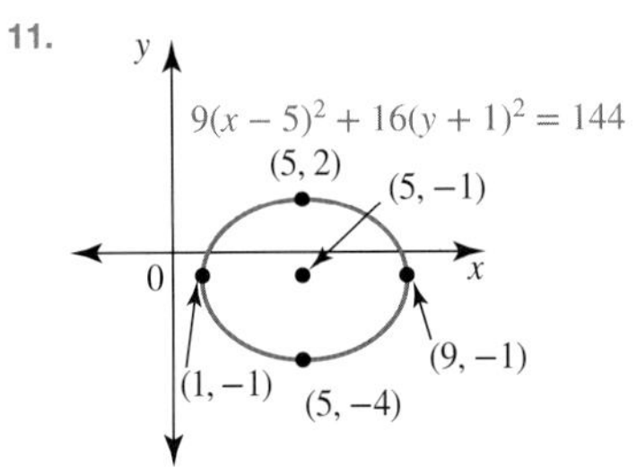

12.

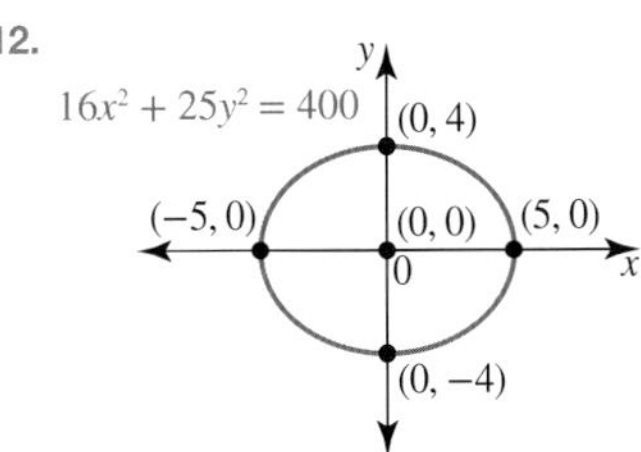

13.

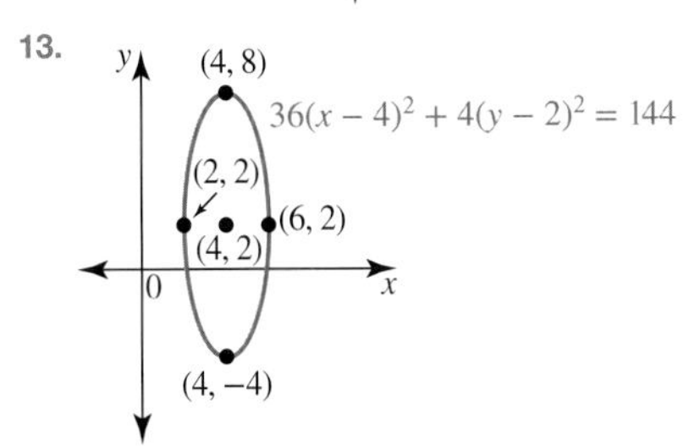

14.

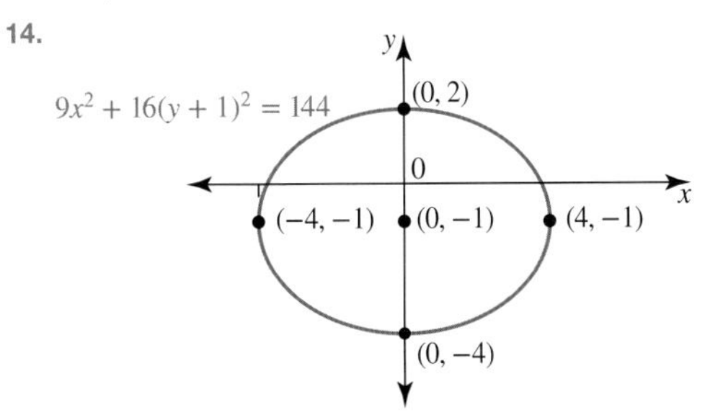

15. a. $x=3$ or $x=-3$ b. $y=5$ or $y=-5$

16. a. $x=4$ or $x=0$ b. $y=3$ or $y=-5$

17. a. $x^2+y^2=4$ b. $x^2+y^2=25$

18. a. $(x-2)^2+(y-3)^2=4$
b. $(x-2)^2+(y+1)^2=25$

19. a. $x=2$ b. $y=2$

20. a. $x=1$ b. $y=4$

12.7 Exam questions

Note: Mark allocations are available with the fully worked solutions online.

1. a. $(x-4)^2+(y-1)^2=9$

b. $\frac{(x-3)^2}{49}+\frac{(y-2)^2}{16}=1$

2. $(x-2)^2+(y+3)^2=25$

Domain $[-3,7]$, range $[-8,2]$

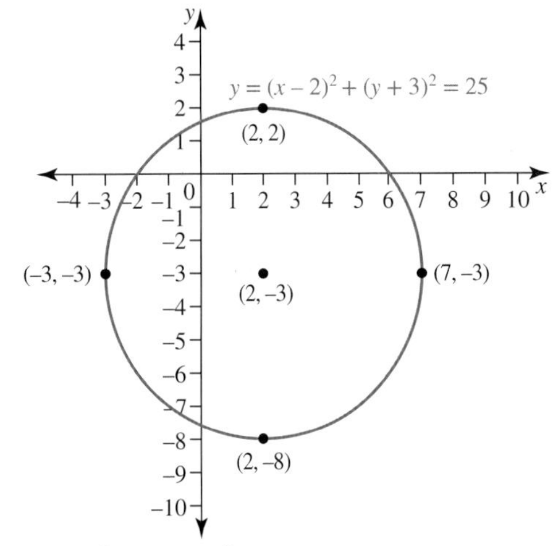

3. $\frac{(x+3)^2}{25}+\frac{(y-4)^2}{49}=1$

Domain $[-8,2]$, range $[-3,11]$

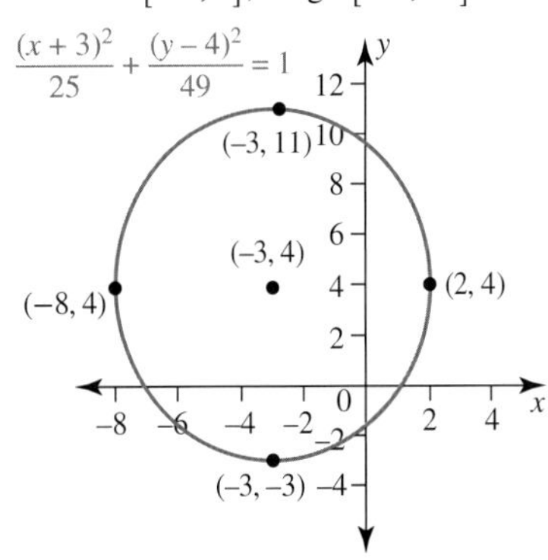

12.8 Hyperbolas

12.8 Exercise

1. a.

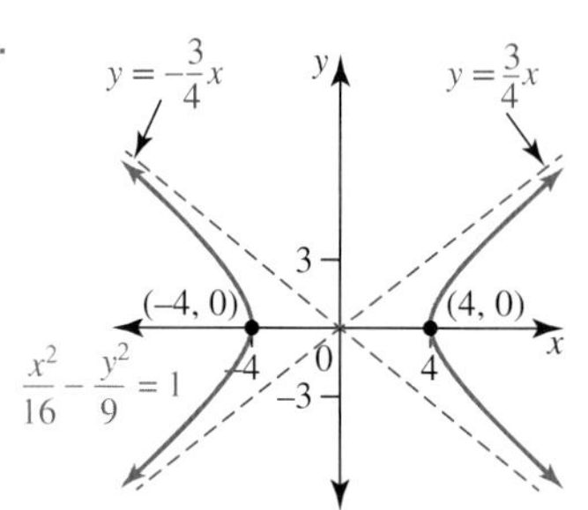

b.

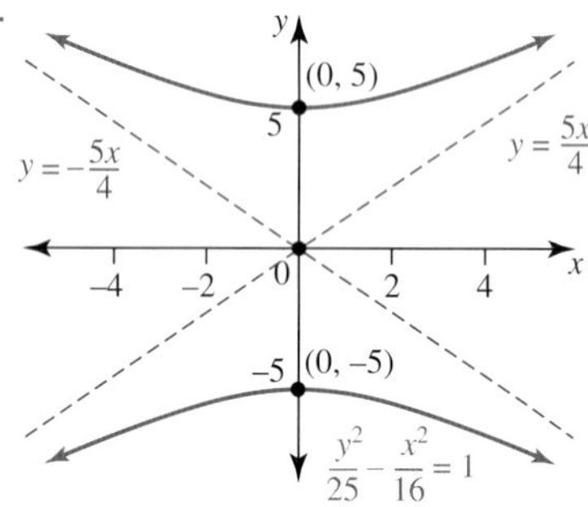

2. a.

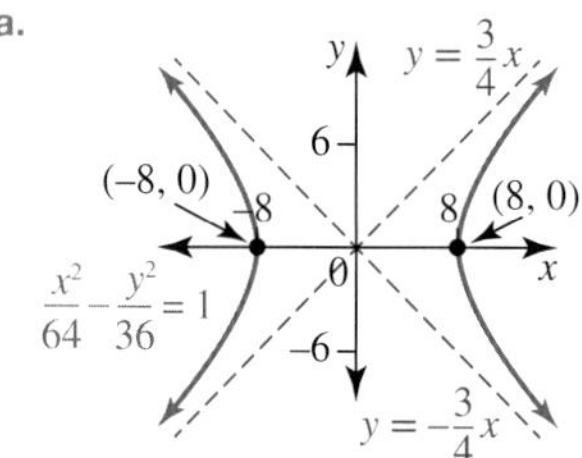

b.

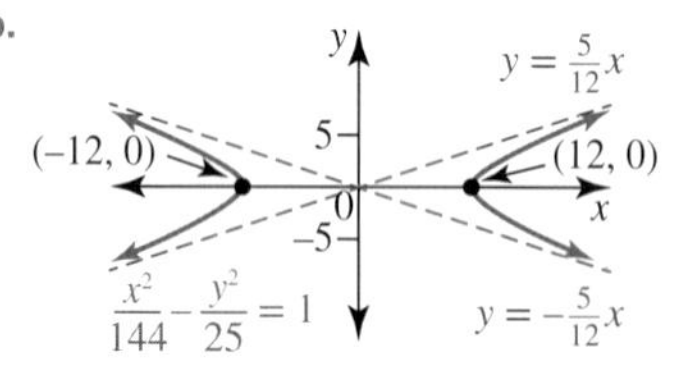

3. a.

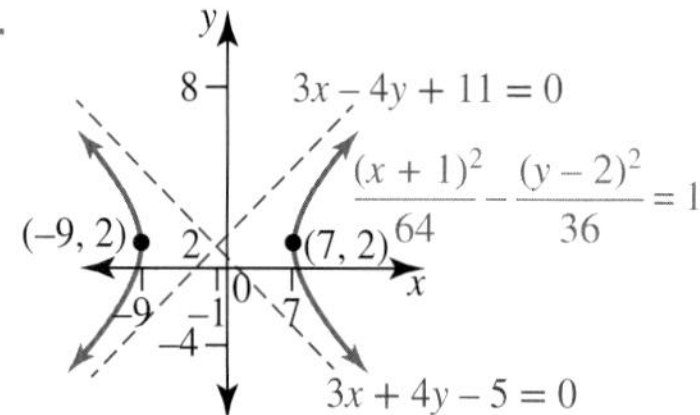

b.

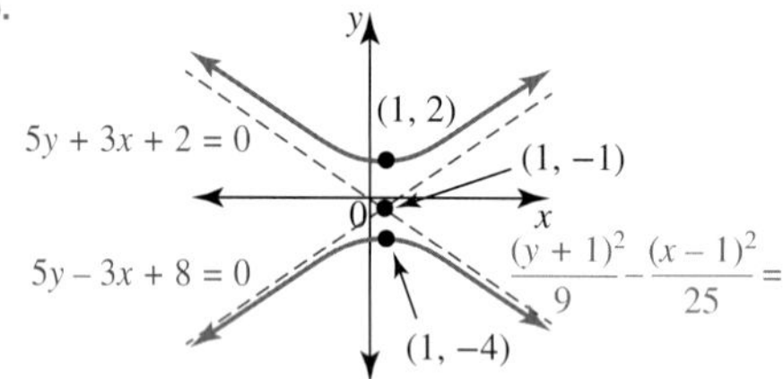

4. a.

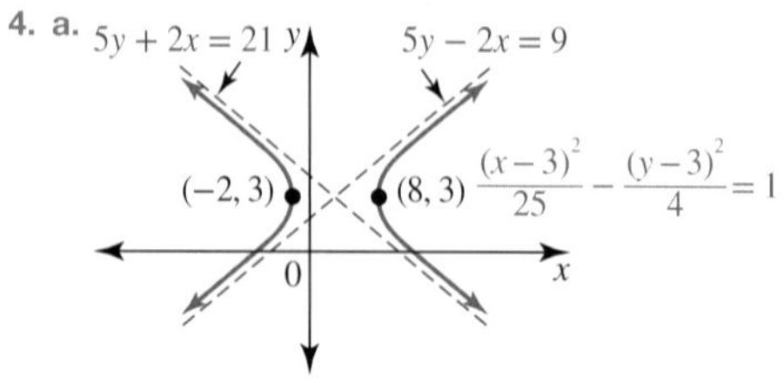

b.

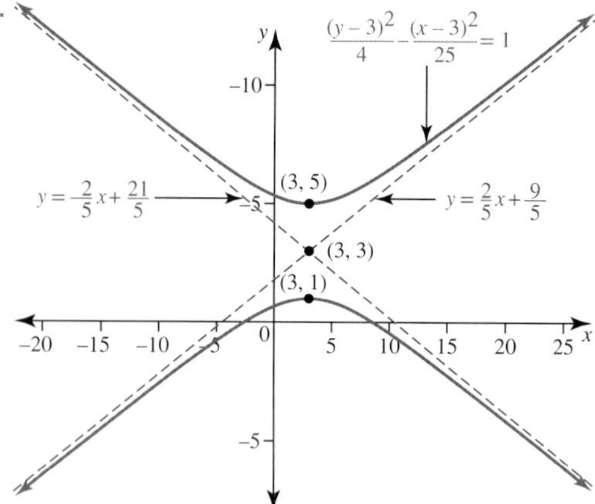

5. a.

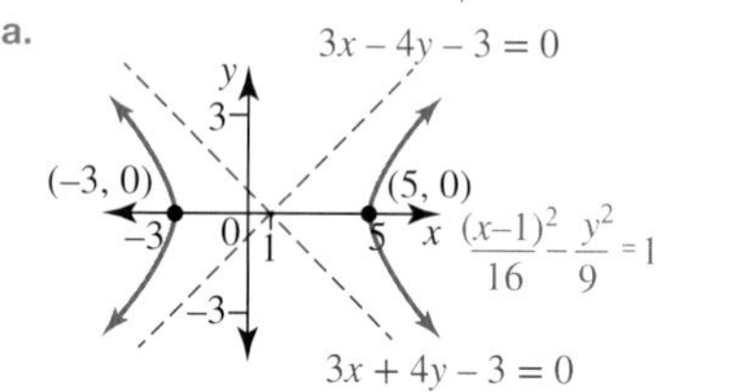

b.

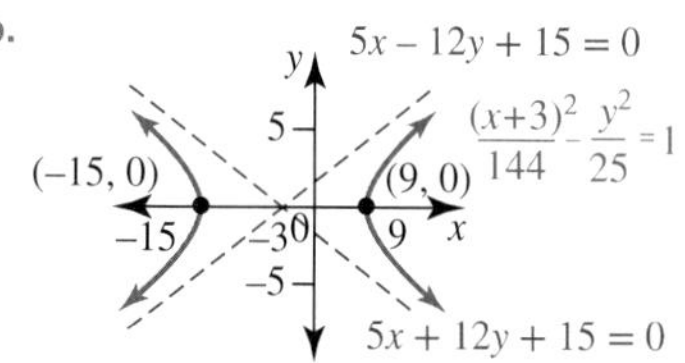

6. a.

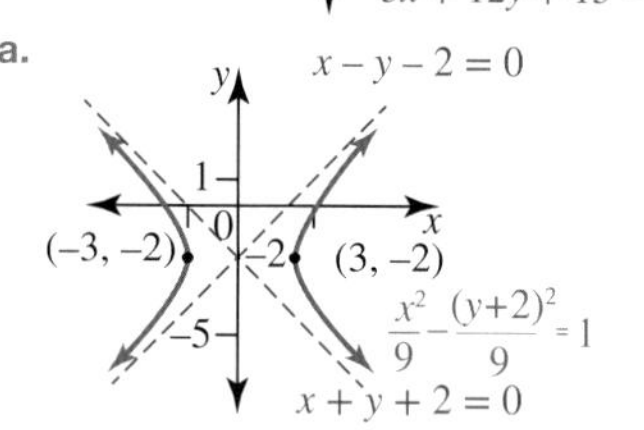

b.

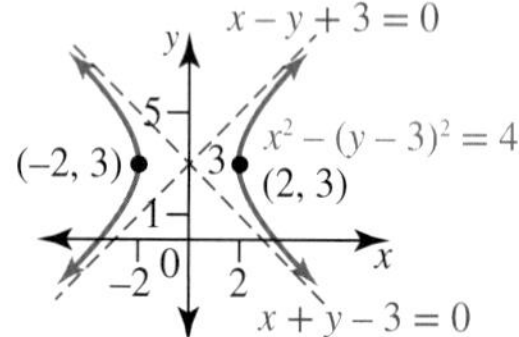

7. a.

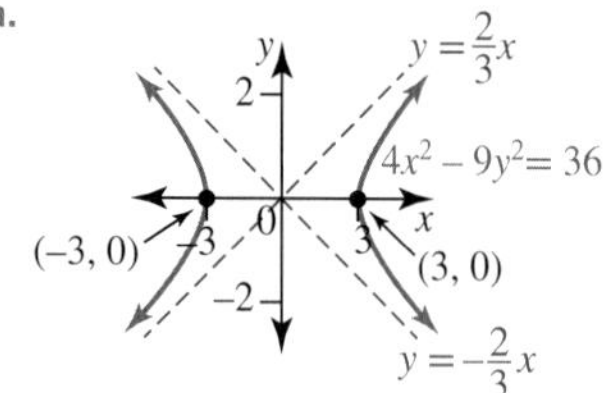

b.

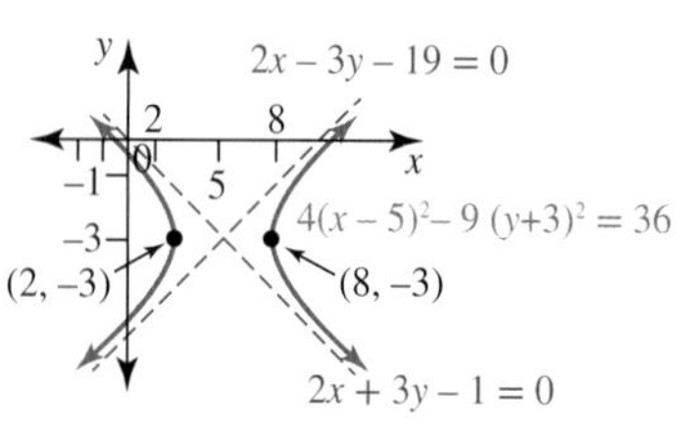

8. a.

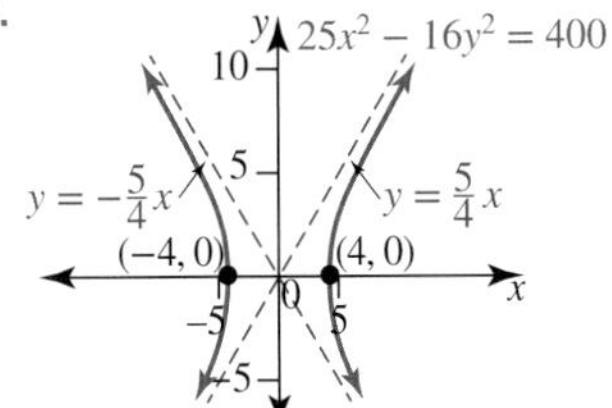

b.

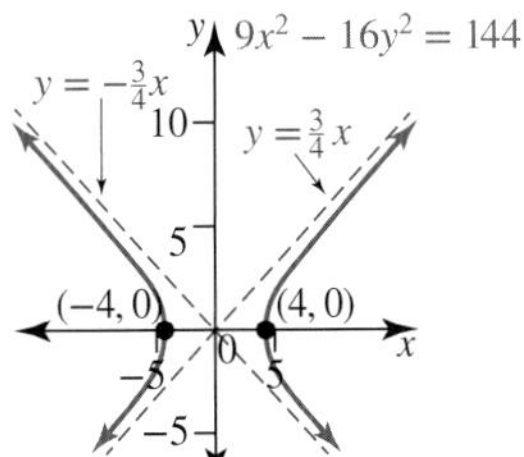

9. a.

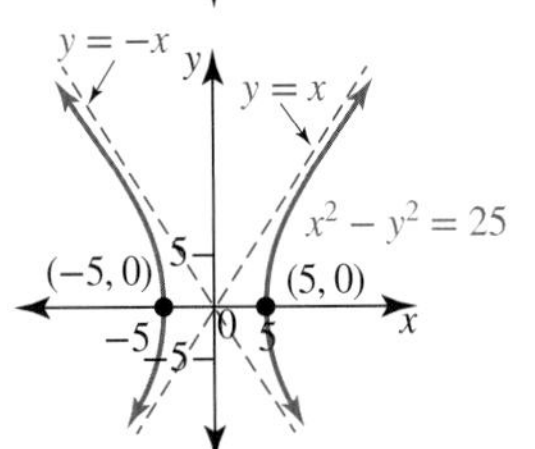

b.

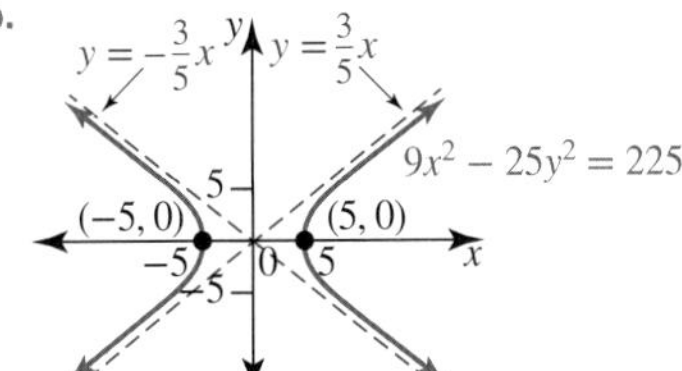

10. a.

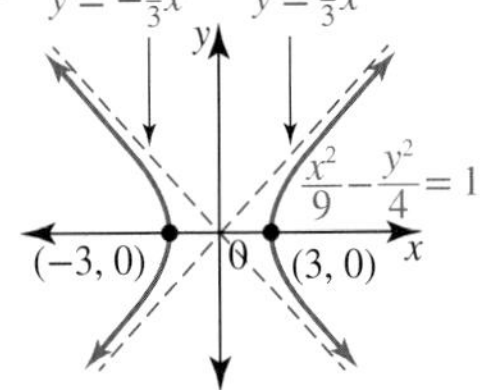

b.

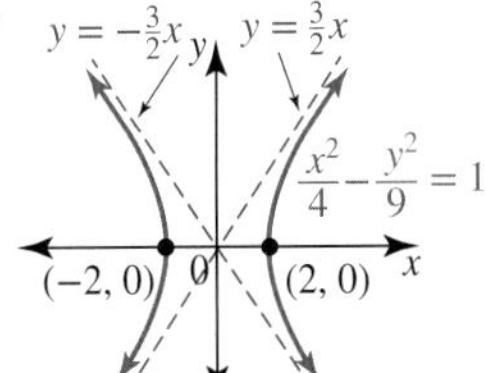

11. a.

$y = \frac{4}{3}x$

$y = -\frac{4}{3}x$

$\frac{x^2}{9} - \frac{y^2}{16} = 1$

(−3, 0) 0 (3, 0) x y

b.

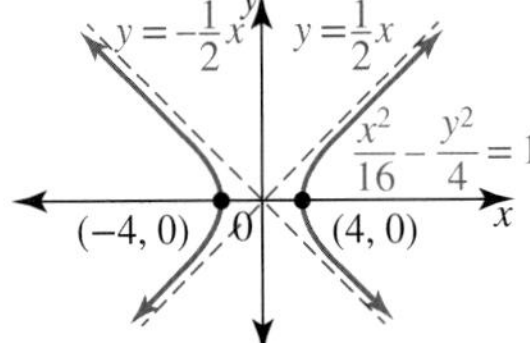

12. a.

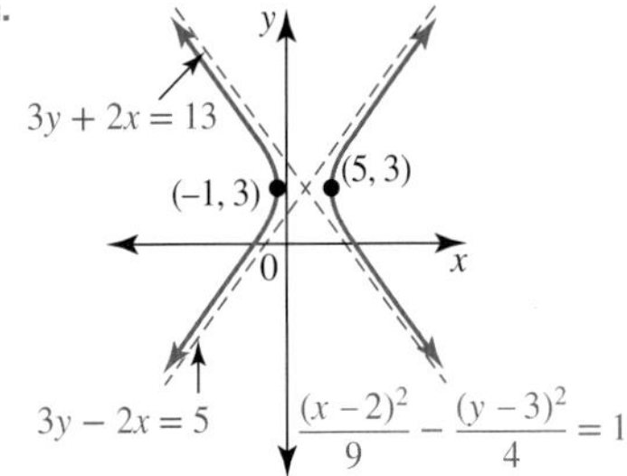

b.

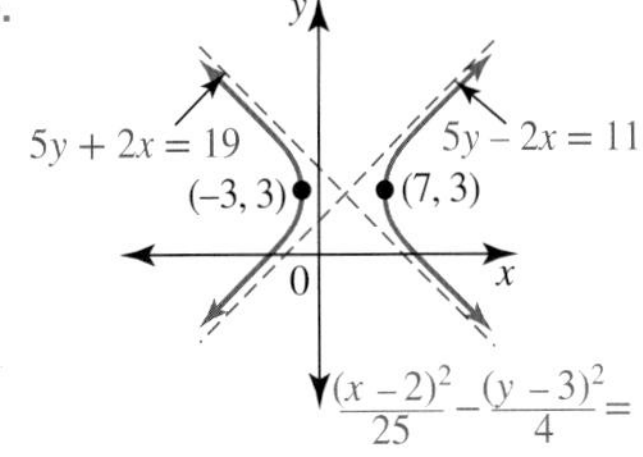

13. a.

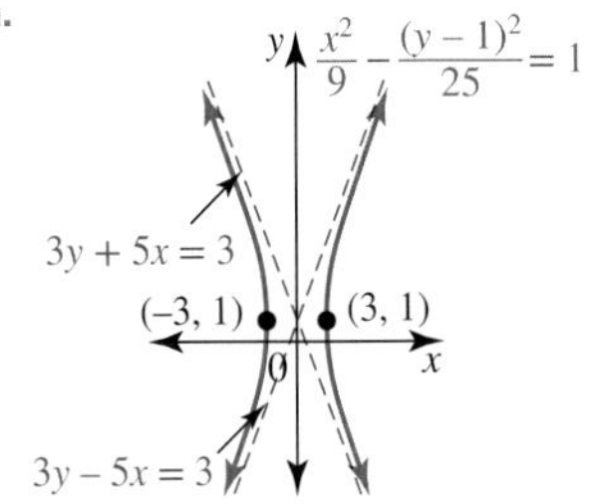

b.

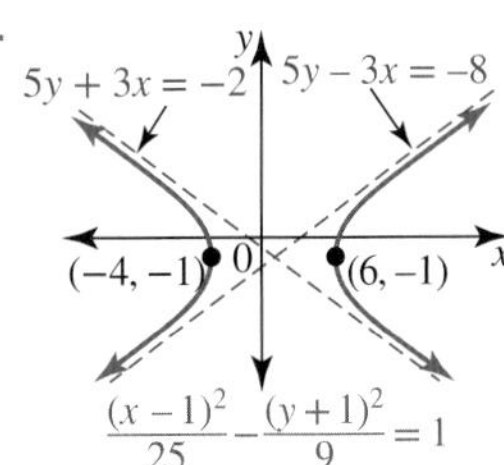

14. a.

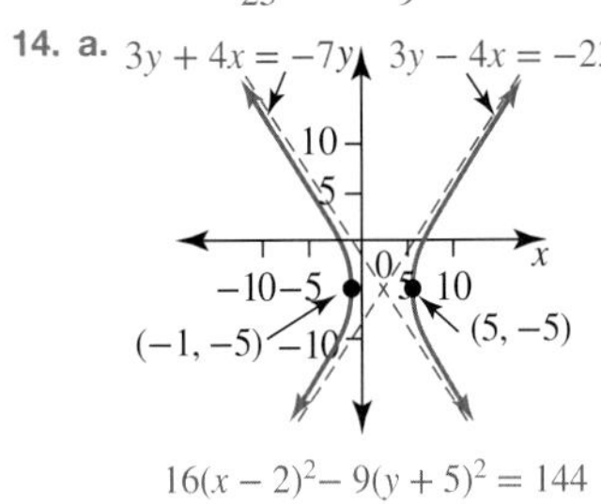

b.

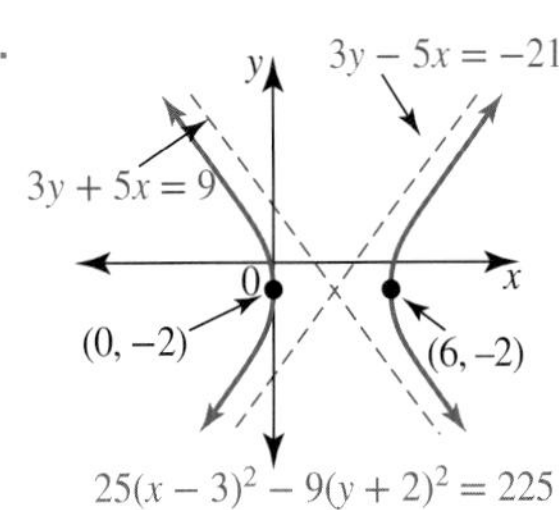

15. a.

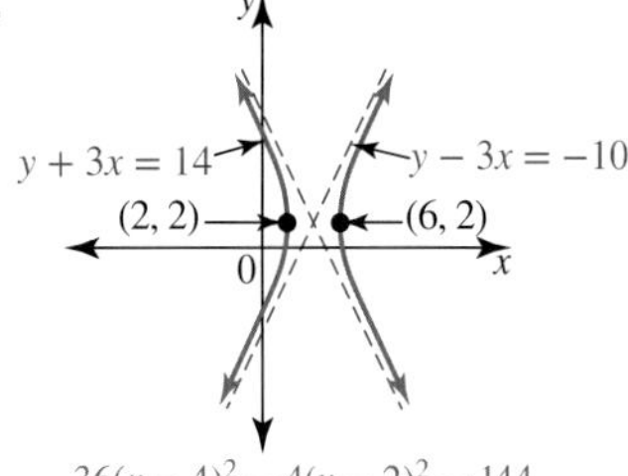

$36(x-4)^2 - 4(y-2)^2 = 144$

b.

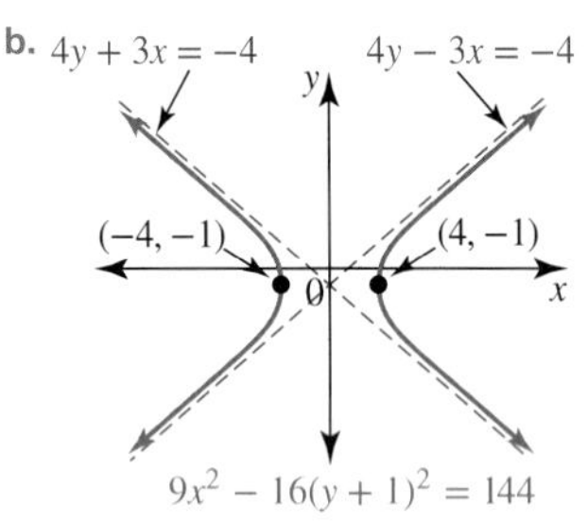

$9x^2 - 16(y+1)^2 = 144$

16. $\dfrac{x^2}{400} - \dfrac{y^2}{500} = 1$

12.8 Exam questions

Note: Mark allocations are available with the fully worked solutions online.

1. $\dfrac{x^2}{16} - \dfrac{y^2}{4} = 1$

2. $h = -2,\ k = -3,\ a = 2,\ b = 3$

3. $\dfrac{(x-3)^2}{16} - \dfrac{(y+2)^2}{9} = 1$

Centre $(3, -2)$

Domain $(-\infty, -1] \cup [7, \infty)$, range R

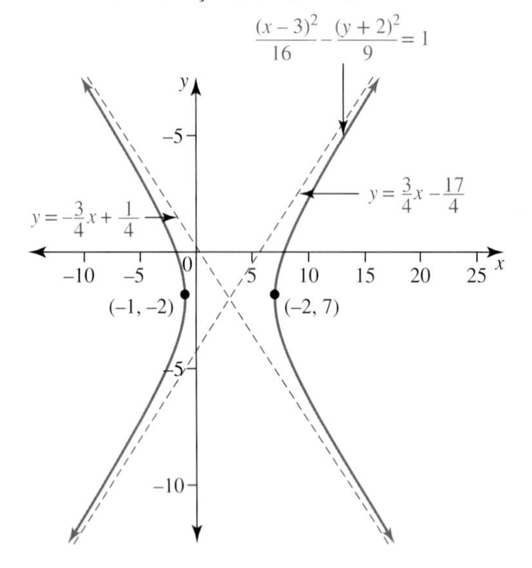

12.9 Polar coordinates, equations and graphs

12.9 Exercise

1.

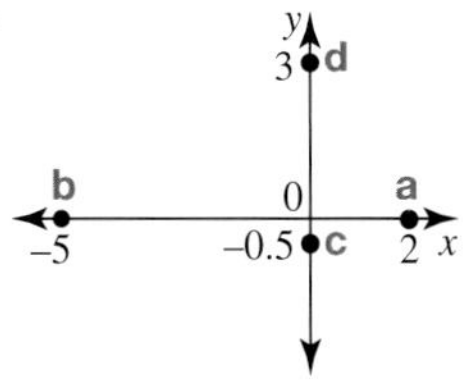

2.

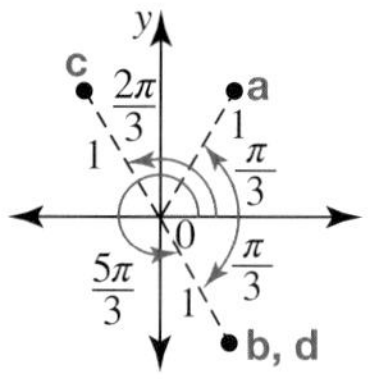

3. a. $(\sqrt{2}, \sqrt{2})$ b. $\left(\frac{5\sqrt{3}}{2}, 2\frac{1}{2}\right)$ c. $\left(1.5, \frac{3\sqrt{3}}{2}\right)$

d. $(0, 2.7)$ e. $\left(-0.75, \frac{3\sqrt{3}}{4}\right)$ f. $(-6\sqrt{3}, -6)$

4. a. $(0, 2.6)$ b. $(-7.8, 0)$

c. $(5, 5\sqrt{3})$ d. $\left(4.55, \frac{-91\sqrt{3}}{20}\right)$

5. a. $[5, 0°]$ b. $[4.3, 90°]$

c. $[30, 180°]$ d. $[9, 270°]$

6. a. $[13, 4.32^c]$ b. $[10, 5.36^c]$ c. $\left[2, \frac{2\pi}{3}\right]$

d. $\left[2\sqrt{2}, \frac{5\pi}{4}\right]$ e. $\left[4, \frac{11\pi}{6}\right]$ f. $[7.81, 0.88^c]$

7. a. $r = \dfrac{3}{\cos(\theta)}$ b. $r = \dfrac{2}{\sin(\theta)}$

c. $r = 3$ d. $r = 6$

8. a. $\theta = 78.1°$ b. $\theta = \dfrac{\pi}{4}$

c. $r = \dfrac{1}{3\cos(\theta) - 4\sin(\theta)}$ d. $r = \dfrac{7}{5\cos(\theta) + \sin(\theta)}$

9. a. $x^2 + y^2 = 4$ b. $x^2 + y^2 = 25$

c. $x^2 + (y-3)^2 = 9$ d. $(x-1)^2 + y^2 = 1$

10. a. $y = 3x$ b. $y = -4x$

c. $y = x$ d. $y = -x$

11. a.

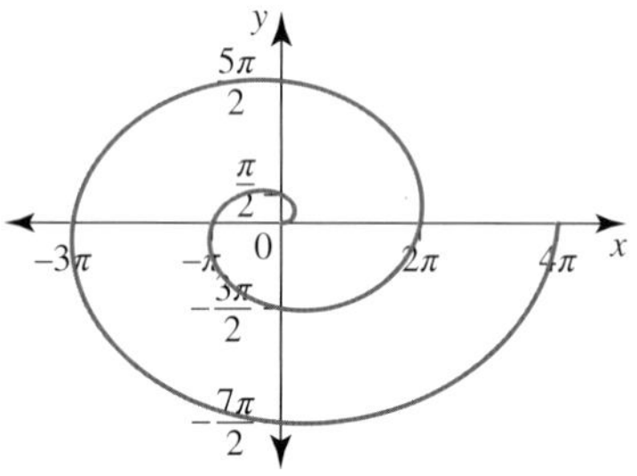

b.

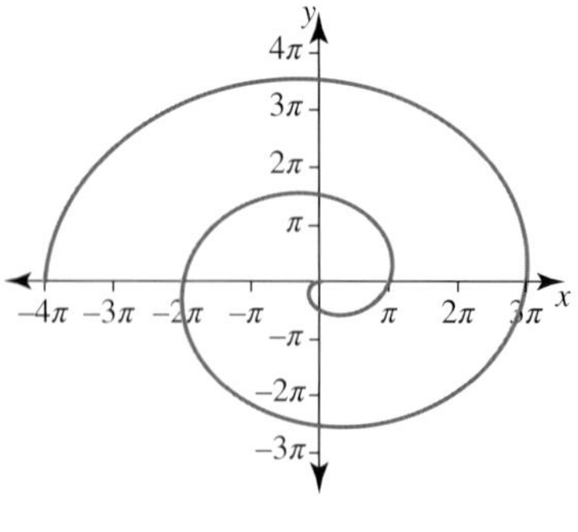

12. a.

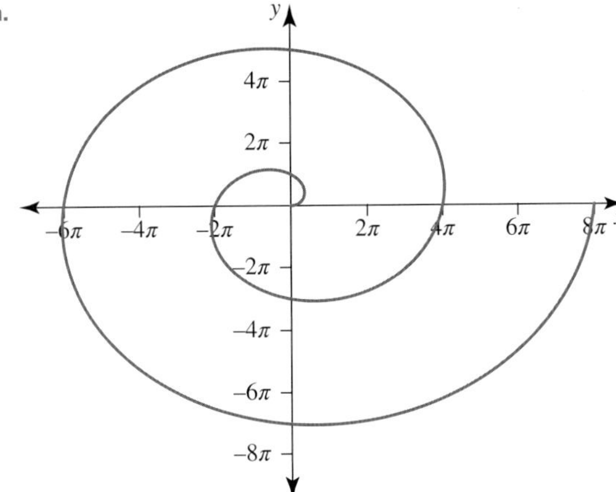

b.

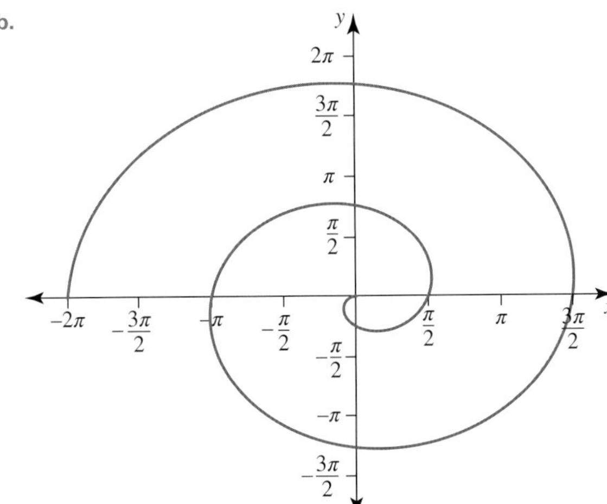

13. a.

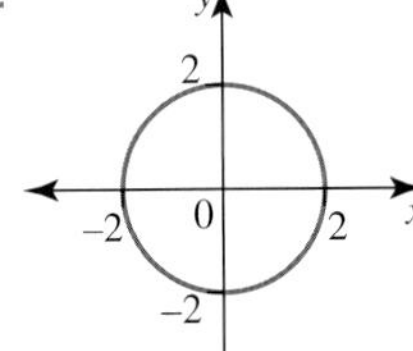

b.

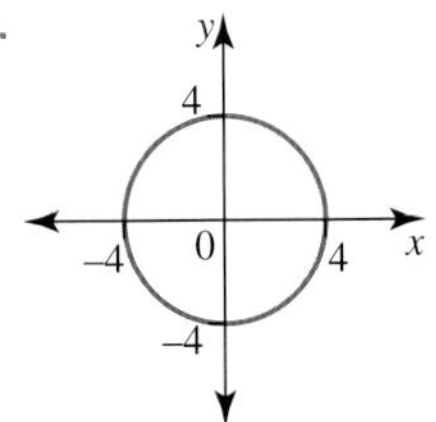

14. a.

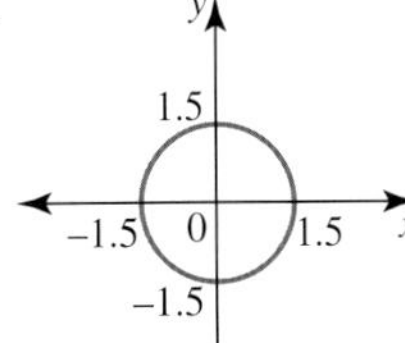

b.

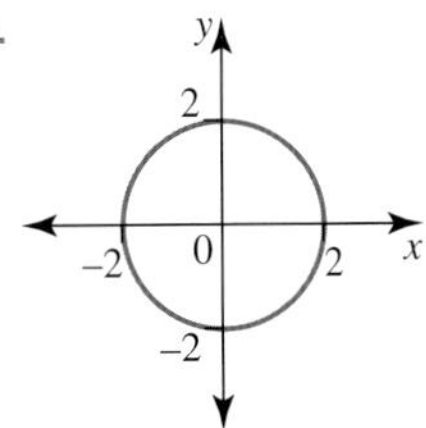

15. a.

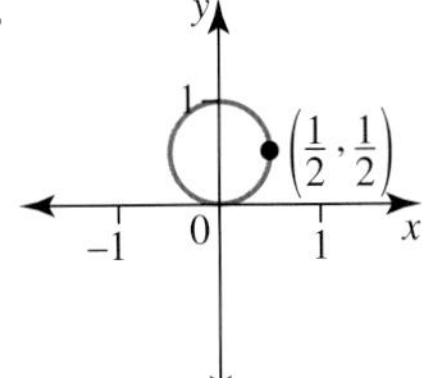

b.

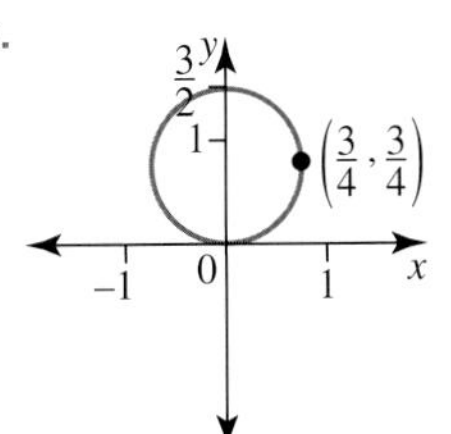

16. a.

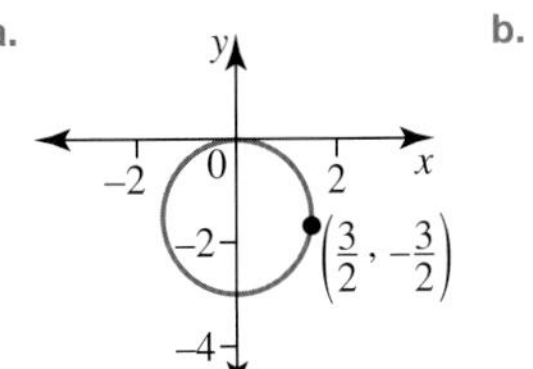

b.

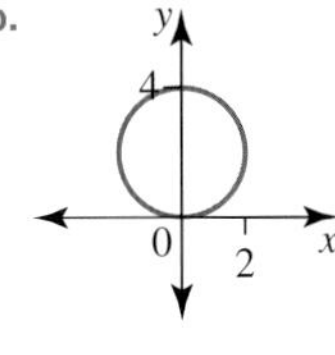

17.

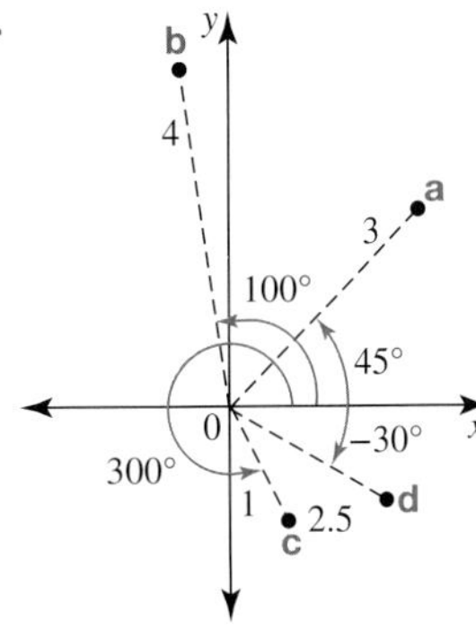

18. a. $r = 10\cos(\theta) - 6\sin(\theta)$

b. $r = -6\cos(\theta) - 8\sin(\theta)$

c. $r = 12\sin(\theta)$

d. $r = 2\cos(\theta)$

e. $r^2 = \dfrac{36}{4\cos^2(\theta) + 9\sin^2(\theta)}$

f. $r^2 = \dfrac{100}{25\cos^2(\theta) + 4\sin^2(\theta)}$

19. a. $x = 4$

b. $y = -1$

c. $x^2 + y^2 + 2x - 4y = 0$

d. $x^2 + y^2 - 8x - 6y = 0$

e. $y = \dfrac{9 - x^2}{6}$

f. $x = \dfrac{y^2 - 16}{8}$

20. a.

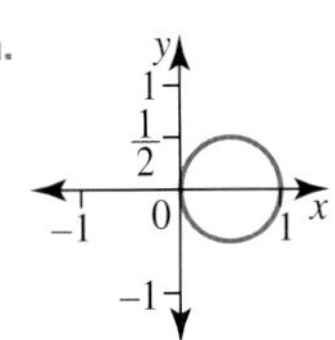

b.

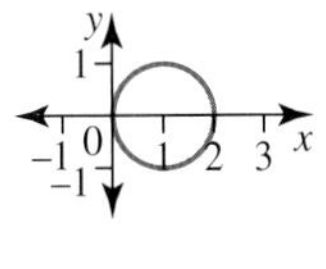

c.

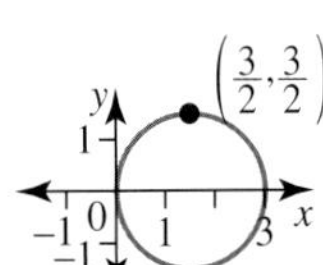

d.

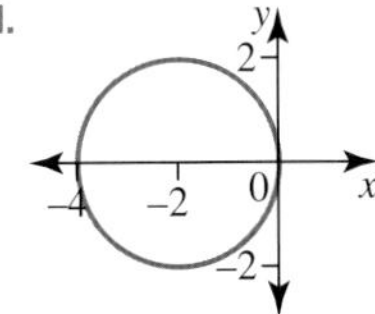

e.

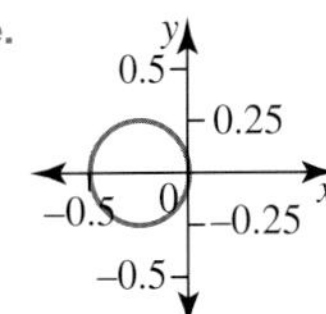

21. a. 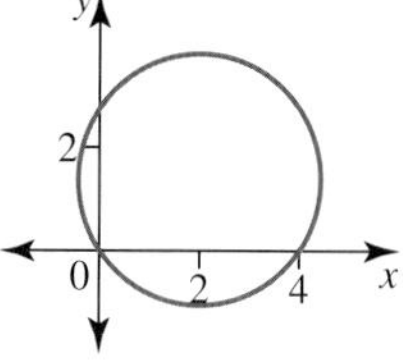

A circle

b. i. 0, 3

ii. 0, 4

iii. 5

iv. 2.5

v. (2, 1.5)

vi. $(x-2)^2+(y-1.5)^2=6.25$

22. a. 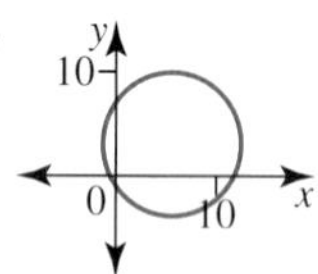

b. A circle

i. 0, 5

ii. 0, 12

iii. 13

iv. 6.5

v. (6, 2.5)

vi. $(x-6)^2+(y-2.5)^2=42.25$

12.9 Exam questions

Note: Mark allocations are available with the fully worked solutions online.

1. a. i. $\left[2, \frac{2\pi}{3}\right]$ ii. $\left[2, \frac{-5\pi}{6}\right]$

b. i. $(-4, -4)$ ii. $\left(-4, 4\sqrt{3}\right)$

2. a. $r=7$, circle

b. $r=\dfrac{13}{5\cos(\theta)-12\sin(\theta)}$, line

c. $r^2=\dfrac{1225}{25\cos^2(\theta)+49\sin^2(\theta)}$, ellipse

3. a. $\left(x^2+y^2\right)^{\frac{3}{2}}=14xy$

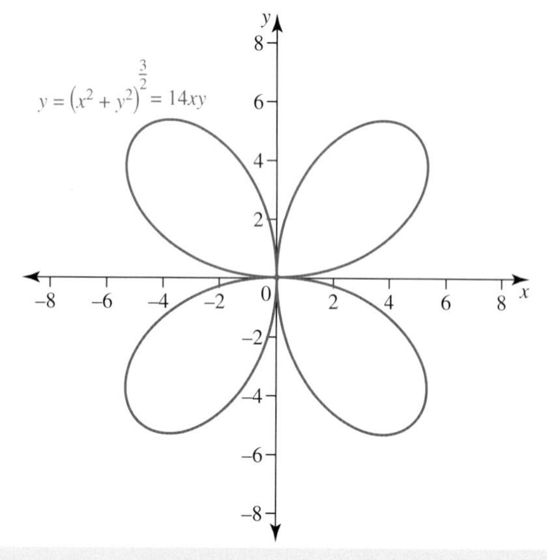

b. $y=-x$

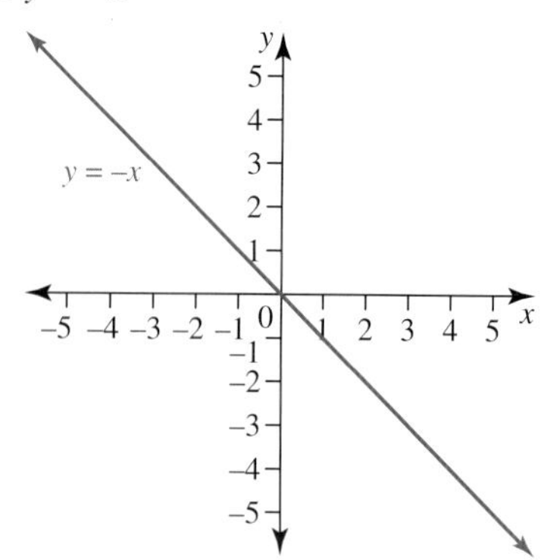

c. $x=\dfrac{9-y^2}{6}$

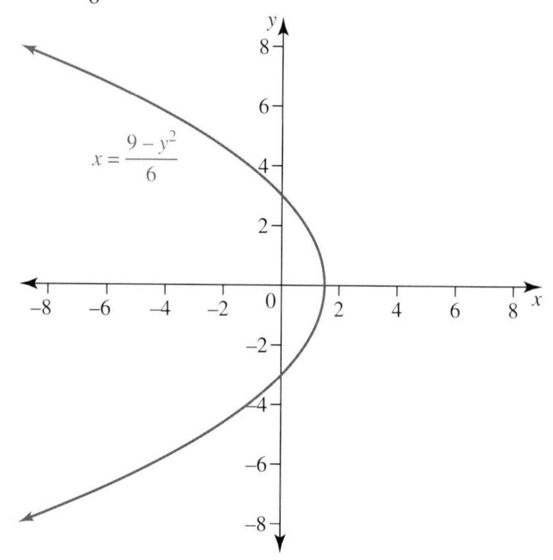

12.10 Parametric equations

12.10 Exercise

1. a. $y=x-7$ b. $y=x^2-2x+5$

2. a. $xy=1$ b. $y=x^{\frac{2}{3}}$

3. a. $x^2+y^2=1$

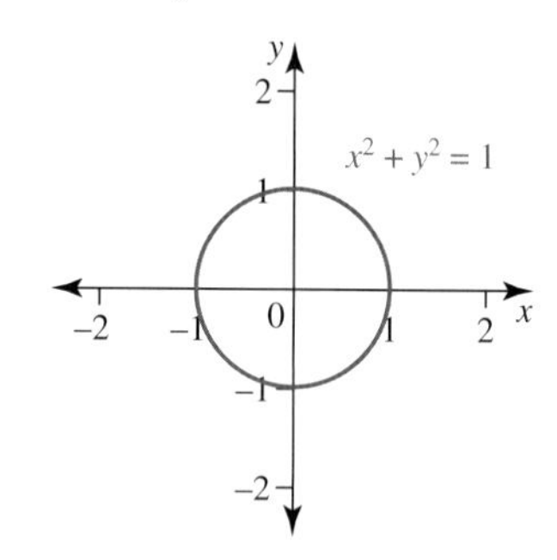

b. $\dfrac{x^2}{16} + \dfrac{y^2}{4} = 1$

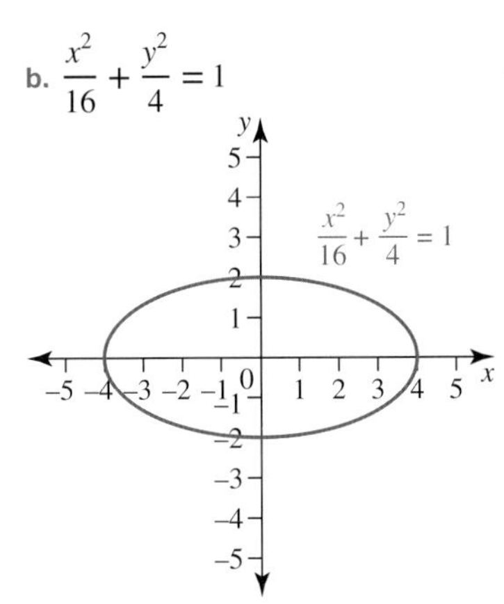

4. a. $x^2 + y^2 = 1$ b. $\dfrac{x^2}{4} - y^2 = 1$

5. $y = x^2 - 6x$

6. $(x - y)^2 = 2(x + y)$

7. $x^2 + y^2 = 1$

8. $x^2 - y^2 = 1$

9. $x = 1 - 2y^2$

10. $(x - 1)^2 + \dfrac{(y - 3)^2}{4} = 1$

11. $y = 2x\sqrt{1 - x^2}$

12. $x = y^2$

13. $x^2 + y^2 = 1$

14. $x^{\frac{2}{3}} + y^{\frac{2}{3}} = 2^{\frac{2}{3}}$

15. a. $x^2 + y^2 = \left(\tan^{-1}\left(\dfrac{y}{x}\right)\right)^2$

b.

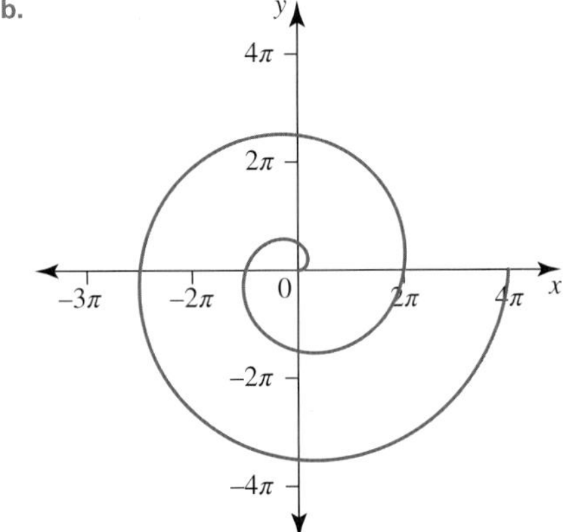

16.

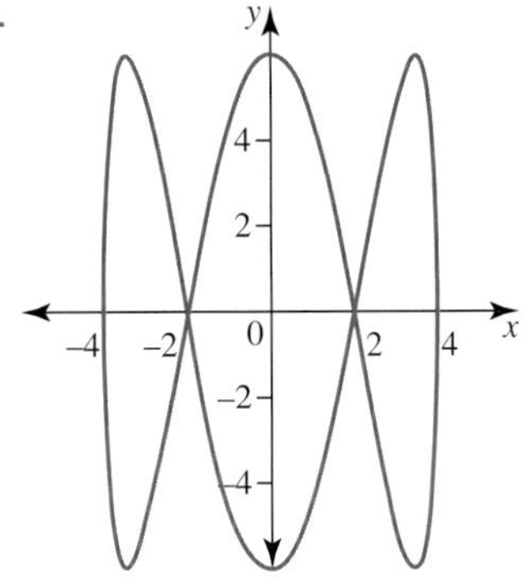

12.10 Exam questions

Note: Mark allocations are available with the fully worked solutions online.

1. $y = 2x^2 - 1$ part of a parabola

Domain $[-1, 1]$, range $[-1, 1]$

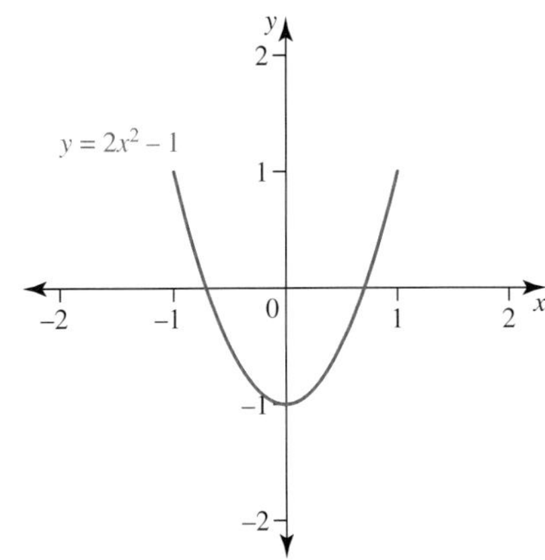

2. $\dfrac{x^2}{16} + \dfrac{y^2}{9} = 1$ ellipse

Domain $[-4, 4]$, range $[-3, 3]$

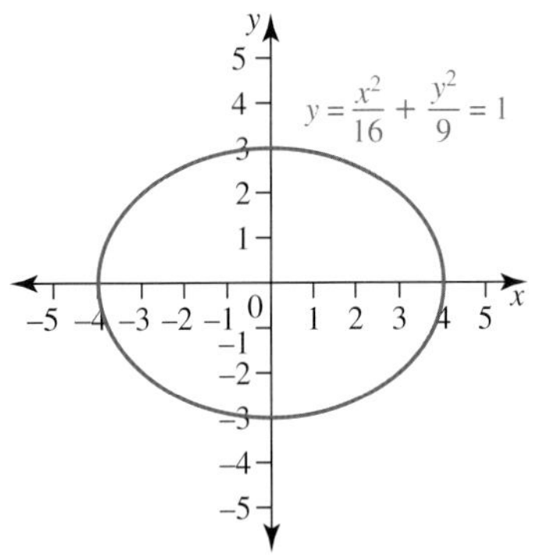

3. $\dfrac{x^2}{9} - \dfrac{y^2}{16} = 1$ ellipse, asymptotes $y = \pm\dfrac{4x}{3}$

Domain $(-\infty, -3] \cup [3, \infty)$, range R

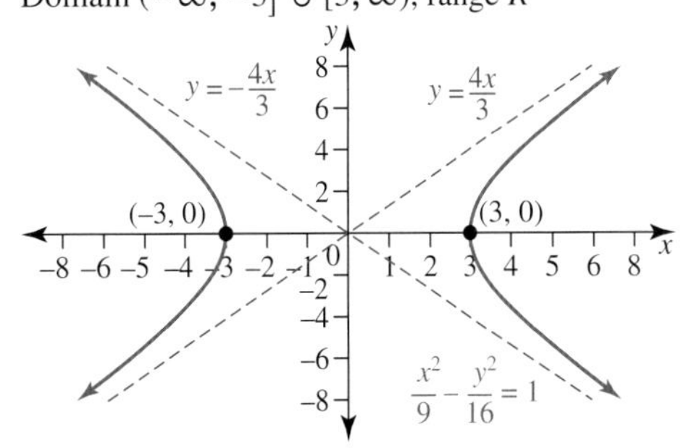

12.11 Review

12.11 Exercise

Technology free: short answer

1. a.

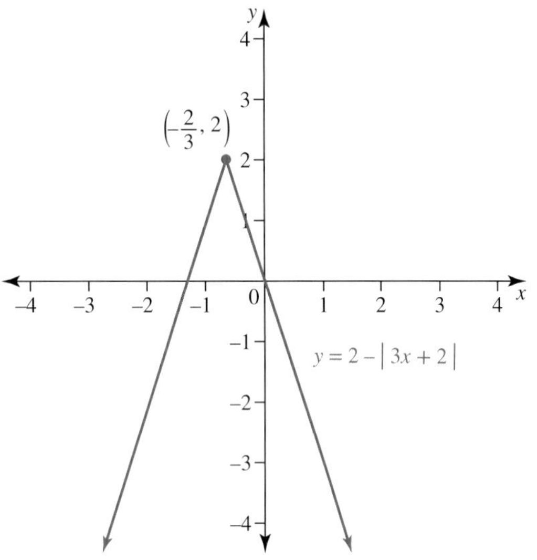

Range $(-\infty, 2]$

b.

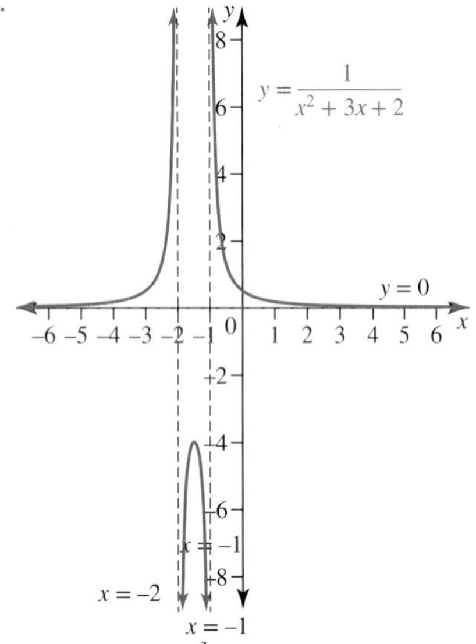

Range $(-\infty, -4] \cup (0, \infty)$

c.

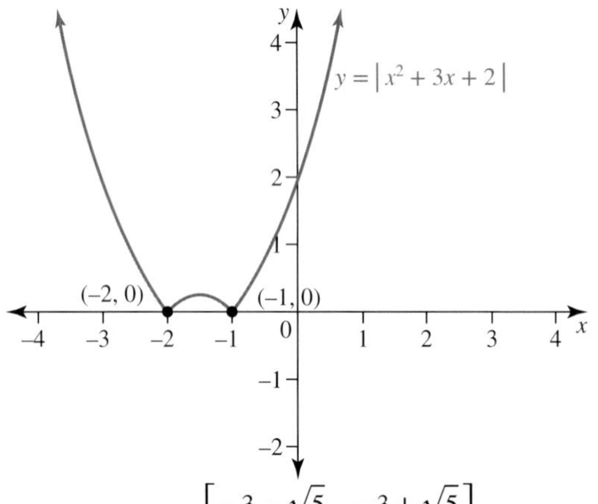

Range $[0, \infty)$ $x \in \left[\frac{-3 - \sqrt{5}}{2}, \frac{-3 + \sqrt{5}}{2}\right]$

2. a.

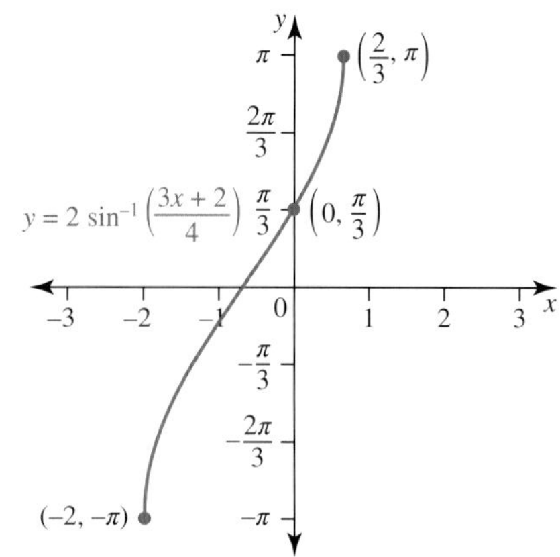

b.

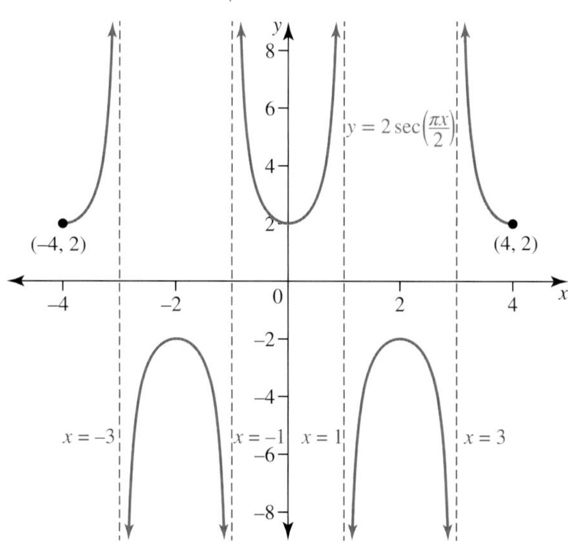

c. See the image at the bottom of the page**

3. a.

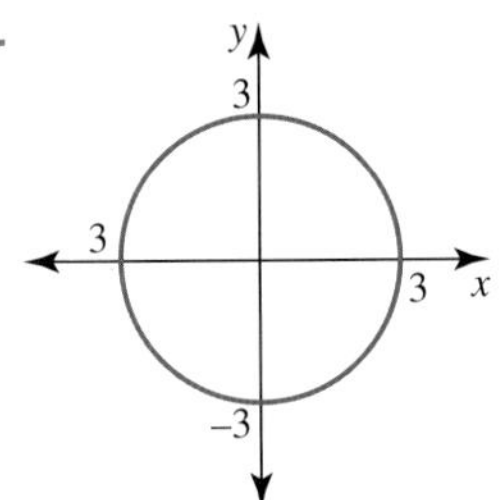

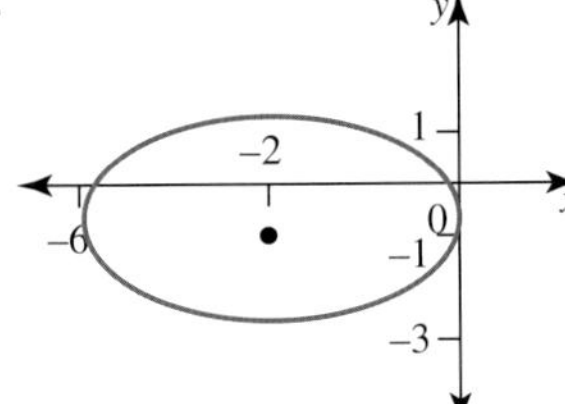

c.

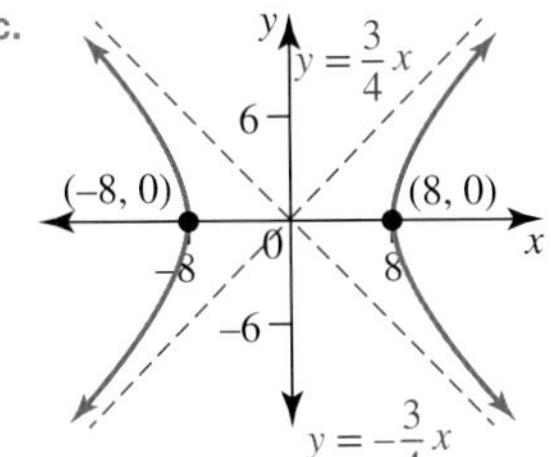

4. a. **i.** 10

ii. 8

iii. $(\pm 5, 0), (0, \pm 4)$

b.

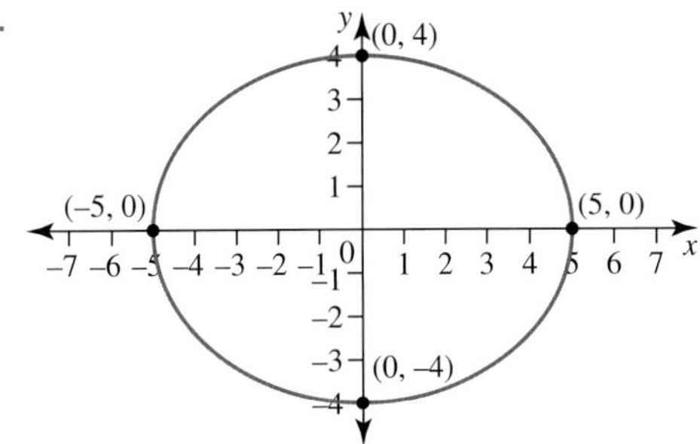

5. a. $x^2 + y^2 = 4$ **b.** $y = -x$

c. $x = 6$ **d.** $y = \dfrac{9 - x^2}{6}$

6. a. $xy = 25$ **b.** $y = x^2$ **c.** $y = 3\sin^{-1}\left(\dfrac{x}{2}\right)$

Technology active: multiple choice

7. E

8. A

9. B

10. A

11. E

12. C

13. C

14. B

15. D

16. D

Technology active: extended response

17. a. 1 and − 1

b. $y = x$ and $y = -x$

c. $a = b$

d. $\dfrac{x^2}{2} - \dfrac{y^2}{2} = 1$

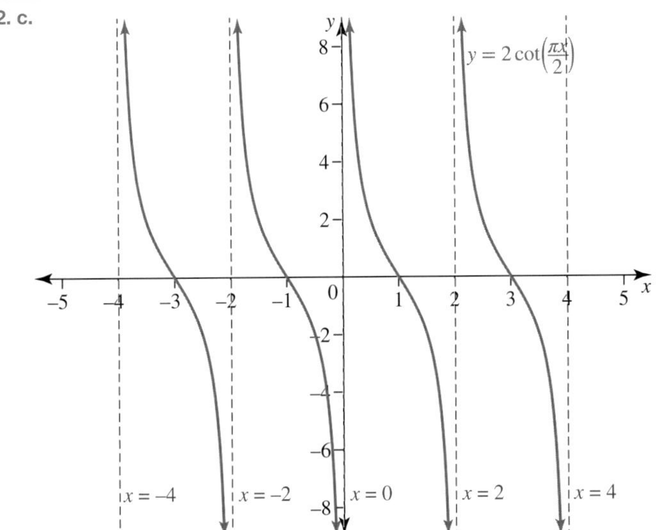

18. a. 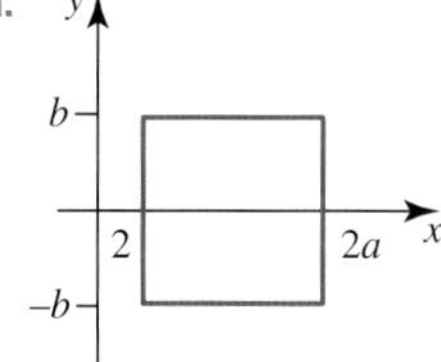

b. The equation is $\dfrac{(x-a-1)^2}{(a-1)^2}+\dfrac{y^2}{b^2}=1$.

19. a. The graph is confined to the first 2 quadrants.

b. Sample responses can be found in the worked solutions in the online resources.

c. The graph consists of an n-sided figure where θ step $=\dfrac{\pi}{n}$ and $n>2$.

d. θ step is infinitely small $(\theta \text{ step} \to 0)$.

20. Sample responses can be found in the worked solutions in the online resources.

12.11 Exam questions

Note: Mark allocations are available with the fully worked solutions online.

1. $y^2=1-2x$
2. $y=\left(x^2-2\right)^3$
3. B
4. C
5. Sample responses can be found in the worked solutions in the online resources.

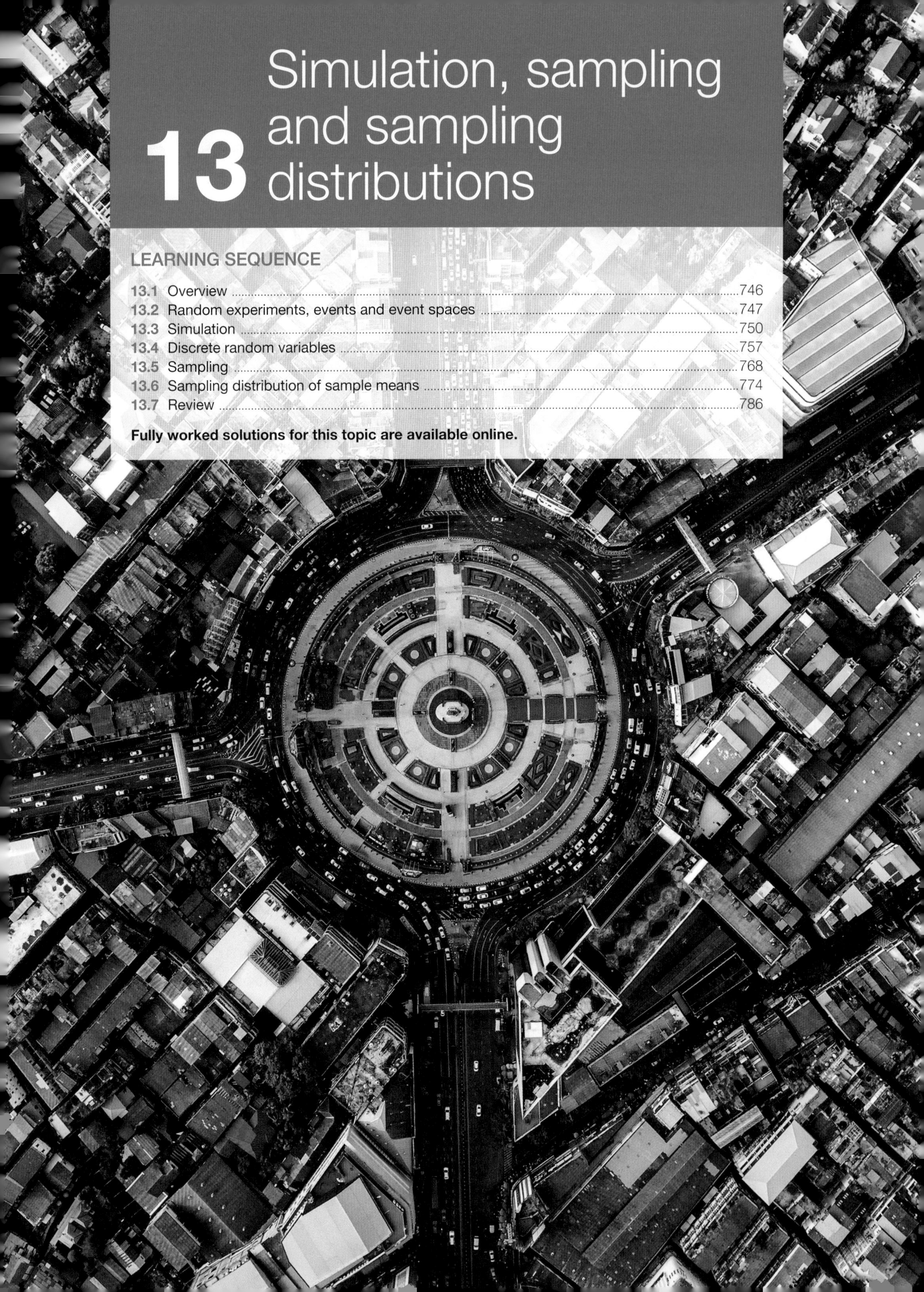

13 Simulation, sampling and sampling distributions

LEARNING SEQUENCE

Fully worked solutions for this topic are available online.

13.1 Overview

Hey students! Bring these pages to life online

Watch videos

Engage with interactivities

Answer questions and check results

Find all this and MORE in jacPLUS

13.1.1 Introduction

Statistics is a very useful tool for uncovering trends in large groups without having to analyse massive quantities of data. In this topic we will be learning about how it is possible to make estimates about large populations by sampling just a small fraction of the population.

When taking a sample, great care must be taken to select a representative sample of the population. For example, if a political party wanted to know voting intentions of the Australian population they would need to sample a selection of the entire (voting age) population. They should not simply send a survey out to signed-up members of their political party as this would give a false (biased) result.

An ice cream company wanting to know the most popular flavour amongst primary school aged children should only include primary school aged children in its sample. To include residents of a retirement home would give a false (biased) result that could lead to incorrect conclusions about the preferences of flavour amongst primary school aged children.

Once a representative, random sample has been collected, predictions can be made about the population as a whole. As we will learn throughout this topic, the predictions that can be drawn from analysing a sample increases in accuracy as the size of the sample increases.

KEY CONCEPTS

This topic covers the following key concepts from the VCE Mathematics Study Design:

- the mean, variance and standard deviation of a discrete random variable X
- the distribution of the sum of n identically distributed independent discrete random variables
- comparison of the distribution of $2X$, where X is a discrete random variable and the sum of 2 independent discrete random variables that are each identically distributed as X
- random experiments, events and event spaces
- use of simulation to generate a random sample
- the distinction between a population parameter and a sample statistic and the use of a sample statistic
- (sample mean $\bar{x}$) as an estimate of the associated population parameter (mean μ)
- the concept of a sampling distribution
- the distribution of sample means considered empirically including comparing the distributions of different size samples from the same population in terms of centre and spread
- display of variation in sample means through dot plots and other displays and considering the centre and spread of these distributions
- consideration of the mean and standard deviation of the distribution of sample means and the effect of taking larger samples.

13.2 Random experiments, events and event spaces

LEARNING INTENTION

At the end of this subtopic you should be able to:

- identify the elements in the sample space and of events in the sample space.

In this topic we explore how to use modelling and sampling to inform us about the likelihood of events occurring.

13.2.1 Basic terminology and concepts

Below is a summary of terminology covered in previous years that will be used throughout this topic.

Key word	Definition
Experiment	An activity or situation that occurs involving probability; for example, a die is rolled.
Trial	The number of times an experiment is conducted; for example, if a coin is tossed 20 times, we say there are 20 trials.
Outcome	The results obtained when an experiment is conducted; for example, when a die is rolled, the outcome can be $1, 2, 3, 4, 5$ or 6.
Equally likely	Outcomes that have the same chance of occurring; for example, when a fair coin is tossed, the two outcomes (Heads or Tails) are equally likely.
Frequency	The number of times an outcome occurs.
Variable	The characteristic measured or observed when an experiment is conducted or an observation is made.
Random variable	A variable that has a single numerical value, determined by chance, for each outcome of a trial.

13.2.2 Random experiments, events and event spaces

Random experiments deal with situations where each outcome is unknown until the experiment is run. For example, before you roll a die, you don't know what the result will be, although you do know that it will be a number between 1 and 6.

The observable outcome of an experiment is defined as an **event**.

A **generating event** is a repeatable activity that has a number of different possible outcomes (or events), only one of which can occur at a time. For example, tossing a coin results in a Head or a Tail; rolling a single die results in a 2 or 6; drawing a card from a standard deck results in a specific card (e.g. ace of hearts, 8 of clubs). These can be described as the generating event of tossing a coin, the generating event of rolling a die, and the generating event of drawing a card. In each case there is no way of knowing, in advance, exactly what will happen (what the event, or observable outcome, will be), making it a random experiment.

An **event space** (or sample space) is a list of all possible and distinct outcomes for a generating event. This list is enclosed by braces { } and each element is separated by a comma. For example, if a coin is tossed once, the event space can be written as {Head, Tail} or abbreviated to $\{H, T\}$. The list of all possible outcomes for a generating event is given the symbol ξ (xi). If a coin is tossed once, $\xi = \{H, T\}$. If a single die is rolled, $\xi = \{1, 2, 3, 4, 5, 6\}$. Capital letters are used to name other events. For example, $A = \{H\}$ means that A is the event of a Head landing uppermost when a coin is tossed once.

The event space

The event space is the set of all possible outcomes for a generating event. The event space is denoted by ξ.

For example, the event space for rolling a die is $\xi = \{1, 2, 3, 4, 5, 6\}$.

WORKED EXAMPLE 1 Listing the elements in an event space and a subset (1)

A die is rolled. List the elements of the event space and list the elements of *X*, the event of rolling an odd number.

THINK	WRITE
1. List the elements of the event space for rolling a die; that is, list all possible outcomes.	$\xi = \{1, 2, 3, 4, 5, 6\}$
2. List the elements of *X*, the event of rolling an odd number.	$X = \{1, 3, 5\}$

WORKED EXAMPLE 2 Listing the elements in a event space and a subset (2)

Two dice are rolled. List the elements of the event space and list the elements of *Y*, the event of the same number appearing on each die.

THINK	WRITE
1. List the elements of the event space for rolling two dice; that is, list all possible outcomes.	$\xi = \{(1,1),(1,2),(1,3),(1,4),(1,5),(1,6),$ $(2,1),(2,2),(2,3),(2,4),(2,5),(2,6),$ $(3,1),(3,2),(3,3),(3,4),(3,5),(3,6),$ $(4,1),(4,2),(4,3),(4,4),(4,5),(4,6),$ $(5,1),(5,2),(5,3),(5,4),(5,5),(5,6),$ $(6,1),(6,2),(6,3),(6,4),(6,5),(6,6)\}$
2. List the elements of *Y*, the event of the same number appearing on each die when two dice are rolled.	$Y = \{(1,1), (2,2), (3,3), (4,4), (5,5), (6,6)\}$

WORKED EXAMPLE 3 Listing the elements in a event space and a subset (3)

A card is selected from a deck of 52 playing cards. List the elements of the event space and list the elements of *K*, the event of selecting a king or a spade.
***Note:* The joker is not included as one of the 52 playing cards.**

THINK	WRITE
1. List the event space for choosing a card from the deck; that is, list all the possible cards in the deck. Use the abbreviations A, 2, 3, ... 10, J, Q, K for the numbers and S, C, H, D for the suits (spades, clubs, hearts, diamonds). *Note:* A, J, Q and K are abbreviations for the ace, jack, queen and king respectively.	ξ = {AS, 2S, 3S, 4S, 5S, 6S, 7S, 8S, 9S, 10S, JS, QS, KS, AC, 2C, 3C, 4C, 5C, 6C, 7C, 8C, 9C, 10C, JC, QC, KC, AH, 2H, 3H, 4H, 5H, 6H, 7H, 8H, 9H, 10H, JH, QH, KH, AD, 2D, 3D, 4D, 5D, 6D, 7D, 8D, 9D, 10D, JD, QD, KD}
2. List all the cards that are a king or a spade. Note that KS is listed only once.	*K* = {AS, 2S, 3S, 4S, 5S, 6S, 7S, 8S, 9S, 10S, JS, QS, KS, KC, KH, KD}

Worked example 3 introduced the idea of multiple events, such as selecting a king or a spade from a deck of 52 playing cards. In mathematics, the terms 'or' and 'and' have very specific definitions, which we covered back in topics 3 and 4.

Let A and B be two events (for example, A is the event of drawing a king and B the event of drawing a spade). Then:

1. A OR B means that either A or B happens, or that both happen. You can get a king or a spade or the king of spades.
2. A AND B means both must happen. You must get a king and a spade; that is, the king of spades only.

A **random event** is one in which the outcome of any given trial is uncertain but the outcomes of a large number of trials follow a regular pattern. For example, when a die is rolled once, the outcome could be either a $1, 2, 3, 4, 5$ or 6, but if it is rolled several thousand times it is evident that about 1 in 6 rolls will turn up a 6. If the die was weighted so that a 6 was more likely than a 2 to land uppermost, then the experiment would be biased and no longer random.

13.2 Exercise

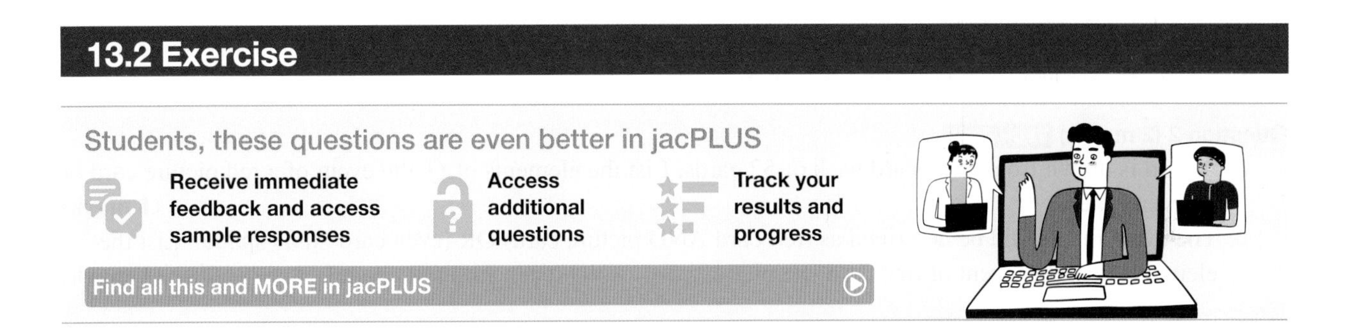

Technology free

1. WE1 A die is rolled. List the elements of the event space and list the elements of Y, the event of a number greater than 4 appearing uppermost.
2. A card is drawn from a deck of 52 playing cards and the suit is noted. List the elements of the event space and list the elements of Z, the event of a black card being selected.
3. WE2 Two dice are rolled. List the elements of the event space and list the elements of Z, the event of a total of greater than 4 appearing on the dice.
4. A die is rolled and a coin is tossed. List the elements of the event space and list the elements of X, the event of a Head appearing with an even number on the die.
5. WE3 A card is selected from a deck of 52 playing cards. List the elements of the event space and list the elements of P, the event of selecting a jack or a spade.
6. A card is selected from a deck of 52 playing cards. Use the list of elements of the event space from question **5** to list the elements of Q, the event of selecting a jack and a spade.
7. A card is drawn from a deck of 52 playing cards and the face value is noted. List the elements of the event space and list the elements of W, the event of a picture card (king, queen or jack) being selected.
8. A coin is tossed and a card selected from a standard pack, with the suit noted. List the elements of the event space, and list the elements of X, the event of a Tail appearing with a red card.
9. Two dice are rolled. List the sample space of D, the event of an odd number being rolled on each die.
10. Two dice are rolled. List the sample space of F, the event of both dice showing a number greater than or equal to 4.

11. A die is rolled and a coin is tossed. List the elements of X, the event of a Head appearing or an even number on the die.

12. A coin is tossed and a card selected from a standard pack, with the suit noted. List the elements of Y, the event of a Tail appearing or a red card.

13. A coin is tossed three times. List the elements of the event space and the elements of D, the event of exactly two coins having the same result.

14. An urn contains four balls numbered 1 to 4. A ball is withdrawn, its number noted, and is put back in the urn. A second ball is then drawn out and its number noted. List the elements of the event space, and list the elements of Q, the event of the second ball having a value greater than the first ball.

13.2 Exam questions

Question 1 (2 marks) TECH-FREE

The three hearts picture cards are drawn at random. List the elements of the event space and list the elements of K, the event of a king being drawn first or a jack being drawn last.

Question 2 (2 marks) TECH-FREE

a. A card is drawn from a standard pack of 52 cards. List the elements of Q, the event of a red picture card or an even spade. **(1 mark)**

b. The event in Q could be described as (red card AND picture card) OR (even card AND spade). List the elements of R, the event of (red card OR picture card) AND (even card OR spade). **(1 mark)**

Question 3 (2 marks) TECH-ACTIVE

An urn contains four balls numbered 1 to 4. A ball is withdrawn and its number noted. A second ball is then drawn out and its number noted (without replacement of the first ball). List the elements of the event space, and list the elements of R, the event of an odd number on both balls.

More exam questions are available online.

13.3 Simulation

LEARNING INTENTION

At the end of this subtopic you should be able to:

- simulate random events using a CAS calculator.

13.3.1 Applications of simulation

Simulation is an activity employed in many areas, including business, engineering, medical and scientific research, and in specialist mathematical problem-solving activities, to name a few. It is a process by which experiments are conducted to model or imitate real-life situations. Simulations are often chosen because the real-life situation is dangerous, impractical, or too expensive or time consuming to carry out in full. An example is the training of airline pilots.

In this section we examine the basic tools of simulation and the steps that need to be followed for a simulation to be effective. We also look at various types of simulation.

13.3.2 Random numbers

Simulations often require the use of sequences of random numbers (most commonly these numbers will be integers in this course).

Consider the following result when rolling a die 12 times:

$$5, 3, 1, 4, 4, 3, 6, 1, 6, 4, 1, 3.$$

This sequence of numbers can be treated as a set of **random numbers**, whose possible values are 1, 2, 3, 4, 5, 6.

A deck of playing cards could be numbered from 1 to 52; by drawing a card, then replacing it in the deck, shuffling the deck and drawing another card, a sequence of random numbers between 1 and 52 would be generated. There are several other ways of generating random numbers; can you think of any?

There are three general rules that a set of random numbers must follow.

Rule 1 The set must be within a defined **range** (for example, whole numbers between 1 and 10, or decimals between 0 and 1). The range does not need to have a definite starting and finishing number, but in most cases it does.

Rule 2 All numbers within the defined range must be equally likely possible outcomes (for example, using a die but not counting 4s is not a proper sequence of random numbers between 1 and 6).

Rule 3 Each random number in a sequence is independent of any of the previous (or future) numbers in the sequence. (This is difficult to prove but is assumed with dice, spinners and so on.)

These rules can make it extremely difficult to obtain a proper set of random numbers in practice. Furthermore, to generate many random numbers, say 1000, by rolling dice could take a long time! Fortunately, computers can be used to generate random numbers for us. (Technically, they are known as pseudo-random numbers, because rules 2 and 3 above cannot be rigorously met.)

WORKED EXAMPLE 4 Simulating a sequence of randoms numbers using a die

Use a sequence of 20 random numbers to simulate rolling a die 20 times. Record your results in a frequency table.

THINK

1. Using CAS, go to the random integer function.
2. Select 20 random integers between 1 and 6 inclusive and store them in a list.
3. Use CAS to count the number of times each element appears in the list.
   Note: Naturally, the values shown may differ each time this sequence is repeated.
4. Enter this information in a table as shown.

WRITE

Die value	1	2	3	4	5	6
Frequency	2	4	6	2	5	1

TI | THINK	DISPLAY/WRITE	CASIO | THINK	DISPLAY/WRITE
On a Calculator page, select MENU 5 Probability 4 Random 2 Integer and complete the entry as shown.	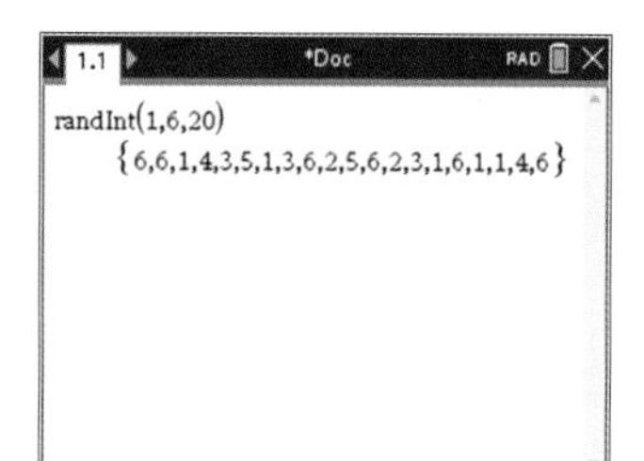	On a Main screen, create a list of 20 random integers between 1 and 6 by completing the entry as shown. randList(can be found in the Catalog in the soft keyboard.	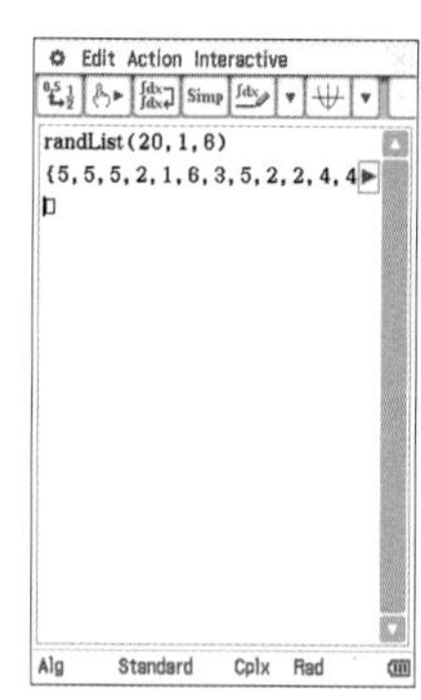

13.3.3 Basic simulation tools

The simulation option that you select may depend on the type and number of events or on the complexity of the problem to be simulated.

1. Coin tosses can be used when there are only two possible, equally likely, events — for example, to simulate the results of a set of tennis where the two players are equally matched. In this case each toss could represent a single game, the simulation ending when either 'H' or 'T' has enough games to win the set.
2. Spinners can be used when there are three, four, five or more possible, equally likely, events. Provided that your spinner is a fair one, where each sector has an equal area, this is an effective simulation device.

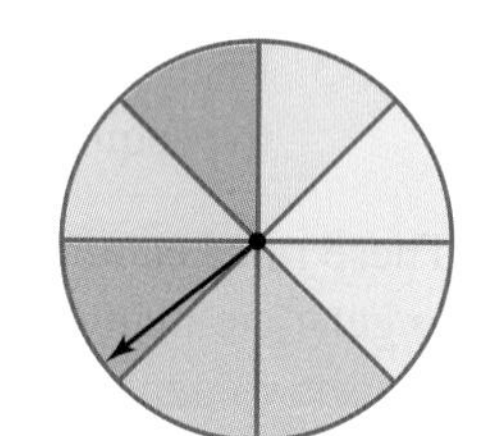

3. Dice — One die can be used to simulate an experiment with six equally likely outcomes. One die can also be used to simulate three equally likely outcomes by assigning a 1 or a 2 to the first outcome, a 3 or a 4 to the second outcome and a 5 or a 6 to the third outcome. Two or more dice can be used for more complex situations. For example, two dice can be used to simulate the number of customers who enter a bank during a 5-minute period.
4. Playing cards can be used to simulate extremely complicated experiments. There are 52 cards, arranged in 13 values (A, 2, 3, ..., J, Q, K) and in 4 suits (spades, clubs, hearts, diamonds); they can be defined to represent all kinds of situations.
5. Random number generators are the most powerful simulation tool of all. Computers can tirelessly generate as many numbers in a given range as you wish. They are available on most scientific or graphics calculators (randInt), spreadsheets (randbetween) and computer programming software.

Steps to an effective simulation

Step 1 **Understand the situation being simulated and choose the most effective simulation tool.**
Step 2 **Determine the basic, underlying, assumed probabilities, if there are any.**
Step 3 **Decide how many times the simulation needs to be repeated. The more repetitions, the more accurate the results.**
Step 4 **Perform the simulation, displaying and recording your results.**
Step 5 **Interpret the results, stating the resultant simulated probabilities. Sometimes you can repeat the entire simulation (steps 3 and 4) several times and compute 'averages'.**

WORKED EXAMPLE 5 Simulating the result of a tennis match using a coin

Simulate the result of a best-of-3-sets match of tennis using simple coin tosses. (*Note:* Each set contains a maximum of 13 games. To win a set you must win 6 games and be 2 or more ahead of your opponent. If someone reaches 6 games and the opponent is on 5 then the set continues. If the score becomes 7 – 5, then the person on 7 wins and if it becomes 6 – 6 then one more game is played, and the result is a 7 – 6 win to whoever wins that last game.)

THINK

WRITE

1. Understand the situation. The winners of games are recorded, not the winners of points.

We will need, at most, 13 coin tosses per set for, at most, 3 sets, assuming that if a set reaches 6 – 6 it will be resolved by a tie breaker.

2. Determine the probabilities.

Assume evenly matched players:
Player A = 'H', Player B = 'T'
(H represents Player A winning; T represents Player B winning.)

3. Decide how many times to simulate.

We will simulate one best-of-3-sets match.

4. Perform the simulation.

Toss	Result	Score	
H	A	1 – 0	
T	B	1 – 1	
H	A	2 – 1	
H	A	3 – 1	
H	A	4 – 1	
T	B	4 – 2	
H	A	5 – 2	
T	B	5 – 3	
H	A	6 – 3	Player A wins Set 1.
T	B	0 – 1	
T	B	0 – 2	
T	B	0 – 3	
T	B	0 – 4	
T	B	0 – 5	
T	B	0 – 6	Player B wins Set 2.
T	B	0 – 1	
H	A	1 – 1	
H	A	2 – 1	
T	B	2 – 2	
T	B	3 – 3	
H	A	4 – 3	
T	B	4 – 5	
H	A	5 – 5	
T	B	5 – 6	
H	A	6 – 6	Tie breaker required
T	B	6 – 7	Player B wins match.

5. Interpret your results.	28 coin tosses were required to simulate 1 match. Player A won 12 games, Player B won 16; thus we could say that Player A has a probability of winning of $\frac{12}{28} = 0.4286$ based upon this small simulation. However, since coins were used, 'true' probability $= 0.5$.

TI | THINK	DISPLAY/WRITE	CASIO | THINK	DISPLAY/WRITE
On a Calculator page, select MENU 5 Probability 4 Random 2 Integer and complete the entry as shown.	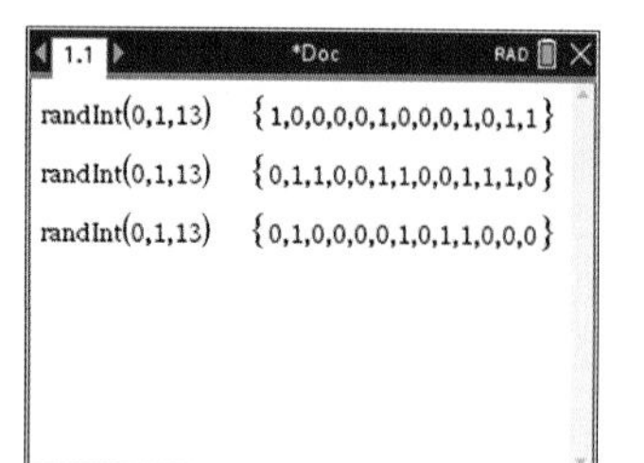	On a Main screen, create 3 lists of 13 random integers between 0 and 1 by completing the entry as shown.	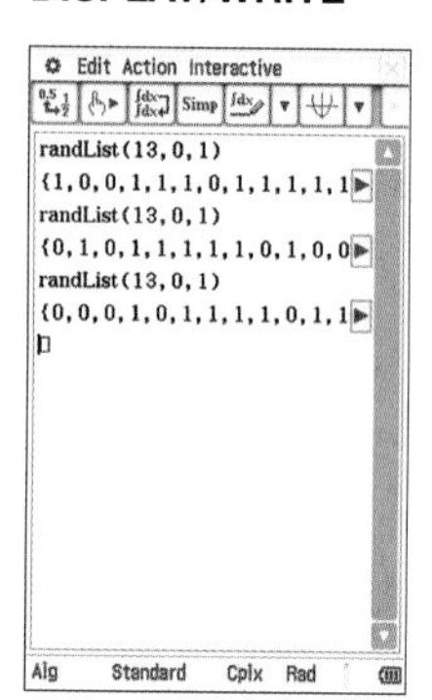
Before creating these random lists decide whether 0 will represent a winning game for player A or player B.	In this simulation, 0 represents player A. Player A wins the first set 6 – 2. (The remaining random numbers in the first list can be ignored.) Player A loses the second set 5 – 7. Player A wins the third set 6 – 2. Player A wins the match by 2 sets to 1.	Before creating these random lists decide whether 0 will represent a winning game for player A or player B.	In this simulation, 0 represents player A. Player A loses the first set 3 – 6. (The remaining random numbers in the first list can be ignored.) Player A loses the second set 2 – 6. The third set is not required, player B has already won the match by 2 sets to 0.

13.3 Exercise

Students, these questions are even better in jacPLUS

Receive immediate feedback and access sample responses

Access additional questions

Track your results and progress

Find all this and MORE in jacPLUS

Technology active

1. WE4 Use a sequence of 6 random numbers to simulate rolling a die 9 times. Record your results in a frequency table.
2. Generate 100 numbers between 10 and 20. Record your results in a frequency table.
3. WE5 Simulate the result of a best-of-5-sets tennis match using a simple coin toss.

4. Use a die to simulate the two-player game described by the table shown. Perform enough simulations to convince yourself about which player has the better chance of making a profit.

Die value	Outcome
1	Player A wins $1
2	Player A wins $2
3	Player A wins $4
4	Player A wins $8
5	Player A wins $16
6	Player B wins $32

5. a. Suggest how a graphics calculator could be used to simulate the tossing of a coin.
   b. Generate 10 'coin tosses' using this method.
   c. Sketch a frequency histogram for your results.
6. Repeat question 5 for 100 coin tosses. State what you notice about the histogram.
7. A mini-lottery game may be simulated as follows. Each game consists of choosing two numbers from the whole numbers 1 to 6. The cost to play one game is $1. A particular player always chooses the numbers 1 and 2. A prize of $10 is paid for both numbers correct. No other prizes are awarded.
   a. Simulate a game by generating two random numbers between 1 and 6. Do this 20 times (that is, 'play' 20 games of lotto). State how many times the player wins.
   b. Determine the player's profit/loss based on the simulation.
8. Simulate 20 tosses of two dice (die 1 and die 2). State how many times die 1 produced a lower number than die 2. (*Hint:* Generate two lists of 20 values between 1 and 6.)
9. A board game contains two spinners shown.
   a. Simulate 10 spins of each spinner.
   b. State how many times the total (that is, both spinners' numbers added) is equal to 5.
   c. State how often there is an even number on both spinners.
   d. State how often the highest possible total occurs.

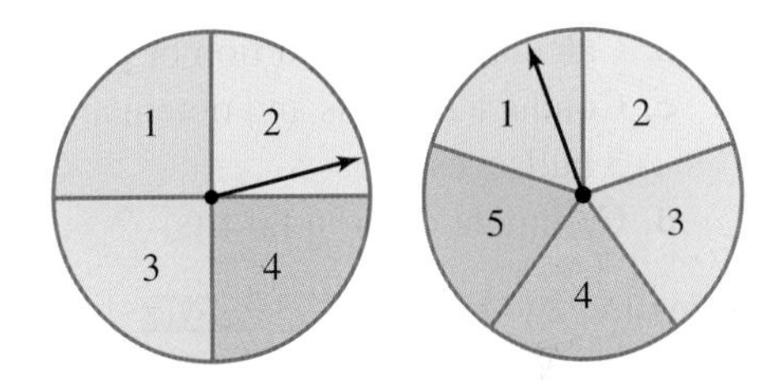

10. Use dice, a spreadsheet or other means to generate data that simulate the arrival and departure of customers from a bank according to the rules in the table shown (assume there are 10 customers to begin with).
    Each total represents what happens over a 5-minute interval. Perform enough simulations to cover at least 2 hours' worth of data. Discuss your results.

Dice total	Outcome
2	1 customer leaves
3	1 customer arrives
4	2 customers leave
5	2 customers arrive
6	3 customers leave
7	3 customers arrive
8	4 customers leave
9	4 customers arrive
10	5 customers leave
11	5 customers arrive
12	6 customers leave

11. Design a spreadsheet to simulate the tennis match from Worked example 5. Create a version for a best-of-3-sets match and a best-of-5-sets match.

12. In the Smith Fish & Chip shop, the owner, Mary Jones, believes that her customers order various items according to the probabilities shown in the table.
Set up a spreadsheet to simulate the behaviour of 100 customers. Determine how many orders of fish only and how many orders of chips only Mary should prepare for 100 customers.

Order	Probability
Chips only	0.20
1 fish and 1 chips	0.15
2 fish and 1 chips	0.26
1 fish only	0.14
2 fish only	0.11
3 fish and chips	0.14

13. Use a spreadsheet to simulate the tossing of 4 coins at once. Perform the simulation enough times to convince yourself that you can predict the probabilities of getting 0 Heads, 1 Head, ..., 4 Heads. Compare your results with the 'theoretical' probabilities.

13.3 Exam questions

Question 1 (4 marks) TECH-ACTIVE

The data below show the number of bullseyes scored by 40 dart players after 5 throws each.

1	4	0	3	4	2	1	5	4	2
3	0	4	5	2	1	2	3	2	1
0	2	1	4	3	5	3	2	4	4
0	2	1	0	3	5	4	2	3	1

a. Explain how a calculator can be used to obtain the range of numbers given in the table. **(1 mark)**
b. Calculate the proportion of players that scored at least 3 bullseyes. **(1 mark)**
c. Conduct 20 trials and obtain another possible value for the proportion of players who scored at least 3 bullseyes. **(1 mark)**
d. Comment on your results. **(1 mark)**

Question 2 (4 marks) TECH-ACTIVE

Use a spreadsheet to simulate the following situation. A target consists of 3 concentric circles and darts are randomly thrown at the target. Let the smallest circle have a radius of r_1 cm, the next r_2 cm and the largest r_3 cm. They sit on a square board $y \times y$ cm $(y \geq 2r_3)$. Assume that all darts hit the target or the square board outside the target.

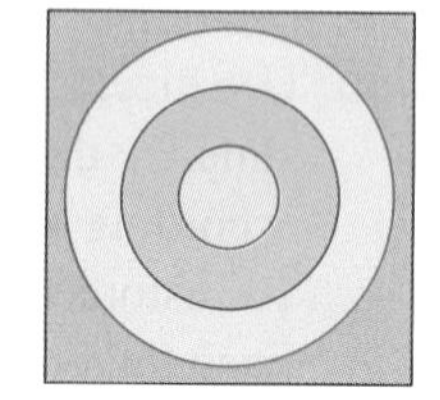

Start with some numeric examples, say $r_1 = 5$, $r_2 = 10$, $r_3 = 15$ and $y = 32$.
If you 'hit' a circle, you score points: the inner circle, 5 points; the middle circle, 3 points; the outer circle, 1 point; and you score no points for hitting the board outside the target.

a. Perform enough simulations so that you can predict the 'expected' number of points per throw. **(1 mark)**
b. Experiment with different values of the radii and board size, and tabulate your results. **(1 mark)**
c. Experiment with different scoring systems and tabulate your results. **(1 mark)**
d. Calculate the 'theoretical' probabilities and expected scores for your different values of radii and board size. **(1 mark)**
Hints on setting up simulation:
1. Generate two random numbers which represent x- and y-coordinates with the target at (0, 0).
2. For each random pair, calculate the distance from the origin and compare it to r_1, r_2 and r_3.
3. Allocate points appropriately.

Question 3 (4 marks) TECH-ACTIVE

A football league has 8 teams. Each team plays all the other teams once. Thus there are 28 games played in all.

a. Simulate a full season's play, assuming that each team has a 50 : 50 chance of winning each game. **(1 mark)**

b. Modify the probabilities so that they are unequal (*hint:* sum of probabilities = 4) and simulate a full season's play. Determine whether the better teams reached the top of the ladder. Discuss your results with other students. **(3 marks)**

Hint: If Team 1 has a probability of winning of 0.7 and Team 2 has a probability of winning of 0.6, then when they play against each other, the probability of Team 1 winning is $\frac{0.7}{0.7+0.6}$.

More exam questions are available online.

13.4 Discrete random variables

LEARNING INTENTION

At the end of this subtopic you should be able to:
- calculate the mean, variance and standard deviation of discrete random variables
- understand the difference between distributions $2X$ and $X+X$.

13.4.1 Introduction to random variables

A random variable is one whose value cannot be predicted but is determined by the outcome of an experiment. For example, two dice are rolled simultaneously a number of times. The sum of the numbers appearing uppermost is recorded. The possible outcomes we could expect are $\{2, 3, 4, 5, 6, 7, 8, 9, 10, 11, 12\}$. Since the possible outcomes may vary each time the dice are rolled, the sum of the numbers appearing uppermost is a random variable.

Random variables are expressed as capital letters, usually from the end of the alphabet (for example, X, Y, Z) and the value they can take on is represented by lowercase letters (for example, x, y, z respectively).

The above situation illustrates an example of a discrete random variable since the possible outcomes were able to be counted. Discrete random variables generally deal with number or size.

A random variable that can take on any value is defined as a continuous random variable. Continuous random variables generally deal with quantities that can be measured, such as mass, height or time.

WORKED EXAMPLE 6 Classifying variables as discrete or continuous

Classify the following variables as discrete or continuous.
a. The number of goals scored at a football match.
b. The height of students in a Specialist Mathematics class.
c. The number of shoes in a closet.
d. The number of girls in a five-child family.
e. The time taken, in minutes, to run a distance of 10 kilometres.

THINK	WRITE
Determine whether the variable can be counted or needs to be measured.	
a. Goals can be counted.	a. Discrete
b. Height must be measured.	b. Continuous
c. The number of shoes can be counted.	c. Discrete
d. The number of girls can be counted.	d. Discrete
e. Time must be measured.	e. Continuous

13.4.2 Discrete probability distributions

When dealing with random variables, the probabilities associated with them are often required.

WORKED EXAMPLE 7 Evaluating probabilities of outcomes of discrete random variables

Let X represent the number of tails obtained in three tosses of an unbiased coin. Draw up a table that displays the values the discrete random variable can assume and the corresponding probabilities.

THINK

1. Draw a tree diagram and list all of the possible outcomes.

WRITE/DRAW

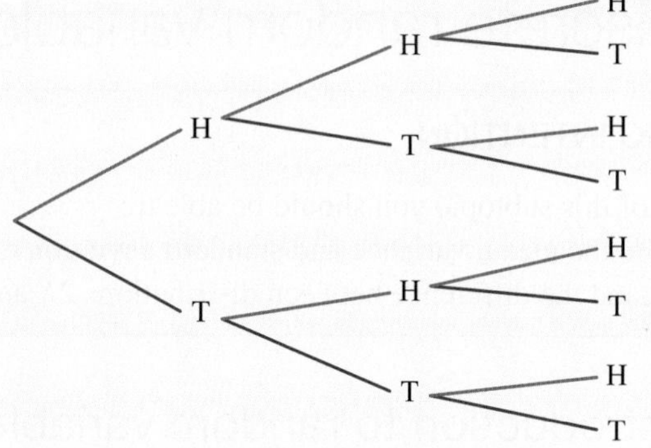

HHH, HHT, HTH, HTT, THH, THT, TTH, TTT

2. Draw a table with two columns: one labelled number of tails, the other probability.
3. Enter the information into the table.

Number of tails (x)	Probability Pr (x)
0	$\frac{1}{8}$
1	$\frac{3}{8}$
2	$\frac{3}{8}$
3	$\frac{1}{8}$

The table above displays the probability distribution of the total number of tails obtained in three tosses of an unbiased coin. Since the variable in this case is discrete, the table displays a discrete probability distribution.

In Worked example 7, X denoted the random variable and x the value that the random variable could take. Thus the probability can be denoted by $p(x)$ or $\Pr(X = x)$. Hence the table in Worked example 7 could be presented as shown below.

x	0	1	2	3
$\Pr(X = x)$	$\frac{1}{8}$	$\frac{3}{8}$	$\frac{3}{8}$	$\frac{1}{8}$

Close inspection of this table shows important characteristics that satisfy all discrete probability distributions.

Properties of discrete probability distributions

1. **Each probability lies in a restricted interval $0 \leq \Pr(X = x) \leq 1$.**
2. **The probabilities of a particular experiment sum to 1, that is,**

$$\sum \Pr(X = x) = 1$$

If these two characteristics are not satisfied, then there is no discrete probability distribution.

13.4.3 Measures of centre of discrete random distributions

The expected value of a discrete random variable, X, is the average value of X. It is also referred to as the mean of X or the expectation of X.

The expected value of a discrete random variable, X, is denoted by E(X) or the symbol μ (mu). It is defined as the sum of each value of X multiplied by its respective probability; that is,

The expected value, E(X)

$$E(X) = x_1 \Pr(X = x_1) + x_2 \Pr(X = x_2) + \ldots + x_n \Pr(X = x_n) = \sum_{\text{all } x} x \Pr(X = x)$$

Note: The expected value will not always assume a discrete value in the sample space, it will usually fall somewhere in between the discrete values.

WORKED EXAMPLE 8 Calculating the expected value of a discrete random distribution

Calculate the expected value of a random variable that has the following probability distribution.

x	1	2	3	4	5
$\Pr(X = x)$	$\frac{2}{5}$	$\frac{1}{10}$	$\frac{3}{10}$	$\frac{1}{10}$	$\frac{1}{10}$

THINK	WRITE
1. Write the rule for the expected value.	$E(X) = \sum_{\text{all } x} x \Pr(X = x)$
2. Substitute the values into the rule.	$E(X) = 1 \times \frac{2}{5} + 2 \times \frac{1}{10} + 3 \times \frac{3}{10} + 4 \times \frac{1}{10} + 5 \times \frac{1}{10}$ $= \frac{2}{5} + \frac{2}{10} + \frac{9}{10} + \frac{4}{10} + \frac{5}{10}$
3. Evaluate.	$= 2\frac{2}{5}$

WORKED EXAMPLE 9 Determining unknown probabilities given the expected value

Determine the values of a and b of the following probability distribution if $E(X) = 4.29$.

x	1	2	3	4	5	6	7
$\Pr(X = x)$	0.1	0.1	a	0.3	0.2	b	0.2

THINK	WRITE
1. Write an equation for the values of a and b using the knowledge that the sum of the probabilities must total 1. Call this equation [1].	$0.1 + 0.1 + a + 0.3 + 0.2 + b + 0.2 = 1$ $0.9 + a + b = 1$ $a + b = 1 - 0.9$ $a + b = 0.1 \quad [1]$

2. Write the rule for the expected value.	$\text{E}(X) = \sum_{\text{all } x} x \Pr(X = x)$
3. Substitute the values into the rule.	$4.29 = 1 \times 0.1 + 2 \times 0.1 + 3 \times a + 4 \times 0.3 + 5 \times 0.2$ $+ 6 \times b \times 7 + 0.2$
4. Evaluate and call this equation [2].	$= 0.1 + 0.2 + 3a + 1.32 + 1 + 6b + 1.4$ $4.29 - 3.9 = 3a + 6b$ $3a + 6b = 0.39$ [2]
5. Solve the equations simultaneously. Multiply equation [1] by 3 and call it equation [3]. Subtract equation [3] from equation [2]. Solve for b. Substitute $b = 0.03$ into equation [1]. Solve for a.	$a + b = 0.1$ [1] $3a + 6b = 0.39$ [2] $3 \times (a + b = 0.1)$ $3a + 3b = 0.3$ [3] $[2] - [3]: 3b = 0.09$ $b = 0.03$ $a + 0.03 = 0.1$ $a = 0.1 - 0.03$ $= 0.07$
6. Answer the question.	$a = 0.07$ and $b = 0.03$

13.4.4 Measures of variability of discrete random distributions

Variance

Variance is an important feature of probability distributions as it provides information about the spread of the distribution with respect to the mean. If the variance is large, it implies that the possible values are spread (or deviate) quite a distance from the mean. A small variance implies that the possible values are close to the mean. Variance is also called a measure of spread or dispersion.

The variance is written as Var(X) and denoted by the symbol σ^2 (sigma squared). It is defined as the expected value (or average) of the squares of the spreads (deviations) from the mean.

The rule for variance is given by:

$$\begin{aligned}\text{Var}(X) &= \text{E}(X - \mu)^2 \\ &= \sum (X - \mu)^2 \Pr(X = x).\end{aligned}$$

Although this rule clearly demonstrates how to obtain the variance, performing the calculation is quite a lengthy process. Hence an alternative rule is used for calculating the variance:

$$\begin{aligned}\text{Var}(X) &= \text{E}(X - \mu)^2 \\ &= \text{E}(X^2 - 2\mu X + \mu^2) \\ &= \text{E}(X^2) - \text{E}(2\mu X) + \text{E}(\mu^2) \\ &= \text{E}(X^2) - 2\mu\text{E}(X) + \text{E}(\mu^2) \\ &= \text{E}(X^2) - 2\mu^2 + \mu^2 \qquad \text{since E}(X) = \mu \text{ (the mean)} \\ &= \text{E}(X^2) - \mu^2 \\ &= \text{E}(X^2) - [\text{E}(X)]^2\end{aligned}$$

Variance

$$\textbf{Var}(X) = \textbf{E}(X^2) - [\textbf{E}(X)]^2$$

WORKED EXAMPLE 10 Calculating the expected value and variance

Calculate the expected value and variance of the following probability distribution table.

x	1	2	3	4	5
$\Pr(X=x)$	0.15	0.12	0.24	0.37	0.12

THINK

1. Write the rule for the expected value.
2. Substitute the values into the rule.
3. Evaluate.
4. Calculate $E(X^2)$.
5. Calculate $[E(X)]^2$.
6. Calculate $\text{Var}(X)$ using the rule for variance.

WRITE

$$\begin{aligned} E(X) &= \sum_{\text{all } x} x \Pr(X=x) \\ &= 1\times0.15+2\times0.12+3\times0.24+4\times0.37+5\times0.12 \\ &= 0.15+0.24+0.72+1.48+0.6 \\ &= 3.19 \end{aligned}$$

$$\begin{aligned} E(X^2) &= \sum_{\text{all } x} x^2 \Pr(X=x) \\ &= (1^2)\times0.15+(2^2)\times0.12+(3^2)\times0.24 \\ &\quad +(4^2)\times0.37+(5^2)\times0.12 \\ &= 1\times0.15+4\times0.12+9\times0.24+16\times0.37 \\ &\quad +25\times0.12 \\ &= 0.15+0.48+2.16+5.92+3 \\ &= 11.71 \end{aligned}$$

$$\begin{aligned} [E(X)]^2 &= 3.19^2 \\ &= 10.1761 \end{aligned}$$

$$\begin{aligned} \text{Var}(X) &= E(X^2)-[E(X)]^2 \\ &= 11.71-10.1761 \\ &= 1.5339 \end{aligned}$$

Standard deviation

Another important measure of spread is the standard deviation. It is written as $\text{SD}(X)$ or denoted by the symbol σ (sigma). The standard deviation is the positive square root of the variance. It is defined by the rule:

$$\begin{aligned} \text{SD}(X) &= \sqrt{\text{Var}(X)} \\ &= \sqrt{\sigma^2} \\ &= \sigma \end{aligned}$$

Variation and standard deviation are used extensively in many real-life applications involving statistics.

Analysis of data would be useless without any information about the spread of the data.

WORKED EXAMPLE 11 Calculating the expected value, variance and standard deviation

A random variable has the following probability distribution.

x	0	1	2	3
$\Pr(X=x)$	$\frac{1}{4}$	$\frac{3}{8}$	$\frac{1}{8}$	$\frac{1}{4}$

Calculate the expected value, the variance and the standard deviation.

THINK / **WRITE**

1. Calculate the expected value.

$$\begin{aligned}E(X) &= 0\times\frac{1}{4}+1\times\frac{3}{8}+2\times\frac{1}{8}+3\times\frac{1}{4}\\ &= 0+\frac{3}{8}+\frac{2}{8}+\frac{3}{4}\\ &= 1\frac{3}{8}\end{aligned}$$

2. Calculate $[E(X)]^2$.

$$\begin{aligned}[E(X)]^2 &= \left(1\frac{3}{8}\right)^2\\ &= 1\frac{57}{64}(\approx 1.890\,625)\end{aligned}$$

3. Calculate $E(X^2)$.

$$\begin{aligned}E(X)^2 &= 0^2\times\frac{1}{4}+1^2\times\frac{3}{8}+2^2\times\frac{1}{8}+3^2\times\frac{1}{4}\\ &= 0+\frac{3}{8}+\frac{4}{8}+\frac{9}{4}\\ &= 3\frac{1}{8}\end{aligned}$$

4. Calculate $\text{Var}(X)$.

$$\begin{aligned}\text{Var}(X) &= E\left(X^2\right)-[E(X)]^2\\ &= 3\frac{1}{8}-1\frac{57}{64}\\ &= 1\frac{15}{64}(\approx 1.234\,375)\end{aligned}$$

5. Calculate the standard deviation.

$$\begin{aligned}SD(X) &= \sqrt{1.234\,375}\\ &= 1.1110\end{aligned}$$

6. Round the answer to 4 decimal places.

13.4.5 Comparison of the independent discrete random variables $2X$ and $(X+X)$

Discrete random variables can be added together, multiplied by a scalar or a combination of both. In this section we will consider the differences between the distributions of $2X$ and $X+X$. The best way to understand the difference between these distributions is with an example. Consider the distribution X being the result of rolling a fair die. The possible outcomes (sample space) of the distribution are $1, 2, 3, 4, 5$ or 6 and the probability distribution is shown below.

x	1	2	3	4	5	6
Pr $(X = x)$	$\frac{1}{6}$	$\frac{1}{6}$	$\frac{1}{6}$	$\frac{1}{6}$	$\frac{1}{6}$	$\frac{1}{6}$

The distribution $2X$ is simply the result of multiplying all of the outcomes in the distribution X by 2, or two times the result of rolling a fair die. The probability distribution of $2X$ is therefore:

$2x$	2	4	6	8	10	12
Pr $(2X = 2x)$	$\frac{1}{6}$	$\frac{1}{6}$	$\frac{1}{6}$	$\frac{1}{6}$	$\frac{1}{6}$	$\frac{1}{6}$

The distribution of $X + X$ is the result of rolling a fair die once, and then rolling the same die again, and adding up the results of both rolls. The outcomes for this distribution are therefore all of the integers between 2 and 12 inclusive, which is very different to the outcomes of $2X$.

The probabilities of the outcomes depend on the number of possible combinations that result in a sum of that outcome. For example, the probability of the outcome 2 in $X + X$ is $\frac{1}{6} \times \frac{1}{6} = \frac{1}{36}$ as the only way the outcome can be 2 is by rolling a 1 and then a 1 again. By contrast, the probability of the outcome 7 is $\frac{1}{6}$ since there are 6 combinations (out of a total of 36 possible combinations) of rolls of two dice that sum to 7. They are: (1,6), (2,5), (3,4), (4,3), (5,2), (6,1). Alternatively, you may notice that regardless of the result of the first roll, there is one particular result of the second roll that will cause the sum to be 7. So the probability of obtaining a sum of 7 is simply $\frac{1}{6}$.

The probability distribution of $X + X$ is shown below.

$x + x$	2	3	4	5	6	7	8	9	10	11	12
Pr $(X + X = x + x)$	$\frac{1}{36}$	$\frac{1}{18}$	$\frac{1}{12}$	$\frac{1}{9}$	$\frac{5}{36}$	$\frac{1}{6}$	$\frac{5}{36}$	$\frac{1}{9}$	$\frac{1}{12}$	$\frac{1}{18}$	$\frac{1}{36}$

By looking at these probability distributions we can see that $2X$ and $X + X$ are very different things!

Note that although the means of the distributions are equal, their variances are not equal. This can be explained best using another, very simple example.

x	0	1
Pr $(X = x)$	$\frac{1}{2}$	$\frac{1}{2}$

$2x$	0	2
Pr $(2X = 2x)$	$\frac{1}{2}$	$\frac{1}{2}$

$x + x$	0	1	2
Pr $(X + X = x + x)$	$\frac{1}{4}$	$\frac{1}{2}$	$\frac{1}{4}$

Just by inspection you should notice that the distribution $2X$ has a larger variance than $X + X$. Although both distributions are centred around the mean of $E(X) = 1$, the distribution $2X$ is split either side, with zero probability of the outcome being the mean. In the distribution of $X + X$ on the other hand, the most likely outcome is the mean!

Calculating the variance for each distribution shows this disparity in variance clearly.

$$\begin{aligned}\text{Var}(X) &= E(X^2) - [E(X)]^2 \\ &= \left[(0)^2 \times \frac{1}{2} + (1)^2 \times \frac{1}{2}\right] - \left[\frac{1}{2}\right]^2 \\ &= \frac{1}{4}\end{aligned}$$

$$\begin{aligned}\text{Var}(2X) &= \sum (2x)^2 \times \Pr(2X = 2x) - [\text{E}(2X)]^2 \\ &= \left[(0)^2 \times \frac{1}{2} + (2)^2 \times \frac{1}{2}\right] - [1]^2 \\ &= 1\end{aligned}$$

$$\begin{aligned}\text{Var}(X + X) &= \sum (x + x)^2 \times \Pr(X + X = x + x) - [\text{E}(X + X)]^2 \\ &= \left[(0)^2 \times \frac{1}{4} + (1)^2 \times \frac{1}{2} + (2)^2 \times \frac{1}{4}\right] - [1]^2 \\ &= \frac{1}{2}\end{aligned}$$

Notice that $\text{Var}(X + X) = 2\text{Var}(X)$ and $\text{Var}(2X) = 4\text{Var}(X)$. This result can be summarised more generally as follows.

$$\begin{aligned}&\text{Var}(2X) = 2^2\text{Var}(X) & & \text{Var}(X + X) = 2\text{Var}(X) \\ &\text{Var}(3X) = 3^2\text{Var}(X) & \text{and} \quad & \text{Var}(X + X + X) = 3\text{Var}(X) \\ &\text{Var}(4X) = 4^2\text{Var}(X) & & \text{Var}(X + X + X + X) = 4\text{Var}(X)\end{aligned}$$

The pattern continues for all positive integers.

The expected values follow a simple pattern as shown below.

$$\begin{aligned}&\text{E}(2X) = 2\text{E}(X) & & \text{E}(X + X) = 2\text{E}(X) \\ &\text{E}(3X) = 3\text{E}(X) & \text{and} \quad & \text{E}(X + X + X) = 3\text{E}(X) \\ &\text{E}(4X) = 4\text{E}(X) & & \text{E}(X + X + X + X) = 4\text{E}(X)\end{aligned}$$

The sum of n independent random variables X, compared to nX

Consider the distribution nX, where X is a random variable and $n \in N$.

$$\text{E}(nX) = n\text{E}(X) \text{ and } \text{Var}(nX) = n^2\text{Var}(X)$$

Consider the distribution $X + X + ... + X$ (n times), where each X is an independent random variable.

$$\text{E}(X + X + ... + X) = n\text{E}(X) \text{ and } \text{Var}(X + X + ... + X) = n\text{Var}(X)$$

WORKED EXAMPLE 12 Determining the mean and variance of a multiple of a distribution

A discrete random variable B has a mean of 20 and a variance of 5. Determine the mean and variance of $2B$.

THINK	WRITE
1. Recall the formula for the mean, $\text{E}(2X) = 2\text{E}(X)$.	$\text{E}(2B) = 2\text{E}(B)$ $= 2 \times 20$ $= 40$
2. Recall the formula for the variance, $\text{Var}(2X) = 2^2\text{Var}(X)$.	$\text{Var}(2B) = 2^2\text{Var}(B)$ $= 4 \times 5$ $= 20$
3. State the answer in words.	The distribution $2B$ has a mean of 40 and a variance of 20.

WORKED EXAMPLE 13 Determining the mean and standard deviation

A discrete random variable M has a mean of m and a variance of v. Determine the mean and standard deviation of $M+M+M+M+M$.

THINK	WRITE
1. The distribution is the sum of 5 of the original distributions. Recall the formula for the mean, $E(X+X+X+X+X)=5E(X)$.	$E(M+M+M+M+M)=5E(M)$ $=5m$
2. Recall the formula for the variance, $Var(X+X+X+X+X)=5Var(X)$.	$Var(M+M+M+M+M)=5Var(M)$ $=5v$
3. Recall that the standard deviation is the square root of the variance.	$\sigma(M+M+M+M+M)=\sqrt{5v}$
4. State the answer in words.	The distribution $M+M+M+M+M$ has a mean of $5m$ and a standard deviation of $\sqrt{5v}$.

13.4 Exercise

Technology free

1. WE6 Classify the following variables as discrete or continuous.
   a. The number of people at a tennis match.
   b. The time taken to read this question.
   c. The length of the left arms of students in your class.
   d. The number of shoes owned by twenty people.
   e. The weights of babies at a maternity ward.
   f. The number of grains in ten 250-gram packets of rice.
   g. The height of jockeys competing in a certain race.
   h. The number of books in Melbourne libraries.

2. a. WE7 If X represents the number of heads obtained in two tosses of a coin, draw up a table that displays the values that the discrete random variable can assume and the corresponding probabilities.
   b. Draw a probability distribution graph of the outcomes in part a.

3. WE8 Determine the expected value of a random variable that has the following probability distribution.

x	0	3	6	9	12
$Pr(X=x)$	0.21	0.08	0.19	0.17	0.35

4. Determine the expected value of a random variable that has the following probability distribution.

x	-2	-1	0	1	2	3	4
$\Pr(X=x)$	$\frac{1}{18}$	$\frac{1}{3}$	$\frac{1}{18}$	$\frac{2}{9}$	$\frac{1}{6}$	$\frac{1}{18}$	$\frac{1}{9}$

5. WE9 Evaluate the unknown probability, a, and hence determine the expected value of a random variable that has the following probability distribution.

x	1	3	5	7	9	11
$\Pr(X=x)$	0.11	0.3	0.15	0.25	a	0.1

6. Evaluate the unknown probability, a, and hence determine the expected value of a random variable that has the following probability distribution.

x	-2	1	4	7	10	13
$\Pr(X=x)$	$\frac{5}{18}$	a	$\frac{1}{9}$	$\frac{5}{18}$	$\frac{1}{18}$	$\frac{2}{9}$

Technology active

7. WE10 Calculate the expected value and variance of the following probability distribution table.

x	1	2	3	4
$\Pr(X=x)$	0.2	0.4	0.3	0.1

8. The cost of a loaf of bread is known to vary on any day according to the following probability distribution.

x	\$1.20	\$1.25	\$1.30	\$1.35	\$1.60
$\Pr(Y=y)$	0.05	0.2	0.1	0.25	0.4

Determine:

a. the expected cost of a loaf of bread

b. the variance of the cost, rounded to 2 decimal places.

9. WE11 A random variable has the following probability distribution.

x	1	2	3	4
$\Pr(X=x)$	$\frac{1}{4}$	$\frac{1}{3}$	$\frac{1}{4}$	$\frac{1}{6}$

Calculate the expected value, the variance and the standard deviation rounded to 4 decimal places where appropriate.

10. A random variable has the following probability distribution.

x	1	2	3	4
$\Pr(X=x)$	$\frac{2}{15}$	$\frac{7}{15}$	$\frac{1}{3}$	$\frac{1}{15}$

Evaluate the following, writing your answers as a decimal correct to 3 decimal places.

a. $E(X)$

b. $E(4X)$

c. $\text{Var}(X)$

d. $\text{Var}(4X)$.

11. A discrete random variable J is given by the following probability distribution.

j	0	1	2
$\Pr(J=j)$	0.3	0.2	0.5

Determine the mean and standard deviation of $J+J+J+J$, correct to 3 decimal places where appropriate.

12. Lucas contemplates playing a new game which involves tossing three coins simultaneously. He will receive \$15 if he obtains 3 heads, \$10 if he obtains 2 heads and \$5 if he obtains 1 head. However, if he obtains no heads he must pay \$30. He must also pay \$5 for each game he plays.

 a. Calculate Lucas's expected gain per game that he plays, ignoring the \$5 fee for playing each game.
 b. Now, considering the \$5 per game playing fee, determine whether he should play the game. Explain your answer.
 c. State if this is a fair game. Explain your answer.

13. In order to encourage carpooling, a new toll is to be introduced on the International Gateway. If the car has no passengers, a toll of \$2 applies. Cars with one passenger pay a \$1.50 toll, cars with two passengers pay a \$1 toll and cars with 3 or more passengers pay no toll. Long-term statistics show that the number of passengers follows the probability distribution given below.

x	0	1	2	3
$\Pr(X=x)$	0.5	0.3	0.15	0.05

 a. Construct a probability distribution of the toll paid.
 b. Determine the mean toll paid per car.
 c. Calculate the standard deviation of tolls paid.

14. The probability distribution of X is given by the formula, $\Pr(X=x)=\frac{x^2}{35}$ where $x=1,3,5$. Determine:

 a. the probability distribution of X as a table
 b. the expected value of X
 c. the standard deviation of X, to 4 decimal places

15. WE12 A discrete random variable F has a mean of 9 and a variance of 4. Determine the mean and variance of $3F$.

16. A discrete random variable W has a mean of 214 and a variance of 15. Determine the mean and variance of $4W$.

17. The mean and variance of a discrete random variable Q are unknown, however it is known that $E(5Q)=2100$ and $\text{Var}(5Q)=275$. From this information, determine the mean and variance of Q.

18. WE13 For a random variable, X, the mean is 5 and the variance 4. Determine the mean and variance of $X+X+X+X$.

19. For a random variable, X, the mean is m and the variance v. If the random variable X is added together p times i.e. $X+X+X+........+X+X$, having p terms, determine the mean and variance of this sum of p terms.

20. There exists a random variable, Y. When the random variable Y is added together n^2 times i.e. $Y+Y+Y+........+Y+Y$, having n^2 terms, the resulting variance is v. Determine the variance of Y.

13.4 Exam Questions

Question 1 (2 marks) TECH-FREE

Casey decides to apply for a job selling mobile phones. She receives a base salary of \$200 per month and \$15 for every mobile phone sold. The following table shows the probability of a particular number of mobile phones, x, being sold per month. Determine the expected salary that Casey would receive each month.

x	50	100	150	200	250
$\Pr(X=x)$	0.48	0.32	0.1	0.06	0.04

Question 2 (2 marks) TECH-FREE

The table below represents the probability distribution of the number of accidents per week in a factory.

x	1	2	3	4	5	6	7	8	9
$\Pr(X = x)$	0.02	0.22	0.18	0.16	0.14	0.07	0.13	0.03	0.05

Given that $\mu = 4.36$ and $\sigma = 2.105$ evaluate $\Pr(\mu - 2\sigma \le X \le \mu + 2\sigma)$.

Question 3 (3 marks) TECH-ACTIVE

The probability distribution of X is given by the formula:

$$\Pr(X = x) = \frac{x^2}{54} \text{ where } x = 2, 3, 4, 5$$

Determine:

a. the probability distribution of X as a table **(1 mark)**

b. the expected value of X, correct to 4 decimal places **(1 mark)**

c. the standard deviation of X, correct to 4 decimal places. **(1 mark)**

More exam questions are available online.

13.5 Sampling

LEARNING INTENTION

At the end of this subtopic you should be able to:

- calculate sample statistics and use them to make predictions about the parameters of a population.

13.5.1 Introduction to sampling

Many of the statistics about life in Australia are drawn from the Census of Population and Housing conducted by the Australian Bureau of Statistics every five years. A census is simply a survey of the population of a country. It collects general information such as gender, age and occupation as well as more detailed data such as ethnicity and housing situation.

It is quite impractical and sometimes impossible to analyse the entire population for a particular situation, so a sample is usually taken. The sample needs to be representative of the population, and multiple samples should be taken where possible. The study of statistics allows us to gauge the confidence we can have in the validity of our samples for an investigation before we make any predictions or assumptions.

Suppose you were interested in the percentage of Year 12 graduates who plan to study Mathematics once they complete school. It is probably not practical to question every student. There must be a way that we can ask a smaller group and then use this information to make generalisations about the whole group.

13.5.2 Samples and populations

A **population** is a group that you want to know something about, and a **sample** is the group within the population that you collect the information from. Normally, a sample is smaller than the population; the exception is a census, where the whole population is the sample.

The number of members in a sample is called the **sample size** (symbol n), and the number of members of a population is called the **population size** (symbol N). Sometimes the population size is unknown.

WORKED EXAMPLE 14 Identifying population and sample sizes (1)

Cameron has uploaded a popular YouTube video. He thinks that the 133 people in his year group at school have seen it, and he wants to know what they think. He decides to question 10 people. Identify the population and sample size.

THINK	WRITE
1. Cameron wants to know what the people in his year at school think. This is the population.	$N = 133$
2. He asks 10 people. This is the sample.	$n = 10$

WORKED EXAMPLE 15 Identifying population and sample sizes (2)

A total of 137 people volunteer to take part in a medical trial. Of these, 57 are identified as suitable candidates and are given the medication. Identify the population and sample size.

THINK	WRITE
1. 57 people are given the medication. This is the sample size.	$n = 57$
2. We are interested in the group of people who might receive the drug in the future. This is the population.	The population is unknown, as we don't know how many people may be given this drug in the future.

13.5.3 Statistics and parameters

A **parameter** is a characteristic of a population, whereas a **statistic** is a characteristic of a sample. This means that a statistic is always known exactly (because it is measured from the sample that has been selected). A parameter is usually estimated from a sample statistic. (The exception is if the sample is a census, in which case the parameter is known exactly.)

The relationship between populations and samples

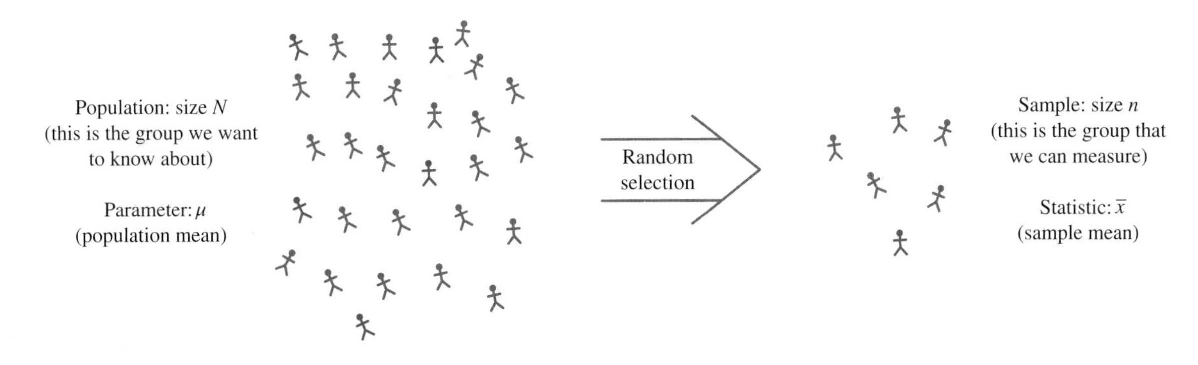

WORKED EXAMPLE 16 Classifying sample statistics and population parameters

Identify the following as either sample statistics or population parameters.

a. **Forty-three per cent of voters polled say that they are in favour of banning fast food.**

b. **According to Australian Bureau of Statistics census data, the average family has 1.7 children.**

c. **Between 18% and 23% of Australians skip breakfast regularly.**

d. **9 out of 10 children prefer cereal for breakfast.**

THINK	WRITE
a. 43% is an exact value that summarises the sample asked.	a. Sample statistic
b. The information comes from census data. The census questions the entire population.	b. Population parameter
c. 18%–23% is an estimate about the population.	c. Population parameter
d. 9 out of 10 is an exact value. It is unlikely that all children could have been asked; therefore, it is from a sample.	d. Sample statistic

13.5.4 Random samples

A good sample should be representative of the population. If we consider our initial interest in the proportion of Year 12 graduates who intend to study Mathematics once they finished school, we could use a Specialist Mathematics class as a sample. This would not be a good sample because it does not represent the population — it is a very specific group of students, who are much more likely to study Mathematics upon finishing school than the majority of year 12 students.

In a **random sample**, every member of the population has the same probability of being selected. The Specialist Mathematics class is not a random sample because students who don't study Specialist Mathematics have no chance of being selected; furthermore, students who don't attend that particular school have no chance of being selected.

A **systematic sample** is almost as good as a random sample. In a systematic sample, every kth member of the population is sampled. For example, if $k = 20$, a customs official might choose to sample every 20th person who passes through the arrivals gate. The reason that this is almost as good as random sample is that there is an assumption that the group passing the checkpoint during the time the sample is taken is representative of the population. This assumption may not always be true; for example, people flying for business may be more likely to arrive on an early morning flight. Depending on the information you are collecting, this may influence the quality of the data.

In a **stratified random sample**, care is taken so that subgroups within a population are represented in a similar percentage in the sample. For example, if you were collecting information about students in Years 9 to 12 in your school, the percentages of students in each year group should be the same in the sample and the population. Within each subgroup, each member has the same chance of being selected.

A **self-selected sample**, that is one where the participants choose to participate in the survey, is almost never representative of the population. For example, television phone polls, where people phone in to answer yes or no to a question, do not accurately reflect the opinion of the population.

WORKED EXAMPLE 17 Determining the subgroups in a stratified random sample

A survey is to be conducted in a middle school that has the distribution detailed in the table. It is believed that students in different year levels may respond differently, so the sample chosen should reflect the subgroups in the population (that is, it should be a stratified random sample). If a sample of 100 students is required, determine how many from each year group should be selected.

Year level	Number of students
7	174
8	123
9	147

THINK	WRITE
1. Calculate the total population size.	Total population $= 174 + 123 + 147 = 444$
2. Calculate the number of Year 7s to be surveyed.	Number of Year 7s $= \frac{174}{444} \times 100$ $= 39.2$ Survey 39 Year 7s.
3. Calculate the number of Year 8s to be surveyed.	Number of Year 8s $= \frac{123}{444} \times 100$ $= 27.7$ Survey 28 Year 8s.
4. Calculate the number of Year 9s to be surveyed.	Number of Year 9s $= \frac{147}{444} \times 100$ $= 33.1$ Survey 33 Year 9s.
5. There has been some rounding, so check that the overall sample size is still 100.	Sample size $= 39 + 28 + 33 = 100$ The sample should consist of 39 Year 7s, 28 Year 8s and 33 Year 9s.

13.5.5 Using technology to select a sample

If you know the population size, it should also be possible to produce a list of population members, whereby each member of the population is assigned a number (from 1 to N). Use the random number generator on your calculator to generate a random number between 1 and N. The population member who was allocated that number becomes the first member of the sample. Continue generating random numbers until the required number of members has been picked for the sample. If the same random number is generated more than once, ignore it and continue selecting members until the required number has been chosen.

On the TI-Nspire this can be done on a Calculator page by navigating to Menu, 5 Probability, 4 Random, 2 Integer and then entering the lower bound, upper bound and the number of random integers that you would like to generate (separated by commas). The following screen shows how to create a list of 10 random integers between 1 and 50.

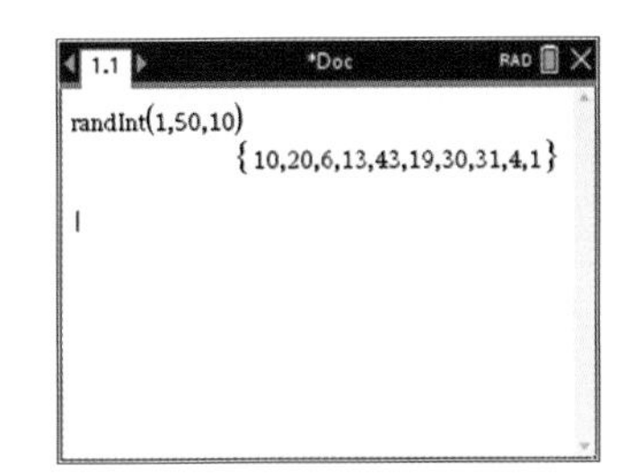

On the CASIO ClassPad this can be done on a Main screen by entering the command randList() and then entering the number of random integers that you would like to generate, lower bound and upper bound (separated by commas). The following screen shows how to create a list of 10 random integers between 1 and 50.

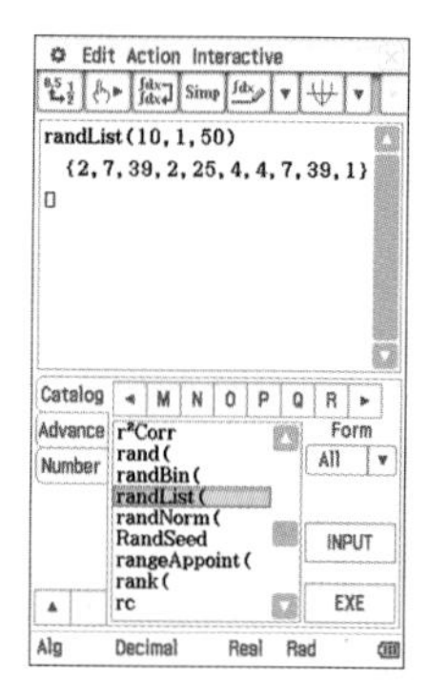

13.5 Exercise

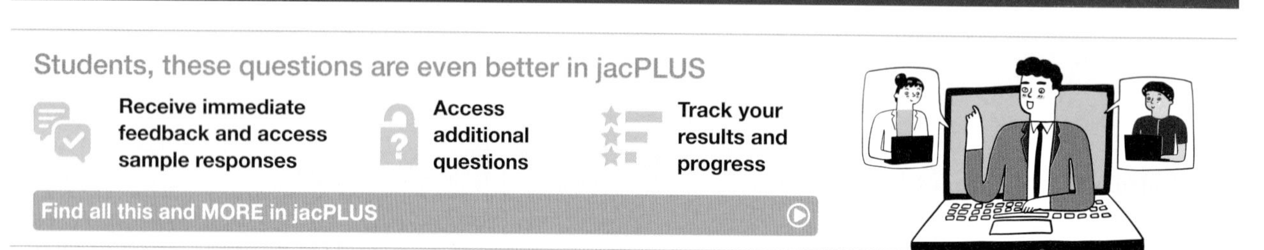

Technology free

1. **WE14** On average, Mr Parker teaches 120 students per day. He asks one class of 30 about the amount of homework they have that night. Identify the population and sample size.
2. Bruce is able to sew the hems of 100 shirts per day. Each day he checks 5 to make sure that they are suitable. Identify the population and sample size.
3. **WE15** Ms Lane plans to begin her statistics class each year by telling her students a joke. She tests her joke on this year's class (15 students). She plans to retire in 23 years' time. Identify the population and sample size.
4. Lee-Yin is trying to perfect a recipe for cake pops. She tries 5 different versions before she settles on her favourite. She takes some samples to school and asks 9 friends what they think. Identify the population and sample size.
5. **WE16** Identify the following as either sample statistics or population parameters.
   a. Studies have shown that between 85% and 95% of lung cancers are related to smoking.
   b. About 50% of children aged between 9 and 15 years eat the recommended daily amount of fruit.
6. Identify the following as either sample statistics or population parameters.
   a. According to the 2021 census, the ratio of male births per 100 female births is 106.3.
   b. About 55% of boys and 40% of girls reported drinking at least 2 quantities of 500 mL of soft drink every day.
7. **WE17** A school has 523 boys and 621 girls. You are interested in finding out about their attitudes to sport and believe that boys and girls may respond differently. If a sample of 75 students is required, determine how many boys and how many girls should be selected.
8. In a school, 23% of the students are boarders. For this survey, it is believed that boarders and day students may respond differently. To select a sample of 90 students, determine how many boarders and day students should be selected.

9. You are trying out a new chocolate pudding recipe. You found 40 volunteers to taste test your new recipe compared to your normal pudding. Half of the volunteers were given a serving of the new pudding first, then a serving of the old pudding. The other half were given a serving of the old pudding first and then a serving of the new pudding. The taste testers did not know the order of the puddings they were trying. The results show that 31 people prefer the new pudding recipe.

   a. State the population size.
   b. State the sample size.

10. You want to test a new flu vaccine on people with a history of chronic asthma. You begin with 500 volunteers and end up with 247 suitable people to test the vaccine.
   a. State the population size.
   b. State the sample size.

11. In a recent survey, 1 in 5 students indicated that they ate potato crisps or other salty snacks at least four times per week. State if this is a sample statistic or a population parameter.

12. Around 25% to 30% of children aged 0 to 15 years old eat confectionary at least four times a week. State if this is a sample statistic or a population parameter.

13. According to the Australian Bureau of Statistics, almost a quarter (24%) of internet users did not make an online purchase or order in 2012–13. The three most commonly reported main reasons for not making an online purchase or order were: 'Has no need' (33%); 'Prefers to shop in person/see the product' (24%); and 'Security concerns/concerned about providing credit card details online' (12%). Determine if these are sample statistics or population parameters.

14. According to the most recent census, there is an average of 2.6 people per household. State if this is a sample statistic or a population parameter.

15. A doctor is undertaking a study about sleeping habits. She decides to ask every 10th patient about their sleeping habits.
   a. State what type of sample this is.
   b. Explain if this is valid sampling method.

16. A morning television show conducts a viewer phone-in poll and announces that 95% of listeners believe that Australia should become a republic. Comment on the validity of this type of sample.

17. Tony took a survey by walking around the playground at lunch and asking fellow students questions. Explain why this is not the best sampling method.

Technology active

18. A company has 1500 staff members, of whom 60% are male; 95% of the male staff work full time, and 78% of the female staff work full time. If a sample of 80 staff is to be selected, identify the numbers of full-time male staff, part-time male staff, full-time female staff and part-time female staff that should be included in the sample.

19. Use your calculator to produce a list of 10 random numbers between 1 and 100.

20. Use your calculator to select a random sample from students in your Specialist Mathematics class.

13.5 Exam Questions

Question 1 (1 mark) TECH-ACTIVE

MC A survey is given to 50 students randomly selected from the Year 12 students at your school. The population is

A. the 50 selected student
B. all year 12 students at your school
C. all students at your school
D. all Year 12 students at in the state
E. all people who live within the school catchment zone

Question 2 (1 mark) TECH-ACTIVE

MC A standard deviation is known as a parameter if it is calculated from the

A. mean
B. sample
C. population
D. statistic
E. variance

Question 3 (1 mark) TECH-ACTIVE

MC The different between $\bar{x}$ and μ is that

A. $\bar{x}$ is a parameter and μ is a statistic
B. $\bar{x}$ is calculated from a population and μ is calculated from a sample
C. $\bar{x}$ is calculated from a sample and μ is calculated from a population
D. $\bar{x}$ is the average value and μ describes the spread of the data
E. μ is the average value and $\bar{x}$ describes the spread of the data

More exam questions are available online.

13.6 Sampling distribution of sample means

LEARNING INTENTION

At the end of this subtopic you should be able to:

- understand how the sample size affects the distribution of sample means
- represent the distribution of sample means graphically.

13.6.1 Populations and samples

In statistics, objects or people are measured. For example, we could measure the heights of a group of 30 Year 11 students from a particular school. We could then calculate the average height of the students in the group.

However, we could measure the heights of the same group of students and for each student ask 'Is this student over 1.6 m tall?'. The answer to this question, although height was measured, is actually 'Yes' (that is, taller than 1.6 m) or 'No' (that is, shorter than 1.6 m). This sort of measurement is called an **attribute**, and subjects either have an attribute or do not have it. By counting those who have the attribute, we can determine the **proportion** of the group with the attribute. In our example, we could determine the proportion of Year 11 students who are over 1.6 m tall.

It is important to distinguish between the **population**, which covers, in some way, everything or everyone in a grouping, and the **sample**, which comprises the members of the group that we actually measure for the attribute. In the above example, all the Year 11 students in the school make up the Year 11 population. The 30 students whose heights were measured represent a sample from that population.

When analysing data, we must be careful in drawing conclusions. It is crucial that the sample comes from the true population that it presumes to represent. For example, it is no good taking a sample of students from one school only and then making statements about all Australian students. Unless the sample is the entire population (which is very rare), knowledge about the population remains unknown. Suppose we measure the heights of 100 students in Victoria and determine that 45 are over 1.6 m tall. We cannot then make statements about the true proportion of all Australian students with this attribute, or even all Victorian students; we can say only that the percentage obtained in our sample (45%) may be close to the true proportion. We can make positive statements only about the sample that we took, and hope that it is representative of the population. When the sample is selected properly and is of a significant enough size it usually *is* representative of the population.

WORKED EXAMPLE 18 Calculating sample proportions

A scientist wishes to study pollution in a river. She takes a sample of 42 fish to see what proportion have mercury poisoning and finds that 12 are poisoned. Calculate the proportion of fish which have mercury poisoning, as a percentage correct to 1 decimal place.

THINK	WRITE
1. Write down the rule for proportion.	$\text{Percentage} = \dfrac{\text{number who have an attribute}}{\text{total number sampled}} \times 100\%$
2. Write down the given values for the total number of fish sampled and the number of fish poisoned.	Total number of fish sampled = 42 Number of fish poisoned = 12
3. Substitute the given values into the rule.	$\text{Percentage} = \dfrac{12}{42} \times 100\%$
4. Evaluate and round to 1 decimal place.	$= 28.6\%$
5. Answer the question.	Approximately 28.6% of the fish were poisoned.

WORKED EXAMPLE 19 Calculating sample proportions

The Department of Health wishes to know how many patients admitted to the hospital have private health insurance and decides on the following sampling technique. For each of 10 hospitals, H, in a city, a sample of 12 patients, P, will be taken. The number of these patients with private insurance, I, will be counted and recorded (see the table below).
Estimate the percentage of the population that has private insurance, using the information in the table, as a percentage correct to 2 decimal places.

	P1	P2	P3	P4	P5	P6	P7	P8	P9	P10	P11	P12
H1	I	N	I	N	I	N	N	N	N	N	I	I
H2	I	N	I	N	N	N	N	N	N	N	I	I
H3	N	N	I	N	I	N	N	I	N	I	N	N
H4	I	N	N	I	N	N	N	N	N	N	I	N
H5	N	I	N	N	I	I	I	I	I	N	N	N
H6	N	N	N	N	I	I	I	N	N	I	N	N
H7	N	I	N	N	N	I	I	N	N	I	N	N
H8	N	N	N	N	I	I	I	I	N	N	N	N
H9	N	N	N	N	N	N	N	N	I	I	I	N
H10	N	N	N	N	I	N	N	I	N	N	N	I

THINK	WRITE
1. Count the number of Is from each of the 10 hospitals.	Hospital 1 count = 5; Hospital 2 count = 4 Hospital 3 count = 4; Hospital 4 count = 3 Hospital 5 count = 6; Hospital 6 count = 4 Hospital 7 count = 4; Hospital 8 count = 4 Hospital 9 count = 3; Hospital 10 count = 3
2. Calculate the percentage of patients that have private health insurance across all 10 hospitals.	$\frac{5+4+4+3+6+4+4+4+3+3}{120} \times 100\%$ $= \frac{40}{120} \times 100\%$ $= 33.33\%$
3. Answer the question.	Of the patients sampled, 33.33% had private health insurance. An estimate of the percentage of people who have private health insurance is 33.33%.

13.6.2 Sampling distribution of sample means

A sampling distribution is created when we take multiple samples of the same size, n, from a population of size N.

Each sample will almost certainly be different from any other and has a sample mean denoted by $\overline{x}_i$, where i is the identifying sample number i.e. 1st sample, 2nd sample, etc.

If we plot the values of $\overline{x}_i$ on a dot plot or the like we can create a visual representation of the spread of values of the sample means.

A sample proportion or sample mean may or may not be close to the population mean, μ. Theory states that if we get many sample statistics then the average of these statistics will be much closer to the population parameters. Why? The answer is quite simple: by repeating the sampling we are effectively taking one large sample. That is, 10 samples of 12 objects each is roughly equivalent to a single sample of 120 (assuming that the population is significantly larger than 120 and that all sampling is random).

Consider a distribution of sample means.

It is also possible to find a distribution of sample means. In this case, the mean of the sample distribution is the same as the mean of the population. Why is this true? Some samples will have means higher than the population, but other samples will have means lower than the population. When taking the distribution of all possible samples, these cancel each other out, meaning that the mean of the distribution, μ_x, is the same as the population mean, μ.

$$\mu_x = \mu$$

The standard deviation of the distribution of sample means can be found using $\sigma_{\overline{x}} = \frac{\sigma}{\sqrt{n}}$. Why is this true? The population standard deviation is σ. This means that the variance of the random variables in the population is σ^2. Each sample is a selection of n of these variables. The total of the variables in the sample will have a variance found by adding the variance of the variables together. This means that the variance of the sample total is $n\sigma^2$.

$$\left(\sigma_{\Sigma x}\right)^2 = \sigma^2 + \sigma^2 + \sigma^2 + ... + \sigma^2 \quad (n \text{ times})$$

$$\sigma_{\Sigma x}{}^2 = n\sigma^2$$

$$\begin{aligned}\sigma_{\Sigma x} &= \sqrt{n\sigma^2} \\ &= \sigma\sqrt{n}\end{aligned}$$

The sample mean is found by dividing the total by n. This means that the standard deviation of the sample means is the standard deviation of the sample totals divided by n.

$$\begin{aligned}\sigma_{\bar{x}} &= \frac{\sigma_{\Sigma x}}{n} \\ &= \frac{\sigma\sqrt{n}}{n} \\ &= \frac{\sigma}{\sqrt{n}}\end{aligned}$$

Note: Larger sample sizes will not change the mean of the distribution of sample means, but will reduce the standard deviation, meaning that larger samples will have results closer to the population parameters.

For purposes of understanding, we will examine a very small population and an even smaller sample. Consider the following data points: 23, 42, 12, 21 and 11. The average of these points is $\mu = 21.8$ and the standard deviation is 11.16.

Consider all of the different samples of size 2 from this data set. The mean of a sample is used to give some indication of the likely population mean. The sample mean is given the symbol $\bar{x}$.

A population of sample means is created.

Data points	$\bar{x}$
23, 42	32.5
23, 12	17.5
23, 21	22
23, 11	17
42, 12	27
42, 21	31.5
42, 11	26.5
12, 21	16.5
12, 11	11.5
21, 11	16

As you can see, there is a lot of variety in the values of $\bar{x}$. This variability would be reduced by selecting larger samples. If you calculate the average of all of the sample $\bar{x}$ values, you will find it is equal to 21.8. Notice that this is the same as the population mean. Although the means of individual samples will vary, the mean of the sample means will be the same as the population mean. That is, $\mu_{\bar{x}} = \mu$.

The standard deviation of the sample means can be found using

$$\begin{aligned}\frac{\sigma}{\sqrt{n}} &= \frac{11.16}{\sqrt{2}} \\ &= 7.89\end{aligned}$$

As the standard deviation is divided by the square root of the sample size, this measure of variability becomes smaller as the sample size increases: $\sigma_{\bar{x}} = \frac{\sigma}{\sqrt{n}}$.

For example if we take 10 samples of size 3, such as $(23, 42, 12)$; $(23, 42, 21)$; $(23, 42, 11)$, etc., we obtain values of $\bar{x}$ of $25.7, 28.7, 25.3, \ldots\ldots$.

Forming a new population of sample means $(\bar{x})$, giving a value of $\mu = 21.8$.

But the standard deviation is $\frac{11.6}{\sqrt{3}} = 6.70$. The sample means are less spread out.

If we take 5 samples of size 4, such as $(23, 42, 12, 11)$; $(23, 42, 21, 11)$; $(23, 42, 21, 12)$, etc., we obtain yet another population of sample means $(\bar{x})$ i.e. $22, 24.3, 24.5, \ldots\ldots$, again giving a value of $\mu = 21.8$.

The standard deviation is $\frac{11.6}{\sqrt{4}} = 5.80$. The sample means are even less spread out.

It would appear that the larger the sample size the closer the average (mean) of the means of the samples are to the mean of the population.

Rounded to the nearest whole number the sample means for all possible samples of size 2, 3 and 4 are shown below. Notice that the spread of the sample means decreases as the size of the samples increase. For example, the means of the samples of size 4 are more clustered than the means of the samples of size 2.

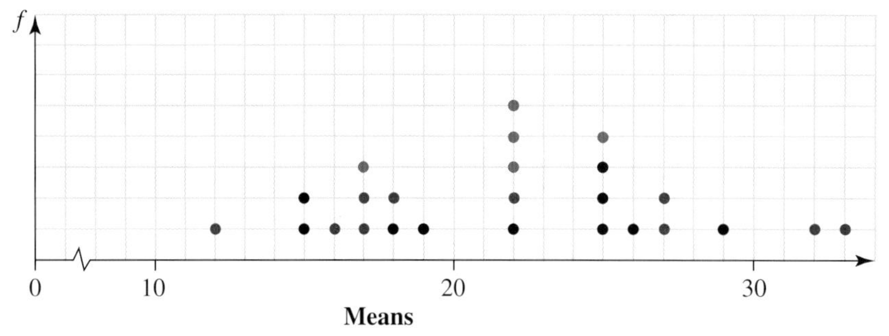

WORKED EXAMPLE 20 Calculating the mean and standard deviation of the distribution of sample means

A total of 500 students are studying a statistics course at Parker University. On a recent exam, the mean score was 73.4 with a standard deviation of 20.

a. **Samples of 5 students are selected and the sample means found. Determine the mean and standard deviation of the distribution of $\bar{x}$.**

b. **State what effect increasing the sample size to 30 would have on the mean and standard deviation of the distribution of sample means.**

THINK	WRITE
a. 1. The mean of the distribution of sample means is the same as the population mean.	a. $\mu_{\bar{x}} = \mu$ $= 73.4$
2. Write down the formula for standard deviation of the distribution of sample means.	$\sigma_{\bar{x}} = \dfrac{\sigma}{\sqrt{n}}$
3. Calculate the standard deviation.	$\sigma_{\bar{x}} = \dfrac{20}{\sqrt{5}}$ $= 8.94$
b. 1. The mean is not dependent on sample size.	b. $\mu_{\bar{x}} = \mu$ $= 73.4$
2. Calculate the standard deviation using $n = 30$.	$\sigma_{\bar{x}} = \dfrac{\sigma}{\sqrt{n}}$ $= \dfrac{20}{\sqrt{30}}$ $= 3.65$
3. Write your conclusions.	Increasing the sample size does not change the mean of the distribution, but the standard deviation is reduced.

13.6.3 Estimating population parameters

In statistics, the entire underlying set of individuals of a group is called the population. In mathematics a population refers not just to people, but to any group. The distribution of a population can be summarised by specific values known as parameters. If the distribution of data is X, the parameters include:

1. The expected value of the distribution, $E(X)$, which corresponds to the mean. That is, $E(X) = \mu$.

2. The standard deviation, σ, of the distribution. This parameter gives information about how the data are spread out from the mean. The standard deviation and mean have the same units making them excellent parameters for analysing the distribution.
3. The variance, σ^2, of the distribution can also be used to describe the spread of a distribution. The variance is the square of the standard deviation and therefore has different units compared to the mean and standard deviation. The variance, also denoted by Var(X), plays a central role in many areas of statistics.

The parameters of mean and standard deviation or variance are most frequently chosen to describe the population.

For very large populations, or where data for the entire population is very difficult and/or expensive to obtain, samples can be used to estimate the population parameters. A sample is a set of individuals or events selected from a population for analysis to give estimates of parameters of the whole population. In this section, we are concerned with determining the mean of different samples. As we will see, the size of the sample has an impact on the accuracy of the predictions that can be made. In practice, if the sample size is greater than 30, it is considered a large sample.

WORKED EXAMPLE 21 Estimating the population parameter from a sample mean

Peter has always had an interest in the length of the Furry Wooknuk which roams the back paddock of his parents farm near Warburton.
Over time Peter collects 12 samples, each of 10 Wooknuks, and carefully measures their length in centimetres.
Calculate the means of each of the 12 samples and hence estimate the mean length of the Wooknuks of Warburton.

Sample	Wooknuk 1	Wooknuk 2	Wooknuk 3	Wooknuk 4	Wooknuk 5	Wooknuk 6	Wooknuk 7	Wooknuk 8	Wooknuk 9	Wooknuk 10
1	60	55	60	60	60	5	55	30	30	30
2	60	30	30	60	35	130	35	60	30	35
3	85	30	50	55	60	60	55	25	55	30
4	30	30	30	30	30	30	30	30	30	30
5	30	30	30	30	30	25	25	25	25	25
6	60	60	60	60	60	60	60	30	30	60
7	30	60	60	60	90	60	35	35	35	30
8	55	55	60	35	35	55	55	55	50	55
9	85	60	60	30	30	60	60	60	60	60
10	30	30	30	45	45	45	45	45	45	45
11	70	70	130	35	35	70	70	35	35	70
12	60	105	60	120	60	60	60	60	60	105

THINK	WRITE
1. Recall the formula for the sample mean.	$\bar{x} = \frac{\sum x_i}{n}$

2. Determine the mean for each sample.

Sample 1: $\bar{x} = \dfrac{445}{10} = 44.5$ Sample 2: $\bar{x} = \dfrac{505}{10} = 50.5$

Sample 3: $\bar{x} = \dfrac{505}{10} = 50.5$ Sample 4: $\bar{x} = \dfrac{300}{10} = 30$

Sample 5: $\bar{x} = \dfrac{275}{10} = 27.5$ Sample 6: $\bar{x} = \dfrac{540}{10} = 54$

Sample 7: $\bar{x} = \dfrac{495}{10} = 49.5$ Sample 8: $\bar{x} = \dfrac{510}{10} = 51$

Sample 9: $\bar{x} = \dfrac{565}{10} = 56.5$ Sample 10: $\bar{x} = \dfrac{405}{10} = 40.5$

Sample 11: $\bar{x} = \dfrac{620}{10} = 62$ Sample 12: $\bar{x} = \dfrac{750}{10} = 75$

3. Calculate the average $\bar{x}$.

$$\text{Average } \bar{x} = \frac{44.5 + 50.5 + 50.5 + \cdots + 75}{12} = \frac{591.5}{12} \approx 49.3$$

4. Answer the question.

An estimate for the mean length of the Furry Wooknuks is 49.3 cm.

TI | THINK	DISPLAY/WRITE	CASIO | THINK	DISPLAY/WRITE
1. On a Home page, select: 1 New 4 Add Lists & Spreadsheet.		1. From a Main screen, select the Statistics application.	
2. Enter the List 1 data by completing the entry lines 60 55 60 30 Continue entering the entire list.	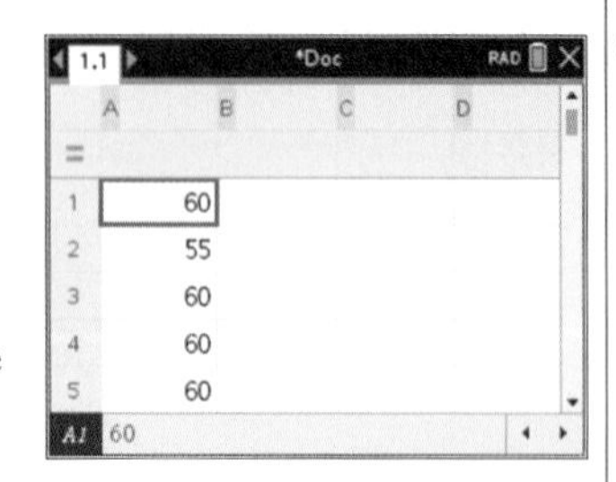	2. Enter the Sample 1 data by completing the entry lines 60 55 60 30 Continue entering the entire list.	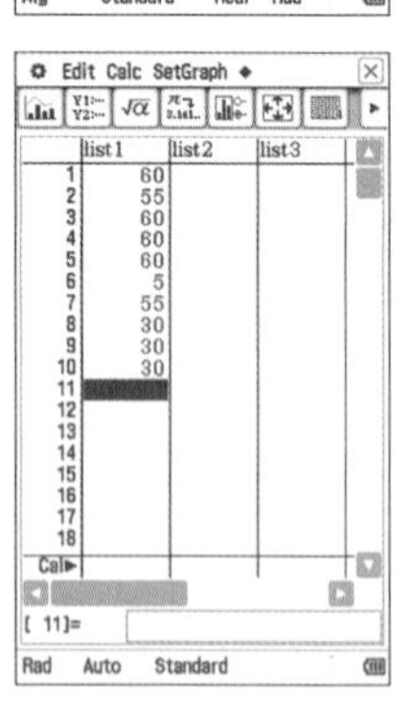

3. Select MENU
   4 Statistics
   1 Stat Calculations
   1 One-Variable Statistics…

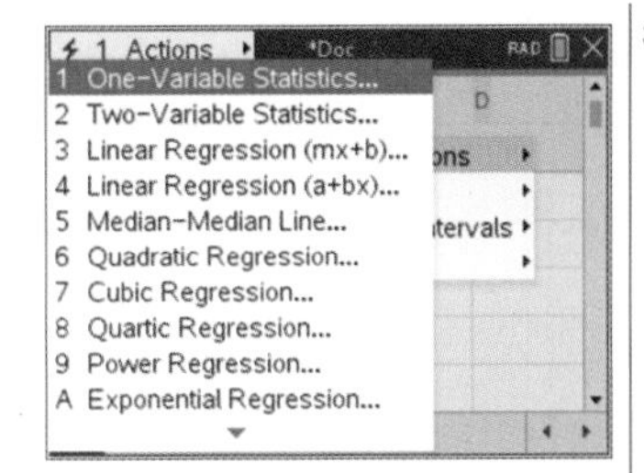

4. Set the number of lists as 1 by pressing the OK button.

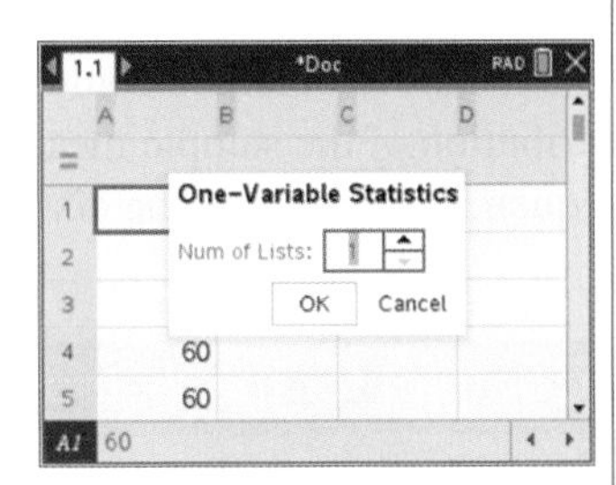

5. Complete the entry lines
   X1 List: a[]
   Frequency List: 1
   1st Result Column: b[]

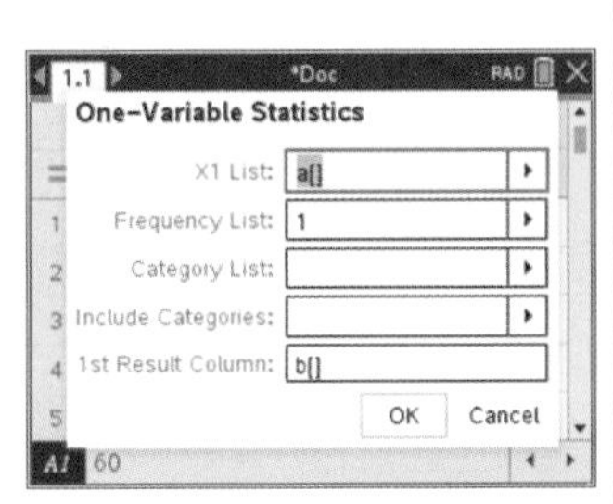

6. Press the ENTER button. The answer appears on the screen.

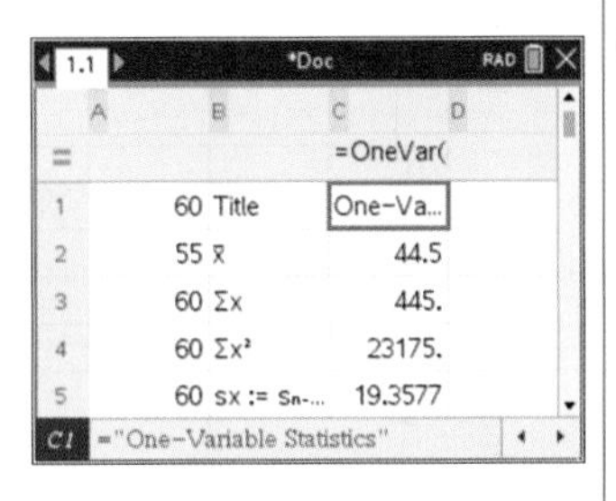

7. Repeat this process to determine the mean for each of the 12 samples.
   To calculate an estimate of the population mean enter the means in a new list as shown.

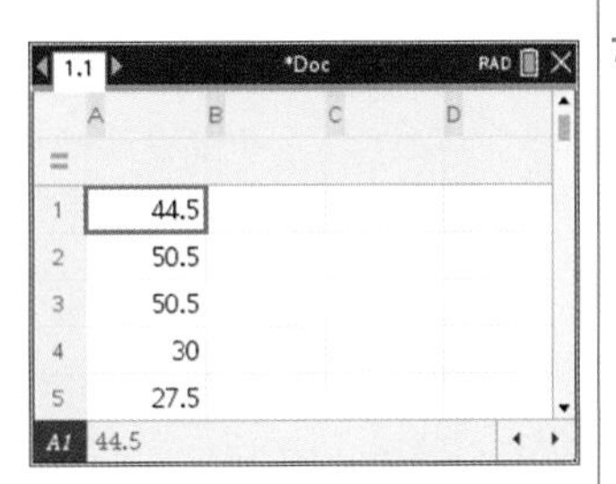

3. Select
   Calc
   One-Variable

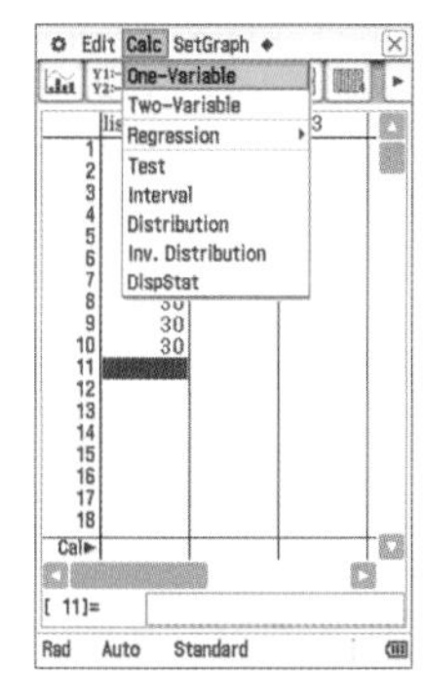

4. XList: list1
   Freq: 1
   Press the OK button

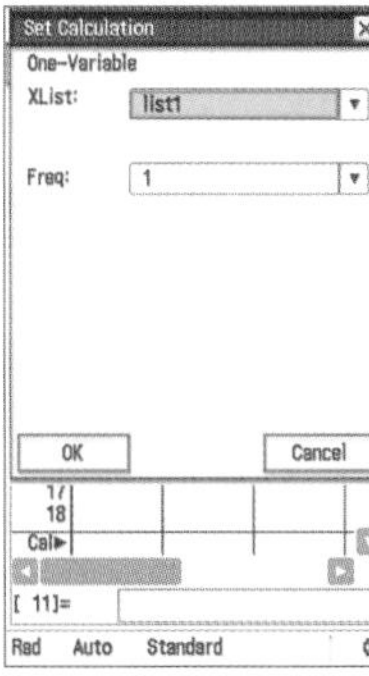

5. The summary statistics appear on the screen.

6. Repeat this process to determine the mean for each of the 12 samples. To calculate an estimate of the population mean enter the means in a new list as shown.

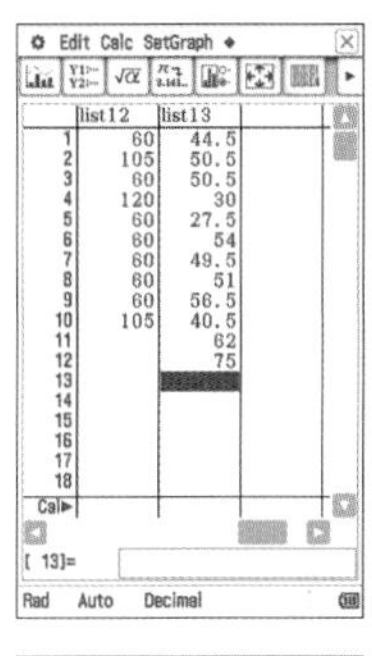

7. Select
   Calc
   One-Variable
   XList: list13
   Freq: 1
   Press the OK button.
   The answer appears on the screen.

8. Press the ENTER button. The answer appears on the screen.

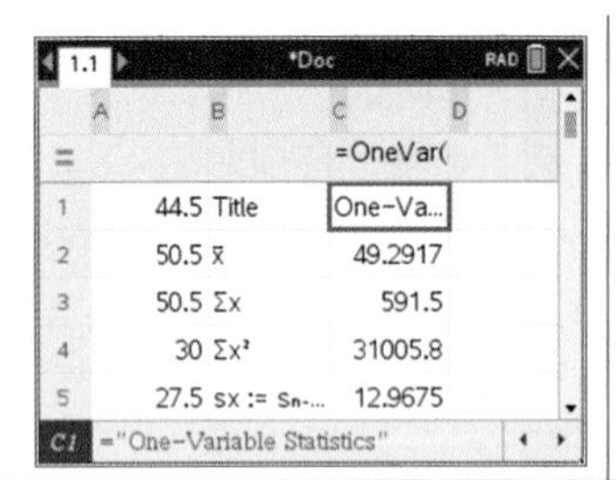

13.6.4 Frequency dotplots

Consider the data from Worked example 21. Each sample mean we obtained could have been graphed using a dotplot. This would give us a pictorial view of the distribution of the sample means. The pictorial view is called a **frequency dotplot**. In a similar fashion, the frequency dotplots for the distribution of values of $\bar{x}$ can be constructed.

WORKED EXAMPLE 22 Creating a frequency dotplot for the distribution of sample means

Consider the data regarding the lengths of furry wooknuks from Worked example 21. Construct a frequency dotplot for the distribution of values of the sample mean.

THINK

1. Set up axes for the frequency dotplot with frequency on the vertical axis and sample mean on the horizontal axis.

WRITE

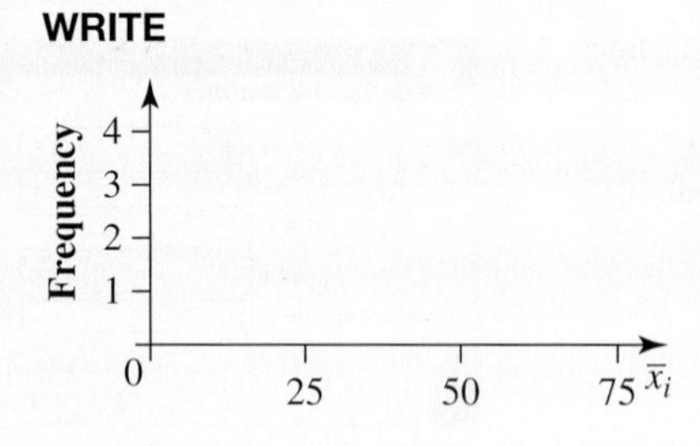

2. Plot the first sample mean (44.5).

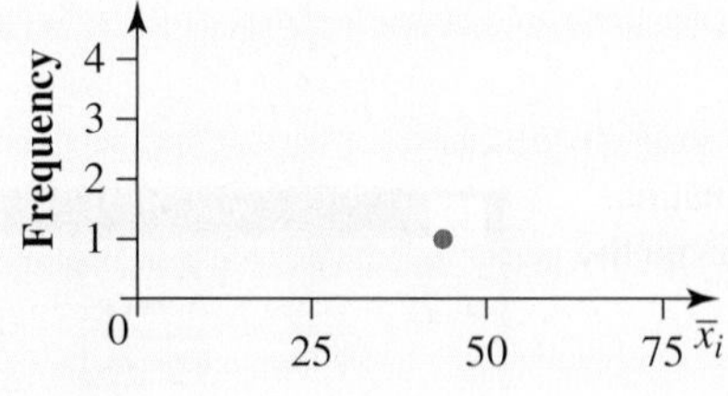

3. Repeat for each of the remaining 11 sample means.

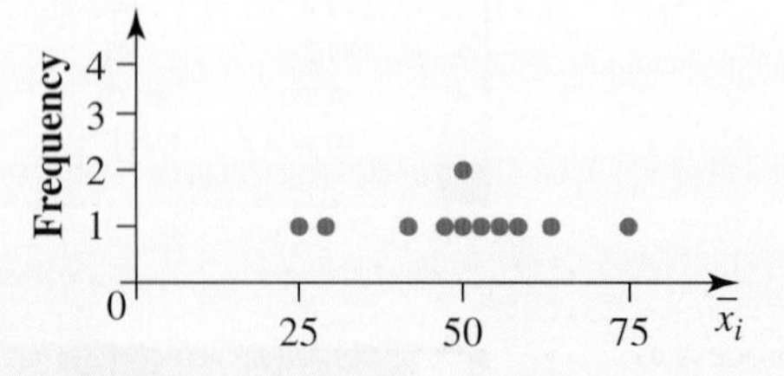

13.6 Exercise

Students, these questions are even better in jacPLUS

Receive immediate feedback and access sample responses

Access additional questions

Track your results and progress

Find all this and MORE in jacPLUS

Technology active

1. WE21 Millicent wants to know the mean movie length for children's movies. She records the lengths for 8 movies every day for a week. Her results are shown in the table below and are recorded in minutes. Estimate the population mean movie length, correct to 1 decimal place.

Day	Movie 1	Movie 2	Movie 3	Movie 4	Movie 5	Movie 6	Movie 7	Movie 8
Monday	115	95	105	95	115	100	90	95
Tuesday	95	85	90	90	105	95	75	95
Wednesday	110	95	80	110	95	90	105	80
Thursday	95	100	90	95	105	100	90	85
Friday	105	95	90	100	105	100	90	105
Saturday	90	85	90	110	80	100	90	90
Sunday	105	100	100	95	90	90	90	110

The following information relates to questions 2 and 3.
The following table shows a simulation of the total obtained when a pair of dice were tossed. There were 9 people who each tossed the dice 8 times.

	Toss 1	Toss 2	Toss 3	Toss 4	Toss 5	Toss 6	Toss 7	Toss 8
Player 1	7	6	10	7	11	2	10	8
Player 2	8	9	8	6	6	11	4	10
Player 3	3	4	2	10	6	8	8	6
Player 4	8	5	5	6	11	7	6	2
Player 5	10	12	6	8	10	8	3	4
Player 6	4	10	9	5	3	6	5	5
Player 7	7	6	5	6	9	10	4	2
Player 8	11	8	9	8	9	9	6	10
Player 9	4	7	10	10	7	4	12	8

2. Determine an estimate of the population mean, to 2 decimal places, and compare it to the theoretical mean.
3. Construct a dotplot for the distribution of values of $\bar{x}$, the sample mean rounded to the nearest integer.
4. Consider a container with 20 balls and $\frac{1}{4}$ of them are red. Samples of 4 balls are drawn. Determine the probability of each different possible sample being drawn, writing your answers as fractions.

5. WE20 Every year 500 students apply for a place at Maccas University. The average enrolment test score is 600 with a standard deviation of $\sqrt{300}$.

a. Samples of 10 students are selected and the means found. Determine the mean and standard deviation of the distribution of $\bar{x}$.

b. State what effect increasing the sample size to 20 would have on the mean and standard deviation of the distribution of sample means.

6. A lolly factory produces lollies of 3 different colours. Each day it produces 100 000 lollies, 50% of which are green. Samples of size 25 are taken from the population of 100 000 and the percentage of green lollies is recorded.

a. Determine the mean and standard deviation of the distribution of sample means.

b. State what effect increasing the sample size to 40 would have on the mean and standard deviation of the distribution of sample means.

The following information relates to questions 7 and 8.

Consider a population with a mean of 67 and a standard deviation of 15.

7. If samples of size 20 are selected, determine the mean and standard deviation of the distribution of sample means.

8. Determine what sample size would be needed to reduce the standard deviation of the distribution of sample means to less than 2.

9. Use the random number generator on your calculator to simulate the following situation: if the number is between 0.25 and 0.47 inclusive, it represents a 4-wheel drive vehicle; otherwise it represents a 2-wheel drive vehicle.

a. Perform the simulation 50 times, calculate $\bar{x}$, the percentage of 4-wheel drive vehicles, and comment on your result.

b. Compare your results with those of your classmates. State how close were they to the theoretical result of 22%.

c. Determine the average of $\bar{x}$ for your class. Comment on how close is it to 22%.

10. A population of 100 has a mean of 30 and a variance of 22.5.
Calculate the mean and standard deviation of the distribution of sample means of samples of size 10.

11. A population of 200 has a mean of 100 and a variance of 25.

a. Calculate the mean and standard deviation of the distribution of sample means of samples of size 10.

b. If the sample size was increased to 20, state how this changes the mean and standard deviation of the distribution of sample means calculated in part a.

12. The following histogram shows a distribution of sample means generated by 9 random samples, each of 100 items.
Estimate the population mean and the variance.

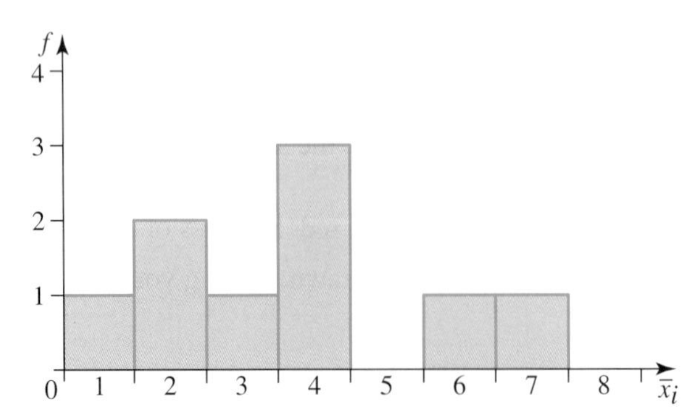

13.6 Exam Questions

Question 1 (3 marks) TECH-ACTIVE

It was stated that to accurately determine the estimate of the population mean from a set of sample means, the size of each sample must be the same. This is not strictly true. Consider the following example of the sample means $\bar{x}$ medical cases requiring hospitalisation at a group of medical clinics. (Note that the hospitals have used different-sized samples.)

Clinic	$\bar{x}$	Sample size
Abbotsford	3	7
Brunswick	5	13
Carlton	3	10
Dandenong	5	15
Eltham	4	10
Frankston	2	8
Geelong	8	21
Hawthorn	3	11
Inner Melbourne	2	15
N. Melbourne	3	13
S. Melbourne	5	17
E. Melbourne	3	7
W. Melbourne	5	14
St Kilda	3	8

a. Convert the sample means of each hospital to the equivalent sample mean of a sample of size 1. **(1 mark)**
b. Calculate an estimate of the population mean if the population is 10 000 cases. **(1 mark)**
c. State whether a dotplot would be an appropriate way to display the spread of the sample proportions. **(1 mark)**

Question 2 (3 marks) TECH-ACTIVE

Consider a population with $N = 600$, $\mu = 180$ and $\sigma^2 = 121$ and samples of size 60 are taken from it.

a. Calculate the mean of the distribution of $\bar{x}$. **(1 mark)**
b. Calculate the standard deviation of the distribution of $\bar{x}$. **(1 mark)**
c. Graph the distribution of $\bar{x}$. **(1 mark)**

Question 3 (1 mark) TECH-ACTIVE

MC A population of N items has a mean of M and a variance of V.

p samples, each of size q, are drawn from the population.

The mean and standard deviation of the distribution of sample means are respectively

A. $M, \sqrt{V}$
B. $\dfrac{M}{q}, \sqrt{\dfrac{V}{p}}$
C. $M, \sqrt{\dfrac{V}{q}}$
D. $\dfrac{M}{p}, \dfrac{V}{\sqrt{q}}$
E. $M, \dfrac{V}{\sqrt{p}}$

More exam questions are available online.

13.7 Review

13.7.1 Summary

doc-37932

Hey students! Now that it's time to revise this topic, go online to:

Review your results

Watch teacher-led videos

Practise past VCAA exam questions

Find all this and MORE in jacPLUS

13.7 Exercise

Technology free: short answer

1. A die is rolled and a coin is tossed. List the elements of the event space and list the elements of Z, the event of obtaining an odd number and a Tail.
2. a. Use a sequence of 6 random numbers to simulate rolling a die 20 times. Record your results in a frequency table.
   b. Produce a histogram of your results.
3. A distribution for $\bar{x}$ has a standard deviation of 2. If the sample size was 150, determine the population variance.
4. a. P is an independent random variable with mean of 4 and variance of 15. Determine the mean and variance of $P + P + P + P + P + P$.
   b. X is an independent random variable with mean of m and variance of z. Determine the mean and variance of $2X$.
5. The table below represents the number of red lollies in bags of 100 coloured lollies. Determine the best estimate of the mean number of red lollies in a jar of 2000 lollies.

34	36	37	31	30	27	31	35	38	33
31	45	41	38	29	26	28	31	26	29

6. This frequency dotplot represents the mean number of second-hand cars requiring warranty repairs out of each sample of 20 cars sold.
   a. Determine how many samples were taken.
   b. Calculate the mean of the distribution of sample means.
   c. Estimate the total number of cars which require repairs if the population contains 300 000 cars.

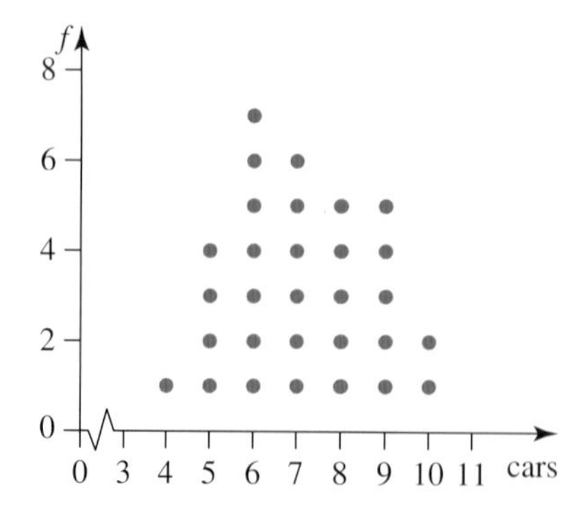

Technology active: multiple choice

7. **MC** Mary has a spinner with 8 equally likely sectors. She wishes to use it to simulate a process where there are 6 equally likely outcomes (numbered from 1 to 6). Select the method which describes how she can do this.

   A. Label the spinner with the numbers 1, 1, 2, 2, 3, 4, 5, 6.
   B. Label the spinner with the numbers 1, 2, 3, 4, 5, 6, 7, 8.
   C. Label the spinner with the numbers 0, 0, 1, 2, 3, 4, 5, 6.
   D. Use either **B** or **C**.
   E. Use either **A**, **B** or **C**.

8. **MC** In a group of 80 Year 11 students it was found that 18 played football, 15 played cricket, 10 played hockey, 7 played volleyball, 12 played tennis and the remainder participated in no sport. The sample percentage of students who did not participate in a sporting activity is

   A. 62.5% B. 77.5% C. 29% D. 71% E. 22.5%

9. **MC** Select which of these sentences is true.

   A. The sample mean may equal the population mean.
   B. The sample mean is always less than the population mean.
   C. The sample mean is always greater than the population mean.
   D. For small values of the sample size, n, the sample mean equals the population mean.
   E. None of the above.

10. **MC** In a survey of 1000 Year 12 students, 600 said that their favourite flavour of ice-cream was chocolate chip. Out of every 10 students questioned, the number who favoured chocolate chip is estimated to be

    A. 600 B. 60 C. 60% D. 6 E. 0.6

11. **MC** In a survey of 1200 Year 10 students, 800 said they favoured a national examination in Year 12. Out of every 30 students questioned, the number who is estimated to favoured a national examination in Year 12 was

    A. 200 B. 20 C. 2 D. 20% E. 0.20

12. **MC** There were exactly 100 lollies in each of three boxes of N & N sweets. The number of blue sweets was 6, 8 and 9 respectively. The best estimate for the proportion of blue sweets in a box is

    A. 0.23 B. 7.67 C. 0.08
    D. 0.0767 E. none of the above

Questions 13 and 14 refer to the following frequency dotplot.

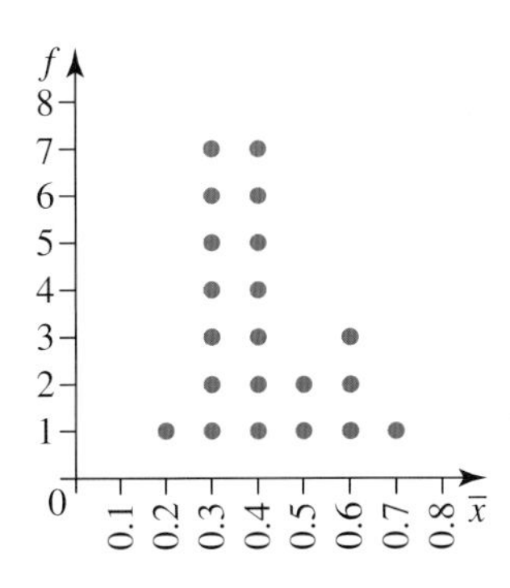

13. **MC** The number of samples is

    A. 17 B. 20 C. 19 D. 18 E. 21

14. **MC** The best estimate of the true population mean is

    A. 0.2 B. 0.3 C. 0.4 D. 0.5 E. 0.35

Questions 15 *and* 16 *refer to the dotplot below.*

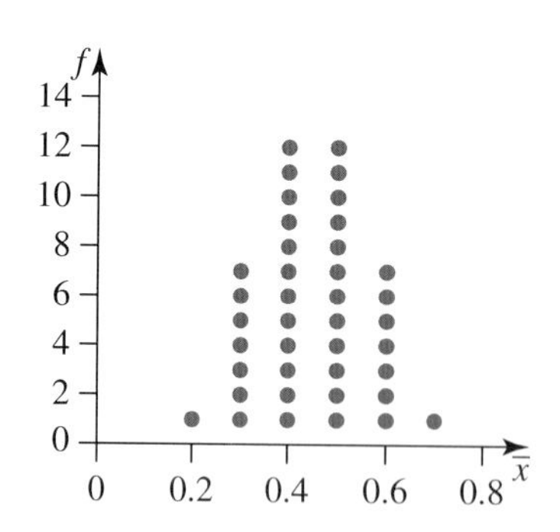

15. **MC** The number of samples is

A. 60 B. 50 C. 40 D. 38 E. 36

16. **MC** The best estimate of the true population mean is

A. 0.25 B. 0.35 C. 0.4 D. 0.45 E. 0.5

Technology active: extended response

17. Use an appropriate simulation tool to simulate the following game, which involves rolling a pair of fair, 6-sided dice.

Dice total	2	3	4	5	6	7	8	9	10	11	12
Outcome	win $1	lose $2	win $3	lose $4	win $5	lose $6	win $7	lose $8	win $9	lose $10	win $11

Play the game enough times to decide whether it is a fair game or not. Comment on your results.

18. The following table represents a sample of 50 cyclists. The letter L indicates those whose bicycles have lights.

a. Determine the percentage number in the sample with lights, correct to the nearest percent.
b. If, instead, 200 cyclists had been sampled, calculate how many you would expect to have lights.

L	N	L	L	N	L	L	L	L	L
N	L	L	L	N	L	L	L	L	L
L	L	N	L	L	N	N	L	N	N
L	N	L	L	N	L	L	L	L	L
N	L	L	L	L	L	L	L	L	N

19. The data below show the number of bullseyes scored by 40 dart players after 5 throws each.

2	2	4	0	1	1	3	2	4	0
2	3	2	1	2	1	1	4	3	1
2	3	3	0	1	2	0	5	3	4
1	5	4	3	3	5	1	4	2	0

a. Determine the percentage of players that scored at least 2 bullseyes.
b. Using a die (or by some other means) conduct 30 trials and obtain another possible value for the percentage of players who scored at least 2 bullseyes.
c. Comment on your results.

20. The following table represents 20 samples of size 10 each of trucks fitted with a speed limiter (L) or not (N).

Sample 1	N	N	N	N	N	L	N	L	L	N
Sample 2	L	N	N	L	N	N	N	L	L	N
Sample 3	N	L	L	L	L	L	L	N	L	L
Sample 4	L	N	L	L	N	N	L	L	N	N
Sample 5	L	N	L	N	L	L	L	N	N	N
Sample 6	N	L	L	N	L	L	L	L	N	L
Sample 7	N	L	N	L	L	L	L	N	L	N
Sample 8	N	N	L	N	L	N	L	N	L	L
Sample 9	N	L	N	L	N	L	N	N	L	N
Sample 10	N	N	N	L	L	N	L	L	N	N
Sample 11	N	N	N	N	N	L	N	L	L	L
Sample 12	N	L	N	L	L	L	N	N	N	N
Sample 13	L	N	N	L	L	N	L	N	L	L
Sample 14	L	L	L	N	N	N	L	L	L	L
Sample 15	N	L	N	N	L	N	L	L	L	N
Sample 16	L	N	N	L	L	L	N	L	L	L
Sample 17	L	L	N	N	N	N	L	N	L	L
Sample 18	L	L	N	L	N	L	N	L	N	N
Sample 19	N	L	L	N	L	N	L	N	L	N
Sample 20	N	L	L	N	L	N	N	N	L	L

a. Determine the mean of trucks with a speed limiter for each sample, correct to the nearest whole number.
b. Display these sample means on a frequency dotplot.
c. Estimate the population percentage with speed limiters, correct to the nearest percent.
d. Calculate the standard deviation of the sample means, correct to 1 decimal place.

13.7 Exam Questions

Question 1 (1 mark) TECH-ACTIVE

MC Which of the following statements best describes the population, as used in statistics?

A. The population is the total number of people living in a country.
B. The population is the total number of people who will participate in a research study.
C. The population consists of people only.
D. The population is the group about whom conclusions are drawn on the basis of a random sample of that group.
E. The population is a subgroup selected from a larger group of research interest.

Question 2 (2 marks) TECH-FREE

Approximately 40% of all Trogfins in the world's population of 1 million Trogfins are tartan with a variance of 250 000. A random sample of 1600 Trogfins is taken. Determine the mean and variance of this sample.

Question 3 (2 marks) TECH-ACTIVE

Using a calculator simulate the outcomes of a game whereby a coin and a six-sided die are tossed simultaneously.

From your simulated results, calculate the probability of tossing a head and rolling a six. Compare your result to the theoretical result.

Question 4 (3 marks) TECH-ACTIVE

Igor is playing a game of Pingo which involves throwing 3 darts at a dart board while blindfolded. The dartboard comprises three sections marked by circles. The inner circle, in black is known as the bull. There is a middle yellow section between the bull and the outer ring which is red. All three sections are of equal area.

Assuming that all three darts hit the target, simulate the throwing of three darts 20 times, and from your results calculate the probability that one dart lands in each of the three coloured areas. Compare this to the theoretical probability of this occurrence.

Question 5 (4 marks) TECH-ACTIVE

The Specialist Mathematics students at Perfect High School perform very well on exams, with a mean score of 91% and a standard deviation of 3.5%.

a. Samples of the percentages gained by 35 students are taken. Calculate the mean and standard deviation of the distribution of sample means. **(2 marks)**

b. If the sample sizes were decreased to 15, state the effect this would this have on your previous answers. **(2 marks)**

More exam questions are available online.

Answers

Topic 13 Simulation, sampling and sampling distributions

13.2 Random experiments, events and event spaces

13.2 Exercise

1. $\xi = \{1, 2, 3, 4, 5, 6\}$, $Y = \{5, 6\}$
2. $\xi = \{\text{spade, club, heart, diamond}\}$, $Z = \{\text{spade, club}\}$
3. $\xi = \{(1,1)\ (1,2)\ (1,3)\ (1,4)\ (1,5)\ (1,6)$
   $(2,1)\ (2,2)\ (2,3)\ (2,4)\ (2,5)\ (2,6)$
   $(3,1)\ (3,2)\ (3,3)\ (3,4)\ (3,5)\ (3,6)$
   $(4,1)\ (4,2)\ (4,3)\ (4,4)\ (4,5)\ (4,6)$
   $(5,1)\ (5,2)\ (5,3)\ (5,4)\ (5,5)\ (5,6)$
   $(6,1)\ (6,2)\ (6,3)\ (6,4)\ (6,5)\ (6,6)\}$
   $Z = \{(1,4)\ (1,5)\ (1,6)$
   $(2,3)\ (2,4)\ (2,5)\ (2,6)$
   $(3,2)\ (3,3)\ (3,4)\ (3,5)\ (3,6)$
   $(4,1)\ (4,2)\ (4,3)\ (4,4)\ (4,5)\ (4,6)$
   $(5,1)\ (5,2)\ (5,3)\ (5,4)\ (5,5)\ (5,6)$
   $(6,1)\ (6,2)\ (6,3)\ (6,4)\ (6,5)\ (6,6)\}$
4. $\xi = \{\text{H1, H2, H3, H4, H5, H6, T1, T2, T3, T4, T5, T6}\}$
   $X = \{\text{H2, H4, H6}\}$
5. $\xi = \{\text{AS, 2S, 3S, 4S, 5S, 6S, 7S, 8S, 9S, 10S, JS, QS, KS,}$
   $\text{AC, 2C, 3C, 4C, 5C, 6C, 7C, 8C, 9C, 10C, JC, QC, KC,}$
   $\text{AD, 2D, 3D, 4D, 5D, 6D, 7D, 8D, 9D, 10D, JD, QD, KD,}$
   $\text{AH, 2H, 3H, 4H, 5H, 6H, 7H, 8H, 9H, 10H, JH, QH, KH}\}$
   $P = \{\text{AS, 2S, 3S, 4S, 5S, 6S, 7S, 8S, 9S, 10S, JS, QS, KS,}$
   $\text{JC, JD, JH}\}$
6. $Q = \{\text{JS}\}$
7. $\xi = \{\text{A}, 1, 2, 3, 4, 5, 6, 7, 8, 9, 10, \text{J, Q, K}\}$, $W = \{\text{J, Q, K}\}$
8. $\xi = \{$H Heart, H Diamond, H Spade, H Club, T Heart, T Diamond, T Spade, T Club$\}$
   $X = \{$T Heart, T Diamond$\}$
9. $D = \{(1,1), (1,3), (1,5), (3,1), (3,3), (3,5), (5,1), (5,3), (5,5)\}$
10. $F = \{(4,4), (4,5), (4,6), (5.4), (5,5), (5,6), (6,4), (6,5), (6,6)\}$
11. $X = \{\text{H1, H2, H3, H4, H5, H6, T2, T4, T6}\}$
12. $Y = \{$H Heart, H Diamond, T Heart, T Diamond, T Spade, T Club$\}$
13. $\xi = \{\text{HHH, HHT, HTH, HTT, THH, THT, TTH, TTT}\}$
    $D = \{\text{HHT, HTH, HTT, THH, THT, TTH}\}$
14. $\xi = \{(1,1), (1,2), (1,3), (1,4), (2,1), (2,2), (2,3), (2,4), (3,1),$
    $= (3,2), (3,3), (3,4), (4,1), (4,2), (4,3), (4,4)\}$
    $Q = \{(1,2), (1,3), (1,4), (2,3), (2,4), (3,4)\}$

13.2 Exam questions

Note: Mark allocations are available with the fully worked solutions online.

1. $\xi = \{\text{KQJ, KJQ, QKJ, QJK, JKQ, JQK}\}$
   $K = \{\text{KQJ, KJQ, QKJ}\}$
2. a. $Q = \{\text{JH, QH, KH, JD, QD, KD, 2S, 4S, 6S, 8S, 10S}\}$
   b. $R = \{\text{2H, 4H, 6H, 8H, 10H, 2D, 4D, 6D, 8D, 10D, JS, QS, KS}\}$
3. $\xi = \{(1,2), (1,3), (1,4), (2,1), (2,3), (2,4), (3,1), (3,2), (3,4),$
   $(4,1), (4,2), (4,3)\}$
   $R = \{(1,3), (3,1)\}$

13.3 Simulation

13.3 Exercise

1–3. Sample responses can be found in the worked solutions in the online resources.
4. Player B
5. a. One way is to use randInt (0, 1, 10) to generate 10 values that are either equal to 0 or 1, and let 0s represent Heads, and 1s represent Tails.
   b. Sample responses can be found in the worked solutions in the online resources.
   c. Sample responses can be found in the worked solutions in the online resources.
6. Generally, the histogram for 100 tosses will be more even than that for 10 tosses, approximately symmetrical.
7. a, b The player can expect to win about once every 15 games, spending \$15 to win \$10 (a loss of \$5).

8–13. Sample responses can be found in the worked solutions in the online resources.

13.3 Exam questions

Note: Mark allocations are available with the fully worked solutions online.

1. Sample responses can be found in the worked solutions in the online resources.
2. Sample responses can be found in the worked solutions in the online resources.
3. Sample responses can be found in the worked solutions in the online resources.

13.4 Discrete random variables

13.4 Exercise

1. a. Discrete b. Continuous c. Continuous
   d. Discrete e. Continuous f. Discrete
   g. Continuous h. Discrete
2. a.

x	0	1	2
$\Pr(X = x)$	$\frac{1}{4}$	$\frac{1}{2}$	$\frac{1}{4}$

   b.

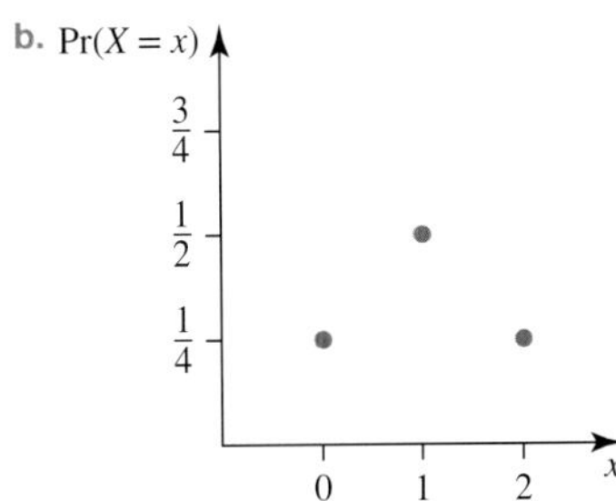

3. 7.11
4. $\frac{13}{8}$
5. $a = 0.09, \text{E}(X) = 5.42$
6. $a = \frac{1}{18}, \text{E}(X) = 5\frac{1}{3}$
7. $\text{E}(X) = 2.3,\ \sigma^2 = 0.81$

8. a. \$1.42 b. 0.02

9. $E(X) = 2\frac{1}{3}$, $\sigma^2 = 1.0556$, $s.d = 1.0274$

10. a. 2.333 b. 9.333 c. 0.622 d. 9.956

11. $E(X) = 4.8$, $\sigma = 3.487$

12. a. \$3.75
    b. No, because although his excepted gain is \$3.75 per game, he must pay \$5 to play each game. Therefore his loss per game will be \$1.25.
    c. No, because the expected gain is not equal to the initial cost of the game.

13. a.

x	2	1.5	1	0
$\Pr(X=x)$	0.5	0.3	0.15	0.05

b. \$1.60
c. \$0.51

14. a.

x	1	3	5
$\Pr(X=x)$	$\frac{1}{35}$	$\frac{9}{35}$	$\frac{5}{7}$

b. $4\frac{13}{35}$
c. 1.0443
d. $\frac{34}{35}$

15. $E(3F) = 27$
    $\text{Var}(3F) = 36$

16. $E(4W) = 856$
    $\text{Var}(4W) = 240$

17. $E(Q) = 420$
    $\text{Var}(Q) = 11$

18. $E(X) = 20$
    $\text{Var}(X) = 16$

19. $E(X + X + X + ... + X) = pm$
    $\text{Var}(X + X + X + ... + X) = pv$

20. $\text{Var}(Y) = \frac{v}{n^2}$

13.4 Exam questions

Note: Mark allocations are available with the fully worked solutions online.

1. \$1595
2. 0.95
3. a.

x	2	3	4	5
$\Pr(X=x)$	$\frac{2}{27}$	$\frac{1}{6}$	$\frac{8}{27}$	$\frac{25}{54}$

b. 4.1481
c. 0.9509

13.5 Sampling

13.5 Exercise

1. $N = 120$, $n = 30$
2. $N = 100$, $n = 5$
3. $n = 15$, population size is unknown
4. $n = 9$, population size is unknown
5. a. Population parameter
   b. Sample statistic
6. a. Population parameter
   b. Sample statistic
7. 34 boys and 41 girls
8. 21 boarders, 69 day students
9. a. The population size is unknown.
   b. 40
10. a. The population is people who will receive the vaccine in the future. The size is unknown.
    b. 247
11. Sample statistic
12. Population parameter
13. Sample statistics
14. Population parameter
15. a. A systematic sample with $k = 10$.
    b. Yes, assuming that the order of patients is random.
16. The sample is not random; therefore, the results are not likely to be random.
17. It is probably not random. Tony is likely to ask people who he knows or people who approach him.
18. Full-time male staff: 46
    Part-time male staff: 2
    Full-time female staff: 25
    Part-time female staff: 7
19. Use the random number generator on your calculator to produce numbers from 1 to 100. Keep generating numbers until you have 10 different numbers.
20. First, assign every person in your class a number, e.g. 1 to 25 if there are 25 students in your class. Decide how many students will be in your sample, e.g. 5. Then use the random number generator on your calculator to produce numbers from 1 to 25. Keep generating numbers until you have 5 different numbers. The students that were assigned these numbers are the 5 students in your random sample.

13.5 Exam questions

Note: Mark allocations are available with the fully worked solutions online.

1. B
2. C
3. C

13.6 Sampling distribution of sample means

13.6 Exercise

1. 95.9 minutes
2. $\bar{x} = 6.93$, theoretical mean $= 7$
3. 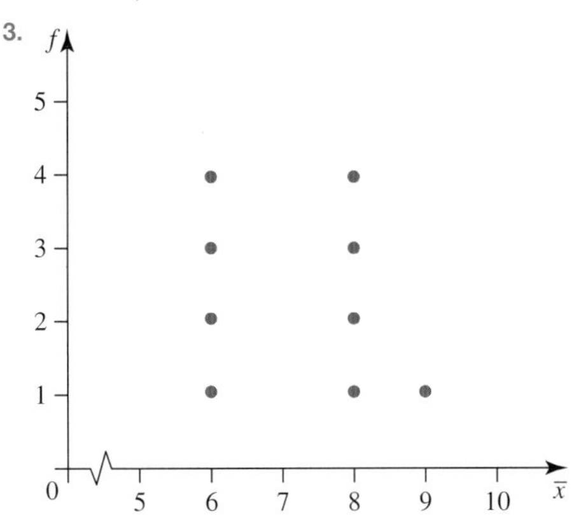

4. Let N represent the number of red balls drawn in the sample of 4.

N	Probability
0	$\frac{1365}{4845}$
1	$\frac{2275}{4845}$
2	$\frac{1050}{4845}$
3	$\frac{150}{4845}$
4	$\frac{5}{4845}$

5. a. $\mu_{\bar{x}} = 600, \sigma_{\bar{x}} = \sqrt{30}$
   b. $\mu_{\bar{x}} = 600, \sigma_{\bar{x}} = \sqrt{15}$. No effect on mean; the standard deviation is reduced.
6. a. $\mu_{\hat{p}} = 0.5, \ \sigma_{\hat{p}} = 0.1$
   b. $\mu_{\hat{p}} = 0.5, \ \sigma_{\hat{p}} = \frac{\sqrt{10}}{40}$. No effect on the mean; the standard deviation is reduced.
7. $\mu_{\bar{x}} = 67, \ \sigma_{\bar{x}} = \dfrac{3\sqrt{5}}{2}$
8. 57
9. Sample responses can be found in the worked solutions in the online resources.
10. $\bar{x} = 30, \ \sigma_{\bar{x}} = 1.5$
11. a. $\bar{x} = 100, \ \sigma_{\bar{x}} = \sqrt{2.5}$
    b. It would have no effect on the mean, but would decrease the standard deviation to $\sqrt{1.25}$.
12. $\mu = 3.30$
    $\sigma^2 = 589$

13.6 Exam questions

Note: Mark allocations are available with the fully worked solutions online.

1. a. Abbortsford: $\bar{x}_1 = \frac{3}{7} \approx 0.43$

   Brunswick: $\bar{x}_1 = \frac{5}{13} \approx 0.38$

   Carlton: $\bar{x}_1 = \frac{3}{10} = 0.3$

   Dandenong: $\bar{x}_1 = \frac{5}{15} = \frac{1}{3} \approx 0.33$

   Eltham: $\bar{x}_1 = \frac{4}{10} = 0.4$

   Frankston: $\bar{x}_1 = \frac{2}{8} = 0.25$

   Geelong: $\bar{x}_1 = \frac{8}{21} \approx 0.38$

   Hawthorn: $\bar{x}_1 = \frac{3}{11} \approx 0.27$

   Inner Melbourne: $\bar{x}_1 = \frac{2}{15} \approx 0.13$

   N. Melbourne: $\bar{x}_1 = \frac{3}{13} \approx 0.23$

   S. Melbourne: $\bar{x}_1 = \frac{5}{17} \approx 0.29$

   E. Melbourne: $\bar{x}_1 = \frac{3}{7} \approx 0.43$

   W. Melbourne: $\bar{x}_1 = \frac{5}{14} \approx 0.36$

   St Kilda: $\bar{x}_1 = \frac{3}{8} = 0.375$

   b. 3264
   c. Yes. A dotplot is very effective, rounding to one decimal place.
2. a. $\mu_{\bar{x}} = 180$
   b. $\sigma_{\bar{x}} = 1.42$
   c. 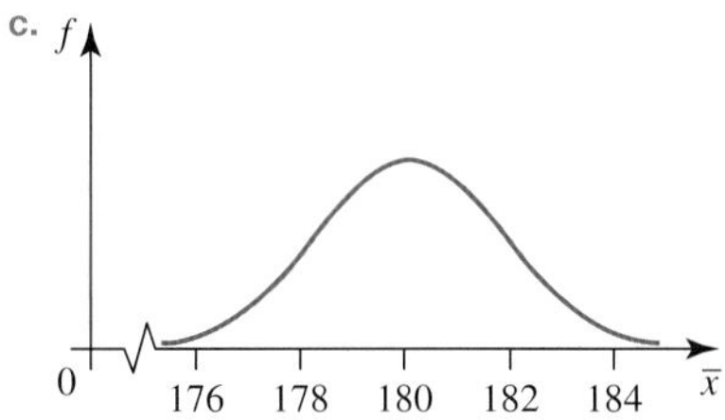

3. C

13.7 Review

13.7 Exercise

Technology free: short answer

1. $\xi = \{H1, H2, H3, H4, H5, H6, T1, T2, T3, T4, T5, T6\}$, $Z = \{T1, T3, T5\}$

2. a.

Random number	Frequency
1	3
2	2
3	1
4	5
5	8
6	1

(*Note:* Answers will vary. This is only one possible answer.)

b. 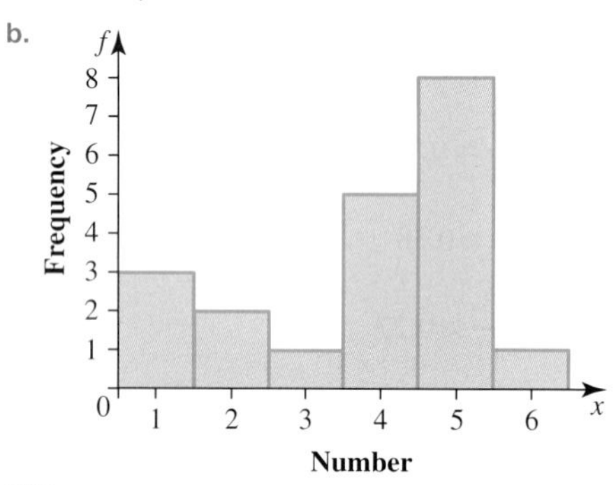

3. 600

4. a. $E(P+P+P+P+P+P)=24$
$\text{Var}(P+P+P+P+P+P)=90$

b. $E(2X)=2m$
$\text{Var}(2X)=4z$

5. 32.8%

6. a. 30 b. 7.1 c. 106 500

Technology active: multiple choice

7. D

8. E

9. A

10. D

11. B

12. B

13. E

14. C

15. C

16. D

Technology active: extended response

17. It is a fair game.

18. a. 74% b. 148

19. a. 65%

b.

Random number	Frequency
0	8
1	7
2	4
3	3
4	5
5	3

(*Note:* Answers will vary. This is only one possible answer.)

$\frac{15}{30}=50\%$

c. The results do not compare well in this case (they differ by 15%). However, each time numbers are generated the probability will vary. Performing more trials or repeating the simulation will provide better estimates of the probability.

20. a. See table at the bottom of the page*

b. 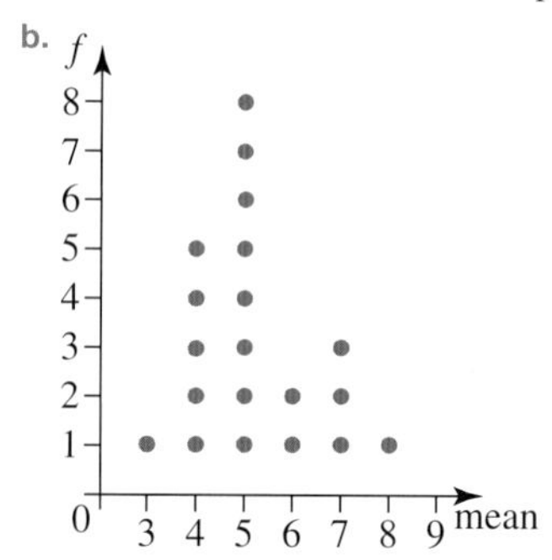

c. 52%

d. 3.95

13.7 Exam questions

Note: Mark allocations are available with the fully worked solutions online.

1. D

2. $\bar{x}=640$
Sample variance $=156.25$

3. Sample responses can be found in the worked solutions in the online resources. The theoretical probability of tossing a head and rolling a 6 is $\frac{1}{12}$.

4. Sample responses can be found in the worked solutions in the online resources.

5. a. $\mu_{\bar{X}}=91\%$
$\sigma_{\bar{X}}=0.5916\%$

b. The mean of the sample means is unchanged.
The standard deviation of the sample means will increase to 0.9037%.

*20 a.

Sample	1	2	3	4	5	6	7	8	9	10
Proportion, $\hat{p}$	3	4	8	5	5	7	6	5	4	4
Sample	11	12	13	14	15	16	17	18	19	20
Proportion, $\hat{p}$	4	4	6	7	5	7	5	5	5	5

GLOSSARY

absolute value $|x|$, the magnitude of a number
addition principle if there are n ways of performing operation A and m ways of performing operation B, then there are $m + n$ ways of performing A *or* B
additive inverse when two matrices $A + B = O$, matrix B is an additive inverse of A
adjacency list a set of unordered lists, whereby each unordered list states the vertices that are adjacent to a particular vertex
adjacency matrix a matrix that represents the vertices and that connect vertices in a graph
affirming the consequent an error in formal logic where if the consequent is said to be true, the antecedent is said to be true, as a result
algorithm a set of rules or instructions used to solve a particular problem
angle difference formulas $\sin(\alpha - \beta) = \sin(\alpha)\cos(\beta) - \cos(\alpha)\sin(\beta)$, $\cos(\alpha - \beta) = \cos(\alpha)\cos(\beta) + \sin(\alpha)\sin(\beta)$
angle of depression the angle measured down from the horizontal line (through the observation point) to the line of vision
angle of elevation the angle measured up from the horizontal line (through the observation point) to the line of vision
angle sum formulas $\sin(\alpha + \beta) = \sin(\alpha)\cos(\beta) + \cos(\alpha)\sin(\beta)$, $\cos(\alpha + \beta) = \cos(\alpha)\cos(\beta) - \sin(\alpha)\sin(\beta)$
antecedent the first statement in a compound statement
argument a series of statements put forward in the study of logic; a standard argument usually consists of two premises and a conclusion
arithmetic sequence a sequence where there is a common difference between successive terms
Argand diagram plotting a complex number $x + yi$ as a point (x, y) on the complex plane with real and imaginary axes (the Argand plane)
Argand plane Geometric representation of complex numbers $x + yi$ with real and imaginary axes. The axes are denoted by 'Re(z)' and 'Im(z)' respectively.
argument of z Arg(z); the angle measurement anticlockwise of the positive real axis;
$\arg(z) = (\theta)\ \sin(\theta) = \dfrac{y}{|z|}, \cos(\theta) = \dfrac{x}{|z|},\ \tan(\theta) = \dfrac{y}{x}$
arrangement a way of choosing things where order is important; *see* permutation
attribute a particular condition related to members of a population
axiom A proposition that is assumed to be true; from Greek, meaning 'agreed starting point'. Every area of mathematics has its own basic axioms.
binomial theorem a rule for expanding expressions of the form $(x + y)^n$:
$(x + y)^n = x^n + {}^nC_1 x^{n-1}y + {}^nC_2 x^{n-2}y^2 + ... + {}^nC_r x^{n-r}y^r + ... + y^n$
Boolean algebra the rules of logic
bridge an edge in a connected graph that, if removed, would cause the graph to become disconnected
Cartesian form of a complex number $x + yi$; consists of an ordered pair of numbers (x, y) which can be plotted on the complex plane or Argand plane
combination the number of ways of choosing r things from n distinct things where order is not important;
${}^nC_r = \dfrac{n!}{r!\ (n-r)!}$, also written as $\dbinom{n}{r}$
complex conjugate roots the complex solutions for quadratic equations with $\Delta < 0$ (i.e. those that have no real solutions); expressed in the form $a \pm bi$
complex number $z = a + bi$; a real number added to a multiple of the imaginary unit i, when a and b are real numbers; $\text{Re}(z) = a$, $\text{Im}(z) = b$
conjugate of a complex number The conjugate, $\bar{z}$, of a complex number z has the opposite sign of the imaginary component. Multiplication or addition of a complex number and its conjugate results in a real number.
Cartesian form of a vector from the origin to point (x, y), the Cartesian form of vector $\underset{\sim}{u}$ is $\underset{\sim}{u} = x\underset{\sim}{i} + y\underset{\sim}{j}$
categorical statement (or proposition) the first premise in an argument

closed trail a trail beginning and ending at the same vertex
closed walk a route taken through a graph that starts and ends at the same vertex
column matrix a matrix with only one column; a vector matrix
combined transformation a combination of two or more transformations
complete bipartite graph a bipartite graph in which every vertex is connected by an edge to each vertex in the other group
complete graph a graph in which every vertex is connected to all other vertices by an edge
conclusion the reason for the argument; it usually follows words such as, 'therefore', 'accordingly', 'it follows that' or 'hence'
conditional in a compound statement when the second statement is conditional on the first statement
conjunction when 'and' is used as a connective
connected graph a graph in which it is possible to visit every vertex by traveling along edges
connective A word used to join two statements into a compound statement. The main connectives are 'and', 'or' and 'not'.
consequent the second statement in a compound statement
contradiction In proof by contradiction, the opposite of what you are trying to prove is assumed to be true. Eventually, the proof will reach a statement that cannot be true, meaning that the initial assumption must be false.
contrapositive A statement that contains the negation of both terms of the original statement but in the reverse order. For example, the contrapositive of 'if p, then q' is 'if not q, then not p'. The contrapositive of a true statement is also true.
conventional (or compass) bearings Directions measured in degrees from the north–south line in either a clockwise or anticlockwise direction. To write the compass bearing we need to state whether the angle is measured from the north or south, the size of the angle and whether the angle is measured in the direction of east or west; for example, N27°W or S32°E.
converge a string of numbers that get closer to a certain fixed value
converse the reverse of a compound statement; for example, the converse of 'if it is raining then I bring my umbrella' is 'if I bring my umbrella then it is raining'
coordinate matrix a matrix that represents the coordinates of points as columns in the matrix
cosecant function the reciprocal trigonometric function of the sine function
cosine rule $a^2 = b^2 + c^2 - 2bc\cos(A)$, is used to solve a side or angle of a triangle when given two sides and the included angle
cotangent function the reciprocal trigonometric function of the tangent function
counter example an example which shows a statement is not true
cycle a path within a graph which stats and ends at the same vertex
deductive argument a logical case containing at least two premises and a conclusion
definition the agreed meaning of a term
degenerate graph *see* null graph
degree in graph theory, the degree of a vertex is the number of edges connecting it
de Moivre's theorem $z_1 \times z_2 = r_1 \times r_2 \text{cis}(\theta_1 + \theta_2)$ and $\left(r\,\text{cis}(\theta)\right)^n = r^n \text{cis}(n\theta)$, where $\text{cis}(\theta) = \cos(\theta) + i\sin(\theta)$; useful for multiplying and dividing complex numbers
determinant of a matrix a value associated with a square matrix, evaluated by multiplying the elements in the leading diagonal and subtracting the product of the elements in the other diagonal; for a matrix $A = \begin{bmatrix} a & b \\ c & d \end{bmatrix}$, $\det(A) = \Delta = |A| = \begin{vmatrix} a & b \\ c & d \end{vmatrix} = ad - bc$
dilation a linear transformation that enlarges or reduces the size of a figure by a scale factor λ parallel to either axis or both
directed graphs graphs in which it is only possible to move along the edges in one direction, where the edges contain arrows
disjunction when 'or' is used as a connective
diverge a string of numbers that grow further and further apart

dot product or scalar product; multiplication of two vectors which results in a scalar

double angle formulas $\sin(2\alpha) = 2\sin(\alpha)\cos(\alpha)$, $\cos(2\alpha) = \cos^2(\alpha) - \sin^2(\alpha)$
$= 2\cos^2(\alpha) - 1$
$= 1 - 2\sin^2(\alpha)$

elements The members of a set; $a \in A$ means a is an element of, or belongs to, the set A. If a is not an element of the set A, this is written as $a \notin A$.

elements of a matrix the numbers in the matrix

equidistant at the same distance from a given point or position

equivalent statements statements p and q are equivalent if the converse of $p \rightarrow q$ is true; that is, $p \rightarrow q$ and $q \rightarrow p$; symbol $\leftrightarrow$

Eulerian graphs a connected graph where you can start at a vertex and move along each edge only once and be able to return to the vertex at which you began

Eulerian trail the trail that is created when you travel around an Eulerian graph

event A set of favourable outcomes in each trial of a probability experiment

event space A list of all the possible outcomes obtained from a probability experiment. It is written as ξ or S, and the list is enclosed in a pair of curled brackets { }. It is also called the sample space.

existential quantifier There exists; symbolically, $\exists$. For a propositional function, consider if there exists a value for the variable which will make the statement true.

faces enclosed areas created by edges and vertices

factorial $n!$, multiplying each of the integers from n down to 1; $0! = 1$

fallacies a deductive argument that follows the correct structure, but while the premises may be true, the conclusion does not follow logically

finite a fixed number

fixed point a point of the domain of a function which is mapped onto itself after a transformation; an invariant point

frequency dotplot a special type of statistical graph where each element in the data set is represented by a dot above a scale

functional definition the general formula for an arithmetic sequence

generating event a means of creating a set of outcomes for a particular experiment

geometric sequence a sequence where there is a common ratio between consecutive terms

graph a collection of points and lines (vertices and edges) that can be used to represent connection in various settings

half angle formulas $\sin(\alpha) = 2\sin\left(\frac{\alpha}{2}\right)\cos\left(\frac{\alpha}{2}\right)$, $\cos(\alpha) = \cos^2\left(\frac{\alpha}{2}\right) - \sin^2\left(\frac{\alpha}{2}\right)$
$= 2\cos^2\left(\frac{\alpha}{2}\right) - 1$
$= 1 - 2\sin^2\left(\frac{\alpha}{2}\right)$

Hamiltonian cycle the trail that is created when you travel around a Hamiltonian graph

Hamiltonian graphs a graph where you can start at a vertex and travel to each vertex only once and be able to return to the vertex at which you began

Hungarian algorithm used to achieve optimal allocation when choosing between multiple options in assignment problems

identical vector a vector with the same magnitude and direction as the original

identity a relationship that is true for all possible values of the variable or variables

identity matrix I; a square matrix that has 1s down the leading diagonal and 0s on the other diagonal; for example, the 2×2 identity matrix is $I = \begin{bmatrix} 1 & 0 \\ 0 & 1 \end{bmatrix}$

iff the condition 'if and only if'; when all stated conditions must apply

image a point or figure after a transformation

imaginary number i; the root of the equation $x^2 = -1$; that is, $x = \pm\sqrt{i^2}$

implication when the first statement in a compound statement implies the second statement; for example, 'if it is raining then I bring my umbrella'
inclusion–exclusion principle for two sets S and T: $n(S \cup T) = n(S) + n(T) - n(S \cap T)$, for three sets S, T and R: $n(S \cup T \cup R) = n(S) + n(T) + n(R) - n(S \cap T) - n(T \cap R) - n(S \cap R) + n(S \cap T \cap R)$
inductive step a step in a proof by induction in which you assume that a formula is true for $n = k$, and use this assumption to show that it must also be true for $n = k+1$
inference another name for an argument
infinite an indefinite number
initial statement a statement that holds true for the smallest integer value in the given range
integers all of the positive and negative numbers and zero: $\ldots -3, -2, -1, 0, 1, 2, 3 \ldots$
invariant point a point of the domain of a function which is mapped onto itself after a transformation; a fixed point
invariants properties which are unchanged by a transformation
intersection the set containing the elements common to both A and B; denoted as $A \cap B$
irrational numbers real numbers not including the rational numbers; that is, real numbers that cannot be expressed in the form $\frac{a}{b}$, where $b \neq 0$
isolated vertex a vertex which is not connected by edges to any other vertex
isomorphic Two graphs are isomorphic if they are structurally equivalent. That is, their vertices and edges differ only by the way in which they are named.
limit the number towards which a convergent sequence tends
linear transformation A transformation of a vector in which the origin does not move. A linear transformation can be represented by a 2×2 matrix $\begin{bmatrix} a & b \\ c & d \end{bmatrix}$. Linear transformations include rotations around the origin, reflection in lines passing through the origin and dilations. Translations are not linear transformations.
locus the path traced by a point that moves according to a condition or rule
logic gate a building block of a digital circuit
logistic equation a model of population growth
loop a sequence of instructions that is continually repeated until a certain condition is reached
magnitude $|x|$; the absolute value of a number
magnitude (of a vector) $|z|$ or r; the length of any directed line segment representing the vector z
major axis the longer axis for an ellipse
mathematical statement A mathematical sentence that is either true or false; a proposition. For example, '5 is a prime number' is a true mathematical statement.
matrix a rectangular array used to store and display data
minimum spanning tree the spanning tree in a weighted graph which has the lowest total weight
minor axis the shorter axis for an ellipse
minors the determinants of a smaller square matrix formed by deleting one row and one column from a larger square matrix
modulus the absolute value of a number
multiple angle formulas $\sin(3\alpha) = 3\sin(\alpha) - 4\sin^3(\alpha)$, $\cos(3\alpha) = 4\cos^3(\alpha) - 3\cos(\alpha)$, $\sin(4\alpha) = \cos(\alpha)\left(4\sin(\alpha) - 8\sin^3(\alpha)\right)$, $\cos(4\alpha) = 8\cos^4(\alpha) - 8\cos^2(\alpha) + 1$
multiplication principle if there are n ways of performing operation A and m ways of performing operation B, then there are $m \times n$ ways of performing A *and* B
multiplicative inverse a number multiplied by its multiplicative inverse equals 1, that is, $z^{-1} = \frac{1}{z}$, $zz^{-1} = 1$
natural numbers the counting numbers $1, 2, 3, \ldots$
negation The opposite of a mathematical statement. If the statement is p, the negation is 'not p'; symbolically, $\neg p$.
network a collection of points and lines (vertices and edges) that can be used to represent connection in various settings
null graph a graph consisting on vertices only, no edges
null matrix a square matrix that consists entirely of '0' elements; a zero matrix

order of a matrix the size of a matrix
oscillating sequence a sequence whose terms fluctuate between two or more values
parameter is a varying constant used to describe a family of polynomials or relates two other variables
Pascal's identity a theorem dealing with binomial coefficients: ${}^{n}C_{r} = {}^{n-1}C_{r-1} + {}^{n-1}C_{r}$ for $0 < r < n$
Pascal's triangle a triangle formed by rows of the binomial coefficients of the terms in the expansion of $(a+b)^n$ with n as the row number
path a walk in which no vertices are repeated, except possibly the start and finish
permutation the number of ways of choosing r things from n distinct things when order is important; ${}^{n}P_{r} = \dfrac{n!}{(n-r)!}$
pigeon-hole principle if there are $(nk+1)$ pigeons to be placed in n pigeon-holes, then there is at least one pigeon-hole with $(k+1)$ pigeons in it
planar graph a graph which can be drawn with no intersecting edges
point of rotation the fixed point around which a rotation occurs
polar coordinates an alternative form of locating a point on a plane, $[r, \theta]$, where r is the distance from a point to the origin and θ is the angle made with the positive direction of the x-axis
polar form (of a complex number) the complex number z expressed in terms of θ; $z = r\,\text{cis}(\theta)$
polar form (of a vector) a vector expressed in the form (r, θ) where r is the magnitude of the vector and $\theta = \tan^{-1}\left(\dfrac{y}{x}\right)$, which is the direction of the vector (the angle the vector makes to the positive direction of the x-axis)
population the whole group under consideration
population proportion the number of elements of a population found to have a certain attribute divided by the total population
population size the total number of distinct objects being considered, it may be unknown
position vector a vector that defines a point by magnitude and direction relative to the origin
postulate in geometry, a postulate is a statement that is assumed to be true without proof (this is usually called an axiom in other areas of mathematics)
pre-image the original figure before a transformation
premise the statement/s used to justify the conclusion in an argument
Prim's algorithm a set of steps that can be used to identify the minimum spanning tree for a weighted connected graph
projection a vector $\underset{\sim}{v}$ can be defined by a projection acting in the direction of vector $\underset{\sim}{u}$ and a projection acting perpendicular to $\underset{\sim}{u}$
proof demonstration that a statement is always true using definitions, postulates and previously proven statements in a formal sequence of steps
proof by mathematical induction the process of proving that a certain property or formula is true for every natural number greater than a certain value
proportion a part or amount considered in relation to a whole
probability the long-term proportion or relative frequency of the occurrence of an event
proposition another name for a statement
propositional function a proposition that includes variables
Pythagorean identities for any θ, $\sin^2(\theta) + \cos^2(\theta) = 1$, $\tan^2(\theta) + 1 = \sec^2(\theta)$ and $\cot^2(\theta) + 1 = \text{cosec}^2(\theta)$
quantifiers used with a propositional function to give information about the scope of the function's variables
radian the size of an angle subtended at the centre of a circle by an arc equal in length to the radius is one radian
random event an event where the outcome cannot be pre-determined
random number a number whose value is governed by chance and cannot be predicted in advance
random sample a randomly selected sample in which every member of the population has the same chance of being selected
range the difference between the largest and smallest values

rational numbers real numbers that can be written in the form $\frac{a}{b}$ where a and b are integers, the only common factor between a and b is 1, and $b \neq 0$
real numbers numbers that can be represented on a number line
recursion defining a problem in terms of itself, this is a very powerful tool in writing algorithms
reflection a transformation of a point, line or figure defined by a line of reflection, where the image point is a mirror image of the pre-image point
regular graph a graph in which every vertex has the same degree
roots of an equation the solutions of an equation
rotation a transformation of a point, line or figure where each point is moved a constant amount around a fixed point
row matrix a matrix with only one row
sample part of a population chosen to give information about the population as a whole
sample size the number of data points selected in a sample
scalar a quantity that has only magnitude, no direction
scalar resolute $\hat{\underset{\sim}{u}} \cdot \underset{\sim}{v}$; the magnitude of vector $\underset{\sim}{v}$ acting in the direction of vector $\underset{\sim}{u}$
secant function the reciprocal trigonometric function of the cosine function
selection the number of ways of choosing things when order is not important; *see* combination
self-selected sample a sample in which participants choose to participate in a survey
semi-Eulerian graphs connected graphs where you can start at a vertex and move along each edge only once without being able to return to the vertex at which you began
semi-Eulerian trail the trail that is created when you travel around a semi-Eulerian graph
semi-Hamiltonian cycle the trail that is created when you travel around a semi-Hamiltonian graph
semi-Hamiltonian graphs graphs where you can start at a vertex and travel to each vertex only once without being able to return to the vertex at which you began
sequence a list of numbers in a particular order
set a collection of elements
simple cycle a cycle in a graph with no repeated vertices
simple graph a graph in which pairs of vertices are connected by one edge at most
sine rule $\frac{a}{\sin(A)} = \frac{b}{\sin(B)} = \frac{c}{\sin(C)}$ is used to solve a side or angle of a triangle when given two sides and an angle or one side and two angles one of which is an opposite angle
singular matrix a matrix that has a zero determinant and hence has no multiplicative inverse
spanning trees graphs that are formed from part of a larger graph that include all the vertices of the original graph
square matrix a matrix with an equal number of rows and columns
statement a sentence that is either true or false
statistic a characteristic of a sample
stratified random sample a random sample in which subgroups of the population are represented in a similar percentage in the sample
subgraph a graph whose vertices and edges are all contained within another graph
surd A root of a number that does not have an exact answer. Surds are irrational numbers. Surds themselves are exact numbers, for example $\sqrt{6}$; their decimal approximations are not.
systematic sample a sample in which every kth member of the population is sampled
theorem a mathematical statement that can be shown to be true using a proof
trace of a matrix $\text{tr}(A)$ for matrix A; the sum of the leading diagonal elements
trail a walk in which no edges are repeated
transformation a geometric operation that may change the shape and/or the position of a point or set of points
translation a transformation of a point, line or figure where each point in the plane is moved a given distance in a horizontal or vertical direction
translation matrix the representation of a translation by a matrix
tree a simple, connected graph which contains no circuits

trigonometric function sine (sin), cosine (cos) or tangent (tan). On a unit circle, $\cos(\theta)$ is the x-coordinate of the trigonometric point $[\theta]$; $\sin(\theta)$ is the y-coordinate of the trigonometric point $[\theta]$; and $\tan(\theta)$ is the length of the intercept that the line through the origin and the trigonometric point $[\theta]$ cuts off on the tangent drawn to the unit circle at (1, 0).

true bearings directions that are written as the number of degrees (3 digits) from north in a clockwise direction, followed by the word 'true' or 'T'; for example, due east would be 090° true or 090°T

truth value Relates to the truth or falsehood of a proposition. Has only two possible values: true or false.

undirected graphs graphs in which edges are not directed

union the set of all elements in any one set and in a combination of sets; $A \cup B$ contains the elements in A or in B or in both A and B

unit vectors the vectors $\underset{\sim}{i}$ and $\underset{\sim}{j}$ with magnitudes of 1 unit, which allow a vector to be resolved into its components; $\underset{\sim}{u} = x\underset{\sim}{i} + y\underset{\sim}{j}$

universal quantifier For all; symbolically $\forall$. For a propositional function, consider if the function is true for all possible values of the variable.

universal set the complete set of objects being considered; usually represented by ξ

vector a quantity that has both magnitude and direction

vector matrix a matrix with only one column

vector resolute the direction of vector $\underset{\sim}{v}$ acting parallel or perpendicular to the direction of vector $\underset{\sim}{u}$; $\underset{\sim}{v}_{||} = \underset{\sim}{u} - (\underset{\sim}{u} \cdot \hat{\underset{\sim}{v}})\,\hat{\underset{\sim}{v}}$; $\underset{\sim}{v}_{\perp} = (\underset{\sim}{u} \cdot \hat{\underset{\sim}{v}})\,\hat{\underset{\sim}{v}}$

walk any route taken through a graph, including routes that repeat edges and/or vertices

weighted graphs Graphs in which every edge has a numerical 'weight' attached to it. Often used to represent some numerical property such as distance or time.

zero matrix a square matrix that consists entirely of '0' elements

INDEX

D

E

F

G

H

I

T

U

V

W

Z